Shifting Cultivation Policies

Balancing Environmental and Social Sustainability

———————————

Shifting Cultivation Policies

Balancing Environmental and Social Sustainability

_____

Edited by

Malcolm Cairns

with the assistance of Bob Hill and Tossaporn Kurupunya

CABI is a trading name of CAB International

CABI	CABI
Nosworthy Way	745 Atlantic Avenue
Wallingford	8th Floor
Oxfordshire OX10 8DE	Boston, MA 02111
UK	USA
Tel: +44 (0)1491 832111	Tel: +1 (617)682-9015
Fax: +44 (0)1491 833508	E-mail: cabi-nao@cabi.org
E-mail: info@cabi.org	
Website: www.cabi.org	

A catalogue record for this book is available from the British Library, London, UK.

Library of Congress Cataloging-in-Publication Data

Names: Cairns, Malcolm, editor.
Title: Shifting cultivation policies : balancing environmental and social
    sustainability / [edited by Malcolm Cairns].
Description: Boston, MA : CABI, [2017] | Includes bibliographical references
    and index.
Identifiers: LCCN 2017004396 | ISBN 9781786391797 (hbk : alk. paper) |
    ISBN 9781786391810 (epub)
Subjects: LCSH: Shifting cultivation--Government policy--Pacific Area. |
    Environmental policy--Pacific Area.
Classification: LCC S602.87 .S545 2017 | DDC 338.1/62--dc23 LC record
    available at
https://lccn.loc.gov/2017004396

ISBN-13: 978 1 78639 179 7

Commissioning editor: Alex Hollingsworth / David Hemming
Production editor: Tim Kapp

Printed and bound in the UK from copy supplied by the editors by CPI Group
(UK) Ltd, Croydon, CR0 4YY

Dʀ Haʀold C. Conklin with Pugguwon Lupaih,
working on maps of the Ifugao rice terraces, in
Baynīnan, northern Luzon, the Philippines.

In Remembrance of Harold C. Conklin

On 18 February 2016, Harold C. Conklin passed away in Hamden, Connecticut, a few miles from Yale University, where he taught for 34 years. Professor Conklin's interests spanned an impressive range of topics, including language, literacy, folk taxonomies, colour categories, ethnobotany and ethnobiology. But perhaps his greatest impact upon the world stemmed from his pioneering work in the study of swidden agriculture, which was universally reviled in the policy worlds of the mid-20th century when Professor Conklin began his career. His first contribution to this field was his Yale dissertation (1954), *The Relation of Hanunóo Culture to the Plant World*, based on his fieldwork among the Hanunóo of Mindoro Island in the Philippines. His discovery that they named over 1600 plants in their lands, and used over 90% of these, changed forever our understanding of the value of the tropical rainforest and also our understanding of the knowledge systems of swidden cultivators like the Hanunóo, who understood this value in a way that we did not. Professor Conklin's second major contribution to swidden studies was his monograph, *Hanunóo Agriculture: A Report on an Integral System of Shifting Cultivation in the Philippines* (1957/1975). Published by and thus winning the imprimatur of the United Nations Food and Agriculture Organization, this became the first definitive defence of swidden agriculture. Professor Conklin made his case not by emotional pleas, but by the most exacting documentation of agricultural practices imaginable. *Hanunóo Agriculture* contains a six-page description of rice planting and a 14-page description of rice harvesting. This was subtly but profoundly political; only people that 'matter' get 20 pages devoted to their planting and harvesting technologies. The Hanunóo responded in kind, adopting the word 'Conklin' into their language – as can be found today in Hanunóo glossaries – as a word for 'things having to do with knowledge'. This present volume continues very much in that tradition of collaborative production of knowledge of swidden agriculture, whose place in the 21st century continues to demand this kind of inquiry.

Michael R. Dove
Yale University

"Scientists don't know how to do that"

"I used to think the top environmental problems were biodiversity loss, ecosystem collapse and climate change.

I thought that with 30 years of good science we could address these problems.

But I was wrong.

The top environmental problems are selfishness, greed, and apathy…

…and to deal with those, we need a spiritual and cultural transformation

…and scientists don't know how to do that."

Professor Gus Speth[*]

[*] Professor James Gustave Speth, formerly Dean of the Yale School of Forestry and Environmental Studies, is a Professor at Vermont Law School and a Distinguished Senior Fellow at Demos, a nonpartisan public policy research and advocacy organization. He co-founded the Natural Resources Defense Council, was founder and President of the World Resources Institute and served as Administrator of the United Nations Development Programme. He is the author of six books, including the award-winning *The Bridge at the Edge of the World: Capitalism, the Environment, and Crossing from Crisis to Sustainability* and *Red Sky at Morning: America and the Crisis of the Global Environment.*

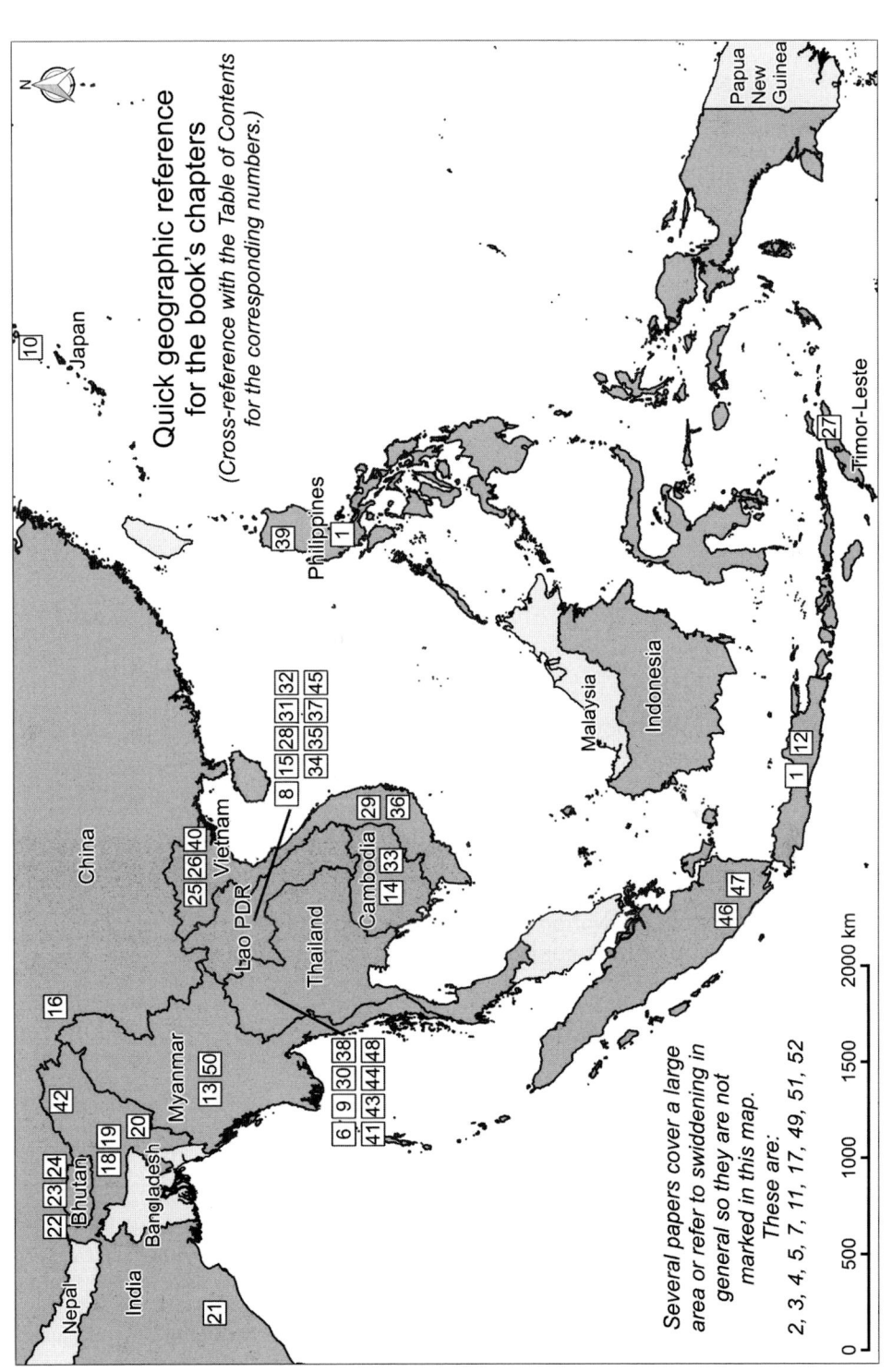

Quick geographic reference
for the book's chapters

(Cross-reference with the Table of Contents
for the corresponding numbers.)

Several papers cover a large
area or refer to swiddening in
general so they are not
marked in this map.
These are:
2, 3, 4, 5, 7, 11, 17, 49, 51, 52

CONTENTS

I. Introductory section

A. Setting the stage

B. Historical overviews from Southeast Asia

II. The impact of state policies on shifting cultivation

B. Removing the 'shifting' from 'shifting cultivation'

III. Policy lessons that we should be learning

A. From research by graduate students

B. From highland development projects

C. From efforts to mediate the role of farmers in the Indonesian forest

D. From the merger of national and local

IV. Concluding section

Addendum

The following chapters add substantial support to the learned arguments presented in the chapters above. They are therefore recommended as additional reading on the broad subject of official policy as it affects shifting cultivation in the Asia-Pacific region. That these chapters are offered for digital download rather than published within these pages is a result of overwhelming contributions and limited space. They can be found at:

www.cabi.org/openresources/91797

A1. Assessing shifting cultivation policies through the lens of
systems thinking
Mai Van Thanh and Tran Duc Vien

ACRONYMS

ACIAR	Australian Centre for International Agricultural Research
ACF	*Action Contre la Faim*, a French NGO
ADAB	Australian Development Assistance Bureau
ADB	Asian Development Bank
ADC	Autonomous District Council, India
AIPP	Asia Indigenous People's Pact
ALMSD	Agricultural Land Management and Statistics Department, Myanmar
AMAN	Alliance of the Indigenous Peoples of the Archipelago (*Aliansi Masyarakat Adat Nusantara*), Indonesia
AMFU	All Mizoram Farmers' Union, India
ANRI	*Arsip Nasional Republik Indonesia* (National Archives of Indonesia)
ANZ Bank	Australia and New Zealand Banking Group
ASEAN	Association of Southeast Asian Nations
ASB	Alternatives to Slash and Burn project
BINLEA	Bureau for International Narcotics and Law Enforcement Affairs, US Department of State
CARP	Comprehensive Agrarian Reform Programme, the Philippines
CBIK	Centre for Biodiversity and Indigenous Knowledge
CBNRM	community-based natural resources management
CCHR	Cambodian Centre for Human Rights
CDM	clean development mechanisms
CGIAR	Consultative Group for International Agricultural Research
CHT	Chittagong Hill Tracts, Bangladesh
CIA	Central Intelligence Agency, USA
CIDA	Canadian International Development Agency
CIFOR	Center for International Forestry Research
CLV-DTA	Cambodia, Laos and Vietnam Development Triangle Area
COHCHR	United Nations Cambodia Office of the High Commissioner for Human Rights
CP	Charoen Phokphand (agribusiness), Thailand
CSEAS	Centre for Southeast Asian Studies, Japan
DANCED	Danish Cooperation for Environment and Development
DANIDA	Danish International Development Assistance
DARD	(provincial) Department of Agriculture and Rural Development, Vietnam
DAS	days after sowing
DENR	Department of Environment and Natural Resources, the Philippines
DNP	Department of National Parks, Wildlife and Plant Conservation, Ministry of Natural Resources and Environment, Thailand
DPC	Dak Lak People's Committee, Vietnam
EED	*Planète Enfants et Développement*, a French NGO
FAO	Food and Agricultural Organization of the United Nations
FCA	Forest Conservation Act, India
FCB	Food Corporation of Bhutan
FIDH	International Federation for Human Rights
FORDA	Forestry Research and Development Agency, Indonesia

FRA	Forest Rights Act, 2006, India
GFB	Gongshan County Forestry Bureau,Yunnan, China
GIM	Green India Mission, India
GIS	geographic information system
GIZ	*Deutsche Gesellschaft für Internationale Zusammenarbeit*, a German federal enterprise for international cooperation
GMO	genetically modified organism
GOI	Government of India
GOO	Government of Odisha, India
GPS	global positioning system
GSO	General Statistics Office, Vietnam
GTZ	*Gesellschaftfürtechnische Zusammenarbeit*, the former German technical cooperation agency, now known as GIZ
GY	grain yield (g/sq. m)
HAMP	Highland Agriculture and Marketing Project, Thailand
HASD	Highland Agricultural and Social Development project, Thailand
HRDI	Highland Research and Development Institute, Thailand
HTD	Hilltribe Development and Welfare Division, Department of Public Welfare, Thailand
ICAR	Indian Council of Agricultural Research
ICEF	India–Canada Environment Facility, India
ICESCR	International Covenant on Economic, Social and Cultural Rights
ICIMOD	International Centre for Integrated Mountain Development
ICRAF	International Center for Research in Agroforestry, (The World Agroforestry Centre)
IDRC	International Development Research Centre, Canada
IDS-Nepal	Integrated Development Society – Nepal, an NGO based in Kathmandu
IEG	Independent Evaluation Group, World Bank
IFAD	International Fund for Agricultural Development
IFM	indigenous fallow management
IIRR	International Institute of Rural Reconstruction, an NGO based in the Philippines
IKAP	Indigenous Knowledge and People, an NGO network
ILO	International Labour Organization
IOM	International Organization for Migration
IPAD	Integrated Pocket Area Development project, Thailand
IPBES	Intergovernmental Science-Policy Platform on Biodiversity and Ecosystem Services
IPCC	Intergovernmental Panel on Climate Change
IRD	*Institut de Recherche pour le Développement*, a French government research organization that replaced the former ORSTOM.
IRRI	International Rice Research Institute
ISFP	Integrated Social Forestry Program, the Philippines
IUCN	International Union for Conservation of Nature
IWGIA	International Working Group for Indigenous Affairs
JFM	Joint Forest Management, India
JICA	Japan International Cooperation Agency

KDTI	*Kawasan Dengan Tujuan Istimewa*, a new forest category created for the Krui agroforests, Indonesia
KfW	*Kreditanstalt für Wiederaufbau*, a German, government-owned development bank
Lao PDR	Lao People's Democratic Republic
LATIN	*Lembaga Alam Tropika Indonesia*, an NGO in Indonesia
LCDC/LNCDCS	Lao (National) Commission for Drug Control and Supervision
LCG	Land Core Group, a group of NGOs working in Myanmar
LFAP	Land and Forest Allocation Programme, Lao PDR
LSUAFRP	Lao-Swedish Upland Agriculture and Forestry Research Program, Lao PDR
MAF	Ministry of Agriculture and Forestry, Lao PDR
MARD	Ministry of Agriculture and Rural Development, Vietnam
MK35	Constitutional Court Ruling 35/PUU-X/2012, Indonesia
MLIPH	Meghalaya Livelihoods Improvement Project for the Himalayas, India
MNRE	Ministry of Natural Resources and Environment, Thailand
MMSEA	Mainland Montane Southeast Asia
MOP	Ministry of Planning, Cambodia
NAFRI	National Agriculture and Forestry Research Institute, Lao PDR
NAP	National Action Programme 2009, Timor-Leste
NCA	Norwegian Church Aid
NCDD	National Committee for Sub-National Democratic Development, Cambodia
NEC	North Eastern Council, India
NEPED	Nagaland Environmental Protection and Economic Development project, India
NERCORMP	North Eastern Region Community Resource Management Project, India
NERLP	North East Rural Livelihood Project, India
NESDB	National Economic and Social Development Board, Thailand
NGO	non-governmental organization
NGP	National Greening Program, the Philippines
NIAPP	National Institute of Agricultural Planning and Projection, Vietnam
NIPAS	National Integrated Protected Area System, the Philippines
NLMA	National Land Management Authority, Lao PDR
NLUP	New Land-Use Policy, India
NPV	net present value
NRM	natural resource management
NSO	National Statistics Office, Thailand
NSSC and PPD	National Soil Services Centre, Policy and Planning Division, Minisry of Agriculture and Forests, Bhutan
NTFPs	non-timber forest products
ONCB	Office of the Narcotics Control Board, Thailand
ORSTOM	*Office de la Recherche Scientifique et Technique Outre-mer* (French development agency, now replaced by IRD)
PCAIV	Principal Component Analysis with Instrumental Variables
PCO	Project Coordination Office, Nagaland, India
PES	payment for ecosystem services
PET	potential evapotranspiration
PGS	Participatory Guarantee System, India
PLUP	participatory land-use planning

P3AE-UI	*Program Penelitian dan Pengembangan Antropologi Ekologi Universitas Indonesia*, an NGO in Indonesia
PNG	Papua New Guinea
R-factor	cropping intensity (Ruthenberg factor, 1980)
RECOFTC	Regional Community Forestry Training Centre, Bangkok
REDD/REDD+	Reducing Emissions from Deforestation and forest Degradation, plus conservation/sustainable management/enhancement of carbon stocks
RFD	Royal Forest Department, Thailand (now part of the Ministry of Natural Resources and Environment)
RGC	Royal Government of Cambodia
RMI	The Indonesian Institute for Forest and Environment, an NGO based in Bogor, Indonesia
RRI	Rights and Resources Initiative, an NGO based in Washington, DC
RRL	Reforestation and Land Rehabilitation, Indonesia (a department within the Indonesian Ministry of Forestry and Environment (KLHK - *Kementrian Lingkungan Hidup dan Kehutanan*)
SALT	sloping agricultural land technology
SESMAC	Strengthening Economic and Social Management Capacity project, Lao PDR
SIDA	Swedish International Development Agency
SLCP	Sloping Land Conversion Programme, China
SMHDP	*Sam Muen* Highland Development Project, Thailand
SNV	*Stichting Nederlandse Vrijwilligers*, a Dutch NGO
TABI	The Agro-Biodiversity Initiative (project) of the Lao PDR
TAHAP	Thai–Australian Highland Agricultural Project, Thailand
TA-HASD	Thai–Australian Highland Agricultural and Social Development project, Thailand
TDRI	Thai Development Research Institute, Thailand
TEC	total ecosystem carbon
TFAP	Tropical Forestry Action Plan (FAO and World Bank)
TRC	Tribal Research Centre, later Tribal Research Institute, Thailand
UNDCP	United Nations International Drug Control Programme
UNDP	United Nations Development Programme
UNEP	United Nations Environment Programme
UNESCO	United Nations Educational, Scientific and Cultural Organization
UNFDAC	United Nations Fund for Drug Abuse Control
UNODC	United Nations Office on Drugs and Crime
UN REDD	The United Nations Collaborative Programme on Reducing Emissions from Deforestation and forest Degradation in Developing Countries, Geneva
UNWMP	Upper Nan Watershed Management Project, Thailand
USAID	United States Agency for International Development
VFV	Vacant, Fallow and Virgin Lands Management Law 2012, Myanmar
VND	Vietnamese dong (currency)
WATALA	'Friends for Nature and Environment', an NGO in Indonesia
WRI	World Resources Institute, an NGO based in Washington, DC
WTG	weight of 1000 grains (rice)
WTO	World Trade Organization

J. Scott Cairns
(16 February 1895 to 25 November 1973)

DEDICATION

On the rare occasions during my overseas career when I've had enough time and money to fit in a home leave, it has always been my ritual to visit the North Bedeque cemetery on Prince Edward Island (PEI), Canada, with my father, to pay respects to our ancestors. One of the main ancestors to be remembered during these visits was my late paternal grandfather, J. Scott Cairns (pictured above).

This book is dedicated to J. Scott Cairns (1895-1973) — a man who I grew up knowing as 'Grampie', but who most others knew as 'Scott'. When I was a toddler, many of my earliest and warmest memories were of sitting on this man's lap as he smoked his pipe and watched television in the evenings. I was immensely fortunate to grow up in an extended family situation, with my grandparents living in one side of a rambling old farmhouse and within easy access.

Scott Cairns was born on February 16th of 1895, as the first-born child of William (1864-1932) and Thirza Picketts Cairns (1868-1931). He was born on the farm, in Lower Freetown, Prince Edward Island, on which he was to spend his entire life. Scott encountered grief at an early age when his only brother, William Cecil (1898-1911), succumbed to a deadly combination of scarlet fever and measles, at the tender age of 12.

As a young man, Scott developed into an accomplished musician, and played the fiddle and bones at the Saturday night dances that were a central part of rural social life at that time. It was at such a dance in Wilmot Valley that Scott met a young, newly arrived teacher named Georgie, who he was to make his wife. Scott and Georgie

Plate 2: Scott relaxes on the lawn with his first-born, William.
It was from the union of their names that Willscott Farm took
its name.

married at a home wedding in Summerside in 1927. Their family expanded quickly
in the following years: William (1928), Helen (1929), Winnifred (1931), Georgina
(1933), Louise (1935) and Amy (1936).

As Scott and Georgie began their young family, it coincided with the Great Depression
of the 1930s when nearly everybody was extremely poor and when a dollar was a
large sum of money. He was a mixed farmer who produced cattle, hogs, sheep and

Plate 3: Scott Cairns (right) and fellow farmers admire silver
trays that they were awarded for their cream production at an
annual meeting of the Dunk River Dairying Company.

poultry, as well as an acreage of potatoes that was regarded as large in those times. He was one of the top cream producers selling to the Dunk River Dairying Company and served a term as the company's President.

In 1940, Scott took the bold step of buying a tractor, a Massey Harris 101 Junior Twin Power (probably about 30 hp). This tractor was one of the first in that part of the country and is still in working condition today, now in storage on the farm.

In 1949, Scott helped to organize and finance an electric-light line to his area in Lower Freetown. This was before electricity was available to much of rural PEI and before the Rural Electrification Programme. In many ways, he was a leader in his time.

In the earlier days, all field work was done by heavy draft horses and horses were also the only means of local transportation. Narrow rural roads often had so much snow on them in winter that not even a horse could get through. As a result, a lot of winter travelling was done on winter trails through the fields and sometimes people took a short cut and travelled part way over the ice, which usually covered the local bays and estuaries. There were times when a horse would lose its way during these ice crossings, and would stray further and further away from land, until the ice thinned to the point where it could no longer support the horse's weight, and it would break through.

Under these conditions, the six Cairns children would bundle into a sleigh every morning, pull a buffalo-skin rug around them for warmth, and take a horse a mile and a half down the road to the one-room Lower Freetown school. My father sometimes

recalled times when winter storms would blow in so fiercely that Scott would become worried about whether the horse could find its way home through the storm. Scott would pull on his heavy raccoon-fur coat and, bent into the wind, would set off on foot for the school, soon disappearing into the swirling snow. Upon arriving at the school, Scott would load his brood into the sleigh and then guide the horse back home through the storm. People were rugged in those times, and Scott clearly didn't want to leave anything to chance when it came to his children's safety.

Plate 4: This old photo of Mrs Scott Cairns, on her way out of the house lane, gives an indication of snow levels during what folks now call 'old-fashioned winters', before climate change became so pronounced.

In 1832, when our ancestors emigrated from Scotland to Prince Edward Island (which was known as *Abegwet* (cradle on the waves) to the local Mi'kmaq Indians), they faced a much colder climate, and a shorter growing season. When I was a child, my mother grew a large vegetable garden every summer. We always had to keep a wary eye on the weather as autumn approached every year, lest an unexpected frost might freeze what remained unharvested in the garden. If frost was forecast, then the choice was either to hurriedly harvest everything remaining before that nightfall, or try to cover it with plastic sheets. This, of course, didn't apply to everything in the garden; root crops and cole crops would not be harmed by a frost. Most folk listened to the weather forecast on the radio or television to learn if there would be frost – but we had a more reliable source. We would ask Scott.

I retain one vivid and specific childhood memory of this daily ritual in late autumn, of asking Scott to forecast if the ground would be whitened with frost that night. He was on his front steps, probably on his way in for supper, before going back out to feed his pigs. Scott accepted the question wordlessly, leaned forward against the step railings, and peered into the evening sky, doubtlessly noting the temperature, the cloud cover, and whether there was any wind. I don't recall what the verdict was on that particular evening, but I do know that it would have been given with certainty. If frost caught us by surprise, it would have meant the loss of a good deal of my mother's garden vegetables. But I never knew him to be wrong. This was the indigenous knowledge of a pioneering son of the soil in eastern Canada.

Although the red loam soils of Prince Edward Island might seem far removed from shifting cultivation in the tropics, let me pause to tell a quick story to demonstrate how interconnected they really are, and how small the world truly is. Among the stories passed down in our family was one about a long-past year in which summer never arrived, and there was great famine and suffering. With the modern advent of Google and Wikipedia, it is now easy to confirm that the year was 1816, and the cause of the missing summer was a combination of the 1815 eruption of Mount Tambora on the island of Sumbawa in Indonesia, and a historic low in solar activity. This reminds us that we all share the same biosphere. While shifting cultivators on the slopes of Mount Tambora undoubtedly had to flee and lost their crops, farmers on the other side of the world lost their entire growing season. While settlers on PEI reported a year of famine the following year in 1816, the Cairns ancestors were still in Scotland at that point, and so would have experienced this year of hardship there.

As a pioneering farmer settling a new land in those times, Scott cleared quite a bit of forest for agriculture in his own lifetime. But it was not an indiscriminate clearing. Very importantly, the woodlots to the west of the homestead were carefully left intact, to protect against the prevailing winter winds that blow from that direction. On both the east and west extremes of the property, the land slopes down to small, meandering brooks. These sloped approaches to waterways were also left forested. There is a scraggly old oak tree that has stood alone, on one side of our barnyard, for all of my life. My father tells a story about when his grandfather and great grandfather were working together clearing forest around the area that would later develop into that barnyard. One pointed to a particular young tree and warned the other, "*Don't cut that one; that's an oak.*" Oak trees were highly valued for the quality of their timber. And the towering old pines that were spared the axe all those years ago now provide the annual nesting sites for the bald eagles that have claimed our farm as their own.

As is the case for many of the farmers discussed in this book, money was tight during my grandfather's time and he sometimes turned to the forest to generate additional income. My father tells a story of one year when Scott had decided to cut firewood for sale to earn some extra income. But he first needed to buy a circular-saw blade to cut the logs into stove lengths. After surveying a number of shops in Summerside, he finally found what he was looking for at Strong's store on west Water Street.

When the old man who was tending the store that day told Scott the price of the saw blade, though, Scott's face must have fallen, because he had to admit that he didn't have that much money. "*Well,*" the old shopkeeper intoned, "*if that's the case, then I guess that you had better get your saw blade elsewhere.*" Scott nodded silently and was halfway to the door when the old man called after him, "*Are you John Cairns's son?*" Scott paused at the doorway and replied, "*No, I'm his grandson.*" "*Oh, in that case then,*" the old shopkeeper offered, "*I guess that you'd better take the saw blade.*" That's how business was often done on Prince Edward Island, and probably still is to a large extent. People dealt with you on the basis of who you were, and your word was as good as your name.

Plate 5: Probably around 1930, Scott Cairns pauses to enjoy a light-hearted moment with a farm dog, at his sister's farm in Chelton, PEI.

As another example of this, my father sometimes tells a story about when he and his father bought a neighbouring farm from R. Louis Cairns (a descendent of the second Cairns brother, Robert, who immigrated to PEI a few years later than the original Cairns settler, John Glen, and arrived on the shores of Prince Edward Island in 1840). In 1970, by then becoming elderly and crippled, Louis had made a decision to retire and sell the farm that he had worked all his life (see Plate 3 in the dedication for Volume III in this trilogy of books). Someone had even made him an offer for it. On his way to Summerside one day, Louis dropped by neighbouring Willscott Farm (the ancestral farm jointly operated by my father and grandfather) to tell his neighbours of his decision to sell, and asked if they might be interested in buying him out. A nearby property had already been sold earlier that year, and so Louis was able to use its selling price as a yardstick to gauge the market value of his own land. William and Scott were indeed interested in buying this adjoining acreage – but they undoubtedly had to discuss a bit between themselves about how they could manage the deal financially. With that understanding, Louis continued on his way to Summerside. While in town that very day, Louis met with lawyers and had ownership of his entire farm (roughly 125 acres) transferred to his neighbours' names.

Virtually everything that Louis owned in this mortal world, including the house that he lived in, was legally transferred to others with not a nickel having changed hands and nothing on paper - but only the expressed interest of his neighbours in buying his land. Such was the level of trust between honourable men.

Many years later, when Scott had reached his early 70s and I was a youngster, I had an opportunity to witness why that old shopkeeper had been so confident in extending him credit to buy the saw blade. Scott was seriously ill in the Summerside hospital, and unsure if he would survive, he summoned his only son, William (my father), to discuss settling his affairs. Like every farmer, at any given time, Scott had both debts and credits around the countryside where he did business. Scott's biggest concern in that meeting was to explain to his son where his outstanding debts were, so that in the event that he never made it out of the hospital alive, William would settle his debts and nobody would be left with an unpaid bill. Such was his honour, that he was determined that no bill would remain unpaid, even in death.

Money, of course, was worth a lot more in earlier times. In the years before he had his own phone installed, if Scott needed to make a phone call, he would have to go down the road and pay a neighbour for the use of his phone. A call outside the local Bedeque exchange cost ten cents at that time – about the same price that a 75-pound bag of potatoes sold for. So it cost the equivalent of 75 pounds of potatoes to make a phone call to nearby Summerside! Furthermore, calls placed after working hours incurred a hefty penalty of 25 cents, or the equivalent value of 187.5 pounds of potatoes! My father recalls Scott saying that he did a lot of thinking at that time before he made a phone call. It was the early 1950s when Scott finally got his own phone line – and, as was the case with the arrival of electricity, he and a neighbour made it happen by building the line themselves.

Grampie left this world during the first winter storm of 1973, probably only a few hundred metres from where he had entered it. He had fitted a lot of farming into his 78 years, and as he was going into cardiac arrest, his last action was to glance at the wall clock and comment that it was almost time to feed the chickens. His wish to die with his boots on had been granted. I take the opportunity provided by this dedication to salute a great man, the late J. Scott Cairns! I am proud to call him my grandfather.

FOREWORD

Carol J. Pierce Colfer*

In 2007, Malcolm Cairns published his first book, *Voices from the Forest: Integrating Indigenous Knowledge into Sustainable Upland Farming*, which reported the results of an earlier workshop on fallows in shifting cultivation systems, held in Bogor, Indonesia in June of 1997 – a sort of 'prequel' to this trilogy of volumes. It went against the dominant perspective within academic and development circles, particularly within the biophysical sciences. This perspective, which has endured since colonial times, saw swiddeners as black sheep, deforesting and raping their environments.[1] Instead, Cairns' book highlighted the value, complexity and rationality of many kinds of swidden agriculture, particularly focusing on fallows, in Southeast Asia. This was a song I'd been singing for some time, but I sang only of the few cases that I knew well. Here was a book that included examples from all over the region, many of which were consistent with my own view of such systems as bursting with insights – if conventional scientists would just listen.

Shortly after finishing this huge and wonderful volume, Cairns began again, assembling even more analyses from the same region, digging deeper, surveying more widely. The second book, *Shifting Cultivation and Environmental Change: Indigenous People, Agriculture and Forest Conservation* (2015), is also filled to the brim with a variety of analyses, ranging from in-depth ethnographies to strictly quantitative studies to remembrances from those who grew up in such systems. While the book series began with a focus on the fallows in such systems, many authors in this new volume provided holistic understandings of the whole sequence of events in swidden systems, from clearing to cultivation to usually productive fallows and sometimes beyond.

Cairns' original call for papers for the 2015 volume elicited a remarkable 150 papers, clearly too many to include in one book, yet too rich a collection to ignore. So Cairns, ever-creative, adapted his plans, managed to find funding and agreeable publishers, and made the 2015 volume the first of a trilogy. This volume, *Shifting Cultivation Policies: Balancing Environmental and Social Sustainability*, is the second volume in that trilogy and the third, *Farmer Innovations and Best Practices by Shifting Cultivators in Asia-Pacific*, is forthcoming.

---

* Dr Carol J. Pierce Colfer is an anthropologist serving as a senior associate at the Center for International Forestry Research (CIFOR) in Bogor, Indonesia; and as a visiting scholar in Cornell University's Southeast Asia Program in Ithaca, NY.

The enthusiastic response to his call for papers is encouraging for those of us who consider swidden agriculture to be one of many viable approaches to managing and living from, and in, forested environments. There is a deep irony as we consider the disdain with which swiddeners tend to be held; many analysts appear to be unconscious of the fact that most of the world's tropical rainforests have survived and flourished in a symbiotic relationship with swidden agriculture (as practised under conditions of low population density). Permanent cultivation, large-scale logging and plantation development: these are the 'innovations' that have resulted in the massive and speedy deforestation in many areas of the tropics. Population growth and in-migration have added to the complexity, of course. But in many areas, local population growth has been augmented substantially by government policies to move people into 'empty' (in many cases read, 'fallow') forest areas.

In the 1990s, there was a lull in the study of swidden agriculture, after it had been studied extensively during earlier eras. But the global interest in climate change and particularly in REDD+ (Reducing Emissions from Deforestation and Forest Degradation) re-ignited the interest of many, including myself. We recognized that our studies showing the complexity of many swidden systems had not been seen, acknowledged, or incorporated into policy-making; the old views and narratives about those who practised 'slash and burn' remained dominant, despite so much evidence to the contrary (in many circumstances). We feared that REDD+ might be the perfect rationale for ignorant planners and corrupt government officials to more seriously implement the common laws against swiddening, endangering the livelihoods and cultures of swidden communities. REDD+ represented a potential source of funding as well; we worried that such funds might make policies that had earlier been impossible to implement due to financial exigencies at least feasible and possibly dangerous. Ziegler et al. (ch 11 of this volume) also 'recognize the irony that some REDD approaches seek to protect a forested landscape by eliminating the long-fallow systems that have inherently helped to shape the landscape in its present form'.

Several strengths are visible in all of Cairns' work: respect for the voices of those with field-based experience (whether academicians, development workers, policy-makers or farmers of either sex); coverage of a wide range of shifting cultivation systems; a balanced offering, in the sense of no particular ideology or commitment to any specific narrative in his selection of cases; and an open-minded view of the various kinds of research methods that authors have used (from experimental to survey to ethnographic and beyond). This volume shares these traits, adding the concern with the variety of policies that have affected those practising shifting cultivation and those contexts affected by it.

Although there is not space here to thoroughly review all of the chapters, I highlight some of the issues that are addressed in this book and that I consider particularly important for us – the global research and policy communities – to think about.

- The first pertains to the variation *in shifting cultivations systems, contexts, practices and effects:* variations that require differentiated policies. We know about the view that shifting cultivation can have bad effects, as argued by almost all governments that have populations who practise it. And there are cases where the effects are adverse: McArthur et al. (ch 27) describe an environmentally damaging form of shifting cultivation resulting in invasions of the noxious weeds *Imperata cylindrica* and *Chromolaena odorata*, in post-conflict Timor Leste (however, see Dove and Kammen (2015), who describe the same kinds of prejudices against *Imperata* as those discussed here for swiddens). A crucial factor, and a common one with adverse effects in shifting cultivation systems, is population growth. The dilemma of what to do as population and other pressures on land increase is of key importance as we consider appropriate policies (and the more common policy failures). It was this inescapable dilemma that led to Cairns' research interest in fallow management in the first place – as a potential entry point in land-use intensification.

  The number of studies that have shown the ecological rationale of shifting cultivation under conditions of low population density is legion; and well represented in all of Cairns' books. Fukushima's analysis in chapter 41 (like many in Volume 1 of this series) builds on conventional ethno-botanical methods to assess differences in plant regrowth in recent, still-used fallows vis-a-vis longer-term, semi-abandoned fallows in northern Thailand. She found different repertoires, with more plants used by people (food, tools, medicines) in the more active swidden areas, and more of a variety of plants sensitive to slashing and burning in the longer fallows. Her conclusion is that we need a variety of approaches (practices and policies) if we want to safeguard biodiversity. Siebert and Belsky (ch 24) make a case for learning from the principles involved in Bhutan's now-abandoned versions of swidden. These authors refer to the 'co-production' of landscapes as people and their habitats interact, with shifting cultivation having played an important ecological role. They suggest consideration of recreating aspects of such practices to maintain their previous ecological functions of small forest-gap creation and 'intermediate-level disturbance'.

- *Existing local management systems* are increasingly being recognized as having under-utilized value. Several chapters discuss indigenous management systems, such as *adat* in Indonesia (de Royer et al., ch 12) or *Tara Bandu* in Timor Leste (McArthur et al., ch 27). Riba's contribution (ch 19) emphasizes the importance of shifting cultivation to the cultural practices and beliefs of people in northeastern India (Arunachal Pradesh). One related present-day policy dilemma involves the clash (cultural, economic and political) between these dynamic but traditional, often somewhat communal, systems on one hand, and the 'modern' world of privatized land and capital, on the other. The study by Hoare (ch 48) represents one partial approach to addressing this dilemma. He reports on a resource-management project that began with a study of existing local management rules and regulations in 45 villages in Thailand's upper Nan watershed – a procedure he suggests all such projects should undertake. Such recognition can also be an excellent starting point for collaborative efforts to manage local resources between local people and outsiders of various hues (see e.g. Colfer, 2005; Evans et al., 2014).

- The significance of *secure land tenure* has also been increasingly bandied about in academic circles of late, and a similar debate has occurred among policymakers around the world. In Indonesia, a Constitutional Decision affecting land tenure for swiddeners has serious implications for making some people's land more secure (Minarchek, ch 47), and that of others less so (de Royer et al., ch 12). Protected-area management has also spelled problems for swiddeners' access to land. Using the conceptual lens of legal pluralism, Prill-Brett (ch 39) provides clear evidence of inter-sectoral conflict and overlapping tenure and authority near two protected areas (Mt. Data and Mt. Pulag) in the Cordillera area of the Philippines. She also discusses the adverse ecological impacts of a shift from traditional shifting cultivation to commercial vegetable production in these areas.

- A big part of the differences from place to place in such systems derives from *historical trajectories*, which are also examined with unusual thoroughness in this book. Useful historical sections often highlight colonial policies, some of which have continued to the present day under national governments. Forsyth and Walker (ch 9) point out saliently that 'environmental policy is often couched in unseen political histories, and that policy debates shape complex trends into convenient, but simplistic, narratives'. Phuntsho et al. (ch 17) provide brief summaries of the histories of swidden–related policies, country by country, in the eastern Himalayas (including the Chittagong Hill Tracts of Bangladesh, eastern Bhutan, north east India, northern Myanmar and parts of Nepal). They find that '…none of the countries of the eastern Himalayas has a policy that aims to support and improve shifting cultivation and recognizes its benefits'. Tripathi et al. (ch 20) document the frequency with which the policies of Mizoram's government related to shifting cultivation (though always denigrating) have changed in historical and more recent decades (in India).

- That '*the will to improve*' (Li, 2007) can also spell dangers is clear from this collection. The funding available at the moment for REDD+ is worrying; Angelsen et al. (2012) report the hopes and fears of both observers and participants. In this volume, Ziegler et al. (ch 11) outline the dangers and possibilities of REDD+ for swiddeners and de Royer et al. (ch 12) identify troublesome implications of REDD+ programmes, as currently implemented, in terms of inter-ethnic conflict and equity in Indonesia. Linking back to the discussion of governments' ubiquitous efforts to eradicate swiddens, Ducourtieux et al. (ch 32) report the adverse livelihood effects of attempts to persuade ethnic minorities in the mountains of Laos to abandon both opium and swidden production.

One topic that has yet to be addressed here is gender, although it was addressed quite thoroughly in the first volume of this trilogy, *Shifting Cultivation and Environmental Change: Indigenous People, Agriculture and Forest Conservation* (2015). I highlight this because of the very common involvement, and not infrequently dominance, of women's labour in swidden systems (Colfer et al., 2015a, 2015b). Indeed, in some areas of Africa, it is considered basically a women's agricultural system; and even in Southeast Asia, women usually have key roles (e.g. Colfer and Dudley, 1993; Colfer et al., 1997; Colfer, 2008; Sutlive, 1991). If, within the fields of agriculture and natural resources, there has been a growing recognition of gender issues, studies at the intersection of this field concerning governance, and particularly policy studies, are far behind.

One function of any book is to help us identify gaps; and this is a gap crying to be filled. The authors in this book are not wholly to blame. It is harder (and less interesting) to write about what governments or policy-makers have not done – in this case, their failure to attend to gender in policies relating to shifting cultivation. What has not been done could be the main topic of a study of gender and governance at larger scales. Neither speaking to, nor thinking about the implications for key actors are sure recipes for policy failure. The tendency to ignore both the differences in men's and women's roles, interests and goals, on one hand, and the power dynamics between them, on the other,[2] is one likely element in the problems governments have had in trying to eradicate or, more optimistically, transform shifting cultivation.

But this one failing aside (and one hardly unique to this collection), this book is an excellent compilation of analyses of policies on shifting cultivation throughout South and Southeast Asia. It provides in-depth knowledge about specific policies and historical trajectories relating to shifting cultivation that have not hitherto been available. Recognizing, as many of these authors do, that policies represent only part of the story, they nonetheless have real impacts in the field – even if moderated by their interaction with local realities (Tsing, 2005). This is one of the extraordinarily valuable aspects of Cairns' series of books: he recognizes that local realities, policies and implementation all have a role to play, and this is so clearly recognized in this three-part series. These books will be welcomed by anyone studying shifting cultivation systems and the people who depend on them.

References

Angelsen, A., Brockhaus, M., Sunderlin, W. and Verchot, L. V. (eds) (2012) *Analysing REDD+: Challenges and Choices*, Center for International Forestry Research, Bogor, Indonesia

Colfer, C. J. P., Peluso, S. and Chin See Chung (1997) *Beyond Slash and Burn: Building on Indigenous Management of Borneo's Tropical Rain Forests*, New York Botanical Gardens, NY

Colfer, C. J. P. (2005) *The Complex Forest: Communities, Uncertainty, and Adaptive Collaborative Management*, Resources for the Future Press/Center for International Forestry Research, Washington, DC

Colfer, C. J. P. (2008) *The Longhouse of the Tarsier: Changing Landscapes, Gender and Well Being in Borneo* (vol. 10), Borneo Research Council, in cooperation with the Center for International Forestry Research and UNESCO, Phillips, ME

Colfer, C. J. P. and Dudley, R. G. (1993) *Shifting Cultivators of Indonesia: Managers or Marauders of the Forest? Rice Production and Forest Use among the Uma' Jalan of East Kalimantan*, Food and Agriculture Organization of the United Nations, Rome

Colfer, C. J. P., Alcorn, J. B. and Russell, D. (2015a) 'Swiddens and Fallows: Reflections on the Global and Local Values of "Slash and Burn"', in M. F. Cairns (ed.) *Shifting Cultivation and Environmental Change: Indigenous People, Agriculture and Forest Conservation*, Earthscan, London

Colfer, C. J. P., Minarchek, R. D., Aier, A., Cairns, M. F., Doolittle, A., Mashman, V., Odame, H. H., Roberts, M., Robinson, K. and van Esterik, P. (2015b) 'Gender analysis: Shifting cultivation and indigenous people', in M. F. Cairns (ed.) *Shifting Cultivation and Environmental Change: Indigenous People, Agriculture and Forest Conservation*, Earthscan, London

Dove, M. R. and Kammen, D. M. (2015) *Science, Society and the Environment: Applying Anthropology and Physics to Sustainability*, Routledge, London

Evans, K., Larson, A., Mwangi, E., Cronkleton, P., Maravanyika, T., Hernandez, X., Müller, P., Pikitle, A., Marchena, R., Mukasa, C., Tibazalika, A. and Banana, A. (2014) *Field Guide to Adaptive Collaborative Management and Improving Women's Participation*. Center for International Forestry Research, Bogor, Indonesia

Li, T. (2007) *The Will to Improve: Governmentality, Development, and the Practice of Politics*, Durham University Press, Durham, UK

Sutlive, V. H. (ed.) (1991) *Female and Male in Borneo: Contributions and Challenges to Gender Studies*, Borneo Research Council, Inc., Williamsburg, VA

Tsing, A. L. (2005) *Friction: An Ethnography of Global Connection*, Princeton University Press, Princeton, NJ

Notes

1. See, for instance, Phuntsho et al. (ch 17), who say 'There is a popular perception that shifting cultivation is a destructive and undesirable practice. This is based on the mistaken belief that untouched forest is being destroyed to make way for low-productivity agriculture. In fact, shifting cultivators are using a practice that is best described as rotational agroforestry, involving planned cyclical use of defined forest plots to grow food while retaining a forest cover'.

2. Not to mention the very important differences that can exist among both women and men, based on other social markers such as age, caste, class, wealth, religion and ethnicity. These also need attention.

PREFACE

*Malcolm Cairns**

An inspirational milieu

In the interests of full and proper disclosure, I should begin with an admission that my credentials as a dispassionate researcher of shifting cultivation deserve to be questioned. My roots in agriculture are too deep to pretend otherwise. My family has been farming in Canada for at least seven generations and, for an unknown number of generations in Scotland before that. It was sufficiently long, in any case, for our lineage of the Cairns clan to already carry with them a reputation for ability in horticulture and agriculture when they arrived on the shores of Prince Edward Island in 1832.

It is interesting to ponder how early childhood experiences influence the paths that we choose later in life. My early childhood years marked me clearly as a hunter-gatherer. Even as a toddler, I spent many warm summer days picking the plentiful wild blueberries and strawberries that grew along the lanes on my parents' farm, and raspberries from my mother's garden. My mother would collect the blueberries that I gathered until she had enough for a pie or cake. There was no tastier dessert than fresh-from-the-oven blueberry pie with a scoop of ice cream on top! By the age of ten, I had joined the garden project of the local 4-H club and was growing a large garden to augment what my mother already planted. I was thus contributing to the food on my family's table from the time that I was a child.

As I grew older, the berry-picking pails were replaced by a fishing rod, a tackle box, and a can of worms and, between farm chores, I spent my summers coursing the brooks that ran through our farm and the nearby Wilmot River, in search of trout. Fishing was a summer pleasure, when the fishing season was open. As autumn approached, and the night-time temperatures dropped below freezing, the fishing rod

---

* Dr Malcolm Cairns is the Editor of this volume. His academic background ranges from Animal Science through a Master's Degree in Environmental Studies, to a Doctorate in Anthropology from the Australian National University, awarded for a study of the cultural ecology of the Angami Nagas in northeast India (Cairns, 2013). The first volume bearing his name as Editor was *Voices from the Forest: Integrating Indigenous Knowledge into Sustainable Upland Farming* (2007, RFF Press, Washington, DC). It arose from a ground-breaking conference held at Bogor, Indonesia, and marked his publishing debut in a scientific field that has become his signature and passion. This volume is the second in a planned series bearing his name as Editor. The first, *Shifting Cultivation and Environmental Change: Indigenous People, Agriculture and Forest Conservation* was published by Earthscan in January, 2015. The third, *Farmer Innovations and Best Practices by Shifting Cultivators in Asia-Pacific*, is now in production. Dr Cairns is a freelance researcher who was most recently a Fellow at the Centre for Southeast Asian Studies (CSEAS), Kyoto University, Kyoto, Japan.

over my shoulder was then replaced by traps and a .22 rifle, and I continued to spend my time in the woods and along the river, now working outside of school hours to run a trap line for fur-bearing animals. So maybe it is not surprising that, years later, my academic interests were drawn to study people who live off the land, much as I had.

This shared background served me well in my later work with indigenous communities in Asia. Many years later, I recall one conversation in particular with an Agta villager in the Philippines. We were taking turns scratching diagrams in the dirt and discussing how we erected branch barriers to channel wildlife into our snares. He seemed puzzled that I would know anything about this, but was eager to trade stories.

In between these periods, I managed Willscott Holsteins, which was to become one of the highest-producing dairy herds on Prince Edward Island. Although not immediately obvious, I feel sure that this experience, of being a herdsman, stood me in good stead for my present task of assembling these books. Managing a herd of 70 milk cows (at that time) called for an ability to remember the individual needs of each and every one of them, depending on where they were in their life and lactation cycles. A first-lactation heifer might need a magnet down her throat, to lodge in her stomach, and capture any hardware that she might accidentally ingest in the coming years. An early-lactation cow that was working hard and mobilizing a lot of calcium from her blood might require a bottle of calcium administered intravenously or subcutaneously (depending on the urgency) to stave off milk fever (hypocalcaemia). Another cow that had already been milking for a few months might need to be bred, if she was properly in heat. A cow walking with obvious discomfort probably needed her hooves trimmed, and any foot rot drained. And yet another, at the end of her lactation, might need dry-cow treatment infused into her udder, to clear up any lingering mastitis problems.

It may be stretching a point, but (with the greatest respect) managing a large number of authors, all at different stages in their writing cycles, requires a similar attention to detail as managing a herd of dairy cows; there are the numbers involved, as well as the need to treat each one as an individual, independent of the rest, but joined by a common purpose. At the very earliest stages, some researchers still need to be invited to participate in the volume. Others, having already submitted their draft chapters, await feedback on how they might be improved or revised to better fit with the volume's theme. Some authors may need to be consulted about what botanical sketches they want included in their chapters; others may be asked to supply maps. And like newly freshened cows being placed into a milking line-up, the newly edited chapters are carefully organized into a Table of Contents – so perhaps our contributing authors should count themselves lucky that they weren't bred, jabbed or dewormed during the long process of getting this volume to press!

I trust readers will forgive my levity in explaining some of the distant background that has led to the birth of this volume.

The length of investment is also almost the same. It takes two to three years of rearing before a dairy heifer gives birth to her first calf and is ready to enter the milking string – and each of these swidden books have also needed almost three years of steady work before they were ready to send to the printer, and a new volume was born.

Setting the stage for conflict between shifting cultivation and conservation

Efforts to protect remaining wildlands and wildlife are often focused on remote areas where human populations and road infrastructure remain sparse, and the environment remains at its most pristine (Figure 1). In the tropics, these tend to be the very same areas where indigenous peoples depend on some variation of shifting cultivation to feed themselves (Figure 2). Governments commonly regard shifting cultivation as a primitive and inefficient form of agriculture that destroys forests, causes soil erosion and robs lowland areas of water supplies.

FIGURE 1: The forested mountains of Asia are often the ancestral homelands of shifting cultivators.

In an earlier draft of this preface, I had speculated that shifting cultivation's poor reputation might be a legacy from the colonial era, when it was poorly understood. In reviewing my draft, Dr Michael Dove rightly objected to this: 'I don't think shifting cultivation was simply misunderstood then, or now. I think the misunderstanding was an active construction, because shifting cultivators are hard to govern and to profit from.' (Dove, 2016; also see Scott, 2009). Professor Dove was absolutely right. I stand corrected.

These negative perceptions have since been called into question by research. Whilst many who lack sufficient understanding of shifting cultivation have condemned it as a threat to sustainable resource use, a growing body of evidence suggests quite the opposite (Cairns, 2015). Essentially, shifting cultivation is not nearly the dire problem that authorities believe it to be.

This growing realization means that we need to revisit our earlier misconceptions that saw shifting cultivation and sustainable resource use as mutually antagonistic, and consider the policy implications. Should policy really be designed to discourage shifting cultivation, or would a more enlightened policy try to support shifting cultivators and help to improve their standard of living? Research findings lead us to

FIGURE 2: Small fields carved into vast forests – the ideal conditions under which rotational shifting cultivation has historically been sustainable. After a brief period of cultivation, the forest rapidly reclaims the fields and the whole cycle is little more than a fleeting disruption to the forest ecosystem.

believe that official policy towards shifting cultivation needs to change away from the former and move towards the latter (Mertz and Bruun, ch 2).

The farmers to whom I refer – along with their families representing up to 200-million people in Asia alone (Karki, ch 49) – have relied for countless generations on shifting cultivation to feed themselves. Their farming systems are a pan-tropical land use that has evolved in Asia, Africa and Latin America, wherever indigenous peoples are living in forest environments.

Perhaps it should not be surprising that indigenous peoples, who generally try to avoid confrontation, have largely retreated into the same remote wilderness as wildlife populations. Both are attempting to escape contact with rapacious lowland dwellers and both have retreated further into the mountains. In the same way that we would preserve a tract of forest to protect an endangered animal, should we not equally set aside tracts of forest land for shifting cultivators as an endangered people – where their culture and relationship with their natural environment can also be protected? Could this not be regarded as a worthy goal for protected forest areas? Is cultural diversity within our own species not equally worthy of protection as the biological diversity between other species? Are shifting cultivators not worthy of saving? They certainly seem to provide a better example of mankind living in harmony with nature than the rest of humanity can muster! Or is this coming too close to relegating indigenous peoples to the role of 'museum pieces'? Some would argue that we should not be contemplating the 'preservation' of shifting cultivation as an antiquated museum piece, but should be helping it to evolve in response to growing populations, market stimuli and other pressures. But surely, its demise should not be hastened by hostile policies, and shifting cultivators should be assisted in adapting to a fast-changing world.

Confusion in terminology

As rotational shifting cultivation systems have come under closer scientific scrutiny over recent decades and the multiple functions of the 'fallow' phase of the shifting cultivation cycle have become clearer (Cairns, 2007b), we've become increasingly aware of the fallacy of referring to fallows as 'wasteland' or land that has been 'abandoned'. Fallows were long regarded as such by many people, but given our present knowledge, we can't even pretend that this is remotely true. To add to the confusion, any policies that issue edicts about 'shifting cultivation' or how 'fallows' should be used, immediately become mired in confusion over terminology. The variability of this age-old system of agriculture often defies easy definition.

I've probably contributed to this confusion myself in the *Voices from the Forest* volume, by conceptualizing 'jungle rubber' as an economic fallow (Cairns, 2007a). This has now been stretched to include oil palm plantations (Mertz, 2015), and I even read of 'fallows' being planted to ginger or opium poppies. Since both are clearly short-term cash crops and make no claims to rehabilitate the soil or perform other fallow functions, there seems to be no justification for calling them 'fallows'. In most cases, I assume that this is unintentional and simply a consequence of using English as a second or third language.

Inherent in this discussion is whether and when a planted fallow becomes a crop. If it regenerates spontaneously, is weedy in appearance, and has little commercial value, then we have little trouble in identifying it as a fallow. But at what point does an improved fallow – be it economically or biologically improved – become a crop in its own right? And by extension, when does land use become permanent and cultivation cease to shift? This is obviously not just a matter of simple semantics; it may be critical to survival in a policy environment that insists that agricultural land use must be permanent. Does vegetation actually have to be planted in order to be considered a crop, or will simply encouraging its spontaneous re-establishment in the fallow suffice?

An example comes to mind. By managing *Alnus nepalensis* in their dryland fields, the Angami Nagas of Nagaland, northeast India, provide a compelling example of how shifting cultivators can use a soil-building tree to intensify land use from shifting to permanent. I remember that during my days as a PhD student undertaking fieldwork in Nagaland, one of my advisory panel members at the Australian National University sometimes informed me gruffly that the dryland system I was studying was *not* 'shifting cultivation'. I never debated his contention because, first, I didn't disagree with him and, second, I saw it as a moot point. Regardless of how the system may now be classified, it certainly had its origins in shifting cultivation and it provided a fascinating case study of farmers who had successfully intensified their land-use system without a major disruption to their culture or economy. That was why it captured my interest, regardless of what it may have been called.

FIGURE 3: The same terminology can invoke quite different understandings among different discussants.

FIGURE 4: Linguistic profusion adds to the complexity.

This confusion reigns within the English language that serves as the *lingua franca* among many Asian countries (see Figure 3). The picture becomes considerably more tangled when the national and minority languages of these countries are added (Figure 4).

The confusion is further multiplied by the forces of 'political correctness' that attempt to give new names to old practices that state governments regard with heavy prejudice. In Thailand, for example, many NGO colleagues who work closely with shifting cultivators are now adamant that the old term '*rai luan loi'* (shifting cultivation) should be avoided and the practice should now be referred to as '*rai mun wian'* (rotational cultivation) instead (Trakansuphakon, 2013). But this new term, of course, could easily mislead people into believing that we are talking about rotating crops, instead of rotating land.

Almost in synchrony, similar changes in terminology are being adopted in Indonesia (Colfer, 2014), and for similar reasons. There, NGOs are avoiding use of the terms '*perladangan berpindah'* or '*ladang berpindah',* as these are viewed as having negative connotations. They instead prefer the terms '*ladang gulir balik'* and '*perladangan'* in Bahasa Indonesia, or when communicating in English, they favour 'the (name of minority group) indigenous farming system' or '(name of minority group) rotational farming' (Bamba, 2015). It can get complicated.

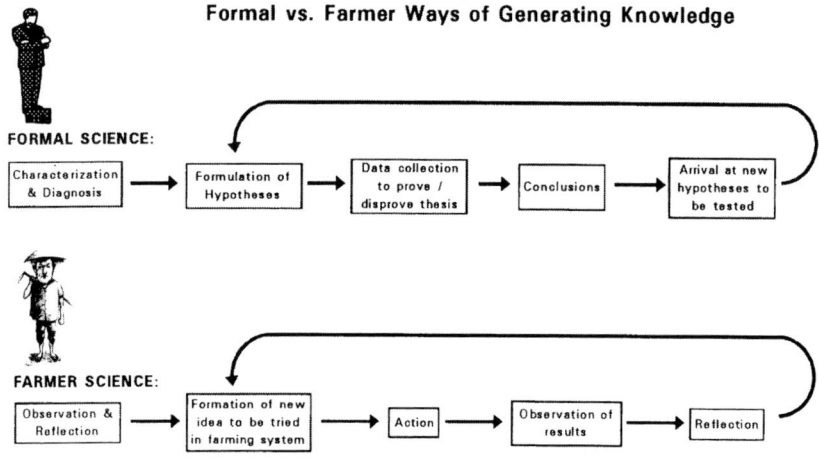

FIGURE 5:   There are clear parallels in how farmers and researchers learn.

The idea seems to be that if shifting cultivation is against the law or attracts instant prejudice, then just call it something else! Of course, similar tensions afflict English speakers. Most of us working with shifting cultivation try to avoid the 'slash-and-burn' moniker and all the negative imagery that it conjures up. Most researchers stick to 'shifting cultivation' or 'swiddening' as less value-laden terms.

Something else we hear far too often is that shifting cultivators need to adopt 'scientific farming methods', as if there is something inherently illogical or unscientific about what they do. This is symptomatic of the common lack of respect for the indigenous knowledge that has been developed by these farmers over centuries of trial and error (Figure 5). It may also reflect an attitude that in order to show themselves as progressive, farmers should grow their crops in nice straight rows of monocultures, and that the usually chaotic appearance of a swidden just can't be right. This view is, of course, consistent with the belief that shifting cultivators are senseless 'marauders of the forest', laying waste to the environment around them (Colfer and Dudley, 1993). These attitudes have spawned the often misguided policies under which shifting cultivators have suffered for so long.

Other problematic terms that are often heard in disparaging shifting cultivation are 'abandoned', 'idle' and 'wasteland'. The narrative here is that 'irrational' shifting cultivators regularly 'abandon' their fields, and leave this land 'idle' or as 'wasteland'. These labels, as misleading as they are, seem to be aimed at justifying governments' wishes to discredit shifting cultivators and expropriate fallowed lands. It helps them to obscure the obvious point that the fallow phase is fundamental to the centuries of proven sustainability of shifting cultivation. Resting this land is just good management, and there is nothing 'abandoned' about it. Nor is it 'idle' or even remotely 'wasteland'. Yet this mistaken notion lies at the heart of much of the conflict between shifting cultivators and policy-makers.

Moving on to other terms, 'agroforestry' is often touted as a wonderful, cutting-edge technology, while 'shifting cultivation' is condemned as a remnant of the Stone Age. Agroforestry is hot; shifting cultivation is not. People seem to forget that shifting cultivation is probably one of the oldest forms of agroforestry. And it is the one form of agroforestry that has probably been most tested over the centuries and across continents by countless ethnic groups, and has proven itself to be most acceptable to farmers. In trying to tap into the positive image of 'agroforestry', some advocates of swiddening prefer instead to refer to it as 'rotational agroforestry', which, of course, it clearly is. But we are continuing to play word games, instead of recognizing shifting cultivation for its own merits. I find it rather ironic that many resources have been invested in extending an agroforestry technology such as contour hedgerows (also known as 'sloping agricultural land technology' (SALT)), but despite its wide promotion, very few farmers have adopted it (Fujisaka, 1991). It remains largely confined to projects and the grounds of national agricultural research stations. Farmers have 'spoken clearly with their feet'.

FIGURE 6: Not many farmers would feel comfortable putting their own perspectives on paper for policy-makers to read. The next-best thing can be accomplished by researchers willing to work closely with farmers and to give them a voice in the policy debate.

In contrast, shifting cultivation is an agroforestry technology that has evolved independently around the world in the hands of hundreds of millions of farmers in both temperate and tropical zones and has been refined over centuries of trial and error. Rather than recognizing the wisdom and practicality of this age-old practice, the official policy response has largely been one of ignorance and repression. The problem of policy misfit is undoubtedly exacerbated in that politicians and policy-makers are typically from the politically dominant lowland groups and have little personal experience with upland peoples or the conditions under which they farm. With their cultures, food security and very survival under threat, swidden farmers have steadfastly refused to give up shifting cultivation and accept replacement technologies that outside scientists have tried to 'ram down their throats', including contour hedgerows. We need to look closely at policies that impact on shifting cultivation, and decide whether they are based on empirical evidence, or on stereotypes, a tinge of racism and prejudices passed down from colonial times. We owe it to farmers to get this right; they are the intended beneficiaries of our work. It is in their names that we receive funding to undertake our research (Figure 6).

The potential role of indigenous peoples in providing solutions to global problems

At the COP21 Climate Summit in Paris in 2015, one speaker after another emphasized how we need to look to indigenous peoples to learn how to take better care of our planet and its environment. Many of the world's foremost experts would argue that shifting cultivators should be viewed in this same light – as allies in the fight to protect the environment. Noam Chomsky is quoted as remarking, 'It's phenomenal all over the world that those who we call 'primitive' are trying to save those of us who we call 'enlightened' from total disaster.' In a similar vein, Dr Jane Goodall recently posed the question: 'Why is the most intelligent being ever to walk on Earth destroying his only home?' Dr Goodall was not referring to the habits and practices of indigenous forest-dwellers. As Noam Chomsky alluded, indigenous peoples are increasingly being looked to as examples of sustainable co-existence of humanity within our natural environment; as examples of low-impact livelihoods that have much to teach wider society. It is ironic that those who draft policies critical of shifting cultivation are often those who live most disconnected from nature, in the large urban centres that are a far more serious environmental problem. They never have to go further than the nearest air-conditioned supermarket for their food needs.

Since the COP21 meetings recently riveted the world's attention on climate change, I should also stress the vital role of shifting cultivation fallows in sequestering carbon from the atmosphere, and the fact that soils farmed under shifting cultivation contain far more soil organic matter (and thus, carbon) than most agriculture growing annual crops. In fact, recent data have shown that newly regenerating forests, such as those that spring up in swidden fallows, can absorb 11 times more carbon from the atmosphere than old-growth forests (Poorter et al., 2016). This shows that regenerating forests that are nurtured by shifting cultivators could play a critical role in mitigating climate change. Although shifting cultivators have been practising this for centuries, it has only recently become fashionably known as 'carbon farming'.

We also know that a landscape farmed under shifting cultivation retains far more tree cover and biodiversity and offers higher-quality ecosystem services than one that is permanently cultivated. Conserving forests and using them economically thus should be very attractive for the anti-climate change movement, so this places this book at the cutting edge of the global environmental agenda!

A combination of distance, mountains, cultural and language barriers makes it practically impossible for shifting cultivators to make their voices heard when governments are formulating policy, so this volume attempts the next-best thing. We have asked researchers working closely with shifting cultivators to speak on their behalf and present their perspectives in this book, within the confines of scientific objectivity.

The urgent need to re-examine policies that marginalize shifting cultivators

Whilst writing my PhD dissertation, I listened repeatedly to CDs of folk music from eastern Canada that reminded me of my roots. I recall the lyrics of one song in particular that paid tribute to the pioneers who had cleared the forests of Cape Breton and laid the foundation for society there as we presently know it. My dissertation was about shifting cultivators, so I couldn't help wondering why nobody was paying tribute to them in song. After all, here was a people who had played a similar pioneering role in their own countries, but instead of being praised in song, they are vilified and marginalized by state policies that see them only as a problem. This is the paradox that lay at the genesis of this book.

Across the Asia–Pacific region, shifting cultivation tends to be largely practised by ethnic minority groups in the uplands, and they are amongst the most marginalized of communities. Their lives are often made even more difficult because of the misguided policies that governments try to enforce on the shifting cultivation practices that they depend on for their livelihoods. In northern Thailand, for example, farmers have even been arrested and jailed for clearing their fields, in preparation for planting (Trakansuphakon, 2013).

Policy-makers in the Asia–Pacific region

FIGURE 7: The image of trees being felled by farmers undoubtedly plays a role in the bad reputation of shifting cultivation. But this is only a transient phase. Managed properly, these trees will regenerate back into secondary forest during the next fallow period and will sequester as much carbon as was lost into the atmosphere when the trees were felled and the field re-opened.

almost universally regard shifting cultivation as a problem that needs to be solved (see Figure 7). Shifting cultivators are often demonized as being responsible for some of the most important environmental problems of our times, including tropical deforestation, biodiversity loss, climate change and lowland flooding.

The policies that governments develop strongly reflect that attitude. This has reached a point where farmers continuing the land-use practices of their ancestors are now in danger of being arrested and incarcerated. Perhaps even worse, state governments are often inclined to legislate their ancestral lands out of their possession, or connive in it being 'grabbed' by state or corporate interests. The literature on shifting cultivation is full of condemnations of state policies that are badly wrong. Neighbouring countries often repeat the mistakes of the government next door, even when it is obvious that the neighbour's policies are not working. It is therefore not surprising that researchers working with shifting cultivation often focus on policy issues. It is an issue that sits squarely on the nexus of humanitarian needs and forest conservation, and invokes strong elements of climate change, food security, indigenous rights and social justice.

Although shifting cultivation has always been viewed negatively by government policy-makers, this attitude is becoming much more extreme in an era of climate change, when the release of greenhouse gases by shifting cultivators burning their fields is believed to add to the problem. This crisis appears to be reaching a climax. What has long been needed is a regional synthesis of this research, so that a bigger picture comes into focus.

In the swidden volumes that we have produced so far, we have not seen or felt such passion as expressed by many of our authors about the deplorable manner in which shifting cultivators have been and are being treated. The concealed atrocities that have been committed in the name of national development are appalling and the duplicity of political leaders shameful. We think that many of our authors have watched all of this happening to countless innocents and have felt powerless to speak out because of the normally clinical nature of scientific reporting. They now have the opportunity to reveal – not individually, but as a group – what has been happening, and to call governments to account for their actions and to allow common sense and humanity to solve the problems for which shifting cultivators stand wrongly accused.

The intended role of this volume in rectifying the problem

Whether speaking directly with shifting cultivators or reading academic papers about their land-use systems, the discussion often turns quickly to the problem of wrong policies, and how these not only add to the woes of shifting cultivators, but also ignore the proven positive attributes of swidden farming and instead act upon myth, prejudice, greed or simple misunderstanding.

In considering the darkness of this governmental landscape and, within it, the spectre of the atrocities that have been committed against shifting cultivators under the twin names of conservation and development, there seemed to be potential for great value in bringing together policy analyses from across the region within a single book cover, in an effort to tease some of the lessons out of our many mistakes. And plentiful those lessons should be, if we claim the capacity to learn from our errors. Therefore, this book attempts to gather together a wide swathe of policy experiences from many countries, and shares them across borders so that the entire Asia–Pacific region can learn from them. We also anticipate that there will be a great deal of extrapolation potential of lessons to other regions in the world where shifting cultivation is widely practised, whether they are in Asia–Pacific, Africa, or Latin America, so the benefits should actually be global. This is a social and economic challenge that is common to many developing countries.

From our project base in Chiang Mai, northern Thailand, we have reached out to the whole Asia-Pacific region, seeking analyses of experiences with shifting cultivation policy in different countries. Chapters have been contributed from Bhutan in the northwest to Timor-Leste in the southeast and most parts in between. Thus, we are dealing with a wide swathe of indigenous peoples throughout the region, and the scientists who are studying them.

This project began with extensive networking across the Asia-Pacific region with researchers, university professors, NGOs, projects, consultants, field-workers, and indigenous groups that work intensively with shifting cultivation. We urged them to contribute their voices as we tried to pull together the aggregate picture of what the region's experience has been with shifting cultivation policy. We invited prospective authors with data, insights and writing skills to contribute chapters on their experiences and research findings.

This volume represents the work of hundreds of serious on-the-ground experts, of whom no fewer than 127 joined in the actual writing. They are generally experts in their fields, many of them known throughout the world as leaders in their academic disciplines. They include agronomists, foresters, anthropologists, sociologists, soil scientists, economists, geographers, historians, development experts, land-use planners and others, all drawn by the opportunity to contribute their knowledge and experience about the effects of official policy on traditional shifting cultivators, and their suggestions for change. This is akin to a regional consultation that uses existing networks among scientists for collaborative learning, ranging from the reasons behind the making of policies, through implementation, and more often than not, to coping with the resulting mess.

The need for this analysis is made more urgent by the growing pressures weighing on policy-makers, ranging from climate change through loss of forests and land-use change, to market demands for expansion of biofuels. The seriousness of these issues is often and easily employed to sweep aside the legitimate interests of shifting cultivators who are occupying and using their ancestral lands. The consequent human suffering has been, and continues to be, extreme. The violation of their indigenous rights invites comparison with the situation of the North American native tribes who gathered to stand in solidarity with the Standing Rock Sioux in North Dakota, USA, to protest against the construction of the Dakota Access Pipeline (DAPL) across their ancestral lands that was under way at the time this preface was being finalized in September, 2016.

Hopefully, more enlightened policies will retreat from the marginalization of both shifting cultivators and the resource base on which they depend. They may not always receive *carte blanche* to cultivate fragile slopes, because well-founded restrictions that protect both their landscapes and their livelihoods seem well-advised. These would probably build upon traditional land-use regulations that shifting cultivators have themselves long imposed on their own activities.

One of this volume's clearest messages is that the same issues and challenges cut across most nations within the Asia-Pacific region, and solutions should be sought collaboratively. There is no need for each country to 'reinvent the wheel'. Any country can learn from its neighbours' experiences in striving for the right policies for shifting cultivation. Gathering these experiences from the pens of leading scientists and experts, and making them readily accessible in an attractive and readable fashion is what this volume aims to achieve.

This book will facilitate information transfers to researchers, development agencies, policy–makers, NGOs, and farmers in developing countries. We are hoping that the availability of these policy analyses will help governments to formulate better informed policies towards shifting cultivation, that will ultimately contribute to both poverty reduction amongst upland peoples, and improved conservation of the forests on which they depend for their livelihoods. This will be particularly important for the remote uplands of the region, where isolation and limited market access mean that most indigenous farmers continue to be heavily reliant on shifting cultivation.

This is an issue of vital importance to several hundreds of millions of the world's poorest farmers who rely on shifting cultivation to feed their families. At present, antagonistic state policies often deny indigenous upland communities the security to occupy and manage their ancestral lands in the ways that have long been their custom.

Getting policies on shifting cultivation right is vitally important, because this age-old system is much more than just a means of producing food and managing forests. Shifting cultivation is often central to the very cultures and identities of its practitioners. It is who they are. Their creation myths, agricultural calendars, festivals, tools, crops, ancestral lands, sense of identity, knowledge systems, world views, rituals, and other interactions with the spirit world are all inextricably interwoven with their shifting cultivation practices. To prohibit their practise of shifting cultivation is akin to banning the practice of their very culture – a kind of cultural genocide. It is now widely accepted that diversity leads to stability and a wide range of cultures, each supported by its own knowledge base and practices, is probably a healthy thing for this planet and its sustainable stewardship.

Part of this cultural loss is the loss of invaluable indigenous knowledge accumulated over successive generations of farming in the forest. If wrong policies are allowed to bring an end to shifting cultivation, the loss will not be limited to injuring the poor; it will mean the loss of hugely valuable indigenous knowledge and crop varieties. The swidden fields, that were once a cornucopia of agrodiversity (see Figure 8), are being replaced by monocultures of oil palm, rubber and GMO maize. Many traditional crop varieties, of which shifting cultivators were the stewards, now no longer exist (see Shen et al., ch 16). This impoverishment of both indigenous knowledge and crop varieties is a serious setback for all of humanity and our quest for food security.

FIGURE 8: Throughout the history of agriculture, swidden fields have probably often been the 'petri dishes' into which wild landraces have crept from surrounding environments and became domesticated into new crop cultivars.

A recurring theme in many swidden studies is the heavy participation of women in shifting cultivation (Colfer et al., 2015). This is to the extent that many swidden communities have local terminology that refers to shifting cultivation as 'women's cultivation'. This means that policies that marginalize shifting cultivation are disproportionately marginalizing women. These women are the custodians of some of humankind's most precious genetic resources – the planting materials for the rich variety of traditional crops that are planted in swidden fields (see Figure 8). This is not only a vital issue to the livelihoods of indigenous peoples – but is a gender issue as well (Figure 9). This raises the stakes for getting the policies right. It is not only an issue of protecting upland environments – but also one vital to the welfare of some of our planet's most vulnerable peoples.

There is a wide consensus today that shifting cultivation is declining, and with it, the livelihoods of hundreds of millions of the world's poorest people. That loss may lead to heightened economic and social disparity and displacement. Volume I – the predecessor to this book in a planned trilogy – presented overwhelming evidence that argued that the loss of shifting cultivation would be regrettable.

Governments have been struggling with the question of how to deal with shifting cultivation since colonial times, so what have we learned from these long years of trial and error? There is a vast amount of experience, spread across time and many nations in the Asia-Pacific region, to provide lessons for the future. We cannot let this be wasted, or we will be condemned to keep repeating the same mistakes. Assuming that a policy-maker is thumbing through this book in search of guidance, what are the

FIGURE 9: Firewood is an important by-product from opening fields for shifting cultivation, especially in areas where higher altitudes and/or latitudes necessitate wood for heating, not only cooking. Gathering firewood is usually the task of women and children.

lessons that we want him or her to take away? What have been the success stories, and what have been the failures? Do we have scientific data to support these policies? Have some countries been more successful with their policies than others? Are there some general, overall lessons to be learned? What are the major obstacles to getting effective policies in place? Are most countries sincere in their efforts to move away from old prejudices and adopt policies based on empirical evidence? What vested interests stand in the way of this process?

Aside from policy-makers, this book should be a treasure trove for the large community of NGOs, indigenous rights groups, project administrators, research scientists, academics, and students who work intensively with or study shifting cultivators and are concerned about environmental sustainability, food security, inequality and the protection of farmer livelihoods.

I find an argument made by Dr Oliver Springate-Baginski (ch 13) to be particularly persuasive, and not often articulated. He points out that the opening of forest patches for shifting cultivation is likely to be less damaging than comparable patches opened for either timber harvest or conversion to sedentary agriculture. The rotational tree plantation system of the forester is not very different than the land use of rotational shifting cultivators – except that the shifting cultivator includes a short cropping phase, usually of one or two years, between the opening of the forest patch and allowing it to regrow. And yet, foresters are among the most vocal critics of shifting cultivation. This begs the question of why the forester's management is considered acceptable – but the shifting cultivator's is not? What accounts for this hypocrisy? The most plausible explanation is that the forester simply views the farmer as an unwelcome encroacher on what should be his sole domain – rather than anything inherently wrong with shifting cultivation as a land-use practice *per se*. And, of course, governments favour the forester's world view because they want the tax revenues that timber sales provide. Both a strength and a misfortune of shifting cultivators is that their ancestral homelands are often rich in valuable natural resources – resources that the state wants. For fans of James Cameron's epic science fiction film, '*Avatar*', the plot should feel eerily familiar.

Regulating how land is used is not an attack on farmers

Fortunately, the choices open to policy-makers are wider than simply permitting or prohibiting; they may ask themselves under what restrictions shifting cultivation might be permitted. To impose controls on the use of land is completely sensible, if the benefits of a land use are to be enhanced and its drawbacks minimized.

As a native of Prince Edward Island (PEI) in eastern Canada, an analogy comes easily to mind. The red sandy-loam soils of Prince Edward Island have proven ideal for growing Irish potatoes (*Solanum tuberosum* L.), to the extent that potatoes have outgrown all other crops and are now the backbone of the Island's economy. Any Island politician who thought to give potato growers a hard time would be committing certain political suicide. Yet the provincial government has seen the need to impose restrictions on how and where potatoes and other row crops are grown. The following examples are taken from the Environmental Protection Act, the Agricultural Crop Rotation Act and the Pesticides Control Act of Prince Edward Island. (The precise wording that imposes each of the following laws can be found in Appendix A.):

1. planting of potatoes or other row crops is not permitted on land with a slope steeper than 9%;
2. land must not be cultivated closer than 15 metres from a water course;
3. potatoes should not be planted more than once every three calendar years on any plot of land greater than one hectare;
4. spraying of potatoes or other row crops is not allowed any closer than 25 metres from a water course or body of water; and
5. spraying of pesticides is prohibited on windy days.

Furthermore, each spring, when the frost of winter leaves the ground, weight restrictions are enforced on PEI's roads, so that heavily laden potato trucks do not break them up.

In later years, when one large agri-business gained an unhealthy dominance of the industry and threatened to convert PEI into one large potato field, policy-makers again had to intervene by limiting the land acreage that any single individual or corporation could own. This helped to keep the benefits of the potato industry in the hands of many smaller farmers – and not solely the domain of a large agribusiness. One could easily draw parallels with the large oil palm or rubber interests that are presently 'grabbing' land across Asia, often evicting shifting cultivators in the process.

Much like the government of Prince Edward Island did with potatoes and other row crops, governments in Asia-Pacific have the task of sorting out what restrictions might need to be imposed on shifting cultivation, if any at all. It's not an all or nothing equation.

Shifting cultivation is not a 'bogeyman' that needs to be eliminated by hostile policies. Conversely, reasonable policy restrictions should not be viewed as an unwarranted attack on shifting cultivators. The challenge is how to refine these policy instruments so that they don't impose undue hardships on shifting cultivators and so that there are trade-offs in tangible benefits in exchange for whatever restrictions are imposed.

Shifting cultivators often impose their own customary regulations on how and where swiddening is practised, backed up by a system of sanctions (Hoare, ch 48). These are akin to village-level policies and show that practitioners are themselves aware that if done wrongly, shifting cultivation can cause adverse impacts on their village and its resource base. These traditional regulations are based on generations of observations and have been closely tailored to the conditions of particular villages. Policy-makers, sitting in national capitals on the other side of the country, don't have that luxury of either time or long experience, and need to forge policies that are more widely applicable.

It would seem eminently sensible to begin by cataloguing village-level restrictions that shifting cultivators have themselves put into place, as a starting point to identify which of them should be scaled up to the national level and included in national policy. This would be an effective way of including farmers' voices and wisdom in the national policy debate.

After so much has been written about shifting cultivation, how can it still be so badly misunderstood? I am reminded of a personal story that helps to illustrate how the same evidence can lead different observers to drastically different conclusions. This event took place in 1995, during my first visit to Khonoma, an Angami village in Nagaland that is widely known for its intricate management of *Alnus nepalensis* (Himalayan alder) in its dryland fields.

I was standing near the village centre awaiting my return drive to the state capital, Kohima, and I was filled with excitement about what I had just witnessed in the village's dryland fields, and viewed as the 'holy grail' of all fallow-management systems that I had ever encountered! The lower mountainsides surrounding Khonoma are cultivated under shifting cultivation, locally known as *jhum*, and the *jhum* fields are scattered with the pollarded alder trees that the farmers of Khonoma have learned to manage so well (Figure 10). Nowhere else in Asia had I encountered such skillful management of a soil-building tree that it allowed such a dramatic intensification of land use! Khonoma's farmers reported that the alder trees allowed them to shorten their fallow periods to as little as two years, without appreciable drops in productivity! The ingenuity of this system had captured my imagination so strongly that I would devote the next several years of my life to studying it (Cairns, 2013). But as I stood there, lost in thought, I became aware that the village had another visitor, standing nearby, and I could overhear his conversation. He was a documentary film-maker from Delhi. Stroking the stubble on his chin thoughtfully, his eyes ranged over the alder *jhums* on the slopes around the village and he finally lamented, 'You can see where the natives are damaging the forest!'

I must have blinked a few times as his words registered in my mind. The very same alder landscape that was so admirable to me was something that he found deserving of condemnation! I saw agricultural fields, in which farmers had carefully nurtured alder trees to improve the soil and provide firewood. He saw a forest that was not as lush and dense as he expected, and so he concluded that the villagers needed to be stopped! I later pondered how easily this visitor could have used his film-making skills to make a damnatory documentary that could rally public opinion against the villagers who were managing the alders. And how very wrong it would have been!

FIGURE 10: Khonoma is well known for its intricate management of the alder trees that are scattered across its dryland fields.

This type of conflict is likely to become more frequent as urban 'environmentalists' increasingly venture out of their cities and into rural areas, and try to impose their views on rural populations. Farmers all over the world are struggling with this problem, in one form or another. The urban visitors know the least, and yet they are the most vocal in making their opinions heard. Much of the condemnation suffered by shifting cultivators is likely to come from this source.

This small experience in Khonoma has remained in my memory as a vivid example of how two different sets of eyes can look at exactly the same thing and see something completely different. Maybe this was symptomatic of why there are such divergent views about shifting cultivation!

States enact their policies by passing supportive legislation. These policies can be tremendously influential – in either a positive or negative sense – so it is imperative that the state gets them right. As an example, I recall that when I first arrived in Thailand in 1986 to work in dairy extension, the fledgling Thai dairy industry was being crippled by competition from cheap milk powder imported from New Zealand. It wasn't until the Thai government stepped in and passed legislation to force local manufacturers of dairy products to buy a certain percentage of their milk locally that the Thai dairy industry became viable. This was a happy story of the state getting a policy right, and the Thai dairy industry has never looked back.

We need to keep in mind, of course, that agricultural and forestry policies are not made in a vacuum, but in a world of vested interests, intense lobbying and often outright corruption. The political community in Washington DC, for instance, seems to have an unhealthy measure of support for a certain giant agribusiness that has poured a lot of money into politicians' pockets. In the Asian context, such powerful agri-businesses, with close ties to government, may have a vested interest in seeing shifting cultivators forced off their land, to become cheap labour for their plantations. Logging interests may regard the machetes of shifting cultivators as competition for their chainsaws, and with their wealth and cozy political connections, they may see it as in their interests to nurture a negative political attitude towards shifting cultivation. Or, most likely, Forestry Departments may try to expand their domains by arguing that swidden fallows should be classified as forests, and their use for agriculture prohibited. All of these powerful lobbying forces are likely to be acting against the interests of shifting cultivators. For their part, shifting cultivators tend to be poor, unorganized and parochial. They live on the fringes of society, geographically, economically and politically, and are unlikely to have either the political connections or the united voice needed to effectively lobby their case or protect their own interests.

We need to be cognizant of the signals, even unintended ones, that policies are sending to farmers and other land owners. I am reminded of a personal experience that is instructive. It took place during a time when I was living in Trang province, in southern Thailand. A local man had come to see me to inquire if I might be interested in buying a sizeable coastal property that he owned. He had fallen on hard times and was in need of cash. He was offering virtually beach-front property at a bargain price. I wasn't particularly interested in buying land, but was nonetheless curious enough to go and take a look at what he was offering. As I recall, his ownership documents for the land were flimsy. He did not have a clear land title – but he had documentation to show that he had been paying tax on the land for quite a number of years. In the absence of a proper land title, evidence of paying taxes on a parcel of land may be taken as tacit proof that the state accepts the claimed ownership of the land. Only scattered trees remained on the property. As we walked through the coarse grasses and bushes, the owner explained that the forest had been destroyed by repeated wildfires in recent years, thought to have been set by local hunters trying to flush out game.

As we walked the land, I pondered what I would possibly do with the property if I did buy it. I certainly had no interest in building a beach resort for over-pampered tourists. The one thought that did appeal to me was to rehabilitate the area's ecology, and to bring back the forest. But no sooner did this thought enter my head, than I knew that it would be exactly the wrong thing to do, given the lack of a proper land title. If I did succeed in restoring the forest to its former glory, then I had little doubt that the Department of Forestry would almost certainly reclaim the land as its own, and declare me an unlawful squatter on state land. Only if this land remained barren and degraded, could I be assured of keeping it. I've wondered many times since then how many shifting cultivators had undergone exactly the same thought process and acted accordingly. Such can be the perverse and unintended effects of policies.

Opium eradication projects as an extreme example of the expanding reach of the state

Although the search for effective policies towards shifting cultivation is an old one, the ability of states to enforce their regulations is much more recent. Shifting cultivators have traditionally operated in the most remote and inaccessible corners, often beyond the reach of state authorities. This probably changed most dramatically in northern Thailand with opium-eradication projects, when shifting cultivators had to get used to the idea that satellites could peer into their most remote fields and see what was planted there. Identification of a poppy (*Papaver somniferum*) field, by someone sitting at a computer monitor in Chiang Mai, would mean that the village could expect a visit from a helicopter full of soldiers, who would disembark long enough to slash the poppy crop to the ground. This new 'airborne pest' thus had to be factored into farmers' management plans. Proving yet again their ingenuity and responsiveness to new problems, some farmers began a strategy of relay-planting a second crop of poppies under the first, just as a PEI farmer might underseed a barley crop with clover. The soldiers, busily slashing the overstorey of poppies, would not notice the younger stand concealed beneath. With the over-storey of poppies reduced to slash, the soldiers would then file back into their helicopter and return to base. The person monitoring the satellite photos back in Chiang Mai would carefully cross that offending field off his satellite photo, as already having been eliminated, and with any luck, it might then fall outside his sphere of surveillance. Meanwhile, the understorey of poppies would eventually push through the slashed first crop and grow to maturity. How is it even remotely credible to say that these farmers are static and unchanging?

It is thus fitting that this volume includes a section of chapters specifically examining official policies to eradicate poppy cultivation – the notorious cousin of shifting cultivation – in Laos and Thailand (Suwannarat, Cohen, Ducourtieux et al., in ch 30, 31 and 32, respectively), and their impacts on former growers.

Now that transportation and communications infrastructure have penetrated into even the most remote corners and states are actually able to enforce their will on shifting cultivators, it is doubly important that states formulate sound policies that are underpinned by solid research and that the growing gravity of policy impacts on complex issues like food security, deculturation, deforestation, biodiversity loss and climate change are thoroughly understood. Now that state policies are finally enforceable, this book will strive to ensure that policy-makers see their responsibilities in a new light of modern science blended with age-old wisdom.

Acknowledgements

The period of fund-raising for this volume was a troubled time – when the world was trying to come to grips with a new scourge called 'ISIS', when the drums of war were beating loudly and when Syrian refugees flooded across Europe in search of sanctuary. The global economy was in the doldrums, widespread heat and drought seemed to herald the arrival of climate change in earnest, and the international squabble for land, water, and other scarce resources gathered pace. In short, the world seemed to be going into a tailspin, and there were far more problems than there was money to address them. It was an extremely difficult environment in which to be searching for funding. For this reason, we were particularly blessed to have several loyal supporters who had supported our work in the past and wanted to see it continue:

- the Australian Centre for International Agricultural Research (ACIAR), Australia;
- the World Agroforestry Centre (ICRAF) – Vietnam programme; and
- Willscott Farm, Prince Edward Island, Canada.

The support from both ACIAR and ICRAF went directly to the publisher as publishing subsidies. Willscott Farm, of course, is the Editor's home farm, and its support was a bit different. It provided monthly support that was used to pay for the project's ongoing costs. Together, it was the support of these three organizations that enabled the birth of this volume. In recognition of this, their logos are displayed on the volume's back cover.

One of the personal challenges in working on this volume has been in coping with the partial paralysis left by a devastating stroke back in 2009. Damage to the right side of my brain at that time has left the left side of my body paralyzed until today, and probably permanently. For starters, my formerly impressive keyboarding skills had gone out the window, and I was left tapping lamely at the computer keyboard with only one hand. Paralysis, of course, has posed a challenge to not only working on such an ambitious book project – but even to just surviving from day to day and trying to take care of myself. This brief explanation is necessary in order to understand the importance of the next several folk who I wish to acknowledge and thank. Chanchira Rattanamanachai (Fa) has played an important role in the development of this volume by working as my care-giver during the project. Her most important duty was showing up at my condo door every morning with several cups of coffee in hand. As a hopeless addict, caffeine was absolutely necessary to get my brain booted up in the mornings. As time has passed, Fa's duties have steadily evolved away from looking after me to spending more time running errands around town related to production of this book. This has helped me to maximize the amount of time spent at my computer – definitely not good for my recovery, but necessary for completing the book.

When I needed to emerge from my 'cave' to run my own errands, Yasin Norasri (Montree) was my faithful driver. Montree is from Amnat Charoen province in Thailand's impoverished Northeast region, and I've never known him to be in a bad mood. He always has a ready smile.

An even more important person, in terms of keeping my mind and body fit for work, is Suthida Chantamanas (Ae). Khun Suthida has been my physiotherapist, and even more importantly, a steadfast friend, during a dark period of my life when I needed both badly. Her constant good cheer and radiant smile could lift the darkest mood. She is a natural-born healer. Since I didn't have a chance to stray from my computer very often, Khun Suthida would even sometimes bring along her two beautiful daughters, Gracie (9 years) and Faith (almost 6 months), to visit me so that I could enjoy their company, watch them grow, and be a doting uncle. Gracie provides lots of warm hugs that make one forget about the world's troubles. Faith (now 6 months old) seems to reckon that if she tugs my beard enough, it may stimulate some good ideas that will improve the book. Both are delightful.

Through the combined efforts of Khun Suthida and Fa, I was able to persevere in working on this volume for the almost three years that it has taken to complete it.

It would be dishonest for me, as Editor, to claim too much personal credit for this trilogy of books about shifting cultivation. I can only point humbly to the exceptional team that I have had working beside me. If I may use a metaphor, the plough was far too heavy for a single horse to pull – and a crippled one at that!

One of our major advantages in working on this volume has been that we've had almost the same team of horses in the stable as for the previous two. It was familiar ground and we had ploughed it before. Straining mightily into their traces, abreast of me, were my two main collaborators in this work, Bob Hill and Tossaporn Kurupunya. The uniform readability of this volume, from beginning to end, is owed completely to the painstaking copyediting of Bob Hill, as he made his way through the daunting pile of chapters, one at a time. That the formatting of the book looks so neat and impressive showcases Tossaporn's talents. She did much of her computer work in the dark of night, as she waited for her rubber trees to finish dripping their latex into the collection cups. The rumble of thunder signalling approaching rain would trigger a quick shut-down of Tossaporn's computer and send her dashing out into the darkness to try to collect her rubber before the rain destroyed her night's production. Such is the life of a rubber farmer; it was between her periods of work in her plantation that Tossaporn switched on her computer to work her formatting magic on these pages. I can only describe the efforts of Bob and Tossaporn, in working on this book, as nothing short of heroic.

Khun Tossaporn's home province of Trang, in southern Thailand, was historically one of the pioneers in rubber production in Asia, but given the expansion of rubber plantations documented by many of the chapters in this volume, there is now a glut of rubber on the market, and the price of rubber has now fallen to a point where it is hardly worthwhile anymore for farmers to get out of bed in the middle of the night to tap their trees. But a little income is better than no income at all, and so most rubber farmers continue to work for smaller and smaller returns.

Although many reports speak glowingly about how the expansion of rubber cultivation has brought increased incomes to farmers in new areas where rubber has been introduced, they seldom note the impoverishment that the resulting price collapse has brought to other areas where rubber cultivation has long been a tradition. This was entirely predictable because all of the work went into expanding the supply, while nothing was done to increase demand. It was hardly rocket science. And that's to say nothing of the environmental impacts of converting such large areas of natural forest, rich in biodiversity, into rubber monocultures.

The dedication with which both Bob and Tossaporn have invested their time and talents in this book cannot be explained by the meagre payments they received. They had adopted the project as their own and were determined to stand beside me in making it as successful as possible. This is the main factor that made this trilogy of books about shifting cultivation exceptional. Money couldn't buy the talents and dedication that I had working beside me.

This volume has been a continuation of our efforts to mix strong chapters with striking artwork. The artwork is of two main styles, from two very talented artists. Again, most of the artwork comes from the same artists who contributed to past volumes. The striking charcoal sketches that appear on the volume's front cover and as section-openers within the book are primarily the work of Paradorn Threemake, probably the most talented of the many Thai artists who base themselves at Chiang Mai's Night Bazaar. This time, though, Paradorn's work was supplemented by a young, up-and-coming artist, Suvarin Suponsin (Musy), from Jom Tong in Chiang Mai province. She stepped in to assist briefly when Paradorn wasn't able to complete the artwork at the pace that we needed.

The second style of artwork is the ink line-drawings of key plants that are discussed in the chapters, to help familiarize readers with their appearance. Here, we again called on the talents of an old friend from my days with the World Agroforestry Centre (ICRAF) in Bogor, Indonesia. Working from afar, and fielding our sketch requests by email, Pak Wiyono's was the patient hand behind all of the botanical sketches. Wiyono's artwork helps to bring alive the botanical components of the chapters, and make them more accessible to readers without training in botany.

Besides assisting readers to visualize what plants are being discussed, we also wanted them to be able to locate the geographic focus of each chapter's research at a glance. We therefore tried to make sure that every chapter had a professionally drawn map to show its research site(s). We were fortunate to recruit a talented Shan (Thai Yai) cartographer by the name of Nang Zarm Moun Hseng (nickname: Ying Tzarm), from Shan State in Myanmar, to join our production team. Ying works for The Border Consortium, a non-governmental organization in Thailand, and fitted her cartography work for us in between her other duties.

As we neared the end of the project, though, Ying's duties with her NGO employer left her with little time to make maps for us, and we were forced to find a replacement. This 'dark cloud' proved to have a substantial 'silver lining' however, because it led us to a very capable replacement, Peter Elstner. Peter is a long-term resident of northern Thailand, and had previously worked with '*The Uplands Program: Research for Sustainable Land Use and Rural Development in Mountainous Regions of Southeast Asia*'. This was a research collaboration between Hohenheim University in Germany and several universities and research institutes in Thailand and Vietnam, which ran from 2000 to 2012. Peter has been a consummate professional in completing the volume's cartographic needs, and has been a genuine pleasure to work with.

In my efforts to give credit and thanks where they are due, I attempted to reflect back on the long path travelled in producing this book, and to give special mention to those contributors who went above and beyond the call of duty in making it a success. Those who came most readily to mind were a number of contributors who came out of retirement to share their experiences. Most of these conscripted retirees were key actors during the busy period of highland development projects that took place in northern Thailand in the 1980s and 1990s, and now live in Chiang

Mai. So, to Dr Gary Suwannarat, Peter Hoare (two chapters) and Garry Oughton, I would like to say an extra special 'thanks!' This volume is much richer for having received their participation. The next generation needs to learn from their wisdom and experience before it is lost in the quiet of retirement.

Among this elite group, one Australian in particular deserves to be singled out for special gratitude. Almost unbelievably, Garry Oughton wrote his chapter without the benefit of vision! Garry, God bless him, relies on Mac-based computer software for the blind called 'Jaws', to help him read and write, and what he lacks in eyesight, he more than makes up for in dedication and perseverance.

In acknowledging outstanding contributions, I would be remiss not to mention our most prolific author. Dr Olivier Ducourtieux, an Associate Professor at Agro Paris Tech in France, is one of the world's foremost experts in the field of shifting cultivation study, particularly as it is practised in the Lao PDR, and has shared his research findings in no less than three compelling chapters in this volume (ch 15, 32 and 37).

Another group of authors who deserve special commendation are those who wrote the opening and concluding chapters. Theirs was the all-important job of both setting the scene and synthesizing the final remarks. We employed two chapters for either task, to ensure that the book has a robust beginning and end. We owe special thanks to Drs Gerard Persoon and Jan van der Ploeg, both at Leiden University (ch 1), and to Drs Ole Mertz and Thilde Bech Bruun, both at the University of Copenhagen (ch 2), for setting the stage so effectively with their introductory chapters. The most challenging task was probably that of our two concluding authors, who had to read and synthesize the volume's entire contents before they could formulate their own chapters. It was a special treat to work with Dr William Found because he had been a favourite professor during my time as a student at York University in Toronto. These synthesis chapters are vital because many readers who are too busy to read the entire volume will flip directly to the concluding chapters to try to pick up some of the main points. For these reasons, we owe immense appreciation to Dr Ken MacDicken (ch 51), most recently of the Food and Agriculture Organization (FAO) of the United Nations, in Rome, and to Dr William Found (ch 52), Professor Emeritus at York University, in Toronto, for their very capable jobs of anchoring the volume. They have the final word.

References

Bamba, J. (2015) Personal communication between John Bamba, Chair of the Consortium of Pancur Kasih Empowerment Movement (GPPK) in Pontianak, West Kalimantan, and the Editor

Cairns, M. F. (2007a) 'Conceptualizing indigenous approaches to fallow management: A road map to this volume', in M. F. Cairns (ed.) *Voices from the Forest: Integrating Indigenous Knowledge into Sustainable Upland Farming*, Resources for the Future Press, Washington, DC, pp16-36

Cairns, M. F. (ed.) (2007b) *Voices from the Forest: Integrating Indigenous Knowledge into Sustainable Upland Farming*, Proceedings of a regional workshop held in Bogor, Indonesia, 23-27 June 1997, Resources for the Future Press, Washington, DC

Cairns, M. F. (2013) 'The Alder Managers: The Cultural Ecology of a Village in Nagaland, N.E. India', PhD dissertation to Anthropology Department, Research School of Pacific and Asian Studies, Australian National University, published online as part of the Digital Himalayas Rare Books and Manuscripts collection, http://www.digitalhimalaya.com/collections/rarebooks/

Cairns, M. F. (ed.) (2015) *Shifting Cultivation and Environmental Change: Indigenous People, Agriculture and Forest Conservation*, Earthscan, London

Colfer, C. J. P. and Dudley, R. G. (1993) *Shifting Cultivators of Indonesia: Managers or Marauders of the Forest? Rice Production and Forest Use among the Uma' Jalan of East Kalimantan*, Food and Agriculture Organization of the United Nations, Rome

Colfer, C. J. P. (2014) Personal communication between Dr Carol J. Pierce Colfer, Senior Associate at the Center for International Forestry Research (CIFOR) in Bogor, Indonesia and Visiting Scholar in Cornell University's Southeast Asia Program in Ithaca, NY, and the Editor

Colfer, C. J. P., Minarchek, R. D., Cairns, M. F., Aier, A., Doolittle, A., Mashman, V., Odame, H. H., Roberts, M., Robinson, K. and Van Esterik, P. (2015) 'Gender Analysis: Shifting cultivation and indigenous people', in M. F. Cairns (ed.) *Shifting Cultivation and Environmental Change: Indigenous People, Agriculture and Forest Conservation*, Earthscan, London, pp920-957

Dove, M. (2016) Personal communication between Editor Cairns and Dr Michael Dove, an anthropologist teaching at the Department of Anthropology at Yale University in New Haven, CT, on 28 September, 2016

Fujisaka, S. (1991) 'Thirteen reasons why farmers do not adopt innovations intended to improve the sustainability of upland agriculture', in *Evaluation for Sustainable Land Management in the Developing World*, IBSRAM Proceedings 12(2), International Board for Soil Research and Management (Thailand), Bangkok, pp509-522

Mertz, O. (2015) 'Oil palm as a productive fallow? Swidden change and new opportunities in smallholder land management', in M. F. Cairns (ed.) *Shifting Cultivation and Environmental Change: Indigenous People, Agriculture and Forest Conservation*, Earthscan, London, pp731-741

Poorter, L., Bongers, F., Aide, T. M., Zambrano, A. M. A., Balvanera, P., Becknell, J. M., Boukili, V., Brancalion, P. H. S., Broadbent, E. N., Chazdon, R. L., Craven, D., de Almeida-Cortez, J. S., Cabral, G. A. L., de Jong, B. H. J., Denslow, J. S., Dent, D. H., DeWalt, S. J., Dupuy, J. M., Duran, S. M., Espirito-Santo, M. M., Fandino, M. C., Cesar, R. G., Hall, J. S., Hernandez-Stefanoni, J. L., Jakovac, C. C. et al. (2016) 'Biomass resilience of neotropical secondary forests', *Nature* 530, pp211-214

Scott, J. C. (2009) *The Art of Not Being Governed: An Anarchist History of Upland Southeast Asia*, Yale University Press, New Haven, CT

Trakansuphakon, P. (2013) Personal communication between Dr Prasert Trakansuphakon, Regional Director of the Indigenous Knowledge and People's Network, and the Editor

Appendix A: Excerpts from the relevant statutes of Prince Edward Island, Canada

1. '*Prohibiting on a provincial parcel of land, the cultivation of one or more hectares of a row crop on any area of that parcel which has a slope greater than 9%.*' – Environmental Protection Act: http://www.gov.pe.ca/law/statutes/pdf/e-09.pdf

2. '*No person shall, without a license or a Buffer Zone Activity Permit, and other than in accordance with the conditions thereof, alter or disturb the ground or soil within 15 metres of a watercourse boundary or a wetland boundary.*' – Environmental Protection Act: http://www.gov.pe.ca/law/statutes/pdf/e-09.pdf

3. '*No grower shall plant and no landowner shall permit regulated crops to be planted on any area of land greater than 1.0 hectares at any time for more than one calendar year in any three consecutive calendar years.*' – Agricultural Crop Rotation Act: http://www.gov.pe.ca/law/statutes/pdf/a-08_01.pdf

4. '*No person shall permit or cause any sprayer or other equipment used to apply pesticides to be (a) filled, discharged, washed or flushed out within 25 metres of the water's edge of any open body of water.*' – Pesticides Control Act: http://www.gov.pe.ca/law/regulations/pdf/P&04G.pdf

5. '*No person shall apply a pesticide, (a) in a dry formulation; or (b) in a liquid formulation that is under pressure, when the wind speed, measured at the point of application, exceeds 20 km/h.*' – Pesticides Control Act: http://www.gov.pe.ca/law/regulations/pdf/P&04G.pdf

I. INTRODUCTORY SECTION

A Naga shifting cultivator returns from her field, her basket a virtual cornucopia of fresh produce. High agrodiversity is one of the many positive aspects of shifting cultivation that should not be lost.

Sketch based on a photo by Malcolm Cairns.

A. Setting the stage

There is typically a strong spiritual dimension in shifting cultivation. Many shifting cultivators believe that their environments are controlled by guardian spirits that must be appeased before the forest or land can be disturbed. Immediately after clearing her new swidden, for example, this shifting cultivator in Bukidnon province, the Philippines, built this altar to perform rituals that are meant to seek the spirits' permission to use the field. It is fear of retribution from the spirit world that compels many shifting cultivators to adopt conservation-oriented practices and try to avoid any serious damage to the environment. Shifting cultivators from the Philippines lowlands are generally Christians, and are not restrained in this way.

Sketch based on a photo by Malcolm Cairns.

1

FIGMENTS OF FIRE AND FOREST

Shifting cultivation policy in the Philippines and Indonesia

*Jan van der Ploeg and Gerard A. Persoon**

FIGURE 1-1: Azryl Alejandrino.

Luzon, the Philippines

A woman walks along a slope under the hot sun. She methodically thrusts a pointed stick into the blackened, bare soil. Tree stumps stand scattered across the field. There are smells of smoke and sweat. In the distance there are towering blue mountains and the call of a hornbill is carried on the still air from the nearby forest.

These are seemingly timeless images, reminiscent of our origins as shifting cultivators. But look more closely, for nothing is as it seems. Azryl Alejandrino is a school teacher at Diwagden, a remote hamlet in the northern Sierra Madre mountain range on the island of Luzon, in the Philippines. After her young pupils have left for the day, she goes to the field to help her husband planting yellow corn. To purchase the hybrid seeds, fertilizer and herbicides, Azryl took a loan from a corn buyer in the town in exchange for the right to procure her crop. After the harvest, the dried kernels will be transported to town in old logging trucks, and used in the production of feedstock for pigs. The local government has constructed a farm-

* Dr Jan van der Ploeg, WorldFish, Solomon Islands, formerly Lecturer in Environmental Anthropology, Institute of Cultural Anthropology and Development Sociology, Leiden University, The Netherlands; Dr Gerard A. Persoon, Professor of Environment and Development and Scientific Director, Institute of Cultural Anthropology and Development Sociology, Leiden University, The Netherlands.

to-market road to facilitate the transport of farmers' produce, and a flurry of cultivation has accelerated the clearing of Diwagden's remaining forest.

Look again. The felled tree is not a native species, but *gmelina*, a fast growing exotic tree used for reforestation. Twenty years ago Azryl's parents applied for a so-called stewardship contract, a tenure instrument issued by the Department of Environment and Natural Resources on the condition that the area remained forested. The foresters urged them not to expand their swiddens

Gmelina arborea Roxb. [Lamiaceae]

Used in an early reforestation programme to replace swidden farming. When harvested, farmers made barely enough from selling the timber to cover their costs.

and instead to plant *gmelina*. But when the trees were big enough to harvest, timber prices were so low that they barely covered the costs. Thereafter, Azryl's parents spurned the conditions of the stewardship contract and, like most of their neighbours, cleared their land to cultivate upland rice and cassava. When Azryl got married her parents gave her the land as a wedding gift. She realizes that the land is not legally hers, but she is not very concerned. After all, most farmers in the municipality do not have land titles and if a land dispute flares up, official papers will make little difference to the outcome.

Look at Azryl. Her parents were landless peasants from Ifugao province who settled in Diwagden in the early 1970s. They bought land from an Agta family, the indigenous hunter-gatherers of the Sierra Madre. The thumb-marked contract is kept by Azryl's mother in a waterproof container under the bed, together with the vaccination records and college diplomas of her seven children. As a young girl, Azryl was sent back to live with her grandparents in Ifugao province, where she went to school. After graduating from college and getting married, she returned to Diwagden. The local government had constructed a school building in the village and Azryl was offered a temporary contract. She considers herself lucky: finding a job as a graduate is not easy.

Think about Azryl's circumstances, and realize that forest policies mean very little to her. The prohibition on shifting cultivation never inhibited Azryl's parents from burning their *kaingin* – a word used in the Philippines for swidden farming, and also for the swidden itself. Community–based forest management, often heralded as a paradigm shift in how the state views and treats swidden farmers, made little difference in her life. Azryl does not know that the forests around the village were officially proclaimed as the Northern Sierra Madre Natural Park, and she is unaware of government plans to issue a Certificate of Ancestral Domain Title to the Agta in

Diwagden. Take a close look at all of these laws, regulations and presidential decrees and realize that here, on this remote sun-burnt slope, they are not what they seem to be.

Sumatra, Indonesia

Another image, another country. A few men and boys sit lethargically alongside a road. The children, barefoot, scabby and skinny, listen to the tales of their grandfather. They are an isolated, destitute group of shifting cultivators in a

FIGURE 1-2: Maritua (second from the right) and some of his relatives.

wasted land. Little grows in the parched soil. Burned hardwood lies scattered around. In the background is a remnant of the once rich forest, destroyed by ignorance and poverty.

But look closely and a different narrative emerges. The man with the wristwatch is Maritua, an *Orang Rimba*. His people, once known as the *Kubu*, but these days more often referred to as the *Orang Rimba* (forest people), are hunter-gatherers in the rainforests of central Jambi province on Sumatra. Maritua no longer hunts pigs and deer. As the forest disappeared he had to look for alternatives. He and his family now live mainly on handouts from companies, government agencies and non-governmental organizations. Just before the photograph, Maritua arrived on a new motorbike he bought with money he received from Sinar Mas, the company that planted rubber and oil palm on his land. He thinks it was a good deal: he can now afford *Gudang Garam* cigarettes, instead of cheap brands. Every month he travels on his motorbike to the company's barracks, where he collects cash. Arriving home, he quickly sheds his trousers and relaxes back into his familiar loincloth.

The landscape is not a swidden, but an industrial-scale tree plantation. Bulldozers have scraped the land and Javanese migrant labourers have made large piles of the debris. It is left to rot because the use of fire is forbidden by the Ministry of Forestry. Small rubber seedlings are planted between the piles. Tractors pass by daily with laborers on their way to the edge of the forest to clear new areas and plant more rubber. Sometimes, Maritua sits alongside the road waiting for traders to collect the rattan fruit called *jernang*, which are used to produce dragon's blood resin. In exchange he gets cigarettes, coffee and sugar. Maritua harvests the fruit in the Bukit Dua Belas National Park. Although it is a protected area, the park is also a *cagar budaya*, or cultural reserve, and the *Orang Rimba* are entitled to harvest non-timber forest products there. On the edge of the park, Maritua and his relatives have opened new plots and planted rubber. He knows it is illegal, but he says he has no choice because

Daemonorops draco (Willd.) Blume
[Arecaceae]

Seeds from this rattan species are used to
produce 'dragon's blood' resin.

wild tubers from the forest no longer yield sufficient food for his family. That is only one side of the story: he also burns the forest to lay claim to the land.

Maritua's situation illustrates the complexity and contradictory character of policies related to shifting cultivation in Indonesia. For a long time, forest-dwellers like Maritua were considered squatters on state-owned land, responsible for forest fires and erosion. To prevent forest loss the government tried to resettle the so-called *orang liar* (wild people). But Maritua was among those who did not accept the offer of a new home. He prefers to live on what the forest provides: tubers, fruit, honey, meat, rattan and nowadays also rubber, oil palm and handouts. 'The forest is my home,' he says. At the same time, the government encourages the conversion of forest into industrial-scale rubber and oil palm plantations. The *Orang Rimba* must now survive in the few remaining forest patches. Maritua's new motorbike conceals the seriously impoverished state of indigenous forest dwellers on Sumatra. Maritua has lost his home.

These two images portray daily life on the forest frontier in two different countries. The stories reveal rapidly changing land-use patterns, social networks and livelihood strategies. They tell of two people confronted with complex and contradictory government programmes on what should be done in, and with, the forest. This chapter describes the history of shifting cultivation policies in the Philippines and Indonesia, and its impact on ordinary people like Azryl and Maritua.

Shifting cultivation policy in the Philippines

Shifting cultivation was the predominant land use in the pre-colonial Philippines. Communities worked collectively on their swiddens, in ritually sanctioned cyclical systems that involved clearing and burning forest vegetation and cultivating multiple crops for two or three years, followed by a prolonged fallow period. Technology was rudimentary: farmers mainly used axes, dibble sticks and knifes. Large trees were often left standing on the fields: the name *kaingin* literally means to cut branches and small trees (Scott, 1994, p199). Millet, taro, yams, rice, bananas and in some areas sago, were important crops. Fields were usually held in usufruct: households could cultivate the land until the soil was exhausted. More permanent rights existed for fields that required more substantial labor investments: irrigated fields and areca palms could, for example, be alienated, leased or inherited. With low population densities and large tracts of residual forest remaining, the environmental impacts of these farming systems were small (Mazoyer and Roudart, 2006).

Colocasia esculenta (L.) Schott
[Araceae]

Taro was a prominent swidden crop in pre-colonial Philippines.

In 1521, Ferdinand Magellan claimed all the lands of the Philippine archipelago for the Spanish Crown. By the mid–17th century, plants from the New World, such as sweet potatoes, corn, cassava, cacao and tobacco had become dominant crops on swiddens throughout the country (Scott, 1994). In theory, the Laws of the Indies, which governed Spanish possessions in the Philippines, explicitly recognized indigenous land-use practices and rights (Lynch, 1982). But in reality the *reducción* system forcibly resettled people in villages in the lowlands in order to facilitate the conversion of, and extraction of tributes and labour from, the *indios*, as the native inhabitants of the Philippine Islands were called by the Spanish *conquistadores*. In the 18th century, large tracts of land were appropriated by religious orders and indigenous elites, mainly for

the production of tobacco, sugar cane, copra or other plantation crops. In the lowlands, most people were in effect reduced to tenants working on the *haciendas* (Larkin, 1993). People who resisted colonial rule fled to the forested mountains, thereby creating the artificial divide between shifting cultivators in the uplands and rice farmers in the lowlands that has haunted policy-makers in the Philippines ever since.

In the 19th century, commercial logging and agricultural development, fuelled by a rapidly growing population, resulted in extensive

Metroxylon sagu Rottb. [Arecaceae]

Sago was an important pre-colonial swidden crop that was discouraged in favour of rice because of its low mainstream cultural status.

deforestation in Luzon, Mindoro and the Visayas (Bankoff, 2004). Uncontrolled timber extraction and shifting cultivation caused massive erosion, siltation, flash floods and changes in precipitation. In response, the Spanish colonial government founded a forestry service in 1863 called the *Inspección General de Montes*. Its primary task was to regulate timber harvesting. The Spanish colonial foresters regarded clearing of forests for *kaingin* as a waste of valuable timber and the primary cause of forest destruction, and aimed to resettle the *kaingineros* in the lowlands. The 1889 Definite Forest Laws and Regulations explicitly prohibited swidden farming in the uplands. But a structural lack of manpower, low morale, petty corruption, institutional conflicts and a growing civil insurgency inhibited the implementation of this policy by the colonial foresters – a pattern that would often repeat itself in the 20th century (Bankoff, 2009).

In 1898, the Spanish ceded the Philippines to the United States, and the new administration formed the Insular Bureau of Forestry, which was tasked with encouraging the rational exploitation of the colony's forests and regulating the conversion of forest into farmland.[1] The colonial foresters estimated that at the turn of the century more than 40% of the archipelago was covered by grasslands, and attributed this largely to *kaingin* (Whitford, 1911). Although the Kaingin Law of 1901 (Act 274) reaffirmed the ban on shifting cultivation, the American foresters lacked the ability to effectively enforce the law (Bankoff, 2013). The priority of the bureau was to facilitate the harvest and export of timber by American logging corporations. American colonial attitudes towards swidden farming were rooted in a mix of racism and evolutionism, which saw shifting cultivation as a primitive stage in societal progress. For example, Dean Worcester, the influential Secretary of the Interior for the Insular Government, wrote (1914, p609):

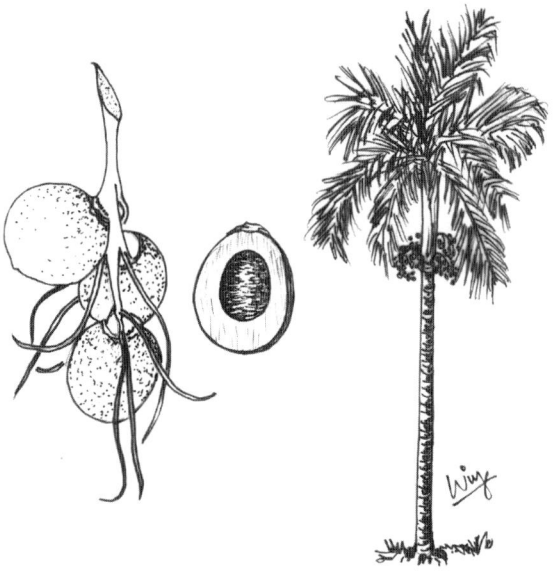

Areca catechu L. [Arecaceae]

Areca (betel nut) palms were one of the early bases for permanent ownership rights. They were increasingly integrated into subsistence systems as a permanent tree crop.

> One thing which renders it difficult to deal with some of the Filipinos [...] is that in its more remote districts they are showing a marked tendency to scatter out into the forests where they make *caiñgins*,

or forest clearings, and live in tiny huts. Little by little they are gravitating back to the barbarism from which they originally emerged.

These views legitimized colonial policies that aimed to resettle shifting cultivators from public forest lands. Paradoxically, the American colonial government also actively stimulated the study of the livelihoods and customs of these primitive peoples: anthropologists such as Merton Miller (1912), John Finley (1913), Fay-Cooper Cole (1913, 1922), Laura Benedict (1916), Ray Barton (1919) and John Garvan (1931) greatly contributed to our understanding of shifting cultivation systems in the Philippine archipelago. In many upland areas, livelihoods then consisted of a mix of *kaingin*, wet-rice cultivation, fishing, hunting and trade in forest products.

After Philippine independence was recognized in 1946, the prejudice towards swidden agriculture remained a persistent characteristic of the country's forest policy. But scientists increasingly challenged the conceived wisdom that shifting cultivation was by definition unproductive and a cause of environmental degradation. For example, Harold Conklin's report on the Hanunóo in the uplands of Mindoro (1957) demonstrated that shifting cultivators had both remarkably high returns to labour and detailed knowledge about their environment. A national conference on the '*kaingin* problem' in 1965 brought together representatives from government, academic institutions, civil society organizations and multilateral donor agencies, and led to the Forest Occupancy Management Programme in 1971 and eventually to the passage of the Revised Forestry Code in 1975 (Magno, 2001). It marked a policy shift, from ejecting farmers who were using public land towards regulating their presence:

> [*K*]*aingineros*, squatters, cultural minorities and other occupants who entered into forest lands […] shall not be prosecuted: provided that they do not increase their clearings [and] that they undertake […] activities to be imposed upon them by the Bureau in accordance with the management plan calculated to conserve and protect the forest resources. (Presidential Decree No. 705: section 53)

In the 1960s, the Philippines became the largest exporter of timber in Southeast Asia. Cronyism and corruption characterized the corporate logging industry during these 'years of plunder' (Broad and Cavanagh, 1993). Forest cover declined rapidly, from 45% in 1957 to 22% in 1987, with the remaining forest severely degraded. The Forest Management Bureau consistently attributed forest loss to shifting cultivation, at least partly to conceal its failure to regulate the logging industry (Kummer, 1992). Rapid population growth, land scarcity and the construction of logging roads led to migration to the remaining forest frontiers on Luzon, Mindanao and Palawan, thereby fundamentally transforming land use and social relations in these remote areas. Although these settlers practised the same slash-and-burn techniques to clear forest vegetation and also maintained long fallows to restore soil fertility, they were much more an integral part of market and governance structures than the indigenous shifting cultivators they replaced.[2]

Growing civil insurgency in the country forced the Marcos regime (1965 to 1986) to expedite the reclassification and redistribution of public land and further reform forest policies. The Integrated Social Forestry Program (Letter of Instruction 1260) was initiated in 1982 to 'democratize the use of public forests and to promote more equitable distribution of the forest bounty'. The programme allowed farmers to cultivate deforested areas, on the condition that forest fires would be suppressed, erosion minimized and tree growth encouraged. But despite its rhetoric on upland development, reforestation and participation, the Bureau of Forest Development still

Pennisetum glaucum (L.) R. Br. [Poaceae]

Pearl millet was a prominent crop in forest clearings in pre-colonial Philippines.

saw 'forest occupants' as destructive squatters whose actions should be controlled and monitored (Pulhin, 2002). Not that these views mattered much under Martial Law: most forest areas of the country were under the functional control of Maoist or Moro rebels.

The fall of the dictatorship in 1986 initiated a period of radical policy reforms that emphasized democratization and decentralization. Under President Corazon Aquino (1986 to 1992) a new constitution was adopted in 1987, which explicitly recognized the rights of indigenous peoples to their ancestral lands. The Local Government Code of 1991 (Republic Act 7160) transferred some of the powers of the national government to provinces and municipalities, among them the responsibility to enforce environmental legislation and the right to collect taxes on real property. These reforms led to a surge in logging and encroachment in the remaining forests as local officials semi-legally sanctioned these activities, a situation that would repeat itself in Indonesia after 1998. In 1991, tropical storm Uring dumped huge amounts of rain on the deforested landscape of Leyte Island, and the resulting flash flood destroyed Ormoc City and took more than 7000 lives. After this disaster, public pressure forced the Department of Environment and Natural Resources to suspend the issuance of licenses to logging companies, close down illegal sawmills and ban timber harvesting in old growth forest (Persoon and van der Ploeg, 2003).

Significant policy reforms continued under President Fidel Ramos (1992-1998). The Community-Based Forest Management Program (Department Administrative Order No 2), initiated in 1995, had the dual goals of 'ensuring social justice and the sustainable development of the country's forest resources' (Magno, 2001). The programme, actively supported by multilateral donors, reflected international policy discourses on sustainable development, indigenous peoples' rights and community-based resource management that had become fashionable in the early 1990s. It highlighted a remarkable turn-around in relations between the state and shifting cultivators: upland communities were no longer regarded as destructive illegal occupants, but as 'stewards of the environment' and 'guardians of the forest' (e.g. Poffenberger, 1990; Walpole et al., 1993; Lynch and Talbot,1995). The provision of land rights to upland communities became a core task of the Department of Environment and Natural Resources. Peoples' organizations could apply for so-called Community-Based Forest Management Agreements, under which they were granted the right to use forest land for 25 years. Indigenous peoples could apply for a Certificate of Ancestral Domain Title on lands that they had occupied since 'times immemorial'. The customary rights of indigenous peoples were further strengthened with the passage of the Indigenous Peoples' Rights Act in 1997 (Republic Act 8371). For some time, legal challenges hampered the implementation of the newly formed National Commission on Indigenous Peoples, but it was deemed to be constitutional in 2001.

But changing things on the ground proved to be more difficult. The Department of Environment and Natural Resources was ill-equipped for its new development task. It lacked political support and continued to be plagued by corruption. Its staff, primarily foresters trained to maximize timber extraction, continued to see shifting cultivation as a primitive, destructive and illegal practice, and often thwarted reforms at the local level (van den Top, 1998). Complex bureaucratic procedures created almost insurmountable obstacles for upland communities. For example, to obtain a Certificate of Ancestral Domain Title communities had to form an association and submit a so-called Ancestral Domain Sustainable Management and Development Plan to the department for approval. This 'co-management approach' required technical expertise and English-writing skills, and therefore depended largely on the support of consultants or non-governmental organizations. As a result most management plans reflected more the agendas of the department and donors than the needs of upland communities: for example, swidden farming was often explicitly prohibited in the ancestral domains, disregarding everyday practices.

The people-centered forest policies of the 1990s thus projected an idealized image of traditional, forest-dwelling, ecology-wise, sustainable and harmonious communities that was increasingly at odds with social realities. In doing so it ignored a growing body of anthropological literature that was documenting rapid change in swidden-based societies (e.g. Olofson, 1981; Headland, 1986; Eder, 1987; Brosius, 1990; Wiber, 1991; Fujisaka and Wollenberg, 1991; Cairns, 1997, 2015). Despite its noble intentions, the Community-Based Forest Management Program foundered in red tape, political interference and ingrained prejudices against shifting cultivators. Likewise, the Indigenous Peoples' Rights Act ultimately made little difference in the lives of marginalized, indigenous communities. The policy reforms were further undermined during the presidency of Joseph Estrada (1998 to 2001). Government officials often used the rights and needs of upland communities as a pretext for inaction and incompetence. The promises of the People Power Revolution faded into a huge disappointment (e.g. Utting, 2000; Bryant, 2000; McDermott, 2000; Li, 2002; Aquino, 2005; Dressler, 2005; Perez, 2010; van der Ploeg et al., 2011; Minter et al., 2012, 2014).

Under President Gloria Macapagal-Arroyo (2001 to 2010), many of the reforms that devolved power and authority over forest resources to upland communities were reversed, and the influence of civil society organizations was drastically curtailed. The issuance of Community-Based Forest Management Agreements was suspended. The Department of Environment and Natural Resources and the National Commission on Indigenous Peoples were tasked primarily with facilitating resource extraction by private companies, especially in the mining sector. Upland communities were again regarded as an obstacle for rapid development. The Wildlife Resources Conservation and Protection Act of 2004 (Republic Act 9147) testifies to this resurgent modernist, technocratic and repressive view: 'squatting or otherwise occupying any portion of critical wildlife habitat', in other words, living in forests, became punishable by 12 years in prison. It demonstrated a complete ignorance of, and disregard for, the livelihoods of shifting cultivators in the Philippine uplands (van der Ploeg and van Weerd, 2004). Unsurprisingly, the law has proven to be another dead letter.

FIGURE 1-3: A signboard of the Department of Environment and Natural Resources alongside a road in the Central Cordilleras of Luzon aptly illustrates the difficulties of shifting cultivation policies in the Philippines. Local election banners divert attention away from a rather impractical message. The *kaingin* in the background underlines its futility.

President Benigno Aquino, who took power in 2010, made the fight against corruption the highest priority of his government, prosecuting high-level officials and reorganizing government agencies. After the devastating floods in Luzon in October 2010, which the media attributed to massive deforestation in the uplands, Aquino signed Executive Order 23, which banned the cutting of timber in natural forests, and Executive Order 26, which initiated the National Greening Program. The latter aimed to plant '1.5 billion trees covering 1.5 million hectares in six years', an unrealistic objective that risked repeating the mistakes of the reforestation programmes of the 1970s. Indeed, its rhetoric betrayed a view of shifting cultivation reminiscent of the discriminatory and repressive policies of those days:

> Decades of mismanagement turned a big chunk of our forestlands (about 8 million hectares) unproductive, denuded or degraded. Thus, the National Greening Program (NGP) was conceived to bring back the lush vegetated cover of the uplands by involving the citizenry, particularly the students and government employees. [...] NGP advocates for increased and sustainable supply of forest-based raw materials; increased productivity of idle lands (value creation); increased economic activity in the uplands; and optimized utilization of upland resources, while avoiding *kaingin*-making or the so-called slash and burn agriculture and other destructive practices (DENR, 2015).

With the National Greening Program, Philippine policies on shifting cultivation seem to have come full circle, from seeing *kaingineros* as squatters to stewards, and back. Meanwhile, a growing demand for agricultural commodities such as cassava, corn, sugar cane and oil palm fuels the rapid conversion of grasslands and forests into farmland.

Shifting cultivation policy in Indonesia

The first specific policy on shifting cultivation in Indonesia dates back to 1874. Through the so-called *Ontginningsordonnantie* (Clearance Ordinance 1874/79) it was decided that inhabitants of Java and Madura were not allowed to open fields on *woeste grond* (wild lands) without prior permission of the authorities.[3] The ordinance was announced to stop the *ongebreidelde roofbouw* (unbridled slash-and-burn cultivation), to prevent erosion and to halt the loss of valuable timber in the Dutch East Indies.[4] People who opened up new fields in the forest could be punished. Earlier, the Dutch East Indies Company had tried to stop the destruction of Java's teak forests, but attempts to regulate land clearing and commercial logging failed. During the first period of the so-called *Cultuurstelsel* (Cultivation System, 1830 to 1870), most fertile lands were claimed for the cultivation of export crops such as coffee, tea, cocoa and sugar cane. To grow subsistence crops such as upland rice, corn and cassava, farmers on Java had no other choice but to clear new land in the forests (van Hall and van der Koppel, 1946).[5]

In the second half of the 19th century the Dutch colonial administration saw the need for a professional forestry service. German foresters were invited to educate Dutch officials and assist them in setting up a forest management system. A number of administrative measures were soon issued, specifically aimed at reducing forest clearing by shifting cultivators (Safitri, 2010). In 1896, the *Ontginningsordonnantie* was replaced by a new ordinance (1896/44) that detailed conditions under which permission could be given for forest clearance. Mention is often made in the colonial archives of the great reluctance of traditional leaders to actually implement regulations seeking to reduce swiddening (Bezemer, 1921, p364). Nonetheless, the colonial government enforced the law: in 1916, for example, 4862 farmers were sentenced to forced labour for illegally opening up wild land (Staten Generaal, 1918). Denial of access to forests by the colonial government occasionally led to resistance, in particular on Java (Peluso, 1992).

The *Domein Verklaring* (Agrarian Act) of 1870 officially designated all wild lands as state land, but at the same time entitled farmers to clear land in village forests, according to local customs. The fact that the *Domein Verklaring* and the Clearance Ordinances of 1874 and 1896 were highly contradictory was made clear in the writings of so-called *adat* scholars, who studied customary rules and regulations throughout the archipelago.[6] For example, the legal scholar Cornelis van Vollenhoven documented the village-law systems that regulated access to land and highlighted the ignorance of the '*bureauheeren*' – 'office gentlemen' who made decisions in the East Indies that would be totally unacceptable in the Netherlands (van Vollenhoven, 1919, p11).[7] Van Vollenhoven was certainly not the only one who wrote about the injustice of colonial forest policies: on one hand the state was denying farmers the right to clear land, while on the other it was permitting large-scale conversion of forest for the cultivation of export commodities. Indeed, the most famous literary work in Dutch colonial history, *Max Havelaar, of de koffieveilingen der Nederlandsche Handelsmaatschappij* (Max Havelaar, or the coffee auctions of the Dutch Trading Company) pointed out how colonial policies reserved land for the cultivation of cash crops, thereby impoverishing local farmers and forcing them to move to marginal lands (Multatuli, 1860).

At the start of the 20th century, forest conservation became a growing concern for the Dutch colonial government. For example, the resident of Banten, W. C. Thieme, wrote in 1930:

> Given the extensive forest reserves, and the desire of the population to preferably practice *huma* cultivation, it is necessary to strengthen the forest police in order to prevent ever larger clandestine reclamations. […] To clamp down on unbridled slash-and-burn farming that is detrimental to the nation's forest [it is essential that] offenders should be disciplined with severe corporal punishment and not with fines. The clandestine reclamation is still not regarded by the people as an offense against the law. (ANRI, 1976, pp16-17, authors' translation).

The first protected area, Ujong Kulon in West Java, was established in 1921, followed by Lorentz Mountain in New Guinea (present-day Papua) in 1923. At the same time, there was growing interest in the different ethnic groups of the archipelago and their modes of livelihood. Scientists, colonial administrators and missionaries such as Pieter Johannes Veth (1881), Julius Jacobs and Johannes Meijer (1891), Anton Willem Nieuwenhuis (1904), Bernhard Hagen (1908) and Albert Kruyt (1923) wrote detailed ethnographies about the agricultural practices of shifting cultivators (see also Koentjaraningrat, 1975). In the minds of many colonial officials, these primitive peoples had to be protected against the negative effects of Western civilization (Jepson and Whittaker, 2002).

In contrast, after independence was proclaimed in 1945 the government saw shifting cultivators primarily as backward and isolated, and in urgent need of development. In the 1950s and 1960s, the Ministry of Social Affairs was put in charge of developing and civilizing forest-dwelling communities. It formulated a resettlement programme that aimed to integrate 'isolated communities' into Indonesian mainstream society, initially on a small scale (targeting 40,000 people in 1966) but increasingly larger (almost 1.7 million people in 1986).

Under the New Order Regime (1966 to 1998) large-scale logging concessions were granted throughout Indonesia, causing much friction with local forest dwellers. In response, the Ministry of Agriculture, which hosted the Forestry Department, developed a resettlement policy for the *perladangan liar* (wild shifting cultivators) (Departemen Kehutanan, 1985). The policy aimed to safeguard the three functions of the Indonesian forests: production of timber, protection of watersheds and conservation of biodiversity, which were, in the view of the foresters, incompatible with slash-and-burn farming (Team Pusat Resetelmen Penduduk, 1980, p108, authors' translation):

> With the transfer of the shifting cultivators to particular locations, which can economically be developed and where the shifting cultivators can become sedentary farmers, the government will focus on the following aims: preventing the loss of valuable timber; facilitating logging operations; preventing disturbances to waterways and reducing erosion; protecting the fertility of the soil and preventing fires; providing a labour force for logging companies; increasing the standard of living of shifting cultivators; raising the level of education; and facilitating the development of the area.

The extent to which the resettlement programme and the logging industry were intertwined is best illustrated by the fact that the funding for resettlement came directly from taxes on timber production (the *Iuran Hasil Hutan Tambahan*). Shifting cultivators would be relocated to areas where they would have to practice permanent agriculture. They would be put in small houses suitable for single nuclear families and served by roads, schools, mosques or churches, polyclinics and other public facilities (Departemen Kehutanan, 1991).[8] Hundreds of resettlement villages were built over

FIGURE 1-4: Two posters, printed by the Ministry of Social Affairs and distributed in isolated communities throughout Indonesia in 1980, exemplify how policy-makers in Indonesia thought, and in many cases, still think, about shifting cultivation. The first urges people to change from shifting cultivators to sedentary farmers, and make a transition from hunting and gathering to animal husbandry, from a barter economy to incorporation into the market, and from living in a hut in the forest to a house with furniture, all under the slogan 'let's develop together' (left). The second equates swidden farming with forest loss, erosion and flooding, and urges 'protect our forest, water and soil' (right).

the years across Indonesia, but in many cases people continued practising shifting cultivation: old swiddens were often not abandoned, or people returned to their old settlements after a few years. The combined effects of shifting cultivation, logging operations and transmigration caused rapid and large-scale deforestation (Davis, 1988; MacKinnon et al., 1996).

In the 1980s, stimulated by the need to identify those responsible for tropical deforestation, anthropologists and social geographers generated a wealth of information on shifting cultivators across the Indonesia archipelago. Among the peoples studied were the Nuaulu on Ceram (Ellen, 1978); the Sahu on Halmahera (Visser, 1984); the Kantu Dayak and the Maloh in West Kalimantan (Dove, 1985; King, 1985); the Buginese in East Kalimantan (Vayda and Sahur, 1985), the Baduy in West Java

Elaeis guineensis Jacq.
[Arecaceae]

Oil palm plantations are blamed
for much of Indonesia's recent
forest loss.

(Iskander, 1992); and the Mentawaians on Siberut (Persoon, 1994). These studies documented the complexity and variability of swidden agriculture, demonstrated that these farming systems could achieve high productivity and efficiency in terms of returns to labour, and highlighted the negative effects of sedentarization and development projects. But the perspective of policy-makers remained largely the same: in their view, swidden farming destroyed valuable timber, contributed to erosion and signified poverty, backwardness and isolation.[9]

In 1997 and 1998, massive forest fires in Kalimantan and Sumatra led to heated debates about who should be held responsible: shifting cultivators, industrial-scale oil palm plantations or governmental agencies that granted the concessions but failed to enforce the law (Glover and Jessup, 1999). This led to more stringent regulations pertaining to the use of fire in forests, but in practice fire continued to be used to clear vegetation. The rapid loss of forests in Indonesia in recent years can, to a large extent, be attributed to the establishment of oil palm plantations (Koh and Wilcove, 2008; Sheil et al., 2009). Policy-makers see oil palm as a kind of miracle crop, generating higher returns than any other type of land use, so land conversion is actively encouraged. Alongside industrial-scale plantations, shifting cultivators are often tempted to begin planting oil palms, in much the same process as that which spawned smallholder rubber plantations at the end of the 19th century (Rist et al., 2010). Sometimes these smallholders aim to produce palm oil themselves, but in most cases they clear the land in order to sell it to companies. In this sense, the promotion of oil palm plantations by government agencies has greatly stimulated slash-and-burn farming in Kalimantan and Sumatra (Potter, 2009).

In May 1998, President Suharto was forced to step down. It implied the end of decades of centralized government, with Jakarta being the center of power. A wave of political reforms swept through the country centered on reformation (*reformasi*) and regional autonomy (Colfer and Resosudarmo, 2002). Numerous new laws allowed for the devolution of authority over natural resources to provinces, districts, subdistricts and villages. For example, provincial governments were given the power to issue logging concessions covering up to 10,000 hectares. Lucrative networks were soon formed to facilitate illegal logging activities around the country (Colchester, 2006).[10] However, not everything changed. In September 1999, a

Presidential Decree (no. 111) was issued regarding the social welfare of the *komunitas adat terpencil* (traditional remote communities) (Departemen Sosial, 2003a). It laid out a programme to improve the welfare of forest-dwelling communities based on resettlement and development of economic activities beyond self-sufficiency, under the responsibility of the Ministry of Social Affairs (Departemen Sosial, 2003b). Despite the minor differences in terminology, the overall policy towards shifting cultivators therefore remained largely unchanged.

The ownership of forest land in Indonesia remains a deeply contested issue. The Forestry Law of 1999 re-defined the relationship between state forest and *hutan adat* (customary forests) and provoked a long and bitter conflict on the status of these customary forests. Since the *Domein Verklaring* of 1870, the state has claimed legal authority to allocate forest land for whatever purpose – for example, for logging concessions, industrial tree plantations, protected areas or transmigration projects. In this view, shifting cultivators living in and cultivating these forest lands were considered to be squatters (Departemen Kehutanan, 2004). The *Aliansi Masyarakat Adat Nusantara* (AMAN), a non-governmental organization of various ethnic groups once classified by the state as 'isolated communities', filed a case at the Constitutional Court challenging the Forestry Law.[11] On 16 May 2013, the Constitutional Court in Jakarta declared that customary forests were not a part of state forests. The landmark ruling, covering 188 pages, implied that customary communities were legally recognized as the owners of their forests – a status they did not enjoy in the past (Rachman, 2014). The historic decision by the Constitutional Court is bound to have a major impact on forest management and the position of shifting cultivators in the Indonesian archipelago.

Conclusion

Over the past 150 years, policy-makers in Indonesia and the Philippines have issued a seemingly endless flow of ordinances, acts, memoranda, decrees, regulations and administrative orders aimed at bringing an end to shifting cultivation. But, for people like Azryl and Maritua, it all meant very little: they continue to use fire to clear vegetation and fertilize their fields. This does not imply that government is absent in the remote forests of Luzon and Sumatra. On the contrary, government policy has drastically changed the land and lives of people like Azryl and Maritua – but not in the way that the scientific literature often assumes that it does.

Much has been written about the almost universal prejudice of the state towards 'unproductive', 'destructive' and 'disorderly' shifting cultivators, and discriminatory policies bent on prohibiting swidden farming (e.g. Dove, 1983, 2015). However, the fact that most of these policies have never been executed has received much less attention. The discrepancy between rule and reality is one of the most striking characteristics of shifting cultivation policies in Indonesia and the Philippines. Since colonial times, government officials at the local level have been simply unwilling or incapable of effectively implementing legislation banning shifting cultivation.

FIGURE 1-5:   A Department of Environment and Natural Resources checkpoint, set up to control illegal logging operations, exemplifies the difficulties of law enforcement in remote rural areas of the Philippines. The building is closed; the forest guards are absent. However, the clandestine chainsaw repair shop next door is full of activity.

They have pointed out that the strict enforcement of regulations risks aggravating rural poverty and fuelling civil unrest, therefore they turn a blind eye to *huma* or *kaingin* making. But that is only one side of the story: an overly complex legal framework, an ineffective judicial system, institutional conflicts, political interference, low staff morale and rent-seeking have also inhibited the implementation of shifting cultivation policies.[12]

The forestry departments of both countries have long played an equivocal role: on one hand prohibiting shifting cultivation in order to curtail forest loss, while on the other facilitating timber extraction by companies and issuing permits for industrial-scale plantations.[13] Corporate logging and plantation agriculture, and the associated construction of roads, remain primary causes of deforestation in insular Southeast Asia (Geist and Lambin, 2002). The expansion of infrastructure has also allowed an influx of migrants looking for arable land, thereby paradoxically encouraging shifting cultivation. Seen in a broader perspective, the dualistic character of government policies becomes even clearer: whereas forestry departments try to stop swidden farming on public land, other departments actually encourage settlers. In Indonesia, district governments and line agencies provide schools, polyclinics and roads to illegally established settlements and village heads actively invite people to settle in their territory. In the Philippines, local governments aim to increase agricultural productivity and provide seeds and fertilizer to constituents in the uplands.

The disconnection between policies regulating shifting cultivation and everyday reality becomes particularly clear in contemporary discussions on community forestry. Here, policy-makers focus on traditional, forest-dwelling, subsistence-oriented indigenous communities. This image is in sharp contrast with daily life on the forest frontier. The cultivation of cash crops has been an integral part of these farming systems for a long time. Shifting cultivators, many of them relatively recent migrants or the children of newcomers, have been eager to adopt new crops,

Jatropha curcas L. [Euphorbiaceae]

A popular biofuel crop among Indonesia's modern shifting cultivators, belying the stereotypical image of primitive and unchanging subsistence farmers.

such as oil palm, jatropha, rubber and sugar cane, and new technologies to increase their productivity, such as herbicides, chainsaws, fertilizers and pesticides (Li, 2002; Masipiqueña et al., 2003; Löffler et al., 2014). As a result of these false images, community forest policies fail to counter the driving forces behind deforestation.

While our understanding of shifting cultivators has greatly improved over the past 150 years, our comprehension of how state bureaucracies actually operate is still inadequate.[14] There is an urgent need to look beyond the mere letter of forest policies pertaining to shifting cultivation and to understand the impact of tangled state interventions on the lives of ordinary people like Maritua and Azryl.

References

ANRI (1976) *Memori Residen Banten (W. Th. Thieme) 1921-1930*, Arsip Nasional Republik Indonesia (National Archives of Indonesia), Jakarta

Aquino, D. M. (2005) 'Resource management in ancestral lands: the Bugkalots in northeastern Luzon', PhD thesis, Leiden University, Leiden

Bankoff, G. (2004) 'The tree as the enemy of man: Changing attitudes to the forests of the Philippines, 1565-1989', *Philippine Studies* 52(3), pp320-344

Bankoff, G. (2009) 'A month in the life of José Salud, forester in the Spanish Philippines, July 1882', *Global Environment* 3, pp8-47

Bankoff, G. (2013) '"Deep forestry": Shapers of the Philippine forests', *Environmental History* 18, pp523-556

Barton, R. (1919) 'Ifugao law', *American Archaeology and Ethnology* 15(1), pp1-186

Benedict, L. W. (1916) *The Study of Bagobo Ceremonial, Magic, and Myths*, Brill, Leiden

Bezemer, T. J. (ed.)(1921) *Beknopte Encyclopedie van Nederlandsch-Indië* (Short Encyclopaedia of the Dutch Indies), Brill/Kolff & Co, Leiden and Batavia

Borras, S. M. (2006) 'Redistributive land reform in 'public' (forest) lands? Lessons from the Philippines and implications for land reform theory and practice', *Progress in Development Studies* 6(2), pp123-145

Broad, R. and Cavanagh, J. (1993) *Plundering Paradise: The Struggle for the Environment in the Philippines*, University of California Press, Berkeley, CA

Brosius, J. P. (1990) *After Duwagan. Deforestation, Succession and Adaptation in Upland Luzon, Philippines*, Centers for South and Southeast Asian Studies, University of Michigan, Ann Arbor, MI

Bryant, R. L. (2000) 'Politicized moral geographies: Debating biodiversity conservation and ancestral domain in the Philippines', *Political Geography* 19(6), pp673-705

Cairns, M. F. (1997) 'Ancestral domain and national park protection: mutually supportive paradigms? A case study of the Mt. Kitanglad Range National Park, Bukidnon, Philippines', *Philippine Quarterly of Culture and Society* 25(1), pp31-82

Cairns, M. F. (ed.) (2007) *Voices from the Forest: Integrating Indigenous Knowledge into Sustainable Upland Farming*, Resources for the Future Press, Washington, DC

Cairns, M. F. (ed.) (2015) *Shifting Cultivation and Environmental Change: Indigenous People, Agriculture and Forest Conservation*, Earthscan, London

Colchester, M. (2006) *Justice in the Forest. Rural Livelihoods and Forest Law Enforcement*, Forest Perspectives 3, Center for International Forestry Research (CIFOR), Bogor, Indonesia

Cole, F-C. (1913) *The Wild Tribes of the Davao District, Mindanao*, Publication 170, Field Museum of Natural History, Chicago

Cole, F-C. (1922) *The Tinguian. Social, Religious and Economic Life of a Philippine Tribe*, Publication 209, Field Museum of Natural History, Chicago

Colfer, C. J. P. and Resosudarmo, I. A. (eds)(2002) *Which Way Forward? People, Forests and Policy making in Indonesia*, Resources for the Future Press, Washington DC

Conklin, H. C. (1957) *Hanunoo Agriculture. A Report on an Integral System of Shifting Cultivation in the Philippines*, Forestry Development Paper 2(12), Food and Agriculture Organization of the United Nations, Rome

Conklin, H. C. (1961) 'The study of shifting cultivation', *Current Anthropology* 2(1), pp27-61

Davis, G. (1988) 'The Indonesian transmigrants', in J. S. Denslow and C. Padoch (eds) *People of the Tropical Rainforest*, University of California Press, Berkeley, CA, pp143-153

Departemen Kehutanan (1984) *Pola pengedalian perladangan berpindah di Indonesia*, Departemen Kehutanan, Jakarta

Departemen Kehutanan (1985) *Pengendalian perladangan berpindah melalui program transmigrasi*, Departemen Kehutanan, Jakarta

Departemen Kehutanan (1991) *Indonesia Forestry Action Programma*, vol. 1-3, Departemen Kehutanan, Jakarta

Departemen Kehutanan (2004) *Himpunan peraturan perundang-undangan Republik Indonesia tentang kehutanan dan illegal logging*, Departemen Kehutanan/Nuansa Aulia, Jakarta

Departemen Sosial (2003a) *Keputusan President Republik Indonesia No. 111, Tahun 1999 tentang pembinaan kesejahteraan sosial komunitas adat terpencil*, Departemen Sosial, Jakarta

Departemen Sosial (2003b) *Keputusan Menteri Sosial RI No. 06/PEGHUK/2002 tentang pedoman pelaksanaan pemberdayaan komunitas adat terpencil*, Departemen Sosial, Jakarta

Departemen Sosial (2004) *Atlas nasional persebaran komunitas adat terpencil*, Departemen Sosial, Jakarta

DENR (2015) *Saving Philippine Forest*, Department of Environment and Natural Resources, Manila, http://www.denr.gov.ph/index.php/component/content/article/597.html, accessed 31 March 2015

Dove, M. (1983) 'Theories of swidden agriculture and the political economy of ignorance', *Agroforestry Systems* 1, pp85-99

Dove, M. R. (1985) *Swidden Agriculture in Indonesia. The Subsistence Strategies of the Kalimantan Kantu*, Mouton, Berlin

Dove, M. R. (2015) 'The view of swidden agriculture by the early naturalists Linnaeus and Wallace', in M. F. Cairns (ed.) *Shifting Cultivation and Environmental Change: Indigenous People, Agriculture and Forest Conservation*, Earthscan, London

Dressler, W. H. (2005) 'Old thoughts on new ideas: Tagbanua forest use and State conservation measures at Puerto Princesa Subterranean River National Park', PhD dissertation, McGill University, Montreal

Eder, J. F. (1987) *On the Road to Tribal Extinction: Depopulation, Deculturation and Adaptive Well-being among the Batak of the Philippines*, University of California Press, Berkeley, CA

Ellen, R. F. (1978) *Nuaulu Settlement and Ecology. An Approach to the Environmental Relations of an Eastern Indonesian Community*, Verhandelingen KITLV 83, Martinus Nijhoff, The Hague

Finley, J. (1913) *The Subanu: Studies of a Sub-Visayan Mountain Folk of Mindanao*, Carnegie Institution, Washington, DC

Fujisaka, S. and Wollenberg, E. (1991) 'From forest to agroforest and logger to agroforester: A case study', *Agroforestry Systems* 14, pp113-129

Garvan, J. (1931) *The Manóbos of Mindanáo*, Memoirs of the National Academy of Sciences, Washington, DC

Geist, H. J. and Lambin, E. F. (2002) 'Proximate causes and underlying driving forces of tropical deforestation', *BioScience* 52(2), pp143-150

Glover, D. and Jessup, T. (1999) *Indonesia's Fires and Haze. The Cost of Catastrophe*, Institute of Southeast Asian Studies, Singapore

Hagen, B. (1908) *Die Orang Kubu auf Sumatra*, J. Baer & Co., Frankfurt, Germany

Headland, T. N. (1986) 'Why foragers do not become farmers: A historical study of a changing ecosystem and its effect on a Negrito hunter-gatherer group in the Philippines', PhD dissertation, University of Hawaii, Honolulu

Holleman, J. F. (ed.)(1981) *Van Vollenhoven on Indonesian Adat Law*, Translation Series 20, Royal Netherlands Institute of Southeast Asian and Carribean Studies (KITLV), Martinus Nijhof, The Hague

Iskander, J. (1992) *Ekologi perladangan di Indonesia. Studi kasus dari daerah Baduy Banten Selatan, Jawa Barat*, Penerbit Djambatan, Jakarta

Jacobs, J. and Meijer, J. J. (1891) *De Badoej's*, Martinus Nijhof, The Hague

Jepson, P. and Whittaker, R. J. (2002) 'Histories of protected areas: Internationalization of conservationist values and their adoption in the Netherlands Indies (Indonesia)', *Environment and History* 8, pp129-172

King, V. T. (1985) *The Maloh of West Kalimantan: An Ethnographic Study of Social Inequality and Social Change among an Indonesian Borneo People*, Verhandelingen 108, Royal Netherlands Institute of Southeast Asian and Carribean Studies (KITLV), Foris Publications, Dordrecht

Koentjaraningrat (1975) *Anthropology in Indonesia. A bibliographic Review*, Bibliographical Series 8, Royal Netherlands Institute of Southeast Asian and Caribbean Studies (KITLV), Martinus Nijhof, The Hague

Koh, L. P. and Wilcove, D. S. (2008) 'Is oil palm agriculture really destroying tropical biodiversity?' *Conservation Letters* 1(2), pp60-64

Kruyt, A. C. (1923) 'De Mentawaiers', *Indian Journal of Linguistics and Anthropology* 62, pp1-188

Kummer, D. M. (1992) *Deforestation in the Postwar Philippines*, Ateneo de Manila University Press, Manila

Larkin, J. A. (1993) *Sugar and the Origins of Modern Philippine Society*, University of California Press, Berkeley, CA

Li, T. M. (2002) 'Engaging simplification: Community-based resource management, market processes and state agendas in upland Southeast Asia', *World Development* 30(2), pp265-283

Löffler, H., Afiff, S. A., Burgers, P., Govers, C., Heeres, H., Manurung, R., Vel, J., Visscher, S. and Zwaagstra, T. (2014) *Agriculture Beyond Food: Experiences from Indonesia*, Netherlands Organization for Scientific Research/WOTRO Science for Global Development, The Hague

Lynch, O. J. (1982) 'Native title, private right and tribal land law: An introductory survey', *Philippine Law Journal* 57, pp268-306

Lynch, O. and Talbot, K. (1995) *Balancing Acts: Community based Forest Management and National Law in Asia and the Pacific*, World Resources Institute, Washington, DC

McDermott, M. H. (2000) 'Boundaries and pathways: Indigenous identity, ancestral domain and forest use in Palawan, the Philippines', PhD Dissertation, University of California Press, Berkeley, CA

MacKinnon, K., Hatta, G., Halim, H. and Mangalik, A. (1996) *The Ecology of Kalimantan (Indonesian Borneo)*, The Ecology of Indonesia series vol. III, Periplus, Singapore

Magno, F. (2001) 'Forest devolution and social capital: State-civil society relations in the Philippines', *Environmental History* 6(2), pp264-286

Masipiqueña, A., Persoon, G. A. and Snelder, D. J. (2003) 'The use of fire in northeastern Luzon (Philippines): Conflicting views of local people, scientists and government officials', in R. Ellen (ed.) *Indigenous Environmental Knowledge and its Transformations: Critical Anthropological Perspectives*, Harwood Academic Publishers, Amsterdam, pp177-212

Mazoyer, M. and Roudart, L. (2006) *A History of World Agriculture: From the Neolithic Age to the Current Crisis*, Earthscan, London

Mikkelsen, C. (ed.)(2014) *Indigenous World 2014*, International Work Group for Indigenous Affairs, Copenhagen

Miller, M.L. (1912) 'The Mangyans of Mindoro', *The Philippine Journal of Science* 7(3), pp135-156

Minter, T., de Brabander, V., van der Ploeg, J., Persoon, G. A. and Sunderland, T. (2012) 'Whose consent? Hunter-gatherers and extractive industries in the northeastern Philippines', *Society and Natural Resources* 25(12), pp1241-1257

Minter, T., van der Ploeg, J., Pedrablanca, M., Sunderland, T. and Persoon, G. A. (2014) 'Limits to indigenous participation: The Agta and the Northern Sierra Madre Natural Park, the Philippines', *Human Ecology* 42(5), pp769-778

Multatuli (1860) *Max Havelaar of de koffieveilingen der Nederlandsche Handelsmaatschappij*, Elsevier, Amsterdam

Nieuwenhuis, A. W. (1904) *Quer durch Borneo: Ergebnisse seiner Reisen in den Jahren 1894, 1896–97 und 1898-1900*, Brill, Leiden

Olofson, H. (ed.)(1981) *Adaptive Strategies and Change in Philippine Swidden-based Societies*, Forest Research Institute, Laguna, Philippines

Otto, J. M. (2009) 'Rule of law promotion, land tenure and poverty alleviation: Questioning the assumptions of Hernando de Soto', *Hague Journal on the Rule of Law* 1, pp173-194

Peluso, N. L. (1992) *Rich Forests, Poor People: Resource Control and Resistance in Java*, University of California Press, Berkeley, CA

Peluso, N. L. (1995) 'Whose woods are these? Counter-mapping forest territories in Kalimantan, Indonesia', *Antipode* 27(4), pp383-406

Perez, P. L. (2010) 'Deep-rooted hopes and green entanglements: Implementing indigenous people's rights and nature-conservation in the Philippines and Indonesia', PhD Dissertation, Leiden University, the Netherlands

Persoon, G. A. (1994) 'Vluchten of veranderen. Processen van verandering en ontwikkeling bij tribale groepen in Indonesië (Fleeing from change: Processes of change and development in tribal groups in Indonesia)', PhD dissertation, Leiden University, Leiden

Persoon, G. A. and van der Ploeg, J. (2003) 'Reviewing the projected future of San Mariano, a boomtown at the Sierra Madre forest fringe', *Philippine Studies* 51(3), pp451-473

Poffenberger, M. (ed.)(1990) *Keepers of the Forest: Land Management Alternatives in Southeast Asia*, Kumarian Press, West Hartford, CT

Potter, L. (2009) 'The oil palm question in Borneo', in G. A. Persoon and M. Osseweijer (eds) *Reflections on the Heart of Borneo*, Tropenbos Series no. 24, Tropenbos International, Wageningen, the Netherlands, pp69-90

Pulhin, J. M. (2002) 'Trends in forest policy of the Philippines', in *Policy Trend Report*, Forest Conservation Project, Institute for Global Environmental Strategies, Kanagawa, Japan, pp29-41

Rachman, N. F. (2014) 'Masyarakat hukum adat adalah bukanpenyandang hak, bukansubjek hukum, dan bukanpemilik wilayah adatnya' (Customary communities are not legal personalities, not legal subjects and not the owners of adat territories), *Wacana* 33, pp25-50

Rist, L., Feintrenie, L. and Levang, P. (2010) 'The livelihood impacts of oil palm: Smallholders in Indonesia', *Biodiversity and Conservation* 19, pp1009-1024

Safitri, M. A. (2010) 'Forest tenure in Indonesia. The socio-legal challenges of securing communities' rights', PhD dissertation, Leiden University, Leiden

Scott, W. H. (1994) *Barangay: Sixteenth-century Philippine Culture and Society*, Ateneo de Manila University Press, Quezon City, the Philippines

Sheil, D., Casson, A., Meijaard, E., Van Noordwjik, M., Gaskell, J., Sunderland-Groves, J., Wertz, K. and Kanninen, M. (2009) *The Impacts and Opportunities of Oil Palm in Southeast Asia: What do we Know and What do we Need to Know?* Center for International Forestry Research (CIFOR), Bogor, Indonesia

Spencer, J. E. (1966) *Shifting Cultivation in Southeastern Asia*, University of California Press, Berkeley, CA

Staten Generaal (1918) *Koloniaal verslag*, Hoofdstuk J, afd. IV (Colonial Report, Chapter J, Div. IV), www.statengeneraaldigitaal.nl/, search for Koloniaal verslag 1918, page 46, accessed 13 July 2017

Team Pusat Resetelmen Penduduk (1980) *Program Resetelmen*, Departemen Kehutanan, Jakarta

Thrupp, L. A., Hecht, S. B. and Bowder, J. O. (1997) *The Diversity and Dynamics of Shifting Cultivation: Myths, Realities and Policy Implications*, World Resources Institute, Washington, DC

Utting, P. (2000) 'An overview of the potential and pitfalls of participatory conservation', in P. Utting (ed.) *Forest Policy and Politics in the Philippines: The Dynamics of Participatory Conservation*, Ateneo de Manila University Press, Quezon City, the Philippines, pp171-215

van den Top, G. M. (1998) 'The social dynamics of deforestation in the Sierra Madre, Philippines', PhD dissertation, Leiden University, Leiden

van der Kolff, G. H. (1929) 'European influence on native agriculture', in B. Schrieke (ed.) *The Effect of Western Influence on Native Civilisations in the Malay Archipelago*, G. Kolff & Co., Batavia, pp103-125

van der Ploeg, J. and van Weerd, M. (2004) 'Devolution of natural resource management and crocodile conservation: The case of San Mariano, Isabela', *Philippine Studies* 52(3), pp345-382

van der Ploeg, J., Masipiqueña, A. B., van Weerd, M. and Persoon, G. A. (2011) 'Illegal logging in the Northern Sierra Madre Natural Park', *Conservation and Society* 9(3), pp202-215

van der Woud, A. (2007) *Het lege land. De ruimtelijke orde van Nederland 1798–1848* (The Empty Land: The Spatial Order in the Netherlands 1798 to 1848), Contact, Amsterdam

van Hall, C. J. J. and van der Koppel, C. (1946) *De landbouw in den Indischen Archipel* (Agriculture in the Indonesian Archipelago), W. van Hoeve, 's-Gravenhage

van Vollenhoven, C. (1919) *The Indonesiër en zijn grond* (The Indonesian and his Land), E. J. Brill, Leiden

Vayda, A. P. and Sahur, A. (1985) 'Forest clearing and pepper farming by Bugis migrants in East Kalimantan', *Indonesia* 39, pp93-110

Veth, P. J. (1881) *Midden-Sumatra reizen en onderzoekingen der Sumatra-expeditie, uitgerust door het Aardrijkskundig Genootschap, 1877–1879* (Travels in Central Sumatra and Discoveries of the Sumatra Expedition of the Geographical Society, 1877 to 1879), Brill, Leiden

Visser, L.E. (1984) 'Mijn kind is mijn tuin. Een antropologische studie van de droge rijstteelt in Sahu (Indonesië) (My child is my garden. An anthropological study of dry rice cultivation in Sahu (Indonesia))', PhD dissertation, Leiden University, Leiden

von Benda-Beckmann, F. and von Benda-Beckmann, K. (2011) 'Myths and stereotypes about adat law: A reassessment of van Vollenhoven in the light of current struggles over adat law in Indonesia', *Bijdragen tot de taal-, land- en volkenkunde* 167(2-3), pp167-195

Walpole, P. (2010) *Figuring Forest Figures*, Environmental Science for Social Change, Manila

Walpole, P., Braganza, G., Ong, J. B., Tengco, G. J. and Wijanco, E. (1993) *Upland Philippine Communities: Guardians of the Final Forest Frontiers*, Research Network Report 4, University of California Press, Berkeley, CA

Whitford, H. N. (1911) *The Forests of the Philippines*, Bureau of Printing, Manila

Wiber, M. G. (1991) 'Levels of property rights, levels of law: A case study from the northern Philippines', *Man* 26(3), pp469-492

Worcester, D. C. (1914) *The Philippines: Past and Present*, Macmillan Publishers, New York

Notes

1. The Public Land Act of 1903 adopted a land classification system that still forms the cornerstone of Philippine land law. Only land classified as 'alienable and disposable' can be privately owned; all other lands remain the property of the state. Public lands are assumed to be uninhabited and uncultivated, and thus forested (Borras, 2006). In fact, it is estimated that nowadays more than 25 million people reside in and cultivate these public lands. Over the past century, large tracts of public land have been deforested, but that does not automatically imply that the legal classification of the land has also changed. Hence the paradox in official statistics on land classification and forest cover in the Philippines: more than 50% of the total land area is classified as public land, but forest cover amounts to less than 23% (Walpole, 2010). Land classified by the government as forest land is by definition public land, but not all forests are on public land: timber plantations, coconut groves and citrus orchards are usually privately owned. The term 'upland' is used by policy-makers to refer to hilly or mountainous areas, in contrast to 'lowlands.' By law all uplands are defined as areas 'with slopes steeper than 18% and above 100m', and are classified as public lands.

2. Increasingly concerned with the rapid deforestation in the country, the government also initiated several reforestation programmes in the uplands, among them the Communal Tree Farming Programme and the Family Approach to Reforestation. The idea, patterned on the Burmese *taungya* forestry system, was to enlist shifting cultivators in reforestation: logging companies could first harvest timber after which people could grow crops for a few years and simultaneously plant and maintain tree seedlings, which could then be harvested by companies in a 20-year cycle (Thrupp et al., 1997). Perhaps unsurprisingly, these social forestry programmes failed to enthuse upland communities. Most tree plantations ended up being burnt.

3. The Dutch concept *woeste grond* is often translated in English as 'waste land'. This is incorrect. *Woeste grond* is 'wild land', in the sense of uncultivated, undomesticated and empty. Another (old) Dutch word for these lands was also *onland*, or 'unland'. Cultivation of these unreclaimed lands in the Netherlands (swamps, heath lands, sand dunes, floodplains, etc.) was a priority for the Dutch government in the 19th century. This perspective on wild lands was also of importance in the forest policies in Indonesia (van der Woud, 2007).

4. The Dutch word *roofbouw* for shifting agriculture is often translated in English as 'robber economy', and explained in terms of competition for forest resources between shifting cultivators and the (colonial) state. This is not correct. The term *roofbouw* is not about people 'competing', 'robbing' or 'stealing' something from someone, but refers to the unwise or even reckless use of land. In Dutch, *roofbouw* is often used metaphorically in the sense of exhausting your body, for instance through excessive use of alcohol.

5. Colonial plantation agriculture also influenced traditional shifting cultivators. Commercial crops like coffee, cocoa, tea, pepper, areca nuts and rubber were increasingly integrated into subsistence systems (Van der Kolff, 1929, pp116-117). As a result fallows were shortened, and ever larger fields were cleared for the cultivation of upland rice and cassava. Over the years, these *humas* were transformed into forests dominated by a variety of perennial commercial crops.

6. *Adat* is a generic term to indicate the morality, customs, traditions, legal institutions and dispute-resolution mechanisms of ethnic or cultural groups in Indonesia (Otto, 2009; von Benda-Beck-mann and von Benda-Beckmann, 2011).

7. Holleman (1981) provides an interesting summary of the writing of van Vollenhoven in English.

8. The Ministry of Forestry's 'wild shifting cultivators' were not necessarily the same people as the Ministry of Social Affairs' 'isolated communities' (Departemen Sosial, 2004). Some groups like the *Orang Bajau* were considered 'isolated communities', but clearly not classified as shifting cultivators. Conversely, not all shifting cultivators were classified as isolated communities, often because they practised a recognized religion. Many shifting cultivators in fact were relatively recent migrants. However, some ethnic groups, such as the *Mentawaians*, living on a chain of islands off the west coast of Sumatra, became the target for resettlement projects of both ministries (Persoon, 1994). There is still much disagreement about the number of shifting cultivators in Indonesia. The Ministry of

Forestry classifies six million people as forest-dwellers (Departemen Kehutanan, 1984). But scientists give much higher numbers: for example, Poffenberger (cited in Peluso, 1995) makes an estimate of 30 to 40 million people. Many non-governmental organizations, such as Aliansi Masyarakat Adat Nusantara (AMAN), use even higher figures (Mikkelsen, 2014).

9. One interesting example is a government effort to replace sago with rice as a staple food, largely because of the higher cultural status of rice in mainstream society.

10. Eventually, in order to combat illegal logging, efforts were made to recentralize some decision-making processes in Jakarta. But it turned out to be extremely complicated to take back the newly acquired powers of local authorities.

11. Interestingly, the case was filed by AMAN and two of its local organizations, one from Riau province and one from Lebak, the setting for Multatuli's famous *Max Havelaar*.

12. The few cases in which forest policies have actually been implemented represent a scale far too small to have adequately tackled the perceived problem, for example, the resettlement projects of the Indonesian Ministries of Agriculture and Social Affairs in the 1970s and 1980s. Effectiveness has also been blunted in the case of policies aiming to serve very different objectives, for example, the 1970s resettlement projects in the Philippines that aimed to isolate Maoist guerillas.

13. In colonial times, shifting cultivation policies in the Philippines and Indonesia were formulated by foresters without significant experience in the tropics, and for whom shifting cultivation was a relatively unfamiliar phenomenon. However, when colonial officials realized the economic potential of tropical forests, they claimed them as state property. Shifting cultivation practices were condemned as being an unproductive and wasteful form of agriculture. Remarkably, there was never much interaction between the Netherlands and the United States on how tropical forests should be managed in their neighbouring colonies. But they independently developed similar policies.

14. In general, there is now a wealth of scientific information on shifting cultivation systems in the Philippines and Indonesia (Conklin, 1961; Spencer, 1966; Cairns, 2007, 2015). However, this knowledge seems to have had little impact on policy. Much can be gained by making this scientific information more easily accessible by street-level bureaucrats, for example through training sessions, briefings and pilot projects.

2

SHIFTING CULTIVATION POLICIES IN SOUTHEAST ASIA

A need to work with, rather than against, smallholder farmers

*Ole Mertz and Thilde Bech Bruun**

Introduction

In 1995, when the first author was doing field work in Sarawak, he was invited to join in the planting of upland rice in one of the shifting cultivation fields that was the subject of his research. The field was very steep and the vegetation had not burned very well because the rains had never really stopped during that dry season. As tradition prescribed, the men were dibbling, or poking holes in the soil for the rice seeds, which would later be planted by the women. The job was like a hurdles race across charred tree trunks and debris of all sizes. The farmers were not happy with the outlook, and were talking about the risk of poor yields despite all their hard work. After having worked a half day, one of the author's good friends remarked, during a rest break: 'Ole, would you happen to know any smarter ways that we can do agriculture in a place like this? Any experiences from your country?'

The author was unable to give a good answer, but he reflected long on the question. It illustrated a concern that we are convinced most shifting cultivators share: shifting cultivation is not so much a choice, but often the only feasible option, and there are few barriers to adopting alternative practices if they offer a real and sustainable alternative. It is therefore puzzling why many countries in Southeast Asia have found it necessary to make shifting cultivation either directly or indirectly illegal, as well as devising a large palette of policies – often donor driven – to curb what is considered an environmentally destructive activity that reinforces poverty, rather than trying to understand what alternatives would be viable from the shifting cultivators' perspective. Fox (2000) made these points very clear 15 years ago, but

* Dr Ole Mertz is Professor of Geography at the Department of Geosciences and Natural Resource Management, Geography Section, University of Copenhagen, Denmark; Dr Thilde Bech Bruun is Associate Professor of Geography at the Department of Geosciences and Natural Resource Management, Geography Section, University of Copenhagen, Denmark.

much has happened in the uplands of Southeast Asia since then. In many areas shifting cultivation has either completely gone, is fast disappearing or is in a state of transformation, to become a small part of a predominantly permanent land-use system (see e.g. Padoch et al., 2007; Mertz et al., 2009; Cramb, 2015; Mertz, 2015; Potter, 2015). Based on case studies in China, Laos, Thailand, Malaysia and Indonesia, Fox et al. (2009) identified six trends that have been shaping shifting cultivation in recent decades: (1) classifying shifting cultivators as ethnic minorities within nation states; (2) dividing the landscape into forest and permanent agriculture; (3) expansion of forest departments and the rise of conservation; (4) resettlement; (5) privatization and commoditization of land and land-based production; and (6) expansion of markets, roads and other infrastructure and the promotion of industrial agriculture. All of these trends have been backed directly or indirectly by policies and policy debates that ruled out shifting cultivation as a viable form of future land management in Southeast Asia, despite numerous studies highlighting the positive role being played by shifting cultivation and mosaic landscapes in terms of environmental management, when compared to many alternative land uses (Padoch and Pinedo-Vasquez, 2010; Ziegler et al., 2011; Mertz et al., 2012; McNicoll et al., 2015).

Policies have indeed been identified by many case studies as a major driver of shifting cultivation change in the region (van Vliet et al., 2012) and shifting cultivation has recently been brought back into the policy arena by global initiatives such as REDD+ (Reducing Emissions from Deforestation and forest Degradation – 'plus' the role of conservation, sustainable management of forests and enhancement of forest carbon stocks in developing countries) (Mertz et al., 2012; Dove, 2015). REDD+ has also resulted in a revitalization of what Dressler (2015) aptly refers to as 'the reproduction of out-dated conservation debates about forests and swidden agriculture in rural localities, leading farmers to 'internalize' beliefs that swidden is illegal and destructive, requiring that it be abandoned'. Before REDD+ entered the scene, many other policies also targeted shifting cultivation. These ranged from top-down land-allocation policies restricting the use of fallow land in Vietnam (Castella et al., 2006; Meyfroidt and Lambin, 2008; Ankersen et al., 2015) and Laos (Castella et al., 2013; Broegaard et al., 2017) to more indirect policies that make shifting cultivation legally impossible, e.g. through a ban on open burning in agriculture in Sarawak, Malaysia (Mertz et al., 2013). Therefore the question for this chapter – and for this book in general – is to what extent land-use policies in South and Southeast Asia have affected not only shifting cultivation practices, but also the transition to other agricultural practices, since this is driven in part by responses to those same policies. We review the policies in each country and then, based on the existing literature, assess how these policies have affected the practice of shifting cultivation.

Policies on shifting cultivation and their effect in South and Southeast Asia

An overview of how shifting cultivation is directly or indirectly affected by legislation in different countries is provided in Table 2-1. There is a large diversity in policy approaches, but a majority of countries have legislation that in some way, directly or indirectly, is designed to inhibit shifting cultivation. The responses of shifting cultivators to these policy pressures have also varied and range from armed conflict in Bangladesh to voluntary abandonment in Sarawak – in the latter case probably driven more by new economic activities than by the policies. Evidence available in the literature for different regions is also very disparate and hence the following analysis does not give equal coverage to shifting cultivation policy and its consequences in every country.

Phuntsho et al. (ch 17) provide a comprehensive overview of shifting cultivation policies in countries that are part of the eastern Himalayas: Bangladesh, Bhutan, China, India, Myanmar and Nepal. As they state, 'none of these countries has a policy to develop shifting cultivation as an agricultural system on its own merits'. This suggests that since colonial times, land and agricultural policy-makers and planners have tried determinedly to make shifting cultivators convert to more permanent

TABLE 2-1: Contemporary policies on shifting cultivation (SC) in South and Southeast Asia.

Country	SC Illegal by law	SC de facto illegal by indirect legislation	SC legal, but restricted to certain areas and/or short fallows	Agricultural and forestry policies aim to stop SC	SC legal and without restrictions
Bangladesh	X			X	
Bhutan	X			X	
Brunei				X	
Cambodia			X		
China				X	
India	X		X	X	
Indonesia				X	
Laos			X	X	
Malaysia		X		X	
Myanmar				X	
Nepal		X			
Papua New Guinea					X
Philippines	X	X	X	X	
Sri Lanka			X	X	
Thailand	X	X		X	
Timor-Leste				X	
Vietnam			X	X	

Source: See text below on each country.

forms of agriculture. The South Asian countries of Bangladesh, India and Bhutan all prohibit shifting cultivation, but manifestations of the laws vary.

Bangladesh

Colonial laws that ban shifting cultivation remain in force and, as the Chittagong Hill Tracts – the only area where shifting cultivation is practised to any significant extent – has been engulfed in violent conflict for decades, there has been little effort to provide good alternatives. The result has been a decline in shifting cultivation areas and shortened fallows due to in-migration, creation of conservation areas and military interventions. However, the conflict remains and there is little or no support for changing farming practices (Phuntsho et al., ch 17), so people are left with few other choices. Moreover, environmental degradation is being exacerbated by involuntary shortening of fallows by shifting cultivators, and by century-old efforts to privatize forest lands and establish annual cropping, horticulture and plantations (Rasul, 2007).

Bhutan

Shifting cultivation is also illegal in Bhutan, although a number of laws exist that aim to secure the tenure of (former) shifting cultivators if their land is registered under different types of land-use systems (Phuntsho et al., ch 17). The result of these policies has been a sharp decline in shifting cultivation (Siebert et al., 2015). However, this can also be attributed to rural-to-urban migration and better opportunities for cash crop cultivation.

India

According to law in India, shifting cultivation is illegal. The approach to it has been described as 'fence and fine' (Blaikie and Muldavin, 2014), but in some states it is considered a privilege and generally tolerated as long as it is practised outside forests reserved for other purposes (Phuntsho et al., ch 17). There has been considerable pressure from government programmes to convert shifting cultivation to permanent agriculture (Phuntsho et al., ch 17), but the system is still widespread in the northeastern states (Cairns and Brookfield, 2011).

Nepal and Sri Lanka

These two countries stand out among others in South Asia by not having legislation directly prohibiting shifting cultivation. However, in Nepal various laws related to private land registration essentially make fallow land susceptible to 'confiscation', because shifting cultivators do not have the means to register their land and pay the associated taxes (Phuntsho et al., ch 17). There is little evidence of how this has affected shifting cultivators in practice. In Sri Lanka, shifting cultivation was regulated as long ago as 1855 by the colonial government (De Zoysa, 2001) and is nowadays only illegal in protected areas (Gunasena and Pushpakumara, 2015).

China

Until recently, shifting cultivation has been practised mainly in Yunnan province, but even here it has largely disappeared, not because it is prohibited by law, but because it has been made impossible by China's Sloping Land Conversion Programme (SLCP). This programme aims to reforest all areas with more than 25% inclination. It has effectively reduced annual cropping in many upland areas despite criticism for its top-down approach and plantation programmes that have been

Hevea brasiliensis (Willd. ex A. Juss.) Müll. Arg. [Euphorbiaceae]

Smallholders have now become the main players in the rubber-growing boom that has spread across southern Yunnan province in China.

neither suitable nor desired by the local people (He et al., 2014; He and Sikor, 2015; Phuntsho et al., ch 17). However, much of the transition away from shifting cultivation has also been driven by expansion of cash crops, for example the dramatic expansion of rubber cultivation in southern Yunnan (Ziegler et al., 2009; Fox et al., 2014; Zhang et al., 2015). This was initiated by the government on state farms, but its economic potential was soon seen by smallholders and external investors, who have since become the main rubber growers (Cao and Zhang, 2007).

In Southeast Asia, shifting cultivation is rapidly transforming and in some countries it barely exists anymore. In mainland Southeast Asia, although there is some diversity in policy approaches to shifting cultivation, most countries are aiming at what has almost been achieved in Thailand: a complete abandonment of shifting cultivation. Many of these policies have roots in past colonial legislation, wars against insurgents that happen to be shifting cultivators, and, not least, the fight against poppy cultivation for opium production (Ducourtieux et al., ch 32).

Thailand

Shifting cultivation was officially banned in Thailand in 1960 (Tomforde, 2003), but the ban was rarely enforced until after the 1980s, when an era with strong political focus on establishment of forest conservation areas began. According to official statistics, Thailand lost 50% of its forest cover in the years between 1970 and 1988 (Sato, 2000). The official narrative is that the 'hill tribes' and their traditional shifting cultivation systems were largely responsible for this forest loss – as clearly stated by the Thai Development

Research Institute: 'Much highland deforestation can be laid directly at their [the hill tribes'] door' (TDRI, 1987). In response to the forest loss, which was believed to be directly related to flooding in the lowlands, new forest-conservation policies established forest reserves, national parks and protected watersheds in upland areas, most of which were already inhabited by shifting cultivators (Hares, 2009).

Since the 1980s, the ban on shifting cultivation in Thailand has been enforced with increasing stringency. The ban is implemented in various ways in different locations, depending on the protection status of the area and on the local context. In some protected areas, shifting cultivation is tolerated by officials, but in other cases shifting cultivation communities living inside protected areas have been resettled and farmers have been fined or imprisoned for practising shifting cultivation (Hares, 2009). In northern Thailand, enforcement of the ban on shifting cultivation – and a long-established general ban on open burning during the field-preparation season – have been tightened in recent years with more regular inspections and more severe penalties (Bruun et al., 2017).

Papaver somniferum L.
[Papaveraceae]

Campaigns to rid Southeast Asian countries of opium poppies became the source of many policies in the region opposed to shifting cultivation.

Myanmar

There is very little research in the former Burma on shifting cultivation and related policies. Shifting cultivation is mainly practised in states that are dominated by ethnic minorities and these states have a long history of armed conflict between the government and ethnic-minority opposition groups. These conflicts are complex and although they are not directly related to shifting cultivation per se, control over land that is used for shifting cultivation is certainly a part of the problem. The Forest and Protection of Wildlife and Conservation of Natural Areas Law prohibits shifting cultivation in protected areas and in other areas supports the conversion of land to permanent farming, e.g. through the community forestry programme, which aims to convert shifting cultivation to permanent agroforestry (Phuntsho et al., ch 17). Most land under shifting cultivation – locally known as *taungya* – falls under the Vacant, Fallow and Virgin Lands Management Law. However, shifting cultivation is not specifically addressed or even mentioned in this law, which has been criticized by human–rights organizations for deliberately avoiding mention of *taungya* and for restricting tenure rights. Shifting cultivators rarely have formal titles to fallowed land and there is some evidence that they have little protection against losing their land to large scale private land developers or to projects undertaken 'in the interest of the state' (Kean, 2012; Guest et al., 2015).

Cambodia

There is also little research available on shifting cultivation policies in Cambodia, where it is generally perceived as an unwanted practice (Andersen et al., 2007). According to the Land Law, from 2001 it may be practised on land with collective ownership, a form of land tenure designed specifically for indigenous groups and recognized by the authorities. However, the process for registration of collective land is lengthy and very complicated, with 11 steps and the involvement of three ministries (CCHR, 2013). As of 2013, fewer than 5% of communities that had applied for collective ownership had completed the full process. Moreover, many shifting cultivation areas are found to exist in a legal 'grey zone': it is not clear if they fall under the Land Law or the Forestry Law, and these laws overlap and are partly contradictory (CCHR, 2013). According to the Forestry Law, shifting cultivation is not allowed in areas that are designated as permanent forest reserves. These reserves are currently being demarcated, but in many cases they include areas already used for shifting cultivation (Andersen et al., 2007). In order to address this issue, a sub-decree on shifting cultivation is being developed under the Forestry Law (Andersen et al., 2007). However, the practical implementation of this sub-decree has yet to be tested. The legal grey zone, the tedious and complicated process of applying for collective ownership and difficulties associated with separating old fallow land from forests all make land under shifting cultivation in Cambodia vulnerable to land acquisition (CCHR, 2013).

Laos and Vietnam

Similar legislation in Laos and Vietnam does not directly prohibit shifting cultivation, but it has been renamed as 'rotational agriculture'. It is a system wherein cultivation can only take place on specifically demarcated lands and with limited fallow periods. This is part of a policy process known as forest land allocation, in which forested areas typically become classified as either protection or production forest or other categories, but in all classifications, shifting cultivation is excluded (while different forms of plantation agriculture are not necessarily excluded). In Vietnam, this has led to a near-complete abandonment of shifting cultivation in many areas, partly because of the prohibitive nature of the legislation, but probably mostly because of associated agricultural intensification programmes that have supported farmers' transition to rely mainly on permanent agriculture (Sikor, 2001; Castella et al., 2006; Jakobsen et al., 2007; Ankersen et al., 2015). There are still large areas under shifting cultivation in Laos (Messerli et al., 2009; Heinimann et al., 2013) and although land has been allocated for short-fallow rotational agriculture in the course of several consecutive land–allocation policy processes (Lestrelin et al., 2011; Broegaard et al., 2017), empirical evidence on fallow lengths and shifting cultivation areas shows that there is limited compliance with these laws (Vongvisouk et al., 2014). Nonetheless, a strong policy vision that the government calls 'turning land into capital' implies rapid large-scale agricultural development and introduction of contract-farming schemes (Fox and Castella, 2013; Vongvisouk et al., 2016). These are likely to cause shifting cultivation to decline further in the years to come.

Papua New Guinea

Policies on shifting cultivation in the countries of insular Southeast Asia are rather different, but the result is more or less the same: a strong push to eradicate shifting cultivation has been on the agenda for decades. One striking exception is Papua New Guinea, where there seems to be no legislation constraining shifting cultivation, and it is believed to be practised by more than five million people (Allen and Filer, 2015). Shifting cultivation is recognized as a traditional practice and there are no governmental policies aiming to change it. Moreover, shifting cultivators account for a large majority of the customary landowners who have legal tenure rights to more than 97% of the country (Allen and Filer, 2015).

Brunei

Brunei also does not – to our knowledge – have prohibitive legislation, but a combination of general policy restrictions on forest use and high availability of alternative livelihoods are likely reasons why there has been no shifting cultivation in the country for at least two decades (FAO, 2000) and probably longer.

The Philippines

Shifting cultivation has been illegal by law in the Philippines since 1873 (Fox et al., 2009). The banning of shifting cultivation – locally known as *kaingin* – was based on colonial forestry principles that sought to maximize timber yields by reducing the threat from shifting cultivation-induced deforestation (Büscher and Dressler, 2012). Controls over shifting cultivators were tightened after independence in 1946, when stricter laws and regulations were issued, and under the Marcos dictatorship after 1975 many upland farmers were sent to jail or fined for practising shifting cultivation (Büscher and Dressler, 2012). Contemporary policies in the Philippines aim to stabilize shifting cultivation by promoting sedentary commercial agriculture. This is done, for example, by supporting the planting of fruit trees in fallow fields and promoting other simultaneous agroforestry practices – albeit with limited success (Dressler, 2009; Dressler and Roth, 2011).

Malaysia

Shifting cultivation is not directly illegal in Malaysia despite decades of condemnation, but in Sarawak and Sabah, the only states where shifting cultivation is still practised to any significant degree, laws on open burning in agriculture make shifting cultivation legally impossible (Mertz et al., 2013). This law was mainly intended to outlaw the burning of cleared vegetation on land destined for new plantations, and it goes generally unnoticed in shifting cultivation communities that do not face legal consequences for burning cleared fallow land. On the other hand, the government push for – and direct involvement in – large-

Elaeis guineensis Jacq. [Arecaceae]

The rapid expansion of oil palm plantations in Sarawak and Sabah has brought strong pressure to bear on shifting cultivation.

scale land development, especially for oil palm plantations, has put strong pressure on shifting cultivators to lease their land to companies and, in practice, lose the customary control of their land for as long as 60 years (McCarthy and Cramb, 2009; Cramb, 2011). General economic development in Malaysia and the profitability of plantation agriculture will probably in themselves make prohibitive legislation on shifting cultivation redundant in coming decades. Many rural people are either turning to smallholder cash-crop plantations (McCarthy and Cramb, 2009; Mertz et al., 2013) that may or may not integrate elements of shifting cultivation (Mertz, 2015) or they are abandoning farming altogether (Cramb, 2007, 2015) and leaving forests to regrow, while considering how they may benefit from the REDD+ mechanism.

Indonesia

Being the largest country in the region, Indonesia also holds the largest diversity of shifting cultivation systems. Colonial legislation made shifting cultivation illegal in 1874, and laid out the foundation for the country's current policies on the practice (Potter, 2003; Li, 2007). There is no direct ban on shifting cultivation, but contradictory legislation has made the tenure situation in shifting cultivation areas precarious. The Agrarian Law recognizes that customary land belongs to communities, but until recently the Forest Law stated that 'customary forest is state forest which lies within the lands of customary communities', making it very difficult for communities to gain official recognition that forest (or in practice, fallows) belong to them (van Noordwijk et al., 2008). However, in 2013 a Constitutional Court ruling removed 'state' from the sentence in the Forest Law to make it read 'customary forest is forest which lies within the lands of customary communities'. According to Minarchek (ch 47) this has opened the door for recognition of 40% of hitherto state-owned forests being returned to local communities. There has been limited implementation of the changed law so far (Minarchek, ch 47) and it remains to be seen whether this can reduce the many conflicts over shifting cultivation land (Peluso, 1995, 2005). It will be especially interesting to see whether there will be diminished claims on community lands from the rapid expansion of oil palm plantations, mining, paper-pulp production and many other large-scale land uses that are also contributing to pressure on shifting cultivators. As in Malaysia, these pressures became more complex in Indonesia after decentralized policies allowed provinces and districts to allocate land concessions and the question is whether these concessions will be modified if local communities claim their forest lands according to the new court ruling. Other recent legislation, such as the 2012 moratorium on deforestation, is also potentially important for shifting cultivators, who, in principle, are not allowed to clear fallow land because of this law. However, its effect on shifting cultivation is not yet clear, and neither is its overall effect in reducing deforestation rates in general (Margono et al., 2014).

Timor-Leste

It has not been possible to find detailed information on legislation affecting shifting cultivation in Timor-Leste, except that it is forbidden in protected areas and discouraged by integrated watershed-management programmes (McArthur et al., ch 27).

Ways forward for policy-making on shifting cultivation in South and Southeast Asia

Past land-use policies in South and Southeast Asia have mainly been aimed at prohibiting or actively replacing shifting cultivation with other systems, and as the practice of shifting cultivation is indeed declining rapidly in the region (Padoch et al., 2007; Mertz et al., 2009), this could be evidence that the policies are working. However, as we have shown, it is very difficult to establish the real causes of decline in shifting cultivation, although in those areas where its disappearance is most rapid, economic incentives to shift to other livelihood strategies appear to have more weight. The classical arguments for these policies have been to protect the environment, particularly to avoid deforestation, and to enhance the well-being of smallholders in remote areas. But the outcomes have been very mixed. Even in places such as Vietnam and China, where reforestation has occurred (Meyfroidt et al., 2010) and where economic growth has created improved livelihoods in rural areas, some rural people who were formerly engaged in shifting cultivation have been left with fewer and more risky livelihood options (Weyerhaeuser et al., 2005; Jakobsen et al., 2007) while others have managed to benefit from new opportunities and are making better incomes than hired labourers (Sturgeon, 2010). In other countries, deforestation continues unabated (Tong, 2009; Harris et al., 2012; Margono et al., 2014) despite the rapid decline in shifting cultivation. Governments and the private sector – either independently or in collaboration – are engaging in large-scale land development of plantation crops and mining, both of which cause complete removal of forest cover in areas where shifting cultivation once maintained mosaic landscapes with forest, fallow and fields. Such developments may, in some cases, result in shifting cultivators losing their land, but they also offer opportunities for entrepreneurial smallholders who are able to withstand pressures from external actors. An example is smallholders who join the plantation-crop race and engage in oil palm cultivation in Sarawak. They do so essentially on their own terms and sometimes in defiance of government interests that favour large-scale businesses, but of course they are dependent on the palm oil infrastructure that is established by large private companies (Mertz et al., 2013; Mertz, 2015).

In places where shifting cultivation persists, official policies tend to have further marginalized shifting cultivators and, in some cases, driven them to modifications of the system that seem difficult to sustain. For example, the expansion of contract farming of hybrid maize in Thailand and Laos in the 2000s has led to permanent cultivation or shifting cultivation of upland fields with very short fallows. This has led to increasing debt problems for farmers and complete deforestation of much larger areas than were earlier used for shifting cultivation (Vongvisouk et al., 2016; Bruun et al., 2017; Yap et al., 2017). In general, shifting cultivators have rarely been given the option to develop their farming practices on their own terms, as advice from extension services has been focused on stopping shifting cultivation and switching to permanent farming with annual or perennial crops (Mertz, 2002; van Noordwijk et al., 2008; Phuntsho et al., ch 17). As stated by van Noordwijk et al. (2008) for Indonesia:

> The Indonesian government's early focus on jump-starting intensive permanent cropping shifted to supporting tree crop monocultures. It would be better to support the gradual evolution of swiddens and the agroforestry systems derived from it in accordance with local expectations.

We began this chapter with a request from a shifting-cultivator friend in Sarawak to find better ways to cultivate upland rice, but it is quite clear that most governments and extension services in the region do not listen to such requests. Nevertheless, resourceful farmers are good at seeking out the best alternatives in their own areas and navigating within the local land-development context. Oil palm smallholders in Sarawak are one example, but some farmers have also discovered that the use of limited amounts of fertilizers and herbicides for upland rice – sometimes taken from the cash-crop subsidies provided by governments – can supplement limited nutrients and alleviate labour-intensive weeding tasks when fallows are too short. For example, limited nitrogen fertilization has led to documented yield increases in shifting cultivation (Mertz et al., 2008).

Advice from shifting cultivators is also rarely sought when it comes to government planning of land uses in their village territories, although some countries have made efforts to include them in such exercises. For example, the most recent forest land allocation process in Laos used participatory land-use planning exercises in which farmers and government officials jointly established zoning within village areas, including decisions on where shifting cultivation (or 'rotational agriculture' as it is called) could take place (Bourgoin et al., 2012; Castella et al., 2014). However, as mentioned earlier, this may be just another planning exercise that once again redefines land use within a village and maybe changes the village boundaries, but the decisions from which are unlikely to be followed by either the farmers or the government (Broegaard et al., 2017). In Malaysia and Indonesia, much of the large-scale land development is done according to contractual agreements with communities, but once again these seem to reinforce state power rather than empower communities (Cramb, 2011). They create more conflict, both between communities and companies (Cramb, 2013) and within communities, where benefits and development are often not equally distributed (Andersen et al., 2016). The new Constitutional Court ruling in Indonesia changing customary forests from state- to community-owned forests (Minarchek, ch 47) is a positive step in the right direction, but its impact has yet to be seen.

Conclusion

We have shown that most countries in South and Southeast Asia have some form of direct or indirect prohibitive legislation against shifting cultivation. It would be an admirable gesture for governments in this region to revise their policies and work with, rather than against, shifting cultivators to improve their well-being and the overall management of the environment. As it is now, policies seem mainly to be poorly concealed ways to promote business interests in land development and show little care for the fact that many shifting cultivators are, themselves, eager to develop and enjoy better lives. This does not mean that shifting cultivation should be promoted or given any specific priority in land-development policies, because many shifting cultivators are in fact looking for alternatives. But criminalizing it clearly will not help the stated goals of governments in South and Southeast Asia, to achieve better environmental management and alleviate poverty in rural upland areas. It is time to give the smallholders who practise shifting cultivation recognition for the benefits of their farming systems and grant them the help necessary to improve their agricultural practices or to move on to other activities, according to their wishes.

References

Allen, B. and Filer, C. (2015) 'Is the bogeyman real? Shifting cultivation and the forests, Papua New Guinea', in M. F. Cairns (ed.) *Shifting Cultivation and Environmental Change: Indigenous People, Agriculture and Forest Conservation*, Earthscan from Routledge, London

Andersen, A. O., Bruun, T. B., Egay, K., Fenger, M., Klee, S., Pedersen, A. F., Pedersen, L. M. L. and Suárez Villanueva, V. (2016) 'Negotiating development narratives within large-scale oil palm projects on village lands in Sarawak, Malaysia', *Geographical Journal* 182, pp364–374

Andersen, K. E., Soporn, S. and Thornberry, F. (2007) 'Development of a sub-decree on shifting cultivation under Article 37 of the Forestry law (2002), Cambodia', International Labour Organization, Phnom Penh

Ankersen, J., Mertz, O., Fensholt, R., Castella, J. C., Lestrelin, G., Nguyen, T. D., Grogan, K. and Rasmussen, K. (2015) 'Vietnam's forest transition in retrospect: Demonstrating weaknesses in business-as-usual scenarios for REDD+', *Environmental Management* 55, pp1080-1092

Blaikie, P. and Muldavin, J. (2014) 'Environmental justice? The story of two projects', *Geoforum* 54, pp226-229

Bourgoin, J., Castella, J.-C., Pullar, D., Lestrelin, G. and Bouahom, B. (2012) 'Towards a land zoning negotiation support platform: "Tips and tricks" for participatory land use planning in Laos', *Landscape and Urban Planning* 104, pp270-278

Broegaard, R. B., Vongvisouk, T. and Mertz, O. (2017) 'Contradictory land use plans and policies in Laos: Tenure security and the threat of exclusion', *World Development* 89, pp170-183, http://dx.doi.org/10.1016/j.worlddev.2016.08.008

Bruun, T. B., de Neergaard, A., Burup, M. L., Hepp, C. M., Larsen, M. N., Abel, C., Aumtong, S., Magid, J. and Mertz, O. (2017) 'Intensification of upland agriculture in Thailand: Development or degradation?', *Land Degradation and Development* 28, pp83–94, doi: 10.1002/ldr.2596

Büscher, B. and Dressler, W. (2012) 'Commodity conservation: The restructuring of community conservation in South Africa and the Philippines', *Geoforum* 43, pp367-376

Cairns, M. F. and Brookfield, H. (2011) 'Composite farming systems in an era of change: Nagaland, Northeast India', *Asia Pacific Viewpoint* 52, pp56-84

Cao, G. and Zhang, L. (2007) 'Rubber plantations as an alternative to shifting cultivation in Yunnan, China', in M. F. Cairns (ed.) *Voices from the Forest: Integrating Indigenous Knowledge into Sustainable Upland Farming*, Resources for the Future Press, Washington, DC, pp600-613

Castella, J.-C., Boissau, S., Hai Thanh, N. and Novosad, P. (2006) 'Impact of forestland allocation on land use in a mountainous province of Vietnam', *Land Use Policy* 23, pp147-160

Castella, J.-C., Lestrelin, G., Hett, C., Bourgoin, J., Fitriana, Y. R., Heinimann, A. and Pfund, J. L. (2013) 'Effects of landscape segregation on livelihood vulnerability: Moving from extensive shifting cultivation to rotational agriculture and natural forests in northern Laos', *Human Ecology* 41, pp63-76

Castella, J.-C., Bourgoin, J., Lestrelin, G. and Bouahom, B. (2014) 'A model of the science–practice–policy interface in participatory land-use planning: lessons from Laos', *Landscape Ecology* 29, pp1095-1107

CCHR (2013) *Cambodia: Land in Conflict. An Overview of the Land Situation*, Cambodian Center for Human Rights, Phnom Penh

Cramb, R. A. (2007) *Land and Longhouse: Agrarian Transformation in the Uplands of Sarawak*, NIAS Press, Copenhagen

Cramb, R. A. (2011) 'Re-inventing dualism: Policy narratives and modes of oil palm expansion in Sarawak, Malaysia', *Journal of Development Studies* 47, pp274-293

Cramb, R. A. (2013) 'Palmed off: Incentive problems with joint-venture schemes for oil palm development on customary land', *World Development* 43, pp84-99

Cramb, R. A. (2015) 'Busy people, idle land: The changing roles of swidden fallows in Sarawak', in M. F. Cairns (ed.) *Shifting Cultivation and Environmental Change: Indigenous People, Agriculture and Forest Conservation*, Earthscan from Routledge, London, pp770-793

De Zoysa, M. (2001) 'A review of forest policy trends in Sri Lanka', *Policy Trend Report* 2001, pp57-68

Dove, M. R. (2015) 'The view of swidden agriculture by the early naturalists, Linnaeus and Wallace', in M. F. Cairns (ed.) *Shifting Cultivation and Environmental Change: Indigenous People, Agriculture and Forest Conservation*, Earthscan from Routledge, London, pp3-24

Dressler, W. (2009) *Old Thoughts in New Ideas: State Conservation Measures, Livelihood and Development on Palawan Island, the Philippines*, Anteneo de Manila University Press, Quezon City, The Philippines

Dressler, W. (2015) 'Governmental pressures on swidden landscapes on Palawan Island, the Philippines', in M. F. Cairns (ed.) *Shifting Cultivation and Environmental Change: Indigenous People, Agriculture and Forest Conservation*, Earthscan from Routledge, London, pp877-890

Dressler, W. and Roth, R. (2011) 'The good, the bad, and the contradictory: Neoliberal conservation governance in rural Southeast Asia', *World Development* 39, pp851-862

FAO (2000) 'Brunei Darusalaam', in *Asia and the Pacific National Forestry Programmes*, Update 34, Food and Agriculture Organization of the United Nations, Rome, pp33-38

Fox, J. (2000) 'How blaming "slash and burn" farmers is deforesting mainland Southeast Asia', *Asia Pacific Issues* 47, pp1-8

Fox, J. and Castella, J-C. (2013) 'Expansion of rubber (*Hevea brasiliensis*) in mainland Southeast Asia: What are the prospects for smallholders?', *The Journal of Peasant Studies* 40, pp155-170

Fox, J., Castella, J-C., Ziegler, A. D. and Westley, S. B. (2014) 'Rubber plantations expand in mountainous Southeast Asia: What are the consequences?', *Asia Pacific Issues* 114, pp1-8

Fox, J., Fujita, Y., Ngidang, D., Peluso, N. L., Potter, L., Sakuntaladewi, N., Sturgeon, J. and Thomas, D. (2009) 'Policies, Political-Economy and Swidden in Southeast Asia', *Human Ecology* 37, pp305-322

Guest, D., Bowman, V. and Wachenfeld, M. (2015) *Land: Myanmar Centre for Responsible Business briefing paper*, Myanmar Centre for Responsible Business, Institute for Human Rights and Business and Danish Institute for Human Rights, Yangon

Gunasena, H. P. M. and Pushpakumara, D. K. N. G. (2015) 'Chena cultivation in Sri Lanka: Prospects for agroforestry interventions', in M. F. Cairns (ed.) *Shifting Cultivation and Environmental Change: Indigenous People, Agriculture and Forest Conservation*, Earthscan from Routledge, London, pp199-220

Hares, M. (2009) 'Forest conflict in Thailand: Northern minorities in focus', *Environmental Management* 43, pp381-395

Harris, N. L., Brown, S., Hagen, S. C., Saatchi, S. S., Petrova, S., Salas, W., Hansen, M. C., Potapov, P. V. and Lotsch, A. (2012) 'Baseline map of carbon emissions from deforestation in tropical regions', *Science* 336, pp1573-1576

He, J., Lang, R. and Xu, J. (2014) 'Local dynamics driving forest transition: Insights from upland villages in Southwest China', *Forests* 5, pp214-233

He, J. and Sikor, T. (2015) 'Notions of justice in payments for ecosystem services: Insights from China's Sloping Land Conversion Program in Yunnan Province', *Land Use Policy* 43, pp207-216

Heinimann, A., Hett, C., Hurni, K., Messerli, P., Epprecht, M., Jorgensen, L. and Breu, T. (2013) 'Socio-economic perspectives on shifting cultivation landscapes in northern Laos', *Human Ecology* 41, pp51-62

Jakobsen, J., Rasmussen, K., Leisz, S., Folving, R. and Quang, N. V. (2007) 'The effects of land tenure policy on rural livelihoods and food sufficiency in the upland village of Que, North Central Vietnam', *Agricultural Systems* 94, pp309-319

Kean, T. (2012) 'No protection for taungya farmers in bylaws: Experts', *Myanmar Times*, 22 October 2012

Lestrelin, G., Bourgoin, J., Bouahom, B. and Castella, J-C. (2011) 'Measuring participation: Case studies on village land use planning in northern Lao PDR', *Applied Geography* 31, pp950-958

Li, T. M. (2007) *The Will to Improve: Governmentality, Development, and the Practice of Politics,* Duke University Press, Durham, NC

McCarthy, J. F. and Cramb, R. A. (2009) 'Policy narratives, landholder engagement, and oil palm expansion on the Malaysian and Indonesian frontiers', *Geographical Journal* 175, pp112-123

McNicoll, I. M., Berry, N. J., Bruun, T. B., Hergoualc'h, K., Mertz, O., de Neergaard, A. and Ryan, C. M. (2015) 'Development of allometric models for above- and below-ground biomass in swidden cultivation fallows of northern Laos', *Forest Ecology and Management* 357, pp104-116

Margono, B. A., Potapov, P. V., Turubanova, S., Stolle, F. and Hansen, M. C. (2014) 'Primary forest cover loss in Indonesia over 2000-2012', *Nature Climate Change* 4, pp730-735

Mertz, O. (2002) 'The relationship between fallow length and crop yields in shifting cultivation: A rethinking', *Agroforestry Systems* 55, pp149-159

Mertz, O. (2015) 'Oil palm as a productive fallow? Swidden change and new opportunities in smallholder land management', in M. F. Cairns (ed.) *Shifting Cultivation and Environmental Change: Indigenous People, Agriculture and Forest Conservation*, Earthscan from Routledge, London, pp731-741

Mertz, O., Wadley, R. L., Nielsen, U., Bruun, T. B., Colfer, C. J. P., de Neergaard, A., Jepsen, M. R., Martinussen, T., Zhao, Q., Noweg, G. T. and Magid, J. (2008) 'A fresh look at shifting cultivation: Fallow length an uncertain indicator of productivity', *Agricultural Systems* 96, pp75-84

Mertz, O., Padoch, C., Fox, J., Cramb, R. A., Leisz, S. J., Nguyen, T. L. and Vien, T. D. (2009) 'Swidden change in Southeast Asia: Understanding causes and consequences', *Human Ecology* 37, pp259-264

Mertz, O., Müller, D., Sikor, T., Hett, C., Heinimann, A., Castella, J-C., Lestrelin, G., Ryan, C. M., Reay, D., Schmidt-Vogt, D., Danielsen, F., Theilade, I., van Noordwijk, M., Verchot, L. V., Burgess, N. D., Berry, N. J., Pham, T. T., Messerli, P., Xu, J., Fensholt, R., Hostert, P., Pflugmacher, D., Bruun, T. B., de Neergaard, A., Dons, K., Dewi, S., Rutishauer, E. and Sun, Z. (2012) 'The forgotten D: Challenges of addressing forest degradation in complex mosaic landscapes under REDD+', *Geografisk Tidsskrift - Danish Journal of Geography* 112,, pp63-76

Mertz, O., Egay, K., Bruun, T. B. and Colding, T. S. (2013) 'The last swiddens of Sarawak', *Human Ecology* 41, pp109-118

Messerli, P., Heinimann, A. and Epprecht, M. (2009) 'Finding homogeneity in heterogeneity: A new approach to quantifying landscape mosaics developed for the Lao PDR', *Human Ecology* 37, pp291-304

Meyfroidt, P. and Lambin, E. F. (2008) 'The causes of the reforestation in Vietnam', *Land Use Policy* 25, pp182-197

Meyfroidt, P., Rudel, T. K. and Lambin, E. F. (2010) 'Forest transitions, trade, and the global displacement of land use', *Proceedings of the National Academy of Sciences* 107, pp20917-20922

Padoch, C., Coffey, K., Mertz, O., Leisz, S., Fox, J. and Wadley, R. L. (2007) 'The demise of swidden in Southeast Asia? Local realities and regional ambiguities', *Geografisk Tidsskrift - Danish Journal of Geography* 107, pp29-41

Padoch, C. and Pinedo-Vasquez, M. (2010) 'Saving slash-and-burn to save biodiversity', *Biotropica* 42, pp550-552

Peluso, N. L. (1995) 'Whose woods are these? Counter-mapping forest territories in Kalimantan, Indonesia', *Antipode* 27, pp383-406

Peluso, N. L. (2005) 'Seeing property in land use: Local territorializations in West Kalimantan, Indonesia', *Geografisk Tidsskrift - Danish Journal of Geography* 105, pp1-15

Potter, L. (2003) 'Forests versus agriculture: Colonial forest services, environmental ideas and the regulation of land-use change in Southeast Asia', in L. Tuck-Po, W. de Jong and A. Ken-ichi (eds) *The Political Ecology of Tropical Forests in Southeast Asia: Historical Perspectives*, Kyoto University Press, Kyoto, Japan, pp29-71

Potter, L. (2015) 'Where are the swidden fallows now? An overview of oil palm and Dayak agriculture across Kalimantan, with case studies from Sanggau, in West Kalimantan', in M. F. Cairns (ed.) *Shifting Cultivation and Environmental Change: Indigenous People, Agriculture and Forest Conservation*. Earthscan from Routledge, London, pp742-769

Rasul, G. (2007) 'Political ecology of the degradation of forest commons in the Chittagong Hill Tracts of Bangladesh', *Environmental Conservation* 34, pp153-163

Sato, J. (2000) 'People in between: Conversion and conservation of forest lands in Thailand', *Development and Change* 31, pp155-177

Siebert, S. F., Belsky, J. M., Wangchuk, S. and Riddering, J. (2015) 'The end of swidden in Bhutan: Implications for forest cover and biodiversity', in M. F. Cairns (ed.) *Shifting Cultivation and Environmental Change: Indigenous People, Agriculture and Forest Conservation*, Earthscan from Routledge, London, pp546-558

Sikor, T. (2001) 'The allocation of forestry land in Vietnam: Did it cause the expansion of forests in the northwest?', *Forest Policy and Economics* 2, pp1-11

Sturgeon, J. C. (2010) 'Governing minorities and development in Xishuangbanna, China: Akha and Dai rubber farmers as entrepreneurs', *Geoforum* 41, pp318-328

TDRI (1987) *Thailand: Natural Resources Profile*, Thai Development Research Institute (TDRI), Bangkok

Tomforde, M. (2003) 'The global in the local: Contested resource-use systems of the Karen and Hmong in northern Thailand', *Journal of Southeast Asian Studies* 34, pp347-360

Tong, P. S. (2009) *Lao People's Democratic Republic Forestry Outlook*, Study APFSOS II/WP/2009/17, Vientiane, Laos

van Noordwijk, M., Mulyoutami, E. and Sakuntaladewi, N. (2008) *Swiddens in Transition: Shifted Perceptions on Shifting Cultivators in Indonesia*, World Agroforestry Centre (ICRAF), SEA Regional Office, Bogor, Indonesia

van Vliet, N., Mertz, O., Heinimann, A., Langanke, T., Pascual, U., Schmook, B., Adams, C., Schmidt-Vogt, D., Messerli, P., Leisz, S., Castella, J-C., Jørgensen, L., Birch-Thomsen, T., Hett, C., Bruun, T. B., Ickowitz, A., Vu, K. C., Fox, J., Cramb, R. A., Padoch, C., Dressler, W. and Ziegler, A. (2012) 'Trends, drivers and impacts of changes in swidden cultivation in tropical forest-agriculture frontiers: A global assessment', *Global Environmental Change* 22, pp418-429

Vongvisouk, T., Mertz, O., Thongmanivong, S., Heinimann, A. and Phanvilay, K. (2014) Shifting cultivation stability and change: Contrasting pathways of land use and livelihood change in Laos', *Applied Geography* 46, pp1-10

Vongvisouk, T., Broegaard, R. B., Mertz, O. and Thongmanivong, S. (2016) 'Rush for cash crops and forest protection: Neither land sparing nor land sharing', *Land Use Policy* 55, pp182-192

Weyerhaeuser, H., Wilkes, A. and Kahrl, F. (2005) 'Local impacts and responses to regional forest conservation and rehabilitation programs in China's northwest Yunnan province', *Agricultural Systems* 85, pp234-253

Yap, V. Y., de Neergaard, A. and Bruun, T. B. (2017) '"To adopt or not to adopt?" Legume adoption in maize-based systems of northern Thailand: Constraints and potentials', *Land Degradation and Development* 28(2), pp731-741, doi:10.1002/ldr.2546

Zhang, L., Kono, Y., Kobayashi, S., Hu, H. B., Zhou, R. and Qin, Y. C. (2015) 'The expansion of smallholder rubber farming in Xishuangbanna, China: A case study of two Dai villages', *Land Use Policy* 42, pp628-634

Ziegler, A. D., Fox, J. M. and Xu, J. (2009) 'The rubber juggernaut', *Science* 324, pp1024-1025

Ziegler, A. D., Fox, J. M., Webb, E. L., Padoch, C., Leisz, S. J., Cramb, R. A., Mertz, O., Bruun, T. B. and Vien, T. D. (2011) 'Recognizing contemporary roles of swidden agriculture in transforming landscapes of Southeast Asia', *Conservation Biology* 25, pp846-848

3

POLICIES IMPACTING SHIFTING CULTIVATION

Getting them right

*John Lindsay Falvey**

Policy context

What policy concerning shifting cultivation would be 'right'? This essay discusses the subject by eliciting what has been learned and offering a context for engagement with policy-makers. Making the 'right' policies is a worthy ideal, but only worthy so long as we recognize that ideals are goals and not destinations. Having such awareness, we can accept the inevitable compromises involved in policy formulation. However, where single-issue advice influences policy, unforeseen and unwanted outcomes can result.

As I have discussed elsewhere, such single-issue advice can be regarded as bias (Falvey, 2013a). I offer just two examples from the past. In the first case some development agencies resisted funding dairy development in tropical nations because of the misguided belief that milk production in the tropics could not be efficient and that milk products were best purchased from temperate dairy nations. Noting that the world's largest dairy producing and consuming nation was India, informed experts were invited to contribute to a book ultimately published by the International Livestock Research Institute, a member centre of the Consultative Group for International Agricultural Research (Falvey and Chantalakhana, 1999). In concert with other forces, this book helped to widen the development agenda away from counterproductive Western conceptions of efficiency and market development.

In the second example, smallholder farmers have been marginalized by policies that favour large-scale agriculture, often expressed in modernizing terms that mimic Western development (Falvey, 2010). This bias continues, despite small farmers in poor nations providing an overall economic benefit that exceeds that possible from their indiscriminate replacement by large-scale farming. Such policies displace

---

* Professor John Lindsay Falvey ftse, Faculty of Veterinary and Agricultural Sciences, University of Melbourne, Australia, and Chair, International Livestock Research Institute, CGIAR.

small farmers who feed themselves and their families from less than two hectares of marginal land and force them into urban environments with concomitant additional demands for welfare and food.

The world has about two billion small farmers, including an unspecified number who practise shifting cultivation. However, these families are spread across nations, often in remote border regions and mountainous terrain, and the policies that affect them are made in the overall interests of the nations in which they reside. In addition, the pace of economic change in most nations has accelerated, with new innovations producing a context of continuous change. For these reasons, the first and rather obvious point to be made is: *policies that affect shifting cultivators have not been aimed specifically at their welfare or their environment.*

What is different today?

Diverse experiences from past decades, countless research papers and development conclusions have provided us with detailed information about particular environments. It is essential to review such information. However, reviews of area-, technology- or culture-specific studies on their own can ignore the overall development context that policy engenders. Even in mainstream societies, people in cities and large-scale agricultural areas are impacted by national and development policies not aimed specifically at their livelihood, comfort or environment. So we might accept it as routine that policy impacts on shifting cultivators over the past 50 years have been a normal part of living in society (Fox et al., 2009). Such 'rain falls on the just and unjust alike', even if a centuries-long time scale is considered.

In past centuries, shifting cultivators may have been remote from centres of civilized power. Nevertheless, they were affected by these centres, at least to the extent that there were established barriers to integration with them. Today's forces of homogenized development and culture, based on communication and infrastructure technologies, may be seen as a final stage in the centuries-old intention of central powers to 'close the frontier'. In the past, this might have been seen as subduing barbarians, securing buffer regions or simply extending the realm. Today's parlance uses such terms as expansion of modern agriculture, reforestation, catchment management or reservation of national parks. Given that such terms were generated by central authorities, there is little conceptual difference to past forces.

Of course it would be disingenuous to argue that nothing is substantially different from the past, but today's differences relate to mechanisms and tipping points. Reduced availability of land often makes shifting cultivation unviable; opportunities have widened and aspirations have blossomed. These are points that concern many commentators about shifting cultivators and their environment, but we should ask ourselves: are they matters that require specific policies separate from, for example, national policies?

Regardless of the response, allowing the silent demise of shifting cultivation carries the danger of wasting a knowledge resource and possibly even damaging natural resources. And when we consider our current knowledge, it becomes clear that our knowledge bank is patchy; we do not understand much about the biological dynamics of shifting cultivation, nor do we know whether permanent agriculture or forest will improve all areas. Therefore, I submit that criticizing the inexorable creep of commercialized agriculture (Fox et al., 2009) or portraying national policies as cultural hegemony are unhelpful in these discussions. We can do better than simply repeating Western value judgements about other cultures and environments. We may elicit as a second point for further consideration that *it is useful to take an historical perspective of change, in order to accept the way things are as the context for policy and to define both policy and shifting cultivation.*

Defining policy and shifting cultivation

If we define policy as a continually adjusted process for implementation in accordance with agreed basic principles, then there are myriad policies that impact on shifting cultivators and their environment. And where multiple policies apply, unintended outcomes are inevitable. Such outcomes fuel criticism of governments for the policies they have imposed. But to assume that governments today have complete freedom in policy-making overlooks the influence of global forces that can compound diverse policies. The influence of the developed world often requires national development actions to conform to international protocols. In the case of shifting cultivation, externally financed economic-development and climate-change agreements may compromise otherwise well-meaning national policies. Sound policies recognize their dynamic situation and accept essential compromises. So we may make the point that *policy is a continually adjusted process implemented in accordance with agreed basic principles and complicated by a range of forces, both internal and external.*

Shifting cultivation also requires definition. As argued elsewhere, the history of civilization can beneficially be portrayed in agricultural terms, since it was the first agricultural settlements that established the conditions for civilization (Falvey, 2013b). Such a perspective of the history of agriculture helps in reconciling some ongoing misunderstandings about shifting cultivation. Just as the early confrontations between pastoralists and settled farmers depicted in early writings (Genesis 4: 1-15) continue into present day denigration of 'primitive nomadic tribes' in some regions, so shifting cultivation is maligned as a primitive form of agriculture. Yet, surprisingly, nomadic rangeland management has proven to be sustainable in areas where fenced ranching and farming have not (Falvey, 2015). One reason for that surprise was a failure to recognize wise management systems that rested lands through sensitive periods by following complex systems of usage rights. So it is with shifting cultivation, with its managed forest fallows that are the longest-lasting environmental component of that sophisticated management system. Each agricultural land use has its place in the rural landscape. Figure 3-1 is a general representation of the history of rural landscapes, indicating a continuing place for shifting cultivation, although admittedly in specific circumstances.

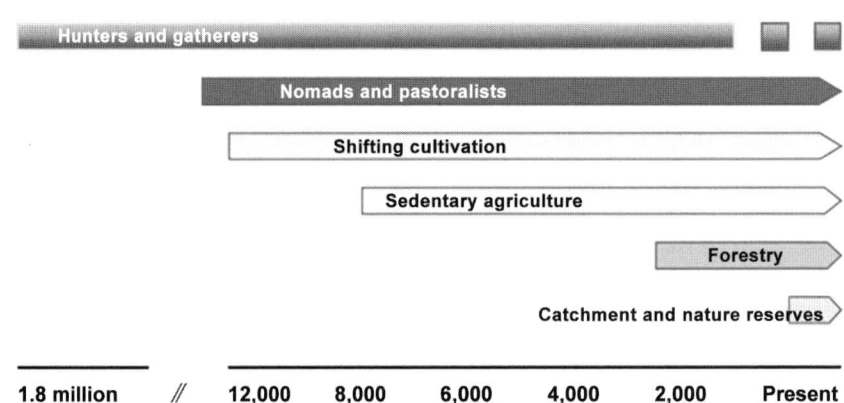

FIGURE 3-1: A general history of rural landscapes.

Shifting cultivation embodies a range of efficient natural farming practices that have evolved over the past 12,000 years (Spencer, 1966). Its benefits escape modern agricultural consideration, possibly for the same reasons that it has proven so durable. Piecemeal considerations of modern agriculture can overlook the integration of agroforestry, animal husbandry, forest management, fallow management, horticulture, food-grain cropping and cash cropping that are all part of shifting cultivation. It has similarly escaped modern environmental consideration when it is pejoratively referred to as 'slash-and-burn'. In fact, common diagrammatic representations of shifting cultivation can create an unintended reaction when presented in a conventional form, as in the left-hand image in Figure 3-2, which can imply that the forest-fallow period is only one-third the duration of the cycle. The right-hand adaptation might be more accurate.

From this, we might consider that rather than dismissing shifting cultivation as a sunset practice, we would do better to define it as *a poorly understood traditional*

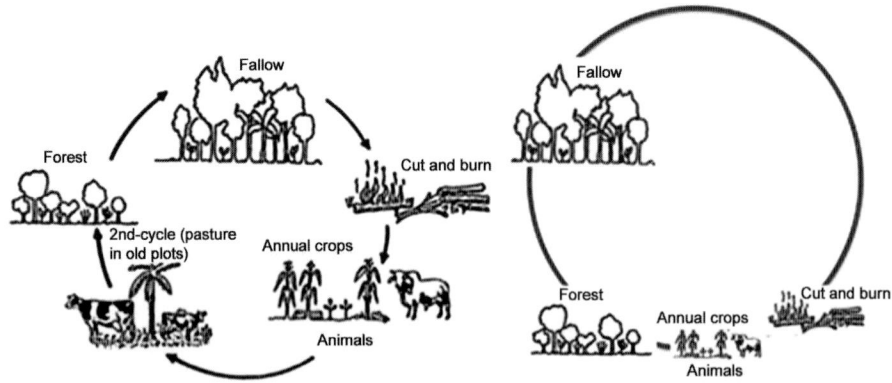

FIGURE 3-2: Correcting conventional diagrammatic representations of shifting cultivation, with conventional (left, Dubois, 1990) and more accurate (right).

farming system integral to various cultural identities. It persists because of continuous change to its land-use system of long forest fallows that are cleared and burned to allow short cropping phases.

Who is a shifting cultivator?

We are blessed with myriad anthropological studies describing the rites, beliefs and myths of individual groups, and in many cases these include pioneering descriptions of shifting cultivation cycles and practices. Agricultural, ecological and demographic studies have built on these works. Thus, it has been estimated that about 300 to 500 million shifting cultivators (Kleinman et al., 1996) live in 40 to 50 countries (Mertz, 2009), mainly in mountainous and upland regions. Is such information useful for the promulgation of policy? We are not able to answer unequivocally, because we must also consider that another estimate puts the number of shifting cultivators living in Southeast Asia – where the majority is said to live – somewhere between 14 and 34 million (Mertz et al., 2009a). Such a lack of clarity is an indication of the alien nature of the practice and the lifestyles of shifting cultivators in the worldview of policy-makers who are reliant on data.

There is also, at least for many east of the Himalayas, uncertainty about the origins of shifting cultivators. They may be historical tribes, refugees from the burdens of civilization, or qualify for that internationally sensitive description: indigenous. It might be argued that this justifies further anthropological research, to better inform policy. Others may claim that the location-specific nature of such research and the biases introduced by site selection and funding sources would limit the utility of any research findings. If assumptions behind policy-oriented research are of national homogeneity, then the results in such circumstances might well support one origin over another (Persoon et al., 2004). This readily becomes an area of confrontation between national aspirations and international ideals of cultural rights, expressed by non-governmental organizations and externally funded development projects. Lack of information about the numbers and origins of shifting cultivators confuses national policies in the face of international protocols, agreements and conventions to which national governments subscribe as global citizens. This implies that we should acknowledge that *national policies influenced by global forces usually take precedence over specific policy proposals oriented towards shifting cultivators and their environment.*

Shifting cultivators or 'highlands'?

Assuming that the majority of shifting cultivators are in montane Southeast Asia (Mertz et al., 2009a), even though their numbers are apparently in decline (Padoch et al., 2007), policy discussions will necessarily focus on highland regions. While highlands or mountains receive some specialized attention, they are a poor cousin to lowland agriculture and cultures. This is one of the reasons for their recognition in Chapter 13, Agenda 21 of the 1992 United Nations' Earth Summit and the designation of 2002 as the International Year of Mountains. With 12% of the world's population, or more than 700 million people, estimated to live in these regions and another 40% relying on mountain watersheds, their common issues are routinely defined as inaccessibility, environmental fragility, political marginality and cultural and environmental diversity (von Braun, 2005). From this perspective, it would appear that *'highlands' is a policy grouping preferred by shifting cultivators and is a policy area particularly influenced by global agendas.*

I have written previously that 'we all know that highlands are vulnerable ecosystems, that they are home to large numbers of people, [and] that they service water and biodiversity while supplying food and landscapes. As nation-building projects around the world now move to fully integrate these once marginalized regions, there is a need to consider our collective experience with the hill peoples and their lifestyles. National perspectives commonly define hill populations as poor, exploitive of natural resources and alienated from education and health services. Past hill-dwellers may well have valued these aspects of their lifestyles and viewed themselves as enjoying freedom from the excessive demands of civilization. But today's communication technologies have probably ended that era. This does not mean that past "civilizing" programmes for the highlands are now vindicated and can be widely implemented. Our science, including the social sciences, suggests that it is an opportune time to reflect and consolidate knowledge for future development advice' (Falvey, 2015). In the context of policy improvement, I must now qualify this statement to note that *this does not mean that we should continue with the same research and projects of past decades.*

With most of the world living in Asia, it is doubly important to recognize that the majority of the world's highlands are also in Asia, as indicated in Figure 3-3 (Huddleston et al., 2003). The International Centre for Integrated Mountain Development (ICIMOD), an intergovernmental centre serving eight countries in the Himalayan region, extends in the east to cover Myanmar, but is otherwise uninvolved in Southeast Asia. Integrating institutions from its eight member nations, ICIMOD aims to 'economically and environmentally … improve the living standards of mountain populations and to sustain vital ecosystem services … '. ICIMOD's work is relevant beyond its mandate, but it is constrained by the interests and influences

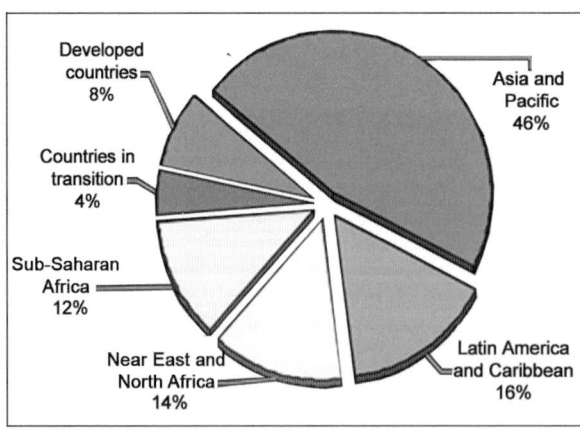

FIGURE 3-3: Location of the world's highlands.

of donors, and so reflects their worldviews in its recommended development policies (ICIMOD, n.d.).

Similarly, lessons are offered by more distant experiences. For example, in the European Alps, subsidized agriculture may be seen as contributing to the cultural heritage, even though its productivity may be constrained by the topography. In that situation, its 'environment-friendly … management depend(s) primarily on the farmer's professionalism and/or an intensity of operation adapted to the location' (Ruffini and Streifeneder, n.d.). To achieve a desirable environment, we would do well to note that *some national policies that are successful probably depend on the strength of national capacity, autonomy and wealth.*

In looking at the Southeast Asian highlands as a policy focus, lessons from elsewhere must be one source of information, and *there are diverse sources of relevant knowledge and experience around the globe.*

Shifting cultivation in Southeast Asia as a policy focus

With the largest share of the world's highlands and most of its shifting cultivators, Southeast Asia – including highland Yunnan province in China – has also experienced most of the research, development projects and angst that accompanies the subject. It has advocates for agriculture, markets, forests, cultures, individual rights, access to national welfare, economic development, relocation, tourism, watershed management, research museums and more.

Experience has been diverse and contradictory. For example, efforts to compare the livelihood vulnerability of shifting cultivators and their impacts on the ecosystem with alternative uses of the same lands have beneficially highlighted the complexity of shifting agricultural systems (Dressler et al., 2015). Experience has indicated the poor record of relocation programmes, such as that in Laos, where implementation assumed governmental omnipotence (Yokoyama et al., 2006), but separated beliefs, traditions and rites from the physical environment and so compromised attitudes to new locations (Tanaka, 1993). More positively, in Yunnan experience has shown that, rather than leading to permanent loss of forest, shifting cultivation can produce more biologically diverse forms of forest succession. This contrasts with past government policies supporting permanent food production, which led to loss of forest cover. In the midst of all this intervention, study and analysis, one policy that produced a beneficial outcome was aimed at neither shifting cultivators nor even the highlands. It was China's Household Responsibility System, which permitted freedom of decision-making within a related national policy of increased agricultural intensification and forest conservation (Xu et al., 1999).

It was a curious distinction from aid-related projects, that the Chinese experience did not overtly promote poverty alleviation and environmental protection, although these were among the outcomes. It is even more curious that experience from aid projects also indicates that more effective development arises from communities managing their own assets than prescriptive programmes that aim, for example, to modernize agriculture and marketing. One study in Vietnam found that improvements within existing shifting cultivation systems, combined with new enterprises, produced greater benefits of poverty alleviation and environmental protection than government-sponsored 'modernization' (Tran et al., 1999). In that case, cattle enterprises integrated well with existing systems and became the primary source of income; yet development programmes had excluded cattle in favour of mainstream lowland enterprises of pig and rice production. In addition, household food security was consistently greater when shifting cultivation was maintained.

Evaluation of such projects is one source of information on which policy must rely, for which the Yunnan Sustainable Highland Agriculture in Southeast Asia Project (Subedi et al., 2009) provides a useful example. It noted the risks of the 'general tendency to consider the completion of project activities as the full achievement of project objectives'. Its analysis claimed scientific and technical outcomes from short-term improvements in crop productivity and reductions in soil and water losses. There was an heroic list of suggestions based on 'good practice(s) for planning/designing', which included (I paraphrase): participatory research; stakeholders' engagement; involving local research and development networks; participatory project planning; realistic objectives; ensuring continuity post-project; and baseline surveys for later evaluation comparisons. In addition, it suggested that technologies should be chosen that had shown 'rapid returns and longer-term benefits' without demanding more labour or costs. The analysis was a laudable attempt to improve development projects and each of the suggestions was no doubt useful, but I question whether such conclusions can be drawn from one development project of limited duration. In any case, the list is too idealistic for policy formulation and it perpetuates the suspect assumption that immutable project plans can foresee the future. From a development project viewpoint, as discussed elsewhere (Falvey, 2015), *plans cannot be practical unless they are sufficiently flexible to allow change during implementation.*

Further constraints include the limited duration of aid interventions, with their burdens of multiple and competing objectives. This may not be the perception of funding agencies or even recipient governments, and if this is the case it might be a reflection of the distance between such interventions and local aspirations. Where projects have a consistent focus over many decades and accord with the general direction of change, enduring benefits may occur, such as in the Royal Projects in the Thai highlands (Angkasith, 2015).

A wide review of development experience has collated these and other experiences into key failures that may serve to inform policy formulation. These include:

- defining shifting cultivators as tribals outside the nation-building definitions of civil society, thereby denying them land-use rights;
- planning land-use to favour forests, permanent agriculture, commercial plantations and logging contractors;
- interpreting environmental conservation as reforestation with increased state control;

- relocating shifting cultivators to new lands; and
- orienting infrastructure and incentives to commercial agricultural developers.

And rather than seeking to redress each failure, the review observes that 'the conditions necessary for swiddening, both the availability of land and the aspirations of people, simply no longer exist in many parts of Southeast Asia' (Fox et al., 2009). This demise of shifting cultivation may turn out to be the pertinent fact for policy development.

The summaries in this section are a taste of the information base that is available for policy formulation. It does not contradict my earlier implication that we have insufficient knowledge about shifting cultivators and their environment, despite hundreds of anthropological, geographical and agricultural studies. Formulation of agricultural, environmental and social policies requires knowledge applicable over large regions and numbers of people. Attempts to provide such information now utilize aerial mapping to classify shifting-cultivation areas into agricultural, degraded or forested lands, according to the phase of the cultivation cycle when the maps are made. However, the timing of mapping and the small size of many cultivated fields can lead to them being overlooked by such remote sensing (Ramankutty and Foley, 1999). As already pointed out, policy cannot be relevant when neither the population nor the area concerned is known. It may even unintentionally marginalize such peoples and areas. Contrary to our well-intentioned desires, *all of the experience mentioned above suggests the merit of avoiding specific policy provisions for shifting cultivation areas and swidden cultivators.*

Unintended marginalization

Some policies in Southeast Asia have inadvertently accelerated the demise of shifting cultivators. These include policies that have classified them as ethnic minorities within nation-states, defined rural highlands as either forested or agricultural, promoted single-minded conservation ideals, assigned land tenure based on permanent land use and increased accessibility (Fox et al., 2009). Yet it is generally agreed that these objectives provide the greatest benefit to the majority in a nation. However, on the basis of United Nations' and other ideals, today's arguments also include rights to protection from collateral policy damage.

Such ideals are wider than those of human rights. An appraisal of more than 100 peer-reviewed studies conducted over the past 20 years has concluded that shifting agriculture in Southeast Asia is attracting increased attention as a result of global concern about carbon emissions (Peng Li et al., 2014). We may leave aside the question of whether this is an equitable application of global initiatives and applaud the fact that it has provided a useful fillip to better define shifting cultivation and forest areas from several decades of Landsat imagery (Woodcock et al., 2008). In reality, however, both the new information and carbon-related policies have, like predecessor global concerns such as opium production, further marginalized shifting cultivation.

Knee-jerk reactions to the burning phase of shifting cultivation as a source of carbon emissions is reminiscent of forest–conservation policies from at least colonial times. Such scientism readily infuses development policy and has produced such aims as seeking to 'reduce emissions from deforestation and forest degradation' by outlawing shifting cultivation (Griffiths, 2008) even as intensification of industrial agriculture is shown to be the sector's major source of carbon emissions (FAO, UNDP, UNEP, 2008). In fact, routine shifting cultivation practices contribute to carbon sequestration by encouraging the regeneration of secondary forest by trimming trees to allow them to re-sprout; by maintaining long fallows with short cultivation periods and creating higher levels of soil organic matter than is normal in continuous cropping. This has now been noted by the Intergovernmental Panel on Climate Change, in the terms of 'forest clearing for shifting cultivation releases less carbon than permanent forest clearing because the fallow period allows some forest regrowth' (IPCC, 2006). But we must wonder what impact such a realization, buried in one paragraph of a long report, will have amid other ideas and detailed information in a politically charged agenda.

When we consider such things, it might be argued that we have not progressed in our understanding beyond the 1957 description of shifting cultivation by the Food and Agriculture Organization of the United Nations as 'a backward type of agricultural practice … [and] … of culture' (FAO, 1957). I would rather state that we have become more aware of the resilience of shifting cultivation systems, and research that once sought to replace such systems is now seeking to 'understand this complex and evolving form of land use' (Mertz et al., 2009b). Is this too little too late? Probably.

Such pessimism appears to be supported by persistent misconceptions of shifting cultivation as a primitive step towards commercial agriculture; uniform and unchanging; exclusive of other occupations; low in productivity; environmentally unsustainable; reliant on primitive technologies and understanding; and devoid of land-use rights (Thrupp et al., 1997). For the sake of clarity, let us acknowledge the counters to each misconception, viz: shifting cultivation is a viable agricultural system in its own right; it is dynamically adaptive; it is practised in concert with other enterprises; it is efficient in its own terms; it has been sustainable across centuries; it utilizes sophisticated and specific practices; and it has produced its own land-use and management systems.

In a related essay, I used a simple definition for activities in the highlands: 'sustainable development of the highlands simply means guiding enhanced long-term use in a manner that causes minimal negative impact' (Falvey, 2015). It aimed to escape from advocacy of one production system over another. We might widen the discussion to note that a closer observation of much shifting cultivation reveals it to be a dynamic approach that adapts to changing environments, and in many cases might otherwise be classified under the more acceptable terminologies of rotational cropping (Padoch et al., 2007) or even agroforestry. As an ICIMOD conference summarized: 'if we change our label from slash-and-burn to rotational agroforestry or agroforestry with a burn cycle, or a form of forest gardening, then we start to use positive words that focus on the growth cycle rather than the cutting cycle. As studies show, farmers spend many more years growing trees and crops rather than burning them – protecting soil, restoring nutrients, fallowing and resting' (Kerkhoff and Sharma, 2006).

The ICIMOD conference recommended that policies that discouraged shifting cultivation should be replaced by addressing land tenure, along with credit access, research and extension, market development and commercialization of new niche products; strengthening customary institutions; and coordination of government agencies. Such observations seem to be supported by a longer-term study in Laos, where changes in land cover, land use and livelihoods indicated a relationship between the vulnerability of highlanders and policies that aimed to increase forest cover, eradicate shifting cultivation and stimulate intensive and commercial agriculture. Rather than designating areas specific to agriculture or forests, the study suggested that the focus be shifted to means of maintaining diverse livelihoods and economic opportunities (Castella et al., 2013). Similar experience in Kalimantan suggests that external guesses about suitable cash crops to grow in combination with food crops are seldom ecologically stable, although in that case locally known rattan production seemed to be an exception (Weinstock, 1996).

Bias and inadequate science persist (Erni, 2009). Such insightful developments as community-based forest management and agroforestry, both incorporating shifting cultivation, may be positive products of such experience, so far as we can see at present. While this may be the prevailing direction of global and national development thought, it contrasts with arguments that rely on the United Nations' Declaration on the Rights of Indigenous Peoples, when the declaration is interpreted to mean that indigenous peoples should be preserved in a museum-like lifestyle. Such interpretations unnecessarily fuel arguments in which people are pitted against the environment.

People, forests or knowledge?

Individuals have rights, but individual rights bow to group rights, whether the group is defined democratically or autocratically. It is simple for a more powerful group to make, for example, blanket recommendations that highlands should be locked up as mountain forests. At present, the world's highlands are said to cover about nine million square kilometres. They provide fresh water for domestic use and for hydropower, industrial and transportation purposes, and they absorb some of the impact from lowland carbon dioxide emissions. To argue for one group over another is to ignore the lessons of experience from multifunctional approaches, such as agroforestry and community forest management. Multifunctional approaches are *ipso facto* more complex than single-focus development and rely on objective use of all available knowledge, including that from stakeholders and well-conducted research.

I have written elsewhere about a long-past project that provided an example of the benefits of an open-minded approach. I wrote: 'the two ingredients of stakeholder inputs and research formed the basis of the Thai-Australia Highland Agricultural Project … In many ways that project was ahead of its time, for it was based on conducting and publishing both social and technical research and it matched well-supervised enthusiastic young scientists with generous budgets. It differed from many of its 1970s contemporaries and 1980 successors in generating knowledge rather than abiding within a log-framed plan of development conceived during a "project design" phase' (Falvey, 2015). Rather than relying on highlanders being 'passive beneficiaries of trickle-down development or technology transfer' (Leach and Scoones, 2006), the Thai-Australia Highland Agricultural Project avoided the agent-projection of 'bottom-up' and the arrogance of 'top-down' approaches by integrating social and technological research to assemble knowledge and build upon it.

There was concern at the time the project began that legislation to protect forests would prove to be ineffective, so forest research formed part of the research base, and this led to such social experiments as payment of hill dwellers to manage forests in recognition of the slow growth rates of trees and the environmental and economic compatibility of grazing ruminants and other forms of agroforestry. The alternative, subsequently witnessed in Laos and Myanmar, was even less control over logging and greater displacement of residents. The conclusions from the 1970s research are being reached again today (Price et al., 2011), which is fortunate since we now have active forest-related research from the Centre for International Forestry Research (CIFOR, www.cifor.org) and the World Agroforestry Centre (ICRAF, www.worldagroforestry. org), both members of the Consultative Group for International Agricultural Research (CGIAR).

Before considering the specific country example of Thailand, it is appropriate to recap the counter-productivity of arguing that shifting cultivators have greater rights than the majority in their nations, or that environmental protection should be ranked above historical practices. Instead, we might say that *integrated approaches may be inevitable and that research could be beneficially oriented to realistic policy rather than a single ideal.*

The case of the Thai highlands

The Thai highlands are one part of a large contiguous region that extends beyond Southeast Asia. These are highlands above an altitude of 300m, spanning Bangladesh, Cambodia, China, India, Laos, Myanmar, Thailand and Vietnam: the Southeast Asian Massif (Michaud, 1997)(Figure 3-4). The definition was sociological more than geographical, like the name Zomia, given to the same highlands but with the addition of a vast part of the Himalayan massif (van Schendel, 2002). Despite the social definition, the area has never been ruled as a single entity and, in political terms, it has been used primarily as buffer space between major powers. For this reason, the peoples of the region have sometimes been studied as isolated tribes and ascribed specific cultural traits and original myths that have later proven to be shared and

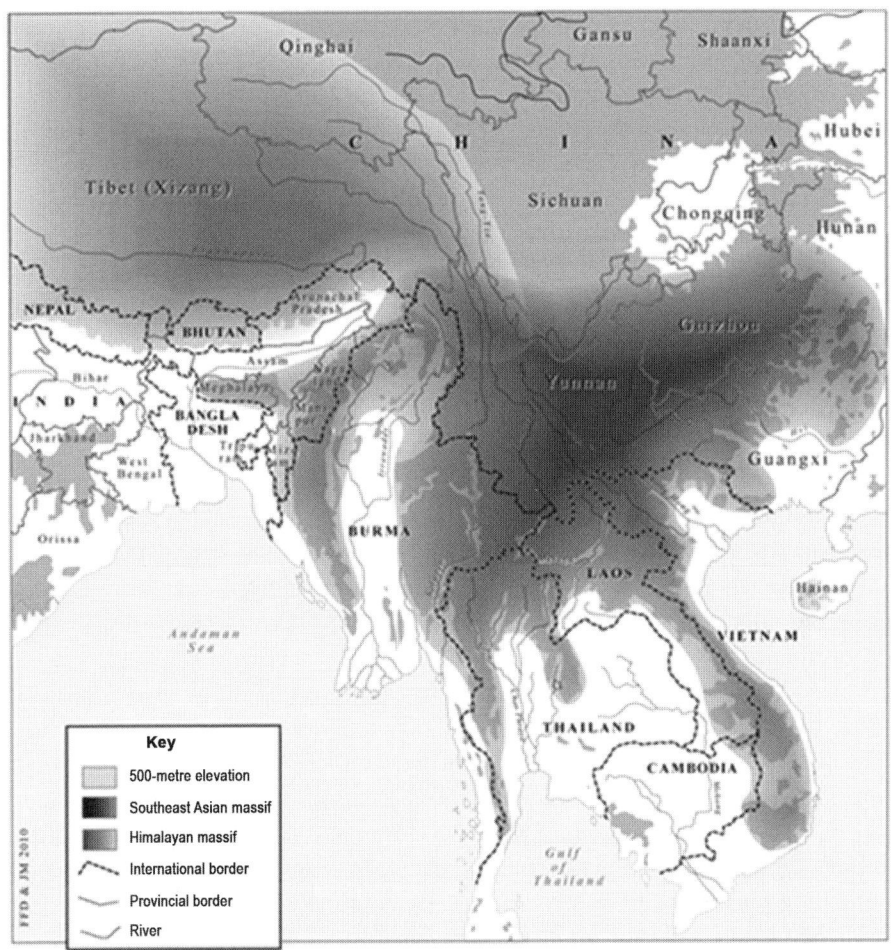

FIGURE 3-4:   The highlands of Southeast Asia.

Source: Michaud (2010).

interchangeable with those of other highland groups and related to lowland cultures. One thesis is that these anomalies are explicable if hill dwellers are understood as self-disciplined refugees from the burdens of the civilized lowlands (Scott, 2009).

In Southeast Asia itself, the highlands are mainly steep north–south ranges between 500 and 2000m in altitude, separated by tributaries of medium to large rivers: the Chao Phya system and the Irrawaddy, Mekong, Red and Salween rivers. Distinct wet and dry seasons and cooler temperatures than lowland regions have produced montane rainforests. Today, much of that forest has been replaced by secondary forests, forest plantations, *Imperata* grasslands and agriculture. In the past, the lifestyle of highland residents was based around shifting cultivation of food crops, opium and feed crops above about 1000m. From the 1960s, policies arose to replace opium through aid-funded interventions. These began in Thailand due to its relations with the West and spread to Laos, Vietnam and Myanmar.

Beginning more than 40 years ago, bilaterally and multilaterally funded projects in the Thai highlands included specific opium-replacement objectives, to which were variously added other development components, including infrastructure, sustainable agriculture, education and health. Variable in impact, these projects usually focused on a specific pre-defined series of interventions for a short period, often less than a decade. And while their objectives were clear in the minds of external agencies, they were inevitably understood as subordinate to Thailand's national security policies and parallel to ongoing arrangements that were often opaque to foreign advisers. A

Papaver somniferum L.
[Papaveraceae]

Historically a major part of highland subsistence in northern Thailand, development projects began about 40 years ago to seek an end to opium growing.

jest of the 1970s was that some agricultural interventions might be seen as projects to improve opium, rather than replace it.

Through the 1950s, Thailand's National Economic Development Plans began to emerge, with the first plan produced in 1961. Mainstreamed from 1972, when the country's National Economic and Social Development Board was moved to the Prime Minister's Office, the third plan, spanning the years from 1972 to 1976, was used to coordinate government actions related to development. For the highlands, the plan introduced a land-capability approach to designate zones in which the administration of diverse government agencies could be coordinated under the Department of Public Welfare, within the Ministry of Interior. The third plan specifically noted in its introduction that 'national plans are extremely important in establishing the economic development policy of the country' (Government of Thailand, 1972).

Thus, the policy environment for development was communicated, including recognition of the 'unique social and development problems' of the highlands. Noting the ethnic and linguistic differences of highlanders and their propensity for producing opium, it was stated that the Border Control Police and the Army were already working to register and educate 'hill tribes people' as part of road-building and agricultural-diversification programmes. Consideration was given to leasehold or titling of land for those who 'intensify their agricultural activities' so that 'land thus released from shifting cultivation would then be forested'. In parallel with these interventions, research into crop diversification and livestock development were to be initiated. It has long been observed that these policies aimed to integrate remote communities into the nation (Tapp, 1979).

This was the context for the foreign-assisted projects of the 1970s and, to a large extent, for the Royal Projects. Succeeding economic development plans continued these intentions with an increasing political spin, such that the eighth, 10th and the current 11th plan attempted to use self-sufficiency as a central ethic, within which hard commerce was promoted. The main focus sees the highlands as watersheds responsible for unmanageable flooding in the lowlands, a problem to be addressed by community forest management and agroforestry. Shifting cultivation is not mentioned, although burning is noted as an air pollution issue. The document might even go as far as misleading those new to the subject into believing that highlanders are part of Thai Buddhism and have long spoken a Thai dialect! (Government of Thailand, 2012).

The Thai example does not differ markedly from others in Southeast Asia, and serves to highlight the reality of the policy environment. Minority rights, even as part of pan-regional programmes, have not been a national priority. With the benefit of hindsight, it is sometimes asked, somewhat cynically, whether foreign projects could ever have been viable when based on single ideals, rather than on the hierarchy of Thai Government policy and the plans of the National Economic and Social Development Board. The Thai experience thus adds the point that *any highland development will be within the overall national interest, which is centred on matters of security and nation-building and served by economic development.*

Utility of development projects

We have seen that the impact of a project may be felt far beyond its evaluation phase and in manners unforeseen by either implementers or evaluators. In the case of the Southeast Asian highlands, impacts from a national perspective will ultimately be judged against first-level national interests. In 1970s Thailand, this was the realm of the Army and the Border Control Police. The roles were further substantiated in the 1980s paranoia about communists and uncontrolled highland borders.

Highland lifestyles have changed remarkably since those times. Dirt-floored tubercular huts are now scarce; accumulating cattle for wealth and ritual sacrifice or for pack animals (Falvey, 1980) has become old-

Imperata cylindrica (L.) Raeusch. [Poaceae]

A species accused of constraining forest regrowth. Colonizing exhausted shifting cultivation fields in an impenetrable sward, *I. cylindrica* is invigorated by regular fires.

fashioned; raising pigs for fat in the absence of other oils (Visitpanich and Falvey, 1981) is no longer necessary; corn-cob tea has been replaced by coffee from a thriving new industry (Angkasith and Warrit, 1999); and growing opium to trade for rice is almost unheard of. I am most familiar with the agricultural work of the highlands, and note the beneficial outcomes of many projects. Perhaps those that provided the most were oriented towards research, rather than development, with the exception of the Royal Projects (HRDI, n.d.), which owe their success to alignment with the highest-level national objectives and consistent implementation over more than 40 years.

The 1970s foreign-aid interventions, spurred by rising drug problems in the West, potentially confused Thailand's national security imperative. Notwithstanding such a milieu, these projects generated useful information and achieved some development outcomes that might not otherwise have occurred as soon. Today, many projects that aimed to replace opium with commercial crop and livestock enterprises may seem to have been naïve, even disingenuous. Nevertheless, they widened options for livelihoods and development and improved nutritional standards. Nowadays, it seems to me that much that has been done is lately being repeated, which suggests that *short-term foreign interventions may be better focused on reputable research in support of national objectives, because institutional memories in aid environments can be unproductively short.*

I have summarized the outcome of the 1970s Thai-Australia Highland Agricultural Project in another essay (Falvey, 2015), and so I only need mention it here to emphasize the research aspect of the above conclusion. This small project produced research outcomes that have been applied in diverse environments beyond their intent. In contrast to contemporary projects, research spanned social, economic and technical fields across crops, livestock and human well-being. The number of research publications may not indicate much, but the application of their findings does, for as was concluded: *development relies on real information from research, and development research relies on an understanding of the socio-economic values and constraints in order to design its technical experiments.*

So development projects have been of use in the implementation of policy, especially when it is realized that the policy context is usually wider than project personnel realize. An integrated approach to highland development is essential today, amid such diverse influences as modern lifestyles, forest replanting, high-value horticulture, off-farm work and mechanization, all served by improved communication systems. One suggestion is that 'in order to solve those problems that are mainly of an agro-technical nature, a sound research-and-extension strategy is needed. But to solve problems that are more of a socio-economic nature, appropriate policy regulations should be worked out' (van Keer et al., 1998). If this implies respect for national objectives as the basis of policy, I see it as a step towards making the outcomes of development projects more useful. It is likely to be more productive than lecturing national governments to develop specific policies for a minority, whether these are policies for land titles, citizenship or cultural diversity, all of which fall under broad national objectives.

Conclusion

There is much to learn from research and development experience to assist our understanding of policies related to shifting cultivation. But since policy must be as dynamic as the environment to which it applies there can be no single 'right' policy on a subject as diverse and marginalized as shifting cultivation. Perhaps it can be said that poor policy has reduced areas for shifting cultivation and so intensified the practice by shortening fallows, fractured communal responsibility, increased dependency on external markets and threatened sensitive environments. Such observations are widely argued to mean that policies should recognize shifting cultivation as dynamically constructive for both the environment and livelihoods because it manages agricultural fallows as medium-term forests. Further, the argument goes, it should be supported by relevant research linked to farmer innovation and rewards for roles of biodiversity conservation, fertility management and pest control through controlled burning, and the development of niche products should be encouraged. However, as this essay has pointed out, policy is overwhelmingly national and the causes of small numbers of people or small areas are likely to be subsumed into higher-level priorities.

In considering development-oriented programmes in the highlands, I have concluded that plans must be flexible, must acknowledge that impacts can often be felt well beyond those planned, and must learn from the experience of past research, including socio-economic values and constraints (Falvey, 2015). I have also concluded that, despite a wide variety of experiences, highland development has been a marked success over the past 40 years. This is a positive outcome, but many of the successes that we might claim for our own specific disciplines or projects have in fact relied on higher-level policies, for example, to stabilize highland regions politically and to integrate their peoples into a nation.

In terms of policy, the 15 italicized points made throughout this discussion may now be extracted and assimilated. The points may overlap, but are repeated individually in the order they appeared in the text. Overall, we may conclude that:

- policies that affect shifting cultivators have not been aimed specially at their welfare or their environment;
- an historical perspective of change means accepting the way things are as the context for policy;
- policy is a continually adjusted process implemented in accordance with agreed basic principles and complicated by a range of forces;
- shifting cultivation is a poorly understood traditional farming system integral to cultural identity and persists as it continuously changes its land-use system of long forest fallows that are cleared and burned to allow short cropping phases;
- national policies influenced by global forces usually take precedence over specific policy proposals oriented to shifting cultivators and their environment;
- 'highlands' is a policy grouping preferred by shifting cultivators and is a policy area particularly influenced by global agendas;
- we should not continue with the same research and projects of past decades;

- some national policies that are successful probably depend on strong national capacity, autonomy and wealth;
- there are diverse sources of relevant knowledge and experience around the globe;
- plans cannot be practical unless they are sufficiently flexible to allow change during implementation;
- there is merit in avoiding specific policy provisions for shifting cultivation areas and swidden cultivators;
- integrated approaches may be inevitable;
- research could be beneficially oriented to realistic policy more than a single ideal;
- short-term foreign interventions may be better focused on research in support of national objectives, because institutional memories in aid environments can be unproductively short; and
- development relies on real information from research, and development research relies on an understanding of the socio-economic values and constraints in order to design its technical experiments.

These points, in combination with the preceding discussion, suggest that it is imperative to acknowledge that major social and technological transformations are continuing as a result of both domestic and global influences, particularly in highland Southeast Asia. Just as changing global realities affect the viability of many family farms in parts of the developed world, so the changing reality in rapidly developing Southeast Asia continues to affect the viability of shifting cultivation. This does not mean that current policies are adequate or even correct, but it does imply that no specific policy for shifting cultivation is likely, and that wider approaches such as well-managed agroforestry may be a more appropriate approach to policy.

We might summarize the implications of this for consideration in future policy formulation – not for specific policies in themselves, but to integrate agricultural, commercial, environmental, political and social factors into research and policy formulation. This includes acknowledging local knowledge from shifting cultivators and their environment to minimize the risk of uniformity being imposed from majority cultures, as well as learning from development experience.

These domestic and global factors can be accommodated within the eight-fold cycle described for effective policy, viz: issue identification, policy analysis, policy instrument development, consultation, coordination, decision, implementation and evaluation, then back to issue identification, etc. (Althaus et al., 2007). This approach may be critiqued as simplistic and divorced from lobbying bodies. Nevertheless, the nations in which shifting cultivation is practised are not always subject to the same non-governmental pressures that have become the norm in more developed economies. So, if we are to 'get policy right', we must understand high-level national priorities and work within them.

References

Althaus, C., Bridgman, P. and Davis, G. (2007) *The Australian Policy Handbook* (fourth edition), Allen & Unwin, Sydney

Angkasith, P. (2015) 'The role of the Faculty of Agriculture and Chiang Mai University in highland agricultural development', paper delivered at the First International Highland Conference, Chiang Mai University, January 2015, http://www.asiahiland2014.agri.cmu.ac.th/uploads/Highland%20%20Sustainable%20Agriculture%20Development.pdf

Angkasith, P. and Warrit, B. (1999) *Highland Arabica Coffee Production*, Mingmoeng Publishing, Chiang Mai

Castella, J-C., Lestrelin, G., Hett, C., Bourgoin, J., Fitriana, Y., Heinimann, A. and Pfund, J-L. (2013) 'Effects of landscape segregation on livelihood vulnerability: Moving from extensive shifting cultivation to rotational agriculture and natural forests in northern Laos', *Human Ecology* 41, pp63-76

Dressler, W., Wilson, D., Clendenning, J., Cramb, R., Mahanty, S., Lasco, R., Keenan, R., To, P. and Gevana, D. (2015) 'Examining how long-fallow swidden systems impact upon livelihood and ecosystem services outcomes compared with alternative land-uses in the uplands of Southeast Asia', *Journal of Development Effectiveness* 7(2), pp210-229

Dubois, J. (1990) 'Secondary forests as a land use resource in frontier zones of Amazonia', in A. Anderson (ed.) *Alternatives to Deforestation*, Columbia University Press, New York

Erni, C. (2009) 'Shifting the blame? Southeast Asia's indigenous peoples and shifting cultivation in the age of climate change', paper presented at a conference on Adivasi/ST Communities in India: Development and Change, 27-29 August, New Delhi

Falvey, J. L. (1980) *Cattle and Sheep in Northern Thailand*, Thai-Australia Highland Agricultural Project, Chiang Mai, www.researchgate.net/publication/236684445_Cattle_and_Sheep_in_Northern_Thailand, accessed 21 September 2015

Falvey, J. L. (2010) *Small Farmers Secure Food: Survival Food Security, the World's Kitchen and the Critical Role of Small Farmers*, Thaksin University Press, Songkhla, Thailand, https://www.researchgate.net/publication/236684402_Small_Farmers_Secure_Food_Survival_Food_Security_the_World%27s_Kitchen__the_Crucial_Role_of_Small_Farmers, accessed 21 September 2015

Falvey, J. L. (2013a) *Beliefs that Bias Food and Agriculture: Questions I'm Often Asked*, Institute for International Development, Adelaide, Australia, https://www.researchgate.net/publication/261357446_Beliefs_that_Bias_Food__Agriculture-_10_Questions_I'm_Often_Asked, accessed 19 September 2015

Falvey, L. (2013b) 'Musing on agri-history', *Asian Journal of Agri-History* 17(2), pp183–191, http://asianagrihistory.org/vol-17/musing-on-agri-history.pdf, accessed 19 September 2015

Falvey, L. (2015) 'Sustainable development in the Thai highlands: The experience of the Thai-Australian Development Project', paper delivered at the First International Highland Conference, Chiang Mai University, Chiang Mai, January 2015, http://www.asiahiland2014.agri.cmu.ac.th/uploads/Sustainable%20Development%20in%20the%20Thai%20Highlands%20Prof%20Falvey%20Keynote%20Paper.pdf

Falvey, L. and Chantalakhana, C. (1999) *Smallholder Dairying in the Tropics*, International Livestock Research Institute, Nairobi, https://www.researchgate.net/publication/236737261_Smallholder_Dairying_in_the_Tropics, accessed 21 September 2015

FAO (Food and Agriculture Organization of the United Nations) (1957) 'Shifting cultivation', *Unasylva* 11, pp9-11

FAO, UNDP, UNEP (2008) *UN Collaborative Programme on Reducing Emissions from Deforestation and Forest Degradation in Developing Countries (UN-REDD)*, Framework Document, Food and Agriculture Organization of the United Nations, United Nations Development Programme, United Nations Environment Programme

Fox, J., Fujita, Y., Ngidang, D., Peluso, N., Potter, L., Sakuntaladewi, N., Sturgeon, J. and Thomas, D. (2009) 'Policies, political-economy, and swidden in Southeast Asia', *Human Ecology* 37, pp305-322

Genesis ch4 verses 1-15, The Bible

Government of Thailand (1972) *Third National Economic and Social Development Plan, 1972-76*, National Economic and Social Development Board, Office of the Prime Minister, Bangkok

Government of Thailand (2012) *Eleventh National Economic and Social Development Plan, 2012-2016*, National Economic and Social Development Board, Office of the Prime Minister, Bangkok

Griffiths, T. (2008) *Seeing 'REDD'? Forests, Climate Change Mitigation and the Rights of Indigenous Peoples and Local Communities*, Update for the 14th session of the Conference of the Parties to the United Nations' Framework Commission on Climate Change, Poznan, Poland, http://www.forestpeoples.org/sites/fpp/files/publication/2010/08/seeingreddupdatedraft3dec08eng.pdf, accessed 25 April 2017

Huddleston, B., Ataman, E. and d'Ostiani, L. (2003) *Towards a GIS-based Analysis of Mountain Environments and Populations*, Environment and Natural Resources Working Paper No. 10, Food and Agricultural Organization of the United Nations, Rome

HRDI (n.d.) *Thailand's Royal Project*, Highland Research and Development Institute, Chiang Mai, www.hrdi.or.th/en/, accessed 25 April 2017

ICIMOD (n.d.) 'About ICIMOD', http://www.icimod.org/?q=122, accessed 17 September 2015

IPCC (2006) 'Land Use, Land-Use Change and Forestry', Paragraph 1.4.1, Intergovernmental Panel on Climate Change, http://www.grida.no/publications/other/ipcc_sr/?src=/climate/ipcc/land_use/008.htm#s6-1, accessed 18 September 2015

Kerkhoff, E. and Sharma, E. (2006) *Debating Shifting Cultivation in the Eastern Himalayas: Farmers' Innovations as Lessons for Policy*, International Centre for Integrated Mountain Development, Kathmandu

Kleinman, P., Bryant, R. B. and Pimentel, D. (1996) 'Assessing ecological sustainability of slash-and-burn agriculture through soil fertility indicators', *Agronomy Journal* 88, pp122-127

Leach, M. and Scoones, I. (2006) *The Slow Race: Making Technology Work for the Poor*, Demos, London

Mertz, O. (2009) 'Trends in shifting cultivation and the REDD mechanism', *Current Opinions in Environmental Sustainability* 1, pp156-160

Mertz, O., Heinimann, A., Rerkasem, K., Thiha, Dressler, W., Pham, V. C., Vu, K. C., Schmidt-Vogt, D., Colfer, C. J. P., Epprecht, M. and Paddoch, C. (2009a) 'Who counts? Demography of swidden cultivators in Southeast Asia', *Human Ecology* 37, pp218-289

Mertz, O., Padoch, C., Fox, J., Cramb, R., Leisz, S., Lam, N. T. and Tran, D. V. (2009b) 'Swidden Change in Southeast Asia: Understanding Causes and Consequences', *Human Ecology* 37, pp259-264

Michaud, J. (1997) 'Economic transformation in a Hmong village of Thailand', *Human Organization* 56, pp222-232

Michaud, J. (2010) 'Editorial: Zomia and beyond', *Journal of Global History* 5(2), pp187-214

Padoch, C., Coffey, K., Mertz, O., Leisz, S. J., Fox, J. and Wadley, R. L. (2007) 'The demise of swidden in Southeast Asia? Local realities and regional ambiguities', *Geografisk Tidsskrift* (The Danish Journal of Geography) 107, pp29-41

Peng Li, Zhiming Feng, Luguang Jiang, Chenhua Liao and Jinghua Zhang (2014) 'A review of swidden agriculture in Southeast Asia', *Remote Sensing* 6, pp1654-1683

Persoon, G., Minter, T., Slee, B. and van der Hammen, C. (2004) *The Position of Indigenous Peoples in the Management of Tropical Forests*, Tropenbos Series 23, Wageningen, The Netherlands

Price, M. F., Gratzer, G., Duguma, L. A., Kohler, T., Maselli, D. and Romeo, R. (eds) (2011) *Mountain Forests in a Changing World: Realizing Values, Addressing Challenges*, Food and Agriculture Organization of the United Nations and the Swiss Agency for Development and Cooperation, Rome

Ramankutty, N. and Foley, J. A. (1999) 'Estimating historical changes in global land cover: Croplands from 1700 to 1992', *Global Biogeochemical Cycles* 13, pp997-1027

Ruffini, F. V. and Streifeneder, T. (n.d.) Agriculture in the Alps: A Challenge for Europe and Society, Commission Internationale pour La Protection des Alpes (CIPRA), http://alpsknowhow.cipra.org/background_topics/mountain_agriculture/mountain_agriculture_introduction.html, accessed 17 September 2015

Scott, J. (2009) *The Art of Not Being Governed: An Anarchist History of Upland Southeast Asia*, Yale University Press, New Haven, CT

Spencer, J. E. (1966) *Shifting Cultivation in Southeastern Asia*, University of California Press, Berkeley, CA

Subedi, M., Hocking, T., Fullen, M., McCrea, A. R. and Milne, E. (2009) 'Lessons from participatory evaluation of cropping practices in Yunnan province, China: Overview of the effectiveness of technologies and issues related to technology adoption', *Sustainability* 1, pp628-661

Tanaka, K. (1993) 'Farmers' perceptions of rice-growing techniques in Laos: "Primitive" or "thammasat"?' *Southeast Asian Studies* 31, pp132-141

Tapp, N. (1979) 'Thailand government policy towards the hill-dwelling minority peoples in the north of Thailand, 1959-1976', Master's Thesis, School of African and Oriental Studies, London

Thrupp, L., Hecht, S. and Browder, J. (1997) *The Diversity and Dynamics of Shifting Cultivation: Myths, Realities, and Policy Implications*, World Resources Institute, Washington, DC

Tran, D. V., Leisz, S., Nguyen, T. L. and Rambo, A. T. (1999) 'Using traditional swidden agriculture to enhance rural livelihoods in Vietnam's uplands', *Mountain Research and Development* 26, pp192-196

van Keer, K., Comtois, J., Turkelboom, F. and Ongprasert, S. (1998) *Soil Fertility Options for Soil and Farmer Friendly Agriculture in the Highlands of Northern Thailand,* Tropical Ecology Support Program (TÖB), Eschborn, Germany

van Schendel, W. (2002) 'Geographies of knowing, geographies of ignorance: Jumping scale in southeast asia', Environment and Planning D, *Society and Space* 20, pp647-668

Visitpanich, T. and Falvey, J. L. (1981) 'Nutrition of highland swine. IV: A comparison of local carbohydrate sources', *Thai Journal of Agricultural Science* 14, pp123-128, https://www.researchgate.net/publication/269875557_NUTRITION_OF_HIGHLAND_SWINE_IV._A_COMPARISON_OF_LOCAL_CARBOHYDRATE_SOURCES, accessed 21 September 2015

von Braun, J. (2005) 'Agricultural development in the highlands', International Food Policy Research Institute presentation, 24 July, Lhasa, Tibet, China

Weinstock, J. (1996) 'Rattan: A complement to swidden agriculture in Borneo', *Journal of Southeast Asian Social Science and Humanities* 48, pp51-61, http://ejournals.ukm.my/akademika/article/view/3114, accessed 18 September 2015

Woodcock, C. E., Allen, R., Anderson, M., Belward, A., Bindschadler, R., Cohen, W., Gao, F., Goward, S. N., Helder, D., Helmer, E., Nemani, R., Oreopoulos, L., Schott, J., Thenkabail, P. S., Vermote, E. F., Vogelmann, J., Wulder, M. A. and Wynne, R. (2008) 'Free Access to Landsat Imagery', *Science* 320, pp1011-1012

Xu, J., Fox, J., Lu, X., Podger, N., Leisz, S. and Ai, X. (1999) 'Effects of swidden cultivation, state policies and customary institutions on land cover in a Hani village, Yunnan, China', *Mountain Research and Development* 19, pp123-132

Yokoyama, S., Tanaka, K. and Phalakhone, K. (2006) 'Forest policy and swidden agriculture in Laos', paper presented to SEAGA Conference, 28-30 November, Singapore

4

TRENDS IN SHIFTING CULTIVATION POLICY

Four decades of efforts to intensify land use in the shifting cultivation tracts of mainland Southeast Asia

Garry Oughton[*]

Introduction

This chapter is not academic, but rather a heartfelt rendition compiled from almost 70 years of successes and failures in interacting with rural landscapes, their cultivators and administrators, in both the southern and northern hemispheres. The first two decades of my life were spent in the farms, forests and mountains of the Tasmanian countryside. Thereafter my professional career was outside Australia, initially among the coastal reefs and foothill farms of the Pacific Islands and subsequently in the hilly and mountainous regions of Thailand, the Lao PDR and Indonesia (with short interludes in Burma and Bangladesh). In all of those places various forms of shifting cultivation had long been practised and in some places it still is. My main professional task was to recruit, train and manage indigenous counterparts and to assist local and national governments, whenever responsive, in persuading villagers to adjust their traditional practices to meet the pressures of population growth and economic necessity.

This chapter is an account of my experience of traditional swidden cultivation in Montane Mainland Southeast Asia from the 1970s on, and of government policies

[*] GARRY OUGHTON, Bachelor of Agricultural Science (Melbourne); Land Use Analyst and retired Managing Director, EcoLao Rural Development Consultants, Vientiane, among many other positions: Technical Adviser, Tribal Research Centre, Department of Public Welfare, northern Thailand; inaugurated the Tribal Data Centre and the Thai-Australian Highland Agronomy Project; Natural Resources Agronomist, Northern Region Agricultural Development Centre, Chiang Mai (UNDP/FAO); Preparation Consultant, Northern Region Highland Development Loan Project (World Bank/FAO); Irrigation Agronomist, Mae Taeng Multipurpose Hydropower Project, northern Thailand (Asian Development Bank); Project Preparation Consultant, Thai-German Highland Development Project, Chiang Rai, Thailand (GTZ); Irrigation Agronomist, Nam Ngum Pump Irrigation Project, Department of Irrigation, Ministry of Agriculture and Forestry, Lao PDR; Agroforestry Consultant, Nam Ngum Watershed Conservation Project, Ministry of Agriculture and Forestry, Lao PDR.

relating to it, including attempts on my part to influence these. In my conclusion I contemplate the future of agriculture as I see it, and refer to the approaches that my consulting firm EcoLao adopted as it attempted to ameliorate the transition from ancient to modern.

About shifting cultivation

As a group, shifting cultivators lie within the category of rural-dwelling primary producers who derive their incomes from direct contact with the soil and/or the vegetation and animals it supports, utilizing traditional, unsophisticated technologies. Much of the shifting cultivation in Montane Mainland Southeast Asia was – and still is – practised by ethnic minorities living in the remoter tracts of the countries.

In Thailand and the Lao People's Democratic Republic I had dealings with Karen and Shan (moving in gradually from Burma), and Hmong, Lahu, Yao, Lisu and Akha (originating from China). In Indonesia's transmigration programme, I also worked with West Papuans and the former headhunters, the Dayaks of East Borneo. In this chapter I am concerned only with interactions with shifting cultivators in Montane Mainland Southeast Asia.

The practice of shifting cultivation, also known as 'swiddening' or (pejoratively) 'slash-and-burn' or 'fire-farming', has been with us for millennia. A 'swidden' is a slashed and burnt forest clearing and traditional practice involves two distinct categories of shifting cultivation: (1) 'pioneer' shifting cultivation, the practitioners of which slash and burn the forest vegetation, then plant and harvest the same fields repetitively, year after year. When the topsoils of all the lands within reach of their hamlets are exhausted, they move to pioneer fresh territories, sometimes far away, perhaps in other provinces or even in neighbouring country; and (2) 'cyclic' or 'rotational' swiddeners, who crop their fields only once before leaving the forest to regrow for a decade or so to recover; they rotate their use of the fields around their hamlets.

Pioneer swiddening was frequently to be found at cool altitudes above 1000 metres. It was used in the production of subsistence crops of upland rice and maize and the cash crop, opium poppy, which was introduced to Southeast Asia by British colonists in India about 200 years ago. Ethnic minority farmers (Hmong, Yao, Akha, Lahu and Lisu) slashed, burned and then cultivated in one tract of land repeatedly until they had exhausted the soil and were forced to move on, leaving behind a treeless wilderness of annually burnt grassland on bald-headed mountain tops. Such practitioners did no terracing or bush-fallowing and might deeply hoe and sow steep, formerly forested fields twice a year until they became infested with weeds. This unsustainable form of land use damaged or eliminated forests and caused permanent loss of topsoil from the steep upper catchments of rivers watering the irrigated lowland plains.

Cyclic or rotational swiddeners, on the other hand (Karen, Lua, Shan and Khamu), took great care to avoid exhausting the soil. A plot would be slashed, burned and cultivated for one season only and after the crops were harvested the forest was allowed to regenerate for a decade or so of secondary forest regrowth termed 'bush fallow'. The field was then reopened for cropping again and the cycle began anew. This swiddening system has proven, in fact, to be the only sustainable method available for continuously producing food crops on lands too steep for terracing.

Under this regime, the burned fields were not ploughed or hoed: the crop seeds, mainly upland rice together with a mixture of herbs and vegetables, were planted by women into small holes punched into the ash-covered soil by men armed with long, springy dibbling poles. The crops were protected temporarily from invasion by large animals (including village livestock) by an encircling fence constructed from sturdy, partially burnt branches woven into a barricade. Throughout the wet season, the crops were weeded using small knives with L-shaped blades. These weeding knives pruned the roots of weeds rather than killing them, allowing for swift recovery of secondary forest after the crop was harvested. This practice had the effect of minimizing soil erosion.

In the dry season, following the harvest, the fences were chopped up for firewood, which was carried back to the hamlet, bundle by bundle, as the workers returned from the fields each evening. The firewood was stacked under the houses to be used for heating, cooking and boiling gruel for pigs. Each settlement was eventually surrounded by a mosaic of fields, some of them under crops, and the rest in various stages of recovery from cultivation under regenerating secondary forest.

In this chapter I am concerned with both these types of shifting cultivation and the government policies relating to them. My account begins in the late 1960s in Thailand.

Shifting cultivation in Thailand

From 1968 until 1980, I worked in Thailand in various advisory roles relating to land use. These years corresponded with part of the reign of the Thai monarch, King Bhumipol Adulyadej, whose technological expertise in agricultural irrigation technology was legendary, as was his concern for his people, of whatever ethnicity. Thailand was extremely fortunate to have a monarch who was benevolent by nature and had a strong interest in science and technology.

During the 1970s, I valued opportunities to follow in the footsteps of the King who, when young, studied for a time in Switzerland before ascending the throne. Like His Majesty, who had enjoyed walking and talking with hill farmers in the foothills of the Swiss Alps, I also delighted in interacting with, and learning from, stabilized hill farmers in Switzerland, the Black Forest and Northern Italy. Some years later I had the unforgettable opportunity to share recollections of those days with the King himself during visits by helicopter to some of the Royal Hill Tribe Development Projects in the northern Thai borderlands.

Some of the agricultural issues of the 1970s in Thailand actually originated 30 years earlier when, in the 1940s, the Chinese communist forces led by Mao Tse Tung drove remnants of the anti-communist Kuomintang army out of Yunnan to seek refuge, first in Myanmar, then later in northern Thailand. Many Akha and Haw Chinese, some Red Lahu, as well as a few Lisu tribespeople, were forced out with them and settled on a 30-kilometre-wide stretch of hilly jungle paralleling the international border at Mae Salaep in Thailand's Mae Chan district.

With the agreement and assistance of the Thai military, the refugee Kuomintang soldiers made a fortified encampment on Doi Mae Salong, the highest peak in the vicinity, and set about establishing terraced tea plantations on the heights. The non-military Akha and Lahu, on the other hand, established hamlets on the surrounding hillsides and proceeded to slash and burn the trees and then to hoe the cleared swiddens with great gusto, planting upland rice and maize. After their rice and maize harvests were gathered at the end of each wet season, the Akha travelled from Mae Salaep to the high-altitude territories of relatives living in other provinces in order to plant poppies to produce opium.

There were soon virtually no trees or associated wildlife remaining anywhere for miles around the Akha and Lahu hamlets. Everything that was not planted to upland rice or maize, or was not the traditional small patch of spirit sanctuary forest adjacent to each village, was a sea of blady grass (*Imperata cylindrica*). The Akha would spend the dry-season months, after their return from the opium harvest, deeply hoeing the land to kill the uprooted grass rhizomes by exposing them to direct solar radiation. Under this Akha system of grass-and-hoe fallow cultivation, the torrential rains at the start of the wet season washed tons of the roughly hoed topsoil downhill, and after a few years the exposed soils became thinner and stonier. Much of the fertile component of the soil took the river journey down the Mekong to enrich the Delta or to disappear forever under the South China Sea.

The labour-intensive repetitive-hoeing strategy resulted in reasonable yields of food crops for a few years, but without any precautions for erosion control it was questionable how long that regime could be sustained. In the 1970s, therefore, the Hill Tribes Division of Thailand's Department of Public Welfare eventually set out to persuade the hoe-wielders to convert to terraced farming.

Imperata cylindrica (L.) Raeusch. [Poaceae]

'Blady grass' (this chapter) is an invasive pioneer species that covers entire landscapes in Asia-Pacific. It takes over when forests are disturbed, spreading with dense underground rhizomes that secrete a toxin to repel competing plant growth. Its coarse leaves are used extensively for roof thatching.

The Akha are arguably the most obstinately traditional and averse to change of all the tribal groups. Although the Chinese at nearby Doi Mae Salong were busily terracing the mountainsides for tea plantations, the message did not seem to get through to the Akha tilling the slopes below them. Presumably because of their eons-old 'move-on' mentality, the Akha people seemed not to care about the irreversible loss of topsoil, or if they did care, they didn't know what to do about it. Fortunately for them in the short-term, the soils of that particular Mae Salong-Mae Salaep tract are derived from a calcareous mudstone that decomposes to a fertile and deep solum, so that this year's erosion exposes next year's cropland.

It took many years and the combined efforts of the Departments of Public Welfare, Land Development, Forestry and the Royal Hill Tribes Project, with technical advice from the Faculty of Agriculture at Chiang Mai University, before the Akha of Mae Salaep and surrounding villages dispensed with opium growing and settled down to an existence of terraced cultivation of rice, maize, peanuts, soybeans, fruit trees and coffee (Thong-ngam et al., 1995).

My first years in Thailand: early field visits

Australian assistance to the Tribal Research Centre at Chiang Mai in northern Thailand began in 1964. The Centre was staffed by several anthropologists, each specializing in socio-cultural studies of a particular highland-dwelling ethnic minority group. The Royal Thai Government was anxious to learn more about these people whose activities, such as river catchment damage through slash-and-burn forest destruction, opium production, and their potential for affiliation with communist cadres in China and neighbouring Laos, were a cause for official concern. In fact, the government had already made several attempts to resettle highland ethnic-minority communities into lowland *nikhoms* – but without success. The re-locatees soon ran away back to the mosquito-free mountains.

Some aspects of highland agricultural and forestry land use were unfamiliar to the paddy-oriented lowland Thai community, and beyond the academic training of the anthropologists at the Tribal Research Centre. Presumably because of my experience with the Tasmanian Department of Agriculture in opium poppy cultivation (for legitimate medical morphine and codeine extraction), I was chosen by the Australian Department of Foreign Affairs to advise on the replacement of poppies by other crops under their foreign aid programme in northern Thailand. I was accordingly appointed in 1968 as Technical (Agro-Forestry) Adviser to the Tribal Research Centre to assist in untangling the riddles of shifting cultivation.

After 40 hours of lessons in the multi-tonal northern Thai dialect (the *lingua franca* in the hills), I spent the next few months of the assignment living with one anthropologist after another in their traditional houses in the study villages, familiarizing myself with the land-use techniques employed by various ethnic groups.

The first field visit that I made in northern Thailand in 1968 was to San Pa Kia, a long-established Hmong village perched high on a ridge below the limestone massif of Doi Chiang Dao. The village was surrounded by a sea of blady grass and the villagers were walking long distances to find fresh forests to fell, even ascending another 1000 metres above the level of their village to cultivate opium poppies amongst the ragged limestones peaks of Doi Chiang Dao.

Blady grass can withstand fire but cannot tolerate shade, so the government set out to reverse this montane desertification by having the villagers graze heavily with their cattle, and then plant pine-tree seedlings, with strict fire control thereafter. Commencing in 1975, these villagers were employed by the Royal Forest Department to plant pine seedlings in the grasslands.

In the course of the ensuing 40 years of fire control, those pine forests have been replaced by re-emergence of the original dry evergreen forest ecotype, germinating from wind-borne seeds. When I revisited in 2013, the San Pa Kia villagers had become lychee and orange orchardists under the sponsorship of the Royal Hill Tribes Programme.

On the other hand, Australian-aided attempts by the Tribal Research Centre to convert *Imperata*-infested grasslands to legume-rich pastures met with little success. While such a conversion was agronomically feasible, it failed to gain sociological momentum in Thailand, although there were some successes across the border in the Lao PDR.

Another of my early visits, late in 1968, was to Dong Luang, an isolated Pwo Karen hilltribe village in the Salween catchment of Mae Hong Son province in northwestern Thailand, where the Tribal Research Centre was conducting an anthropological study and resources survey. The villagers, living far from the nearest road access, practised sustainable rotational swiddening to provide food for their 40 households. Under their land-use regime, one half of the village territory was rotated between cropland and bush-fallow, and the other half was left to remain as tall primary community forest. Edible non-timber forest products gathered from that community forest provided an essential supplement to the villagers' basic diet of rice and tubers.

On my first morning in Dong Luang the cacophony of barking dogs, squealing pigs and rice pounders woke me somewhat earlier than I would have desired after my long walk in from the highway. After breakfast we recruited the services of the headman and, armed with binoculars, sketch maps and aerial photographs, headed off around the village territory seeking out hilltops from which we could identify and map the land-use boundaries.

The headman told us that each household laid claim to about ten plots of secondary forest, one of which was slashed, burned and planted to hill rice and vegetables each year, the remainder being allowed to rejuvenate for a decade or so, one by one. When each plot was to be cleared, the young men slashed the undergrowth and climbed up selected tall seed trees with impressive agility, lopping the side branches to admit the sunlight, but leaving some of the top foliage so as not to kill the trees. The felled vegetation was sun-dried throughout the dry season, then burnt, after perimeter firebreaks were cleared to prevent fire spreading into the adjacent secondary forest regrowth, upon whose integrity the ensuing years' rice harvests depended. Rice seeds, mixed with the seeds of miscellaneous vegetables, were then dibbled into the ash-covered topsoil.

We paced the boundaries of a few cleared plots and found them to be about one hectare each in extent. A hectare of upland rice, grown without fertilizer, will yield about two tons of grain-in-the-husk per harvest. The average six-member village family consumed about two tons of grain per year, so there was little, if any, surplus.

Assisted by our aerial photographs and starting from the highest ground, we mapped the boundary of the entire village territory. The secondary forest 'bush fallows' were separated from the communal forest that the villagers kept unburned and used for hunting and collecting food reserves and natural medicines. The bottom boundary was the steep-sided gorge of the Mae Rit river, where the inhabitants sometimes climbed down to trap fish and catch shrimps and crabs. So deep and sheltered from drying winds was the valley bottom that the jungle there had the characteristics of the equatorial rainforests that mainly occur much further to the south. Sago and a myriad of other types of palm were interspersed with huge emergent buttressed dipterocarps, draped with lianas and prickly rattan vines. Everything was locked in a furious but eerily silent struggle for the available sunlight.

According to the headman, the population of the village was growing and he was worried that they were on the threshold of a land shortage. There was no way of expanding the village's farming area because their lands were hemmed in on all sides by the territories of neighbouring Skaw Karen and Lua communities. All of the Mae Rit tributaries were too steep-sided for the construction of paddy fields. Moreover, the Hmong who grew poppies on hilltops above the village were sneaking their cultivation downhill into some of the Pwo Karen bush fallows to grow their rice (which does not grow well on lands higher than 1000 metres above sea level). Needless to say, this creeping 'land-grab' was not conducive to harmonious inter-tribal relations.

Another cause for concern was that the neighbouring Hmong were offering paid employment to some of the young Karen men to work in their poppy fields. They started by providing meals and paying cash wages, but soon got them addicted to smoking opium and thereafter paid only in opium.

Years later, at Tee Char, another Pwo Karen village nearer to 'civilization', Chiang Mai University researchers reported that the villagers had been able to maintain productivity of upland rice and swidden crops under shortened cultivation cycles by enriching fallows with the fast-growing tree species *Macaranga denticulata* (Youpensuk et al., 2005). *Macaranga* associates with arbuscular mycorrhizal root fungi ('truffles'), which are active in 'fixing' atmospheric nitrogen and excreting some of it into the surrounding topsoil to be stored in the humus to fertilize future foodcrops. The shortened, but still sustainable, *Macaranga*-facilitated swidden cycle enabled the enduring support of 25 to 30 people per square kilometre of village territory (Figure 4-1).

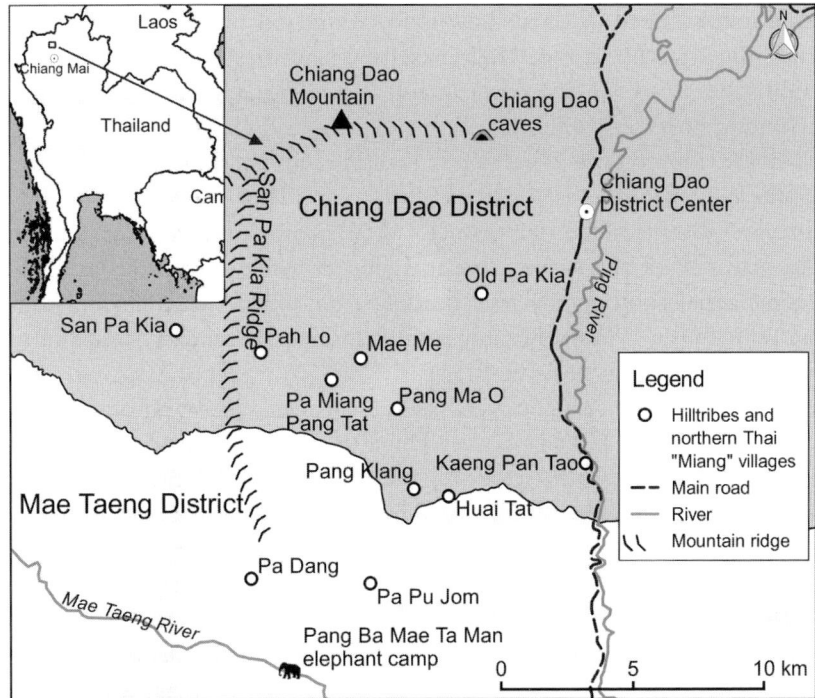

FIGURE 4-1:   The 'crowded' landscape of the small part of northern Thailand that was the scene of the author's main activities, as it was in about 2000.

The ethnicity of the villages or significance of locations shown in Figure 4-1 is as follows:

Pa Mae Ta Man, Khon Muang; **Old Pa Kia,** Hmong; **Pah Lo,** Lisu; **Huai Tat,** Lahu; **Pa Pu Jom,** Hmong; **Pa Dang,** Karen; **Pang Klang, Pang Ma O** and **Pa Miang Pang Tat,** all 'Khon Miang' identifying with lowland Thais; **San Pa Kia,** Hmong; **Kaeng Pan Tao,** a police post; **Mae Me,** a hilltribe orphanage.

Aerial mapping for land-use planning

As I became familiar with conditions in the hills it became evident that increasing population pressure was making formerly sustainable land-use regimes unsustainable. It was also leading to conflicts concerning access to forest and farmland between pioneer and rotational shifting cultivators on one hand, and between highlanders and lowlanders at their territorial margins on the other. The rate of deforestation was also cause for alarm and land tenure issues were becoming increasingly fractious. Even in the lowlands, land tenure was customary rather than legalized and certified. Across the region, traditional land use was failing to meet subsistence requirements and economic needs, so modifications were required.

Following submission of the Tribal Research Centre's preliminary socio-cultural reports, the Director-General of Public Welfare commented: 'So far, so good, but as an administrator I need to know how many Akha and Lisu and Lahu and Hmong and Karen there are up in those hills, exactly where they are and how much forest is remaining. How are we, the government, to rectify inappropriate land use? What will it cost and how long will it take?'

Land-cover mapping was a first step in meeting some of these needs and answering some of these questions.

Before the advent of remote sensing (aerial photography and satellite imagery) to make maps, all of the surveying had to be done from ground level. But after stereoscopic aerial photography was released by the military to civilian cartographers in Asia-Pacific in the 1960s, it became possible to map the extent of various forms of land cover much more expeditiously.

The maps that were available were assembled from aerial photography taken by the US-supported Thai Military Survey Department using camera-equipped aircraft flying in a grid pattern. They were still very basic. To make them more useful for land-use planning, therefore, the Tribal Research Centre embarked on a campaign of ecotype analysis in a typical highland tract, in collaboration with the Faculty of Forestry at Kasetsart University, funded by the Asia Foundation. Specialist technical advice was provided by Dr Heng L. Thung of the Advanced Research Projects Agency of the US Department of Defence (see his memoirs entitled *The Pigeons and the Witch Doctor*) and frequent reference was made to two masterful and detailed works by Dr Tem Smitinand of the Royal Forest Herbarium: *A Botanical Ascent of Doi Inthanon* and *A Botanical Ascent of Doi Chiang Dao*.

In 1973 I was snapping aerial photographs for land-use planning from a Swiss-made Air America Pilatus Porter STOL aircraft when the pilot took the opportunity to land briefly to inspect the condition of a border-security emergency airstrip at Ban Wat Chan, an ancient Skaw Karen village halfway between Pai district and the Mae Hong Son provincial seat in northwestern Thailand. In those days, Wat Chan village was two days' walk from anywhere in any direction.

After a couple of low-level passes to frighten off the cattle and buffaloes grazing on the sinuous cleared ridge-top that served as a landing strip, we touched down in a cloud of choking red dust whipped up by the turboprop at full throttle in reverse thrust, to help to stop us before the aircraft collided with houses at the end of the incredibly short runway. The noise panicked dozens of pigs and chickens, which stampeded away, colliding with the streams of curious villagers who, roused from their siestas, were running towards the strip. It was high noon on a hot windless day as the pilot and I walked into the village to find a cup of tea to wash down our mixture of runway dust and luncheon sandwiches.

Suddenly a tall, thin column of smoke wafted skywards from a patch of bush fallow on a hillside below the village. Almost immediately there was a staccato tattoo on the village fire-alarm drum. All the able-bodied young men, each pausing only to cut a leafy branch to use as a fire-fighting beater, raced downhill towards the patch of secondary forest from which the smoke was rising. After swiftly smothering the flames they returned to the airstrip to watch us take off again in another cloud of dust to resume our photographic mission. Playing to an enthusiastic audience, the pilot advised me to hang on tight and immediately after takeoff dived down the hillside to gain airspeed then 'looped-the-loop' while still in full view of the village. I am sure the audience enjoyed the free air show, but my stomach definitely did not.

Government policy and opium poppy production

In late 1969, I was delegated by the Australian Embassy to advise the first United Nations Fund for Drug Abuse Control (UNFDAC) meeting in Thailand. I soon found that the views of the UN team did not coincide with my recommendations concerning land-use stabilization and crop replacement, formulated on the basis of my field research.

I had proposed: (a) embracing all available opportunities for rice irrigation on hillside terraces; (b) reclaiming low-producing highland grass 'deserts' with fenced leguminous pastures on lower slopes; and (c) encouraging forest regrowth on the higher slopes with the trees being owned by the villagers to compensate them for relinquishing pioneer shifting cultivation and, coincidentally, opium production. My recommendations were rejected as being too mundane to attract UN funding.

I was told that what was desired was exotic crops like persimmon and cut flowers. But the real stumbling block was the government reluctance to grant mountain-top peasants any shareholding in the country's forest and timber resources. My only success was in convincing the UN mission that the non-erosive subsistence activities of rotational swiddeners were not a cause for immediate government concern, as opposed to the environmental depredations of pioneer swiddeners.

In the worldwide uproar over the production, consumption, smuggling and suppression of narcotics, I believed that one important aspect was continually overlooked: Mother Nature does not care *what* crop is grown on a particular tract of land, or *who* grows it; Nature is concerned only about *how* it is grown. Are

Diospyros kaki L. f. [Ebenaceae]

Persimmon was one of the first 'new' fruit crops introduced to Thailand's opium-growing areas at the behest of the Chairman of the Royal Project Foundation, Prince Bhisatej Rajani. It is a popular fruit in lowland markets. The Foundation now has 29 extension centres with 27,680 highland farmers growing more than 350 types of horticultural crops.

the poppies being cultivated on steep erodible hillsides or on erosion-proof hillside terraces? Will the topsoil still be in place for next year's crop, whatever that crop may be?

In the mid-1970s, Thailand's 13th Prime Minister, M. R. Kukrit Pramoj (the aristocrat who established diplomatic ties with communist China), decreed that rural development funds be distributed to every rural subdistrict (*tambon*) in the country. In highland *tambons* most of the funds were devoted to construction of small irrigation systems and quite substantial concrete access roads.

Not long after the skeletal access routes were installed in the highlands, some of the more prosperous Hmong opium-growers purchased motorcycles and then second-hand pickup trucks. A garage at Mae Rim, near Chiang Mai, began to specialize in beefing-up the suspension of those mountain pickups, including replacing standard rear axles with heavier units from wrecked two-ton Japanese trucks. Few of those four-wheeled packhorses were registered, and the Hmong chauffeurs did not bother with such troublesome formalities as a driver's licence. To avoid police prosecution on trips to town it was sufficient to wear full tribal dress whilst driving and to profess no knowledge of the local Thai dialect. It was a fascinating sight in 1975: a mountain pickup, with an overflowing load of female passengers in colourful tribal dress standing precariously in the tray, happily proceeding the wrong way up a one-way Chiang Mai business street!

The improved access roads had another unexpected effect. They also reinforced government attempts to outlaw opium production: they negated the previously superior 'value-for-weight' characteristic of opium resin. Bulkier, more perishable products such as cabbages, lychees, pomeloes, oranges and even strawberries became economically viable for the first time.

Hitherto, opium resin had been the perfect commodity. The poppies could be sown as a second crop in the cool dry season following the maize harvest simply by sprinkling the seeds on the ground without heavy soil preparation. The crop

Papaver somniferum L.
[Papaveraceae]

Opium poppies grow best at cooler altitudes above 1000 metres in Southeast Asia. Shifting cultivators would sow poppies in swiddens in November, after their food crops were harvested, so they would mature in February. The raw opium resin was gathered from slashes in the seed pods.

did not require irrigation. Women and children could do the light work of seeding, thinning, weeding and scarifying the capsules and scraping off the congealed resin. Post-harvest treatment consisted merely of boiling the raw opium to produce a stable resinous product that could be stored underground, away from prying eyes. Boiled opium resin was high in value for weight and bulk and thus economically viable to transport to lowland markets by packhorse caravans, which visited the mountain villages annually to buy the product. The farmers did not even have to waste time transporting their own product to lowland markets. Payment was usually in silver coins, which could be melted and forged into neck rings that could be worn both day and night, for the sake of security.

In contrast to opium, highland cabbages or potatoes, for instance, were much more difficult to harvest and transport to remote markets. While packhorse trails were self-building and required no maintenance, highland feeder roads were expensive to construct and maintain. The packhorses also found their own 'fuel', by grazing along the trails and around the villages at night.

There was no doubt that opium resin was the perfect commodity for highland producers. But it could be converted into addictive heroin and sold for outrageous profits in urban areas. At the urging of concerned foreign governments, opium growing was outlawed in Thailand in the 1960s and substitute crops were introduced.

Subsequent efforts to replace opium poppies with cabbages and potatoes, supported by the United Nations, led to aggravated soil erosion when these crops were planted in rows aligned up-and-down steep hills. But red kidney beans, whose seeds were contributed by the Mexican Ambassador and airlifted to several Hmong villages, were a success story.

Critics had initially grumbled that it would be impossible to change the entrenched dietary preferences of traditional ethnic minorities, but they did not reckon on the ingenuity of the King's cousin, Oxford-educated Mom Chao Bhisatej Rajani, who was manager of the Royal Hill Tribes Programme. The monocled, helicopter-commuting 'Mom Bhee', as he became affectionately known throughout the hill tracts, arranged a crash course in domestic science for young Hmong women, teaching them how to combine the beans with chillies and mix them with the basic rice and maize ingredients of their traditional cuisine. Within a year, red kidney beans had become a regular Hmong dietary item and most families came to cultivate a plot in their gardens or mixed them amongst their swidden rice crops.

These days, considerable investment has been made in the terracing of steep, densely inhabited hillsides, and crops such as strawberries, blueberries, passion fruit, dragon fruit, oranges, pomeloes, lychees, rambutan, zapotes, grapes and even cut flowers have become popular. With the expansion and maintenance of tourist-oriented highland feeder roads and the strong support of the Royal Hill Tribes Project, most Hmong and other former pioneer swiddeners in Thailand's Chao Phya watersheds converted to sedentary production of high-value horticultural crops on terraces and the plantation of fruit trees. When they relocated their hamlets after abandoning eroded, unproductive fields, the forests slowly re-established, but it sometimes took

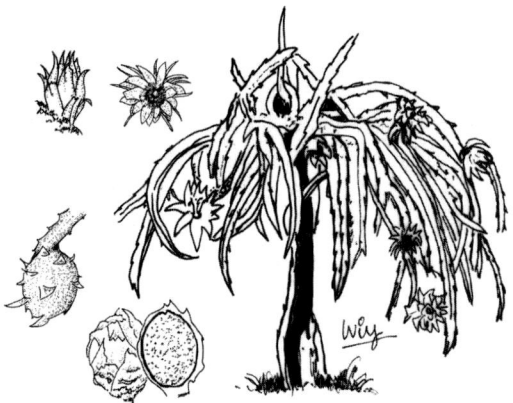

Hylocereus undatus (Haw.) Britton and Rose [Cactaceae]

Dragonfruit is thought to have been introduced to Vietnam from South America by French colonists. It has since spread throughout Laos and Thailand. It is a sprawling cactus that climbs by using aerial roots, and can be grown on extremely steep terrain without risk of soil erosion.

30 years to recover previous levels of biodiversity.

The situation is vastly different in the river catchments of Mae Hong Son province, where the waters flow down the Pai river, through Myanmar into the Salween river, and ultimately exit through its delta on the Andaman Sea. In the Pai catchment there has been no concern about erosion or catchment integrity, only a vigorous campaign to substitute opium poppy with maize.

The campaign is financially supported by the United Nations Office for Drugs and Crime (UNODC) and supplies a greedy urban market fed by one of the world's largest industrial conglomerates, Charoen Phokphand. This Thai company operates in more than 20 countries in agribusiness and food (retail and distribution) and telecommunications. CP, as it is known, plays a role in promoting legitimate agriculture in the northern Thai highlands by contracting farmers and supplying them with hybrid seeds, fertilizers and herbicides. But nowadays in the Salween catchments, for mile after mile, instead of former poppy swiddens covered in grass, there is a vista of Hmong-cultivated maize fields, green in the wet season and yellow in the dry. A few clumps of straggly trees remain on land that is too eroded or rocky to cultivate. It begs the question: what will happen when the topsoil is all gone and maize-growing is no longer economically viable?

Village life in the hills

Over the months and years I became more familiar with, and more admiring of, the skills of villagers in the hills. I found that, at family level, the head of each household acquired the knowledge and the leadership skills to organize his close relatives to erect and repair their dwelling and their elevated rice barn; also to execute all of the various tasks that had to be performed throughout the farming year. At the same time the youngsters were guided in their experiential learning. They accompanied their parents to the swiddens and forests and learned how to survive and thrive in their remote rural environment. They learned which jungle plants were edible, which curative and

Vaccinium corymbosum L. [Ericaceae]

Blueberry is a recent introduction from North America to the mountains of northern Thailand. It is believed to have been cultivated by native Americans for thousands of years. As a replacement for shifting cultivation in Thailand, the shrubs produce a sweet and popular fruit that is an ice-cream and cake-shop favourite.

which poisonous. (When visiting the Karenni rebels in Burma's Kayah State in 1976 they told me how they inflicted psychological warfare on the oppressive Burmese military with curare-tipped blowpipe darts, causing those they ambushed to die in cries of anguish, frightening the remaining invaders.) The tribal children also learned how to perpetuate their ethnic identity through music, dances, customs, rituals and modes of dress that distinguished one group from another and reinforced their social awareness.

Using primitive looms, the women wove a coarse type of homespun cloth made from swidden-grown cotton. The warps were tensioned by attaching them to a belt around the waist of the operator, who was seated on the floor. Cotton seeds were removed from the lint using a wooden-geared mangle. A hand-turned spinning wheel made of bamboo and cord produced the thread for the loom. The woven bolts of cloth were coloured by boiling them in natural dyes gathered in the forest.

I saw how, even without literacy or the use of maps, the headmen could effectively adjudicate in disputes concerning selection of the current year's rotational farming plots. Without written calendars, these wise men could interpret the trajectory of the sun, the phases of the moon, the positions of the stars and the direction of the prevailing winds. These were used to predict in advance precisely when the trees of the bush fallow should be felled in order to be dry enough to burn completely before rice-planting time and to predict the date for sowing the seeds so they would be ready for the onset of the rains. They also knew which sectors of the village territory should be left untouched for watershed protection and as biodiversity reserves.

Most significantly, the village elders knew how, at harvest time, to select vigorous seed heads of the four or five different varieties of upland rice that were grown each year. They ranged from a fast-growing but low-yielding variety, harvested to provide calories before the end of the rainy season, through a couple of later maturing but disease-resistant varieties, to the main crop, which was heavier-yielding and harvested last. The selected panicles were cut by sickle and hung in bundles under the houses to air-dry before being suspended from the rafters to await the next planting season, preserved from the ravages of weevils by smoke from the open fireplace, which was mounted on a square clay platform on the elevated, woven split-bamboo floor. The smoke escaped to the atmosphere through the thatched-grass roof, helping to preserve it from insect attack en route.

Population growth and control

Before the advent of family planning, the scarcity of land for rotational swiddening meant that slowing population growth by controlling adolescent sexuality was of constant concern to those tribal elders whose villages had limited farmland. They had to resort to many cultural and religious strategies: the Karen tribes, for instance, dressed their unmarried girls in a long, pale-coloured ankle-length smock, while after marriage the women changed to a more accessible, shorter red-coloured skirt and blouse.

Manilkara zapota (L.) P. Royen [Sapotaceae]

Sapodilla is a native of Mexico and Central America, believed to have been introduced to Southeast Asia by Spanish colonists in the Philippines. It has been used as an alternative to shifting cultivation in northern Thailand on slopes below 600 metres. The popular fruit, known locally as *lamut*, has a soft edible skin and very sweet flesh.

By contrast, pioneer swiddeners, many of whom were polygamous, encouraged early marriage and large family sizes. More hands for weeding and tapping opium poppies meant bigger areas under crops and more wealth for the household. The Akha people, for instance, had several social mores promoting early marriage. One of these was the construction of the 'lan kod sao' (literally 'arena hug girl') platform on a stream bridge near the village, where it was the approved custom for young men to meet and cuddle young women in the romantic cool of evenings on the way home after work in the fields. The Akha version of a beauty pageant at festival time was to load a wooden ferris wheel at the village gate with four young women, each on a separate swing, and the wheel would be rotated by the young men to the accompaniment of great hilarity and feasting on dog meat and rice whisky.

New Year festivities in the poppy-growing Hmong villages similarly included a day-long event where a row of boys in their best clothes stood in line facing a row of teenage girls. A cloth ball was tossed back-and-forth between the boys and the girls and partners were changed and exchanged until all were compatibly paired-off. Nine months later, new members for the household workforce ensued.

Shifting cultivation in the Lao PDR

For most of the years from 1990 to 2011 I worked in Laos, and although Laos and Thailand were politically different, the problems with shifting cultivation were similar. As in Thailand, natural resources were still plentiful in the mountains until the 1950s and the use of montane land, water and forest was of little interest to the urban-based governing cliques. But until 1954, Laos was a French colony.

After independence from France, Laos was embroiled in 21 years of civil war between the Lao Government and the armed forces of the communist Pathet Lao. At the same time, the country was impacted by the spill-over from the Vietnam War. Considerable sections of the famous Ho Chi Minh trail, the vital supply line for the North Vietnamese forces invading the South, ran through parts of eastern Laos, and this area was mercilessly bombed by US forces, and many communities of shifting cultivators were forced to shelter in caves or flee and relocate. Many of the younger men were inducted into the Pathet Lao army. In 1975, as US forces withdrew from the war in neighbouring Vietnam, the communist Pathet Lao gained control of the whole country and it became the Lao People's Democratic Republic.

Across the Mekong to the southwest, the Thai Government was having some success in curbing both destructive pioneer shifting cultivation and the production of opium, and at the same time was providing farmers with alternative livelihoods to shifting cultivation. However, the Lao Government was neither as benevolent nor as successful in persuading their highland farmers to adopt another way of life.

Government policy and shifting cultivation

When Laos became communist, American personnel were replaced for a while by Russian advisors. A Soviet-financed oil pipeline was built to deliver fuel to the Mekong valley from Russian tankers berthed at the Vietnamese port of Vinh. When the access road to the pipeline was upgraded, miners and loggers moved in, the latter capitalizing on the lucrative timber trade. In fact, this trade became the major source of wealth for members of the politburo and the Central Government military elites in the Lao PDR as well as in Vietnam.

As a result of the timber trade, an unpublicized, hidden agenda evolved: all of the swidden-cultivating ethnic minorities were to be cleared out of the hills and resettled along the sides of lowland roads. This would allow the timber-rich forests in the mountains to recuperate, not as biodiversity conservation or watershed protection reserves *per se*, but as unofficial timber-rich tracts to be harvested for the ultimate benefit of the families and descendants of the communist elites. Most well-meaning non-governmental organizations and other rural-development workers in the 1990s were quite unaware of this unspoken policy.

One fallout of this hidden policy, under the guise of opium-poppy eradication, was to deprive would-be opium producers of access to the cool highlands, free of malaria, where their poppies could thrive. Thousands of Akha died when army and police forces, financed by the United Nations Office for Drugs and Crime, forcibly moved them out of the highlands without proper protection against malaria and dengue fever.

Lao Government policy was to resettle all of those who practised shifting cultivation. The level of environmental damage from pioneer swidden cultivation had the unfortunate effect of sullying the reputation of its benign and sustainable rotational counterpart. After all, both systems were labelled 'slash and burn' and studying the vital difference between them required more attention than many at official level were prepared or equipped to give.

Another issue was also clouding the thinking of Lao policy-makers. Until relatively recently, most rotational swidden farmers were non-monetized and primarily subsistence-oriented. Although they might have interacted with the ecosystem in ways that did not threaten environmental or ecological integrity, they did not generate much, if any, saleable produce or revenue that was of use to urban dwellers, who faced the need to accumulate money to purchase their food instead of growing it themselves. There was a tendency to regard subsistence shifting cultivators as parasites who contributed nothing to the rest of the country and who burnt potentially valuable timber instead of allowing it to grow to maturity.

The Lao Government did not find it easy to eradicate swidden farming, nor to resettle the villagers. Anti-swidden and anti-opium policies and targets proclaimed at national level were often not sufficiently realistic or morally right and fair to be respected at community level. To this day, some remote mountain-top Hmong, Akha and addicted Lahu still follow the poppy-growing tradition. In northern Myanmar, it is the militant Kachin who persevere.

In May 2009, *The Vientiane Times* reported:

> The Lao PDR Ministry of Agriculture and Forestry (MAF) is persisting with its commitment to end shifting slash-and-burn cultivation by 2010, even as Prime Minister Bouasone Bouphavanh last week reported difficulties with the policy to the National Assembly. The Prime Minister said that there were no detailed plans for providing alternative jobs or re-locating villagers to a permanent living place if they had to give up their traditional farming methods.
>
> Development of agriculture, which is one of the driving forces behind the Lao economy, remains slow and poverty reduction is not fully associated with the Government's directive to resettle villages and establish village development groups. The MAF official in charge of the issue, Mr Boualy Phameuang, said slash-and-burn cultivation could not be addressed while the provision of alternative work was not dealt with, and the ministry was aware of the problem. He said land area under [pioneer] shifting slash-and-burn cultivation had dropped to between 6000 and 8000 hectares, mostly in mountainous provinces. 'With the decline in recent years, I think we are capable of ending shifting cultivation by 2010,' he said.
>
> But rotating (cyclic) slash-and-burn cultivation may not cease entirely by that time due to the challenges of providing alternative work for people who live in rural mountainous regions. Mr Boualy said that between 60,000 and 70,000

hectares of land was currently farmed by the rotating slash–and–burn method and authorities hoped to end this form of farming by 2015 or 2020. A major challenge is the lack of funding and competent officials to work in these areas. The ministry has pledged to set up a special section to deal specifically with the matter so that more progress is made. Mr Boualy said another important issue was land allocation for villagers, who needed enough land to productively cultivate commercial crops to avoid encroachment into protected forest areas. 'It will take several years to know whether the provision of alternative jobs is working effectively. Land surveys are important to identify which areas are best suited to growing which crops. We also need to build access roads so that villagers can transport their products to markets,' he said. 'Eliminating slash–and–burn cultivation relates to several sectors such as education, health and information, so cooperation between all of these sectors is needed to effectively address the issue.'

Previously, the Party and Government had named the year 2005 as the time when this form of farming would be phased out. But this target was not reached due to continuing poverty and the lack of development in rural mountainous regions. In 2005, the Party Congress set a further target of 2010, aiming to focus on rural development and poverty reduction.

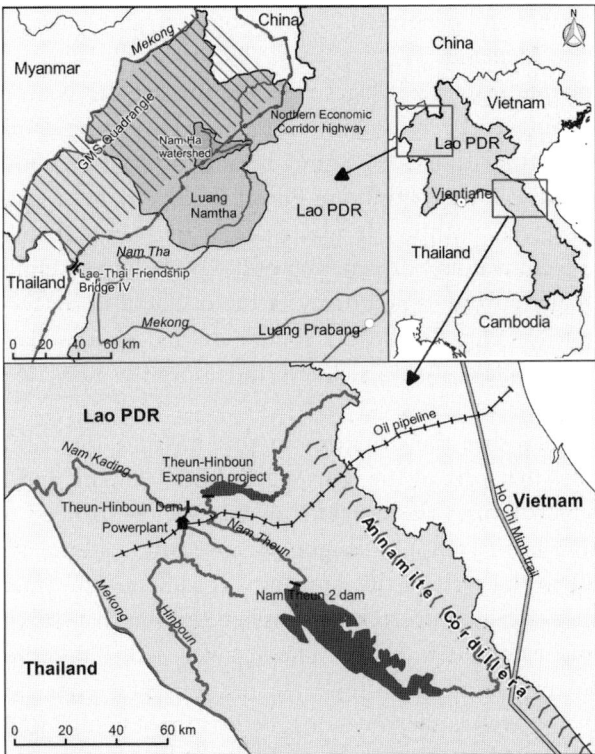

FIGURE 4-2:   The Lao PDR, showing physical features mentioned in this chapter.

Several other writers also recounted the effects of various aspects of the Lao Government's actions concerning the suppression of narcotics-related shifting cultivation, e.g. Cohen (ch 31) and Ducourtieux et al. (ch 32).

Two government infrastructure projects

When major infrastructure projects are undertaken, the lives of villagers on the sites can be seriously impacted. During my time in Laos I was involved in working with shifting cultivation villagers who found themselves in this situation. One such occasion was in 1998, when the Lao Government attracted a considerable level of critical scrutiny from environmental and human rights organizations around the world over the Theun-Hinboun Hydropower Project in the country's central provinces (Figure 4-2). The original design of this project gave priority mainly to economic issues and insufficient thought to the human and land-use implications of its construction and operation.

The project was initiated as a joint venture between a victorious but bankrupt communist Lao government and the wealthy Thai private sector, and saw a US$260-million dam erected across the Theun river. The construction funds were ultimately augmented by a consortium of commercial and international banks, amongst them the Asian Development Bank and, later on, ANZ bank. Stage One of the project was commissioned in 1998 and generated 210MW of electricity, most of which was sold across the border to Thailand.

About 6000 people from 25 villages near the project site were impacted in one way or another, some to be resettled, leading to changes in their social context, lifestyle and agricultural practices. Some heavily impacted households received compensation from the Theun-Hinboun Power Company, but most project-affected people initially received nothing. It was claimed by critics that for thousands of subsistence shifting cultivators, construction of the Nam Theun-Hinboun project was a life-changing tragedy from which many of them would never recover. According to the International Rivers Network, 'some resettled groups have suffered from declining nutritional intake, rising sickness and mortality rates, loss of dialect and deculturation'.

However, I believe that the criticism in this instance was unfairly harsh. The Theun river crosses that part of the map where both Vietnam and Laos are at their narrowest and which houses the corridor of the Soviet-financed oil pipeline. The Ho Chi Minh trail from Hanoi to Saigon traversed the eastern edges of the Nam Theun catchment to the west of the Annamite cordillera. The International Rivers Network and other critics completely overlooked the fact that the cannabis-growing populations in the catchments of the Theun river had, for decades before, been severely disrupted by fallout from the Vietnam war. The poor quality of life the critics described was well underway long before the Theun-Hinboun developers came upon the scene. Moreover, the war effort and the exodus of the intelligentsia after the communist takeover bankrupted the Lao Government, financially, culturally and technically. Until Theun-Hinboun came along there were no funds or personnel to provide more than illusory health and educational services to the outlying communities for whom relocation was by no means a novelty.

Belatedly, at the end of 1999, the Asian Development Bank responded to humanitarian criticism by commissioning the author's consulting firm, EcoLao, to investigate the extent of the socio-environmental impact and to recommend necessary restitution measures. The first measure recommended and ultimately implemented by EcoLao technicians was the installation of pumped irrigation systems for villages along streams whose maize and vegetable gardens had been inundated by the head pond reservoir. (They had already, de facto, been compensated temporarily by illegal lumbering opportunities in the few remaining forests in the reservoir catchment.)

Also in response to the concerns of the Asian Development Bank, Stat Craft – a Norwegian quasi-governmental agency and one of the eventual Theun-Hinboun shareholders – hired the Norwegian consulting firm Norplan to conduct comprehensive socio-environmental impact surveys. These culminated in a long term livelihood restoration plan which was duly inaugurated and continues to the present day.

While much should be said about responsibility for the human cost of infrastructure development, it is also worthwhile to acknowledge what was eventually done, both for the affected swidden and paddy farmers and for the Lao PDR. First, the affected villagers were provided with effective access to adequately staffed education, family planning and human and animal health facilities. Additionally, they were provided with efficient irrigation systems wherever feasible and guided in the exploitation of the project catchments in ways that both maximized retention of rainfall in the underground aquifer and minimized soil erosion, thus benefiting both farmlands and the reservoir.

Another development project, this time with humanitarian motives, was undertaken in Luang Namtha province in 2013 to offset possible hardship caused by highway construction. The 'Project for Poverty Alleviation, Land-use Stabilization and Environmental Protection in the Nam Ha Watershed, Northern Economic Highway Corridor (ADB Route 3)' was known for short as the Nam Ha Catchment Project and was funded by the Asian Development Bank. It was implemented by the non-governmental organizations Adventist Development and Relief Agency and German Agro-Action, with technical support from the Wildlife Conservation Society and EcoLao Consultants. The latter were advising the Luang Namtha Provincial Government on both resettling shifting cultivation communities displaced by construction of the Greater Mekong Subregion Quadrangle Highway (as it was called) and stabilizing land use and land tenure in the corridor.

That highway was the last remaining undeveloped link in a regional 'economic corridor' connecting Singapore and Thailand through Bokeo and Luang Namtha provinces in the Lao PDR to Kunming, in China's Yunnan province. In 2013 the road link was finally completed with the opening of a bridge across the Mekong between Bokeo and Chiang Khong district in Thailand's Chiang Rai province. The highway provided Yunnan province, once a remote corner of southwestern China, with convenient road access to trade opportunities via seaports in Thailand and land links as far south as Singapore. Quite suddenly, with the commencement of highway traffic, the previously isolated and very traditional village economies of the Luang Namtha denizens were jerked into the 21st century and exposed to the full gamut of both the good and evil influences of the outside world.

The Luang Namtha Provincial Government and the Central Government of Laos did not always see eye-to-eye on policy issues, and face-to face discussions required a five-hour door-to-door journey by Lao Aviation (when it was flying) or a 24-hour torment by unsealed roads between the provincial and national capitals. The more leisurely riverboat alternative could take up to four days and one had to wear a crash helmet when negotiating the rock-strewn whirlpools. The Luang Namtha Vice-Governor (a former schoolteacher) accordingly authorized the project team to 'implement first and argue with Vientiane later'.

The initiative incorporated the village territories of nine shifting cultivation communities, seven of them rotational swiddeners of Khamu extraction, except for one paddy-growing village of Tai Lue ethnicity, and two of them – one Akha and one Hmong – opium-producing pioneer swiddeners. The project area was to be divided into government forests and village forests, plantations and farmland, on the basis of each family's landholding being adequate for self-sufficiency in rice and protein sources, and with enough forested or plantation land for an adequate annual income. It was a project that aimed to take care of both people and land. The rationale was to provide an incentive to villagers to take care of the land for themselves and for future generations.

Experiences with Lao swidden farmers

My involvement in government projects in Laos meant that I also became acquainted with Lao swidden farmers. I saw at first-hand how pioneer shifting cultivators, particularly the Sino-Tibetan Yao (who were literate, using Chinese-style characters) and the Hmong, could be quite ingenious.

In 2001, I visited a Hmong hamlet amongst the limestone crags on the ridges above the Luang Prabang provincial capital. The soil between the megaliths was of good quality, but there was no surface water supply except for a large sinkhole in front of a limestone cave beneath a ridge-top crag. I was surprised to see that it had been fenced off, but that the cattle were fenced into the pond area, and not fenced out as one would expect.

The villagers told me that the cattle manure needed to be regularly trodden into the sandy bottom of the pond to seal it against leaking and losing the ponded water between rain events. The murky water was bucketed out of the pond and the sediment allowed to settle, and then it was trickled through a filter on the front porch of each house, comprising a suspended clay-filled basket draining into another bucket on the floor beneath. The filtered water was then boiled before consumption. The people told me that it was very tasty!

Presumably as a bonus from there still being some cash-producing opium fields nestled amongst the cliffs and crags, many of the houses in that hamlet were equipped with solar panels powering radios and television sets. Their favourite viewing was the uncensored, no-words-barred Thai Parliamentary broadcasts of the day – '*peun da kun muan*' (them abusing each other is enjoyable). It was certainly more invigorating than the monotonous and repetitive lecturing of the Lao Politburo, as well as useful training for them in the trans-Mekong dialect.

Halfway down the mountain towards the provincial capital was another Hmong village whose headman had retired early from a senior government position in Vientiane. Instead of the usual scenario of the pigs wandering freely through the mud between the village houses, he had organised his people to fence off a 20-hectare enclosure to contain the entire pig population, within which they proceeded to cultivate and manure the soil without mercy. Each family had a pig-feeding trough in the enclosure for administering the evening meal, which was a gruel of chopped banana stems, taro and maize-meal. Somewhat incomprehensibly, each pig knew which family trough to go to when called. In the absence of wandering cattle and pigs, there was a small fruit orchard and flower and vegetable garden planted in front of each residence, and there was a kiosk in the village for selling blooms, fruit and vegetables to tourists and tradesmen who drove up the steep feeder road from the town. After a year or so of pig-rooting, the 20-hectare enclosure was denuded of vegetation and thoroughly manured without any need for human labour, and was subsequently planted with coffee. Meanwhile, a new swine feed-lot was fenced-off in an adjacent tract.

Further south, in the wildlife-rich shifting cultivation areas of central Laos, the Vietic villagers set snares and erected low woven-bamboo fences around their fields (Chazée, 2015). Gaps at intervals along these fences were booby-trapped with triggered drop-guillotines to capture rodents and other small edible wildlife to bolster family protein intake. Rice field rats were considered a delicacy!

Each rotational swidden community had at least one blacksmith equipped with the wherewithal to make bellows for his forge and charcoal to fuel it. He fashioned the steel machetes, weeding knives, dibbling sticks and sickles required by his fellow villagers for their farming pursuits.

An improved quality of life in remote mountain villages sometimes had unexpected consequences. After the introduction of antibiotics into remote areas by itinerant 'injection quacks' (long before the arrival of birth-control awareness and technology), population growth in many mountain villages escalated, leading to the production of a surplus of both cannon fodder (e.g. young Hmong militants on both sides of politics in Laos during the Vietnam war) and factory fodder (young, semi-educated and lowly paid workers who were both docile and nimble, for distant construction industries, garment factories and fishing boats).

Up until the late 1950s, there was still a profusion of wildlife roaming the steep-sided valleys and remote hillside jungles and savannahs of Southeast Asia. Turtles, pangolins, civet cats, bears, gibbons, wild pigs, white-winged ducks and all manner of now rare and threatened fauna were still common. This was particularly so near the Annamite cordillera on the Lao–Vietnam border, where the limestone cliffs, crags and caves had been a non-glaciated genetic refugium during the Ice Age. Species such as rhinoceros, gaur, kouprey, banteng, dhole, tigers and leopards, smaller and larger deer, antelope and the 'unicorn goat' (saola) were once common, but they are now virtually extinct. Even elephant and gibbon numbers are in decline. Snakes though, both large and small, are still quite plentiful, and bats and swallows still crowd the walls of caves.

Initially, much of the decline in wildlife numbers was due to the predations of Hmong tribespeople, armed with crossbows, and the Lahu (known locally by their Burmese name, Muser, meaning 'hunter'), armed with spears. The latter were introduced to gunnery by the sons of Baptist Christian missionaries in Burma (see *Tracks of the Intruder*, by Gordon Young). Both of these tribes inhabited the high country, combining hunting with slash-and-burn cultivation of maize and opium poppy, and moving ever deeper into the hilltop jungles as the wildlife population declined and the fertile forests were transformed into exhausted and barren grasslands. This led to the aforementioned phenomenon of vast stretches of bald-headed mountains, when seen from afar.

The ethnic and linguistic origins of the Lahu people are Tibeto-Burman, and they relish their folklore. A hunter once advised me that if I was charged by a leopard, I should squat down, angle my spear in the direction I estimated it would jump from, and thrust the butt of the spear firmly into the ground. The hunter claimed that if I calculated the angle and range correctly, the beast would impale itself on my spear instead of impaling me on its claws, and I would get a valuable pelt for my trouble. Don't try it with a tiger though – 'that cat is too heavy and will only break your spear!' Another of his pearls of indigenous wisdom concerned being charged by a gaur: as it thunders toward you, throw yourself prone directly across its path and it will daintily lift its feet as it prances through.

The Lahu hunter cautioned our team that if, when walking through the forest in wild elephant territory, we stumbled across a herd, we should not make the potentially fatal mistake of heading for the hills, as elephants can climb uphill with amazing agility. However, probably due to their ponderous weight, they are very reluctant to run downhill and a human can easily outstrip an elephant in that direction. I never had the mischance to test the hunter's good advice. But I did notice that he had deep scars on his face and arms – just the kind that might have been inflicted by a leopard.

The future of highland agriculture and EcoLao

In 1971, our team from the Tribal Research Centre stood on a smoky, deforested hilltop with Dr Thalerng Thamrongnavasawat, then Secretary-General of Thailand's

National Economic and Social Development Board. As we took in the panorama of bare, fire-ravaged grasslands, punctuated here and there by a few straggly remnants of the verdant rainforest that once clothed the slopes, I explained to him our recommendations for restoring both the ecosystem and the local economy by means of government-financed reforestation and fire-control; also by incorporating villagers' ownership of the trees, to replace the income they made from cultivating the opium poppies responsible for the ecological devastation before our eyes. After listening closely to our technical recommendations, Dr Thalerng commented: 'You have convinced me. Now I need seven years to convince my colleagues in parliament!'

Having an advisory role on land use was not always rewarding. Sometimes our recommendations to authorities were received and acted upon, sometimes not, perhaps due to short-term political or budgetary constraints. Too often, the solution for enforced abandonment of opium poppy farming was for the farmers to enter contract agreements with agri-business corporations and to plant maize repeatedly on the steep lands, burning the bulky crop residue after every harvest to make way for the next planting. (Hence, the heavy smoke-haze problem that, in recent years, has begun to occur ever earlier in the dry season across northern Thailand.)

Contract farming is a money-spinner for all concerned, but it demonstrates a shameful lack of awareness on the part of local economists that cash-flow projections are not accompanied by environmental-impact projections. Where such projections do exist, they cost money. If implemented they reduce the exorbitant profits made by corporations, which doubtless dampens their enthusiasm for responsibility as world citizens.

More people populate the earth today than ever before, and the number of urban dwellers outnumbers those who produce the food and fibres to feed them. Consequently, the urban understanding of rural livelihoods is declining. This is a serious issue because urban-based institutions are the wellsprings of policies that guide society's behaviour in relation to the resources it consumes to support life. Hence, governments flood farmlands to

Morus indica L. [Moraceae]

Mulberry has many cultivars, some used to produce fruit, others to produce leaves to feed silkworm caterpillars. The rich red fruit are eaten raw or made into jam. The leaves are fed to silkworm caterpillars that are kept indoors. The cocoons in which they later pupate are unravelled to be spun into silk thread.

create hydropower or irrigation reservoirs; the corporate sector campaigns to grab previously farmed land under fallow to mine it; or to establish broadacre plantations of industrial crops like cassava, sugar cane, coffee, tea, rubber, palm oil and *Eucalyptus*.

Once-remote mountain areas are not as remote as they used to be and shifting cultivators are just one of many types of rural dwellers who are being forced by the worldwide tide of so-called 'development' to alter their traditional modes of land use. They are being required to deliver more products of nutritional and economic importance per unit of area, per worker, and per unit of time.

If current trends persist, shifting cultivation – a centuries-old means of producing life-supporting foodstuffs and fibres, often on steeply sloping and remote forested and agricultural landscapes – is doomed as a way of life. Its productivity per unit of area and time is too low to survive the widespread clamour for cultivable land.

I established the rural development consulting firm called EcoLao in 1997, at the urging of some of the post-Communist, younger generation of internationally trained Lao Government officials, and I managed it until my 75th birthday in 2011. The firm trained and employed young Lao technicians to address, along with villagers and outposted officials, the land-use constraints and inter-ethnic conflicts arising out of infrastructure developments, many of them internationally funded. These developments included highways, hydropower schemes, irrigation systems and mining, as well as the land-use intensification associated with expanding agroforestry plantations, such as those producing cassava, pulpwood, sugar, rubber, coffee, agar-wood and firewood.

In the years that I managed EcoLao, my policy was to place particular emphasis on the conservation and management of the soils of once-forested steep slopes, partly as a consequence of seeing with my own eyes in my Tasmanian birthplace what happens to soil productivity when slopes that are too steep are repeatedly cultivated and grass is grazed too heavily. After all, 75% of the surface area of the Lao PDR is mountainous, and so is more than 50% of Vietnam. My approach was an attempt to focus on intensifying land use in tracts utilized by shifting cultivators. I constantly emphasized, to those who would listen, the potential long-term effects on ecosystems in general, and the soils and watersheds in particular, of on-going practices and planned future actions. Modern approaches, if adoptable, must be truly sustainable with particular regard to long-term impacts on soils and waterways.

At a regional level, I constantly recommended that the planning of future land use should be carried out on a holistic, river-catchment by river-catchment basis; and that in every instance, planning should incorporate calculations of sustainable population carrying capacity. This was because what is done to watersheds in highland catchments governs what happens lower down: whether downstream 'rice bowls' enjoy ample flood-free water, or are starved of stream-flow by excessive upstream extraction, or subjected to flash floods and siltation due to soil erosion in the highlands.

At every opportunity, I attempted to alert farmers, on one hand, and policy-makers and budget-holders on the other, to the potential outcomes – both positive and negative, both technical and sociological – of alternative, sustainable ways of utilizing land and water for their subsistence or economic pursuits.

Sometimes the interventions I led were successful. But often, due to government policy inadequacies, they were not followed up. However, the land is still there and the people still need to be fed. It is my fervent hope that in a couple of decades – after the loggers have taken their trees, the land-grabbers have despoiled the land, the plantations have over-matured, the hydro dams have silted up and the mining sites have been exhausted and abandoned while the dogs of war have exterminated one another – the policy-makers will be more inclined to support analysts, planners and technicians, such as those trained by EcoLao, in rehabilitating the landscape and its farming communities for the benefit of all. Forever the optimist, I am devoting my remaining years to the preparation of a training manual for land-use planners in the Asia–Pacific region.

Acknowledgements

I am most grateful for assistance in compiling and refining this chapter from Liz Hinton, who became an expert on the Karen people and shifting cultivators in northern Thailand in general while at the side of her late husband, Dr Peter Hinton. The Hintons lived with and studied the Pwo Karen at Dong Luang in northern Thailand during 1968 and 1969, while Peter was employed by the Tribal Research Centre. Liz later published a translation of Karen folk tales entitled *Oldest Brother's Story*. Her help in sifting through my voluminous memoirs and recollections to produce this chapter was a superhuman effort for which she deserves my deepest gratitude. I am also indebted to the assistance of Peter Hoare, whose service to development projects in northern Thailand made him a contemporary of mine. Peter gave constructive feedback on earlier versions of this chapter, including advice on how to write a 'scientific' paper, help in digging out some references and guidance in developing the maps.

References

Chazée, L. (2015) 'Valuation and management of forest ecosystem services: A skill well exercised by the forest people of Upper Nam Theun, Lao PDR', in M. F. Cairns (ed.) *Shifting Cultivation and Environmental Change: Indigenous People, Agriculture and Forest Conservation*, Earthscan from Routledge, London

Thong-ngam, C., Shinawatra, B., Healy, S. and Trébuil, G. (1995) 'Farmers' resources management and decision-making in the context of changes in the Thai highlands', Proceedings of a Regional Symposium on Montane Mainland Southeast Asia in Transition, 12-17 November 1995, Chiang Mai, Thailand, pp462-487

Youpensuk, S., Rerkasem, B., Dell, B. and Lumyong, S. (2005) 'Effects of arbuscular mycorrhizal fungi on a fallow enriching tree (*Macaranga denticulata*)', *Fungal Diversity* 18, pp189-199

5

TENURE AND SHIFTING CULTIVATION

*Joseph A. Weinstock**

Introduction

In the course of human settlement of most countries in the world, the fertile plains and valleys best suited for agriculture were settled by farmers, while coastal areas and places with easy access were occupied by ports, cities and industry. Remote inland forest lands were considered of minimal value, so they were left for people to practise shifting cultivation. As populations increased and the best lands became increasingly scarce, more people moved into forest areas. Governments and private investors constructed new housing estates around cities, developed new settlements and industrial zones and established resorts in formerly undeveloped forest lands. At the same time, timber and other forest products increased in value in local and global markets and demand rose for tree crops such as cacao, rubber and palm oil. These forces have increased the pressure on forest lands and on communities dependent on shifting cultivation throughout the world, from the Amazon and the Congo to Southeast Asia. This has amplified the importance of recognizing and protecting the rights of shifting cultivators to the lands they occupy and the crops they grow.

Tenure and rights to use land

Shifting cultivation often evokes images of nomadic people slashing and burning the jungle to grow crops, before moving on in a couple of years to slash and burn a new section of jungle, in an endless cycle. Nothing could be further from the

* Dr Joseph A. Weinstock, conducted field research among shifting cultivators in Kalimantan (Indonesian Borneo) between 1979 and 1981 for his doctorate at Cornell University. He later spent 13 years as Senior Sustainability Specialist, Senior Environment Specialist and Focal Point for Disaster Rehabilitation at the Asian Development Bank. He is currently an Environment and Natural Resources Management Advisor and Climate Change Specialist working on the projects and programmes of various development agencies.

truth. While the fields for cultivation shift every couple of years, the people do not. Shifting cultivators are not nomadic; they live in stable villages that have existed for decades, if not centuries, in the same location. Primary forest would have been cleared when the village was first established, but slashing and burning primary forest is hard work. After its use to grow crops, the land is abandoned to regrow secondary forest, enabling it to once again sustain food crops, typically after 10 to 15 years. It is far easier to slash, burn and reuse an old site than it is to cut down the huge trees of a virgin jungle. The key concept is sustainable food production. Unlike grassland ecosystems, in which the majority of nutrients lie in the soil, where they have built up over time through the decay of the roots and vegetation of annual plants, the majority of nutrients in forest ecosystems are locked-up in the trees and other perennial vegetation, with relatively few nutrients retained in the soil. To sustain food-crop production in a forest ecosystem, the trees and standing vegetation need to be slashed, dried and burned to release the nutrients needed to grow annual food crops. Fifteen- or twenty-year-old secondary forest is not only easier to clear than primary forest, but it also dries faster after being cut down and burns more fully to provide the nutrients to grow rice, corn or other food crops. Hence the secondary forest of former shifting cultivation fields is valuable property for people practising shifting cultivation.

Problems arise when people outside the forest fail to understand the tenure systems of shifting cultivators living in the forest. They cannot see the ploughed fields and fences of annual-crop farms, or the neat rows of tree-crop plantations, so outsiders wrongly assume that no-one owns the forest, or has a claim to it. The secondary-forest fallow of shifting cultivation is often considered to be unowned land when it seems to have been abandoned for a decade or more. Government planners and policy-makers regard a secondary-forest fallow of shifting cultivation as empty government land that can freely be given as a concession to an investor wanting to establish a plantation of timber or tree crops such as oil palm (*Elaeis guineensis*), coffee (*Coffea* sp.) or cacao (*Theobroma cacao*). Such land may also be provided to developers for housing or industrial estates or given to settlers wanting to establish new farms and villages.

There are two types of land tenure in traditional shifting cultivation communities. One is a hierarchical system of use rights to individual parcels of land; the other is group or community ownership of a large area of land. The first system

Koompassia excelsa (Becc.) Taub. [Leguminosae]

Ownership of plants, rather than land, is important in swidden communities. These 'bee trees', or common homes for wild honey bees, are often owned by individuals.

vests ownership of a specific parcel of land in the individual who first cleared it from primary forest and burned the vegetation in order to grow crops. He is recognized by others in the community as having all future rights to use this parcel of land, even if it is 10, 15 or 20 years later. In the future, should the original cultivator no longer wish to use this land, or is no longer alive, the rights to use the land pass to his children. Bilateral kinship tends to dominate among shifting cultivators, so both male and female children inherit use rights to their parents' land.[1] The general pattern is that the oldest child has superior use rights to his or her parents' parcel of land, but should he or she not wish to use it, then the second-eldest child, then the third, and so forth, can use it. After another generation, the grandchildren of the original pioneer have secondary rights after their parents. Once several generations have passed, all, or nearly all, members of a village may have varying degrees of use rights to this parcel of land. Besides the hierarchy of generational use rights, the identity of the last user of the land comes into play. An example is the case of two people who descended from the original cultivator and have roughly equal use rights to a field that has been fallow for 12 years. One individual used the land 25 years ago, while the other person has never used it. The latter person's use rights will prevail as it will now be his turn to use the land.[2] A weakness in the hierarchical system of use rights arises when there is a threat of expropriation of land by outsiders. With many people having varying degrees of use rights to each parcel of land, it is often difficult, or at least complicated, to organize communities to fend off an attempt by a developer or a government authority to commandeer a tract of land comprising multiple parcels of shifting cultivation land.

The other system of land tenure in shifting cultivation occurs where there is a strong tribal, ethnic or communal culture, and the ethnic group or village claims ownership of a large block of land surrounding the community. The village leader or council of elders gives individuals or family groups the right to use individual parcels of land for shifting cultivation for one or two years. Priority may or may not be given to the person who originally cleared the forest or his descendants. After the crops have been harvested the land returns to the pool of communal land. When it is ready to be used again, in 10 to 15 years, the right to use this parcel of land may be granted to any member of the community. Examples of this land-tenure system can be found in Papua New Guinea and in the Indonesian provinces of Papua. One of the strengths of communal tenure is group cohesion in protecting the land from potential incursion by outsiders, be they investors or government authorities.

Ownership of economically valuable plants

Another difference between the two systems of land tenure in shifting cultivation occurs with respect to valuable perennials. Valuable trees are held as the private property of an individual or his heirs. Ownership of valuable perennials, either planted or found in the wild, is separate from ownership of, or the right to use, the land on which they are growing. In the first mentioned hierarchical tenure system, valuable perennials belong to the individual who planted them or who was the first to discover them in the wild. From my experience in Indonesia, valuable wild perennials include fruit trees such as durian (*Durio zibethinus*); wild rattans (*Calamus* sp.); 'bee trees' (*Koompassia excelsa, Alstonia* spp. and *Ficus* spp.), which are favoured for nesting by wild honey bees; and other economically significant species. Valuable perennials planted in swiddens in Indonesia may include rattans (typically *Calamus caesius* and *Calamus trachycoleus*); rubber (*Hevea brasiliensis*); cacao (*Theobroma* cacao); bamboo (*Bambuseae* genera); fruit trees

Calamus caesius Blume [Arecaceae]

Rattan: planted and owned by individuals regardless of who is using the swidden land on which it is growing.

(both intentionally planted or which spring up in the refuse of a field hut); and others. Ownership of these trees and their fruit belongs to specific individuals or families, and this must be respected by whoever uses the land on which they grow for future shifting cultivation. These trees must also be protected from fires that clear slashed vegetation, and payment must be made to the owner should they be damaged. Harvesting of these trees or their fruit is the right of the person who owns them, and not the user of the land.

Where communal tenure prevails over shifting cultivation land, ownership of valuable wild perennials lies with the community. Planted perennials are likewise owned by the community as a whole, so there is little incentive for individuals to plant and take care of valuable trees. Decisions to plant fruit trees, make a rattan or cacao garden, or plant any other valuable perennial species, must therefore be made by the community as a whole. The planting of economically valuable trees by individuals is often actively discouraged, out of fear that individuals will make future claims to the trees or their fruit, effectively reducing communal lands and communal values. This is why tree crops are less common in areas of communal tenure, such as Papua New Guinea, than they are where individual ownership of perennials exists, such as in Kalimantan (Weinstock and Vergara, 1987).

Where communal tenure exists, the community as a whole, be it an ethnic group or a village, can stand up to government planners and policy-makers to assert their ownership or rights over a block of land. Where a hierarchical system exists, with many different people claiming rights to various individual parcels, it is difficult to convince government planners and policy-makers that fallow shifting cultivation fields are 'owned'. This is where individual ownership of economically valuable perennials is important. When a government official sees a swathe of secondary forest as unowned 'government land', shifting cultivators can point out rattan gardens, fruit trees, clumps of bamboo and other valuable perennials in the secondary forest that are their individual property. In Indonesia, ownership of economic plants is recognized by the government as separate from the land on which they are growing. Most shifting cultivators live in the forest, far from government offices, and haven't registered their property. But in less remote parts of Indonesia it is not unusual for ownership of trees to be registered at a local level.[3]

Conclusion

Disputes over property rights, be they related to land, to trees or to other natural resources, often lead to the failure of agricultural or rural development projects and programmes. At worst, conflict over property rights can lead to fights and even armed clashes. Lack of clarity or transparency in tenure is not limited to shifting cultivation, but it is made more serious in the case of shifting cultivation by unmarked field boundaries and intermittent uses of the land, often separated by a decade or more of fallow. Reliance on oral histories and shared use rights makes land tenure in shifting cultivation both complex and contrary to formal legal codes in most countries where property rights are based on demarcation and individual ownership.

It is important for government planners and policy-makers to understand that where there is secondary forest that has not resulted from a natural forest fire, it is probably the result of shifting cultivation and therefore subject to a claim of ownership. Before declaring secondary forest to be empty land and handing it over to an investor for a plantation or a new settlement, care needs to be taken to investigate whether anyone claims the right to use the land or if anyone owns economically valuable perennials growing in the forest. To protect the rights of shifting cultivators to use forest lands and the natural resources contained in them, national legal codes and policies covering land, forest and natural resources need to be restructured.

Central to such legal reformation would be recognition of group or communal tenure/ownership of land and forests, so that intermittent use rights can be protected and environmentally sustainable shifting cultivation maintained. One way this might be achieved would be for legal title to blocks of land or forest to be given to tribal or ethnic groups, communities or villages. This might require such groups to organize

Theobroma cacao L. [Malvaceae]

The tree from which chocolate is produced may belong to the person who planted it, no matter who is using the swidden land on which it grows.

themselves into registered legal entities, such as village corporations, to which formal deeds or land titles to demarcated blocks of land or forest could be issued. It would then become the responsibility of each of these legal entities to define the rules by which they would operate, including a definition of its membership and the rights each member has to use the group's property. Shifting cultivation and shifting cultivators can only survive increasing competition for land, forests and natural resources if ways are found in which traditional land-tenure systems in shifting cultivation can be incorporated into national laws and legal codes.

References

Weinstock, J. and Vergara, N. T. (1987) 'Land or Plants: Agricultural Tenure in Agroforestry Systems', *Economic Botany* 41(2), pp312–322

Notes

1. The terms 'he' and 'his' are used throughout this chapter, but individuals holding land rights can be either male or female.
2. This system of land-use rights prevails throughout Borneo, including Indonesia's Kalimantan provinces, the Malaysian states of Sarawak and Sabah, and Brunei. Similar systems probably prevail elsewhere.
3. Several years ago I bought 113 teak trees (*Tectona grandis*) that were planted by my father-in-law in the late 1970s on land in Lampung (Sumatra) that was owned by his sister. My ownership of the trees was registered in the village, while ownership of the land remained with my father-in-law's sister.

B. Historical overviews from Southeast Asia

An old Khmer woman looks pensive, as if gathering her memories of life in upland forests at a time when official policy towards shifting cultivation was dictated by the foreign administrators of French Indochina.

Sketch based on a photo by Malcolm Cairns.

6

THE GEOPOLITICS OF SHIFTING CULTIVATION IN THAILAND

A brief history of the 'hill tribe problem'

*Katharine McKinnon**

Introduction

One of the mysteries of shifting cultivation is how this form of land use has come to acquire such a bad reputation. Agricultural development programmes across Southeast Asia and into the Pacific seem to universally push people away from shifting cultivation, often based on ill-informed assumptions about its ecological, economic and social impacts. Such programmes ignore research that shows how shifting cultivation is a sustainable form of land use; one that promotes biodiversity and provides a sufficient livelihood for many communities, for whom industrialized agricultural production based on commodity (rather than subsistence) crops is not a realistic option. In my work in northern Thailand I traced the history of how shifting cultivation came to be demonized in that setting. It occurred alongside the complex geopolitics of domestic nation-building and international concerns about the Cold War, accompanied by what can only be described as discriminatory policies directed at highland minority groups. This chapter considers the role played by power and politics in creating particular understandings about swidden systems and shaping the policies that control agricultural development.

Until the 1950s, the highlands of northern Thailand were mostly inaccessible and highland villages had very little contact with the Thai authorities. Although they had been drawn into Thai territory half a century before, little had been done to

This chapter is an edited excerpt from McKinnon, K. (2011) *Development Professionals in Northern Thailand: Hope, Politics and Power*, ASAA Southeast Asia Publications Series, National University of Singapore Press, University of Hawaii and Nordic Institute of Asian Studies, Singapore. Reprinted with the permission of NUS Press.

* DR KATHARINE MCKINNON, Community Planning and Development, Department of Social Inquiry, College of Humanities and Social Sciences, La Trobe University, Bendigo, Victoria, Australia.

extend formal Thai control. Highland groups were allowed to live independently and autonomously, as they had for centuries, building rich cultures based on shifting cultivation livelihoods.

Why, then, did highlanders become the object of development? The answer is not that they were desperately impoverished and in need of outside assistance in order to thrive. In the 1960s most highland communities were still subsistence-based economies, with very little cash income to provide a buffer against a bad season or to enable the purchase of modern luxuries such as a radio, or for those communities close to town, modern medicines. Many highland villages were remote from main roads and cities, accessible only on foot, and health indicators like life expectancy and infant mortality were certainly much lower than they are today. Nevertheless, highland villages were not isolated. Most communities maintained close contact with neighbouring groups and villages throughout the mountains traded with lowland towns. The northern Thai dialect was widely spoken.[1] Furthermore, highland communities maintained a rich set of cultural practices tied to the seasons, the growing cycle of rice crops, and important events in people's lives, such as births, marriages and deaths. Villages had functioning leadership systems, social and economic inequalities were minimal, and shamans, healers and spirit mediums worked to ensure the health and well-being of all. In other words, while highland communities were materially poor, according to most early anthropological accounts they were also healthy, functional and culturally and spiritually rich in many other ways. Thus the need for development assistance did not emerge from any sense of need within mountain communities or any self-perceived problem with swidden farming systems.

The imperative for development came instead from the political interests of the Thai state in securing its borders and putting in place a modern form of sovereignty over all Thai territory. Development has long been entangled with questions of national sovereignty and the political complexities of progress (Sidaway, 2007). The conditions through which development programmes were introduced to the mountains were no different. When development professionals first ventured into highland communities in the 1960s and 1970s, they entered a political landscape being reshaped by the geopolitical movements of previous decades. Since the early 1900s, the way highland villages were conceptualized in relation to the state authorities had changed dramatically. The mapping of state borders had drawn the highlands clearly inside Thai territory; a new order had been introduced based on European models of the nation-state; and a new Thai national subject was being constructed, one which excluded highland groups from national belonging (Renard, 2000; Thongchai, 1994, 2000). By the time of the first development programmes, Thailand's population was exploding, putting increasing pressure on land (Hirsch, 1997). Furthermore, the rise of communism and fears that a 'red wave' was about to engulf Southeast Asia had brought the Cold War to the region, with conflicts in Korea and Vietnam and communism gaining power in Laos and Cambodia. There were fears that highlanders might join a communist rebellion in Thailand and thus the people of the borderlands could become a potential threat to the newly emerging Thai nation-state.

In the space of just a few years highlanders went from being ignored by the Thai administration to being considered the highly problematic objects of extensive development programmes. In this chapter, I trace how the highlands were reconfigured as a problematic space inhabited by problematic populations of 'hill tribes'. Tied up with both these movements was the introduction of a new conceptualization of shifting cultivation as a part of the 'problem': at best an inefficient use of productive land, at worst a destructive mode of land use that facilitated undesirable production of opium and dangerous patterns of mobility.

Modern Thai territoriality

The highlands would not have been re-imagined as problematic but for the introduction of the modern nation-state system. Southeast Asian historians such as Thongchai Winichakul (1994) and Ron Renard (2000) have shown that the nation and territory of Thailand emerged only over the past 150 years, and took its shape largely in response to the influence of colonial powers on the geopolitics of the region.

Thongchai (1994) argues that the process of mapping introduced an entirely new idea of state and territory. When the British and French were seeking to establish a precise border between territories they had to contend with local perceptions of what a border was. For both the Siamese and the Burmese, the mountains and forests were understood as natural markers of the spatial divisions of their states. Wide tracts of mountains and forests were effectively ungoverned, and were non-state spaces that formed a buffer between lands controlled by the Siamese on one hand and the Burmese on the other. The colonial authorities, however, pushed for a boundary. The exact location of this cartographic line was not finally decided until the early 1900s. Even then, laying claim to the territory was not the same thing as being able to govern every inch inside the national boundary. That would come much later.

Introducing Thai nationalism

Within the new state boundaries a new nation also started to take shape. For the people of the borderlands, the mountains and the forests, these changes meant that their position in relation to the state also shifted when the places they inhabited were re-imagined as state (rather than non-state) spaces. The earliest evidence of this appears to be in the rise of amateur ethnologies undertaken by Bangkok elites in the Siamese hinterlands, pointing to a fundamental transformation of the *tai–kha* relationships of old (Thongchai, 2000). In pre-colonial Siam, relationship to the state was not defined on the basis of descent. Instead, subjects were labelled according to whether they were seen to be part of a system of state allegiances and known as *tai*, or whether they existed outside it and were known as *kha* (Renard, 2000). In place of the fluid *tai–kha* relationship, difference was newly presented in terms of essentialist notions of culture and race. As in Europe, where domestic ethnology was one of the many new forms of knowledge enabling modern modes of rule (Godelier 1997), the ethnological writings of Bangkok elites helped to make the highland subject, hitherto largely absent from a Bangkok-centric view of the nation, a known and recognized subject. In these new discourses, highlanders entered the consciousness of Bangkok elites as racialized 'others'; examples of primitive peoples who were juxtaposed with the sophistication and civilization of the Siamese. Over time, as European ideas of the nation-state became more pervasive, highlanders became more and more sharply defined against an emerging Thai nation and their 'otherness' was increasingly set in ethnic and racial terms.

It was not until the 1950s that the deliberate exclusion of highlanders from the Thai nation became enshrined in law. In 1956, household registration was introduced as a first step towards creating comprehensive and centralized registration of all Thai citizens. This was the first piece of legislation that 'set down in law the procedure for classifying between Thai and non-Thai people' (Chainarong and Suppachai, 1999). Because of the inaccessibility of many mountain communities, the majority of highlanders were not registered. While the 'wild' and 'backward' subjects of turn-of-the-century ethnologies were a recognized and documented presence within the Kingdom, no formal steps were taken to identify these subjects as either 'insiders' or 'outsiders'. Anthropologist Peter Hinton suggested that this could have been due to a lack of resources and the difficulty of navigating the mountainous terrain, where there were still few roads (Hinton, 2002) (Figure 6-1). It could also have been that geopolitical circumstances did not give rise to any urgent need for the Thai authorities

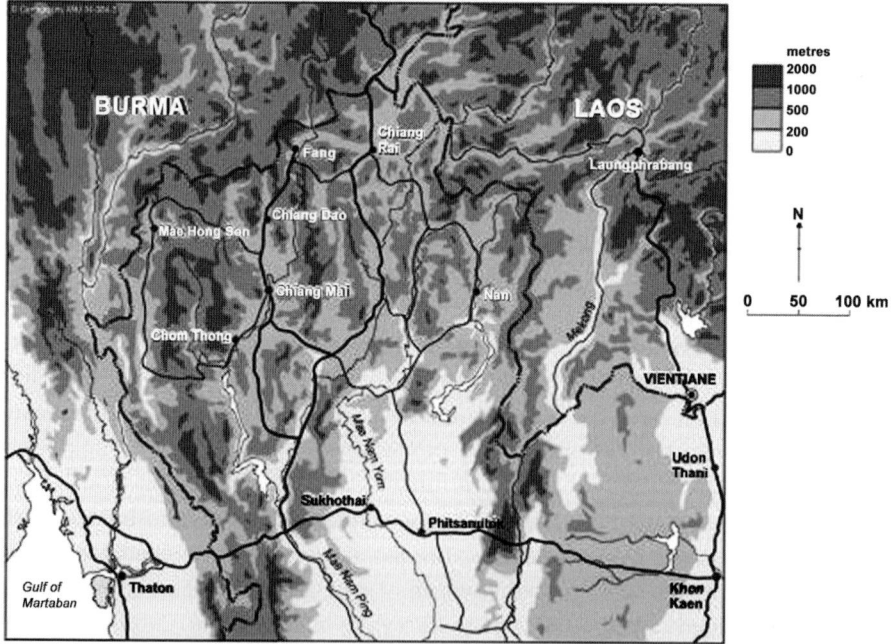

FIGURE 6-1:   Northern Thailand: The highlands of the Thai nation-state.

Source: Reprinted from McKinnon (2011).

to gain access to the highlands. Whatever the reason, while lowland Thais were being registered, highlanders remained documented subjects, but were positioned beyond the regulating (and legitimizing) gaze of the state.

Re-imagining the highlands as a threat

Thai nationalism and modern Thai territoriality put highlanders in a contradictory position: they were within the boundaries of the Thai state, but ostensibly excluded as legitimate subjects. For a time the contradiction did not matter because most communities remained inaccessible and had little to do with state authorities. For all intents and purposes, the borderlands remained non-state space; beyond the regulatory authority of the Thai state. In the late 1950s and early 1960s the Thai government began to bring the border territories under its control. There was an emerging belief that the highlands were a problematic zone, and highlanders themselves were re-imagined as a potential threat to the Thai kingdom. Shifting cultivation, the predominant land use of highlanders, became a central part of how that threat was defined.

The shift came with changed geopolitical circumstances and the growing fear of a communist takeover in Southeast Asia. With communist governments being established in China, North Korea and Vietnam and the beginnings of the war against communist forces in South Vietnam, it was feared that Thailand would be the next 'domino' to fall. The borderlands came to be seen as vulnerable to communist infiltration. At last, the Thai state began to act upon the vision of a nation-state as a contiguous and uniformly governed territory within its existing boundaries.

In the highlands, national and international concerns over the region's vulnerability to communist takeover translated into concerns that communist forces would forge alliances with highlanders. It was recognized that highlanders had an intimate knowledge of the borderlands and that exceptional communication channels existed between scattered mountain settlements across state borders. According to Saihoo, a lecturer in social anthropology at Chulalongkorn University in Bangkok, in the absence of a strong sense of 'national loyalty', it was feared that

> … given adequate support and encouragement from outside, the hill tribes in a particular country may easily engage themselves in subversive activities to further their own ends or to put up resistance to the national authority which seeks to impose some control contrary to their interests (Saihoo, 1963, pp16–17).

Highlanders thus came to be seen as a potential threat to national security because they had not yet been wholly incorporated into the Thai state, and had not been included in the process of national-identity building that had occurred over preceding decades in mainstream Thai society.

The concern that the highlands could be a focal point for communist insurgency against the Thai government did eventually seem to be justified by events in eastern parts of the northern province of Nan. In 1967, a sizeable contingent of Hmong rose there in open rebellion against the state (Hanks and Hanks, 2001). The Thai government was quick to identify it as a communist uprising, although the Hmong involved in the events that precipitated the 'uprising' made no such ideological claim. The Thai authorities saw the conflict as a '[Hmong] conspiracy cooked up by outside communists and directed from headquarters in Laos' (Cooper, 1979, pp325-326). Anthropologists Jane and Lucien Hanks (2001) conducted several extensive surveys among highlanders in the region between 1969 and 1979, and claimed that the insurrection was a response to repeated attempts by Thai officials to exact fines for cutting trees without permission. Robert Cooper, another anthropologist who worked with Hmong communities in the region, provided a different account. He wrote that 'police patrols were only ambushed after they had destroyed Hmong villages which refused to pay an increase in the unofficial tax that is levied on opium cultivation in some areas' (Cooper, 1979, pp325-326). Regardless of the underlying causes, the Thai government responded as if it was a communist-inspired uprising. The Thai army began to clear out Hmong and Yao villages in the area and the air force began bombing raids in the mountains. According to Tapp (1989), 40% of the upland population had been made homeless by 1968. The Thai Cabinet declared the area a free-fire zone and it remained so into the 1970s. All villages in the free-fire zone were evacuated and anyone remaining in the area was considered a communist. At some point during this time, the insurrection became a reality and Thai army units in the hills were targeted by rebels. Even the provincial governor of Chiang Rai and his police chief were killed in an ambush on a mountain road (Hanks and Hanks, 2001). The Hanks' interviews with Hmong who lived in the region at the time made it clear that the Pathet Lao – the eventually triumphant communist faction in the civil war in Laos – had begun recruiting among Hmong across the Thai border and supporting the insurrection with arms and ammunition (Hanks and Hanks, 2001, p196). If the rebellion had not started out with an ideological component, it had certainly acquired one along the way.

At the same time as the highlands became the focus of national security, the region also began to draw attention in relation to opium production, a practice which had been made illegal in 1958. Opium had been grown in the region for centuries for its medicinal qualities. After the two Opium Wars of 1839–1842 and 1858 between Britain and China and the Chinese government's subsequent legalization of the trade, opium became an important cash crop in China. Hmong and Yao farmers cultivated it in China's remote southern highlands of Yunnan. As these people moved south to escape the fighting that erupted after the first Opium War, they took their crop with them and introduced cash cropping of opium to Thailand (Renard, 2001). Opium cultivation was legalized in Thailand in 1855 and the supply was regulated by the Royal Opium Department. The processed opium was sold at exorbitant prices in licensed opium dens. While a profitable black-market business thrived, official trade in opium earned a good revenue for the state (Renard, 2001). When the sale and consumption of opium was finally banned, it was in response to international pressure and only took place after many years of internal negotiation within the ranks

Papaver somniferum L.
[Papaveraceae]

Opium provided good revenue for the Thai state until international pressure had it banned in 1958. It was some years later before it was finally eradicated.

of leading military and police personnel (McCoy, 1972). According to Ajhan Chupinit Kesmanee of Srinakharinwirot University in Bangkok, Field Marshall Sarit Thanarat launched a campaign in 1957 to make opium illegal across the board. This was shortly after Sarit became Thailand's 11th Prime Minister by seizing power in a military coup. However, he was persuaded that revenue from the opium trade was vital to securing the ongoing loyalty of his subordinates, so the ban was shelved (Chupinit, 2001). By the time opium was finally made illegal in 1958, the opium poppies, which were ideally suited to the difficult terrain and touchy soils of the highlands, were well established as an important cash crop, supplementing the livelihoods of many highland communities (Geddes, 1976).[2] The continuing production of opium despite its illegality thus became an additional problem in the view of the Thai authorities.

As well as concerns about national security and opium production, the highlands began to draw attention because of land-use practices that were thought to be causing environmental damage and deforestation. At that time the dominant form of agricultural practice in the highlands was various forms of swidden farming. Also known as 'slash and burn' or shifting cultivation, swiddening is commonly used in tropical upland areas where irrigation systems cannot be built due to unreliable water supply or hilly terrain. Fields are cleared in the forest, the dried vegetation burned and the land planted in rice and vegetables. When yields start to decline, the fields are either left fallow for up to 12 years so that the soil may recover its nutrient value or abandoned altogether, leaving the forest to reclaim the fields. As illustrated in Figure 6-2, the resulting landscape is a pattern of mixed land use in which primary or second-growth forests, fallow fields and cultivated swiddens dot the hillsides. This land-use system was characterized in the case of Akha groups in northern Thailand and southern China by 'landscape plasticity', a term referring to responsiveness to changing conditions of production and the ongoing need to sustain both village households and the land (Sturgeon, 2005).[3] The problem was that the system was thought to compromise the health of the soil, thereby stunting forest regeneration, and to require a degree of clearing that was quickly depleting Thailand's precious rainforest reserves and exacerbating seasonal flooding in the lowlands.

FIGURE 6-2: Highland swidden landscape at Mae Suai in Chiang Rai province, 2005.

Photo: K. McKinnon.

In fact, the data upon which these assumptions rested were not particularly strong. Very little good research had yet been conducted on the environmental impacts of swidden farming. The assumption that swidden agriculture was a bad thing was probably based firstly on the norms set by dominant forest-management paradigms, as well as the links that were being made between swiddening and other concerns about opium and national security.

According to historian Ronald Renard, anti-swiddening prejudices may have been introduced to Thailand from British colonial authorities in Burma (Renard, 2001). The Royal Forest Department of Siam was established in 1896 under the leadership of Herbert Slade, an Englishman trained in German forestry (Usher, 2009). The Royal Forest Department was responsible for managing the forested highlands, which were territories officially owned by the state. Its ethos was to produce more teak – the wonder wood for ship building. The approach was to create carefully managed single-species plantation forests that were clearly separated from agriculture, in the belief that the proper place for agricultural production was in lowland valleys. The existence of agricultural lands scattered through the forests was an affront to the German system of clearly delineated and carefully managed forests. When teak production was the aim, any clearance of forest by swidden farmers was considered to be a wasteful use of resources (Saihoo, 1963). According to Saihoo, early ethnologies from the highlands remarked on the extent of forest destruction (Saihoo, 1963). For example Blofield, who published his account in 1955, described the highlanders' method of agriculture as wasteful, entailing the systematic destruction of valuable jungle. Blofield did not consider the matter particularly serious at the time 'in view of the enormous area of jungle in Thailand, which has hitherto been put to no use whatever' (Blofield, quoted in Saihoo 1963, p21). As Scott (1998) contends, the complex usage of forest resources by swiddeners was simply not 'legible' to the dominant perspectives of the time, which remained enthralled by the orderly and exact production of plantation forests under the German forestry system.

As the priority for the Royal Forest Department shifted from teak production to forest preservation, the perception of the practice of swiddening shifted from wasteful to destructive. By 1960, when the first highland Land Settlement Projects were established, so-called 'slash-and-burn' agriculture was seen to be such a damaging practice that it was outlawed (Manndorff, 1967, p533). Four years later, the Department of Public Welfare identified the major cause of forest and watershed destruction as 'the shifting cultivation practised by the hill tribes' (Department of Public Welfare, 1964). Despite a lack of accurate data, the Department estimated that hill tribe activity had destroyed two thirds of the forest in Chiang Mai and Lamphun provinces. These claims have since been strongly contested.

While the concern in the 1950s and early 1960s focused on the risk that shifting cultivators posed to primary forest, by 1969 broader environmental concerns began to emerge, many of which were closely related to the cultivation of opium. Although, as mentioned above, the cultivation of opium poppies was banned in 1958, little direct action had been taken against opium producers, who continued to grow it both for domestic use and as a cash crop. Opium poppies could be grown on marginal land which was either very steep or had poor soils. Even on otherwise productive land, opium was an excellent crop when soil fertility fell below that required for producing food. In these cases, planting an opium crop was an effective way to continue to generate an income from depleted land. As a crop that could be planted in otherwise unproductive sites, opium worked well with the swiddening system and provided cash income to supplement a subsistence farming livelihood. From an environmental point of view, however, the cultivation of opium on very steep land was believed to cause erosion and subsequent siltation of streams, which in turn was affecting lowland water supplies (Grandstaff, 1980). In addition, because opium was able to thrive in soil that had already supported several rice crops and was no longer fertile enough for rice, the further use of the land to grow poppies was thought to severely deplete soil fertility. As a result, the only plant that would grow when the fields were eventually abandoned was the rhizomatous grass *Imperata cylindrica*, which was thought to inhibit regeneration of the forest (Keen, 1972). There was little or no scientific evidence; nevertheless the dominant view emerged that opium production was leaving swathes of the highlands denuded of precious rainforest.

Finally, swidden farming supported the movement of highland communities through the hills. Opium production was believed to be a key element in the livelihoods of 'pioneer swiddening' groups – those communities that would periodically relocate to find new land rather than rotating the same fields through a production cycle. The pioneer swiddeners – Hmong, Akha, Lisu and Lahu – were considered to be the main opium producers and their periodic relocation added to national security concerns. This movement was unregulated by state authorities and often led communities across state borders. Under the modern form of territoriality, such unauthorised movement violated state sovereignty and the integrity of national borders. In the 1960s and 1970s, this was also believed to carry the added risk that communist ideologies might travel alongside farming communities. Put together, these concerns shaped the negative stereotypes that would influence much of the Thailand government's policies towards highlanders and the practice of shifting cultivation.

Imperata cylindrica (L.) Raeusch. [Poaceae]

This plant grows readily on soil exhausted by crops of rice and opium. It was then thought to inhibit forest regrowth.

The threefold 'hill tribe problem'

By the early 1960s, the three issues of national security, opium and swidden farming had been rolled together to become 'the hill tribe problem'. It was under the rubric of the 'hill tribe problem' that the Thai state sought increasing engagement with highland communities. In the process, highlanders themselves became redefined by the governing gaze of the state.

The first important act of redefinition came with the introduction of the term 'hill tribe' (*chao khao*) to describe highland peoples. The term was adopted as the Thai state began formally to engage with highland communities after the establishment in 1959 of the Hill Tribe Development and Welfare Programme of the Department of Public Welfare. From this point on, 'hill tribe' became the term most widely used to refer to highland groups. This had the effect of grouping together as a single population a diverse range of peoples, while disguising the wealth of ethnic, cultural and linguistic diversity among them. The 10 officially recognized groups were Karen, Hmong, Mien, Akha, Lisu, Lahu, H'tin, Lu, Yao and Khamu. Furthermore, the 'hill' designation obscured the extent of interrelationships between these peoples and valley dwellers, well recognized since Leach's ground-breaking research in the 1940s (Leach, 1954; Toyota, 1998, 2003; Jonsson, 2005). As 'hill tribes', highlanders became part of a state discourse as a single entity that was defined by its distinction from lowland Thai groups. Highland peoples thereby entered both popular Thai consciousness and the attentions of state authorities not as a range of culturally rich additions to the nation, but as the 'other' to an emerging Thai national identity. 'Hill tribes' were defined as strange, dirty and primitive, and were imagined to be far distant from the sophisticated, civilized and modern Thais.

Official definitions of 'hill tribes' are exemplified by the following extract from a Department of Public Welfare brief on the Hill Tribe Development and Welfare Programme (1964). It is an example that has not dated:

> It is estimated that there are between 200,000 − 300,000 hill peoples living interspersedly in the densely forested hill ranges of Northern Thailand. These people belong to various tribes having their own distinguished languages, cultures, traditions, beliefs, etc. Most of them raise and sell opium, practise shifting cultivation and always keep on moving to hunt for new pieces of land for cultivation which have greatly resulted in the forest and watershed destruction. These hill tribes are generally illiterate, have ill health and are economically deprived and could become the victims of [communist] infiltration so easily. These lead to the problems of social economic development, administration and political security of the nation which therefore demand the most urgent solution (Department of Public Welfare, 1964, p1).

In this picture, the 'hill tribes' are identified by their primitiveness, their illiteracy, ill health and economic deprivation, and by their destructive agricultural practices, for which they must 'always keep on moving'. This discourse on the 'hill tribe' subject attributes a single identity to diverse and widespread communities on the basis of a shared 'problem' they present to the Thai state. By defining highlanders in this way, a homogeneous identity was brought into being that was fundamentally abnormal and, like the development subjects discussed by Escobar (1995), could be treated and reformed through the development process.

The Thai state had a clear interest in representing highlanders in this way. Long-standing prejudices aside, the new discourse on 'hill tribes' made perfect sense in the context of the emerging need to be able to govern the highlands more effectively. Being rendered a singularity had one convenient effect: the 'hill tribes' no longer appeared to be mobile, diverse and complex peoples, but a single population that could become the object of policy-making and state policing. The discourse created, in theory at least, a population that could become a manageable group and the object of special legislation. It created a knowable and, at the same time, governable subject, and was part of the state's attempt to make highland societies 'legible' in order to control them (Scott, 2009).

The first efforts to govern highlanders established a pattern that would characterize highland-state relationships in the coming decades. The terms 'welfare' and 'development', embedded in the title of the Hill Tribe Development and Welfare Programme, suggested a rather benign approach. But apart from the stated aim of promoting socio-economic standards, the programme's objectives built on negative stereotypes about highlanders and approached them from the outset as being a problem. As stated in 1964, its objectives were:

1. To promote and develop the socio-economic standard of the hill tribes by way of promoting their occupation, education and health as well as helping to develop their own communities.
2. To prevent forest and watershed destruction by way of introducing stabilized farming.

3. To abolish opium production by way of introducing other occupations to replace opium raising.

4. To guarantee the public safety in border provinces by way of promoting mutual understanding and loyalty (Department of Public Welfare, 1964).

These objectives held true to popular beliefs about highlanders at the time but had no basis in any extensive or critical research. As anthropologist Saihoo (1963, p15) recognized, although there was no 'complete data or results of specific studies' in 1963, the three interrelated issues of swidden, national security and opium were well entrenched as a 'hill tribe problem' that needed urgently to be addressed. Saihoo characterized these three issues (quoted verbatim) as follows:

> It is by now generally agreed among persons interested in Thailand hill tribes that they invite careful consideration in three important respects:
>
> 1. Their shifting cultivation which involves the destruction of extensive areas of the forests on the mountains and moving the villages in search of new fields with possible consequences of soil erosion and damage to watersheds which would affect the supply of water for the lowland Thai cultivators.
>
> 2. Their little recognition of international boundaries and national authority and control with possible consequences of border insecurity, especially in the present world political situations, for the country in which they reside.
>
> 3. The production of raw opium of some tribes which supplies the country and the world with an illegal and harmful product in various forms.

In the absence of hard data, trust was placed in the 'reliable observations of those who are in well-qualified positions' (Saihoo, 1963, p15). In practice, such knowledge could be traced back to observations by colonial explorers such as Pendleton and Blofield, supplemented by the preliminary results of the first broad social survey in the highlands. The first comprehensive survey was carried out by a United Nations team led by anthropologist Hans Manndorff (see Manndorff, 1965). The survey itself was not designed to question the basis for the threefold 'hill tribe' problem, but to gather data describing the extent of the issues. On this tenuous basis, the 'hill tribe problem' became the cornerstone of development and research policy in the 1960s and it remains a powerful driving force behind much state policy in the present day.

The beliefs that highlanders posed a serious security threat, were supplying narcotics and threatening precious old-growth forests also shaped the development strategies of the international community. From the early 1970s, Thai efforts to address the 'hill tribe problem' were bolstered by foreign development assistance that also focused on issues of opium production, national security and preservation of forests.

Why was the 'hill tribe problem' so compelling despite the absence of any foundation of substantive research? The most plausible explanation has little to do with the welfare or development of highland peoples. What seems most likely is that there was an accepted geopolitical imperative for Thailand to secure its borders. Development interventions were one conduit towards achieving this while appearing to act in a benign and altruistic way. The core 'problems' for which aid was required coincided with national and international geopolitical concerns of the day. The possibility that highlanders could threaten national security fitted neatly with Cold War paranoia about a 'red wave' of communism sweeping through Asia. Concerns with opium production coincided with rising rates of heroin addiction in Western cities and increasing attention being given to strategies to limit, if not eliminate, supply (McCoy, 1972). At the same time there was burgeoning interest in the value of old-growth forests and the need to preserve them. As highland development programmes were aimed at addressing all three of these overlapping concerns, few were motivated to find out if the investment of aid money was, in fact, justified.

Conclusion

This chapter has traced the ways in which highland development in Thailand, and its focus on transforming the shifting cultivation practices of highland minorities, was bound up with processes of nation building and the management of international geopolitics. The introduction of agricultural-development programmes to the highlands coincided with a belated push to realize a vision of a Thai nation-state in which control spread from the centre at Bangkok out to its edges at the cartographic line that formed its boundaries. As part of this process, the highlands and highland peoples came to be understood in terms of the problems they presented and the need they had for the ministrations of the development industry. Development was thus introduced to the highlands as a tool for transforming remote mountains into accessible Thai territory and making it possible for the state to administer the highland population.

The discourse of the 'hill tribe problem' was a key element in this transition. It effected a negative reconstitution of the highlands in terms of their lawlessness, distance from the state and damaging 'otherness' to the Thai nation. Thus constructed as a peripheral and disadvantaged space, inhabited by 'dangerous' and 'deficient' subjects, the highlands became a space of 'problems' that could then be 'fixed' through outside intervention. In the process, the highlands became known and governed in new ways, transforming remote and largely ungoverned communities into governable spaces and governable subjects. The discourse of the 'hill tribe problem' thus found a place for highlanders within the wider hegemonic discourses of the nation-state. On the basis of erroneous evidence, and with very little correspondence to the rich lives and livelihoods of highland communities, 'hill tribes' become known as a problematic and potentially dangerous 'other' to the Thai national identity.

By taking advantage of international geopolitical concerns, the Thai state gained the cooperation of foreign governments in bilateral development programmes in the mountains. But the process of defining highlanders on the basis of their

Litchi chinensis Sonn. [Sapindaceae]

Lychees are among the monoculture cash crops that have replaced swidden systems in the mountains of northern Thailand.

'problems' in effect demonized the foundations of highland livelihoods and introduced the belief that shifting cultivation was inefficient and damaging. Despite research since then having shown just the opposite, shifting cultivation continues to be understood as a problematic form of land use in northern Thailand. Although the geopolitical concerns that were bound up with its characterization as part of the 'problem' in the 1950s and 1960s have changed, an almost superstitious hostility to highland agriculture remains in place. Nowadays, however, the public narratives associated with the problem of highland agriculture tend to focus more on risks to lowland communities that come from permanent agriculture. Fertility in mountain soils is quickly depleted and the traditional method for dealing with that problem was to leave the fields fallow for long enough to allow the forest to regenerate and soils to regain their nutrients. Pressure on highland resources, along with the outlawing of swidden farming methods, placed increasing pressure on highland farmers to find alternative production methods. In many communities subsistence farming has given way to monocultures of new crops like lychees, cut flowers, coffee and cabbages. Tourism interests, commercial forestry and lowland farmers looking for new land have placed increasing pressure on highland resources. To achieve continuous cultivation, highland farmers often rely on chemical fertilizers and are now accused of polluting waterways due to runoff from their fields. At the same time, lowland flooding is often blamed on deforestation that continues to be attributed to highland farmers despite plentiful evidence to the contrary. While the issues have changed, there remains a dangerous pattern of basing agricultural policy on poor (or non-existent) science, and an accompanying failure to question the foundations of assumptions. It is important to remember that current debates around highland agricultural practices are set in the context of decades of discriminatory and exclusionary politics, and the 'knowledge' that underlies policy is seldom immune to the biases arising from that history.

References

Chainarong, S. and Suppachai, J. (1999) 'Citizenship, ethnic identity and state policy: Thai or non-Thai for hilltribe people?' paper presented at the 7th International Thai Studies Conference, Special Round Table, Amsterdam

Chupinit, K. (2001) Personal communication between the author and Ajhan Chupinit Kesmanee of Srinakharinwirot University, Bangkok

Cooper, R. G. (1979) 'The tribal minorities of northern Thailand: Problems and prospects', in *Southeast Asian Affairs* 1979, Institute of Southeast Asian Studies, Singapore, pp323–332

Department of Public Welfare (1964) *A Brief on Hill Tribe Development and Welfare Program in Northern Thailand*, Department of Public Welfare, Ministry of the Interior, Bangkok

Escobar, A. (1995) *Encountering Development: The Making and Unmaking of the Third World*, Princeton University Press, Princeton, NJ

Geddes, W. R. (1976) *Migrants of the Mountains: The Cultural Ecology of the Blue Miao (Hmong Njua) of Thailand*, Clarendon Press, Oxford, UK

Godelier, M. (1997) 'American anthropology as seen in France', *Anthropology Today* 13(1), pp3–5

Grandstaff, T. B. (1980) *Shifting Cultivation in Northern Thailand: Possibilities for Development*, Resource Systems Theory and Methods Series (No. 3), United Nations University, Tokyo

Hanks, J. R. and Hanks, L. M. (2001) *Tribes of the North Thailand Frontier*, Yale Southeast Asia Studies, New Haven, CT

Hinton, P. (2002) Personal communication between the author and anthropologist Dr Peter Hinton

Hirsch, P. (ed.) (1997) *Seeing Forests for Trees: Environment and Environmentalism in Thailand*, Silkworm Books, Chiang Mai

Jonsson, H. (2005) *Mien Relations: Mountain People and State Control in Thailand*, Cornell University Press, Ithaca, NY and London

Keen, F. B. G. (1972) *Upland Tenure and Land Use in North Thailand*, Tribal Research Centre, Chiang Mai

Leach, E. (1954) *Political Systems of Highland Burma: A Study of Kachin Social Structure*, London School of Economics and Political Science, London

McKinnon, K. (2011) *Development Professionals in Northern Thailand: Hope, Politics and Power*, ASAA Southeast Asia Publications Series, National University of Singapore Press, University of Hawaii and Nordic Institute of Asian Studies, Singapore

McCoy, A. (1972) *The Politics of Heroin in Southeast Asia*, Harpers and Row, New York

Manndorff, H. (1965) *The Hill Tribe Program of the Public Welfare Department, Ministry for the Interior, Thailand*, Division of Hill Tribe Welfare, Bureau of Self-help Land Settlement, Department of Public Welfare, Bangkok

Manndorff, H. (1967) 'The hill tribe program of the Public Welfare Department, Ministry of the Interior, Thailand: Research and socio-economic development', in P. Kunstadter (ed.) *Southeast Asian Tribes, Minorities and Nations*, Princeton University Press. Princeton, NJ, pp525–552

Renard, R. D. (2000) 'The differential integration of hill people into the Thai state', in A. Turton (ed.) *Civility and Savagery: Social Identity in Tai States*, Curzon Press, Richmond, UK, pp63–83

Renard, R. (2001) Personal communication between the author and historian Ronald Renard, June 2001

Saihoo, P. (1963) *The Hill Tribes of Northern Thailand: A Study,* Southeast Asia Treaty Organization Cultural Programme, Bangkok

Scott, J. (1998) *Seeing Like a State: How Certain Schemes to Improve the Human Condition have Failed*, Yale University Press, New Haven, CT and London

Scott, J. (2009) *The Art of Not Being Governed: An Anarchic History of Upland Southeast Asia*, Yale University Press, New Haven, CT

Sidaway, J. (2007) 'Spaces of postdevelopment', *Progress in Human Geography* 31(3), pp345–361

Sturgeon, J. (2005) *Border Landscapes: The Politics of Akha Land Use in China and Thailand*, University of Washington Press, Seattle, WA

Tapp, N. (1989) *Sovereignty and Rebellion: The White Hmong of Northern Thailand*, Oxford University Press, Oxford, New York and Toronto

Thongchai, W. (1994) *Siam Mapped: The History of the Geo Body of a Nation*, University of Hawaii Press, Honolulu

Thongchai, W. (2000) 'The others within: Travel and ethno-spatial differentiation of Siamese subjects, 1885-1910', in A. Turton (ed.) *Civility and Savagery: Social Identity in Tai States*, Curzon Press, Richmond, UK, pp38-62

Toyota, M. (1998) 'Urban migration and cross-border networks: A deconstruction of the Akha identity in Chiang Mai', *Southeast Asian Studies* 35(4), pp803-829

Toyota, M. (2003) 'Contested Chinese identities among ethnic minorities in the China, Burma and Thai borderlands', *Ethnic and Racial Studies* 26(2), pp301-320

Usher, A. D. (2009) *Thai Forestry: A Critical History*, Silkworm Books, Chiang Mai

Wanat, B. (1989) 'Government policy: Highland ethnic minorities', in J. McKinnon and B. Vienne (eds) *Hill Tribes Today*, White Lotus, Bangkok, pp5-31

Notes

1. Northern Thai, or *Kham Meuang*, is closely related to central Thai and Lao and is spoken widely in the provinces of Chiang Mai, Chiang Rai and Mae Hong Son.

2. Opium was made illegal in Thailand by Proclamation of the Revolutionary Party No. 37, on 9 December 1958 (Wanat, 1989, p13).

3. Sturgeon's study involved two case-study villages, one located in southern Yunnan province in China and the other in Mae Fa Luang district in northern Thailand. From her results she extrapolates to Akha land use in general.

7

THE FRENCH COLONIAL ADMINISTRATION VERSUS SWIDDEN CULTIVATION

From political discourse to coercive policies in French Indochina

*Mathieu Guérin**

Introduction

The French agronomist, botanist and anthropologist Jacques Barrau defined swidden cultivation, or *essartage,* in Southeast Asia as a mode of culture 'on land cleared by fire with long fallow in the forest' (Barrau, 1972, p100). This general definition has the merit of sticking to the reality of practice while remaining broad enough to incorporate local variations. Swidden cultivation has long been practised in the highlands of the southern part of the Annamite Range, which extends through Laos, Vietnam and into northeastern Camdodia, by native people, including the Mnong, Jarai, Brao, Stieng, Bahnar, Rhades, and so on. These people are considered to be the indigenous occupants of this region: they have languages, customs and social organisation distinct from the Viet, Lao or Khmer peoples of the plains. Because of their strong connection to the forest, they can be viewed as 'people of the forest'.

In the second half of the 19th century, France set foot on the peninsula and built the far-eastern component of its colonial empire: French Indochina. It included the territories of Vietnam, Cambodia and Laos, and also the highlands that were once a buffer zone between lowland states. This chapter aims to describe the process that led to the emergence of a deprecatory discourse against swidden cultivation; a discourse that eventually led to the drafting of regulations and policies. The fight against swidden carried out by the French colonial administration appeared to be one of the many facets of France's 'civilizing mission' in the region. It had severe consequences for the highland peoples and for the preservation of Indochina's forests.[1]

* Dr Mathieu Guérin, Member of the Southeast Asia Centre (UMR 8170, CNRS EHESS INALCO), Associate member of the Research Center for Quantitative History (UMR 6583) and a member of the Rural Studies Group at the Maison de la Recherche en Science Sociale, Caen, France.

Swidden in the southern highlands of Indochina

A farmer in the highlands of Indochina who wanted to cultivate a field in a forested area would chop down the trees on the land he had decided to farm, and then set the resulting slash on fire before growing his crops. He then used the field for one to three years before moving to a new plot, where he renewed the operation. Swidden cultivation could not be regarded as a permanent clearing technique, because the forest would regenerate on the abandoned field. The farmer would return after a variable time, usually 10 to 20 years, and the operation would be repeated. The forested territory of a village could be considered as a large space of fallow, potentially available for cultivation. Knowing that crop yields would be higher on swiddens cleared from dense forests, it was in the interests of the farmer to leave the forest to recover sufficiently before a plot was cleared again. However, when they had to clear areas too far from their village, they would move closer to their fields.

The agricultural calendar began at the end of the dry season in February and March (Condominas, 1983, pp21-28). Plots for swiddens were selected on the basis of expected yields, assessed by examining the slope, to allow rainwater to flow, and the quality of the soil and the forest cover. There were also religious considerations. For instance, the presence of a cobra on the ground to be cleared was often considered a sign that the divinities opposed the use of the land, and another location had to be chosen. When they found a suitable plot, the trees that covered it were chopped down, but the largest stumps were left in place. Some of the slashed branches were used to build a fence around the field to protect it from predators, and the rest of the slash was left to dry on site. At the time of the first rains in April, the dried slash was burned. Sowing took place shortly after. In most villages, the men would move forward in line and make planting holes in the ground with sticks. Behind them, women poured some rice seeds into each hole. As well as rice, vegetables, corn and tobacco, plants were grown for their textile uses, their dyes and their medicinal properties. The extreme variety of crops on a swidden was one of the features of the system. A small house was built on the field to allow farmers to continuously supervise their crops. These houses, on cleared land well hidden by the surrounding forest, could even serve as shelters in cases where people were driven from the main village by some form of danger. In June, when the rice panicles were beginning to form, the oversight of the fields was tightened. Vegetables began to be collected. The corn was harvested in July, and then came the early rice. In November, at the end of the rainy season, late rice was harvested, and this was the community's most important harvest.

A cropping system seen as unproductive and destructive by the French

One of the first detailed descriptions of swidden cultivation by a French traveller was provided by the explorer Henry Mouhot in 1859 after he spent three months in a Catholic mission established in the Stieng area in eastern Cambodia. Mouhot's account carries no value judgment on swidden cultivation as an agricultural practice.

He chose to describe it because it was exotic to him. He considered that it 'is very different from [the techniques that] our farmers use in their wheat or oats fields' (Mouhot, 1868, p155). Mouhot's description has long been an exception in the tenor of comments on swidden farming (Figure 7-1).

During the installation of the French colonial presence in Vietnam, Cambodia and Laos, a deprecatory discourse was constructed on swidden cultivation. As early as 1853, the Reverend Combes, who was a founder of the Catholic mission in Kontum, among the Bahnar people, wrote:

> The land seems fertile enough in Banhar territory, although it is of much lower quality than in some neighbouring tribes' territories. Several villages reap a hundredfold, and the least fortunate still collect another fifteen to twenty for one. Such results lead to the conclusion that here the savages do not know the hardships of hunger, and yet it is rare that there is enough rice to fill the gap from a harvest to another without suffering. This is due to several causes, and especially the agricultural system. When a Banhar cuts down a forested area and burns it to make a field, the land acquires a strength that indeed compensates the fatigue necessarily attached to such a difficult work, for two or three years only. However, the tools available to him do not allow him to maintain this fertility: everything he does and all he can do is to deliver its seeds to the ground with a sharp stick, and later in pulling weeds, with a small pickaxe only three inches wide. He cannot stir the earth; or rather he doesn't have the means nor the force to do it. So it exhausts very fast, and after three crops, it is abandoned and returned to the forest, which rises and falls and to fall and rise again, without ever disappearing. (Dourisboure, 1922, pp309-310)

Labour et semailles chez les sauvages stiéngs. — Dessin de E. Bocourt d'après M. Mouhot.

FIGURE 7-1: A mid-19th century illustration of swidden farmers at Stieng in Cambodia.

Source: Mouhot (1868).

Combes' description deserves a critical analysis, especially because of its contradictions. It describes both exceptional returns and scarcity of food – if not famine – experienced by swidden cultivators. Part of his explanation involves the decline in soil fertility after two or three years, while reporting that Bahnar farmers then changed the location of their fields, and part of it refers to the simplicity of the tools and cropping processes. For the Reverend Combes, yields were insufficient because he perceived Bahnars as primitive people; and in his view 'savages' couldn't possibly have efficient production techniques.

The ethno-historian Oscar Salemink showed the impact of representations made by missionaries in the 19th century, as they were regarded as the best experts on the highlanders (Salemink, 2003, pp40-50). Combes' letter was published in Pierre Dourisboure's 1873 book, *Les Sauvages Ba Hnar*, which was reprinted more than a dozen times between 1875 and 1929. It certainly helped to impose the idea that a swidden, then known as *ray*, was unable to feed those who cultivated it. This assertion was repeated in most books on the subject published during the colonial period. Thus, the monograph on Darlac province, published on the occasion of the Colonial Exhibition at Vincennes in 1931, says 'the performance of upland rice is, however, much lower than the rice paddy, the *Moi* [savages] also grow corn' (Monfleur, 1931, p45).

Swidden was also accused of destroying the forests of Indochina, especially for turning areas of dense forest into savannah or open forest. In a seminal work written in the 1930s, the forest engineer Maurand, from the Institute of Agricultural Research and Forestry of Indochina, who was then chief of the forestry station of South Indochina, began a chapter on deforestation by writing: 'The clearing of forests is caused by the creation of permanent crops, the *rays* [swiddens], forest fires and wildfires' (Maurand, 1938, p35).

He went on to describe swidden as 'one of the causes of the ruin of forest stands'. Eugene Teston and Maurice Percheron drew a similar conclusion in their famous *L'Indochine Moderne, Encyclopédie Administrative, Touristique, Artistique et Économique*. In a chapter on forests in this impressive and luxurious book, which was designed as an exhaustive introduction to French Indochina for a large audience, we read:

> The fires started by the natives and the foolish exploitation of forest, in which they were engaged before the organisation of the French Forest Service, are the plague of Indochinese forests. Indeed, in all parts of the colony, and especially in the vast interior forests, wandering lands of the *Moi* tribes, the only practice of this technique causes incalculable losses to the forest and we could foresee the gradual disappearance of the finest stands. Thanks to the expert, patient and tenacious efforts of French foresters, who limited as much as they could these traditional devastations, Indochina will preserve its admirable wooded area. (Teston and Percheron, 1931, p931)

The Forest Service Department of Indochina believed that the forest had been in significant decline since the colonial conquest. The regression of the forest at that time was confirmed by historical studies (Thomas, 2000). Members of the Forest Service Department, who were considered the main experts in this field by the leaders of the colonial administration, blamed swidden as the main cause of deforestation.

Moreover, swidden was seen as encouraging a nomadic lifestyle, allowing its practitioners to elude the control of the state. The location of villages could change, making it difficult to assert administrative control over their inhabitants.

A perception of swidden, in the colonial rhetoric

This vision of swidden as archaic, unproductive and destructive fitted perfectly into the evolutionary discourse in vogue throughout the colonial period; a discourse that was one of the justifications of French colonization, especially during the Third Republic, whose motto was *Liberté, Égalité, Fraternité*. It was considered a duty of France, as a part of its civilizing mission, to bring progress to colonized peoples. This was its 'white man's burden'. Within this mindset, swidden farmers in the highlands of Indochina appeared as a boon for advocates of colonial action. They were the perfect incarnation of 'savages': they lived in the forest, they dressed with a simple strip of cotton wrapped around the waist, men and women went topless on a hot day, some of them filed down their front teeth, they did not know how to write, they refused the authority of the state and their religious beliefs were considered to be superstitions by Christians. The French chose to call swidden farmers in the highlands *Moi,* a Vietnamese term meaning 'savages'. Therefore, the inferiority of these 'savages' and the need to bring them some progress seemed obvious. An outspoken defender of French colonialism, Paul Lechesne, was moved to write:

> We have the rare chance to face a kind of *tabula rasa,* and nothing and no one will distort our action: no ancient civilization as in Annam, no valid organisation, no formidable resistance, no contiguous influence other than ours, simply Nature, and untouched! [...] We don't find the *Moi* in a respectable evolution but in an inveterate stagnation, morbid, and precarious state of single embryo of "Human Capital". How dare say, "Respect this!" How not to seek the formula of Progress, which can only be administrative and economic, an element supporting and justifying the other? (Lechesne, 1925, pp64–65)

For the French administration, the implementation of the 'civilizing mission' therefore meant replacing swidden cultivation with techniques perceived as being more productive; in line with the European perception of what must be modern. Thus, the Resident Superior of Annam (a former subdivision of French Indochina, now the central region of Vietnam) Pierre Pasquier, issued an order on 30 July 1923 entitled 'Guidelines for the Administration of the *Moi* Provinces'. These provided a general framework for action in the highlands. Its first economic goal was the eradication of swidden cultivation.

An inaudible rehabilitation of swidden

A few voices rose against the dominant and official discourse on swidden cultivation. The Resident administrator of Kratie, in northeastern Cambodia, Adhémard Leclère, who had studied the local Mnong people for some years, described their swidden-farming system as a fairly productive one (Leclère, 1898, p204). Leclère's analysis was confirmed early in the 20th century by the work of agronomists and geographers who tried to rehabilitate the swidden system, with long forested fallows. In 1910, the director of Cambodia's Agricultural and Commercial Service wrote after a visit to the highlands of Kratie:

> This mode of cultivation may seem primitive and even barbaric. However, it is quite rational, if one considers the very special conditions under which these people are [...]. Upland rice has the same requirements as cereals in Europe, that is to say that it needs decomposed minerals to grow, while ashes from the incineration of the forest gives precisely the necessary minerals and when they are exhausted, harvests cannot be reaped anymore and the land must be abandoned. (Magen, Director of the Service of Agriculture and Trade in Cambodia, [Tournée dans la Province de Kratié, Agrologie], report to the Résident Supérieur du Cambodge, 1910, RSC 4147, National Archives of Cambodia, Phnom Penh)

Pierre Gourou, a geographer who specialized in Indochinese rural areas, drove the point further in his seminal work *L'Utilisation du Sol en Indochine Française* (Land use in French Indochina):

> The system of *ray* is not awkward, it is indeed practising a very long fallow revolution [...]. The system of *ray* does not allow a high density of population, but it protects the soil from erosion, since there is no ploughing of the land on which [in] the second year grow some shrubs, and since the reconstitution of the forest, hampers laterisation. (Gourou, 1940, p347)

Gourou's analysis was based mainly on the work of the agronomist Yves Henry, who had been studying Indochinese soils and ways to preserve them (Henry, 1931).

Blaming deforestation on swidden was also quite unconvincing. While all research indicated that high-quality timber quickly disappeared from easily accessible forest areas in the decades following the arrival of the French, most French colonists accused swidden cultivation of being the main cause, a view that was supported by the Forest Service of French Indochina, even if swidden cultivators had been using their system for several centuries, if not millennia. In fact, the timber trade was an important element of the economy in early colonial times. As well, the author Frédéric Thomas was able to highlight the fact that the majority of forest officers – those who became experts in the field for the colonial administration and the general public – had very low technical competencies (Thomas, 1999, pp14-16).

After World War II, the work of anthropologists embracing participant observation, such as Georges Condominas and Jean Boulbet, contradicted the most established assertions about swidden cultivation (Condominas, 1957; Boulbet, 1966, 1975). They calculated that swidden yields could be very high. The comparison between wet paddy and swidden yields in Stung Treng in northeastern Cambodia in the early 20th century, drawn from colonial sources, confirms their writings. Swidden had the highest yields in this area at that time (Guérin, 2001).

Anthropologists also showed that only in exceptional circumstances would swidden cultivators clear dense or primary forest, because this was believed to host very powerful divinities. Jean Boulbet went further. For him, the ability to maintain forest cover was an essential part of swidden cultivation. 'Only those who can manage their sylvan capital in a way as to retain its agricultural and regeneration potential can be called *paysans de la forêt* (forest farmers)' (Boulbet, 1975, p8).

Georges Condominas also explained that shifting cultivators were not nomads, since, except in times of crisis such as wars or epidemics, they moved their villages only within the boundaries of their territory. However, the work of Condominas and Boulbet was published too late to influence colonial policies.

A rehabilitation sanctioned by recent studies

Studies conducted at the end of the colonial period or shortly afterwards, in line with the work of Harold Conklin on swidden cultivation in the Philippines, confirmed that another approach to swidden was needed (Conklin, 1954, 1957, 1961). In former French Indochina, the routine of blaming swidden for deforestation was significantly cooled by the work of Jean Vidal (Vidal, 1957-1961). More recently, Jefferson Fox, comparing sets of aerial photographs of the highlands of Cambodia between 1952/1953 and today (Figure 7-2), was able to show that where swidden cultivation was practised, forest cover was maintained, even if there was an observable decline in the quality of the forest. But where perennial crops were grown, the forest had disappeared (Fox, 2002).

Moreover, studies conducted by Jacqueline Matras, Frédéric Bourdier and more recently Ian Baird, have shown that production of crops other than rice was significantly underestimated in the French colonial era (Matras-Troubetzkoy, 1983; Bourdier, 1995; Baird, 2008). During his stay among the Mnong Gar in 1949, Georges Condominas collected a herbarium of 27 species of food plants, in addition to different varieties of rice, tobacco, cotton, indigo, 'magical' plants for rituals and medicinal plants. Ian Baird observed that the Brao of Northeastern Cambodia could grow more than 180 different plants (Condominas, 1983, pp38, 53; Baird, 2008, p93).

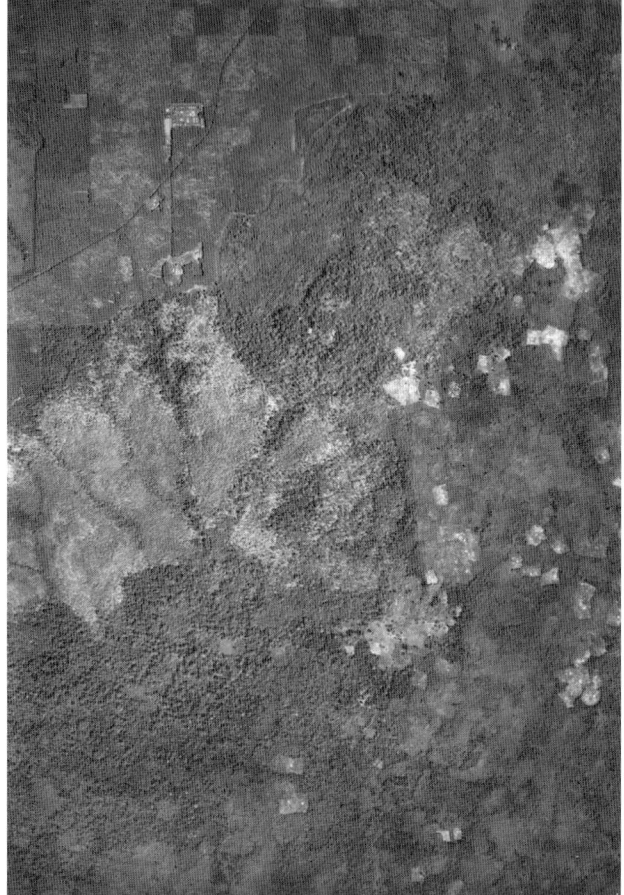

FIGURE 7-2:   An aerial photograph taken in 1952–1953
reveals swiddens southeast of the rubber plantation at
Memot, in Cambodia.

Source: Service Géographique de l'Indochine,
Institut Géographique National.

Studies in the post-colonial period confirmed the conclusions of those who, in the
early 20th century – and against the dominant ideology based on the superiority of
homo occidentalis – began seeking to rehabilitate the practice of swidden cultivation.
The major conceptual breakthrough in reassessing swidden was to take into account
the environment, soil quality and forest cover, population densities, actual production,
contributions in input and labour and cultural practices. As Georges Condominas
noted, this rehabilitation was primarily the result of the work of researchers who
took the time – often years – to study the system of swidden in depth (Condominas,
1983, p53).

It also appeared that swidden was complementary to other forms of agriculture,
including wet-rice production, thanks to harvest calendars and an opposite sensitivity

to natural hazards: wet rice was particularly vulnerable to floods as young plants could rot if submerged, while upland rice was very sensitive to drought. Trade networks between swidden and wet-rice farmers reduced the risk of famine in the region (Condominas, 1957, pp203-204; Guérin, 2001).

However, despite the writings of agronomists, the work of Pierre Gourou, who, in 1948, introduced the concept of *Civilisation du Végétal* (Gourou, 1948), and the writings of the anthropologists, French administrators never changed their perception of swidden. Efforts to rehabilitate swidden were a clamour barely heard. Indeed, to challenge official thinking on swidden was to question the dogma of progress – a part of the justification of the colonial action. To look more closely at this retrospective vindication, I return to the highlands of French Indochina at the beginning of the 20th century, when the colonial administration launched a campaign against shifting cultivation.

Policies implemented to fight swidden cultivation

Prohibition

The first step taken by the colonial administration against shifting cultivation was to ban it. This was aimed primarily at protecting forests, since these had been considered a resource in the colonial-development scheme since the late 1860s. By decree of the President of the French Republic dated 9 January 1895, practising shifting cultivation in Indochina became punishable by six months in prison and a fine of 25 to 100 piastres, to be paid by the village. This was the heaviest penalty imposed for a forest crime. Similar bans were soon introduced into local laws.

It soon became clear that such laws were impossible to implement because they were disconnected from the capacity of villages. Moreover, local French administrators and residents were among the first to take side with the people under their jurisdiction, even opposing the Forest Service staff. In 1902, the Resident administrator of Kratie received complaints from the residents of three Mnong villages following the application of regulations of the Department of Agriculture, Forestry and Trade. Eventually, the Résident Supérieur of Cambodia ruled that swidden cultivation would be tolerated in the open forest of northeastern Cambodia (Letters between the head of the forest service in Cambodia, the Résident Supérieur of Cambodia and the Résident of Kratié, July-August 1902, RSC 758, National Archives of Cambodia, Phnom Penh). In 1907, a new decree lowered the penalties. Shifting cultivators were nevertheless still subject to one to three months' imprisonment and their villages liable to fines of 20 to 250 francs.

In fact, prohibition could only be enforced in areas where the administration was strong enough to be aware of breaches of the law and to impose court decisions. This wasn't the case in most of the highlands (Carton, 1951, p455; Guérin, 2003, pp226-241). Moreover, administrators whose main task it was to keep order in their jurisdiction were very reluctant to impose unpopular orders that could lead to general uprisings, like those that occurred between 1901 and 1936 on the Boloven plateau of southern Laos or between 1912 and 1935 in the Mnong territories of Cambodia (Moppert, 1978; Guérin, 2008).

Modernization of land law, concessions and expropriation

The modernization of land laws was also used as a tool to fight swidden cultivation. Individual private ownership of land, as conceived in the West, was gradually introduced into all countries of Indochina. In Cambodia, a royal decree of 24 January 1908 stated that land belonged to those who worked it for at least three consecutive years. Article 15 stipulated that land abandoned for more than five years returned to the public domain. These provisions made land ownership inaccessible to shifting cultivators who cultivated their fields for one to three years before shifting to another plot, and let the land rest and the forest fallow for several years, even decades (Letters between the Résident of Stung Treng and the French Administrator of Veunsai, 1908, RSC 26315, National Archives of Cambodia, Phnom Penh). On 17 October 1921, a royal ordinance introduced property taxes on land, cultivated or not. This was partly justified by the fight against swidden cultivation. However, because the administration didn't have the means to carry out a comprehensive cadastral registration of land, this ordinance could never be enforced (Forest, 1980, pp244-253).

Another measure that had a tremendous impact on swidden in these areas was the provision of large areas of land to colonial capitalist companies. When such concessions were granted, swidden cultivators were denied access to the land. These domain concessions were instruments by which land belonging to the state was alienated to an individual or a company under the condition that the land was then developed. The concession was at first temporary, but became final after the authorities duly recognized the implementation of a development project. Only land that belonged to the state – meaning unfarmed land – could be granted as a concession. However, the fallow forest areas of shifting cultivators were considered public land and therefore available for alienation (Deroche, 2004, pp36-40). In the early 20th century, with the development of coffee, tea, and above all, rubber cultivation, French investors started to become very interested in the fertile red earth of southern Indochina, which was found mainly in the territories of swidden cultivators, who could harvest very high yields from these soils.

With the arrival of large rubber companies, many villages were expropriated. This is what happened to the Stieng villages in the Loc Ninh region of Cochin-China (the southern part of Vietnam) and those in the Memot region of Cambodia. The concessions could cover thousands of hectares. Memot provides a good example of the setting up of a concession and its consequences for forest farmers. In 1929, the Memot concession extended over 9500ha on a red-earth plateau covering 24,000ha, on which lived 8763 inhabitants, mostly swidden cultivators. Due to a relatively high population density – 36 people per square kilometre – the swidden farmers had shortened the length of their fallows to about four years, and cleared swiddens from bamboo thickets. In the farmers' view, there was no land available. A confrontation arose between the villagers and the owners of the concession, and the administration acted as an intermediary. Finally, the people received an 'eviction allowance *(indemnité de déguerpissement)*' for their plots, crops and huts, not even equivalent to half the value of a yearly rice harvest. In addition, 9500ha of the plateau was declared a reserve for the people. This area appeared to be insufficient for swidden cultivation as the rotation period would be too short to allow time for the forest, or even a thicket of bamboo, to regenerate. The villagers had no option but to leave or to change their cultivation techniques with the support of the administration (Guérin, 2008, pp291–294).

Education

To help swidden cultivators displaced by the establishment of large plantations, and also because of the failure of police enforcement, the French administration tried to set up educational programmes to teach intensive-farming techniques to shifting cultivators.

The desire to inculcate the techniques of wet-rice farming or commercial cropping into forest farmers appeared early in the 20th century, in conjunction with the implementation of the decrees of 1895 and 1907 prohibiting swidden. Schools that opened in the highlands integrated the teaching of wet-rice cultivation into their curricula. The results were very disappointing for French administrators. In 1921, for example, the school of Veunsaï, Cambodia, had only 18 students, including seven highlanders. Young Brao and Jarai, although good students, didn't show any interest in the study of wet-rice farming. After being forcibly closed by lack of resources, the school reopened in 1938, and alternative modes of production were then at the core of the curriculum: 'We have to inculcate the habit of growing rice without resorting to the disastrous technique of *ray*, and to pull off the earth a much larger variety of production than that known by *Kha* people.' (Order of the Résident Supérieur du Cambodge, 19 May 1938, *Bulletin Administratif du Cambodge*, 1938, pp997-1003)

The results remained almost nil. The French never understood that they were trying to teach agricultural techniques that were less diversified and less efficient than swidden. The highlanders didn't change their cultivation methods because, contrary to the beliefs of the French and Khmer leaders, it was not in their interests to do so.

Promotion of alternative crops also required education campaigns aimed directly at farmers. Leopold Sabatier, administrator of the Darlac province in the highlands of Annam (central Vietnam) between 1914 and 1926, was a pioneer in this field. He was also one of the main characters in the history of the highlands of Indochina. He married the daughter of the prestigious Rhade chief, Khun Jonob Thou of Ban Don. He drafted the 1923 order on 'the Administration of *Moi* Provinces'. He imagined a specific policy toward shifting cultivators, which he implemented before being ruled out of Indochina in 1926 by colonial circles that looked longingly at the land that fell under his jurisdiction. Sabatier had sought to encourage the people of his constituency not to totally abandon swidden, but to practise it in conjunction with other agricultural techniques, especially wet rice and fruit-tree plantations. He made buffalo and seeds available to those who wanted to plough wet-rice fields.

Sabatier's approach served as an example in Cochin-China and Cambodia, particularly in the red-earth areas where the land of swidden farmers was being expropriated for the establishment of rubber plantations. A school was opened at Snoul, in northeastern Cambodia, in 1931, half of whose students were Stieng. They were taught wet-rice cultivation. As well, the Agricultural Services of the Protectorate supported former shifting cultivators in introducing a three-fold crop-rotation system with corn, cotton and cassava. It also tried to introduce coffee. Seeds and fertilizer were distributed in the villages. The results were encouraging at Snoul, where after a few years, local governors witnessed the development of wet-rice cultivation:

> I had the satisfaction to see during my last tour conducted in the district of Snoul on the 2nd to 6th of December that many people in this district have abandoned the practice of ancestral *ray* for permanent rice fields. In the commune of Sre Char everybody started to grow wet rice. Of course for this year, there is no sufficient space for each family. (Governor of Kratié province, monthly report to the Resident of Kratié, November 1938, RSC 380, National Archives of Outre-Mer, Aix-en-Provence)

This success, in the view of the administration, could be explained in part by a local environment favourable to this type of culture, but also by the fact that large population densities existed on lands that remained after the installation of plantations, which left Stieng farmers in the region with no other choice. In Memot, where the quality of the land was different, the situation appeared to be much more difficult for the Stieng and many left for Vietnam in the 1940s and 1950s. However, swidden didn't disappear in the area, as shown clearly in aerial photographs taken in 1952-1953 by the Geographic Service of Indochina (Figure 7-2).

The failure of French colonial policy vis-à-vis swidden

Neither the prohibitions and penalties, nor expropriation of land, nor teaching other forms of cropping were successful in eradicating swidden in Indochina during the French colonial domination. This failure of the French administration can be explained by its weakness in the highlands, but also by the very low understanding of what swidden cultivation really was. The French decision-makers never understood, even after the publication of the first ethnological works, those of the Reverend Kemlin in 1909, or later those of Georges Condominas, that swidden was not only a method of agricultural production, but it was also part of the civilization of the peoples of the highlands, with implications for their religious and social patterns (Condominas, 1957; Kemlin, 1998). They failed to acknowledge that swidden agriculturalists relied on the forest not only for production of food and to supply life's needs, but also as the ever changing framework of their existence; past, present and future. The land and the forest, left fallow for years or decades, were more than a reminiscence of their history; they were the canvas on which it was written (Condominas, 1957; Padwe, 2011, pp58-101).

The policy against swidden maintained by the French colonial authorities, and then taken up by Cambodia, Laos and Vietnam after the collapse of French Indochina, has been a source of pain and hardship for shifting cultivators. Swidden farmers suddenly found themselves regarded as outlaws; they were under strong pressure from the authorities to change their cultures and therefore their ways of life and sometimes their land was expropriated. On top of this, false representations about the farmers of the forest led authorities to introduce absurd methods of taxation, with regard to the agricultural calendar and the financial capacity of villagers to pay (Guérin, 2001, pp48–52). This policy was not based on serious studies, but simply on the way that decision-makers perceived highlanders; this is where we find the atavistic distrust of modern states for the people of the forest (Figure 7-3).

The prohibition of swidden has to be understood in the broader context of the colonial administration's willingness to strongly regulate use of the forest by villagers in all of French Indochina (Thomas, 2000). When France established its domination over Indochina, it aimed at imposing the paramountcy of a modern European state. Forests not only had to be protected, but also controlled, and shifting cultivators were very hard to control. They were seen as part of a wild and strange world, thus were considered as 'savages' in need of civilization. This will to control what was perceived

FIGURE 7-3: A Mnong house on a swidden in Mondulkiri province, Cambodia.

Photo: Mathieu Guérin.

as the 'wilderness' had tremendous consequences for the forest, for the flora and fauna of Indochina, and for the people who were most closely connected to it. Coercion was used to bring highlanders under the authority of the state and its regulations. Animals considered as pests or vermin were slaughtered against bounties: these included not only rice-field rats and crabs, but also tigers, panthers and crocodiles (Guérin, 2010). Hundreds of thousands of hectares of forest were felled for plantations, timber, roads and perennial agriculture. While official discourse claimed that the fight against swidden was necessary to protect Indochinese forests, the main threat to forests at that time was the colonial system of domination itself. Then, it gave the new states of Vietnam, Cambodia and Laos the rhetoric and the intellectual tools to take over the job.

Post-colonial situation

Over the past 70 years, the publication of many research reports has supported the rehabilitation of swidden in Southeast Asia, yet the same policies are still enforced, with a persistent background in which swidden is negatively represented. Working in Indonesia, Michael Dove has shown that the denigration of swidden provides a basis for state control; for agricultural settlers or logging companies to grab land from the farmers of the forest (Dove, 1983; Dove and Kammen, 1997). This analysis, together with the concept of 'a civilizing mission', helps to better understand the politics of the French colonial administration vis-à-vis shifting cultivators in Indochina, even though authoritative views were, at that time, challenging the dogma of swidden cultivation being unproductive and destructive. It seems that the French decision-makers never understood that forest in the vicinity of swidden cultivators' villages was not free land, but land left fallow for years to allow the forest to regenerate. They never understood that swidden combined with low densities of population made it possible to maintain a forest cover along with agricultural practices.

The parallel that can be drawn between the colonial era and the situation that exists today is actually striking. Supporters of European colonial expansion in the 19th and early 20th centuries justified the grabbing of others' land and resources by claiming the ability to more efficiently exploit those resources. Some even claimed a global right of expropriation, in the name of the need to access natural resources: 'The human community has, as each State, the right to expropriate. No property is inviolable when public necessity requires its sacrifice' (Pelletier and Roubaud, 1936, p33). Today, forgetting that highlanders have been incorporated very reluctantly into the modern states of the peninsula (Scott, 2009), Vietnam, Cambodia and Laos still give to local and foreign companies and individuals the right to take the land of swidden cultivators in the name of development. The concession granted in 2007 to the former French colonial rubber company SOCFIN and its Cambodian partner KCD at Bu Sra, in northeastern Cambodia, (see Leemann and Nikles, ch 33) appears as a strong reminder of what happened in Memot in the 1920s and 1930s. Abuses took place when the concession was established and attempts to take into account the specificity of local cultures have been dismissed by top management (Slocomb, 2007; FIDH, 2011;Vogel, 2011). The founding of an efficient modern agro-industry is leading to the impoverishment of local farmers and the destruction of the forest cover, all in the name of development, or *mise en valeur* (enhancement), as it was called by the French-expansion supporter Jules Ferry at the end of the 19th century.

References

Baird, I. (2008) 'Various forms of colonialism: The social and spatial reorganisation of the Brao in Southern Laos and Northeastern Cambodia', PhD dissertation, University of British Columbia, Canada

Barrau, J. (1972) 'Culture itinérante, culture sur brûlis, culture nomade, écobuage ou essartage? Un problème de terminologie agraire', *Etudes Rurales,* no. XXX, January-March, p99-103

Boulbet, J. (1966) 'Le miir, culture itinérante avec jachère forestière en pays maa, région de Blao – bassin du fleuve Daa'Dööng (Dông Nai)', *Bulletin de l'Ecole Française d'Extrême-Orient,* vol. LIII, p77-98

Boulbet, J. (1975) *Paysans de la Forêt,* École Française d'Extrême-Orient (EFEO), Paris

Bourdier, F. (1995) *Connaissance et Pratique de Gestion Traditionnelle de la Nature dans une Province Marginalisée du Cambodge,* Association des Universités Partiellement ou Entièrement de Langue Française – Université des Réseaux d'Expression Française (Aupelf-Uref), Phnom Penh

Carton, P. (1951) 'Les prospections poursuivies en Indochine en vue de la mise en valeur des hautes régions incultes', *Comptes-rendus des Séances de l'Académie des Sciences Coloniales,* vol. XI, p443-472

Condominas, G. (1957) *Nous Avons Mangé la Forêt de la Pierre Génie Gôo,* Mercure de France, Paris.

Condominas, G. (1983) 'Aspects écologiques d'un espace social restreint en Asie du Sud-Est, Les Mnong Gar et leur environnement', *Etudes Rurales,* January-September, no. 89-90-91, p11-76

Conklin, H. C. (1954) 'An ethnoecological approach to shifting agriculture', *Transactions of the New York Academy of Sciences,* 17(2), p133-142

Conklin, H. C. (1957) *Hanunoo Agriculture: A Report on an Integral System of Shifting Cultivation in the Philippines,* Forestry Development Papers no. 12, Food and Agriculture Organization of the United Nations, Rome

Conklin, H. C. (1961) 'The Study of Shifting Cultivation', *Current Anthropology* 2(1), pp27-61

Deroche, A. (2004) *France Coloniale et Droit de Propriété. Les Concessions en Indochine,* L'Harmattan, Paris

Dourisboure, P. (1922) *Les Sauvages Bahnars (Cochinchine Orientale), Souvenirs d'un Missionnaire,* Missions Etrangères de Paris, Paris

Dove, M. (1983) 'Theories of swidden agriculture, and the political economy of ignorance', *Agroforestry Systems* 1, pp85–99

Dove, M. and Kammen, D. (1997) 'The epistemology of sustainable resource use: Managing forest products, swiddens, and high-yielding variety crops', *Human Organization* 56(1), pp91-101

FIDH (the International Federation for Human Rights) (2011) 'Cambodia land cleared for rubber, rights bulldozed. The impact of rubber companies in Bousra, Mondulkiri', http://www.fidh.org/IMG//pdf/report_cambodia_socfin-kcd_low_def.pdf, accessed 29 February 2012.

Forest, A. (1980) *Le Cambodge et la Colonisation Française, Histoire d'une Colonisation sans Heurts (1897-1920),* L'Harmattan, Paris

Fox, J. (2002) 'Understanding a dynamic landscape: Land use, land cover, and resource tenure in Northeastern Cambodia', in S. Walsh and K. Crews-Meyer (eds) *Linking People, Place, and Policy: A GIScience Approach,* Kluwer Academic Publishers, Boston, pp113-130

Gourou, P. (1940) *L'Utilisation du Sol en Indochine Française,* Centre d'Études de Politique Étrangère, Paris

Gourou, P. (1948) 'La civilisation du végétal', *Overdruck mit Indonesië* 5, pp385-396

Guérin, M. (2001) 'Essartage et riziculture humide, complémentarité des écosystèmes agraires à Stung Treng au début du XXᵉ siècle', *Aséanie* 8, pp35-55

Guérin, M. (2003) 'Des casques blancs sur le Plateau des Herbes. La pacification des aborigènes des hautes terres du Sud-Indochinois, 1859-1940', PhD dissertation, Université Paris 7-Denis Diderot, France

Guérin, M. (2008) Paysans de la Forêt à l'Époque Coloniale: La Pacification des Aborigènes des Hautes Terres du Cambodge, 1863-1940, *Bibliothèque d'Histoire Rurale,* AHSR, Caen, France

Guérin, M. (2010) 'Européens et prédateurs exotiques en Indochine: le cas du tigre', in J-M. Moriceau and P. Madeline (eds) *Repenser le Sauvage Grâce au Retour du Loup. Les Sciences Sociales Interpellées,* Presses Universitaires de Caen, Caen, France, pp211-224

Henry, Y. (1931) *Terres Rouges et Terres Noires Basaltiques d'Indochine, leur Mise* en Culture, Gouvernement Général de l'Indochine, Hanoi

Kemlin, Emile (1998) *Les Reungao, Rites Agraires, Songes et Alliances,* P. Le Roux (ed), École Française d'Extrême-Orient (EFEO), Paris

Lechesne, P. (1925) 'Roland Dorgelès chez les Moïs', *Revue Indochinoise,* July–August, pp64-65

Leclère, A. (1898) 'Les Pnongs, peuple sauvage de l'Indochine', *Mémoires de la société d'ethnographie,* Paris, pp137-208

Matras-Troubetzkoy, J. (1983) *Un Village en Forêt, l'Essartage chez les Brou du Cambodge,* SELAF, Paris.

Maurand, P. (1938) *L'Indochine Forestière, Rapport au VIIe Congrès International d'Agriculture Tropicale et Subtropicale,* Imprimerie d'Extrême-Orient, Hanoi

Monfleur, A. (1931) *Monographie de la Province du Darlac,* Imprimerie d'Extrême-Orient, Hanoi

Moppert, F. (1978) 'Mouvement de résistance au pouvoir colonial français de la minorité protoindochinoise du plateau des Bolovens dans le sud-Laos : 1901-1936', PhD dissertation, Université Paris 7-Denis Diderot, France

Mouhot, H. (1868) *Voyages dans les Royaumes de Siam, de Cambodge, et du Laos et Autres Parties Centrales de l'Indochine,* Hachette, Paris

Padwe, J. (2011) 'Garden Variety Histories: Postwar Social and Environmental Change in Northeast Cambodia', PhD dissertation, Yale University, New Haven, CT

Pelletier, G. and Roubaud, L. (1936) *Empires ou Colonies?,* Plon, Paris

Salemink, O. (2003), *The Ethnography of Vietnam's Central Highlanders: A Historical Contextualization,* 1850-1990, University of Hawaii Press, Honolulu

Scott, J. (2009) *The Art of Not Being Governed: An Anarchist History of Upland Southeast Asia,* Yale University Press, New Haven, CT

Slocomb, M. (2007) *Colons and Coolies: the Development of Cambodia's Rubber Plantations,* White Lotus, Bangkok

Teston, E. and Percheron, M. (1931) *L'Indochine Moderne, Encyclopédie Administrative, Touristique, Artistique et Économique,* Librairie de France, Paris

Thomas, F. (1999) *Histoire du Régime et des Services Forestiers Français en Indochine de 1862 à 1945: Sociologie des Sciences et des Pratiques Scientifiques Coloniales en Forêts Tropicales,* The Gioi, Hanoi

Thomas, F. (2000) 'Forêts de Cochinchine et "bois coloniaux", 1862-1900', *Autrepart,* no. 15, pp49-72

Vidal, J. (1957-1961) *La Végétation du Laos,* Faculté des Sciences, Toulouse, France

Vogel, S. (2011) *Aspects de la Culture Traditionnelle des Bunoong du Mondulkiri,* Tuk Tuk Editions, Phnom Penh

Note

1. This chapter is based on a lecture given at a seminar on rural studies at the Centre for Research in Human Studies in Caen, France, and published in *Enquêtes Rurales* 12 (2009).

8

LAO SWIDDEN FARMERS

From self-initiated mobility to permanent-settlement trends imposed by policy, 1830 to 2000

*Laurent Chazée**

Introduction

This chapter aims to achieve a better understanding of the process of evolution faced by, or undertaken by, Lao swidden communities during the period from 1830 to 2000, which led them from mobility to a condition now known as 'sedentarization', or more permanent forms of settlement. Traditionally, mobility was a key element of the lifestyle of ethnic minorities involved in swidden farming. This mobility was either decided by the farmers or faced by them reluctantly, depending on the situation. In 'ordinary' cases, mobility was decided mainly on the basis of their indigenous practices and the characteristics of their territory and its natural resources. In 'extraordinary' cases, mobility was something to be faced as a response to beliefs or external forces.

Studies conducted in 700 villages in all provinces of the Lao PDR between 1989 and 2000 identified 24 key drivers to explain the mobility of swidden farmers during the period from 1830 to 2000. Official policies began to intervene in the self-initiated mobility decision-making processes of swidden farmers in 1950. Between 1990 and 2000, the impact of centrally conceived policies began to result in high social costs, impoverishment and loss of social and human capital. Moreover, medium-term prospective analysis suggested that this was only the beginning.

When policy, either through the communist Politic Bureau or government line, became the main driver of change in swidden farming communities, some elements of mobility were documented in the case of some ethnic swidden groups. However, in 2000 there was still no certainty that the politically driven territorial approach was

* Dr Laurent Chazée is currently working on the protection of Mediterranean wetlands at Tour du Valat, in the south of France, and is an expert in local development and decentralized governance at Al Hoceima, in Morocco. He is also involved in the publication of books and articles on his earlier research work in Southeast Asia and Africa.

socially accepted by swidden farmers, given the unbalanced emphasis on territorial and technological issues compared to the human dimensions of the changes. Indeed, lessons learned around the world indicate that when social forces are underestimated or ignored in the process of development decision-making, there may be physical results in the short term, but neither the processes nor the results may be sustained in the medium term. A more serious consequence of ignoring the human and social capital built up over countless generations is the possible loss of cultural values, solidarity mechanisms, indigenous safety nets, environmentally friendly practices, social organizations and religious beliefs and rules against the overuse of natural resources. Once lost, these things are very difficult to restore. Prior to 1991, indigenous human capital was of the highest value in the Lao PDR. Since then, there has been compelling and growing evidence of negative outcomes from the neglect of social and human capital in the course of development policy-making.

Historical background

Before the Lane-Xane Kingdom (14th to 19th centuries), the territory of the current Lao PDR was occupied by several ethnic groups, mostly animist swidden farmers from the Mon-Khmer linguistic family. Starting in the 11th century, several groups of Tai speakers came from China and Vietnam. They included Buddhists and animists, and most were experienced in permanent and irrigated rice cropping using draft animals. They often incorporated swidden farming into their agricultural systems. During this period, the country's demographic density was very low, ranging from 1 to 10 people per square kilometre, depending on the district. Swidden farmers could practise their agriculture, based on production of glutinous rice, unrestricted by limits on territory or natural resources. The French protectorate began in 1893 and ended in 1954. The First Indochina War in Vietnam and Laos lasted for eight years and

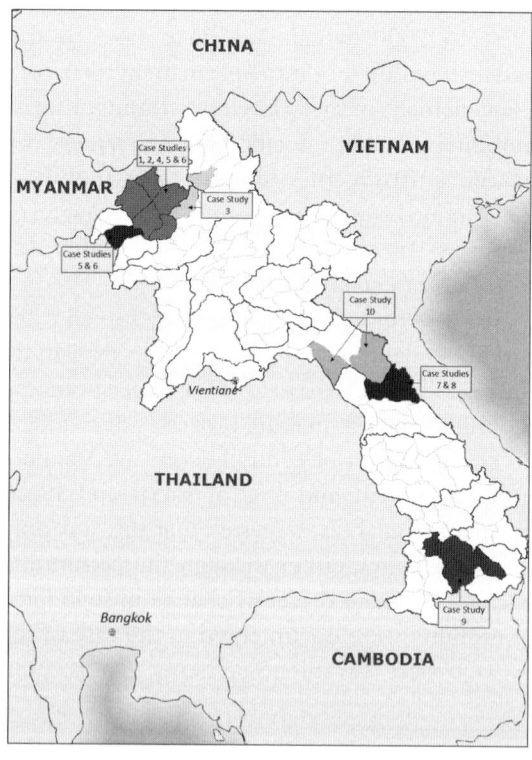

FIGURE 8-1:   Lao PDR, showing the location of case studies mentioned in this chapter.

ended in 1954 with the battle at Dien Bien Phu, the final clash between the forces of the Viet Minh and the Pathet Lao against the French Army. Earlier, during the 19th century and early 20th century, several forest-based communities of Miao-Yao and Tibeto-Burman speakers emerged out of China, Vietnam and Burma to settle in northern Laos. Most of these groups practised pioneering agriculture at higher altitudes, producing maize, glutinous and non-glutinous rice and opium. After the departure of the French came the Second Indochina War – known in the West as the Vietnam War – from 1964 to 1975, between the United States and the Pathet Lao and the Viet Minh. Many people were killed in the two Indochina wars; indigenous communities were displaced and people fled the country or were resettled elsewhere within Laos. During this period, warfare became the major external driver impacting the mobility of swidden communities, especially those living along the now-famous Ho Chi Minh trail, in Khammouane province and in all of northern Laos.

In 1975, a communist government took power and created the Lao People's Democratic Republic (the Lao PDR). After a period of collectivist production from 1976 to 1980, which was aimed at ensuring national rice sufficiency, the country began to open its economy in 1986 and then opened its borders with Thailand and China in 1990. In 1991, national policy in the Lao PDR launched a large-scale rural development programme that would have a significant impact on the future life of swidden communities.

Swidden farmers in the Lao PDR during the period 1990 to 2000

It is not the intention here to describe swidden farmers in Laos. However, some characteristics are important as background information, in order to understand and analyse their evolving mobility and settlement trends.

At the end of the 1980s and beginning of the 1990s, the forests of the Lao PDR were used and managed by about 300,000 households of swidden farmers, comprising 1.75 million people. According to Boupha (1995) the figure was about 220,000 families or 1.3 million people, representing about 40% of the national population. Swidden farming was the backbone of their household food sufficiency while exchanges, barter trade and sales of surpluses produced some cash income.

About 380,000 hectares of forest and fallow land were slashed and burned every year to prepare swiddens (locally known as *rây*) to grow rice, maize, cassava, cotton, tobacco, sesame, soya, opium poppies and various vegetables and spices. The average length of the fallow cycle in this period was estimated to be five and a half years, but this varied from three years to 20 years depending on the area, population density, soil quality and ethnic agricultural practices. Shifting cultivation was then associated with frequent uncontrolled fires, and it was estimated that about 2.3 million hectares of forest and fallow land were burnt in the course of the average five-and-a-half-year cycle. The area of land used for shifting cultivation represented 9.7% of the country's total area and 20.6% of its forests (Chazée, 1998). About 80% of household energy consumption was based on firewood that was collected from swidden land before burning. Shifting cultivation communities also took substantial advantage of forest resources and benefited from gathering non-timber forest products (NTFPs), including medicinal plants, and from hunting, trapping and fishing.

Between 1990 and 2000, shifting cultivation communities accounted for more than one third of the country's 4.6-million inhabitants. They belonged to four major linguistic families, represented by the Lao and by at least 131 ethnic minorities and sub-groups which could be further divided into numerous clans and lineages, whether genealogical or totemic, with various kinship systems (Chazée, 1999).

In 1995, Austro-Asiatic speakers – the first inhabitants of the Lao PDR – were represented by 59 ethnic and sub-ethnic minorities of the Mon-Khmer sub-family, totalling 30% of the national population. Among them, the Khmu, making up 38% of the Austro-Asiatic group, were the most numerous in the north. They were followed by the Katang, Makong and Suay in the south. Generally, they were less well organised socio-politically than the Tai, Miao-Yao and others of the Tibeto-Burman sub-family, so they were less protected against social, economic and political turmoil. By tradition, they preferred to settle in or near middle-range forested areas to practise shifting cultivation. Most groups managed rotational farming with a fallow cycle of more than eight years. Glutinous rice, maize, cassava and vegetables were their main crops, but they also grew sesame, cotton, Job's tears, peas, beans and sweet potatoes. Some also had private gardens (*souan*), usually near rivers, while others, influenced by Tai communities, adopted permanent rice fields in lowland areas, sometimes with irrigation. They also kept goats and poultry and sometimes cattle or buffaloes, practised basket work, gathered forest products and hunted and trapped forest wildlife.

The Tai linguistic family made up about 60% of the population of the Lao PDR during this period, and was a clear majority. There were 27 Tai groups living in the country that had come from China and Vietnam in a migration that began in the 11th century. The Lao were the largest community, representing about 35% of the total population and 58% of the Tai family. In the same linguistic family, the Phutai, Phouane, Tai Nyo, Lue, Tai Deng and Tai Khao together represented another 20% of the total population. The remaining 20 Tai ethnic minorities constituted 4% to 6% of the national population. While most Lao, Lu, Tai Deng, Tai, Yang and Yuane cultivated only permanent ricefields and private gardens on the plains and in valleys, most other Tai speakers were swidden farmers, especially in the eastern mountainous areas of the country. As well as cropping and livestock husbandry, they were also involved in weaving and other handicrafts, fishing, trading, small businesses and other services.

The Miao-Yao speakers were represented by four ethnic and sub-ethnic minorities living in the northern provinces, from the southern boundary of Bolikhamxay province northwards. They were Hmong Khao, Hmong Lay, Mien (Yao) and Lantene (Lao Huay). Most of these groups entered the Lao PDR from China and Vietnam between 1830 and 1950. In 1995, they represented between 6% and 10% of the population. About 80% of them were Hmong. They were socio-politically well organized, gathered in clans and notions of survival and freedom were their everyday priorities. Only the Yao were dispersed throughout northern Laos without much contact between different communities; others remained grouped in several villages. During the two decades from 1960 to 1980, most Lantene communities accepted resettlement in lowland and valley areas, but other Miao-Yao groups were more reluctant to move from the mountains. Their traditional livelihood was based on the production of maize, glutinous and non-glutinous rice, opium, pigs and cattle, and hunting and gathering.

Sino-Tibetan speakers were quartered in the extreme north of Laos, represented by 33 ethnic and sub-ethnic minorities. Although in the majority in Long and Phongsaly districts, they comprised only 3% to 4% of the national population. Except for the Ho, all Sino-Tibetan speakers living in Laos belonged to the Tibeto Burman Loloish. Among them, the Kho (Akha), with about 60,000 people in 1994, was the major group. Most came to Laos between 1800 and 1900, but some were still entering Laos from Vietnam, China and Upper Burma until 1930. Their livelihood was similar to that of the Miao-Yao, but their beliefs, spiritual pantheon and community organization were very different. By choice, most of them preferred to live in remote mountainous areas far from local administrations, in settlements of 15 to 40 households.

Main drivers of mobility decided or faced by swidden farmers from 1830 to 2000

Traditionally, mobility is a key lifestyle element of ethnic minorities who practise shifting cultivation. This mobility is decided or faced by the farmers, depending on their situation. In 'ordinary' cases, farmers opt for mobility on the basis of their indigenous farming practices and the characteristics of their territory and its natural resources. In 'extraordinary' cases, mobility is a reluctant response to beliefs or external forces. Field studies conducted by the author between 1989 and 2000, assisted by S. Syphanravong in 1999 and 2000, covered 700 villages in all provinces of the Lao PDR. These studies indicated 24 key drivers explaining the mobility of swidden farmers during the period from 1830 to 2000 (Chazée and Gehin, 2017). Table 8-1 lists these key drivers and they are later discussed in greater detail.

The mobility principle of swidden farming

The need to divide a village as a consequence of population growth or because of beliefs based on bad omens and natural disasters are three historical, self-initiated drivers of mobility in shifting cultivation communities. These drivers were internal social forces built into the agrarian functioning and social, cultural and religious references of swidden farmers. They were managed in a precise community and with individual decision-making processes specific to each ethnic group. These mobility drivers were common and freely utilized until 1950, after which they became more controlled, then almost stopped after 1995.

TABLE 8-1: Key drivers of mobility decided or faced by swidden farmers of the Lao PDR between 1830 and 2000.

Drivers of mobility	Regions or areas and groups most concerned or impacted	Period in which this had higher occurrence
1 The mobility principle of swidden farming, based on forest-and-fallow rotational cropping or pioneering system.	All areas where swidden farming is the backbone of the production system, especially in forested mountainous areas.	Until 1992 with some mobility until 2000 in remote and inaccessible areas. Some small groups such as Mlabri in Sayabury province and Temarou and Atel in Khammouane province, who are not attached to official villages, continue to follow an aboriginal way of life.
2 Need to divide the village because of insufficient land after population increase, or due to internal conflict, including conflict over land.	All areas where swidden farming is the main production system, especially in mountainous areas.	Until 1992, with some more recent cases in remote areas such as Attapeu, Phongsaly, Houaphan and Oudomxay provinces.
3 Traditional beliefs: human epidemics (cholera, leprosy), animal epidemics, successive bad harvests, fire in the village, natural disasters (flooding, drought, invasion by rats), successive illnesses.	Everywhere, with more intensity in northern provinces and Attapeu and Xekong provinces in the South. The strength of beliefs in the need to move villages is especially strong among Tibeto-Burman ethnic groups and some Austro-Asiatic groups of central and southern regions.	A free practice until 1950, then more controlled by the Pathet Lao for security reasons and in the fight against superstition. Starting from 1992, mobility is controlled through village decree, land allocation and focal site development policy.
4 Balance of power between ethnic groups sharing the same territory, involving cases and degree of mobility with progressive assimilation, syncretism, acculturation, insertion, adoption, integration.	Everywhere, with more intensity of impact and influence among less organized ethnic groups found among Austro-Asiatic speakers.	This was a continuous phenomenon at all times.

TABLE 8-1 (cont.): Key drivers of mobility decided or faced by swidden farmers of the Lao PDR between 1830 and 2000.

Drivers of mobility	Regions or areas and groups most concerned or impacted	Period in which this had higher occurrence
5 Employment opportunities involving mobility.	Mobility of swidden farmers had been taking place since 1855, following the Bowring Treaty between Siam and England. Since 1990, mostly in provinces bordering Thailand. Economic migration from Bokeo, Luang Namtha, Sayabury, Oudomxay for the logging industry in northern Siam.	Opportunities took place during the colonial period (French and English) in line with commercialization and industrialization. This was limited from 1975 to 1990 but started again after 1990 when borders with Thailand were opened.
6 Attacks by tigers and leopards.	In all mountainous areas, especially in the Annamite Range, northern mountains and Sayabury forest.	Most occurrences were before 1940. A few cases up to 1970 in Sayabury province (Mlabri group) and Nakai forest (Khammouane province), Attapeu and Xekong.
7 Deportation to Siam (temporary slavery).	Central region, mostly in the Annamite Range. Ethnic groups such as Sek, Phong and Brou So suffered from this 'deportation'.	1835 to 1890.
8 Ethnic and religious rebellion in China, and Chinese policy (including the Cultural Revolution).	Northern regions bordering China, especially Phongsaly, Oudomxay and Luang Namtha.	1800 to 1970.
9 Intrusion of armed groups coming from China to organize pilferage and aggression (in particular black flag, red flag and yellow flag).	Northern provinces, from the Chinese border to Xieng Khouang and Luang Prabang. High intensity in the watershed of the Nam Ou river, crossing Phongsaly, Oudomxay and Luang Prabang provinces.	1865 to 1930.
10 Corvées (unpaid labour for the state) during the French period.	All rural areas, especially where French were settled and where access roads were constructed: Luang Prabang, Muang Sing, Boten, Champassak, and so on.	1900 to 1936 A revolution against corvées took place between 1900 and 1910 on the Boloven Plateau in Champassak province.

TABLE 8-1 (cont.): Key drivers of mobility decided or faced by swidden farmers of the Lao PDR between 1830 and 2000.

Drivers of mobility	Regions or areas and groups most concerned or impacted	Period in which this had higher occurrence
11 Mobility linked to the general state of security during the two Indochina wars, including political refugees and displaced people.	Everywhere, but most common in zones more affected by bombing: Xieng Khouang, Viangxay, Nam Bak, Savannakhet, Xekong, Attapeu.	Between 1940 and 1987. High insecurity between 1960 and 1974 in several rural areas; fleeing the country between 1958 and 1962, 1973 and 1974, 1975 and 1987.
12 Occupation of frontline areas between armies during wars, with ethnic groups in political positions (areas from which people were evacuated or escaped).	Oudomxay, Houaphan, North of Vientiane, Xieng Khouang.	1950 to 1985.
13 Frontline areas where ethnic groups were in political positions (with the Pathet Lao, Lao Issala or US army (refuge areas).	Bokeo, Vientiane, Thailand.	1955 to 1985.
14 Occupation by newcomers of land that had been abandoned (due to slavery, deportation or wars), using existing swiddens and ricefields.	Annamite Range. Everywhere, especially in Oudomxay, Luang Namtha, Phongsaly, the north of Luang Prabang, Khammouane and Saravane.	1850 to 1890 (Deportation to Siam). 1970 to 1985 (War).
15 Return of refugees and displaced people to their ancestral land (refugee resettlement).	Returns took place in different provinces, especially in Vientiane municipality and in the provinces of Vientiane, Bokeo, Sayabury, Champassak, Luang Prabang and Khammouane.	1976 to 1979. 1985 to 1997.
16 Control and dismantling of political opposition networks in non-secured zones, with forced displacement of population at the end of the second Indochina war and during subsequent attempts to destabilize the country's political situation.	Northern and western parts of Sayabury, southern Oudomxay, Bokeo, Luang Namtha, Luang Prabang, Xieng Kouang, Houaphan and Xyasomboun were most impacted.	1970 to 2000.

TABLE 8-1 (cont.): Key drivers of mobility decided or faced by swidden farmers of the Lao PDR between 1830 and 2000.

Drivers of mobility	Regions or areas and groups most concerned or impacted	Period in which this had higher occurrence
17 Collective production period (cooperatives).	All provinces, with greater intensity in Vientiane, Savannakhet, Luang Prabang, Champassak, Saravane.	1976 to 1980 along with dismantling of the last cooperatives in 1989.
18 Displacement as a side-effect of repeated campaigns against superstition organized by the government and the army.	Mountainous areas, with more intensity among ethnic groups that were not on the Pathet Lao side (high intensity in Xieng Khouang, Houaphan, Luang Prabang, Luang Namtha).	1950 to 1986.
19 Displacement of population in line with construction of large public infrastructures such as hydroelectric dams, roads, irrigation schemes, mines and wood, gold and sapphire concessions, including resettlement in line with reservoir and watershed protection.	In several areas such as Muong Home (dam), Nam Theun, Hinboun (dam), Nam Theun II, and so on. Provinces of Bokeo, Sayabury, Vientiane, Khammouane, Bolikhamxay and Champassak.	1970 to 2000.
20 Outcome of rural development policies in mountainous areas, in particular those against swidden farming including opium poppy (decrease of fallow cycle, encouragement of sedentarization and settlement along roads or in accessible lowlands, and so on).	Everywhere in mountainous areas.	Began in 1970, organized at larger scale after 1990.
21 Outcome of the village decree encouraging the grouping of small rural communities, mostly remote swidden farmers.	Everywhere in mountainous areas, especially in remote areas with small ethnic hamlets.	1992 to 2000.
22 Rural development policy by priority development area (Focal sites)[a].	Focal sites in each province.	Established in 1994, operational in 1995 and developed after 1997.

TABLE 8-1 (cont.): Key drivers of mobility decided or faced by swidden farmers of the Lao PDR between 1830 and 2000.

Drivers of mobility	Regions or areas and groups most concerned or impacted	Period in which this had higher occurrence
23 Land reform, land allocation and land titling.	Country-wide after a test period in selected provinces.	Since 1993.
24 Increased urbanization[b].	Everywhere, with more intensity in large cities bordering Thailand, such as Vientiane, Savannakhet, Pakse.	Since 1980, with increasing trend after 1990, following the opening of the border with Thailand.

Notes: [a]This policy targeted the least-favoured zones with high incidence of poverty, insecurity and opium production. The willingness of the political party to speed-up the development of these communities was assumed to involve the adoption of a permanent settlement model, control of the population and their political education, as well as development activities. A National Rural Development Committee (NRDC) was created with direct links to the central Party, and a Rural Development Committee (RDC) was set up in each province under the direct authority of the Governor. [b]Between 2005 and 2010, as a consequence of rural-urban migration, the urbanization rate in the Lao PDR reached 5.6%, a rate higher than the rate of national population increase (estimated at 1.8%), and much higher than the 0.1% population increase in rural areas (Bouté, 2011).

In the case of the first driver of mobility – over-population – some ethnic groups such as the Hmong, Yao and Akha had a clear historical preference for pioneering swidden farming. They cut and burned old forests and moved their village setting every 10 to 25 years. Other groups such as the Khmu stayed as long as 20 to 40 years in one place and used a rotational farming system with a fallow cycle of 8 to 20 years. When the number of households and people became difficult to support and the distance to new agricultural land was too time-consuming, they were obliged to divide the village. In some cases, such a split occurred after a conflict between two lineages, usually involving land, power and attitude issues. Departing households selected a propitious new village site, usually not far from the original village.

For animist ethnic swidden communities, the explanation for extraordinary events often arose from beliefs in the supernatural and guardian spirits. In some cases, such events were attributed to, or could impact upon, the future prospects of the village and its swidden farming area. Such events identified in the field studies included village fire, flooding, an epidemic, repeated accidents or illnesses, natural disasters and destruction of crops by rodents or insects (see plate 62, Coloured Plates section 2). If the consequences of these events were judged to be of prime concern because of their serious effects on human health and security and swidden production, then traditional healers, sorcerers and priests, after performing their rituals, could decide that the community should abandon the village to find a more propitious place. Among the Tibeto-Burman and Miao-Yao communities involved in this study, belief in bad omens was the main driver of community mobility, up until 1950. There was always a major event that obliged them to move before their village reached a state of over-population. Hence, until 1950, some communities moved every 8 to 10 years, on average, because of bad omens.

Balance of power

The balance of power between ethnic groups has long been a continuous influence on community mobility. Relationships between different groups has sometimes deteriorated into conflict. This driver of mobility is mostly an external social force that includes push and pull elements. Since the 11th century, when ethnic groups first began to find refuge in Laos after fleeing China and Vietnam, the balance of power between them has usually resulted in peaceful coexistence. Land and natural resources were sufficient for all and production territory could be selected without much competition. Indeed, before 1990, there were only a few examples of cultivated forest territories of swidden communities actually bordering each other. However, in some areas where communities were concentrated, usually as an outcome of war or rebellion, the balance of power between communities sparked conflict, often involving social and customary issues, misunderstandings and fights, especially in central and northern mountainous areas where the ethnic mosaic was most diversified and concentrated. Such conflict provoked displacements and escape into deep forest. Historically, this also led to some major documented events, such as the Khmu Cheuang revolt in Xiengkhouang between 1875 and 1980, which arose following the 16th-century intrusion and settlement of Burmese in Vieng Poukha, in what is now Luang Namtha province. Another such event followed an intrusion from neighbouring Siam (now Thailand) in 1873.

In other cases, a balance of power was achieved naturally through the recognized supremacy of one ethnic group by its neighbours. This usually came after a long and mutual social, organizational, linguistic and technological exchange that could end up with changes to agricultural territory and practices and even the merger of two groups into one village. This exchange could translate into differing degrees of adoption, assimilation, syncretism, integration and sometimes acculturation. In the period from 1990 to 2000, these influences between groups at linguistic, handicraft, clothing, house architecture and agricultural levels were clear. For instance, the Lu group of Luang Namtha and Oudomxay provinces had influenced the agricultural practices, irrigation management and house construction of the Khmu Lu, the Yang and the Tai Dam. The Tai Youane group had influenced the religion, architecture and agricultural technologies of the Kouene and Nguan Austro-Asiatic groups in Luang Namtha and Bokeo provinces.

Employment opportunities

In the Lao PDR, economic opportunities usually create an external (pull) factor affecting the mobility of ethnic swidden communities through seasonal, semi-permanent and permanent migration. This factor was documented as long ago as the 19th century, when Kouen and Nguan groups from Luang Namtha migrated to the Lanna Kingdom in what is now northern Thailand following the signing of the Bowring Treaty (1895-1930) for the timber industry. Another such migration was prompted by the mining industry, specifically the extraction of gold and sapphires at Huay Xay and Xieng Khong. Employment opportunities during the colonial period included work on commercial clearing of land, commercial plantations, the timber industry and large infrastructure projects. After 1990, when the border with Thailand was opened, formal and informal seasonal and semi-permanent migration became commonplace from Lao provinces bordering Thailand, especially Khammouane, Savannakhet, Champassak and Sayabury, with migrants seeking work in agribusiness value chains.

Tigers and leopards

Until about 1940, tigers and leopards were quite common in mountain forests and when there were repeated attacks on domestic animals or humans, villagers sometimes chose to move to a less dangerous place. This external push factor of mobility was reported by several swidden communities living in remote eastern parts of Attapeu, Xekong, Saravane and Khammouane provinces, as well as in Northern Oudomxay and Sayabury. After 1940, the problem declined due to the introduction of firearms for warfare and hunting of big cats (tigers, leopards, clouded leopards and golden cats) for the Chinese market – a trade that lasted until about 1995. The most recent attacks by tigers or leopards were reported by Austro-Asiatic groups living in the southern provinces of Attapeu and Xekong and in the eastern mountains of Saravane and Khammouane, and by the Mlabri Group living in the Phieng Forest in Sayabury province.

'Deportation' to Siam

The 'deportation' of labour to neighbouring Siam (Thailand) was, in reality, a form of temporary slavery and debt-bondage (Cruikshank, no date). Labourers were taken from villages in the central mountains of Laos between 1835 and 1890, before the abolition of slavery in both Siam and Laos in 1905, and set to work in the central plains of what is now Thailand. In fear of such 'deportation', the people often fled. This external push factor mostly affected small mountain tribes living in the Annamite range, especially in the Nam Poui area of Khammouane province. According to these people, oral history passed down by their ancestors indicated that the business was organized by Siamese, Vietnamese and some Lao people, sending forest people to work on the plains near Bangkok. The swidden communities in Khammouane most affected were the Phong, Sek, Brou So, Maleng, Malang, Atel and Arao ethnic groups, and the resulting mobility took two forms: they were either forcibly deported to Siam or they fled into remote forest. Either move involved severe trauma and fear. Very few of the forcibly deported ever returned; it was assumed that they were assimilated into the Thai population. Those that hid in the forest were excluded from the modern word until between 1992 and 2000, when they were discovered in the course of a social survey preceding construction of the Nam Theun II hydropower project.

Rebellion and new policies in China and armed incursions

These two influences (external push factors) have a prominent place in the history and memory of swidden farming communities. Ethnic and religious rebellions in China and official policies imposed in that country had effects in communities across the border in Laos between 1800 and 1970. Attacks on villages and pilferage by armed groups occurred between 1865 and 1930. Some villagers still pass the ancestral memories of this period down to younger generations. Ancestors of both the Khmu and Lu communities currently living in the Nam Ou watershed fled and abandoned their villages several times because of attacks and pilferage by so-called 'Black Flag' fighters. Tibeto-Burman groups (Mounteun, Akha, Lolo, Ho, Sila, Hani, and so on) who settled in Phongsaly province escaped the influences of both the rebellions and the cultural revolution in China.

Government corvées

Corvées, or forced labour exacted in lieu of taxes, occurred during the French period, and took place mostly in areas where roads were being constructed. Even though the impact of this driver was relatively short (1900 to 1936) and localized, it was an external push factor that provoked fleeing, village abandonment by swidden farmers and migration to more remote areas, sometimes into China and Vietnam.

General wartime insecurity, scorched-earth policy and steady bombing

During and after the two Indochina wars, the key drivers of mobility (external push and pull factors) took place between 1945 and 1987. General insecurity resulted in displacement, resettlement and fleeing from threatened territory, while scorched-earth policies and steady bombing led to abandonment of their villages by people in areas north of Vientiane and west of Xieng Khouang. On the other side of displacements, there were influxes of people, including several swidden communities, seeking refuge in secured areas such as Vientiane municipality, Huay Xay in Bokeo province and in Thailand. This mobility was linked to insecurity, bombing, scorched-earth strategies and the political orientation of the rural communities. In total, it affected more than one third of the lowland and mountain population of Laos.[1] During these four decades, much agricultural land was abandoned; some communities took refuge in the forest, in towns, in Thailand and in China. Others were accepted in the United States, France, Australia and other countries.

Others take over abandoned land

Agricultural and forest land abandoned because of insecurity has often attracted, and been taken over by, other communities. This applies not only to land abandoned during the wars (1970 to 1985), but also that affected by earlier 'deportation to Siam' (1850 to 1890). This driver of mobility was usually a pull factor. However, in some cases, local authorities under the Pathet Lao regime forced landless farming communities and swidden farmers to settle on abandoned farm land, such as in certain areas of Oudomxay, Luang Namtha, Luang Prabang (particularly Nam Bak district), Xieng Khouang, the Viangxay district of Houaphanh province, Savannakhet, Xekong and Attapeu provinces. There were several cases where swidden farmers adopted abandoned rainfed or irrigated rice fields: Khmu Lu, Tai dam and Yang farmers took the land of the Lu group in Oudomxay and Luang Namtha provinces, and Brou-So people took over the ricefields of the Sek in Navang village, Nakai district, Khammouane province.

The return of displaced people and refugees

Between 1976 and 1997, many thousands of displaced people and refugees returned to Laos, most of them refugees from camps in Thailand and China. Their return occurred in two periods: 1976 to 1979 and 1985 to 1997, and was assisted by the UN High Commissioner for Refugees, the European Union, German aid and international non-governmental organizations. This was a push and pull factor of mobility. The return was sometimes associated with high tensions on the land between the original and the new occupants of both ricefields and forestland. For returnee swidden farmers, especially those of the Austro-Asiatic groups, the return was also motivated by the importance of reconnecting with ancestral territories, including ancestors' spirits, territorial spirits (*Phi Muang*) and related guardians that ensured their protection. Most of the time, ricefields were given back to returnees. In some cases, Austro-Asiatic farmers who had adopted permanent rice cultivation between 1960 and 1985 returned to shifting cultivation (Kouene, Nguan and Khmu in Luang Namtha province). In the negotiation process, local authorities usually encouraged landless swidden communities to accept accessible lowland areas for permanent settlement and agriculture, or gave them no choice in the matter. Most of the time, these lowland areas were along access roads. These decisions affected several swidden-farming ethnic minorities, especially in the country's northern and central provinces (Kui Lung, Kui Sung, Khmu, Hmong, Yao, Lantene, Kouene, Nguan, Lamet, and so on). The result was cases of impoverishment, high mortality, psychological shock and begging on urban streets, at least during the first years of settlement.

Official control and dismantling of political opposition networks

This was an external push factor that was poorly documented and communicated beyond confidential political and army files. However, these actions were clearly recalled by the villagers involved. This mobility factor first appeared in 1970 and was implemented with more intensity between 1985 and 1997, following attempts in Sayabury (Chazée, 2000), the south of Oudomxay, Bokeo, and the Xaysomboun special zone to destabilize the country. While national security was said to be the key objective for dismantling opposition networks, officials often used double standards. Some mountain ethnic groups were more affected by having, in the past, been on the wrong political side or having settled in territory controlled by the colonial army. Most affected were mountain-dwelling shifting cultivators, particularly Hmong and Yao in Luang Prabang, Xieng Khouang and Vientiane provinces; Kui Lung and Kui Sung in Luang Namtha; and Akha in Luang Namtha and Oudomxay provinces. They were 'controlled' and displaced outside of their ancestral ethnic and cultural domains into other districts and provinces. Some Hmong were sent from the northern provinces to the country's southern provinces.

Between 1975 and 1987, security continued to be dealt with by full-time control and assimilation of mountain minority groups, and by introducing Tai households into ethnic villages. Their role was to channel political education, to encourage permanent agriculture and to report on the mobility of the ethnic groups. The areas most targeted by this policy were the mountain districts of Muong Fuang, Vang Vieng and Kasi in Vientiane province; the districts of Sayabury, Phiang, Xanakham, Ngeun and Hongsa in Sayabury province; the districts of the Xaisomboun special zone; the districts of Vieng Poukha and Long in Luang Namtha Province; Viengthong district in Houaphanh province and along the Chinese border in Luang Namtha and Oudomxay provinces.

The collective production period

The collective production period, between 1976 and 1980, was driven by the Lao government's top agenda of that time: national food security. In other words, rural farmers were urged to provide food for urban and army people. The establishment of cooperatives involved some movement of swidden farmers. As a driver of mobility, this was an external push factor. The first effects of the cooperative period were felt by lowland communities already involved in permanent agriculture. However, about 20% of cooperative members were believed to be of Austro-Asiatic origin, some resettled in the valleys and plains of Vientiane, Luang Prabang, Champassak and Saravane provinces. After the dismantling of the cooperatives between 1980 and 1989, some Austro-Asiatic groups stayed in the lowlands while others returned to their traditional mountainous territories.

Campaigns against superstition

Between 1950 and 1985, repeated campaigns against superstition resulted in mobility of swidden farmers, and they thus became an external push factor. All non-Buddhist communities were potentially impacted by this action, with the most intense campaigns targetting animist Austro-Asiatic, Miao-Yao and Tibeto-Burman ethnic groups in Xieng Khouang, Houaphan, Luang Prabang and Luang Namtha provinces.

Public infrastructure construction and granting of land concessions

The displacement and resettlement of large numbers of people, including a high proportion of swiddening communities, took place directly or indirectly in line with the construction of large public infrastructures and the granting of land concessions. As drivers of mobility, these arose from external push factors, with some cases of induced pull factors.

These hydroelectricity dams and road construction projects involved resettlement in line with reservoir and watershed protection, roads, irrigation schemes, and land concessions for timber and mining companies, including those mining gold and sapphires. In most of these cases, 'strongly encouraged' or forced displacement or internal resettlement provoked fear, psychological shock, resistance and sometimes rebellion or fleeing into deeper forest. In some cases, resettlement was claimed to be 'voluntary' or 'negotiated'. In reality, village studies found that the 'spontaneous village-initiated dynamic of displacement' because of a project almost never took place during project preparation. The use of the term 'voluntary' referred only to the way in which projects asked villagers how they preferred to be 'cooked', or resettled.

The construction of large hydropower schemes such as Nam Ngum, Nam Theun Hinboun and Nam Theun 2 had the greatest impact, in terms of displacement and changes of livelihood, on swidden farmers living in river watersheds or future flooded reservoirs. Between 1994 and 1998, construction of the Huay Ho hydro-electric project on the Boloven plateau was responsible for the involuntary displacement of several swidden groups, especially the Nyaheune people. Of all these dam projects, only Nam Theun 2, which was constructed in the years from 1992 to 2010, adopted a transparent and global approach to its effects on people living within the project area, with continuous socio-economic and environmental feasibility studies, monitoring of implementation and updated communications via a website.

In some cases, during or after the construction of road, irrigation and hydropower schemes, swidden communities changed their minds. They realized that the new socio-economic environment was sufficiently attractive for them to adopt a new way of life with improved access to social and economic facilities and opportunities. This situation arose during the implementation of access-road and hydropower schemes on the Boloven plateau, irrigation schemes in Vientiane, Savannakhet, Sayabury and Luang Namtha, and road-rehabilitation projects between Takhek and Pakse, between Vang Vieng and Luang Prabang and between Saravane, Xekong and Attapeu. The settlement of villages along the Vang Vieng to Luang Prabang road between 1992 and 1997, for instance, was intended to reduce insecurity and was supported by incentives such as access to electricity, water supplies and health and education facilities.

Political orientation, strategies and tactics

The period between 1990 and 2000 was marked by a series of policy and strategic actions aimed at consolidating permanent agriculture and protecting the natural environment, especially the forest. Political orientations, strategies and tactics introduced under the umbrella of 'rural development' launched a large-scale, institutionalized and legalized process towards a permanent-agriculture model involving sedentarization of swidden farmers. Three main elements impacted swidden farmers: rural development policies as they related to shifting cultivation, village restructuring and a focal-site approach to rural development. These drivers of mobility were external push factors, with an induced pull effect in the process. The rural development model was a communist-party-driven approach based on permanent agriculture and resettlement. It was aimed mostly at swidden farmers. It involved steady and large-scale strategic leverage aimed at settling and controlling shifting cultivators in continuity of actions begun in the 1970s. The programme set out to achieve a rapid and permanent sedentarization of mountain communities, institutionalized by an integrated and mutually reinforcing set of policies, laws, decrees, directives and strategies. All of these actions were launched or were tested in early 1990. Implementation increased in 1997, when it incorporated foreign budget assistance under the umbrella of international commitment to poverty reduction and environmental protection.

Finally, urbanization

Before 1990, Laos was a very rural country by international standards. This was a consequence of poor urbanization and industrialization during the French Protectorate and long periods of warfare; of an urban and rural population of political refugees prone to fleeing en masse in fear of conflict both during the wars and after they ended in 1975; and because of the autocratic production-based economic situation that existed without private-sector development between 1975 and 1990.

Urbanization really began as recently as 1990. After opening its borders with Thailand and China and under the new influences of a market economy and industrialization, the Lao PDR opened to the world. In the heady atmosphere of globalization and new partnerships, the country became a member of the Association of Southeast Asian Nations (ASEAN) in 1997. Among the pull factors influencing the mobility of ethnic minority groups were the attraction of towns in terms of job opportunities, quality of education and health services, access to economic services and a higher quality of life. Among the push factors were the traditional patrilinear and patriarchal system in rural areas, the problems of inheritance or access to new agricultural land for new generations, forced displacements in 'unacceptable' rural areas, poor living conditions and social services in rural areas and sometimes natural disasters such as flooding. Permanent resettlement of swidden communities along access roads and in focal sites certainly played a role in paving the way for increased rural-to-urban migration. Indeed, permanent settlement patterns and improved access to social facilities were elements catalysing human concentration and urbanization. Similarly, impoverishment and food insecurity among some resettled swidden farmers who were unable to pursue and replicate their farming systems in rural areas may have pushed them to survive in urban settings. They became a part of the urban poor and may now be called the swidden-policy-induced displaced people.

Case studies: The impact of recent policies and development programmes on ethnic minority groups

In 1995, when rural development policies became more detailed, it became clear that mountain communities would face important changes to their traditional way of life. Large-scale, top-down attempts to impose territorial and technological integration and social assimilation could, at that stage, be firmly associated with mass social acculturation and possible alienation. Experiences elsewhere had already shown that underestimating or ignoring social forces such as cultural values, beliefs and social organization in development policies resulted in unbalanced and unsustainable outcomes (Chazée and Gehin, 2012). In the Lao PDR, where indigenous and cultural values were still strongly anchored in the daily way of life of swidden communities, this simplistic approach was – as early as 1993 – being considered a backward step in the evolving international concept of sustainable development (Chazée and Gehin, 2017).

The traditional decision-making processes of most swidden communities were seriously challenged by involuntary displacement driven by administrative 'grouping' into larger villages or by sedentarization of agriculture in lowland areas. Rural development policies were insensitive to their criteria and beliefs in mobility management and territorial selection processes. In new territories, swidden farmers suddenly lost control of the spirit guardians that were supposed to protect them; they were confronted by new languages, different crop varieties and tastes, different religions and different cultural values. Social vulnerability due to such change was sometimes associated with economic and psychological poverty, for which no kind of development effort could compensate. Beginning in 1990, precipitate reactions to development policies and social deviation became unavoidable. These included spontaneous migration, rural exodus, increased numbers of beggars and marginalization of ethnic groups wandering along roads and through urban areas. These displacements were the outcome of swidden-farming policies and strategies, including integrated rural development projects. These policies and strategies were the root cause of a high rate of mortality involving thousands of children and adults, with greater intensity in the northern and central mountainous regions of the Lao PDR. Particularly affected were Lantene, Hmong, Yao, Khmu, Poussang, Kui Lung and Kui Sung communities. The minority groups of Attapeu, Xekong and Saravane provinces in the south also suffered from internal resettlement. This high rate of mortality was the most severe negative outcome of the early years of resettlement into lowland areas, and was perceived by swidden farmers as a severe punishment for having offended the spirits by breaking with generations of tradition.

Case study 1: the Kui Lung of Luang Namtha and Bokeo provinces

The Kui Lung are Tibeto-Burman speakers. In 2001 their population was estimated to be 5000, living in 22 villages, including three villages with Kui Sung and des Mousseu Dam ethnic groups.

Most Kui Lung entered Laos between 1850 and 1900. They settled in the northwestern mountainous and forested areas of the country, between 600 and 1200 metres above sea level, not far from water sources. The geomorphologic conditions were conducive to the growing of rice, maize and opium poppies. The high altitude also provided natural protection against mosquitoes and consequently malaria, and against the polluted water and dust that were common problems in the lowlands. These criteria were vital for the Kui Lung, for whom good health was synonymous with the strong labour force required to ensure food sufficiency and sustain household well-being. At the end of their southern and eastern migrations, they now live in mountain territories shared by the Mousseu, Lamet, Hmong, Khmu and Akha ethnic groups. Between 1976 and 1978, they were encouraged by local authorities to settle in lowland areas. The Kui Lung households of Photong village were resettled in Dong Yeng, near a district town, but returned to the mountains in 1983 after five years, having lost more than a quarter of their population to malaria, dysentery and other illnesses. Faced with similar resettlement, the community of Phonsaysavang fled from Long district in Luang Namtha province in 1976 and settled in Huay Xay district in Bokeo province. Between 1985 and 1995, recognizing the failure of this approach to ethnic minority integration, the authorities allowed the remaining Kui Lung villages to remain in mountain areas until the implementation of the focal site approach to rural development. Beginning in 1995, almost all villages in Luang Namtha and Bokeo provinces were incorporated into focal sites, with a very high proportion of internal resettlement of shifting cultivation households. A high level of dissatisfaction built up in 1996 and 1997, and an ethnic movement in Luang Namtha evolved into local guerrilla bands. In response, the government strengthened village militias and provincial authorities attempted to channel most of the available foreign-aid funds into the affected areas to boost development and speed-up benefits. The foreign aid came from the French non-governmental organization Action Contre la Faim (ACF), NCA (Norwegian Church Aid), the former German technical cooperation agency GTZ (Gesellschaftfürtechnische Zusammenarbeit), the European Commission and EED (Enfants et Développement). Based on EU project evaluation in 1997, villages displaced along roads after 1995 suffered 15 to 25% mortality in the two first years.

Case study 2: the Kui Sung of Luang Namtha and Bokeo provinces

The Kui Sung are Tibeto-Burman speakers. In 2001, their population was estimated to be 3500, living in 12 villages, including five villages in association with other ethnic groups such as the Kui Lung, Mousseu and Yao, in Luang Namtha and Bokeo provinces.

Kui Sung people from Kui Portong village in Luang Namtha province's Vieng Poukha district say they came from Upper Burma in the 19th century. About 150 households settled first in Muong Sing district, near the border with China, and stayed there for about 70 years. In 1945, an epidemic forced them to move and create a new village called Ban May, near the border between Long and Vieng Poukha districts. In 1970, following a serious land conflict with a Yao group, they moved to Long district. In 1988 the need for new agricultural land led them to move again, to settle at Ban Nam Pe, in Vieng Poukha district, near the border between Luang Namtha and Bokeo provinces. In 1992, 18 families returned to Vieng Poukha district and established Kui Portong village, where agricultural land was available. However, in 1997 district authorities asked them to move nearer to a new road under construction between Vieng Poukha and Long, into a village called Nam Kab. This was near Ban May village, which they abandoned in 1970. Since 1997, along with other Mousseu Dam villages, they have been living in one of three priority focal sites in Luang Namtha province.

Case study 3: the Akha Djepia of Phongsaly and northern Oudomxay provinces

The Akha Djepia are Tibeto-Burman speakers. In 2002 they had an estimated population of 3400 in 17 villages in Phongsaly and Oudomxay provinces.

The Akha Djepia have been living in Phongsaly and Oudomxay provinces since the end of the 19th century. Until recently, most village displacements affecting these people have been due to epidemics, village fires, or the need for new agricultural land. For example, the community of Chaluang Li in Phongsaly's Bountai district lived in mountainous areas about one hour's walk from their current village until 1972. Then, at the insistence of local authorities, they moved to a location near the Nam Lan river to practise permanent agriculture. They stayed until 1982, but were unable to develop rice fields sufficient for their needs because of inadequate means and limited land. The regular occurrence of malaria forced them to return to the mountains in 1982 and a village fire in 1995 led them to their current village site.

Another Akha Djepia community called Chaluang May moved from its mountain village in 1997 to settle near the Nam Ngen river after repeated requests from district authorities. They were part of a commercial sugar cane plantation project aiming to provide raw material to sugar factories in China's Yunnan province. This project came to an end in 1999, after the failure of commercial production. The district then asked them to move again, to settle along a new road to be built between Bountai district in Phongsaly province and Namo district in Oudomxay. In 2001, construction on the road had not begun and the villagers had yet to move.

Case study 4: the Mousseu Dam (Black Lahu) group of Luang Namtha province

The Mousseu Dam are Tibeto-Burman speakers. In 2000, their estimated population was 5500, spread across 31 villages, including three villages in association with Kui

Sung, Mousseu Khao and Hmong people. Some households of Mousseu are divided into other villages in Luang Namtha and Bokeo provinces.

The first Mousseu arrived in Laos between 1820 and 1840, trekking through Muang Sing and Long districts of Luang Namtha province from Yunnan, upper Burma or northern Thailand. The Black Lahu settled in Sing and Long districts during the war years between 1964 and 1973, and were sometimes supportive of either the American forces or the communist Pathet Lao Army. Thus, some pro-communist villages in the Sobloi and Buakbo areas of northern Long district were bombed by US planes. Other villages in the Siengkok area of Luang Namtha province supported the US side and managed guerrilla fighters against the communist troops. In 1974 they were subjugated by the Pathet Lao army and 'punished' with an authoritarian resettlement programme. Some households succeeded in fleeing into Thailand or Burma, while others organized themselves into guerrilla groups (Iu Mien, Khmu, Kui and Akha). The communist authorities took until 1978 to bring this Black Lahu area under their control. Sporadic rebellions arose even until 1992 in the old sub-district of Tchongka.

The Black Lahu community at Tchayi village in Luang Namtha's Long district arrived from Yunnan through Upper Burma in 1940, crossing Muang Sing district. They originally created two villages, Khalang and Phasi, in the search for sufficient agricultural land. In 1975, they came under the control of the Lao army and some were killed while trying to escape to Burma. In 1993, local authorities asked them to move to lowland areas and practise permanent agriculture, so the Khalang community separated into two groups and created Tchayi and Tchador villages. Households were resettled in 1995. In 1996, 25 new households from the Mousseu and Kui Lung ethnic groups arrived from three other areas to join their community, so that in 1998, it comprised 51 houses, 57 households and 230 persons. By June 2000, only 13 houses and 64 persons remained in the community. Some had gone back into the mountains; others had died or joined other villages.

Case study 5: the Kouene from Luang Namtha and Bokeo provinces

The Kouene are Austro-Asiatic speakers. In 2002 their population was estimated to be 11,000, living in about 30 villages in the provinces of Luang Namtha and Bokeo.

This Austro-Asiatic ethnic group has long been settled in the Namtha, Nalae and Vieng Poukha districts of Luang Namtha province. These swidden farmers stayed together despite pilferage by armed groups, epidemics and wars.

During the Second Indochina War (the Vietnam War), Vieng Poukha district was the scene of successive military events, particularly between 1960 and 1975. Bombing in the years between 1962 and 1967 forced a number of families to seek refuge in the forest. After the takeover of power by the communists, the priority for local authorities was securing their area. One strategy was to resettle and control the rural population known to include people who supported the other side. This tactic was maintained until 1994. Similarly to the Nguan minority, Kouene swidden farmers were displaced between 1975 and 1985 from mountainous areas to accessible areas alongside roads. It is estimated that about 1000 people from Vieng Poukha district and a similar number from Namtha district were internally displaced during this period. Between 1985 and 1995, local authorities tried to resettle other swidden farmers into a lifestyle of permanent residence and agriculture, in line with the national agricultural and forestry policies. Between 1995 and 2000, the territory occupied by Kouene communities was slowly developed as a result of road rehabilitation between Namtha and Houay Xay, in Bokeo province, and a rural development project was supported by the European Union.

Case study 6: the Nguan from Luang Namtha and Bokeo provinces

The Nguan are Austro-Asiatic speakers. In 2002, their population was estimated to be 20,500, living in 54 villages in the provinces of Luang Namtha and Bokeo.

This ethnic group originates from Luang Namtha province, and in particular its Namtha, Nalae and Vieng Poukha districts. During the second Indochina war (the Vietnam war), Vieng Poukha, in the southeast of Luang Namtha, suffered a succession of military events, in particular between 1960 and 1975. When Lao Issala troops took over the province in 1962, several communities, including Nguan families, sought refuge in Houay Xay, the capital of neighbouring Bokeo province, which remained under US army occupation. After the Communist takeover, local authorities secured the area by resettling and controlling opposition groups, including the Nguan. But guerrilla activity continued in Vieng Poukha, Long and Nalae districts. The authorities decided to remove all communities from mountainous areas and resettle them by force along roads between Namtha and Vieng Poukha. This resettlement strategy, along access roads in lowland areas, continued until 1985. At the same time, Tai-speaking people who had been temporary refugees in Houay Xay during the war returned to their land to reclaim their ricefields. Some Nguan farmers were obliged to give back this occupied land and they returned to the uplands to resume shifting cultivation. It is estimated that at least 5000 Nguan people (3000 in Vieng Poukha and 2000 in Nalae) were displaced between 1975 and 1985. After 1986, the Nguan faced another episode of internal resettement as an outcome of the Village Decree and the national policy and action plan to increase permanent agriculture and encourage the abandonment of swidden farming, including the growing of opium poppies. About 2000 Nguan, half from Vieng Poukha and half from Nalae, may have been displaced between 1985 and 1995. Then, between 1995 and 2000, the Nguan faced a third wave of displacement and sedentarization as the country's rural development policy shifted to focal sites and land allocation, supported by foreign aid.

In 1999, Kamphone was a lowland village situated along road No. 3 between Vieng Poukha capital town and Huay Xay, in Bokeo province. The community originated from this mountainous region of Laos and traditionally based its production system on shifting cultivation. During the American period between 1960 and 1970, the villagers launched some commercial activities and undertook an education programme. The current village alongside road No. 3 resulted from the grouping of different highland villages beginning in 1975 under the direction of local authorities. In 1985, the village had about 55 families and a population of 350. In 1991, numerous cases of illness and an epidemic resulted in 54 deaths, sparking off a lengthy period of migration and movement. At first, several families moved to Nongkham, Namthcaling, Namseua and Théo villages. Then another group of families established a new village site, also called Kamphone. Between 1992 and 1996, some of those who had fled the original Kampone returned. Overall, the community of Kamphone had a long history of migration and permanent settlement before a European Union rural development project arrived in 1995 to give them five years of assistance.

Case Study 7: the Atel of Khammouane province

The Atel are Austro-Asiatic speakers from the Vietic branch of the language family. In 2002 their population was estimated to be just 50 people from 11 families with a single parent of Atel origin. They were based in Nakai district, Khammouane province. The new generation is 100% Atel-Arao.

In 1979, a commander from Maka village, in the upper Nam Noy valley in Nakai district, Khammouane province, was given the responsibility of organizing the grouping together of Atel households from the Gnot Nam Xot and Kanil areas, and of Malang groups from Boungkong and Thamuang villages. The mission, requested by Pathet Lao authorities, aimed at better controlling the itinerant forest people and reducing the risk of guerrilla activities in an area that had been heavily affected by the war. The commander was of a different language group to the people he was 'organizing'. Eighteen Atel families from Kanil and 12 from Kouay were displaced into Thamuang village, occupied by Arao people. Another group of Atel was forced to resettle in a village called Nam Houay, with tragic consequences (see plate 61, Coloured Plates section 2). All of these Atel families died within a few years of resettlement from malaria, fever, dysentery, stress, fear of no longer being protected by their guardian spirits and general psychological shock. In Thamuang, most resettled Atel families died for the same reasons, especially during the first three years. Arao, Malang and Atel people in the village confirmed this high rate of mortality during field studies in 2001, but the subject had become taboo. The basic details of between 30 and 60 deaths from illness between 1979 and 1990 were gathered from various sources. The Atel attributed the high number of deaths to the fact that their guardian spirits, which they believed protected them in their Kanil ancestral domain, did not accept the Thamuang territory and remained in Kanil. The situation did not improve after 1982, when the Atel, Arao and Malang ethnic groups of Thamuang village erected a new spiritual pillar for their territorial spirit (Phi Din) (Figure 8-2). Since 1982, even though Atel households have built houses in Thamuang to please the

FIGURE 8-2: Spiritual pillars erected in new villages of forced settlement, seeking the protection of guardian spirits left behind.

Photo: Laurent Chazée.

authorities, they go regularly to their ancestral domain at Kanil to perform their rituals, seek protection and connect again to their forest-based way of life. By the end of 2000, only one single Atel household could be found in Thamuang village. Others were living on swidden farms in the northern part of Thamuang territory, half way to their ancestral domain at Kanil.

Case Study 8: the Temarou of Khammouane province

The Temarou are Austro-Asiatic speakers from the Vietic branch of the language family. In 2003 there were just 30 people of Temarou origin remaining. They were living in 19 households with partners of other origins in Nakai district, Khammouane province.

The traditional Temarou livelihood involved rotational shifting cultivation in mountainous areas based on a forest regeneration cycle. At the end of the 1960s the Temarou were told by local authorities to join the Brou-So villages of Vangchang and Songlerk. The field study covered the village of Vangchang in 2002, when it had 39 houses, 40 households and 208 people of Brou-So ethnic origin. The villagers told of about 35 to 40 Temarou households settling in their village. In fact, the Temarou preferred to live in the vicinity of the village, in the Nyalark and Prryang areas, rather than within the settlement. About half of them died in the first 10 years after resettlement, several of them from fevers in 1970 and 1971. Starting from 1992, the remaining Temarou households built shelters in the northern quarter of Vangchang to please the authorities, but continued their life in the forest. Their Brou-So neighbours told of how, for several years, the bodies of the Temarou were 'imprisoned in the new territory they had never adopted', but their souls continued to wander in the forested swidden areas of their ancestral domain. The surviving Temarou households tried to stay in Vangchang for a few months per year, then spent the rest of the year in the forest, hoping they would not fall foul of the authorities. The chief of Vangchang village kept the habits of the surviving Temarou to himself and reported nothing amiss to the authorities. During a study conducted in the course of the Nam Theun II project in the years from 1996 to 2000, and since there were no project activities in Vangchang village, the remaining Temarou decided to quit Vangchang and Songlerk and settle again in their traditional homes in the forest. Seven households selected a mountain area called Huay Nyalark, two-and-a-half hours' walk northwest of Vangchang and the remaining four households opted for a forest home on the bank of the Nam Theun river, near where two streams called Koran and Prryang join the bigger waterway.

Case Study 9: the Nyaheune of Champassak province

The Nyaheune are Austro-Asiatic speakers. In 2002 it was estimated that their population numbered 5400, living in 27 villages in Champassak and Attapeu provinces.

The Nyaheune originated from the Boloven plateau, in southeastern Laos, from the Paksong, Bachiang, Xaisetha and Sanamxay districts of Champasak and Attapeu provinces where, until about 1960, they were able to freely pursue shifting cultivation. They were accustomed to living at altitudes between 800 and 1200 metres above sea level. Between 1960 and 1973, they were caught up in military events during the second Indochina war (the Vietnam war). Prior to this, Nyaheune communities had conducted an annual ritual between October and November at the junction of Houay Ho and Sé Nam Noi rivers, but in 1970, this traditional ceremony was forbidden by the Pathet Lao.

Following an agreement between the Lao PDR and Thailand concerning sales of electricity, the South Korean Daewoo company began a series of feasability studies for the construction of a hydropower dam on the Huay Ho river, in the middle of Nyaheune territory. Construction of an access road began in 1994, and a year later two Thai companies were contracted to log trees in a 32-square-kilometre area that would later be flooded. The resulting displacement of Nyaheune communities lasted until 1998. It affected nine out of the 11 Nyaheune villages and involved about half of the entire Nyaheune population of Lao PDR. However, as revealed by Khamin (2000), just one of these villages was actually in the area to be flooded for the future reservoir and the large resettlement programme was not among the recommendations of the Environmental Impact Assessment mission. The first four villages – Thong Nyao, Nam Han, Nam Tiang and Nam Ngaw – were resettled on an involuntary basis in the Chat San area at the end of 1994. Sé Nam Noi and Lassassin villages were then displaced in 1995, followed by Nam Leng and Keo Koung Muang in 1996, and finally by part of Nam Kong and Houay Soy villages in 1998. In 2000, Khamin found that the Nyaheune villagers resettled at Chat San, as well as suffering the trauma of the forced resettlement, were unable to find enough agricultural land to ensure their food sufficiency. The land given to them by the government was not only infertile fallow land, but some of it was also claimed by neighbouring farmers (Khamin, 2000).

The new location for Nyaheune people displaced from Lassassin village was selected by provincial and district authorities without consulting the people involved. Villagers found that its natural resources were of poor quality and insufficient to support them. However, the hydro-electricity project financed the transport of the population (except for seven families that arrived later than the rest) and new houses were built, along with an access road, dispensary and school. The villagers later claimed that the government failed to meet an important part of the resettlement agreement. They said they were promised 20 kg of rice per person per month for three years to compensate them for lost production while new farms built up capacity. In reality, the rice supply was irregular. Each person received between 1 and 10kg of rice per month and the supply ended after only two years.

Case Study 10: the Tai Meuiy of Bolikhamxay, Khammouane and Xieng Khouang provinces

The Tai Meuiy are Tai speakers. In 2003, their population was estimated to be 24,000, living in 60 to 70 villages, mostly in Bolikhamxay, Khammouane and Xieng Khouang provinces.

This group came from the north and northeast of Laos. Some of them claim to come from Son La, West of Hanoi in Vietnam. At the beginning of the 19th century, they migrated to Xieng Khouang province's Kham district, which was then known as Meuang Meuiy. In 2002, there were still some Tai Meuiy families living in Kham district, in particular in Sopmar village. Until 1975, the Tai Meuiy farmers in Sopmar were mostly shifting cultivators. However, with continuous encouragement from local authorities, some farming households developed permanent agricultural systems on flat land. Since 1989, 26 families have managed ricefields as well as cultivating swiddens to ensure their food sufficiency. In 2001, other Tai Meuiy households still relied totally on shifting cultivation (Figure 8-3).

Other Tai Meuiy communities, following the example of Tai minorities such as Tai Laan, Tai Mène and Tai

FIGURE 8-3: A makeshift forest house for a Tai Meuiy family displaced by local authorities without assistance. Bolikhamxay province, 2000.

Photo: Laurent Chazée.

Pao, migrated southwards to escape regular pilferage and armed conflict, reaching Bolikhamxay and Khammouane provinces. But the fighting followed them. In 1959 Pathet Lao troops and the Royal Lao Army joined in battle in Bolikhamxay and Khammouane. By 1964, each army was struggling to occupy new territories, displacing rural communities in the conflict zone. Phatong district – today called Vienthong – was divided between the Pathet Lao, the Royal Lao Army and Vang Pao troops. In 1967, as the Pathet Lao began to gain land previously occupied by the other armed factions, United States forces began bombing areas of Bolikhamxay and Khammouane inhabited by Tai Meuiy communities. Several groups found temporary refuge in nearby forest while others fled the area altogether. Daily bombing by US aircraft intensified, and continued until 1973 – even after the capitulation of troops in Vientiane in 1969 and 1970. Some Tai Meuiy families from Khamkeut district in Bolikhamxay were evacuated to camps around Vientiane but returned after 1975. Other families from Viengthong village in Khamkeut district were settled in Nam Sang village in Bolikhamxay's Pakkadin district with assistance from the United Nations High Commissioner for Refugees.

References

Boupha, P. (1995) *Shifting Cultivation Extermination Program Up To Year 2000, Through Permanent Occupations in Lao PDR*, Vientiane

Bouté, V. (2011) *Nouveaux phénomènes migratoires au Laos: l'exode rural vers les villes.* Thématique A :Dynamiques migratoires, enjeux postcoloniaux. 4ème Congrès du Réseau Asie&Pacifique, 14-16 September 2011, Paris

Chazée, L. (1998) *Evolution des systèmes de production ruraux en république démocratique populaire du Laos*, L'Harmattan, Bonchamp-Lès-Laval, France

Chazée, L. (1999) *The People of Laos, Rural and Ethnic Diversities*, White Lotus, Bangkok

Chazée, L. (2000) Xayabury province. History, Actors, territories and Resources for further socio-economic development, photocopy report, Xayabury, Lao PDR

Chazée, L, and Gehin, E. (2012) *Say, femme Poussang. Peuple de la forêt. De la montagne à la plaine, au Laos* (Women from the Poussang minority: People of the Forest, from Mountain to Plain, in Laos), BuchetChastel, Paris

Chazée, L. and Gehin, E. (2017) *Peuples, territoires et moyens d'existence du Laos rural, de 1990 à 2005* (Peoples, Territories and Livelihoods of Rural Laos, 1990-2005), draft in progress. Tentative publication years: Vols 1 and 2, 2018; vol 3, 2019; vol 4, 2020.

Cruikshank, R.B. (no date) *Slavery in Nineteenth Century Siam*, http://www.siamese-heritage.org/jsspdf/1971/JSS_063_2j_Cruikshank_SlaveryIn19thCenturySiam.pdf

Khamin, Nok (2000) 'More trouble for the Heuny', *Indigenous Affairs* 4, pp22-29

Note

1. From 1958 to 1962, the failure of the first government of coalition led to 250,000 people fleeing the country; from 1973 to 1974, the failure of the second government of coalition saw 500,000 people leave the country for Thailand, the United States and France; from 1975 to 1987, about 400,000 people escaped after the Communists took control of the country. Most went to Thailand, France, the United States and Australia. Some returned after 1990.

C. The wider context in which shifting cultivation takes place

From battlefield to swidden field. A derelict army tank huddles beneath encroaching vegetation near the old Ho Chi Minh trail in southern Laos, a reminder of the battles that raged there during the Vietnam War. The land has now reverted to agricultural use, but the ground remains littered with unexploded ordinance. The heat of fires often detonates hidden bombs when shifting cultivators burn the slash on newly-cleared fields, prompting tongue-in-cheek researchers to label the practice 'slash-and-boom' agriculture. This twisted hulk was a relic of Operation Lam Son 719, a disastrous attack into Laos by the South Vietnamese Army, aimed at cutting the Ho Chi Minh trail.

Sketch based on a photo by Peter Xenos.

9

ROMANTICIZING AND VILLAINIZING SHIFTING CULTIVATORS WITHIN NATIONAL POLICIES

Co-producing ethnic politics and resource-use legitimacy in Thailand's community forestry debate

*Tim Forsyth and Andrew Walker**

Introduction

Many studies of shifting cultivation have focused on the traditional cultivation practices of different ethnic groups. These studies have often emphasized how agriculture has allowed vulnerable smallholders to produce food in challenging locations, and how knowledge about cultivation is an important part of ethnic identity. In northern Thailand, for example, much research has highlighted how different upland cultivators, such as the Hmong, Karen, or Iu Mien, have distinctive styles of shifting cultivation, and how this, in turn, impacts upon the types of livelihoods adopted by these people and on the agricultural landscapes that result (Kunstadter et al., 1978; A. R. Walker, 1995).

It is clear that there is now a need to see shifting cultivators in a wider context. Many of these traditional practices have changed as a result of greater permanency of farming settlements, and new agricultural techniques and livelihoods are now available. Moreover, shifting cultivators are increasingly targeted by national and international policies that seek to regulate their impacts on forests, biodiversity and climate change (Myers, 1980; Palm et al., 2005). As a result of these trends, traditional studies of ethnicity and shifting cultivation need to be seen side-by-side with studies of how shifting cultivators are portrayed within national and international policy debates.

This chapter seeks to define how shifting cultivators have been portrayed in national policy debates in Thailand, in order to demonstrate new and important challenges for representing them fairly.[1] 'Getting policies right' for shifting cultivation

* Dr Tim Forsyth, Department of International Development, London School of Economics and Political Science, London; and Dr Andrew Walker, Department of Political and Social Change, School of International, Political and Strategic Studies, College of Asia and the Pacific, Australian National University, Canberra.

is no longer a matter of pointing to historic studies of indigenous knowledge: increasingly, it requires an understanding of how and why different shifting cultivators are represented in policy. This objective requires looking at how policy processes represent shifting cultivators in varying ways, and identifying means of making these processes more transparent and equitable.

This chapter analyses the differing representation of the Karen and Hmong ethnic groups in Thailand, especially in the context of the ongoing debate about community forestry as an appropriate means of managing the use of local resources. Over time, there has been a tendency for policy-makers and popular writers to portray Karen ethnic groups as 'forest guardians' on account of their traditional land-use practices, but Hmong ethnic groups have been labelled as 'forest destroyers.' These representations persist even though there is immense diversity and uncertainty in many of the beliefs about environmental cause-and-effect that are used to justify them (Forsyth and Walker, 2008).

In order to explain these trends, we use two concepts from social science that have been used to analyse how social norms and scientific explanations become interwoven. The first concept is environmental narratives, which refers to the use of convenient summaries of environmental cause-and-effect to explain how shifting cultivation allegedly impacts on watersheds and environmental quality (Roe, 1991). The second concept is co-production, which refers to how authoritative knowledge and visions of social order evolve together, usually allocating a specific role to different social actors (Jasanoff, 2004). These concepts show how different shifting cultivators such as the Karen and Hmong have been placed in different roles. The concepts also offer the possibility for making these social processes more transparent, and therefore more likely to lead to better approaches to community forestry and more equitable treatment of upland agriculturalists.

To start, we briefly summarize the debate about environmental narratives and co-production, and how shifting cultivation in Thailand bears many similarities to other environmental controversies in different locations. We then apply these concepts to the debate about shifting cultivation in Thailand, and how shifting cultivation has become increasingly subsumed into the policy process of debating community forestry. In particular, the policy debate about community forestry has tended to highlight a state-centred view that seeks to centralize forest management for fear of damage by agriculture versus a more people- or development-oriented perspective that argues that shifting cultivators and other smallholders can be trusted to protect forests.

Our key argument, however, is that both of these perspectives are simplified positions. Rather than asking which side is right, 'getting policies right' requires analysts to inquire how different perspectives create narratives to justify their own positions and attribute social roles to different social actors. Representing shifting cultivators more equitably usually means looking at diverse evidence and examining uncertainties about their alleged impacts on their environments, and seeking to address concerns about livelihoods and development needs more inclusively.

Environmental narratives and co-production

Various studies of environmental degradation, especially in developing countries, have highlighted how environmental policy is often underlain by convenient summaries of cause-and-effect that are both simplistic and frequently unhelpful. For example, in the drylands of Africa, the term 'desertification' has often been used to propose that land will become permanently degraded as the result of overgrazing or intensive agriculture. But this proposition has been widely criticized for avoiding the complex causes of drought and a more useful analysis of the social origins of vulnerability to drought. Indeed, some analysts have also argued that belief in the simplistic model of desertification can actually be harmful to attempts to manage both drought and the impacts of land degradation (Ribot, 2011). In West Africa, other analysts have also argued that much policy-making is based on the assumption that smallholders have pushed back large areas of closed forest through the use of fire and agriculture. But research has indicated that various aspects of climate and ecological dynamics control the border between savannah and closed forest, and that smallholders can actually help to restore forest (Fairhead and Leach, 1996).

In both cases, the mainstream 'received wisdom' of how specific actors impact on environmental degradation are known as environmental narratives (Forsyth, 2003). Narratives, or 'storylines,' are succinct summaries of environmental cause-and-effect that are seen as factual within popular and formal policy debates, but which are rooted in historical social influences on the formation of knowledge. According to two analysts:

> Storylines are devices through which actors are positioned, and through which specific ideas of 'blame' and 'responsibility' and 'urgency' and 'responsible behaviour' are attributed (Hajer, 1995, pp64-65).

and:

> Development narratives tell scenarios not so much about what should happen as about what will happen according to their tellers – if the events or positions are carried out as described (Roe, 1991, pp288).

The purpose of analysing narratives is to demonstrate how common beliefs are based on assumptions that are considered 'factual,' and then to link these assumptions to the contexts or networks that created them. By making these assumptions transparent, it should be possible to make environmental policy more inclusive by demonstrating how received wisdom often predefines problems and solutions, and the identities of actors linked to each.

The second concept, co-production, is a more recent attempt by social scientists to analyse how the generation of scientific knowledge and politics come together. According to one definition:

> Co-production is shorthand for the proposition that the ways in which we know and represent the world (both nature and society) are inseparable from the ways in which we choose to live in it (Jasanoff, 2004, p2).

Whereas the concept of narratives refers to the social ordering of complex facts, co-production is an analysis of how facts and social norms are linked together. A co-productionist framework highlights the role of social processes and institutional arrangements first in driving, and then in settling political controversies about matters of uncertainty. In the case of shifting cultivation, co-production might indicate how different knowledge systems or evaluations of agriculture might drive different claims about the impacts or benefits of cultivation.

Shifting cultivation and northern Thailand

Northern Thailand is a familiar location for scholars of shifting cultivation. It is without doubt one of the most researched locations in Asia, largely because of its status for decades as a geopolitically sensitive zone near the borders of capitalist Thailand and communist China, Vietnam, Laos and Burma (Figure 9-1). Added to this, many social scientists have been attracted by the breathtaking diversity of ethnic groups and the mosaic of cultures juxtaposed in a complex landscape of valley bottoms, hill slopes and mountain ridges. Many of these groups practise (or once practised) forms of shifting cultivation.

Shifting cultivators, too, have been represented in simple and sometimes insensitive ways. There is a long record of upland minorities in northern Thailand being called 'hill tribes,' and for many years being considered as marginal to Thai lowland society. The Thai expression for 'hill tribe' – *chao khao* – has a double meaning that underlines this 'othering' of the uplands. *Khao* means both 'mountain' and 'they', so the *chao khao* are both 'mountain people' and 'they people'. The problems of their upland existence are thus symbolically set apart from the concerns of 'we' Thai (Pinkaew Laungaramsri, 2001, pp43–44).

FIGURE 9-1:   The provinces of northern Thailand.

Shifting cultivation was part of this concern. Van Roy (1971, p198) reported that as early as the 1950s 'the belief gathered force that the form of swiddening practised by the hill tribes is destructive of watersheds, that it causes flash flooding as well as altered climatic and rainfall conditions,' making upland shifting cultivators 'scapegoats of northern Thai public opinion.' Indeed, one government report in 1966 claimed that 40% of deforestation in Thailand was the result of shifting cultivation (Boonserm Weesakul, 1966, p13). These negative sentiments were summarized by one of the Royal Forest Department's leading upland catchment researchers:

> There are almost 700,000 shifting cultivators in the highlands. ... As a result the hill evergreen forest has been widely destroyed by shifting cultivation. The farmers have cut and burnt the forest with slash and burn systems to grow annual crops without using soil and water conservation methods. ... This practice has caused deforestation and destruction of the ecosystem of catchment areas and decreased biodiversity in the hill evergreen forest. ... As a result, there are now large areas of wasteland on the mountains and a lack of water in the summer, with floods and landslides occurring in the rainy season (Pornchai Preechapanya, no date, section 1.4).

A dominant narrative, therefore, is that upland shifting cultivation has contributed to problems such as water shortages, deforestation and lowland sedimentation. But there is also a tendency to divide responsibilities between different types of shifting cultivation. Much classic research on shifting cultivation has identified two types: rotational (or the cultivation of land after regular periods of fallow around semi-permanent villages) and pioneer (or the cultivation of all land around villages for 10 to 20 years, before relocating the village) (Grandstaff, 1980). These are similar to the older categorizations of *Waldhackbauern* (forest swiddeners) and *Berghackbauern* (hill swiddeners) (Credner, 1935).

The Karen are perhaps the best known example of an ethnic group that, at least historically, practised the rotational form of shifting cultivation. Although comprising different groups,[2] the Karen as a whole are the largest 'hill tribe' group in northern Thailand and number more than 350,000 (Social Welfare Department, 1995). The history of Karen settlement in the region is uncertain, but many analysts believe they have lived in northern Thailand longer than the *khon muang* (northern Thais), especially in rugged or remote areas. Most Karen villages are located in the middle zones of the uplands (between about 600 and 1000 metres) although there are also numerous Karen communities in the lower reaches of intermontane valleys. Traditional Karen shifting cultivation is known as *rai mun wian* (rotating upland fields). This form of shifting cultivation classically adopted a seven-year fallow cycle in order to allow the soil to recover its fertility. Practitioners also left tree stumps in the ground rather than completely removing trees to facilitate fast forest recovery (where they were known as 'relict emergents') (Schmidt-Vogt, 1998).

The Hmong, on the other hand, are often portrayed as a typical example of pioneer shifting cultivation. They are the second-largest upland ethnic minority in northern Thailand, numbering, according to government statistics, about 73,000 in the provinces of the far north.[3] Hmong farmers have been migrating into northern Thailand, predominantly from Laos, since the latter decades of the 19th century and their villages are now scattered throughout the region. A substantial in-migration occurred after the 1975 communist victory in Laos – many Hmong having sided with the ousted Royal Lao Government – and while most refugees ultimately settled overseas, a significant number moved into established Hmong communities in the northern Thai uplands. Classically, Hmong shifting cultivation was characterized by the relocation of villages every 10 to 20 years, and the intensive cultivation of surrounding land without fallow periods. Often, Hmong farmers continued to use land near old settlements for fruit crops such as lychees. During the middle of the 20th century, Hmong farmers lived near the highest peaks, at about 1000 metres above sea level or more. Since then, they have increasingly moved downslope in pursuit of paddy fields, fruit orchards, tourism and other enterprises (Geddes, 1976; Cooper, 1984). The Hmong have the reputation of being more commercially adventurous than the Karen, and for hiring Karen farmers to work as labourers on Hmong farms (Cohen, 1984).

Representations of the Karen and Hmong

Various narratives characterize discussions about environmental problems in northern Thailand. A common theme is that shifting cultivation in general has caused deforestation, and linked to this, it has also caused dry-season water shortages and wet-season flooding in the lowlands; the pollution of water through excessive agrochemical use; and soil erosion that has damaged both upland agricultural productivity and lowland ricefields by depositing sediment. A sub-theme is that rotational shifting cultivators, especially the Karen, are examples of appropriate upland land use, while the Hmong are responsible for the worst cases of land mismanagement.

While it is clear that any analysis of watersheds needs to understand the impacts of upland agriculture, there is a growing appreciation that the simple narratives employed in Thailand are perhaps too simple (Forsyth and Walker, 2008). Statements such as 'deforestation causes water shortages', 'upland cultivation causes erosion' and 'commercialized cash cropping is environmentally destructive' organize complex biophysical processes into assumptions of linear cause-and-effect in ways that can place too much emphasis on upland agriculture, and too little analysis of pre-existing non-anthropogenic factors or changes in the lowlands, such as increased demand for water or land-cover transformations. For example, hydrological research has proposed that there is more evidence linking changes in rainfall to lowland water supply than to changes in land cover (Alford, 1992; A. Walker, 2003). International research has also shown that it is difficult to make linear connections between lowland water supply and upland agriculture (Calder and Aylward, 2002; Bruijnzeel, 2004).

The implications of these studies are that environmental policies will not address underlying environmental problems if they adopt simplistic narratives of cause-and-effect.

There is also a need to look critically at the representations of Karen and Hmong within these narratives. A co-productionist analysis considers how far different explanations of environmental cause-and-effect are made hand-in-hand with social perceptions of different ethnic groups. This analysis does not deny that the Karen and Hmong forms of ethnic identity and historic styles of using shifting cultivation are strongly different. However, it does consider how the representations of these differences reflect wider social concerns within Thailand. It also asks how far policy proposals based on these differences might work against environment and development objectives.

In particular, how do wider political trends influence how different shifting cultivators are portrayed? According to A. Walker (2001), there has been a distinct trend since the 1990s for popular writing and advocacy groups from non-governmental organizations to portray the Karen *rai mun wian* approach to shifting cultivation as environmentally appropriate. A. Walker (2001) calls this the 'Karen consensus' because, he argues, it forms a politically convenient way for pro-development non-governmental organizations to champion the rights of minorities without offending environmentalists, and because of wider social trends that are beginning to place extra attention on Thailand's forest margins as attractive places.[4] Sometimes Karen have used these views. In 1997, a *Bangkok Post* article stated:

> The life of [the] Karen is closely linked to nature. Our traditions, beliefs and tales all reflect our respect for nature. … By reviving our folk wisdom and practices that enhance forest conservation, the hill communities won't have to be uprooted and the forest can thrive. The Karen's respect for nature is evident in their farming-related ceremonies year round to pay homage to the spirits of fire, water, soil, mountain, trees and ancestors to protect their crops and bless them. The Karen's legends, which are often told at bedtime to children, are rich in stories concerning nature and wildlife with a stress on the necessity for humans and nature to live in harmony. According to the Karen's customs, newly-wed couples are prohibited from shooting birds and the husbands of pregnant women cannot hunt animals or cut down trees. It is believed such acts may harm the babies. Many trees and animals in their folktales cannot be harmed either, for fear of bad luck (Karnjariya Sukrung, 1997).

In 2003, the Thai Airways in-flight magazine carried an article, lavishly illustrated with Karen in traditional dress, entitled 'A love of the land' (Mecir, 2003). It concluded:

> Thus, to the Karen, the dictates of conservation and those of the world beyond merge: a tree cannot be cut down because it embodies a person's soul, a patch of forest in an environmentally sensitive watershed area must not be cleared

because the resident spirit would surely be angry. And as an old Karen poem says, all things under the sky are bound together.

But whereas the Karen have come to represent the romantic and sometimes tragic heroes of upland environmental narratives, the Hmong are undoubtedly the villains. One of the main reasons for this is the association of the Hmong with the sporadic insurgencies and military conflicts that occurred in northern Thailand in the 1960s and 1970s. The Hmong are often portrayed as outsiders whose loyalty to the nation is questionable. Another important reason is their reputation as opium cultivators. As documented by Geddes (1976), poppy cultivation was once the 'mainstay' of Hmong farmers' engagement with external commercial systems. He described an upland agricultural system in which poppies were growing in more than 80% of the fields (often planted after the harvest of a prior maize crop) and the subsistence rice crop occupied about 17% of fields. In a striking contrast to recent material on the Karen, Geddes (1976, p131) argued that '[d]irect subsistence agriculture is only a small part of the [Hmong] economy. The major part is the production of opium as a cash crop'. Opium was, in many respects, an ideal crop for this upland system. As Renard (2001, p3) pointed out:

> Poppy grew well in the hills despite the poor tropical soils there. It required no advanced production technology, nor did it need agricultural inputs such as chemical fertilizers or pesticides. … Furthermore, opium as a crop had advantages in terms of marketing and handling. No cold storage or sophisticated protection against spoilage was required.

The pioneer shifting cultivation with which the Hmong and opium are associated has also been considered akin to ecological vandalism. One classic representation is:

> The communities remove most of the trees, dig out rootstock and stumps and burn the entire swidden areas. They then farm these fields for two to three years, or until the soil's fertility is badly depleted and weeds such as *Imperata* grass invade. Abandoning the plots in an ecologically disturbed, inhibited state of succession, upland tribals rarely attempt to return one day to reutilize them. Instead they move on to clear new vulnerable uplands, repeating the process of forest overexploitation, degradation and conversion to grassland (Poffenberger and McGean, 1993).

A Thai sociologist has also argued that this agricultural system is the antithesis of the irrigated paddy fields of the lowland *muang* because it depends on mobility. Pinkaew Laungsamiri (2001) argues that the Thai term *rai luan looy* (literally, sliding and floating upland cultivation) conveys the sense that this is an ungovernable, unstable, directionless and disordered form of agricultural activity and this imagery resonates with broader security concerns about the loyalty of minority upland groups.

The Thai community forestry debate

These narratives and representations of shifting cultivation are repeated in the debate about community forestry in Thailand. Community forestry is a form of land management where local communities share governance of forested landscapes with the central government. It is usually considered to be a form of decentralized forest policy. It also comprises forms of agroforestry, or the combination of agriculture and forest management. But, as with other countries, community forestry in Thailand is not simply a technical debate about agroforestry or forest management; it is also an arena in which roles of appropriate citizenship, blame and responsibility are allocated.

The debate about community forestry in Thailand is usually dated to the national logging ban, the law for which was passed in 1988, to come into effect in 1989. The logging ban was a response to public concerns about rapid rates of deforestation during the 1980s, and forbade both the cutting of trees and the planting of tree plantations, by agents other than those working under official state policies. The campaign for so-called community forestry was an attempt to allow a more flexible form of access to forestland for rural communities, without losing the ability to control exploitative or unsanctioned deforestation. Since 1989, various bills and proposals for community forestry have been considered within legal and parliamentary debates, but there has been no agreement on a formal framework. In 2003, it was estimated that between one and two million people in Thailand lived in zones bordering official protected areas such as national parks and wildlife sanctuaries, but these people had no official land certification or recognition under the law (Sato, 2003).

Over time, two different approaches to community forestry have emerged. One perspective emphasizes centralized control over forest resources, usually on the grounds of national security and environmental protection. This approach is often proposed by government agencies such as the Royal Forest Department, and it is also supported by some conservation-based non-governmental organizations such as the Seub Nakhasathien Foundation. The other argument proposes the decentralization of authority to communities, and reflects wider discussions of democratization. This viewpoint is usually supported by pro-development non-governmental organizations (NGOs) such as the Foundation for Ecological Recovery.

Proponents of centralized forest control have often invoked the language of scientific certainty to make public statements about appropriate forest use. In 1999, the director of the Royal Forest Department, Plodprasob Surasdi, declared: 'A virgin forest is an untouched forest, but that's a utopian notion, so we have to find a way to mingle the two [forests and human occupation] with minimum impact. But please don't ever say [that] we need humans in the forest to protect it. That's a lie' (Uamdao Noikorn, 1999).

Unsurprisingly, the government – and especially the Royal Forest Department – made various statements during the 1990s that community forestry should be adopted on a highly centralized basis, where specially sanctioned communities could be allowed to administer forestland in accordance with the department's directives and advice. However, the Forest Department and other government agencies changed their positions over time and under different governments. Under the temporary military regime of the early 1990s, the *Khor Jor Kor* reforestation programme sought to establish eucalyptus and pine plantations on officially sanctioned 'forest land', where villages in the northeast of Thailand had settled or grown crops for decades. Unsurprisingly again, villagers resisted these proposals by marching on Bangkok (Pye, 2005). Later, in 1995, under democratically elected governments, more than 2000 hill tribe farmers from six northern provinces marched in protest against the Forest Department's desire to evict them from forest reserves (Bangkok Post, 1995). A similar protest occurred in May 1999, when about 5000 lowland and hill tribe farmers formed a 'rally for rights' in front of the Chiang Mai provincial hall, demanding better land tenure, less state-sponsored reforestation and better access to citizenship (Figures 9-2 and 9-3). This, and many similar protests, were organized partly by NGOs such as the Assembly of the Poor, which was created from various pre-existing networks of NGOs and trade unions in 1995 to represent poorer farmers and workers in informal Thai politics (Missingham, 2003).

Meanwhile, the proponents of a more decentralized vision of community forestry portrayed the debate as a testing ground for appropriate forms of democracy in Thailand. The 1997 Constitution of Thailand mandated local participation in natural resources policy, which, in turn, was supported by the Decentralization Act of 1988 (Anan Ganjanapan, 1997). These political statements were considered to be hard-

FIGURE 9-2: A large poster dominates a protest gathering in Chiang Mai, showing farmers from ethnic groups in Maoist stance, but with an ordained tree in the background.

Photo: Tim Forsyth.

FIGURE 9-3: The sign reads 'Akha people', above villagers in traditional costumes at the same Chiang Mai protest meeting.

Photo: Tim Forsyth.

won indicators of democratization after years of military rule in Thailand. They have since been under challenge by more recent constitutional changes. One editor of the English-language newspaper *Bangkok Post* described the Royal Forest Department as comprising 'gun-toting rangers, at the invitation of forest authorities' who sought to evict 'peasants from their ancestral homes' (Sanitsuda Ekichai, 1999). Some analysts have also argued that community forestry continues a longer-term trend of national security policy by defining citizenship and land rights on the basis of racial identity – a process alleged to exclude recent migrants to Thailand such as the Hmong (Vandergeest, 2003).

A so-called 'people's version' of the Community Forestry Bill was proposed in 1998 by various members of non-governmental and national social-advisory organizations. It emphasized the ability of local communities to govern forest land without strong intervention by the state (Anonymous, no date) (Figure 9-4). The National Legislative Assembly passed a weaker version of this Bill in 2007 (RECOFTC, 2014).

Narratives and shifting cultivators in community forestry

Both the centralized and decentralized perspectives on shifting cultivation employ a central narrative that emphasizes the need to protect upland forest, along with stereotypical representations of shifting cultivators as either forest guardians or forest destroyers. This narrative simplifies the options for livelihoods and environmental protection in the uplands. It also demonstrates how proponents of both perspectives on shifting cultivation participate in co-producing representations of shifting cultivators.

In particular, the 'people's version' of the Community Forestry Bill – despite its emphasis on community rights – carried implicit definitions of both appropriate communities and appropriate forests. For example, there was no discussion of agriculture within community forests and instead the main emphasis concerned planting trees and maintaining forest areas. Indeed, Article 34 of the 2007 Bill stated that community forests should be defined as protected forest that allowed

FIGURE 9-4:   A multi-stakeholder meeting in an Akha village near Doi Mae Salong in Chiang Rai province in 2014 demonstrates the organizing abilities of local ethnic communities. Although there is a predominance of men, there is one traditionally dressed woman. The men include a western consultant and various military advisers.

Photo: Tim Forsyth.

local communities to collect non-timber forest products (RRI, 2008). The Bill also required eligible communities to be 'original local communities ... that live together as a society in the same area and pass down their culture together' (Anonymous, no date). These statements present an image of tradition and isolation within villages that avoids the migration and transience of citizens between villages, or the levels of modernization that have already occurred in many villages.

Statements like these suggest that community forestry (or participation in natural resources management) can only be for appropriate communities that act and live in an appropriate fashion. Some Thai lawyers and journalists also see this implication: one NGO representative said the Bill 'aims to make us responsible for protecting nature in our communities. It doesn't allow a person or a group of people to live in, or to make a living in the forest' (Supara Janchitfah, 2002). An academic advocate of community forests argued that the Bill 'gave local communities the right to manage the forests, not to occupy forest land' (Kultida Samabuddhi, 2002).

These simplifications arise partly because the policy debate about community forestry focuses on the differences between the centralized and decentralized visions, rather than on areas of agreement between them. For example, both sides adopt the 'Karen consensus.' One activist opposed to centralized forestry wrote: 'If plantation forestry is a logical extension of colonial sustained-yield logging, then conservation forestry is its mirror opposite' (Usher, 2009, p10). But this activist also portrayed the

Karen as fragile traditionalists in order to illustrate this point. She described how a Karen elder in the Hot district of Chiang Mai province opposed state-led pine plantations because: 'If they remove the [original] trees from the watershed, the rivers will run dry, the soil will lose its fertility and we won't be able to grow rice. How will we eat?' (Usher, 2009, p110).

However, for A. Walker (2012, p192) this kind of representation 'relies on an imagery of local cultural identity, self-sufficient agriculture and ecologically-friendly lifestyles … that is largely disconnected from the livelihood aspirations of Thailand's commercially connected middle-income peasantry.' Clearly, there is much evidence to indicate that monoculture forest plantations can reduce water supply and biodiversity, as well as increase soil erosion under certain circumstances (Calder, 1999; Bruijnzeel, 2004). But representing plantation forestry only in terms of its threat to traditional Karen lifestyles takes attention away from these hydrological concerns and romanticizes the Karen.

Meanwhile, the Hmong ethnic group is typically vilified in NGO reports and the popular press, especially for being involved in large-scale cash cropping. In the words of another *Bangkok Post* article:

> Hmong tribespeople who have resettled in Phop Phra district [in Tak province] are destroying forests and dealing in drugs, the government claims. Ladawan Wongsriwong, deputy labour minister, said problems were reported at 19 villages among those which had developed on a 125,000-rai [21,000 hectares] area of deteriorated forest land along the Phop Phra-Umphang highway since 1987. … the Hmong were resorting to their 'old tricks' at their new home, she said. … Mrs Ladawan said the locals had destroyed their land with excessive use of chemical fertilisers and pesticides and in their search for new fertile plots for cultivation had destroyed forest areas around their villages. State agencies must act swiftly to protect the forest which was the villages' only source of water, she said (Kasem, 2001).

Statements like this need to be taken alongside other research that, for example, highlights how supposedly 'monoculture' upland crops such as cabbages contain patches of diverse vegetation, especially along streamlines and ridge tops, that are actively managed by farmers 'for growing a variety of traditional crops and local vegetables formerly grown in swidden fields for household consumption' (Kanok Rerkasem, 2003, p296). Similarly, research has also suggested that monocropped cash crops have raised income and substantially reduced land pressure in areas of population growth, where rice production is insufficient. According to one analyst, 'various studies of Hmong agriculture indicate that newly adopted cash crops, such as cut-flowers, yield the highest average gross margin per unit of area' (Waranoot, 2002, p107).

Moreover, it is worth noting that the location in Tak province referred to above became notorious during the 1960s as a site of Hmong insurgency, and indeed a road built to increase military access was so plagued by attacks it was called 'the road that cost nine men per mile' (The Nation, 1973). In the late 1980s, 157 Hmong families from this region were moved to lowland sites on official grounds of protecting the headwaters of local rivers and maintaining the integrity of a newly declared wildlife sanctuary (Pratya Sawetrimon, 1987).

Statements like these do not imply that upland agriculture has no impacts, or that environmental policy is misplaced. But they do indicate that environmental policy is often couched in unseen political histories, and that policy debates shape complex trends into convenient, but simplistic, narratives. A critical approach to these narratives requires assessing how they were made, which perspectives were excluded and how a more inclusive and effective environmental policy can be made.

Conclusion

This chapter has asked: 'How can the right policies be formulated to represent shifting cultivators in national environmental planning?' The main argument is that traditional studies of ethnic practices of shifting cultivation are valuable ways of understanding how marginal people achieve their livelihoods and food security. But there is also a need to analyse how shifting cultivators are represented in national policy debates in order to work towards achieving a fair and helpful treatment of shifting cultivation in environmental policy.

The experience of Thailand has shown that national policy-making does not reflect local studies of shifting cultivation. Rather, stereotypical images of shifting cultivators are presented, often in tandem with simplified representations of scientific cause-and-effect, in order to justify other political concerns. In particular, Thailand's debate about community forestry has shown tensions between organizations and activists who seek to maintain a strong state control over resources and national security and those who see forest management as a case study of democratization and uneasiness about modernization. Despite the deep divisions between these two perspectives, there is a need to see how both sides also share, and support, stereotypical and overly simplified representations of shifting cultivation and upland agriculture as a common theme in their discussions. The Karen, in particular, are portrayed as forest guardians because they are alleged to represent a benign way of combining tradition and ecological wisdom. The Hmong, by contrast, are labelled as forest destroyers. Yet, these simplified visions do not help to address environmental problems such as water shortages or declining soil fertility; nor do they acknowledge the potential benefits of diverse systems of shifting cultivation, or of some shifting cultivators adopting modernized agriculture.

A short-term solution to this problem is to acknowledge the significance of so-called narratives in environmental policy and the emergence of convenient, but inaccurate, summaries of cause-and-effect. Similarly, there is a need to understand how political objectives – or visions of social order – are co-produced simultaneously with knowledge about environmental problems and the people linked to them. In essence, knowledge about shifting cultivation in Thailand is implicitly linked to the national security concerns that followed the Vietnam War; the predominance of population and policy-makers in the lowland plains beneath the northern highlands; and the concerns of modern Thai society about modernization and democratization. A longer-term objective, however, is to diversify the objectives of, and participation in, environmental planning in order to acknowledge a broader range of development needs in the uplands and the greater role of lowland water demand in managing national problems such as water shortages.

References

Alford, D. (1992) 'Stream flow and sediment transport from mountain watersheds of the Chao Phraya basin, northern Thailand: A reconnaissance study', *Mountain Research and Development* 12(3), pp237-268

Anan Ganjanapan (1997) 'The politics of environment in northern Thailand: Ethnicity and highland development programs', in P. Hirsch (ed.) *Seeing Forests for Trees: Environment and Environmentalism in Thailand*, Silkworm Books, Chiang Mai

Anonymous (no date) *Draft Community Forest Laws: Differences between the People's Version and the Government Versions* [Thai language] Project for Ecological Recovery, Bangkok

Bangkok Post (1995) 'Hilltribe eviction protest tops 2000', *Bangkok Post,* 13 May 1995

Boonserm Weesakul (1966) 'Preliminary socio-economic survey of the hill tribes in Thailand, 1965-66', in Central Bureau of Narcotics (ed.) *Preliminary Socio-economic Survey of the Hill Tribes in Thailand, 1965-66*, Ramin Press, Bangkok, pp13-19

Bruijnzeel, L. A. (2004) 'Hydrological functions of tropical forests: Not seeing the soil for the trees?', *Agriculture, Ecosystems and Environment* 104(1), pp185-228

Bupho (1997) *My Pakakoeyo Life* (Thai language), Samnakphim Sarakhadi, Bangkok

Calder, I. (1999) *The Blue Revolution: Land Use and Integrated Resource Management*, Earthscan, London

Calder, I. and Aylward, B. (2002) *Forests and Floods: Perspectives on Watershed Management and Integrated Flood Management*, Food and Agriculture Organization of the United Nations and the University of Newcastle, Rome and Newcastle, UK

Cohen, P. (1984) 'Opium and the Karen: A study of indebtedness in northern Thailand', *Journal of Southeast Asian Studies* 15(1), pp150-165

Cooper, R. G. (1984) *Resource Scarcity and the Hmong Response: Patterns of Settlement and Economy in Transition*, Singapore University Press, National University of Singapore, Singapore

Credner, W. (1935) *Siam, das Land der Tai. Eine Landeskunde auf Grund eigener Reisen und Forschungen* (Siam, the Land of Thai. A Regional Study based on His Own Travels and Research), J. Engelhorns nachf, Stuttgart

Fairhead, J. and Leach, M. (1996) *Misreading the African Landscape: Society and Ecology in a Forest-Savanna Mosaic*, Cambridge University Press, Cambridge, UK

Forsyth, T. (2003) *Critical Political Ecology: The Politics of Environmental Science*, Routledge, London and New York

Forsyth, T. and Walker, A. (2008) *Forest Guardians, Forest Destroyers: The Politics of Environmental Knowledge in Northern Thailand*, University of Washington Press, Seattle

Geddes, W. (1976) *Migrants of the Mountains: The Cultural Ecology of the Blue Miao (Hmong Njua) of Thailand*, Clarendon, Oxford, UK

Grandstaff, T. (1980) *Shifting Cultivation in Northern Thailand: Possibilities for Development*, United Nations University, Tokyo

Hajer, M. (1995) *The Politics of Environmental Discourse*, Clarendon Press, Oxford, UK

Jasanoff, S. (2004) 'The idiom of coproduction', in S. Jasanoff (ed.) *States of Knowledge: The Co-production of Science and the Social Order*, Routledge, London, pp1-12

Kannika Phromsao and Bencha Silarak (1999) *The Seven Layer Forest: From the Words of Headman Joni Odochao* (Thai language), Local Wisdom Foundation, Bangkok

Kanok Rerkasem (2003) 'Thailand', in H. Brookfield, H. Parsons and M. Brookfield (eds) *Agrodiversity: Learning from Farmers across the World*, United Nations University Press, Tokyo

Karnjariya Sukrung (1997) 'The fight for the forests', *Bangkok Post*, 19 June 1997

Kasem, S. (2001) 'Hmong not doing what they should: Queen had warned of forest destruction', *Bangkok Post*, 2 August 2001

Kultida Samabuddhi (2002) 'Academic slams senators over ban', *Bangkok Post*, 29 March 2002

Kunlawadi Bunphinon (1993) 'Karen lifestyle: A lifestyle in harmony with nature' (Thai language), *Niwet* 20(3), pp43–56

Kunlawadi Bunphinon (1997) 'Collective resource management: Mechanisms for forest protection in the Karen village of Sanephong' (Thai language), *Niwet* 24(3), pp19–31

Kunstadter, P., Chapman, E. C. and Sabhasri, S. (eds) (1978) *Farmers in the Forest: Economic Development and Marginal Agriculture in Northern Thailand*, University Press of Hawaii, Honolulu

Mecir, A. (2003) 'A love of the land', *Sawasdee*, January 2003 issue, pp36–43

Missingham, B. (2003) *The Assembly of the Poor in Thailand: From Local Struggles to National Protest Movement*, Silkworm Books, Chiang Mai

Myers, N. (1980) *Conversion of Tropical Moist Forests. A Report prepared for the Committee on Research Priorities in Tropical Biology of the National Research Council*, National Academy of Sciences, Washington, DC

Palm, C., Vosti, S., Sanchez, P. and Ericksen, P. (eds) (2005) *Slash and Burn Agriculture: The Search for Alternatives*, Columbia University Press, New York

Pinkaew Laungaramsri (2001) *Redefining Nature: Karen Ecological Knowledge and the Challenge to the Modern Conservation Paradigm*, Regional Center for Social Science and Sustainable Development, Chiang Mai University, Chiang Mai

Poffenberger, M. and McGean, B. (1993) *Community Allies: Forest Co-management in Thailand*, Southeast Asia Sustainable Forest Management Network, Bangkok

Pornchai Preechapanya (no date) *Indigenous Highland Agroforestry Systems of Northern Thailand*, Chiang Dao Watershed Research Station, Chiang Mai

Pratya Sawetrimon (1987) '5000 tribespeople to be moved from forest reserve', *The Nation*, 9 March 1987

Pye, O. (2005) *Khor Jor Kor: Forest Politics in Thailand*, White Lotus, Bangkok

RECOFTC (2014) *Community Forestry Adaptation Roadmap to 2020 for Thailand*, Regional Community Forestry Training Centre, Bangkok

Renard, R. (2001) *Opium Reduction in Thailand, 1970-2000: A Thirty-year Journey*, United Nations International Drug Control Programme and Silkworm Books, Chiang Mai

Ribot, J. (2011) 'Vulnerability before adaptation: Toward transformative climate action', *Global Environmental Change* 21, pp1160-1162

Roe, E. M. (1991) 'Development narratives, or making the best of blueprint development', *World Development* 19(4), pp287-300

RRI (2008) *Thailand's Community Forestry Bill*, Rights and Resources Initiative, Washington, DC

Sanitsuda Ekichai (1999) 'An unholy alliance lays siege to Karens', *Bangkok Post*, 13 May 1999

Sato, J. (2003) 'Public land for the people: Institutional basis of community forestry in Thailand', *Journal of Southeast Asian Studies* 32(2), pp329-346

Schmidt-Vogt, D. (1998) 'Defining degradation: The impacts of swidden on forests in northern Thailand', *Mountain Research and Development* 18(2), pp135-149

Social Welfare Department (1995) *Residence of Highland Communities in Thailand*, Hill Tribe Welfare Division, Social Welfare Department, Ministry of Labour and Welfare, Bangkok

Supara Janchitfah (2002) 'Falling trees have long shadows', *Bangkok Post*, 24 March 2002

The Nation (1973) 'The road that cost nine men per mile', *The Nation*, 1 December 1973

Uamdao Noikorn (1999) 'Tough and in the thick of things', *Bangkok Post*, 7 June 1999, p10

Usher, A. D. (2009) *Thai Forestry: A Critical History*, Silkworm Books, Chiang Mai

Van Roy, E. (1971) *Economic Systems of Northern Thailand: Structure and Change*, Cornell University Press, Ithaca, NY

Vandergeest, P. (2003) 'Racialization and citizenship in Thai forest politics', *Society and Natural Resources* 16(1), pp19-37

Walker, A. (2001) 'The "Karen Consensus": Ethnic Politics and Resource-use Legitimacy in Northern Thailand', *Asian Ethnicity* 2(2), pp145-162

Walker, A. (2003) 'Agricultural transformation and the politics of hydrology in Northern Thailand', *Development and Change* 34(5), pp941-964

Walker, A. (2012) *Thailand's Political Peasants: Power in the Modern Rural Economy,* University of Wisconsin Press, Madison, WI

Walker, A. R. (1995) 'From the mountains and the interiors: A quarter of a century of research among Fourth World peoples in Southeast Asia (with special reference to northern Thailand and peninsular Malaysia)', *Journal of Southeast Asian Studies* 26(2), pp326–365

Waraalak Ithiphonorlan (1998) *Rotational Shifting Cultivation: Mother of Plant Varieties* (Thai language), Project for the Development of Northern Watersheds by Community Organisations, Chiang Mai

Waranoot Tungittiplakorn (2002) 'Limitations of subsistence agriculture in the highlands', in P. Dearden (ed.) *Environmental Protection and Rural Development in Thailand: Challenges and Opportunities*, White Lotus, Bangkok

Notes

1. This chapter draws upon Forsyth, T. and Walker, A. (2008) *Forest Guardians, Forest Destroyers: The Politics of Environmental Knowledge in Northern Thailand*, University of Washington Press, Seattle, WA and Forsyth, T. and Walker, A. (2014) 'Hidden Alliances: Rethinking environmentality and the politics of knowledge in Thailand's campaign for community forestry,' *Conservation and Society* 12(4), pp407–417.

2. There are two major sub-groups of Karen: the Sgaw and the Pwo. Sgaw villages are located mainly within the Ping river catchment area while the Pwo tend to be located further west, closer to Mae Sariang.

3. The Hmong are frequently classified into White and Blue Hmong sub-groups. The White are found mostly in the far northeast of Thailand, including Nan, Phayao and eastern Chiang Rai provinces; the Blue usually live further west, in western Chiang Rai and in Chiang Mai provinces.

4. For example, Bupho (1997), Waraalak Ithiphonorlan (1998), Kannika Phromsao and Bencha Silarak (1999) and Kunlawadi Bunphinon (1993, 1997).

10

CONSERVATION AND RESTORATION OF TRADITIONAL GRASSLANDS IN THE MOUNT ASO REGION OF KYUSHU, JAPAN

The role of collaborative management and public policy support

*Yoshitaka Takahashi, Andreas Neef and Hiroshi Yokogawa**

Introduction

Shifting cultivation was a common practice in mountainous regions of Japan until the early 1960s, when modernization of agriculture and rural-urban migration started to bring profound changes to the social and economic fabric of rural societies in remote upland areas (Oki, 1993; Sekido, 1994). In some areas of Ishikawa and Yamagata prefectures on Japan's largest island, Honshu, farming with fire (*yakihata*, literally 'burned field' agriculture) has survived even until today and is still part of a diverse cultural landscape (Kurata, 2013). Remarkably, these traditional practices have undergone a re-evaluation by academics and policy-makers in recent years and are increasingly regarded as important cultural and historic treasures that warrant public efforts for their preservation and even revival.

On the southern island of Kyushu, the grasslands around the Mount Aso volcanic landscape in Kumamoto prefecture (Figure 10-1) have been maintained historically by human activities such as mowing, burning and grazing. They have been embedded in a mosaic landscape pattern of agricultural fields, grassland and forests (commonly referred to as *satoyama* in Japanese) and have provided habitat for a variety of grassland plants and animals, including continental element species (northern China-Korea tertiary relict species) and Red Data Book species. In recent times, however, because burning, mowing and grazing have been largely abandoned, the grassland

* Dr Yoshitaka Takahashi, NARO Western Region Agricultural Research Center, Ohda, Shimane, Japan; Professor Andreas Neef, Development Studies, School of Social Sciences, Faculty of Arts, University of Auckland, New Zealand; and Professor Hiroshi Yokogawa, Kyushu Kyoritsu University, Kitakyushu, Fukuoka, Japan.

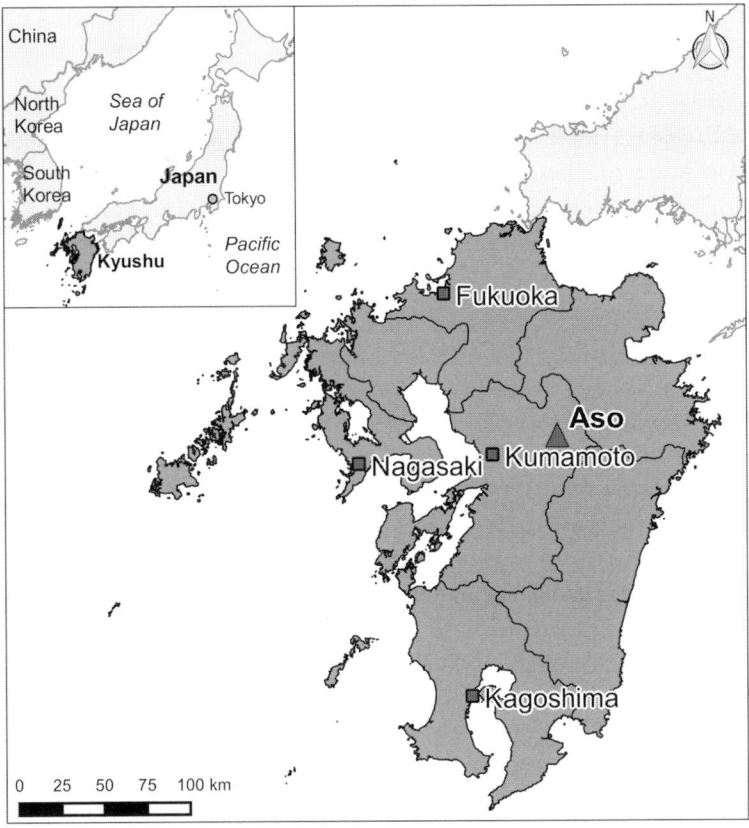

FIGURE 10-1:   Location of the Aso grassland on the island of Kyushu, Japan.

area has declined and the number of grassland species has fallen dramatically as a consequence. The Aso Grassland Restoration Committee was established in 2005 to address these problems and to conserve and restore the grassland. In this chapter, we argue that traditional grassland management practices need to be modified to address the current condition of the grassland and the realities of modern lifestyles. We also demonstrate that long-term collaborative management of such multifunctional landscapes can be successful through the cooperation of various stakeholders and strong public policy support.

The case of the Aso grassland restoration may offer a surprising model for the preservation of swidden systems in Southeast Asia and the Pacific, and more importantly, the biodiversity that has long been nurtured by these systems. Despite wide differences in climate and plant species, at grass-roots level, the principles are the same.

Human intervention and natural processes combine to create a cultural heritage

The Aso grasslands are 22,000 hectares in extent, and are the largest area of grassland in Japan. The Engishiki book of laws (901–923AD), compiled during the Heian Period, refers to two horse ranches, Futae and Hara, in Higo province (today's Kumamoto prefecture), so the Aso grassland became known as the 'grassland of a thousand years' (Takahashi, 2009). However, recent analyses of plant opal content in the soil has confirmed that the grassland extended across the area more than 10,000 years ago (Miyabuchi and Sugiyama, 2006, 2012; Miyabuchi et al., 2012; Kawano et al., 2012).

The Aso grasslands have been developed and maintained by a successful combination of natural and human factors, including volcanic activities (Aso Grassland Restoration Committee, 2007), highly permeable geology (Hayakawa, 1981) and burning, mowing and grazing. These factors have prevented the area from shifting into forest (Figure 10-2) (Yamauchi and Takahashi, 2002; Takahashi, 2009). Grasslands such as those at Aso, which have been maintained for sustainable use through a range of natural and artificial disturbances, are referred to as semi-natural or secondary grasslands.

The role of grasslands has changed according to the needs of various eras. Prior to World War Two, they provided roof-thatching materials and fodder for war horses; immediately thereafter, they provided fodder for cattle and horses. Today, the grasslands attract local and foreign visitors as they are part of a cultural landscape formed by the interaction of people, livestock and nature. The grassland landscape that is part of the caldera landform of the Aso region is both a natural and historical heritage that is a source of national identity and pride.

FIGURE 10-2: The grassland landscape is created by human activity. Top: burning eliminates shrubs but herbaceous plants survive to regenerate the following season. Middle: grasses gain value by being mowed; and bottom: the cycle of grazing and excretion creates a pastoral landscape.

Photos: Aso Green Stock Foundation (top);
Norio Otaki (middle);
Yoshitaka Takahashi (bottom).

A treasure of diversity

An estimated 1600 different species of plants, representing about 70% of all the species found within Kumamoto prefecture, are distributed around Aso. More than 600 of these species grow in the grasslands, including continental element species (northern China-Korea tertiary relict species) such as *Viola orientalis, Echinops setifer, Silene sieboldii* and *Campanula glomerata* var. *dahurica*, which advanced southward during the glacial age when the island of Kyushu was part of the Asian continent (Okubo, 2002; Takahashi, 2009). Many species of birds and butterflies are also found in the area because the natural environment comprises both forests and grasslands (Kumamoto Prefecture, 1998).

People used the grasslands for livestock grazing in late spring and summer, mowing for fresh feed in summer to early autumn (when grazing was prohibited in certain areas), and mowing for hay and thatch in early to late autumn (Figure 10-2, middle). The vegetation has formed a mosaic in the grasslands, albeit seemingly uniform in character and composed of similar plants. The dominant species in the mowed fields (meadows) are *Miscanthus sinensis* and *Pleioblastus chino* var. *viridis*. When the fields are mowed and the growth of these dominant species is reduced, an environment is created where various species of plants can coexist. The custom of placing wildflowers on tombs during the Bon Festival, as flowers for Bon (Figure 10-3), coincides with the time at which many Red Data Book plant species blossom in the mowed areas.[1]

Conversely, artificial grassland in which temperate (cool season) grasses, such as *Lolium multiflorum* and *Dactylis glomerata*, have been introduced have only a few species, although they may appear to be diverse at first glance. Also, since Japanese cedar trees have been planted in many mowing areas, the natural habitats that formerly supported Red Data Book plant species have been lost (Kumamoto Prefecture, 1998; Sei, 2006; Takahashi, 2009). Wetlands found in the depressed hollows and valleys of the grassland have also deteriorated because of the inflow of fertilizers and soils from surrounding artificial grassland and vegetable fields.

FIGURE 10-3:   The custom of offering wildflowers for the Bon Festival is a form of grassland culture. The flowers bloom in the meadows.

Photo: Shingo Kaneko.

Thus, while the Aso region contains some of the most abundant grassland plants and animals in Japan, the unregulated expansion of forested sites, artificial grassland and cultivated areas is leading to the extinction of native grassland, natural heritage plants and animals from the global cold age and is compromising the traditional relationship between the grassland and humans. To maintain the semi-natural grassland, human-induced management activities, including burning in the spring and mowing in the autumn, are important for species conservation and biodiversity (Sei, 2006; Takahashi, 2009).

The Aso grassland and society in crisis

After having been conserved for more than 1000 years, the Aso grasslands have recently reached a point of crisis. Historically, they were preserved by various organizations on the basis of shared territorial bonds within *shuraku* or hamlets. This arrangement involved equal distribution of resources, access to grass resources and benefit sharing among *iriai* (rights holders), or people who held the land in common and shared the labour of *noyaki* (burning) and *wachigiri* (creation of mown firebreaks). After World War Two, local grassland associations were organized in many places by livestock-holding farmers. However, these farmers comprised only a portion of commonage holders and grassland management was later undertaken at district level or by all commonage holders in a collaborative effort.

Relatively recent changes in farming and lifestyles have made this type of management unsustainable. Beginning in the late 1960s, livestock-rearing was influenced by fluctuating economic growth. Labour shortages in farming communities, an ageing population and recently expanded imports of low-priced farm products brought pressure to bear on an increasing number of livestock farmers to sell their animals and abandon farming. The result is a serious shortage of labour for *wachigiri* and *noyaki* (Yamauchi and Takahashi, 2002).

The total area of grassland in the Aso region that is no longer burned, mowed or grazed has already reached several thousand hectares, and this is likely to increase rapidly because of an expected decrease in membership of the grassland associations and the advancing age of the community. In addition, *wachigiri* – making mown firebreaks – is a dangerous activity undertaken during the intensely hot late summer and early autumn seasons and on complex landforms such as steeply sloping ground. *Noyaki*, or burning, is often abandoned because of the effort involved (Yamauchi and Takahashi, 2002).

Changes threatening flora and fauna

Dramatic changes in the use and management of the Aso grasslands over the past several years have meant that the landscape structure has also changed, and many plants and animals are now facing extinction (Figure 10-4).

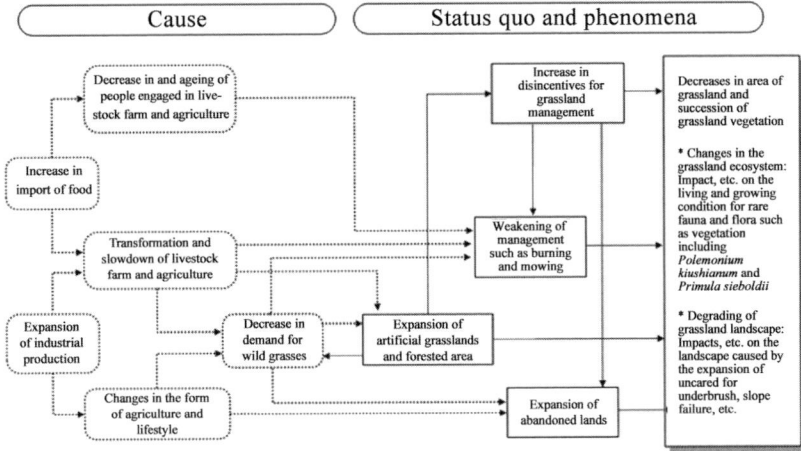

FIGURE 10-4:   Causes and the existing state of changes in the Aso grassland.

Of the approximately 600 species of plants growing in the Aso grasslands, 72 are listed on the Ministry of the Environment's Red List of endangered and threatened species (Ministry of the Environment, 2014). One species faces a high risk of extinction in the very near future and is listed as 'Threatened IA' (or CR: Critically Endangered). Sixteen species face a slightly lesser risk of extinction in the near future and are categorized as 'Threatened IE' (or EN: Endangered); 43 species face an increasing risk of extinction, and are regarded as 'Threatened II' (or VU: Vulnerable), and 12 species are having trouble maintaining a viable population, and are listed as 'Quasi-Threatened' (or NT: Near Threatened) (Table 10-1). In total, about 10% of all Aso grassland plant species are facing extinction, and this gives the area one of the highest concentrations of species threatened with extinction in Japan.

Polemonium kiushianum is an endemic species that grows only in the Aso grassland. The Ministry of the Environment has designated it a domestic rare species of fauna and flora under the provisions of the Law for the Conservation of Endangered Species of Wild Fauna and Flora. It once grew naturally in 46 areas (Sei, 2006), but today it is found in only a few areas. Many of the continental element species (northern China-Korea tertiary relict species) that are concentrated in the Aso region of Kyushu, such as *Echinops setifer*, *Silene sieboldii* and *Campanula glomerata* var. *dahurica*, have also lost their habitat and are threatened with imminent extinction (Sei, 2006; Takahashi, 2009).

The grassland no longer functions as a habitable environment, and this is affecting not only plants but also small animals and insects. For example, *Sophora flavescens*, the food plant of the butterfly *Shijimiaeoides divinus*, is not eaten by cattle or horses because of its toxicity. The plant is relatively plentiful in grazing areas and is intentionally left unmown in the grassland meadows. However, when the grassland is abandoned and burning, grazing and mowing are neglected, other plant species grow more vigorously. As a result, the butterfly has begun to decline along with the

TABLE 10-1: Threatened species of plants in the Aso grassland.

Threatened IA (CR)	Threatened IE (EN)	Threatened II (VU)	Semi-threatened (NT)
Polemonium kiushianum	Aconitum ciliare	Moehringia trinervia	Adonis multiflora
	Hedyotis chrysotricha	Silene kiusiana	Epimedium grandiflorum var. grandiflorum
	Lithospermum erythrorhizon	Silene sieboldii	Primula sieboldii
	Trigonotis radicans var. sericea	Pulsatilla cernua	Swertia pseudochinensis
	Euphrasia multifolia var. multifolia	Ranunculus ternatus var. ternatus	Vincetoxicum pycnostelma
	Veronica linariifolia	Thalictrum simplex var. brevipes	Mosla japonica var. hadae
	Campanula glomerata var. dahurica	Paeonia obovata	Utricularia uliginosa
	Saussurea japonica	Hypericum ascyron var. longistylum	Achillea alpina subsp. subcartilaginea
	Tephroseris flammea subsp. flammea	Potentilla discolor	Artemisia stolonifera
	Asparagus oligoclonos	Lespedeza tomentosa	Carex kujuzana
	Lilium callosum var. callosum	Geranium soboliferum var. kiusianum	Pecteilis radiata
	Lilium concolor	Osbeckia chinensis	Pogonia japonica
	Schoenoplectus mucronatus var. ishizawae	Bupleurum scorzonerifolium var. stenophyllum	
	Habenaria dentata	Pterygopleurum neurophyllum	
	Herminium lanceum	Lysimachia barystachys	
	Liparis odorata	Mitrasacme indica	
		Vincetoxicum amplexicaule	
		Vincetoxicum atratum	
		Trigonotis radicans var. radicans	
		Ajuga ciliata var. villosior	
		Leonurus macranthus	

TABLE 10-1 (cont.): Threatened species of plants in the Aso grassland.

Threatened IA (CR)	Threatened IE (EN)	Threatened II (VU)	Semi-threatened (NT)
		Scrophularia buergeriana	
		Veronica kiusiana var. *kiusiana*	
		Veronicastrum sibiricum var. *zuccarinii*	
		Codonopsis ussuriensis	
		Platycodon grandiflorus	
		Artemisia rubripes	
		Aster maackii	
		Aster tataricus	
		Cirsium sieboldii subsp. *Austrokiushianum*	
		Echinops setifer	
		Inula linariifolia	
		Ixeris chinensis subsp. *strigosa*	
		Leucanthemella linearis	
		Ligularia sibirica	
		Saussurea pulchella	
		Iris rossii	
		Arisaema heterophyllum	
		Fimbristylis dichotoma subsp. *podocarpa*	
		Fimbristylis pierotii	
		Schoenoplectus mucronatus var. *mucronatus*	
		Schoenoplectus gemmifer	
		Habenaria sagittifera	
1 species	16 species	43 species	12 species

Notes: CR: Species facing a very high risk of extinction in very near future; EN: Species facing a risk of extinction in near future, though not as high as those listed in category IA; VU: Species facing an increasing risk of extinction; NT: Species facing difficulties in maintaining a viable population.

Source: Red List 2012, Ministry of the Environment, Japan.

decline of its food source. Other butterflies including *Fabriciana nerippe, Leptidea amurensis* and *Maculinea teleius daisensis* are also facing extinction because of the degeneration and disappearance of grassland caused by discontinued mowing and grazing and changes in land use (Takahashi, 2009; Yamauchi and Takahashi, 2002).

Public efforts to maintain and restore the Aso grassland

Various activities were developed to maintain and restore the multifunctional landscape of the Aso grassland in 1997, including volunteer burning programmes. In December 2005, the Aso Grassland Restoration Committee was established under the provisions of the Law for the Promotion of Nature Restoration (Aso Grassland Restoration Committee, 2007). The challenge has been to reorganize and modernize traditional management and use of the grassland, including burning, mowing and grazing, that is not only sustainable, but also considers species conservation. Four strategies have been adopted:

1. Collaborative work by local and urban residents

Miscanthus sinensis Andersson
[Poaceae]

This perennial grass grows up to two metres tall, in dense clumps, and is widely cultivated around the world as an ornamental plant. Its purplish flowers are held above the foliage. It is a dominant species in the Aso grasslands and is mowed to allow other plants to grow and flower.

The largest-scale activity related to the conservation of the Aso grassland is the recruitment of volunteers to support burning. In the 20 years since volunteers first began burning the grassland, these people have become an indispensable asset to the local grassland associations. About 2000 volunteers participate in a coordinated and cooperative burning programme for one third of all common grassland in which burning, including *wachigiri*, is practised (Aso Green Stock Foundation, 2008). Volunteers are required to undertake training before they take part. They are made aware of the dangers of the task and the need for consistent concentration and they undertake 'hands-on' burning practice on small-scale grassland before being admitted to the programme.

Volunteers who take part often express feelings of goodwill and gratitude to Mount Aso, which consistently provides enjoyment to sightseers and others. During a collective grassland burning programme in March 2015, one volunteer said: 'I consider my involvement in *noyaki* as a way of honouring the hard work that earlier generations put into the creation of this cultural landscape. Yet, when I participate in this collective effort, it fills me with such joy and energy that it becomes a reward in itself.' By showing that they are conscientious and sincere in their work, the volunteers have earned the trust of local residents. They visit the Aso region on a frequent basis and can be regarded as ideal examples of responsible citizens. The relationship of mutual trust between local residents and urban visitors has contributed greatly to the work of the Aso Grassland Restoration Committee. Burning of the abandoned grassland, which provides habitat for rare species but has long been neglected and overgrown with bushes, has therefore resumed. The results, in terms of restoring the grassland, have been far better than originally anticipated.

As well, a Restoration Movement of Grassland with High Species Diversity has been established, in which non-profit organizations purchase abandoned grassland to burn and mow. This movement seeks to restore the former mowing fields of Aso (called *hana-no*, literally 'flower fields'), where many flowering herbaceous plants once grew, and to conserve and re-establish valuable plant species (Sei, 2006; Takahashi, 2009). Organizations recreate the historical uses and management of the grassland, such as burning, mowing and collecting grasses, under the guidance of local farmers. The grasses are sold to local tea cultivators who use them as compost and as 'tea grass' (mulch for tea fields) (Takahashi, 2009). In some neighbouring grasslands, areas of artificial forest have been cut down, revitalizing rare species such as *Polemonium kiushianum*. This has inspired plans to return other forested areas to semi-natural grassland in the future.

2. Linking the concept of grassland conservation to production of food

People can become involved in grassland conservation not only by direct participation in its management, but also through economic activities or payment of subsidizing fees to promote the coexistence of agriculture, forestry and conservation.

The Aso Green Stock Foundation, which organizes the volunteers who take part in the burning activities, also promotes the consumption of beef produced on the grassland as a means of supporting the protection of the area (Yamauchi and Takahashi, 2002). The foundation began a unique programme named after a region-specific cattle breed called *Aka-ushi* (Japanese red cattle). Under this scheme, urban residents can become owners of *Aka-ushi* cattle, to both increase the breed's numbers and expand the consumption of beef. Other organizations are also engaged in similar schemes, such as the certification of agricultural products cultivated with wild-grass compost and experimental activities that use wild-grass biomass, to promote awareness of grassland conservation (Takahashi, 2008).

In 2004, the Ministry of the Environment launched a programme that provides Grassland Restoration labels for agricultural products that are cultivated with wild-grass compost (Figure 10-5). A Grassland Restoration Label Producers' Association has now been established with vegetable farmers comprising the main membership, and a manual has been produced to promote the use of wild-grass compost (Takahashi, 2009). The activities of the association were expanded further following aid grants in fiscal year 2007 from the Ministry of Agriculture, Forestry and Fisheries' Pilot Program for Rural Landscapes and Natural Environment Conservation and Restoration.

By linking the mowing and use of wild grasses to the growth of plants and animals associated with food, grasslands become directly associated with essential human consumption. The dissemination of such information can enhance a sense of belonging in the community and generate more diversified partnerships. Relationships between producers and consumers, as well as an awareness of the conservation of grasslands and grassland species, ensure both sustainability and future capacity for expansion.

3. Promoting the use of grassland biomass

Grasses can be used for feed, fibre, fertilizer and fuel. However, much of the existing grassland is burned only in spring and is rarely mowed. It is proposed that if the grasslands are mowed for intended purposes and use of the grass biomass is promoted, this could result in the creation of a diversified grassland ecosystem (Nakabo, 2006; Takahashi, 2008).

Moreover, biomass from native grassland can be used as a clean and renewable energy resource. According to a survey conducted in Europe, the ratio of energy output to energy input for cultivated crops of the herbaceous perennial grass *Miscanthus sinensis* is as high as 30 or even higher, so it can be regarded as a crop that is able to provide alternative energy to fossil fuel (Takahashi, 2008). This energy value surpasses that for both wheat and rapeseed, and is far higher than the energy value of mown temperate (cool-season) grasses introduced into artificial grassland in Japan.

FIGURE 10-5: There is popular demand for vegetables grown in composted grassland cuttings.

Photo: Yoshitaka Takahashi.

The government of the city of Aso has embraced the cause of conserving the extensive Aso grassland and has launched an experimental project to develop a grass-based biomass-energy system. Established on a business footing, the system uses grass biomass resources in a way that conserves the grassland culture, lifestyle and ecosystem. Grasses have never before been used as an energy source in Japan, so data from the project are expected to establish a basis for the future use of the country's grass resources.

4. Developing next-generation leaders

As the grassland restoration activities have expanded, some associated problems have arisen. For instance, there is sometimes excessive reliance upon external volunteers, to the point where they become responsible for the entirety of the work. Volunteer leaders, who are concerned about the potential problems inherent in this trend, have proposed the development of a geographical information system that includes data such as work procedures, techniques and points to be noticed on landforms or wind conditions for each grassland area. Since fiscal year 2006, this information has been compiled into *noyaki* and *wachigiri* work charts (Aso Green Stock Foundation, 2008), which can be used by volunteers working in unfamiliar grassland and as an effective tool for fostering the development of new local leaders.

In addition, since fiscal year 2005 the local community, with the support of the Ministry of the Environment, has been investigating plants and locations identified by the grassland associations (Aso Grassland Restoration Committee, 2007). This allows the local community to review the status of the grassland and its environment and to reaffirm its value. Currently, the Ministry of the Environment investigates three or four common grassland areas each year. These investigations have already led some grassland associations to begin managing grassland on the basis of conserving native wild-grass fields. There is now a need to develop a feedback system to compare and assess the results of these activities in order to support their use in future land-management procedures.

Continuity of the current Aso grassland environment will undoubtedly require the sustained participation of present and future generations, since the people currently managing the grassland are growing older. For this reason, the Ministry of the Environment and key local museums are working with local grassland associations to implement programmes that allow children and other residents, both inside and outside the Aso region, to study and experience the grassland environment. The grasslands are also being developed as a tourism resource, for ecotourism and environmental education through exchange programmes with urban districts. Members of local grassland associations are continually seeking new opportunities for such educational programmes with a view to promoting voluntary conservation of grassland environments.

Strengthening cooperation among civil society, government and business groups

There is a continuing need to educate the Japanese public, particularly inhabitants of Kumamoto prefecture, regarding the ecological importance of the Aso grassland, and for this reason, there is also a need to establish an external advisory body or supporters group and cooperation across regions. This will not only help cooperation among residents and awareness of the importance of water conservation, but will also help to secure funds to support grassland restoration and preservation by people and enterprises for generations to come.

1. Creation of effective support groups

In October 2010, the Aso Grassland Restoration Committee launched a fundraising event attended by government and business leaders, citizens' groups and the media, mostly from the Kumamoto prefecture, which established the Millennium Board for Aso Grassland Restoration (Takahashi, 2012; Kumamoto Nichinichi Newspaper

Primula sieboldii E. Morren
[Primulaceae]

Japanese primrose is a species endemic to East Asia. It grows in damp meadows, its creeping rootstock building large clumps that cover the earth with blossoms in late spring. Although this plant has been bred and cultivated as a garden flower from at least the 16th century in Japan, it faces difficulties in maintaining a viable wild population on the Aso grasslands.

Company, 2013). The Millennium Board was formed to uphold the principle that the Aso grassland should be conserved by the Japanese people and residents of Kumamoto prefecture, who enjoy its many benefits.

The board has since supported the Aso Grassland Restoration Committee in its conservation activities, promotional events, campaigns to raise awareness of the crisis being faced on the grasslands and on going restoration efforts. It has also supported the Aso Grassland Restoration Fundraising Campaign, which aims to raise ¥100 million (US$838,212).[2] This campaign is not only promoting a 'private-sector environmental direct payments' programme, but is also laying the foundation for future agri-environmental strategies and securing permanent funding (Takahashi, 2012).

A new three-year second stage of support for the restoration of the Aso grasslands began in 2013. In this stage, business groups and media organizations from Fukuoka

– the biggest city on Kyushu – were included, to broaden the activities of the Millennium Board. Nineteen delegates, including the Governor of Kumamoto Prefecture, Ikuo Kabashima, attended the board's first meeting in August 2013. At this meeting it was decided that all of Kyushu – the southernmost of Japan's four main islands – would support the Aso grasslands because the area fostered assets such as tourism and water supplies. As well, fundraising efforts by working groups were directed towards securing permanent funding and supporting efforts to list the grasslands as a World Cultural Heritage Site.

2. Resolution to support grassland restoration

As part of the Aso Grassland Restoration Fundraising Campaign, cooperation was obtained from numerous enterprises and groups both within and outside Kumamoto prefecture. Fundraising products were developed, including a charitable prepaid card, time-deposit accounts and sponsored drinks-vending machines. However, natural disasters, including the devastating Great East Japan Earthquake, intervened and fundraising activities had to be suspended. Nevertheless, more than ¥70 million was raised in three years (Aso Grassland Restoration Committee, 2013).

The funds were used for several activities, including the introduction of Japanese red cattle to farmers, re-introduction of burning of abandoned grassland, management of volunteer deployment and grassland environmental training. A system was established to manage the funds, including the involvement of fundraising committee members chosen from the public, to offer advice.

Responses to a questionnaire completed by donors suggested that more than 80% of respondents intended to continue donating and even indicated their interest in increasing their donations. This showed that, when people were unable to participate directly, for example, by volunteering, fundraising could be an effective strategy to obtain indirect support.

The people of Kumamoto prefecture contributed more than 90% of the amount raised in the first fundraising event. The second three-year fundraising event sought support from the entire Kyushu region. This campaign focused on Northern Kyushu – including the city of Fukuoka – and was made possible by cooperation and support from the Millennium Board.

3. Integrated collaboration between the public and private sectors

Organizations such as the Aso Grassland Restoration Committee and the Millennium Board have become established centres for discussions on grassland restoration and dissemination of information. They also pursue various activities ranging from volunteering to fundraising. The Kumamoto prefecture and local municipalities have also shown interest in grassland restoration. In 2012, Kumamoto prefecture announced the 'Kabashima Initiative,' a strategy for regional development by promoting the restoration and use of grasslands. In doing so, it suggested that instead of its existing indirect support via tourism, environmental administration and agriculture and livestock breeding, it would provide direct support for grassland restoration.

In 2013, Kumamoto prefecture established the Grassland Restoration Vision, integrating the relevant measures required for grassland restoration (Kumamoto Prefecture, 2013). At the same time, a 'General Strategy for Revitalization of the Aso Region Using the Grassland of a Thousand Years' was announced, outlining measures to be undertaken by local government organizations and groups in the area (Aso City et al., 2013). In addition, approval was given by the Cabinet Office for declaration of the eight municipalities of the Aso region as a 'Comprehensive Special Zone for Creative Use and Inheritance of Grassland with a Thousand Years' History (the Special Zone of Grassland)'. These programmes are now part of a broad framework for the Aso Grassland Restoration Committee. Since associations between the committees and programmes have been clarified (Figure 10-6), an integrated public and private approach to grassland restoration has been achieved, involving close cooperation between government, private sector and citizens' groups.

In May 2013, the Aso region, comprising seven municipalities, was listed as a Globally Important Agricultural Heritage System, certified by the Food and Agriculture Organization of the United Nations, and in September 2014, the Aso region, comprising eight municipalities, became a member of the Global Geopark Network. Efforts are now being made to have the grassland listed as a World Cultural Heritage Site. It is hoped that these regional restoration measures, which involve a reassessment of the value of the grassland, will facilitate their sustainable use and conservation.

Formation of a national consensus

The conservation of grasslands, whether in Aso or other areas of Japan, cannot be successful unless the grasslands are used in a sustainable way. If profit is generated by cutting and collecting the grasses or by allowing grazing, the grassland will inevitably

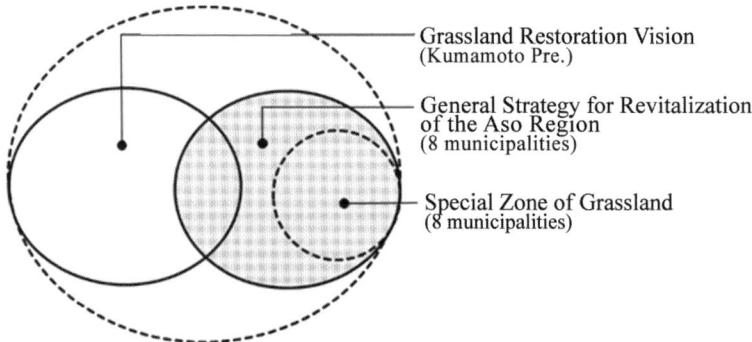

FIGURE 10-6:   Overall plan for Aso grassland restoration.

Source: Aso Grassland Restoration Committee (2013).

be regarded as valuable and protected as a base for industry. However, if grassland use is not economically feasible, discussion of a system of payments for environmental services may be useful, while gaining broad public understanding.

Services provided by biodiversity can be considered a public good, and every citizen may express his or her intention to conserve important living creatures and landscapes, or to pay money for that purpose, even if he or she does not enjoy direct benefits from grasslands. In order for it to gain public support, the idea needs to be presented as an easily understandable concept. Only then is it possible to establish sufficient support systems, e.g. through direct payments or a certification system for environmentally sustainable land-use practices.

A system for direct payments to farmers who agree to adopt practices that allow the coexistence of wildlife has not yet been developed in Japan. However, the biodiversity strategy of the Ministry of Agriculture, Forestry and Fisheries', released in July 2007, clearly states that the role and responsibility of agriculture is to 'give rise to a diversity of living things' (Ministry of Agriculture, Forestry and Fisheries, 2007). The third phase of the National Biodiversity Strategy was also formulated in that year (Ministry of the Environment, 2008). In both of these strategies, grasslands are expected to play an important role as refuges for wildlife and as buffer zones in mountain and village areas to promote the preservation of Japan's biodiversity through the recycling and reuse of grass biomass resources.

Whereas many countries and regions in Europe and the United States have established environmental guidelines for grassland management, Japan has not yet followed suit. A serious constraint is the low degree of conservation awareness at local level. In accordance with the compartmentalized public administration system in Japan, grassland management is under the jurisdiction of the animal husbandry administration of the Ministry of Agriculture, Forestry and Fisheries. However, grasslands provide a wide range of benefits and services with far more than livestock-raising value (Figure 10-7). There is a need for public relations strategies and activities designed to persuade a wide variety of beneficiaries and local residents of the vital importance of preserving biodiversity through grassland conservation. With that goal in mind, simple and objective biodiversity indices should be developed at field level. This will

Viola orientalis W. Becker [Violaceae]

A pansy-like plant of the violet family, this is one of the northern China-Korea tertiary relict species growing at Aso. Its bright yellow blooms contribute to the grasslands' profuse show of flowering plants.

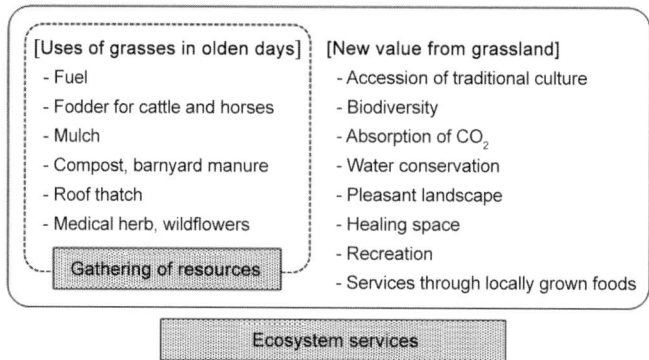

[Uses of grasses in olden days]
- Fuel
- Fodder for cattle and horses
- Mulch
- Compost, barnyard manure
- Roof thatch
- Medical herb, wildflowers

Gathering of resources

[New value from grassland]
- Accession of traditional culture
- Biodiversity
- Absorption of CO_2
- Water conservation
- Pleasant landscape
- Healing space
- Recreation
- Services through locally grown foods

Ecosystem services

FIGURE 10-7:   Diversified ecosystem services provided by grassland.

be essential if a system of direct payments or certification is to be established for agricultural products that help to preserve biodiversity.

Conclusions and policy implications

In order to manage the Aso grasslands, coordinators are needed who can foster agreements between urban residents (volunteers) and farmers (local residents). Fortunately, the volunteers who have been supporting burning activities to date have attained a high level of approval and trust from local residents. Accordingly, the government will be expected not only to rely on the efforts of the public sector and NGOs, but also to encourage participation by more diversified stakeholders and thereby play an important role in management and operations. A major factor in future success will be how local government authorities can use the various organizations associated with the Mount Aso region to gather public financial support for sustaining the management activities.

The Aso Grassland Restoration Committee is a partnership of diversified members whose ability to develop an overall concept despite their differing opinions and attitudes is noteworthy (Aso Grassland Restoration Committee, 2007, 2014). In order to sustain restoration efforts, the leaders of conservation activities establish firm roots in local communities and coordinate a division of responsibilities among all members on the basis of collective agreements. The scope of the Aso grassland restoration also extends to the Aso-Kuju National Park, so the restoration effort is inseparably related to the management and operation of the National Park. The knowledge gained thus far regarding the cooperation of diversified participants is expected to evolve further and become a concrete system of park management in the near future.

We believe that the case of the Aso grassland restoration has important policy implications for the wider Asia–Pacific region. In many parts of South and Southeast Asia, shifting cultivation and – more generally – the use of fire in land management has been largely dismissed or even criminalized as an environmentally destructive practice of primitive upland dwellers. However, the Aso case study demonstrates that regular and controlled burning can actually enhance biodiversity and help to maintain multifunctional landscapes. The re-evaluation of these traditional management practices in Japan has garnered increased public efforts and triggered the formation of multi-stakeholder partnerships for the preservation and restoration of grasslands. Similar public movements to preserve swidden environments may emerge in other parts of the Asia-Pacific region, when the wider public in these countries can be convinced that traditional farming practices – which have created unique cultural landscapes and sustained agro-biodiversity over many decades – still have an important role to play in a rapidly modernizing world.

Acknowledgements

This chapter is an outcome of the research project 'Ecosystem Services from the Aso Landscape – Their Foundations, Economic Valuation and Support Programs' (04/2013 to 03/2016), funded by the Japanese Society for Promotion of Science (JSPS) and headed by Hiroshi Yokogawa (JSPS Grants-in-Aid for Scientific Research (C), Project No. 25450349). Constructive comments from the editor of this volume, Malcolm Cairns, on an earlier version of this chapter are gratefully acknowledged.

References

Aso city, Minami-Oguni town, Oguni town, Ubuyama village, Takamori town, Minami-Aso village, Nisihara village, Yamato town, Aso Green Stock Foundation and Aso Design Center (2013) *General Strategy for Revitalization of the Aso Region Using the Grassland of a Thousand Years* (Japanese language), Aso Design Center, Aso, Japan

Aso Grassland Restoration Committee (2007) *General Plan of Aso Grassland Restoration* (Japanese language), Aso Grassland Restoration Committee, Aso, Japan

Aso Grassland Restoration Committee (2013) *Annual Activity Report 2012 of Aso Grassland Restoration* (Japanese language) Aso Grassland Restoration Committee, Aso, Japan

Aso Grassland Restoration Committee (2014) *General Plan of Aso Grassland Restoration* (Japanese language), Aso Grassland Restoration Committee (Stage 2), Aso, Japan

Aso Green Stock Foundation (2008) *Protecting Aso Grassland of 1,000 Years: Assorted Reports of Activities by Volunteers Assisting with Burning* (Japanese language), Aso Green Stock Foundation, Aso, Japan

Hayakawa, Y. (1981) 'Establishment of subclimax grassland and the number of farmers feeding beef cattle in western Japan: 1. Aso area' (Japanese language with English summary), *Bulletin of the Kyushu Agricultural Experiment Station* 21, pp273–288

Kawano, T., Sasaki, N., Hayashi, T. and Takahara, H. (2012) 'Grassland and fire history since the late-glacial in northern part of Aso Caldera, central Kyusyu, Japan, inferred from phytolith and charcoal records', *Quaternary International* 254, pp18–27

Kumamoto Nichinichi Newspaper Company (2013) *Crisis of Grassland* (Japanese language), Kumamoto Nichinichi Newspaper Company, Kumamoto, Japan

Kumamoto Prefecture (1998) *Wild Species of Fauna and Flora in Kumamoto Prefecture that are Important for Protection: Red Data Book Kumamoto* (Japanese language) Environment Conservation Section, Kumamoto Prefectural Government, Kumamoto, Japan

Kumamoto Prefecture (2013) *Aso Grassland Restoration Vision* (Japanese language), Kumamoto Prefecture, Kumamoto, Japan

Kurata, T. (2013) *Farming with Fire: Revaluing a Japanese Agricultural Tradition*, available online at http://ourworld.unu.edu/en/yakihata, accessed 20 October 2015.

Ministry of Agriculture, Forestry and Fisheries (2007) *MAFF Biodiversity Strategy* (Japanese language), Ministry of Agriculture, Forestry and Fisheries, Tokyo

Ministry of the Environment (ed.) (2008) *The Third National Biodiversity Strategy* (Japanese language), Ministry of the Environment, Tokyo

Ministry of the Environment (ed.) (2014) *Red Data Book 2014: Threatened Wildlife of Japan* (Japanese language), vol. 8: Vascular Plants, GYOSEI Corporation, Tokyo

Miyabuchi, Y. and Sugiyama, S. (2006) 'A 30,000-year phytolith record of a tephra sequence, east of Aso caldera, southwestern Japan' (Japanese language with English summary), *The Quaternary Research* 45, pp15-28

Miyabuchi, Y. and Sugiyama, S. (2012) 'Holocene vegetation history based on phytolith records in Asodani Valley, northern part of the Aso caldera, Japan', *Quaternary International* 254, pp73-82

Miyabuchi, Y., Sugiyama, S. and Nagaoka, Y. (2012) 'Vegetation and fire history during the last 30,000 years based on phytolith and macroscopic charcoal records in the eastern and western area of Aso Volcano, Japan', *Quaternary International* 254, pp28-35

Nakabo, M. (2006) 'A report from Aso: Cascade use of grassland biomass' (Japanese language), *Resources and Environment Measures* January 2006, pp86-90

Oki, A. (1993) 'Forms of forest exploitation: A comparative study of shifting cultivation in Java and Japan' (Japanese language with English summary), *Southeast Asian Studies* 30(4), pp457-477

Okubo, K. (2002) 'The present state in the study of biological diversity on semi-natural grassland in Japan' (Japanese language with English title), *Japanese Journal of Grassland Science* 48, pp268-276

Sei, S. (2006) 'Current status of grassland plants in Aso', in *Plants of Kyushu are Threatened* (Japanese language), proceedings of a public symposium at the 70th Plenary Meeting of the Botanical Society of Japan, Kumamoto, Japan, pp13–20

Sekido, A. (1994) 'The social-spatial organization of forest-lands and the subjective classification of its lands in a shifting cultivation village: A case study of Onaru village, Omogo-mura, Ehime prefecture', *Jimbun Chiri (Human Geography)* 46(2), pp144-165

Takahashi, Y. (2008) 'Possible utilization of native grass resources other than as a feed for livestock' (Japanese language with English title), *Japanese Journal of Grassland Science* 53, pp318-325

Takahashi, Y. (2009) 'Management and restoration of grassland landscape for species conservation: A case from Aso grassland' (Japanese language with English title), *Landscape Research Japan* 72, pp394-398

Takahashi, Y. (2012) 'New movement in collaborative activities for conservation and restoration of Aso grassland' (Japanese language with English summary), in The Folklore Society of Kyushu (ed.) *Aso and Grassland*, Koumyakusha, Miyazaki, Japan, pp5-27

Yamauchi, Y. and Takahashi, Y. (2002) 'Citizens' participation in conservation activities of Aso grassland' (Japanese language with English title), *Japanese Journal of Grassland Science* 48, pp290-298

Notes

1. The Bon Festival is a Japanese custom honouring the spirits of the ancestors. It has been celebrated for more than 500 years.
2. Japanese yen converted to US dollars at October 2015 exchange rate.

D. The complexities of implementing REDD+

The basic concept of REDD+ is to create a financial value for the carbon stored in tree biomass and to offer incentives for developing countries to reduce emissions from forested lands. The spectre of shifting cultivators cutting down trees to reopen fallows is therefore viewed with concern that forest carbon banks are being diminished and the carbon released into the atmosphere by burning.

Sketch based on a photo by Keith Barney.

11

THE VIABILITY OF SWIDDEN AGRICULTURE

and its uncertain role in REDD+

*Alan D. Ziegler, Daniel Borstein and Jia Qi Yuen**

Introduction

In the first part of this chapter, we summarize how swiddening, also called shifting agriculture and slash-and-burn agriculture, may still be a viable and relevant form of livelihood and food production in some locations and situations (Padoch and Pinedo-Vasquez, 2010; Ziegler et al., 2011). While our focus is largely on Southeast Asia, many of the ideas we discuss may be relevant elsewhere, such as in the tropical regions of Africa, Asia and the Americas, where swiddening is now transforming (van Vliet et al., 2012). In many parts of the world, shifting cultivation is often viewed negatively as a form of substantial forest degradation, which therefore contributes to anthropogenic climate change through the emission of greenhouse gases during burning (Cramb et al., 2009; Fox et al., 2009; van Vliet et al., 2012). Thus, swidden landscapes are increasingly being identified as potential sites of entry for projects designed around the United Nations' collaborative programme Reducing Emissions from Deforestation and Forest Degradation (REDD). This is an effort to create a financial value for the carbon stored in forests, offering incentives for developing countries to reduce emissions from forested lands and invest in low-carbon paths to sustainable development. 'REDD+' goes beyond deforestation and forest degradation, and includes the role of conservation, sustainable management of forests and enhancement of forest carbon stocks (UN REDD, 2015).

Many proponents envision that REDD+ incentives, based on a financial value for carbon stored in forests, could tempt forest-frontier communities away from swidden agriculture and towards other systems that potentially reduce emissions and/or increase carbon sequestration (Ziegler et al., 2012). We question the viability

* Dr Alan D. Ziegler, Professor, Geography Department, National University of Singapore; Daniel Borstein, a student at Dartmouth College, Hanover, NH, USA; Jia Qi Yuen, completed MSc studies in geography at the National University of Singapore in 2015.

of such programmes, largely because of the great uncertainty regarding the carbon outcomes of many of the main land-use transitions involved. This issue is explored in the second part of this chapter. In moving forward, we believe that a critical stage has been reached in establishing the nexus between swiddening and REDD+. There is an urgent need to advance understanding of the extent to which swiddening fits into the causes and cures of climate change, not just in terms of carbon emissions and sequestration, but in terms of how swiddeners are affected by the process. Therefore, the final part of this chapter explores unanswered questions with regard to REDD+ and swiddening.

Recognizing the contemporary role of swidden agriculture in transforming the landscapes of Southeast Asia

Rapid transitions in land use occurring in the humid tropical uplands of Southeast Asia often involve replacing shifting cultivation (Mertz et al., 2009). Swiddening was once the dominant agricultural system practised on sloping forested lands in the region (Figure 11-1), but it is now generally denounced by governments whose sole regard for the practice is to associate it with deforestation and the degradation of soil and water resources (Fox et al., 2009).

FIGURE 11-1:   Upland rice growing in a swidden in northern Laos. Swiddening was once the dominant agricultural system practised on sloping lands in the region.

Photo: Alan Ziegler.

Land uses that are replacing swidden include extensive, long-term cultivation of annual crops (e.g. cabbages, onions, beans, carrots, tea), tree crops grown in monoculture plantations (particularly rubber and oil palm), greenhouse-based horticulture (especially cut flowers) and livestock grazing (Schmidt-Vogt et al., 2009). These commercial systems are often favoured by antagonists of swiddening and agriculture entrepreneurs who claim they are more productive and cause fewer environmental problems than shifting cultivation (Ziegler et al., 2009b). However, we believe that in some situations swiddening is still a productive agricultural system that has a role to play in preserving biodiversity, conserving soil and water and mitigating climate change.

Criticisms of swiddening were commonly voiced by colonial governments during the last century, when there was increasing awareness of high rates of deforestation in the tropics (Fox et al., 2000). Concurrently, land degradation as a result of accelerated soil erosion and landslides was observed in areas where traditional long-fallow swiddening was being replaced by more intensive cultivation systems (Ziegler et al., 2009b). In the past it was nearly impossible to accurately identify those who practised swidden agriculture or distinguish between landscapes temporarily deforested for swidden and those permanently deforested for other reasons (Schmidt-Vogt et al., 2009; Mertz et al., 2009; Messerli et al., 2009). Thus, the total extent of land degradation deemed to be the result of swiddening was probably overestimated, contributing to widespread negative public sentiment (Fox et al., 2000). Regional governments responded by restricting swiddening. Upland areas were divided into forest and agricultural land-use categories based largely on slope class. Forest departments expanded to enforce bans on tree cutting and to limit cultivation on steep slopes. Ethnic minorities were resettled or, if nomadic, were encouraged to settle in one location; and most were encouraged (or forced) to adopt the cultivation of permanent cash crops, for example, cabbages in northern Thailand (Fox et al., 2009).

Land-use changes on former swidden lands continue to be driven by the emergence of regional markets for highland agricultural products and subsequent expansion of commercial agriculture (Fox et al., 2009). The customs of shifting cultivation communities are often closely linked to seasonal swidden activities; therefore substantial cultural changes are occurring as closer ties are forged between rural livelihoods and urban economic forces (Cramb et al., 2009). While we agree that commercial agriculture allows many former swiddeners to earn higher incomes, specialization in one product, for example, rubber production in Yunnan (China) or oil palm in Borneo, also increases the vulnerability of resource-poor farmers who are dependent on the weather and distant market conditions (Sturgeon, 2005; Cramb et al., 2009). Pursuit of income solely through commercial agriculture can also leave farmers, such as those producing coffee in the central highlands of Vietnam, with large debts (Cramb et al., 2009).

Efforts to increase agricultural productivity on sloping lands have led to permanent deforestation as well as contributing to other types of land degradation. For example, permanent agriculture can result in relatively high rates of erosion and increase the probability of landslides (Sidle et al., 2006; Ziegler et al., 2009b). Agricultural surfaces often generate overland water flow that accelerates erosion on sloping lands (Ziegler et al., 2001, 2004b, 2006, 2007; Turkelboom et al., 2008; Valentin et al., 2008). Moreover, the greatest contributors to high loads of sediment in rivers are often the road systems that make commercial agriculture possible (Ziegler et al., 2004a). Water quality in highly modified landscapes is degraded by the use of fertilizers and pesticides in areas where commercial agriculture has replaced swiddening (Ziegler et al., 2009b) (Figure 11-2). Extraction of surface and ground water for irrigation of annual crops is an increasing cause of stream desiccation (Forsyth and Walker, 2008). There is also the recognized possibility of reduced water availability across entire regions as a result of uncontrolled expansion of monoculture plantations of non-native tree species, such as rubber (Guardiola-Claramonte et al., 2008, 2010), which have spread across former swidden areas at relatively high latitudes and elevations in mainland Southeast Asia (Ziegler et al., 2009a).

Swidden agriculture requires relatively large forested landscapes and provides lower short-term economic returns per unit of area than intensive cropping systems, but it provides several benefits that permanent agriculture does not. Soil fertility, plant biomass and richness of plant species typically decline over time in swidden systems, yet the maintenance of lengthy fallow periods and short cropping periods

FIGURE 11-2:   Cultivated row crops growing on former swidden land in northern Thailand.

Photo: Alan Ziegler.

not only helps to slow this decline, but also fosters the regeneration of diverse secondary forests (Lawrence, 2004; Finegan and Nasi, 2004; Rerkasem et al., 2009; Lawrence et al., 2010). Swiddening is also particularly good at maintaining a diversity of human food sources (Padoch and Pinedo-Vasquez, 2010). For example, Conklin (1957) found 280 food crops – including more than 90 rice varieties – growing naturally and planted in the swiddens of the Hananoo people on Mindoro Island, in the Philippines. More recently, nearly 370 plant species were identified in the swiddens, home gardens, paddy fields and forests surrounding a Karen community in Thailand (Rerkasem et al., 2009). While swidden fallows have a substantially different structure than old-growth forest, they nonetheless provide essential habitats for wildlife within a managed landscape (Naughton-Treves et al., 2003; Zhijun and Young, 2003). Therefore, landscapes that contain swidden systems rather than those that are extensively converted to permanent agriculture could be expected to facilitate the conservation of many varieties of useful, endemic plants (e.g. rice varieties and medicinal plants), as well as wildlife populations (in the absence of hunting).

Transitions from traditional swiddening to many intensive cropping systems also reduce total carbon stocks. Above-ground carbon may decline by more than 90% when long-fallow swidden systems are replaced by rotational systems with short fallows, or by continuous cycles of annual crops (Bruun et al., 2009). Soil organic carbon is reduced by about 10% to 40% as a result of conversion to continuous annual cropping, with the largest declines associated with mechanically established plantations (Bruun et al., 2009). Uncertainty about how much carbon is sequestered in swidden systems, compared to other tree-based or biofuel plantations, is now an important obstacle for the implementation of mechanisms to Reduce Emissions from Deforestation and forest Degradation (REDD+) in Southeast Asia (Mertz, 2009).

Uncertainty about carbon-stock changes related to swidden transitions

Two recent meta-analyses reporting above- and below-ground carbon estimates for different land-use types highlighted the great uncertainty in the net total ecosystem carbon (TEC) changes that can be expected from many transitions, including the replacement of various types of shifting agriculture with oil palm, rubber and some other types of agroforestry systems (Ziegler et al., 2012; Yuen et al., 2013). Estimates of changes in total ecosystem carbon associated with these land covers are reported in Table 11-1. These analyses highlight the great uncertainty about the carbon stocks associated with many important land covers in the tropics, including swiddening. This uncertainty, which is greatest for below-ground carbon sources, including root masses and soil organic carbon, currently prevents accurate and reliable estimates being made of the carbon changes associated with many land-cover transitions, except for radical changes such as permanent forest removal (so-called deforestation) and forest restoration.

TABLE 11-1: Estimated ranges of above-ground carbon (AGC), below-ground carbon in the roots (BGC), soil organic carbon in the upper 2m (SOC), and total ecosystem carbon (TEC = ACG + BGC + SOC) for important vegetation types involved in land-cover conversions in Southeast Asia (Mg/ha).

Land covers [*]	AGC		BGC		SOC		TEC	
	min	*max*	*min*	*max*	*min*	*max*	*min*	*max*
FOR	40	400	11	74	75	225	126	699
LOF	30	210	5	26	68	205	103	441
OTP	15	200	5	33	65	196	85	429
RP	25	143	5	32	65	196	95	371
LFS	25	110	3	16	64	191	92	317
AGF	15	100	3	16	61	182	79	298
OP	17	69	4	22	65	196	86	287
IFS	4	50	3	16	62	187	69	253
GPS	3	35	2	4	66	198	71	237
SFS	2	22	3	16	59	178	64	216
PC	2	15	1	5	53	158	56	178

Notes: [*] Land covers are: FOR = forest; LOF = logged-over forest; OTP = orchard and tree plantation; RP = rubber plantation; LFS = long fallow swiddens; AGF = non-swidden agroforest; OP = oil palm plantation; IFS = intermediate fallow swiddens; GPS = grassland, pasture or shrub land; SFS = short fallow swiddens; PC = permanent crop land.

Forests have by far the highest estimated range for total ecosystem carbon of all land covers considered – up to about 700Mg of carbon per hectare (Table 11-1). The maximum values for logged-over forests (secondary forests) and various other tree plantations are lower (429 to 441Mg/ha), but still higher than for rubber plantations, oil palm plantations, non-swidden agroforests and long-fallow swiddens (287 to 371Mg/ha). At the bottom end of these estimates is permanent cultivation (178Mg/ha), with short-fallow swiddening, grasslands, pastures, shrublands and intermediate-fallow swiddening being slightly higher (216 to 253Mg/ha). Again, these values are maximum estimates for individual land covers. Estimates of total ecosystem carbon for any individual land cover often involve a three- or four-fold increase from the lowest to highest figures, not only because of variability, but also uncertainty in the estimates and variations in methods of determination.

Despite the uncertainty shown in the range of total ecosystem carbon estimates for the land covers listed in Table 11-1, we believe the following transitions should produce positive carbon outcomes (adapted from Ziegler et al., 2012): (1) abandonment of any agricultural system to allow permanent forest regeneration; (2) regeneration of logged forest into high-biomass 'primary' forest; (3) conversion of permanent croplands and short-fallow swidden systems to other land uses, including tree-based plantations, orchards, various agroforests and intermediate- to long-fallow swidden systems; (4) transition from intermediate-fallow swidden to long-fallow swidden systems, other agroforestry systems, rubber and other tree plantations; (5) regeneration of grasslands or pastures – particularly if they are degraded – or their replacement with orchards, rubber and timber plantations; and (6) conversion from oil palm plantations to rubber plantations, orchards, or other tree-based plantations.

In contrast, the data also suggest the following transitions could produce negative carbon outcomes: (1) logging of high-biomass primary forest; (2) conversion of primary or other high-biomass forest into any type of agriculture or plantation; (3) conversion of any land-cover type, except short-fallow swiddening, to permanent croplands; (4) intensification of any type of shifting agriculture by shortening the fallow period and/or increasing the length of the cropping period; (5) replacement of long-fallow swiddening by permanent croplands (this may also apply to some intermediate-fallow swiddens); and (6) conversion of rubber to oil palm, permanent croplands, or short- and intermediate-fallow swidden systems and probably to grasslands, as well (Ziegler et al., 2012).

The data lead us to believe that many other land-cover or land-use transitions would produce uncertain or potentially neutral carbon outcomes: e.g. (1) transitions between short-fallow swidden systems and permanent croplands; (2) land-cover changes between or among long-fallow swidden, other agroforestry systems, and rubber; and (3) land-cover changes between or among intermediate-fallow swiddening, grasslands, pastures, shrub lands and oil palm plantations (Ziegler et al., 2012).

Many of these transitions that are now underway throughout the tropics lie at the heart of REDD+ debates. Of interest to this chapter are swidden systems with fallow periods that are long enough to allow (natural) forest regeneration. These systems are probably more beneficial than was assumed in the past, particularly when compared with permanent types of cultivation or plantation agriculture where entire plots are cleared for extended periods of time (Fox et al., 2013). Therefore, programmes that encourage land-cover conversion away from long-fallow shifting cultivation to other, more cash-crop-oriented systems producing ambiguous carbon-stock changes – including oil palm and rubber plantations (Figure 11-3) – may be counter-productive with respect to climate change mitigation (Ziegler et al., 2012). In some cases, lengthening fallow periods may increase the carbon benefits of an existing shifting cultivation system (Ziegler et al., 2012). Here, we recognize the irony that some REDD+ approaches seek to protect a forested landscape by eliminating the long-fallow systems that have inherently helped to shape the landscape in its present form.

Both meta-analyses mentioned above found that the paucity of data currently in existence highlights the need for additional field investigations to improve the precision of carbon-stock estimates (Ziegler et al., 2012; Yuen et al., 2013). These data are crucial for making informed recommendations or policy decisions regarding which land uses optimize or increase carbon sequestration. As some land-use

FIGURE 11-3: Rubber is now planted extensively on moderately steep hill slopes on former swidden lands in Xishuangbanna, Yunnan province, China.

Photo: Alan Ziegler.

transitions may negatively impact other ecosystem services, as well as food security and local livelihoods, the entire carbon and non-carbon benefit streams should be taken into account before prescribing transitions when the carbon benefits are, at best, ambiguous (Ziegler et al., 2012). The present inability to do this represents a major uncertainty in the viability of REDD+ programmes that attempt to include swidden agriculture systems, either by preserving them as is or by transforming them into more productive stationary systems.

Key questions at the intersection of swiddening and REDD+

Smallholder intensification, which can be summarized succinctly as farming more productively on a lesser land area, is a strategy that is gaining momentum for halting pressure on forests and potentially creating a 'win-win' outcome in the stand-off between development and conservation in forest-frontier areas (Ziegler et al., 2012; Byerlee et al., 2014). The key differences between the extensive and intensive land uses in question – in a general sense – are that swidden systems are spread across the landscape, use burning to clear plots of land and involve fallowing to restore nutrients and reduce weed pressure. Intensive systems, on the other hand, are stationary and often require the application of agrichemicals, use of hybrid or genetically modified seeds and irrigation to maintain productivity over time (Ziegler et al., 2009b). While smallholder intensification has long been a development strategy (Pirard and Belna, 2012; Byerlee et al., 2014), only recently is this approach being driven explicitly by the objective of sparing forests to combat climate change. This fully warrants an examination of how strategies for reduced forest degradation may be converging with calls for a new 'green revolution' – a model predicated on the expansion of new seeds and fertilizers (see Bezner-Kerr (2012) for a good overview). The potentially greater impacts associated with intensive practices, summarized earlier in this chapter, lead us to question agricultural intensification as an effective REDD+ instrument.

We believe there are several questions that should be asked of proponents of REDD+ programmes that affect swidden systems. Moreover, these questions should be answered before such programmes are implemented. For example, will the fact that various environmental discourses deny the legitimacy of swiddening in some landscapes affect its treatment in a REDD+ programme? We refer here to the myriad negative perceptions of the purported environmental impacts of swiddening that are commonly held by most governments and forest conservation agencies, while at the same time they seem to accept the impacts of permanent agriculture as a part of the development process (Ziegler et al., 2012; Fox et al., 2009). The impacts of roads are accepted in much the same way (Sidle and Ziegler, 2012). In the application of proposed REDD+ programmes, how will forest farmers be compelled or coerced to depart from their long-standing practices in favour of new crop technologies that ostensibly reduce land degradation? This evokes questions of social justice.

In general, there is a need for more research that focuses on the politics of knowledge production about degraded forests, to understand how intensification functions in the context of payments for carbon-sequestration schemes. Partner countries in the United Nations' REDD+ programme have been working on developing viable monitoring, reporting and verification systems, which are necessary to produce hard and accurate evidence of reductions in carbon emissions (Samek, 2011). The potential for local stakeholder participation in this process is also mentioned as a vital component in country proposals, but such processes do not necessarily leave room for local interpretations of what should be valued in a forest, i.e. biodiversity, rural livelihoods, and so on (Gupta et al., 2012). Likewise, instituting local management of forests may do little to allow for local autonomy in decision-making about how land is used (Beymer-Farris and Bassett, 2012).

Another question relates to how participatory approaches entrust farmers with the task of monitoring carbon stocks. This places value on their knowledge of forest carbon while diminishing the value of the knowledge that has long guided their particular swiddening practices inside the same forests. We find clear relevance in Goldman's (2005) emphasis on how global market-based environmental discourses depend upon the participation of a diverse array of national and local actors, ranging from political elites and technical experts to the tillers of the land.

To give this production of knowledge a wider application, we should bring science and policy together to make degraded forests part of a global scheme on payment for ecosystem services. Countries seeking to receive payments for reduced emissions must first ascertain present-day carbon stocks as a baseline, known as the 'reference emission level', against which to assess future changes in forest carbon under a REDD+ regime (UN REDD, 2014). As pointed out earlier, carbon stocks in almost all relevant land covers are difficult to measure, not only because of the difficulty in accessing below-ground components, but also because the systems as a whole are frequently changing, even as the total forested area on a landscape may remain the same (Houghton, 2005).

There is also a need to debate another fundamental issue: to what extent do we value the ability to sequester carbon over other environmental and socio-cultural benefits?

Conclusion

From both economic and environmental perspectives, swiddening may in some instances still be the most viable land use for farmers in forests (Fox et al., 2009; Padoch and Pinedo-Vasquez, 2010; Ziegler et al., 2011). Although government policies and market forces will almost certainly push shifting cultivators towards stationary and permanent agricultural practices, swiddening should be encouraged in areas where it is beneficial for the preservation of ecosystem services, as well as cultural identity (Ziegler et al., 2011). We believe REDD+ could play a role by encouraging the maintenance or rehabilitation of traditional long-fallow swidden systems that allow the regeneration of mature secondary forests, thereby producing both carbon-sequestration and livelihood benefits. These ideas seem to run counter to some emerging visions that see smallholder intensification, and subsequent sparing of forests, as viable REDD+ options. Currently, there are many uncertainties that hinder the development of viable "REDD" frameworks that involve swidden agriculture and the livelihoods of its practitioners. The initial obstacle is the development of reasonable standards of accuracy for carbon monitoring for use in global payment schemes. Another obstacle is agreeing on schemes that address issues other than climate change, such as preserving cultures, livelihoods, biodiversity and limiting negative environmental impacts associated with intensified agriculture. The key to achieving this will be viewing swidden landscapes at appropriate scales in space (landscapes) and time (decades), and drawing from the informed inputs of physical scientists, social scientists, conservation officials, policy-makers and affected local peoples.

Acknowledgements

The first part of this chapter is taken from 'Recognizing the contemporary roles of swidden agriculture in transforming landscapes of SE Asia', published in *Conservation Biology* in 2010 (Ziegler et al., 2011) and reproduced with the permission of the publisher, John Wiley and Sons. The second part borrows from ideas published in the papers 'Carbon outcomes of major land-cover transitions in

SE Asia: great uncertainties and REDD+ policy implications', published in *Global Change Biology* (Ziegler et al., 2012) and 'Uncertainty in below-ground carbon biomass for major land covers in Southeast Asia', published in *Forest Ecology and Management* (Yuen et al., 2013). We thank the following co-authors for their contributions to the original articles: Jefferson Fox, Edward Webb, Christine Padoch, Stephen Leisz, Rob Cramb, Ole Mertz, Thilde Bruun, Tran Duc Vien, Deborah Lawrence, Casey Ryan, Wolfram Dressler, Unai Pascual and Lian Pin Koh.

References

Beymer-Farris, B. and Bassett, T. (2012) 'The REDD Menace: Resurgent protectionism in Tanzania's mangroves forests', *Global Environmental Change* 22, pp332-341

Bezner-Kerr, R. (2012) 'Lessons from the old green revolution for the new: Social, environmental, and nutritional issues for agricultural change in Africa', *Progress in Development Studies* 12, pp213-229

Bruun, T. B., de Neergaard, A., Lawrence, D. and Ziegler, A. D. (2009) 'Environmental consequences of the demise of swidden agriculture in Southeast Asia: soil nutrients and carbon stocks', *Human Ecology* 37, pp375-388

Byerlee, D., Stevenson, J. and Villoria, N. (2014) 'Does intensification slow cropland expansion or encourage deforestation?' *Global Food Security* 3, pp92-98

Conklin, H. C. (1957) *Hanunoo Agriculture: A Report on an Integral System of Shifting cultivation in the Philippines*, Food and Agriculture Organization of the United Nations, Rome

Cramb, R. A., Colfer, C. J. P., Dressler, W., Laungaramsri, P., Quang, T. L., Mulyoutami, E., Peluso, N. L. and Wadley, R. L. (2009) 'Swidden transformations and rural livelihoods in Southeast Asia', *Human Ecology* 37, pp323–346

Finegan, B. and Nasi, R. (2004) 'The biodiversity and conservation potential of shifting cultivation landscapes', in G. Schroth, G. A. B. da Fonseca, C. A. Harvey, C. Gascon, H. L. Vasconcelos and A-M. N. Izac (eds) *Agroforestry and Biodiversity Conservation in Tropical Landscapes*, Island Press, Washington, DC, pp153-198

Forsyth, T. and Walker, A. (2008) *Forest Guardians, Forest Destroyers*, University of Washington Press, Seattle

Fox, J., Truong, D. M., Rambo, A. T., Tuyen, N. P. and Cuc, L. T. (2000) 'Shifting cultivation: A new old paradigm for managing tropical forests', *BioScience* 50, pp521-528

Fox, J., Fujita, Y., Ngidang, D., Peluso, N., Potter, L., Sakuntaladewi, N., Sturgeon, J. and Thomas, D. (2009) 'Policies, political-economy and swidden in Southeast Asia', *Human Ecology* 37, pp305-322

Fox, J., Castella, J-C. and Ziegler, A. D. (2013) 'Swidden, rubber and carbon: Can REDD+ work for people and the environment in Montane Mainland Southeast Asia?' *Global Environmental Change* 29, pp318-326, DOI: 10.1016/j.gloenvcha.2013.05.011

Goldman, M. (2005) *Imperial Nature: The World Bank and Struggles for Social Justice in the Age of Globalization*, Yale University Press, New Haven, CT

Guardiola-Claramonte, M., Troch, P. A., Ziegler, A. D., Giambelluca, T. W., Vogler, J. B. and Nullet, M. A. (2008) 'Local hydrologic effects of introducing non-native vegetation in a tropical catchment', *Ecohydrology* 1, pp13-22

Guardiola-Claramonte, M., Troch, P. A., Ziegler, A. D., Giambelluca, T. W., Durcik, M., Vogler, J. B. and Nullet, M. A. (2010) 'Hydrological effects of the expansion of rubber (*Hevea brasiliensis*) in a tropical catchment', *Ecohydrology* 3(3), pp306-314, DOI: 10/1002/eco.110

Gupta, A., Lovbrand, E., Turnhout, E. and Vijge, M. J. (2012) 'In pursuit of carbon accountability: The politics of REDD+ monitoring, reporting, and verification systems', *Current Opinion in Environmental Sustainability* 4, pp726-731

Houghton, R. A. (2005) 'Aboveground forest biomass and the global carbon balance', *Global Change Biology* 11, pp945-958

Lawrence, D. (2004) 'Land-use change, biodiversity and ecosystem functioning in West Kalimantan', in G. Gerold, M. Fremery and E. Guhardja (eds) *Land Use, Nature Conservation and the Stability of Rainforest Margins in Southeast Asia*, Springer-Verlag, Berlin, pp253-268

Lawrence, D., Radel, C., Tully, K., Schmook, B. and Sneider, S. (2010) 'Untangling a decline in tropical forest resilience: Constraints on sustainability of shifting cultivation across the globe', *Biotropica* 42, pp21-30

Mertz, O. (2009) 'Trends in shifting cultivation and the REDD mechanism', *Current Opinion in Environmental Sustainability* 1, pp156-160

Mertz, O., Leisz, S.J., Heinimann, A., Rerkasem, K., Thiha, Dressler, W., Pham, V. C., Vu, K. C., Schmidt-Vogt, D., Colfer, C.J.P., Epprecht, M., Padoch, C. and Potter, L. (2009) 'Who counts? Demography of swidden cultivators in Southeast Asia', *Human Ecology* 37, pp281-289

Messerli, P., Heinimann, A. and Epprecht, M. (2009) 'Finding homogeneity in heterogeneity: A new approach to quantifying landscape mosaics developed for the Lao PDR', *Human Ecology* 37, pp191-304

Naughton-Treves, L., Mena, J. L., Treves, A., Alvarez, N. and Radeloff, V. C. (2003) 'Wildlife survival beyond park boundaries: The impact of slash-and-burn agriculture and hunting on mammals in Tambopata, Peru', *Conservation Biology* 17, pp1106-1117

Padoch, C. and Pinedo-Vasquez, M. (2010) 'Saving swidden to save biodiversity', *Biotropica* 42, pp550-552

Pirard, R. and Belna, K. (2012) 'Agriculture and deforestation: Is REDD+ rooted in evidence?' *Forest Policy and Economics* 21, pp62-70

Rerkasem, K., Lawrence, D., Padoch, C., Schmidt-Vogt, D., Ziegler, A. D. and Bruun, T. B. (2009) 'Consequences of swidden transitions for crop and fallow biodiversity in Southeast Asia', *Human Ecology* 37, pp347-360

Samek, J. (2011) *REDD/REDD+ Monitoring Reporting and Verification Reference Material*, Sumernet Project, Stockholm Environment Institute, Bangkok

Schmidt-Vogt, D., Leisz, S. J., Mertz, O., Heinimann, A., Thiha, Messerli, P., Epprecht, M., Cu, P. V., Chi, V. K. and Hardiono, M. (2009) 'An Assessment of trends in the extent of swidden in Southeast Asia', *Human Ecology* 37, pp269-280

Sidle, R. C., Ziegler, A. D., Negishi, J. N., Abdul Rahim, N., Siew, R. and Turkelboom, F. (2006) 'Erosion processes in steep terrain: Truths, myths and uncertainties related to forest management in SE Asia', *Forest Ecology and Management* 224, pp199-225

Sidle, R. C. and Ziegler, A. D. (2012) 'The dilemma of mountain roads', *Nature Geoscience* 5, pp437-438

Sturgeon, J. (2005) *Border landscapes: The politics of Akha land use in China and Thailand*, University of Washington Press, Seattle

Turkelboom, F., Poesen, J. and Trébuil, G. (2008) 'The multiple land degradation effects by land-use intensification in tropical steeplands: A catchment study from northern Thailand', *Catena* 75, pp102-116

UN REDD (2014) *Emerging approaches to Forest Reference Emission Levels and/or Forest Reference Levels for REDD+*, REDD Web Platform: Forest Reference Levels and Forest Reference Emission Levels, United Nations Framework Convention on Climate Change

UN REDD (2015) *About REDD*, The United Nations Collaborative Programme on Reducing Emissions from Deforestation and Forest Degradation in Developing Countries, Geneva. Switzerland, http://www.unredd.net/about/what-is-redd-plus.html, accessed 25 April 2017

Valentin, C., Agus, F., Alamban, R., Boosaner, A., Bricquet, J. P., Chaplot, V., de Guzman, T., de Rouw, A., Januea, J. L., Orange, D., Phachomphonh, K., Do, D. P., Podwojewski, P., Ribolzi, O., Silvera, N., Subagyono, K., Thiebaus, J. P., Tran, D. T. and Vadari, T. (2008) 'Runoff and sediment losses from 27 catchments in Southeast Asia: impact of rapid land use changes and conservation practices', *Agriculture Ecosystems and Environment* 128, pp225-238

van Vliet, N., Mertz, O., Heinimann, A., Langanke, T., Pascual, U., Schmook, B., Adams, C., Schmidt-Vogt, D., Messerli, P., Leisz, S., Castella, J-C., Jorgensen, L., Birch-Thomsen, T., Hett, C., Bech-Bruun, T., Ickowitz, A., Vu, K. C., Yasuyuki, K., Fox, J. M., Padoch, C., Dressler, W.

and Ziegler, A. D. (2012) 'Trends, drivers and impacts of changes in swidden agriculture on tropical forest frontiers: A global assessment', *Global Environmental Change* 22(2), pp418-429

Yuen, J. Q., Ziegler, A. D., Casey, R. and Webb, E. L. (2013) 'Uncertainty in below-ground carbon biomass for major land covers in Southeast Asia', *Forest Ecology and Management* 310, pp915-926

Zhijun, W. and Young, S. S. (2003) 'Differences in bird diversity between two swidden agricultural sites in mountainous terrain, Xishuangbanna, Yunnan, China', *Biological Conservation* 110, pp231-243

Ziegler, A. D., Sutherland, R. A. and Giambelluca, T. W. (2001) 'Acceleration of Horton overland flow and erosion by footpaths in an agricultural watershed in Northern Thailand', *Geomorphology* 41, pp249-262

Ziegler, A. D., Giambelluca, T. W., Sutherland, R. A., Nullet, M. A., Yarnasarn, S., Pinthong, J., Preechapanya, P. and Jaiaree, S. (2004a) 'Towards understanding the cumulative impacts of roads in agricultural watersheds of montane mainland southeast Asia', *Agriculture Ecosystems and Environment* 104, pp145-158

Ziegler, A. D., Giambelluca, T. W., Tran, L. T., Vana, T. T., Nullet, M. A., Fox, J. M., Tran, D. V., Pinthong, J., Maxwell, J. F. and Evett, S. (2004b) 'Hydrological consequences of landscape fragmentation in mountainous northern Vietnam: Evidence of accelerated overland flow generation', *Journal of Hydrology* 287, pp124-146

Ziegler, A. D., Tran, L. T., Giambelluca, T. W., Sidle, R. C. and Sutherland, R. A. (2006) 'Effective slope lengths for buffering hillslope surface runoff in fragmented landscapes in northern Vietnam', *Forest Ecology and Management* 224, pp104-118

Ziegler, A. D., Giambelluca, T. W., Plondke, D., Leisz, S., Tran, L. T., Fox, J., Nullet, M. A., Vogler, J. B., Dao, M. T. and Tran D. V. (2007) 'Hydrological consequences of landscape fragmentation in mountainous northern Vietnam: Buffering of accelerated overland flow', *Journal of Hydrology* 337, pp52-67

Ziegler, A. D., Fox, J. M. and Xu, J. (2009a) 'The rubber juggernaut', *Science* 324, pp1024-1025

Ziegler, A. D., Bruun, T. B., Guardiola-Claramonte, M., Giambelluca, T. W., Lawrence, D. and Lam, N.T. (2009b) 'Environmental consequences of the demise in swidden agriculture in Southeast Asia: Hydrology and geomorphology', *Human Ecology* 37, pp361-373

Ziegler, A. D., Fox, J. M., Webb, E. L., Padoch, C., Leisz, S., Cramb, R. A., Mertz, O., Bruun, T. B. and Tran, D. V. (2011) 'Recognizing contemporary roles of swidden agriculture in transforming landscapes of Southeast Asia', *Conservation Biology* 25, pp846-848

Ziegler, A. D., Phelps, J., Yuen, J. Q., Lawrence, D., Webb, E. L., Bruun, T. B., Ryan, C., Leisz, S., Mertz, O., Pascual, U., Koh, L. P., Dressler, W., Fox, J. M. and Padoch, C. (2012) 'Carbon outcomes of major land-cover transitions in SE Asia: Great uncertainties and REDD+ policy implications', *Global Change Biology* 18, pp3087-3099

12

INVOLVING ALL LOCAL STAKEHOLDERS AND HOLDERS OF LAND-USE RIGHTS IN REDD+

Indigenous people and/or local communities in Indonesia

Sébastien de Royer, Leontine E. Visser, Meine van Noordwijk, Gamma Galudra and Ujjwal Pradhan*

Introduction

In the years since it was launched in 2008, the United Nations collaborative programme Reducing Emissions from Deforestation and Forest Degradation (REDD) has not only gathered substantial momentum, but has also stirred up a global debate that ranges from land tenure and land grabbing to the ability to prove carbon stocks and the rights of indigenous people.

In essence, the programme aims to create a financial value for the carbon stored in forests, offering incentives for developing countries to reduce emissions from forested lands and invest in low-carbon paths to sustainable development. 'REDD+' goes beyond deforestation and forest degradation, and includes the role of conservation, sustainable management of forests and enhancement of forest carbon stocks. By June 2014, total funding to support national REDD+ readiness efforts in Africa, Asia-Pacific and Latin America totalled nearly US$200 million (UN REDD, 2015). Organizations representing indigenous peoples and local communities see REDD+ as an opportunity to strengthen their territorial control and have demanded respect for their rights, territories and autonomy, along with adequate consultation processes. It is a field fraught with conflict, with contestants ranged behind very valid and substantial issues. For example, it is easy to state that any discussion of new plans for sustainable development should include all local stakeholders, and that planning should respect

---

* Sébastien de Royer, researcher on the Human Dimension of Climate Change, World Agroforestry Centre (ICRAF), Bogor, Indonesia; Dr Leontine E. Visser, Professor of Rural Development Sociology, Department of Social Sciences, Wageningen University, the Netherlands; Dr Meine van Noordwijk, Chief Science Advisor, World Agroforestry Centre (ICRAF), Bogor, Indonesia; Gamma Galudra, Policy and Tenure Specialist, World Agroforestry Centre (ICRAF), Bogor, Indonesia; and Dr Ujjwal Pradhan, Regional Coordinator, ICRAF Southeast Asia.

and support the purposes of all holders of land-use rights (Galudra et al., 2014a). Among the stakeholders, rights-holders deserve a special place. But who determines which rights are to be recognized? And who has the moral or legal entitlement to decide that a claim to rights can be ignored? These questions can be easily answered within the legal framework (constitution, laws) of a nation-state, but in many parts of the world, the historical process of formation of nation-states has been based on military and economic power and political domination, and many groups have fallen by the wayside. These groups still retain a sense of identity from their specific history and heritage and adhere to an internal system of rights, responsibilities and conflict resolution. Many of them have historically practised shifting cultivation with long fallow periods, intensive agroforestry and natural resource extraction – all of it associated with rituals and strong spiritual values.

Internationally, groups that do not fully identify with the dominant nation-states in which they live have found mutual support under the description 'indigenous people'. In some parts of the world the term indigenous (and its derivative noun indigeneity) refers to pre-colonial history and groups whose ancestors inhabited various territories before the arrival of European settlers. However, as it is currently understood, human history has been one of constant movement; of groups taking over from and blending in with those who came before, so European colonial expansion is just one of many such events and not a basis for an operational dichotomy between 'indigenous' and 'non-indigenous'. The primary criterion accepted by international convention is that of self-identification as being different. But being different from the dominant nation-state does not mean that indigenous groups are similar. Current procedures that are meant to recognize and involve all stakeholders in planning for sustainable development are challenged by the diversity of cases. These procedures tend to impose a certain preconceived idea of what indigeneity (the specific character of indigenous groups) is, or even what it should be.

This chapter is a short version of an analysis of indigeneity in the context of Indonesia, an archipelago of nearly 18,000 islands, rich in ethnic diversity, languages and religions, and with a complex pre-colonial, colonial and post-colonial history. [1] Recent efforts to reduce emissions from deforestation and (forest) degradation (REDD+) brought fresh impetus to the need to clarify indigenous peoples' rights. This also revealed some major challenges to these efforts, where the diversity of local groups did not match the existing paradigm of indigeneity. In this chapter, we will discuss:

- general aspects of being indigenous in Indonesia;
- a case study in West Kalimantan (Indonesian Borneo);
- historical concepts of *adat*;
- recent changes in the legal recognition of *adat*;
- evidence that groups are still falling through the cracks;
- REDD+ social safeguards and shifting cultivation; and
- lessons learnt that may be applicable elsewhere.

Being indigenous in Indonesia

The narrative discourse about who owns and controls the forest and the role and position of customary people's laws (*adat*) within state law in Indonesia began during the Dutch colonial era, continued after Independence and persists up to the present day (Thiesenhusen et al., 1997; Peluso and Vandergeest, 2001; McCarthy, 2005; Galudra and Sirait, 2009; Von Benda-Beckmann and Von Benda-Beckmann, 2011). During President Suharto's New Order regime (1967 to 1998) the government designated 120 million hectares of forest (roughly two-thirds of the country's land area) as a state forest zone (*kawasan hutan*) under the jurisdiction of the Ministry of Forestry (Contreras-Hermosilla et al., 2005). Under this regime, forest communities suffered strong pressure and marginalization in the name of national interest. Government encroachment on the village commons was perceived by customary communities as an illegal infringement of their *adat* rights (Von Benda-Beckmann and Von Benda-Beckmann, 2011). Indigenous agriculture was referred to as 'shifting agriculture' (*perladangan berpindah-pindah*), a pejorative term dating back to the colonial period. Traditional swidden agriculture based on dry land rice was perceived by government officials as being inefficient, unable to raise subsistence living standards, damaging to the environment, a source of forest fires and a major cause of deforestation (Peluso, 1995; Setyawan, 2010).

In the wake of the reform era (*Reformasi*) since 1998 and the subsequent democratization and decentralization processes that have taken place in Indonesia, there have been efforts to reinstate *adat* as an alternative source of meaning and legitimization for local governments and non-governmental organizations across the archipelago (Acciaioli, 2008; Bakker, 2008; Von Benda-Beckmann and Von Benda-Beckmann, 2013).

In 2001, the Indonesian Parliament decreed the reform of natural resources and land-tenure laws and policies in accordance with principles that recognize, respect and protect the rights of *adat*-law communities (*masyarakat hukum adat*). However, it did not give any concrete answer to the historical narrative discourse about the precise role of *adat* in land-tenure and natural resources management, its status as customary law or *adat* law (*adatrecht*), and the relation of *adat* institutional orders to the state order (Widiyanto and Mary, 2012). It was only in May 2013 that the Constitutional Court (decision no 35/PUUU-x/2012) removed state claims on *adat* forests, which are now referred to as customary forests (*hutan adat*). The decision implied that indigenous people were to be considered legal entities and customary forests declassified from being state forests. It opened up the necessary space to clarify and settle historical discourse and disputation about customary rights, as different from state rights, over forested land, especially in relation to the recent issue of carbon rights and benefit sharing in the context of REDD+ (Lyster et al., 2013).

By adopting the UN Declaration on the Rights of Indigenous People, REDD+ social safeguards (Box 12-1) focus on efforts to ensure that mitigation programmes related to deforestation and emission reductions from the forest do not harm local communities and indigenous peoples' rights, their access to land or their livelihood

---

BOX 12-1: REDD+ social safeguards

The global discussion on the implementation of REDD+ has recognized the need for social safeguards. The concept is being developed and set in operation by global, governmental and non-governmental or private actors who claim to act on behalf of indigenous and vulnerable communities by protecting them from infringements of their rights, knowledge and aspirations, while at the same time protecting non-carbon forest values. Financial institutions such as the World Bank have referred to measures to prevent and mitigate undue harm from investment or development activities, while the United Nations Framework Convention on Climate Change has recognized that REDD+ could exacerbate social injustice and environmental challenges. These concerns led to approval in 2009 of the Cancun Agreement, which specified the adoption of social and environmental safeguards to prevent adverse consequences (Jagger et al., 2012). The safeguards are a direct response to critiques and protests from civil-society organizations representing indigenous peoples and forest-dependent communities about the potential social risks arising from REDD+. The safeguards have been influenced by and have built upon available international instruments, especially the UN Declaration on the Rights of Indigenous Peoples. They are expected to generate long-term social co-benefits, prevent negative outcomes, ensure better livelihoods and avoid harm to local communities (Steni et al. 2010; McDermott et al., 2012, Chhatre et al., 2012; Visseren-Hamakers et al., 2012).

---

aspirations. Unfortunately, the single focus on indigenous peoples generates another problem in today's multi-ethnic societies, namely the potential exclusion of in-migrants whose rights derive from a different source and who cannot claim historical or *adat* rights to the land on which they depend for their livelihood. Recent recognition of indigenous rights and territories by the Constitutional Court therefore underscores the need to seriously consider the impact of the conceptualization of indigeneity in relation to land tenure and territorial conflict in Indonesia (Bakker and Moniaga, 2010; Hall et al., 2011).

Questions like 'who is in and who is out?' or 'who is considered indigenous and who is not?' are fundamental in the context of REDD+ and cannot be ignored. There could be the danger of puritanism in defining indigeneity as the single prerogative of customary law communities, at the expense of in-migrating forest dwellers (Bakker, 2008, p156; Fisher and Lyster, 2013, p189). As in the colonial era, the mapping and control over land again make indigeneity and indigenous land rights critical issues to be addressed, particularly in setting the REDD+ social safeguards into operation and effect. Identifying who is and who is not 'indigenous' (Box 12-2), and therefore entitled to claim rights over forested land, is a delicate exercise in an era when decentralization is triggering the revitalization of *adat* and ethnic-identity claims over land rights.

A group that identifies itself as indigenous does not earn a natural or given status, but a 'positioning which draws upon historically sedimented practices, landscapes and repertoires of meaning, and emerges through particular patterns of engagement and struggle' (Li, 2000, p151). This approach relates to the concepts of articulation and positioning developed by Hall (1996). Hall suggests in his articulation theory that collective identities, common positions or a shared interest such as self-identification as an indigenous group need to be seen as provisional and non–permanent. He argues that cultural identities always come from somewhere and have histories, but are not externally fixed in some essentialized past. Articulation is not just a connection, but a process of simplification and boundary–making, as well as creating connections (Slack, 1996). Identities are subject to the continuous 'play' of history, culture, power and positioning. Li uses this theory to demonstrate how certain groups in Indonesia identify themselves as indigenous during moments at which global and local agendas have been conjoined in a common purpose and, presented within a common discursive frame, realign the ways these groups connect to the nation, the government, and their own unique tribal place. These, she says, are the contingent products of agency and the cultural and political works of articulation.

A case study from West Kalimantan (Indonesian Borneo)

Self-identified customary communities are claiming legitimate rights over land and territories in many parts of Kalimantan, as they are elsewhere in Indonesia. People are contesting boundaries based on different definitions of indigeneity. The following case study shows that this is happening even between communities contesting each other's ethnic boundaries.

A recent study was conducted in two villages located in the Kapuas Hulu regency in the province of West Kalimantan, bordering Betung Kerihun National Park (Figure 12-1). The two villages, Menua Sadap, home to Iban Dayak communities

---

BOX 12-2: Self-identification

Indigeneity and self-identification set indigenous peoples apart from other local communities. The concept of indigeneity was framed in the UN Declaration on the Rights of Indigenous Peoples in general terms:

> Indigenous communities, peoples and nations are those which, having a historical continuity with pre-invasion and pre-colonial societies that developed on their territories, consider themselves distinct from other sectors of the societies now prevailing on those territories, or parts of them. They form at present non-dominant sectors of society and are determined to preserve, develop and transmit to future generations their ancestral territories, and their ethnic identity, as the basis of their continued existence as peoples, in accordance with their own cultural patterns, social institutions and legal system (Karoba,2007, pp106-107).

This definition clearly connects the concept of indigeneity to a specific place or territory. The definition is linked to efforts to draw boundaries on maps, associating claims with places. However, the current location of indigenous peoples and the mobility of immigrants from elsewhere mean that narrow interpretations of place and territory don't match current identities (Galudra et al., 2014b). In the Indonesian context, where people's mobility, especially in upland forested areas, has been the rule rather than the exception, this territorialized definition of indigeneity can easily lead to conflicts and competing claims among groups who may all have lived in a certain place at different points in time.

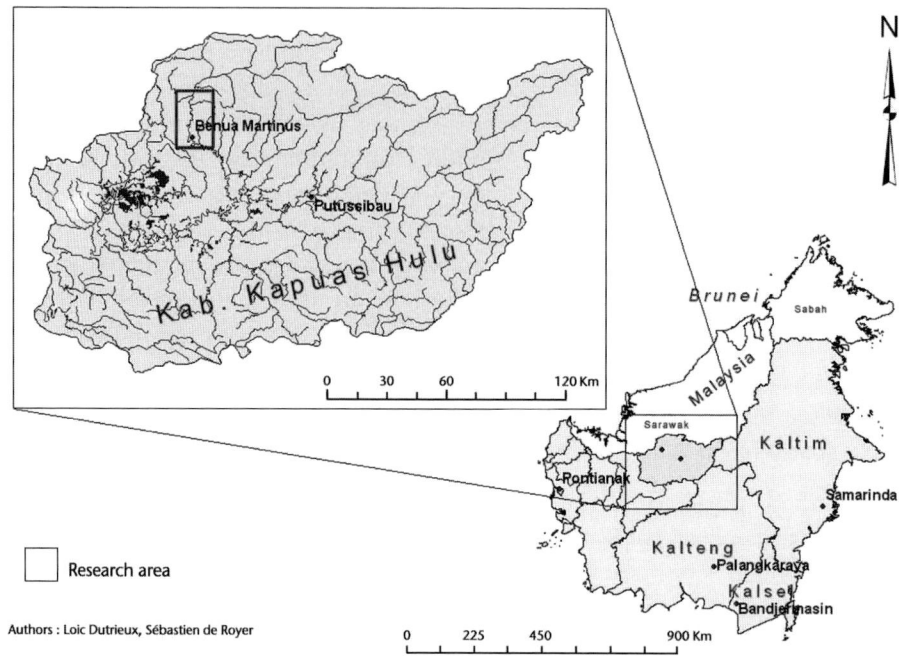

FIGURE 12-1: The study sites in Kapuas Hulu district, West Kalimantan, Indonesian Borneo.

and Pulau Manak, populated by Embaloh Dayak groups, are located on the upstream reaches of the Embaloh river, a tributary of the Kapuas river. Administratively, both villages are divided into hamlets (*dusun*), each of which consists of a longhouse (*rumah betang*) with its own forested territory to which the longhouse community holds exclusive customary land rights. A hamlet's territory is usually distinguished from that of the neighbouring longhouse by natural features. The people practise swidden cultivation to produce rice and form landscape mosaics consisting of natural forest, artificial forest, land in fallow and agricultural fields. Dependency on rice is high and its production is associated with strong rituals.

Both communities of longhouse people have enjoyed peaceful relations for generations, based mainly on trade and intermarriage. Both villages are also home to 'outsiders' who have settled in the villages more recently through marriage. These newcomers may be either Dayak or non-Dayak, and usually do not live in the longhouses. The staple foods of both communities are rice and forest products, while their main source of cash income is tapping of rubber from agroforests and seasonal collection of *tengkawang* nuts (*Shorea* ssp). Both villages are targeted for implementation of a pilot REDD+ project.

The Embaloh Dayak, or Tamanbaloh of Pulau Manak, belong to a sub-division of a broader ethnic group known in the anthropological literature as the Maloh Dayak, most of whom have inhabited the upper Kapuas region for at least 20 generations (King, 1976, p54). Embaloh Dayak live in nine communities on flatlands bordering the Embaloh river. They practice rice cultivation on the fertile river plains and claim

Shorea macrophylla (de Vriese) P. S. Ashton [Dipterocarpaceae]

Producer of *tengkawang* nuts, and from them Illipe oil, a dominant product from Kapuas Hulu forests.

to be the original inhabitants of the watershed. The good quality of the soil allows them to crop repeatedly on the same land after short periods of fallow. Because their land is flat and with easy access to watercourses, they have turned to irrigated wet rice agriculture and their swiddens have been converted into paddy fields. The Embaloh usually prefer to clear secondary forests rather than primary forest and concentrate their efforts on low-lying areas.

The other village is populated by Iban Dayak people who are spread all over northwestern Borneo, inhabiting large portions of the Malaysian state of Sarawak with a smaller population along the border in the province of West Kalimantan. The Iban living in our study site are descendants of migrants from Iban communities in the upper reaches of the Batang Lupar river in Malaysian Sarawak, who settled in the upper Kapuas region about 100 years ago. This migration of Iban people from the British-ruled province of Sarawak to West Kalimantan (then Dutch Borneo) began around the end of the 19th century (Wadley, 2003). King (1976) describes hostilities in which the upriver Iban of Batang Lupar were engaged with downriver Iban and other Dayak groups. Raiding Iban took advantage of the political boundary between Sarawak and Dutch Borneo to engage in forays then seek refuge on the other side of the border. King also advances the theory that Iban land and vegetation resources in their homeland were hard-pressed and there was a need to expand into new areas.

Unlike the Embaloh Dayak, Iban livelihood is based on swidden cultivation of rice in the uplands, usually in hillside swiddens using a long fallow cycle. They are usually located further upriver, where the land is higher and more sloping. The Iban are usually described as integral swidden farmers (Wadley, 2007), meaning that they engage in integral swidden systems including pioneer cultivation (i.e. cultivating large portions of old-growth forest, often integrated into traditional social, economic and ritual life) and cropping in established swidden fields (largely in secondary forest at various stages of maturity).

Most villagers are small landholders. Shifting cultivation is still commonly practised, implying that households move from one swidden plot to another every one to three years, generally returning to previously cultivated sites after a fallow period of variable length. For a long time the agricultural system was based on a pure shifting and land-clearing system with a cropping frequency varying between 20 and 25 years, regulated by customary-law leaders. However, recent demographic changes have accelerated competition for land, shortened fallow periods and increased the frequency of rotation and burning of forest. Pressure on the land over the past few decades has been accompanied by the introduction of pesticides and fertilizers, and households have been pressed to adapt their farming practices and cultivate fewer plots of land in a rotational system (*ladang gulir*).

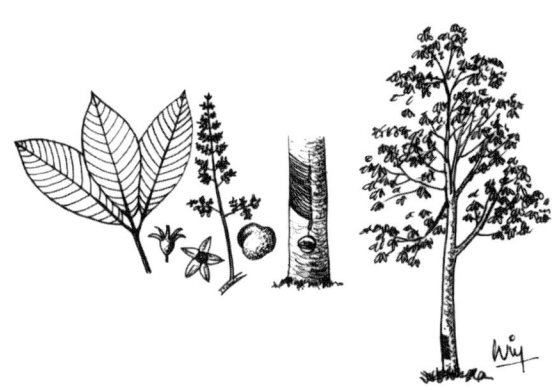

Hevea brasiliensis (Willd. Ex A. Juss.) Müll. Arg. [Euphorbiaceae]

Managed swidden fallows, or agroforests, often include rubber among fruit- and oil-producing trees.

Hillside swiddens and swidden fallows have also been converted to other uses, most frequently into rubber or mixed rubber-and-fruit gardens.

Besides being skilled cultivators, both communities are highly dependent on access to forest and forest gardens for secondary occupations and livelihood. Since a complete ban on illegal logging was implemented in the area in 2004, the economic situation has deteriorated drastically. Employment is almost non-existent and many villagers have to seek wage-labour opportunities as labourers in oil palm plantations, on the other side of the border in Sarawak. This is in keeping with their history of using their ethnic identity to enable circular labour migration across the international border into Sarawak, which dates back to pre-colonial times (Eilenberg and Wadley, 2009; Ishikawa, 2010).

Due to the lack of employment opportunities, people are highly dependent on subsistence strategies, and these make them even more reliant on access to their forests, since forest products provide them with their daily needs. Dayak communities actively manage the growth of both timber and non-timber products in swidden fallows of different ages and in other types of forest. Swiddening, for Dayak groups, is a form of rotational agroforestry, practised on a relatively broad regional basis, involving the management not only of swiddens but also of swidden fallows in various stages of regrowth, including standing forests (Peluso, 1995; Colfer et al., 1997).

Historical concepts of *adat*

Adat is a broad term commonly used throughout Indonesia to describe a social entity incorporating morality, customs, traditions, rituals and rules or practices of social life, as well as the underlying legal institutions. It is often translated as 'custom' or 'customary law', with conflicting implications that can be traced back to interpretations by different scholars during the Dutch colonial era, and which still apply (Von Benda-Beckmann and Von Benda-Beckmann, 2011).

In relation to land and natural-resources management, *adat* is a generic concept that comprises a wide understanding of customary and local practices by which people organize access to land and its resources for agriculture, hunting and the collection of forest products. Before the arrival of Europeans, many such customary practices existed throughout the various landscapes of the archipelago. *Adat* law (*adatrecht*), as described by Dutch scholars at the end of the 19th and early 20th centuries, represented only a limited part of the full scope of *adat*, comprising only those rules that were recognized, defined and codified by the colonial authorities and their collaborators, including legal anthropologists led by Van Vollenhoven (Von Benda-Beckmann and Von Benda-Beckmann, 2008, pp5–6). Van Vollenhoven and his fellow scholars from Leiden, in a substantial ethnological documentation enterprise, divided the archipelago of the Dutch Indies into 19 legal regions (*adatrecht* territories) and described them in 45 volumes (*Adatrechtbundels*). Dutch Borneo was one of Van Vollenhoven's *adatrecht* territories, known as the Dayak territory (*Adatrechtbundel* Vol. XLIV: Borneo).

Van Vollenhoven argued that the *adat* legal communities (*adat rechtsgemeenschappen*) with which he was concerned were autonomous communities that claimed to exercise a right of allocation over land (Burns, 1989; Von Benda-Beckmann and Von Benda-Beckmann, 2011). However, this recognition of local sovereignty over land control and access was challenged in 1870 by the Agrarian Law and the underlying Domain Declarations (*Domeinverklaringen*), which directly addressed the issue of land rights. The Agrarian Law proclaimed that all land was state owned, apart from land that was previously alienated to private holders under the civil code. All other land was considered to be state land (Peluso and Vandergeest, 2001).

As a result of *adat* codification, the Dutch administrators legalized the formation of territorial institutions called 'villages' and the notion of 'village land', over which a village head had the authority to allocate use rights. In that exercise, an *adat* law community needed to be associated with a circumscribed territory within the colonial schema. Territorialization therefore involved the mapping of the administrative territory (*landschap*). This demarcation process of forestland marked the beginning of a long forest delineation process throughout Indonesia, which is often still being challenged and which may threaten the implementation of REDD+ and the land-rights security of forest dwellers today. By fixing landscapes on maps, both the colonial government (Peluso and Vandergeest, 2001; Burns, 1989) and modern

Indonesian governments imposed a stereotyped notion of local group identity on a variety of social realities.

Recent changes in legal recognition of *adat*

In the early years of the Indonesian nationalist movement, *adat* was used as an ideological weapon in the struggle to make independence embrace the entire territory of the archipelago. After independence, President Sukarno relied on *adat* to support his ideology of national unity. During Suharto's New Order regime, the official discourse was that of a nation where all native Indonesians (*pribumi*) were, in a sense, indigenous, so there was no need for an 'indigenous people' concept. Native Indonesians were differentiated from *non-pribumi* (Chinese and other migrants) and cultural differences were accepted as long as they complied with the unification motto 'unity in diversity'. Claims to *adat* were strongly discouraged and efforts were made to restrict *adat* to the cultural domains of marriage customs, kinship and art.

The fall of the Suharto regime in 1998 was followed by an era of rapid reform promoting decentralization and aiming at democratization. This wave of reforms opened up the opportunity of negotiation for many indigenous peoples. An *adat* movement began to resurface throughout the archipelago (Davidson and Henley, 2007; Acciaioli, 2007). Indigenous peoples had the opportunity to recover from the injustices and dispossessions of the past. Decentralization and political freedom have since been a fertile ground for revitalization of local claims to political authority and natural resources on the basis of *adat* law, or *adat* societies. With the establishment of the Alliance of the Indigenous Peoples of the Archipelago (AMAN) in 1999, *adat* communities expressed their wish for the state to respect their traditional affiliation to their respective lands (Li, 2001). This wave of claims was supported and strengthened by a discourse on indigeneity in activist circles, which in turn was influenced by international indigenous peoples' movements and ideas imported by non-governmental organizations (NGOs) and other international advocacy groups. The creation of an indigenous identity in modern Indonesia was the result of NGO imagination stimulated by international laws and treaties (International Labour Organization Convention 169, the UN Declaration on the Rights of Indigenous Peoples, the Convention on Biological Diversity, and so on). In the evolution of international legal instruments seeking to define and recognize indigenous peoples, there has been an increasing trend to use self-identification.

In 2003, the request for a special law on indigenous peoples was formulated for the first time in Indonesia. Ten years later, in April 2013, the Bill on Recognition and Protection of the Rights of Indigenous Peoples (RUU PPHMHA) was presented to the government and adopted as draft law. The draft law has yet to be approved by the parliament, and public consultations are continuing throughout the country. However, the term that has been chosen to describe indigenous peoples is again *masyarakat hukum adat*, and not *masyarakat adat*, despite its colonial connotation, with emphasis on *adat* legal communities:

> A group of people who have been living in a certain geographical area for generations in the territory of the Republic of Indonesia because of the ancestral connection and a special relationship with the land, territory and natural resources, who own a customary governance system and *adat* law order on their territory.

The clear reference to *adat* law (*hukum adat*) and affiliation to territory in the draft law is consistent with the AMAN definition, which stipulates that indigenous people are:

> Communities that live on the basis of their hereditary ancestral origins in a specific customary territory, that possess sovereignty over their land and natural resources, whose socio-cultural life is ordered by customary law, and whose customary institutions manage continuity of their social life (AMAN, 1999).

Evidence that groups are still falling through the cracks

Access and rights to forest lands and village territory in the Kapuas Hulu villages have for generations been arranged according to *adat* institutional methods, which can be related to territorial *adat* rules and practices. These arrangements provide privileged access for historically well-established households to community resources while excluding others who have been integrated into the community more recently.

Villages also have designated forest areas where rules of access and use of forest resources, such as timber, are strongly controlled and monitored under village *adat* rules. These forest patches (*pulau*) fall into two categories. The first relates to the recognition by the community of the environmental services provided by tree cover in certain areas surrounding villages, such as water catchment and prevention of soil erosion. The second kind of *pulau* is related to religious and spiritual values and can be defined as sacred forests. These include a variety of sites that are either places of human deaths, burial sites, old settlements or areas inhabited by non-human spirits (Wadley and Colfer, 2004). These forest patches commemorate important historical and mythical events and provide villages with meaningful connections to the landscape. While felling trees and farming are prohibited in

Nephelium lappaceum L. [Sapindaceae]

Rambutan is a common component of managed swidden fallows in the study villages.

these *pulau*, they may provide important forest products such as fruits, leaves, medicinal plants and game.

During early migration movements in the Upper Kapuas watershed area (the study sites), Iban Dayak migrated to ecological niches that were left unoccupied by the Embaloh Dayak, further upstream. The Embaloh already possessed areas of fertile alluvial lowland along the Embaloh river, while the Iban were more interested in exploiting uninhabited tracts of old mature forests for their swiddens. The Iban opted for sites where the hills were accessible and large areas of primary forest were still

Arenga pinnata (Wurmb) Merr. [Arecaceae]

Commonly found in swiddens and fallows, this species is important as the source of aren palm wine.

available. Their territorial choice appeared to be complementary to the territorial preference of the Embaloh. This enabled the development over decades of non-hostile relationships between the two communities that were formalized by various *adat* laws and by-laws (King, 1976, p56).

However, during fieldwork in 2011 we found that, when it came to forest boundaries and ownership over forest resources, the two groups did not speak with one voice. Their different interests and the friction between them were kindled by recent discussions about the possibility of obtaining financial or economic profits from standing forest stocks of carbon through the possible implementation of REDD+. As soon as the topic of potential credit payments from village forests and surrounding agroforests was discussed, tensions rose. It is now commonplace for Embaloh to state that Iban Dayak are only guests and that they are not the legitimate owners of the forested land, due to the history of their migration into the area. In addition, Embaloh often accuse Iban of squatting on more land than permitted by customary agreements and of depleting forest resources by using unorganized farming systems. This criticism is similar to historical accusations that the Iban, by practising pioneer shifting cultivation, are depleting land resources. They are depicted as *mangueurs de bois*, or 'forest eaters' because unlike more sedentary Dayak groups, their pioneer shifting cultivation practices have led to massive forest degradation. Likewise, Dutch colonial officials described Iban swidden farming as 'plunder farming' due to their rapid depletion of forest and soil fertility (Padoch, 1982; Wadley, 2007).

Conversely, Iban often claim that Embaloh Dayak owe them eternal recognition for their readiness to protect Embaloh interests by engaging in conflicts with Melayu groups. In other words, both communities are reconstructing and reinventing history related to migration and settlement in order to justify their present and future claims over forest lands and resources, particularly in view of their new value in terms of potential carbon benefits. Dominant, socially well-structured and established groups with strong agricultural practices – such as the Embaloh– reaffirm their superiority over more marginalized groups with more extensive practices.

REDD+ social safeguards and shifting cultivation

In the debate about the benefits of REDD+ to local or indigenous communities, land tenure and secured access to forest products, as a positive outcome of avoided deforestation, is often seen as a priority (Mertz, 2009). But in many cases – such as in our Kalimantan case study – forest products may not be sufficient for shifting cultivators who are dependent on swidden production of staple food such as rice. Food production under shifting cultivation, despite its low overall productivity, has indeed been shown to be competitive with alternative income sources. So one of the challenges under the current discussion is to what extent REDD+ will address the complex reality of forest-dependent shifting cultivators, in terms of both securing reduced emissions and ensuring their livelihoods. Although little or no mention of swiddens has been made in the existing documents and literature of the United Nations Framework Convention on Climate Change, available references suggest that reducing emissions from degradation and deforestation are among the leading categories into which swiddens might fall because swiddens are seen in some cases as a cause of deforestation and in other instances as a cause of degradation (Mertz et al., 2012; Van Noordwijk et al., 2015). This categorization might affect poor swidden communities as it may provide a further reason for governments to ban shifting cultivation altogether. But the managed tree-system fallows practised by Dayak groups may in fact be beneficial and it would be appropriate if the REDD+ mechanism could reward the maintenance of long fallow systems rather than just focusing on natural forest conservation. However, it will be important to pay attention to the variety of shifting cultivation systems, knowing that some pioneer shifting systems are releasing more carbon dioxide than those systems that are more rotational.

Durio zibethinus L. [Malvaceae]

Durian agroforests, planted over many generations, engender a strong feeling of identity.

Lessons learnt that may be applicable elsewhere

Ethnic classification and territorialization were common practices during the colonial era, but independent Indonesia has not moved beyond them; they have not been abolished. In (old) Dutch Borneo, the identification of multiple legal systems associated with different ethnic groups created social units on the basis of customary law, and this added to the fixation and territorialization of identity. The creation of these legal categories has resulted in the granting of privileges to certain local practices and institutions perceived as indigenous or 'native' customary rights and certain lands as 'native land' under the authority of particular longhouses (Peluso and Harwell, 2001; Peluso and Vandergeest, 2001).

The colonial administrative practice of essentializing the social structuring of the landscape ignored the fact that in the colonial era as well as today, the social and territorial boundaries of ethnic groups are permeable and dynamic; they are distorted by changing social–political interactions that create contention and conflict (Gunawan and Visser, 2012). Ethnic groups presently attached to a particular territory or perceived by colonial authorities as 'original inhabitants' often originated elsewhere and have their histories rooted in migration, intermarriage, seasonal shifting cultivation practices, trade or armed conflict that forced them to move and settle anew, as in the case of the Iban Dayak.

Such essentialized, narrow descriptions of territorialized indigeneity are widely encountered in the current international discourse on indigenous rights and policy frameworks trying to define indigenous peoples. Indigenous identities are commonly defined by their strong ties to specific territories, such as they are in the UN Declaration on the Rights of Indigenous Peoples. While this notion is further strengthened by contemporary popular media, it is regularly countered by modern anthropological studies that highlight the social and geographical mobility of many indigenous peoples and oppose old-fashioned views of ethnic identity as being rooted in static place and territorial boundaries (Li, 2000; Davidson and Henley, 2007; Acciaioli, 2008; Persoon and Osseweijer, 2008; Visser and Adhuri, 2010; Hauser-Schäublin, 2013). One problem with this line of thinking is that concepts of state ownership and management rights are also 'rooted in static place and territorial boundaries', pushing anyone concerned about indigenous people's welfare to adopt the same approach.

In the wake of *adat* forest recognition in Indonesia, it will be important to move away from the colonial concept of identifying 'tribes' and 'native' lands with specific, culturally and geographically bounded peoples by adopting a more holistic approach to customary communities. In the Indonesian context, this will be particularly relevant since very few groups have inhabited a similar space throughout history. The problem with current advocacy is that indigenous groups, like 'the Dayak', are too often constructed according to a colonial discourse as homogeneous and harmonious swidden agriculturalist groups.

The importance of safeguarding social attributes has been recognized as a critical element for the successful implementation of REDD+. In the discourse on social safeguards under REDD+, indigenous peoples' advocates and international NGOs have supported the recognition of tenure security as a major issue to be addressed in order to protect vulnerable indigenous groups. It is seen as a way to demarcate clear boundaries and resolve conflicts over contested land (Chhatre et al., 2012). However, our study shows that the concept of boundaries may itself be an important part of the problem. The present situation is like a timebomb, where indigenous people as well as modern migrants are taking advantage of unclear and contested boundaries to try to appropriate more land and to justify their legitimate ownership rights over it. If these issues are underestimated and are left unattended by local and central governments, as well as international donors and NGOs, it may have a strongly negative effect on the success of any REDD+ project and prevent social safeguards being effectively set in operation.

References

Acciaioli, G. (2007) 'From customary law to indigenous sovereignty: Reconceptualizing *masyarakat adat* in contemporary Indonesia', in J. Davidson and D. Henley (eds) *The Revival of Tradition in Indonesian Politics: The Deployment of Adat from Colonialism to Indigenism*, Routledge, London and New York, pp295-318

Acciaioli, G. (2008) 'Strategy and subjectivity in co-management of the Lore Lindu National Park (Central Sulawesi, Indonesia)', in N. S. Sodhi, G. Acciaioli, M. Erb and A. K.-J. Tan (eds) *Biodiversity and Human Livelihoods in Protected Areas: Case Studies from the Malay Archipelago*, Cambridge University Press, Cambridge, UK, pp266-288

Adatrechtbundels (1952) *Part XLIV: Borneo*, Royal Netherlands Institute of Southeast Asian and Caribbean Studies (KITLV), Leiden

AMAN (1999) '*Atatan Hasil Lokakaryadalam Rangka Kongres Masyarakat Adat Nusantara I* (Notes on the Outcome of a Workshop on the Framework of the First Congress of Indigenous Peoples), Alliance of the Indigenous Peoples of the Archipelago, 14 March, Jakarta

Bakker, L. (2008) 'Politics or tradition: Debating *Hak Ulayat* in Pasir', In G. A. Persoon and M. Osseweijer (eds) *Reflections on the Heart of Borneo*, Tropenbos Series 24, Tropenbos International, Wageningen, the Netherlands, pp141-158

Bakker, L. and Moniaga, S. (2010) 'The space between: Land claims and the law in Indonesia', *Asian Journal of Social Science* 38(2), pp187-203

Burns, P. (1989) 'The myth of *adat*', *The Journal of Legal Pluralism and Unofficial Law* 21(28), pp1-127

Chhatre, A., Lakhanpal, S., Larson, A. M., Nelson, F., Ojha, H. and Rao, J. (2012) 'Social safeguards and co-benefits in REDD+: A review of the adjacent possible', *Current Opinion in Environmental Sustainability* 4(6), pp654-660

Colfer, C. J. P., Peluso, N. L. and Chin, S. C. (1997) *Beyond Slash and Burn: Building on Indigenous Management of Borneo's Tropical Rain Forests*, New York Botanical Garden, Bronx, NY

Contreras-Hermosilla, A., Fay, C. and Effendi, E. (2005) *Strengthening Forest Management in Indonesia through Land Tenure Reform: Issues and Framework for Action*, Forest Trends, Washington, DC

Davidson, J. S. and Henley, D. (eds) (2007) *The Revival of Tradition in Indonesian Politics: The Deployment of Adat from Colonialism to Indigenism*, Routledge, London and New York

Eilenberg, M. and Wadley, R. L. (2009) 'Borderland livelihood strategies: The socio-economic significance of ethnicity in cross-border labour migration, West Kalimantan, Indonesia', *Asia Pacific Viewpoint* 50(1), pp58-73

Fisher, R. and Lyster, R. (2013) 'Land and resource tenure: The rights of indigenous peoples and forest dwellers', in R. Lyster, C. Mackenzie and C. McDermott (eds) *Law, Tropical Forests and Carbon: The Case of REDD+*, Cambridge University Press, Cambridge, UK, pp187-206

Galudra, G. and Sirait, M. (2009) 'A discourse on Dutch colonial forest policy and science in Indonesia at the beginning of the 20th century', *International Forestry Review* 11(4), pp524-533

Galudra, G., de Royer, S., Agung, P. and Pradhan, U. (2014a) 'Planning for social justice', in J. Chavez-Tafur and R. J. Zagt (eds) *Towards Productive Landscapes*, ETFRN News 56, Tropenbos International, Wageningen, the Netherlands, pp212-219

Galudra, G., van Noordwijk, M., Agung, P., Suyanto, S. and Pradhan, U. (2014b) 'Migrants, land markets and carbon emissions in Jambi, Indonesia: Land tenure change and the prospect of emission reduction', *Mitigation Adaption Strategies for Global Change* 19(6), pp715-732

Gunawan, B. I. and Visser, L. E. (2012) 'Permeable boundaries: Outsiders and access to fishing grounds in the Berau Marine Protected Area', *Anthropological Forum* 22(2), pp187-207

Hall, S. (1996) 'On postmodernism and articulation: An interview with Stuart Hall, edited by Lawrence Grossberg', in D. Morley and K-H. Chen (eds) *Stuart Hall: Critical Dialogues in Cultural Studies*, Routledge, London, pp131-150

Hall, D., Hirsch, P. and Li, T. (2011) *Powers of Exclusion: Land Dilemmas in Southeast Asia*, Singapore University Press and Hawaii University Press, Singapore and Honolulu

Hauser-Schäublin, B. (2013) 'The power of indigeneity: Reparation, readjustment and repositioning', in B. Hauser-Schäublin (ed.) *Adat and Indigeneity in Indonesia: Culture and Entitlements between Heteronomy and Self-Ascription*, Göttingen Studies in Cultural Property (vol. 7), Universitätsverlag Göttingen, Germany, pp5-15

Ishikawa, N. (2010) 'State-making and transnationalism: Transborder flows in a borderland of western Borneo', in W. de Jong, D. Snelder and N. Ishikawa (eds) *Transborder Governance of Forests, Rivers and Seas*, Earthscan, London, pp31-50

Jagger, P., Lawlor, K., Brockhaus, M., Gebara, M. F., Sonwa, D. J. and Resosudarmo, I. A. P. (2012) 'REDD+ safeguards in national policy discourse and pilot projects', in A. Angelsen, M. Brockhaus, M. Kanninen, E. Sills, W. D. Sunderlin and S. Wertz-Kanounnikoff (eds) *Analysing REDD+*, Center for International Forestry Research (CIFOR), Bogor, Indonesia, pp301-316

Karoba, S. (2007) *Declaration on the Rights of Indigenous Peoples*, Ndugu Ndugu Research and Publishing Foundation, Yogyakarta

King, V. T. (1976) 'Some aspects of Iban-Maloh contact in West Kalimantan', *Indonesia* 21, pp85-114

Li, T. M. (2000) 'Articulating indigenous identity in Indonesia: Resource politics and the tribal slot', *Comparative Studies in Society and History* 42(1), pp149-179

Li, T. M. (2001) '*Masyarakat adat*, difference, and the limits of recognition in Indonesia's forest zone', *Modern Asian Studies* 35(3), pp645-676

Lyster, R., Mackenzie, C. and McDermott, C. (eds) (2013) *Law, Tropical Forests and Carbon: The Case of REDD+*, Cambridge University Press, Cambridge, UK

McCarthy, J. F. (2005) 'Between *adat* and state: Institutional arrangements on Sumatra's forest frontier', *Human Ecology* 33(1), pp57-82

McDermott, C. L., Coad, L., Helfgott, A. and Schroeder, H. (2012) 'Operationalizing social safeguards in REDD+: Actors, interests and ideas', *Environmental Science and Policy* 21, pp63-72

Mertz, O. (2009) 'Trends in shifting cultivation and the REDD mechanism', *Current Opinion in Environmental Sustainability* 1(2), pp156-160

Mertz, O., Muller, D., Sikor, T., Hett, C., Heinimann, A., Castella, J., Lestrelin, G., Ryan, C.M., Reay, D.S., Schmidt-Vogt, D., Danielsen, F., Theilade, I., van Noordwijk, M., Verchot, L.V., Burgess, N. D., Berry, N. J., Pham, T.T., Messerli, P., Xu J. C., Fensholt, R., Hostert, P., Pflugmacher, D., Bruun, T. B., Neergaard, A., Dons, K., Dewi, S., Rutishauer, E. and Sun, Z. L. (2012) 'The forgotten D: Challenges of addressing forest degradation in complex mosaic landscapes under REDD+', *Geografisk Tidsskrift* (Danish Journal of Geography) 112(1), pp63-76

Padoch, C. (1982) *Migration and its Alternatives among the Iban of Sarawak*, Verhandelingen van het Konginklijk Instituutvoor de Taal-, Land- en Volkenkunde 98, Martinus Nijhoff, S'Gravenhage

Peluso, N. L. (1995) 'Whose woods are these? Counter-mapping forest territories in Kalimantan, Indonesia', *Antipode* 27(4), pp383-406

Peluso, N. L. and Harwell, E. (2001) 'Territory, custom, and the cultural politics of ethnic war in West Kalimantan, Indonesia', in N. L. Peluso and M. Watts (eds) *Violent Environments*, Cornell University Press, Ithaca, NY, pp83-116

Peluso, N. L. and Vandergeest, P. (2001) 'Genealogies of the political forest and customary rights in Indonesia, Malaysia, and Thailand', *Journal of Asian Studies* 60(3), pp761-812

Persoon, G.A. and Osseweijer, M. (eds) (2008) *Reflections on the Heart of Borneo*, Tropenbos Series 24, Tropenbos International, Wageningen, the Netherlands

Setyawan, Dwi A. (2010) 'Review: Biodiversity conservation strategy in a native perspective; A case study of shifting cultivation at the Dayaks of Kalimantan', *Nusantara Bioscience* 2(2)

Slack, J. D. (1996) 'The theory and method of articulation in cultural studies', in D. Morley and K-H. Chen (eds) *Stuart Hall: Critical Dialogues in Cultural Studies*, Routledge, London, pp112-127

Steni, B., Indarto, G. B., Surya, M. T. and Indradi, Y. (2010) 'Beyond carbon: Rights-based safeguard principles in law', *Huma*, Jakarta

Thiesenhusen, W., Hansted, T., Mitchell, R. and Rajagukguk, E. (1997) *Land Tenure Issues in Indonesia*, Rural Development Institute (RDI), for USAID, on file at RDI, Seattle.

UN REDD (2015) About the UN REDD Programme, http://www.unredd.net/about/what-is-redd-plus.html, accessed 28 April 2017

van Noordwijk, M., Minang, P. A. and Hairiah, K. (2015) 'Swidden transitions in an era of climate-change debate', in M. F. Cairns (ed.) *Shifting Cultivation and Environmental Change: Indigenous People, Agriculture and Forest Conservation*, Earthscan, London, pp261-280

Visser, L. E. and Adhuri, D. S. (2010) 'Territorialization re-examined: Transborder marine resources exploitation in Southeast Asia and Australia', in W. de Jong, D. Snelder and N. Ishikawa (eds), *Transborder Governance of Forests, Rivers and Seas*, Earthscan, London, pp83-98

Visseren-Hamakers, I. J., McDermott, C., Vijge, M. J. and Cashore, B. (2012) 'Trade-offs, co-benefits and safeguards: Current debates on the breadth of REDD+', *Current Opinion in Environmental Sustainability* 4(6), pp646-653

Von Benda-Beckmann, F. and Von Benda-Beckmann, K. (2008) 'Traditional law in a globalising world. Myths, stereotypes, and transforming traditions', Van Vollenhoven Lecture 2008, Leiden

Von Benda-Beckmann, F. and Von Benda-Beckmann, K. (2011) 'Myths and stereotypes about *adat* law: A reassessment of Van Vollenhoven in the light of current struggles over *adat* law in Indonesia', *Bijdragen tot de Taal-, Land- en Volkenkunde* 167(2-3), pp167-195.

Von Benda-Beckmann, F. and Von Benda-Beckmann, K. (2013) *Political and legal transformations of an Indonesian polity. The nagari from colonization to decentralization*, Cambridge University Press, UK

Wadley, R. L. (2003) 'Lines in the forest: Internal territorialization and local accommodation in West Kalimantan, Indonesia (1865–1979)', *South East Asia Research* 11, pp91-112

Wadley, R. L. (2007) 'Slashed and burned: War, environment, and resource insecurity in West Borneo during the late nineteenth and early twentieth centuries', *Journal of the Royal Anthropological Institute* 13(1), pp109-128

Wadley, R. L. and Colfer, C. J. P. (2004) 'Sacred forests, hunting and conservation in West Kalimantan, Indonesia', *Human Ecology* 32(3), pp313-338

Widiyanto, S. and Mary, R. (2012) 'Refleksi perjalanan kembali Tap MPR N. IX Tahun 2001 tentang pembaruan agraria dan pengelolaan sumber daya alam' (Reflections on the way back, MPR No. 9 of 2001 on agrarian reform and natural resource management), *Huma*, Jakarta, 27 December 2011

Note

1. Please see De Royer, S., Visser, L. E., Galudra, G., Pradhan, U., and van Noordwijk, M. (2015) 'Self-identification of indigenous people in post-independence Indonesia: A historical analysis in the context of REDD+', *International Forestry Review* 17(3), pp282-297, for a more detailed version of this chapter.

II. THE IMPACT OF STATE POLICIES ON SHIFTING CULTIVATION

Shifting cultivators, like this woman in southern Laos, depend on sufficiently long fallow periods to suppress weeds during subsequent cropping. If the weed-suppression function of the fallow is lost, there is a steep increase in the need for labour in weeding and crop yields plunge. The government of Laos guaranteed the decline of shifting cultivation systems when it adopted a misdirected policy restricting fallows to no more than three years in duration.

Sketch based on a photo by Malcolm Cairns.

A. Shifting cultivation in the crosshairs of state policy

Most governments in Asia have had shifting cultivators in their crosshairs for many generations and have formulated policies aimed at suppressing their practices.

Sketch based on a photo by Laurent Chazée.

13

RETHINKING SWIDDEN CULTIVATION IN MYANMAR

Policies for sustainable upland livelihoods and food security

Oliver Springate-Baginski[*]

Introduction

Swidden cultivation, also known as shifting cultivation or, in Myanmar, *Shwe Pyaung Taung-ya*, describes a range of agroforestry-based cropping practices prevalent in the country's uplands. It contributes to the livelihoods of millions of citizens, for whom it is a solution to food security.

During the colonial era, forest departments, attracted by timber, competed for control of the uplands with swiddeners, and sought to restrict them. Even after independence, colonial-era prejudices and hostile assumptions persisted. But in recent years, perceptions have become more sympathetic, mainly due to scientific studies that confirm that under conditions conducive to swiddening, these systems are efficient, productive, sustainable and environmentally beneficial. Shifting cultivation is also an agricultural system that is particularly resilient to climate change.

However, conditions conducive to stable swidden systems are declining in many areas, particularly due to an adverse policy environment in which customary tenure is being undermined, fallow areas are being treated as 'wasteland' and 'land grabbing' is thereby facilitated.

The present-day policy challenge is how to support swidden cultivators in adapting their livelihoods to the changing conditions. This will require:

- acceptance and endorsement of swidden cultivation practices;
- assurance of secure tenure by (a) reinforcing customary authorities; (b) revising national land legislation; (c) handing back community forests; and (d) protecting shifting cultivators from land grabbing;

[*] Dr Oliver Springate-Baginski is a Senior Lecturer at the School of International Development, University of East Anglia, UK.

- providing technical support for sustainable intensification by building on local technical knowledge and innovations; and
- promoting rural enterprises for jobs and cash incomes.

What is swidden cultivation?

Swidden or shifting cultivation, also known as 'rotational agroforestry' and 'long-fallow forest cultivation', is a traditional agricultural practice from which many other farming systems have developed. The specific nature of upland environments, characterized by fragility, fed by rain, with lower soil fertility and slopes making cropping more laborious, presents different challenges than farming on the plains, and so it leads to location-specific adaptations (Figure 13-1). Swidden cultivation involves cultivating a series of plots, one after the other, sequentially. Abandoned plots are left fallow for several years, typically long enough for pioneer tree growth. Shifting cultivation is thus more complex than either permanent cultivation or tree farming. It is not nomadic, as the farmers return to abandoned areas, and it is not frontier 'slash and burn', where sites are subsequently put under sedentary or permanent cultivation, or tree crops.

There is currently a spectrum of shifting cultivation practices prevalent in upland Asia. It is known as *jhum* in Bangladesh and north east India, *shwe pyaung taung-ya* in Myanmar, *Khoriya* in Nepal, *Lunxi* di in China, and various local names in Indonesia and Papua. Across the Asia-Pacific region about 15% of the population is considered 'forest dependent'. Many of these people are shifting cultivators. In

FIGURE 13-1:   A swidden landscape in Chin State, western Myanmar.

Photo: U Kyaw Moe Aung.

Myanmar, estimates of the number of shifting cultivators range from two million to 20 million, possibly as many as half of Myanmar's upland population, cultivating and fallowing at various levels of intensity perhaps as much as a quarter of the country's land area (Win, 2009). However, with no nationwide studies we must still rely on coarse estimates.

How does swidden cultivation work?

The cultivating household returns to a fallowed patch after, in some cases, as long as 20 years or more, and clears the regrowth. Sometimes trees are only lopped or pollarded, and tree stumps may be left to help later regeneration. After the debris has dried, controlled burning clears the debris, and serves a number of beneficial functions. It releases nutrients (especially potassium), improves the soil structure, making it easier for planting, and decreases soil acidity. Burning also eliminates pest insects, plant diseases, soil microbial pathogens and weeds. Fire control may involve community mobilization to prepare fire lines and patrol. In some areas bio-charing is practised (i.e. covering the fuel materials with soil to reduce the presence of oxygen during combustion, leading to charred residues that are excellent for the soil structure).[1] In longer-fallow regimes, there is a lesser need for weed suppression, and sometimes there may be mulching of the debris rather than burning.

The cultivation phase may range from just one to several years. Cultivation practices can be highly complex, involving variously intercropping (other crops cultivated in the spaces between the main crop), relay cropping (a second crop started in between the plants of the first), green-manure cover crops, low or zero tillage (dibbling), inclusion of trees and so on. Shifting cultivation commonly involves a wider range of crops than permanent cultivation. As many as 50 crop species have been reported in one swidden in areas of northeast India.

After a patch has been cultivated for a few years, the soil fertility declines and weeds become more persistent. This presents a dilemma for all cultivators. Where land is the primary limiting factor, cultivators typically increase labour inputs to cycle nutrients and tackle the weeds. But where labour is the limiting factor and alternative land is available – conditions typical in hill areas – the most efficient use of labour is to move to another plot, leaving the abandoned plot fallow.

During the fallow period, the soil gradually recovers its condition (Figure 13-2). The fallow phase can involve active management interventions, including promotion of specific tree and/or shrub species for a range of purposes, and there is an important balance to be struck between the effectiveness of fallows (i.e. for fertility recovery) and productivity. Interventions can promote more effective fallows (e.g. nitrogen fixing, soil stabilization and soil organic-carbon promotion); or more productive fallows (e.g. fruit or wood products, or fodder); and can also serve other purposes such as hydrological, ecological or religious functions.

Swidden systems produce a distinctive patchy mosaic landscape, containing cultivated plots and regenerating fallows of different ages. Within swidden systems there are also typically a range of soil- and water-conservation measures at swidden sites as well as at landscape level. These can include contour bunds, stone walls, ridge and gulley protection or conservation of forest; zero tillage by planting with dibble sticks; nitrogen-fixing alder and other species, and so on. The balance between the cropping and fallow phases is critical: it determines whether or not the system suffers a net loss of nutrients and erosion.

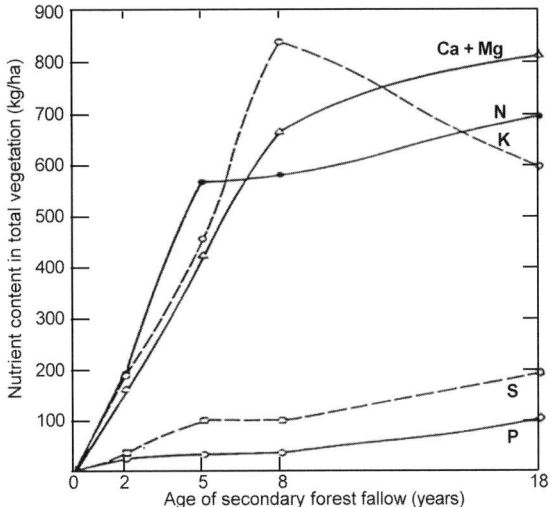

FIGURE 13-2:   Soil nutrient accumulation under shifting cultivation: Data from a secondary forest fallow in Yangambi, Zaire.

Source: Bartholomew et al. (1953).

If the fallow period is shortened there will be less time in which the soil recovery processes and vegetation successions can take place. If the fallow period continues to be reduced, an observable change occurs in fallow vegetation: secondary forests may be reduced to shorter, thinner-stemmed vegetation with fewer woody bush or jungle species. Other critical changes that are not so directly observable also occur in the soil, often resulting in declining crop yields (Figure 13-3).

Cultural integrity, social capital and social security

Swidden cultivation is typically performed by indigenous ethnic groups whose cultural practices include customary tenure institutions and authority structures, collective and reciprocal work, and indigenous technical-knowledge systems, including ethno-botanical knowledge.

Village-level customary institutions are essential to

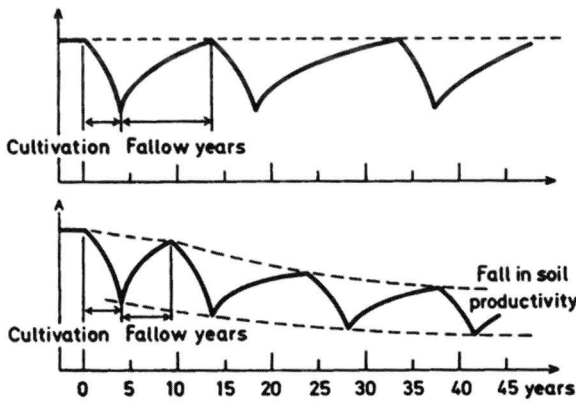

FIGURE 13-3:   The relation been fallow duration and soil productivity in shifting cultivation, illustrating the decline in productivity resulting from shorter fallow periods.

Source: Guillermin (1956).

regulate a swidden farming system as a community. These institutions involve shared norms, values and cultural traditions. A particularly important institution is common property land tenure. Land is typically held not as an 'open access' resource, but under regulated access, whereby portions are allocated to different households and their fallow areas are protected against squatting by others. Customary authorities also act to resolve conflicts over land. Reciprocal labour sharing is another important social institution. The social organization of labour increases efficiency particularly for land clearing and collective fire management. In some areas, patches are cultivated by the whole community, but apportioned out to individual households.

In these ways, cultural practices relating to swidden produce a material cultural heritage, including indigenous technical skills passed on to the next generation, and also a unique 'cultural landscape' involving a unique and rich biological diversity.

Swidden cultivation is found in upland areas across the World. In Asia, it is found across the mid-altitudes of the eastern Himalaya region and widely in upland Southeast Asia, as illustrated by the 'upland intensive mixed (4)' and 'highland extensive mixed (5)' farming-system categories in Figure 13-4.

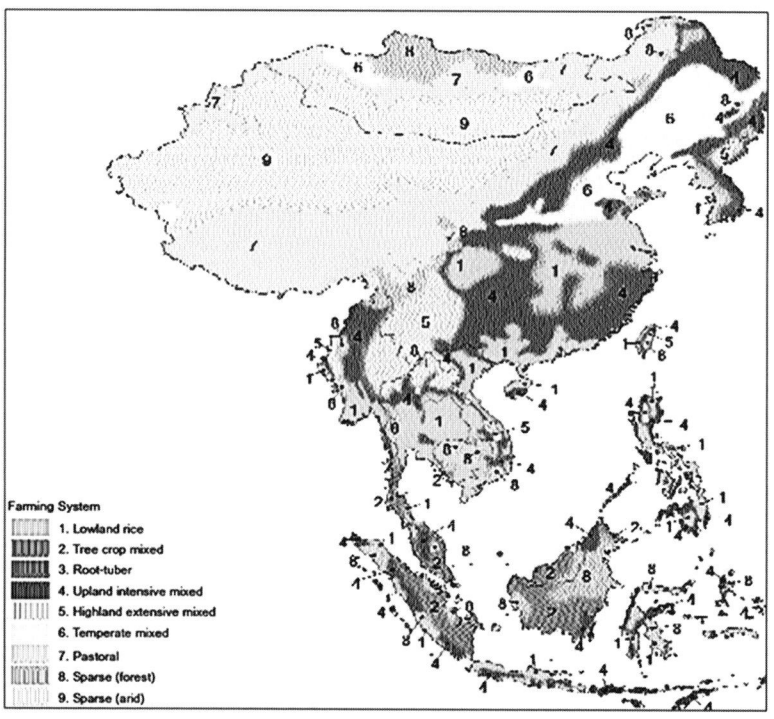

FIGURE 13-4: Farming systems in Southeast Asia.

Source: FAO (2010).

Is swidden cultivation bad? Science and logic

It was not until the mid-19th century that swidden cultivation began to be seen as destructive. As colonial forest departments entered into competition with shifting cultivators for control of the land, they began to characterize swidden practitioners as primitive, unproductive, even 'pre-agricultural', in contrast to the so-called 'modern' and 'scientific' management of foresters. Governments introduced policies to obstruct the practice and remove ethnic populations from valued forests. The colonial administrations imposed 19th-century euro-centric assumptions on the legitimate use of land. Many of these policies, and the hostile assumptions and attitudes on which they were based, persist to this day. Even though they are rarely based on more than a subjective preference, they have been selectively employed to unfairly quash the legitimacy of cultivators' rights in ways that would not be acceptable for lowland cultivators.

Myanmar's Forest Policy (1995) includes the standpoint:

> '[D]iscourage shifting cultivation practices which are causing extensive damage to the forests ....'

However, this seems to be superseded by Article 354 of the 2008 Constitution of Myanmar, which states:

> 'Every citizen shall be at liberty in the exercise of the following rights, ... (d) to develop their … culture they cherish ... and customs ...'

Despite this apparent softening of official policy, Myanmar has seen lengthy government efforts to control shifting cultivation. But the practice remains prevalent over large areas of the country.

Scientific studies of swidden cultivation

Any objective evaluation of swidden, by comparing it with other land uses, would need to compare specific criteria such as biodiversity impact, livelihood benefits and labour and land-use efficiency (bearing in mind that generic comparisons are difficult due to the diverse spectrum of location-specific practices). There are also surprisingly few comprehensive studies of shifting cultivation. However, several rigorous research studies in the tropics have pointed to the strength and resilience of many of these systems, the high returns to labour they offer and the species enrichment and biodiversity conservation they achieve.[2]

Comparing environmental impacts

Foresters often blame swidden cultivation for deforestation. But is shifting cultivation relatively more or less destructive of forests at site level than forestry harvesting or clearing of forests for permanent, sedentary agriculture? Probably less: a long-fallow swidden system is very similar to a forester's rotational tree-plantation system, but with an additional few years of food-crop cultivation between felling and replanting. It involves a more diverse mix of tree and other species. The Global Forest Resource Assessment conducted by the Food and Agriculture Organization of the United Nations included a category for 'temporarily unstocked forests' to encompass lands under forest management after felling and before regrowth (FAO, 2016). Swidden cultivated areas and early fallows could reasonably be categorized as 'temporarily unstocked forests'. Since foresters do not object on environmental grounds to their own tree felling it seems illogical that they should object to others doing the same.

Swidden cultivation undoubtedly has a greater impact on natural forests than rigorously implemented practices of low-impact selective logging within continuous-cover forestry. But low-impact logging, in Myanmar as elsewhere, has remained more a noble intention than a reality, as the Myanmar Timber Enterprise and local sawmills have been accused of widespread unregulated felling of forests.

In comparison to permanent, sedentary cultivation, swidden has lower environmental impacts: the agricultural ecology is far more biodiverse, in terms of genetic, crop and species diversity, especially during the fallow period, when forest regrowth provides diverse habitats (Table 13-1). Moreover, virtually no toxic external inputs are used in swidden cultivation, such as pesticides, herbicides and synthetic fertilizers, which are damaging to the soil and water systems across large areas. In recent years, agribusiness plantations have been spreading across hill areas of Asia. From an environmental point of view these are among the most damaging of all land uses, far more so than swidden: exotic monocultures provide no biodiversity habitat and often involve the heavy use of toxic inputs. In fact, agribusiness plantations in Asia are heightening the risk of species extinctions.

On a global scale, plantation agriculture is now the single greatest driver of deforestation (Foley et al., 2010), and in comparison, shifting cultivation has a relatively minor impact. Compared to permanently cultivated areas and even to tree plantations, a stable long-fallow shifting cultivation system provides higher forest cover.

Elaeis guineensis Jacq. [Arecaceae]

Oil palm agribusiness is one of the main drivers of land-grabbing in the highlands of Southeast Asia. So-called 'green deserts' of oil palm have very low biological diversity. But one hectare of oil palms produces about 20 tonnes of fruit that yields four tonnes of oil and 750kg of seed kernels, which in turn produce about 500kg of palm kernel oil and kernel meal that is used as livestock feed.

TABLE 13-1: Comparing the impacts of different land uses.

Land use	Initial impact of land-use change from primary forest	Biological diversity	Productivity and beneficiaries
Natural primary forest (e.g. 'protected area')		√ Very high	Range of ecosystem services and NTFPs to a variety of beneficiaries
Swidden cultivation	Reduced forest density to secondary forest	√ High	Food products, wood and NTFPs to swidden farming families
Settled agriculture	× Deforestation	× Low to very low	Food production benefitting farmers
Timber extraction from natural forests	√ Low degradation in MSS theory; medium to high degradation in MTE practice due to over-extraction	√ High to medium (variable)	Timber benefitting foresters and merchants
Tree plantations	× Deforestation and replacement with monocultures	× Very low	Timber and poles to benefit foresters and merchants
Agribusiness plantations	× Deforestation	× Very low	Commodities to benefit merchants

Notes: NTFPs=non-timber forest products; MSS=Myanmar Selection System; MTE=Myanmar Timber Enterprise.

In terms of biodiversity, studies suggest that, in some cases, swidden systems have higher plant- and wildlife-biodiversity levels than permanent forest (due to edge effects), and certainly have higher agro-biodiversity than modern farming systems. Swidden secondary forests and trees also provide valuable biodiversity corridors around protected areas.

Internationally, traditional shifting cultivation systems are not necessarily a major cause of forest loss. This suggests that greater attention needs to be given to other causes of deforestation, including timber trading and land conversion

Swidden systems also maintain relatively high carbon levels, unlike permanent agriculture and industrial plantations.

Livelihoods

A well-functioning swidden system can be a highly efficient use of labour that is more productive in hill areas than alternatives, and can lead to very good food security. Rotational agroforestry also supports, through customary institutions, social security and cultural integrity. Swidden cultivation is one of the most intensive and productive ways farmers can manage uplands, and provides substantial labour opportunities when compared, for instance, to agribusiness plantations, the profitability of which depends on minimizing labour costs, thereby creating unemployment and social dislocation.

As rural economies become increasingly incorporated into the cash economy, swidden production is gradually adapting through incorporation of some cash crops, such as elephant's foot yam (*Amorphophallus paeoniifolius*), corn and vegetables.

The current status of, and stresses on, Myanmar's shifting cultivation systems

Their nature and extent

It is not known how many swidden cultivators there are in Myanmar, or how widespread swidden is. Based on several approximate measures, a rough estimate of the swidden population is about seven million, suggesting that about half of the upland region's population of about 17 million are involved, although until a proper survey is conducted, this remains a guess.

There seem to be clear variations in the practice of swiddening by area. In Chin state, fallows remain for a relatively long time, as long as 10 to 15 years, and customary authorities there are robust. In Shan state, the long border with China has led to commercial initiatives that have changed cultivation incentives in more accessible areas, for instance to production of maize.

The main factors under which shifting cultivation systems were stable and sustainable, and which now give rise to adaptation, include low population density, extensive land and remoteness, especially from markets. These conditions have been changing rapidly across Southeast Asia in recent decades due to a range of stresses, and in Myanmar, these stresses are more recent and more rapid: population growth, enclosure of remaining wild lands for protected areas, the strengthened reach of the state to induce the adoption of permanent agriculture and the threat and actual emergence of 'land grabbing' for agribusiness. All of these factors are stressing shifting cultivation systems

Amorphophallus paeoniifolius
(Dennst.) Nicolson [Araceae]

Elephant's foot yam is a traditional subsistence crop planted by shifting cultivators in Myanmar. Access to markets has now led some swidden farmers to make this species a cash crop. The plant gives off a putrid smell, but the corms are an important part of Indian cuisines. It is also said to have a broad range of medicinal properties.

and pushing cultivators to lengthen the cultivation phase and shorten fallows. Also, partly due to the hostile political environment, there is an inter-generational decline in interest in shifting cultivation from many in the younger generation.

These stresses are leading to poverty in some shifting cultivation areas. However, we must be careful not to equate shifting cultivation with poverty *per se*, as the highest poverty incidence in Myanmar is in the area of highest agricultural productivity – the lower Ayeyarwady (UNDP, 2011).

Demographic change

Despite the lack of a reliable census, there seems to be gradually increasing numbers of people in Myanmar's uplands, even after taking outmigration into account. This is leading to generational splitting of family lands and overall hunger for land. There are no simple solutions, and intensification of cultivation may be needed in many areas. There may be an increase in outmigration, mainly of men, resulting in the feminization of agricultural labour. Comparing the number of people per unit of area employed by swidden and those supported by plantations, swidden clearly supports many more, as most plantation crops (e.g. oil palm and rubber) employ few people, so they are not a reasonable alternative for generation of employment.

Reduced access to land and growing tenure insecurity

Access to land has become increasingly competitive in recent years. Factors driving this include growing populations, expansion of protected areas and land grabbing by agribusinesses. These factors mean that the area of land available for cultivation has reached a physical and administrative limit, bringing about the so called 'end of the hinterland', which is now a common issue around the world (RRI, 2011).

Swidden cultivators have historically relied on customary tenure to manage their land. Yet with the decline in conflict within Myanmar, the state has been increasingly able to impose central governance and statutory regulation on previously remote areas (Table 13-2). Statutory ownership is vested with the Agriculture Department and administered by the Land Records Department, and recent reforms to land laws have made it easier for land to be allocated to business interests.

Weak or insecure tenure acts as a disincentive for cultivators to invest in sustainable land management. Thus, secure tenure is an important priority. The President's Office recently declared that occupiers of land in Myanmar would get tenure security – a very laudable statement of intent. However, swidden cultivators occupy each patch of land only temporarily, so it remains to be seen whether this tenure security will cover their fallows.

Market access, monetization and incorporation into the global economy

As the cash economy expands, rural people have an increasing need for cash incomes. With improving access to and from remote areas, there are also increasing commercial

TABLE 13-2: Property rights and swidden.

Laws and regulations	Conditions
Land law	• Most swidden cultivation fallows are considered 'VFV' land under Department of Agriculture jurisdiction. • Fallow lands are particularly vulnerable to reallocation. • There is currently no indigenous land law or endorsement of customary land systems.
Land-use regulations	• Customary management systems. • Protected Area regulations. • Community forestry – but this currently discourages land uses not considered 'forest', and only offers a 30-year lease.
Recent changes	• Reduced state-ethnic conflict and democratization, increasing the state's reach into remote areas. • Continuing land-law revision process. • Lobby from agribusiness for land allocation, threatening to strip assets from swidden cultivators.

Note: VFV=Virgin, fallow and vacant land.

opportunities for niche products and benefits from local comparative advantages. However, there is also increasing commercial interest in access to areas of land used by shifting cultivators, and in several cases, willingness by governments to sacrifice the interests of swidden communities to secure international investment.

Promotion of a statutory legal system that overrules customary structures can transform the common property regime, in which everyone has a share, into private or state property, leading to landlessness and poverty. It can also lead to increased dependence on market vagaries for incomes and food security.

Adverse policy environment

Problems in swidden cultivation systems are often brought about by counter-productive policies rather than inappropriate land-use practices. Governments typically oppose shifting cultivation in their desire for the revenue benefits and administrative simplicity of permanent forests or agriculture, and have implemented extensive programmes to achieve this. Even today, government staff typically lack understanding of, or interest in, the basic livelihood realities of upland cultivators, and there is a prevalence of hostile departmental policies relating to forests, agriculture, hydropower, rural development, and so on. The policies of different departments are often conflicting, and hostile policies are often promoted by lobbies of vested interests such as the 'timber mafia', settlers and commercial enterprises.

These things have proven to be counterproductive for social development: a hostile policy environment can force cultivation systems into suboptimal practices by undermining customary-authority structures and shortening fallows, or obstructing them altogether, leading to poverty and conflict, alienating cultivators and resulting in degraded land. Areas labelled as 'vacant' or 'wastelands' are typically productive or regenerating fallows and mislabelling them often leads to their allocation to other uses.

In some countries, agriculture departments have promoted chemical inputs, particularly synthetic fertilizers, which have had a seriously deleterious effect on the soil and nutrient cycling. Moreover, regulatory frameworks for farm products and uniformity norms for commodities are often biased against local landraces that are adapted to input-free shifting cultivation systems.

Stresses lead to deteriorating performance

In the absence of mitigating adaptations, farmers may be forced to shorten fallow periods and extend the cropping phase, pushing the systems beyond their ecological resilience and into a degradation cycle in which weeds proliferate and yields decline, with declining returns to labour and increasing food insecurity (Cairns and Garrity,1999). Also, as productivity and economic returns are not assured, investment in land management is increasingly risky.

Supporting sustainable livelihoods for Myanmar's swidden cultivators

There is growing international recognition of the principle that a hostile policy bias against shifting cultivation practices is both unscientific and unfair. Instead, policy-makers should be supporting the interests of their citizens by ceasing to obstruct their livelihoods and instead helping to improve and adapt livelihood systems in Myanmar's rapidly evolving context.

Take, for example, The Shillong Declaration, made in 2004 by government agencies, farmers, international bodies, non-governmental organizations, academics, science and research institutions, local institutions, international donors and development-assistance agencies from the eastern Himalayan countries. It said, in part:

> Shifting cultivation must be recognised as an agricultural and an adaptive forest management practice which is based on sound scientific and ecological principles ... Regional, national and local policies ... need to be reappraised and ... reformulated (ICIMOD, 2004).

Policy recommendations

The following policy recommendations emerge from this discussion. Although they relate directly to conditions in Myanmar, they could apply equally to other parts of Southeast Asia and, indeed, all tropical regions:

A. Reorient the hostile policy and administrative environment to one supportive of improvements to swidden systems and swiddeners' livelihoods

Government policy should focus on sustainable poverty alleviation by working with citizens to support their aspirations and needs. This would mean:

1. Removing policies that are explicitly opposed to shifting cultivation (e.g. hostile policy statements in the Forestry Master Plan, Forest Department restrictions and land re-allocation measures).
2. Strengthening the implementation of existing beneficial policies (e.g. agricultural development extension, marketing support, on-farm timber production and marketing) and improving the provision of services, such as health and education, to remote areas.
3. Strengthening and capacitating swiddeners' customary institutions for resource management and governance.
4. Encouraging coordination between government agencies at all levels (national, state, township, village tract). There is also a need for some sort of role for local government in land-use planning.

B. Ensure tenure security and customary mechanisms for shifting cultivation

The central issue for swiddeners is tenure security, covering both their swiddens under cultivation and their fallows, because land is their prime productive asset. Improving tenure security improves the incentive for sustainable land management. Steps towards this include:

1. Customary resource-management institutions should be given statutory authority for tenure management and arbitration. Traditional conflict-resolution mechanisms must be empowered. However, customary institutions are not perfect and they may need to be supported towards greater equity and fairness (e.g. more equal land access for the land poor and more gender equity).
2. Statutory legal instruments are needed to protect swidden land tenure, which typically overlaps formal land categories for forest and agriculture. At present, the best possibility is for regional government to develop new sub-laws to permit and regulate the use of swidden land, perhaps under the 2012 Farmland Law.
3. Village forest and agroforest lands should be handed over to community control, possibly under Community Forestry Instruction, where communities request this. Community Forest Instruction should also be strengthened and converted into a law, and modified to accommodate sustainable swidden cultivation practices. The 30-year lease period currently governing community forests should become perpetual.
4. Asset stripping of swiddeners by re-allocating land plots under cultivation or fallow to outside agribusiness land grabbers should be prohibited. Land-use

planning is needed to endorse current use practices. If agribusinesses seek land, there should be institutional mechanisms through which they can negotiate with farmers to purchase plots, although non-ethnic ownership of hereditary ethnic-domain lands should be restricted

C. *Support technical innovations for sustainable intensification and build upon indigenous innovations*

Governments, academic bodies, development agencies and non-governmental organizations must invest in on-farm participatory action research to build upon and enhance traditional swidden practices and identify local innovations. There is a need to develop pathways for more sustainable intensification by building on traditional skills and working with leading farmers as innovators and extension agents within and between communities. Many aspects of swidden livelihoods could benefit from the following:

- Intensifying the swidden cultivation system (e.g. improving production by enhancing soil management and nutrient cycling for improved yields, involving both the cultivation and fallow phases).
- Intensifying and expanding into other components of the farming system, such as home gardens, livestock rearing and sedentary cultivation.
- Improving post-harvest technologies and processing.
- Addition of value to the production of traditional handicrafts.
- Introducing labour-saving technologies (e.g. draught power and mechanization).
- Improving landscape management, including upgraded fire management and forest and water-channel protection.

A wide range of indigenous practices have already been identified around upland Asia that go a long way towards achieving these goals, although the techniques involved must be transferred and locally adapted. As noted by Cairns and Garrity (1999) 'Land use in shifting cultivation communities is often transitional: Land use intensification is a near-universal process'.

A range of existing technical innovations for sustainable intensification can be tried out across the four phases of swidden. These include traditional practices, recent indigenous innovations and international innovations. For instance, improved fallowing methods (e.g. Franzel, 1999); improved burning-off (e.g. 'bio-char'); and nutrient cycling (e.g. advanced composting like Activated Compost Tea). Development approaches that build on existing capacities and potential are likely to be more acceptable to farmers and therefore more achievable (Kerkhoff and Sharma, 2006).

D. Promote livelihood diversification and enterprise development

It is increasingly important to help households to adapt to societal changes by moving beyond subsistence livelihoods, taking advantage of opportunities for additional income, spreading their risks and avoiding indebtedness. The following measures could augment incomes and diversify livelihoods:

1. Market development should be encouraged, based on local niche products. Commercial niche products can contribute to economic development that is adjusted to local circumstances and builds on existing potential.
2. Alternatives are needed to informal money lending at high interest rates, if intractable indebtedness is to be avoided. Saving and credit schemes may help.
3. Schemes that provide payments for ecosystem services may help, although international experience so far indicates that such payments are very small and have strings attached, such as required changes to land use that compromise existing practices. Nevertheless, opportunities may emerge, e.g. catchment management or carbon sequestration.

Acknowledgements

The author expresses his gratitude to Professor San Win, Dr Salai Kungliang Thawng and Ms. Julia Fogerite for their very helpful comments on early drafts of this chapter.

References

Bartholomew, W. V., Meyer, J. and Laudelout, H. (1953) *Mineral Nutrient Immobilization under Forest and Grass Fallow in the Yangambi (Belgian Congo) Region*, INEAC série scientifique no. 57, The National Institute for Agronomy in Belgian Congo, Brussels

Cairns, M. F. and Garrity, D. P. (1999) 'Improving shifting cultivation in Southeast Asia by building on indigenous fallow management strategies', *Agroforestry Systems* 47(1), pp37-48

FAO (2010) *Farming Systems and Poverty: Improving Farmers' Livelihoods in a Changing World*, Food and Agriculture Organization of the United Nations, Rome, ftp://ftp.fao.org/docrep/fao/003/y1860e/y1860e.pdf, accessed 12 September 2016

FAO (2016) *Global Forest Resource Assessment 2015*, Food and Agriculture Organization of the United Nations, Rome

Foley, J. A., Ramankutty, N., Brauman, K. A., Cassidy, E. S., Gerber, J. S., Johnston, M., Mueller, N. D., O'Connell, C., Ray, D. K., West, P. C., Balzer, C., Bennett, E. M., Carpenter, S. R., Hill, J., Monfreda, C., Polasky, S., Rockstrom, J., Sheehan, J., Siebert, S., Tilman, D. and Zaks, D. P. M. (2010) 'Solutions for a cultivated planet', *Nature* 478(7369), pp337-342

Franzel S. (1999) 'Socioeconomic factors affecting the adoption potential of improved tree fallows in Africa', *Agroforestry Systems* 47, pp305-321

Guillemin, R. (1956) 'Evolution de l'agriculture autochtone dans les savanes de l'Oubangui', *Agronomie Tropicale* 11, pp143-176

ICIMOD (2004) *The Shillong Declaration on Shifting Cultivation in the Eastern Himalaya*, International Centre for Integrated Mountain Development, Kathmandu

Kerkhoff, E. E. and Sharma, E. (2006) *Debating Shifting Cultivation in the Eastern Himalayas: Farmers Innovations as Lessons for Policy*, International Centre for Integrated Mountain Development (ICIMOD), Kathmandu

RRI (2011) *The End of the Hinterland*, Rights and Resources Initiative, Washington, DC

UNDP (2011) *Myanmar Poverty Assessment*, United Nations Development Programme, Yangon

Win, San (2009) *Investigation on Shifting Cultivation Practices Conducted by the Hill Tribes for the Development of Suitable Agroforestry Techniques in Myanmar*, International Tropical Timber Organization, Yokohama

Notes

1. This fact has been known to cultivators around the globe for at least five centuries, but has been recognized by agronomists for its soil enhancement benefits only in the last few decades.
2. The two most significant international studies being the ongoing global ASB Partnership for the Tropical Forest Margins, which began as the 'Alternatives to Slash and Burn' project, and is coordinated by the World Agrofestry Centre (ICRAF) in Nairobi, and a recent project of the International Centre for Integrated Mountain Development (Kerkhoff and Sharma, 2006).

14

SWIDDEN AGRICULTURE UNDER THREAT

The case of Ratanakiri, northeast Cambodia:
Opportunities and constraints from the national
policy environment

Jeremy Ironside, Gordon Paterson and Anne Thomas*

Introduction and context

Historically isolated, northeast Cambodia's previously abundant forest and rich volcanic soils have long been home to a culturally diverse mix of upland swidden farmers. A long period of insecurity and conflict, lasting from the 1960s to the late 1990s, added to the region's isolation and meant that traditional swidden systems still function in the more remote villages. Over the past decade or more, broad socio-economic changes have dramatically impacted the natural ecology and traditional agriculture systems of this region. This chapter examines the processes of rapid land-use change impacting on swidden systems in Ratanakiri province. Building on this, it will also outline the possibilities and some of the changes needed for swidden agriculture to survive as a viable land use in this region.

Changes occurring in Ratanakiri and other parts of northeast Cambodia are similar to processes reported from many parts of Southeast Asia and elsewhere (see Salemink, 2003; Guerin et al., 2003; Shiva, 2003; Nevins and Peluso, 2008). Some of the factors causing extreme stress on the traditional agro-ecosystem in this area include:

- The Cambodian government's vision of the country's northeastern region becoming a pillar of national economic development;[1]

---

* Dr Jeremy Ironside, has been working in Ratanakiri province since 1996, completing a PhD in communal land management based on these experiences. He is presently a consultant for The McKnight Foundation's Southeast Asia programme and an independent researcher; Gordon Paterson is an independent consultant who has worked for 24 years in Cambodia and has been instrumental in setting up two organizations working on land and natural-resource rights for indigenous communities; Anne Thomas is an independent consultant who has worked in Ratanakiri province since 1996. She assists indigenous peoples to gain access to health and education services and works to protect their traditional lands and natural resources.

- Plans for a Cambodian, Lao and Vietnamese 'Development Triangle' to economically integrate the isolated regions of southern Laos, central Vietnam and northeast Cambodia; and
- With support from the Asian Development Bank, development of an economic corridor that will pass through Ratanakiri, eventually linking Bangkok by road with Vietnam and China.

Notwithstanding these plans, Cambodia has a reasonably favourable policy environment towards swidden agriculture. Article 23 of Cambodia's 2001 Land Law recognizes and permits swidden cultivation 'according to customary rules of collective use', and forest fallows are recognized as an integral part of the system.[2] The Land Law also allows indigenous communities to register their traditional agricultural lands under a collective title. A key challenge we consider in this chapter is whether the Land Law will be implemented rapidly enough, and on a scale sufficient to provide some protection for indigenous communities and their swidden agriculture systems against the rapid changes that are planned and currently being implemented. In Cambodia, as in many similar countries, implementation of any policy that supports the poor is impeded by the lack of a functioning governance structure, a virtual absence of political will and the absence of an effective grass-roots civil society able to claim its legal entitlements and hold the government accountable.

We begin this discussion of the dynamics of economic change and the adaptations underway in traditional management systems with an introduction to Ratanakiri and its indigenous swidden farmers. In particular we consider the predominant discourse, from lowland parts of the country, of the need to bring 'economic development' to the people and the region. We then explore the traditional swidden systems found in Ratanakiri, and argue that the communal-property arrangements practised by indigenous groups are fundamental to the continuation of the swidden agriculture system. Villages that have been able to retain their communal lands have also retained their swidden agriculture practices. Because of this, we briefly consider the progress being made in strengthening communities to secure their land, as well as some possibilities arising from the communal model for both preserving swidden systems and building on these to develop diverse land uses and economic opportunities.

Ratanakiri province and its indigenous peoples

Ratanakiri province has long been seen by Cambodians of the lowland plains as a remote, wild and forested area, populated by people with distinctly different cultural practices (Meyer, 1979; Bourdier, 1995). Located in the northeastern corner of the country, bordering Vietnam to the east and Laos to the north (Figure 14-1), it covers 10,782 sq. km, with a relatively low population density of 17 people per sq. km (MOP, 2013). Large areas in the north and south of the province have few or no inhabitants. Settlement and land use, therefore, is concentrated along main roads, around the main centres, along the two main rivers and on the fertile 2000 sq. km basalt plateau in the centre of the province.

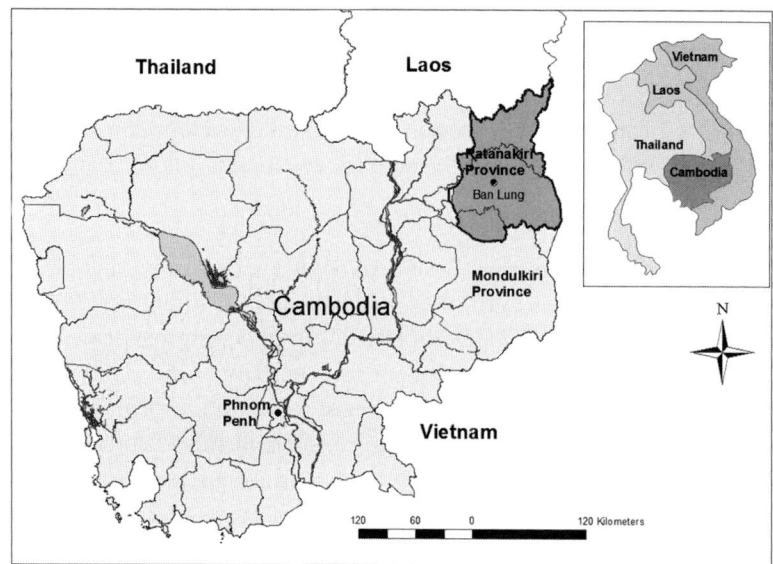

FIGURE 14-1: Ratanakiri province, northeast Cambodia.

Ratanakiri, and Mondulkiri immediately south of it, are the only two of Cambodia's 24 provinces where indigenous peoples make up about half of the population. While there are no precise figures for the population of indigenous groups in Cambodia,[3] the 2008 census identified 23 different minority mother tongues, spoken by a total of 179,215 indigenous people, or 1.4% of the population (MOP, 2009). Figure 14-2 shows the distribution of indigenous groups found in Ratanakiri and Mondulkiri provinces.

Indigenous groups found in Ratanakiri range in size from a few hundred to the largest group, the Tampuan, with a population of 30,888 (MOP, 2009).[4] Despite their different languages, these groups share common cultural practices and livelihood systems based on swidden agriculture and collection of forest products. The story of Cambodia's indigenous communities is one of adaptation to their forest environment over considerable periods of time. Their belief systems allow for the coexistence of natural and supernatural forces and include communal land-management that allows for rotation of agricultural plots around a village's land area to ensure forest regrowth. Each community traditionally governs its affairs autonomously, including the resolution of conflicts, using systems of community law that dispense justice in the interests of maintaining community harmony (Backstrom et al., 2006). Only recently have pan-village organizations begun to develop. While they are now adopting cellphones, motorcycles and television, most indigenous peoples in Ratanakiri live without running water or grid electricity and rely almost entirely on agriculture for their livelihoods.

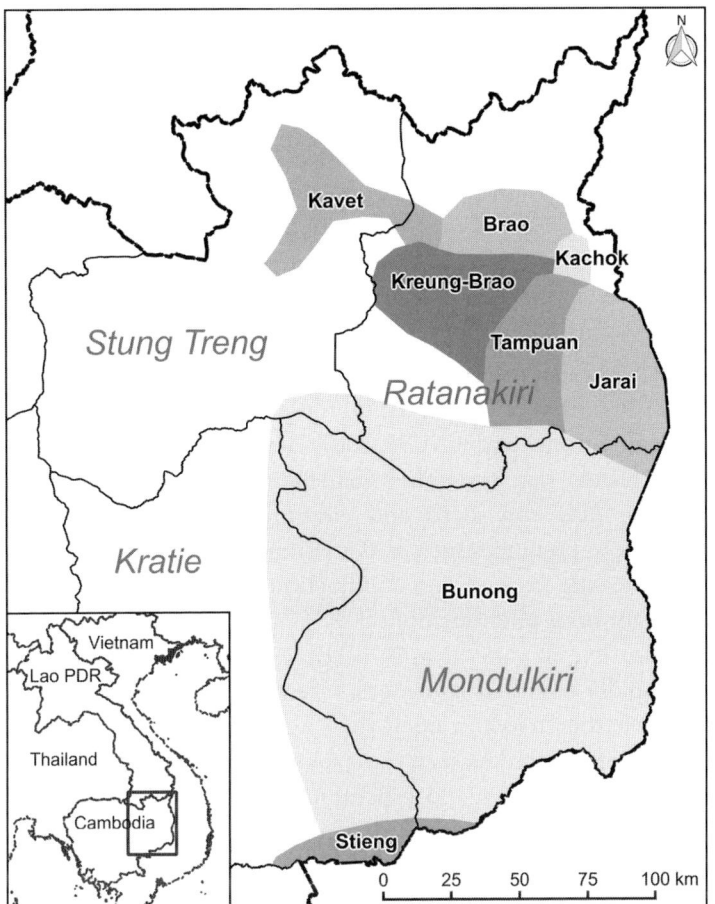

FIGURE 14-2:   Distribution of indigenous groups in northeastern Cambodia.

Source: Colm (1996).

The context of socio-economic development in Ratanakiri

Apart from trading contacts, Ratanakiri's isolation persisted up to the middle of the 20th century (Meyer, 1979).[5] Local indigenous groups were able to maintain their autonomy despite efforts by the French and later by the post-independence Cambodian government in the 1960s to establish control over Ratanakiri and Mondulkiri provinces. Thirty years of civil war, which finally ended in 1998, further helped to maintain the region's isolation. However, the end of hostilities marked the start of a rapid opening-up of areas such as Ratanakiri and significant exploitation of their richness in land and resources. In less than a generation, local indigenous groups had to adapt to a market-driven economic transformation.

The increase in Ratanakiri's population since the 1998 census shows the rapidity of change. From 1998 to 2008 the population grew at an annual rate of 4.65%, from

94,243 to 150,466 (MOP, 2009). From 2008 to 2013 the annual population growth slowed slightly to 3.99%, to reach a total population of 183,699 – still the third highest provincial growth rate in the country (MOP, 2013). Immigrants are attracted particularly by the availability of cheap and fertile soils for cash cropping.

Increased central control of Ratanakiri and other highland provinces by Cambodia's political elite is now being accomplished by free-market economic policies and the opening up of the area for industrial-agricultural investment, both implemented through neo-patrimonial networks.[6] A key part of the free-market economic reforms, which were implemented in Cambodia in the 1990s, is the promotion of the private sector as an 'engine of economic growth' to reduce poverty (RGC, 2005b, p19). In 2003, Cambodia became the first so called 'lesser developed country' to join the World Trade Organization, and since then there has been a significant opening of its markets to international trade (MOP, 2007; Beresford et al., 2004).

Government claims that these free-market reforms have achieved significant poverty reduction hide the social and environmental costs being paid by the inhabitants of remote provinces such as Ratanakiri. Communities in these areas are under the most pressure from resource-extraction activities. Poverty in Cambodia is an overwhelmingly rural phenomenon; 80% of the population is rural and 92% of the country's poor live in rural areas (UNDP, 2010b).[7] However, poverty has been reduced most rapidly in urban and more accessible rural areas (UNDP, 2010b). It is questionable whether free and open trade will really benefit indigenous groups living in marginal provinces. As seen in development plans for the northeast of the country, Vietnamese merchants and processing factories stand to reap most of the trade benefits from border areas such as Ratanakiri.

Partly because of its high percentages of indigenous peoples, Ratanakiri has one of the highest poverty rates in the country. As shown in Table 14-1, Ratanakiri is ranked at the low end of the country's 24 provinces in terms of achieving Cambodia's Millennium Development Goals. The province's position at the low end of the country's socio-economic indicators gives an idea of the problems faced by Cambodia's indigenous peoples. Low levels of literacy further show the significant problems indigenous groups face in dealing with Khmer (the national language) and lowlander-dominated institutions. Effectively, literacy is negligible among indigenous women.

Economic-development plans for Ratanakiri province

The Cambodian government's economic plans for its northeastern provinces include mining, agro-industry and eco-tourism as the key drivers of growth (COHCHR, 2007). As part of this, a 'Development Triangle' has been created by joining five provinces in Vietnam and four each in Cambodia and Laos (CLV-DTA, 2004).[8] Plans for this Cambodia, Laos and Vietnam Development Triangle (CLV-DTA) involve making use of the 'under-utilized economic potential' in what are considered to be the 'least developed territories' of each of the countries (CLV-DTA, 2004, p10).[9]

TABLE 14-1: Rankings of Ratanakiri province in Cambodia's Millennium-Development-Goal statistics.

Cambodian Millennium Development Goals	Ratanakiri's ranking (out of 24 Provinces)	Cambodian Millennium Development Goals (May 2015)*	Key statistics
Food and Income Security	22nd	Off track	Poverty rates of 36.2% in 2012, (RGC, 2014).
Education	23rd	Slow	39.37% of Ratanakiri's children 6-11 years old are not attending school (MOP/UNDP, 2010)
Gender	21st	Slow	
Infant Mortality	22nd	On track	Up to 180:1000 mortality for children under five (UNDP, 2010a)
Maternal Mortality	23rd	Off track	
HIV/AIDS, Malaria, TB	19th	On track	Malaria is endemic in Ratanakiri (Hubbel, 2007)
Land and Forests	22nd	Off track	

* *Source:* http://www.khmerdigitalpost.com/cambodia-millennium-development-goals-cmdgs-to-be-reached-during-2015/, accessed 5 September 2016.

A reading of the Development Triangle Master Plan, however, suggests that Cambodia and Laos will more likely become suppliers of raw materials to Vietnam and China. As well as plans for hydroelectricity, infrastructure development, eco-tourism and so on, the Master Plan is heavily focused on agricultural and forestry development to create a 'market-oriented commodity-producing economy' (CLV-DTA, 2004, p92). In particular, the development of high-value cash-cropping ventures on the red basalt soils of Ratanakiri and Mondulkiri is envisaged, including coffee, rubber, cashews, pepper and raising livestock. Despite these plans having been in existence since 1999, Ratanakiri's indigenous inhabitants have never been informed of them.[10] Consequently, the social and environmental impacts of the plans have never been assessed. This lack of consideration of the CLV-DTA's impacts means that there will be little recognition of existing settlement and land-use patterns or the land rights of indigenous communities. These plans, therefore, seem to 'sanction and foreshadow further alienation of indigenous land' for planting crops such as cashews and rubber (COHCHR, 2007, p16). This highlights the insecurity felt by indigenous villagers, who are constantly told by officials and land brokers that their land belongs to the state, and that the state will take it away from them if they don't accept a minimal payment of money and acquiesce to whatever plans powerful outsiders have for it. Fallow management in these circumstances is extremely difficult.

Further justification for displacement and dispossession, as well as a disregard for local peoples' rights and management systems, can be seen in comments within the CLV-DTA Master Plan, such as 'ethnic minority groups' … practice of forest burning for cultivation of land and inconsiderate forest exploitation has rapidly decreased the forest area [and has had] serious negative impacts on the ecological environment' (CLV-DTA, 2004, p96). The plan suggests that 'land reserve' areas be identified 'that can be potentially exploited for agricultural and forestry development … in order to plan for population distribution along the lines of permanent cultivation and fixed settlement … with an aim to stabilize the lives of the ethnic minority people', purportedly in order to mitigate deforestation and protect the environment (CLV-DTA, 2004, pp148-149).

The 'Development Triangle' therefore aims to transform groups of self-employed subsistence swidden farmers into producers for the market economy and workers for large-scale commodity production.[11] 'Ethnic minority groups' are portrayed in the Master Plan as in need of 'modernization' and 'civilization', since their 'production customs and practices remain backward' (CLV-DTA, 2004, p129). Planned interventions in education aim to train 'the Cambodian and Laotian people, and the Vietnamese ethnic-minority people, to familiarize [them] with, and [help them] adapt to, the market economy and market-oriented commodity production' (CLV-DTA, 2004, p111).

Traditional land-use systems in Ratanakiri

As mentioned earlier, traditional swidden systems have survived in isolated parts of Ratanakiri, or where communities have strongly defended their lands. Because of this we argue that defence of communal-property arrangements is crucial for the survival of swidden farming. In this section we describe the functioning of these swidden systems.

Despite the changes mentioned above, swidden systems are still integral to the livelihood of indigenous people in many villages in Ratanakiri. The 2000sq. km basaltic plateau in the middle of the province has attracted swidden communities over the centuries because of the fertility of the soil and rapid fallow regrowth. In flatter areas, these soils are able to be cropped for three to four years before being fallowed for 10 to 15 years (Ironside, 1999). However, as mentioned earlier, these soils are increasingly in demand for a variety of cash crops. In other parts of the province, where the soils are less fertile, swidden cycles have much shorter periods of cropping (one or two years) and regeneration periods of between six and 20 years, depending on the rate of fallow regrowth. Along the mountain streams in the very north of the province, indigenous farmers developed an effective short-rotation swidden system using bamboo as the predominant fallow species (Box 14-1).[12]

Swidden systems operated by indigenous groups in Ratanakiri have been shown to be well-developed and based on an intimate understanding of the local ecosystem (Bourdier, 1995; Fox, 2002; Ironside and Baird, 2003; Fox et al., 2009). Up until the recent past, overall forest cover (primary and secondary forest) was maintained at around 80% for several centuries (Bourdier, 1995; Fox, 2002). As with all swidden systems, a relatively low population density permitted the maintenance of this forest

---

BOX 14-1: The swidden system in Virachey National Park

The Brao and Kavet people living in the north of Ratanakiri Province (see Figure 14-2) practised a swidden system that consisted of moving along rivers and streams and cutting the predominantly bamboo forest along their edges. Bamboo forest once stretched along all of the large tributaries in the northern part of Ratanakiri.

Clearing swiddens close to streams meant easy access to water and fishing grounds. Depending on the area, people would spend 5 to 10 years moving up one side of a stream, and then they would change sides and start working their way back down, or they would change stream valleys and start working their way down a new valley. When the village centre got too far away from the swiddens, it would be moved and rebuilt in front of the current swidden areas. It took about 10 to 20 years to return to the same location. People would distribute themselves widely in different stream valleys on their village land in groups of around 10 families each. About every five years the whole village would congregate and build their village houses in one place before dispersing once more to cultivate their swiddens.

Following the cropping period, the bamboo fallow would regenerate quickly – generally within six years and sometimes within three or four. One elder said that a swidden could be cut again if necessary after only a few years of bamboo regrowth. Every 6 to 20 years, all the bamboo would drop its seeds and die. This system is now a limited practice because the area was declared a national park in 1993 and the communities now live south of the park, where there are poorer soils and open deciduous forest that is not suitable for swidden cultivation. Large areas have been taken over by *Imperata cylindrica* grass. Villagers have been appealing to the national park administration and the Ministry of Environment to be allowed to resume swidden agriculture in the bamboo areas within their 10,000ha community-protected areas inside the park. Their request remains under discussion.

Source: Ironside and Baird, 2003.

Bambusa polymorpha Munro
[Poaceae]

One of the most useful species found in bamboo forests growing alongside rivers and streams in northern Ratanakiri.

---

cover. However, aerial photographs taken in 1953 show intensive swidden farming on the most productive soils, such as those on the basalt plateau (Figure 14-3).

This maintenance of forest cover and ecosystem function over a long period has been demonstrated by longitudinal land-use studies (see Fox, 2002; Fox et al., 2009). In one study, comparison of 1953 and 1996 land cover showed that over this 43-year period, overall tree cover (forest and secondary forest) remained constant at between 77% and 96% of the landscape, depending on the area studied (Fox, 2002). This was despite the fact that between 50% and 81% of the landscape was being used for swidden farming over this period, again depending on the area.[13] Fox (2002) further points out that 'while 77% to 96% of the landscape remained under forest or secondary growth in different stages of regeneration, land cover on any particular plot may have changed several times'.

A further feature of indigenous land use in Ratanakiri was the dispersed distribution of villages over the landscape, minimizing impact on the forest and allowing for sufficient time for recovery (Bourdier, 1995; Fox, 2002). A study in a forested district of Ratanakiri, for example, showed that villages maintained a fairly constant population density of 30 people per sq. km within their agricultural boundaries (Fox, 1997). The larger a village's population, the larger was the village's land area (agriculture and forest), in a fairly constant ratio. This pointed to a balanced demographic impact on the forested environment and reinforced the importance given in traditional law to preserving this even dispersal. For example, it is prohibited to encroach onto the territory of another village.

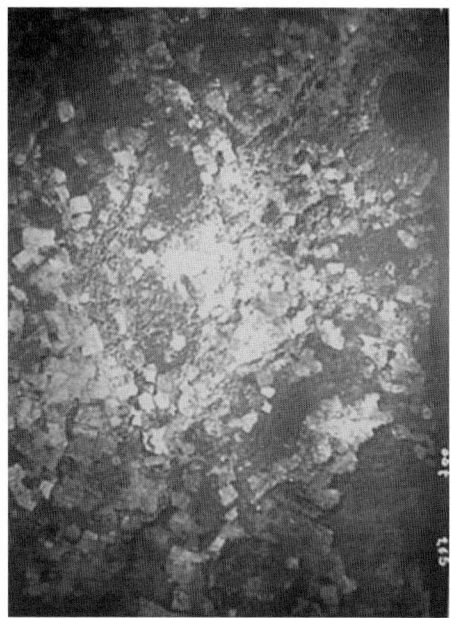

FIGURE 14-3: Intensive land use for swidden farming near the present-day site of Ratanakiri's provincial capital, Ban Lung, revealed by an aerial photograph taken in 1953. The 800m-wide Yeak Laom volcanic crater lake, 5km east of Ban Lung, can be seen in the top right of the photo. The area in the southern part of this photo was planted in rubber in the 1960s.

Traditional law also maintained and regulated a flexible system of land-user rights that allowed the swidden system to function. In the past, but less so now, individual rights over a piece of land applied during periods of use, but the land reverted to community ownership when the field was left fallow and the forest grew back. As discussed above, the village land area has always been large enough to accommodate the scattered family fields and the periodic movement of the village site. Within this village area, land was traditionally considered the property of the forest spirits who lived there. It was only with the agreement of these spirits, contacted through ceremonies and rituals, that the field could be temporarily cleared, with the intention always that the forest would be allowed to regenerate after cultivation finished.

Villagers therefore traditionally depended on a given land area to supply their food, medicine, building materials and other resources. This required careful management, as Cupet (1898, in Ironside and Baird, 2003, p52) pointed out from his travels in the region in 1890:

As a consequence of secular fighting, the inhabitable territory has been divided up between the villages ... Within it, the inhabitants mark out their fields as they see fit, fish and hunt as they please. The smallest incursion into neighbouring territory brings about a conflict. ... The different peoples are, consequently, more or less immobilised where they are established.

Cupet may have overlooked the shared use of forest areas, but his point that each village was constrained from expanding its agriculture domain because of its neighbours is correct. Immobilization, however, did not mean sedentarization (in the anthropological sense), as agricultural fields and even village sites were rotated for fertility regeneration and to avoid or evade disease build-up. This arrangement of dispersed and essentially independent villages lasted well into the 20th century.

The key to these land-use systems is diversity. A single family oversees plots of different ages and different stages of use or regeneration. Different rice varieties are planted in the plots, depending on the number of years the field has been in use, soil type and so on. A great variety of crops is planted in swidden fields along with rice. For example, in 1999, Baird (2013) documented 145 different species or varieties of non-rice crops and 36 varieties of rice growing in the swidden fields of two villages (one Kavet and one Kreung) in the northern part of Ratanakiri. Different fields growing different rice varieties allows for lengthening harvest times and shortening hunger gaps. As well, it exploits agrobiodiversity and spreads the risk of seasonal variations. In younger fallows, longer-term crops such as bananas, papaya, fruit trees and plants that have self-seeded are harvested and these areas are also used for grazing. Older forested fallows are used for collecting forest products.

A mix of ecological niches therefore allows for a range of products from agriculture, hunting and collection, all of which are needed for subsistence and commerce. Villagers reported that before the fighting (in the 1960s), people usually had sufficient rice and other food to eat and there were plenty of animals and other resources in the forest when they were short (Ironside, 2009). Local people said they were often able to grow enough rice to be able to sell surpluses. With this income they could buy domestic animals, ceremonial gongs, ceramic urns for rice beer and so on. However, despite their self-sufficiency, many other aspects of life in their forested environment was, and continues to be, difficult.

Swidden systems and social cohesion

Swidden systems in Ratanakiri show the importance of a wider social organization that provides for exclusive individual-use rights, along with community rights and responsibilities to protect and encourage fallow regrowth. Social harmony, maintained by traditional systems of conflict resolution, is essential for managing these systems of user rights to allow for alternating periods of cultivation and fallow. Social cohesion, for example, permits negotiation and agreement between land users over which pieces of land they intend to use, and what rights a particular family might have to a particular site.[14] This maintenance of social cohesion, therefore, is the basis of communal land management, and this is a key lesson from indigenous management systems in Ratanakiri. This section explores the importance of supportive social organization for the ongoing functioning of swidden systems.

Village spatial arrangements

As well as providing methods for resolving conflicts, traditional management systems in Ratanakiri also foster social cohesion by fixing the arrangement of village houses in a circular pattern around the central village meeting house. The meeting house is the centre of both the social and spiritual life of the village. Since the Prince Sihanouk government in the 1960s, many villages have been persuaded by the authorities to break up these circular layouts in the name of 'development' (Bourdier, 1995). Villagers have been required to arrange their houses in a linear fashion along roads and rivers in 'street villages', in the classic Cambodian style (Matras-Troubetzkoy, 1983, p26). Whereas previously all of the families were within hearing distance, several villages now stretch a half a kilometre or more along a road, seriously affecting communication and village solidarity. The result is a change in the community's emphasis, from the village and its land as the central focus, to the street and trading relations with the outside.

Community harmony and swidden ceremonies

Further examples of the close association between swidden agriculture and social organization are the traditional ceremonies conducted to ensure harvests will be plentiful and the community prosperous. The swidden system is founded on a number of *saen* (ceremonies) for planting and harvesting. The *samaki* (cooperation and solidarity; social capital) developed as a result of these ceremonies is in turn fundamental to sharing the communal lands.

The effectiveness of traditional governance systems in managing change

Krola Village, of Poey commune in Ratanakiri's O Chum district, is one of the diminishing number of villages that has been able to maintain its traditional swidden system. With 208 families and a population of 857 people of Kreung/Brao ethnicity, Krola is a good example of the importance of maintaining traditional governance systems as the basis for the continuation of a viable swidden system. Residents feel that the village has maintained its strong solidarity because conflicts are resolved according to traditional means. Land is still used communally and village leaders affirm that there is still enough land and fallow areas. However, the demand to plant cassava and cashew trees for cash cropping means there is less land for swiddens. As a result, fallows have been shortened to about two years instead of the previous five. A Land Management Committee established by the community oversees the village's collective land and natural resources. Community leaders explain that the villagers simply do not permit the village chief to authorize the sale of any land or forest, even though outsiders are continually seeking to buy it.

The effectiveness of community land management in Krola village was confirmed by a study examining land-use changes between 1989 and 2006 in three villages (Fox et al., 2009). Over this period, the overall forest cover in Krola Village was reduced by only 0.86% per year (Fox et al., 2009; Figure 14-4). This forest reduction was

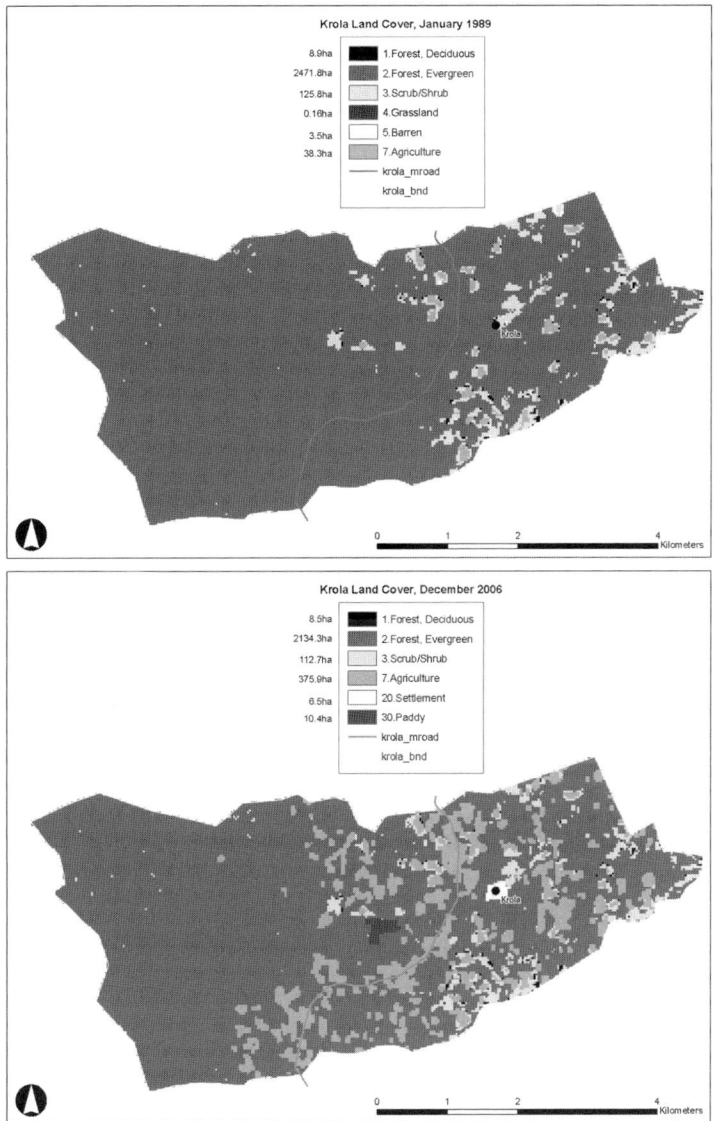

FIGURE 14-4: Land-cover change in Krola Village, Poey commune, O Chum district, Ratanakiri province, 1989 (upper) and 2006 (lower).

Source: Fox et al. (2009).

mainly due to expansion of villagers' cashew plantations. However, village regulations (instituted by the Land and Natural-Resources Management Committee in 1998) restricted the area of cashew trees to a maximum of five hectares per family. This, along with controls over land sales, has meant that traditional agricultural systems have been maintained, largely intact.

The situation of the other two villages in the study differed from Krola village: they were more accessible, there was greater pressure on the land and community management was less effective. The conversion of land to cashews, as well as an influx of outsiders buying up land for cash cropping, resulted in widespread deforestation and the demise of swidden systems (Fox et al., 2009). This highlights the difference between villages that have been able to act in unison versus those in which community authority has been fragmented by land privatization. One of the villages studied, Tuy village of Tingchak commune, in Ratanakiri's Borkeo district, is situated along the main road through the province. Here, the village authority had fragmented and pressure to sell and clear land had been intense. Deforestation rates were found to be 4.88% per year over the 17-year period of the study (Fox et al., 2009; Figure 14-5). In the third study village, Leu Khun village of Keh Chong commune in Bokeo district, annual deforestation rates were 1.63% (Fox et al., 2009).

Recent changes in the swidden system

The above study gives a good indication of the different ways villages are being impacted by the changes described earlier in this chapter. The example of Krola village and others like it also points to the importance of traditional communal management systems for allowing the flexibility and mobility needed for the land and fallow management of properly functioning swidden systems. In this section we explore some of the problems that need to be dealt with if swidden agriculture is to survive and adapt to growing pressures.

The individualization of communal lands

As discussed, mass land insecurity, and because of this the inability to protect fallow areas from outside pressures, is a major factor affecting the future viability of traditional communal swidden-agriculture systems in Ratanakiri. Partly as a way of claiming ownership of their land and partly out of a desire for cash income, swidden farmers have been planting their fields in cashew trees.[15] Their swiddens have thus been taken out of the swidden-fallow cycle. In addition, both villagers and outsiders are clearing forest land and older fallows in response to increasing demands for land for cash cropping.

A further major problem, also related to mass land insecurity, has been the granting of economic land concessions, often over the swiddens and fallow fields of indigenous farmers. Publically available information indicates that 32 companies have been granted more than 230,000ha of land in Ratanakiri.[16] However, it is

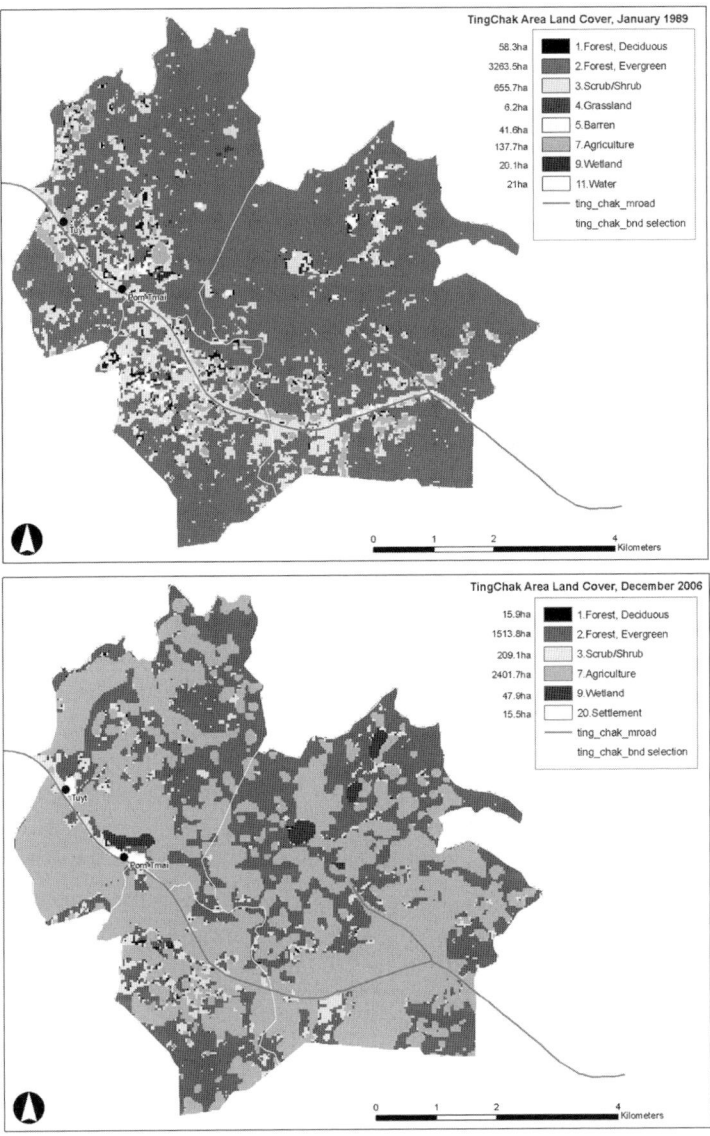

FIGURE 14-5: Land-cover change in Tuy Village, Tingchak Commune, Borkeo district, Ratanakiri province, 1989 (upper) and 2006 (lower).

Source: Fox et al. (2009).

impossible to know the exact figure and how much of this land is actually being converted to industrial agricultural crops, mainly rubber. Again, fallow areas are given no recognition when excising indigenous community land from these concessions, despite their legal status in the 2001 Land Law.[17] Only land that is currently in use is subject to any legal-possession rights that exclude it from land concessions. All land that is lying fallow thus becomes fair game for concessionaires. This represents a serious threat to the future of swidden agriculture.[18]

Villagers complain that the result of recognizing only those fields that are in use is like 'putting them in a cage like chickens', and a ploy to make poor people sell their land to companies and rich people (Ironside and Nuy, 2010). Villagers have difficulty accessing small patches of land when it is surrounded by a rubber plantation; they are unable to expand their swidden fields and their complaints are generally never heard or acted upon. In these situations, families often give up, or are coerced into selling their land cheaply to the company. This is indicative of a lack of respect for the rights of indigenous people, despite laws and policies intended for that purpose. Being 'caged like chickens' conjures up a worrisome future for people finding themselves within the 'Development Triangle'.

This trend of recognizing the ownership of only those fields that are in use, and not fallow areas or communal land rights, was further reinforced during implementation of a country-wide land titling campaign (Order 01) in 2012-2013, in villages whose lands overlapped with economic land concessions. Launched on 9 July 2012, this Order 01 campaign was intended to dampen down conflicts between communities and land-concession companies and reduce widespread anger over land alienation in the run up to the 2013 general election. Thousands of student volunteers were deployed throughout Cambodia to demarcate these conflicted lands.[19]

A study of 79 villages in Ratanakiri, in an effort to understand the impact of the implementation of Order 01, found that 26 of these villages were being impacted by land concessions and had some or all of their land privately titled as part of this campaign (Rabe, 2013). In these 26 villages, communities were forced to choose between accepting private land titles that were immediately available for fields that were in use or waiting indefinitely for a communal title. Twenty five of the 26 villages said they were unsatisfied because the land titling did not allow them to register their land as communal (Rabe, 2013). The study further found that 71 of the 79 surveyed villages were in varying stages of communal land titling before this Order 01 was implemented, and 40 of the 79 villages had problems with 26 land and mining concession companies (Rabe, 2013).

Perhaps predictably, the study also found that the land privatization process had increased land loss in villages. For example, many of the villagers' newly titled landholdings from the Order 01 process were surrounded by company lands and the companies denied them access to these landholdings and forced villagers to sell this land to them (Rabe, 2013). The implementation of policies, ostensibly intended to improve land security and reduce land conflicts, actually increased land loss and livelihood insecurity. In many of the villages impacted by the Order 01 process, villagers stated that their communities had been 'broken' because of the privatization of their communal lands (Rabe, 2013). The private land-titling process also identified community fallow lands that were not able to be registered and were easier for companies to take over (Rabe, 2013). This increasing insecurity is not a strong base upon which to build livelihood improvements and adaptations to existing swidden agriculture systems.

In addition to these processes of land loss to concession companies and through the land-titling Order 01 campaign, there has also been widespread and generally illegal buying and selling of indigenous community land. The buyers have mostly been outside cash-crop farmers and speculators, both large and small. Villagers have been driven to sell their land for many reasons, but most relate to their feelings of land insecurity, and what one villager called 'the new era of money'. Destitution, increased debt (partly due to land loss as a result of economic land concessions), land insecurity and other economic pressures mean that communities are now more vulnerable and at risk of selling their land to powerful and coercive companies and outsiders. As more villagers are driven to sell their land, collective community efforts to resist outside encroachment are weakened.

One of the consequences of this widespread land alienation in indigenous villages and the economic changes underway has been a change in gender roles. Women are now under greater stress as heads of households and are more and more becoming the main cultivators of the swidden fields. This has resulted from the increasing mobility of men, many of whom have motorcycles, and the increased availability of labouring work outside the village. Women's decreasing dependence on men for livelihood support and their increasing role as breadwinners through traditional livelihood activities means that women are becoming increasingly responsible for heavy labour along with decision-making in the marketing of their agricultural products.[20] As men are more exposed to outside economic pressures they also come under increased pressure to sell land. Recent research into land alienation has shown that the idea of selling community land has come overwhelmingly from men, rather than from women (Ironside, 2011). To a large extent, women are becoming the guardians of swidden farming, because it has always guaranteed food security for their families.

Given the important role of swidden agriculture in indigenous communities in Ratanakiri, not only for livelihoods but also in the social, cultural and spiritual life of the people, these recent changes are forcing a fundamental redefinition of cultural meaning in indigenous communities. Evidence of several companies illegally bulldozing, claiming and clearing burial and spirit forests (Figures 14-6 and 14-7) illustrate a much deeper cultural clash over how the land will be used, and who will live

FIGURE 14-6: Several rice-wine jars signify a burial forest in Kanat Thom village, Ratanakiri. Forest clearing can be seen on the edge of this area, all of which was subsequently cleared for rubber.

on it. A report to the United Nations Human Rights Commissioner points out:

> Land is the repository of memory and keeps traces of the past in the absence of a strong written tradition … place names, old roads, legends and stories attached to places. For local people, bulldozing the landscape is seen as erasing their history and disturbing social organization and traditions (COHCHR, 2004, p27).

FIGURE 14-7: A rubber-concession company's office, built in the spirit forest of Pu Rapet village, Mondulkiri province.

An article about land clearing for a rubber plantation by a Khmer and French joint venture in neighbouring Mondulkiri province describes the machines that are ploughing villagers' fallow areas, swidden fields, graves, spiritual sites and forests as 'weapons of mass destruction' (Rith and Strangio, 2009, p1). In this sense, the survival of swidden systems in Ratanakiri also needs to be seen as part of a wider cultural struggle and fundamentally linked with revaluing indigenous culture, including maintaining communal land-management arrangements.

Several contradictions arise from this conflicting dynamic between plantation agriculture, intensive cash cropping and swidden systems. Little consideration is given to the consequences of converting large areas of land from food crops and diverse secondary-forest regrowth to monocultures of luxury and non-food crops such as cashews, rubber and cassava for biofuel. A new cultural perspective is transforming Ratanakiri's landscapes, and is fundamentally revaluing its land and natural resources. In this context, without the establishment of concrete land-tenure security, traditional swidden systems stand little chance of surviving the blatant abuses discussed above.

Securing communal lands in Ratanakiri to ensure a future for swidden farming

Experiences in other tropical upland areas suggest that there is a certain inevitability about the processes of land-use change; that they will lead to individual land holdings, cash cropping and large numbers of landless indigenous people. However, we argue that there is another possible scenario. We now explore the option and potential of communal-land titling as a basis for allowing the swidden system to continue and to adapt to new circumstances.

As mentioned earlier, adoption of Cambodia's new Land Law in August 2001 made it possible for indigenous communities to register their customary agricultural land (including swidden fallows) under a collective title. The provisions of this law, as it relates to collective land title for indigenous communities, resulted from a very timely and unprecedented advocacy effort that mobilized local communities, non–governmental organizations, academics and legal groups (Williams, personal communication). See Box 14-2 for details of how the law came about.

The legal framework to protect and maintain indigenous community land, therefore, is already in place. The challenge is implementing and enforcing it. In view of the powerful economic forces driving privatization, cash cropping and land concessions, there is a strong need for a vibrant, grass-roots civil society together with a functioning legal system, in order to claim and defend the legal entitlements of swidden cultivators (such as communal title).

Under the 2001 Land Law and the procedures that were subsequently formulated through a sub-decree in 2009, obtaining a communal land title entails a three-

---

BOX 14-2: How communal-land titling in Cambodia came about

Provisions for communal titling of land for indigenous communities were included in Cambodia's 2001 Land Law because of a combination of well-timed advocacy by community representatives and non-governmental organizations (NGOs), donor pressure, assistance from the King of Cambodia, and a measure of goodwill towards indigenous minorities from some Members of Parliament and government officials.

The campaign for communal land titling grew out of consultations with community leaders during the mid-1990s. Around this time, political manoeuvrings in Phnom Penh ironically allowed a certain amount of leverage to lobby the Government to recognize indigenous peoples' land rights. In July 1997, factional fighting ousted the elected first Prime Minister, Norodom Ranariddh.[21] When several important donors, including the United States, cut their development assistance as a result of the fighting, the Government found itself forced to improve its image to regain international legitimacy.

In 1998, the Government was forced to undertake a review of the 1992 Land Law as a condition for a loan from the Asian Development Bank for agricultural development (Baird, 2011). An Oxfam-funded land-study project, together with local and international NGOs, found that they were able to lobby for land reforms favouring poor and indigenous people, and these calls were supported by upland communities. Advocacy and support from the Ratanakiri Provincial Governor strengthened the call to include communal land titling for indigenous communities into the draft of the new Land Law.

Initially the Ministry of Land Management rejected the idea of including provisions for communal land titling, but pressure from NGOs and donors, helped by community-level consultations and support from key government officials finally resulted in agreement from the Government. Announcement of the Government's concurrence came directly after a meeting of Cambodia's international donors in Tokyo. It enabled comprehensive community consultations to take place, resulting in a draft section of the new Land Law covering the collective rights of indigenous communities.

A June 2000 draft of the new law not only included eight articles outlining provisions for indigenous communities to claim their land as a communal title, but it also legitimized rotational swidden agriculture as a land use. The Council of Ministers threatened to excise the articles that both defined an indigenous community and recognized the rights of these communities to fallow swidden lands. It was only through media publicity and a personal plea from the King of Cambodia that the Prime Minister finally ordered that the articles be retained. Section three of the Land Law, allowing for communal land ownership for Buddhist pagodas and indigenous communities, was passed by 71 votes to 15 in Cambodia's National Assembly on July 6, 2001 (RGC, 2004).

stage process.[22] However, even though more than a decade has passed since the Land Law came into effect, only 12 communities have achieved a communal title. Nevertheless, a significant number of indigenous villages are at different stages in the titling process. As of July 2016, a further six villages are in the final stages of getting their title. An additional 14 villages have submitted their land title applications to the Ministry of Land Management. The Ministry plans to issue 10 communal land titles per year nationally, although donor support ended for this work in mid 2016 and communal land titling is now likely to be slower than this. In total, 107 villages throughout the country (including the villages mentioned above) have been recognized by the Ministry of Interior as legal entities, allowing them to be issued a communal title once land management regulations and the land demarcation processes are completed. Altogether 117 villages, also throughout the country (again including the villages mentioned above), have completed the first step in the three stage process of communal land titling by applying and becoming approved as an indigenous community by the Ministry of Rural Development.[23] The process, however, has been far too slow for many villages, which have, in the meantime, lost large parts of their land, given the speed of land-use changes. For many villages throughout Cambodia's northeast, it has become a case of villages trying to secure what land remains.

Given the importance of communal land tenure for the survival of the swidden system, it is useful to summarize some of the lessons learned from efforts to secure it both in legislation and on the ground. Advocacy in this regard has been ongoing in Ratanakiri, by indigenous communities and on their behalf, for the past 18 years. Some of these lessons can be summarized as follows:

1. Strong and informed communities are essential. In many cases, communities have shown surprising resilience in the face of attempts by outsiders to encroach on their lands, or attempts by government officials to intimidate them and weaken their resolve. The value of legal-registration processes, therefore, is in the community building that is part of this. Selection of community representatives to serve on Land Management Committees, for example, has been most effective when broad community involvement and consensus has ensured their accountability.

2. Thorough negotiations between neighbouring villages are essential to resolve boundary issues. Granting official community tenure where there are disputed village boundaries has proven to be an opportunity for land brokers to foment conflicts, leading to the disenfranchised community selling off the disputed land.

3. A pan-indigenous voice is essential to keep the government accountable and to claim legal entitlements. This also needs to link with wider 'rural poor' and international indigenous peoples' lobbies.

4. Support from non-governmental organizations (NGOs) for local communities has been most effective when the partnership is initiated by the community,

and the NGO acts within agreed parameters in response to community needs and requests. Community organizing and strengthening has also proven most effective in times of crisis, such as when the community is facing illegal logging or illegal land encroachment.

5. Working in land-security activities has been challenging. Many NGO and community leaders have faced intimidation and violence. NGOs and leaders have had to find a balance between arguing against land rights abuses and demanding that rights are recognized, and finding a cooperative arrangement with government authorities. However, in many cases the balance has gone too far in the direction of attempting to reduce confrontation with those in power (often out of valid concern for personal security). The result has been that the focus on land security has been lost.

Building on existing swidden systems in Ratanakiri

Given the pace of change that is underway, sustained and concerted advocacy by swidden farmers will not be enough to see the system survive on any scale; a significant level of adaptation will also be required. This should include, for example, creatively accessing markets and cooperative processing of swidden and fallow products, as well as intensifying fallows, and the processing and marketing of products from the swidden system. As seen from the example of Krola village, economic initiatives by individuals and families can still flourish within a framework of traditional communal tenure. The rapid adoption of cashews as a mainstay economic crop by indigenous farmers highlights the flexibility of the system if it is properly managed, and also highlights the need for the survival (and adaptation) of the traditional system, to accommodate changes in land use and economic activity.

In the current economic environment, community solidarity and cultural identity will remain resilient only if there is some shared economic interest. The forced-labour cooperatives of the Pol Pot era (1975 to 1979) and the Vietnam-backed State of Cambodia (1979 to 1991) have left a bad taste in people's mouths.[24] However, communities will need to organize themselves in some way in order to get the most out of production, processing and marketing their swidden products. Some form of cooperative organization within (and between) communities, that yields a net economic (livelihood) benefit to all members, is necessary to galvanize action to protect the remaining land and resources. Promoting artisanal specialization and family enterprises would help communities to adapt as livelihood niches are created by the new processing and marketing activities. New skills are needed for this kind of community adaptation, and we strongly recommend that NGOs refocus their efforts in this area, rather than continuing with the currently prevalent 'service delivery' approach. This would complement and greatly enhance the efforts of some groups to promote cultural identity and solidarity within and between indigenous communities. In the current political environment, direct advocacy has been almost impossible. Economic advocacy may prove to be more successful, and NGOs could have a strategic role in providing the necessary skills for this. For instance, several NGOs are engaged in providing community-managed education in both local and national languages for adults and children. This addresses the immediate and long-term needs of the communities. Lack of education is a major contributing factor to the indigenous peoples' increasing marginalization; people are not well prepared for the changes, and this leaves them with a feeling of powerlessness. Ideally, indigenous peoples' movements should be taking on more and more of this work in the future.

Swidden agriculture has a number of distinct and potentially marketable advantages over cash-crop monocultures. The biggest advantage is that communities can remain intact, make a contribution to the national economy and provide a national service by maintaining the landscape in an ecologically fit state. However, changing the mindset of decision-makers and promoting an alternative vision is required. Such a vision would see strong indigenous communities functioning as viable land-management and economic-production units as an alternative to multinational business corporations.

It is important now to develop alternatives to the conversion of biodiverse forested areas – such as swidden landscapes – to monoculture plantations. The significant decrease in rubber prices over the past two to three years has called into serious question the government's economic strategy over the past decade or more, of converting forest and swidden land to industrial concessions to plant rubber. Instead, we propose that it would make economic sense for the government to support the development of indigenous communities and their capacity as production managers on securely tenured community land. Ultimately, a system of biodiverse, multi-storeyed agroforest-type agriculture may be the most logical end-point in the process of adapting the swidden system. Such a system would maintain the ecosystem as well as produce food and income throughout the year. Communal swidden lands could be more a mix of cropping, agroforestry and fallows. Cooperative technical services, processing and marketing (or local private businesses that provide these services) would greatly facilitate the transition. Several possibilities that maintain forest systems while generating economic return are possible if there is secure communal tenure. These kinds of interventions have not yet been widely promoted or adopted. However, as the pressure on land increases, there will be increasing incentives for farmers to invest in sustainable adaptations. Local farmers need to demonstrate the productiveness of their fallows while they still have them, so that they are not lost to concessions and speculators.

Conclusions

Swiddening is an agricultural system that is still alive and widely practised by indigenous communities in Ratanakiri. However, it has been greatly modified in the past decade due to severe pressure on fallowed land and the rapid trend towards cash cropping. The entire upland swidden-agriculture system, and indeed the whole livelihood, lifestyle and identity of indigenous communities in Cambodia, is now very much under threat due to unmitigated land speculation, land concessions and in-migration from the lowlands. If current trends continue, the onslaught on swidden systems will eventually breakdown social solidarity at family and community levels, increasing landlessness and food insecurity.

In this chapter we have argued that the continued survival of swidden systems requires attention to the wider socio-ecological, political and cultural contexts in which it is embedded. At the local level, this means maintaining communal-property arrangements and the socio-cultural arrangements of which swidden systems are a part. However, land use in Ratanakiri is increasingly dominated by central-level decision-makers. For this reason, we argue that attention is needed at this level to ensure that as much land as possible is officially recognized as the communal property of indigenous communities. The only future for swidden systems is proper recognition of land rights and implementation of existing laws.

Communal management appears to play an important role in promoting diversity in landscapes, and this is essential for sustainable swidden systems. The overall productivity, covering a range of different products, of diverse agro-ecosystem landscapes that are able to adapt to changing climatic conditions needs much more serious analysis. As Plant and Hvalkof (2001, p27) point out in reference to South America, individual titling of indigenous lands freezes farmers in one location and reduces the 'flexibility of the individual production unit to the detriment of productivity'. They argue that 'communal titling in tropical-forest environments, apart from the social arguments, also proves to be the most viable approach for enhancing the productivity and full economic potential of the individual producer' (Plant and Hvalkof, 2001, p27).

Swidden will survive, in some form at least, in the small number of villages that manage to register their collective land and then defend it strongly. There is a greater chance of building on swidden systems for the long-term social, environmental and economic well-being of indigenous communities, and to provide a model of sustainable land use in the uplands, if as many villages as possible have their land rights recognized, and if local people are empowered to protect their land. If this does not happen, then swidden/fallow systems in northeastern Cambodia will not survive the present onslaught of land-use change that is underway in the region.

References

Aymonier, E. (1895) *Voyage dans le Laos* (Travel in Laos), Leroux, Paris

Backstrom, M., Ironside, J., Paterson, G., Padwe, J. and Baird I.G. (2006) *Case Study of Indigenous Traditional Legal Systems in Ratanakiri and Mondulkiri Provinces,* United Nations Development Programme (UNDP) and Ministry of Justice, Phnom Penh

Baird, I. G. (2011) 'The construction of "indigenous peoples" in Cambodia', in Leong Yew (ed.) *Alterities in Asia: Reflections on Identity and Regionalism*, Routledge, London, pp155-176

Baird, I. G. (2013) 'The ethnoecology of the Kavet peoples in northeast Cambodia', in M. Poffenberger (ed.) *Cambodia's Contested Forest Domain: The Role of Community Forestry in the New Millennium,*' Ateneo de Manila University Press, Manila, pp155-186

Beresford, M., Ngoun, S., Rathin, R., Sisovanna, S. and Ceema, N. (2004) *The Macroeconomics of Poverty Reduction in Cambodia*, Asia-Pacific Regional Programme on the Macroeconomics of Poverty Reduction, United Nations Development Programme, Kathmandu and Ponleu Khmer Printing House, Phnom Penh

Bopha, P. and Marks, S. (2009) 'Forbes lists Cambodia in top 10 for corruption', *Cambodia Daily* 26 March 2009, Phnom Penh

Bourdier, F. (1995) *Knowledge and Practices of Traditional Management of Nature in a Remote Province*, Report of a research mission on the theme of environment in Cambodia, sponsored by AUPEL/UREF (October 1994 - July 1995), unofficial English translation by Dr Carol Mortland, East-West Centre, Honolulu

CLV-DTA (2004) *Socio-economic Development Master Plan for the Cambodia-Laos-Vietnam Development Triangle Area* (unpublished), CLV-DTA, Hanoi (See also: http://clv-development. org/en/, accessed 28 April 2017)

COHCHR (2004) *Land Concessions for Economic Purposes in Cambodia: A Human Rights Perspective*, United Nations Cambodia Office of the High Commissioner for Human Rights, Phnom Penh

COHCHR (2007) *Economic Land Concessions in Cambodia: A Human Rights Perspective*, United Nations Cambodia Office of the High Commissioner for Human Rights, Phnom Penh

Colm, S. (1996) *The Highland Minorities and the Khmer Rouge in Northeastern Cambodia 1968-1979*, Document Center of Cambodia, Phnom Penh

Fox, J. (1997) 'Customary boundaries in Ratanakiri: A study of three villages in Poey Commune', (unpublished) East-West Centre, Honolulu

Fox, J. (2002) 'Understanding a dynamic landscape: Land use, land cover, and resource tenure in Northeastern Cambodia', in S. Walsh and K. Crews-Meyer (eds) *Linking People, Place, and Policy: A GIScience Approach*, Kluwer Academic Publishers, Boston, pp113-130

Fox, J., Vogler, J. and Poffenberger, M. (2009) 'Understanding changes in land and forest resource management systems: Ratanakiri, Cambodia', *Southeast Asian Studies*, 47(3), pp309-329

Guerin, M. (2001) 'Essartage et riziculture humide: Complementarite des ecosystems agraires a Stung Treng au debut du XX siecle' (Slash-and-burn and wet rice: The complementarity of agrarian ecosystems in Stung Treng at the beginning of the century), *Aseanie* 8, pp35-56

Guerin, M., Hardy, A., Chinh, N. and Hwee, S. (2003) *Des montagnards aux minorities ethniques: Quelles integration nationale pour les habitants des hautesterres du Vietnam et du Cambodge?* (Of mountain ethnic minorities: What national integration for people in the highlands of Vietnam and Cambodia?), Institut de Recherche sur l'Asie du Sud Est Contemporaine (IRASEC), Bangkok

Hubbel, D. (2007) 'Indigenous people and development in northeast Cambodia', *Watershed* 12(2), pp33-42

Ironside, J. (1999) *Culture and Agri-culture. Hill Tribe Farming Systems from an Agroecological Perspective: A Case Study of Yeak Loam Commune, Ratanakiri Province, Cambodia*, IDRC/CARERE, Ban Lung, Ratanakiri, Cambodia

Ironside, J. (2009) 'Poverty reduction or poverty creation? A study on achieving the MDGs in indigenous communities in Cambodia', in F. Bourdier (ed.) *Development and Dominion: Indigenous Peoples in Laos, Vietnam and Cambodia*, White Lotus Press, Chiang Mai, pp79-113

Ironside, J. (2011) 'Competition for the communal lands of indigenous communities in Cambodia', paper presented to an International Academic Conference on Global Land Grabbing, 6-8 April 2011, Institute of Development Studies, University of Sussex, Brighton, UK

Ironside, J. and Baird I.G. (2003) *Wilderness or Cultural Landscape? Settlement, Agriculture and Land and Resource Tenure in Virachey National Park, Northeast Cambodia*, Biodiversity and Protected Areas Management Project, Ministry of Environment and World Bank, Phnom Penh

Ironside, J. and Nuy, B. (2010) *Development with Identity: Assessment of the Impact of Tenure Security from Legal Entity Registration in Indigenous Communities in Cambodia (Mondulkiri and Ratanakiri Provinces)*, Danish International Development Agency (DANIDA), Phnom Penh

Maitre, H. (1912) *Mission Henri Maitre (1909-1911) Indochine Sud-Centrale: Les Jungles Moi*, Emile Larose, Libraire-Editeur, Paris

Marks, S. (2010) 'Governments, investors to map Ratanakiri investment', *Cambodia Daily*, 12 March 2010, Phnom Penh

Matras-Troubetzkoy, J. (1983) *Un village en foret: L'essartage chez les Brou du Cambodge* (A Village in the Forest: Swidden Cultivation among the Brou of Cambodia), Society for Linguistic and Anthropological Studies of France (SELAF), Paris (Unofficial English translation by C. Mortland, East-West Centre, Honolulu, December 1995)

Meyer, C. (1979) 'Les nouvelles provinces: Ratanakiri – Mondolkiri (The new provinces, Ratanakiri and Mondolkiri)', *Revue Monde en développement* 28, pp682-690

MOP (2007) *Progress in Achieving Cambodia Millennium Development Goals: Challenges and Opportunities*, 2007 annual ministerial review of the high-level segment of ECOSOC, Geneva, 2-4 July 2007, Ministry of Planning, Phnom Penh

MOP (2009) *2008 Cambodia Population Census*, CELADE - Population Division, ECLAC and National Institute of Statistics, Ministry of Planning, Phnom Penh

MOP (2013) *Intercensal Population Survey*, National Institute of Statistics, Ministry of Planning, Phnom Penh

MOP/UNDP (2010) *Commune database and the implementation of Cambodia Millennium Development Goals at the sub-national level* (Powerpoint presentation), D&D and Seth Koma working group, Ministry of Planning, Phnom Penh

Naren, C. (2012) 'Cambodia surges full speed ahead with land concessions', *Cambodia Daily*, 23 March 2012, Phnom Penh

Nevins, J. and Peluso, N. (eds) (2008) *Taking Southeast Asia to Market: Commodities, Nature, and People in the Neoliberal Age*, Cornell University Press, Ithaca, NY

Plant, R. and Hvalkof, S. (2001) *Land Titling and Indigenous Peoples*, Inter-American Development Bank, Washington, DC

Rabe, A. (2013) *Directive 01BB in Ratanakiri Province, Cambodia: Issues and impacts of private land titling in indigenous communities*, Asia Indigenous Peoples Pact, Chiang Mai

Ratanakiri Department of Agriculture (2009) *Provincial Agricultural Statistics* (unpublished), Ban Lung, Ratanakiri, Cambodia

RGC (2001) *Land Law*, Royal Government of Cambodia, Phnom Penh

RGC (2004) *National Assembly Debates*, Chapter 3, Land Law - Collective Ownership, Transcripts (English translation), Royal Government of Cambodia, Phnom Penh

RGC (2005a) *Subdecree on Economic Land Concessions,* Urban Planning and Construction, No. 146 ANK/BK, Ministry of Land Management, Royal Government of Cambodia, Phnom Penh

RGC (2005b) *National Strategic Development Plan 2006-2010*, (Unofficial translation from the Khmer version), Royal Government of Cambodia, Phnom Penh

RGC (2014) *Annual Progress Report 2013: Achieving Cambodia's Millennium Development Goals*, Ministry of Planning, Royal Government of Cambodia, Phnom Penh

Rith, S. and Strangio, S. (2009) 'Villagers curse Mondulkiri plantation', *Phnom Penh Post*, 19 June 2009, Phnom Penh

Salemink, O. (2003) 'Enclosing the highlands: Socialist, capitalist and protestant conversions of Vietnam's Central Highlanders' (draft), Vrije Universiteit, Amsterdam

Shiva, V. (2003) 'Globalization and the war against farmers and the land', In N. Wirzba (ed.) *The Essential Agrarian Reader: The Future of Culture, Community, and the Land*, University Press of Kentucky, Lexington, KY, pp121-139

Sturrock, T. (2010) 'Cambodia ranked as second most corrupt in the region', *Cambodia Daily*, 10 March 2010, Phnom Penh

UNDP (2010a) *Current Status of Cambodian Millennium Development Goals (CMDG)* (draft), United Nations Development Programme, Phnom Penh

UNDP (2010b) *What We Do: Poverty Reduction*, United Nations Development Programme, Phnom Penh, www.un.org.kh/undp/what-we-do/poverty-reduction, accessed 6 June 2013

Vrieze, P. and Chancy, C. (2009) 'Cambodia scores low on International Hunger Index', *Cambodia Daily*, 3 December 2009, Phnom Penh

Vrieze, P. and Naren, C. (2012) 'Carving up Cambodia: One concession at a time', *Cambodia Daily*, 10-11 March 2012, Phnom Penh

Williams, S. (personal communication) Shaun Williams was coordinator of the Oxfam Land Study Project that facilitated most of the advocacy on the Land Law with the central Government between 1998 and 2000.

Notes

1. The government sees the northeast of the country becoming one of four 'pillars' of economic development for the country by 2015. The other three 'pillars' are the capital Phnom Penh; Siem Reap, where the temples of Angkor Wat are located; and Sihanoukville, the country's main port (COHCHR, 2007).

2. Article 25 of the Land Law states that 'the lands of indigenous communities include not only lands actually cultivated but also include reserved land necessary for the shifting of cultivation which is required by the agricultural methods they currently practice and which are recognized by the administrative authorities' (RGC, 2001).

3. The 2008 census counted only those people who spoke a mother tongue different from Khmer. Indigenous groups face significant difficulty in being accurately accounted for in national statistics.

4. The 2013 inter-census population survey puts the number of Tampuan mother-tongue speakers at 56,800 (MOP, 2013). However this was based on a sample and is thus perhaps an overestimation. All of these ethnic groups, except the Jarai, speak an Austroasiatic language in the Bahnaric branch of the Mon-Khmer family. The Jarai speak an Austronesian language related to that of the Cham, who originate from what is now central Vietnam. It is also related to Indonesian (Ironside and Baird, 2003).

5. Highland villages traded elephants, buffaloes, pigs, chickens, tobacco, rice, gongs, ceramic urns, iron, brass, ivory jewelry and other items between themselves and with neighbouring Lao lowland villages (Guerin, 2001). Depending on the particular political configuration in their area, individual villages also sold goods to Thai, Lao, Khmer, Vietnamese, Burmese and Chinese traders (Aymonier, 1895; Maitre, 1912).

6. Other highland, or upland, provinces with significant indigenous populations that are being similarly affected include Preah Vihear, Kratie, Stung Treng and Mondulkiri.

7. A UNDP study (2010b) found that one-third of Cambodians still live below the poverty line. Cambodia is one of 36 countries with high rates of child undernutrition and has been grouped with 33 countries with 'alarming' levels of hunger and undernutrition (Vrieze and Chancy, 2009; UNDP, 2010b). Both Transparency International and Forbes Magazine have repeatedly listed Cambodia as one of the most corrupt countries in the world (Bopha and Marks, 2009; Sturrock, 2010).

8. The CLV-DTA provinces are Mondulkiri, Ratanakiri, Stung Treng and Kratie in Cambodia; Attapeu, Saravane, Sekong and Champassak in Laos; and Dak Lak, Dak Nong, Gia Lai, Kon Tum and Binh Phuoc in Vietnam.

9. The primary sources of funds for the 'Development Triangle Area' are the governments of Japan, China and Vietnam, along with private businesses (Marks, 2010).

10. A meeting involving all three countries, to discuss implementation and coordination of the Development Triangle plan, was held at Ban Lung, Ratanakiri, in March 2010. Local villagers said they had heard nothing of the plans and were not included in discussions or in making important decisions concerning the future of their area (Marks, 2010).

11. Despite prices for Vietnamese and Cambodian rubber that are 7% to 8% lower than those for rubber of the same category from Thailand and Malaysia, the CLV Triangle region is seen as having a price advantage due to 'favourable land conditions and cheaper labour costs' (CLV-DTA, 2004, p95).

12. These bamboo areas have now been incorporated into Virachey National Park.

13. Fox reached his conclusion that 81% of the landscape was being used for swidden agriculture by analyzing 1953 aerial photographs. These showed that 21% of the area was active swidden and 60%

was secondary forest (fallow land that was part of long-term rotations). In other words, 81% (21% + 60%) of the land area was being used for swidden agriculture in different stages of the cycle.

14. Matras-Troubetzkoy's (1983, p45) anthropological study in the 1960s of a Brao-Kreung village highlights the social negotiation involved in sharing land. The study carries the comment that 'the leisure time of the dry season, which favours long conversations before a fire or a jar of rice wine, have allowed the inhabitants of the village to share their intentions on this subject [the choice of land for the new agrarian cycle] and to discuss it with their neighbours'. In the more remote villages of Ratanakiri, this practice of negotiating and planning the sharing of land continues, during drinking and eating after religious ceremonies.

15. In the 10 to 15 years up to 2009, 20,170ha of cashew trees were planted in Ratanakiri, mainly by indigenous farmers. Total production in 2009 was 7093 tonnes (Ratanakiri Department of Agriculture, 2009).

16. See http://www.opendevelopmentcambodia.net/company-profiles/economic-land-concessions/ and http://www.licadho-cambodia.org/concession_timelapse/ (both accessed 17 February 2015). In Cambodia as a whole, Vrieze and Naren (2012) reported that as of the end of 2011, 2,036,170 hectares had been leased in Cambodia for economic land concessions to 227 private companies. Naren (2012) reported that the total area was 2.5 million hectares and that 300,000 hectares had been given away in less than three months in 2012. These figures represent roughly 12% of the country's land area.

17. Article 23 of the 2001 Land Law states that '… while waiting for legitimate recognition of the community by-laws, the groups actually existing at present shall continue to manage their community and immovable property according to their tradition and shall abide by the provisions of this law'. Further, Article 25 states that 'The lands of indigenous communities are those lands where the said communities have established their residences and where they carry out traditional agriculture. The measurement and demarcation of boundaries of immovable properties of indigenous communities shall be determined according to the factual situation as asserted by the communities, in agreement with their neighbours' (RGC, 2001).

18. Article Four of the 2005 Subdecree on Economic Land Concessions states that 'An economic land concession may be granted only on land that … has been registered and classified as state private land' (RGC, 2005a). Article Four states that 'The [government] Contracting Authority shall ensure that there will not be involuntary resettlement by lawful land holders and that access to private land shall be respected' (RGC, 2005a). Article 30 of the 2001 Land Law allows for a person with five years' uncontested possession of a piece of land to claim ownership of that land (RGC, 2001). This means that after five years, the land becomes privately owned and can no longer be considered state land. Therefore, fields that villagers have been using for five years or more must by law be excluded from concession areas.

19. According to statistics from the Ministry of Land Management, up to mid-December 2014 the Order 01 campaign surveyed more than 710,000 plots in 357 communes throughout Cambodia and issued 610,000 titles. As of 17 December 2014, the campaign had resulted in the reclassification of about 1.2 million hectares of land that was previously regarded by the Government as state land. This included 270,000ha from 17 forest concessionaires, 380,000ha from 134 economic land concession companies and 530,000ha from other types of state land and forest areas.

20. The great majority of indigenous women not only have limited marketing skills, but they also have limited use of the Khmer language. They are thus ill-equipped to obtain fair prices from traders who arrive at swidden fields on motorcycles to buy cashews, dried cassava chips, sesame, corn, small animals and so on. Often, families must depend on their primary-school children to count the money from cashew sales. However, some progress is being made through various empowerment, education and land initiatives that are increasingly using the women's own languages. Use of local languages helps women to become actively involved in addressing the issues affecting their livelihoods.

21. From 1993 to 1997, Cambodia's two main political parties shared power in a coalition government to avoid the potential break-up of the country.

22. The process requires official confirmation of a community's indigeneity from the Ministry of Rural Development's Department of Ethnic Minorities; registration of the community as a legal entity

with the Ministry of Interior; and approval of the community's land-management regulations, surveying, demarcation and titling of the community's land by the Ministry of Land Management. Registration as a legal entity is necessary for the village to hold a communal land title. Since the passing of the General Policy on Highland Peoples Development in 2009, the Department of Ethnic Minorities Development, within the Ministry of Rural Development, now verifies that a community is 'indigenous' before it can begin the legal-entity registration process. This is because the communal-titling provisions in the Land Law apply only to indigenous communities.

23. Ratanakiri province has a total of 241 villages, although only 60% to 70% of them are indigenous minority villages.

24. The Khmer Rouge actually controlled Ratanakiri and other northeastern provinces from 1970 to 1979.

15

THE GROWING VOICE OF THE STATE IN THE FALLOWS OF LAOS

*Olivier Ducourtieux**

Introduction

The voices of states often growl in the forest, overpowering the voices of villagers living there. For decades – even centuries – states have been trying to push shifting cultivators out of the forest fallows that are part of their farming systems. To review these processes and ascertain the causes, which have evolved over the years, this chapter examines the situation in Laos, and more precisely, in the villages of the Phunoy people in Phongsaly province (Figure 15-1).

According to Bouté (2011), the origin of the Phunoy ethnicity lies in the mid-19th century, tied to the duty of defending the borders of the Lane Xang kingdom, a Tai state based in Luang Prabang. The Lao nation had not yet been born (Ivarsson, 2008), so these communities were not dedicated to Laos as we know it today. The core task undertaken by the Phunoy came from direct allegiance to a political leader (the king), in a social contract rewarding them with land in return for part-time military duty. This was during political convulsions associated with the decline of the Chinese empire, expansion of the Siamese kingdom and the Vietnamese empire, and the first French and English colonial ambitions in the region. Phunoy society was therefore not defined by opposition to the State (Clastres, 1989), but rather by an innate intimacy with it. However, this early link remained for years a distant contract, with a limited impact on the day-to-day life of the villagers.

In tales of the origin of Komen village, told by its former head, the State appears twice. First, in an initial contact, the distant king in Luang Prabang rewarded the founders of the village with permission to settle in the Phongsaly area. The local state representative, the *Chao Muong*, then specified precisely where they could settle.

_____

* DR OLIVIER DUCOURTIEUX is an Associate Professor in the Comparative Agriculture Unit/UMR PRODIG at AgroParisTech, in France. He worked on rural development projects in Laos from 1993 to 2007.

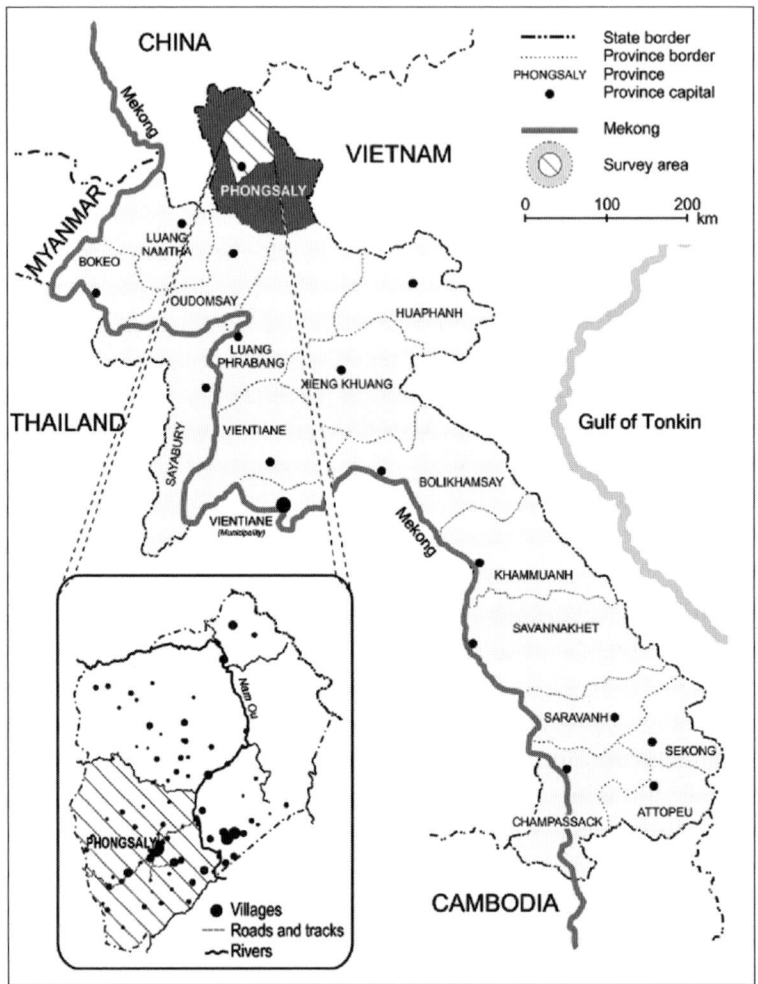

FIGURE 15-1: Phongsaly city, the capital of Phongsaly province, and surrounding villages in northern Laos.

Second, there were subsequent yearly contacts, when the villagers had to pay a tribute for the land to the *Chao Muong*. Apart from this, they were left alone to live as they wished, practising shifting cultivation and managing the land and the fallows in accordance with community regulations and internal social relations. This autonomy continued until the end of the 1960s, and for most villages, even into the 1990s.

Phunoy livelihood has been based on shifting cultivation for many generations. The nuclear family performed most of the farming activities (Ducourtieux, 2006). Sometimes, compelled by the need for a larger workforce or driven by the inexorable timing of farm operations, households would exchange labour. Some farm operations were managed at a community level, but never beyond the village. For example, the council of elders would meet after the villagers harvested the rice to discuss and select the area of secondary forest that would be cleared for the next swidden cycle.[1] This piece of land, which would have been cleared and cropped many times since the foundation of the village, would comprise a plot for each household (Ducourtieux, 2009). The council of elders would also guide the process of burning the slashed vegetation by deciding the date and organizing the process on a village level. However, after the cleared land was fenced, the community would not interfere further in the cultivation cycle. Village land was managed at different levels, but there was never a role to be played by the State. Even expansion into new territories involved the State only marginally. Natural demographic growth caused the number of households per village to grow, and this increased the need for land for shifting cultivation. When land-management mechanisms could no longer bear the tensions of villagers' increasing needs and demands, some families would leave to settle another village in a new zone. Often, the new villages were named after the old ones, with the adjectives 'big, small, new, old, south or north' appended.[2] The voice of the State was very weak – if it was ever heard at all – in the Phongsaly fallows. However, by the early 20th century, Laos had become a part of colonial French Indochina.

Progressive building of the Lao State's aversion for shifting cultivation

Uneasy French acceptance

As the British and French colonial empires expanded during the 19th century, explorers, military officers, executives and scientists discovered the wide scale of shifting cultivation in their tropical territories. It was a farming practice that had already vanished from the European mainland. The colonial administrators were quick to fix their minds against swidden agriculture and built up a discourse advocating its elimination (Grove, 1994). For example, Hugh Cleghorn, while managing the colonial Forest Department in Madras, India, wrote:

> 'Shifting cultivation ought not to be tolerated except in a very wild and unpeopled country. […] It leads to unsettled habits and takes away from the regular cultivation of a fixed spot. It is carried on by a set of savages who would be more profitably employed on public works or coffee plantations (Cleghorn, 1861).'

Subsequently, in 1860, shifting cultivation was officially banned in the Madras Presidency (Das, 2005). Similar discourses were also common in Indochina – the pearl of the French empire. Clovis Thorel, physician and botanist of the Mekong Exploration Commission from 1866 to 1868, reported:

> 'In terms of forestry, all peoples of Indochina, including the Chinese, only know ways of destroying forests. Everywhere they burn forests, whether to grow forest rice, maize, and cotton; to clear land so they can more easily

move about and hunt animals; or, as we saw so many times, to simply distract themselves (Thorel, 2001, p185). The second mode of rice cultivation, which is practised in forests, is a barbarian, transitory method that is destined to disappear with the progress of civilization. [...] It is practised from Saigon to China, but more frequently in Cambodia and Laos, where civilization is still much more backward and where forests are more extensive.' (Thorel, 2001, p79).

The text encapsulates the French colonizers' position towards shifting agriculture. It was seen as an archaic, unproductive and degrading practice throughout the 90 years of French administration in Indochina. The condemnation took its origins from the very sources of Western colonization: racism, the 'civilizing mission' and greed. Racism was commonplace among administrators, scientists and colonists alike in Indochina, as well as their supporters in France (Gunn, 2003). For example, Major Georges Aymé wrote of the population of Phongsaly:

'[The Phunoy] is by temperament rather intelligent, frank, gay and at times naïve; he is not belligerent, indeed perhaps not even very courageous. He contents himself to be submissive, devoted and industrious (pp36–37). [...] Temperamentally, the Ho are quite the least interesting tribe of the Territories, to which they are of little benefit: hypocritical and deceitful, [...] living from the sale of alcohol and contraband [...] the Ho should be displaced [...] to a location where they could languish in the mud they value so much, without disturbing anybody (Aymé, 1930, p57)'

Such examples are endless, given how colonial publications unanimously and continuously reiterated the same clichés. Colonial racism was used to control Indochina, underpinning the systematic division of tasks and powers to be left in the hands of local people (Vann, 2003). The different ethnic groups were played off against one another to prevent an anti-colonial alliance that mainland France could not have controlled, given the weak military deployment in Indochina. Before becoming a Marshall of France after World War I, Colonel Joseph Gallieni was in charge of military and security issues in Tonkin (the northernmost part of Vietnam). In discussing the 'pacification' of Tonkin in 1895, he wrote:

'Every agglomeration of individuals, race, peoples, tribes or family, is the sum of its shared or opposing interests. [...] there are rivalries we should know how to defuse, or to use for our gain by playing them off against one another.' (Ferro, 2003)

One of the motivations for colonization, proclaimed by the Third French Republic, was the so-called 'civilizing mission' (Cleary, 2005). Travel writers regularly reflected upon France's moral responsibility to raise people from their supposed archaism and lead them towards 'progress'. Progress was the main point of the argument put forward in a debate in the National Assembly in 1885 over the allocation of funds for the conquest of Indochina. Left-wing representative and promoter of public education in France, Jules Ferry, challenged opponents of Indochinese colonization with the question: 'Provocative? Civilisation, when it seeks to open up barbarian lands?' Ferry was also president of the council that would defend Indochinese colonization. Such messianic expressions conveyed blatant paternalism towards the local populations. Earlier, when reporting on agriculture and ethnobotany following the Mekong Exploratory Commission in 1868, Clovis Thorel established a discourse of superiority and promoted France's guiding role:

> 'Particularly in the south of Indochina, agriculture can greatly benefit from improved methods, and European influence [...]. This foreign influence would not only work upon their farming methods but also on their social and territorial organization, to which the defective state of agriculture greatly owes. [...] we have little to learn from the farming methods of the Indochinese' (Thorel, 2001, pp2-3).

However, the 'generous impulse' of 'bringing civilization' concealed a motivation that had been a recurring theme for the French colonial empire, ever since its origins in the 16th century: economic exploitation. The founder of a pro-colonial lobby called the French Colonial Union, representative Eugène Etienne, declared in 1894: '[The colonial empire seems necessary] to assure the future of our country in new continents, to preserve an opening there for our goods and to find raw materials for our industries'. Colonial investors were a powerful lobby active in the sphere of influence of the Bank of Indochina and formed powerful financial cartels, influencing even the political authorities in Hanoi and Paris (Gunn, 2003). The collection of resources from the colonies entailed rejecting any local techniques that failed to profit the colonial power, and took two forms that concerned rural populations in Indochina:

- A poll tax to extract revenue from farmer labour; indirect taxation through company monopolies (salt, alcohol, mineral oils, tobacco and opium); tenant farming for the rice plantations of Cochinchina (the southern third of Vietnam); and taxes on the trading of forest-harvested products, such as sticklac, benzoin and cardamom.
- Plantation economies (notably rubber, coffee and tea) using salaried employees and controlled by private financial groups, situated mainly in the highlands of Cochinchina, Annam (central Vietnam) and Cambodia (Daniel et al., 1992; Murray, 1992), that owed their existence to land conquests made at the expense of local shifting cultivators.

Relations between the colonial plantation holders and the local smallholders were often strained. Early antagonism resulting from expropriation grew due to the colonizers' insatiable desire for land and their attempts to extend their franchises into village territories. Criticism of shifting agriculture served as a perfect argument for expropriating land, after villagers were accused of degrading the environment. It was even more blatant with regard to the forests and their exploitation. The colonial administration constantly stressed the economic and social damage of deforestation caused by shifting agriculture. For example, in 1932, Colombon, chief of staff for the Resident Representative in Annam, signed a circular stipulating:

> 'My attention has been drawn on several occasions to the frequency of forest fires and the damages such practices cause to the province's economy. The population has to be made aware of the serious disturbances caused by the destruction of the forests to the running water and rain systems. The possible immediate benefits of these practices for a few ignorant or unscrupulous inhabitants risk compromising a whole region's prosperity in the near future.'[3]

In 1941, the economic service of the General Inspection of Agriculture, Livestock and Forests of the General Government published a memorandum entitled 'Measures against Deforestation'. This concluded that 'deforestation through *rays* [shifting cultivation] and fire is clearly the true cause of intensive deforestation, particularly in mountainous territories'.[4] Even in Laos, where officials were unable to lament about deforestation – given the huge forest cover at the time – they nonetheless criticized shifting cultivators for their thoughtless handling of the forest and the subsequent loss of fiscal resources:

> 'The dense forest of Laos is quickly losing its best stock due to frequent gratuitous tree felling, and the abusive farming. […] It is important to fight such creaming off of the forestry assets in order to save timber for the colonization of fertile regions, for the booming furniture industry, and for a more lucrative form of exportation' (letter no 40 from the Agricultural and Forestry Senior Technical Advisor in Laos to the Superior Resident in Laos, 11 February 1939).[5]

The colonial administration wanted to substitute local uses of the forest, which were condemned as destructive and archaic, with a rational operation based on European exploitation (Cleary, 2005). Forestry reserves were created in all the colony's different territories. This was not an environmental measure. These reserves acted as forestry exploitation estates, to the exclusion of the local population. To facilitate and hasten forestry exploitation in line with the principles of 'colonial development', a regulatory framework was established when colonization began in Cochinchina in the 1860s, and it was later extended throughout Indochina. The Forestry Code, up to the final version in 1930, strengthened the colonial hold over controlled territories, in particular by denying villagers' customary rights to officially establish land ownership. The only forest lands not regulated by the colonial forest administration were those ceded against payment, such as logging concessions. Shifting agriculture was officially forbidden (Forestry Code, article 92).[6] These increasingly restrictive regulations betrayed the colonial administration's opposition to shifting agricultural practices. They would have created an ideal situation for forest companies seeking concessions, had the authorities not been forced to make gestures of lenience in the face of peasant resistance. The colonial administration was never able to fully apply its forestry policy because it lacked economically credible alternatives to convince the mountain-dwelling people to abandon shifting agriculture, and forcibly controlling shifting cultivators would have required colossal political and military resources, dispersed over vast, hard-to-access territories. The administration was thus forced to be flexible with shifting cultivators and their practices, ignoring them as long as they did not interfere with 'colonial development' operations: agricultural plantations, forestry concessions, mines, road and rail operations, and so on.

I shall now return to Phongsaly to assess the local impact of colonial rule. French colonization did little more than touch the region before the end of the 19th century. It was not until 1917-1918 that Phongsaly came entirely under control, after the 5th Military Territory was created and power transferred from civil to military authorities (Aymé, 1930). Despite the martial implications of this change, colonial power in Phongsaly remained subtle. With disturbed times in China and with limited means,[7] local administrators gave priority to border control over domestic matters. The Resident in Vientiane constantly complained about the limited nature of the province's contribution to the colony's budget,[8] but the 5th Military Territory commander answered: 'Only once a sufficiently solid and precise framework is created for defence and then administrative requirements, will we consider filling it in with the necessary tools for economic development'.[9] Of 52 reports drafted by the territorial administration from 1921 to 1932,[10] 42 dealt with Chinese concerns such as political events in Yunnan and the border police. No colonists settled in Phongsaly to invest in agriculture, forestry or mining. The military authority, limited to only a handful of French officers, relied on the royal administration to represent the State for the villagers. The local administration ignored farmers' practices, as long as tax was paid, forced labour was carried out and supplies delivered. As was the case elsewhere in Indochina (Izikowitz, 1951), the tax system evolved from a collective one (a tribute in kind paid by the village) to an individual one (capitation, or payment per head, in cash). Local military administrators were reluctant to apply the State's new financial rules, because this risked alienating the highlanders and jeopardising colonial control over the newly settled and unstable border. The State was weak in remote regions like Phongsaly; it had to reach a compromise with villagers to obtain their support or, at least, their neutrality. The administration bargained the peace provided by military control in the troubled area for some taxes and intelligence.

Colonial rule had a limited direct impact on Phongsaly. Between 1918 and 1940, only two villages were forcibly displaced by a few hundred metres and charged with maintaining the horse tracks and the telegraph line. Another village was created in the study area for the same purpose (see Figure 15-1). Villagers continued to rely on shifting cultivation and to manage their fallows according to customary regulations.

However, colonial rule significantly affected village communities, not directly, but by the monetization of levies, leading farmers to slightly change their economic practices. Priority was increasingly given to commercial production, with crops such as opium poppies (Trocki, 1999; Chouvy, 2009), and villagers developed commercial ties with Phongsaly town. The colonial administration did not demand changes at village and household levels; it simply modified the economic context and the Phunoy villagers attuned their economic practices to the new reality.

The Phongsaly case illustrates the impact of colonization on shifting cultivators and their fallows. The State's voice rose to condemn shifting cultivation, but it was not loud enough to reach the upland villages. Colonialism was equivocal: on one hand, there was the outdated mission of 'bringing progress', and on the other, a greed for natural resources and wealth that lay waiting to be seized. The mix was a corrosive one for shifting cultivators: for the first time, the State was claimed loudly to have the right and the power to rule their economic livelihood. Since that time, the voice of the State has never been still in the fallows of Laos.

Turbulent times: war and revolution in Phongsaly

From 1940 to 1975, Laos went through turbulent times (Brown and Zasloff, 1986; Stuart-Fox, 1996, 2001; Evans, 2002). First there was Japanese occupation, followed by an attempt to restore colonial rule that was eventually aborted by decolonization (1946-1954). There followed a civil war lasting from 1956 to 1975. The State vanished as a unified, central power. It was replaced by several military parties aiming to control the villages for their resources – opium, labourers and rice – all of them invaluable to sustaining the war effort (McCoy, 2003) (Figure 15-2). Phongsaly remained at the margin of the conflicts. Open fighting and bombing were very limited, but the social impact of the conflict was nevertheless obvious, and is still remembered by villagers. Taxes and requisitions were common, sometimes from two sides within 24 hours – the Royal Lao Government in the daytime[11] and the Pathet Lao at night (Gunn, 1998; Ducourtieux, 2009).[12] Military service became a frequent calling for young Phunoy males and this continues to the present day. Conscription gave the Pathet Lao army a large Phunoy contingent; some of the soldiers, when demobilized, settled close to their barracks with their families, rather than return to Phongsaly. They formed the early wave of a diaspora that is still a sizeable movement today, to the urban centres of Oudomsay, Luang Namtha, Luang Prabang and Vientiane (see Figure 15-1). Of the Phunoy who joined the army, some were promoted and today are members of the provinces' administrative corps (Bouté, 2006, 2007). In the early stages of the civil conflict, the eastern part of Phongsaly quickly went over to the control of the Pathet Lao, and remained one of its staging grounds for the whole period. The Pathet Lao progressively expanded to the West. In 1964, the royal forces in Phongsaly, cut off from their base after a rightist coup in Vientiane, joined the Pathet Lao. Thus, Phongsaly was at peace 11 years earlier than most other provinces of the country. It was the time of a 'socialist revolution' in the Phunoy villages.

In a country with no workers – because there was no industry – it was ideologically complex for the Pathet Lao to promote a proletariat dictatorship. Therefore, the movement's leader, Kaysone Phomvihane, wrote:[13]

'The Party and the working class must educate the peasants, call them to unite and fight for socialist collectivization, help them raise their cultural level, introduce science and technology for production, progressively raise the material and technological base of agriculture and thus consolidate the peasant-farmer bloc alliance across various fields: politics, economics and culture, thereby strengthening the people's union and enforcing working people's right to collective governance of the whole country' (Phomvihane, 1980, pp234-235).

Such rationale reduced the peasantry to the raw material of the revolution. They were seen as ignorant and backward and

FIGURE 15-2:   Sowing rice in a freshly-burnt swidden in the Phunoy village of Samlang, Phongsaly – 'Military service became a frequent calling for young Phunoy males...'.

Photo: Olivier Ducourtieux (2003).

thus to be shaped, formed and 'educated' towards socialist modernity. As for shifting cultivation, the Pathet Lao had to juggle their rhetoric in order to win the support of upland minorities while maintaining a discourse on the socialist modernization of the country:

'Another no less important task is to rescue the absolute majority of the Laotian people, its working peasants, from poverty and backwardness. […] half of the Laotian population consists of minority ethnic groups who live scattered across highly remote forest and mountain areas, with no cultural, scientific or technological contact with the outside world. […] Economically speaking, half of the country still lives from slash-and-burn farming. Not only are hundreds of thousands of hectares of forest devastated and burnt each year, destroying inestimable quantities of precious fuel, […] but rice yields decline year upon year, and food shortages are becoming ever more frequent and serious' (Vongvichit, 1968).

This kind of discourse and reasoning closely echoed those of the colonial administrators: shifting agriculture was condemned for economic reasons; it was perceived as outdated and preventing the country from exploiting its forestry resources.

In Phongsaly in 1968, the new authorities initiated an ambitious mobilization of the population, especially in the Phunoy villages around the provincial capital. The aim was firstly to promote the socialist revolution, and secondly to sustain the war effort – to 'liberate' other provinces (Brown and Zasloff, 1986; Gunn, 1998). This required an increase in agricultural production and a flow of produce out of remote Phongsaly villages. The provincial and district administrations organized compulsory roadworks to open a network of driveable tracks from the villages to the main town. Each village in the district was assigned a section of road according to the size of its population, and had to provide labour over a period of two months. This involved most of the workforce of the communities, under the supervision of district officers and Vietnamese technical advisors. A revolutionary rhetoric window-dressed this unpaid forced labour, which was carried out in a similar fashion to the forced-labour corvées of the former colonial administration. The villagers' reluctance to perform this task led to a low level of labour productivity, and this exhausted the administrative commitment. The project lasted for only two dry seasons (1968 and 1969), and barely 30km of road was completed in three sections (Figure 15-3). The driveable tracks reached only five villages relatively close to Phongsaly and their economic impact was limited. The wartime shortage of motorized vehicles and fuel meant that they

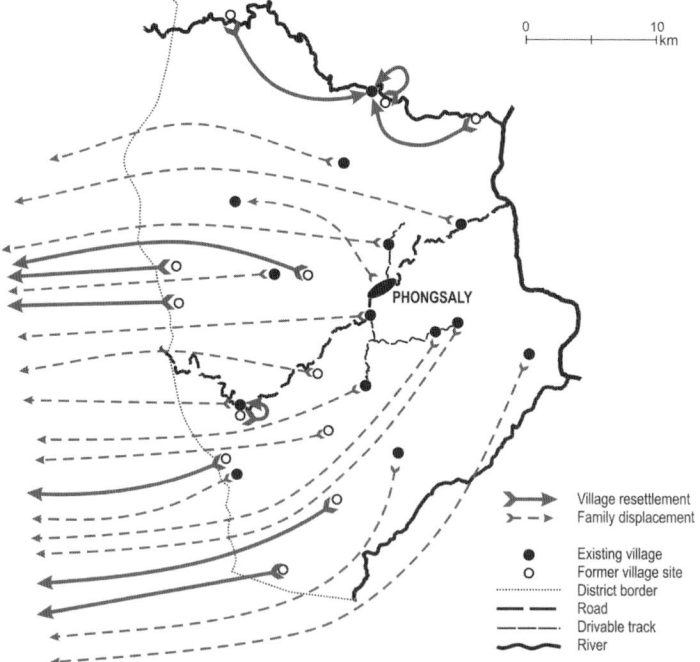

FIGURE 15-3: Population resettlement in Phongsaly district during the Paddy Field Movement.

were used by only a few rare administrative vehicles. Road construction was probably more politically than economically motivated; it was a way to prove the power of the socialist revolution to the villagers.[14] Another provincial programme from 1968 to 1970, the Paddy Field Movement, was further evidence supporting the hypothesis that the local administration was motivated by revolutionary zeal.

In order to boost agricultural production, the provincial administration launched a programme in 1968 aiming to convert shifting agriculture into paddy cultivation. In the district of Phongsaly, each farmer had to develop 0.2ha of land into irrigated terraces. Villagers unable to meet this target were forced to relocate to new villages in the neighbouring districts of Boun Neua and Bountay, which had lowlands for conversion into paddy fields (Laffort and Jouanneau, 1998). The programme replicated a policy for settling mountain-dwelling populations that originated in the Socialist Republic of Vietnam (Kerkvliet, 2005; Castella et al., 2006), and the Phongsaly administration benefited from the support of technical consultants from North Vietnam. In 1969, a large political meeting of more than 500 shifting cultivators was held in Boun Neua to launch the programme. In two years, all of the villages in the southwest of Phongsaly district were affected by this imperious project. Six villages were completely displaced, while in 14 other villages, a proportion of families left (Figure 15-3). In total, close to 1200 families left their land, representing 40% of the Phunoy population in 1967. It was the biggest population movement in living memory. The population density dropped brutally from 14 to 8 inhabitants per square kilometre (Figure 15-4).

In Phongsaly, village households developed almost 500ha of rice terraces, but never gave up shifting cultivation. In 1996, fewer than 200ha of terraces remained under cultivation; more than half of the terraced land had been abandoned. Were the Phunoy too 'backward' to adopt modern agriculture? Phunoy villagers from Komen (Figure 15-5), Samlang and Bakanoy gave credible and rational explanations in denial of this

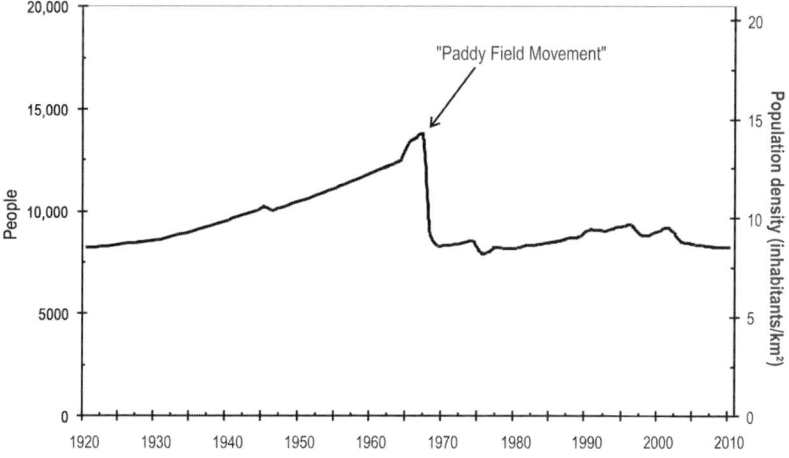

FIGURE 15-4:   Demographic effect of the Paddy Field Movement in Phongsaly district.

assertion: they built the terraces to distract the attention of the local administration and to stay in their villages. Some would have been interested in cultivating the new paddy fields, but cement and pipes were not available to build a working irrigation system. Besides, the district's officials quickly realized a negative consequence of the programme: with mass emigration, tax revenue fell and the district lost already scarce resources to its neighbours. The programme was suspended in 1970 and officially abandoned in 1973 at a 'self-criticism' political meeting in Boun Neua that reversed the decisions of 1968, but failed to reverse the displacements.

The Paddy Field Movement was the first state intervention in Phongsaly's agricultural sector and, more generally, the first such direct intrusion into the day-to-day life of the Phunoy at a household level. However, the programme did not intend to punish the Phunoy, who represented an important human resource to both the Pathet Lao army and the civil administration. It was presented to villagers along enthusiastic and paternalistic lines: the revolution would bring progress and prosperity, rewarding the people who supported the Pathet Lao (Bouté, 2006). With hindsight, the villagers' recollections are more circumspect. The short-lived Paddy Field Movement was considered a failure by the Phongsaly district administration, but the State had nevertheless affected the Phunoy livelihood at a speed and a scale without precedent. The State ceased to be a distant murmur that was more or less harmless; it became an imperious voice over the fallows, dictating direct orders that were to be executed at once. After the intrusion was withdrawn, the local administration ceased direct interventions in the villages until the mid-1990s, apart from a very limited and cautious test of collectivism for shifting cultivation in the years from 1979 to 1981, which was implemented in only five villages in Phongsaly district and then aborted. This marked the end of revolutionary voluntarism in Phunoy villages, and the interventions of the local administration were pared down to a predictable minimum: tax collection. At that time, the central State in Vientiane gave only limited attention to provinces with weak economic potential, such as Phongsaly, even though it was the historical base of the Lao Revolutionary Party. It was only well after the market-economy reform launched in 1986 – the so-called New Economic Mechanisms (Bourdet, 2000) – that Phongsaly was brought back to light.

At the sixth Party Congress in 1994, the leaders of the northern provinces explicitly demanded State assistance for their territories, in the name of the historical debt they said the regime owed them. So after 25 years at a low ebb, the State was back on track – intervening against shifting cultivation and yelling at Phunoy villagers.

Eradication of swidden for the sake of environment and poverty alleviation

Responding to new international paradigms

In the second half of the 1980s, a new concern rose up in the world's scientific community, involving civil society, the media and political networks: how to prevent the degradation of the environment. In 1985, the Food and Agriculture Organization of the United Nations and the World Bank launched the Tropical Forestry Action

Plan, a move joined by bilateral agencies[15] and environmental non-governmental organizations (NGOs).[16] In the founding document of the plan, Laos was one of the priority countries, leading to an increased allocation of international aid. For the first time, the plan aimed to stop tropical deforestation on environmental grounds, and of course, shifting cultivation was again presented as the main cause of deforestation. Even the General Assembly of the United Nations joined the movement at the 1992 Earth Summit in Rio: among the objectives of Agenda 21 was a recommendation to 'limit and aim to halt destructive shifting cultivation by addressing the underlying social and ecological causes' (Articles 11 to 13).

In 1986, postponing communist objectives with no appointed date for their resumption, the Party committed the country to a 'socialist market economy'.[17] Private ownership of the means of production and free enterprise became new principles for prosperous development. In 1996, the 6th Party Congress set the goal of removing Laos from the list of 'least advanced' countries by 2020. In 2001, the 7th Congress reinforced that position and enshrined its national aims in the United Nations' Millennium Development Goals. The Party set quantified objectives of reducing poverty by half by 2005 and eradicating it by 2010, based on three pillars: economic growth, socio-cultural development and environmental protection. Later Party congresses (2006 and 2011) reiterated the objectives, but had to successively postpone the completion terms. The government now has the target of reaching its goals by implementing the National Growth and Poverty Eradication Strategy, along with the 6th and 7th Socio-economic Development Plans, covering the years from 2006 to 2015. The Ministry of Agriculture and Forestry takes part in the policy by promoting modern, permanent and intensive agriculture, which must generate substantial quantities of raw materials to supply the domestic market, contribute to growth of exports and support the emergence of a 'national agro-industrial fabric'.[18]

Shifting cultivation is perceived as an obstacle that is impossible to bypass, and as one of the main causes of rural poverty. Lao officials say that demographic growth in swidden agricultural regions tends to reduce forest areas, which leads to the reduction of income by the families involved, who get poorer while burdening the country's future development with the destruction of natural resources. The vicious circle is complete and poverty is self-maintained (Dasgupta et al., 2005). The solution seems obvious: farmers practising 'slash-and-burn' must be converted to permanent cropping or to non-agricultural activities, making it possible to interrupt the process and therefore eliminate poverty.

An increasing policy focus on shifting cultivation

While eliminating shifting cultivation was one of the founding goals of the Lao Revolutionary Party when it was founded in 1975, almost no effective action was taken until the 1990s. As with the Paddy Field Movement in Phongsaly, the national solution was to displace swidden cultivators to valleys, but fewer than 15,000 people – less than 5% of the population involved in shifting cultivation – were resettled between 1975 and 1985 (Souvanthong, 1995). At the Party's 5th congress in 1991, it was explained that the transition from subsistence to a market economy would require the abandonment of shifting agriculture (Goudineau, 1997), and in 1993, the National Assembly voted to eradicate swidden cultivation by 2000. However, in 2000, the objective was postponed until 2020,[19] before being brought back down to 2010 in the current National Growth and Poverty Eradication Strategy. Although some provinces were declared 'free from slash-and-burn' before 2010, the countrywide goal had to be postponed once again to 2015.

Increasingly, the elimination of shifting cultivation has become a recurring theme in the Lao political discourse and in the national press, which is controlled by the Party. The expression 'shifting cultivation' appears regularly in official documents, for example 28 times in the National Growth and Poverty Eradication Strategy (2004); 30 times in two recent National Socio-Economic Development Plans (2006 and 2011); 18 times in the Ministry of Agriculture and Forestry's strategy (2010);[20] and twice in the Party's Decree on Village Organization (2011).[21] In 2010, the Ministry of Agriculture and Forestry published a special decree banning shifting cultivation,[22] and in 2012, the Prime Minister issued a decree reorganizing the Ministry and, giving the elimination of shifting cultivation top priority:

> 'The main role of the Ministry of Agriculture and Forestry [MAF] is to act as a secretariat of the government in macro-management of agriculture and forestry development, aiming to ensure food security; supplying raw materials for processing industries and sustainable, clean and modern commodity productions; creation of permanent jobs for ethnic groups in order to stop shifting cultivation and eradicate poverty across the country' (Decree on MAF Organization and Function, Prime Minister's Office [262/PM 28/06/2012]).

Presented at first as a necessary step to reach the foremost goals of environmental protection and poverty alleviation, the eradication of shifting cultivation has become one of the government's highest aims.

To reach the Party's objectives, government bodies at national and local levels have designed and implemented different programmes aimed at helping to bring an end to swidden cultivation in the Lao PDR. Budgetary resources were limited in the 1990s and 2000s, so the government favoured administrative measures based on reorganization of rural areas rather than investment programmes or agricultural subsidies. It focused on three different levers: resettlement, land reforms and cash-crop cultivation.

Removing shifting cultivators from their fallows was a classical approach in colonial times and in the early revolutionary years. More recently, cloaked in new terminology – 'stabilizing shifting cultivation', 'providing villagers with fixed occupations', and so on – resettlement was revived in the 1990s and is still in effect. The principle involves moving people from uplands towards lowlands, where they can change their way of life and access public services and markets. It was revived in the mid-1990s by the

Focal Zones Programme (Goudineau, 1997;Vandergeest, 2003; Baird and Shoemaker, 2007), but this was replaced in 2006 by the more politically correct Village Cluster Development Programme.[23]

Based on the Lao PDR's 1991 constitution, which states that land belongs to the State, the Land Allocation Programme aims to produce sustainable land use in the uplands, and provides for distributing land to villagers, even though they may already have been managing that land. The programme's objective is twofold: to increase land tenure security that will encourage and enable farmers to invest in their land and to encourage village communities to protect the forest (Ducourtieux et al., 2005; Lestrelin et al., 2012). Under the land-allocation process, the village territory is zoned according to existing vegetation and past use by villagers. Farmland is defined as areas of permanent cultivation while forestland, by default, is the rest of the territory. In a bid to encourage farmers to change their practices, tenure of swidden land is deliberately limited, as is the land area, because shifting agriculture is said to 'consume too much space'.[24] Land allocation is finalized after a quick 10-day implementation exercise led by district officers. An agreement is signed between the village and the State, and a map is published illustrating the village territory. The agreement has all the characteristics of a land lease in which the State is the owner and the farmers are borrowers. Through this agreement, farmers formally acknowledge the rights of the State over land.

Aid institutions supported the widespread implementation of the land reforms. Most rural-development projects launched in Lao PDR after 1995 included land allocation as a key component (Evrard and Goudineau, 2004). Proponents often present the programme as a model of collaboration between village communities and the State; the term 'participatory' is used systematically by both the administration and international-aid projects, for example, 'participatory land-use planning' and 'participatory integrated land-use management' (Lestrelin et al., 2012). 'Participation' is a telling term because it encompasses contrasting concepts of the role to be given to civil society in defining and implementing development actions. Indeed, some authors even consider the term tyrannical (Cooke and Kothari, 2001); it is compared with the similar lack of options for villagers 'participating' in the forced-labour corvées of the former colonial administration (Ribot, 1999). Land allocation is part of this double standard: the generosity of the participatory concept masks a hierarchical top-down programme (Vandergeest, 2003; Lestrelin and Giordano, 2006). District services have implemented a generic approach with little thought for the farmers and their micro-level land-management practices (Ducourtieux et al., 2005).Villagers do, indeed, participate in the programme, but the consensus among them is that they are obliged to do so. An impressive list of the lands 'allocated' in each village is published – usually within a few months – and multi-coloured zoning maps are placed proudly and conspicuously at the village entrance.

Promotion of cash crops is another standard tool for replacing shifting cultivation. In Laos, the authorities promote either perennial plantations (rubber, coffee, cassava, jatropha, eucalyptus, tea, and so on) or annual crops (mainly maize), cultivated either by local smallholders or by foreign investors with salaried staff.

I now return to Phongsaly to review the local implementation of these programmes and to measure their impact on local shifting cultivators and their fallows.

Impact of the policies in Phongsaly

Three state programmes designed in the early 1990s have been implemented in Phongsaly, aimed at eliminating shifting cultivation: resettlement of villages, land reform and mandatory cultivation of commercial crops.

Increased state involvement began in Phongsaly with the Focal Zones Programme. From 1993 to 1997, the administration ordered the displacement of four Phunoy villages to a selected priority zone along the main road from Phongsaly to Boun Neua (Figure 15-6). In 1996, the district authorities made plans for a second, more ambitious stage, with a decree for the displacement of 21 villages. However, in the absence of the necessary means and any real will power, the decree went unheeded. Alongside the Focal Zones Programme, the Phongsaly administration enacted a governmental decree that fixed the minimum size for a village in an upland zone at 20 households.[25] Although the decree did not explicitly impose the grouping together of households at one site, it was strictly applied in Phongsaly: four villages were merged with larger neighbouring villages and some families moved to the town of Phongsaly. In total, more than 300 families were displaced between 1992 and 1998 due to state policy. After a pause between 2002 and 2006, the district authorities revived the resettlement issue when the government launched its Village Cluster Development Programme, which was an expanded version of the Focal Zones Programme, but with a new title.

In 1997, the Phongsaly administration launched land allocation in the villages. The first communities affected were in the town of Phongsaly and in 28 rural villages, most of them Phunoy. The villages involved were those with the best access to administrative services. After a timid start, the process picked up speed in 2000, with the support of a European Union-funded rural-development project. However, land reform in the district remains incomplete and land registration (the titling of private property) has so far been ignored; contrary to the political objectives of the reform, land allocation in Phongsaly has neither secured nor commercialized land tenure (Ducourtieux et al., 2005). The 'reform' has been limited to the reorganization

FIGURE 15-5:   The ridge-top Phunoy village of Komen is dwarfed by its mountainous setting of forest fallows.

Photo: Olivier Ducourtieux (1999).

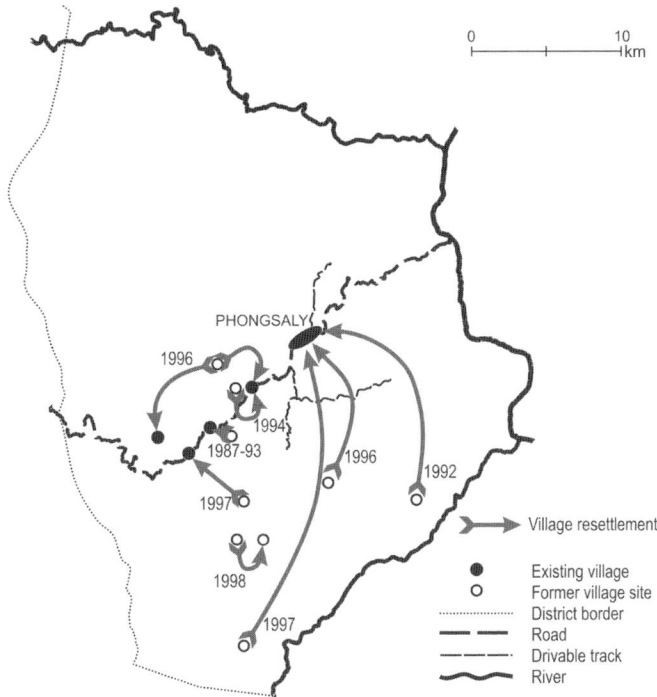

FIGURE 15-6: Villages resettled for the Focal Zones Programme.

of village territory and reducing village lands with a view to eradicating shifting agriculture. In villages where land allocation has been completed, the average size of a farm has fallen by a quarter, from 17ha to 13ha, including a drastic reduction of fallow lands, from 14ha to 7ha per household (Figure 15-7). This had led to an abrupt shortening of the swidden rotation cycle, in which the average length of fallow has plunged from 12 years to six. Shorter rotation causes increased weed competition and declining yields, which can only be stopped with harder, longer work: the result is a drop in labour productivity and in food security (Ducourtieux, 2006). Access to land has been drastically transformed. The private type of ownership arising from the former community-controlled Phunoy land system has given way to temporary state allocation of land. Family ownership of fallow areas was wiped out in a few days, meaning that the sacrifice endured by generations of villagers who emigrated in an effort to preserve the forest as a staple resource for shifting cultivation was for nothing.

To complement land allocation, the Phongsaly administration promoted the expansion of commercial cropping. Alongside resettlement, the Focal Zones Programme made sugar cane cultivation for export to China a further compulsory measure for Phunoy villages along the main road. Peaking at 700ha in 1998, but with an objective of 3000ha, the project collapsed within two years due to the default of the Chinese contractor. Following the demise of sugar cane in Phongsaly, the local

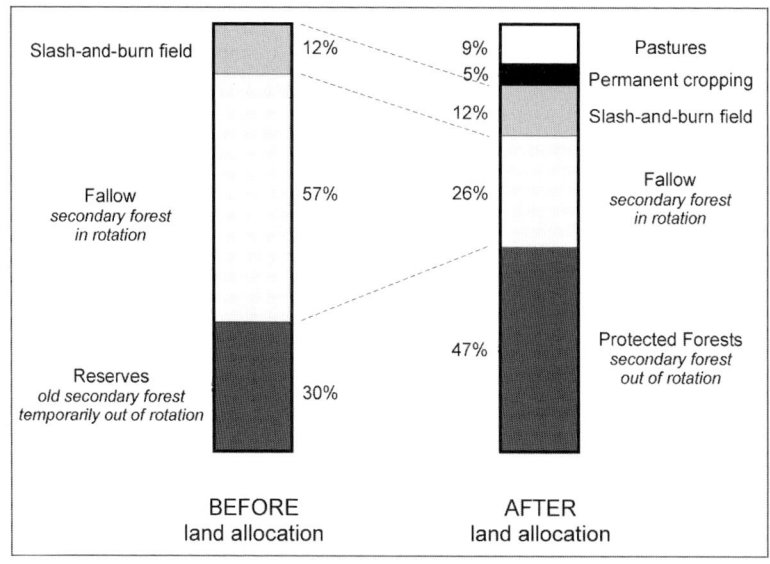

FIGURE 15-7: Land allocation and reduction of fallow area in Phongsaly.

Source: A survey of 12 Phunoy villages in Phongsaly district, 2004.

administration turned its attention to the cultivation of tea, fruit (citrus) and galangal. The principle was simple: farmers from selected villages were required to plant a minimum area per worker. Forty-five per cent of Phunoy rural households were involved in tea cultivation; 13% grew galangal; and 5% grew citrus fruit (Figure 15-8). For tea, the cropping perimeter was determined by processing constraints: all the villages within one hour of the road from the drying plants in Phongsaly were involved. In a bid to maximize production, the local authorities displaced three villages. However, some of the displaced villagers chose not to move to designated sites, and shifted instead into Phongsaly town (Figure 15-8). While the authorities decided the surface area to be cultivated under different crops, each household was left with the burden of purchasing the plants. In the case of tea plantations, for example, the plants represented an investment of nearly US$75, or more than 60% of the annual cash income for an average Phunoy household (Ducourtieux, 2006). As well as promoting cash cropping in the villages, the administration granted foreign investors, mainly from China, the right to operate three tea-drying plants in Phongsaly town, and rubber plantations along the Nam Ou river. The land for the plantations of *Hevea brasiliensis* came from former village fallows confiscated during the land allocation procedure. This land was instantly transformed from 'protected areas' to monoclonal rubber plantations. To an extent, Phongsaly district was spared a wider incursion by perennial plantations because of its lack of access and slope limitations. In many other upland provinces, the prevalence of rubber, eucalyptus, jatropha and cassava plantations has placed Laos at the forefront of the land-grabbing controversy (Manivong and Cramb, 2008; Mann, 2009; Ziegler et al., 2009; Zoomers, 2010; Deininger et al., 2011).

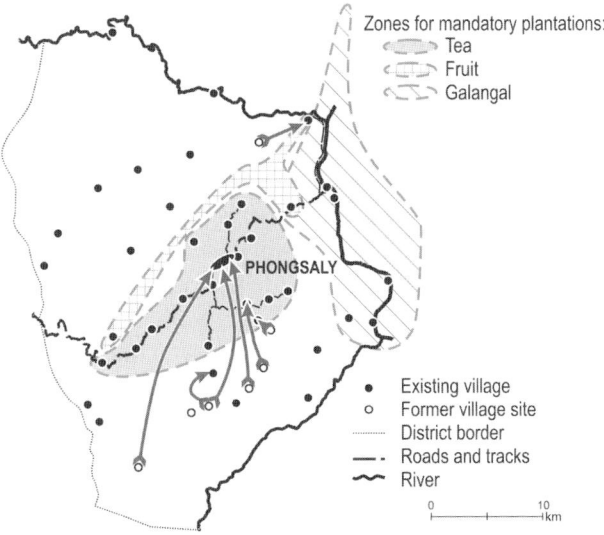

Zones for mandatory plantations:
Tea
Fruit
Galangal

PHONGSALY

• Existing village
○ Former village site
········· District border
—·— Roads and tracks
〜〜 River

0   10
⊢————⊢————⊣km

FIGURE 15-8: Compulsory cash-crop programme and further induced migrations in Phongsaly.

Less than 15 years ago, Phunoy farmers used to manage environmental resources and economic activities as they wished, under the customary control of their peers in the village community. Since that time, the Lao State has shifted from this laissez-faire policy in the uplands to a multifaceted interventionism in the everyday economic life of the villagers. In 2003, a study compared two Phunoy villages – Samlang and Yapong (Ducourtieux, 2006). Prior to 1995, they were very similar in terms of ecosystems, culture, social organization, economic practices and results. Then, Yapong became involved in the state intervention process and received the whole bundle: resettlement in 1996, land allocation in 1999, compulsory cash crops with sugar cane from 1997 to 1998, then tea since 2000; a ban on hunting in 2000; and a ban on shifting cultivation in 2005. Access to the other village, Samlang, was more difficult, so it managed to stay outside the process and mostly continued with its former organization and economy. An in-depth economic interview was conducted with each of the households in the two villages, dealing with the family, farming practices and their results, as well as other economic activities, such as gathering, fishing, hunting, handicrafts and trade over the five years prior to 2003. The survey in Samlang was used as a counterfactual case to appraise, by comparison with Yapong, the exclusive impact of state policies in the Phongsaly villages.

Land allocation had a direct impact on swidden cultivation in Yapong. Forest reserves were taken out of the swidden rotation, then the area of fallow land available for swidden cultivation was reduced. The length of the fallow period fell as a consequence from 10 to just three years. That directly impacted harvests, with yields limited to 600kg/ha of paddy rice, compared to 1300kg/ha in Samlang, or a 54% reduction for Yapong. In an attempt to maintain rice production, Yapong farmers developed a strategy of increasing the area under cultivation within the limits of the land allocation, with two to three successive years of cultivation, compared to one year in Samlang. Extending the area under cultivation meant that households had to face the crucial problem of weed control. Due to a lack of resources, farmers were not able to devote more time to weeding. The Yapong villagers compensated for the lack of an available workforce with massive use of herbicides: the consumption of weedkiller per ton of rice produced was 20 times higher in Yapong than in Samlang. The herbicide, of Chinese origin, was poorly identified and used, with probable effects on public health and the environment. With an increasing workload, a falling yield and production costs that were rising, the work productivity of villagers fell drastically. Production per family dropped as well, which increased problems of staple-food shortages. Rice shortages became the norm in Yapong. Sixty per cent of its families suffered rice shortages for an average of three months every year. In Samlang, such shortages remained a rarity. Twenty per cent of its families ran short of rice for about half of one month per year in 2000 to 2003. The impact of state interventions was conspicuous: in the state-targeted village, the average per capita income was only half that of the other village (Figure 15-9). In the name of poverty reduction, the intrusion of the State in household economies had contributed to the impoverishment of Phunoy families and to increasing socio-economic differentiation within the village communities.

Contrary to earlier state interventions, the impacts of state policies over the past 15 years will probably persist because the environment of the villages has been disrupted by substantial migrations, whether forced, voluntary or induced (Vandergeest, 2003; Evrard and Goudineau, 2004; Bouté, 2006; High, 2008). Of 55 villages in the southwest of the Phongsaly district in the mid 1960s, only 39 remained by 2008,

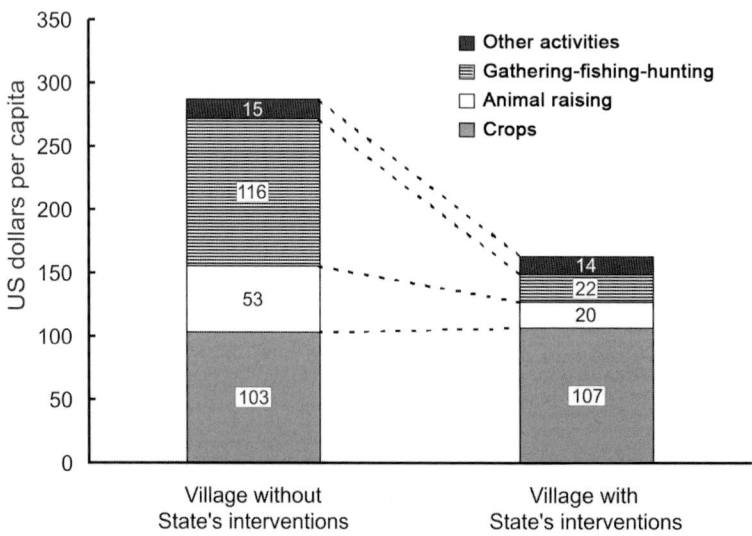

FIGURE 15-9: Comparison of average income in two Phunoy villages.

including eight that were displaced outside their former territory (Figure 15-10). According to current trends and in the opinions of the villagers, there will be only 35 left by 2020. During the Paddy Field Movement between 1968 and 1970, 1200 families were resettled on the plains. Between 1995 and 2008, the State moved 830 households and more than 400 households left of their own accord (Ducourtieux, 2009). Although the local administration reduced the pressure for resettlement after 2005, the trend of depopulation has continued, and villages continue to vanish. A prime example is the 'counterfactual' village surveyed in 2003. Although it was not directly affected by state interventions, Samlang comprised 144 people in 2003. Only 70 remained in 2005 and in 2006, the village site was abandoned, all of its inhabitants having migrated to Phongsaly town or to other regions of Laos. These spontaneous departures are due to a disintegrating social fabric, the growing impact of predation from wild forest animals and the threat to their future from the proclaimed ban of shifting cultivation. With the reduction in the number of villages, the average distance between communities has increased (Figure 15-10); social relations have become more difficult and less frequent, whether for economic, religious or leisure activities (including encounters of young people, marriages and celebrations). Socializing is limited to the village neighbours, which can encourage young people to leave the community, further impoverishing exchanges in a self-amplifying vicious circle. The workforce is the limiting factor in farming systems in Phongsaly, so the demographic drop automatically brings about a reduction in cultivated area and an increase in forest. The expansion of forest areas, along with the confiscation of firearms for hunting, has allowed wildlife to proliferate. This has brought increased predatory pressure on

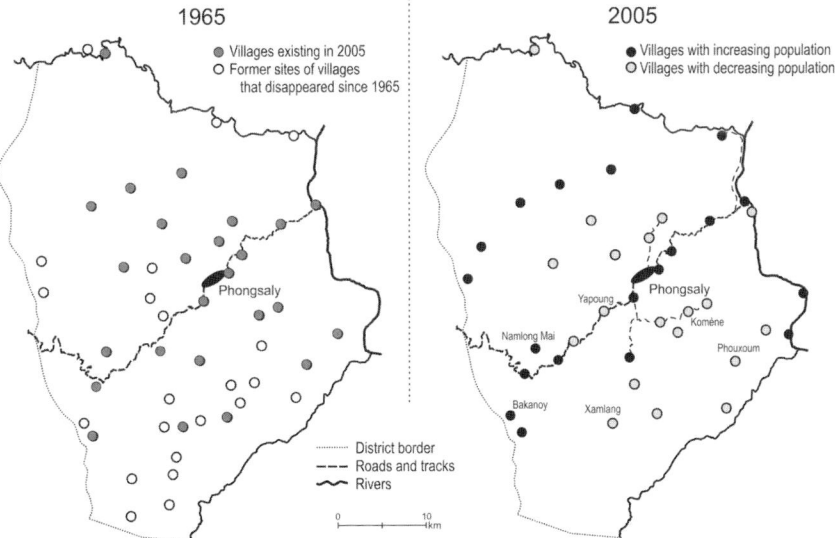

FIGURE 15-10: Evolution of Phunoy villages in the southwest of Phongsaly district, 1965 to 2005.

crops and livestock herds from elephants, boars, bears, felines, wild dogs (dholes) and rodents.

Ongoing political pressure to eradicate shifting agriculture has radicalized during the last three congresses of the Lao Revolutionary Party. Local officials repeat political slogans, as does the state-controlled media: the state voice carries weight in the Phongsaly fallows. The general threat of the ban on shifting cultivation has led many families to question their future. During survey interviews, fear of the ban was regularly discussed. Some elders said: 'If we stop shifting cultivation, what else can we do? We will have to die', or 'how will our children be able to live in the village? They should leave'. Better-off families prefer to sell their cattle to buy paddy fields in neighbouring districts or to move to urban areas and set up as traders. Poorer families emulate them by drawing on existing family networks in immigration zones. In Phunoy villages, departure has long been a means of regulating demographic and land pressures that tend to shorten the rotation of fallow land. Since the late 20th century, departures have tended towards the tertiary sector in Vientiane or other cities in northern Laos; particularly into the army for young men, and more recently, the clothing industry or prostitution for young women. Village departures are a community-recognized fact facilitated, even encouraged, by the Phunoy diaspora outside Phongsaly district. Social bonds survive these departures; the migrant maintains contact and a place in his or her lineage or clan (Bouté, 2006). The tendency to depart is reinforced by schooling. Young people who receive a secondary education are more attracted by urban life than by the difficult village life, which is seen as backward and offering no future (High, 2008). Since the foundation of the Lao PDR, schooling drives have been considerable in Phongsaly district and have been particularly effective in the Phunoy villages. More than half of the young people under 18 are educated to high-school level. After several years of urban life, few of these young people are interested in returning to set up a household in their village of origin. While the rural population tends to decrease (Figures 15-4 and 15-10), the provincial capital of Phongsaly is growing. From 1995 to 2005, the number of inhabitants of rural villages fell by 10%, while the urban population grew by 17%. On a restricted regional level, we are witnessing a rapid deruralization process. But is the integration of migrants through urbanization guaranteed in a small agglomeration like Phongsaly, whose activities are limited to administrative services, trade and shaky tourism? In the town, recent rural migrants live more off family solidarity networks (including subsidies from relatives still in rural villages) than from their own economic activities, in contrast to their previous situation in their villages of origin.

Conclusion

The State has roared and its voice goes on roaring in the fallows of Laos, banning shifting cultivation and criticizing upland farmers who have no option but to live by swidden farming. The condemnatory discourse is based on a trinity: (1) Progress:

shifting cultivation is an archaic practice to be replaced by modern practices; (2) Economic development: shifting cultivation destroys valuable wood resources and hampers the expansion of profitable commercial plantations, mining operations and hydropower; and (3) Environmental protection: shifting cultivation is the main cause of tropical deforestation. The trinity has been progressively built up over the past century, beginning in the colonial empires. Following decolonization, the new states took it up without hesitation or guilty conscience. Shifting cultivation is presented as a poverty trap for upland minorities and a direct threat to forests. Thus, the State, as the guarantor of the general interest, assumes the right to judge shifting cultivators, whose pile of handicaps includes cultural, language and lifestyle differences, along with a lack of political representatives. The points of view expressed by lowland elites far outclass the interests of the upland minorities (Menzies, 2007).

In the lowland towns, it is easy for the media, politicians and many scientists to disgrace shifting cultivators by simply displaying pictures of recently burnt fallows. Conversely, the economic and ecological rationale of shifting cultivation is too discreet and complex: the long period of fallow, wherein healthy soils develop, biomass accumulates and biodiversity-rich secondary forests grow is too complicated to fit into quick grabs of media argument. The environmental and economic services of shifting cultivation are therefore overlooked. In this book, the assumption that shifting cultivation is the main cause of tropical deforestation has been vigorously refuted. New voices must override those of the different lobbies aiming to expel shifting cultivators and deny them free access to their fallows. These new voices must not only be heard in the fallows, but also in places of power and decision-making, seeking new policies under which shifting cultivators will be paid for their environmental services. Alas, time is running out and the voices of shifting cultivators themselves are vanishing, even from the fallows.

References

Aymé, G. (1930) *Monographie du Vème Territoire militaire*, Imprimerie d'Extrême Orient, Hanoï

Baird, I. G. and Shoemaker, B. (2007) 'Unsettling experiences: Internal resettlement and international aid agencies in Laos', *Development and Change* 38(5), pp865-888

Bourdet, Y. (2000) *The Economics of Transition in Laos: From Socialism to Asean Integration*, Edward Elgar Publishing, Cheltenham, UK

Bouté, V. (2006) 'Empowerment through acculturation: Forgetting and contesting the past among the Phunoy in Northern Laos', *South East Asia Research* 14(3), pp431-443

Bouté, V. (2007) 'Political hierarchical processes among some highlanders of Laos', in F. Robinne and M. Sadan (eds) *Social Dynamics in the Highlands of Southeast Asia: Reconsidering Political Systems of Highland Burma*, Brill Academic Publishers, Leiden

Bouté, V. (2011) *En miroir du pouvoir : les Phounoy du Nord Laos, ethnogenèse et dynamiques d'intégration,* EFEO, Paris

Brown, M. and Zasloff, J. (1986) *Apprentice Revolutionaries: The Communist Movement in Laos 1930-1985,* Hoover Press/Stanford University, Stanford, CA

Castella, J-C., Boissau, S., Thanh, N. H. and Novosad, P. (2006) 'Impact of forestland allocation on land use in a mountainous province of Vietnam', *Land Use Policy* 23(2), pp147-160

Chouvy, P-A. (2009) *Opium: Uncovering the Politics of the Poppy*, I. B. Tauris, New York

Clastres, P. (1989) *Society against the State: Essays in Political Anthropology,* Zone Books, New York

Cleary, M. (2005) 'Managing the forest in colonial Indochina c.1900-1940', *Modern Asian Studies* 39(2), pp257-283

Cleghorn, H. (1861) *Forests and Gardens of South India*, H. Allen, London

Cooke, B. and Kothari, U. (eds) (2001) *Participation: The New Tyranny?* Zed Books, New York

Daniel, E. V., Bernstein, H. and Brass, T. (eds) (1992) *Plantations, Proletarians, and Peasants in Colonial Asia*, Library of Peasant Studies no 11, Frank Cass Publishers, London

Das, P. (2005) 'Hugh Cleghorn and forest conservancy in India', *Environment and History* 11(1), pp55-82

Dasgupta, S., Deichmann, U., Meisner, C. and Wheeler, D. (2005) 'Where is the poverty–environment nexus? Evidence from Cambodia, Lao PDR and Vietnam', *World Development* 33(4), pp617-638

Deininger, K., Byerlee, D., Lindsay, J., Norton, A., Selod, H. and Stickler, M. (2011) *Rising Global Interest in Farmland: Can It Yield Sustainable and Equitable Benefits?*, Agricultural and Rural Development, World Bank, Washington, DC

Ducourtieux, O. (2006) 'Is the diversity of shifting cultivation held in high enough esteem?' *Moussons* 9-10, pp61-86

Ducourtieux, O. (2009) *Du riz et des arbres : l'interdiction de l'agriculture d'abattis-brûlis, une constante politique au Laos*, Hommes Et Sociétés, Karthala/IRD, Paris

Ducourtieux, O., Laffort, J-R. and Sacklokham, S. (2005) 'Land policy and farming practices in Laos', *Development and Change* 36(3), pp499-526

Evans, G. (2002) *A Short History of Laos: The Land in Between,* Short History of Asia Series, Silkworm Books, Bangkok

Evrard, O. and Goudineau, Y. (2004) 'Planned resettlement, unexpected migrations and cultural trauma in Laos', *Development and Change* 35(5), pp937-962

Ferro, M. (ed.) (2003) *Le livre noir du colonialisme : XVIe-XXIe siècles, de l'extermination à la repentance*, Hachette Littératures, Paris

Goudineau, Y. (ed.) (1997) *Resettlement & Social Characteristics of New Villages: Basic Needs for Resettled Communities in the Lao PDR*, Vol 1, United Nations Development Programme, Vientiane

Grove, R. H. (1994) 'A historical review of early institutional and conservationist responses to fears of artificially induced global climate change: The deforestation–desiccation discourse, 1500-1860', *Chemosphere* 29(5), pp1001-1013

Gunn, G. C. (1998) *Theravadins, Colonialists and Commissars in Laos*, White Lotus, Bangkok

Gunn, G. C. (2003) *Rebellion in Laos: Peasant and Politics in a Colonial Backwater*, White Lotus, Bangkok

High, H. (2008) 'Implication of aspirations: Reconsidering resettlement in Laos', *Critical Asian Studies* 40(4), pp 531-550

Ivarsson, S. (2008) *Creating Laos: The Making of a Lao Space between Indochina and Siam, 1860-1945*, Nias Monograph Series, NIAS Press, Copenhagen

Izikowitz, K. G. (1951) *Lamet: Hill Peasants in French Indochina*, Etnologiska Studier 17, Etnografiska Muséet, Göteborg

Kerkvliet, B. T. J. (2005) *The Power of Everyday Politics: How Vietnamese Peasants Transformed National Policy,* Cornell University Press, Ithaca, NY

Laffort, J-R. and Jouanneau, R. (1998) *Deux systèmes agraires de la Province de Phongsaly: deux systèmes agraires contrastés d'une province montagneuse du Nord Laos,* CCL, Paris

Lestrelin, G. and Giordano, M. (2006) 'Upland development policy, livelihood change and land degradation: Interactions from a Laotian village', *Land Degradation & Development* 18(1), pp55-76

Lestrelin, G., Castella, J-C. and Bourgoin, J. (2012) 'Territorialising sustainable development: The politics of land-use planning in Laos', *Journal of Contemporary Asia* 42(4), pp581-602

Manivong, V. and Cramb, R. (2008) 'Economics of smallholder rubber expansion in northern Laos', *Agroforestry Systems* 74(2), pp113-125

Mann, C. C. (2009) 'Addicted to rubber', *Science* 325(5940,) pp564-566

McCoy, A. (2003) *The Politics of Heroin*, third edition, Lawrence Hill, Chicago

Menzies, N. K. (2007) *Our Forest, Your Ecosystem, Their Timber: Communities, Conservation, and the State in Community-Based Forest Management,* Columbia University Press, New York

Murray, M. J. (1992) "White gold' or 'white blood?' The rubber plantations of colonial Indochina, 1910-40', *Journal of Peasant Studies* 19(3-4), pp41-67

Phomvihane, K. (1980) *La Révolution Lao*, Editions du Progrès, Moscow

Ribot, J. C. (1999) 'Decentralisation, participation and accountability in Sahelian forestry: Legal instruments of political-administrative control', *Africa* 69(1), pp23-65

Souvanthong, P. (1995) *Shifting Cultivation in Lao PDR: An Overview of Land Use and Policy Initiatives*, IIED Forestry and Land Use Series No 5, International Institute for Environment and Development, London

Stuart-Fox, M. (1996) *Buddhist Kingdom, Marxist State: The Making of Modern Laos*, Studies in Southeast Asian History, White Lotus, Bangkok

Stuart-Fox, M. (2001) *Historical Dictionary of Laos*, Asian-Oceanian Historical Dictionaries Series No 35, second edition, The Scarecrow Press, Lanham, MD

Thorel, C. (2001) *Agriculture and Ethnobotany of the Mekong Basin,* Translated by W. E. J. Tips, The Mekong Exploration Commission Report 1866-1868, Volume 4, White Lotus, Bangkok

Trocki, C. A. (1999) *Opium, Empire and the Global Political Economy,* Asia's Transformations, edited by Mark Selden, Routledge, London

Vandergeest, P. (2003) 'Land to some tillers: Development-induced displacement in Laos', *International Social Science Journal* 55(1), pp47-56

Vann, M. G. (2003) 'The good, the bad, and the ugly: Variation and difference in French racism in colonial Indochina', in S. Peabody and T. Stovall (eds) *The Color of Liberty: Histories of Race in France*, Duke University, Durham, NC, pp187-205

Vongvichit, P. (1968) *Le Laos et la lutte victorieuse du Peuple Lao contre le néo-colonialisme américain*, Neo Lao Haksat, Vientiane

Ziegler, A. D., Fox, J. M. and Xu, J. (2009) 'The rubber juggernaut', *Science* 324, pp1024-1025

Zoomers, A. (2010) 'Globalisation and the foreignisation of space: Seven processes driving the current global land grab', *Journal of Peasant Studies* 37(2), pp429-447

Notes

1. The council of elders comprises male villagers who are deemed successful: they have established and maintained a family, they maintain cordial relationships and they participate actively in the community. Membership of the council is informal, marked by an invitation to attend meetings. Generally, members are household heads who are more than 50 years old (Bouté, 2007; Ducourtieux, 2009).
2. For example, Khounsouk Luang and Khounsouk Noy (the great and small villages called Khounsouk), and Nampongsang Kao and Nampongsang May (the old and new villages called Nampongsang), and so on.
3. This quote taken from Overseas Archives (CAOM, Aix-en-Provence), reference INDO/RSL/L17.
4. From Overseas Archives (CAOM), ref. INDO/GGI/SE/2687.
5. From Overseas Archives (CAOM), ref. INDO/GGI/SE/2647.
6. From Overseas Archives (CAOM), ref. INDO/GGI/SE/2647.
7. Major Georges Aymé, who was in charge of the province in the 1930s, wrote that 'The Fifth Military Territory is the poor child of Laos, which stands alone in its deprivation among the countries of the [Indochinese] Union' (Aymé, 1930, p131).
8. From Overseas Archives (CAOM), ref. INDO/GGI/40465 and INDO/RSL/Q2.
9. Letter from the commander, 5th Military Territory, dated 4 January 1917, in Overseas Archives (CAOM), ref. INDO/RSL/L17.
10. From Overseas Archives (CAOM), ref. INDO/GGI/40423 to 40475.
11. This was based at Luang Prabang and was generally supported by Western countries.
12. This was the Lao Patriotic Front, including the dominant Lao Communist Party, supported by the USSR, China, the Viet Minh – a post World War II guerrilla force opposed to the French – and later, North Vietnam.
13. Kaysone Phomvihane was general secretary of the Communist Party from 1955 until his death in 1992, Prime Minister from 1975, and President of the Lao PDR from 1986 to 1992.

14. Perhaps the road construction was also a sign of loyalty to the Pathet Lao from the new governor, Khammuan Bupha, a former royalist officer commanding the newly joined royal forces, who soon became a general in the Pathet Lao army, before taking on ministerial positions in the successive governments of Lao PDR (Stuart-Fox, 2001).

15. Notably, those in Germany, Finland, Norway, Sweden and the United Kingdom.

16. Notably, the World Wide Fund for Nature (or World Wildlife Fund), the International Union for Conservation of Nature and the World Resources Institute.

17. In the Lao PDR, the Lao Revolutionary Party is constitutionally the superior body of political power.

18. The name given to the Government's strategic vision for the agricultural sector, according to the Ministry of Agriculture and Forestry in 1999.

19. According to the Ministry of Agriculture and Forestry's 2000 Framework of Strategic Vision on Forest Resources Management to the Year 2020.

20. The Ministry of Agriculture and Forestry's 2010 Strategy for Agricultural Development 2011 to 2020: Sector framework, vision and goals.

21. The Central Committee's instruction 03/PBP, 30/05/2011: Transformation of villages to be development units.

22. The Ministry of Agriculture and Forestry's Ministerial Instruction 22 on shifting cultivation (02/2010).

23. The Focal Zones Programme focused too explicitly on resettling the population and this worried some international aid institutions. The Cluster Development Programme (Prime Minister's Decree no 9, 7/2007) emphasizes decentralization and village organization, which are more likely to attract foreign support. However, it includes a minimum village size, which is the 'back door' to justify resettlement of upland communities, which are generally much smaller than lowland ones.

24. Personal communication between a high official of the Ministry of Agriculture and Forestry and the author (6/2001).

25. Prime Minister's Decree no 162 (7/1993), completed in 2004 with the Law on Local Administrations.

COLOURED PLATES* ...Part I (Plates 1-31)

Recording the bountiful agrodiversity nurtured by shifting cultivators at Khonoma, in Nagaland, northeast India, an Angami Naga research assistant leans close to photograph a specimen of *Solanum kurzii*, a member of the Solanaceae family.

Sketch based on a photo by Malcolm Cairns (2001)

* Refer to the accompanying regional map, on the next page, to see where each photo was taken. Please note that these photographs relate to rotational shifting cultivation, which has been shown to be sustainable, and not to pioneering shifting cultivation, which would be much harder to defend. The References section can be found at the end of Part II of the Coloured Plates section.

Every effort has been made to contact copyright holders for third party images, but in some instances this has not been possible. If you believe you are the copyright holder of a particular image, please contact either the Editor or the publishers.

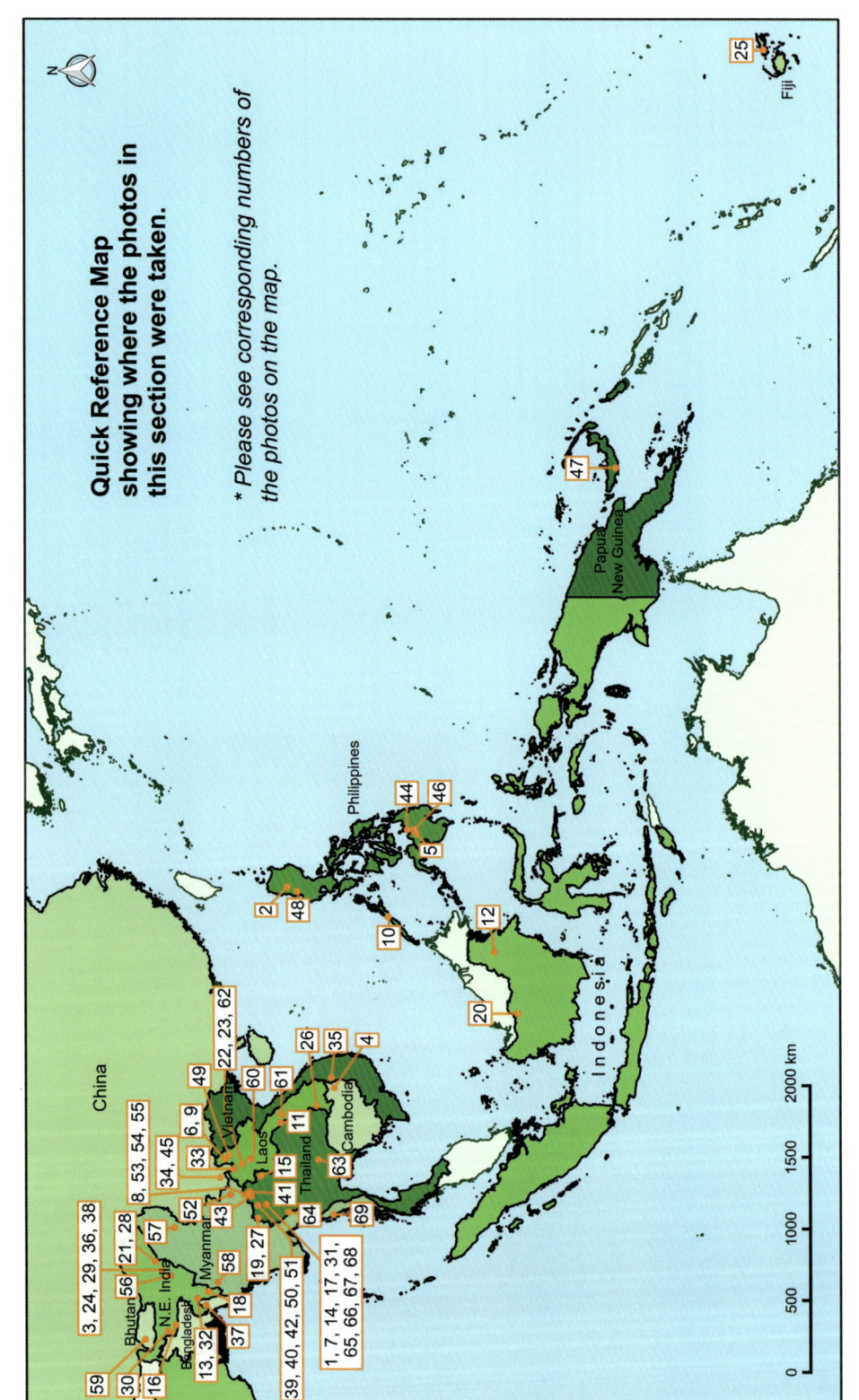

Quick Reference Map showing where the photos in this section were taken.

* Please see corresponding numbers of the photos on the map.

1. THIS IS A GENDER ISSUE: SHIFTING CULTIVATORS ARE MOSTLY WOMEN

Official policies that marginalize shifting cultivators have a disproportionately negative effect on women, who perform a large part of the labour, management and decision-making in swidden agriculture (Colfer et al., 2015).

1. This Lisu woman's betel-stained smile belies the arduous labour that typified her life in her village swiddens near Huai Thung Choa, about 60km northwest of Chiang Mai in northern Thailand, the site of early efforts to find crops to replace opium poppies.

Photo: Jack Ives (1979)

It is generally the case that men are more likely to cut down the trees in re-opening fallows, and women are more likely to weed the crop that follows. Pressures on the use of land, often arising from official policies, force the shortening of fallows. Shorter fallows mean fewer and smaller trees but escalating weed populations. Thus, shorter fallows generally imply an increasing burden on the women who weed the crops and a lighter task for the men who clear the land.

A vital policy recommendation is that governments should formulate policies and regulations that encourage local decision-making, including that by women.

2. The worn features of an old Igorot woman at Sagada, in the Philippines' Mountain province, tell of a tough existence in the upland *kaingin* (swiddens).

Photo: Jochim Voss (1978)

3. Amongst the Naga subgroups of northeastern India, Angami women are renowned for their hard work in agriculture.

Photo: Malcolm Cairns (2001)

Such is the extent of women's participation in shifting cultivation that it is often known as 'women's cultivation'.

Naga children generally get their first exposure to shifting cultivation from their mothers' backs.

2. EASY ACCESS TO CONTRACEPTION: A VITAL FACTOR IN THE USE OF LAND THAT POLICY-MAKERS OFTEN OVERLOOK

4. A young Kavet woman carrying her baby and a heavy basket of rice arrives at Kongnok village, in Cambodia's Ratanakiri province, after walking for several hours. Various pressures, including population growth, have resulted in greater distances between these people and their swiddens.

Photo: Anne Thomas (2008)

5. The Editor recalls that when he visited this Tala-andig village in the buffer zone of the Mount Kitanglad National Park in Mindanao, the Philippines, the women were desperate for some way to control their fertility.

Photo: Malcolm Cairns (1995)

Family planning could limit the number of children who inherit ancestral land, forcing its subdivision into ever-smaller plots in which a sustainable fallow period may become impossible. For women, it could also provide a balance between productive and reproductive roles (Colfer, 2015).

With no off-farm jobs or ability to intensify their farming practices, there seems little alternative for these villagers in the Philippines but to clear more fields from the forest.

3. DO FORESTS REALLY NEED TO BE PROTECTED FROM SHIFTING CULTIVATION?

One of the great ironies is that shifting cultivation has been one of the major factors in forming the very forested landscapes that policy-makers now want to protect from shifting cultivation (Ziegler et al., ch 11).

For centuries, forests and shifting cultivation have occupied the same parts of the map. Clearly, shifting cultivation has not destroyed these forests. Instead, it may have produced more biological diversity in forests (Falvey, ch 3). It would be a different, more destructive story if the same areas were occupied by permanent agriculture.

6. Not the primary rainforest it would seem. This is a 15-year-old fallow in Phongsaly province, northern Lao PDR.

Photo: Olivier Ducourtieux (1998)

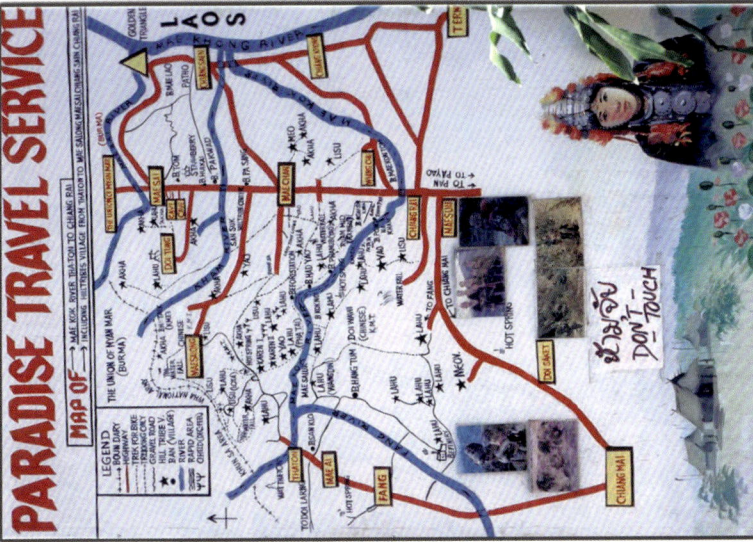

7. A shop selling 'hill tribe' treks in Chiang Mai, northern Thailand, offers visits to a wide variety of ethnic groups that have long been portrayed as primitive or socially backward.

Photo: Tim Forsyth (1990)

8. A map of villages and ethnic groups in northern Thailand, displayed in Doi Mae Salong, Chiang Rai province. Many tourists trekked the highlands to see villagers dressed in traditional clothing (like the Akha woman in the picture), and to smoke opium.

Photo: Tim Forsyth (1991)

Its detractors say that shifting cultivation is a 'primitive' practice. They ignore the fact that it is backed by centuries of trial, error and experimentation that has created a remarkable legacy of accumulated indigenous knowledge.

5. A MELTING POT, OR DISTINCT SOCIETIES?

Most upland shifting cultivators belong to ethnic minority groups. This suggests the need for broadly-based social policies that opt either to absorb these groups into wider society, with all the assistance that might require, or to respect them as distinct societies and encourage them to retain their cultures, with shifting cultivation as a central feature.

9. Ho children at school in Phongsaly province, northern Lao PDR, where they learned the national language, but still wore their ethnic costumes.

Photo: Olivier Ducourtieux (2000)

6. WHAT IS IT ABOUT SHIFTING CULTIVATION THAT CONCERNS OFFICIALDOM?

A. Many observers regard the clearing of secondary forest to reopen fallowed land as environmental vandalism. They fail to inquire about the long term.

10. A migrant Cuyonon farmer on Palawan, in the Philippines, fells a forest giant prior to planting rice and later, coconuts, bananas and fruit trees in a settled agroforestry system.

Photo: James Eder (1971)

11. This Tai Bor swidden farmer in Khammouane province, central Lao PDR, was anxious to assert his use rights over village fallow land that had been zoned as 'degraded forest' by government authorities and allocated to a private company for a *Eucalyptus* plantation.

Photo: Keith Barney (2006)

Shifting cultivators generally avoid large trees, like the one pictured above, because felling them is far too much work with their rudimentary tools. They simply work around them. This man might have been a logger, but as the caption reveals, he is a permanent cultivator.

The photograph on the right is far more typical of the secondary forests that shifting cultivators clear when reopening their fallows.

B. Shifting cultivators' use of fire attracts a lot of criticism. Concerns about haze and climate change have increased pressure on swidden farmers to stop burning.

12. Kenyah Dayak communities in the Apo Kayan region of East Kalimantan, Indonesian Borneo, use a cyclical system of fallow and fire to grow rice and to create a succession of forest environments that provide timber and non-timber species, including medicinal plants.

Photo: Danna Leaman (1988)

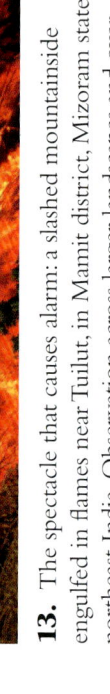

13. The spectacle that causes alarm: a slashed mountainside engulfed in flames near Tuilut, in Mamit district, Mizoram state, northeast India. Observation across larger landscapes and over longer periods creates a better appreciation of shifting cultivation.

Photo: Zakhuma (2014)

Fire continues to be the most practical way of releasing plant nutrients from slashed biomass and clearing fields in preparation for planting. However, the burning phase of shifting cultivation causes concern as a source of carbon emissions. Sarawak recently banned open burning for agriculture and in doing so, indirectly banned shifting cultivation (Mertz and Bruun, ch 2).

C. As haze has become a problem of increasing severity across Southeast Asia, this has built enormous political pressure on governments to try to minimize its sources, including shifting cultivation.

15. Maize stover (leaves, stalks, husks and cobs) remaining in permanent cultivation fields after maize harvest on unregistered land in Pua District, Nan Province. The stover will be burnt *in situ* during the hot-dry season of March–April in preparation for new maize plantings. Extensive burning of maize stover contributes significantly to the haze problem in northern Thailand in the hot-dry season.

Photo: Soontorn Khamyong (December, 2014)

14. Chiang Mai city is in a valley surrounded by mountains. There is little wind in the dry season from November to April. The air quality declines as burning from forest ground fires and the burning of agricultural residues between February and April is added to the daily urban pollution. By mid March, Air Quality Indexes commonly exceed levels dangerous to human health.

Photo: Marisa Marchitelli (March, 2016)

It is largely contract farming that has fuelled the expansion of maize monocultures. It is unlikely that the value of the maize is sufficient to pay for the environmental damage caused by the hunger for corn and profits.

The number of respiratory complaints treated by the local hospitals skyrockets during the burning season.

D. Along with concerns about deforestation and burning come worries about loss of wildlife and biodiversity.

17. The pelt of a forest cat is offered for money to buy food. A young hunter risks indignation at the slaughter of wildlife in a Lisu village near Huai Thung Choa, in northern Thailand.

Photo: Jack Ives (1979)

Hunting is generally a part of the livelihood strategy of farmers living in the forest. But the sanctity of wildlife was always guarded, until poverty began to bite, the value of beautifully-cured pelts became known and development projects brought outside markets.

16. This elephant was spotted in an area of swiddens, wet-rice fields and fallowed land in the South Garo Hills, Meghalaya, northeast India. The area has a large number of elephants and they often destroy swidden crops as they roam the landscape.

Photo: Varun Goswami (2011)

Rather than being a threat, swidden fallows attract many kinds of wildlife. For example, abandoned food crops can be an important source of food for elephants (Gunasena and Pushpakumara, 2015). The pioneering biogeographer, Alfred Russel Wallace, found … that of all available landscape features, patchy swidden areas were the richest in biodiversity (Dove, 2015).

E. Another of the main criticisms levelled at upland shifting cultivation is that steeply sloped land is very vulnerable to soil erosion.

18. Don't slip or fall! When the only arable land is precipitous, shifting cultivators may have no choice but to farm it. A swidden near Hnahthial town, in Lunglei district, Mizoram, northeast India.

Photo: David Vanlalfakawma (2013)

19. Children watch from above as adults and teenagers cultivate a swidden cleared after a nine-year fallow in northern Thailand's Mae Hong Son province.

Photo: Carrie Sedlak (2009)

To minimize erosion, it is standard procedure for shifting cultivators to dibble seeds directly into the undisturbed soil, a practice that also allows regrowth to fairly quickly cover the soil again. MacDicken (ch 51) suggests that restrictions commonly used for commercial harvesting of timber may usefully be applied to swiddening. For example, no cultivation within 100m of a watercourse or on slopes greater than 35%.

F. Shifting cultivators are gaining access to modern technologies. Does this mean that their traditional balance with nature may be lost and they may become more destructive?

21. A Konyak Naga farmer in Mon district, Nagaland, northeast India, sprays a salt solution to kill broadleaf weeds that became a problem because of shorter fallow periods.

Photo: Malcolm Cairns (2001)

Shifting cultivators have long 'hitched a ride with nature', but as the ecological foundations of their farming systems are increasingly undermined, it is likely that they will turn to the technological 'fixes' that are commonly used in permanent cultivation, to overpower nature. For example, it may prove to be a small step for them to move from spraying salt solutions to using commercial herbicides.

20. With his modern chainsaw, this Iban villager in the northwest corner of the Danau Sentarum National Park in West Kalimantan, Indonesia, was cutting up logs felled on his swidden territory for a Ministry of Agriculture project encouraging rubber planting in an area regarded by the Ministry of Forestry as a wildlife reserve.

Photo: Carol Colfer (1993)

Shifting cultivators, working with axes, may be unable to seriously damage the forest, but what may happen if they take up chainsaws? Do the tools used by shifting cultivators need to be regulated?

G. Another big reason for the vilification of shifting cultivation: most of the opium produced in the 'Golden Triangle' came from swiddens.

23. Raw opium is gathered from poppy pods in a Hmong Lay village called Nampik, in Xay district, Oudomxay province, Lao PDR. The crack-down on opium growing was devastating for many shifting cultivators, who were offered no alternative livelihoods (Cohen, ch 31).

Photo: Laurent Chazée (1994)

Opium was the ideal cash crop for many shifting cultivators because it was highly valued, low in volume and not very perishable. Poppies also fitted perfectly into the cropping calendar and thrived in poor upland soils.

22. A Khmu Lu woman tends to a crop of opium poppies in Oudomxay province, Lao PDR. Shifting cultivators were the main target in the fight against opium production in Laos.

Photo: Laurent Chazée (1999)

H. Governments tend to regard it as a sign of poverty and backwardness that shifting cultivators grow crops for their own needs and not for sale. Of course, this also means that they don't provide taxable revenue. Another historical concern was the fear in some countries that remote shifting cultivators could become communists.

25. A Fijian shifting cultivator digs up a yam (*Dioscorea alata*) from a yam mound during the annual harvest on Yadua Island, in the northwestern Fiji Islands.

Photo: Randolph Thaman (2012)

24. An Angami Naga harvests her swidden rice in Khonoma village, Nagaland, northeast India

Photo: Malcolm Cairns (2001)

Across most of Asia, upland rice is grown in swiddens as the main staple crop .....

.... but as we move further south, into the Pacific, the rice disappears and is replaced by root crops.
  Government antipathy towards cultivators who do not contribute to tax revenue is probably universal. One of the world's most famous naturalists, Carl Nilsson Linnaeus, struck trouble with Swedish authorities when writing in support of shifting cultivation in the 1750s. This prompted a leading scholar of Linnaeus' work to comment: *"For that reason [the lack of tax revenue] it was politically expedient for the authorities to see to it that burn-beating [shifting cultivation] was not especially encouraged ... On the other hand, the destructive effect on the forests and the soil was emphatically pointed out."* (Weimarck, 1968, cited by Dove, 2015, p12)

I. Powerful interests often try to push shifting cultivators off their ancestral land so that they can claim it as their own. They include state governments, investors, corrupt politicians, logging companies, military groups and international organizations.

27. The rise of eco-tourism has seen hotels and resorts, such as this one in Mae Hong Son province, northern Thailand, spring up in pristine forest environments and even on farm land.

Photo: Malcolm Cairns (1992)

26. A mission from Canada's International Development Research Centre (IDRC) pauses to admire some logs in southern Laos. Loggers usually harvest the very best trees, and their seeds are then lost from the forest's gene pool. This is the destructive opposite to selective breeding.

Photo: Anonymous (1991)

J. The legal systems of many countries do not recognize the communal land tenure systems that are used by many shifting cultivators. This leaves them vulnerable to 'land grabbing'.

The lack of secure land tenure is by far the most frequently cited constraint to the evolution of shifting cultivation into more sedentary forms of agriculture (MacDicken, ch 51).

Clearing an entire hillside and temporarily dividing it into household allotments, as can be seen in this photo, offers significant benefits to shifting cultivators (Yaden, 2015):

- Everyone shares the same access path, so it is easier to keep it cleared;
- Fencing is reduced and it is easier to guard crops against damage by wildlife and stray livestock;
- When flocks of birds plunder the crops, the losses are shared and no single field is completely wiped out;
- Since everyone is working in the same general area, the arrangement is good for communal labour;
- Traditional village leadership finds large swidden blocks easier to manage;
- The communal nature of the fields adds a social element to working in swiddens. It is 'more fun' to work in groups;
- In Nagaland's head-hunting days, working together offered greater security. There was safety in numbers.

There are strong arguments for adopting customary land tenure for shifting cultivators, but governments are generally uninterested or unable to allow for a set of diverse customary land-tenure arrangements (MacDicken, ch 51).

28. Typical swiddening in Mon district, Nagaland, northeast India. The hillside would be owned by an entire village or kinship group and the huts mark its division into family plots for one swidden cycle.

Photo: Malcolm Cairns (2001)

7. GOVERNMENT POLICY HAS ALWAYS SOUGHT TO REPLACE SHIFTING CULTIVATION WITH SOMETHING ELSE

A. Cash crops

State policy-makers want upland agriculture to contribute to tax revenue, so they encourage a move away from subsistence agriculture to cash crops and plantation monocultures.

30. This man is a member of a self-help group cultivating a half-hectare plot of turmeric at Joropura village, East Garo Hills, Meghalaya, northeast India, as a source of cash income. His family still depends on shifting cultivation for its food.

Photo: Sanat Chakraborty (2010)

Another major problem is getting crops to markets. Shifting cultivators tend to live in remote areas with no farm-to-market roads. Farmers find themselves at the mercy of buyers, in terms of prices received for their crops.

Photo: Malcolm Cairns (2001)

29. An Angami farmer in Khonoma village, Nagaland, northeast India, inspects his crop of ginger.

A worrying number of swidden farmers across the Himalayan foothills have turned away from their high-agrodiversity systems to plant ginger, turmeric and a very narrow range of other cash crops. This threatens to leave them just one market crash or disease outbreak away from disaster.

31. In the long campaign to eradicate opium in northern Thailand, cabbages spread to cover entire mountainsides, like this field in Chiang Mai province.

Photo: Malcolm Cairns (1992)

Cabbages were one of the crops introduced in northern Thailand as a replacement for opium poppies. When deluged by monsoon rains, chemical inputs applied to the crops are likely to be washed downslope and into the nearest stream.

In contrast, swidden crops are generally organically produced and there is no danger from chemicals to either the environment or consumers.

The subject '7. GOVERNMENT POLICY HAS ALWAYS SOUGHT TO REPLACE SHIFTING CULTIVATION WITH SOMETHING ELSE' continues in Part II of the Coloured Plates Section.

16

SWIDDEN AGRICULTURE AND SLOPING LAND CONVERSION IN CHINA'S DULONG VALLEY

Impact and adaptation

Shen Shi-Cai, Li Di-Yu, Zhang Fu-Dou, Xu Gao-Feng, Andreas Wilkes, Yin Lun and Jin Gui-Mei*

Introduction

Land degradation has become a serious problem in China because of environmental conditions and inappropriate human management. The main causes include a rising population and low availability of land per capita, resulting in increasing intensity of land use and degradation. Ninety per cent of poor people live in areas of moderate to severe land degradation. Living standards have risen rapidly in urban areas, increasing the demand for meat and livestock products, and this encourages the overuse of grasslands and increased use of chemical fertilizers. As well, policies are applied in a 'top-down' process with no respect for local conditions; the regulatory environment for dealing with land degradation is inadequate; and there are poor financial incentives for conservation (Wang et al., 1999; Li, 2002).

In 1999, the Chinese government announced its Sloping Land Conversion Programme (SLCP), otherwise known as 'Grain for Green'. Its objective was to reduce land degradation, especially by increasing the vegetative cover and reducing soil and water loss (Bennett, 2008). It provided that all cropland on slopes exceeding 25 degrees should revert to forests and trees should be planted on grasslands. For this, farmers would be provided with grain subsidies or cash for five to eight years. Additionally, the programme was structured in the hope of alleviating poverty and introducing

* Dr Shen Shi-Cai, Li Di-Yu, Zhang Fu-Dou, Xu Gao-Feng and Jin Gui-Mei are all from the Agricultural Environment and Resource Research Institute, Yunnan Academy of Agricultural Sciences, Kunming, China; Author Shen is also affiliated with the Centre for Biodiversity and Indigenous Knowledge, Kunming, China; Dr Andreas Wilkes is from the Centre for Mountain Ecosystem Studies, Kunming Institute of Botany, the Chinese Academy of Sciences, Kunming, and the World Agroforestry Centre's China Programme in Beijing; Dr Yin Lun is from the Institute of Ethnic Literature, Yunnan Academy of Social Sciences, Kunming, China.

ecologically sound practices to China's farmers. Since the programme began in 1999, there have been many research studies and findings. However, much of this research has been focused on the advantages of the programme and its success in economic, environment and ecological terms. Only a few studies have examined its negative impacts, such as the conflicts between rural livelihoods and implementation of the Sloping Land Conversion Programme (Weyerhaeuser et al., 2005; Xu et al., 2005). The programme's potentially negative impacts on indigenous people and its disregard for indigenous knowledge have rarely been mentioned (Xu and Ribot, 2004; Xu and Wilkes, 2005).

In 2002, the programme was implemented in Dulong valley, in the northwestern corner of China's Yunnan province. The valley is the home of the Dulong people, one of the least populous ethnic groups in Yunnan (Long et al., 2004). Apart from retaining their wet-rice fields, permanent-cultivation fields and vegetable gardens, the Dulong found that all of their remaining arable land and rotational arable fields were incorporated in the conversion programme. This brought traditional cultivation to an end, threatened the survival and continuity of Dulong crop-species diversity and traditional culture, and transformed livelihoods in Dulong valley. This chapter discusses the impacts of the Sloping Land Conversion Programme on land use and agrobiodiversity, livelihoods and ethnic culture in Dulong valley. It also describes a series of activities that have raised awareness of the value of agrobiodiversity and have supported conservation efforts by communities and local governments.

Methods

Dulong valley is situated in the northwestern corner of Yunnan province ($98°–98°\ 5'$ E, $27°5'–28°$N), bordering Myanmar to the west, Tibet to the north, and the Salween valley to the east (Figure 16-1) (Chen et al., 2006). The area is inhabited by fewer than 4052 members of the Dulong (also known as Drung) ethnic group (Gao, 2003; Luo, 2006; Wilkes and Shen, 2007). The valley's location makes it an area of extremely high natural biodiversity (Chen, 1993), and its traditional agriculture system generates a significant level of agrobiodiversity.

In view of the importance of traditional agriculture in Dulong livelihoods and culture and the potential uniqueness of agrobiodiversity in Dulong valley, the Centre for Biodiversity and Indigenous Knowledge (CBIK), a non-governmental organization based in Yunnan, undertook a number of surveys and action-research studies on the impacts of the SLCP (Xiao, 2005; CBIK, 2006; Wilkes and Shen, 2007). This chapter draws on those studies. As well, an additional survey of the status of traditional crop cultivation was conducted in two villages in 2009, and data provided by the local government were analysed. The two villages involved in the 2009 survey were Dizhengdang, in the upper reaches of the valley, about 65km from the township seat of government and the local market, and Kongdang, in the central area, only about 2km from the township seat and the local market. These communities were selected to represent different degrees of access to the market and government support, as well

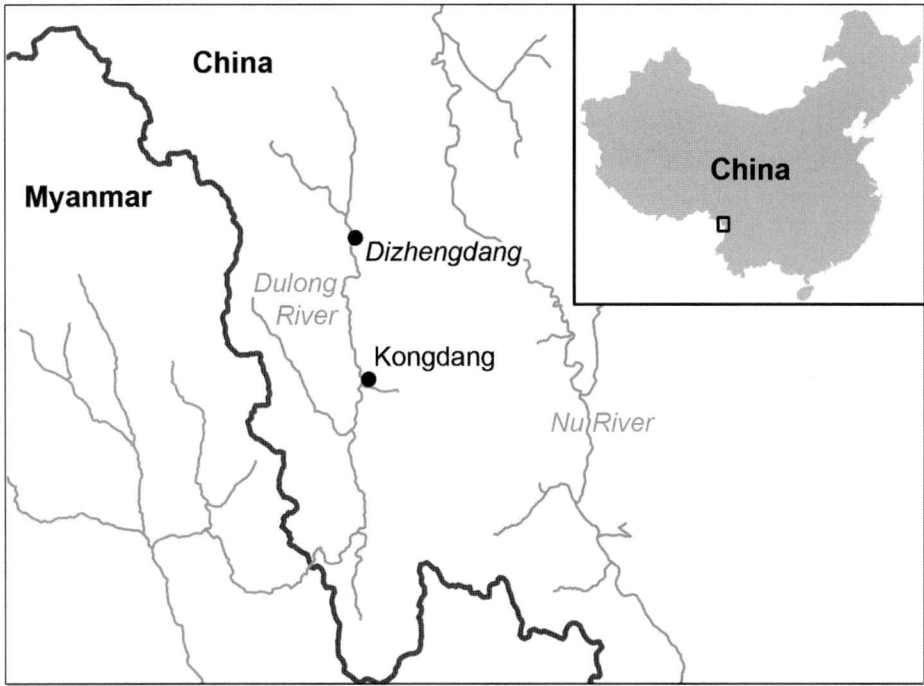

FIGURE 16-1:   The study area, Dulong valley, Gongshan county, Yunnan province.

as for their different natural-resource endowments. Dizhengdang (61 households) has more abundant forest and cropland resources per household, while Kongdang (73 households) has limited cropland per household. Our survey sampled 25 households at random in each village. In addition, 15 unstructured individual interviews and three focus-group discussions, which involved 45 elderly villagers and women, were held in the two villages to investigate householders' diverse perceptions and responses to the SLCP. Interviewees were selected to represent diverse social groups.

Dulong agriculture and the SLCP

Conversion of swidden agriculture

Historically, the Dulong people depended mainly on traditional shifting agriculture, along with hunting, fishing and gathering. Dulong valley had five kinds of agricultural land-use systems (Yin, 2001; CBIK, 2006).

1. Long-fallow swidden land: This was one of the main rotational-farming systems, and was usually the greatest distance from the village. Choosing a piece of forest dominated by *Alnus nepalensis* or pines, the farmers cut the trees and burnt the dry slash. Crops were then grown for one or two years, followed by more than six years of fallow. Wood ash was used to improve soil nutrients, and wood from the cut trees was also used as firewood and for house construction.

When the cropping phase ended, *A. nepalensis* was planted in the swiddens and forest cover regenerated quickly. This kind of land had no property rights. Families were able to clear as much land as their labour resources could manage.

2. Short-fallow swidden land: This was also used for swidden cropping. Dominated by *A. nepalensis* or other mixed shrubs, it was closer to the village and was used more frequently than the long-fallow land. The use of this land and the fallow period depended on the grain grown by the farming family. In general, it was cropped for one to three years and left fallow for two years. Whoever planted the land with *A. nepalensis* had clear ownership rights. *A. nepalensis* was routinely planted at the end of cultivation.

3. Maize land: This was privately owned and permanently cropped, and was nearby the family home. Normally, mixed crops were planted every year under contract. An example was growing maize for two years followed by one year of taro. Every family had a finite plot of cultivated land to which it held clear ownership rights.

4. Paddy-rice land: This was created in a small number of locations in the 1950s, and consisted of bench terraces made by Dulong villagers in the lower reaches of the valley, relying on manpower, sticks, soil and stones. The local government encouraged construction of the terraces, and families held clear ownership rights. The rice varieties and growing techniques came from special technicians, brought from outside. The bench terraces were flat, located close to villages in the lower reaches of the valley, and had higher soil nutrients and higher temperatures. During rice growing, the terraces were fertilized with pig and cattle manure. Although the paddy-rice land played an important role in some villages, wet-rice cropping was halted when the conversion programme began because the programme undertook to supply the villagers with rice.

5. Vegetable gardens: These were close-by each farmer's house, and were privately owned. These gardens boasted the highest soil nutrients, and mostly grew vegetables such as maize, taro and yams.

Swidden agriculture, including both long- and short-term fallows, occupied about 85% of the total agricultural land in Dulong valley, and it was central to the traditional livelihoods and culture of the people (Yin, 2001; Gao, 2003; Qi, 2006). Traditionally, primary or secondary forest was cleared for swiddens and crops were cultivated for one to three years before cropping was abandoned and the land left fallow. During the cropping period a number of agroforestry practices were followed. Sites that were dominated by *A. nepalensis* – a fast-growing, nitrogen-fixing tree – were the most preferred areas for swidden agriculture. During clearing, larger *A. nepalensis* trees were retained, but pollarded (Figure 16-2). Smaller trees and other undergrowth were cut, dried and burnt. Burning the dried vegetation accelerated decomposition, released useful nutrients for crop production and killed weeds and pests. Some months after burning, *A. nepalensis* saplings previously collected from nearby locations were transplanted into the cleared sites. Pollarding the larger, retained *A. nepalensis* trunks

FIGURE 16-2: Larger *Alnus nepalensis* trees are retained and pollarded by Dulong villagers.

Photo: Andreas Wilkes, 2002.

reduced shading impacts on crops. The thin branches that subsequently grew were sometimes also cut and burned in the second and third years of cropping, to maintain soil fertility. The cropping phase lasted up to three years.

During the cropping phase, a variety of annual rotations were used, involving several crop types, for example, maize, millet, buckwheat, upland rice, finger millet, beans and taro. The crops chosen depended on household needs, the fertility of the soil at each site and the perceived site requirements of each crop (Yin, 2001). The fallow that followed the cropping phase was, to some extent, managed. The retained and planted *A. nepalensis* trees and other trees that occurred at the site were allowed to regenerate for at least five or six years. *A. nepalensis* trees were regarded as very important because they had nitrogen-fixing properties that benefited soil fertility for future cultivation (Li et al., 2006). Fallows were also exploited for other purposes, such as livestock grazing, hunting, and gathering of non-timber forest products (NTFPs) and timber. When the site was cleared again, transplanted trees were cut along with the other vegetation, unless they were selected to be retained and were then pollarded. A scheme partially comparable with the *A. nepalensis* improved-fallow system of the Dulong can be found in Nagaland, northeast India, and has been described by Cairns et al. (2007). Such systems support agrobiodiversity because they create and depend upon landscape and habitat diversity, while also meeting multiple livelihood needs.

Poverty, as measured by the government, has always been both widespread and deep in Dulong valley. In 1995, the average annual net per capita income (including the imputed value of agricultural produce) was just 344 yuan (US$41.38). From 1995 to 2001, per capita incomes rose to 684 yuan ($82.63), bringing average-income levels for the whole valley to just above the national poverty line. Since the 1960s, the government has tried to discourage swidden or rotational agriculture by providing relief grain to Dulong villages at subsidized low prices. The SLCP was the first official effort to introduce specific implementation measures, using funding from the central government to supply sufficient grain, to bring Dulong villagers' grain consumption levels up to the poverty line. Thus, in 2002, implementation of the SLCP began in Gongshan county, and most of the first year's conversion quota was allocated to Dulong valley and special implementation measures were established. All

Dulong villagers were required to take part in the programme and all of their swidden land and most permanent arable land had to be converted. Most land on slopes of more than 25 degrees on both sides of the Dulong river was taken for conversion to forest. In return, farming households were given grain subsidies covering eight years. The subsidy was paid annually. At the end of 2002, 14,000 mu or 924 hectares (1 mu=0.066ha) of cropland in the Dulong valley was returned to forest. This included all of the swidden land (9943 mu) of the Dulong people and 78% of their permanent arable land, such as maize land and paddy-rice land (GFB, 2009). The national guidelines for implementation of the sloping-land conversion programme stipulated that grain subsidies should be given on the basis of the area of land converted. However, given the large area converted in Dulong valley and the low rates of grain self-sufficiency among farming households over many years past, the local government decided to allocate the subsidy on a per capita basis, with all rural inhabitants (adults and children) receiving 180kg of rice per year. According to national-government policies, grain subsidies under the SLCP should have ended in 2011. However, in early 2012 the central government launched a new SLCP policy under which afforested households were given grain subsidies for a further eight years, but the subsidy was cut in half. As well as the grain subsidy provided by the local forestry department, villagers were also given rural-subsistence allowances of 50 to 60 yuan per person every month. Up to 2009, total cash subsidies provided by local governments amounted to more than 1.86 million yuan (US$272,727) (GFB, 2009). Nowadays, more than 98% of Dulong villagers are paid cash subsidies.

Loss of agrobiodiversity

Since the implementation of the SLCP, agricultural land use and crop varieties have changed greatly. In 2002, the cultivated area under swidden agriculture accounted for 67% of all cropland in Dulong valley. By 2009, all land in the cultivation phase of swidden agriculture and a large proportion of permanent maize fields had been converted to forest. Surveys in Dizhengdang and Kongdang both showed similar trends in land-use change. In 2002, the total area of cropland of surveyed households in both villages was 68.7ha, of which cultivated swiddens accounted for 71%; maize land, 22%; and vegetable gardens, 7%. In 2009, the surveyed households had less than 11ha of permanent maize fields and vegetable gardens, although the average land area in Dizhengdang was higher than that in Kongdang.

There was no baseline survey of crop diversity before 2002, although Long et al. (2004) reported the results of a rapid assessment that identified more than 20 crop species planted in Kongdang village in 2002. Interviews and informal discussions with villagers in both Kongdang and Dizhengdang identified 12 different types of crops (19 varieties) that were commonly planted by most households before 2002. Of these, seven crop types were 'under-utilized species', such as foxtail millet (*Setaria italica* [L.] P. Beauv.), pearl millet (*Setaria italica* var. *germanica* [Mill.] Schrad.), an unknown variety of millet (*Echinochloa* sp), amaranth (*Amaranthus paniculatus* L.), and cocoyams (*Colocasia esculenta* [L]. Schott). Because of the valley's location as a link between Chinese and Indo–Burmese genetic resources (Chen, 1993), some varieties were likely to have been unique to the area. The status of two common crops and seven under-utilized crops planted by households interviewed in the two study villages in 2009 are shown in Table 16-1. In 2009, most households no longer cultivated the traditional crop varieties. Compared with Dizhengdang, however, a larger proportion of households in Kongdang continued to cultivate traditional crops because of the proximity of the township seat and market demand for these crops and their products, for example, flour made from buckwheat (*Fagopyrum esculentum* Moench.) and wine made from foxtail millet (*Setaria italica* [L.] Beauv.).

Land-use change has been the main reason for discontinuing the cultivation of traditional crops. Specifically, few farmers continue to plant traditional crops because the yields are lower when they are planted in permanent fields rather than in swiddens. Some varieties can be grown only on swidden land. Many crop seeds can be stored for only one to three years (Table 16-1), so maintenance of genetic diversity requires that these varieties continue to be cultivated. Survey results show that many households believe that other farmers are retaining traditional-crop seeds and expect that if, in the future, they want to plant these crops, then they will be able to obtain seeds through seed exchanges. But, in fact, only a few farmers continue to grow some of the traditional species.

Livelihood change

Implementation of the SLCP has changed the pattern of livelihood activities and the structure of incomes in the Dulong valley (Table 16-2). In 2002, the main income sources were agriculture, forestry and, to a lesser extent, livestock husbandry. Forestry

TABLE 16-1: Cultivation of crop species in two study villages in 2009.

Crop type	Farmers cultivating crops in		Maximum seed storage time (y)
	Dizhengdang (n = 25)	*Kongdang (n = 25)*	
Maize: *Zea mays* L.	13	21	1
Soybean: *Glycine max* (L.) Merr.	13	15	2–3
Bitter buckwheat: *Fagopyrum tataricum* (L.) Gaertn.	9	15	1
Sweet buckwheat: *Fagopyrum esculentum* Moench	9	11	1
Finger millet: *Eleusine coracana* (L.) Gaertn.	8	9	1–2
Millet (variety unknown): *Echinochloa* sp.	0	7	5–6
Foxtail millet: *Setaria italica* (L.) Beauv.	0	5	3–4
Pearl millet: *Setaria italica* var. *germanica* (Mill.) Schrad.	0	2	1–2
Amaranth: *Amaranthus paniculatus* L.	0	3	3

income was largely derived from gathering and marketing NTFPs. In addition, both agriculture and forestry supported livestock production by providing feed and grazing land. Few households engaged in transportation or trading activities because the Dulong valley was sparsely populated and far from the county administration centre. Before the SLCP, most households were unable to meet their grain needs from their own agricultural production (Long et al., 2004). However, supplementary cash income from hunting, NTFP gathering and livestock husbandry enabled them to procure grain from other sources.

In 2009, the main sources of household income were forestry and state transfers (cash subsidies). Some of the increased income from forestry was from cultivation of cardamom in the understorey of secondary forests that were not converted as part of the SLCP. This agroforestry system has been promoted since 2007 by the local government in the Dulong valley. However, most of what is classified as 'forestry income' in government statistical reports is direct cash payments made in lieu of grain subsidies under the SLCP. Some households in Kongdang, because they are close to the township seat, have developed off-farm income sources (including transportation and trade). Because of restrictions on all uses of forests converted under the SLCP, as well as on uses of a nature reserve created in the 1980s that covers forests more distant from residential areas, many traditional activities, such as hunting, NTFP gathering, timber collection and livestock grazing, had ceased by 2009. State income subsidies now comprise about 25% to 30% of income in the two study villages, and cash payments made in lieu of grain subsidies under the SLCP account for a further 18% to 27% (Table 16-2).

Changes in land use and grain production have also had an impact on livestock husbandry. Before 2002, livestock husbandry was important for meeting household food-security needs, mostly through sales of animals on the market. Animals raised included pigs, chickens, goats, sheep, cattle, and some horses, mules or donkeys. In 2009, most livestock were raised to meet household consumption needs. Conversion of cultivated swidden land and SLCP restrictions on the use of this land after

TABLE 16-2: Structure of per capita income in two study villages in Dulong valley, Yunnan (2002, 2009).

Village income sources	Dizhengdang		Kongdang	
	2002	2009	2002	2009
Total annual income (yuan)	789.00	2301.00	1156.98	2734.00
Forestry (%)	24.65	34.21	28.05	31.01
Cropping (%)	54.21	26.03	41.05	17.76
Livestock and poultry (%)	9.80	4.50	13.95	4.92
Transportation (%)	0	1.01	2.85	6.15
Trade (%)	0	3.57	4.09	14.41
Subsidies (%)	11.34	30.68	10.01	25.75

Note: US$1=8.27 yuan in 2002; 6.82 yuan in 2009.
Source: Local government statistical reports (2002, 2009).

TABLE 16-3: Average livestock holdings per household in two study villages.

Livestock	Dizhengdang		Kongdang	
	2002	*2009*	*2002*	*2009*
Pigs	6.2	2.3	5.5	2.1
Cattle	4.2	1.6	1.0	0.9
Chickens	14.2	5.6	14.9	7.8
Goats/sheep	3.9	1.3	4.7	1.4
Horses, mules and donkeys	0.5	0.3	0.2	0.1

Source: Local government statistical reports (2002, 2009).

conversion meant that cattle, sheep and goats were no longer allowed to graze in these areas, so the scale of livestock husbandry has declined (Table 16-3). Some villagers reported giving up livestock husbandry because they had insufficient grain to use as feed.

Traditional culture change

Traditional rotational agriculture relied on knowledge about the characteristics of swidden sites, such as vegetation cover, slope, aspect, soil and so on, as well as knowledge relating to the treatment of different forest resources and the use of fire. Special farming tools were used to minimize soil erosion caused by cultivation on steep slopes, and there was also a lot of knowledge related to the production and use of these tools. Traditionally, Dulong hamlets were based around one patrilineal clan, and elders had a great deal of influence over the use of forest resources, such as the choice of land plots for swidden cultivation. In the process of cultivation, there were all sorts of joint-cultivation arrangements between households that were based on traditional social ties. And for those Dulong who had not converted to Christianity, cultivation had to be preceded by rituals to appease the spirits. Thus, traditional agriculture was a core element of Dulong culture, relating not only to ecological knowledge, but also to religion and social organization. In northern parts of the valley that had not converted to Christianity, clearing of forests for swidden agriculture was preceded by religious rituals performed by a *namusa*, a kind of shaman (Box 16-1). The purpose of the rituals was to seek the support of spirits for a good harvest. Such rituals ceased with the end of swidden cultivation in 2002. Some older Dulong villagers are concerned that the knowledge and abilities required to perform these rituals will now disappear, particularly because younger people are not interested in learning them.

Agrobiodiversity is also central to ethnic cuisine. For many older people, food is a basic expression of culture. Traditional Dulong food does not include paddy rice, although this is the standard grain provided by subsidies under the SLCP. Most villagers have therefore not been able to eat traditional varieties of grain since the SLCP was implemented. Crops other than paddy rice are referred to as 'ethnic food', and elder villagers insist that in their cultural views, mixed grains other than rice are

---

BOX 16-1: A religious ritual performed by a *namusa*

A table was brought into the centre of the square. Into the square stepped two old Dulong men, one of whom was one of the village leaders. The other was a *namusa*. They were dressed in regular clothes - one was wearing a pink long-sleeved sweater, a dark-grey hat and blue Adidas-style trek pants rolled up over his thighs. They were draped in a woven cloth, which was white with multi-coloured length-wise stripes. It covered their shoulders and backs. Covering their ankles were similar pieces of white cloth, but with fewer stripes. The village leader wore a long-bladed knife in a sheath attached to a piece of string hanging over his shoulder. From his shoulder and crossing his torso was a white fish-net bag; he also carried a long bamboo stick.

The ritual-performance began when a male attendant entered the square and placed a bowl on the table in front of the two men, together with a small twig from a pine tree. The village leader then unsheathed his knife and placed it on the table, while the other man started chanting and reciting words in the Dulong language. He then dipped the twig in the bowl, and started moving it around, sprinkling alcohol on the ground and towards the sky. The village leader picked up the knife again and started moving it around in circles, while joining in the chanting. This continued for a while, until the movements and the chanting suddenly stopped; the 'performance' was over. The whole process had taken 40 to 60 minutes.

---

better for human health. For example, when mothers rest for a month after giving birth, they are usually given these mixed, traditional grains to eat (CBIK, 2006). Finger millet (*Eleusine coracana* [L.] Gaertn.) (Figure 16-3) is also a widely known cure for diarrhoea.

Besides changes in diet, many older people worry about the overall loss of their distinctive ethnic culture (CBIK, 2006; Wilkes and Shen, 2007). They say: 'If young Dulong people are not good in school and are unable to find work, and they no longer understand ethnic food and don't know which wild vegetables to eat and how to cultivate ethnic foods', then they will be severely lacking in the usual social and ethical standards. Other villagers ask: 'If you don't know these things, then are you still a Dulong?' This illustrates the important part played by swidden agriculture and traditional foods in what it means to be Dulong. Nowadays, many younger villagers say that, even when the SLCP ends, they do not want to return to traditional practices, because they are labour-intensive and physically demanding.

Responses to the impacts of the SLCP

Since 2002, the local government has introduced several measures to support implementation of the SLCP and its goals – as they are understood by local officials. These include promoting alternative energy sources (e.g. biogas and hydro-electricity) to reduce dependence on firewood in areas near the township seat in Kongdang and helping to bring building materials from outside the valley to reduce demand for locally sourced timber. Alternative sources of income have also been established, such as another eight years of grain subsidy for the SLCP, annual payments of long-term rural-subsistence allowances and cash cropping in secondary forest, to reduce poverty and improve local livelihoods. The crops include *Amomum tsao-ko,* different kinds of mushrooms and traditional medicinal plants. The conservation of agrobiodiversity and related aspects of Dulong culture were not previously a priority of local government. However, under pressure from the Center for Biodiversity and Indigenous Knowledge, local government has been taking a new interest in both agrobiodiversity conservation and protecting Dulong cultural inheritance.

FIGURE 16-3: Finger millet (*Eleusine coracana* [L.] Gaertn.) grown in the Dulong valley.

Photo: Shen Shi-Cai (2006).

In the course of promoting conservation of agrobiodiversity and preservation of traditional culture, the CBIK has been working with community members and local government officials to convene workshops and other community events to raise awareness and provide support for conservation activities. Since 2008, traditional cultural-exchange meetings have been convened in different communities in the Dulong valley. The main activity has been the staging of 'seed fairs', in which villagers bring traditional seed varieties and exchange knowledge about seeds, as well as exchanging the germplasm itself (Figure 16-4). The fairs have provided an opportunity for local government officials to understand the range and characteristics of traditional seed varieties and their current conservation status. The fairs also provide an occasion for cultural shows and competitions, and local 'superstars' are selected and awarded prizes in various categories, such as traditional crop-variety preservation, traditional food making, song and dance and traditional costume making. Between 2006 and 2009, 563 villagers from 11 communities in the Dulong valley have been involved in these meetings. At the end of 2009, incomplete

FIGURE 16-4: A seed-exchange gathering in the Dulong valley.

Photo: Shen Shi-Cai (2006).

statistics collected to evaluate the impact of these workshops showed that at least 21 additional households had renewed their cultivation of traditional crops (Figure 16-5). In 2010, the county agricultural bureau and township government took over from CBIK the task of convening traditional cultural-exchange meetings.

The county forestry bureau and township government also agreed in 2008 to the use of 1ha of land for swidden cultivation. A plot of land was selected in the upper part of the Dulong valley and eight

FIGURE 16-5: A Dulong farmer who is voluntarily cultivating a traditional crop species on his own land in the Dulong valley.

Photo: Andreas Wilkes (2006).

traditional crops were cultivated using traditional methods. The crops were cocoyams (*Colocasia esculenta* L. Schott.), bitter buckwheat (*Fagopyrum tataricum* L. Gaertn.), sweet buckwheat (*Fagopyrum esculentum* Moench.), finger millet (*Eleusine coracana* L. Gaertn.), millet (*Echinochloa* sp), foxtail millet (*Setaria italica* L. Beauv.), pearl millet (*Setaria italica* var. *germanica* Mill. Schrad.), and amaranth (*Amaranthus paniculatus* [L.]). These crops were chosen by villagers primarily on the basis that they did not grow well in permanent fields. Five villagers and one township official were nominated as responsible for the cultivation, but many others volunteered to work in the special plot. From 2005, CBIK and local government officials have been collecting agricultural resources in Dulong valley, especially traditional crops that do not grow well in permanent fields. In 2010, more than 400 accessions of agricultural biological resources, including germplasm from 42 species or subspecies of 18 plant families, were collected and stored in the Biotechnology and Germplasm Resource Research Institute of Yunnan Academy of Agricultural Science. These agricultural biological resources consist of food crops, vegetables and medicinal plants, and if the need arises, they can be made available for the Dulong people to use once again.

With support from CBIK, a government official has also documented the process of swidden cultivation, including ritual observances (Figure 16-6), using a digital video camera. These activities serve to preserve agrobiodiversity and to document traditional agricultural practices and culture. When the video is completed, a Dulong-language version will be used to educate school children about traditional agriculture and related culture. Chinese-language versions may also be useful to support future dialogues with policy-makers. Some villagers have now begun to package traditional seeds and crop produce for sale to tourists, and local-government agencies are assessing the feasibility of developing processed products using traditional crop

FIGURE 16-6:   The oldest Dulong wizard capable of performing traditional rituals in Dizhengdang village.

Photo: Shen Shi-Cai (2006).

varieties. With local-government support, three tourism centres were created in 2010 in Dizhengdang, Kongdang and Qinlangdang villages in Dulong valley. These centres organized local villagers to grow traditional crops, collect wild medicines and cash plants, and make traditional costumes for markets. When tourists arrived, traditional seeds, crop products, traditional handicrafts and traditional cultural shows were provided and enjoyed. This was regarded as a very important strategy for strengthening the linkage between agrobiodiversity conservation, the preservation of traditional culture and livelihood development in Dulong valley.

Discussion and conclusions

For many mountain peoples in the eastern Himalayas, swidden agriculture is an integral part of not only natural-resources management and genetic-resources conservation, but also of ethnic identity and biocultural heritage. In areas rich with natural biodiversity, the persistence of swidden cultivation is mostly seen as a 'problem' that obstructs the achievement of conservation objectives, and policies originating from a forest-conservation perspective often seek to eradicate swidden agriculture. However, traditional rotational shifting cultivation contributes to the maintenance of diversity of both crop and animal genetic resources, and these play important roles in maintaining cultural identity (Wilkes and Shen, 2007).

Since the compulsory conversion to forest of all areas in the Dulong valley that were used for swidden cultivation, many farmers have incorrectly assumed that others are preserving traditional seed varieties, when the danger is that nobody is doing so. The time for which seeds remain viable in storage is often brief. Given the centrality of these genetic resources to Dulong cultural identity, as well as uncertainty over extension of the SLCP beyond 2011, the preservation of genetic resources with which to maintain traditional farming systems is an urgent task. Activities supported by CBIK in the Dulong valley have shown that communities and local governments, including officials who are sometimes of local ethnic origin, can reassess the value of traditional agricultural practices, even in the midst of a dominant policy environment oriented towards achieving forest-conservation objectives. More and more villagers and local government officials have become aware of the potential value of maintaining agrobiodiversity. Renewed implementation of the SLCP policy, along with the interventions introduced by the local government and the CBIK, may explore a pathway to sustainable development that promotes the conservation of agrobiodiversity and the preservation of traditional culture while supporting livelihood development in Dulong valley. However, there is a need for further study of how the Dulong people measure the impacts of the SLCP on their health and future food security and how they assess the interventions implemented by local government and the CBIK.

Acknowledgements

This research was supported by the World Wide Fund for Nature, Canada's International Development Research Centre, and the World Agroforestry Centre's Rewards for, Use of and Shared Investment in Pro-poor Environmental Services project.

References

Bennett, M. T. (2008) 'China's sloping land conversion programme: Institutional innovation or business as usual?' *Ecological Economics* 65, pp699-711

Cairns, M. F., Keitzar, S. and Yaden, A. (2007) 'Shifting forests in northeast India: Management of *Alnus nepalensis* as an improved fallow in Nagaland', in M. F. Cairns (ed.) *Voices from the Forest: Integrating Indigenous Knowledge into Sustainable Upland Farming*, Resources for the Future Press, Washington, DC pp341-378

CBIK (2006) *Consultations on Agrobiodiversity Loss and Conservation in the Dulong Valley, Yunnan. China*, unpublished report, Center for Biodiversity and Indigenous Knowledge, Kunming, Yunnan

Chen, Z. L. (1993) *China's Biodiversity: Current Status and Protective Measures*, Science Press, Beijing

Chen, Z. M., Pan, X. F., Kong, D. P. and Yang, J. X. (2006) 'Fish biodiversity and its distributional characters during winter in the Dulong River Basin, Yunnan, China', *Zoological Research* 27(5), pp505- 512

Gao, Y. X. (2003) *Upland Agricultural Minorities' Ecological Economy Research*, Yunnan Science and Technology Press, Kunming

GFB (2009) *The implementation scheme of sloping land conversion programme in Dulong township*, unpublished report, Gongshan County Forestry Bureau, available from Gongshan County Forestry Bureau and Dulong Township People's Government

Li, S. M., Long, C. L. and Dao, Z. L. (2006). 'An effective way to improve soil fertility in traditional agroforestry: Planting *Alnus nepalensis*', *Acta Phytoecologica Sinica* 30(5), pp878-886

Li, Z. (2002) *Survey of the Linkage between Land Degradation and Poverty*, ADB TA 3548 PRC paper, Asian Development Bank, Manila

Long, C. L., Li, R., Dao, Z. L., Liu, Y. T. and Li, H. (2004) 'Natural resource management in the Dulong ethnic community' (in Chinese with English abstract), *Acta Botanica Yunnanica* 15(Suppl), pp34-41

Luo, R. F. (2006) 'Dulong agriculture', in Jiaji Guo (ed.) *Sustainable Development and Yunnan's Ethnic Minorities*, Yunnan Minorities' Press, Kunming, China, pp75-83

Qi, Y. F. (2006) 'Dulong forest indigenous knowledge: Rotational agriculture research', in Q. H. Xiong and X. C. Shi (eds) *Ethnic Minorities and Biodiversity Research in the Gaoligongshan*, Science Press, Beijing, pp145-167

Wang, Z., Ye, X. and Li, G. (1999) *Chinese Ecological Agriculture and Intensive Farming Systems*, Environmental Science Press, Beijing

Weyerhaeuser, H., Wilkes, A. and Kahrl, F. (2005) 'Local impacts and responses to regional forest conservation and rehabilitation programmes in China's northwest Yunnan Province', *Agricultural Systems* 85, pp234-253

Wilkes, A. and Shen, S. C. (2007) 'Is biocultural heritage a right? A tale of conflicting conservation, development, and biocultural priorities in Dulongjiang, China', *Journal of Policy Matters* 15, pp76-82

Xiao, J. W. (2005) *Preliminary Survey of the Impact of Sloped Land Conversion on Agro-biodiversity in Dulongjiang Township*, CBIK Livelihoods Programme Working Paper 24, Centre for Biodiversity and Indigenous Knowledge, Kunming

Xu, J. C, Ai, X. H. and Deng, X. Q. (2005) 'Exploring the spatial and temporal dynamics of land use in Xizhauang watershed of Yunnan, Southwest China', *International Journal of Applied Earth Observation and Geoinformation* 7(4), pp299-309

Xu, J. C. and Ribot, J. (2004) 'Decentralization and accountability in forest management: Case from Yunnan, Southwest China', *European Journal of Development Research* 14(1), pp153-173

Xu, J. C. and Wilkes, A. (2005) 'State simplifications of land use and biodiversity in the uplands of Yunnan, eastern Himalayan Region', in U. Huber, H. Bugman and R. Mel (eds) *Advances in Global Change Research. Global Change and Mountain Regions: A State of Knowledge Overview*, Kluwer Academic, Dordrecht, The Netherlands, pp541-550

Yin, S. T. (2001) *Man and Forest*, Yunnan University Press, Kunming

17

POLICIES ON SHIFTING CULTIVATION IN THE COUNTRIES OF THE EASTERN HIMALAYAS

Karma Phuntsho, Gopal S. Rawat, Golam Rasul and Wu Ning[*]

Introduction

Shifting cultivation originated as an agricultural system during the Neolithic Period of human technological development (Das, 2007) and is still widely practised in tropical and subtropical parts of the world. At one time an estimated 400 million people practised shifting cultivation across tropical and subtropical Asia (Spencer, 1996; Ma, 1999; Kerkhoff and Sharma, 2006) and it is still common in the eastern Himalayas and adjacent upland regions, including the Chittagong Hill Tracts of Bangladesh, eastern Bhutan, northeast India, northern Myanmar, and parts of Nepal. In 2006 it was estimated that 10 million hectares of land in the region were under shifting cultivation (Kerkhoff and Sharma, 2006).

There is a popular perception that shifting cultivation is a destructive and undesirable practice. This is based on the mistaken belief that untouched forest is being destroyed to make way for low-productivity agriculture. In fact, shifting cultivators are using a practice that is best described as rotational agroforestry (Kerkhoff and Sharma, 2006), involving planned cyclical use of defined forest plots to grow food while retaining a forest cover. Short periods of cultivation are followed by long forest fallows, when the forest is enriched by useful cropping species. Traditional shifting cultivation is characterized by unique ecological, economic and social features which make it markedly different to both arable agriculture and forestry. These characteristics are related to the specific physical and climatic features of the land, particularly in areas with steep terrain and high rainfall where irrigated and mechanized agriculture is not feasible. Shifting cultivation is a low-productivity system that involves cultivation of extensive areas of land to secure both basic food subsistence and a wide range of other products, including medicine, firewood, fodder, agricultural implements,

[*] KARMA PHUNTSHO, DR GOPAL S. RAWAT, DR GOLAM RASUL and DR WU NING are all from the International Centre for Integrated Mountain Development (ICIMOD), Kathmandu, Nepal.

utensils, construction materials and ornaments (Tiwari, 2003). Essentially, it enables forest land to be used for growing food while retaining the advantages of forest cover. When practised in its original form, farmers spend many more years growing trees and crops than burning them, with fallowing used to protect the soil and restore its nutrients. The fallow phase harbours more diverse floral species than climax forests and the cropping phase is endowed with a diverse agro-biodiversity conserved in situ by the shifting cultivators, based on their traditional knowledge (Kerkhoff and Sharma, 2006).

Shifting cultivation is environmentally sustainable as long as the duration of the fallow is sufficient to regenerate the vegetative cover and the productive capacity of the soil (Ramakrishnan, 1984; Warner, 1991). Shortening the length of the fallow reduces yields (Boserup, 1965) and impacts overall productivity. Several authors have suggested that shortening the length of the fallow is the major cause of decline in crop yields (Mishra and Ramakrishnan, 1981; Toky and Ramakrishnan, 1981a; Maikhuri and Ramakrishnan, 1991; Ramakrishnan, 1993; Tawnenga and Tripathi, 1996, 1997a, b; Wenjuan and Yin, 2007). The mean annual total economic yield over a full rotational cycle from shifting cultivation systems cultivated with a longer fallow time of 10 to 30 years is twice that from areas cultivated with a shorter fallow time of five years (Toky and Ramakrishnan, 1981a). There are a number of reasons for this. Soil loss from shifting cultivation land is higher during the cultivation years than during the fallow years (Borggaard et al., 2003). As a result, surface run-off and sediment loss is about 40% higher in shortened (five-year) fallow systems than in those with a fallow period of 10 to 30 years (Toky and Ramakrishnan, 1981b). Consequently, concentrations of organic carbon and nitrogen in the top 7cm soil layer are about 25% lower (Ramakrishnan and Toky, 1981) and extractable and pooled phosphorus and potassium concentrations two-to-three times lower in fields with a shortened fallow period (Ramakrishnan and Toky, 1981; Tawnenga and Tripathi, 1996, 1997b). However, it is difficult to determine the precise impact of fallow length on yield as a single factor because yield levels are influenced by a wide range of biophysical, socio-economic and cultural factors (Mertz, 2002). Short fallow lengths also tend to reduce the availability of non-crop resources such as forage and firewood (Dalle and de Blois, 2006). Enforced shortening of the fallow length is now considered to be the major cause of environmental degradation in shifting cultivation systems, which manifests in weed invasion, habitat fragmentation, biodiversity loss, soil erosion and hydrological imbalance (Mishra and Ramakrishnan, 1981; Toky and Ramakrishnan, 1981a; Maikhuri and Ramakrishnan, 1991; Ramakrishnan, 1993; Tawnenga and Tripathi 1996, 1997a,b). Short fallows tend to reduce the abundance of mid- and later-seral and animal-dispersed plant species (Dalle and de Blois, 2006) and is associated with rampant weed growth (Wenjuan and Yin, 2007). In long fallows, on the other hand, species diversity, above-ground biomass, net primary productivity and leaf litter production all increase as vegetation succession proceeds (Toky and Ramakrishnan, 1983).

In its original form, shifting cultivation is able to contribute to biodiversity conservation while maintaining agricultural and forest productivity. It is a form of agriculture that is well suited to steeply sloping forested lands with heavy seasonal (monsoon) rainfall, such as that found in subtropical South Asia. Nevertheless, it has been considered a destructive and primitive method of farming since the colonial era (Dove, 1983) and most conventional studies have emphasized its environmental ill-effects and economic inefficiency. It has been described as an economically inefficient and wasteful practice that prevents development and keeps people in poverty (AIPP/IWGIA/IKAP, 2009). Various authors have also reported that it is able to support only small populations because of its inherently low return per unit of land and labour (Atal and Bennagen, 1983; Sanchez, 1994). The number it is capable of supporting has been variously estimated at fewer than 32 per sq. km (Nye and Greenland, 1960; Watters, 1971), three to nine per sq. km (Das, 2007) and six per

Alnus nepalensis D. Don [Betulaceae]

This fast-growing tree is a traditional fallow species that is valued for its role in rehabilitating land in many parts of the eastern Himalayas. It has root nodules that fix nitrogen, contributing to rapid soil-nutrient replenishment following shifting cultivation. It also yields valuable firewood.

sq. km (Ganguly, 1968; Saha, 1970). After an estimated cost of inputs (mainly labour) of US$380 per hectare per year, total output was reportedly only US$360 per hectare per year (Borggaard et al., 2003). Development agencies have also regarded shifting cultivation as outdated and destructive (Brady, 1996; Thrupp et al., 1997; O'Brien, 2002).

Some authors have stressed the need to improve present shifting cultivation systems, especially with fallow management (Cairns and Garrity, 1999), while others consider that population pressure will eventually bring about a collapse of shifting cultivation systems (Mertz, 2002), even though there is currently little evidence of this. A recent study by the International Centre for Integrated Mountain Development (Phuntsho et al., 2015) showed that where cultivators in the eastern Himalayas gained access to markets, they tended to transform their traditional plots to horticulture and agroforestry. In more remote areas, cultivators have also innovated, adapted and improved their traditional systems in response to external forces. Given the suitability of shifting cultivation for these marginal lands and the subsistence and livelihood needs of shifting cultivators, there is a strong case to be made for supporting and promoting development of an improved form of the traditional system that exploits modern approaches. This could be the conversion of shifting cultivation to forms of agroforestry and horticulture that retain the advantages of the traditional system.

For decades, governments across South Asia have adopted policies and legislation aimed at discouraging or stopping shifting cultivation in the name of conservation and development (Kafle, 2011). In promoting alternatives, they have implemented plans to transform these lands to permanent forest, sedentary agriculture, horticulture, cash-crop plantations or other agroforestry systems, in a bid to improve economic efficiency and reduce environmental degradation. Where proper funding has been provided for a switch from shifting cultivation to agroforestry or horticulture, economic and environmental benefits have accrued to the shifting cultivators. However, where policies have focused on rooting out the practice, or have not been properly implemented, they have sometimes had the opposite effect to that intended: livelihood options have been reduced, erosion and soil degradation has increased, forest cover has diminished and the vulnerability of fragile marginal lands has been compounded.

Converting shifting cultivation into monoculture forms of sedentary agriculture, horticulture and commercial plantation leads, among other things, to a loss of biodiversity that is likely to increase vulnerability to climate change. Such conversion is rarely accompanied by a workable property-rights arrangement and has entailed disruption to customary tenure, tenure insecurity and inequitable access to land and associated resources. It fails to recognize the value and benefits of traditional knowledge associated with shifting cultivation and has generally failed to provide the intended economic benefits. In light of this, participants at the 'Shifting Cultivation Regional Policy Dialogue Workshop for the Eastern Himalayas', held at Shillong in Meghalaya in 2004, adopted the Shillong Declaration, which made concrete policy recommendations and underlined the need to improve shifting cultivation in its own right (Kerkhoff and Sharma, 2006). The participants included representatives of government agencies, farmers, international bodies, non-government organizations, and others. However, in general, policies have not yet changed.

This chapter reviews government policies in the countries of the eastern Himalayas and how they have affected shifting cultivation (Figure 17-1). The review is based on secondary information, comprising peer-reviewed research articles and progress reports on shifting cultivation development projects. It focuses largely on qualitative aspects as quantitative information is limited. The policies, implementation, and outcomes in individual countries are presented first, followed by a discussion of gaps and weaknesses across the region and some suggestions for improvements.

Country policies

Bangladesh

Shifting cultivation in Bangladesh is found mainly in the Chittagong Hill Tracts, in the east of the country, bordered by the Indian states of Tripura and Mizoram, and

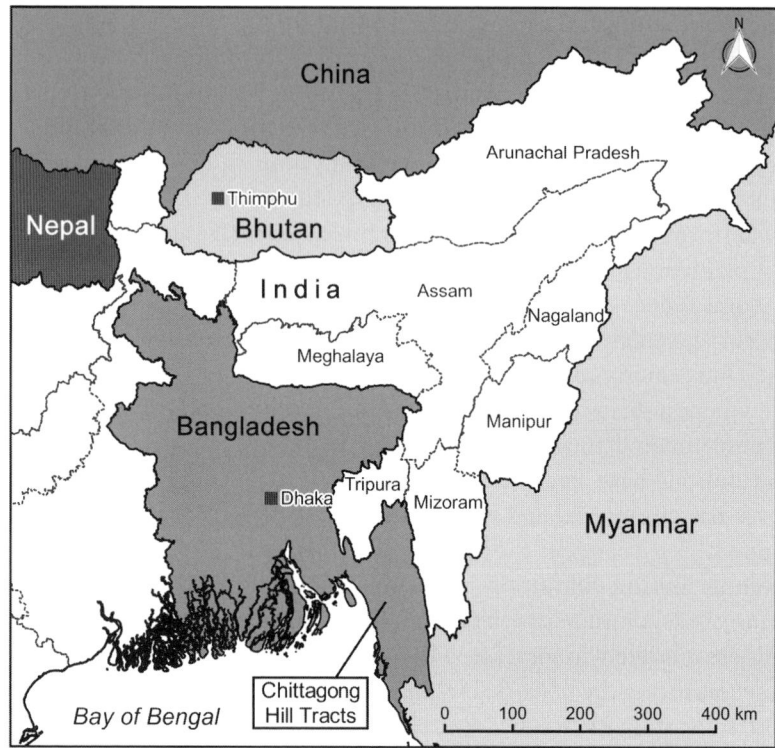

FIGURE 17-1: The countries of the eastern Himalayas region, including the states of northeast India.

Myanmar. Laws and policies to control shifting cultivation were first introduced in the British colonial period (1860 to 1947) and have not changed since. In 1860, shifting cultivation was the only form of agriculture practised in the Chittagong Hill Tracts (Lewin, 1869), but the colonial government considered it to be a primitive and destructive use of land (Lewin, 1869) and decided to ban it (Van Schendel et al., 2000). Between 1867 and 1900, several policy measures were introduced to control the practice and promote sedentary agriculture. The government leased inheritable land-use rights to local residents for plough cultivation and to establish villages (Rasul, 2009). While tax was levied on shifting cultivation, financial support was given for adopting plough cultivation. Livestock were also integrated into plough agriculture (Khan and Khisa, 1970). Priority was given to cash crops such as tea and tobacco. The forests were nationalized in 1871 to support commercial exploitation, and one-third of the Chittagong Hill Tracts was declared reserved forest and the rest became unclassed state forest (Rasul et al., 2004; Rasul, 2009; Phuntsho et al., 2015). Villagers were permitted to collect timber and minor forest products from the unclassed state forest, provided it was genuinely for their personal use. The CHT Regulation 1900 restricted immigration of people from the plains into the Chittagong Hill Tracts; prohibited the allotment of land to non-residents; and empowered customary institutions to take part in the administration of leasing of land-use rights for plough agriculture (Phuntsho et al., 2015).

During the Pakistan era (1947 to 1971), industrial development received priority over agriculture. As well as creating reserved forests, some forests were declared 'protected forest', and shifting cultivation and collection of forest products in them were prohibited (Rasul, 2007). The British policy of commercial extraction of forest resources continued and a corporate entity was created for commercial extraction to support the industrial use of forest products (Rasul, 2007). The government resettled Muslim refugees from Assam into the Chittagong Hill Tracts in the 1950s and abolished the region's special status in order to integrate it into the national mainstream. The role of customary institutions in administering the leasing of land-use rights was reduced.

After independence in 1971, the Government of Bangladesh lifted the restriction on leasing of land to non-residents of the Chittagong Hill Tracts. It also lifted restrictions on immigration from the plains and resettled lowland people in the Chittagong Hill Tracts at the end of the 1970s. Initially, about 25,000 Bengali families were resettled on government-owned fallow land. In reality, this was community land that had been used by the hill tribes for generations (Barua, 2001). The ceiling on the amount of land leased to local residents for plough agriculture and horticulture was reduced from 25 acres (a little more than 10ha), as fixed by the Chittagong Hill Tracts Regulation 1900, to five to 10 acres (Phuntsho et al., 2015). The leasing of land-use rights has since been suspended. In 1992, the government declared a further 50,000ha of forest as reserved forest (Rasul, 2009). It leased a considerable area of unclassed state forest to private entrepreneurs for rubber plantations and the Forest Department increased its control of shifting cultivation (Rasul, 2007). In 1989, the Chittagong Hill Tracts Act came into force to ensure, among other things, the transfer of responsibility for land administration from the Deputy Commissioner to the Hill District Councils. This was followed by the Chittagong Hill Tracts Peace Accord, which was signed in 1997 to address, among other things, socio-economic and political problems related to shifting cultivation, land use and rights to land.

Policy implementation and impact

During the British period, about one-third of the land in the Chittagong Hill Tracts was declared as reserved forest (Rasul, 2009; Phuntsho et al., 2015). With an increasing population and falling crop yields under shifting cultivation, sedentary agriculture (plough cultivation) increased from nothing in 1860 to 8097ha by 1908 (Khan and Khisa, 1970). The local people also grew about 21,000 tonnes of cotton annually for export (Lewin, 1869). The colonial government then introduced the *taungya* system, a new forest-management system originating in neighbouring Burma, in which annual agricultural crops were grown together with tree saplings until the tree canopy closed and cropping was no longer possible. Emphasis then fell upon growing the trees to maturity while the cultivators of annual crops shifted to another site. This system was used to create teak plantations. Creation of reserved forest, promotion of sedentary cultivation and prohibition of shifting cultivation in reserved forest, all reduced the land available for shifting cultivation. This led to conflict over land tenure between the Forest Department and shifting cultivators.

During the Pakistan era, the area under plough cultivation increased further, from 8097ha in 1908 to 40,486ha in 1960 (Khan and Khisa, 1970). But by 1964 the area had shrunk to 25,354ha, as the Kaptai dam submerged about 40% of the arable land in the Chittagong Hill Tracts, displacing about 100,000 people, 55% of them settled (plough) cultivators (Rasul, 2009). Most of the displaced cultivators were not compensated by the government. Some returned to shifting cultivation in the upper parts of the hills, with others encroaching into reserved forests as there was no other appropriate forest land. The resettlement of Muslim refugees from Assam in the early 1950s also resulted in the loss of unclassed state forest land that had traditionally been used for shifting cultivation. Removal of the special status of the Chittagong Hill Tracts in the 1960s led to large-scale migration of lowland people to the Hill Tracts; the population of lowland Bengalis increased from 26,000 to 119,000 in the 10 years between 1951 and 1961 (Gain, 1998; Shelley, 1999). There were no alternative sources of livelihood, so shifting cultivation expanded. The prohibition of use of protected forests infringed the shifting cultivators' customary forest rights. The combined effect of all these policy measures was to markedly reduce the area available for shifting cultivation, and this resulted in a reduction of the fallow period, from 15 to 20 years at the end of the 19th century to just three to four years by the 1960s (Forestal, 1966; Khan and Khisa, 1970).

After independence, the Government of Bangladesh continued leasing rights to use land for plough cultivation and cash plantations. There is no reliable data on land use, but it is estimated that the area of land under plough cultivation increased again, from 25,354ha in 1964 to 42,450ha in 1990 (Khan and Khisa, 1970). At the same time, the area under horticulture and vegetable and root crops increased from 6860ha and 5610ha, respectively, in 1970 to 16,780ha and 13,620ha in 1990 (Uddin et al., 2000). According to Phuntsho et al. (2015), the government had, by then, leased land for plough cultivation and plantations to about 40% of the Chittagong Hill Tracts' resident population. However, leasing has now been suspended, which means that about 60% of shifting cultivators in the region cannot switch to plough cultivation or cash-crop plantation, even if they want to.

For those who leased rights to use land in the Chittagong Hill Tracts, the desire to make more money, market demand for horticultural products and improved transport facilities invited a diversification of land use. Rasul (2009) reported six types of land use in the Hill Tracts: *jhum* (shifting cultivation), valley farming, annual cash cropping, commercial horticulture, agroforestry and timber plantation. Phuntsho et al. (2015) described five types of cash-crop plantation: annual cash crops (turmeric, ginger, aroids and others); quick-growing fruit (pineapples, bananas, papaya and others); mixed-fruit gardens with perennials (mangoes, lychees, jujubes, citrus, guavas, lemons and others); perennial cash crops (cashews, coffee, and others); and forest-tree plantations.

At present, there is no specific policy aiming to improve shifting cultivation. At the same time, support such as agricultural research and extension, marketing, value addition, enterprise development and general funding for plough cultivation, horticulture, or cash-crop plantations is either inadequate or absent in shifting cultivation areas. With increased control of shifting cultivation, leasing of land for plantations by non-residents of the Chittagong Hill Tracts, resettlement of lowlanders in the Hill Tracts, expansion of reserved forests, occupation of lands for military purposes and creation of conservation areas, the land available for shifting cultivation has shrunk and conflict between the Forest Department and shifting cultivators has escalated. The administration of land has yet to be transferred to the district councils in accordance with the Chittagong Hill Tracts Act 1989 and the Peace Accord of 1997, and there are no signs that the conflict is abating.

Bhutan

Policy-makers in Bhutan began to pay formal attention to shifting cultivation in 1957, when the National Assembly, a legislative body established in 1953, passed a law called the *Thrimzhung Chenmo*. This prohibited shifting cultivation in areas where the age of the forest fallow exceeded 12 years.

A first cadastral survey was carried out between 1957 and 1974, and shifting cultivation lands were measured and entered into the *Ma-thram Chem* (main land record). During the 30th session of the National Assembly in 1969, a resolution was passed aiming to persuade shifting cultivators to convert to growing cash crops through conversion to *chhuzhing* (irrigated land) and *kamzhing* (rainfed land). The tax on land thus converted was to be levied at the rate for shifting cultivation land – until the first crop was harvested. The National Assembly also passed the Forest Act, which created government reserved forests and confined shifting cultivation to the areas it occupied prior to the Act coming into effect. The Land Act of 1979 legitimized legal ownership of shifting cultivation land, but it also stipulated that shifting cultivation land with a fallow period of less than 12 years should be converted to irrigated or rainfed agriculture or cash-crop plantation, while shifting cultivation lands with a fallow period of more than 12 years should become government reserved forest. However, in 1980 the National Assembly, at its 52nd session, acknowledged that the government was not ready to enforce the Land Act 1979. Farmers were permitted to continue with shifting cultivation if their livelihood depended solely upon it. Then, in 1983, the 58th session of the National Assembly allowed owners of shifting cultivation land to sell it to the government, at the government's rate, for conversion to government reserve forest, if the land was geophysically unsuitable for conversion to irrigated, rainfed or plantation agriculture. As a significant number of farmers depended solely on shifting cultivation for their livelihoods, selling the land to the government was not practical (Upadhaya, 1994). The decision was then changed to granting substitute lands in lieu of cash compensation, but farmers were equally unwilling to do this. In 1996, the 74th session of the National Assembly decided that the government should also pay landowners for shifting cultivation lands with a fallow period exceeding 12 years and convert this to government reserved forest. The Land Act, 1979 was amended accordingly, but the decision was not implemented due to a lack of funds.

Contrary to the Land Act, 1979 and the resolutions of the National Assembly, a second cadastral survey (carried out from 1980 to 1997) simply converted 11,553 acres of shifting cultivation lands with a fallow period exceeding 12 years (including some rainfed cultivated land) to government reserve forest (MoA, 2006). This meant it could no longer be used for shifting cultivation. Then the Land Act of 2007 formally deleted shifting cultivation as a category of legal land use, implying that it could not be practised and the land should be registered as *chhuzhing*, *kamzhing* or land for cash cropping. Finally, a third cadastral survey was conducted from 2008 to 2013 and the government considered granting ownership of shifting cultivation land with a fallow period of more than 12 years to the owners, if the land was included in the *thram* (legal land-ownership document) arising from the second cadastral survey. A Royal Decree was issued granting owners the choice of converting such land to private forest, in accordance with the Forest and Nature Conservation Rules, 2006.

Implementation status

It was not possible to carry out a detailed critical review of the outcome of policy implementation because of a lack of systematic information. However, available information confirmed that during the Fifth and Sixth Five-Year Plans, spanning the years from 1982 to 1991, the Department of Agriculture provided cash incentives for terracing and contour bunding in order to convert shifting cultivation land to rainfed and irrigated cultivation, and funded research and extension support for establishing cash-crop plantations. During the Seventh Five-Year Plan (1992 to 1997), the department promoted agroforestry as an option; during the Ninth Five-Year Plan (2002 to 2006) it invested in increasing road

Elettaria cardamomum (L.) Maton
[Zingiberaceae]

This herbaceous member of the ginger family is grown on shifting cultivation lands in many parts of the eastern Himalayas region. The spice cardamom, which comes from its seed, is widely used in Indian and other Asian cuisines.

access to rural areas, including shifting cultivation areas, to improve market linkages; and in the Tenth Five-Year Plan (2008 to 2013) it continued to grant cash incentives for the conversion of shifting cultivation land to rainfed agriculture, cash-crop plantations and horticulture. So far, about 1800ha of shifting cultivation land has been terraced and converted to rainfed farming (Phuntsho et al., 2015). Orange and cardamom orchards have replaced shifting cultivation in subtropical areas, triggered by demand from India, Bangladesh and global markets. The number of orange trees increased from about 1.7 million in 2000 to 3.2 million in 2007 (DoA, 2011). Since being granted legitimacy by the Royal Decree, private forest is now likely to emerge as an alternative to shifting cultivation.

China

Shifting cultivation has been practised for centuries, if not millennia, in Yunnan and Sichuan provinces in southwest China, and in parts of Tibet Autonomous Region. Prior to the 1950s, farmland was owned by farmers and landlords and used for growing perennial and annual crops (Guo and Padoch, 1995). The Land Reform Law of 1950, enforced between 1950 and 1952, nationalized farm and forest land, which was then collectivized between 1952 and 1956 and people's communes were formed in 1958 (Xu and Ribot, 2004). The collectivization policies, particularly during the 'great leap forward', emphasized grain production to achieve food self-sufficiency. Much land under shifting cultivation (Xu et al.,1999) was turned into state-owned farmland (Shirasaka, 2006). Before this (up to 1978) shifting cultivation with a fallow period of up to 15 years had prevailed (Thang et al., 2007).

Generally, an open economic policy came into force after 1978. The Household Responsibility System of 1979 legalized the leasing of use rights to agricultural land on the plains and in valley floors to shifting cultivators. In 1982, a policy called *liangshanyidi* (freehold and contracted forest lands and swidden land) was implemented which gave shifting cultivators the use rights to their shifting cultivation lands (Xu and Ribot, 2004). However, this was accompanied by considerable encouragement to convert shifting cultivation land to agroforestry. After the 1980s, the government viewed shifting cultivation as a cause of deforestation and environmental degradation (Yin, 2001). In 1981, forestry-sector reform was launched and this moved the responsibility for forest management from the state to local communities and individuals. A new programme in 1981 supported tree planting with a subsidy of 50 yuan (then about US$30) per mu (1/15 hectare) for procuring seedlings and saplings. Financial compensation was provided for five years for establishing timber or fruit plantations (economic forests) and eight years for trees with a primary ecological function (ecological forests). Funds were also provided to forestry agencies to cover technical support and programme design. This was followed by the Forest Law (1984), which prohibited denudation and deforestation, thus discouraging shifting cultivation. Some shifting cultivators from upland regions were relocated to designated farmlands in valley regions (Liang et al., 2008). The government also supported intensive agriculture in irrigated land, thereby reducing the need to produce food from shifting cultivation and triggering the development of agroforestry.

Efficient use of community or collective forests was promoted in order to reduce economic dependence on state-run forests (ZLN, 1992). After 1985, the government forest research establishment supplied improved Chinese fir (*Cunninghamia lanceolata*) for agroforestry plantations. The Natural Forest Protection Programme, which was launched in 1998, banned the cutting of natural forests and promoted replanting and afforestation.

In 1999, the government launched a Sloping Land Conversion Policy that was intended to convert farmlands with a slope of more than 25 degrees –

Cunninghamia lanceolata (Lamb.) Hook. [Cupressaceae]

One of the most valued timber-tree species in China, Chinese Fir (although it is not actually a fir), produces a fragrant and highly durable wood. The trees may grow to 50m in height.

including shifting cultivation land – to forest or grassland with the aim of reducing soil erosion and runoff and increasing forest cover on marginal lands. Under this policy, the government spent a total of 2370 billion yuan (about US$313 billion) in compensation to participating farmers for their lost food-grain production (Wenjuan and Yin, 2007), and a further 785 million yuan in subsidies to farmers in compensation for the conversion of their land. The government also gave farmers 20 yuan per mu per year to support access to health and educational services.

Implementation status

The changes in policy, combined with population growth, changes in markets and scientific innovations, led to a considerable transformation of traditional shifting cultivation land to other land uses, as well as to different forms of agriculture based on shifting cultivation principles (Wenjuan and Yin, 2007). According to Thang et al. (2007), the changed use was distributed between arable agriculture (5%), agroforestry (20%), tree-dominated fallows (55%, see below), orchards (10%), and commercial forests (10%). At the same time, population growth meant that the average area of traditional shifting cultivation land per capita decreased from 1.45mu (0.1ha) in 1980 to 0.97mu (0.06ha) in 2005 (Thang et al., 2007). This led to a shortening of the fallow period. In cases of change to arable agriculture, use rights to arable land suitable for irrigated cultivation were contracted to shifting cultivators for intensive agriculture. High-yielding hybrid rice varieties were widely promoted. Improved management of agricultural crops, irrigation, nutrient inputs, control of pests and diseases and access to markets all led to higher agricultural productivity, which in turn enabled farmers to reduce their dependence on shifting cultivation for food security (Liang et al., 2008).

Currently, there are about 220 types of agroforestry systems, grouped into 82 different forms, in Yunnan province alone (Guo et al., 2007). The traditional shifting cultivation system in this area involves a rotation between upland rice and alder or other trees. This has been modified to systems such as those based on tea, pine trees and shellac host trees (Guo and Padoch, 1995). In these modified systems, the traditional principle of rotating crops and trees has been maintained, but new crops such as maize, soybeans, peanuts and pineapples have been introduced and new trees are being planted, such as pine (*Pinus yunnanensis, Taiwania cryptomerioides*), fir (*Cunninghamia lanceolata*), birch (*Betula alnoides*) and *Macadamia integrifolia* (Liang et al., 2008). The use of fertilizers and improved varieties for higher crop yields is common in the modified systems. When shifting cultivators have enough flat land for agriculture, they forego the cultivation of rice in the agroforestry part of their land and instead rear livestock to diversify their cash income. They also grow maize as fodder for the livestock in a maize-fir rotation system that was developed to replace the upland rice-alder system (Liang et al., 2008).

Another agroforestry system features tree-dominated fallows. Shifting cultivators grow short-term crops together with saplings of long-term trees (Liang et al., 2008). They stop growing annual crops once the tree canopy closes and the forest-fallow phase is then managed intensively for timber production. In these cases, the cropping and fallow phases are longer (3 to 4 years and 20 to 30 years, respectively). Unlike the seed broadcasting method used in traditional systems, tree seedlings are raised in nurseries or purchased from suppliers. Forest fires and pests are actively controlled, and pruning and selective cutting or thinning is done to improve the forest growth. This approach often

Toxicodendron vernicifluum (Stokes) F. A. Barkley [Anacardiaceae]

Among those species now grown in managed-tree fallows in Yunnan province, the Chinese lacquer tree is tapped for its highly toxic sap. This makes a highly durable lacquer for Chinese, Japanese and Korean lacquerware.

leads to a reduction in biodiversity and promotes monoculture cropping. Many combinations of crops and trees have arisen in recent years. Economic species traditionally grown during the fallow phase include *Cassia siamea*, *Gmelina arborea*, *Cajanus cajan* and *Alnus nepalensis* (Liang et al., 2008). However, changing market demand has meant that the trees commonly grown today are pine, birch, Chinese fir, *Macadamia integrifolia*, *Toxicodendron verniciffuum* and rubber (*Hevea brasiliensis*). This has led to a shorter fallow period of 15 to 20 years as these trees mature faster. In some cases, tea gardens developed on shifting cultivation lands have been abandoned and the associated shade trees are being managed as forests for firewood and timber production (Xu, 2007). Some shifting cultivation lands have been converted to plantations of medicinal plants such as *Amomum tsao-ko*, *Dendrobium nobile* and *Coptis teeta* (Huang and Long, 2007).

The area of land under orchards increased by 40% in rural China between 1980 and 1993, with fruit and nut trees accounting for 80% of the growth (Rozelle et al., 2000). The development of orchards reflected China's transition from a planned to a market-driven economy. They included oil-bearing, fruit, nut and rubber trees, as well as tea, medicinal herbs and other cash-producing non-timber tree species. The Sloping Land Conversion Policy in 1999 and the Natural Forest Protection Programme also triggered the conversion of some degraded shifting cultivation fallows to commercial forests. Natural fallows remained (Guo et al., 2002), which enabled people to meet increasing demands for timber as well as non-timber forest products such as pine needles, pine nuts and wild mushrooms (Xu et al., 2005).

India

National policies and legislation that govern shifting cultivation in India came into existence during the colonial era (1865 to 1947). The Indian Forest Act of 1865 promulgated reserved forest and legitimized the government as the *de jure* owner of these forests, within the bounds of which it prohibited the practise of shifting cultivation. However, no blanket restrictions were imposed. The fact that shifting cultivation was practised was recorded during the process of creating reserved forests in order to settle compensation for the rights. Shifting cultivators were granted the right to practise their farming systems outside of reserved forests. The National Forest Policy of 1894 reaffirmed the intent of the Indian Forest Act, 1865. The Indian Forest Act of 1927 deemed shifting cultivation to be a privilege, subject to control, restriction and abolition by state governments (Maithani, 2005).

During the colonial era there were also state-level regulations. In Assam, the Assam Forest Regulation, 1891 (applicable to Meghalaya, Arunachal Pradesh and Mizoram), recognized the jurisdiction of customary laws that governed shifting cultivation (Darlong, 2004), and the Sylhet *Jhum* Regulation, 1891, restored the right to practise shifting cultivation in the districts of Sylhet and Cachar, which had earlier been denied. The Garo Hills Regulation, 1882, introduced irrigated rice cultivation in place of shifting cultivation and legalized land titles and private ownership in the Garo Hills.

The Nagaland *Jhum* Land Regulation, 1946, treated shifting cultivation as a privilege, and not a right. The Tripura Tenant and Landlord Act, 1886, guided the allotment of land to shifting cultivators who were changing to settled agriculture (Ganguli, 1990), while the Forest Reservation Act, 1887, lent legitimacy to the creation of reserved forests in which shifting cultivation was prohibited (Maithani, 2005).

After independence the situation changed. Article 371 and Schedule VI of the Constitution granted safeguards to protect the rights of shifting cultivators to their land as well as preserving their customary tenure and institutions. As a result, 92% of the land in Nagaland, 73% in Meghalaya, 70% in Manipur, and 61% in Mizoram is under customary tenure (Darlong, 2004). As in the Indian Forest Act of 1927, the National Forest Policy, 1952, suggested that shifting cultivation should be regulated rather than controlled, by integrating forest regeneration into the system. However, the National Forest Policy of 1988 regarded shifting cultivation as having harmful effects on forest and suggested improving its agricultural practices in order to contain any spread into new areas. In essence, both policy and legislation recognized shifting cultivation, but sought a workable alternative to replace it. Schedule VI of the Constitution also gave autonomous district councils the right to regulate shifting cultivation.

Other state-level regulations affecting shifting cultivation were introduced after independence. In Arunachal Pradesh, the Balipara Sadiya/Triap Frontier Tract *Jhum* Land Regulation, 1947, recognized the customary ownership of shifting cultivation lands (Dollo, 2004; Darlong, 2004). In Meghalaya, the Garo Hills District (*Jhum*) Regulation, 1954, established rules for district council village heads and encouraged fallow enrichment, terrace cultivation and horticulture. The Garo Hills District (Forest) Act, 1958, prohibited shifting cultivation near water sources, public roads, and in sal (*Shorea robusta*) forests. The Manipur Acquisition of Chiefs (Rights) Act, 1966, empowered the government to acquire certain rights and interests held by village chiefs over shifting cultivation lands. The Manipur Forest Rules, 1971, recognized shifting cultivation as a privilege, subject to control and supervision by the Forest Department, while the Perspective Plan of the Forest Department (Vision 2020) envisioned reducing the pressure on forests from shifting cultivation. In Mizoram, the Lushai Hills (*Jhumming*) Regulations, 1954, empowered village councils to regulate tenure in communal shifting cultivation lands. The Mizoram District (Forest) Act, 1955, restricted the right to practise shifting cultivation in unclassified district council forest; the New Land Use Policy, 1984, prescribed replacement of shifting cultivation by settled agriculture; and the Soil Conservation Act, 1987, specified terracing for soil conservation and promotion of settled agriculture. The Nagaland *Jhumland* Act, 1970, repealed the Nagaland *Jhum* Land Regulation, 1946, lent legitimacy to customary laws and recognized the rights of people to regulate the gathering of forest products from shifting cultivation lands. The Tripura Forest Rules of 1950 prohibited shifting cultivation in reserved forest, but permitted it on protected forest lands and in unclassed state forests; the Shifting Cultivation Resettlement Guidelines of the 1950s proposed directions for resettlement of shifting cultivators; the Tripura Land Revenue and Land Reforms Act, 1960, regarded land that was not owned by any entity or person as state property; and the Forestry Perspective Plan (Vision 2010) envisioned economic rehabilitation of shifting cultivators and ethnic groups.

Implementation status

India's central government has designed and funded numerous programmes and projects to encourage a change from shifting cultivation to settled agriculture, horticulture and cash–crop plantations. The main programmes and projects are listed in Table 17-1.

A lack of systematic information has made it impossible to carry out a proper critical review of the outputs and impact of various programmes. The following review is limited to a description of what took place in the different states. Maithani (2005) provided details of many of the programmes. In Arunachal Pradesh, the programmes covered a combined total of 4313 families and 2893ha of shifting cultivation land, with the Watershed Management for Shifting Cultivation Areas programme covering 15 micro-watersheds. The total budget outlay was about 114 million rupees. During the Fifth Five-Year Plan the focus was on controlling shifting cultivation, but it was later realized that the mountain topography limited the scope for settled agriculture and the focus changed to raising plantations of coffee, tea and cardamom on abandoned shifting cultivation land. Permanent irrigated rice fields were established on valley floors. By the end of the Sixth Five-Year Plan, 1400ha of plantations had been established and 1613 families were settled in irrigated rice cultivation (Maithani, 2005). Horticulture gardens were created on blocks of land measuring 25, 50 and 100ha in nine districts. Each of the 2767 families taking part in the scheme was promised ownership rights over 2ha of the horticulture gardens. However, although the families took part in establishing the plantations, they did not wish to replicate the switch to horticulture on their own land, and they continued with shifting cultivation, as before (Sachidananda, 1984).

In Assam, programmes focusing on cultivators covered about 2600 families and about 2280ha of land, while the Watershed Management for Shifting Cultivation Areas programme covered 15 micro-watersheds. The total budget outlay for all three programmes was about 100 million rupees. In the 1970s, the Soil Conservation Department and the Assam Plantation Crops Development Corporation developed coffee and rubber plantations in the districts of Karbi Anglong and North Cachar Hills (Dima Hasao) and handed them over to shifting cultivators (Das, 2007), while an interdisciplinary 'composite project' developed terraces for irrigated rice cultivation and supported the modernization of agricultural management and tree plantation. However, the interdisciplinary project was implemented without piloting and was unable to integrate sectoral schemes. It was unable to transfer the titles of developed lands to families and did not involve the beneficiaries in project design and implementation, so there was a mismatch between the project interventions and the local conditions and endowments. Based on lessons learned from this project, the

TABLE 17-1: Central government support for the implementation of shifting cultivation policy in India.

Five-year plan period	Programme/project	States	Budget (Indian rupees x 100,000)
1950s and 1960s	Alternative land use: • Cultivation of cashew nuts, rubber, black pepper and coffee.	Assam (Meghalaya and Mizoram were part of Assam).	not available
	Soil conservation: • Afforestation and terracing programme.	Assam, Arunachal Pradesh, Manipur, Nagaland and Tripura.	
Fifth and Sixth Five-Year Plans (1974-1975 to 1985-1986)	Pilot Project on Control of Shifting Cultivation (13 project units covering 1300 families and 2600ha of shifting cultivation land): • Settled agriculture – irrigated and non-irrigated – and cash-crop plantation (terminated in 1979 and transferred to state governments).	Assam, Meghalaya, Arunachal Pradesh, Mizoram, Tripura, Manipur and Nagaland.	105.6
	River Basin Scheme for Control of Shifting Cultivation (covering 25,000 families and 60,000ha of land): • Soil conservation (in Assam); • Settled agriculture and cash-crop plantation; • Forest and rubber plantations (Tripura).	All states	4896.16
Seventh Five-Year Plan (1986-1987 to 1990-1991)	Model Scheme for Control of Shifting Cultivation (covering 18,500 families): • Primary sector – agriculture, horticulture, cash-crop plantation, livestock, pisciculture, sericulture and others for the older generation. • Non-farm sector – assistance for household and cottage industries for young people.	All states	552.5

TABLE 17-1 (cont.): Central government support for the implementation of shifting cultivation policy in India.

Five-year plan period	Programme/project	States	Budget (Indian rupees x 100,000)
Eighth Five Year Plan (1991-1992 to 1995-1996)	Watershed Management for Shifting Cultivation Areas (covering 272 micro-watersheds): • Watersheds used as a basis for planning and implementation of socio-economic development and ecological restoration.	All states	5582.6 (released); 4490.4 (spent)

Note: INR100,000 = about US$2222 in 2005.
Source: adapted from Maithani, 2005.

Integrated *Jhumia* Development Programme was piloted for five years to the end of the 1980s, followed by the Compact Area Development Programme, for three years from 1989. The Watershed Management for Shifting Cultivation Areas programme was implemented in many parts of Karbi Anglong and North Cachar Hills (Dima Hasao) later in the 1990s. By 2007, close to 20% of the sampled population had given up shifting cultivation and switched to alternatives (Das, 2007).

In Manipur, programmes focusing on cultivators covered 4369 families and 1379ha of shifting cultivation land, while the Watershed Management for Shifting Cultivation Areas programme covered 84 micro-watersheds. The total budget was about 280 million rupees. An Intensive Valley Development Programme was implemented with the aim of replacing shifting cultivation with settled agriculture, horticulture and cash-crop plantations. In the late 1980s, an attempt was made to transform about 20,000ha of shifting cultivation land to terraced cultivation, plantations and horticulture. Efforts continued in the 1990s under the watershed management programme to convert 29,694ha of shifting cultivation lands to settled agriculture, horticulture and cash-crop plantations under a state programme, and a further 7876ha under a centrally sponsored scheme (Tiwari, 2003).

In Meghalaya, programmes focusing on cultivators covered 779 families and 2893ha of shifting cultivation land, while the watershed management programme covered 12 micro-watersheds. The total budget was about 119 million rupees. Shifting cultivators were offered a range of options and activities including settled agriculture, horticulture, silvopasture, forestry, fishery, sericulture, piggery, poultry and duck raising, water harvesting, soil conservation and land management, tailoring, carpentry, and others. There was a gradual decrease in shifting cultivation (Government of Meghalaya, 2003), from 41% of the rural population engaged in shifting cultivation in 1971 to 13% in 2003. The area under cash crops increased by about 30% between 2001 and 2008 (Leduc and Choudhury, 2012).

In Mizoram, programmes focusing on cultivators covered about 3550 families and 2785ha of shifting cultivation land, while the watershed management programme covered 33 micro-watersheds. Cash-crop plantations of coffee, rubber, black pepper, cloves, oranges, cardamom and cashews were established over the years from 1956. Terracing was carried out after 1972 to promote both irrigated rice and rainfed cultivation to improve food self-sufficiency. However, a review of soil conservation work in 1979 revealed that of 10,000 families that took part in the terracing programme, only 300 of them ceased shifting cultivation (Maithani, 2005). A Department of Forest programme from 1984-1985 to 1989-1990 under the New Land Use Policy 1984 (which aimed to phase out shifting cultivation within 10 years) supported 6086 families in establishing plantations of teak, oranges, rubber, and other species on 11,410ha of land (Singh, 1996). Land that was not owned by any individual or entity was notified as natural and protected forest. The state government took over from village councils the task of leasing land long-term for settled agriculture, horticulture, dairy farming, piggery, plantation crops and similar pursuits. However, the programme did not reflect the needs of the families it set out to assist and the long development period of plantations delayed returns from investment. In 1990-1991, the policy was modified to overcome these weaknesses. Its scope was expanded to incorporate a wide range of both farm and non-farm activities. Beneficiaries were limited to those practising shifting cultivation and those below the poverty line. Financial assistance of 30,000 rupees per family was provided over three years. A total of 33,911 families benefited from the modified policy over a period of four years. However, a study of three areas found a decline in the incidence of shifting cultivation in one area but an increase in the other two (Singh, 1996).

In Nagaland, programmes focusing on cultivators covered 5302 families, while the watershed management programme covered 104 micro-watersheds. The total budget outlay was about 262 million rupees. In the late 1980s, the focus was on terracing, farm forestry, orchards and cash crops. A model scheme was implemented in 51 villages covering 4818 families, but input supplies, logistics, administrative support and arrangements for marketing of produce proved to be inadequate. In the 1990s, the focus shifted to appropriate land-use planning and water and soil conservation (Anonymous, 1999). Despite these efforts, the area under shifting cultivation in Nagaland increased more than three-fold in the 10 years from 1983 to 1993 (Maithani, 2005).

Activities in Tripura began with a government scheme to settle shifting cultivators, adopted during the First Five-Year Plan from 1951 to 1956. Programmes that focused on cultivators, implemented under the Fifth to Eighth Five-Year Plans, covered around 2500 families. The total budget outlay was about 90 million rupees. In the mid-1980s, a separate scheme was implemented that sought to settle shifting cultivators by setting up colonies. The government and the Rubber Board of India employed shifting cultivators to establish rubber plantations in shifting cultivation areas. After six to seven years, the rubber plantations were handed over to the cultivators. Each family was given 2ha of land – 1.5ha of it under rubber trees and 0.5ha as homestead land for multiple uses. This scheme resettled a total of 45,000 families in 28 colonies (Maithani, 2005). To support the scheme, the Tripura Land Revenue and Land Reforms Act declared that customary use of land by tribal farmers was no longer authorized (Ganguli, 1990). In the late 1980s, the focus moved from settlement to conversion to horticulture, crop plantation, animal husbandry and raising of fish. In the early 1990s, a 'Restoration Assistance Scheme' was launched to restore abandoned shifting cultivation lands. At the same time, a scheme called 'Purchase of Land for Rehabilitation of *Jhumias* and Landless Tribals' was launched, under which the government provided grants of 50% of the total cost of land purchased by shifting cultivators with loans covering the remainder of the cost. The government also set up autonomous district councils to empower tribal populations engaged in shifting cultivation in the government of their development programmes.

Following the end of the Eighth Five Year Plan in 1996, there have been further policy-related activities in the states of northeast India.

In Nagaland, the Nagaland Environment Protection and Economic Development Project (NEPED) was implemented from 1994 to 2011 to improve the relevance of interventions in shifting cultivation. It focused on planting commercial tree species interspersed with diverse food crops in order to check soil erosion and improve the productivity of shifting cultivation lands. From the start of the first phase (1994 to 2000), the project's operations, from planning to implementation, were decentralized to village councils and village boards. Village elders from various indigenous ethnic groups formed local expert teams to support the project with their traditional knowledge. Women were empowered to solve development and gender-related problems. During the first phase, the project set up 18,084 agroforestry test plots in 854 villages and more than seven million trees were planted on 5500ha of land (Anonymous, 1999). The project's interventions were replicated in other villages, and as a result about 33,000ha of shifting cultivation land was brought under agroforestry (Anonymous, 2013). Test plots were allocated to women who received finance to set up tree nurseries. During the second phase (2000 to 2005), village councils and village development boards were empowered to function as grassroots institutions to manage micro credit, support processing and marketing of cash crops and

Passiflora edulis Sims [Passifloraceae]

Passion fruit, a South American native that is now grown throughout the world, was proposed by the NEPED project (this page) as a cash crop to improve the productivity of shifting cultivation land in Nagaland.

turn subsidy-reliant project funding into self-reliant credit-based financing. Some of the 'best practices' and farmer innovations observed in the first phase of the project were demonstrated in a model demonstration plot. Cash crops such as cardamom, ginger, black pepper, betel vines and passion fruit were promoted. Small business plans were developed to guide investment of money lent by the project. Self-help groups and marketing boards were formed to improve production and marketing of the farm produce. Women were given a quarter of the project funds to promote economic activities. For the first time, they were allowed to own land by purchase or long term lease, and were permitted to use their share of grant-in-aid funds for such purchases. Eighteen women's self-help groups purchased land. District forest committees were mandated to ensure sustainable management of the plantations developed by the project. During the third phase (2006 to 2011) the micro-credit scheme was expanded to include enterprise development in animal husbandry, horticultural crops, cottage industries, blacksmithing and carpentry. The impact of NEPED has not yet been systematically evaluated.

Another programme, the North Eastern Region Community Resource Management Project (NERCORMP), was implemented between 1999 and 2008 in the Karbi Anglong and North Cachar Hills (Dima Hasao) districts of Assam, the West Khasi Hills and West Garo Hills districts in Meghalaya, and the Senapthi and Ukhrul hill districts in Manipur. The focus of the project was to improve the livelihoods of vulnerable groups through improved management of their natural resource base (Phanbuh et al., 2008). It was based on the rationale that the beneficiaries needed to be empowered institutionally, socially, technically and financially and enabled to develop the capacity to plan, manage and sustain their livelihoods based on enterprises supported by local natural resources. The project recognized the need to combine the application of indigenous and scientific knowledge and to attract the participation of the beneficiaries in planning, implementing, funding and sustaining its activities. Expert teams were formed, with experts

Areca catechu L. [Arecaceae]

Betel vines (*Piper betle* L.) are a recommended cash crop in Nagaland. Their leaves are used to wrap the nuts from this palm, for chewing together. Betel nuts contain alkaloids which, when chewed, are intoxicating and slightly addictive. They are also carcinogenic. Habitual chewing stains the teeth black.

from local communities and government technical departments. Natural resource management groups, involving both men and women, were formed to strengthen traditional organizations in planning, implementing and monitoring of natural resource management plans. The financing of enterprises was supported by micro-credit schemes managed by the beneficiaries, in contrast to traditional grant-giving systems. Self-help groups with men and women members were established, along with group federations, to institutionalize the financing of farm production, marketing and enterprise development based on lending, and not grants. The natural resource management groups and self-help groups received training in natural resource management, agricultural production, marketing, credit, business management and accounting. The project's teams of experts provided technical support to the groups. The impacts of the project have yet to be evaluated.

Myanmar

Shifting cultivation in Myanmar is traditionally governed by customary tenure and institutions (Macqueen, 2013) and is practised in reserved forest, protection forest and unclassed forest (Win, N. R., 2013). A system of shifting cultivation called *taungya* was introduced to control the practice during the colonial era (1824 to 1948) and is now found in 11 of the country's 14 administrative regions (Win, N. R., 2013), although it is mostly confined to the Kachin, Kayah, Kayin, Chin and Shan states. Under this system, the Forest Department employs shifting cultivators to raise commercial tree plantations on shifting cultivation land, and allows them to cultivate food crops among the growing saplings for two to three years, or until the canopy begins to close. Food cropping then ceases, the trees continue to be nurtured for their commercial value, and the cultivators move to another site. The *taungya* system now covers close to 23% of the total area of the country and around two million families depend on it for their livelihood (Win, S., 2004). Nevertheless, the *taungya* system is still not recognized in law.

The first legislation to affect shifting cultivation in Myanmar (then Burma) was the Forest Rules 1856, proclaimed by the British colonial administration to legitimize its proprietary rights over teak forests (Bryant, 1997). The authorities considered shifting cultivation an evil to be stamped out, and shifting cultivators were required to obtain approval from the Forest Department to practise their farming system in any area containing more than 50 teak trees, seedlings included. A breach of the rules could result in a fine or imprisonment. In 1881, the Burma Forest Law prescribed one class of reserved land in which use of the land was limited to those with explicitly recognized rights (Bryant, 1997). While settled agriculturalists in the vicinity of such reserves were selectively granted use rights, shifting cultivators were not (Bryant, 1997). Infraction was subject to a fine or imprisonment. The Act also made no specific provision for common rights in unreserved forest. Flat lands in valleys were allotted to shifting cultivators, but this land turned out to be unsuitable for shifting cultivation because of dense weed growth (Win, S., 2009).

After independence, the Forest Law, 1992, and the Protection of Wildlife and Conservation of Natural Areas Law, 1994, were enacted and the Forest Policy, 1995, adopted. These prohibited shifting cultivation in reserved and protection forests and supported the transformation of shifting cultivation to settled agriculture, commercial plantations and horticulture in unclassed forest. Since 2002, the government has been promoting conversion of shifting cultivation into agroforestry under a community forestry programme introduced in a five-year plan (Win, S., 2004).

Implementation status

The *taungya* system – growing food crops and tree seedlings together and calling a halt to cropping once the tree canopy closes – is still practised, although the government has made efforts to control it. According to S. Win (2004), the colonial authorities raised teak plantations at a very low cost under this system, but the system led to conflicts between shifting cultivators and foresters as the former were exploited by the latter (Win, S., 2004). Where socio-economic and biophysical situations were favourable, shifting cultivators continued to switch to settled agriculture. S. Win (2009) concluded that around 50% of shifting cultivators in his study area had moved to irrigated rice cultivation. Since the 1970s, the government has grouped shifting cultivators together and in some cases has settled them along forest fringes.

Myanmar's Community Forestry Instructions (1995) allow the establishment of agroforestry systems in reserved, unclassed and protected forests. Agroforestry based on community forests is being promoted as an alternative to shifting cultivation, in keeping with successive five-year plans. User groups are formed by shifting cultivators who wish to take part in the community forestry programme. The government leases land to user groups for 30 years in reserved, protected or unclassed forests and waste lands, with a provision for extension. The user groups must follow management plans for community forests, which encourage them to establish home gardens with edible-fruit producing trees, such as coconut palms (*Cocos nucifera*), mangoes (*Mangifera indica*), guavas (*Psidium guajava*) and bananas (*Musa* spp.), as well as vegetables (Yamamoto and Maung, 2008).

Nepal

The Government of Nepal does not have a specific policy for shifting cultivation (Kafle, 2011; Phuntsho et al., 2015). However, the Civil Code, 1854, the Land Act, 1964, the Land Revenue Act, 1977, the Land Acquisition Act, 1977, and the National Forestry Act, 1993, all affect shifting cultivation indirectly (Phuntsho et al., 2015). The Civil Code, 1854, states that 'as long as tenants pay rents, lands cannot be confiscated from them even if they leave the lands barren' and 'land which is not registered in anybody's name belongs to no one'. Given the nature of land use in shifting cultivation, where farmers cultivated their fields in rotation, it was not practically possible to register the land as privately owned and pay tax. As a result, shifting cultivation lands became 'land belonging to no one.' The Land Act, 1964, does not recognize shifting cultivation as a legal land use category, so shifting cultivation land is not registered as community land or private land, except in eastern Nepal, where it is registered as grazing land or rainfed agricultural land (Phuntsho et al., 2015). The National Forest Act, 1993, treats 'land belonging to no one' as part of national forest. Although the Government of Nepal is a signatory to International Labour Organization Conventions 169 and 111, the Land Act, 1964, and the National Forest Act, 1993, do not comply with the provisions of these conventions, which guarantee the right to practise a traditional occupation, as part of the culture that identifies the indigenous people who practise it, and the right to customary land as a resource needed for this.

Swertia chirata Buch.-Ham. ex Wall. [Gentianaceae]

A plant with abundant medicinal properties, so-called *chiraito* is one reason swidden farmers in Nepal are switching to permanent cropping. It is used in Indian Ayurvedic medicine to cure fever, is also said to have hypoglycaemic properties and is used as a tonic for stomach and digestive complaints.

Implementation status

Although the government has no policy governing shifting cultivation, it is a prevalent practice in Nepal. In general, shifting cultivators with larger landholdings and small families are voluntarily transforming to sedentary agriculture, cash-crop plantations and agroforestry. In eastern Nepal, shifting cultivation lands are being converted to agroforestry, even though they are registered as grazing land or land for rainfed agriculture. Cardamom-alder and agroforestry systems based on *chiraito* (*Swertia chirata*) are common. Farmers first introduced the large cardamom agroforestry system from Sikkim (India) at their own expense. However, agricultural development banks now provide loans in support of such systems and District Agriculture Development Offices supply planting materials. In the *chiraito*-based systems, farmers cultivate maize and *chiraito* together or sometimes *chiraito* on its own. *Chiraito* is less prone to pests and diseases than cardamom. In central Nepal (Chitwan, Gorkha, Dhading, Makwanpur and Nawalparasi districts) horticulture-based agroforestry, or agri–horti–silviculture, is common, with fruit trees as an important element. Major species cultivated are bananas, pineapples, ipil ipil, bakaino and broom grass. In some places, shifting cultivators are establishing integrated hedgerows for fruit, fodder, forage, improvement of soil fertility and erosion control. In others (for example Bhumlichok in Gorkha district) good access to markets has triggered intensive vegetable farming. In some areas, the government is promoting leasehold forestry: land is leased to forestry groups to develop silvopastural systems in which goats are reared while forest products are cultivated to generate cash income.

Discussion

Since colonial times, national laws and policies related to shifting cultivation in the countries of the eastern Himalayas have been based on the idea that the practice is primitive, destructive and economically inefficient. For more than a century, governments have been determined to transform land used for shifting cultivation into reserved forest, settled agriculture, horticulture or cash-crop plantations. Essentially, the same legislative and policy intent has persisted since independence in the colonial countries and similar policies have been followed in the non-colonial countries.

Shifting cultivation has yet to receive specific attention in the overall planning framework of any of the countries, and none has a policy that considers developing shifting cultivation as an agricultural system on its own merits, even though the Shillong Declaration acknowledged this need in 2004. The bias against shifting cultivation arises from the mistaken belief that it can be compared with settled cultivation. Essentially, shifting cultivation uses land that is not suitable for settled agriculture to grow food, while maintaining the forest cover on these steep slopes to enrich and protect the soil from degradation (Kerkhoff and Sharma, 2006). Over the millennia, people around the world have established settled agriculture wherever it was able to provide them with sufficient food to survive. Policy-makers rarely consider why people in the eastern Himalayas continue to choose shifting cultivation over settled agriculture despite years of attempts to eradicate the practice. The assumption is that the local people are ignorant and need to be shown a better approach; a conclusion that they may have discovered a sustainable and ecologically sound way to use marginal land is never considered. Shifting cultivation is seen as an inferior and wasteful system compared to settled agriculture (Maithani, 2005), yet studies suggest that it enables both optimum utilization of natural resources and a stable and sustainable form for agriculture in the mountains (Sharma, 1984; Ramakrishnan, 1993). While overall productivity under shifting cultivation may be relatively low (Barkakoti, 1990), it provides varied and sufficient food when this would otherwise be unobtainable from the forest alone. The emphasis on regeneration maintains soil quality, which actually increases yields. Divakar (1990) reported that the mean yield of rice, millet and pulse crops was considerably higher from shifting cultivation than in the plains of Assam. Similarly, Srivastava (1996) reported that the gross value of agricultural produce per hectare in hill districts was higher than that in the plains. When compared to settled agriculture and monoculture cash-crop plantations, shifting cultivation offers a range of benefits. It harbours more agrobiodiversity (Kerkhoff and Sharma, 2006) as well as diverse flora and fauna. The range of crops, mixed cropping, and sequential harvesting ensure the availability of food throughout the year (Maithani, 2005; Choudhury, 2012). Forest fallows, which are often seen by outsiders as abandoned waste lands, are actually carefully managed to produce a wide range of economic products and environmental services. After cropping is finished, shifting cultivation fields are converted into secondary forest gardens. Shifting cultivators plant trees that provide fruit, nuts, resins, fibre, medicinal herbs and building materials. They also introduce nutrient-enriching trees into their fields to enhance the biological efficiency of the fallow in restoring soil fertility, in order to shorten the fallow phase. The forest fallows are more beneficial for hydrology and water resources than horticulture or cash-crop plantations. As well as contributing to the household economy, the fallows also fix more carbon than settled agriculture and help to mitigate climate change. From the perspective of ecosystem-based adaptation, which views conservation of biodiversity and ecosystem services as an adaptation strategy (UNEP, 2012), shifting cultivation offers better possibilities for adaptation to climate change than settled agriculture or cash-crop plantations as it maintains considerably higher levels of biodiversity.

There are two main approaches that may be taken in the formulation of policy regarding shifting cultivation and the land required to make it viable. It is an environmentally appropriate form of land use with many benefits in terms of the ecosystem services it provides and conserves. Therefore, shifting cultivation maybe maintained and promoted, especially in areas with poor access to markets where shifting cultivators are dependent upon it for their livelihood. This means both ensuring that the necessary conditions are in place for the practice to function and supporting appropriate farmer-led adaptations. However, where there is good access to markets, shifting cultivators themselves tend to choose to convert their lands to some form of cash cropping. In this situation, support needs to be provided to ensure that they can carry out conversion in the most appropriate way, and that the forms of cropping they introduce maintain the integrity of fragile lands and do not lead to long-term degradation. At present, however, governments across the region are failing to provide the necessary support for either of these approaches – with the exception, to some extent, of China.

Land tenure

One of the major problems faced by shifting cultivators, whether they follow a traditional system or a modified approach focused on horticulture, is lack of land tenure. Disinterest and misunderstanding have meant that governments have seen fallows as forest, rather than as an essential part of a sophisticated agricultural cycle, and the transitory nature of cultivated plots rules out their recognition as agricultural land. This is one of the fundamental issues in the failure to grant land tenure to shifting cultivators.

Traditionally, shifting cultivation land has been a community-owned resource governed by customary tenure. To be successful and sustainable, shifting cultivation needs community-based management at a landscape level, and local regulatory institutions play an essential role in this. The institutions provide social security, while customary tenurial arrangements ensure equitable access to resources (Kerkhoff and Sharma, 2006). However, conversion of shifting cultivation land to forests, settled agriculture and cash-crop plantations has disrupted both customary tenure and local institutions, as well as reducing the land available (Rasul, 2009). In Bangladesh, the government has replaced customary tenure with leases for land-use rights and taken away the legitimacy of customary institutions. In Bhutan, the government has abrogated shifting cultivation as a legal land use, although the practice continues and the authorities permit it in some places. In India, the Constitution theoretically safeguards the legitimacy of customary tenure and local institutions, but in reality their legitimacy and jurisdiction have been altered (Maithani, 2005). In Myanmar and Nepal, the governments do not recognize either customary tenure or local institutions and in Nepal, the government has only granted legal ownership of shifting cultivation land based on customary tenure to farmers in the eastern part of the country (Kafle, 2011; Phuntsho et al., 2015). In the absence of a strategy to manage tenure issues, shifting cultivators in Bangladesh, Myanmar and Nepal continue to use nationalized forests for shifting cultivation. The tenure conflict that persists between forest departments and shifting cultivators has been identified as the main cause of forest degradation in these areas (Rasul, 2009; Phuntsho et al., 2015).

Government policies and legislation have led to inequitable access to shifting cultivation lands, while disrupting the social and economic security provided by customary tenure. In Bangladesh, the 'administrative plurality' (mixed role) of government agencies and customary institutions in administering land-use rights has resulted in leasing of land-use rights to non-residents of the Chittagong Hill Tracts. The registration of land-use rights has now been suspended and this has deprived around 60% of shifting cultivators of their use rights (Phuntsho et al., 2015). The dilution of customary tenure and local institutions has also disrupted the passing on of shifting cultivation traditions and indigenous knowledge to the younger generation. In northeast India, customary tenure has ceased to apply to shifting cultivation lands converted to settled agriculture or cash-crop plantations, and this has enabled elites and well-off farmers to acquire private ownership over community-owned shifting cultivation land (Maithani, 2005; Choudhury, 2012; Leduc and Choudhury, 2012). In some cases, cash-crop plantations were handed over to farmers without land titles, leading to their failure (Das, 2007). Early settlers who owned permanent fields (irrigated rice fields and orchards) as well as having access to shifting cultivation land have benefited from the policies, while later settlers who only had access to shifting cultivation have been excluded (Maithani, 2005).

Support for traditional shifting cultivation

Shifting cultivation is not only an appropriate land use for fragile mountain slopes, it is also the only option for those who do not have the right to use land for settled agriculture, horticulture or cash-crop plantation. However, the only land available for shifting cultivation in most countries is unclassed government forest, and shifting cultivators continue to face controls from forest departments and are often penalized because the boundary between unclassed and reserved forests is unclear. Restrictions on shifting cultivation in reserved and protected forests, leasing of land-use rights for settled agriculture in unclassed forest and resettlement of people from outside has markedly reduced the land available for shifting cultivation, and this has led to a reduction in the fallow period as farmers strive to produce the same amount of food from less land. At the same time, the demand for shifting cultivation land is growing in line with the growth in population, leading to an escalation of conflicts between foresters and shifting cultivators. Policy-led changes have also resulted in tenure inequity among shifting cultivators: while some have rights to use land for settled agriculture and horticulture as well as access to land for shifting cultivation, others only have access (often restricted) to land for shifting cultivation, and some have converted common shifting cultivation land to private property. Governments need to prepare and administer workable plans to balance consumption and conservation needs and promote equity of access to shifting cultivation lands.

Support for conversion of shifting cultivation land

Even after half a century, government efforts to stop shifting cultivation and convert the land it uses to settled agriculture, horticulture, cash-crop plantations or agroforestry have met with little success, with the exception of China, and even there, it is successful only to some extent. In some cases, this is because the alternatives promoted as replacements for shifting cultivation are simply not appropriate either for the land or for the shifting cultivators. In others it is because government policies and legislation have not been accompanied by the support needed in terms of funding, research and extension. The policies promoted in Bhutan and Bangladesh could not be implemented systematically because of a lack of funds, while in Nepal and Myanmar alternatives for shifting cultivators were not even part of the plans. India's approaches to policy implementation kept changing, while funding was limited to targeted communities, rather than the entirety of the country's vast areas with shifting cultivation.

With modernization, market forces continue to penetrate into shifting cultivation areas, leading to a spread of the monetized economy. Shifting cultivators wish to gain more cash income, but to achieve this they must either accept alternative livelihood options or transform shifting cultivation into a commodified form (Choudhury and Sundriyal, 2003). In developing alternatives to shifting cultivation, governments tend to ignore the role that customary institutions and indigenous knowledge could play, and farmers are not involved in planning for alternative livelihoods. As a result, the promoted alternatives are often inappropriate. For example, the long development period needed for cash-crop plantations means that there is a considerable delay between establishing the plantation and receiving any income, but there is no strategy for ensuring that farmers can survive while they wait for the returns to materialize. Similarly, market linkages for the sale of crops such as coffee on the open market are either absent or weak, and lack of marketing support makes farmers vulnerable to exploitative traders. Storage facilities, processing, value addition and enterprise development are needed for the management of perishable agricultural and horticultural products in order to profit from plantations, but these are far from adequate, if not absent, as are supplies of inputs such as seeds and saplings. Across the region, supplies of inputs, agricultural research and extension support and technical backstopping provided by governments are inadequate (Maithani, 2005; Choudhury, 2012). The change to settled agriculture, horticulture and cash-crop plantations means a change to multiple farming systems (Choudhury, 2012), but farmers are not supported in developing the capacity needed to manage these alternatives. Shifting cultivators also need access to credit for investment and long-term management if they want to convert their land, but they can only obtain credit if they have secure land tenure to provide collateral. Credit policies don't usually recognize the common property rights that govern the tenure of shifting cultivation land as sufficient. There has been little attempt to address the tenure issue in order to enable financing of conversion, and this has acted as a further disincentive to movements away from shifting cultivation.

A further problem is that the alternatives to shifting cultivation promoted by governments generally lack strategies to address gender issues (Leduc and Choudhury, 2012). Women have a major stake in shifting cultivation as they bear a large share of the work. They are involved equally with men in farming and natural-resource management practices, both as users and managers. But women have limited access to decision making at policy, institutional, community and household levels.

Conclusions

Shifting cultivation persists partly because it is one of the most effective and sustainable ways of ensuring subsistence in a challenging landscape. Shifting cultivation can guarantee food and nutritional supplies around the year, while the forest-fallow period ensures a wide range of ecosystem services. However, except in China, none of the countries of the eastern Himalayas has a policy that aims to support and improve shifting cultivation and recognizes its benefits. In areas where access to markets is limited, and shifting cultivation is the most appropriate land use, governments need to formulate, fund and implement policies that recognize the benefits of shifting cultivation and support its development. Shifting cultivators and their culture should be recognized as assets rather than liabilities and seen as part of the solution rather than the problem.

In areas closer to markets, it may be appropriate to convert shifting cultivation land to horticulture, cash-crop plantations or agroforestry (Rasul et al., 2004). However, this must be done in a way that is appropriate for the land and includes proper support for the shifting cultivators. Interaction with markets has already reshaped livelihood strategies and cropping options for shifting cultivators in many areas, but reliance on external markets has made them vulnerable to the vagaries of the pricing system, especially as they are mostly unorganized. Transformation away from shifting cultivation has led to food and nutritional insecurity, as well as a loss of agrobiodiversity, which could be important as a basis for adaptation to a changing climate. So far, products from monocultures have benefited a few innovative shifting cultivators, but the great majority have become more vulnerable.

Certain conditions need to be fulfilled in order to enable the effective adoption by shifting cultivators of alternative farming systems (Rasul et al., 2004). These include ensuring tenure security: providing market access (Rasul et al., 2004); ensuring food security; participation of shifting cultivators in the planning and implementation of plans and programmes associated with alternatives; recognition by policy-makers of customary institutions and improving them where necessary; and funding for alternatives including production, processing, value addition and marketing. There must also be an increase in the capacity of shifting cultivators to manage livelihood alternatives; appropriate technical support to shifting cultivators must combine the technical expertise of scientists with indigenous knowledge; and shifting cultivators must be empowered to direct the planning, implementation, progress reviews, monitoring and evaluation of programmes associated with alternatives. All programmes and activities to facilitate inclusive growth must also include equitable gender goals and perspectives, and mainstream gender considerations.

Acknowledgements

This chapter was partially supported by the core funds of the International Centre for Integrated Mountain Development (ICIMOD), contributed by the governments of Afghanistan, Australia, Austria, Bangladesh, Bhutan, China, India, Myanmar, Nepal, Norway, Pakistan, Switzerland and the United Kingdom.

Disclaimer

The views and interpretations in this publication are those of the authors. They are not necessarily attributable to the International Centre for Integrated Mountain Development (ICIMOD) and do not imply the expression of any opinion by ICIMOD concerning the legal status of any country, territory, city or area within its authority, or concerning the delimitation of its frontiers or boundaries, or the endorsement of any product.

References

AIPP/IWGIA/IKAP (2009) *Shifting Cultivation and Climate Change*, briefing paper for intersessional meeting of the United Nations Framework Convention on Climate Change, Asia Indigenous People's Pact, International Working Group for Indigenous Affairs and Indigenous Knowledge and People, Bangkok

Anonymous (1999) *Nagaland Environmental Protection and Economic Development Project: A Self-Assessment Using Outcome Mapping*, NEPED, Kohima, Nagaland

Anonymous (2013) 'NEPED', *Nagaland Journal*, https://nagalandjournal.wordpress. com/2013/03/23/112/, accessed 22 April 2015

Atal, Y. and Bennagen, P. L. (1983) 'Introduction' in Y. Atal and P. L. Bennagen (eds) *Country profiles: India, Indonesia, Malaysia, Philippines, Thailand* (vol. 2), UNESCO Regional Office for Education in Asia and the Pacific, Bangkok

Barkakoti, S. (1990) 'Alternatives Plan for Jhum Area Development' in D. N. Majumdar (ed.) *Shifting Cultivation in NE India*, Om Sons Publications, New Delhi

Barua, B. P. (2001) *Ethnicity and National Integration in Bangladesh: A Study of the Chittagong Hill Tracts*, Har-anand Publications, New Delhi

Borggaard, O. K., Gafur, A. and Petersen, L. (2003) 'Sustainability appraisal of shifting cultivation in the Chittagong Hill Tracts of Bangladesh', *Ambio 32*, pp118-123

Boserup, E. (1965) *The Conditions of Agricultural Growth*, George Allen and Unwin, London

Brady, N. C. (1996) 'Alternatives to slash-burn: A global imperative', *Agriculture, Ecosystems and Environment* 58, pp3-11

Bryant, L. R. (1997) *The Political Ecology of Forestry in Burma 1824-1994*, University of Hawai'i Press, Honolulu

Cairns, M. F. and Garrity, D. P. (1999) 'Improving shifting cultivation in Southeast Asia by building on indigenous fallow management strategies', *Agroforestry Systems* 47(10), pp37-48

Choudhury, D. (2012) 'Why do *jhumias jhum*? Managing change in shifting cultivation areas in the uplands of northeastern India', in S. Krishnan (ed.) *Agriculture in a Changing Environment: Perspectives on Northeastern India*, Routledge, New Delhi, pp78-100

Choudhury, D. and Sundriyal, R. (2003) 'Issues and options for improving livelihoods of marginal farmers in shifting cultivation areas of northeast India', *Outlook in Agriculture* 32(1), pp17-28

Dalle, S. P. and de Blois, S. (2006) 'Shorter fallow cycles affect the availability of noncrop plant resources in a shifting cultivation system', *Ecology and Society* 11(2), http://www.ecologyandsociety.org/vol11/iss2/art2/, accessed 19 January 2016

Darlong, V. T. (2004) *To Jhum or Not to Jhum: Policy Perspectives on Shifting Cultivation*, The Missing Link, Guwahati, Assam, India

Das, G. N. (2007) 'Shifting Cultivation and Development Programmes in Northeast India with Special Reference to the Hill Areas of Assam', in K. G. Saxena, L. Liang and K. Rerkasem (eds) *Shifting Agriculture in Asia: Implications for Environmental Conservation and Sustainable Livelihood*, Bishen Singh Mahendra Pal Singh, Dehradun, India

Divakar, G. D. (1990) 'Cropping Patterns, Growth Rates and Yield Stability of Different Crops under RI-RIAD and KYNTI and Roytiwari Land Systems', paper presented at the National Seminar on Agrarian Relations in NE India, National Institute for Rural Development, NE Regional Centre, Guwahati, Assam, India (mimeograph)

DoA, (2011) *Agriculture Statistics 2011*, Department of Agriculture, Ministry of Agriculture and Forests, Thimphu, Bhutan

Dollo, M. (2004) *Shifting Cultivation and Conservation Laws of Arunachal Pradesh*, North East Unit, G. B. Pant Institute of Himalayan Environment and Development, Vivek Vihar, Itanagar, Arunachal Pradesh, India

Dove, M. (1983) 'Theories of swidden agriculture and the political economy of ignorance', *Agroforestry Systems* 1, pp85-89

Forestal (1966) *Reconnaissance Soil and Land Use Survey, Chittagong Hill Tracts 1964-65*, Forestal Forestry and Engineering International, Vancouver

Gain, P. (1998) *Bangladesh Environment: Facing the 21st Century*, Society for Environment and Human Development, North Dhanmondi, Dhaka, Bangladesh

Ganguli, J. B. (1990) *The Process of Transition from Communal to Individual Land Tenure System in Tripura*, Tripura University, Agartala, Tripura (mimeograph)

Ganguly, J. B. (1968) *Economic Problems of Jhumias of Tripura*, Bookland Private Ltd, Calcutta

Government of Meghalaya (2003) *Draft Report of the Task Force on Shifting (Jhum) Cultivation*, Government of Meghalaya, Shillong

Guo, H. and Padoch, C. (1995) 'Patterns and management of agroforestry systems in Yunnan: An approach to upland development', *Global Environmental Change* 5(4), pp273–279

Guo, H., Padoch, C., Coffey, K., Aiguo, C. and Yongneng, F. (2002) 'Economic development, land use and biodiversity change in the tropical mountains of Xishuangbanna, Yunnan, Southwest China', *Environmental Science and Policy* 5, pp471–479

Guo, H., Yongmei, X. and Padoch, C. (2007) '*Alnus nepalensis*-based Agroforestry Systems in Yunnan, Southwest China', in M. F. Cairns (ed.) *Voices from the Forest: Integrating Indigenous Knowledge into Sustainable Upland Farming*, Resources for the Future Press, Washington, DC

Huang, J. and Long, C. (2007) '*Coptis teeta*-based agroforestry system and its conservation potential: A case study from northwest Yunnan', *Ambio* 36(4), pp343–349

Kafle, G. (2011) 'An overview of shifting cultivation with reference to Nepal', *International Journal of Biodiversity and Conservation* 3(5), pp147–154

Kerkhoff, E. and Sharma, E. (2006) *Debating Shifting Cultivation in the Eastern Himalayas: Farmers' Innovations as Lessons for Policy*, International Centre for Integrated Mountain Development (ICIMOD), Kathmandu

Khan, F. K. and Khisa, A. L. (1970) 'Shifting cultivation in East Pakistan', *The Oriental Geographer* 14(2), pp24–43

Leduc, B. and Choudhury, D. (2012) 'Agricultural transformations in shifting cultivation areas of northeast India: Implications for land management, gender, and institutions', in D. Nathan and V. Xaxa (eds) *Social Exclusion and Adverse Inclusion: Development and Deprivation of Adivasis in India*, Oxford University Press, New Delhi pp 237–258

Lewin, T. H. (1869) *The Hill Tracts of Chittagong and the Dwellers Therein, with Comparative Vocabularies of the Hill Dialects*, Bengal Printing Company, Calcutta

Liang, L., Shen, L., Yang, W., Yang, X. and Zhang, Y. (2008) 'Building on traditional shifting cultivation for rotational agroforestry: Experiences from Yunnan, China', *Forest Ecology and Management* 257(2009), pp1989–1994

Ma, Q. (1999) *Asia-Pacific Forestry Sector Outlook Study: Socio-Economic and Non-Wood Products Statistics*, Food and Agriculture Organization of the United Nations, Rome

Macqueen, D. (2013) *Myanmar: Could an Unusual Yam Help the March of Community Forestry?*, International Institute for Environment and Development, London, www.iied.org/myanmar-could-unusual-yam-help-march-of-community-forestry, accessed 22 April 2015

Maikhuri, R. K. and Ramakrishnan, P. S. (1991) 'Comparative analysis of the village ecosystem function of different tribes living in the same area in Arunachal Pradesh in North-eastern India', *Agriculture Systems* 35, pp377–399

Maithani, B. P. (2005) *Shifting Cultivation in North-East India: Policy Issues and Options*, Mittal Publications, New Delhi

Mertz, O. (2002) 'The relationship between length of fallows and crop yield in shifting cultivation: A rethinking', *Agroforestry Systems* 55, pp149–159

Mishra, B. K. and Ramakrishnan, P. S. (1981) 'The economic yield and energy efficiency of hill agro-ecosystems at higher elevations in Meghalaya in North-eastern India', *Acta Oecologica (Oecologia Applicata)* 2, pp369–389

MoA (2006) *The Task Force Report on Shifting Cultivation not Registered during the Second Cadastral Survey*, Ministry of Agriculture, Royal Government of Bhutan, Thimphu

Nye, P. H. and Greenland, D. J. (1960) *The Soil under Shifting Cultivation*, Commonwealth Agricultural Bureau, Buckinghamshire, England

O'Brien, W. E. (2002) 'The nature of shifting cultivation: Stories of harmony, degradation and redemption', *Human Ecology* 30, pp483–502

Phanbuh, S., Albano, A. and Darlong, V. (2008) *Increasing Benefits from Forests: Forest-based Interventions of NERCORMP in Meghalaya, India*, North Eastern Region Community Resource Management Project for Upland Areas, International Fund for Agricultural Development and the Government of India, Shillong, Meghalaya

Phuntsho, K., Aryal, K. P. and Kotru, R. (2015) *Shifting Cultivation in Bangladesh, Bhutan and Nepal: Weighing Government Policies against Customary Tenure and Institutions*, working paper 2015/7, International Centre for Integrated Mountain Development, Kathmandu

Ramakrishnan, P. S. (1984) 'The science behind rotational bush fallow agriculture systems (*jhum*)', *Proceedings of the Indian Academy of Science (Plant Science)* 93, pp379-400

Ramakrishnan, P. S. (1993) *Shifting Agriculture and Sustainable Development: An Interdisciplinary Study from Northeastern India*, MAB series 10, Oxford University Press, New Delhi

Ramakrishnan, P. S. and Toky O. P. (1981) 'Soil nutrient status of hill agro-ecosystems and recovery after slash and burn agriculture (*jhum*) in northeastern India', *Plant and Soil* 60, pp41-64

Rasul, G. (2007) 'Political ecology of the degradation of forest commons in the Chittagong Hill Tracts of Bangladesh', *Environmental Conservation* 34, pp1-11

Rasul, G. (2009) *Land Use, Environment and Development Experiences from the Chittagong Hill Tracts of Bangladesh*, AH Development Publishing House, New Market, Dhaka, Bangladesh

Rasul, G., Thapa, G. B, and Zoebisch, M. A. (2004) 'Determinants of land-use change in the Chittagong Hill Tracts of Bangladesh', *Applied Geography* 24, pp217-240

Rozelle, S., Huang, J., Husain, S. A. and Zazueta, A. (2000) *China: From Afforestation to Poverty Alleviation and natural Forest Management*, Evaluation Country Case Study Series, The International Bank for Reconstruction and Development, The World Bank, Washington, DC

Sachidananda (1984) 'Swidden cultivation among the Wanchos of Arunachal Pradesh', in Z. Majid (ed.) *Swidden Cultivation in Asia*, vol. 3 (1985), United Nations Educational, Scientific and Cultural Organization, Bangkok

Saha, S. B. (1970) *Socio-economic Survey of the Noatia Tribes*, unpublished report to the Tribal Welfare Department, Government of Tripura, Agartala, Tripura, India

Sanchez, P. A. (1994) 'Alternative to slash and burn: A pragmatic approach for mitigating tropical deforestation', in J. R. Anderson (ed.) *Agricultural Technology: Policy Issues for the International Community*, CAB International, Wallingford, UK, pp451-479

Sharma, B. D. (1984) 'Shifting cultivators and their development', *North Eastern Hill University Journal of Social Science and Humanities* 2, pp1-36

Shelley, F. (1999) 'Socio-economic status and development of Chittagong Hill Tracts of Bangladesh: An overview', in A. Kamal, M. Kamaluddin and M. Ullah (eds) *Land policies, Land management and Land Degradation in the Hindu Kush Himalayas: Bangladesh Study Report*, International Centre for Integrated Mountain Development, Kathmandu

Shirasaka, S. (2006) 'Shifting cultivation in Xishuangbanna, southwestern China: A vanishing mountain culture', *Global Environmental Research* 10(1), pp21-38

Singh, D. (1996) *The Last Frontier: People and Forest in Mizoram*, Tata Energy Research Institute, New Delhi

Spencer, J. (1996) *Shifting Cultivation in Southeastern Asia*, University of California publication Geography 19, University of California Press, Berkeley, CA

Srivastava, S. C. (1996) *Levels and Structures of Development: An Inter-District Study of North-East India*, NE Regional Centre, National Institute of Rural Development, Guwahati, Assam, India (mimeograph)

Tawnenga, U. S. and Tripathi, R. S. (1996) 'Evaluating second year cropping on *jhum* fallow in Mizoram, north-eastern India: Phytomass dynamics and primary productivity', *Journal of Biosciences* 21, pp563-575

Tawnenga, U. S. and Tripathi, R. S. (1997a) 'Evaluating second year cropping of *jhum* fallow in Mizoram, north-eastern India: Energy and economic efficiencies', *Journal of Biosciences* 22, pp605-613

Tawnenga, U. S. and Tripathi, R. S. (1997b) 'Evaluating second year cropping of *jhum* fallow in Mizoram, north-eastern India: Soil fertility', *Journal of Biosciences* 22, pp615-625

Thang, Y., Lixin S. and Xinkai, Y (2007) 'Shifting agriculture and agrobiodiversity: A case study from western Yunnan Province, Southern China', in K. G. Saxena, L. Liang and K. Rerkasem (eds) *Shifting Agriculture in Asia: Implications for Environmental Conservation and Sustainable Livelihood*, Bishen Singh Mahendra Pal Singh, Dehradun, India

Thrupp, L. A., Hecht, S. B. and Browder, J. O. (1997) *The Diversity and Dynamics of Shifting Cultivation: Myths, Realities, and Policy Implications*, World Resources Institute, Washington DC

Tiwari, B. K. (2003) 'Traditional management of NTFP/MAPs in north east India. A case study of bayleaf and broom grass', Paper presented at a Consultation Meeting on Improving Livelihoods of Mountain Communities through Sustainable Utilization of Non-Timber Forest Products, 18-20 December 2003, International Centre for Integrated Mountain Development, Kathmandu

Toky, O. P. and Ramakrishnan, P. S. (1981a) 'Cropping and yield in agricultural systems of northeastern hill regions of India', *Agro-Ecosystems* 7, pp11-25

Toky, O. P. and Ramakrishnan, P. S. (1981b) 'Run-off and infiltration losses related to shifting agriculture (*jhum*) in northeastern India', *Environmental Conservation* 8, pp313-321

Toky, O. P. and Ramakrishnan, P. S. (1983) 'Secondary succession following slash and burn agriculture in northeastern India.1. Biomass litterfall and productivity', *Journal of Ecology* 71, pp735-745

Uddin, M. S., Kamal, M. S. and Mollah, M. H. (2000) *Hill Farming Systems and Resource Utilization in Chittagong Hill Tracts*, Hill Agriculture Research Station, Khagrachari, Bangladesh Agriculture Research Institute, Joydebpur, Bangladesh

UNEP (2012) *Ecosystem-Based Adaptation Guidance: Moving from Principles to Practice*, working document, United Nations Environment Programme, Nairobi, Kenya

Upadhaya, K. (1994) *Shifting Cultivation in Bhutan: A Gradual Approach to Modifying Land Use Patterns. A Case Study*, Department of Forestry, Ministry of Agriculture, Thimpu, Bhutan

Van Schendel, W., Wolfgang, M. and Dewan, A. K. (2000) *The Chittagong Hill Tracts: Living in Borderland*, White Lotus, Bangkok

Warner, K. (1991) *Shifting Cultivators: Local Technical Knowledge and Natural Resources Management in the Humid Tropics*, Community Forestry Note 8, Food and Agriculture Organization of the United Nations, Rome

Watters, R. F. (1971) *Shifting Cultivation in Latin America*, Food and Agricultural Organization of the United Nations, Rome

Wenjuan, Z. and Yin, S. (2007) 'Swidden agriculture and its evolution in Yunnan, China', in K. G. Saxena, L. Liang and K. Rerkasem (eds) *Shifting Agriculture in Asia: Implications for Environmental Conservation and Sustainable Livelihood*, Bishen Singh Mahendra Pal Singh, Dehradun, India

Win, N. R. (2013) *Shifting Cultivation and Land Management in Myanmar*, Powerpoint presentation dated 4 July 2013, Forest Department, Ministry of Environmental Conservation and Forestry, Napypyidaw, Myanmar

Win, S. (2004) *Investigation on Shifting Cultivation Practices Conducted by the Hill Tribes for the Development of Suitable Agroforestry Techniques in Myanmar*, Forest Department, Ministry of Environmental Conservation and Forestry, Naypyidaw, Myanmar

Win, S. (2009) *Investigation on Shifting Cultivation Practices Conducted by the Hill Tribes for the Development of Suitable Agroforestry Techniques in Myanmar*, National Commission for Environmental Affairs, Ministry of Environmental Conservation and Forestry, Naypyidaw, Myanmar

Xu, J. (2007) 'Rattan and tea-based intensification of shifting cultivation by Hani farmers in southwestern China', in M. F. Cairns (ed.) *Voices from the Forest: Integrating Indigenous Knowledge into Sustainable Upland Farming*, Resources for the Future Press, Washington, DC, pp667-675

Xu, J. and Ribot, J. C. (2004) 'Decentralization and accountability in forest management: A case from Yunnan, southwest China', *European Journal of Development Research* 16(1)

Xu, J., Fox, J., Lu, X., Podger, N., Leisz, S. and Ai, X. (1999) 'Effects of swidden cultivation, state policies and customary institutions on land cover in a Hani village, China', *Mountain Research and Development* 19(2), pp123-132, http://www.jstor.org/stable/3674253, accessed 22 April 2015

Xu, J., Ai, X. and Deng, X. (2005) 'Exploring the spatial and temporal dynamics of land use in Xizhuang watershed of Yunnan, southwest China', *International Journal of Applied Observation and Geoinformation* 7, pp299-309

Yamamoto, M. and Maung, T. M. (2008) 'Exploring the socio-economic situation of plantation villagers: A case study in Myanmar Bago Yoma', *Small-scale Forestry* 7, pp29-48, DOI 10.1007/s11842-008-9039-1

Yin, S. (2001) *People and Forests: Yunnan Swidden Agriculture in Human-Ecological Perspective* (translated by Magnus Fiskesjo), Yunnan Education Press, Kunming, China

ZLN (1986-1992) *Chinese Forestry Yearbook (Zhongguo Linye Nianjian)*, Forestry Publishing House, Beijing

18

REFLECTIONS ON THE IMPACTS OF STATE POLICIES ON SHIFTING CULTIVATORS IN NORTHEAST INDIA

*Vincent Darlong**

Introduction

Shifting cultivation, known as *jhum* in northeast India, continues to be a subject of considerable debate and conflict.[1] In terms of official policy and development planning, it is confronted with numerous obstacles. Its practitioners in northeast India have been given very little space in which to find support, particularly for demonstrating innovations within their cultivation systems. Numerous policies and programmes seeking to change shifting cultivation and replace it with entirely different agro-horticultural systems remain largely unsuccessful. However, shifting cultivators themselves are making changes through their own ingenuity and efforts. Despite official obstruction, *jhumscapes* (shifting cultivation lands comprising current *jhum* fields and secondary-forest fallows of different ages) remain the most dominant feature of the landscape in most upland areas of northeast India (Figure 18-1), interspersed with settled agriculture such as wet rice cultivation, terraced fields on hill slopes, mixed horticultural crops and forest reserves.

Shifting cultivation in India is increasingly trapped between two conflicting paradigms operating at the policy and institutional levels. The dominant perspective is that shifting cultivation is a wasteful and ecologically dysfunctional system, detrimental to forests and soil conservation, and therefore in need of eradication by inducing the shifting cultivators to adopt alternative forms of livelihood. The other emerging paradigm is that shifting cultivation is a legitimate practice that ensures the survival of people living on marginal lands and hence should be allowed to carry on as it is, without external influence (Government of India, 2008).

Since the 1950s, there have been many attempts, through government policies, programmes and interventions, to improve shifting cultivation lands and wean shifting

* Dr Vincent Darlong is Country Programme Officer in the India Country Office of the International Fund for Agricultural Development (IFAD), New Delhi.

FIGURE 18-1:   Northeast India and the states discussed in this chapter.

cultivators away from the practice. Much of the concern has been focused on the fallow period, that part of the *jhum* cycle in which vegetation is left undisturbed in order to restore the soil nutrients lost to cropping. Under pressure from a swelling population, farmers have attempted to intensify their shifting cultivation systems by shortening the fallow, and this has increasingly been linked to ecological and social insecurity for both the shifting cultivators and their environment. Some of the most visible environmental impacts of shortening the *jhum* cycle include soil and nutrient loss, increased runoff and other hydrological shifts, water scarcity, vegetal degradation, loss of floral and faunal biodiversity, increased fire hazards, decreasing crop yields, increasing food and nutritional insecurity and increasing weed infestation of fallow areas. Socio-economic indicators often associated with this malaise are persistent poverty (income, economic, nutritional, human, basic amenities) and continued household burdens, particularly persistent drudgery for women as they struggle to gather drinking water, firewood, wild vegetables and feed for domestic livestock. Numerous policy and programme interventions have resulted from the fact that many of the areas used for shifting cultivation are also suitable for cash crops and horticulture crops, such as rubber, tea, oranges, cashew nuts, areca nut, pineapples, passion fruit, and so on. At the same time, development planners and social scientists have expressed concern that unsustainable shifting cultivation practices due to the shortening of fallow periods, combined with inequity in land access following the introduction of cash crops and horticultural schemes, are straining the socio-cultural fabric of shifting cultivation communities. The overall impression in the context of modern India is that shifting cultivation is an inherently unstable and unsustainable practice that should give way to more efficient forms of farming and more productive land-use systems.

Intensive efforts by the Government of India and state governments in northeast India have benefited large numbers of shifting cultivators by replacing their swidden agriculture with terraced fields, agroforestry, vegetable cultivation, horticulture crops, high-value crops, tree farming and plantations for gathering non-timber forest products (NTFPs) (see figures in this chapter). Over the years, many families across northeast India have abandoned shifting cultivation and adopted settled agriculture and other alternative livelihoods. However, there are still an estimated 443,000 families of shifting cultivators in the region (Government of India, 2008). Various studies have identified both underlying constraints to the adoption of alternative forms of agriculture and the reasons why a large number of shifting cultivators remain either partially or wholly dependent on the ancient practice. Some of the key constraints are the remoteness and poor infrastructures of many shifting cultivation communities, with difficulties in market access as well as problems for government extension agencies in providing effective extension services. Such services are unable to reach out to shifting cultivators with existing schemes, provide new seeds and technologies or give personal support. Other important factors include the traditional land-tenure system, under which land access is mainly for shifting cultivation, and the unsuitability of steep upland landscapes for any form of agriculture other than shifting cultivation.

This chapter is an attempt to explore and understand the impacts of various state policies on shifting cultivators. The broad context of state policies in this chapter encompasses any acts, laws, regulations, notifications, schemes and programmes of the Government of India and the governments of the States, including Autonomous District Councils. Even externally aided projects or programmes implemented by the government with assistance from multi-lateral or bilateral agencies, but addressing issues of shifting cultivation and thereby impacting shifting cultivators in some ways, are grouped in this category. International funding agencies working in northeast India include the International Fund for Agricultural Development (IFAD), the World Bank and the United Nations Development Programme (UNDP). Bilateral agencies that have worked in the region, or are currently working there, include the India-Canada Environment Facility (ICEF), KfW and GIZ of Germany and the Japan International Cooperation Agency (JICA).

Overall state policies, well intended in design, strategies and approaches, have transformed the lives of many shifting cultivators. At the same time, many policies have also brought them new and unintended challenges. Broadly, these opportunities and challenges apply to the social, cultural, economic, environmental, physical and knowledge domains of shifting cultivators. The present study is an appreciation of this wide range of impacts, with reflections on opportunities for ways forward.

Policy landscapes with implications for shifting cultivation

There are a number of natural-resource management policies relating particularly to forest, agriculture and watershed development that have implications for shifting cultivators in northeast India (Darlong, 2004). While many of these policies and programmes have benefited a large section of the swiddening population in the region, many others continue to be impacted adversely by unintended consequences, making them vulnerable and, in a way, distressed. The impacts of state policies on shifting cultivation communities can only be properly appreciated through an understanding of the full range of state policies with implications for shifting cultivation. This full range is outlined below.

Forest policies and programmes

Forest policies and management systems have always tended to be at variance with shifting cultivation, from the time state control of forests was first introduced in India. While shifting cultivators depend on forests for their livelihoods by clearing and cultivating patches of forest and then allowing them to recover, foresters and natural-resource managers have always regarded shifting cultivation as a wasteful practice. As well as cultivating patches of forest, shifting cultivators also collect a variety of edible leaves, fruits, flowers, creepers, roots, tubers, and diverse NTFPs such as honey from primary forests and from fallow areas of secondary forest in various stages of recovery.

To the forest communities, the overriding philosophy is one of shifting the forests (fallow forests) rather than shifting the cultivation.

British colonial forest policies in India attempted to impose supremacy of state interests over those of the people, particularly the people who protected and managed the forests on the basis of their customary practices. Shifting cultivators were invariably the targets. According to Malik (2003),

FIGURE 18-2: A *jhum* plot ready for planting in Dima Hasao district, Assam.

Photo: Vincent Darlong.

a forest officer, one Mr Baden-Powell, made the following statement to a forest conference in Allahabad in 1874:

> The fact is that the system is so wasteful that somehow or the other it must be put a stop to (sic), just like ... any other great evil. It consists in destroying a large and valuable capital to produce a miserable and temporary return. To put a stop to it is only to anticipate by a few years the natural determination of the system, which will happen if the system continues long enough, because there will be no more forest to cut down and burn. The way out is to reserve large areas and prohibit *jhum*. Efforts should be made to change people to permanent agriculture (Malik, 2003).

The first forest law that sought some direct control over shifting cultivation was the Indian Forest Act of 1927. Section 10 of the Act carried certain 'special provisions' related to the treatment of claims by people seeking to practise shifting cultivation. A Forest Settlement Officer would record such claims and deliver an opinion to the state government on whether the practice should be permitted or prohibited, wholly or in part. The state government would then order permission or prohibition, wholly or in part, and if the order involved permission, the Forest Settlement Officer would demarcate and apportion land for the practice. However, the Act specifically declared that 'the practice of shifting cultivation shall in all cases be deemed a privilege subject to control, restriction and abolition by the State Government' (Government of India, 1927).

Even post-independence forest policies in India were not supportive of shifting cultivators, although tribal rights and customary practices were protected in the Sixth Schedule of the Constitution. While admitting that shifting cultivation was a wasteful practice, the Indian Forest Policy of 1952 advocated a 'missionary approach' of persuasion in dealing with the needs of shifting cultivators, so as to wean them away from swiddening. Section 23 of the Policy stated:

> The damage caused to forests by shifting cultivation in certain areas must be guarded against. To wean the aborigines, who eke out a precarious living from axe-cultivation, moving from area to area, away from their age-old and wasteful practices, requires persuasion, not coercion; a missionary, not an authoritarian, approach. Possibilities of regulating shifting cultivation by combining it with forests regeneration (taungya) to the benefit of both should be fully explored. Success in this direction largely depends on enlisting the cooperation of the cultivators and gaining their confidence, and in showing consideration to their needs and wishes (Government of India, 1952).

The National Forest Policy was revised in 1988. While discouraging the continuation of shifting cultivation, the new policy also emphasized the need for alternative sources of income for shifting cultivators. Section 4.7 stated:

> Shifting cultivation is affecting the environment and productivity of land
> adversely. Alternative avenues of income, suitably harmonised with the right
> land-use practices, should be devised to discourage shifting cultivation. Efforts
> should be made to contain such cultivation within the area already affected,
> by propagating improved agricultural practices. Areas already damaged by such
> cultivation should be rehabilitated through social forestry and energy plantations
> (Government of India, 1988).

Following introduction of the national forest policies, some of the states in northeast
India also produced state-specific forest policies. The Assam Forest Policy of 2004,
dealing with shifting cultivation in its Section 4.3.5, declared that an 'Integrated
Area Development Programme, with due regard to local tradition and culture, will
be the mainstay in tackling the problem related to *jhumming'* (sic). The programme
would raise awareness in shifting cultivation communities about the benefits of more
sedentary land-use systems, and forest development works would generate sustained
employment through implementation of short- and long-term projects that would
contain *jhumming,* and shifting cultivation sites would be rehabilitated through
innovative community-based reafforestation, agroforestry and cash-crop plantations
(Government of Assam, 2004a).

Successive five-year plans of the Government of India adopted a variety of
approaches to promoting afforestation programmes. Starting from the 10th Five-Year
Plan (2002 to 2007), the Government merged several existing programmes, such as the
national afforestation and eco-development programme, and initiated the National
Afforestation Programme. This included the 'eco-restoration' of shifting cultivation
areas. More recently, the Report of the Working Group on Forests and Sustainable
Management of Natural Resources for the 12th Five-Year Plan (2012 to 2017)
proposed a variety of schemes and programmes under the National Afforestation
Programme. These included afforestation and eco-restoration of shifting cultivation
areas by adopting local, specific and viable models; soil and moisture conservation,
including water harvesting structures; eco-development of people residing in forests;
and treatment of problem soils (Government of India, 2011b). These programmes
would certainly have impacts on shifting cultivators.

Various contemporary offshoots of the forest policies and programmes of both the
central and state governments in India also have implications for shifting cultivation.
In line with the Joint Forest Management Notification of 1990, which aimed to
encourage peoples' participation in the protection, management, regeneration and
development of forests (Government of India, 1990), many state governments in
northeast India introduced their own joint forest management (JFM) resolutions.
Often, one of the objectives of the state-government JFM policies was to wean
swidden communities away from the practice of shifting cultivation. For example,
the Government of Nagaland's JFM resolution, while aiming to elicit the active
participation of villagers in creating, managing and protecting plantations, also
proposed 'to wean away the land-owning communities from shifting cultivation by

adopting an alternative, i.e. tree farming, and to productively utilize the degraded *jhumland,* thereby checking soil erosion' (Government of Nagaland, 1997).

   The beginning of the New Millennium in India coincided with the prospect of 'gregarious flowering' of *muli* bamboo *(Melocanna baccifera).* It was predicted that this would affect most states of northeast India, particularly during the years 2004 to 2007. *Melocanna baccifera* flowers at an interval of 40 to 50 years in northeast India and this phenomenon is often associated with ecological, economic and social problems (Bipin Behari, 2006). Bamboo flowering and the production of its pear-shaped seeds triggers a rapid increase in the population of rats, which ultimately invade agricultural fields. In the 1960s, destruction of crops by rodents led to famine in the state of Mizoram. To minimize the impacts of bamboo flowering, the central Ministry of Environment and Forests initiated a series of national bamboo action plans and the Ministry of Agriculture initiated the National Bamboo Mission in 2006. Taking advantage of these initiatives, state governments in northeast India launched Bamboo Development Agencies and introduced state-specific bamboo policies to minimize the impacts of bamboo flowering and to optimize rural incomes generated by bamboo-based livelihoods and enterprises. These included the Mizoram Bamboo Policy 2002, the Nagaland Bamboo Policy 2004 and the Assam Bamboo and Cane Policy 2005. These policies joined other official initiatives in promoting the cultivation of high-value bamboo species in *jhum* land. Therefore, they had implications for shifting cultivation and swidden communities. For example, paragraph 3.1(b)(ix) of the Nagaland Bamboo Policy 2004 provided for 'bamboo cultivation to synchronize with existing farming practices, such as *jhumming,* etc, to maximize interim benefits' (Government of Nagaland, 2004). The Bamboo Policy of Mizoram 2002 (section 6.3) envisaged 'regulating bamboo exploitation in *jhum* regrowth and *jhum* areas by involving Village Councils [and] Village Forest Development Committees (VFDCs) and facilitating gradual change over to agroforestry management and practices' (Government of Mizoram, 2002). Thus, the underlying bamboo policy in Nagaland was to replace *jhum* with high-value bamboo cultivation, while that in Mizoram sought to replace *jhumming* with different kinds of agroforestry.

Agriculture and land-use policies

General agricultural policies and programmes in India have not been supportive of shifting cultivation. It is not considered to be a part of mainstream agriculture. In fact, the National Commission on Agriculture in 1976 suggested that shifting cultivation should be banned (Government of India, 1976). However, the National Policy for Farmers issued in 2007 recognized the special needs of tribal farmers engaged in shifting cultivation (Government of India, 2007). The policy acknowledged the fact that tribal farmers, including shifting cultivators, were among the country's most disadvantaged farmers. It emphasized the need to uplift their economic conditions as a matter of government priority, by taking several courses of action. These were updating land records in areas inhabited by tribal farmers; strengthening institutional structures to give tribal farmers a more participatory role in decision making; enabling easy access to institutional credit for all tribal farmers and providing them with *kisan* credit cards (with which Indian farmers may access affordable credit); documenting traditional crops and recording the indigenous knowledge of tribal farmers and creating an economic stake in conserving their crops; and providing appropriate technology and extension services for tribal farmers and relaxing criteria for providing them with inputs such as water, fertilizers, seeds and so on (Government of India, 2007).

FIGURE 18-3: A swidden farmer in northeast India looks over his bumper crop.

Photo: Vincent Darlong.

A number of states in northeast India have introduced policies on agriculture and land use. Notable among them have been the New Agriculture Policy of Arunachal Pradesh 2001, the Assam Agriculture Policy 2004 and the New Land Use Policy of Mizoram 2009. All of them have a broad common objective of discouraging shifting cultivation and weaning shifting cultivators away from *jhumming*. The Assam Agricultural Policy 2004 recognized hill agriculture – another term for shifting cultivation – but only with the aim of replacing it with 'alternative high-income generating avenues like horticulture crops', along with support for value additions. The focus areas were the districts of Karbi Anglong and Dima Hasao, which were formerly the North Cachar Hills (Government of Assam, 2004b).

The New Agriculture Policy of Arunachal Pradesh specifically attempted to address the problems related to shifting cultivation. It stated: 'Special emphasis is to be given on shifting cultivation, ensuring better land management, introducing improved cultivation on sloping land through agroforestry, horticulture and encouraging other household activities. The programme is to be designed in such a way that there would be a simultaneous thrust in weaning the *jhum* farmers towards better cultivation' (Government of Arunachal Pradesh, 2001).

The Arunachal Pradesh policy further quoted the observations of the S. P. Shukla Commission Report on 'Transforming the Northeast', pertaining to *jhum* farming:

> Hill farming in the northeast is largely under *jhum*, though there are some excellent terraces in certain states and expanding patches of wet rice cultivation. *Jhum* farming is becoming less productive with a shrinking *jhum* cycle and has caused erosion and forest regression in certain areas. Not all *jhumias*

resettlement schemes have worked well; nor can *jhumming* be ended all at once. The problem needs to be tackled sensitively, as *jhum* cultivation is also a way of life.

The ICAR [Indian Council of Agricultural Research] has evolved a three-tier hill farming package combining forestry, horticulture or tree farming and terraced cultivation, as one moves down the hills. *Jhum* improvement is advocated by others and can be carried further through appropriate R&D. Nagaland has pioneered an excellent method of upgrading *jhum* by interposing a strong and increasing component of agroforestry through assisted tree planting of selected fast growing economic timber, the menu being a producer of meticulous exercise in biodiversity mapping, documentation and breeding of plant material for widespread propagation (Government of India, 1997).

Mizoram embarked upon a New Land Use Policy in 2009. Its primary objective was: 'to wean farmers away from destructive *jhum* practices and assist the workforce hitherto engaged in *jhumming* to be employed in sustainable economic ventures to create productive assets in each family'. The new policy envisaged the promotion of a 'one village and one crop' concept, by targeting agro-horticultural development for proven markets, growing on a commercial scale and treating villages on a cluster basis. Under this policy, a family of shifting cultivators would typically be provided with about 2ha of land to be used for four different farming systems: mixed agroforestry covering 0.5ha; passion fruit and turmeric over 0.5ha; orange trees and broom grass over 0.5ha; and lowland irrigated crops over 0.2ha (Government of Mizoram, 2009).

For its part, the State of Tripura implemented what it called its Approach to People's Plan in 2000. One of the plan's objectives was the 'economic uplifting' of *jhumias*. It stated:

> All primary sector departments will coordinate with each other to draw up programmes for the economic uplift of *jhumias*. These programmes will have strong market focus and be subsidy based. The Tribal Welfare Department will monitor the proper implementation of the programmes and insist that a significant portion of departmental resources should target *jhumias*. Apart from *jhumias*, there are a large number of tribal cultivators who, while not primarily dependent on *jhum*, live in upland areas. The focus in upland areas will be significantly on forestry, horticulture and animal husbandry. All primary sector departments will target the upland areas through specific programmes framed [while] keeping in mind the special needs of the tribals and the tribal areas (Government of Tripura, 2000).

Watershed development policies

The Watershed Development Project in Shifting Cultivation Areas was first launched during India's Fifth Five-Year Plan, between 1977 and 1982. It was discontinued during 1991 and 1992, but was re-introduced for the states in northeast India in 1994 and 1995. The scheme was aimed at overall development of *jhum* areas on a watershed basis, reclaiming land affected by shifting cultivation and improving the livelihoods of *jhumia* families so as to encourage them to adopt settled agriculture.

The scheme's broad objectives were to protect and develop hill slopes in *jhum*-cultivation areas by adopting different soil and water conservation measures on a watershed basis and to reduce further land degradation; to encourage and assist *jhumia* families to develop their land for productive use with improved cultivation and 'a suitable package of practices' that would lead to settled cultivation; to improve the socio-economic status of *jhumia* families through household and land-based activities; and to mitigate the ill-effects of shifting cultivation by introducing appropriate land-use and improved water-management technologies. In practice, each participating family had to own or have access to at least 1.5ha of land and be willing to adopt a new use for at least one-third of its *jhum* area under plantation (Government of India, 2001).

Regulation of shifting cultivation

Forest and other natural-resource policies introduced by both the colonial and post-colonial administrations in India discouraged shifting cultivation. A number of regulations to contain the spread of shifting cultivation were implemented in northeast India. First introduced by the colonial government, these regulations later came from the post-colonial Government of India, state governments, and more recently, the Autonomous District Councils (Darlong, 2004).

The first *jhum* law was the Sylhet *Jhum* Regulation, 1891, under which the rights of shifting cultivators were revoked, by payment of compensation, once their land was permanently settled. It appears that this Act was more to benefit tea companies, by encouraging the expansion of tea cultivation, than to help and to satisfactorily re-settle shifting cultivators. In contrast, the Balipara/Tirap/Sadiya Frontier Tract *Jhum* Land Regulation, 1947, which still applies in Arunachal Pradesh, appears more progressive in balancing the customary rights and livelihood needs of shifting cultivators with protection of forests. It seeks to establish and protect the customary

FIGURE 18-4: A *jhum* field converted into a tea plantation in West Garo Hills, Meghalaya.

Photo: Vincent Darlong.

rights of shifting cultivators to the forest land that they have traditionally cultivated (Murtem et al., 2008). The Nagaland *Jhumland* Act, 1970, which was notified in 1974, similarly sets out to safeguard and regulate rights to *jhum* land for the benefit of shifting cultivators as part of their customary rights. The act also provides for acquisition of land by the government, particularly for prevention of erosion and protection of forests, or as an award following offences.

The creation of Autonomous District Councils (ADCs), under Articles 244 and 275 of the Constitution of India, made provision for the administration of tribal areas in Assam, Meghalaya, Mizoram and Tripura. Amongst many other functions, the ADCs were given power to regulate shifting cultivation and they subsequently enacted their own *jhum* regulations. These were the Mikir Hills District *(Jhumming)* Regulation, 1954, to regulate *jhum* in the Mikir Hills (present day Karbi Anglong) district of Assam; the Garo Hills District *(Jhum)* Regulation, 1954, to regulate *jhum* in the entire Garo Hills of Meghalaya – presently divided into the West, South, South-West, North and East Garo Hills districts; the Lushai Hills District *(Jhumming)* Regulation, 1954, regulating *jhum* in the erstwhile Lushai Hills district of Assam, which, since 1987, has become the state of Mizoram; and the Pawi Autonomous District *(Jhum Regulation)* Act, 1983, to regulate *jhum* in the Pawi ADC areas in the southern part of Mizoram. While generally, these regulations were to safeguard the rights of shifting cultivators to practise subsistence *jhumming,* they also provided equitable access to land for *jhum,* protected forests from fire, prohibited *jhum* in certain areas, and so on. The regulations, particularly those of the Mikir and Garo Hills, encouraged wet rice farming and other forms of permanent cultivation, wherever suitable.

Other policies and programmes with implications for shifting cultivators

It is interesting to note that many other centrally sponsored and land-based schemes or programmes of the Government of India have implications for shifting cultivators. These include the National Horticulture Mission, the National Bamboo Mission, the Rubber Board, the Spices Board, the Tea Board and the Coffee Board. Many shifting cultivators have availed benefits from these programmes, because a common element of them all is that their activities also aim at shifting cultivation. For example, plantations of cashew nuts, oranges or passion fruit under the National Horticulture Mission, bamboo plantations under the National Bamboo Mission, tea plantations under the Tea Board and rubber plantations under the Rubber Board almost always aim for establishment in *jhum* fallow areas of most states in northeast India. The idea, of course, is to rehabilitate and resettle shifting cultivators with permanent sources of livelihood. The programmes or schemes involve income-generating activities or enterprise development by promoting cash-crop plantations, along with improving the environment, rather than perpetuating the existence of degraded *jhum* fallows.

The Government of India has also initiated the National Mission for a Green India, or the Green India Mission, as one of eight missions under the National Action Plan on Climate Change. The Green India Mission (GIM) has been conceived as a multi-stakeholder, multi-sectoral and multi-departmental mission, recognizing that climate change will seriously affect the distribution, type and quality of natural resources, as well as the associated livelihoods of the people. GIM puts 'greening' into the context of climate-change adaptation and mitigation, and is meant to enhance ecosystem services such as carbon sequestration and storage (in forests and other ecosystems), hydrological services and biodiversity. It also promotes provisioning services like fuel, fodder and small timber through agroforestry, farm forestry and production of NTFPs. One of the primary objectives of the GIM is to increase forest and tree cover through agroforestry, social and farm forestry, and in the context of northeast India, one of the target areas would be land under shifting cultivation (Government of India, 2011b).

Another policy that might have far-reaching impacts on shifting cultivators is the Scheduled Tribes and Other Traditional Forest Dwellers (Recognition of Forest Rights) Act, 2006, popularly known as the Forest Rights Act (FRA). This Act recognises the rights of shifting cultivators to use forest land under their occupation. Rights to as much as four hectares of forest land can be given to individual households, for purposes of cultivation, in the names of both the husband and wife. Community rights can be granted to entire communities. Such land cannot be alienated, but can be inherited. In practice, most families are given rights under the FRA to between 1 and 1.5 acres (0.405ha to 0.607ha) of land, on average. In some states, such as Tripura, indigenous people already had limited concessions for *jhum* cultivation in forest areas. Now, if they are granted land for cultivation under the FRA, they may no longer be allowed the old concessions to practise *jhum* on forest land.

Field assessments have indicated that two other government programmes could have interesting impacts on shifting cultivators, and even on the way in which shifting cultivation is undertaken. The Mahatma Gandhi National Rural Employment Guaranteed Scheme provides 100 days of rural wage employment in a financial year to any poor person – including shifting cultivators – through a 'job card' issued by a competent government authority for a member of a poor household. Under this national flagship programme, the rural poor – if they belong to scheduled tribes – can claim 100 days of employment on land- and water-related activities, including terrace development and plantations of fruit trees and other economic species on private land. In effect, areas currently under shifting cultivation are gradually being transformed into terraced fields or into permanent plantations.

The other programme having implications for shifting cultivators is the Prime Minister's *Grameen Sadak Yojana* – a scheme for improving rural connectivity in remote areas of the country. Such have been the improvements achieved by this programme that transport 'connectivity' in remote hilly areas where shifting cultivation was once predominant has changed local farming patterns. Shifting cultivation has been replaced by cash crops and other horticultural crops that can now be easily transported to markets.

Externally aided projects having implications for shifting cultivation

Externally aided projects in northeast India (Table 18-1) have one common feature: they are working to improve areas under shifting cultivation and to improve the livelihoods of shifting cultivators. The India–Canada Environment Facility (ICEF) has funded NEPED projects in Nagaland to promote tree farming and various agroforestry models on *jhum* land. The NERCORMP and MLIPH projects, financed by the International Fund for Agricultural Development, are striving for *jhum* modification and integrated *jhum* development, primarily to promote more remunerative agro-horticultural crops, or conversion of *jhum* land into community-conserved areas for harvesting NTFPs. NERCORMP has piloted a 'rationalization and optimization' approach to *jhum* clearing, based on available labour for weeding and seeds for

TABLE 18-1: Major externally aided projects in northeast India with implications for shifting cultivation.

Name of projects	Funding agency	States in operation	Period
Nagaland Environment Protection and Economic Development through People's Action (NEPED 1)	ICEF (Canada)	Nagaland	1995-2001
Nagaland Empowerment of People through Economic Development (NEPED 2)	ICEF (Canada)	Nagaland	2001-2006
North Eastern Region Community Resource Management Project for Upland Areas (NERCORMP – I)	IFAD	Assam, Manipur and Meghalaya	1999-2008
North Eastern Region Community Resource Management Project for Upland Areas (NERCORMP – II)	IFAD	Assam, Manipur and Meghalaya	2010-2016
Meghalaya Livelihoods Improvement Project for the Himalayas (MLIPH)	IFAD	Meghalaya	2004-2013
North East Rural Livelihoods Project (NERLP)	World Bank	Mizoram, Nagaland, Sikkim and Tripura	2011-2017
Sustainable Land and Ecosystem Management in Shifting Cultivation Areas of Nagaland for Ecological and Livelihood Security	UNDP	Nagaland	2009-2013
Participatory Natural Resource Management in Tripura	Kfw–GIZ (Germany)	Tripura	2008-2016
Tripura Forest Environment Improvement and Poverty Alleviation Project	JICA (Japan)	Tripura	2008-2016

Source: Compiled by the author from www.necorps.org; www.nerlp.gov.in; www.tripurajica.com; www.tigproject.in; www.in.undp.org

sowing. With technical assistance from the International Centre for Integrated Mountain Development (ICIMOD), NERCORMP has also piloted a 'participatory three-dimensional model' which sets out to achieve more rational and sustainable land-use planning, including the avoidance of certain critical areas, such as water sources, in the selection of *jhum* sites. Similarly, the World Bank-funded North East Rural Livelihoods Project (NERLP), and other projects in Nagaland (funded by the United Nations Development Programme), and in Tripura (funded by KfW-GIZ of Germany and the Japan International Cooperation Agency), all aim to improve areas under shifting cultivation and secure sustainable livelihoods for shifting cultivators.

Impacts of state policies on shifting cultivators

Shifting cultivators have been impacted in a variety of ways by state policies concerning forests, agriculture, horticulture, NTFPs, watersheds, soil conservation and so on – natural-resource management in general – together with laws and regulations banning or restricting shifting cultivation in forest areas. On one hand, shifting cultivators find practically no support for continuing their traditional agriculture because these policies set out to discourage shifting cultivation. On the other hand, a different range of impacts come from expansion of state forest reserves, monoculture plantations of commercial timber species and expanding areas of shifting cultivation land converted to cash crops like rubber, tea and cashew nuts, in the name of rehabilitating degraded *jhum* fallow lands or resettling shifting cultivators. While many shifting cultivators have taken best advantage of government programmes, to emerge from shifting cultivation with improved livelihoods and incomes, there are many others who have yet to gain access to such benefits. There are also instances of unsuccessful experiences and distressed situations arising out of state policies, often with families returning to shifting cultivation.

Policies that create reserved forests and protected areas have led to changes from community control to state control. At the same time, policies that discourage or ban shifting cultivation in reserved forests have resulted in a gradual diminution of customary rights over forest, diluting historical bonds between shifting cultivators and their surrounding forests and bringing a decline in traditional forest conservation and management systems in many shifting cultivation communities. Restriction of shifting cultivation to small and demarcated areas has forced swidden farmers to shorten fallow cycles or to prolong cultivation on designated plots, resulting in further environmental deterioration.

The Scheduled Area and Scheduled Tribes Commission of 1960-1961 (popularly known as the Dhebar Commission) was one of the first to analyse the impacts of forest policies on indigenous communities. It reiterated the importance and value of forests to these communities, while observing the gradual expansion of government authority over forests to the disadvantage of indigenous people and their livelihoods. The Commission also noted that the traditional rights of indigenous communities over forests had changed from 'rights and privilege', in the Forest Policy of 1894, to 'rights and concessions' in the National Forest Policy of 1952. It made the point that the need of swidden communities to practise shifting cultivation in forest areas was an issue the government had to address (Government of India, 1961).

Field assessments have indicated that the impacts of various state policies on shifting cultivators are multi-dimensional, and are felt at individual-household level, at village level and at the level of entire communities. These impacts relate to communities' immediate environment and natural resources, traditional livelihoods and economies, social and cultural traditions, governance, biodiversity and agro-biodiversity, knowledge and skills, adaptation skills to cope with climate change, the status of women and the overall outlook for shifting cultivation. Brief reflections on these impacts are outlined below.

Environmental, livelihood and economic impacts

Expansion of reserved forests, protected-area networks for wildlife sanctuaries and national parks and biosphere reserves in northeast India have led to decreasing per capita availability of community forest land for shifting cultivation. Introduction of cash crops like rubber, cashew nuts and tea by various government programmes in the name of rehabilitating degraded *jhum* land or re-settling shifting cultivators, while aiming for the betterment of shifting cultivators, has resulted in further shrinkage of available land for shifting cultivation. This lack of land, coupled with government programmes aimed at weaning people away from shifting cultivation, has resulted in the per-family requirement for *jhum* land shrinking by as much as 60% in Manipur, and by 22% in Meghalaya (Government of India, 2006). Overall, the per-capita availability of forest land in northeast India has been significantly reduced (Table 18-2) (Poffenberger, 2007).

Meanwhile, with increases in both population and family size, but with decreasing land available to them, shifting cultivators have been forced to intensify their farming practices in an effort to produce more crops from the same land. Competition for land has led to a gradual reduction in fallow periods – the time during which the land is left idle to enable the soil to recover – and in most cases this has been compounded by longer periods of cultivation on the same plots. Fallow periods have been reduced from 12 to 20 years in the 1950s and 1960s to currently average three to five years

TABLE 18-2: Per-capita availability of forest land (hectares).

State	*1981*	*1995*
Arunachal Pradesh	10.88	6.0
Assam	0.13	0.14
Manipur	1.26	0.83
Meghalaya	1.17	0.54
Mizoram	3.68	2.32
Nagaland	1.85	0.71
Tripura	0.26	0.23

Sources: Poffenberger, 2007, p11; Datta Ray and Alam, 2002, p8.

(Table 18-3). Combined, in many cases, with extension of the cropping phase in the same plot from the traditional duration of one year to two to three years, this has seriously disrupted traditional soil-conservation and land–husbandry systems, leading to increasing soil erosion, land degradation, biodiversity depletion and deforestation.

Shifting cultivators have been blamed for all of these negative environmental consequences; in the eyes of most policy-makers, development planners, forest and natural-resource managers, shifting cultivation became a wasteful practice that needed to be done away with immediately. The vicious cycle of the need to continue *jhumming,* shortening fallow periods, extended years of cropping, reducing *jhum* yields, declining land productivity and most of all, the depleting quality of fallow forests and land, has seriously disturbed the very environmental base of shifting cultivators and consequently their livelihoods and economies. The shortening of the *jhum* cycle, as periods of fallow recovery have been reduced, has depleted secondary forests and this has upset local hydrological conditions. Natural springs have dried up and the quality of fallow forests of different ages has suffered. Shifting cultivators have long depended on the health of their fallow forests for NTFPs, including edible leaves, flowers, fruits, roots and tubers and other small trees, bamboo and firewood for households needs and birds and small animals as protein sources. The fabric of traditional shifting cultivation livelihoods has thus been all but destroyed.

The NERCORMP project,[2] supported by the International Fund for Agricultural Development, has found in the West Garo Hills of Meghalaya that many of these impacts can be addressed if shifting cultivators are professionally and technically assisted for 'rationalization and optimization' of *jhum* practices through community-based participatory planning, implementation, monitoring and social auditing of their *jhum* procedures. Prior to NERCORMP's intervention, it was found that almost one-fifth of land cleared for *jhum* was not effectively and productively used, due to

TABLE 18-3: Forest and *jhum* statistics of northeast India.

State	Total forest area(ha)	Unclassed forests(ha)	Annual area under jhum (ha)	% of jhum to total forest area	% of jhum to unclassed forests	Fallow period (years)	Number of jhumia families	Jhum land per family (ha)
Arunachal Pradesh	5,154,000	3,146,600	70,000	1.36	2.22	3–10	54,000	1.29
Assam	2,683,200	896,800	69,600	2.59	7.76	2–10	58,000	1.20
Manipur	1,741,800	11,780	90,000	5.17	7.64	4–7	70,000	1.29
Meghalaya	949,600	837,100	53,000	5.58	6.33	5–7	52,000	1.01
Mizoram	1,671,700	524,000	63,000	0.37	0.12	3–4	50,000	1.26
Nagaland	922,200	862,800	19,000	0.02	2.20	5–8	116,000	0.16
Tripura	629,400	211,700	22,300	0.03	0.10	5–9	43,000	0.51
Total	13,751,900	6,490,780	386,900	15.12	26.37		443,000	

Sources: Forest Survey of India, 2012; Government of India, 2008.

a lack of seeds for sowing or lack of labour for weeding as the crops grew. Moreover, clearings for *jhumming* were sited haphazardly due to the shortage of land and forest-fire prevention measures were minimal. 'Rationalization and optimization' of *jhum* meant essentially that clearing of land was based on an assessment of the availability of labour for weeding and crop seeds for sowing. It also meant observance of the Garo Hills *Jhum* Regulations 1954, which required that 50m of forest be preserved alongside rivers or streams, and that measures be taken to prevent forest fires. Shifting cultivators were also encouraged to exchange seeds with others. Within three years of these interventions, case-study sites in the West Garo Hills showed that 70% of natural springs had been rejuvenated, damage to forests due to *jhum* fires escaping became negligible, *jhum* areas were reduced by 30% to 40% annually, the *jhum* cycle was increased from seven to nine years, and shifting cultivators reported increased crop yields (NERCORMP, 2008). It must be mentioned here that the project also provided additional inputs in the form of livestock (mainly pigs) and home-garden developments. A revolving fund was also established for women's Self-Help Groups and village-level Natural Resource Management Groups. Households could borrow from this fund and repay according to conditions decided by the villagers. Added to these interventions were systematic capacity building and awareness education.

Monoculture plantations promoted under different government programmes have also had different impacts among shifting cultivators. Rubber plantation, which was promoted in Tripura by the Tripura Forest Development Plantation Corporation and the Tripura Rehabilitation Plantation Corporation for rehabilitation of shifting cultivators, reported positive results. However, a citrus rejuvenation programme funded by the North Eastern Council in West Garo Hills, Meghalaya, in 2004 was a disaster, according to a study by Teegalapalli (2008). Under this plan, many shifting cultivators abandoned *jhum* and established orange orchards, but within a year, as many as 51 households in just one study village had returned to traditional *jhum* cultivation. Staff of an IFAD-funded project working in the same district found that the

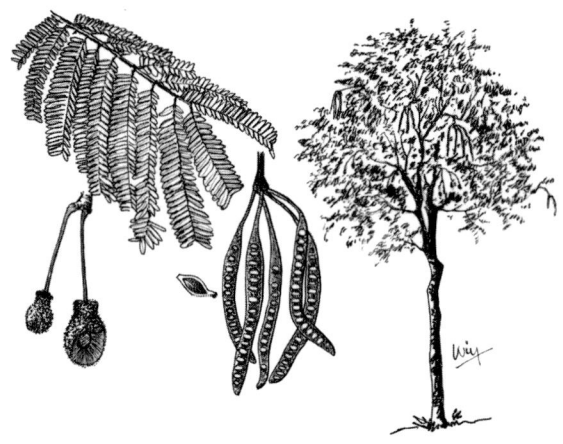

Parkia roxburghii G. Don [Leguminosae] Syn. of Parkia timoriana (DC.) Merr.

The multi-purpose Tree Bean, popular in swidden fallows in northeast India for its edible pods and seeds, its timber, and its soil-improvement, medicinal and insecticidal properties.

programme failed because a lack of timely rain resulted in a very low tree-survival rate. There was also a lack of extension services by programme staff and a lack of support for alternative livelihoods needed by the shifting cultivators while they waited for their orchards to mature. Shifting cultivation, with its characteristic multi-layered and multi-cropping systems, offering both above-ground crops and roots and tubers, ensured some level of food security to the stricken families, even in such an adverse situation.

The impacts of monoculture plantations on shifting cultivators – particularly the negative impacts – have been widely debated in India (Saxena, 1999). The end of the 1980s and the early 1990s saw widespread promotion of monoculture plantations, particularly of teak *(Tectona grandis)* and *gamari (Gmelina arborea),* in many shifting cultivation areas of northeast India, aimed at improving the income of the people. Later field assessments of farmers' benefits from these plantations in Mizoram and Nagaland, made during a preliminary scoping mission for the North East Rural Livelihood Project in 2009, showed that the farmers – most of them shifting cultivators – never received the income they expected from the plantations due to regulations on harvesting, transportation and sale of their trees. Moreover, indigenous people saw little value in the species chosen for plantation. Citing an example from central India, Saxena (1999) observed:

> [Industrial monoculture] plantation offers little of the product range of the old forests. For instance, *mahua (Madhuca longifolia)* is of no significance to the Forest Department, nor have any efforts been made to increase its number in forests. No doubt, *mahua* is also not felled by the Department, but its significance for them is not the same as for tribals. Compared with the Forest Department, tribal involvement in *mahua* is pervasive to a profound degree. In addition to collecting flowers and seeds for sale at the weekly market, or for exchange for salt or cloth, tribals use the wood to support the canopy at wedding celebrations, the dried flower to add bulk to their food or to feed their animals, the seeds and the flowers for preparing liquor and for religious ceremonies.

Social, cultural and governance impacts

Traditionally, the social, cultural and governance systems of the indigenous communities of northeast India were built around the practices and cycles of shifting cultivation. This evolved over many generations because the people had a common need for shifting cultivation and consequently a need to protect and manage both the natural resources of their villages and the well-being of surrounding forests. Social systems evolved to ensure justice and equity in apportioning land for *jhum* to every household. Traditional village institutions, acting on behalf of the community, governed and administered community-controlled forests and other natural resources as custodians and guardians. Lifestyles and livelihoods that evolved around shifting cultivation not only made the communities cohesive, but imbued them with an

FIGURE 18-5: Women in Ukhrul district, Manipur, gather NTFPs as they tend their livestock.

Photo: Vincent Darlong.

overwhelming attitude of action for the common good. The Mizos of Mizoram[3] even have a traditional social ethic called *tlawmngaihna*, which means, literally, 'service above self'. For the Mizos and many similar communities, *tlawmngaihna* is a code of social conduct that compels individuals to be seen as performing good and beneficial deeds, particularly in their dealings with others, without expecting anything in return. It also involves putting the common good of society or the community above personal interests. Similarly, the cultural milieu of indigenous communities has evolved around the practices of shifting cultivation, as depicted in the traditional festivals, dances and songs of communities whose history is closely inter-woven with swiddening.

With the decline in shifting cultivation and many shifting cultivators adopting alternative livelihood systems, the traditional social and cultural fabric of indigenous communities has been significantly distorted. Social-value systems of equity and justice for land access and ownership by all households in a village have changed, with traditional common or community ownership leaning more and more towards individual or privatized title. This privatization and fragmentation of common property resources, particularly as it affects community forests, is regarded as the single most significant change impacting on the lives of shifting cultivators. While, in many cases, the poorest households generally have smaller shares of land, better-off families generally possess larger shares of land because they can make investments that justify their occupation of community land by cultivating cash crops and horticultural crops. This has made the availability of land shrink even further, adversely affecting those families who, either by choice or without any options, continue to practise *jhum*. The consequences of these changes are linked to an increasing incidence of conflicts and disputes over land between households and between neighbouring villages, over traditional village boundaries. Conflicts remain unresolved, particularly where traditional village social-governance systems have become weak or partisan due to the influence of modernization. Leaders are often no longer chosen on the basis of popular consensus, so they may not be the most respected individuals.

Displacement is another negative consequence of state development policies. The commissioning in the 1960s of the 8.6MW Gumti Hydel Project in Tripura submerged 46.34sq. km of land. Officially, the inundation displaced 2558 tribal families. They had land deeds, so were compensated as 'oustees'. About double that number – between 5500 and 6500 *jhum*-dependent families – were ignored because they had no land deeds, and were simply abandoned. According to Fernandes (2004), the only alternative these families had was to practise *jhum* in the catchment area of the new dam, and they were labelled enemies of nature because of their effects on the environment. Interestingly, Fernandes met and interacted with some second-generation descendants of families displaced by the Gumti Hydel project in 2004, in the Gandacherra areas of Dhalai district in Tripura. Many of these families remained landless, poor and dependent on shifting cultivation. The meeting was about two years before the Forest Rights Act came into force, so their occupation of any kind of forest land was seen as illegal.

Shifting cultivators have also felt other kinds of impacts from policy decisions, particularly when land acquisition was involved for development works. The broad-gauge railway-expansion works between Lumding and Silchar, through Dima Hasao district in Assam, affected large areas traditionally used by shifting cultivators. Although the affected people were paid compensation, many families reported hardship due to loss of orchards, gardens and fallow forests, from which they gathered NTFPs to supplement their food and income.[4]

A noticeable social impact of state policies has been the increased visibility of 'inequality' in the otherwise homogenous society of shifting cultivators. Gaps are widening between the poor and the better off because of inequality of opportunity. For example, access to government schemes for promotion of cash crops or horticultural crops is often taken by elite members of indigenous communities, while the illiterate poor have no idea of how to access such opportunities. Consequently, privatization of common village land, through the planting of cash crops or horticulture crops, has in many cases benefited elite families more than the poorer or poorest among shifting cultivators.

Impacts on community forestry stewardship and fire management

State policies have affected shifting cultivators perhaps most visibly in the changes they have inflicted upon traditional community-forestry stewardship and governance systems. In many places, community forests have become fragmented due to privatization and the weakening of community leadership through the exercise of vested interests. Some elements of control over resources have also been unclear. Even Autonomous District Councils, which are mandated to safeguard customary and traditional practices, have levied transit taxes when resources extracted from community forests are transported from place to place. Such systems have led to a decline in the traditional bonds between communities and their forests. Increasingly, forests have been torn between community and government control, in both a physical sense and, perhaps more importantly, in the mindset of the people.

State ownership of forests, coupled with the consequent centralization of forest governance, has alienated the people from their forest land. The result of this forest alienation has been an increasing incidence of local people surreptitiously collecting as much as they can of their needs from government-controlled forests – as long as they are able to get away with it. A similar situation is that in which local people have been employed to illegally remove timber from government-controlled forests. This unplanned and uncontrolled removal of forest resources has caused greater damage to forests. At the same time, local communities and shifting cultivators have long been suspects in the eyes of the authorities, leading to increasing conflicts between the Forest Department and local communities. There have also been instances of heightened tension when communities have refused to recognise government notifications regarding the creation of reserved forests. For example, local communities in the Upper Subansiri district of Arunachal Pradesh refused to allow Forest Department staff into the Langsa reserved forest (covering 37sq. km, and notified in 1984) and the Singen reserved forest (covering 96sq. km, also notified in 1984). The people claimed both forests as their traditional reserves in spite of official notification that they had become state reserved forests (Kemp, 2012).

Another outcome of policy-making has been a decline in community-based fire management practices. Many communities with extensive areas under shifting cultivation in northeast India have always had effective community-based fire management strategies and systems. Burning of slashed vegetation for *jhum* has always been a community affair in any typical village with many shifting cultivators. Preparations included the clearing of fire-lines, notification of burning dates and times, bans on any person travelling outside the village during the burning period and availability of every person in the village for community action in the event of fire spreading outside the *jhum* areas. With the decline in shifting cultivation due to state policies, community-based fire-management strategies and practices have fallen into disregard. Frequent forest fires and the extent of fire damage to forest areas in northeast India could be indicators of growing disfavour for such practices (Darlong and Vanchhong, 2001).

Agro-biodiversity impacts

Traditionally, shifting cultivators have been custodians, managers and conservators of a rich agricultural and forest biodiversity. They also sustained a rich agro-biodiversity in their *jhum* fields and home gardens, by conserving and propagating many species of cereals, pulses, oilseeds, vegetables, spices, cucurbits, cotton, roots and tuber crops. Their fallow lands, with varying ages of vegetation succession, as well as their pristine community forests, harboured a wealth of plant and animal life. State policies seeking to replace *jhum* have seriously undermined the biodiversity resources of shifting cultivators, particularly their agro-biodiversity. The introduction of monoculture tree farming (teak, *gamari*, and so on, as in Nagaland), cash cropping (rubber, tea, cashew nuts, areca nuts, and so on, as in the *jhum* lands of Assam, Meghalaya, Tripura and Mizoram), and the agricultural policy of introducing high-yielding rice varieties into *jhum* fields to replace traditional varieties, have all impacted on shifting cultivators. Even without these introductions, the traditional crop varieties and biodiversity of crops available to shifting cultivators have been dwindling in an environment of soil nutrient and biotic-quality loss because of farmers' needs to intensify their production by extending years of cropping and reducing fallow periods. In this changing *jhum* environment, it appears that only a limited number of crop varieties are responding to the needs of shifting cultivators, to intensify and lift their production on the same area of land. Crops that do not grow well in the poorer conditions fail to provide seed for the following season, and are eliminated. Much research is now needed to help upgrade the quality of seeds for *jhum* crops.

During field assessments for the NEPED project in Nagaland in 2002 and the NERCORMP project in Assam and Meghalaya in 2008 and 2012, shifting cultivators reported a progressive reduction of crop varieties in their *jhum* fields (Table 18-4). These reductions in crop varieties were more significant among shifting cultivators who lived in villages closer to roads and markets; remote villages had retained a greater diversity of crops. Shifting cultivators attributed the changes to a variety of influences, including non-availability of seeds, market demands for certain *jhum* crop varieties, reduction in *jhum* area due to land and labour constraints, and replacing *jhum* with tree farming and cash-crop plantations.

Other causes of shrinking *jhum*-crop diversity are the consequences of state policies to promote horticulture and livestock rearing among shifting cultivators. For example, piggery and poultry promotion in Nagaland has encouraged large-scale cultivation in *jhum* fields of tapioca for pig feed and maize for poultry feed, with an adverse effect on *jhum*-crop diversity. In recent years, government departments have been promoting cabbage and potato cultivation in *jhums*, because of high market demand, bringing another assault on crop diversity (NEPED, 2011). A recent assessment by an IFAD-supported project revealed that although such interventions had brought an increase in cash income, the nutritional security of many families who had just left shifting cultivation could be at risk because of their bulk consumption of carbohydrates (rice) and limited vegetables, along with lifestyle changes. The project has therefore directed its efforts towards promoting home gardens and community-based biodiversity conservation for wild edible plants and *jhum* vegetables.

TABLE 18-4: Changes in the number of crops planted in *jhum* fields in northeast India.

State	*Number of crop varieties*		
	1960s to 1980s	*1990s to 2000s*	*present*
Nagaland	40–60	22–36	12–24*
Assam	36–56	24–36	14–22
Meghalaya	30–50	20–32	10–20

Sources: Field data through key informants from selected villages during NEPED project assessment in 2002 for Nagaland and NERCORMP project assessment in 2008 and 2012 for Assam and Meghalaya. *Nagaland data for this period is from personal communication with Mr Vengota Nakro, of NEPED, on 14 April 2012, but relates mostly to villages with roads and market access. According to a recent NEPED survey, interior villages still have up to 41 crop varieties planted in *jhum* fields. However, in Tuensang district, many farmers cultivate only a single variety of Kolar or Rajma bean on their *jhum* land.

Of all the *jhum* crop varieties, perhaps the crop suffering most from policy-related change has been rice. Cultivation of rice remains the main objective of most *jhum* farming, but there have been rapid changes in the genetic erosion of rice germplasm and reduction in the number of varieties. While shifting cultivation has been the principal means of propagating and conserving upland rice germplasm and maintaining the large number of varieties, Sharma and Hore (1990) say that in Tripura alone a record of 42 indigenous varieties in 1986 fell to 32 by 1990 – a loss to that state's *jhum* farmers of 10 varieties in just four years. The primary reason for this genetic erosion was the modernization of agriculture with the introduction of high-yielding varieties to meet the needs of a growing population. According to the same study, while rice production increased, landraces were lost at rates varying from 12.5% to more than 73% in 1990.

Shifting cultivators have also reported adverse impacts on biodiversity in their fallow land, community forests and village reserve forests. The quality of vegetation and hence the biodiversity in secondary-forest fallows has been significantly depleted, particularly in those fallows where the *jhum* cycle has been shortened to just three to five years. In Karbi Anglong district of Assam, shifting cultivators with a *jhum* cycle of only three to five years reported that their fallows were being dominated by weeds (*Eupatorium* sp.) and small-sized bamboo (mostly *Melocanna* sp.), making the land unproductive and making the job of clearing the dense *Eupatorium* a much longer task.[5] Elsewhere, particularly in West Khasi Hills, Meghalaya, community forests have been harvested to make charcoal, which has then been sold locally at a much more profitable rate than timber (NERCORMP-KCRMS, 2008). This situation arose because sales of charcoal were not restricted, while a whole range of activities for harvesting, processing, transportation and sale of timber were regulated by existing government policies.

Knowledge and skill impacts

As the lifestyles, livelihoods and landscapes of shifting cultivators have been influenced by state policies, their indigenous knowledge and skills have suffered in many ways. Older generations of shifting cultivators once held a significant fund of indigenous knowledge related to land husbandry, cropping technologies, weed and seed management and post-harvest crop management, including, for instance, the ratios of various seeds to be mixed for sowing in *jhum*

FIGURE 18-6:   Women sowing seeds in a *jhum* field in Wokha district, Nagaland.

Photo: Vincent Darlong.

fields – a calculation that depended on soil texture, the slope of the land, its aspects and sequential harvesting. They were also skilled in making animal traps to protect their crops from foraging animals and birds, while at the same time providing meat for the table. They improvised their own tools and implements for dibbling, weeding and harvesting and they wove bamboo baskets for storage of seeds and carrying or transporting crops. They knew the different kinds of bamboo needed for making different types of baskets and mats. The women brewed different beverages from rice, millets and maize to be used in religious rituals and at social events. The women were also experts in preparing food from the *jhum* fields, and weaving colourful traditional cloth. They also knew the kinds of herbs and shrubs to be consumed according to the seasons of the year. Above all, both men and women knew the names of all the various crops and the seasons for planting and harvesting, based on generations of field experience and learning.

Indigenous knowledge associated with shifting cultivation also encompassed the various cultural milieus, rituals and social events interwoven with the seasonal agricultural cycle. Shifting cultivators observed spring festivals (after successfully sowing seeds in the *jhums*), autumn (harvest) festivals and winter festivals, where merry-making helped the selection of sites for next season's *jhums*. Rituals and offerings were associated with major events in the *jhum* cycle, such as site selection, slashing of vegetation, making the fire lines and burning the dried biomass, seed exchange and preparation of seeds for sowing, weeding, guarding against intrusion by animals, harvesting, threshing of *jhum* paddy and storage of seeds for the next season. Each occasion was marked with merry-making, celebration and telling of stories that had been passed down from generation to generation. These were the processes by which community-based knowledge was passed from parents to children and from elders to younger generations in the many shifting cultivation communities in northeast India.

The overall decline of shifting cultivation has brought with it degeneration, if not a loss, of knowledge and skills associated with swiddening. The range of local vocabularies, terminologies and phrases associated with the cycles and events of shifting cultivation are gradually disappearing; so also are the songs, dances, jokes and legends associated with various events in the *jhum* cycle. New generations of shifting cultivators, and those who are practising 'modified' *jhum*, are no longer able to appreciate or value much of the traditional knowledge and skills of traditional practitioners. Should traditional shifting cultivation cease altogether, the consequent loss of priceless knowledge and skills learned over generations – as a negative dimension of state policies – would be immense and irreparable.

Meanwhile, the introduction of new livelihood strategies by way of promotion of cash-crop plantations in *jhum* land is posing new challenges for traditional shifting cultivators, requiring that they adopt new technologies and skills. While many have been able to quickly take advantage of government policies and programmes aimed at replacing shifting cultivation, others have been slow to adjust. In the absence of adequate and appropriate technical and extension support, the need to learn new technologies and skills and cope with changing livelihood strategies could lead many families back into poverty, or at least to return to shifting cultivation. A recent survey by the NEPED project in Nagaland showed that of 116 villages, 65% had reduced shifting cultivation and promoted cash-crop plantations, while 12% of villages had actually increased their areas under shifting cultivation (Nakro, 2012). Some village councils in Nagaland are even encouraging families to continue shifting cultivation in designated areas so as not to lose the traditional knowledge and skills of upland farming, and also to conserve the rich variety of crops, including chillies, vegetables, maize, cucumbers, millets, roots and tuber crops, that the communities have been cultivating for countless years. The increasing popularity of the Slow Food Movement in Nagaland and in northeast Indian generally has brought an eagerness for the return of native food-crop varieties and local cuisines among many shifting cultivators.

Through their indigenous institutions, shifting cultivators in Arunachal Pradesh, Assam, Manipur, Meghalaya, Nagaland and Mizoram have generated a body of knowledge related to the management of natural resources – particularly of forests – for apportioning shifting cultivation land. For example, in Mizoram, forest lands classified as 'safety' and 'supply' reserves were managed by the village council; *anchal* reserves were created in Arunachal Pradesh and placed under the management of the *Anchal Samitis* (village committees). In Meghalaya, various community forests (*raid land, law kyntang, law niam, law ri sumar,* and so on) were officially acknowledged and operated under different community bodies with varying jurisdictions (Poffenberger, 2007). Although this is poorly documented, it is understood that these community bodies gathered a wealth of indigenous forest-management knowledge. Many of the relict forests can still be seen, particularly in Meghalaya, as sacred groves. The Mawphlang sacred grove in East Khasi Hills in Meghalaya is one such relict forest. It is at least 600 years old, and has survived various biotic and human-society pressures to remain a pristine forest, thanks to the indigenous management knowledge and skills of local communities, led by their customary institutions.

Indigenous stewardship of community forests, along with the associated body of knowledge in traditional forest-management practices, is under serious threat. In spite of having safeguards under the Sixth Schedule and Article 371 of India's Constitution, the institutions managing community forests, upon which villagers depend for shifting cultivation and other livelihood activities, are all being transformed, and along with them the knowledge systems used in this management. The effects of government policies and programmes, coupled with growing demands for timber, cultural changes disrupting institutional mechanisms, privatization of community land, demographic expansion, growing internal demands for expansion of 'modern' agriculture and cash-crop plantations and competition from migrant families are all combining to adversely affect shifting cultivators and their resource bases. The weakening of traditional village institutions is also resulting in the degeneration of indigenous-knowledge systems related to natural-resource management, and this is an area in which shifting cultivators once held an advantage.

Gender impacts

Field assessments during participatory rural appraisal exercises for the NERCORMP project in 2011 revealed that perceptions of adverse policy impacts were felt more acutely by women shifting cultivators than by their male counterparts. In typical communities dependent upon shifting cultivation, women have always spent much more time in the *jhum* fields than the men. A recent study in Tripura showed that women farmers typically attended to about 75% of the work related to shifting cultivation (Darlong et al., 2012). Interestingly, although the women spent much longer working in the *jhum* fields, most did not consider it 'a burden'. There is no strict division of labour between men and women in shifting cultivation. While the tasks carried out by women include sowing seeds by dibbling, weeding, watching for birds and harvesting crops, the men generally perform jobs that are physically more demanding and strenuous, such as cutting and clearing vegetation, preparing fire lines and burning slash, construction of *jhum* huts and carrying harvested crops from the *jhum* to the village. Weeding in *jhum* fields is a continuous process. In one annual cropping period, weeding is generally done two or three times. Weeds compete profusely with agricultural crops in the *jhum* during the monsoon months of June to September. In a case study in Tripura, while the women undertook the weeding, the men were collecting bamboo from fallow forests for sale in local markets for additional cash income (Darlong et al., 2012). Even when cash crops were introduced into *jhum* fields, the weeding work and protection of the fields from grazing animals was mostly done by the women.

Promotion of monoculture crops to replace *jhum* appears to affect women shifting cultivators more seriously than the men. Monoculture tree plantations are often termed 'green deserts' by environmental activists because of their adverse effects on the local environment, biodiversity and the livelihoods of forest dwellers (Hance, 2008). To maintain an adequate livelihood, shifting cultivators need leaves, fruits, flowers and many other NTFPs from forests. Monoculture tree plantations do not provide for these needs. Therefore, the replacement of *jhum* with monoculture tree farming, along with decreasing areas for *jhum* and reducing fallow forests, adversely affects women shifting cultivators, who must gather wild vegetables and other NTFPs, including roots and tubers, for their families. Bamboo shoots have practically disappeared from many traditional shifting cultivation areas because bamboo forests have been replaced with cash crops like pineapples, bananas, tea, and so on. Secondary forests that are fallow lands in various stages of regrowth have a diversity of vegetation that has traditionally provided firewood, wild vegetables, roots and tubers and feed for livestock. Families who have taken up horticultural and cash-crop plantations in their *jhum* fields must depend on other natural resources for their firewood and wild vegetables. IFAD-funded projects in northeast India that promoted monoculture (high-value timber species) and cash-crop plantations judiciously planned to also promote protection and maintenance of agroforests, mixed forests, home gardens and even forests specifically for gathering NTFPs. Such a balanced approach reduces negative impacts and ensures both economic and environmental security at a local level.

Impacts on emerging social and environmental activism

Amid the positive and negative effects of state policies on swiddening communities in northeast India, a new social and environmental activism has emerged. The bitter experiences of environmental and ecological degradation, scarcity of drinking water, biodiversity depletion and declining wildlife, coupled with increased awareness of the need for environmental conservation, have led many shifting cultivators to take up the cause of environmental protection through their traditional village institutions, youth groups and other non-governmental organizations. Externally aided projects in northeast India have also focused on community-based biodiversity and forest conservation. The IFAD-funded NERCORMP project promoted community-based biodiversity conservation and succeeded in having more than 1860sq. km of secondary forest fallows protected as community conserved areas in more than 660 villages in Assam, Manipur and Meghalaya (NERCORMP, 2008). In Mizoram, the Young Mizo Association has been in the forefront of forest and wildlife conservation since the 1990s, creating reserved forests in many shifting cultivation-based villages. In recognition of its work in conservation of natural resources, the Young Mizo Association was awarded the Government of India's Indira Gandhi *Paryavaran Puruskar* in 1993 (Government of India, 2011a).

Conclusions

The shifting cultivators of northeast India are witnessing the gradual demise of *jhum*. To these people, *jhum* is both forest management and an agricultural practice. However, neither the contemporary mainstream agricultural policy nor the forest policy in India recognize shifting cultivation as either conventional agriculture or forestry. If neither is the case, opponents ask, then how does one explain the large variety of food crops, including cereals and legumes, fibres, spices, cucurbits, oil seeds, vegetables, roots and tuber crops that shifting cultivators have conserved, managed and perpetuated in their *jhum* fields for countless generations? Equally, how does one ignore the large tracts of fallow forests that shifting cultivators have long sustained with rich biodiversity and environmental services? Arguments against *jhum* have focused on its overwhelming environmental consequences, along with the social consequences of dwindling *jhum* harvests. This latter aspect has been more pronounced over the past few decades in northeast India, due to progressive reduction in fallow periods, once again influenced by a variety of factors, not least among them state policies on natural resources. The tilt against shifting cultivation has been aggravated in recent times by the transformation of traditional, sustainable *jhum* – with reasonable and adequate fallow periods of at least 10 years or more – to 'modified' *jhum*, in which fallow periods tumbled to four or five years in the desperate effort to intensify production from exhausted soil.

Communities of shifting cultivators who have never known any other way of life but *jhum* are in different stages of transition. Caught between various state policies of non-support for *jhum*, while drawn by compelling factors forcing it to change, many of these people endure a life of dilemma and uncertainty. It is a fact that a large number of shifting cultivators and their families have benefited from state policies, particularly those promoting agroforestry, cash cropping and horticulture plantations, to replace *jhum*. These programmes have made significant improvements to the livelihoods of people who no longer depend on shifting cultivation. However, the NEPED project in Nagaland introduced tree farming and agroforestry (cardamom with trees) to show that *jhum* fields could be sustained by innovative land-use strategies that were both acceptable to environmentalists and profitable for farmers (Karunakaran, 2008). This clearly demonstrated that the state, shifting cultivators and the environment could all benefit if appropriate extension and hand-holding services were made available to shifting cultivators.

Experiences gained while working on these issues in northeast India have led inexorably to the belief that state policies are not alone in bringing an end to shifting cultivation. There are various other social, political, ecological and economic factors – both external and internal – that are forcing changes in shifting cultivation, not only in northeast India, but also elsewhere, such as in Southeast Asia, where shifting cultivation was once a prominent livelihood pursuit (Schmidt-Vogt et al., 2009). A modified participatory rural appraisal exercise carried out by the author during review missions in the states of Assam, Meghalaya and Manipur in 2011 and 2012 for IFAD-funded projects showed that more than 92% of young men and women from traditional shifting cultivation families would prefer not to practise *jhum* if they had other alternatives. Only about 8% said that they would still practise *jhum*, even if they had other alternatives. This minority group said *jhum* was a practical farming system best suited to their landscapes; it taught them many practical lessons and gave them the opportunity to remain connected to their forefathers. What's more, food grown in *jhum* fields was different in both flavour and quality.

Transformation of *jhum* is happening very rapidly in communities that now enjoy all-weather road access, including access to credit and markets. Most remote villages and communities with neither roads nor access to markets remain dependent on shifting cultivation, although often not wholly so. In either situation, the influence of state policies and other factors on shifting cultivators are varied. Table 18-5 shows the results of a rapid survey made during a supervision mission in March and April, 2012, for IFAD-funded projects in Assam and Meghalaya. It sought to determine the percentage of shifting cultivators who perceived the effect of various clearly identified impacts on their lifestyle and livelihoods.

Although shifting cultivators are blamed for causing environmental degradation in upland areas, the products of *jhumming* are generally considered to be superior by many people in downstream communities, particularly those who live in small cities and towns in the vicinity of shifting cultivation communities. With increasing demands for clean and nutritious food and awareness of health concerns, *jhum* products are known to be chemical and pesticide-free and are therefore often regarded as superior to vegetables grown in plains areas, where chemical fertilizers and pesticides are used extensively. This could perhaps gradually influence the negative image of *jhum*; a wider acceptance could come from the 'organic' nature of *jhum* and its recognition as a production practice that is chemical and pesticide free.

The global problem of climate change has also drawn shifting cultivation to the attention of rural-development practitioners and agricultural scientists. While many opine that shifting cultivators could be among the worst hit by climate change, their practice of growing many crops in a single plot (at times more than 30 or 40 varieties)

TABLE 18-5: Percentage of shifting cultivators perceiving different impacts.

Impact categories	Perceived by % of shifting cultivators		
	Women	Men	Youth
Shrinking availability of land for *jhum* cultivation	100	100	100
Decreasing size of *jhum* plots per household	100	100	100
Shortening *jhum* cycle	100	100	100
Declining *jhum* productivity and yield	100	100	100
Decreasing crop varieties in *jhum* fields	100	100	100
Declining quality of *jhum* fallow forests	100	100	100
Declining availability of edible plants in fallow forests	100	100	100
Decrease in traditional crop varieties in *jhum*	100	100	100
Decline in overall traditional food basket	100	100	100
Increasing food insecurity	40	30	40
Increasing environmental insecurity (water problem)	25	25	30
Increasing environmental insecurity (firewood problem)	30	30	40
Increasing environmental insecurity (construction materials problems)	30	40	20
Increasing environmental insecurity (problems of soil erosion and soil fertility)	40	50	30
Increasing environmental insecurity (pest problems)	20	10	5
Overall decline in knowledge on traditional *jhum* systems and crop varieties	70	80	80
Changes in community-forest governance systems	20	50	30
Privatization of community forests	90	95	80
Forest-land alienation (sense of disconnect of communities with forest)	40	50	50
Decline in community-based fire-management practices	70	80	90
Increase in inter-village conflicts	10	12	8
Increase in inter-household conflicts	5	5	0
Increase in inter-tribal community conflicts	20	30	10
Increase in conflicts with forest department	20	25	10
Increased adoption of new or diversified livelihoods (other than *jhum*)	80	90	90
Increased *jhum* areas planted with cash crops (tea, broom grass, areca nut)	40	50	70

TABLE 18-5 (cont.): Percentage of shifting cultivators perceiving different impacts.

Impact categories	Perceived by % of shifting cultivators		
	Women	Men	Youth
Increased *jhum* area planted with horticulture crops (oranges, bananas, pineapples)	60	80	70
Increase *jhum* area with tree farming (monoculture trees)	10	20	20
Increased *jhum* areas converted to agroforestry (mixed trees, fruit trees and vegetables)	40	40	40
Increase in drudgery for women and men due to diversified livelihoods	30	45	40
Difficulties due to new crops replacing *jhum*	45	50	30
Inadequate knowledge and skill for new plantation crops	25	20	40
Decline of knowledge about *jhum* practices among youth	50	70	50
Decline in knowledge of culture, songs and dances connected with *jhum*	80	90	90
Decline in number of youth wanting to practise *jhum*	85	95	95

Notes: Rapid field assessment from supervision mission for IFAD-funded NERCORMP and MLIPH projects in Assam and Meghalaya during March and April 2012; average from total sample sizes of 120 women, 110 men and 60 young people from 12 villages of Karbi Anglong and Dima Hasao districts in Assam and the East, South and West Garo Hills districts in Meghalaya.

could actually be one of the best adaptation strategies. Debating the issues affecting shifting cultivation in the Eastern Himalayas, researchers from the International Centre for Integrated Mountain Development (ICIMOD) have called for recognition of farmers' innovations in shifting cultivation areas as a basis for policy dialogue (Kerkhoff and Sharma, 2006). Raman (2000) has pleaded for a balanced approach to development that recognizes the merits of *jhum*, because it plays such an important cultural role in local customs, traditions and practices, as well as offering economic security to farmers in northeast India. The noted French anthropologist Georges Condominas, while delivering a keynote address at a conference entitled 'The Demise of Swidden in Southeast Asia?' in Hanoi in 2008, remarked:

> They (swidden cultivators) cannot be eliminated in a systematic, authoritarian way because their many and varied models are a testimony to the creative richness of humans as social beings. Any human experience that seeks to improve living conditions deserves to be studied very closely in the search for balanced solutions (Condominas, 2009).

While state policies in India and in most other countries discourage or even ban shifting cultivation, a recent contribution to the debate from Roy et al. (2012), presented at the 11th Session of the United Nations Permanent Forum on Indigenous

Issues in New York in May 2012, outlined the importance of shifting cultivation and the socio-cultural integrity of indigenous peoples from the perspectives of their economic, civil and political rights. The study referred to the importance of various traditions, practices and usages of shifting cultivation to the maintenance and protection of indigenous people's socio-cultural integrity, including their identity as distinct peoples, their spirituality, history, traditions, democratic decision-making norms, social

FIGURE 18-7: A former swidden in West Garo Hills, Meghalaya, eight years after conversion to an areca nut plantation.

Photo: Vincent Darlong.

unity, literature, music and dances. The study noted how shifting cultivation was closely related to forest protection, sustainable forest management, the protection of watersheds and the conservation and maintenance of biological and linguistic diversity. It concluded that practising shifting cultivation was a right of indigenous peoples and, as such, needed to be maintained, strengthened and promoted in its sustainable forms (Roy et al., 2012).

Traditionally, shifting cultivation was sustainable because, at any given time, the people cleared only about 20% of their land for *jhum* and the remaining 80% was kept as forest fallows of different ages, allowing the recovery of both the forest and the fertility of the soil. The number of families in the villages was maintained, in line with the availability of land for cultivation. Social and natural-resource governance was exercised by traditional village institutions. Using a participatory three-dimensional model of landscape mapping of a typical shifting cultivation-dominated village, the IFAD-funded NERCORMP project in northeast India attempted to understand a village-level land-cover land-use plan. In consultation with the communities, and mindful of shifting cultivators' indigenous-knowledge systems related to natural-resource management, it was concluded that one of the most practical options was to maintain the village-level land-cover land-use more or less under three broad categories: one-third of the area conserved by the community for protection of critical areas such as water sources and biodiversity-rich pockets, with strict regulations for resource extraction; another one-third as a multi-purpose forest area for community use, with a broad system of regulation for NTFP extraction and other forest-based income-generating activities; and the final one-third to be used for agriculture, horticulture, *jhum*, human habitations, and so on. Initial results from experimental villages in West Garo Hills, Meghalaya, have shown promising opportunities. The key lesson was the need to maintain a healthy land-to-family ratio of about five to seven hectares per family of upland areas. The challenge then was maintaining and sustaining a minimum *jhum* cycle of 10 years, through strong institutional and extension support and involving other external assistance and services for alternative livelihoods and income-generating activities. Only then could *jhum* be sustained, the adverse impacts of state policies cushioned, and the positive impacts more gainfully absorbed.

It is acknowledged, even by shifting cultivators themselves, that *jhum* in its present form cannot be sustained. They also understand that state policies are in some way responsible for the unsustainable practices that they have no option but to pursue. The governments of India and the states of the northeastern region have put years of effort into addressing the problems of shifting cultivation and rehabilitating shifting cultivators, through policy, institutional and programme support (e.g. Government of India, 1981; North Eastern Council, 2008). In spite of this, it is an acknowledged fact that the practice of shifting cultivation will remain in northeast India – for a while. Intensive engagements with shifting cultivators through IFAD-funded projects, as well as evidence from a number of government programmes relating to shifting cultivation, provide ample proof that shifting cultivators are looking for transformation to something more practical and sustainable. However, effective delivery of the packages of changes over the last mile in the path towards acceptance remains the key challenge. It is a sincere conviction of many working in this field that the good intentions of government, for social, economic and financial inclusion of shifting cultivators by way of policy development and programme implementation, can best be achieved through the principle of free prior informed consent. Shifting cultivators need to be taken into confidence and become a part of all the policy-development, programme-implementation and decision-making processes of transformation and change.

Acknowledgements

I would like to thank Dr Malcolm Cairns for his persistence in following up and trusting me to complete this essay. I would also like to thank my friends and colleagues in the IFAD-funded projects in northeast India, particularly the Managing Director of NERCORMP, Mr L. Baite, and the Project Director of MLIPH, Mr Daniel Ingty, for bearing with me during their project-review processes so that I could collect additional information for use in this chapter. The contribution of my friend, Mr Vengota Nakro, of NEPED, in providing updated information on the work of his project, is also gratefully acknowledged.

References

Bipin Behari (2006) 'Status of bamboo in India', in *Compilation of Papers for Preparation of National Status Report on Forests and Forestry in India,* Ministry of Environment and Forests, New Delhi, pp109-120

Condominas, G. (2009) 'Anthropological reflections on swidden change in Southeast Asia', *Human Ecology* 37, pp265-267

Darlong, V. (2004) *To Jhum or Not to* Jhum: *Policy Perspectives on Shifting Cultivation,* The Mission Link, Guwahati, Assam, India

Darlong, V. T. and Vanchhong, R. (2001) 'Community-based fire management: Tools and techniques of the Mizos in India', in IFAD, IDRC, CIIFAD, ICRAF and IIRR, *Shifting Cultivation: Towards Sustainability and Resource Conservation in Asia,* International Institute of Rural Reconstruction, Cavite, Philippines, pp86-92

Darlong, V., Hore, D. K. and Deb Barma, S. K. (2012) 'Gender, food security and rice farming in Tripura' in Sumi Krishna (ed.) *Agriculture and Changing Environment in Northeastern India,* Routledge, New Delhi

Datta Ray, B. and Alam, K. (eds) (2002) *Forest Resources in North East India,* Omsons Publications, New Delhi

Dhakal, S. (2000) 'An anthropological perspective on shifting cultivation: A case study of Khorira cultivation in the Arun Valley of Eastern Nepal', *Occasional Papers in Sociology and Anthropology* 6, pp96-111

Fernandes, W. (2004) 'Forest issues, forest dwellers and emerging situations', paper presented at the National Seminar on Human Origins, Genome and People of India, 22-24 March 2004, Anthropological Survey of India, New Delhi

Forest Survey of India (2012) *State of Forest Report 2011,* Ministry of Environment and Forests, Government of India, Dehradun, Uttarakhand

Government of Arunachal Pradesh (2001) *New Agricultural Policy of Arunachal Pradesh 2001,* Government of Arunachal Pradesh, Itanagar, http://arunachalpradesh.gov.in/nnap.htm, accessed on 7 April 2012

Government of Assam (2004a) *Assam Forest Policy 2004,* Government of Assam, Dispur, http://asbb. gov.in/Downloads/Assam%20Forest%20Policy%202004.pdf, accessed 2 May 2017

Government of Assam (2004b) *State Agriculture Policy, Assam, 2004,* Government of Assam, Dispur, http://assamagribusiness.nic.in/ASAGRI.pdf, accessed 2 May 2017

Government of India (1927) *The Indian Forest Act, 1927,* Ministry of Agriculture, New Delhi, www. envfor.nic.in/legis/forest/forest4.pdf, accessed on 13 May 2012

Government of India (1952) *National Forest Policy 1952,* Ministry of Food and Agriculture, New Delhi, http://circle.forest.kerala.gov.in/tckollam/images/docs/policies/nfp1952forest.pdf, accessed 2 May 2017

Government of India (1961) *Report of the Commission for Scheduled Areas and Scheduled Tribes,* Ministry of Home Affairs, New Delhi

Government of India (1976) *Report of the National Commission on Agriculture,* Ministry of Agriculture and Irrigation, New Delhi

Government of India (1981) *Report on Development of North Eastern Region,* Planning Commission, Government of India, New Delhi

Government of India (1988) *National Forest Policy 1988,* Ministry of Environment and Forests, New Delhi, www.moef.nic.in/downloads/about-the-ministry/introduction-nfp.pdf accessed on 13 May 2012

Government of India (1990) *Joint Forest Management Notification,* Ministry of Environment and Forests, New Delhi

Government of India (1997) *Transforming the Northeast: Tackling Backlogs in Basic Minimum Services and Infrastructural Needs,* Planning Commission, Government of India, New Delhi

Government of India (2001) *Report of the Working Group on Watershed Development, Rainfed Farming and Natural Resource Management for the 10th Five Year Plan,* Planning Commission, Government of India, New Delhi

Government of India (2006) *Report of the Working Group on Forests for the 11th Five Year Plan (2007-2012),* Planning Commission, Government of India, New Delhi

Government of India (2007) *National Policy for Farmers 2007,* Department of Agriculture and Cooperation, Ministry of Agriculture, New Delhi

Government of India (2008) *Report of the Inter-Ministerial National Task Force on Rehabilitation of Shifting Cultivation Areas,* Ministry of Environment and Forests, New Delhi

Government of India (2011a) *Indira Gandhi Paryavaran Puraskar: Fellowships and Awards,* Ministry of Environment and Forests, New Delhi, www.moef.nic.in/citizen/award/igpp.html, accessed on 12 May 2012

Government of India (2011b) *Report of Working Group on Forests and Sustainable Management of Natural Resources for 12th Five Year Plan (2012-2017),* Planning Commission, Government of India, New Delhi

Government of Mizoram (2002) *The Bamboo Policy of Mizoram 2002*, Environment and Forest Department, Government of Mizoram, Aizawl

Government of Mizoram (2009) *The New Land Use Policy 2009*, Government of Mizoram, Aizawl

Government of Nagaland (1997) *Joint Forest Management Resolution,* notification no FOR-153/80 (vol. II), dated 5 March 1997, Government of Nagaland, Kohima

Government of Nagaland (2004) *Nagaland Bamboo Policy 2004,* Office of the Agriculture Production Commissioner, Government of Nagaland, Kohima

Government of Tripura (2000) *Approach to Peoples' Plan in Tripura,* Planning Department, Government of Tripura, Agartala, www.planningtripura.gov.in/Gramoday/Guidelinesofgramoday.pdf, accessed on 7 April 2012

Hance, J. (2008) 'Monoculture tree plantations are 'green deserts' not forests', *Environmental News,* https://news.mongabay.com/2008/09/monoculture-tree-plantations-are-green-deserts-not-forests-say-activists/, accessed 2 May 2017

Karunakaran, N. (2008) '*Jhum* fields forever', *Business Outlook*, http://archive.outlookbusiness.com/article_v3.aspx?artid=101608, accessed 2 May 2017

Kemp, R. (2012) Personal communication between the author and Dr R. Kemp, Chief Conservator of Forests and Principal Secretary, Department of Environment and Forests, Government of Arunachal Pradesh, 12 April 2012

Kerkhoff, E. and Sharma, E. (2006) *Debating Shifting Cultivation in the Eastern Himalayas,* International Centre for Integrated Mountain Development, Kathmandu

Malik, B. (2003) 'The problem of shifting cultivation in Garo Hills of North East India, 1860-1970', *Conservation and Society* 1(2), pp87-115, http://www.conservationandsociety.org/text.asp?2003/1/2/87/49352, accessed 2 May 2017

Murtem, G., Sinha, G. N. and Dopum, J. (2008) 'Jumias' view on shifting cultivation in Arunachal Pradesh', *Bulletin of Arunachal Forest Research* 24(1 and 2), pp35-40

Nakro, V. (2012) Personal communication between the author and Mr Vengota Nakro, of the NEPED project

NEPED (2011) *Status Report of NEPED,* Nagaland Empowerment of People through Economic Development project, Government of Nagaland, Kohima

NERCORMP (2008) *Annual Progress Report* (unpublished), North Eastern Region Community Resource Management Project for Upland Areas, Shillong, Meghalaya

NERCORMP-KCRMS (2008) *Annual Status Report* (unpublished), North Eastern Region Community Resource Management Project for Upland Areas and Khawkylla Community Resource Management Society, West Khasi Hills, Meghalaya, NERCORMP-KCRMS, Nongstoin, Meghalaya

North Eastern Council (2008) *North Eastern Region Vision 2020,* Ministry of Development of North Eastern Region, New Delhi and North Eastern Council, Shillong, Meghalaya

Poffenberger, M. (ed.) (2007) *Indigenous Forest Stewards of Northeast India,* Community Forestry International, California

Raman, T. R. S. (2000) 'Jhum: Shifting opinions', paper delivered at Environment: Reality and Myth – A Symposium Re-evaluating Some Prevailing Beliefs, Mysore, India, www.india-seminar.com/2000/486/486%20raman.htm, accessed on 12 May 2012

Rasul, G. (2005) *State Policies and Land Use in the Chittagong Hill Tracts of Bangladesh,* Gatekeeper Series 119, International Institute for Environment and Development, London, www.pubs.iied.org/pdfs/14511IIED.pdf, accessed on 16 May 2012

Roy, D., Xavier, B. and M'Vidouboulou, S.W. (2012) *Study on Shifting Cultivation and the Socio-cultural Integrity of Indigenous Peoples,* UN Economic and Social Council, Permanent Forum on Indigenous Issues, 11th Session, 7-18 May 2012, New York

Saxena, N. C. (1999) *Forests in Tribal Lives,* Planning Commission, Government of India, New Delhi, www.planningcommission.nic.in/reports/articles/ncsxna/index.php?repts, accessed on 23 April 2012

Schmidt-Vogt, D., Leisz, S. J., Mertz, O., Heinimann, A., Thiha, T., Messerli, P., Epprecht, M., Van Cu, P., Chi, V. K., Hardiono, M. and Dao, T. M. (2009) 'An assessment of trends in the extent of swidden in Southeast Asia', *Human Ecology* 37, pp269-280

Sharma, B. D. and Hore, D. K. (1990) 'Rice germplasm collection in Tripura State', *Indian Journal of Plant Genetic Resources* 3(2), pp71-74

Teegalapalli, K. (2008) 'Shifting livelihood options and changing attitudes of communities in the Garo hills, Western Meghalaya', *Current Conservation* 2(3), pp18-21

Notes

1. The term *jhum* is widely used in the eastern Himalayas, more so in northeast India and the Chittagong Hill Tracts of Bangladesh and notably in some parts of eastern Nepal (Dhakal, 2000). *Jhum* may generally mean both the practice of shifting cultivation and the plot of land, or swidden, where crops are grown. However, in Tripura (northeast India) and in the Chittagong Hill Tracts of Bangladesh (Rasul, 2005), the practice of shifting cultivation is called *jhumming* or *jhuming*, the farmer who practises shifting cultivation is called a *jhumia* and the plot of land where crops are grown is called a *jhum*. The use of the terms *jhumias* and *jhumming* appears to be more common in Bengali-speaking communities.

2. North Eastern Region Community Resource Management Project for Upland Areas (1999-2008), a joint project of the International Fund for Agricultural Development and the Government of India, working in three states of Northeast India: Assam, Manipur and Meghalaya. The project has been refinanced by IFAD and the Government of India as NERCORMP II, in the period from 2010 to 2016.

3. The Mizos are a tribal community living predominantly in Mizoram. However, they are also found in parts of Manipur, Tripura, Assam and Meghalaya.

4. This information was gathered during a NERCORMP project Joint Review Mission in villages of Dima Hasao district, Assam, in March 2012.

5. Information sourced from a supervision field-assessment report of the NERCORMP project in Karbi Anglong district, Assam, 5-6 March 2012.

19

VANISHING SHIFTING CULTIVATION AND LOSS OF TRIBAL CULTURE IN ARUNACHAL PRADESH, NORTHEASTERN INDIA

*Tomo Riba**

Introduction

The state of Arunachal Pradesh is located in the northeasternmost corner of India, and is inhabited by 26 major tribal groups with more than 100 subtribes, all speaking Tibeto-Burman languages (Nyori, 1993). All of these groups practise shifting cultivation on the southward-facing slopes of Eastern Himalaya. Being located in the easternmost part of India, this mountainous state receives the first sunshine in the country. Due to southward-facing slopes with no higher ground between them and the Bay of Bengal, Arunachal Pradesh receives abundant sunshine and heavy rainfall, favouring the growth of luxuriant mixed tropical evergreen forests. The state has international borders with China in the north, Myanmar in the east and Bhutan in the west, with the Brahmaputra river in the plains of India's Assam state to the south (Figure 19-1). Entry to the state by outsiders is restricted to those who obtain an Inner Line Permit from the Indian authorities. Outsiders are also not allowed to settle permanently in Arunachal Pradesh. Thus, the state is inhabited solely by its recognized indigenous peoples.

The altitude of Arunachal Pradesh ranges from 150 metres on the Brahmaputra plains in the south to more than 7000 metres in the north. The land and people of the state are India's least known, even to most of their countrymen. The vegetation gradually merges from tropical rainforest into subtropical, to temperate and ultimately to alpine forest (Taher and Ahmed, 2001).

---

* DR TOMO RIBA, Department of Geography, Faculty of Environmental Sciences, Rajiv Gandhi University, Arunachal Pradesh, India. In his own words: 'I was born in a shifting cultivator's family and as a young boy I spent most of my time off school with my mother in the field, especially after the death of my father when I was 10. I still have an emotional attachment to it; the smells of freshly fallen trees, of the ash just after burning a field, of weeds, ripening paddy, and the call of the birds. I miss them. The taste of nature is original, and much different to what we get today.'

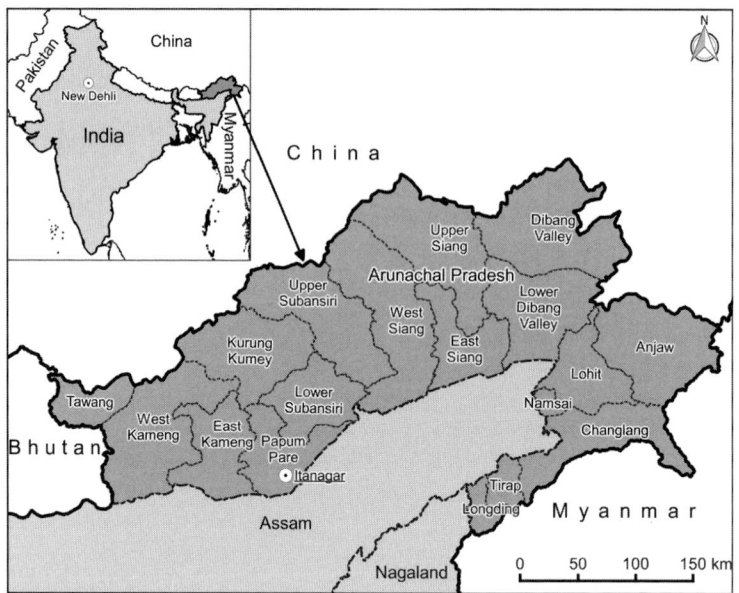

FIGURE 19-1:   Arunachal Pradesh state, northeastern India.

As well as its isolation, rugged topography, heavy rainfall (an annual average of 300cm) and dense vegetation (82% of the area is under forest), Arunachal Pradesh has low technological development and, according to the 2011 census, a low population of 1,382,611. This gives the state a population density of just 17 people per square kilometre. These factors have favoured the continuation of shifting cultivation as the main means of sustenance for the state's many ethnic groups. According to the Government of India's *Jhum* Land Regulations, 1947, the people have a customary right to cultivate *jhums* (swiddens) in the state's forests (T. Riba, 2013).

Centuries of practice has led to improvements in swiddening techniques that have increased production, in terms of both quantity and crop varieties. Along with this, the cultural life of the farmers has also improved substantially over recent decades.

Nowadays, shifting cultivation has become both an institution and a culture for the tribal people of Arunachal Pradesh, blended with economy, anthropology and ecology. The tribal farmers have a richness of traditional ecological knowledge (TEK). They are part and parcel of the forest and feel an emotional attachment to it (T. Riba, 2002). When they are in the forest they feel safe, and constantly feel the presence of unseen living beings around them, watching whatever they do. This sense of being watched restrains them from any acts of violence in the forest. Various deities are believed to own the forests; the farmers are merely tenants, and if they fail to please the forest deities they will never have good harvests. So, at every stage of cultivation, starting from the first day of clearing the forest to harvesting and the first meals of new rice, they stage rituals and feasts to appease and offer gratitude to all of the benevolent and malevolent deities of the forest (T. Riba, 1996). The farmers also believe that the Goddess Prosperity makes regular visits to their fields and will bless them if they perform due rituals in a timely fashion. For this reason, the farmers keep their fields clean and the roads leading to them tidy. The sense of the presence of deities in their fields makes them maintain decency, refrain from using abusive terms to animals and plants and avoid inauspicious comments while in the forest. If they see the footprint of a tiger, they will say:

> 'Oh elder brother, I did not know that you were also here. So, please, take your path and I will take mine. Let us not disturb each other' (T. Riba, 1998).

They strive at all times to please the forest deities, for the sake of their own safety, in pursuit of good production, and to ensure successful hunting, fishing and collection of forest products.

Cultivation techniques

The author belongs to the Galo tribe. The farming techniques of the Galos, along with other tribal groups, do not simply involve slashing and burning the forest, as many scholars suggest. From beginning to end, cultivating a *jhum* is a meticulous undertaking, using effective techniques that the farmers have acquired over generations of practice. They don't clear the forest until it has gathered sufficient litter, the undergrowth is thin and most of the dormant weed seeds have exceeded their germination period. *Jhums* cleared from forests that are too young are less productive because of inadequate humus deposits and more weed growth. Clearing of the forest begins from the bottom of the slope. This is more convenient and leads to less back pain for the farmers. First, the undergrowth is cleared and the bigger trees are left standing. While clearing the undergrowth, not even the smallest plant is left alive; creepers are pulled out, climbers are pulled down, litter is raked, tossed and turned upside down so that the coming fire will sear the topsoil. On a sunny day when the fallen undergrowth has dried sufficiently, the big standing trees are felled. During this operation the farmers take the utmost care to avoid accidents. The

Solanum indicum L [Solanaceae]

The fruit of 'small brinjal' is eaten in Arunachal Pradesh as a subsidiary vegetable, or as a bitter snack with drinks. Also known as Indian nightshade, it is an important medicinal plant whose roots are said to have powerful anti-inflammatory qualities. The fruit is also used to treat respiratory disorders.

position of the trees is carefully studied, along with branches that are entangled with other trees. Before the first axe blow, the farmer performs a small ritual by sacrificing a chicken and chanting:

> 'Oh, Deity of the Forest, today let me fell the big trees, which belong to you. I want to grow crops here. For this, I am offering you this chicken. Let trees fall easily without any obstruction or accident'.

After the trees fall, their raised branches are chopped to the ground so that the fire burns uniformly. About one month later, in March, on a bright sunny day at noon, fires are lit at the bottom of the field. Earlier, the farmer will have prepared a track through the slash so he is able to set fires quickly and avoid accidents. While setting the fires, the farmer calls on the Goddess of Wind to help with strong winds to fan the flames. He chants:

> 'Oh Goddess of Wind, fan the fire like the wings of a quail bird' (T. Riba, 2013).

On the day after the fires, the farmer begins clearing the half-burnt logs. In the first few days, these charred logs remain soft and easy to cut with an axe or machete. But the roasted logs soon become hard and completely dry and cutting them is difficult and takes time. While clearing the debris, the farmer gathers long, straight logs of medium size to build a farm hut. He will use the rest of the smaller debris as a fence around the field, or will keep it for future firewood. Bigger logs are also pegged across the slope to prevent downward movement of topsoil. At different points, heaps of debris are left for dumping weeds pulled from the field. Care is always taken to make these heaps of debris unattractive to mice, which would destroy the crops. However, in the damp and cold of winter, the farmers will trap rats around these heaps.

Mustard seeds are sown while the half-burnt debris is still being cleared, mostly where the ash deposits are deeper. Yams, taro, chillies and beans are also sown in advance of the main rice crop, near to tree stumps. Maize is also sown in advance. If the field has one corner close to uncleared forest, it could become easy prey for birds and wild animals. In such cases, sowing begins in that vulnerable corner, so that sprouting seeds will gain some protection from the constant activity while the debris is cleared from the rest of the field. The sound of wood-cutting, smoke from burning debris and the vigilant farmer will keep birds and animals at bay. Big logs are also kept burning all night. Generally, harvesting also begins from the same corner so that ripening crops are saved from marauding wildlife.

Upland rice is the staple crop. While sowing the other crops, care is always taken to avoid them inhibiting the growth of the rice. Proper gaps are left when sowing corn and millet. Beans, onions, chillies, tobacco, and so on are grown along either side of the farm path and around the farm hut. Pumpkins, gourds, foxtail millet and cucumbers are grown on the periphery of the field. Creeping and climbing crops are avoided in the middle of the field. The rice is sown by using dibble sticks. An expert woman planter can make more than 60 holes per minute and throw the correct amount of seed into the holes from a distance of one metre, after which she covers the holes with quick sweeps of her feet. The dibbling method of sowing has many advantages. Seeds remain intact in the holes, safe from rushing rainwater, wind and birds. Soil moisture helps quick germination and the seed is protected from the drying sun. The growing seedlings can also send roots firmly into steep slopes.

The Nocte, Tangsa and Wancho tribes, who live in the eastern part of Arunachal Pradesh, do not dibble their rice seeds, but broadcast them on to the slopes instead. Their seeds can be swept away by strong winds and rainwater, and can accumulate at the bottom of the field, leaving large areas empty. Exposed seeds are also easily eaten by birds or roasted in the sun. These people also leave certain big trees standing in the middle of their fields, and although this helps the quick regeneration of the forest after cropping is finished, it also provides resting places for hungry birds (Ralangham, 2009).

For many generations, the shifting cultivators of Arunachal Pradesh have grown varieties of rice, maize, millet, yams, taro, chillies, beans,

Zanthoxylum rhetsa DC. [Rutaceae]
Possibly synonymous with *Zanthoxylum limonella* (Dennst.) Alston

Called *onyor* at the study sites, the bark and immature fruit of this tree are used as a spice; the seeds can be a substitute for black pepper and the fruit and young leaves are cooked as a vegetable. The fruit is also used to kill fish when fishing.

ginger, sweet potatoes, cucumbers, gourds, pumpkins, sesame, eggplants, onions, mustard, and so on. They are hesitant to adopt new varieties that they have not tested because of the risk of crop failure. They have confidence in their own seeds from varieties that they have been growing for centuries, so they are quite correctly regarded as preservers of diversity in the region's crop species. Their domestication of animals matches this tradition.

Weeding is the most tedious task in shifting cultivation, but weeding determines the productivity of the crop. It is done in July and August, so the days are long, warm and humid. There is also a shortage of grain in the village granaries. The heat and the smell of the weeds and the crop combine to challenge the strength of the farmers. Weeding is done three times. On the first occasion it is random, removing the fast-growing weeds, shoots of bananas, wild cardamom and so on, with a machete. The second weeding is very systematic; not a single weed is allowed to remain in the field. While weeding, the soil is also scratched and loosened; crops that are growing thickly and competing for space are thinned and transplanted so each has space to grow. The final weeding involves the selective removal of fast-growing plants. Fields that have not been weeded thoroughly yield poorly because of competition from weeds and crop damage from mice and insect pests (B. Riba, 2009).

As the crops ripen, especially the rice, the farmers feel a growing sense of hope and security. Proudly, and with a sense of contentment, they constantly inspect their fields; every plant that falls under its own weight is helped to stand again. However, shifting cultivation occurs in the deep forest, amid wild birds and animals. Unprotected crops are always devoured by wild animals and such attacks can lead to starvation for a farmer's family. In order to protect their crops from hungry wildlife, the farmers have many techniques and equipment made from locally available materials. Bows and arrows are used, and a wide range of traps are made from stone, wood, cane and bamboo. They also use running water and wind to produce sounds to scare the wildlife away from ripening crops. Because fire is employed as the main tool for clearing the slashed vegetation and fallen trees, there is a very low incidence of disease in the crops. However, farmers do not have any measures to save their crops from erratic climatic conditions.

Traditional rituals play a prominent role in harvesting; strict taboos must be followed. Every fallen grain is lovingly picked up. Sometimes a farmer may be seen talking to plants in the field. While harvesting, especially the rice, grasshoppers and green leaves are collected for family consumption, particularly for the children. Harvested grain is carried back to the village and stored in granaries away from the dwellings, in case of an accidental fire. All farmers, especially the women, have their own methods and techniques for selecting, storing and preserving seeds for next season's crops.

Traditional ecological knowledge of shifting cultivators

People who blame shifting cultivation for a variety of environmental problems are those who have the least practical knowledge about it and who have never attempted to understand why farmers have continued to practice it since time immemorial, in spite of its many challenges. A combination of factors has compelled them to continue swidden farming, including their remoteness, mountainous topography, dense vegetation, low population, lack

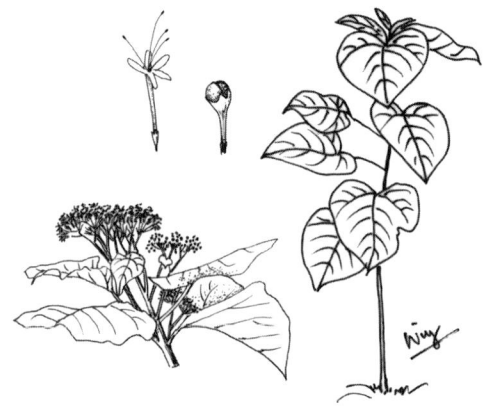

Clerodendrum colebrookianum Walp. [Lamiaceae] Synonym of *Clerodendrum glandulosum* Lindl

The broad leaves of this perennial shrub are eaten as a wild vegetable, mostly during the winter. It is also used in Arunachal Pradesh as a treatment for hypertension, and to combat skin diseases, coughs and dysentery.

of markets and low technological development. Besides these things, agriculture is deeply ingrained in the traditions and culture of shifting cultivators; their food habits, festivals, folk songs and beliefs are intensely agriculture-oriented (Fox, 2000).

Swidden farmers should be seen not only as forest destroyers, but also as forest maintainers and keepers of crop and domestic-animal diversity (ICIMOD, 2004). They should also be recognized as preservers of traditional knowledge about the behaviour of different organisms and the properties of various plants. If the present trend of rural depopulation continues, this traditional knowledge will disappear over the course of time. In the light of this outlook, we should be thankful to those shifting cultivators who, like museum keepers, persevere in the preservation of our oldest traditions.

Rural depopulation and the end of shifting cultivation

Many scholars have proposed alternative practices and livelihoods for shifting cultivators, without thinking of the traditions, culture and religion of these people. Their concern is only for food; they have no suggestions concerning traditions and culture. Another question is: will the alternative occupations they propose have any impact on the forest and the soil? Continuing to urge the discontinuation of shifting cultivation may encourage more young people to migrate to urban areas, leading to rural depopulation and the loss of age-old cultures and rich traditional ecological knowledge (Shirasaka, 2006).

As an evolutionary process, change is inevitable. Shifting cultivation will meet a natural death and disappear, along with most of the tribal culture and traditional ecological knowledge gathered over centuries, such as the behaviour of plants and animals, techniques for hunting and fishing and counting of the seasons. Depopulation has already begun in many villages in Arunachal Pradesh (Figure 19-2). Today, every village has a free primary school and, having experienced elementary education, the children are gradually moving out of the villages to urban centres, where they can find higher education.

In its regard for traditional societies, the Government of India holds contradictory positions. On one hand, it emphasizes the promotion of traditions and culture. On the other, it constantly attempts to persuade farmers to cease shifting cultivation, without accepting that its demise will mean the disappearance of rich and varied traditions and culture. Anything viewed from a distance is always perceived differently from impressions gained in close proximity. Generalized views therefore fail to grasp the underlying intricacies of any social situation. It follows that in policy matters, changes that are imposed from the top down, that is, from the authorities down to the small people, are far less likely to work than social or livelihood changes proposed from the bottom and adopted at the top. And while there is a universal need to constantly adjust the relationship between humankind and nature, it should not be expected that a successful policy in one place will work well in another. For example, conditions within the equatorial belt vary from Malaysia, which has a highly developed rubber plantation industry, to the Congo and Amazon basins, where indigenous ethnic groups remain dependent upon shifting cultivation. Such differences may also be found between adjacent villages. In 2010, the author hosted a visit to his village in Arunachal Pradesh by an Emeritus Professor from Oxford University in the United Kingdom. It was during the rice harvest, and after close observation of the system, she commented, 'Interesting! There is a science

FIGURE 19-2: The traditional tribal village of Diisi, in Arunachal Pradesh's West Siang district, the site of livelihood studies for this chapter.

to it'. This confirmed my conviction that most information regarding shifting cultivation is not based on first-hand observation in the real world, but rather on hearsay, boosted by imagination. This is why the negative aspects of shifting cultivation are commonly exaggerated. Many people make very hasty conclusions from 'outside the door', without understanding what lies inside. One of the interesting aspects of this debate is that people who know little about shifting cultivation do most of the talking; those who are familiar with it don't talk much because they don't find much to say. The tragedy is that most people fail to collect information from the source, but rely on distorted information furnished by people with little experience.

Estimates of areas under shifting cultivation are also inaccurate and rather generalized. The fact is, there is no need for the farmers themselves to know either the size of their fields or their rate of production. Areas reported to be under shifting cultivation are calculated from aerial photographs, which include frost-bitten forest and alpine vegetation in northern parts of the state where there are no settlements.

Nowadays, more than half of the children born in a village will leave. Many do not return, but settle permanently near to where they work and sever their connections with shifting cultivation and the villages of their birth. There are also many families where only the aged parents are left in the village, and they do not perform any agricultural activities. They depend on remittances from their children in urban jobs. There are also many village families who have adopted either terraced or wet-rice cultivation and no longer practise shifting cultivation. Many people who have left the villages are working in government jobs in urban centres. There are even families left in the villages who are in government jobs. For example, two of the author's brothers are teachers in their home village and have given up shifting cultivation. Thirty years earlier in the author's family there were five sons living under a single roof. Today, all five have left shifting cultivation and have government jobs. Three of them, with their eight children, have settled permanently in urban centres. The elderly parents remain in the village, but are too old for agricultural activities. Many families are much the same: all of the children have gone, leaving only the last of the shifting cultivators. When they die, there will be no more shifting cultivation.

Therefore, in the case of Arunachal Pradesh, the actual number of shifting cultivators cannot be calculated from numbers of family members and households. The size of most old villages has either remained static or has shrunk. Many families have shifted to roadsides, nearer to urban centres. Fine details of the status of village populations are shown in Tables 19-1 through 19-4. In 1950, a major earthquake in Assam devastated many villages in the hills. Many village families, frightened of a reoccurrence, migrated to the foothills and adopted terraced cultivation. They are now economically better off, but have lost the traditional culture of the hills.

TABLE 19-1: Reduction in shifting cultivators in five selected villages (February 2015).

Village	Percentage of shifting cultivators		
	In forefathers' time	In father's time	At present time
Diisi, West Siang	100%	Partial	67%
Piira, West Siang	100%	100%	24%
Bomjir East Siang	100%	Partial	14%
Dambuk, East Siang	100%	Partial	0.62%
Kangkong, East Siang	100%	Partial	nil

Source: Compiled by the author.

TABLE 19-2: Status of shifting cultivators at Diisi village, West Siang district, Arunachal Pradesh (February 2015).

House no.	Head of family	Major occupation	Agriculture			Remarks
			Only shifting cultivation	Only terrace cultivation	Both	
1	To:hen Dirchi	Agriculture			√	Children are in jobs
2	Jummar Riba	Service	√			Service in the village
3	Miito Riba	–	√			
4	To:mo Dirchi	–	–	–	–	No prominent occupation
5	Karyom Riba	Agriculture	√			One son stays at home
6	Iri Riba	Agriculture			√	
7	Mo:li Dirchi	Agriculture	√			
8	Mo:mar Dirchi				√	
9	Jumya Dirchi	Agriculture	√			
10	Marto Riba	Agriculture			√	
11	To:ke Riba	Agriculture		√		
12	Dakken Riba	Service		√		Engaged relative
13	Kardak Riba	Agriculture		√		

Source: Collected and compiled by the author.

TABLE 19-2 (cont.): Status of shifting cultivators at Diisi village, West Siang district, Arunachal Pradesh (February 2015).

| House no. | Head of family | Major occupation | Agriculture | | | Remarks |
			Only shifting cultivation	Only terrace cultivation	Both	
14	Dakkar Riba	Agriculture			√	All sons in jobs
15	Chiiken Dirchi	Agriculture			√	One young son at home
16	Imo Riba	Agriculture	√			
17	Miirik Tacha	Agriculture	√			
18	Baarik Riba	Agriculture		√		Children in jobs
19	Jummo Dirchi	Agriculture	√			
20	Jumke Riba	Agriculture	√			
21	Go:rik Riba	Agriculture	√			One young son at home
22	Do:ge Dirchi	Service				No one does cultivation
23	Minya Riba	Agriculture			√	
24	Mo:kar Riba	Agriculture	√			
25	Mo:bom Riba	Agriculture			√	
26	Mo:dak Riba	Agriculture			√	
27	Goken Riba	Service		√		Wife cultivates
28	Go:ri Riba	Agriculture			√	
29	To:ken Riba	Agriculture		√		
30	Baato Riba	Agriculture		√		
31	Karchi	Agriculture	√			
32	Kambu	Agriculture	√			
33	Kiike Kambu	Agriculture			√	
34	To:ba Riba	Service		√		Tenant. Service in the village
35	To:dak Riba	Agriculture			√	
36	Liimi Riba	Agriculture			√	
37	Jumri Kambu	Service	√			Wife cultivates
	Total Agriculture		12	10	13	

Source: Collected and compiled by the author.

TABLE 19-3: Categories of farming households (n=37) in the author's village (February 2015).

Families dependent solely on shifting cultivation	Families dependent solely on terraced rice cultivation	Families who practise both shifting and terraced rice cultivation	Families who do not depend on agriculture	Families without next-generation shifting cultivators
12	10	13	02	08
33%	27%	35%	5.40%	22%

Source: Collected and compiled by the author.

TABLE 19-4: State of village population from a survey of 206 families (2012).

Particulars	Figures
Two-member working families	126 (61%)
Family members frequently visit the village	16%
Family members rarely visit the village	10%
Students in higher schools	16%
Children (including those going to school)	27%
Working population	42%

Source: Collected and compiled by the author.

It can be seen in Table 19-4 that the percentage of shifting cultivators, or the village's working population, has fallen from 100% to 42% in the space of one generation. In most families (61%) there are not more than two working members, mostly a husband and wife only. After a few decades, after the death of these working groups, finding workers to replace them in the swiddens will be very difficult as there is a tendency for children to leave the village, mainly for further studies. Family members who visit the village regularly (16%) will cease visiting upon the death of their parents. Students who are in the village school will join them after a few more years, in search of higher education. Thus, the working population will diminish further. Ultimately it will lead to marginalization of villages, which will become non-functional.

There is also much exaggeration in reports concerning the length of fallow periods. There is always a difference between figures supplied by farmers and the writings of many scholars concerning the age of forest cleared for swidden cultivation. Those who criticize shifting cultivation consider shortening fallow periods to be a strong argument against the continuation of swidden farming. They report that fallow periods have fallen to just two to three years, from 10 to 20 years in the past. It is true that fallow periods have fallen from the days when farmers cleared primary forest for the first time and populations were very small. These very short fallow periods can be found only in certain exceptional locations, because farmers cannot expect good production from swiddens cleared from such young forest. The biomass deposits would be very thin and the growth of weeds very high. Moreover, it is very difficult to clear such forest due to thick undergrowth. Such information comes from forest areas cleared near to settlements or along roadsides. It is not from isolated interior villages. Forests near to settlements are always subject to frequent human interference and are damaged by domestic animals. The trees grow only slowly, even though they may be 'old' by secondary forest standards. Another factor contributing to shorter fallow periods in certain places is not an increase in the number of farmers, but the fact that people don't like to cultivate swiddens in forests far distant from their homes, such as was the case in olden days. Shorter fallow periods can be found mainly in those forests in close proximity to urban centres. In these places people cultivate swiddens either to grow cash crops or as a source of vegetables and firewood. In remote interior areas, the length of the swidden cycle, and in particular the fallow period, has remained almost unchanged. In some cases, farmers are using increased fallow periods. However, if the forest is too old, large fallen logs can occupy much of the field and the thick deposit of biomass leads to over-vigorous growth of crops, with less grain.

Conclusion

As an evolutionary process in human civilization, the traditional practice of shifting cultivation has recently lost its importance in many places. Many factors can be blamed for this, but the cause is mostly rural depopulation. Villages, particularly those in remote interior areas, are experiencing an inability to replace experienced shifting cultivators. As shown in Table 19-1, many villagers moved away from shifting cultivation as long as 50 years ago. Many villages located in river valleys, such as that of the Siang river (Brahmaputra) and its tributaries, have long abandoned shifting cultivation as a major source of sustenance.

Shifting cultivators also consider terraced cultivation to be a better form of agriculture. Families with terraced fields are usually accorded a higher social status. Thus, most gentle slopes with the capacity for irrigation have been converted into terraces. There is also a widespread desire among shifting cultivators for their children to avoid the hardship of work in swiddens, and instead to secure government jobs. For this reason, they believe it is important for their children to gain an education. The day is not far off when these tribal people will have lost most of the traditions and culture that are attached to shifting cultivation (Figure 19-3). Shifting cultivation itself will have disappeared. Another factor contributing to the loss of traditional practices is the conversion of many tribal people to Christianity (T. Riba, 2003). These tribal Christians continue many cultural practices, such as food habits, dress, language and house patterns, but they discontinue many of their customs and religious practices that are associated with cultivation. Traditional feasts and rituals are no longer performed at every stage of cultivation.

There are also other factors that are responsible for the loss of tribal culture. When people return to their village after their education is complete, they have forgotten many of the customs in their prolonged absence. There are fewer and fewer shamans who can perform rituals, and the sheer cost of traditional ritual sacrifices have discouraged many from maintaining indigenous customs.

FIGURE 19-3: A group of men, educated in the 'outside world', make a rare visit to the village of their birth.

There is also another side to this coin: there have been a few recent instances where hillsides adjacent to urban centres have been cleared to grow cash crops. This has provided a lesson in side-effects, including the tastlessness of crops grown with inorganic manure and pesticides. Although they have moved to urban areas, these tribal people still have a preference for vegetables grown in shifting cultivation that are free from chemicals, fertilizers and pesticides. Moreover, they seek out traditional food items that are mostly grown in swiddens. The increasingly popular terraced cultivation is mostly for monocropping of rice as a staple crop; it does not supply vegetables or firewood, and firewood remains

Piper pedicellatum C. DC.
[Piperaceae]

This forest climber of the pepper family is not only a wild vegetable in northeast India, but is also attracting scientific interest for its antifungal and antioxidant qualities.

the only fuel for cooking and warming in rural houses. Therefore, despite the growth of permanent agricultural fields, shifting cultivation continues as a source of fuel and vegetables.

In the foothills adjacent to the Brahmaputra river, tribal people who migrated from the mountains 50 or 60 years ago, especially after the 1950 Assam earthquake, cultivate extensive terraces and wet-rice paddies. Their ability to sell their excess production in markets on the plains has led these people to cultivate swiddens on sunny, southward-facing slopes, mainly for cash crops like chillies, eggplants, taro, yams, ginger, cucumbers, beans and green leafy vegetables.

Another situation has also compelled many of them to return to shifting cultivation as a main means of sustenance. A boundary dispute between the people of the hills and those of the plains has resulted in the newcomers losing their land. According to the people of the Assam plains in the Brahmaputra valley, hills are for tribal people and the plains are for the people of the plains.

A new situation has also developed in many rural areas where the number of educated young people is increasing. Finding no scope for engaging themselves in urban society, many educated but unemployed youths are returning to their villages. Having been away for a long time in pursuit of education and employment, most of them know little about traditional economic activities and fail to take part in gainful activities.

Shifting cultivation has undergone tremendous changes over the past 50 to 60 years, in terms of the area under cultivation, the number of cultivators, techniques and motives for cultivation and crops grown in fields. The number of swidden farmers has fallen considerably. There will be much greater reductions over the next few decades. Many traditional forms of shifting cultivation, along with the traditions and culture attached to them, will disappear. Further depopulation of rural areas will also see the disappearance of certain villages.

References

Fox, J. (2000) 'How blaming "slash and burn" farmers is deforesting mainland Southeast Asia', *Asia Pacific Issues* 47, pp1–8

ICIMOD (2004) *The Shillong Declaration*, Regional Shifting Cultivation Policy Dialogue Workshop for the Eastern Himalayas, Shillong, Meghalaya, 6-8 October 2004, International Centre for Integrated Mountain Development, Kathmandu

Nyori, T. (1993) *History and Culture of Adi*, Omsom Publications, New Delhi

Ralangham, M. (2009) 'The indigenous knowledge system of the Wancho and its relevance for sustainability of the environment', unpublished PhD dissertation to the Faculty of Environmental Sciences, Rajiv Gandhi University, Arunachal Pradesh, India

Riba, B. (2009) 'The relevance of the indigenous knowledge system of the Galos of Arunachal Pradesh for sustainable development of forest resources', unpublished PhD dissertation to the Faculty of Environmental Sciences, Rajiv Gandhi University, Arunachal Pradesh, India

Riba, T. (1996) 'Galos: Their beliefs and ecosystem', *Arunachal University Research Journals* 1

Riba, T. (1998) 'Problems and prospects of modernisation of agriculture in Arunachal Pradesh (Special reference to the Galos of West Siang District), in M. C. Behera (ed.) *Agricultural Modernisation in Eastern Himalaya*, Directorate of Research, Government of Arunachal Pradesh, Itanagar

Riba, T. (2002) 'Tribal belief systems and their ecosystem: An approach towards environmental conservation', *Radiant*, The Indigenous Faith and Cultural Society of Arunachal Pradesh, Itanagar

Riba, T. (2003) *Tribals and Their Changing Environment, with Special Reference to Galos of Arunachal Pradesh*, Himalayan Publishers, Delhi and Itanagar

Riba, T. (2013) *Shifting Cultivation and Tribal Culture of Tribes of Arunachal Pradesh, India*, Center for Southeast Asian Studies, Kyoto University, Kyoto, Japan and Rubi Enterprise, Dhaka, Bangladesh

Shirasaka, S. (2006) *Shifting Cultivation in Xishuangbanna, Southwestern China: A Vanishing Mountain Culture*, College of Tourism, St. Paul's University, Saitama, Japan

Taher, M. and Ahmed, P. (2001) *Geography of North-East India*, P. K. Dutta, Mani Manik Prakash, Guwahati, Assam, India

20

SHIFTING CULTIVATION ON STEEP SLOPES OF MIZORAM, INDIA

Impact of policy reforms

*S.K. Tripathi, David C. Vanlalfakawma and F. Lalnunmawia**

Introduction

Shifting cultivation, a primitive agricultural practice believed to have originated in the Neolithic period around 7000 BC, is widely practised by millions of poor people inhabiting the world's tropical forest areas. This system of agriculture is prevalent in many Indian states, where it is practised by about 60 million nomadic tribespeople. It occurs most commonly in northeast India, where it is locally called *jhum* (meaning 'to group or work together'). It has there become a way of life, and is deeply associated with many cultural activities. In Mizoram, one of the seven sister states of northeast India, about 60% of the population is believed to depend on shifting cultivation for their livelihood (Anonymous, 2004a, pp1-9). Shifting cultivation in Mizoram is characterized by steep slopes; about half of the total land area has slopes of 40 to 100%. It therefore stands out from the other northeast Indian states in the need to perform the many activities of shifting cultivation, like slashing, burning, sowing, weeding and harvesting, on these steep slopes. The state also suffers huge losses, of about 60 tons per hectare, of fertile soils every year through erosion – a problem that requires meticulous scientific understanding.

   In their traditional practice, farmers slash an area of forest and burn the vegetation in situ after it dries. This is followed by manual seeding and cultivation of agricultural crops, mostly for one or occasionally two years, and then abandonment of the land to allow the restoration of soil fertility through natural processes. The farmers then move to cultivate another area of forest. In early times, this system was adequately productive, economically feasible and ecologically balanced because of a prolonged

---

* PROFESSOR S. K. TRIPATHI, Department of Forestry, Mizoram University, Aizawl, northeast India; DR DAVID C. VANLALFAKAWMA, Research Associate, Department of Forestry, Mizoram University, Aizawl, northeast India; and DR F. LALNUNMAWIA, Associate Professor, Department of Botany, Mizoram University, Aizawl, northeast India.

fallow period of 20 to 30 years. However, in recent years, exponential expansion of the human population has forced a drastic reduction in the length of fallow periods, to less than five years. This has led not only to a substantial decrease in soil fertility and crop productivity (Grogan et al., 2012), but also to widespread concern on the part of the government and non-governmental organizations about the sustainability of shifting cultivation. The government has therefore implemented various policies to improve the livelihood of the majority of poor people.

Over the past few decades, a number of policies seeking to replace shifting cultivation with alternative agricultural systems have been implemented in Mizoram. However, none of them has succeeded against traditional shifting cultivation in terms of productivity, economic feasibility and popularity; these policies have failed because of a lack of interest by shifting cultivators (*jhumias*) for one or another reason. Moreover, many of these policies were implemented for only a very short time of less than five years before being replaced by a new one. There were administrative reasons for this, but there is one exception: the recently introduced New Land-use Policy (NLUP).

This chapter highlights the major characteristics of Mizoram state and its dominant traditional *jhum* agriculture. Further, it discusses the policies adopted by the state pertaining to shifting cultivation and their impact.

The evolution of Mizoram

In 1890, the present Mizoram state was known as the Lushai Hills district under the British colonial administration. It was included in Assam state when India attained its independence. In 1952, the Lushai Hills district attained the status of an Autonomous District Council, and after implementation of the North-Eastern Re-organization Act in 1972, it became a Union Territory. In 1986, after three decades of insurgency, the Mizo National Front (the major insurgent group) and the Government of India signed a peace accord and Mizoram Union Territory became the 23rd state of India. Earlier, after attaining Autonomous District Council status in 1952, agriculture was identified as a major

Hibiscus sabdariffa L. [Malvaceae]

Roselle has many uses among Mizoram's *jhumming* communities. Most commonly, a beverage is made from the fleshy red calyces, but the leaves are also eaten as a vegetable and fibre from the stems can be used as a substitute for jute. It is said to be the most popular and widely eaten vegetable in Mizoram's eastern neighbour, Myanmar.

priority area, and the Agriculture Department of Mizoram was established as a fully fledged directorate in 1972. Since then, the department has been the nodal agency in most of the government's land-use policies.

Location and climate

Mizoram is a hilly state lying between 92° 15′ and 93° 29′ E longitude and 21° 58′ and 24° 35′ N latitude. In the northeastern part of India, it is sandwiched between Myanmar and Bangladesh, sharing international borders of 404km and 318km, respectively, with these countries (Figure 20-1). Mizoram is bounded by the states of Assam and Manipur in the north, by Myanmar's Chin state and Arakan Hills in the east and south, by the Chittagong Hill Tracts of Bangladesh in the west and by the state of Tripura in the northwest. The state is divided into eight administrative districts: Aizawl, Champhai, Kolasib, Mamit, Lunglei, Serchhip, Lawngtlai and Saiha. Mizoram's total geographical area is 21,087sq. km. The Government of India census in 2011 found its population to be 1,091,014, consisting of 552,339 males and 538,675 females. The tropic of cancer (23° 30′N latitude) divides the region into two almost equal parts (Pachuau, 1994).

Mizoram enjoys a moderate and pleasant climate. The mean temperature varies from 9°C to 24°C in winter, and in summer from 24°C to 32°C. The entire state comes under the direct influence of the southwest monsoon, receiving an annual average rainfall between 1926mm and 2479mm (Anonymous, 2011).

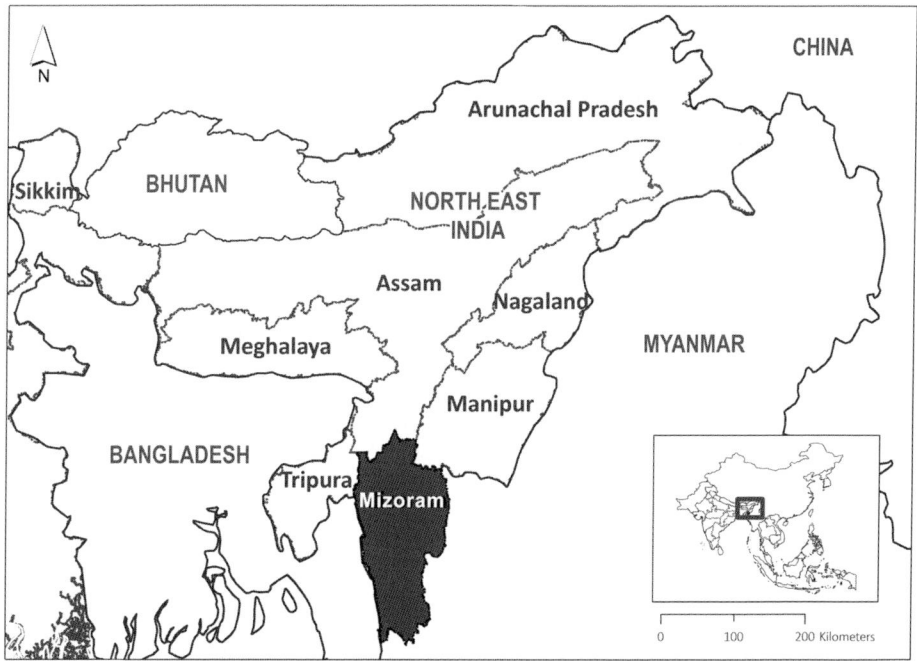

FIGURE 20-1: Mizoram state, in northeastern India.

Physiography of the state

Mizoram is composed of several ranges of hills formed from tertiary rocks. These ranges are separated by narrow, deep river valleys. The elevation ranges from 20m at Tlabung, near the western border with Bangladesh, to 2157m at Phawngpui (Blue Mountain), in the southeast. There are a few small patches of plains, most of them under permanent rice cultivation (Anonymous, 2003, pp1-64). The rest of the state is characterized by slopes ranging in steepness from 4% to more than 100%. The state is drained by a number of rivers, streams and rivulets. All are fed by monsoon rain and their volume is very limited in the dry season.

The soils of Mizoram are dominated mainly by loose sedimentary formation. They are described as young, immature and sandy (Pachuau, 1994). Analysis of soil samples from different parts of Mizoram indicate that manganese, copper and iron are adequately available, but zinc is not. The soil carbon pool of bamboo forest – the dominant natural resource – ranges from 51.91Mg per ha to 84.24Mg per ha (Vanlalfakawma et al., 2014). In most cases the pH and organic C content decrease with depth (Anonymous, 2004b, pp1-3, 42-43).

Forest cover

The state has a subtropical, humid climate that favours the growth and reproductive potential of living organisms. Thus, it harbours diverse flora and fauna with a high degree of endemism. The state is part of the Indo-Burma biodiversity 'hot spot', having about 90% forest cover. The area under forest is 19,117sq. km (SFR, 2011). As a result of anthropogenic activities about 61% of the forests are open, about 29% are moderately dense, less than 1% are dense and about 9% are regarded as non-forest areas (SFR, 2011). The state's forests are governed by the Mizoram (Forest) Act, 1955 and commercial utilization of forest resources is prohibited. However, small-scale felling is permitted in order that bona fide locals can meet their basic needs (SFR, 2006).

Genesis of Mizoram's shifting agriculture

The forefathers of the Mizo people lived a nomadic lifestyle characterized by the practice of shifting cultivation. The practice was passed down from generation to generation. It continues today because of the transfer of knowledge and agricultural tools from earlier generations, providing the opportunity for the pursuit of agriculture with no initial cost. There has also been a lack of productive and economic alternatives. As a result, shifting cultivation has had a deep influence on the socio-culture of Mizo society. Many traditional festivals are named and observed on the basis of various stages of shifting cultivation, and of the 12 months of the lunar calendar, six months are named for different stages of swidden farming.

Many Mizo historians have claimed that shifting cultivation was practised long before the people settled in the area that is now Mizoram, and it was the sole occupation of the Mizo until the late 19th century (Liangkhaia, 1938; Siama, 1953; Lalthangliana, 1998, 2001; Lianthanga, 1999). The advent of the British colonial empire in the late 19th century offered opportunities for new occupations other than shifting agriculture, but a significant proportion of the population still depends upon it for their livelihood (Anonymous, 2009).

Major activities and crops under shifting cultivation

Shifting cultivation, in Mizo called *Lo neih*, is practised throughout Mizoram with only slight modifications from place to place to accommodate different topography and climate. It involves the clearing of forest followed by burning, sowing and reaping (Figure 20-2). Traditionally, village elders select the sites in December and January, then allot the land to the *jhumias* for cultivation. The *jhumias* prepare the land by slashing the forest trees in January and February and burning the slashed vegetation in March and April. Seeds are sown in May prior to the onset of the monsoon. There is nowadays a slight change in the timing of these procedures because the State Government has ordered that all lands allocated for shifting cultivation should be burnt before 15 March. Moreover, a fresh order regarding this timing is issued by the authorities every year.

Weeding takes place three times a year and on each occasion it is given a specific name in the Mizo language –*hnuhlâk*, *hnuhhram* and *pawhchhiat*. Harvesting begins in the month of July and lasts until December, depending upon the crops. In Mizoram, the predominant practice is to crop a *jhum* for one year. Typical shifting cultivation crops are listed later in this chapter. Some of the most common crops are upland rice (*Oryza sativa*), maize (*Zea mays*), chillies (*Capsicum annuum*),

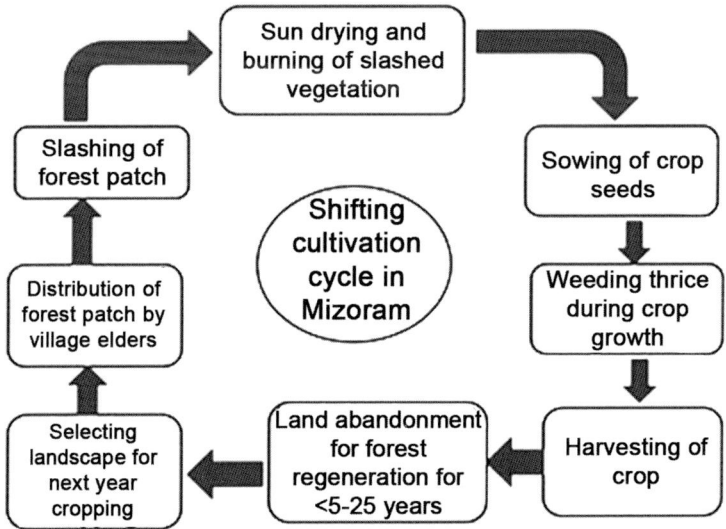

FIGURE 20-2: Flowchart showing the major events of the shifting agriculture cycle in Mizoram.

mustard (*Brassica juncea*), taro (*Colocasia esculenta*), eggplant or 'brinjal' (*Solanum melongena*), cowpea (*Vigna unguiculata*), squash (*Cucurbita maxima*) and Indian nightshade (*Solanum anguivi*). In addition, ginger (*Zingiber officinale*) and turmeric (*Curcuma longa*) are frequently planted in recently burned sites because they grow well on steep slopes and, unlike bananas for example, are high-value crops that store and transport well.

Impact of various land-use policies on shifting agriculture in Mizoram

Since most of Mizoram's people are farmers, the Department of Agriculture has been the nodal agency acting to achieve land-use reform since its establishment in 1972. Every government has attempted to enhance the livelihood of farmers by formulating certain land-use policies based on theoretical principles or by replicating ideas from other places. Regular changes in the political allegiance of various governments has seen a variety of policies introduced from time to time, including those described in the following sections.

Garden colony

This was the first land-use policy implemented in Mizoram under the Integrated Rural Development Programme during the years from 1979 to 1983. The main focus was to effect a move from shifting to permanent cultivation by incorporating horticultural crops like bananas, oranges and pineapples. The Agriculture and Horticulture Department implemented the scheme in agricultural lands around the state capital Aizawl, with a view to creating gardens in compact areas (Lalnunmawia and Lalzarliana, 2013). Under this scheme, farmers were given subsidies of 375 rupees (about US$47.40) per hectare on the purchase of seeds of various agricultural and horticultural crops. In 1984, just five years after the policy was implemented, the ministry lost its power and a new government ended the scheme before any fruitful outcome could be achieved. A new land-use policy was then introduced.

Abelmoschus esculentus (L.) Moench [Malvaceae]

Okra, or Ladies Fingers, is commonly grown in *jhums* in Mizoram. Although the fruit is most often used in local dishes, the whole plant is edible. It is also a perennial, capable of crossing the line between cultivated swidden and fallow.

Land-use Policy

The new scheme introduced in 1984 was called, simply, the Land-use Policy. It was a successor to the previous scheme in terms of its objectivity. Action plans and schemes were drawn, but before the scheme could be implemented, a peace accord between the insurgent Mizo National Front and the Government of India ushered in a new Ministry led by the Mizo National Front in 1987. Like its predecessor, the 1984 Land-use Policy simply faded away without any outcome and the new Ministry initiated another programme.

Jhum *Control*

The new policy in 1987 was called *'Jhum* Control'. Its aim was to achieve settled cultivation in Mizoram by introducing plantations of coffee, oranges and pineapples. It focused intensively on the Aibawk Community Development Block (more recently called Aibawk Rural Development Block), headquartered at Aibawk village, about 30km south of Aizawl city, where it was given the name 'Aibawk *Jhum* Control Project' (Lianzela, 2008). Under this scheme, the entire development block, comprising 22 villages with a total population of 17,128, became involved in the years from 1987 to 1990. The scheme was funded by the North Eastern Council and administered by the state Agriculture and Soil Conservation Department (Lalnunmawia and Lalzarliana, 2013). It was more systematic than its predecessor as the beneficiaries were: (1) supervised and monitored by technical personnel from the department; (2) given financial assistance soon after the completion of one task so that they could proceed with the next; (3) given liberty to choose the trade they preferred and the materials were provided; and (4) informed through awareness campaigns, leaflets and brochures. However, the scheme also had its drawbacks (Lalnunmawia and Lalzarliana, 2013). First, each family was allotted two hectares of land and this was too large for them to manage properly in one year. Thus, only half of the allotted land (i.e. one hectare per family) was used in the first year and the rest was cultivated in the second year, along with the previous year's plots; Second, the financial support (i.e. 375 rupees per hectare) given to the beneficiaries was insufficient for proper management of large plots. These issues meant that the programme met with only partial success. After the 1991 state general election, the ministry changed and the incoming administration implemented yet another new policy.

New Land-use Policy

In 1991, the ministry formulated the New Land-use Policy, which was an updated version of the 1984 Land-use Policy. It envisaged the provision of an alternative land-based permanent occupation to *jhumias*, with a stable income. It also turned out to be the first properly implemented land-use-based policy in the state. Its major aims and objectives were: (1) to wean *jhumias* away from shifting cultivation; (2) to induce each *jhumia* family to adopt an alternative permanent means of livelihood based on either agriculture, industry or animal husbandry on land allotted for that purpose; (3) to undertake wet-rice cultivation in all potential flat lands in order to promote self-sufficiency in food production; (4) to protect and afforest all land other than that earmarked for programmes under the New Land-use Policy; and (5) to provide marketing outlets for farmers' produce. The programme targeted families who depended upon *jhumming* for their livelihood and those with no permanent means of livelihood, but with able-bodied members capable of physical labour (Lalnunmawia and Lalzarliana, 2013).

Moreover, the state government reorganized and enlarged its Rural Development Department to implement the programme and deputed technical personnel for this purpose from various agriculture and allied departments. The programme was operated on a yearly basis in all of Mizoram's 26 Rural Development

Brassica juncea (L.) Czern. [Brassicaceae]

Mustard greens are among the most common plants grown in Mizoram *jhums*. The pungent leaves are used in many local dishes; the seeds are known as brown mustard and yield a widely-used cooking oil.

Blocks, which were to be covered in their entirety by the end of the state's Eighth Five-Year Plan, from 1992 to 1997. In addition, the beneficiaries were given assistance to undertake subsidiary trades for three consecutive years.

The policy offered 31 main trades in the agriculture and allied sector, animal husbandry sector and industrial sector. Most of the trades (28) were in the agriculture and allied sector, but piggery and poultry were offered in the animal husbandry sector and cottage industries in the industrial sector. Apart from the main trades, the beneficiaries were offered five subsidiary trades that could be 'packaged' with a main trade, for example, the main trade of wet-rice cultivation could have subsidiary trades of either piggery or poultry, but for a main trade of hill cattle (mithun) rearing, there were no subsidiary trades offered. Assistance in the form of cash or kind was released in instalments on the basis of verification reports from Block Development Officers and according to a schedule laid down in a calendar of works.

Considering the financial expenditure and the number of beneficiaries in the years from 1990 to 1994 (Table 20-1), it could be safely, but erroneously, assumed that shifting cultivation had been effectively replaced by settled cultivation and other cottage industries by the end of the plan.

TABLE 20-1: Physical and financial progress under the New Land-use Policy (1990-1994).

Year	Expenditure (million Indian rupees)	Number of beneficiaries
1990 – 1991	120	17,159
1991 – 1992	207	26,773
1992 – 1993	222	25,468
1993 – 1994	279.8	31,147
Total	828.8	100,547

Source: Garbyal (1999).

Assessment of the New Land-use Policy

Academicians reviewed the outcome of the New Land-use Policy from time to time. The introduction of tea plantations in Biate and Ngopa villages in 1994 was a landmark alteration in landuse under the policy, but the beneficiaries were unable to reap income benefits because of a lack of processing units in both areas after the plantations reached maturity and their production was wasted. Garbyal (1999) stated that the level of assistance given to beneficiaries was quite insufficient and income earned from trades was too little to support families. Therefore, most *jhumias* who received assistance under the New Land-use Policy were forced to return to *jhuming* and the policy failed to achieve any of its objectives.

Jha (1995) made the point that the New Land-use Policy of 1994 had a poor impact on the economy and farming systems. In assessing the probable reasons for this poor performance and suggesting possible solutions, he said that the beneficiaries found difficulty in accepting imposed technologies because of a poor approach to extension. Acceptance of the programme as a continuing long-term feature should have been ensured by the appointment of extension workers to work with farmers, examine their requirements and modify the technologies (cropping systems) accordingly. Another hurdle was the lack of a sincere desire to motivate the farmers and frequent transfers of officers associated with the New Land-use Policy. Jha also added poor technology as another shortcoming of the policy because

Eryngium foetidum L. [Apiaceae]

Wild or Spiny coriander or culantro is grown worldwide. The leaves are a vital garnishing, marinating, flavouring and seasoning ingredient in food in Mizoram and India generally. Its flavour is said to be a stronger version of that of common coriander (*Coriandrum sativum* L.).

it lacked sound technology that was capable of increasing farm productivity on a sustained basis without degrading the land (Jha, 1995).

The New Land-use Policy aimed to replace *jhumming*, so most of the beneficiaries should have been those engaged in *jhum* cultivation. However, this principle was broadly ignored. Garbyal (1999) interviewed 116 families at Seling village who benefitted from the New Land-use Policy and found that only 28 of them were practising *jhum* cultivation. Sixty-seven of the families were not *jhumias*. According to Lalremsanga (2011), implementation of the New Land-use Policy in the years from 1991 to 1997 had both positive and negative impacts. Some success stories were observed among farmers practising wet-rice cultivation, but most of the policy's beneficiaries failed to withdraw from traditional *jhum* cultivation.

Mizoram Intodelhna Project

In 2002, a new Mizo National Front government came into power and introduced a new policy. This one was called the 'Mizoram Intodelhna Project', meaning in the local dialect, 'Mizoram Self-sufficiency Project'. Its initial aims were to attain self-sufficiency in food grains, pulses, oil seeds, fruits and spices; to assist farmers – particularly *jhumias* – to adopt settled cultivation; and to improve the livelihood of rural communities. After thorough revision of the project's proposed guidelines, it was implemented with the following aims and objectives: (1) to conserve, upgrade and utilize basic resources (land, water, plant, animal and human) in an integrated manner, in order to meet growing demands for agricultural crops; (2) to generate employment both during and after the project period for small and marginal farmers; (3) to improve the environment and promote agricultural production through farmland development and terraces; (4) to promote the conservation of water resources for domestic use as well as constructing rainwater-harvesting dams for small-scale irrigation schemes; (5) to promote the use of quality planting materials in community nurseries for the development of rainfed horticulture; and (6) to promote livestock production to meet growing demands for meat and manure.

In line with the Mizoram Intodelhna Project's guidelines for community organization and institutional arrangement (Lalnunmawia and Lalzarliana, 2013), a Project Implementing Agency was set up in each Rural Development Block, and these were supervised by District Project Committees. Responsibility for the project's overall execution was given to a Mizoram Intodelhna Project Executive Authority, under the chairmanship of the Agriculture Minister. Further, a State-level Project Committee and a High-level Project Committee were set up under the chairmanship of the Chief Minister, the state's paramount authority, to monitor its smooth and effective implementation. In the 2002-2003 and 2003-2004 financial years, a total of 46,812 families were given assistance (Lianzela, 2008). Initially, the proposed assistance to be disbursed to each beneficiary was 50,000 rupees (about US$1033) over a period of five years. This was to be released in instalments, for example, 7500 rupees in the first year, 15,000 in second and third years and the remaining 12,500 rupees in the final year. However, the exact amount disbursed to each of the beneficiaries was a mere 4000 to 7500 rupees (US\$82 to \$153), as a first instalment. The remaining instalments could not be disbursed due to a lack of funds.

Lianzela (2008), concluded that the Mizoram Intodelhna Project had made no significant impact on farming systems in Mizoram because most of the beneficiaries continued *jhum* cultivation even after the scheme was implemented. It also failed to achieve any perceptible improvement in the economic condition of the beneficiaries.

Revised New Land-use Policy

In 2008, another change of political power saw the implementation of a revised version of the New Land-use Policy of 1991. It incorporated several changes that were aimed at improving the livelihood of poor farmers under a flagship government programme. A draft proposal was submitted to the Central Government of India in February 2009 (Lalnunmawia and Lalzarliana, 2013). It was re-submitted four months later after incorporating suggestions made by an expert Central Government panel. Then, the Planning Commission of India sanctioned initial expenditure of 1 billion rupees (about US\$20.88 million) in the 2009-2010 financial year for the launch of the project, and finally, in July 2010, after a thorough study of the proposal by various committees, India's Cabinet Committee on Economic Affairs sanctioned total spending of 28.73 billion rupees under the policy. The State Government organized awareness and training programmes for bureaucrats and technical personnel involved in implementing the policy, and finally, it was launched across the state in January 2011.

The Revised New Land-use Policy aimed to gradually change shifting cultivation practices with a pattern of land use, people empowerment, environmental conservation, commercial utilization of abundant domestic resources and planned marketing of agricultural products. It planned to provide sustainable means of livelihood to 120,000 families in 750 villages over a period of five years. Remarkably, the policy was formulated after careful analyses of the socio-economic status of the people and hence, multi-disciplinary trades were offered to the beneficiaries. Land-based trades were classified according to a land-capability profile. Eight state government departments undertook various developmental activities related to the land-use policy through community participation and utilization of village institutions.

Successful implementation of the project is expected to transform Mizoram socially and economically, in the assumption that *jhumias* will gradually forego shifting cultivation and switch to permanent and sustainable means of livelihood. Overall growth of the state's economy in the next decade is expected to double, from 8% per annum over the past decade to 16%.

Monitoring the Revised New Land-use Policy

The State Government has established monitoring cells at different hierarchal levels to closely examine and supervise various personnel involved in the project, including both beneficiaries and line departments. A State-level Monitoring Cell, headed by the vice chairman of the state-level Revised New Land-use Policy Implementing Board, will oversee reports compiled by village-level Monitoring Committees and submitted through district-level Monitoring cells. The village-level committees function at grass-roots level and are intended to perform on-the-spot verifications, so they are expected to function without bias and maintain a truthful approach to their task. In 2013, the government that introduced the Revised New Land-use Policy was re-elected, so the life of the policy was extended for another five years. Statistical data show that since 2008, the area under shifting cultivation has declined in Mizoram. Rice production from *jhums*, while falling in the years up to 2010, has more recently levelled out. On the other hand, both the area of paddies and production of wet rice are rising (Figures 20-3 and 20-4).

Centrally sponsored schemes

In addition to state government programmes, several schemes sponsored by India's central government have been implemented in Mizoram with the objective of providing farmers with sustainable livelihoods and permanent land-use systems. Of these, the most notable were the Integrated Wasteland Development Programme, the Integrated Watershed Management Project and the National Watershed Development Project for Rainfed Areas.

Integrated Wasteland Development Programme

This programme was launched in 2000 and 2001, and was focused on land and water management and its contribution to sustainable development of natural resources, environmental conservation and improvement of the socio-economic conditions of resource-poor sections of society. It used a participatory approach to achieving its goals. However, the programme's performance was regarded as unsatisfactory and it was terminated due to the state government's failure to achieve certain criteria laid down by the Government of India.

Integrated Watershed Management Project

This programme sought to restore an ecological balance by harnessing, conserving and developing degraded natural resources such as soil, vegetative cover and water. Pior to 2008, the Rural Development Department had implemented three watershed programmes: the Integrated Wasteland Development Programme, the Drought Prone Area Programme and the Desert Development Programme. These programmes were combined under the name Integrated Watershed Management Programme and were implemented under the Common Guidelines on Watershed Development, 2008.

Area under *Jhum* and WRC

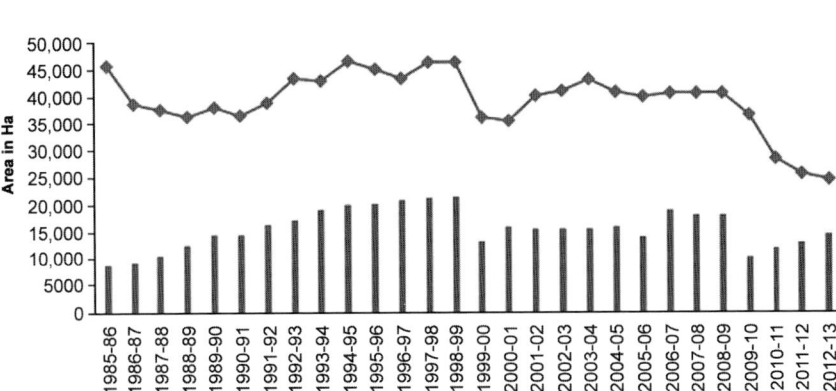

FIGURE 20-3: Area (ha) in Mizoram under shifting cultivation (*jhum*) and wet-rice cultivation (WRC). Line represents *jhum* and bars represent wet rice.

Rice production in *Jhum* and WRC

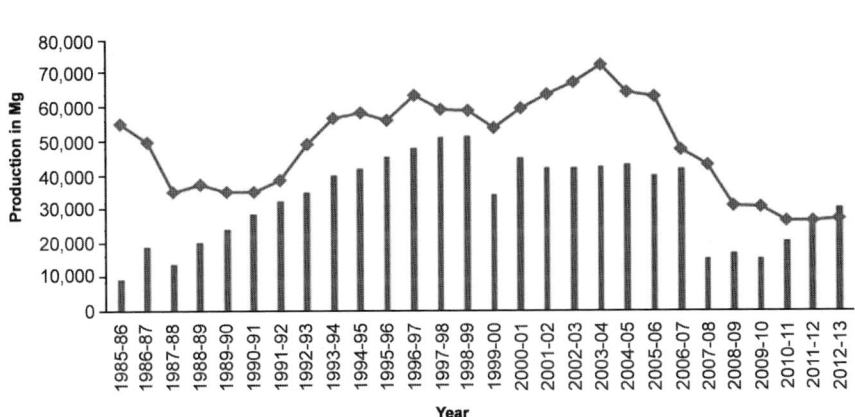

FIGURE 20-4: Rice production in Mizoram from shifting cultivation (*jhum*) and wet-rice cultivation (WRC). Line diagram represents *jhum* and bars represent wet rice.

Note: In rice volume, Mg (Megagrams) equals tonnes.

National Watershed Development Project for Rainfed Areas

This project was implemented during the 11th Five-year Plan, beginning in 2007, as a 'thrust programme' in all of Mizoram's eight administrative districts, with the Agriculture Department as the nodal agency. Development efforts were concentrated on both arable and non-arable lands, and included treatments of natural-drainage lines. High priority was given to sustainable integrated farming systems in rainfed areas on a watershed basis, with emphasis on food production, reducing the regional disparity between irrigated and rainfed areas, increasing employment opportunities, restoring the ecological balance and reducing the need for migration within watersheds. Further, it aimed at in situ conservation of moisture, mainly through vegetative measures, to conserve rainwater, control soil erosion and generate a green cover on both arable and non-arable lands.

Watershed Development Programme In Shifting Cultivation Areas

This was a special central assistance programme for the benefit of *jhumia* families who were living below the poverty line in the northeastern states. The scheme was implemented for five years, from 2006 to 2011. It involved 61 projects with a total treatable area of 30,000ha, and its main focus was on natural-resource management, economic enhancement to alleviate poverty and eco-friendly living.

The National Agriculture Development Programme (Rastriya Krishi Vikash Yojana)

India's slow growth rate in its agriculture and allied sectors prompted the National Development Council to launch a country-level scheme in all states, aimed at (1) stimulating public investment in agriculture and allied sectors; (2) ensuring that all districts and states formulated agricultural plans on the basis of agro-climatic conditions, availability of technology and natural resources; (3) reducing the gap between potential and actual yields in important crops and increasing production in agriculture and allied sectors through focused initiatives; (4) ensuring emphasis on local needs, crops and priorities in these district and state agricultural plans; (5) providing flexibility and autonomy to states in planning and implementing schemes related to their agriculture and allied sectors; and (6) maximizing the incomes of farmers.

Spilanthes acmella var. *oleracea* (L.) C. B. Clarke [Compositae] A synonym of *Acmella oleracea* (L.) R. K. Jansen

Known in some places as the 'toothache plant', the flowering buds have a tingling and numbing effect. *Jhum* farmers in Mizoram harvest the leaves for their strong flavour, but when boiled as a leafy green, the flavour is moderated.

The Green Revolution Mizoram Movement

As recently as November 2014, the All Mizoram Farmers' Union launched a programme known as the Green Revolution Mizoram Movement. It was the outcome of two workshops, one held in September 2010 and called 'Sustainable Agriculture Revolution in Mizoram' and the second in June 2012, called 'Green Revolution in Mizoram'. It is regarded as Mizoram's first non-governmental and non-political land-use based economic policy. To quote the All Mizoram Farmers' Union (AMFU, 2014), the aims and objectives of the Green Revolution Mizoram Movement are as follows:

1. To attain self-sufficiency in food grains, with a special focus on rice as the staple-food crop.
2. To conserve the environment, maintain the ecological balance and promote sustainable use and development of natural resources.
3. To promote the production and marketing of selected crops for better livelihoods.
4. To develop better market linkages for produce.
5. To work hand-in-hand with the state government to establish a partnership with funding agencies, in order to achieve the movement's visions.

The All Mizoram Farmers' Union is the state's biggest and most respected non-governmental organization dealing with land-use based programmes. It was established by a group of farmers in 1994, and has since been working to uplift farmers and develop agriculture. To fulfill the aims and objectives of its new Green Revolution Mizoram Movement, the Union has formulated the following action plans:

1. In administrative reforms, to develop town planning and re-orient the agriculture and allied departments.
2. In land reforms, to restructure the land-ownership system and division of land.
3. In environmental and ecosystem development, to safeguard and conserve the ecosystem by adopting a sustainable approach to utilization and development of natural resources.
4. In economic strategies, the development of infrastructure, crop selection, farming systems, livestock development, financial-support systems, a family-support scheme and a mitigation scheme for victims of natural disasters.
5. Promotion of Changkham technology, a soil- and moisture-conservation technique based on a traditional method.
6. In agricultural marketing, the establishment of market linkages to improve incomes.
7. The development of a Green Revolution Mizoram Movement calendar to earmark the movement's activities.

The organizational structure of the All Mizoram Farmers' Union enhances the potential for achieving its targets, provided it gets the co-operation of the state government. Efforts are being made to involve funding agencies to enhance the chances for success. The organization has great potential in terms of manpower, basic know-how in upland farming, involvement of technical personnel and the workmanship of its members. Since the Green Revolution Mizoram Movement was launched only recently, there has yet to be any significant action. However, the organization has been taking all the necessary steps to create a path for success.

An overview of land-use systems in Mizoram

Mizoram's land-use and land-cover pattern has been classified into six categories: built-up land (Settlement area), agricultural land, forest, shifting cultivation land, scrubland and bodies of water. Moreover, the land has been classified on the basis of a slope and soil survey. The state has 58,638ha of land with slopes of less than 10%. This is considered potential land for cultivation of wet rice and other seasonal crops (Lalnunmawia and Lalzarliana, 2013). A further 1,217,325ha has moderate slopes of 10% to 50%), and this is suitable for terracing, horticulture and plantation crops. A total of 832,770ha of land is steeply sloped at more than 50% and can be used for medium sized plantations of fruit crops such as bananas, citrus, *Parkia*, mangoes, jackfruit, or bamboo. Exceptionally steep slopes are recommended for forest regeneration.

In recent years, policies aimed at persuading *jhumias* to move to permanent agriculture have introduced several exotic trees, such as *Eucalyptus* spp. and teak (*Tectona grandis*), for plantation on government and private land. As a result, a large area of the state is now occupied by these plantations (Vanlalfakawma, 2009). Further, a gregarious bamboo flowering in 2006 was followed by the death of bamboo (*Melocanna baccifera*) over large areas and the state government of that time introduced several exotic bamboo species under the National Bamboo Mission, aiming to restore the plantations affected by the bamboo flowering. In 2008, the state government, under its Revised New Land-use Policy, encouraged large scale plantation of Red Oil Palm (*Elaeis guineensis*). This resulted in

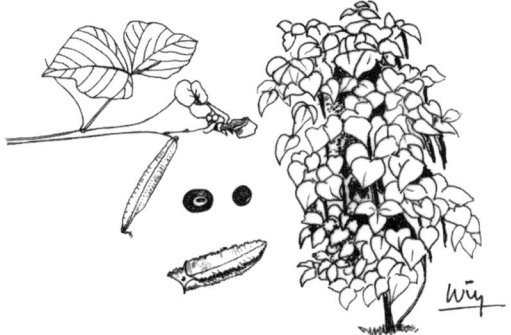

Psophocarpus tetragonolobus (L.) DC. (Leguminosae]

The Winged Bean, a native of New Guinea, is an underutilized species with the potential to be a major food crop. Every part of the plant is edible – pods, leaves, flowers, beans and roots – making this climbing species a popular food source in Mizoram's *jhumming* communities.

reduction of the shifting cultivation area and, as farmers strove to produce the same amount of food from smaller plots of land, the length of the *jhum* cycle suffered and this exerted more pressure on land. It appears that decisions to encourage tree plantations were taken only with revenue collection in mind and the potential impact of these plantations on ecosystems and microclimate was completely ignored.

Over the past two decades, many changes have occurred as a result of six land-use policies, but the area used for *jhum* cultivation still dominates Mizoram's land-use systems (see Table 20-2 for a list of plants commonly cultivated in *jhums*). The amount of rice produced from *jhum* fields still exceeds production from wet-rice cultivation. However, changes in crop selection for *jhum* fields, coinciding with the popular adoption of the System of Rice Intensification (SRI) methodology, resulted in higher production of rice from permanent agricultural land in 2012-2013 (Anonymous, 2013). The low production of rice in the years from 2006 to 2008 was due to an infestation of rodents following the gregarious flowering and seeding of bamboo (*Melocanna baccifera*).

TABLE 20-2: List of plants cultivated in shifting cultivation sites in Mizoram.

Local name (Mizo)	Botanical name	Common name	Family	Parts used
Anka-sa	*Spilanthes acmella* var. *oleracea*	–	Compositae	Leaf
Antam	*Brassica juncea* [**]	Mustard	Cruciferae	Leaf
Anthur	*Hibiscus sabdariffa*	Roselle	Malvaceae	Leaf
Awmpawng	*Luffa acutangula*	Ridged gourd	Cucurbitaceae	Fruit
Bakhawr	*Eryngium foetidum* [*]	Wild coriander	Umbelliferae	Leaf
Bal	*Colocasia esculenta* [**]	Taro	Araceae	Leaf & tuber
Balhla	*Musa* spp.	Banana	Musaceae	Fruit
Bawkbawn	*Solanum melongena* var. *esculentum* [**]	Brinjal/egg plant	Solanaceae	Fruit
Bawrhsaiabe	*Abelmoschus esculentus* [*]	Lady's finger	Malvaceae	Fruit
Behlawi	*Vigna unguiculata* [**]	Cow pea	Papilionaceae	Leaf & fruits
Behliang	*Cajanus cajan*	Lentil/pigeon pea	Papilionaceae	Fruit
Bekang	*Glycine max* [*]	Soybean	Papilionaceae	Fruit/seeds
Bepui	*Lablab purpureus*	Hyacinth bean	Papilionaceae	Fruit
Bepuithlanei/ Bepawr	*Psophocarpus tetragonolobus*	Winged bean	Papilionaceae	Fruit
Berul	*Trichosanthes cucumerina*	Snake gourd	Cucurbitaceae	Fruit
Buh	*Oryza sativa* [**]	Rice	Poaceae	Fruit
Changkha	*Momordica charantia* [*]	Bitter gourd	Cucurbitaceae	Leaf & fruits
Chhawhchhi	*Sorghum cernuum* [*]	White durra/ sorghum	Poaceae	Seeds
Dawnfawh	*Citrullus lanatus*	Watermelon	Cucurbitaceae	Fruit

TABLE 20-2 (cont.):   List of plants cultivated in shifting cultivation sites in Mizoram.

Local name (Mizo)	Botanical name	Common name	Family	Parts used
Fanghma	Cucumis sativus*	Cucumber	Cucurbitaceae	Leaf
Fangra	Canavalia gladiatus	Sword bean	Papilionaceae	Fruit
Fu	Saccharum officinarum*	Sugar cane	Poaceae	Culm
Hmarchate	Capsicum annuum**	Chilli	Solanaceae	Fruit
Hmarchapui	Capsicum frutescens*	Hot long pepper	Solanaceae	Fruit
Hmazil	Cucumis melo var. saccharinus	Honeydew melon	Cucurbitaceae	Fruit
Kawlbahra	Ipomoea batatas	Sweet potato	Convolvulaceae	Tuber
Lakher Anthur	Hibiscus sabdariffa var. sabdariffa	Roselle	Malvaceae	Leaf & fruits
Lengser	Elsholtzia communis*	–	Labiatae	Leaf
Mai	Cucurbita maxima**	Squash	Cucurbitaceae	Leaf & fruits
Maipawl	Benincasa hispida	Ash gourd/melon	Cucurbitaceae	Fruit
Mim	Coix lacryma-jobi	Job's tears	Poaceae	Fruits/seeds
Pangbal	Manihot esculenta*	Cassava/tapioca	Euphorbiaceae	Tuber
Pardi	Trachyspermum roxburghianum	–	Umbelliferae	Leaf
Phuihnam	Clerodendrum colebrookianum	–	Verbenaceae	Leaf
Runhmui	Ocimum americanum	Wild/hoary basil	Labiatae	Leaf
Samtawk	Solanum spp.*	–		Fruit
Samtawk-te	Solanum anguivi**	Indian nightshade	Solanaceae	Fruit
Satinrem	Solanum spp.*	–		Leaf
Sawhthing	Zingiber officinale**	Ginger	Zingiberaceae	Tuber
Thialbal	Maranta arundinacea	Arrowroot	Marantaceae	Tuber
Thingfanghma	Carica papaya	Papaya	Caricaceae	Fruit
Tumthang	Crotalaria juncea	Sunn hemp	Papilionaceae	Leaf
Vaimim	Zea mays**	Maize	Poaceae	Fruit
Zawngtur	Pueraria montana var. chinensis	–	Papilionaceae	Tuber/root
Zo-Purun	Allium hookeri*	–	Liliaceae	Leaf & root
Inaieng	Curcuma longa*	Turmeric	Zingiberaceae	Tuber

Notes: ** = most common species, and * = common species grown in shifting cultivation fields in Mizoram; the plant family Papilionaceae is a subfamily of Leguminosae.

Current trends in land use in Mizoram

As a result of various land-use policies implemented over the past two decades, several trends are now apparent, including: (1) poor forest regeneration due to shortened fallow periods in the *jhum* cycle; (2) reduced agricultural productivity due to decreasing soil fertility; (3) conversion of community land to private ownership; (4) increasing landlessness and social insecurity following the privatization of land; (5) conversion of shifting cultivation land into exotic plantations, including *Eucalyptus* spp., *Tectona grandis*, *Aleurites fordii*, *Jatropha curcas*, several bamboo species, red oil palm (*Elaeis guineensis*), rubber (*Hevea brasiliensis*) and so on, reducing land available for cultivation of subsistence food and cash crops; (6) loss of authority by traditional community institutions; and (7) government encouragement of afforestation of *jhum* lands and consequent migration of the rural poor to urban areas.

Conclusions

There are two general views on shifting cultivation: (1) it is a wasteful and ecologically dysfunctional system that is detrimental to both forests and soil, and therefore needs to be eradicated by motivating *jhumias* to opt for other forms of livelihood; and (2) it is a legitimate practice that ensures the survival of people living on marginal lands and hence should be allowed to continue without external influence. The policies framed in Mizoram were based on the former view and constantly emphasized replacement of *jhumming* with permanent agriculture. This approach was regarded as both patriotic and an example of good governance. Further, a single policy was formulated for the whole state, regardless of the physiography, traditions and customs of the community, with the objective of weaning *jhum* farmers away from their traditional system to adopt a permanent land use system designed for land that was almost flat. And while *jhumias* are still discouraged from practising shifting cultivation on the grounds that it is detrimental to forests, corporations are encouraged to enrich the government treasury in the name of economic development even though they destroy forests. The eminent Indian scientist, Professor P. S. Ramakrishnan, who spent many years researching *jhumming* and *jhumias* in northeast India, has rightly commented: 'A system this old cannot be that bad'. Shifting cultivation continues to evolve after many centuries, and it manages to serve the rural poor. In fact, it is still their only source of food. The ecological services rendered by traditional shifting agriculture in terms of regeneration during the fallow stage are still unrecognized by policy-makers. In Mizoram, bamboo is a pioneer species

Maranta arundinacea L.
[Marantaceae]

Arrowroot is grown in Mizoram's *jhums* for the starch obtained from its rhizomes. This large perennial herb grows in many of the world's rainforest habitats. The roots are harvested when the plants are about one year old.

that occurs abundantly in secondary succession and serves as a multi-purpose species for rural people. Bamboo occupies about 33% of the total geographical area of the state (Anonymous, 2010) in pure stands or mixed with natural forests (Vanlalfakawma et al., 2014). The large extent of bamboo is attributed to its rapid regeneration as a pioneer species in secondary succession on abandoned *jhum* land. The species has tremendous potential to sequester carbon from the atmosphere (Vanlalfakawma et al., 2014) and can generate income from many utility products in markets both in Mizoram and beyond, if properly managed.

Statistical data show that *jhuming* is still being practised by a majority of Mizoram's population, so when formulating any future policy that targets *jhumias*, there should be deep and committed involvement by bureaucrats, educationists from different backgrounds, opposition leaders in the state Legislative Assembly and groups of farmers from different areas of the state. Care should be taken in overall policy implementation to ensure that: (1) it benefits overall society; (2) beneficiaries are selected without nepotism; (3) it is implemented for at least 10 years, with allowance for minor modifications if needed, or if the government changes; (4) it is cost effective; (5) disseminated knowledge is properly provided to farmers; (6) it is suitable to the physiography of the state and the traditions and customs of Mizoram society; (7) it covers the availability of water for irrigation and introduction of high-yielding rainfed crops to farmers; (8) it provides for improvements in transport systems and market avenues; and (9) the impact of alien species on the ecological balance is properly assessed before their introduction.

In the context of Mizoram, shifting cultivation is presently a system that no policy can overcome. Therefore, it is advisable that instead of trying to change the tradition and culture of rural society, policy-makers should accept high-quality research advice and reform certain innovative systems that can be popularly accepted as harmonious supplements to *jhum* cultivation that are both cost-effective and eco-friendly.

Acknowledgement

The authors are grateful for the financial support of the Department of Biotechnology, Government of India.

References

AMFU (2014) *Green Revolution Mizoram Movement: The New Economic Policy 2014*, All Mizoram Farmers' Union, Aizawl, Mizoram

Anonymous (2003) *Mizoram Forest*, Department of Environment and Forest, Government of Mizoram, Aizawl, pp1–64

Anonymous (2004a) *Statistical Handbook of Mizoram 2004*, Directorate of Economics and Statistics, Government of Mizoram, Aizawl

Anonymous (2004b) *Statistical Abstract*, Department of Agricultural and Minor Irrigation, Government of Mizoram, Aizawl

Anonymous (2009) *Statistical Handbook of Mizoram 2009*, Directorate of Economics and Statistics, Government of Mizoram, Aizawl

Anonymous (2010) *Bamboos of Mizoram*, Department of Environment and Forest, Government of Mizoram, Aizawl

Anonymous (2011) *Meteorological Data of Mizoram for the Year 2009*, Department of Economics and Statistics, Government of Mizoram, Aizawl

Anonymous (2013) *Agriculture Statistical Abstract 2012-2013*, Directorate of Agriculture (Crop Husbandry), Government of Mizoram, Aizawl

Garbyal, S. S. (1999) '"*Jhuming*" (Shifting cultivation) in Mizoram (India) and new land use policy: How far has it succeeded in containing this primitive agriculture practice?', *India Forester* 125(2), pp137-148

Grogan, P., Lalnunmawia, F. and Tripathi, S. K. (2012) 'Shifting cultivation in steeply sloped regions: A review of management options and research priorities for Mizoram state, Northeast India', *Agroforestry Systems* 84, pp163-177

Jha, L.K. (1995) *Advances in Agroforestry*, APH Publishing, New Delhi

Lalnunmawia, F. and Lalzarliana, C. (2013) *Land Use systems in Mizoram*, Akansha Publishing House, New Delhi, ISBN 978-81-8370-335-2

Lalremsanga, K. (2011) 'Ecological and economical analysis of different land-use systems in Mizoram', MSc Dissertation (unpublished), Department of Forestry, Mizoram University, Aizawl

Lalthangliana, B. (1998) *Pi Pu Chhuahtlang: Studies in Mizo Culture and Folktales*, Hrangbana College, Aizawl

Lalthangliana, B. (2001) *India, Burma and Bangladesh - a Mizo Chanchin*, Remkungi, Aizawl

Liangkhaia (1938) *Mizo Chanchin*, (fifth edition, 2002) LTL Publications, Mission Veng, Aizawl

Lianthanga, C. (1999) *Hmanlai Mizo Nun*, Mizoram Publication Board, Aizawl

Lianzela (2008) 'Political economy of Mizoram: A Study of MIP', in J. K. Patnaik (ed.) *Mizoram, Dimensions and Perspectives: Society, Economy, and Polity*, Concept Publishing Company, New Delhi

Pachuau, R. (1994) *Geography of Mizoram*, R. T. Enterprise, Aizawl

SFR (2006) *State of Forest Report*, Department of Environment and Forest, Government of Mizoram, Aizawl

SFR (2011) *State of Forest Report 2009*, Forest Survey of India, Ministry of Environment and Forest, Government of India, Dehradun

Siama, V. L. (1953) *Mizo History*, (reprint, 2001), Lengchhawn Press, Khatla, Aizawl

Vanlalfakawma, D. C. (2009) 'Analysis of soil fertility and litter nutrient retranslocation of natural forests and teak plantations of Aizawl district of Mizoram, India', MSc Dissertation (unpublished), Department of Forestry, Mizoram University, Aizawl

Vanlalfakawma, D. C., Tripathi, S. K. and Lalnunmawia, F. (2014) 'Estimation of biomass of three major bamboo species of Mizoram, India', in Lalnuntluanga, J. Zothanzama, Lalramliana, Lalduhthlana and H. T. Lalremsanga (eds) *Issues and Trends of Wildlife Conservation in Northeast India*, Mizo Academy of Science, Aizawl, ISBN 978-81-924321-7-5

Vanlalfakawma, D. C., Lalnunmawia, F. and Tripathi, S. K. (2014) 'Soil carbon pools of bamboo forests of Mizoram, India', *SciVis* 14, pp46-50

21

STATE LAND POLICIES AND SHIFTING CULTIVATION IN ODISHA, INDIA

*Kundan Kumar**

Introduction

Shifting cultivation of different kinds was widely practised in forested areas of India in the pre- and early-British periods. It formed part of a mixed use 'agroforestry landscape that included domesticated forests, grasslands and farms nestled amongst each other' (Sivaramakrishnan, 1999) in the hilly and forested zones of India. However, by the time Independence came, shifting cultivation as an important land use survived only in certain pockets of central India and in the northeastern part of the country. The Eastern Ghats in Odisha and contiguous Andhra Pradesh remain the main shifting cultivation zones in mainland India. Disputes and conflicts related to shifting cultivation are common in these areas.

In this chapter, I examine the processes through which hill slopes customarily used by ethnic communities in Odisha for shifting cultivation were taken over by the state as forest lands or revenue wastelands through deliberate policies aimed at bringing shifting cultivation to an end. Similar processes occurred in other parts of India, but with the important difference that the tenurial transformation in Odisha happened after Independence under a democratic dispensation. Another difference is that in spite of state attempts to stop shifting cultivation, it persists in parts of Odisha. The failure to recognize the rights of shifting cultivation communities in the state to the hill slopes they traditionally cultivated has had serious implications on their access to land and livelihoods.

The process of denying legitimacy to the practice of shifting cultivation has been common around the world, and the experience of Odisha's shifting cultivators fits this mould. However, the dispossession in Odisha was remarkable for a number

---

* Dr Kundan Kumar is Director for Asia, Rights and Resources Initiative, Washington, DC. At the time of submitting this chapter he was an Assistant Professor (CLTA), Faculty of Forestry, University of Toronto, Canada.

of reasons: (1) it took place under a democratic regime which was simultaneously involved in land reforms aimed at providing legal rights over land 'to the tillers'; (2) shifting cultivators formed a sizeable minority of the population; and (3) the process of criminalizing shifting cultivation occurred through a complex conjunction of laws, procedures and practices. Ironically, all of them provided for the settlement of 'tillers' rights over the land they used.

I discuss briefly the widespread occurrence of swidden cultivation in mainland India and the efforts of the colonial rulers to end it. I then discuss the colonial and post-colonial situations in Odisha that led to denial of the legitimacy of shifting cultivation. I end with the impacts of these processes on shifting cultivators and landscapes.

Shifting cultivation in mainland India

In general perception, shifting cultivation in India has been primarily practised in the northeastern part of the country, where it is commonly referred to as *jhum*. However, in the past, shifting cultivation was widely practised in almost all forested parts of India. A number of authors have recently discussed the variety of shifting cultivation practices in different parts of the country and the efforts of the colonial rulers to extinguish them.[1]

As the British expanded their control over forested areas of colonial India in the 19th century, shifting cultivation was seen as a major threat to valuable forest resources. The difficulties of generating land taxes from often rebellious shifting cultivators and the preference for a stable, tax-paying sedentary peasantry was another reason for strongly discouraging shifting cultivation (Jyotishi, 2000; Kumar et al., 2005). Forest and land laws were used to restrict shifting cultivation. The forest laws, i.e. the Madras Forest Act, 1882 and the Indian Forest Act, 1927,[2] explicitly criminalized the felling trees and setting of fires. Land laws often withdrew or limited the legitimacy of shifting cultivation as a valid cultivation practice.

In 1801, Francis Buchanan highlighted widespread *kumri* cultivation (shifting cultivation) in the Western Ghats of Uttara Kannada (Chandran, 1998, p678). Restrictions on *kumri* were imposed from 1848 and became progressively stricter. *Kumri* was also banned in Coorg in 1848 and in Belgaum in 1856 (Stebbing, cited by Rangarajan, 1996). *Dalhi* cultivation practised by Warlis, Thakurs, Katkaris and other *adivasis*[3] of North Konkan and Thane district was brought under restrictions from as early as the 1850s. Attempts to control shifting cultivation in India's central provinces began in 1862 with the passing of forest rules that banned *dhya* by the Gonds, Korkus, Baigas and Bhumias and other hill tribes in well-timbered areas (Rangarajan, 1996). By 1900, shifting cultivation was eradicated in almost all forests under direct British control in the central provinces. In the Dangs of Gujarat, restrictions on shifting cultivation (*khandad*) in teak and sissoo forests were initiated in 1844. By 1909, it could be claimed that *khandad* had completely ceased in the Dangs (Skaria, 1999).

Similar historical narratives outlining the erasure of shifting cultivation are available from other parts of the country (see Chandran, 1998, for Uttar Canara; Saravanan, 1998, for parts of Tamil Nadu). By Independence, shifting cultivation had been more or less stamped out in most of mainland India by the colonial regime and was confined to parts of the Eastern Ghats in Odisha, northern Andhra Pradesh and the Bastar region of Madhya Pradesh. In the remote forested hills of the Eastern Ghats, particularly in Odisha, the British were unable to stop shifting cultivation for various reasons and it continued as a major land-use practice, even after Independence.

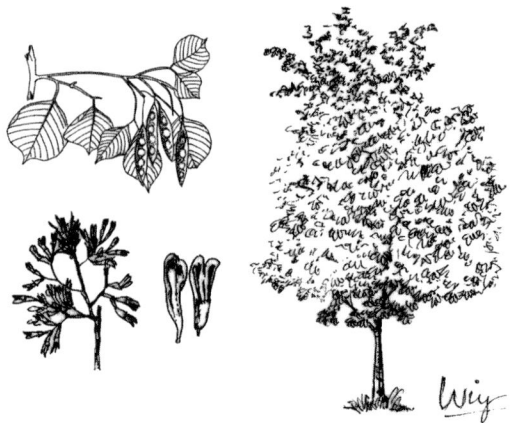

Dalbergia sissoo DC. [Leguminosae]

This species, commonly known as Indian redwood and much sought-after for its fine timber, dominated some forests in 19-century India and was one reason for restricting shifting cultivation.

Shifting cultivation and the colonial state in Odisha

The Eastern Ghats in Odisha are one of the few areas in India where shifting cultivation and its variant practices still persist. Present-day Odisha historically consisted of parts of Madras Presidency, the Central Provinces and Bengal, as well as 24 princely states. A large part of Odisha is hilly, with plains limited to the coastal areas and the western parts of the state. Various major river valleys, such as the Mahanadi and Brahmani, dissect the hilly areas on their way to the Bay of Bengal. Forests range from semi-evergreen on the coastal hills to dry-deciduous in the western part of the state, with sal (*Shorea robusta*)-dominated deciduous forests being the most common. Officially, as many as 62 scheduled tribes comprise almost 22% of the state's population, with Kandhs being the most numerous. Most of the ethnic population is concentrated in the hilly parts of the state. The regions surrounding these mountainous upland tracts have generally been dominated by communities whose social status was fixed by India's caste system, and who practised settled cultivation.

The major communities that have been practising shifting cultivation in the hilly parts of Odisha include the Kandhs, Parojas, Gadabas, Bondos and Saoras in southern Odisha and the Juangs, Bhuiyans and Eranga Kols in the north of the state. The major tracts under shifting cultivation are located in south-central Odisha (Kandhamal and Gajapati in the Eastern Ghats); southwestern Odisha (the erstwhile Koraput district and parts of Kalahandi district, forming part of the Eastern Ghats) and a third area covering parts of Keonjhar, Sundargarh and Angul districts in northern Odisha (Figure 21-1).

Shifting cultivation has long been an intrinsic part of the culture and cosmologies of the ethnic communities. Amongst the Kandhs, for example, a clan group lays claim to 'owning' the earth within a territorially defined area, and the political, social and cultural structure of Kandh communities is built around their relationship with the land (Bailey, 1960). Historically, anyone who wanted to settle within these territories needed to obtain permission to do so from the clans (Bailey, 1960). Similar systems existed for other ethnic communities. They once had complex relationships with neighbouring kingdoms and chiefdoms and often enjoyed a great deal of autonomy. These kingdoms often had mountainous frontiers, and the ability of the rulers to exercise control over these communities was limited. This led to a hybridized mosaic of legitimacies and institutions, and there were many examples of rulers seeking legitimacy from various communities through rituals and ceremonies.

British attempts to exercise control over Odisha faced some of the greatest resistance from these communities, especially from the Kandhs in central Odisha. This group fought against the British for decades before they were 'pacified' in the 1850s (Padel, 1995). Although these wars were waged over the issue of human sacrifice, the real reason was that the Kandhs feared that their lands would be taken away or taxed (Bailey, 1960). By the 1900s, the shifting cultivation areas of Odisha were either under British sovereignty or under the nominal sovereignty of princely states. Special administrative establishments called 'Agencies' were designed by the British for *adivasi* areas directly under their control and normal laws were not generally exercised in these areas. The major concern of the British in these areas was two fold: maintaining peace and control and increasing revenue from land and forests.

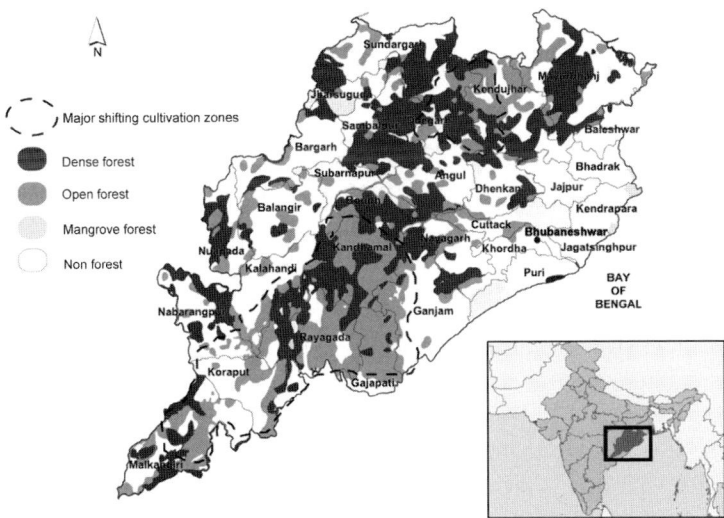

FIGURE 21-1:   Odisha state showing forested areas and major zones of shifting cultivation.

The apparatus for extracting land taxes in these remote and hilly *adivasi* regions remained rudimentary, compared to that in the lowlands, until the 1950s. Survey and settlement and preparation of records of rights were extended to very few upland parts of the country. This was an outcome of the low potential for land revenue from these areas, difficulties faced in carrying out survey and settlement procedures and more important, fear of unrest. It is important to note that in response to the resistance of the Kandhs, the British Government issued a proclamation in 1855 that the Kandhs would be exempt from land-revenue taxation in perpetuity. Areas such as the present Kandhamal district in Odisha were left largely untouched by land settlements, record maintenance and land revenue up to the time of Independence.[4] In other upland areas, especially in the princely states of Kalahandi and Keonjhar, instead of land tax being based on acreage and land category, a simple taxation was assessed on the basis of the number of ploughs or hoes owned, or simply on the basis of an observation. Shifting cultivation was permitted in tribal parts of princely states, e.g. in Juangpirh of Keonjhar and in large areas of Bamra, Bonai, Ranpur and Kalahandi princely states, where shifting cultivators paid a hoe (*kodki*) tax or house tax to the rulers (Ramdhyani, 1947).

The connection between these communities and the higher echelons of state was generally made by intermediary tenure holders, village-heads or others. In most cases these individuals were legally responsible for administering local land at village level. In areas with greater control, especially on the plains, these intermediaries were quite powerful, and could exercise control over the allocation of land and fixing of revenues. In the remote shifting cultivation areas, this power generally tended to be nominal and traditional kinship leaders played this role.

As mentioned earlier, the views of the British administration on shifting cultivation were harsh; it was seen as destruction of valuable timber. This perspective was also echoed in Odisha, where shifting cultivation was seen as a major cause of forest destruction. A 1908 report by the Agency Commissioner of Madras, one Mr. Harries, was quoted verbatim in a 1940 Enquiry Committee Report (GOO, 1940):

> The evils of *podu* (shifting cultivation) are as follows: (i) It causes the springs below the hills to dry up; (ii) causes the soil on *podu* lands to be washed off; (iii) ruins valuable timber for the sake of much less valuable crops or grains; (iv) causes very heavy floods in the rivers below …; (v) causes the hot-weather supply of these rivers to diminish; (vi) brings down heavy silt into tanks and onto fields and destroys the crops

Even though it is very clear that all of these statements can be questioned according to the current scientific understanding of shifting cultivation, the passage illustrates the thinking at that time. This was a widespread perspective that underpinned the administration's opposition to shifting cultivation. Its main strategy was the extension of reserve forests into shifting cultivation areas and enforcing forest laws that prohibited the practice of swidden cultivation.

However, the British were cautious about stopping shifting cultivation because of fear of unrest. As the then Forest Conservator of Orissa, J.W. Nicholson (1938), put it:

> It is essential that the present practice of shifting cultivation should be curtailed and gradually put a stop to… There is certain to be considerable agitation and trouble over further forest reservation in the hill tracts, but the problem is one that will have to be faced boldly.

In parts of present-day Odisha that were under British rule, the extent of forest reservation in shifting cultivation areas was minimal, although plans were made to reserve forests. Thus in Kandhamal, much of the area proposed for reservation as forest was demarcated by the time Independence came, but reservation was never carried out. In other areas used for shifting cultivation, legal forests were created, but no serious efforts were made to curb swiddening and forest laws were not strictly enforced. Forest reservations were established in shifting cultivation areas in the princely states of Kalahandi, Bonai and Bamra, but these remained reservations on paper. The rights and settlement processes were not followed according to the law,[5] and the reservations didn't affect shifting cultivation except in a few pockets. In Keonjhar Princely State, the ruler refrained from notifying the reservation of any forests that were recognized as homes for shifting cultivation communities, such as the Juangs (Juangpirh) and Bhuiyans (Bhuiyanpirh).

Thus, unlike most of mainland India, where shifting cultivation had been stamped out before Independence, swiddening continued in the hilly tracts of Odisha. This was due to fear of unrest, the inability of the colonial state to control remote areas and the difficult nature of the terrain.

The post-colonial period and shifting cultivation

After Independence, the most pressing task for the government of Odisha was to ensure consolidation of the new state, which involved the merger of the 24 princely states and creation of a common administrative system. This included consolidating the land and forest administration and major changes included the abolition of tenure intermediaries and enactment of land reforms laws based on the principle of 'land to the tillers'.

The post-independence government prioritized land administration and began extending survey and settlement procedures into all parts of the state, including the erstwhile untouched remote hilly areas. The forest administration focused on extending its control to cover the vast forest areas in the former princely states and the tracts of forest formerly owned by the aristocratic zamindars. Forest reservation processes were speeded up in the 1960s and 1970s, extending into shifting cultivation zones. These territorialization processes were based on the frameworks and practices bequeathed by the British, with some modifications to take account of land reforms. They had a deep impact on the practice of shifting cultivation and on the communities in Odisha that depended upon it. I will first discuss the scale of shifting cultivation in Odisha in the post-colonial period and then show how state land and forest policies since then have dismantled the territorial nature of swidden farming in the state.

The extent of shifting cultivation in Odisha

There has never been a clear-cut estimate of the extent of shifting cultivation in Odisha that was based on actual measurement, so estimates varying widely. Part of the confusion arises from how shifting cultivation should be defined, and whether fallows should be included in any estimate. In certain areas, due to pressures on land, shifting cultivation has effectively become

Shorea robusta Gaertn. [Dipterocarpaceae]

Sal, as it is known, is one of the most important sources of hardwood timber in India. It dominates many of Odisha's deciduous forests.

short-fallow agriculture, and the fallows are dominated by grasses and bushes. My estimates include both kinds of shifting cultivation fallows, i.e. secondary forests and short fallow areas on hill slopes, within the ambit of shifting cultivation as it is regarded from the perspective of tenure.

Over the past 60 years, various estimates of the area affected by shifting cultivation in Odisha range from 1445sq. km to 32,681sq. km (Table 21-1). The first estimate was made by H. F. Mooney, a former conservator of forests, who used personal observations to claim that about 32,681sq. km (almost 20% of Odisha's total area) was affected by shifting cultivation in 1951. He also estimated that one million people were then dependent on shifting cultivation (GOO, 1958, 1959; Dash, 2006). In 1981, the FAO's Tropical Forest Resources Assessment Project used coarse-resolution Landsat imagery to estimate that 16,580sq. km. of forest in Odisha were affected by shifting cultivation (FAO, 1981).[6] This seems to have been a study that involved some amount of verification and cross-checking. The recent Wasteland Atlas of India shows only about 1445sq. km as being affected by shifting cultivation in Odisha (GOI, 2010), probably because it classifies shifting cultivation fallows and short fallows in the categories of land with open and dense scrub (6828sq. km) and degraded forests (6623sq. km). It is also possible that much of the shifting cultivation land under long-rotation fallow would have been considered as secondary forest. The widely fluctuating estimates make it almost impossible to get a clear picture of the extent of

TABLE 21-1: Different estimates of the area affected by shifting cultivation in Odisha.

Source	Year	Area (sq km)
H F Mooney	1951	32,681
ICAR	1958	8,000
Dhebar Commission	1961	8,333
French Institute Pondicherry & ICAR	1967	30,233
Task Force on Development of Shifting Cultivation Areas, Ministry of Agriculture, GOI	1983	26,490
Tropical Forest Resources Assessment Project, FAO	1981	16,580
Wasteland Atlas of India	2010	1,445

Source: Adapted from Dash, 2006; FAO data from FAO, 1981; Wasteland Atlas of India, 2010 (GOI, 2010).

shifting cultivation in Odisha. However, high resolution Google Earth maps make it clear that in most of the shifting cultivation areas, hill slopes remain under cultivation.

In two areas studied by the author, an analysis of plots under cultivation found that on average a family was farming swiddens covering two to three acres (0.8 to 1.2 hectares) and tended to have two or three patches under rotation. These families also cultivated bunded paddy lands in valleys, although some of them were completely dependent on shifting cultivation. In one of the areas, Kirkicha watershed, there were 121 families, 70 of which were cultivating swiddens on 301 acres (122 hectares). In the other area, Gaurigaon village, there were 29 families who were cultivating 61 acres (24.7ha) on hill slopes (Kumar and Kerr, 2013).

Anecdotal evidence suggests that there has been a clear reduction of area under shifting cultivation over the last few decades, as conservation laws have become stricter in areas defined as legal forests. Efforts by the state to stop shifting cultivation, including aggressive policing and criminalization of shifting cultivation on forest land, have forced many swidden farmers to stop cultivating hill slopes. In certain areas, pressure from the government, along with the availability of alternative income-generating activities through migration and government-led wage-creation schemes, have contributed to the abandonment of hill-slope cultivation. However, the author's recent studies, as well as the maps of Google Earth, show that it remains a major land use in many pockets of Odisha.

There is little acknowledgement of this fact among policy-makers and even less recognition of the rights and claims of shifting cultivators to these tracts of land. As I will show later, the state has erased shifting cultivation, both as a land use and as the basis of claims to rights and livelihoods. This has had major social and ecological consequences. The territorialization processes of the post-colonial state, namely survey and settlement procedures and forest notifications, have been the main instruments for legal dispossession of shifting cultivators' rights to land.

Post-Independence survey and settlement in shifting cultivation areas

The 1940 Enquiry committee report (GOO, 1940), mentioned earlier for its regurgitation of historic beliefs in the 'evils' of shifting cultivation, seems to have formed the basis of government policies on shifting cultivation in Odisha. Official documents from the decades immediately following Independence exemplify the deep bias against both shifting cultivation and *adivasi* communities. They repeat the same notions of shifting cultivation being primitive and destructive and leading to flooding, soil erosion and the drying up of perennial springs. Shifting cultivators are portrayed as being 'indolent and carefree in spirit' and lacking all desire for self-improvement (GOO, 1958). One proposal suggests that swidden communities should be settled in colonies. Such discursive memes are to be found in almost all of the official documents that sought to deal with shifting cultivators in Odisha.

Two major instruments of landscape territorialization – survey and settlement and creation of legal forests – were used to deprive shifting cultivation of its legitimacy. As mentioned earlier, most shifting cultivation areas hadn't undergone survey and settlement procedures prior to Independence, so they were subjected to the territorialization process after 1947. When an area underwent survey and settlement for the first time, certain criteria based on the land laws were used to identify and recognize individual rights over land, and remaining land was recorded as belonging to various categories of state land. Community jurisdiction over landscapes, the practice followed by most ethnic communities, was not recognized in the formal survey and settlement process. This was a legacy of the British concept of property rights. Thus, the first and most critical impact of survey and settlement was a denial of the legitimacy of common control of resources and landscapes.

Survey and settlement was an exhaustive, time-consuming process, carried out by a specialized government department and involving highly skilled land surveying. This meant that the processes of making the landscape legible to the state were totally illegible to the largely illiterate subjects of survey and settlement. At village level, the process involved fixing the boundaries of the village and then conducting a cadastral survey and mapping inside the boundaries. Areas that were seen as remote, or had already been notified as forest under forest laws, were excluded from the village boundaries. Within the village, occupied land was recorded as such and all other areas were classified into different forms of government land. Some general categories of government land within village boundaries were grazing lands, cultivable wastelands, uncultivable wastelands, waterways, playgrounds, and so on. Within the category of wastelands, some areas were categorized as revenue forests.

Within the village, land occupied by families was mapped, categorized and occupancy rights were recorded. There were two main principles, established under various laws, that were used to recognize individual rights over land: evidence of land-tax payments to former regimes and evidence of occupation and cultivation. Given these principles, the survey and settlement process should have recognized shifting cultivation lands as proprietary land, even if community jurisdiction was not recognized. This principle was deliberately overturned, primarily because of a strong bias against shifting cultivation and the unwillingness of the government to hand over large areas of land to ethnic communities.

Settlement in Dongarlas

The farmers of Dongarlas, a hilly southeastern part of Kalahandi princely state, used to pay a 'hoe tax' to the king in order to practise shifting cultivation on hill slopes. The first survey and settlement procedure after Independence in areas occupied by ethnic communities was conducted in Dongarlas. It was the first time that the area had been subjected to survey and settlement. The matter of occupancy rights over shifting cultivation land on hill slopes was referred to the Odisha Government's Board of Revenue in the early 1950s. According to legal precedent, the occupancy rights of the cultivators should have been recognized and recorded in relation to that land for which tax was being paid to the sovereign. However, instead of recording occupancy rights over the shifting cultivation lands, the Government of Odisha established a special category of records for these areas called *dongarkhasra*. The government explicitly stated that the *dongarkhasras* were not to be treated as occupancy rights (Sundarajan, 1963), although it offered no legal reason for this. Based on these instructions, *dongarkhasras* were prepared for 86,000 acres (34,830ha) of shifting cultivation lands. Inexplicably, the government unilaterally withdrew these *dongarkhasras* in the 1980s, and the land was deemed to be unencumbered state land, mostly revenue wasteland.

Koraput settlement

The next major decision regarding the recognition of rights over shifting cultivation land was taken during the survey and settlement procedure in the undivided Koraput district (currently the Koraput, Rayagada, Malkangiri and Nowrangpur districts), in the southeastern part of Odisha. This was the first time that survey and settlement procedures had been applied in this vast area, formerly owned by a zamindar and inhabited mainly by *adivasi* hill communities. Since shifting cultivation was widespread in the area, the Government of Odisha initially adopted the attitude that even though shifting cultivation was harmful to forest growth and soil conservation, it should be regulated rather than completely stopped because the people depended on it for their livelihoods (Behuria, 1965). The government's Board of Revenue decided that occupancy rights would be granted for cultivated land on hill slopes of up to one in ten gradient. Cultivation on steeper slopes was to be regarded as encroachment and the perpetrators liable to be evicted. The physical process of demarcating those areas for shifting cultivation, with slopes up to a gradient of one in ten, began in 1954 and continued in 1955 (Behuria, 1965).

In 1960, the Board of Revenue reversed its earlier decision and issued a letter to the Settlement Officer[7] instructing that no rights should be granted for hill-slope shifting cultivation and all such activities should be treated as encroachments.

Moreover, the land should be recorded as government land. A clever use was made of a provision of the Madras Estate Land Act – the law under which the survey and settlement was being carried out.[8] It stated that 'a person who had occupied *ryoti* [farm] land for a continuous period of 12 years should automatically be considered a *ryot* [farmer] with occupancy rights.' The Board of Revenue decided that 'shifting cultivation couldn't be construed as continuous possession for a period of 12 years' and therefore occupancy rights should not be granted for land under shifting cultivation (Behuria, 1965). The legal and procedural basis for these decisions was not only highly ambiguous, but also a violation of the principle of 'land to the tiller'. The shifting cultivation lands had been very clearly claimed and occupied by the same families for generation after generation and, in any case, continuous occupation of the lands had been claimed by kinship-based communities. It is very clear from this that the government took a highly political decision through a clever interpretation of the law, and that this interpretation, in one stroke, extinguished all rights or claims by shifting cultivators over their ancestral lands on hill slopes.

Other survey and settlement cases

The decision in the Koraput survey and settlement procedure became a precedent, and in all later cases of survey and settlement, shifting cultivation lands were summarily classified as government land. The main shifting cultivation zones affected in this manner were Kandhamal district, the present Gajapati district and Juangpirh and Bhuyanpirh, in Keonjhar district. In Kandhamal, almost half of the area within the village boundaries was classified as state-owned revenue forest, even though this land was commonly used for shifting cultivation.

Legal forests and shifting cultivation after Independence

Forest laws were the second major instrument that affected the rights and claims of shifting cultivators, the main ones being the Madras Forest Act, 1882 (MFA), the Indian Forest Act, 1927 (IFA) and the Odisha Forest Act, 1972 (OFA), which was based on the Indian Forest Act. Various pre-Independence forest laws were also used in the princely states. The Wild Life Protection Act, 1972 (WLPA) and the Forest Conservation Act, 1980, (FCA) were other important post-Independence laws with a major impact on swidden farming. Under these laws, vast areas of reserved and protected forests were created in shifting cultivation zones.

The forest acts (MFA, IFA and OFA) were created to conserve forests, with protection of timber being one of their specific objectives, so they were fundamentally inimical to shifting cultivation. Once an area was declared a forest under these laws, activities that constituted the basis of shifting cultivation, such as 'making fresh clearings', 'setting fires', 'damaging trees' and 'breaking land for cultivation', were explicitly forbidden. Penalties included fines, imprisonment and confiscation of tools. The forest bureaucracy had the power to arrest a person without warrant and impose a fine. Thus, inclusion of an area within a legal forest immediately made shifting cultivation illegal and criminalized the shifting cultivators. As discussed earlier, the criminalization of shifting cultivation and the resultant outrage and protests were among the reasons why the British were cautious about extending the notification of forests into shifting cultivation tracts.

The forest laws provide clear instructions for the settlement of rights in any area before it is notified as either a reserved forest or a protected forest. In the case of reserved forests, the settlement of rights must be completed before an area is finally notified as reserved. However, an area can be declared a protected forest without the settlement of rights, provided that no existing rights of individuals or communities are curtailed as a consequence. But the rights that must be taken into consideration during forest notifications are limited to those rights that have been recognized by earlier survey and settlement procedures or any other law. In cases where there has been no previous survey and settlement procedure, the Forest Settlement Officer will take existing land-use practices and settlements into account before making a decision about excluding an area from the forest. In the pre-Independence period, this often meant that areas where shifting cultivation was being practised would be excluded from the forest, or the government would at least deliberate the strategic advantages of notifying shifting cultivation areas as forests.

After Independence, once the government decided not to recognize rights over shifting cultivation lands in survey and settlement procedures, there was no longer any need to exclude shifting cultivation zones from areas declared as legal forests and large-scale conversion of landscapes took place (Table 21-2). The Government of Odisha used the creation of legal forests as a conscious strategy to establish legal control over areas under shifting cultivation and discourage swidden farming.

There was a second trajectory through which large areas of land were converted into legal forests. The survey and settlement process also provided for the creation within village boundaries of a land category named revenue forests. Almost 6600sq. km of land was categorized as revenue forests by survey and settlement procedures in Kandhamal, Gajapati and undivided Koraput district. These areas were next to settlements and within village boundaries, so they were often used for shifting cultivation or short-fallow farming. Control of these revenue forests was not vested in the Forest Department, but remained with the Revenue Department,[9] with comparatively lax control and regulations. However, this changed after introduction of the Forest Conservation Act in 1980, which effectively banned non-forest land uses in any type of forest, including the revenue forests. This situation was reinforced by the famous Godavarman case interim decision of the Supreme Court in 1996, which defined forest lands as land recorded as any type of forests or land having forest growth (Rosencranz and Lele, 2008; Kumar and Kerr, 2013).

All of the forests notified under the typologies in Table 21-2 were not necessarily areas affected by shifting cultivation. However, empirical case studies, secondary data and anecdotal evidence tell us that large areas of shifting cultivation land were included within legal forests. For instance, in Kirkicha watershed, almost 150 acres

TABLE 21-2: Legal forests created after independence in areas where shifting cultivation is prevalent.

Areas with shifting cultivation	Pre-Independence status	Legal forests at time of Independence	Legal forests at present	Legal forest area created after Independence
Juangpirh and Bhuiyanpirh Keonjhar district	Part of Keonjhar Princely State	No legal forests	One RF (875ha). About 600sq. km more classified as revenue forest*	608sq. km
Kandhamal and Gajapati district	Kandhamal, Baliguda and other Ganjam Agency areas under British rule	About 520sq. km of RF.	Reserved forests: 2426sq. km; DPFs, reserved lands, UDPFs: 2010sq. km; forests created through revenue survey and settlement: 2708sq. km. Total legal forests: 7144sq. km	6624sq. km
Rayagada, Koraput and Malkangiri districts	Jeypore Zamindari	4140sq. km as reserved or protected lands	RF: 2136sq. km; Proposed RFs,★★ Reserved lands, ★★★ UDPFs: 4417sq. km; forests created through revenue survey and settlement: 3948sq. km. Total legal forests: 10,501sq. km	6461sq. km

Notes: RF=reserved forest; DPF=demarcated protected forest; UDPF=undemarcated protected forest; ★ The Census of India, 2001, shows the forest area included within village boundaries. In two blocks of Telkoi and Banspal, which overlap with the Juangpirh and Bhuiyapirh, the total forest area inside the village boundaries adds up to 62,491ha, i.e. 624sq. km; ★★ Proposed Reserved Forests are areas where initial notification for reservation has been issued, but final notification as a Reserved Forest has not yet been issued; ★★★ Reserved lands and Protected Lands are ambiguous legal categories in South Odisha (the area that came from Madras Presidency) because they were specially created under Section 26 of Chapter III of the Madras Forest Act, 1882, in the pre-Independence period. The main reason for notifying these areas as Reserved or Protected Land was to avoid the rights-settlement process that creating Reserved Forests would have entailed (Behuria, 1965).

(61ha) of current shifting cultivation land was within the recently notified Mohangiri Proposed Reserved Forest (Kumar and Kerr, 2012). The entire Mohangiri Reserved Forest is a mosaic of fallows, secondary forests and patches in which cultivation is ongoing. Another 32 acres of swiddens were found to be on land categorized as revenue forests. In Mangara village, Koraput district, the author documented a few existing patches of shifting cultivation persisting deep within reserved forests, and villagers mapped out areas in reserved forests where they had been forced to abandon shifting cultivation lands by the Forest Department since the late 1990s (Kumar et al.,

2004b). In Gaurigaon village, Kandhamal district, about 61 acres of land currently under shifting cultivation by 29 families is located within revenue forests created during survey and settlement procedures in 1982.

It is difficult to estimate the extent of shifting cultivation land that has been converted to legal forests, particularly since abandoned shifting cultivation areas have reverted to secondary forest. Perusal of forest-notification proceedings throws little light on the issue because Forest Settlement Officers don't discuss it. The author's detailed examination of forest reservation proceedings in Kandhamal district turned up just one case in which a Forest Settlement Officer acknowledged that shifting cultivation was being practised in a proposed reserved forest, and then promptly proceeded to reserve the area. Indirect evidence of such actions can be found in various Forest Department documents, including Forest Working Plans. Illustrating this point, extracts from the Working Plan of the Rayagada Forest Division, describing various 'working circles' are shown in Table 21-3.

Similarly, the Balliguda Working Plan, covering part of Kandhamal district, states:

> Large forest areas in the division have been destroyed since time immemorial due to the pernicious practice of *podu* and this has changed the complexion of some of the forest areas completely. Some of the *podu* areas, which once supported lofty trees, are now bereft of vegetation. Extensive forest areas now bear a crop of poles and young saplings due to recent *podu* (GOO, 1989).

There is no concern in these documents about the loss of livelihood and food supplies suffered by the ethnic peasantry. Using legal instruments, the livelihood of vast

TABLE 21-3: Details of shifting cultivation activities in the Working Plans of the Rayagada Forest Division.

Working Circles	Total area	Shifting cultivation area affected	Remarks in the Working Plan
Improvement Working Circle	41,526ha	Part of the circle	All the blocks having congested crops at pole stage, mainly of *podu* origin, were included in this circle
Rehabilitation-cum-soil conservation Working Circle	27,768ha	Part of the circle	It included *podu* areas thoroughly degraded due to repeated hacking
Teak plantation Working Circle	2507ha	2288ha of *podu* area affected	*Podu* areas and existing teak plantations
Protection Working Circle	987ha	Half of Rafukona Reserved Forest (647ha) under shifting cultivation	

Notes: Podu = shifting cultivation.
Source: Rayagada Division Working Plan (GOO, 2006).

numbers of Odisha's most vulnerable people was criminalized. Land claimed and used by shifting cultivators for many generations became state land, over which the ethnic communities no longer had any rights or any voice in its management. Extended periods of contestation and conflict followed the legal criminalization of shifting cultivation and continue up to the present day, as shifting cultivation communities strive to maintain their critically important livelihood and cultural practices. This struggle for survival now shapes both the landscapes and the lives of ethnic farmers in these tracts of Odisha state.

Contested shifting cultivation landscapes

Both survey and settlement procedures and forest notifications have legally converted shifting cultivation lands to state ownership, with laws that specifically criminalized swidden cultivation or short fallows on hill slopes. In areas where survey and settlement procedures occurred just after Independence (the Dongarlas and the undivided Koraput district), most of the hill slopes used for cultivation were categorized as revenue wastelands. In areas that had somewhat more secondary forest, and where survey and settlement occurred later (Kandhamal and Gajapati districts), most of the hillslopes were converted into legal forests, through both forest laws and survey and settlement procedures. These government lands now cover more than 80% of these districts. If only hilly areas are taken into consideration, almost 90% of land in shifting cultivation tracts is now owned by the state. The only legal rights left to the inhabitants relate to that land that was recorded in the name of families and individuals in valley bottom lands.

After examining land ownership and population figures drawn from a number of sources, including census data, it is very clear that the amount of land legally owned by households in these areas is extremely inadequate to provide a livelihood. Most of the households in shifting cultivation communities have very little legal access to land.

There are very clear implications arising from the legal land-tenure systems in these areas. Shifting cultivation was a rotational agroforestry land use that evolved in response to the peculiar topography of these hilly regions and co-evolved with the cultural and social practices of the people. The conversion of these lands to state ownership not only criminalized shifting cultivation, but it left the local agrarian communities with very little legally owned land. The local communities, in order to survive, have continued to practise shifting cultivation, albeit as a 'criminal' activity. Given the state's variable capacities and interest in enforcing the laws, this has led to a contested landscape full of conflict, struggle and dispossession.

The Forest Department, with its jurisdiction over reserved and protected forests, has been quite aggressive in its efforts to stop shifting cultivation. In certain areas, particularly in reserved forests, it has succeeded through the use of force. The suffering that this has caused has made the Forest Department 'enemy number one' in these landscapes. It has also created an underground economy of rents and extractions, whose major costs are borne by the poor and marginalized. The Revenue Department, which administers all government land (excluding reserved and protected forests), has been comparatively less aggressive in stopping shifting cultivation. The issues arising out of this situation are discussed below.

Loss of community control over local landscapes

Most tribal communities exercised de facto or customary territorial claims to the landscapes in which they lived, with both individual and collective rights embedded in these customary systems. Shifting cultivation was an integral part of these dynamic landscape-based systems and unencumbered use of the land was also closely intertwined with the cultural and social practices of various ethnic groups (Mohapatra, 1997). The territorialization processes that imposed simplified private and state property rights over these landscapes completely ignored these complex relationships. In legal terms, the territorialization processes led to the loss by ethnic communities of control over, and rights to, most of their landscapes, including their swiddens, and left them with no say in the management or use of those lands.

Criminalization of life and livelihood itself

The source of a basic livelihood for a large number of Odisha's most vulnerable communities became criminalized. The cultivation of state land was criminalized not only by forest laws, but also be revenue laws such as the Odisha Prevention of Land Encroachment Act, 1972. Penalties included fines, evictions and arrests. In protected areas (National Parks and sanctuaries), simply entering the area without permission could lead to prosecution. Thousands of prosecutions are filed against shifting cultivators every year by the forest department, dragging ethnic farmers into the complex arena of courts and lawyers. These cases drag on for years and often involve prison sentences. For example, in the G. Udaygiri Forest Range in Kandhamal district, out of 87 cases brought before the courts between 2001 and 2006, 47 related to shifting cultivation (Kumar et al., 2008). With dependence on oral traditions and an extremely poor literacy rate, ethnic shifting cultivators find dealing with the official legal system a traumatic experience. Discretionary powers provided to petty officials under the law creates an economy of rent extraction, coercion and humiliation for hapless shifting cultivators.

Losing access to land and livelihoods

The greatest loss of access to land, and hence ability to practise shifting cultivation, has occurred in areas declared as forest land, primarily because of the aggressive approach of the Forest Department. Even in those places where the department has been unable to stop shifting cultivation, a low-level running battle continues between shifting cultivators and department officials. Remoteness, difficulties in monitoring, bribes, local power relationships and an increasing influx of left-wing insurgents all play a key role in these battles. As well as using the coercive tools at its disposal, the Forest Department has consciously used plantation projects as a strategy to evict shifting cultivators. The plantations – including teak plantations – have budgetary provisions for 'watch and ward' covering several years, and the department uses this for continuous monitoring. My studies have found that even those plantations that involve participatory schemes, such as Joint Forest Management, poverty alleviation and ecological rehabilitation schemes, have been used by the department to evict shifting cultivators. There have been cases where the Forest Department has collaborated with local elites, often outsiders, to launch Joint Forest Management

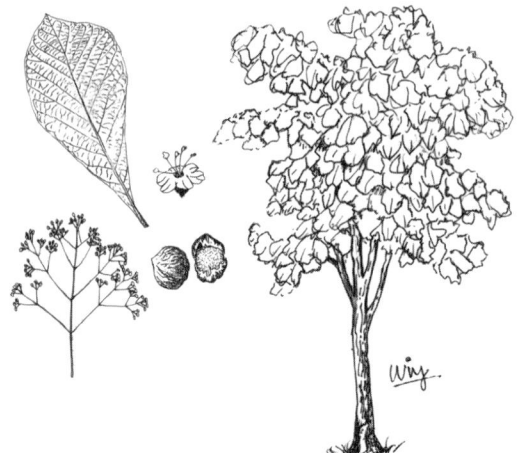

Tectona grandis L.f. [Lamiaceae]

In Odisha, as in many other shifting cultivation territories, teak plantations figured prominently in programmes aimed at 'improving' land use and ousting shifting cultivators.

plantations on shifting cultivation lands, as in Mangara and Podagarha villages, Koraput district (Kumar et al., 2005). In 2009 and 2010, the department introduced Joint Forest Management plantations on customary land cultivated by the Kutia Kandh communities in Kandhamal district (Behera, 2011). In neighbouring Andhra Pradesh state, the Forest Department used the Joint Forest Management programme to evict vulnerable shifting cultivators from 37,000ha of forest land (Blaikie and Springate-Baginski, 2007). The approach to land being used for shifting cultivation is exemplified by the Rayagada Forest Division Working Plan (Table 21-3), which developed plans for 'improving' nearly 73,000ha of land affected by shifting cultivation. Of this, more than 2500ha were to be used for teak plantations (Kumar et al., 2005).

In the case of state land administered by the Revenue Department, law enforcement is not so strict, even though shifting cultivation remains illegal. However, even on these lands, plantations are used by the government as a means of removing shifting cultivators. The first major plantations of cashews and eucalyptus in Odisha, involving the Soil Conservation Department, occupied watersheds of the Machkund reservoir in Koraput district, where shifting cultivation was being practised. There was resistance to this takeover of customary land. B.K. Roy Burman, an eminent anthropologist, documented the burning of plantations by shifting cultivators in his report for the Planning Commission (GOI, 1986). Despite such acts of resistance, cashew plantations on shifting cultivation land have continued and vast areas have been transferred to the state-owned Cashew Development Corporation, which manages them commercially. Large areas of plantations continue to expand into revenue lands, ironically often under tribal development and poverty alleviation schemes, while shifting cultivators are forced off the same lands.

A study in Dekapar village, Koraput district, revealed how the same patch of shifting cultivation land had been taken over by the government four times in the past 30 years for plantations. On each occasion, the take-over resulted in out-migration and the plantation was destroyed by the local people (Kumar et al., 2005). In Kadalibadi village, Keonjhar, 116ha of shifting cultivation land used for swidden farming by extremely vulnerable Juang farmers, was converted to plantations by the Forest

Anacardium occidentale L. [Anacardiaceae]

Odisha has large areas of cashew plantations, the first of which were used to displace shifting cultivators. Some plantations were burned in early conflicts over the taking of shifting cultivation land.

Department (Rath, 2005). Such examples are found in all shifting cultivation areas.

Lack of legal rights means that shifting cultivators are evicted whenever the government needs their land for any other purpose. As well as plantations, other purposes include diversion of land for development, industrial or mining projects and leasing it to private parties. The government has leased shifting cultivation lands to coffee growers (generally rich non-*adivasis*), and major industrial and mining projects have been allocated vast areas of land, including that used by *adivasis* for short-fallow cultivation. An increasing emphasis on production of bio-fuels and sequestering carbon has also given many companies a new regard for the vast tracts of revenue wastelands and forest land in shifting cultivation areas. Once again, this may lead to the de facto loss of these lands to shifting cultivators, without any compensation, and lost access to land means lost livelihoods.

Extraction of local surpluses

The criminalization of subsistence shifting cultivation has made the communities that maintain their traditional farming systems vulnerable to coercion and surplus extraction, or payment of taxes, rents or tributes to chiefs, landowners, officials, or the state. Households are forced to pay fines or bribes to be able to cultivate shifting cultivation lands. Almost every shifting cultivation village visited by the author has had stories of how the people have been forced to pay bribes to forest officials in order to access their land to grow crops. When cases are filed against villagers they have to spend money to engage the legal system or pay fines, often by taking high-interest loans from local moneylenders. For example, in 2002, the people of Dekapar village in Koraput district were fined 20,000 rupees (US$290) for cutting down government plantations that had occupied shifting cultivation land (Kumar et al., 2005). These acts of surplus extraction and coercion link up with other rent-seeking processes to create powerful local networks of exploitation and repression based on money lending and coercion (Viegas, 1991).

Marginalization and poverty

Criminalization of livelihood activities and surplus extraction through bribes, displacement and the loss of access to their ancestral land are important factors underlying extreme levels of poverty in the shifting cultivation landscapes of Odisha. The loss of access to shifting cultivation lands has meant that communities have lost control of their major source of subsistence and nutrition. Swiddens traditionally have a high level of agricultural biodiversity and are a vital source of nutrition. Frequent payment of bribes or fines and occasional imprisonment for cultivating state land have become serious financial liabilities for forest dwellers. The need to take loans from moneylenders at high interest rates often leads to chronic indebtedness. These problems are magnified in remote villages deep inside protected areas or reserved forests. Cases of malnutrition and deaths from preventable diseases have been reported from Nowrangpur district, Simlipal National Park and Satakosiya sanctuary in Odisha.

Conflicts and violence

Most shifting cultivation communities have a history of conflict with the Forest and Revenue Departments, with stories of prosecutions, arrests, fines and violence (Kumar et al., 2008). Rights over shifting cultivation land have long been a sensitive issue in these areas, and the insensitivity of the government to the land issues of shifting cultivators has led to recurrent unrest. It has been a major factor in increasing support in Southern Odisha for left-wing insurgents (Kumar et al., 2005; GOI, 2008).

Land and ecosystem degradation

The lack of legal rights to their ancestral land has implied that shifting cultivators have little incentive to stabilize the hillsides they cultivate. As populations have increased and more hill slopes have been classified as forests and forced out of cultivation, those hillsides remaining under shifting cultivation have begun to degrade because of shorter rotations and less time for soil recovery (Misra-Panda, 1999). Even though many tribal communities have sophisticated knowledge of soil and moisture conservation methods,[10] most swidden farmers hesitate to apply these labour-intensive methods to stabilize shifting cultivation land in the absence of secure tenure.

Government officials also discourage farmers from building permanent structures like terraces and stone bunds on state land. Shifting cultivation areas, particularly those where the rotational system has switched to short fallows due to pressures on availability of land, could easily be converted into stable, fruit tree-based agroforestry systems, but tree planting on state land would be illegal. The cultivators cannot access either state support or bank loans to invest in their farming systems. Moreover, they cannot access benefits from on-going government-subsidized watershed development or land improvement programmes (Kumar et al., 2005). Communities have little incentive to protect watersheds and forests on a collective basis because they have no formal claims to these areas.

In spite of the lack of rights, many communities still protect patches of forests and hill slopes from degradation (Singh, 2002) and this could be scaled up very quickly if communities were confident of state recognition for their land claims (Nayak and Berkes, 2008). But more sustainable management of the hilly landscape of Odisha is confounded by conflicting formal and local legitimacies.

Wastage of development funds

The forest communities that once practised shifting cultivation are some of the poorest people in India, and they live on extremely fragile landscapes that are of immense ecological importance. The Government of India and the state government of Odisha spend millions of dollars every year on various development and poverty-alleviation schemes in these areas. Despite this, poverty seems to be increasing rather than diminishing. The main reasons lie in the political economy of development in these regions and the political marginalization of the poor. The issue of rights over land, including hill slopes, aggravates the wastage of development funds and efforts. Plantations have been regarded as a favoured instrument for the use of development funds because they can be seen as generators of wage labour, improvers of ecological conditions and providers of benefits for the future. But they lead to the eviction of shifting cultivators from their land and are often destroyed over time. Watershed projects have been launched without any consideration of the tenure issue, and they tend to convert land used for shifting cultivation into forest use or plantations. There have been efforts by the government to take account of the farmers' use of hill slopes, but persistent problems arising out of a lack of understanding and appreciation of customary land-tenure systems have made these efforts only partially effective. At the same time, the considerable body of ecological knowledge, good practices and innovations held by shifting cultivators has been ignored by these programmes. There have been examples of shifting cultivators adapting to the increasing pressures on land by switching to terracing, fruit-based agroforestry and the use of innovative soil and moisture-conservation systems. These could have been leveraged by the development programmes, but they have been ignored because of contempt for the 'backward shifting cultivators' and the fact that the hill slopes belong to the state.

State interventions for 'improving' shifting cultivation

Governments at both state and federal levels wanted to stop the practices of shifting cultivators, who were accused of 'destroying' Odisha's forests, and get them to shift to other livelihoods. The criminalization of shifting cultivation and the denial of rights over land for swidden farmers was accompanied by programmes aimed at 'improving' shifting cultivators. Initial efforts included schemes for resettling them in lowland colonies and allocating them agricultural land for settled cultivation. Most of these colonies did not do well (Dash, 2006).

In the 1980s, federally funded poverty-alleviation and development programmes established plantations on shifting cultivation land with the promise of usufructuary rights to cultivators.[11] These schemes were dogged by conflicts between the customary owners of the land and the newcomers, failure to allot the plantations to beneficiaries and poor plantation survival. Another initiative was a small-scale programme funded by the central government named 'Plantation by *Podu* Cultivators'.[12] The scheme envisaged the plantation of commercial fruit trees such as mangoes, tamarind, jackfruit, guavas and oranges by shifting cultivators.[13] The cultivators were to be granted usufructuary rights and intercrops were permitted between the growing trees. The scheme was never upscaled to larger areas.

The issue of rights over hill slopes was taken up by non-governmental organizations in the late 1980s, leading to the inclusion of land-tenure issues in a large watershed-development project in Kashipur block of Rayagada district in the early 1990s, funded by International Fund for Agricultural Development (IFAD). In Kashipur, most shifting cultivation lands had been converted to short-rotation fallows on highly degraded hill slopes. The Government agreed that shifting cultivators could settle on hills with slopes up to 30 degrees.[14] In the course of the project, rights for *dongar* (hills) land were granted to 6837 *adivasi* households in 236 villages, covering a total area of 17,175 acres (7956ha). This was the first time that rights over non-forest hill slopes were recorded in favour of shifting cultivators. According to a later evaluation report of the Orissa Tribal Development Project (OTDP), the allocation of land rights helped to restore the area's agro-ecological balance (IFAD, 1999). However, the author found during his research that the rights-settlement process had not recognized the traditional tenure system. As a result, land customarily owned by one household was often settled by another household, causing local conflicts (Kumar et al., 2004a).

The Kashipur order was extended to all *adivasi* areas of Odisha in 2000, due to pressure from civil society.[15] However, this extension did not apply to legal forests and, until recently, its actual implementation has been almost non-existent. Another government initiative that is attempting to address the issue of land tenure in shifting cultivation areas is the Odisha Tribal Empowerment and Livelihood Project, funded by IFAD and the UK Department for International Development. One of its objectives is to facilitate the settlement of land by landless persons in ethnic areas. It is too early to assess the impact being made by this programme.

Responses of the shifting cultivators: Taking back the land

As ethnic shifting cultivators have become more aware of their situation and the laws under which it has been created, they have sometimes reasserted their claims to their customary landscapes. A strong civil-society movement that exists across India has supported the assertion of customary ownership over common land, and this movement has been active in Odisha. Many ethnic communities have adopted community protection of degraded hill slopes in order to regenerate forests, and as a reassertion of their control over these lands. A report prepared for the Food and Agriculture Organization of the United Nations, using satellite data and ground truthing, found that almost 1000 communities in Kandhamal district were involved in community forest protection covering 37,000ha of land (Singh et al., 2005). Most of this land had formerly been used for shifting cultivation but had regenerated into secondary forests under the protection of the communities. Hundreds of villages in other parts of Odisha are now involved in protecting forests that cover their erstwhile shifting cultivation land. This process has been facilitated by the emergence of new means of livelihood, such as wage labour in government programmes, intensified agriculture, cultivation of cash crops like turmeric and ginger, and so on. This bodes well for the ecological regeneration of shifting cultivation lands.

There has also been a movement to assert control over cashew plantations that were planted on shifting cultivation land and are now managed by the Cashew Development Corporation. In 22 villages in Koraput district, ethnic communities forcefully took over cashew plantations that were leased to the corporation by the Odisha government. This led to conflict and litigation against the *adivasis*. The struggle over the cashew plantations expanded across the landscape, finally resulting in a government order in 2008 that declared that the cashew plantations would be allocated only to ethnic farmers.[16] It is clear that as the *adivasi* communities became more assertive, their efforts to reassert their rights over government-controlled land increased. In areas affected by left-wing armed insurgency, it has been extremely difficult for the state to exercise control over land, and local communities have tended to control what happens within their traditional territories. This is obvious in protected areas such as Kotgarh and Lakhari sanctuaries, where the insurgency prevents forest officials from trying to stop shifting cultivation, even though wildlife laws are normally applied with extreme strictness.

New possibilities following new laws

In 2006, a new forest-rights law was passed in India after prolonged nationwide mobilization of ethnic and forest-dwelling communities (Kumar and Kerr, 2012). Enactment of the Scheduled Tribes and Other Forest Dwellers (Right to Forest) Act, 2006 (FRA) aimed to redress the historical injustices done to *adivasis* and other forest dwellers by the reservation of forests. The Act created the possibility of resolving some of the contradictions regarding shifting cultivation lands categorized as legal forests. Before its enactment, forest lands could not be settled by cultivators without the permission of the central government. The FRA eased the situation by recognizing the right of ethnic cultivators and other forest dwellers to occupy up to 4ha of forest land. This provision also applies to forest land under shifting cultivation.

Unfortunately, the FRA does not recognize the practice of shifting cultivation as a legal right. This means that in secondary forests that have grown on former shifting cultivation slopes, even if they are settled by individual families, the clearing of swiddens from fallow growth may be construed as a violation of the Indian Forest Act, 1927, which forbids the felling of trees. The outcome may be that the only forest land allocated to shifting cultivators for agriculture is that where the vegetation cannot be regarded as a forest because it has already been used for short-fallow agriculture.

Before enactment of the FRA, many of the communities that oversaw the generation of secondary forests on part of their shifting cultivation land had no formal rights over the regenerated forests. The FRA provides for 'community forest resources (CFR)', opening up the possibility of shifting cultivation communities asserting effective control over all of the forest land within their traditional territories, and protecting and benefiting from these lands. This process has begun in earnest in Odisha. CFR titles have been issued to a number of villages in Kandhamal district, many of which are already protecting secondary fallows.

Conclusion

Official denial of shifting cultivators' customary rights has been a common problem around the world (RRI, 2009). Most often this denial has involved the blanket takeover of vast areas of land by the state. In India, re-territorialization of landscapes was undertaken on a massive scale, with intensive survey and settlement procedures and forest notifications creating the present property-rights regimes in forested landscapes. Both of these processes provided for settlement of existing rights, so that the state has been able to claim that the current property regime is legitimate. However, as I have shown, this legitimacy was selective, and in Odisha, the very processes of detailed territorialization and rights settlement dispossessed a large number of shifting cultivation communities. Ethnic communities all over Odisha were left with legal control over only small areas in valleys that had been settled as private property, while most of the land they used for their subsistence and livelihood was taken by the state. This has had major impacts on local socio-ecological systems, leading to impoverishment of ethnic peoples as well as, arguably, degradation of their landscapes.

Shifting cultivation landscapes in India's Eastern Ghats, including those in Odisha, are in environmental and social crisis. Part of the reason for this is the gap between *de jure* and *de facto* tenure over land and forests. The state has taken ownership responsibilities over more than three-quarters of forested landscapes, even though it doesn't have the capacity to govern them effectively and sustainably, much less to address the livelihood needs of the millions of people who live in them. The imposition of forest laws on pre-existing systems of rotational agroforestry has led to dispossession of ethnic communities, conflict and degradation of hillsides.

In the absence of alternatives, many shifting cultivators continue to eke out a living from the same hill slopes, even though they do so illegally. The constructed 'illegality' of shifting cultivation has placed tremendous power in the hands of petty officials, especially forest officials, and has led to a regime of coercion, exploitation and marginalization. Marginalization through criminalization is an integral part of a larger vicious cycle of exploitation and surplus extraction from *adivasis* by a nexus of local elites, moneylenders and officialdom. Since their use of the land is 'illegal', cultivators have neither the incentive to invest time or resources in the land nor the capacity to access government incentives to improve it, thus leading to further degradation.

Only in recent years has civil-society activism and grassroots mobilization led to a shift in the situation. Special protection was provided in the Constitution for the land of ethnic communities, and this has become a key reference point for strategies attempting to untangle the mess created by land and forest laws. The solution lies in providing ethnic communities with more autonomy and control over their landscapes and allowing community governance systems to manage natural resources. Large-scale forest protection and and soil- and moisture-conservation efforts undertaken by ethnic communities in these landscapes has demonstrated the potential that exists in rationalizing land and forest tenure.

The FRA has provided strong legal tools with which to shift control of legal forests to ethnic forest communities, and similar legal initiatives need to be taken in the case of non-forest state land on which people have settled in tribal areas. Existing legislation has the potential to allow such transfers. These initiatives can be supported by participatory development programmes that seek to stabilize hill slopes while they continue to generate income and livelihoods.

Providing secure tenure over hill slopes to communities of shifting cultivators may also enable them to take part in activities such as carbon sequestration by planting trees and regenerating forests, growing commercial crops like cashews on barren slopes and adopting agroforestry systems involving coffee and other plantation crops. These developments could potentially augment household and community incomes and reduce poverty. They would also create incentives for local communities to regenerate forests and move towards sustainable governance of Odisha's hilly landscapes.

References

Bailey, F. (1960) *Tribe, Caste and Nation: A Study of Political Activity and Political Change in Highland Odisha*, Manchester University Press, Manchester, UK

Baker, D. (1991) 'State policy, the market economy, and adivasi decline: The Central Provinces, 1861-1920', *Indian Economic Social History Review* 28(4), pp341-370

Behera, S. (2011) Personal communication between Shricharan Behera, an independent researcher, and the author

Behuria, N. C. (1965) *Final Report on the Major Settlement Operation in Koraput District 1938-64*, Government of Odisha Press, Cuttack, Odisha

Blaikie, P. and Springate-Baginski, O. (2007) *Forests, People and Power: The Political Ecology of Reform in South Asia*, Earthscan, London

Chandran, M. (1998) 'Shifting cultivation, sacred groves and conflicts in the colonial forest policy in the Western Ghats', in R. H. Grove, V. Damodaran and S. Sangwan (eds), *Nature and the Orient: The Environmental History of South and Southeast Asia*, Oxford University Press, New Delhi, pp674–707

Dash, B. (2006) 'Shifting cultivation amongst the tribes of Odisha', *Odisha Review* 2006, http://odisha.gov.in/e-magazine/Orissareview/july2006/engpdf/76-84.pdf, accessed 25 May 2017

FAO (1981) *Tropical Forests Resource Assessment Project: Forest Resources of Tropical Asia* (India brief), Food and Agriculture Organization of the United Nations, Rome, http://www.fao.org/docrep/007/ad908e/AD908E13.htm, accessed 14 February 2009

GOI (1986) *Report of the Study Group on Land Holding Systems in Adivasi Areas*, Planning Commission, Government of India, New Delhi

GOI (2008) *Development Challenges in Extremist Affected Areas*, Planning Commission, Government of India, New Delhi.

GOI (2010) *Wasteland Atlas of India*, 2010, Department of Land Resources, Ministry of Rural Development, Government of India, New Delhi, http://www.dolr.nic.in/wasteland2010/orissa.pdf, accessed 5 December 2015

GOO (1940) *The Partially Excluded Area Enquiry report*, Government of Odisha Press, Cuttack

GOO (1958) *Report of the Administration Enquiry Committee*, Government of Odisha Press, Cuttack

GOO (1959) *Forest Enquiry Report*, Government of Odisha Press, Cuttack

GOO (1989) *Working Plan for Baliguda Division*, Forest Department, Cuttack

GOO (2006) *Revised Working Plan for the Reserved Forests, Proposed Reserved Forests, Demarcated Protected Forests and Compensatory Afforestation Areas of Rayagada Forest Division for the period 2006-07 to 2015-16*, Forest Department, Cuttack

IFAD (1999) *India: Completion Evaluation of Odisha Development Project. Seven Lessons Learned*, International Fund for Agriculture, Rome, http://www.ifad.org/evaluation/public_html/eksyst/doc/agreement/pi/Odisha.htm, accessed 16 January 2009

Jyotishi, A. (2000) *Swidden Cultivation: A Review of Concepts and Issues*, Institute for Social and Economic Change, Bangalore

Kumar, K., Choudhary, P. R. and Kerr, J. (2004a) 'Tenure and access rights as constraints to community watershed development in Odisha, India', paper delivered at The Commons in an Age of Global Transition: Challenges, Risks and Opportunities, the Tenth Conference of the International Association for the Study of Common Property, August 9 to 13, Oaxaca, Mexico, http://pdf.wri.org/ref/kumar_04_tenure.pdf, accessed 25 May 2017

Kumar, K., Giri Rao, Y. and Sarangi, S. (2004b) *Odisha (India) Case Study Report: IUCN's Sustainable Livelihoods, Environmental Security and Conflict Mitigation Project*, (mimeo), Vasundhara, Bhubaneswar, Odisha

Kumar, K., Choudhary, P. R., Sarangi, S., Mishra, P. and Behera, S. (2005) *A Socio-Economic and Legal Study of Scheduled Tribes' Land in Odisha* (mimeo), Vasundhara, Bhubaneswar, Odisha, http://siteresources.worldbank.org/INTINDIA/Resources/Kumar1.pdf, accessed 25 May 2017

Kumar, K., Behera, S., Sarangi, S. and Springate-Baginski, O. (2008) 'Historical injustices: The creation of poverty through forest tenure deprivation in Odisha', in O. Springate-Baginski (ed.) *Understanding Livelihood Impacts of Participatory Forest Management Implementation in India and Nepal*, Overseas Development Group, University of East Anglia, UK

Kumar, K. and Kerr, J. M. (2012) 'Democratic assertions: The making of India's Recognition of Forest Rights Act', *Development and Change* 43(3), pp751-771

Kumar, K. and Kerr, J. M. (2013) 'Territorialization and marginalization in the forested landscapes of Orissa, India', *Land Use Policy* 30(1), pp885-894

Misra-Panda, S. (1999) 'Towards a sustainable natural resource management of tribal communities: Findings from a study of swidden and wetland cultivation in remote hill regions of Eastern India', *Environmental Management* 23(2), pp205-216

Mohapatra, L. (1997) 'Tribal rights to land and the state in Orissa' in P. M. Mohapatra and P. C. Mohapatro (eds) *Forest Management in Tribal Areas: Forest Policy and Peoples Participation*, Proceedings of the seminar 'Forest Policy and Tribal Development', 15-16 February 1994, Koraput, Concept Publishing Co., New Delhi

Nayak, P. K. and Berkes, F. (2008) 'Politics of co-optation: Community forest management versus joint forest management in Orissa, India', *Environmental Management* 41(5), pp707-718

Nicholson, J. W. (1938) A*nnual Progress Report on Forest Administration in the Province of Orissa for the Year 1936-37*, Government of Orissa, Government Press, Cuttack

Padel, F. (1995) *The Sacrifice of Human Being: British Rule and the Konds of Odisha*, Oxford University Press, New Delhi and New York

Ramdhyani, R. K. (1947) *Report on Land Tenures and Revenue Systems of Odisha and Chattisgarh State*, ILP Press, Berhampur, Odisha

Rangarajan, M. (1996) *Fencing the Forest: Conservation and Ecological Change in India's Central Provinces, 1860-1914*, Oxford University Press, New Delhi and New York

Rath, B. (2005) *Vulnerable Adivasi Livelihoods and Shifting Cultivation: The Situation in Odisha*, Vasundhara, Bhubaneswar, Odisha

Rosencranz, A. and Lele, S. (2008) 'Supreme Court and India's Forests', *Economic and Political Weekly* 43(5), p11

RRI (2009) *The End of the Hinterland: Forests, Conflict and Climate Change*, Rights and Resources Initiative, Washington, DC

Saldanha, I. (1998) 'Colonial forest regulations and collective resistance: nineteenth-century Thana district', in R. H. Grove, V. Damodaran and S. Sangwan (eds) *Nature and the Orient: The Environmental History of South and Southeast Asia*, Oxford University Press, New Delhi

Saravanan, V. (1998) 'Commercialisation of forests, environmental negligence and alienation of adivasi rights in Madras Presidency, 1792-1882, *Indian Economic & Social History Review*, 35(2), p125

Saravanan, V. (2004) 'Colonialism and coffee plantations: Decline of environment and *adivasis* in Madras Presidency during the nineteenth century', *Indian Economic &Social History Review*, 41(4), pp465-488

Singh, N. (2002) 'Federations of community forest management groups in Orissa: Crafting new institutions to assert local rights', *Forests, Trees and People Newsletter* 46, pp35-45

Singh, K. D., Sinha, B. and Mukherji, S. D. (2005) *Exploring Options for Joint Forest Management in India*, Forest policy and institutions working paper, World Bank, Worldwide Fund for Nature (WWF) and Ashoka Trust for Research in Ecology and the Environment (ATREE)

Sivaramakrishnan, K. (1999) *Modern Forests: State Making and Environmental Change in Colonial Eastern India*, Oxford University Press, New Delhi

Skaria, A. (1999) *Hybrid Histories: Forests, Frontiers and Wildness in Western India*, Oxford University Press, New Delhi

Sundarajan, S. (1963) *Final report on the Survey and Settlement of the Kashipur, Karlapat, Mahulpatna and Madanpur-Rampur Ex-Zemindaries in the District of Kalahandi,* Odisha Government Press, Cuttack, Odisha

Viegas, P. (1991) *Encroached and Enslaved: Alienation of Tribal Lands and its Dynamics*, Indian Social Institute, New Delhi

Notes

1. Some of the terms used for different forms of shifting cultivation are *bewar* and *dhya* in Madhya Pradesh (Baker, 1991; Rangarajan, 1996); *khandad* in the Dangs, Gujarat (Skaria, 1999); *Ponnakadu* in Tamil Nadu (Saravanan, 1998, 2004); *kumri* in North Kanara (Chandran, 1998); and *dalhi* in Thane district of Bombay Presidency (Saldhana, 1998).

2. The Indian Forest Act, 1927, in section 10, directs that in the Reservation of Forests, claims for shifting cultivation should be submitted to the State Government for approval. In cases where the State Government decides that shifting cultivation can take place, an area may be excluded from the

reservation or certain areas separately demarcated where shifting cultivation is conditionally allowed as a privilege. It also states that 'Shifting cultivation shall in all cases be deemed a privilege subject to control, restriction and abolition by the State Government'.

3. *Adivasi* is an umbrella term for a heterogeneous set of ethnic and tribal groups considered to be the aboriginal population of India.

4. In Kandhamal, a special survey and settlement was undertaken between 1921 and 1925, but only to assess the land of non-*adivasi* plainspeople who had moved into the area to take advantage of not having to pay tax (Kumar et al., 2008)

5. For instance, Ramdhyani wrote about the Kalahandi Princely State: 'Reserve forests are simply those which may be reserved, and the rules do not prescribe any matter to be taken into account while making reservations' (Ramdhyani, 1947).

6. The report said: 'A detailed and extensive field visit of Odisha state was undertaken to supplement and correct the office interpretation. Interpretation carried out in the office was checked in the field and areas marked as closed forest, degraded forest, water bodies, non-forest etc.... Data regarding areas affected by shifting cultivation were obtained from related working plan and local forest officers for the entire state and reported on to 1:250 000 scale, Survey of India topo sheets. The data collected on topo sheets were used in delineating the areas affected by shifting cultivation in closed forest and degraded forest.' (FAO, 1981) However, even this exercise depended on secondary data rather than confirmation on the ground.

7. Letter no. 434-XXI-57/58-L-LRS dated 12 March, 1959.

8. The Madras Estate Land Act, 1908, section 15, subsection 3.

9. There is an intense contest over these revenue forests between the Revenue Department and Forest Department, even though they were created under revenue laws and not under forest laws. The exact legal status remains unclear, but the FCA and the Godavarman case has strengthened Forest Department claims.

10. Most ethnic communities have sophisticated water- and soil-management systems for wetland cultivation along stream beds in narrow valleys. The Saoras and Hill Bondas, and to lesser extent the Paraja and Khond people, use terrace cultivation on steep slopes.

11. Circular no GE(GL)-S-*/81-37565/REV from the Government of Odisha's Revenue Department laid down operational guidelines for the provision of usufructuary rights in plantations undertaken as part of this programme on government land (except reserve forests). Each beneficiary was to be given usufructuary (*Dafayati*) rights.

12. Vide circular no. GE (GL)– S–69/79–3755/Rev, Revenue Department, Government of Odisha, dated 18 January 1980.

13. 'Measures against shifting cultivation allotment of *Podu*-affected Government land for plantation by *Podu* Cultivators - Conferring of usufruct rights on temporary basis', Government of Odisha Revenue Department Letter No. GE (GL) – S – 69/79 – 3755/Rev dated 18 January 1980.

14. Government of Odisha letter no TD-I(IFAD)-18/91/2628/HTW, dated 10 April 1992, issued following a review meeting on the Odisha Adivasi Development Project, sponsored by the International Fund for Agricultural Development under the chairmanship of Odisha's Chief Minister.

15. Government of Odisha letter no. 14643-R-S-60/2000, dated 23 March 2000.

16. Letter no. GU (Cashew) 54/08_32240/Ag, dated 11 November 2008. Minute from a meeting chaired by the Chief Minister, Odisha on 31 July 2008 regarding usufructuary rights over existing cashew plantations on government land.

22

THE DRAGON AND ITS ATTEMPTS TO PUT OUT THE FIRE

Chencho Dukpa*

Introduction

In the first half of the 20th century, the two World Wars were cataclysmic events that all but tore the world apart. Cities were razed, millions lost their lives and global relations were forever changed. Yet the people of the tiny Buddhist kingdom of Bhutan (Figure 22-1) had barely a clue of what was happening in the outside world. Nestled in the eastern Himalayas, Bhutan was in a world of her own, with her doors closed, following a policy of self-isolation: no roads, no telephones, no TV, no postal system, no contact. Then, in 1952, Jigme Dorji Wangchuck became the third King of Bhutan, and the doors squealed open. In 1958, the new king invited the Indian Prime Minister, Jawaharlal Nehru, and his daughter, Indira Gandhi, to make a historic visit, and their entourage inched its way into Bhutan on horses and yaks. A long lasting and friendly relationship was forged, and an era of modernization began.

On the advice of India, Bhutan embarked on a systematic and planned approach to economic development. Its first five-year plan was launched in 1961 and two years later, a road finally reached its capital, Thimphu. The first vehicle to drive into Thimphu was a Jonga jeep, lurching and bumping along the rough road, and people dropped everything they were doing to watch in awe.

Moving history 'fast forward' to 2012, we find that Bhutan is one of the fastest-developing countries in Southeast Asia, with 11.8% growth in gross domestic product in 2010 (National Statistics Bureau, 2011). Agriculture, which was once Bhutan's dominant sector, has slid down to number four, with hydropower, tourism and construction taking the top spots. By 2020, it is expected that – for better or for worse – Bhutan will emerge from the group of Least Developed Countries. Surprisingly, these rapid changes have not come at a cost to the country's culture, environment

* CHENCHO DUKPA is chief research officer, Council for Renewable Natural Resources Research of Bhutan, Ministry of Agriculture and Forests.

or social fabric. All along, a cautious path was chosen and various policies were adopted to guide the country towards sustainable growth. These included policies on education, self-reliance, preservation of culture, health and forest conservation. Implementation of these policies invariably involved bans, and these were not always popular. They included bans on the use of plastics, the sale of tobacco, the export of unsawn timber – and the practice of shifting cultivation. Some critics felt compelled to brand Bhutan 'the Nation of Bans'.

This chapter is concerned with Bhutan's ban on shifting cultivation. It will first present a historical description of events leading to the ban. Then it will look at what happened to the practice of shifting cultivation after the ban came into force and discuss some policy lessons learnt from that.

Events leading to the ban on shifting cultivation

Prior to 1969, shifting cultivation was practised not only on farmers' registered land, but also on common property, or more correctly, in government forests. In those days, it was an era of open access to forests. However, the government realized as early as the 1950s that Bhutan's forests were one of the country's greatest sources of wealth. What's more, those same forests were in need of protection. In 1952, the country's Directorate of Forests was established – perhaps one of Bhutan's first modern institutions. A number of Bhutanese were sent to India to be trained as

FIGURE 22-1: Bhutan – 'the tiny Buddhist kingdom … nestled in the eastern Himalayas'.

professional foresters. By the late 1960s, these educated foresters managed to convince the government to give top priority to forests. They were instrumental in drafting the Forest Act 1969, which was the first formal Act of Bhutan's parliament. This nationalized all forests, made them government reserved forests, and banned shifting cultivation in them forthwith. Thus, shifting cultivation suffered its first major blow. It was thereafter limited to registered private land.

Private land for shifting cultivation had long been registered as *tseri* in Bhutan's subtropical agro-ecological zones and as *pangzhing* in temperate zones. Other categories of registered agricultural land were *kamzhing* (dryland), *chuzhing* (wetland), *tsoesa* (kitchen gardens) and *ngueltho dumra* (land for cash crops). However, the days of shifting cultivation were numbered, even on registered land. The government turned to consider its banishment, even there.

One of the government's first moves to discourage shifting cultivation on registered land came in the form of a taxation concession. Soon after passage of the Forest Act 1969, the National Assembly passed another resolution stating: 'In order to encourage conversion of shifting cultivation lands *(tseri)* to dry or wet lands, all farmers desirous of carrying out such a conversion will be entitled to pay taxes for two or three years at the prevailing rate for *tseri*. However, as soon as the first crops are harvested from the converted land, the farmers will have to start paying taxes according to the type of land in their possession, i.e. dry or wet.'

In those days, the tax rates were 4 to 6 ngultrum (roughly 8 to 12 US cents) per *langdo* of wetland, depending on the class of wetland; 3 to 4 ngultrum per *langdo* of dryland, depending on its agro-ecological location; and 0.75 ngultrum per *langdo* of *tseri* or *pangzhing*. A *langdo*, which translates literally as 'a pair of oxen', is the area of land that can be ploughed in a day by one pair of oxen. It is a measure still used in some rural areas of Bhutan.

Then, the government passed a new Forest Policy in 1974. Among other things, it declared:

> Shifting cultivation has been another cause for the depletion of forest wealth by means of destruction of forests, soil erosion and fall in soil fertility. The practice has, therefore, to be abolished if forests are to be conserved. In order to avoid hardship to the present population dependant on this practice of cultivation within forest areas, appropriate compensation will be given to the dispossessed, who can take up alternative avocations or adopt intensive forms of cultivation. The Agriculture Department will extend necessary assistance and co-operation in this respect.

However, in 1979, a new Land Act was passed and, surprisingly, *tseri* and *pangzhing* were still named as two of the agricultural-land categories. This meant that farmers could legally continue to practise *tseri* or *pangzhing*, and doubts about legality arising from the Forest Policy were removed. But some policy-makers were clearly impatient. In 1980, the then Home Minister told the National Assembly:

Tseri land can be cultivated only once in about 10 years at the earliest. If all the trees grown during that period were cut down indiscriminately for *tseri* cultivation, this would result in a loss to the country. Although *tseri* farmers own a large acreage of land, they cannot cultivate the land at a time to benefit them. The government now plans to resettle them and convert their land to permanent cultivable land. Therefore, it is time the *tseri* practice should be stopped.

His recommendations were shot down by members of the National Assembly, who decided that shifting cultivation should be allowed to continue until such time as alternative arrangements were in place.

The alternative arrangements, as well as resettlement, consisted of further efforts to encourage *tseri* farmers to convert their land into permanent dryland or wetland. In Bhutan's fifth five-year plan, from 1981 to 1986, the government aimed to convert 8800 hectares of *tseri* land into permanent fields and subsidize contour bunding on a further 2250 hectares of steep *tseri* land. It offered subsidies of 300 ngultrum (about US$5.78), subject to a maximum of 600 ngultrum per family, and 100 ngultrum, subject to a maximum of 300 ngultrum per family, for every acre (0.405ha) of sloping land terraced or contour-bunded, respectively. The scheme was continued into the next five-year plan, which ended in 1991. The conversion process was made easier by clear rules and procedures laid down by the Land Act 1979, for the conversion of *tseri* land into dryland or wetland. In 1983, the government introduced a new rule to facilitate conversion of *tseri* land into cash-crop land or pasture.

The government also floated the idea of buying shifting cultivation land from farmers in the 1980s. Normally, shifting cultivation occupies steep slopes and even if a farmer wants to sell such land, there are no eager buyers. So the government announced that any farmers wishing to sell their shifting cultivation land need look no further for potential buyers: the government was ready to make such deals. There are few records to tell how extensively this scheme was implemented, but according to some sources (e.g. Wangchuck and Tashi, 2004), it never got off the ground because of a lack of budget, and because the area of shifting cultivation land available for purchase made the cost of the scheme astronomical.

Nevertheless, it is clear that the 1980s saw a concerted effort by the Bhutan government to phase out shifting cultivation. Market forces were also working in the government's favour, with many *tseri* farmers planting mandarin oranges, cardamom and ginger as cash crops on their own initiative. Trade in these crops had by then gathered momentum, with the encouragement and promotion of the newly formed Food Corporation of Bhutan (FCB). The corporation offered a trusted platform for farmers to market their produce; if they could not fetch good prices in the open market, they could always sell to the FCB at a floor price and the latter would then dispose of the produce, often running at a loss. The signing of a trade agreement between Bhutan and Bangladesh in 1980 further broadened Bhutan's cash-crop horizon, which until then had been limited to markets in India.

Then came the 72nd session of the National Assembly in 1993. It was to be a historic one as far as shifting cultivation was concerned. Speaking on the subject of shifting cultivators and landless people, government officials and the people's representatives stated:

> The people practising *tseri* cultivation are all landless people who have no other means of earning a livelihood. The returns from their slash-and-burn cultivation are very meagre and very often the crops they sow are washed away by floods or destroyed by wild animals. As these families face a hard life and produce barely enough food to feed themselves and their families, they have applied for resettlement. These people should be resettled as soon as possible, wherever suitable land can be found.

The then Planning Minister gave the issue further emphasis:

> Being a mountainous country with limited flat and arable land, many people have resorted to *tseri* cultivation to eke out a living. In this slash-and-burn system of cultivation, converting one acre of *tseri* into cultivable land often results in seven to eight acres of nearby forest being burnt. The heavy toll inflicted on our forests by *tseri* cultivation is well known to all of us. Timely steps to do away with *tseri* cultivation have become very essential to protect our environment and preserve our rich and diverse flora and fauna. The people practising *tseri* cultivation have derived very minimal benefits from the last 30 years of planned development. The main reason for this is that *tseri* cultivators move from place to place in remote and rugged terrain looking for new *tseri* land to slash and burn. As a result it is never feasible for the government to provide them with cost-effective service facilities and infrastructure.

The Home Ministry also reported to the assembly that there were 25,126 households in Bhutan that were totally dependent on shifting cultivation and they were slashing and burning more than 200,000 acres of forestland in their efforts to eke out a hand-to-mouth living. The sentiment in the assembly called for a bold decision, and the timing could not have been better: The seventh five-year plan, covering 1993 through 1997, was about to begin and a target to phase out shifting cultivation by the end of the period seemed appropriate.

The move was given added impetus by other aspects of the seventh five-year plan. A resettlement programme was to be given major emphasis. Plentiful vacant agricultural land had appeared in the south of the country as a result of the southern Bhutan crisis in the early 1990s – a period of violence sparked by cross-border migration and illegal, uncontrolled settlement. Landless people, including the shifting cultivators, were now to be resettled on land vacated by illegal migrants, as well as in other parts of the country. Moreover, an Integrated Horticulture Development Project was to be launched to take the promotion and development of cash crops to a higher level. There was much optimism that the task of ending shifting cultivation, which had eluded the authorities for the past 30 years, would be achieved within the short space of the coming five years.

The National Assembly thus decided that '*tseri* cultivation must be stopped completely by the end of the seventh five-year plan, in the interests of the people and the preservation and protection of the environment'.

Soon afterwards, in 1995, the Forest and Nature Conservation Act was passed, giving the Ministry of Agriculture authority to make rules regarding the conversion of *tseri* to other forms of land use and to set a time frame, at the end of which *tseri* cultivation would be prohibited. In 2006, the Forest and Nature Conservation Rules were passed and these clearly stated that '*tseri* is banned and shall not be permitted'. They further provided for ownership certificates of *tseri* land to be cancelled and the land to revert to government reserved forest if it was left uncultivated for 12 years or more. Although the rules left some level of ambiguity, they essentially meant that land owners had to take proactive measures to convert *tseri* land to other land uses, or part with it; the implied logic being that if the owner could afford to keep *tseri* land fallow for 12 years or more, he or she no longer needed it.

Eventually, the earlier Land Act of 1979 was amended and passed as the Land Act 2007. This time there was no mention of *tseri* or *pangzhing* as agricultural-land categories, reinforcing the message that shifting cultivation had passed into history. Shifting cultivation was no longer recognized as a legal agricultural practice and land registered as *tseri* or *pangzhing* was to be categorized, by default, as dryland.

Current situation and reflections

Nearly two decades have passed since 1997 – the year in which shifting cultivation was supposed to have passed into Bhutanese history. The truth is, shifting cultivation is far from gone. A recent survey (Dukpa et al., 2011), conducted in selected shifting cultivation 'hotspots' across the country, found that 45% of farmers were still practising shifting cultivation. Why does it endure? The following section attempts to throw some light on this question.

First, although the government imposed the ban, it has never enforced it strictly in the field – on humanitarian grounds. Swidden agriculture is practised mainly by poor farmers in very remote and inaccessible villages where it is very hard to provide extension support. Strict enforcement of the ban would have meant sentencing these people to untold hardship. This is why, in some cases, government functionaries even make it easier for farmers to clear swiddens. For instance, in Nabji-Korphu – a remote *Khengpa* community in central Bhutan – farmers wishing to practise shifting cultivation inform forest officials, who then verify that the lands are registered as private and that there are no overgrown trees, and they issue permits allowing the farmers to go ahead. Nabji-Korphu is, in fact, located in the centre of a national park and the forestry rules on clearing and burning activities are even more stringent.

Second, there is a socio-cultural aspect involved. A close look reveals that it is mainly the indigenous communities, for whom shifting cultivation is an age-old occupation, who are keeping it alive. For instance, there are the *Doyaps,* in the Am Mo Chhu river basin in south-western Bhutan, whose distinctive culture and traditions are very different from those of 'mainstream' Bhutanese. Then there are the *Monpas* of central Bhutan, who are sometimes regarded as the country's earliest inhabitants. Along with these indigenous groups, there are similar communities in other parts of Bhutan, and the one thing they have in common is their practice of shifting cultivation. They have been shifting cultivators for generations beyond memory, and they argue that there is nothing wrong with their farming systems – as long as they are practised according to their age-old traditions. They say that with their skills and traditional knowledge, there are seldom problems with soil erosion or fire escaping into state forests. For these groups, shifting cultivation also has sentimental value, because some of the crops grown in their swiddens are vital to their rituals. To these people, the ban on shifting cultivation makes no sense. In any case, they argue, they were never consulted about the proposed ban before it was imposed.

It appears that policy-makers failed to recognize the sometimes subtle differences between various practitioners of shifting cultivation; between the indigenous communities for whom shifting cultivation has long been an integral part of their livelihood, and those who are more opportunistic, such as landless families and migrants who roam the forests in search of new patches to slash and burn. This failure to differentiate led to shifting cultivation being generalized as a primitive and wasteful method of farming, and this prompted the blanket ban. The arguments of the indigenous communities therefore make sense.

Third, the government placed too much hope in its resettlement programme becoming the solution to the problem of shifting cultivation. The resettlement programme had its own share of problems and could not be fully implemented. Although a lot of agricultural land lay vacant in the south, beckoning people – including shifting cultivators – to come and settle, there were only limited takers because of the security issues in the area. As a matter of curiosity, it is perhaps pertinent to ponder whether the government would have gone ahead with its decision to ban shifting cultivation if the early 1990s crisis in southern Bhutan had not happened. Most likely not. If that episode had not happened, there would not have been plentiful vacant land to bring the high expectation of success to a major resettlement programme for shifting cultivators. The connection between a political situation and an agricultural system is, nevertheless, an interesting one.

Even without the security issues in southern Bhutan, it soon became apparent that people were unwilling to leave the place of their birth behind and resettle in an unfamiliar place. In hindsight, perhaps policy-makers were too quick to brand shifting cultivators as landless, or that *tseri* was their only asset. A study by Dukpa et al. (2011) found that shifting cultivators, particularly the indigenous communities, owned not only *tseri* or *pangzhing* land, but other kinds of assets, like livestock, a decent house, land for cash crops, dryland and wetland. In fact, the study found that many of them owned more land assets than average rural Bhutanese households. It was unlikely, therefore, that resettlement of such farmers would succeed.

Fourth, the horticultural alternatives to swiddening developed major problems of their own. Citrus greening, cardamom wilt and ginger rot struck down plantations of mandarin oranges, cardamom and ginger, respectively. The diseases appeared in the late 1990s and to date no lasting solution has been found. Farmers have panicked and many who earlier gave up shifting cultivation have returned to it as a means of ensuring food security.

Conclusion

Shifting cultivation in Bhutan was banned in two stages: firstly in government reserved forests by the Forest Act 1969 and secondly on private registered land by the Forest and Nature Conservation Act 1995, following a National Assembly resolution in 1993. While the ban on swiddening in government reserved forests has proven to be effective, that related to privately registered land has not. Shifting agriculture continues to be practised, particularly by indigenous communities who see it as an integral part of their livelihood and a means of ensuring food security. Their refusal to abandon shifting cultivation shows that a farming practice that has proven to be a reliable tool for producing food from rugged terrain cannot be thrown out the window overnight – especially when the alternatives (cash crops) have yet to prove that they are foolproof.

References

Dukpa, C., Moktan, M. R. and Dorji, R. (2011) *Regional Project on Shifting Cultivation: Land-use Options and Extension Approaches,* International Centre for Integrated Mountain Development (ICIMOD) and International Development Research Centre (IDRC), Thimphu, Bhutan

National Statistics Bureau (2011) *National Accounts Statistics,* NSB, Thimphu, Bhutan

Wangchuck, P. and Tashi, K. (2004) *Shifting Cultivation in Bhutan,* International Centre for Integrated Mountain Development (ICIMOD), Kathmandu, Nepal

23

FROM FARMERS TO FORESTERS?

Response to pine encroachment on former swidden fields in Choekhor Valley, Bumthang district, Bhutan

Laura S. Meitzner Yoder, Sonam Phuntsho, A.J. Conrad, Hannah Doren, Rachel Haney, Christina Johantgen, Katie LeBoeuf, Sarah Miller, Zoe Reich-Aviles, Annalise Ritter and Greg Zegas *

Introduction

The many chapters contributed to this volume amply illustrate how agricultural decisions and transitions occur in dynamic agronomic, socio-economic and policy environments. In this chapter, we provide a snapshot of some of the factors that influence the choices faced by farmers in planting and fallowing land in the Choekhor Valley of Bumthang district, Bhutan.

Over the past few decades, this region has undergone accelerated socio-economic change, reflected in its production of staple foods (Guenat, 1991). Until recently, swidden-cultivation systems with a grass fallow were the dominant form of agriculture, producing mainly upland, rainfed grains including buckwheat, barley and wheat. The region has been transformed into a mixed agricultural area with heavy emphasis on production of potatoes as a cash crop for local and export markets, widespread use of commercial fertilizer, an increase in irrigated production systems and the introduction and establishment of high-altitude wet rice in valley locations. Importantly, the region's land-use systems are still changing. With a decline in swidden production over the past two decades, there has been a widespread increase in blue pine (*Pinus wallichiana*) forest, covering the former swidden land. This trend mirrors a dramatic increase in forest cover across the country, and a concomitant decrease

* Dr Laura S. Meitzner Yoder is Director of the Human Needs and Global Resources programme and Associate Professor in the Environmental Science Department of Wheaton College, Illinois, USA. Sonam Phuntsho is Forestry Officer at the Ugyen Wangchuck Institute for Conservation and Environment (UWICE), Bumthang, Bhutan. Other authors are A. J. Conrad, Hannah Doren, Rachel Haney, Christina Johantgen, Katie LeBoeuf, Sarah Miller, Zoe Reich-Aviles, Annalise Ritter and Greg Zegas, all of whom were student researchers from different colleges and universities in North America affiliated with the 2012 collaborative programme of the School for Field Studies, Beverly, Massachusetts and the Ugyen Wangchuck Institute for Conservation and Environment, Bumthang, Bhutan.

in cultivated land (NSSC and PPD, 2011). Within the context of national net forest-cover increase, as demonstrated by analysis of satellite images, Bumthang district had the greatest amount of land-cover change from non-forest to forest, adding 277sq. km of forest between 1990 and 2010 (Gilani et al., 2015).

Pinus wallichiana A.B. Jacks [Pinaceae]

Blue pine is a shade-intolerant, aggressive pioneer species with broad soil pH tolerance. It is widely distributed and is often

Blue pine is spreading aggressively in the wake of reduced shifting cultivation.

the dominant native species across the mid-altitude band of the Hindu Kush, in the southern Himalayas. It is valued for its timber. It readily colonizes cleared land, including fallowed swidden fields, grazing land, abandoned pastures and even sites newly exposed by the retreat of glaciers or ice sheets (Mong and Vetaas, 2006; Cochrane, 2009). Without any interventions such as planting or care of seedlings, nearly pure stands of pine rapidly become established once land management ceases or other land cover disappears. Around the Choekhor Valley, the recent increase in blue pine is due mainly to abandoned management; a long fallow where blue pine becomes established is often land that is removed from agricultural production for a long time. This study describes some primary considerations at a time of agricultural transition in the Choekhor Valley, and how landowners are making plans for their land once the area becomes overgrown with pine trees.

A team of North American undergraduate students and advisors conducted biophysical and socio-economic surveys in June and July 2012 to determine the extent of change in forest cover, the reasons for the increase in blue-pine forests and farmers' interests and concerns about future land uses in pine-encroachment areas. We established 24 randomly located transects in seven privately owned forested plots, and held semi-structured interviews, participatory mapping sessions and sketch-card selection exercises with 32 households (evenly divided between women and men as primary informants) in three villages in the Choekhor Valley, all of them within 10km of the Bumthang district capital of Jakar. One village (Chamkhar) was located near Jakar town, although farmers' fields were scattered over a wide area. Two rural villages (Pongkhar and Norgang) had both agricultural fields and 20- to 30-year-old pine forests adjacent to the settled area (Figure 23-1).

Crop changes and decline of swidden

Until the 1980s, the predominant food-production system in this region was swidden cultivation followed by a grass fallow, known locally as *pangshing*, described by

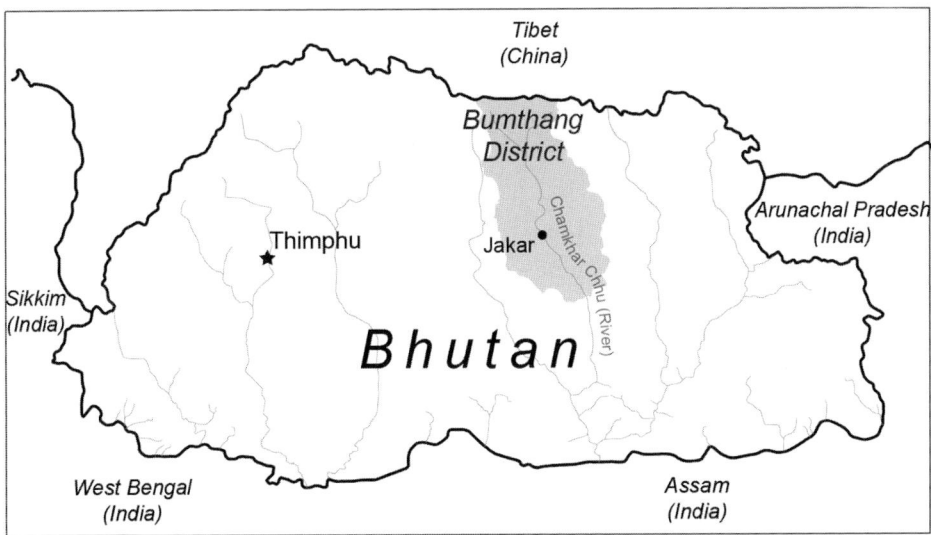

FIGURE 23-1: Bumthang district in Bhutan. The study villages are within 10km of the district capital, Jakar.

Roder et al. (1992), and Norbu et al. (1996). The process involved building mounds of vegetation and soil in fallowed fields, then burning the dried biomass in the mounded area over the course of several days. After letting it rest, the mounds were then spread out across the plot and crops were planted. Planting continued for several years before the plot was left fallow again. In this region, buckwheat and wheat were the core crops of this agricultural system.

The average size of the land held by the surveyed households was 5.5 acres (2.23 hectares), with an average area of forest (among the 23 households who had forested land) of 2.2 acres (0.89 hectares). All of the forested areas we surveyed had been cultivated in *pangshing* grass-fallow swidden systems 15 to 30 years earlier. Nearly all of the farmers interviewed owned multiple non-contiguous plots. This was a typical landowning pattern resulting from past cultivation of upland swiddens that were located to take advantage of different microclimates and the need to rotate fields through periods of fallow. Nearly all continued to cultivate a variety of starch and vegetable crops for home consumption and for sale, and 69% of respondents owned at least one cow, with most owning between three and ten head of cattle. The majority of farmers said that their main cash income came from selling potatoes and milk.

Production for household consumption was still predominant; nearly all farmers cultivated a variety of crops and 72% of them said they still ate at least half of the food they grew (down from 88% a generation earlier). Importantly, landowners in this region still wanted to grow their own food, rather than seeking to abandon a farming livelihood. When asked whether they would prefer to grow their own food or to purchase it, 100% of them stated a preference for growing their own food – even when jobs, age, or other factors prevented them from doing so. They gave varied reasons for this: 83% explained that it was economically beneficial to grow their own food rather than to buy expensive foodstuffs, often imported from India; 17% wanted to produce food for the additional income it made; 10% wanted either to be self-sufficient or to have something to do; and less than 7% in each case opted for having nutritious food in the house, ensuring supply of a preferred crop, or ensuring sufficient food supplies for the winter.

Fagopyrum esculentum Moench [Polygonaceae]

Bumthang is known for its buckwheat, which is now attracting attention as an urban health food.

Despite considerable increases in domestic grain yields per production unit in recent decades, Bhutan's reliance on imported staple grains, produce and processed foods from India remains significant (Duba et al., 2006). The rapid transition from diverse upland crops broadly adapted to montane microclimates to a very narrow genetic range of introduced, input-intensive, often irrigated crops has carried major consequences in terms of lost agrobiodiversity, once characteristic of highland cultivation. The recently introduced high-altitude rice grown in the Bumthang region goes by the name *Paro-China*. It is locally believed to have originated in China, and was first tested in the Paro region of western Bhutan. Many respondents mentioned that repeated efforts to introduce paddy rice to Bumthang had previously been unsuccessful, and they credited the current success to either a warming climate or (less frequently) to 10 years of effort to introduce the *Paro-China* variety. Just three potato varieties (Desiree, *Kufri Jyoti* and *Yesukaap*) were grown in Bumthang, for both local and export markets. By contrast, the landrace varieties of buckwheat and their wild relatives, for which the Bumthang region had become known as a centre of biodiversity, stood to lose ground as rice and potatoes gained prominence (Norbu, 1995; Campbell, 1997; National Biodiversity Centre, 2009). The main types of buckwheat for which Bumthang had become known included both the sweet (*Fagopyrum esculentum*) and more frost-tolerant bitter (*F. tataricum*) landrace varieties and their wild relatives (*F. debotrys* and others). Buckwheat is now undergoing something of a transformation, from a subsistence producers' crop grown on marginal land to a health food, sought after by urban consumers within Bhutan and newly appreciated for its particular nutritional and medicinal benefits.

Dukpa (ch 22) outlines in detail how the state has sought since 1969 to curtail swidden cultivation through a variety of economic and regulatory mechanisms. In contrast to the study by Siebert et al. (2015), the government's policy to discourage swidden was not mentioned by people in our interviews; instead, they indicated that the policy was indirectly felt through programmes supportive of permanent agriculture. On a national level, cultivated land fell from 7.85% of Bhutan's land area in 1995 to 2.93% in 2010 (NSSC and PPD, 2011). A change of this magnitude indicates that there was a simultaneous need for significant intensification of local production and increased food imports. Landowners we interviewed drew attention to the indirect effects of state policy, citing its outcomes, such as crops newly available in national seed-production programmes, subsidized technologies and favourable irrigation hierarchies for paddy production, rather than mentioning any direct state efforts to reduce swidden practices.

In descending order of prevalence, farmers said the following motives had contributed to their decisions to abandon swidden agriculture over the past two decades:

- Increased interest in commercial potential and profitability (mentioned by 32% of landowners, who gave more than one reason). One of the most important shifts in the region's agricultural landscape is a widespread crop change from upland grains (buckwheat, barley and wheat) to high-altitude paddy rice for home consumption, and potatoes for local sale and export to India. Agricultural policies favouring paddy production, use of commercial chemical fertilizers, newly available varieties and adoption of mechanized-cultivation technologies have resulted in rapid transition to different crops. Of potato-growing farmers interviewed, 90% said they sold their potatoes; 52% said they sold them for export to India. Nearly all farmers reported greatly expanding the relative area in which they cultivated potatoes in recent years.

- Wildlife damage (24%). Although wild pigs were by far the main cause of crop destruction in all three villages, 92% of the households in rural settlements adjacent to established blue-pine stands reported crop damage, while it affected only 42% of those in the peri-urban village. In the rural settlements surrounded by forest, 67% of residents perceived an increase in crop damage over the past decade, while most residents of the peri-urban village said there had been no change in the incidence of crop damage in recent years. Increased fencing and nightly crop guarding were not sufficient to prevent damage, and several farmers said the damage caused by wildlife was the primary reason they had abandoned agriculture in fields near forests. These forests, which would include all areas previously under swidden cultivation, now provide wild boar habitat. Several farmers told of how excessive damage from wild boar incursions into their fields prompted them to abandon cultivation of distant plots. After wild boar, bears and deer were identified as the greatest wildlife threats to agriculture in the region. Nearly all farmers cited their following of Buddhism or 'Bhutanese beliefs' as the main reason they were unable to kill or otherwise harm wild animals that destroyed their crops.

- Availability of new technologies (16%). The advent of mechanization, the ready availability of annual-crop seeds and new high-yielding grain and vegetable varieties have increased farmers' production options. In one rural village, farmers described the sudden switch they made from upland swiddens on steep slopes far from their settlement, to planting on irrigated flat land after a large landowner sold them the prime fields adjacent to the village. The sudden availability of irrigated land drove their transition to permanent field cropping and its associated technologies. Three farmers indicated that their choices of crops were bound to the seeds that they found available at planting time.

- Land shortages (16%). Farmers mentioned various causes for this, including the encroachment of blue pine on to former swiddens. This, in turn, was connected to the transition to more intensive, permanent cropping using available land for rice and potato production.

- Changes in staple diets (16%). Growing high-altitude paddy rice is relatively new to the region, but it follows a growing increase in the prevalence of imported (Indian) rice consumption, supplanting to a degree many traditional upland grains in local consumption habits. As electric rice cookers appear in local kitchens following ongoing rural electrification, the processing and preparation of rice is deemed simpler and less laborious than that of other staple starches.

- Labour shortages (12%). Migration from rural areas for work and education in towns and cities is rapidly increasing in Bhutan, drawing young people away from traditional family farms.

- Climate change (12%). Villagers noted that palpably shorter winters and higher average year-round temperatures had increased their ability to grow paddy rice and other crops that previously would not grow in Bumthang's cold climate. Farmers were acutely aware that upstream glacial melting (and associated flood risks), noticeably lighter snowfalls and changes in seasonal rainfall were affecting many factors in their agricultural systems and practices.

Other reasons cited include pine-forest encroachment on to agricultural land, the difficulty of selecting crops that are best adapted to given soil types, changing from cultivated fields to pasture land and the increasing availability of imported foodstuffs.

Might swidden farming continue in long rotations with blue pine? It would be possible, in theory, for landowners to allow the spontaneous expansion of blue pine on to their fallow fields to persist for 20 to 30 years so the trees could reach economically harvestable size, and then clear-cut the land, sell the timber, and start the cycle over again with a season or two of upland-grain crops. However, the long-standing state determination to discourage swidden agriculture is only one of several factors reducing the likelihood of this scenario becoming regular practice. Current law does not support land classification that permits long-swidden rotations for forest and agricultural use. The current generation of young farmers has cultivated more potatoes and wetland rice than buckwheat and barley, so they have diminishing hands-on familiarity with upland grain production. In a generation without cultivation, local and diverse stocks of upland-grain seed will have been depleted and may eventually be difficult to locate in sufficient quantities. Buckwheat, in particular, is known for the low viability of its planting seed, even under optimal planting conditions. Regulatory mechanisms within Bhutan strongly favour permanent forest cover, and present rural-urban migration patterns and government policies on food trade with India may further discourage the resurgence of swidden with a long forest-fallow adaptation of the grass-fallow *pangshing* system.

From forest farmers to foresters: Land management after swidden

Dukpa (ch 22) describes the changing, and sometimes ambiguous or seemingly contradictory, national regulations in Bhutan surrounding land and forest management, as it relates to swidden cultivation. In contrast, the national agenda to conserve a high percentage of forest cover throughout the country is consistently highlighted and clearly articulated. Several older villagers recounted how their village was relocated across the Chamkhar River (*Chamkhar Chhu*) in 1964, and the villagers were employed to plant blue-pine trees on the cleared lands surrounding their former settlement. Now, the blue pine is self-propagating and landowners face the issue of controlling spontaneous pine seedling growth in pastures and agricultural lands.

How do landowners view this blue-pine incursion on to their land? When we asked this question, 81% of our respondent landowners said they were happy to see pine saplings growing on their land. Of those, half said they would like to use the trees for their personal consumption, one third hoped to generate income through timber sales, and the rest cited the environmental benefits of healthy forests. All of those who said that they were unhappy to see blue-pine seedlings on their land had small landholdings of less than 5 acres (2.02ha). We showed the landowners eight sketches depicting various alternatives for how their forested land might be used in the future (Figure 23-2), and asked them to select their preference for what their land should look like in five years' time. Most farmers chose dense forest, followed by pasture. None of them chose the images that showed part forest and part cropland (sketches D and E), possibly because of the inevitable issue of pine seedling encroachment on to their crop land. Figure 23-3 is a diagrammatic representation of the majority view of how their land would be used in five years' time.

Some of the important land-management issues facing landowners in Bhutan are their ability to harvest trees for personal use or for sale and the permissibility of cutting or uprooting pine seedlings that appear in their fields. In the past, landowners have not been allowed to harvest trees from their own land for any reason, without official permission, and the state has maintained the right to harvest trees from private land for distribution to other individuals or to state enterprises. Our tree and stump measurements in transect plots showed that 73% of stumps were less than 16cm in diameter – the minimum size accepted at local sawmills. The cutting of more small-diameter trees than those of commercial size indicated that the trees had been cut

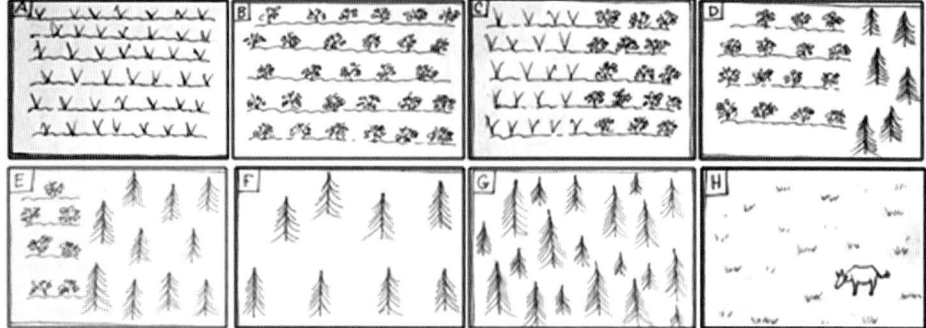

FIGURE 23-2: Sketches from which farmers selected their preferred use status of their forested land in five years' time.

Note: Although the incremental increase in potatoes is small, most farmers indicated that the relative area planted to potatoes had greatly expanded in recent years.

for firewood or other household uses. Additionally, during interviews some farmers said that in the past they had cut trees while they were still small in order to avoid state claims on the trees, or to ensure that a certain plot would remain available for projected uses in the future, such as home construction for children, and would not be gathered up and given state-mandated conservation status by virtue of it having large trees. This situation is changing, and may account for the large percentage of farmers who hoped to see their land covered in dense forest in the future.

Revision of the rules related to forests is an ongoing process, and implementation of past initiatives has been incomplete, leaving the public with a partial or outdated awareness of the changes. As a result, there is some uncertainty among landowners regarding what is allowed and what is not, and the doubt extends to the process required for gaining permission to harvest trees for firewood, timber, or other purposes. We asked villagers to describe their understanding of permissible forest use, and also to name their sources of information about permissible forest use. As shown in Figure 23-4, of the 23 households that owned private forests, 61% believed it was illegal to do anything with their land. In the peri-urban village of Chamkhar, which is closer to government offices, only 20% believed that using their forests for any purpose was illegal, compared to 80% of those interviewed from rural areas. It was striking that 74% of forest owners named forestry officials as an important information source, and that current understanding of the rules among our

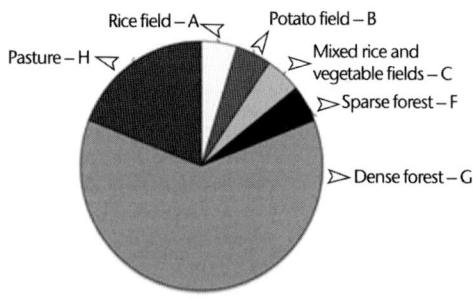

FIGURE 23-3: Landowners' preferred composition of their forested land in five years' time, from sketches depicting various possible uses (n=21).

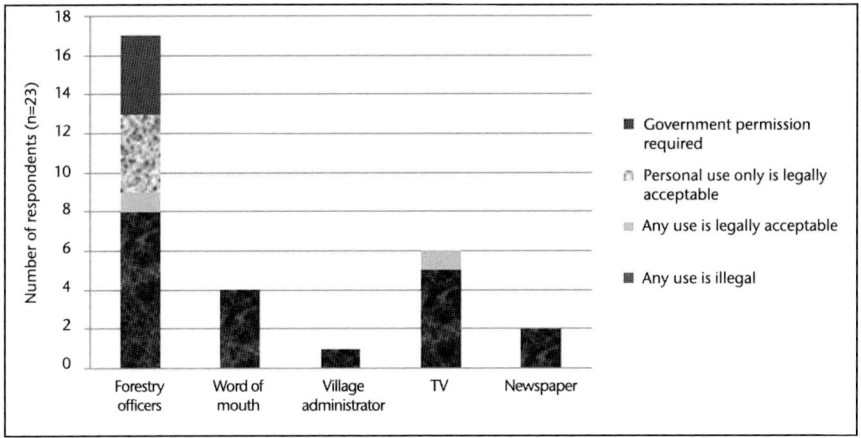

FIGURE 23-4: Correlation between perceptions of permissible use of private forest and sources of information about how landowners may use their forest.

interview subjects varied so widely. Clearly, official information sources played an important role in public information.

Following a mid-1990s increase in attention to community forestry, a new process for registering and certifying private forests was introduced in 2003. It modified the regulations surrounding individual rights to tree use on private land (Social Forestry Division, 2003), and provided greater latitude for landowners to actively manage their forested land (Phuntsho, 2011). Farmers seemed to favour the ability to maintain land with standing trees as forest for their own use and as a source of supplementary cash income. When we asked landowners with forested plots (n=23) how they would prefer to use those forests, if given complete freedom to choose, 83% (19/23) said they would like to harvest their own trees for personal use; 57% (13/23) prioritized use of the trees for their own home construction, and 17% (4/23) mentioned firewood. In addition, 52% (12/23) said they would like to sell trees, but nearly all of these noted that personal use was primary and they would sell only small quantities or excess wood after household needs were met. One respondent spoke of leaving the forest alone; another would opt to clear the forest and convert it to pasture; and just two expressed the desire to clear their forest for farmland, one of them saying that he would plant buckwheat and barley.

Of all the plots we measured, there was an average volume between 1842 and 2343 cubic feet (52.13 and 66.3 cubic metres) of timber, usable for construction and firewood. 'Usable' timber was calculated by subtracting 50% of total volume to account for bark and branches, which is not included in this analysis. Applying the government-regulated price for timber as of July 2012 (280 ngultrum per cubic foot) and after subtracting 20% to account for varying transportation, labour and processing costs, the landowners of the blue-pine forests we measured could potentially make between 412,551 and 524,757 ngultrum (US$7457 and $9486) per acre (about $18,412 to $23,422 per hectare) if they were to clear-cut and sell all of their trees larger than 16cm diameter at breast height. Even if only 10% of their timber were harvested per year, there was a significant income potential that could be managed sustainably with the help of forestry officers. This analysis concurs with Dukpa et al. (2007), who found that returning former *pangshing* lands to blue-pine cover and private forestry was economically favourable. It offered an alternative that required reduced labour and no inputs, but a good annual income – as long as the landowner could wait as long as 30 years for first returns.

In conclusion, this study illustrates some of the reasons farmers give for agricultural transitions. Although none of our respondent farmers explicitly cited government policies opposing swidden cultivation among the reasons for their cropping-system changes, they were taking part in a range of government programmes that supported intensified permanent cropping and production of export commodities. Land that is fallowed for a long time in their mountainous territory tends to become blue-pine forest, which most farmers viewed as a favourable condition, with the expectation that future government regulations would allow them to harvest trees for sale and personal use. The data suggest that income potential from private blue-pine forests on former swidden fields is significant enough to merit additional attention to this land use. Forestry officers are well placed to aid landowners in understanding new regulations and procedures as they evolve.

Acknowledgements

The School for Field Studies, Beverly, Massachusetts, provided support and funding for this research. However, it would not have been possible without the commitment and assistance offered by many members of staff at the Ugyen Wangchuck Institute for Conservation and Environment in Bhutan and the participation of village residents who welcomed us into their homes and showed us their fields and forests.

References

Campbell, C. G. (1997) 'Buckwheat. *Fagopyrun esculentum* Moench.', 19, *Promoting the Conservation and Use of Underutilized and Neglected Crops,* Institute of Plant Genetics and Crop Plant Research and International Plant Genetic Resources Institute, Gatersleben, Germany and Rome

Cochrane, M. A. (2009) *Tropical Fire Ecology: Climate Change, Land Use and Ecosystem Dynamics*, Springer-Praxis, Heidelberg

Duba, S., Gurung, T. R. and Ghimiray, M. (2006) *Assessment of SARD-M Policies in the Hindu Kush Himalayas: The Case of Land Use Policies in Bhutan*, Renewable Natural Resources Research Centre, SARD Mountain Policy Project of the Food and Agriculture Organization of the United Nations, and the International Centre for Integrated Mountain Development (ICIMOD), Bajo, Bhutan

Dukpa, T., Wangchuk, P., Rinchen, Wangdi, K. and Roder, W. (2007) 'Changes and innovations in the management of shifting cultivation land in Bhutan', in M. F. Cairns (ed.) *Voices from the Forest: Integrating Indigenous Knowledge into Sustainable Upland Farming*, Resources for the Future Press, Washington, DC, pp692-699

Gilani, H., Shrestha, H. L., Murthy, M. S. R., Phuntsho, P., Pradham, S., Bajracharya, B. and Shrestha, B. (2015) 'Decadal land cover change dynamics in Bhutan', *Journal of Environmental Management* 148, pp91-100

Guenat, D. (1991) *Study of the Transformation of Traditional Farming in Selected Areas of Central Bhutan: The Transition from Subsistence to Semi-subsistence, Market Oriented Farming,* PhD dissertation to the Swiss Federal Institute of Technology, Zurich, Switzerland

Mong, C. E. and Vetaas, O. R. (2006) 'Establishment of *Pinus wallichiana* on a Himalayan glacier foreland: Stochastic distribution or safe sites?' *Arctic, Antarctic, and Alpine Research* 38(4), pp584–592

National Biodiversity Centre (2009) *Biodiversity Action Plan* 2009, Ministry of Agriculture, Royal Government of Bhutan, Thimphu

Norbu, S. (1995) 'Buckwheat in Bhutan', in T. Matano and A. Ujihara (eds) *Current Advances in Buckwheat Research*, vol. 1, proceedings of the 6th International Symposium on Buckwheat, Shinshu University Press, Asahi Matsumoto City, Japan, p55–60

Norbu, S., Wangdi, D., Roder, W. and Wangdi, K. (1996) 'Traditional practices of Bhutanese mountain farmers to maintain soil fertility in buckwheat systems,' presented at an International Crop Science Congress, New Delhi, November, 1996

NSSC and PPD (2011) *Bhutan Land Cover Assessment 2010* (LCMP-2010), National Soil Services Centre, Policy and Planning Division, Ministry of Agriculture and Forests, Thimphu

Phuntsho, S. (2011) 'Forests, community forestry and their significance in Bhutan', in S. Phuntsho, K. Schmidt, R. Kuyakanon and K. J. Temphel (eds) *Community Forestry in Bhutan: Putting People at the Heart of Poverty Reduction*, Ugyen Wangchuck Institute for Conservation and Environment and Social Forestry Division, Jakar and Thimphu, Bhutan, pp1–3

Roder, W., Calvert, O. and Dorji, Y. (1992) 'Shifting cultivation systems practised in Bhutan', *Agroforestry Systems* 19, pp149–158

Siebert, S., Belsky, J., Wangchuk, S. and Riddering, J. (2015) 'The end of swidden in Bhutan: Implications for forest cover and biodiversity', in M. F. Cairns (ed.) *Shifting Cultivation and Environmental Change: Indigenous People, Agriculture and Forest Conservation*, Earthscan from Routledge, London, pp546–558

Social Forestry Division (2003) *Private Forestry Manual for Bhutan, August 2003*, Department of Forestry, Ministry of Agriculture, Royal Government of Bhutan, Thimphu

24

KEEPING ECOLOGICAL DISTURBANCE ON THE LAND

Recreating swidden effects in Bhutan

*Stephen F. Siebert and Jill M. Belsky**

Introduction

For many centuries, swidden agriculture was the basis of complex, linked socio-ecological systems throughout much of the world (Cairns, 2007; Xu et al., 2009; Padoch and Pinedo-Vasquez, 2010). In South and Southeast Asia, hundreds of ethno-linguistically and culturally unique societies developed and managed tree- or shrub-fallow farming systems that reflected context-specific environmental (e.g. climate, soils, slope and vegetation) and social conditions (e.g. cultural beliefs, governance institutions and socio-economic resources). For centuries these practices provided communities and households with food, fibre, building materials, medicines and other valuable products and, in the process, created and maintained floristically diverse and structurally complex vegetation mosaics across landscapes (Fox et al., 2009; Mertz et al., 2009a; Xu et al., 2009; Siebert and Belsky, 2014). However, in recent decades, swidden systems and the diverse societies that created and maintained them have disappeared or been radically transformed (Kerkhoff and Sharma, 2006; Fox et al., 2009; Mertz et al., 2009a; Xu et al., 2009). The bio-cultural losses associated with modernization and the integration of formerly isolated peoples into nation states and a globalized economy are well documented, as are changes in traditional agricultural knowledge and practices (e.g. replacement of diverse, nutritionally rich subsistence food systems with more risky cash crops and monocultures that are dependent on chemical inputs) (Kerkhoff and Sharma, 2006; Xu et al., 2009; Maffi and Woodley, 2010; Pilgrim and Pretty, 2010; Wangchuk and Siebert, 2013). What

* Dr Stephen F. Siebert, Professor of Tropical Forest Conservation and Management, College of Forestry and Conservation, University of Montana, Missoula, MT; Dr Jill M. Belsky, Professor of Rural and Environmental Social Science, College of Forestry and Conservation, University of Montana, Missoula, MT

is less understood are other implications associated with the cessation of swidden systems, especially their socio-ecological legacies.

In a recent paper we examined the socio-ecological aspects of historical swidden practices in Bhutan and argued that they functioned as intermediate ecological disturbances that contributed to the creation and maintenance of biodiversity (i.e. the composition, abundance, structure and distribution of flora and fauna) (Siebert and Belsky, 2014). In this chapter we suggest that some integral (i.e. what we call 'historical') swidden systems offer principles that may be built upon or adapted to promote biodiversity conservation and to provide goods, income and employment for rural households in the future. Our argument is not that swidden can or should be recreated in Bhutan or anywhere else; that is impossible due to dramatic changes in agrarian conditions, political economy and culture. Rather, we suggest that understanding the socio-ecological attributes and disturbance effects associated with swidden (and other historical land uses) can and should inform the conservation and development policies and programmes of the present day.

Swidden, cultural landscapes and biodiversity

Swidden, or shifting cultivation, was for centuries widely practised in Bhutan and much of the tropical and temperate world (Cairns, 2007; Fox et al., 2009; Padoch and Pinedo-Vasquez, 2010; Dove, 2015; Dukpa, ch 22). The extent of human modification has led to the use of the term 'cultural landscapes' to denote land forms that are the cumulative result or 'co-production' of human activities and environmental conditions (Balée, 2013). Despite aggressive attempts to suppress swidden by colonial regimes, post-independence governments and protectionist conservation approaches, in the early 2000s about 10 million hectares remained under some form of forest farming in the eastern Himalayas, at the hands of between 50 and 200 million swidden farmers (Kerkhoff and Sharma, 2006; Mertz et al., 2009b; Ziegler et al., 2009). The extent of swidden in Bhutan prior to the country's political and economic integration with the outside world in the late 20th century is unknown, but in 1988 swidden fields or fallows were believed to cover about 200,000ha, or 5.2% of the country's land area (Roder et al., 1992) (Figure 24-1).

The effect of swidden and other historical land uses on flora, fauna and biodiversity in general can be evaluated by conceptualizing anthropogenic activities as ecological disturbances and identifying their effects (Siebert and Belsky, 2014). Ecological disturbances can be characterized on the basis of specific attributes, the most important of which are their type, spatial features, temporal characteristics, specificity, intensity and resulting synergisms (Mori, 2011). While disturbance effects vary with soil, topography, climate and other factors, three disturbance parameters are particularly important in forest ecosystems: (1) the return interval (i.e. the time between disturbances); (2) the severity (i.e. amount of vegetation killed and the type and amount of space available for new plant growth); and (3) the landscape-level spatial patterns that are created (Seymour and Hunter, 1999).

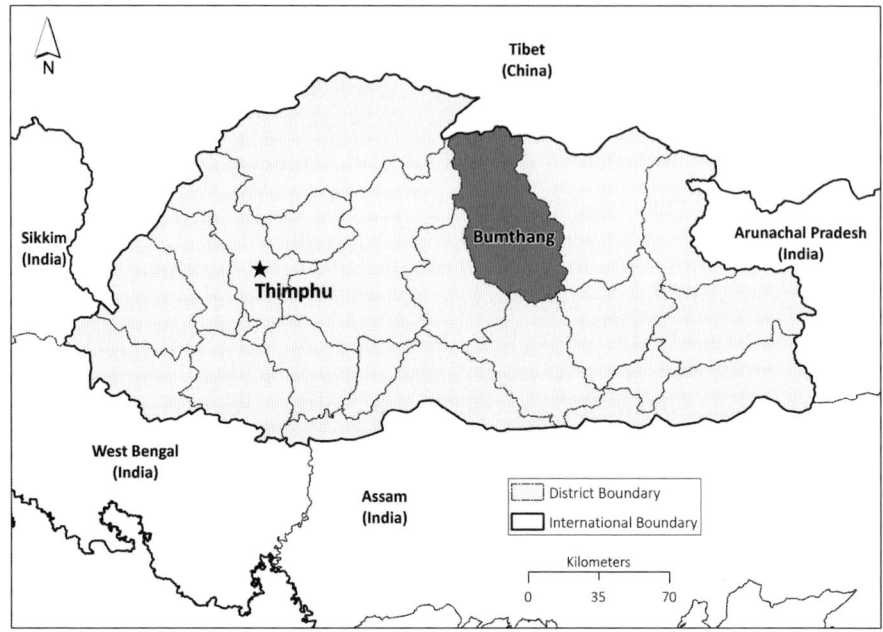

FIGURE 24-1: Bhutan, showing the locations of Bumthang district and the capital, Thimphu.

Disturbance effects associated with traditional Bhutanese swidden farming and two important and widespread natural disturbances (i.e. individual tree mortality and landslides) are summarized in Table 24-1. This comparison suggests that ecological disturbances due to swidden farming held an intermediate position between natural tree-falls and landslides in terms of size, intensity and duration. Another important effect of swidden was the maintenance of complex vegetation mosaics or patchiness, with significantly more early seral vegetation (i.e. grasses, forbs and other plant species that establish following disturbance) and open habitats. The species composition and structure of the flora in swidden fallows was different to that which developed following natural treefalls and landslides. Swidden landscapes also included areas that were not cultivated or managed due to steep slopes, poor soils, difficult access, religious beliefs and local customs that maintained habitats for forest-dependent flora and fauna (Siebert and Belsky, 2014). At the landscape level, the overall effect was to increase plant species richness and structural heterogeneity, and maintain open and disturbed habitats of value to ungulates and carnivores, such as tigers, leopards and wild dogs (Wikramanayake et al., 2011; Post and Panday, 2013; Siebert and Belsky, 2014).

Swidden has largely ceased in Bhutan and the rest of Asia. This has been due to government policies based on modern development and conservation politics that purposefully disregarded or failed to understand the productivity, priorities, sustainability and mutually reinforcing social and ecological services provided by historical livelihoods and land uses. Instead, they pursued sedentary and export

TABLE 24.1: Disturbance effects associated with swidden, treefalls and landslides in Bhutan.

Disturbance attribute	*Disturbance type and effects*		
	Tree-falls	*Swidden* (tseri)	*Landslides*
Size	Small and partial – less than 0.25 ha	Intermediate. 1 to more than 3ha	Large. Tens to hundreds of hectares
Intensity	Low. Plants crushed, pits and mounds	Intermediate. Biomass cut and burned, cultivated, not ploughed	Extreme. Vegetation and soil removed to subsoil or bedrock
Duration	Short. Rapid regrowth	Intermediate. 1 to 2 yrs of cultivation	Long and variable: yrs to decades
Frequency	?	Two-yr to more than 10-yr fallows	?
Landscape pattern	Variable, widespread	Managed mosaic of different species and seral stages	Variable – slopes and deposition areas
Ecological effects	Maintains +/- dense closed forests	Greater percentage of early seral spp. with unused closed forests	Widespread rock and bare ground
Succession	Rapid, native spp.	Secondary fallow spp.	Primary, slow

agriculture, forest exploitation and Western protectionist conservation (Dove, 1983, 2015; Phuntsho et al., ch 17). New market opportunities, cultural changes, rural to urban migration and other dimensions of agrarian transformation have also propelled farmers to replace swidden with different agricultural practices and livelihood activities (Mertz et al., 2009a; Xu et al., 2009; Wangchuk and Siebert, 2013). Consequently, swidden-related, intermediate-scale disturbances no longer provide the ecological and livelihood functions they did in the past. Throughout Bhutan and other areas of the eastern Himalayas, swidden landscapes have transformed into either dense, closed-canopy forests or intensive annual and perennial cash-crop monocultures (Mertz et al., 2009a; Xu et al., 2009; Siebert et al., 2015). As a result, the flora, fauna and rural economy are changing with uncertain, but potentially profound implications for bio-cultural diversity and household livelihoods. We suggest that understanding historical swidden practices and their effects should inform rural livelihood development and biodiversity conservation efforts.

Opportunities for recreating swidden effects in Bhutan

Recreating the intermediate-level ecological disturbances associated with historical swidden practices in contemporary Bhutan will be challenging. It will require: (1) identifying the range of effects associated with historical swidden practices, particularly the size and intensity of the ecological disturbances they create and the resulting landscape patterns; (2) revising the forest policies and regulations of the Royal Government of Bhutan so that the ecological importance of swidden-related disturbances is appreciated and experimentation with novel forest-management practices that replicate swidden disturbance effects is encouraged; and (3) supporting livelihoods and land uses that generate products and income while meeting ecological objectives. We suggest that experimental activities and land uses be context-specific and collaborative, involving resource users, local managers and others to understand available options and to pursue land uses and livelihoods that make sense in the light of local social, economic and environmental conditions and concerns.

We suggest two practices that may maintain some of the ecological-disturbance effects associated with historical swidden farming in Bhutan: timber harvesting by group selection or clear felling and the cultivation of tree crops in open orchards. It is important to emphasize that both practices should occur at intermediate scales and be managed in a patchy or mosaic-landscape pattern to maintain historical landscape-scale disturbance effects. Group selection or clear felling of small stands of blue pine (*Pinus wallichiana*) in cool temperate areas and oak (e.g. *Quercus griffithii*) and other hardwood species in warm temperate zones for timber and firewood could mimic some historical swidden-disturbance effects while generating products for household use or sale. Blue pine, oak and other economically valuable species have grown in former swiddens throughout Bhutan and are widely used for timber and firewood. Blue pine is an aggressive, shade-intolerant pioneer species that rapidly colonizes swidden fields and other open areas in mid-altitude zones throughout the southern Himalayas (Cochrane, 2009). Thousands of hectares of dense, even-aged, blue pine stands now cover much of Bumthang district alone (Siebert et al., 2015), on what are now Royal Government of Bhutan forest lands, following their nationalization (Dukpa et al., 2007) (Figure 24-2). Yoder et al. (ch 23) estimate that harvesting all pines with a diameter at breast height greater than 16cm on a 30-year rotation could generate between US$18,412 and $23,422 per ha in this region. This is an attractive financial return compared to available alternatives, particularly since it requires little labour or capital investment (Dukpa et al., 2007) and would also produce firewood for domestic consumption and sale.

To recreate disturbances to mimic the historical effects of swidden, harvesting could be conducted in small blocks (1 to 3ha), followed by

Pinus wallichiana A. B. Jacks.
[Pinaceae]

Blue pine is an aggressive, shade-intolerant pioneer species that rapidly colonizes swiddens and other open areas in the southern Himalayas.

FIGURE 24-2:   Thousands of hectares of dense, even-aged, blue pine stands now cover much of Bhutan's Bumthang district.

low-intensity burning of slash (i.e. tops, small branches and leaves). Repeated across the landscape, this would create mosaics of open and regenerating pine stands of different ages and sizes, along with unutilized, closed-canopy forests. Seed trees could be retained to facilitate regeneration and to improve the growth, vigour and form of subsequent trees. Parcel sizes, landscape patterns and other disturbance attributes could reflect previous site-specific swidden practices, which vary from one region to another. These practices could potentially be implemented in government reserve forests, including community forests, and on private land, with harvesting activities regulated by the government (Phuntsho et al., ch 17). Harvesting could be conducted by private contractors who pay the government, community forest-management groups or private landowners on the basis of stumpage values.

This same approach could be explored in temperate broadleaf forests dominated by oak, chestnut and other economically valuable species. Unlike blue pine, some hardwood species, such as oak, coppice vigorously, which eliminates the need to replant or to retain seed trees. Potential market opportunities and returns from clear felling of hardwood stands in Bhutan are unknown, but would provide timber for furniture and veneer as well as high quality firewood.

Some historical disturbance effects might also be recreated by cultivating tree crops in relatively open stands. One example is a current project being undertaken by the private Mountain Hazelnut Venture, the Royal Government of Bhutan and private landowners, in which hazelnuts are being grown as a cash crop on private, former swidden lands and degraded slopes between 1600m and 3000m above sea level, particularly in eastern Bhutan and in the Punakha and Bumthang districts, where swidden was once widespread (Mountain Hazelnut Venture, 2014). The company provides assistance with seedlings and planting, and later with processing and marketing. As of 2014, 4400 private landowners had planted two million hazelnut seedlings, with a goal of 20,000 landowners cultivating 10 million trees on 20,000 acres (8100ha) (Mountain Hazelnut Venture, 2014).

Orchards are more open than primary or secondary forests and support understorey growth of grasses and forbs that could be used by wild ungulates. However, orchards will not provide other swidden effects, such as regenerating secondary forests. Importantly, the long-term economic viability, benefits to landowners and sustainability of hazelnut cultivation will depend on the terms of production and exchange under which they are grown and marketed, crop yields, market demand and prices, returns to labour, pest losses and other factors, all of which are presently unknown.

Policy and regulatory changes

The biodiversity effects and economic returns associated with clear felling trees in small parcels for firewood, and tree-crop cultivation, warrant examination under a range of socio-economic and environmental conditions. Field trials should evaluate and monitor variable harvesting and management effects on plant regeneration, residual vegetation, soils, wildlife, economic costs and benefits, labour requirements, marketing and other factors. Including older farmers with experiential knowledge of traditional swidden practices and their ecological effects could help to identify possible management approaches and evaluation criteria. These strategies may require the modification of existing government forestry and agricultural policies and regulations to: (1) fund trials under a range of climatic, soil and vegetation conditions (e.g. cool temperate blue pine and warm temperate oak stands); (2) modify current timber harvesting practices, which emphasize single trees and selective systems, and evaluate group selection or clear felling in small patches; (3) allow increased timber harvesting over a greater proportion of landscapes, particularly around villages where timber and firewood can be readily transported to market, where forest fires pose increasing risks and where swidden was formerly practised; (4) evaluate market demands, returns to labour and opportunities for locally profitable, value-added processing for domestic and export

Quercus griffithii Hook. f.
& Thomson ex Miq. [Fagaceae]

Clear-felling of small stands of oak
is proposed to mimic the beneficial
disturbance effects of swidden.

markets; and (5) revise attitudes towards swidden farmers to consider them potential management partners.

Another challenge to recreating swidden-like disturbances in Bhutan is the need for government officials, elected representatives, environment and development organizations and the general public to recognize that some traditional swidden systems contributed to the creation and maintenance of bio-cultural diversity. Swidden agriculture and the people who managed it have been denigrated for decades in Bhutan as being both primitive and destructive (Dukpa, ch 22). Since the early 1980s, the government has subsidized the conversion of swiddens to permanent wet and dry fields and cash crop plantations (Phuntsho et al., ch 17). In Bhutan, increased forest cover and density are widely assumed to be desirable. Indeed, the constitution provides that at least 60% of the country must remain under natural forest cover, and more than half of the country is in some form of national park or protected area (MoA, 2009; Kuensel, 2013). Despite this, the promotion of market-based enterprises in rural areas has become a development objective in Bhutan (PDP, 2013), but their pursuit in the forestry sector is fraught with challenges. For example, the government recently suspended the designation of new community forests because of concerns over inadequate monitoring and excessive tree harvesting by some community forestry management groups (Dema, 2014).

Attitudes towards traditional swidden systems and expanding forest cover may be changing in Bhutan and elsewhere. The International Centre for Integrated Mountain Development (ICIMOD) has concluded that some swidden systems were not only productive and sustainable, but also preferable to the agricultural practices that have replaced them in the eastern Himalayas (Kerkhoff and Sharma, 2006). Moreover, the governments of Bangladesh, Bhutan, China, India, Myanmar and Nepal have acknowledged that 'shifting cultivation must be recognized as an agricultural and adaptive forest-management practice that is based on scientific and sound ecological principles' (cited in Kerkhoff and Sharma, 2006). This view mirrors those of long-term swidden researchers throughout Asia (Cairns, 2007, 2015; Xu et al., 2009). In addition, swidden maintained open areas and limited the accumulation of biomass over large areas for centuries. Royal Government of Bhutan forestry officials have recently concluded that forest fires are increasing in frequency, size and intensity and have become a serious problem due, in part, to the increasing cover and density of the country's forests (MoAF, 2013).

Conclusion

In Bhutan and much of the eastern Himalayas, historical swidden-based societies created, managed and maintained open habitats and early seral vegetation, along with large expanses of mature, closed forests in complex landscape mosaics. For centuries, these landscapes sustained local food production and cultures, along with diverse flora and fauna (Kerkhoff and Sharma, 2006; Xu et al., 2009; Siebert and Belsky, 2014). In recent decades, the swidden-associated, intermediate-level ecological disturbances of the past have ceased in Bhutan and elsewhere. Government policies, emerging market opportunities and other agrarian changes have led to the establishment of annual and perennial crop monocultures and dense, closed canopy forests across formerly heterogeneous landscape mosaics in Bhutan and other upland areas of Southeast Asia. Flora and fauna that prefer open habitats, such as ungulates and rare and endangered carnivores, may be adversely affected by these land-use changes. The ecological sustainability and livelihood security of households and communities who are dependent upon cash crops are questionable and increasingly vulnerable to economic, social and environmental (e.g. climate) change. Increasing forest cover and density are also resulting in different disturbances, such as larger and more destructive fires, which affect people and their environments.

The conservation of biodiversity and human well-being, particularly the capacity of socio-ecological systems to adapt to dynamic and unpredictable social, economic and climatic change, require the identification and development of new functional links between sustainable livelihoods, culture and biodiversity (Xu et al., 2009). Small-scale clear-felling of forest stands for the production of timber and firewood, followed by burning of slash, could generate income and products for rural households while recreating the intermediate-scale ecological disturbances formerly provided by swidden farming, thereby supporting national development and biodiversity-conservation objectives. Recognizing and building upon historic social-ecological legacies and traditional knowledge could and should inform future conservation and development paths.

References

Balée, W. (2013) *Cultural Forests of the Amazon*, University of Alabama Press, Tuscaloosa, AL

Cairns, M. F. (ed.) (2007) *Voices from the Forest: Integrating Indigenous Knowledge into Sustainable Upland Farming*, Resources for the Future, Washington, DC

Cairns, M. F. (ed.) (2015) *Shifting Cultivation and Environmental Change: Indigenous People, Agriculture and Forest Conservation*, Earthscan, London

Cochrane, M. (2009) *Tropical Fire Ecology: Climate Change, Land Use and Ecosystem Dynamics*, Springer-Praxis, Heidelberg

Dema, T. (2014) 'Community forest formation suspended', http://www.kuenselonline.com/community-forest-formation-suspended/#.U6GBSJRDVJK, accessed 7 January 2014

Dove, M. (1983) 'Theories on swidden agriculture and the political economy of ignorance', *Agroforestry Systems* 1, pp85-99

Dove, M. (2015) 'The view of swidden agriculture by the early naturalists Linnaeus and Wallace', in M. F. Cairns (ed.) *Shifting Cultivation and Environmental Change: Indigenous People, Agriculture and Forest Conservation*. Earthscan, London

Dukpa, T., Wangchuk, P., Rinchen, Wangdi, K. and Roder, W. (2007) 'Changes and innovations in the management of shifting cultivation land in Bhutan', in M. F. Cairns (ed.) *Voices From the Forest: Integrating Indigenous Knowledge into Sustainable Upland Farming*, Resources for the Future, Washington, DC, pp692-699

Fox, J., Fujita, Y., Ngidang, D., Peluso, N., Potter, L., Sakuntaladewi, N., Sturgeon, J. and Thomas, D. (2009) 'Policies, political-economy and swidden in Southeast Asia', *Human Ecology* 37, pp305-322

Kerkhoff, E. and Sharma, E. (2006) *Debating Shifting Cultivation in the Eastern Himalayas*, International Center for Integrated Mountain Development (ICIMOD), Kathmandu, Nepal

Kuensel (2013) 'The fuelwood paradox', http://www.kuenselonline.com/the-fuelwood-paradox/#.
VBsLq_ldWCk, accessed 5 December 2013

Maffi, L. and Woodley, E. (2010) *Biocultural Diversity Conservation: A Global Sourcebook*, Earthscan,
London

Mertz, O., Padoch, C., Fox, J., Cramb, R., Leisz, S., Lam, T. and Vien, T. (2009a) 'Swidden change in
Southeast Asia: Understanding causes and consequences', *Human Ecology* 37, pp259-264

Mertz, O., Leisz, S., Heinimann, A., Rerkasem, T., Dressler, W., Cu, P., Vu, K., Schmidt-Vogt, D., Colfer,
C., Epprecht, M., Padoch, C. and Potter, L. (2009b) 'Who counts? The demography of swidden
cultivation', *Human Ecology* 37, pp281-289

MoA (2009) *Bhutan Biodiversity Action Plan*, Ministry of Agriculture, Thimphu, Bhutan

MoAF (2013) *Forest Fire Management Strategy for Bhutan*, Ministry of Agriculture and Forests,
Thimphu, Bhutan

Mori, A. (2011) 'Ecosystem management based on natural disturbances: Hierarchical context and non-
equilibrium paradigm', *Journal of Applied Ecology* 48, pp280-292

Mountain Hazelnut Venture (2014) *Fact Sheet*, Mountain Hazelnut Venture, Thimphu, Bhutan

Padoch, C. and Pinedo-Vasquez, M. (2010) 'Saving slash and burn to save biodiversity', *Biotropica* 42,
pp550-552

PDP (2013) *Manifesto*, People's Democratic Party, Thimphu, Bhutan, http://pdp.bt/2013-election/
manifesto, accessed 5 December 2013

Pilgrim, S. and Pretty, J. (eds) (2010) *Nature and Culture: Rebuilding Lost Connections*, Earthscan,
London

Post, G. and Panday, B. (2013) 'Comparative evaluation of tiger reserves in India', *Biodiversity
Conservation* 22, pp2785-2794

Roder, W., Calvert, O. and Dorji, Y. (1992) 'Shifting cultivation systems practised in Bhutan', *Agroforestry
Systems* 19, pp149-158

Seymour, R. and Hunter, M. (1999) 'Principles of ecological forestry', in M. Hunter (ed.) *Maintaining
Biodiversity in Forest Ecosystems*, Cambridge University Press, Cambridge, UK, pp22-61

Siebert, S. F. and Belsky, J. M. (2014) 'Historic livelihoods and land uses as ecological disturbances and
their role in enhancing biodiversity: An example from Bhutan', *Biological Conservation* 177, pp82-
89

Siebert, S., Belsky, J., Wangchuk, S. and Riddering, J. (2015) 'The end of swidden in Bhutan: implications
for forest cover and biodiversity', in M. F. Cairns (ed.) *Shifting Cultivation and Environmental
Change: Indigenous People, Agriculture and Forest Conservation*. Earthscan, London, pp546-558

Wangchuk, S. and Siebert, S. (2013) 'Agricultural change in Bumthang, Bhutan: Market opportunities,
government policies and climate change', *Society and Natural Resources* 26, pp1375-1389

Wikramanayake, E., Dinerstein, E., Seidensticker, J., Lumpkin, S., Pandav, B., Shrestha, M., Mishra, H.,
Ballou, J., Johnsingh, A., Chestin, I., Sunarto, S., Thinley, P., Thapa, K., Jiang, G., Elagupillay, S., Kafley,
H., Pradhan, N., Jigme, K., Teak, S., Cutter, P., Abdula Aziz, M. and Than, U. (2011) 'A landscape-
based conservation strategy to double the wild tiger population', *Conservation Letters* 4, pp219-227

Xu, J., Lebel, L. and Sturgeon, J. (2009) 'Functional links between biodiversity, livelihoods, and culture
in a Hani swidden landscape in southwest China', *Ecology and Society* 14(2): 20

Ziegler, A., Bruun, T., Guardiola-Claramonte, M., Giambelluca, T., Lawrence, D. and Lam, N. (2009)
'Environmental consequences of the demise in swidden cultivation in montane mainland Southeast
Asia: Hydrology and Geomorphology', *Human Ecology* 37, pp361-373

25

SHIFTING CULTIVATION IN VIETNAM

Impacts of various policy reforms

*Delia Catacutan, Nguyen Thi Hoa, Do Trong Hoan,
Elisabeth Simelton and Hoang Thi Lua**

Introduction

Traditional shifting cultivation has been practised in Vietnam for many years. It has often been blamed for forest degradation and deforestation. 'Slash and burn' – as one method of shifting cultivation – has also been proven to be a trigger for large-scale forest fires. In 2007, the Ministry of Agriculture and Rural Development in Vietnam reported that shifting cultivation was the cause of 60% to 70% of all reported forest fires and 60% of illegally logged forest areas. It is also often identified as a cause of soil erosion that leads to sedimentation in rice fields on lower slopes and in watershed areas (Tran, 2001). Due to the complexity of these issues, successive governments since the early 1960s have adopted a range of policies aimed at halting or stabilizing the practice of shifting cultivation. Overall, the outcomes of these policy interventions, especially in terms of reducing the area under shifting cultivation, have been laudable. Mixed results have been reported only in specific aspects of policy implementation. For instance, a significant reduction in the area under shifting cultivation was reported by the Ministry of Agriculture and Rural Development in 2007, but some claimed that the national figure masked a myriad of issues and challenges faced by local people. Each new policy brought new challenges that affected shifting cultivators either positively or negatively. However, with new mechanisms such as REDD+ emerging at a global level,[1] it is important to reflect on how shifting cultivators might be more effectively involved with implementation, rather than being construed as either the victims or the causes of failure. Based on literature reviews, recent surveys and interviews with local farmers and provincial officials in the country's northwest region between April

* Dr Delia Catacutan is Country Representative of the World Agroforestry Centre (ICRAF) in Vietnam; Nguyen Thi Hoa is a researcher; Do Trong Hoan, a researcher; Dr Elisabeth Simelton, a scientist; and Dr Hoang Thi Lua, a researcher. All authors work for the World Agroforestry Centre, Vietnam.

and June 2012, this chapter provides an overview of shifting cultivation practices in Vietnam, the key policy interventions and their impacts on livelihoods, land use and the environment, as well as insights into how shifting cultivators could benefit from new incentives such as REDD+.

Shifting cultivation practices in Vietnam: an overview

Shifting cultivation types

There are various shifting cultivation practices in Vietnam (Do, 1994; Tran, 2007), and as this chapter will show, the practices transform according to policy changes. First, 'pioneer' shifting cultivation involves forest clearing for extensive cultivation, normally for a period of three to five years, before the farmers move on to other areas with little or no intention of returning (Figure 25-1). Upland rice, maize and cassava are the main crops planted. This practice was typically carried out by H'mong people in mountainous areas at altitudes of at least 700 metres above sea level (masl), but it is now declining due to increasing population density (Lundberg, 1996) and the allocation of land to individual households.

Second, there is 'rotational' shifting cultivation, in which the land is cultivated for a number of years and then left under fallow for three to five years to allow the natural recovery of vegetation and soil fertility. Upland rice is usually grown for two to three years, followed by three years of cassava or maize, beans and sweet potatoes (Figure 25-2). The practitioners, including the Dao, Ede, Bana and Gie Chieng ethnic groups, open their swiddens by slashing the vegetation from fallowed land and burning it.

Third, the most common shifting cultivation practice at present is 'supplementary' or 'composite' shifting cultivation. This is undertaken on steep slopes surrounding permanently cultivated valleys with small paddy fields at 300 to 700masl. The hillside vegetation is slashed and burned and then planted with upland rice, maize, cassava,

FIGURE 25-1: Typical pioneer shifting cultivation in a mixed–bamboo forest in Dien Bien province, where upland rice is planted for three years after forest clearing.

FIGURE 25-2: Typical rotational shifting cultivation area after crop harvesting in Dien Bien province, where maize is planted after a three-year fallow.

Photos: Nguyen Thi Hoa, ICRAF-Vietnam.

sweet potatoes and edible canna. This system is commonly practised by the Muong, Thai, Tay and Nung ethnic groups (MARD, 2007).

Distribution of shifting cultivation in Vietnam

In the mid-1990s, it was reported that 1.4 million hectares of sloping land was under shifting cultivation (Thai and Nguyen, 2002), but by 2007 this had dropped to 212,271ha (Table 25-1). Of this, 40% was in the northern mountainous region, 57% in the north-central region and 3% in the south-central region. The largest groups of shifting cultivators were the H'mong and Dao people, with an estimated total population of 800,000. The distribution of shifting cultivation varied across the country in 2007. In the northern mountainous areas, it was most common in Lai Chau and Dien Bien provinces, while it was dominant in Thanh Hoa and Quang Ngai provinces in the north-central region (Figure 25-3). In terms of products, 84% of shifting cultivation areas were planted with upland rice, maize, potatoes, cassava, edible canna, beans, sugar cane, cotton and mulberry; remaining areas were planted with perennial trees, grasses and other trees in a fallow cycle (Table 25-1).

Benefits of shifting cultivation

There are both temporal and spatial dimensions to the benefits of rotational and composite shifting cultivation. Long fallow periods reduce soil erosion and improve soil fertility, whilst the integration of nitrogen-fixing trees increases benefits when

TABLE 25-1: Distribution of shifting cultivation by size (ha) and by region (2007).

Location	Total shifting cultivation area	Annual crops	Perennial trees and crops	Plateau	Others
Northern mountains (Bac Giang, Cao Bang, Dien Bien, Ha Giang, Hoa Binh, Phu Tho, Lai Chau, Lao Cai and Tuyen Quang provinces)	86,292	76,149	5,746	358	4,038
North central (Quang Ngai, Quang Tri, Thanh Hoa and Thua Thien-Hue provinces)	120,473	96,928	4,055	35	19,455
South central (Khanh Hoa, Ninh Thuan and Phu Yen provinces)	5,506.7	4,986.2	140	0	381
Total	212,271	178,064 (84%)	9,941 (4.7%)	393 (0.19%)	23,874 (11.11%)

Source: Vietnam Forest Protection Department (2007).

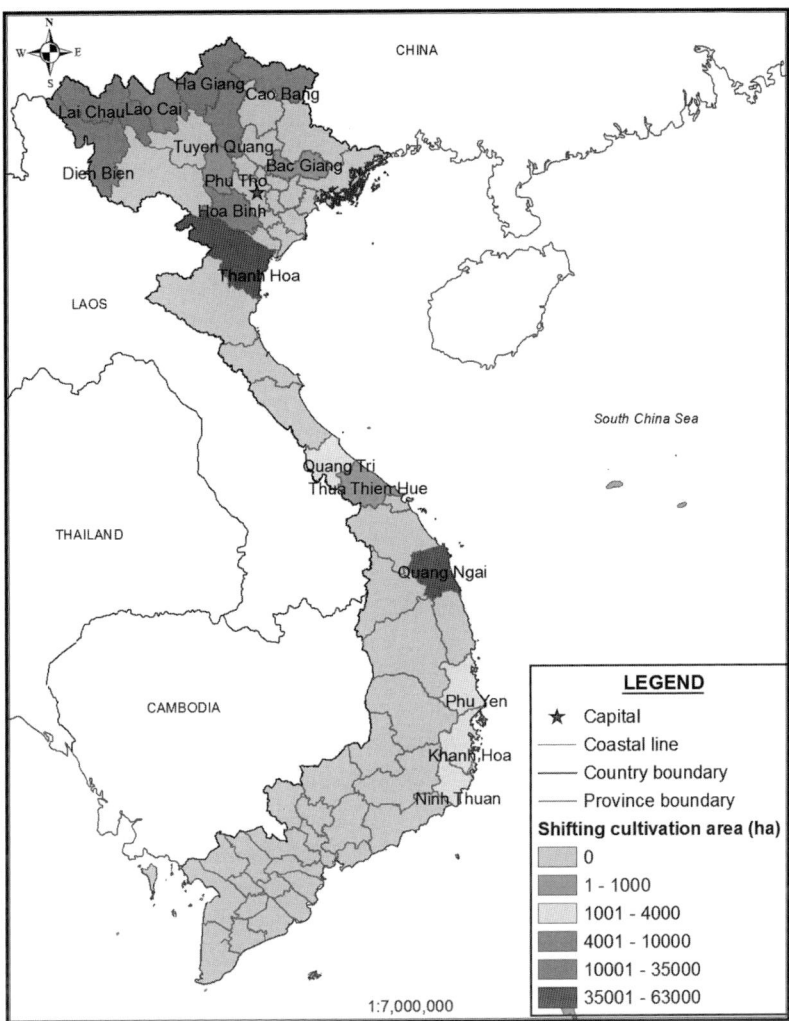

FIGURE 25-3: Distribution of shifting cultivation in Vietnam, 2007.

fallow periods are shortened. In the Northern Highlands, farmers plant *Leucaena leucocephala* during fallow periods to improve soil fertility, which not only shortens the required fallow time but also provides firewood (Tran, 2001). In Bac Giang province, hybrid *Acacia, Leucaena leucocephala, Gliricidia sepium* and *Calliandra calothyrsus* are planted as part of an improved fallow method that provides fodder, fencing materials and additional income (IFAD, IDRC, CIIFAD, ICRAF and IIRR, 2001). Furthermore, burning slashed vegetation saves labour, enriches the soil with ash and increases water infiltration. Fire also plays an important role in removing fungal diseases and noxious insects that may affect crops (FAO, 1984). There are also opportunities for generating income by harvesting products from fallow vegetation and gathering fodder for livestock (IFAD, IDRC, CIIFAD, ICRAF and IIRR, 2001).

Typically, rotational shifting cultivation is an ecologically sound practice where the provision of human needs is integrated with those of the natural environment, and species composition and diversity in the natural forest are preserved by allowing natural regrowth during fallow periods. This cultivation system occupies an established position in the indigenous economy and constitutes a vital part of the culture and livelihood of the Vietnamese highland population (Pham, 1999).

Issues and challenges of shifting cultivation

Over time, shifting cultivation has become a controversial land use. On one hand, it is an important aspect of the traditions and identities of many upland people and an inexpensive means of clearing income-generating land (Pham, 1999). But on the other hand, it is believed to cause deforestation, environmental degradation and forest fires. These problems contrast with the benefits noted across the rotational shifts from burning and cultivation through fallowing. Clearing of forest and replacing it with agricultural crops is a characteristic of pioneer, rather than rotational, shifting cultivation. This causes forest loss. Moreover, while it may be a low-cost and simple way of preparing land for cropping, total clearing of vegetation is known to exacerbate water runoff and soil erosion, which often leads to sedimentation in lower watersheds and paddy fields.

As the government of Vietnam shifted its focus towards forest management in the 1990s, it targeted shifting cultivation areas for reforestation projects. Burning was prohibited, so shifting cultivators were confronted with the need for higher inputs of labour during land preparation. Population increases also put pressure on shifting cultivation. According to Rambo (1997), traditional systems of swidden agriculture in northern Vietnam are sustainable only if population density is well under 40 people per sq. km, demanding a level of food production that allows time for the forest to regenerate sufficiently to maintain the cycle. Nowadays, this is almost impossible, since even the remotest parts of the Central Highlands has a population density of 95 people per sq. km (General Statistics Office, 2010). The population in the northern uplands tripled between 1960

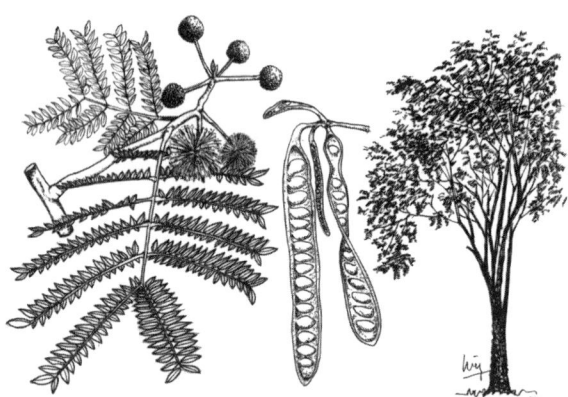

Leucaena glauca Benth. [Leguminosae] Syn. of *Leucaena leucocephala* (Lam.) de Wit

Planted in Vietnam's northern provinces to bolster swidden soil nutrients in the face of shrinking fallow periods.

and 1984 (Sunderlin and Huynh, 2005). As a result, the area available for shifting cultivation diminished, forcing shifting cultivators to shorten fallow periods, cultivate their land continuously without fallow breaks, or sell their land and move deeper into the forest to embark upon destructive pioneer shifting cultivation. Reports of shortened fallow periods are commonplace, and the changed farming practice was confirmed in interviews with farmers in the northwest region. Interviewees disclosed that most farmers now leave their land fallow for only one or two years. Very few of them fallow their land for as long as three or four years.

Shortened fallow periods have led to declining crop yields, which has placed household food security at risk. Average rice or maize productivity in shifting cultivation areas in the northwest region is as low as 1 ton per ha per year, compared to 6 to 7 tons per ha per year in the Red river delta (MARD, 2007). Staple-food production in shifting cultivation areas now meets only 60% to 70% of household needs. In addition, extreme weather events and climate change, particularly water and temperature stress, are further aggravating the problems faced by shifting cultivators.

Despite accusations that shifting cultivation causes deforestation and land degradation, Boucher et al (2011) argue that small-scale farmers and shifting cultivators are not necessarily to blame for most deforestation, particularly in those tropical areas with the highest deforestation rates. Van Noordwijk and Minang (2009) and Trakansuphakon (2010) say that based on the FAO's definition of forest, shifting cultivation may cause degradation over time, but not deforestation, as long as the fallow phase is long enough to achieve a minimum tree height and crown cover. Moreover, in terms of mitigating climate change, shifting cultivation can be expected to be carbon neutral at the plot level if fallow periods are sufficiently long and a 'carbon bank' is established at the landscape level by keeping forests young and growing (Trakansuphakon, 2010). With improved fallow management, deforestation can thus be avoided and shifting cultivation can contribute to mitigation of climate change. More data needs to be collected and examined in Vietnam on the role being played by shifting cultivation in reducing land-based greenhouse-gas emissions. Opportunities obviously exist in which shifting cultivation can be harnessed for its potential to help mitigate climate change, rather than blindly opposing it on the spurious grounds that it causes deforestation.

Policy interventions and their impacts on shifting cultivation

Widespread adoption of shifting cultivation began as the First Indochina War ended in 1954. At that time it was deemed necessary to use shifting cultivation to accelerate economic recovery and address serious food deficits. However, the increase in shifting cultivation brought with it a number of problems that led the government to implement several measures in the early 1960s to stabilize the practice. A literature review suggested that these policy interventions brought significant changes to the livelihoods of shifting cultivators in particular, and to shifting cultivation in general. Interviews with farmers and government officials in the northwest region also revealed that the changes resulting from these early policy interventions affected both farming systems and cultivation techniques. The following sections describe the policy interventions and their specific impacts.

Subsidies, new economic zones and land allocation from the late 1960s to 1993

In 1968, the government launched a major programme aimed at stabilizing shifting cultivation and promoting 'sedentarization' – or the permanent abandonment by farmers of traditional systems of rotating plots between cultivation and fallow, and acceptance of permanent cultivation. The programme included infrastructure development, expansion of irrigated-rice areas, forest rehabilitation and provision of subsidies and loans. Further, in the 1970s, cash and rice incentives were offered to farmers who agreed to plant forest trees on shifting cultivation lands. However, the results were unsatisfactory due to a lack of markets or a limited demand for timber during that period (Tran, 2001). Also in the 1970s, the government launched its 'cooperative approach' to reforestation, and this exacerbated dislike for forest-based timber plantations. The cooperative economy in Vietnam was based on a system in which outputs were distributed to households according to their labour contribution (Fatoux et al., 2002). However, in cooperative reforestation, the returns per person for labour input were insufficient to meet individual household needs, and this forced poor households to clear areas of upland forest to plant maize and cassava to meet their immediate food needs – a move directly opposed to the aims of the programme. In 1981, Vietnam's so-called 'cooperative economy' came to an end with the issuance of Decree 100 (*Khoan* 100) and for the first time, farmers were granted property rights over their cultivated lands (Tran, 2001).

In 1980, prior to land allocation, the government passed Decision 95/CP, which created new economic zones and set in train a major population redistribution through 'organized mass migration' to the Central Highlands. Most families were offered financial support to relocate to the uplands, although some migrated voluntarily, without direct support, as the lowlands were becoming congested and land was becoming more and more scarce. Not surprisingly, the population of the Central Highlands increased rapidly, giving rise to large-scale forest conversion, mainly through shifting cultivation. This was undertaken initially for subsistence, but was later converted to cash–crop production. In the northern regions, migration from the Red River Delta to mountainous areas began earlier, in 1972, when large numbers of people began moving further north as a result of new economic zones being established in the area. The occupation of shifting cultivation lands by new settlers led to an abruptly shorter fallow period in mountainous areas as swidden farmers strove to produce more food from less land (Thai and Nguyen, 2002). Farmers in Lao Cai province told in interviews how fallow periods had been reduced from 10 to 15 years to 6 or 7 years by the end of the 1980s.

Forest land allocation policy, 1993

Forest lands were allocated to farmers under a revised Land Law in 1993 and Decree No 02/CP in 1994. The forest-land allocation policy was considered a milestone in the history of the forestry sector and aimed to provide forest-land tenure and achieve massive reforestation. However, much of the land allocated was fallowed shifting cultivation land, so the ability to open new swiddens was severely restricted, and fallow periods tumbled. Farmers were quick to adapt to the changes, first as a matter of obedience to the 'top-down' style of local governance, and second because they received various incentives: there was both demand for cash crops and a form of land-tenure security that enabled them to invest permanently in the land. Several case studies later revealed the impacts of the policy on shifting cultivation and land-use change in general.

The number and concentration of shifting cultivation fields diminished as settled farming spread across the allocated lands. Data from Lao Cai province, bordering China in Vietnam's far north, revealed a reduction in the area under shifting cultivation after forest lands were allocated in 1993. As more forest land was granted for private use, the fall in shifting cultivation fields became significant between 1998 and 2001. Castella et al. (2006) found a declining trend in the number of shifting cultivation fields at Khuoi Noc, a mountainous village in Na Ri district of Bac Giang province, north of Hanoi. As shifting cultivation areas decreased, fallow periods, in turn, became shorter. In a case study in Que village, in Nghe An province's Con Cuong district, on Vietnam's north-central coast, Jakobsen et al. (2007) reported that fallow periods had fallen from between five and 10 years to a maximum of only two years. They also noted a concentration of shifting cultivation in the allocated area instead of it being widely dispersed across the village territory. This had resulted in Que village's shifting cultivation area shrinking from 92ha in 1991 to 43ha in 2003.

New crops were also being introduced, along with changed farming practices in shifting cultivation areas. In Bac Giang, Thai Nguyen and Quang Ninh provinces, the Dai people quickly changed from shifting cultivation to terraced rice cultivation. Similarly, farmers in Que village shifted from sloping land cultivation to growing paddy rice. New crops such as cinnamon were planted in Yen Bai and Lang Son provinces in the north, while ethnic groups in the Central Highlands started permanent cultivation of perennial trees such as coffee, cashews and rubber, and joined in forest enterprises (Do, 1994). The same pattern was observed in the south-central region: a case study in Dai Lao commune, Lam Dong province, found that tea plantations quadrupled in area between 1994 and 1999. In addition, the area under coffee grew from 300ha in 1994 to 1291ha in 1999 (Dang et al., 2001). This change was attributed to the land allocation programme and a rise in the price of coffee and tea.

Contrary to the objectives of the forest-land allocation policy, it actually placed a strain on household food security, and increasing population density compounded the problem. Interviews conducted in the northern provinces of Son La and Yen Bai revealed that a number of households were suffering from food shortages. Further south in Que village, rice yields were not sufficient to offset the loss in production due to the reduced shifting cultivation area. In 1991 and 1998, rice harvests provided an average of 1840 and 1790kg/household respectively, but this fell to 1100kg/household after forest lands were allocated (Jakobsen et al., 2007). Fortunately, cattle raising and the gathering of non-timber forest products (NTFPs) contributed to livelihoods. In some areas of the northern mountainous region, reduced shifting cultivation areas in upland forests also led to intensification of production in the lowlands, where farmers introduced a second, spring-season rice crop. Farmers adapted to the impacts of the forest-land allocation policy in many different ways. Some accessed credit from neighbours and policy banks, others applied more fertilizer to increase production, rented land to expand their farms, harvested NTFPs, or found seasonal off-farm employment. In Tram Tau district of Yen Bai province, farmers were even regularly contracted to perform forest-protection activities.

Poverty reduction, reforestation and sloping-land policies, from 1995 to 2007

Several policies supporting reforestation and settled farming were adopted between 1995 and 2007. Together with forest-land allocation, these appeared to have positive consequences, not only for farmers' livelihoods, but also in terms of shifting cultivation lands reverting to forest.

Decision 661/QD-TTg, a 12-year reforestation programme, applied to the years from 1997 to 2010 and was also known as the '661 programme'. It took the problems faced by shifting cultivators into account by offering farmers cash and rice supplies as incentives for planting trees on shifting cultivation land (IKAP, 2005). The programme sought to achieve five million hectares of reforestation. Large areas traditionally used for shifting cultivation were converted to production and protection forests. Subsequently, the Ministry of Agriculture and Rural Development reported a significant increase in forest cover, from 26% in 1994 to 38% in 2006 (Dinh and Dang, 2008). However, as large areas of shifting cultivation land were reforested under the 661 programme, farmers on the dwindling number of swiddens had no option but to shorten fallow periods even further.

In 1998, Decision 135/1998/QD-TTg was adopted, with a land allocation and certification component aimed at reducing poverty through sedentarization in mountainous areas. Some farmers were given land by provincial governments, while others settled on land rented from government enterprises. The programme was quite successful: an evaluation in 2005 found that more than 80% of the country's farming households had sufficient agricultural land; more than 90% of registered households had permanent farms, and forest encroachment was down to zero.

Nevertheless, shifting cultivation was still in evidence, and was still seen as a cause of forest degradation. From 2000 to 2005, it was estimated that 2165 hectares of natural forest had been lost, and 60% of this loss was blamed on shifting cultivators (Government of Vietnam, 2007). Consequently, in 2006, Decision 07/2006/ QD-TTg – also known as Programme 135 (2006-2010) – was introduced, with the triple aims of promoting improved farming practices, halting shifting cultivation and reducing poverty. At the same time, Decree 08/TTg/2006 became law, with the aim of preventing illegal forest logging, burning and deforestation. This decree was designed specifically to deal with pioneer shifting cultivators carving swiddens out of protection and special-use forests. Under this programme, households were relocated and provided with agricultural lands and other support.

The ultimate effect of these various policies has been a significant reduction in the area under shifting cultivation in Vietnam (Figure 25-4). The area of cultivated sloping land also fell from 2.68 million hectares in 1993 to about 1.2 million hectares in 2004 (MARD, 2007), and further to 1.18 million hectares in 2007 (Vietnam Forest Protection Department, 2007). Table 25-2 shows the distribution of cultivated sloping lands in different regions of Vietnam in 1993, 2004 and 2007, respectively.

Support for sustainable sloping-land cultivation from 2007 to 2011

In 2007, the Ministry of Agriculture and Rural Development issued Decision no. 2945/QD-BNN-KL (2008-2012), which created a large-scale programme to halt shifting cultivation throughout the country. It focused on 34 out of Vietnam's 63 provinces and municipalities, and aimed to convert all shifting cultivation land – amounting to 360,000 hectares at the end of 2004 – to protection and production forests. In addition, 840,000 hectares of sloping lands were earmarked for intensive cultivation, and fertilizer subsidies, improved seed varieties and livestock were to be provided. Several policy instruments were adopted in support of this major initiative: Decision 147 concerned the development and plantation of protection forests, and adjustments to Decision 100 provided for conversion of shifting cultivation land with slopes equal to or more than 25 degrees to protection forest, and shifting cultivation areas with lesser slopes to production forest.

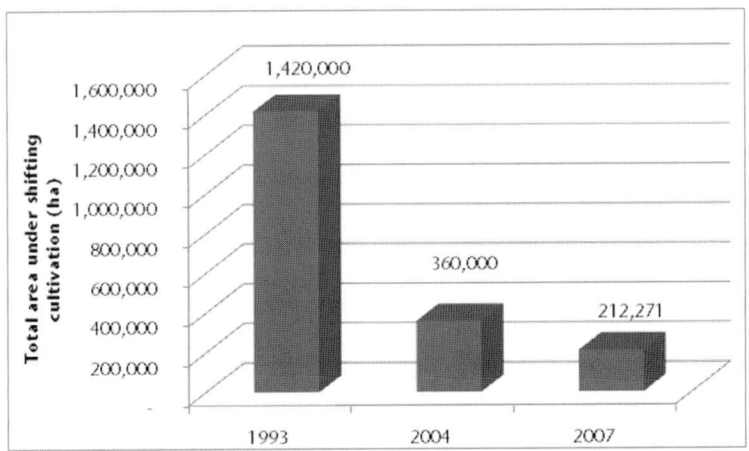

FIGURE 25-4: Total shifting cultivation areas in Vietnam, 1993 to 2007.

TABLE 25-2: Distribution of cultivated sloping lands by region in Vietnam in 1993 and 2004 (hectares).

Region	1993	2004
Northern mountainous	1,257,400	944,190
North Central	305,300	72,500
South Central	195,100	108,020
Central Highlands	375,900	73,340
Southeast	548,900	1,650
Total	2,682,600	1,199,700

Source: Thai and Nguyen (2002) and MARD (2007).

In 2011, the Ministry of Natural Resources and Environment hailed the success of the programme in converting shifting cultivation and sloping land into 'more productive' agricultural lands. In Lao Cai, for example, 7906 hectares of shifting cultivation land had reportedly reverted to forest, while 360 hectares of permanently cultivated sloping land were converted into areas of intensive agricultural production.

Despite these claimed achievements, the state's goal of eradicating shifting cultivation may yet prove to be elusive, if not unattainable. Our 2012 survey found that 15 out of 32 farmers in Son La and Yen Bai provinces were still practising shifting cultivation (Figure 25-5). Furthermore, we found that farmers had employed a variety of strategies in response to policy changes (Figure 25-6). With shrinking availability of land for shifting cultivation and shorter fallow periods, farmers had changed their practices: they were applying more chemical fertilizers in an attempt to counter the constant depletion of soil nutrients.

A number of households said they had switched to settled farming, while others continued with shifting cultivation and suffered the consequences of sharply reduced fallow periods. In Yen Bai province, the fallow phase of the swidden cycle had been reduced to between one and three years, compared with five to six years previously. Farmers reported using nitrogen-fixing species or legumes as planted fallow vegetation. Intercropping, crop rotation and controlled succession of fallow species were becoming common practices. Maize or upland rice was usually intercropped with high-value fruit and industrial tree crops such as *Acacia mangium, Melia* spp, *Dimocarpus longan,* oranges or *Docynia indica,* as well as coffee and tea in Son La and Yen Bai provinces, respectively.

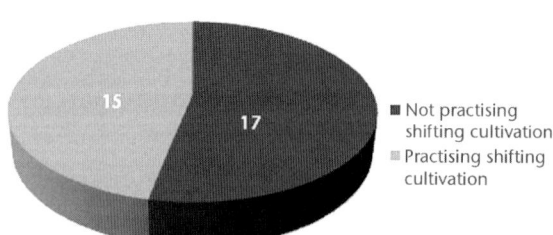

■ Not practising shifting cultivation
■ Practising shifting cultivation

FIGURE 25-5: Proportion of farmers still practising shifting cultivation among 32 interviewees in Son La and Yen Bai provinces, 2012.

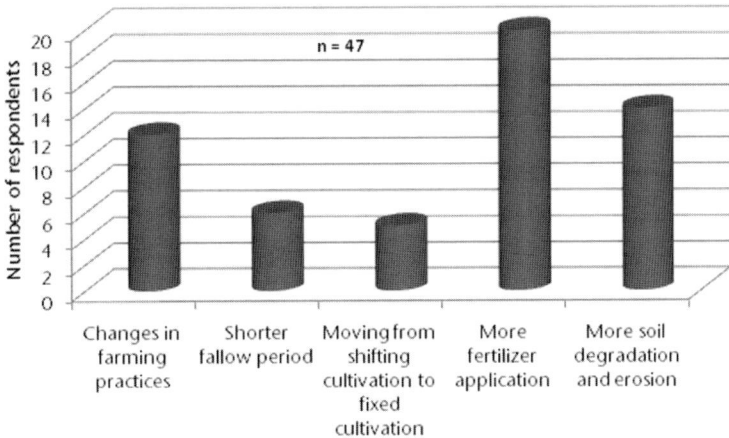

FIGURE 25-6: Impacts of state policies on farmer practices and the environment in Son La and Yen Bai provinces.

Shifting cultivation and REDD+

As a country likely to be seriously affected by climate change, Vietnam has long been engaged in international discussions and preparations to implement REDD+ (Pham et al., 2012). The government expects that as a global mechanism, REDD+ will offer economic benefits to forest-dependent communities – mainly ethnic minorities making up about 9% of Vietnam's total population – that have been practising shifting cultivation (MARD, 2008). REDD+ is also viewed as a means of addressing the drivers of deforestation and degradation, including shifting cultivation. However, shifting cultivation creates distinct opportunities and challenges in relation to REDD+.

From an economic viewpoint, REDD+ incentives can be effective in compensating farmers for accepting restrictions against the practice of pioneer shifting cultivation, since the net present value (NPV) of this system is often lower than that of other sustainable practices, such as agroforestry (Pham, 1999). Do (1994) supports this idea, arguing that pioneer shifting cultivation is not economically sound in the long term. However, although the NPV of pioneer shifting cultivation is low, the returns to labour are high. For example, for husked rice, the return

Melia azedarach L. [Meliaceae]

This member of the mahogany family produces high-quality timber and is commonly used to improve fallows in Vietnam.

to labour is about 6 to 8kg per working day, which is two or three times higher than that of wet-rice cultivation in the Red River and Mekong River Deltas (Nguyen, 1993; Do, 1994; Tran, 2007), making it worthwhile for pioneer shifting cultivators to stick to their practices. In addition, a boom in coffee, pepper and rubber plantations has given rise to land speculation by well-resourced farmers, and this has posed another challenge to attracting shifting cultivators to REDD+. Lucrative offers often tempt shifting cultivators to sell their land to rich farmers, mostly people who have no tradition of shifting cultivation themselves (Pham, 1999; Hoang et al., 2010). Land deals of this nature are quite common in the Central Highlands, where the opportunity cost of converting forests into industrial plantations is much higher than the current carbon price in the global market (Hoang et al., 2010; JOFCA and JFTA, 2011). In such cases, REDD+ incentives will be less attractive to shifting cultivators and consequently will fall short of bringing an end to the practice. Nevertheless, it may still achieve some decrease in shifting cultivation by attracting larger-scale farmers, although this would redirect REDD+ benefits to elite households. It should be noted that pioneer shifting cultivation has been closely linked to the identity and culture of some ethnic minorities (AIPP, 2011). REDD+ projects should offer sustainable land-use alternatives that are not only economically and environmentally feasible, but are also culturally acceptable to ethnic people.

Unlike pioneer shifting cultivation, rotational shifting cultivation has a much smaller direct impact on forests. In some parts of Vietnam, rotational shifting cultivation fields have already been transformed into sedentary farms with very short fallow periods, or none at all. However, these areas are often a source of conflict between the government and indigenous groups. Having lived in an area for generations, indigenous people hold strongly to their rights over surrounding forest resources; they consider patches of rotational shifting cultivation fields to be theirs, even when they are lying fallow, and make claims for recognition of this ownership. Despite the historical basis for their claims, Vietnamese law states that ethnic minorities are, at best, granted only de facto use rights over forest lands, while the state remains the legal owner. Conflicts between the so-called 'legal owner and user' of forest lands (the state) and shifting cultivators will likely heighten once shifting cultivators begin to participate in REDD+ activities. On one hand, REDD+ may inadvertently encourage shifting cultivators to cut more trees for fear of losing their lands to REDD+ agents; but on the other, they will have a greater opportunity to receive REDD+ incentives if tenure conflicts are resolved. REDD+ may compensate farmers for extending their fallow periods or for abandoning their fields so the forest can regenerate. This could be economically attractive, given that the opportunity cost of converting forest to rotational shifting cultivation in many parts of Vietnam is relatively low, and falls below US$5 per tonne of carbon dioxide equivalent (Hoang et al., 2010; Do et al., 2012). An optimistic outlook suggests that formalizing the land-tenure status in this area and making it private agricultural land – with certain requirements for tree cover – would help to sustainably increase carbon stocks and partially satisfy local demand for timber, thus indirectly reducing pressure on forests. It could also be covered by other schemes such as Reducing Emissions from All Land Uses (REALU).[2]

Conclusion

Since the 1960s, successive waves of policy reform have brought significant changes to shifting cultivation in Vietnam, but addressing the latent issues arising from these reforms has not been easy. The establishment of new economic zones in mountainous areas before 1993 led to an increase in shifting cultivation, but key policy reforms such as forest-land allocation, sedentarization, poverty reduction and reforestation programmes have contributed significantly to reducing its scale and changing its character. In many areas, farmers converted shifting cultivation areas into terraced rice fields, or began permanent cultivation of farms with little or no fallow period. Others converted their fields into industrial and perennial tree-crop plantations, or incorporated these trees into their annual cropping systems. The regrowth of trees on shifting cultivation land requires financial capital and supplementary income, because trees take time to mature and harvest and, because they do not ensure short-term food security, there is a risk that farmers with growing tree plantations will revert to shifting cultivation for the sake of their survival. These issues are not straightforward – even more so because these same shifting cultivators have already suffered loss of income, displacement and, in some cases, subtle commercial land grabbing.

Although the results of the dynamic policy changes that have occurred in Vietnam over the past two decades are impressive, shifting cultivation remains deeply rooted in the cultural practices of the country's ethnic peoples. Unless there are further consistent and significant implementation efforts and commitment to financial investments, eradicating shifting cultivation may be unachievable. The most recent intervention, REDD+, can help to transform shifting cultivation into a system of more settled and sustainable farms, as long as they are not targeted by land grabbers. Land-tenure issues have been central to the REDD+ debate, and are even more crucial to the plight of many shifting cultivators. Any effort to halt shifting cultivation, even when backed by REDD+, will continue to face difficulties as long as land-tenure issues remain unresolved and the opportunity cost of converting forests into more profitable land uses remains higher than the incentives on offer.

References

AIPP (2011) *REDD+ Implementation of Indigenous Peoples in Asia and the Concerns of Indigenous Peoples,* Asia Indigenous People's Pact, technical report, Chiang Mai, Thailand

Boucher, D., Elias, P., Lininger, K., May-Tobin, C., Roquemore, S. and Saxon, E. (2011) *The Root of the Problem: What's Driving Tropical Deforestation Today?* Union of Concerned Scientists, Cambridge, UK

Castella, J. C., Boissau, S., Nguyen, H. T. and Novosad, P. (2006) 'Impact of forest land allocation on land use in a mountainous province of Vietnam', *Land Use Policy* 23, pp147–160

Dang, T. H., Pham, H. D. P., Nguyen, N. T., Le, V. D., Pham, T. H., Espaldon, M. V. O. and Magsino, A. O. (2001) 'Impacts of changes in policy and market conditions on land use, land management and livelihood among farmers in the central highlands of Vietnam', http://pdf.usaid.gov/pdf_docs/PNACX449.pdf, accessed 4 October, 2013

Dinh, H. H. and Dang, K. S. (2008) *Land and Forest Allocation in Vietnam: Policy and Implementation,* Institute for Policies and Strategies in Agriculture and Rural Development (IPSARD), Hanoi

Do, D. S. (1994) *Shifting Cultivation in Vietnam: its Economic and Environmental Values Relative to Alternative Land Uses,* International Institute for Environment and Development (IIED), London

Do, T. H., Jindal, R., Hoang, M. H. and Catacutan, D. (2012) *Feasibility Notes for Reducing Emissions from All Land Uses in Bac Kan Province,* World Agroforestry Centre, Hanoi

FAO (1984) *Shifting Cultivation in Bhutan: A Gradual Approach to Modifying Land Use,* Food and Agriculture Organization of the United Nations, http://www.fao.org/DOCREP/006/V8380E/V8380E01.htm, accessed 20 August 2012

Fatoux, C., Castella, J. C., Zeiss, M. and Pham, H. M. (2002) 'From rice cultivation to agroforestry within a decade: The impact of *Doi moi* on agricultural diversification in a mountainous commune of Cho Moi district, Bac Kan province, Vietnam' in J. S. Castella and D. Q. Dang (eds) *Doi moi in the Mountains: Land-use Changes and Farmers' Livelihood Strategies in Bac Kan Province, Vietnam,* The Agricultural Publishing House, Hanoi, pp73-97

General Statistics Office (2010) *Population and Population Density in Vietnam by 2010,* General Statistics Office, Hanoi

Government of Vietnam (2007) *Strategies for Forestry Development in Vietnam from 2006-2020,* Policy document, Hanoi

Hoang, M. H., Do, T. H., van Noordwijk, M., Pham, T. T., Palm, M., To, X. P., Doan, D., Nguyen, T. X. and Hoang, T. V. A. (2010) *An Assessment of Opportunities for Reducing Emissions from All Land Uses: Vietnam Preparing for REDD,* ASB Partnership for the Tropical Forest Margins, World Agroforestry Centre, Nairobi, Kenya, http://asb.cgiar.org/publication/assessment-opportunities-reducing-emissions-all-land-uses-%E2%80%93-vietnam-preparing-redd-final, accessed 15 July 2017

IFAD, IDRC, CIIFAD, ICRAF and IIRR (2001) *Shifting Cultivation: Towards Sustainability and Resource Conservation in Asia,* International Fund for Agricultural Development, International Development Research Centre, Cornell International Institute for Food, Agriculture and Development, International Centre for Research in Agroforestry (World Agroforestry Centre) and the International Institute of Rural Reconstruction, Cavite, Manila, Philippines

IKAP (2005) *Rotational Farming in Vietnam,* Indigenous Knowledge and People's Foundation (IKAP), Chiang Mai, Thailand, www.ikap-mmsea.org/document%20new/IK-BD-RF/RF_Vietnam.pdf, accessed 30 August 2012

Jakobsen, J., Rasmussen, K., Leisz, S., Folving, R. and Nguyen, Q. V. (2007) 'The effects of land tenure policy on rural livelihoods and food sufficiency in the upland village of Que, North Central Vietnam,' *Agricultural Systems* 94, pp309-319

JOFCA and JFTA (2011) *The Economic Feasibility of Reduction of Emissions from Deforestation and Forest Degradation in Vietnam: Case Studies in Forest Conservation, Community Forest Management, Forest Plantation and Rubber Development,* Japan Overseas Forestry Consultants' Association and Japan International Cooperation Agency, Hanoi

Lundberg, M. (1996) *Ethnic Minorities and the State: Conflicting Interest Between Shifting Cultivators and the Governments in Peru and Vietnam,* research report no 7, Environmental Policy and Society (EPOS), Linkoping University, Sweden

MARD (2007) *Support Upland Farmers for Sustainable Cultivation on Sloping Lands from 2008-2012',* proposal attached to Decision 2945/QĐ-BNN-KL, Ministry of Agriculture and Rural Development, Hanoi

MARD (2008) *Readiness Plan Idea Note (R-PIN),* a report offered to the Forest Carbon Partnership Facility, Ministry of Agriculture and Rural Development, Hanoi

Nguyen, Q. H. (1993) *Renovation of Strategies for Forestry Development until the Year 2000,* Ministry of Forestry, Hanoi

Pham, T. M. (1999) 'Socio-economic analysis of shifting cultivation versus agroforestry systems in the upper stream of lower Mekong watershed in Dak Lak province', MA dissertation in economic development, National University, Ho Chi Minh City and Institute of Social Studies, The Hague, Netherlands

Pham, T. T., Moeliono, M. , Nguyen, T. H., Nguyen, H. T. and Vu, T. H. (2012) *The Context of REDD+ in Vietnam: Drivers, Agents and Institutions,* Occasional Paper 75, Center for International Forestry Research (CIFOR), Bogor, Indonesia

Rambo, A. T. (1997) 'Development trends in Vietnam's northern mountain region' in D. Donovan, A. T. Rambo, J. Fox, L. T. Cuc, and T. D. Vien (eds) *Development Trends in Vietnam's Northern Mountain Region,* vol. 1, National Political Publishing House, Hanoi

Sunderlin, S. D. and Huynh T. B. (2005) *Poverty Alleviation and Forests in Vietnam,* Center for International Forestry Research (CIFOR), Bogor, Indonesia

Thai, P. and Nguyen, T. S. (2002) *Sustainable Use of Uplands and Mountainous Areas,* Hanoi Agricultural Publishing, Hanoi

Tran, D. V. (2001) *Experiences in Managing Fallow Land in Vietnam,* workshop proceedings, Hanoi Agricultural Publishing, Hanoi

Tran, D. V. (2007) 'Indigenous fallow management with *Melia azedarach* Linn in Northern Vietnam', in M. F. Cairns (ed.) *Voices from the Forest: Integrating Indigenous Knowledge into Sustainable Upland Farming,* Resources for the Future Press, Washington, DC, pp435–443

Trakansuphakon, P. (2010) *Rotational Farming/Shifting Cultivation and Climate Change,* concept paper, Strategy Workshop on Rotational Farming/Shifting Cultivation and Climate Change, Indigenous Knowledge and People's Foundation (IKAP), Chiang Mai, Thailand

van Noordwijk, M. and Minang, P. A. (2009) *If We Cannot Define It, We Cannot Save It,* policy brief no 15, ASB Partnership for the Tropical Forest Margins, World Agroforestry Centre, Nairobi, Kenya

Vietnam Forest Protection Department (2007) *Data on Forest Allocation, Sloping and Shifting Cultivation Area 2007,* http://www.kiemlam.org.vn/Desktop.aspx/List/Giao-rung-Quan-ly-nuong-ray/So_lieu_co_ban_ve_giao_rung_cho_thue_rung_va_canh_tac_nuong_ray_nam_2007/, accessed 5 May 2017

Notes

1. REDD+ stands for Reducing Emissions from Deforestation and Forest Degradation plus conservation of forest carbon stocks, sustainable management of forests and enhancement of forest carbon stocks. Thus, REDD+ has been expanded beyond the scope of its precursor mechanism, REDD (Reducing Emissions from Deforestation in Developing Countries), which proposed financial incentives for developing countries to reduce emissions from deforestation. Having evolved rapidly during the technical and political discussions of the United Nations' Framework Convention on Climate Change, REDD+ remains controversial, particularly in terms of the ways in which its building blocks are defined and the scope of activities considered eligible for accounting. A firm definition of REDD+ is therefore lacking.

2. Reducing Emissions from All Land Uses (REALU) is an initiative of the World Agroforestry Centre's Alternatives to Slash and Burn Partnership. It enables the development of landscape approaches to REDD+, allowing fair, efficient and effective calculation of emission reductions from land uses in tropical forest margins. These would include, but would not be restricted to, deforestation and forest degradation. The initiative is part of the post-Kyoto regime of the United Nations' Framework Convention on Climate Change. For more information, visit www.asb.cgiar.org.

26

MISINTERPRETING THE UPLANDS OF VIETNAM

How government policies and maps lead to a misunderstanding of swidden and its associated livelihood systems

*Stephen J. Leisz**

Introduction

Since the late 1800s, when European colonial administrators were expanding their control across the uplands of Southeast Asia, there has been a mistaken belief within governments of the region and also among a majority of the public that continues to the present day: that the uplands of Southeast Asia were in the past either covered by forest, or were meant to be covered by forest. This view was enshrined into laws that placed the whole of the uplands of French Indochina under the colonial forest service (Thomas, 1999). A corollary to this view is that the livelihood systems and natural-resource management systems that are practised in the uplands by the people who have historically lived there are destructive of both the land and the natural resources. This view is clearly expressed in both early reports from Southeast Asia's upland regions and present day laws regulating land use in the uplands.

Reports from French officials who ventured into the uplands in the mid-to-late 1800s made it clear that even then, the uplands were not a continuous blanket of forest, or even tree cover. Moreover, the people who lived there cultivated the sloping land using different swidden/fallow agricultural techniques,[1] and depended upon both the cultivated land and the vegetation that regrew in fallow areas for their livelihoods. Given the background of the administrators governing the uplands and their desire to extract timber and other wood products from these areas, they viewed these management practices as destructive. To them, the uplands were forest land and the general view was that the practices of the people in managing the land and the natural resources were destructive. Captain P. Cupet, a member of the Pavie expedition to Indochina from 1879 to 1895, wrote:

---

* Dr Stephen J. Leisz is Associate Professor of Geography at Colorado State University, USA.

These savages are the greatest destroyers of forests I know. One never meets them except in the vicinity of the highest crests, on elongated hilltops offering a great surface of arable land… The terrain is almost everywhere deforested and covered with rai[2] (Cupet, 2000, pp35-40).

The Garde-Générale des forêts, M. Thome, expanded on this view in an 1897 report in which he stated that the belief that Indochina's highlands were at that time dominated by rich forest was an illusion. Instead, he described the uplands as being covered in bush, grasses, lianas, bamboo and regrowing trees (Thomas, 1999). The views of the French colonial era, that the uplands should be covered in trees and forest, were transferred to Vietnam's government institutions during their development in colonial times, and are perpetuated to the present day (examples can be seen in Poffenberger, 1998).

Present day views across Southeast Asia

Swidden/fallow cultivation systems are still practised in many parts of the world, including the uplands of Southeast Asia (see, for example, Fox, 2000; Ickowitz, 2006). As recently as the mid-1980s, it was believed that these systems were the most prevalent form of agriculture in the tropics (Lanly, 1985). Government views of swidden/fallow cultivators and their agricultural systems have not changed much since the late 1800s. Both the people and their practices are viewed by governments and agriculture and forestry administrators as primitive. They are blamed for destroying trees and forests and causing soil degradation, erosion and downstream flooding (Myers, 1984, 1992, 1993; Bandy et al., 1993; ASB, 1994; FAO, 1999; Fox, 2000; Devendra and Thomas, 2002; Vandergeest, 2003; Jorgensen, 2006). As a result, national governments throughout Southeast Asia have tried to stop swiddening by imposing land laws and agricultural policies that are intended to sedentarize upland populations and encourage the practice of fixed, permanent cultivation in the uplands (Dove, 1984; Fox, 2000; Padoch and Coffey, 2003).

The belief behind these policies is that swidden/fallow systems are unproductive and can be replaced

Gossypium hirsutum L. [Malvaceae]

Cotton is an important crop in Vietnam, but local production falls far behind large-scale imports to supply industries. There are constant efforts to expand cotton plantations, posing a threat to swidden land.

by other agricultural systems that, from the point of view of governments and other policy-makers, are more rational and economically productive. Some conservation organizations also hold beliefs similar to these and view swidden/fallow systems as overly destructive and a threat to biodiversity (IUCN, 2006). They believe that an end to swiddening will lead to a 'regreening' of the uplands and a return to primary forest coverage, a situation that many believe would help to increase biodiversity and save rare and endangered plants and animals.

Present-day views in Vietnam

A combination of the two views is commonly the case in Vietnam. The Vietnamese government's rhetoric and policies since independence in 1954 make it clear that it views swidden/fallow as both a primitive form of agriculture that wastes upland resources and a main cause of deforestation and land degradation (Dang, 1991; Jamieson, 1991; Rambo, 1995, 1998; Cuc, 1996; Morrison and Dubois, 1998; Fox et al., 2000). In line with this rhetoric, the government of Vietnam has passed laws and enabling regulations in a bid to limit and ultimately stop the practice of swidden/fallow (see Catacutan et al., ch 25, for a full discussion of Vietnam's laws regarding swidden/fallow). The first of these policies (Tran, 2003) was the Sedentarization and Fixed Cultivation Programme, implemented in the 1960s and 1970s. Other laws with similar underlying motives followed, but have been focused more on land and forest-area allocation (Tran, 2003). Some of these are detailed in Table 26-1.

As well as having a negative view of swidden/fallow as a farming system, the government does not recognize it as a land-use system, so it is not recognized on official land-use maps. Moreover, official government land-use and land-cover maps do not recognize swidden/fallow systems or the vegetation that regrows in fallow areas. Figure 26-1 is an official land-use map for three upland districts of Nghe An province in 2002. The categories on the map for agricultural land are: annual planted land; irrigated land; upland agriculture land; perennial planted land; rice land; and flat wasteland. The categories for forest lands are: artificial forest; natural forest – protected forest land; natural forest – specialized forest land; natural forest – productive; and unused hill and mountain forest land.[3] The categories for other types of land are aquaculture land; built-up land; exploited/mined land; national defence land; other wasteland; river and lake land; rural residential land; salt marsh land; urban areas; and religious/cultural land. Figure 26-1 shows that human settlements are scattered throughout the three districts; that most of these human settlements are located in mountainous areas; and that there is no officially recognized agricultural or cultivated land near most of the settlements. The major land-cover/land-use type found in association with the settlements on the land-use map is unused hill and mountain forest land (*dat doi nui chua su dung*).[4]

Land-cover maps show some cultivated land associated with settlement areas in the mountains, but they show none of the vegetation found in fallow lands and do

TABLE 26-1: Government policies and programmes in the uplands and their objectives since 1983.

Year	Government policies/ programmes	Objectives
1983	Directives 29/7CP	Allocation of forest land not only to forest enterprises and co-operatives, but also to farming households.
1986	Decision No. 1171/QD	Management regimes of production, protection and special-use forest.
1988	Land reform	Allocation of agricultural land directly to land users.
1990	Decision 72 – HDNT	Socio-economic development programme in the uplands.
1991	Forest law	Law on forest protection and management.
1992	327 programme	Re-greening barren hills and making use of wasteland.
1992	1586/QDUB	The province policy on forest-management renovation in state-owned farms and forest enterprises.
1994	Resolution No. 02/CP	Forest land allocation for forestry purposes.
1995	Resolution No. 01/CP	Allocation of contract land for agriculture, forestry and aquaculture.
1998	661 programme	Reforestation of 5 million ha during the period 2000 – 2010.
1999	Decision 245/TTg	State management of forest with recognition of the role of communes.
1999	Circular letter 56/1999/TT/BNN-KL	Guidance related to agreements for protecting and developing forests in native hamlet communities.
2001	Direction 52/2001/CT/BNN- KL	To promote local groups and local people's participation in forest protection and development activities.
2001	Determination 178/2001/QD protected species list	Allowing local people to collect NTFPs, except for those noted on the protected species list.
2004	Determination 04/2004/QD- BNN-LN	Regulating the exploitation of timber and other forest products.

Note: NTFPs = non-timber forest products.

not recognize fallow land. In fact, the largest land-cover category near to settlements is barren land (*dat trong*). This is true for land-cover maps from both 2000 (Figure 26-2) and most recently, 2010 (Figure 26-3). Despite this, the population of these settlements is recorded in official district documents as being overwhelmingly made up of farmers (Ky Son, Tuong Duong, and Con Cuong statistics, unpublished, 2003), and farmers in these areas practice swidden cultivation and their land management system includes fallow land (Leisz and Rasmussen, 2012).

The failure of official maps to acknowledge the existence of swidden/fallow farming systems, fallow areas and associated vegetation cover puts into practice the contentions of Thongchai (1994), that maps anticipate spatial reality rather than representing it, and a map is a model for, rather than a model of, what it

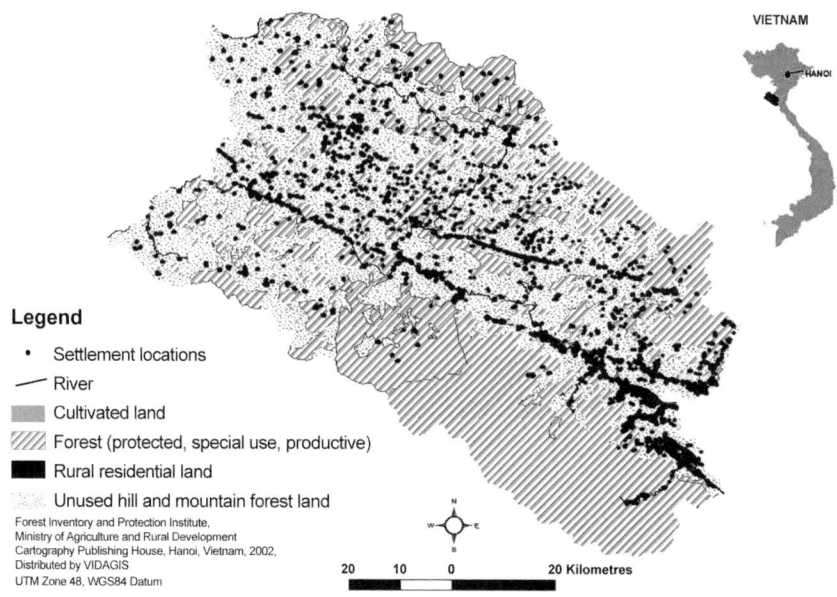

FIGURE 26-1: Location of settlements and official land uses in three districts of Nghe An province in 2002.

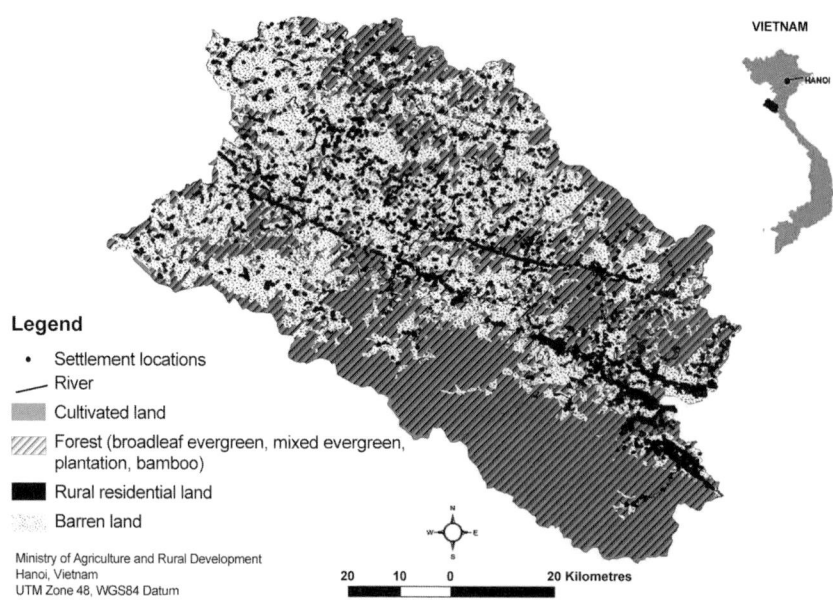

FIGURE 26-2: Location of settlements and official land cover of three districts of Nghe An province in 2000.

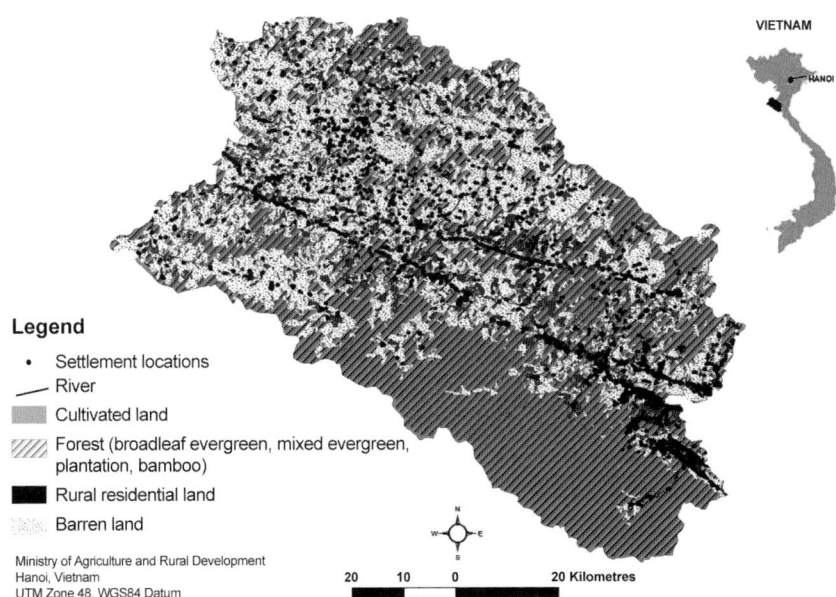

FIGURE 26-3:   Location of settlements and official landcover of three districts in Nghe An province in 2010.

represents. In this case, the maps of the uplands produced by the government of Vietnam show almost all mountainous areas as being covered by some type of forest. This representation of land cover acts to delegitimize the swidden/fallow farming that has historically been practised by most of the mountain population; swidden/ fallow systems and the vegetation associated with them figuratively disappear from the maps. If neither the system nor the vegetation is found on official maps or in official records, figures and statistics, then it is hard, if not impossible, to consider these farming systems when government planning is undertaken or when laws are made to regulate the way areas are governed and their resources officially distributed (e.g. land allocation) and managed.

At the same time as shifting cultivation is being discouraged in the uplands by government policies and delegitimized by exclusion from official maps and figures, a number of specific farming systems and agricultural activities, originally developed for lowland conditions, are being implicitly, and sometimes explicitly, encouraged in the uplands. These include:

- *Irrigated and seasonally inundated paddy-rice cultivation.* This is promoted through the gardens (*vuon*), fishponds (*ao*) and livestock-raising (*chuong*) programme (VAC), the gardens (*vuon*), fishponds (*ao*), livestock raising (*chuong*) and tree crops (*rung*) programme (RVAC) and other credit programmes that provide money for the building of terraces.

- *Animal husbandry.* Many projects promote the raising of pigs in sties (pens) to replace traditional free-ranging pigs. The same programmes promote the construction of fish ponds. Credit programmes are oriented to providing money for pigs and fish and for buying and raising cattle.
- *Permanent agriculture.* The Ministry of Agriculture and Rural Development (MARD) promotes crops grown on permanent upland fields, the most notable of these being hybrid maize (Wezel, 2000; Le and Rambo, 2001). The hybrid maize programme provides inputs (seeds and fertilizer) and credit, as well as marketing channels. Other projects specific to provinces and districts use similar mechanisms to promote permanent cultivation of new crops for markets (see, for example, the project described by Castella and Quang, 2002). Land allocation regulations support these programmes.

Arguments for a counter-reading of the uplands

The dominant position of these negative views of swidden/fallow and its practitioners, held in many government circles, along with the history of these views, raise the question of whether they are warranted. In recent years, this question has challenged accepted interpretations by governments in other countries where traditional farming systems have been, and still are, viewed as destroying the environment. A prime example is the view that was generally held of farmers and dryland farming systems in the West African savannah. This view held that dryland farmers and their farming system were degrading and deforesting the savannah. However, in 1996, Fairhead and Leach showed how this understanding was actually a misreading of the savannah landscape. They demonstrated how, when the landscape was understood from a different perspective, e.g. from the perspective of the local people, it was clear that the farmers' actions were actually enriching and enlarging forest areas, and did not degrade the landscape. They demonstrated through an examination of historical aerial photographs and other records that tree cover on the West African savannah was associated with human habitation. When humans were removed from the savannah, trees did not reproduce and grass dominated the landscape.

Kerkhoff and Sharma (2006) provided a similar re-reading of the landscape of South Asia. Through case studies, they presented evidence that environments where swidden/ fallow cultivation had been practised for hundreds of years had been misread. Their compilation of research conducted by the International Centre for Integrated Mountain Development (ICIMOD) in Bangladesh, Bhutan, India, Nepal and Myanmar showed that while there were cases where swidden/fallow cultivation systems were breaking down, these problems were as much a result of counterproductive government policies as they were of inappropriate land-use practices. Further, they found many examples of long-running swidden/fallow systems that had been sustainable, even in the face of increasing population pressure; that did not lead to land degradation; and that protected biodiversity. They demonstrated that these systems were resilient and adaptive in the face of internal and external pressures. Rather than viewing these areas as degraded forest, they argued that in reality they were examples of 'rotational agroforestry', or 'agroforestry with a burn cycle', or 'forest gardening'. They argued that since their swidden/fallow study areas had not been primary forest for hundreds, if not thousands, of years, they should not be compared, as was usually the case, to areas of primary forest. Rather, these areas should be compared to the alternative of permanent plantation agriculture – the system that was usually suggested as a replacement for it.

Corchorus capsularis L. [Malvaceae]

Production of jute features in Vietnam's agricultural statistics, but it is not only grown for its industrial fibre. At village level its leaves are used as a green vegetable and the unripe fruit and roots are also used in traditional medicine.

The arguments made against swidden cultivation in South Asia are the same as those made against the system in Southeast Asia. Paralleling the findings of Kerkhoff and Sharma are case-study examples found in literature on swidden/fallow cultivation systems in Southeast Asia. For example, Schmidt-Vogt (1998) in Thailand and de Jong (1997) in Indonesia detail how biodiversity found in swidden/fallow systems is of a similar level to that found in primary forest and vastly superior to that found in areas where permanent agriculture and plantation agriculture have replaced swidden cultivation. More recent literature on swidden/fallow systems support these conclusions (Rerkasem et al., 2009; Van Vliet et al., 2012). Recent work has also pointed out that long-term swidden/fallow systems are comparable to primary forest in their ability to sequester carbon, and even medium-term swidden/fallow systems equal or exceed proposed replacement systems in their ability to sequester carbon (Ziegler et al., 2012).

Other claims regarding the negative effects of swidden/fallow systems on the Southeast Asian upland resource base are also being increasingly questioned (Hamilton, 1988; Calder, 1999; Lindert, 2000). Furthermore, research over the past three decades has pointed out an absence of long-term data that are needed to determine trends in land degradation and changes in vegetation cover; a failure to separate the impacts of naturally occurring processes from those caused by human activities; and a failure to distinguish seasonal changes from permanent ones (Hamilton, 1988; Ives and Messerli, 1989; Fox, 2000; Fox et al., 2000; Ives 2004).

One possible reason why swidden/fallow cultivation is still viewed by the Vietnamese and other governments in a negative manner, even in the face of a growing number of studies that detail its benefits, is as much to do with differences in 'landscape visions' as with science (Sturgeon, 2000, 2004). In Sturgeon's case, the differences are shown in the landscapes of northern Thailand and southwestern China, and how these landscapes are viewed by their respective government officials and swidden/fallow cultivators. An analysis of different landscape visions could also be applied to Kerkhoff and Sharma's discussion of swidden cultivation in South Asia, as well as helping to explain why no swidden/fallow areas or vegetation are delineated on the land-cover and land-use maps for Vietnam's Nghe An province. In all of these cases, the reasons why there is a difference in landscape vision between the local people and the government may range from the remoteness of the area from the policy-makers to a misunderstanding of the people who live there and of the land-cover found there. This could apply particularly if the people living there are from ethnic minorities and the policy-makers are from an ethnic majority, or if the people making the policies or interpreting the maps are from lowland areas and have no experience in the uplands.

Another possible explanation is that the view of swidden/fallow cultivation systems as environmentally destructive has become the orthodox way to see them, both at the official level and by the public in Southeast Asia, and orthodox views are hard to change. Forsyth (2003) and Fairhead and Leach (1996) suggest that scientific error, or delays in understanding or accepting scientific advances, may not be the only reasons for environmental orthodoxies. They suggest that scientific facts only hold true given certain epistemic boundaries and that facts may be subverted if they are seen from a different starting point or vantage point. It is easy to see how this could be the case in Vietnam. As mentioned earlier, the first programme aiming to stop swidden cultivation in the northern mountains was the 'Sedentarization and Fixed Cultivation Program' in the 1960s. In tandem with this programme was the explicit government-supported movement of lowland peoples – ethnic-majority Kinh – to the uplands. The programmes stopped in the 1980s, but to the present day, the government continues to encourage lowland people to migrate to the uplands (Hardy, 2003; Winkels, 2011) by changing both the laws related to land rights and the ways in which land is allocated throughout the country, as well as through economic incentives. In order for the government of Vietnam to continue to justify a policy of encouraging lowland migrants to move to the uplands, it is necessary to perpetuate a viewpoint that the current inhabitants do not use upland resources wisely. Thus, the government argues that the uplands are under-populated. Its logic proposes that if resources were used more wisely, the uplands could support more people. Therefore, lowland Kinh people (ethnic Vietnamese) should be allowed to continue migrating to the uplands.

Implications of misreading the uplands

The government's maps (Figures 26-1, 26-2 and 26-3) show most of the uplands of the three districts in Nghe An province labelled as unused or barren land. This provides a basis for the government to promote policies aimed at changing the farming systems and livelihoods of the people living there and promote other policies, such as the 327 and 661 programmes, with the goal of 're-greening the wasteland'. However, this is not the only way that these lands can be classified. Leisz and Rasmussen (2012) presented a different view of the same three districts in Nghe An province. They used a land-cover classification system based on the land-cover types that the people who live in the uplands actually see covering their landscape. The resulting map shows a drastically different picture of the upland area (Figure 26-4). They found the areas that government maps (Figures 26-1, 26-2 and 26-3) designate as 'upland unused land' or 'barren land' are really covered with grass, grass and bush, bush, bamboo and tree cover.

All of the vegetation in these classes is usually found on fallow land and is used productively by the local people. Grass is collected and sold (broom grass) or used as fodder for animals. 'Bush' invariably includes plants that are collected and used domestically or sold for medicinal purposes. These plants are used as fodder, firewood, wild vegetables for cooking or as condiments, and as construction materials. Old-fallow vegetation includes many plants that are used for animal fodder or collected as non-timber forest products (NTFPs) that are used domestically or sold. Tree areas also produce NTFPs that are used as fodder, firewood and construction material. Besides these uses of the fallow vegetation, the fallows themselves provide grazing for large animals such as cattle, buffaloes and free-ranging pigs.

Thysanolaena latifolia
(Roxb. ex Hornem) Honda
[Poaceae]

Broom grass is among the many fallow species gathered by the villagers of Nghe An province. They make brooms for home or market and feed the foliage to livestock.

Table 26-2, which illustrates the points made above, is based on research conducted in 11 hamlets chosen on the basis of distance from a district town and markets, ethnicity and predominant farming systems in the village. The hamlets are spread across the three study districts illustrated in the government maps shown in Figures 26-1, 26-2 and 26-3. In each district, a combination of semi-structured, focused and structured interviews were conducted to investigate the farming systems and livelihood strategies as well as the customary rules and regulations used to manage natural resources. The interviews, following a structured questionnaire, were held with randomly selected households. A total of 240 structured interviews were conducted in the 11 hamlets during 2002, 2003 and 2004. Thirty households were randomly chosen from each

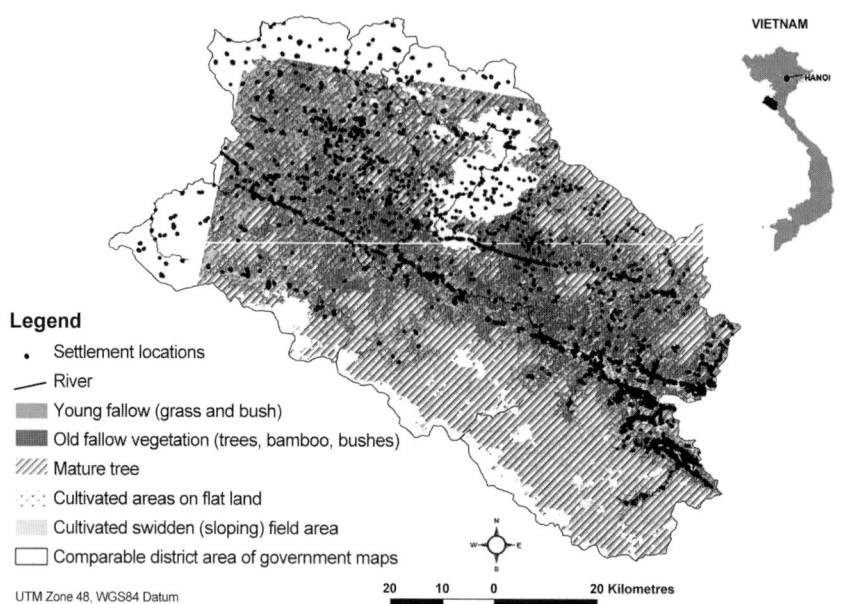

FIGURE 26-4: Location of settlements and land cover of three districts in Nghe An province from a local perspective in 2003.

Source: Based on Leisz and Rasmussen (2012).

of seven hamlets, but in the other four hamlets, time constraints meant that fewer households were randomly chosen and visited. Almost 100% of the households interviewed reported making use of fallow vegetation and fallow areas classified by the government maps as 'unused land' or 'barren land' for at least one of the activities listed. Table 26-3 presents an analysis of the data in order to quantify money made from selling products collected from the fallow areas.

The values shown in Table 26-3 are not insignificant when considered in terms of the average income of farming families in Nghe An province in 2003. According to the Government of Vietnam's Statistical Yearbook for 2003 (2004), the average agriculturally based family had an income equivalent to US$395. On this basis, the NTFPs sold per household alone were

Ipomoea batatas (L.) Lam.
[Convolvulaceae]

Sweet potatoes are among the few major food crops for which Vietnam keeps records of agricultural production. The many products of fallows and non-timber forest products go unrecorded.

TABLE 26-2: Number of households from study sample who made use of fallow vegetation.

Activity dependent on fallow vegetation	No. of households involved in this activity	% of households involved in this activity
Raising small animals (chickens and ducks – fodder for animals)	185	77.0
Raising large animals (buffaloes, cattle, pigs – collecting fodder and grazing animals)	217	90.0
Raising buffaloes	88	37.0
Raising cattle	137	57.0
Raising pigs	191	80.0
Collecting NTFPs for domestic use or sale (including firewood)	239	99.6
Collecting NTFPs for domestic use or sale (not including firewood)	235	97.9
Collecting firewood	237	98.8
Collecting bamboo	183	76.3
Collecting bamboo shoots	204	85.0
Gathering mushrooms	125	52.1
Collecting medicinal plants	123	51.3
Hunting/trapping bush meat	80	33.3
Collecting broom grass	117	48.8
Honey	21	8.8
Orchids	25	10.4
Other products	128	53.3

Note: NTFPs = non-timber forest products.

TABLE 26-3: Estimated annual value of fallow products sold by study households.

Product sold	Total value (VND)	Total value (USD)	Value per household (VND)	Value per household (USD)
All products	807,039,500	53,989.80	3,362,665	224.96
Livestock	699,156,000	46,772.54	2,913,150	194.89
Firewood	1,000,000	66.90	4,167	0.28

Note: Livestock were dependent on fallow vegetation for fodder. All data were collected in 2002, 2003 and 2004. During this time the official exchange rate varied from 14,382 Vietnam dong to US$1 in January 2002 to 15,756 Vietnam dong to US$1 in January 2005. The average during this period was 14,948 Vietnam dong to US$1 (http://www.oanda.com/currency/historical-rates/, accessed 1 December 2015).

equal to 8% of the average farm household's total income. If the value of animals produced using fodder from fallow areas is included, more than 50% of the average farming household's income was dependent on fallow lands. Considering that the government's average household income included lowland farming families, whose incomes could be expected to be as much as two-thirds higher than those in the uplands, it is clear that a large part of the income of upland households came from products collected from fallow land. It is also clear that the value of those products was much higher than shown here, because these data cover only those products that were sold, and do not include the value of products that were used within the household. Since the vast majority of products collected from fallows were reportedly used within households, the true value of the fallow vegetation and fallow areas to communities in the three study districts was much higher. The implication of this analysis is that the true value of products produced in the uplands is most likely not being recorded by the government, since these products are not found in government agricultural statistics.

Another implication of the official land–cover and land-use maps and official statistics is that they provide justification for implementing new land regulations and justify the explicit and implicit encouragement of lowland migration to upland areas.

An important issue that is raised by this misreading of the uplands is that the government is not recording, and may not be aware of, the true value of economic production in the uplands. As pointed out, the area that the government considers 'unused land' and 'barren land' is actually used by the local people as a source of many products that are used within the household or sold. None of these products is noted in the government's General Statistics Office records. The only statistics for agricultural production are for paddy, maize, sweet potatoes, cassava, key annual industrial crops, cotton, jute, sedge, sugar cane, peanuts, soybeans, tobacco, tea, coffee, rubber, pepper, coconuts, and the numbers of livestock, including buffaloes, cattle and pigs. The only statistics for forests note the value of timber output (Statistical Yearbook 2003, 2004). With the exception of some value given to livestock, the products listed in Table 26-2, especially the NTFPs, are not considered in the government's economic statistics. In short, the value of upland products that are dependent on fallow lands and fallow vegetation are missing from the government's statistics. Some of the products that are sold could conceivably be recorded if the right questions were asked upon collection of yearly statistics. The value of the domestic use of these products needs to be known, in order to gain an accurate view of the true productivity of Vietnam's upland areas. These values are far from insignificant, and this knowledge is needed in order to devise more appropriate policies towards agricultural development in the uplands and towards immigration to the uplands from the lowlands.

By ignoring the value of these products, the actual returns from the uplands, as they are currently managed, cannot be accurately known. What is evident, though, is that the current understanding of the uplands undervalues its products. It is worth asking whether the land- and resource-management systems being proposed as replacements in the uplands will produce enough added value to replace that which will be lost as fallow areas and fallow vegetation disappear.

References

ASB (1994) *Alternatives to Slash-and-Burn: A Global Initiative*, International Centre for Research in Agroforestry, Nairobi

Bandy, D. E., Garrity, D. P. and Sanchez, P. A. (1993) 'The worldwide problem of slash-and- burn agriculture', *Agroforestry Today* 5(3), pp2-6

Calder, I. (1999) *The Blue Revolution: Land Use and Integrated Water Resources*, Earthscan, London

Castella, J. C. and Quang, D. D. (2002) '*Doi Moi in the Mountains*', The Agricultural Publishing House, Hanoi

Cuc, L. T. (1996) 'Swidden agriculture in Vietnam', paper presented to a conference 'Montane Mainland Southeast Asia in Transition', Chang Mai

Cupet, P. (2000) *Travels in Laos and among the Tribes of Southeast Indochina*, translated by W. E. J. Tips, White Lotus Press, Bangkok

Dang, Ng. V. (1991) 'La culture sure brulis et le nomadisme', *Etudes Vietnamiennes* 1(99), pp16-28

de Jong, W. (1997) 'Developing swidden agriculture and the threat of biodiversity loss', *Agriculture, Ecosystems and Environment* 62, pp187-197

Devendra, C. and Thomas, D. (2002) 'Smallholder farming systems in Asia', *Agricultural Systems* 71, pp17-25

Dove, M. R. (1984) *Government versus Peasant Beliefs concerning Imperata and Eupatorium: A Structural Analysis of Knowledge, Myth and Agricultural Ecology in Indonesia*, East-West Center, Honolulu

Fairhead, J. and Leach, M. (1996) *Misreading the African Landscape: Society and Ecology in a Forest-Savanna Mosaic*, Cambridge University Press, Cambridge, UK

FAO (1999) *State of the World's Forests 1999*, Food and Agriculture Organization of the United Nations, Rome

Forsyth, T. (2003) *Critical Political Ecology: The Politics of Environmental Science*, Routledge, London

Fox, J. (2000) 'How blaming "slash and burn" farmers is deforesting mainland Southeast Asia', *Asia Pacific Issues* 47(47), pp1-8

Fox, J., Truong, D. M., Rambo, A. T., Tuyen, N. P., Cuc, L. T. and Leisz, S. (2000) 'Shifting cultivation: A new old paradigm for managing tropical forests', *BioScience* 50, pp521-528

Hamilton, L. (1988) 'Forestry and watershed management', in J. D. Ives and D. C. Pitt (eds) *Deforestation: Social Dynamics in Watershed and Mountain Ecosystems*, Routledge, London

Hardy, A. (2003) *Red Hills: Migrants and the State in the Highlands of Vietnam*, University of Hawai'i Press, Honolulu

Ickowitz, A. (2006) 'Shifting cultivation and deforestation in tropical Africa: Critical reflections', *Development and Change* 37, pp599-626

IUCN (2006) *Biodiversity Programme*, available at http://www.iucn.org/places/vietnam/our_work/species/biodiversity.htm (accessed 30 November 2006)

Ives, J. D. (2004) *Himalayan Perceptions: Environmental Change and the Well-Being of Mountain Peoples*, Routledge, London

Ives, J. D. and Messerli, B. (1989) *The Himalayan Dilemma: Reconciling Conservation and Development*, Routledge and United Nations University, London

Jamieson, N. L. (1991) *Culture and Development in Vietnam*, East-West Center, Honolulu

Jorgensen, B. D. (2006) *Development and 'The Other Within': The Culturalisation of the Political Economy of Poverty in the Northern Uplands of Vietnam*, Department of Peace and Development Research, Goteborg University, Goteborg, Sweden

Kerkhoff, E. and Sharma, E. (2006) *Debating Shifting Cultivation in the Eastern Himalayas: Farmers Innovations as Lessons for Policy*, International Center for Integrated Mountain Development, Kathmandu

Lanly, J. P. (1985) 'Defining and measuring shifting cultivation', *Unasylva* 37(147), pp17-21

Le, T. C. and Rambo, A. T. (2001) *Bright Peaks, Dark Valleys: A Comparative Analysis of Environmental and Social Conditions and Development Trends in Five Communities in Vietnam's Northern Mountain Region*, The National Political Publishing House, Hanoi

Leisz, S. and Rassmussen, M. S. (2012) 'Mapping fallow lands in Vietnam's north central mountains using yearly Landsat imagery and a land-cover succession model', *International Journal of Remote Sensing* 33(20), pp6281-6303

Lindert, P. H. (2000) *Shifting Ground: The Changing Agricultural Soils of China and Indonesia*, The MIT Press, Cambridge, MA

Morrison, E. and Dubois, O. (1998) *Sustainable Livelihoods in Upland Vietnam: Land Allocation and Beyond*, Forestry and Land Use Series 14, International Institute for Environment and Development, London

Myers, N. (1984) *The Primary Source: Tropical Forests and our Future*, Norton, New York

Myers, N. (1992) 'Tropical forests: The policy change', *Environmentalist* 12, pp15-27

Myers, N. (1993) 'Tropical forests: The main deforestation fronts', *Environmental Conservation* 20, pp9-16

Paddoch, C. and Coffey, K. (2003) 'Monitoring the demise of swidden in Southeast Asia: Local realities and regional ambiguities', in O. Mertz, R. Wadley and A. E. Christensen (eds) *Local Land Use Strategies in a Globalizing World: Shaping Sustainable Social and Natural Environments*, vol. 1, proceedings of an International Conference, 21-23 August, 2003, Danish University Consortium for Environment and Development and Sustainable Land Use and Natural Resource Management Programme, Institute of Geography, University of Copenhagen. Denmark, pp103-124

Poffenberger, M. (1998) *Stewards of Vietnam's Upland Forests*, Asia Forest Network, Berkeley, CA

Rambo, A. T. (1995) 'Slash-and-burn farmers: Villains or victims?', *Earthwatch* 39, pp10-12

Rambo, A. T. (1998) 'The composite swiddening of the Tay ethnic minorities of the Northwestern Mountains of Vietnam', in A. Patanothai (ed.) *Land Degradation and Agricultural Sustainability: Case Studies from Southeast and East Asia*, Southeast Asian Universities Agroecosystem Network, Khon Kaen, Thailand

Rerkasem, K., Lawrence, D., Padoch, C., Schmidt-Vogt, D., Zeigler, A. D. and Bruun, T. B. (2009) 'Consequences of swidden transitions for crop and fallow biodiversity in Southeast Asia', *Human Ecology* 37, pp347-360

Schmidt-Vogt, D. (1998) 'Defining degradation: The impacts of swidden on forests in northern Thailand', *Mountain Research and Development* 18, pp135-149

Statistical Yearbook 2003 (2004) Statistical Publishing House, General Statistics Office, Hanoi

Sturgeon, J. (2000) 'Practices on the periphery: Marginality, border powers and land use in China and Thailand', PhD dissertation to the Yale School of Forestry and Environmental Studies, New Haven, CT

Sturgeon, J. (2004) 'Border practices, boundaries, and the control of resource access: A case from China, Thailand and Burma', *Development and Change* 35(3), pp463-484

Tran, D. (2003) *The Farm Economy in Vietnam*, The Gioi Publishers, Hanoi

Thomas, F. (1999) Histoire du Regime et des Services Forestriers Francais en Indochine de 1862 a 1945, The Gioi Publishers, Hanoi

Thongchai, W. (1994) *Siam Mapped: A History of the Geo-body of a Nation*, University of Hawaii Press, Honolulu

Van Vliet, N., Mertz, O., Heinimann, A., Langanke, T., Adams, C., Messerli, P., Leisz, S., Pascual, U., Schmook, B., Schmidt-Vogt, D., Castella, J. C., Jorgensen, L., Birch-Thompson, T., Hett, C., Bech-Bruun, T., Ickowitz, A., Vu, K. C., Yasuyuki, K., Fox, J., Dressler, W., Pacoch, C. and Ziegler, A. D. (2012) 'Trends, drivers and impacts of changes in swidden cultivation in tropical forest-agriculture frontiers: A global assessment', *Global Environmental Change* 22, pp418-429

Vandergeest, P. (2003) 'Racialization and citizenship in Thai forest politics', *Society and Natural Resources* 16, pp19-37

Wezel, A. (2000) 'Weed vegetation and land use of upland maize fields in north-west Vietnam', *GeoJournal* 50, pp349-357

Winkels, A. (2011) '"Stretched livelihoods": Social and economic connections between the Red River Delta and the Central Highlands', in T. Sikor, P. T. Nghiem, J. Sowerine and J. Romm (eds) *Upland Transformations: Opening Boundaries in Vietnam*, National University of Singapore Press, Singapore

Ziegler, A. D., Phelps, J., Yuen, J. Q., Webb, E. L., Lawrence, D., Fox, J. M., Bruun, T. B., Leisz, S. J., Ryan, C., Mertz, O., Dressler, W., Pascual, U., Padoch, C. and Koh, L. P. (2012) 'Carbon outcomes of major land-cover transitions in SE Asia: Great uncertainties and REDD+ policy implications', *Global Change Biology* 18(10), pp3087-3099

Notes

1. The term swidden/fallow is used in this chapter in preference to other terms, such as 'slash and burn', that are used to describe cultivation systems where land is cleared of vegetation, burned and cultivated for one to three or four years, then left fallow for a number of years before being cultivated again.
2. Rai in this context refers to an upland field that has been cultivated.
3. All terms are translated from Vietnamese as found in Bui Phung (2002) Vietnamese English Dictionary, Gioi Publishing, Hanoi.
4. Translated from Bui Phung (2002).

27

CHANGING PATTERNS OF SHIFTING CULTIVATION IN TIMOR-LESTE

Harold J. McArthur, James B. Friday and Michael J. Jones*

Introduction

Timor-Leste, or East Timor, officially the Democratic Republic of Timor-Leste, is an island country at the far eastern end of the Indonesian archipelago (Figure 27-1). The young nation, located about 500km north of Australia, comprises the eastern half of Timor Island and includes the enclave of Oecusse (within Indonesian West Timor) and the islands of Atauro and Jaco. The country has a land area of about 15,000sq. km, of which nearly half has steep slopes, often exceeding 40% (Timor-Leste National Action Program, 2008).

---

*  We were greatly saddened to learn of the passing of Dr Harold McArthur during the final editing phases of this chapter. Hal, as he was known to friends and colleagues, worked and taught for more than 30 years at the University of Hawaii and inspired generations of students. With his background in anthropology, Hal strove to understand why farmers and rural people made the decisions they did and how agricultural and development programmes could be improved by understanding rural societies. His career included projects in Timor-Leste, the Philippines, Vietnam, Laos, Indonesia, Papua New Guinea, Mexico, Honduras and Peru. Hal also served as President of the International Farming Systems Association and as Visiting Senior Scientist at the International Center for Living Aquatic Resources Management (ICLARM). Upon his retirement Dr McArthur became an associate professor for the Asian Institute of Management in Manila. His colleagues remember him for his constant cheerfulness and optimism, even in difficult situations, and his steadfast faith in the ability of farmers to make the right decisions. – J. B. Friday

DR HAROLD J. MCARTHUR, Retired Assistant Vice Chancellor for Research Relations, University of Hawaii at Manoa, Honolulu; DR JAMES B. FRIDAY, Associate Specialist, Department of Natural Resources and Environmental Management, College of Tropical Agriculture and Human Tropical Resources, University of Hawaii at Manoa, Honolulu; MICHAEL J. JONES, Independent Consultant and Adjunct Research Fellow, College of Business, Law and Governance, James Cook University, Queensland, Australia.

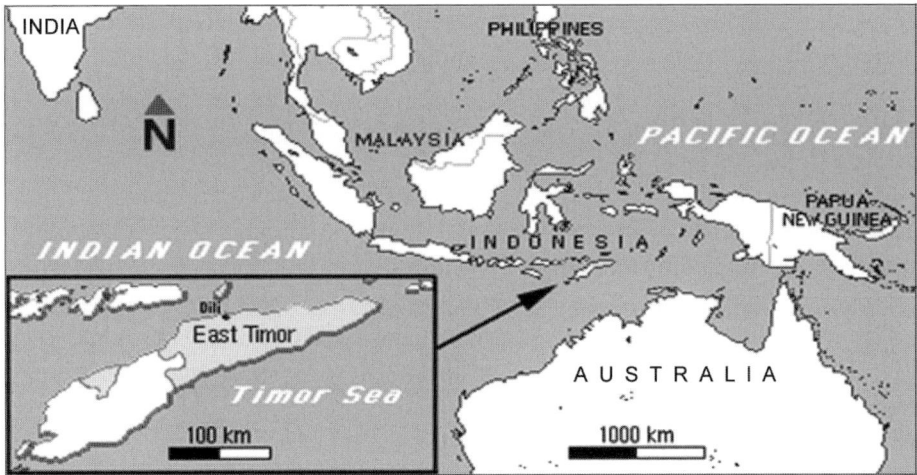

FIGURE 27-1: Location of Timor–Leste, or East Timor.

The island is dominated by a mountain range running east to west and rising to an elevation of 2900 metres above sea level. Most of the population, and the capital Dili, are located on the northern coast, a seasonally dry area with as little as 600mm of rainfall per year and a dry season lasting up to seven months (Fox, 2003). The southern slopes receive more rain, but all areas are seasonally dry. Heavy monsoon rains and steep slopes make flash floods frequent and irrigation and transportation difficult. About 57% of the land area is classified as forest (National Directorate of Forestry and Water Resources, 2004).

Timor-Leste is one of the newest independent countries in the 21st century and some consider it to be in a state of serious ecological crisis as a result of swidden agriculture and population pressure (Pannell, 2011).

During 400 years of Portuguese colonial rule and 25 years of Indonesian occupation the harsh, drought-stricken landscape of Timor-Leste has been cleared, cultivated and ultimately degraded by human activity (McWilliam, 2003). The relationship between society and the forests has been dramatically shaped by politics and civil administration during three distinct periods of the country's history.

Post-conflict nation

In July 1999, under the auspices of the United Nations Secretary-General, a referendum was held and the people voted overwhelmingly for separation from Indonesia. On 20 May 2002, the new post-conflict nation emerged from the ashes and destruction left behind by departing Indonesian civil servants and military personnel. The departing troops and some pro-integration militia members burned government buildings and schools and destroyed much of the country's infrastructure, including bridges, irrigation dams and canals. Crops and much of the livestock were also destroyed. During the 24 years of Indonesian occupation it is estimated that 200,000 people, or about one third of the population, were killed or starved to death. The armed aggression of Indonesia against the Timorese received little media attention and was referred to as the 'forgotten conflict' (Ramos-Horta, 2013-2014). In 1999, The UN Security Council, with support from the United States and Australia, authorized a multinational peacekeeping force to restore peace and order to the war-torn country during its transition to independence.

Although many improvements have been made over the past 15 years, the country remains fragile in terms of political, social and economic stability. Since independence, there have been several instances of civil unrest (in 2003 and again in 2006) that required intervention by UN peacekeeping forces (Brady and Timberman, 2006).

The most significant features of this situation include deep divisions among the country's senior political leaders, institutional weaknesses and rivalries and a propensity for violence among elements of the population. It has been suggested that post-conflict countries such as Timor-Leste can learn useful lessons from other countries with a similar history of conflict (eTN Global Travel Industry News, 2015). There is now a growing literature on strategies adopted by post-conflict countries in order to restore peace and move development forward. For example, the African nation Rwanda is a well-known post-conflict country. Its High Commissioner to South Africa, Vincent Karega, says his country was nearly destroyed by a horrific genocide that was born from internal conflicts. Rwanda's case has many similarities with the conflict in Timor-Leste.

Indigenous agriculture

Traditional crops

Unlike many Southeast Asian and Pacific nations (including the Philippines, Thailand, Cambodia and Vietnam), Timor-Leste's economy still depends largely on subsistence agriculture (Chao, 2009). Maize and rice are the staple crops. Maize was introduced to Timor by the Portuguese. Farmers also cultivate sweet potatoes, cassava, taro, squash, beans, peanuts, tomatoes, ginger, garlic, onions and shallots for home use or for sale in local markets. In some regions, breadfruit is a staple crop along with coconuts, citrus, bananas, papaya and mangoes. Candlenuts (*Aleurites moluccanus*) are harvested from semi-wild groves and several species of palms are cultivated for palm wine, thatch and other products, including sago. Yams, bean, and other wild foods may also be gathered from forest areas in lean seasons. Livestock include

Artocarpus altilis (Parkinson ex F. A. Zorn) Fosberg [Moraceae]

Breadfruit is one of the staple crops in some regions of Timor-Leste.

chickens, pigs, goats, cattle and water buffaloes (Lopes and Nesbitt, 2012).

Forest management and agriculture under the Portuguese

Beginning in 1515, Portuguese traders arrived in Oecusse and quickly began buying up tons of valuable sandalwood (*Santalum album*), which grew throughout what is now Timor-Leste. By the early 1800s, when nearly all of the commercial-grade sandalwood had been extracted, the Portuguese

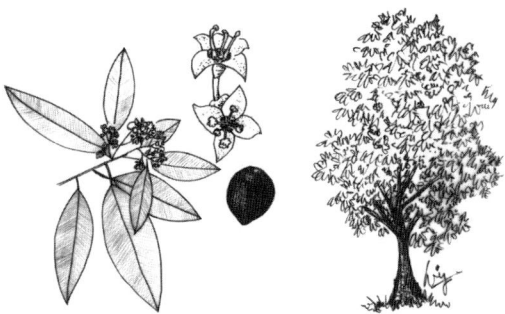

Santalum album L. [Santalaceae]

Such was its value that nearly all commercial-grade Sandalwood was gone by the early 1800s.

established plantations of coffee and cotton (Ramos-Horta, 2013-2014). These crops were planted and managed by Portuguese plantation owners and had little impact on the subsistence-level swidden agriculture of most of the population. Portugal ruled Timor-Leste through the use of indigenous power structures in a manner that left little impact on traditional Timorese society and culture (Ramos-Horta, 2013-2014).

Chao (2009) suggests that apart from trading and the extraction of forest resources, Timor-Leste was a largely neglected colonial outpost of Portugal. Only a small portion of the land was considered suitable for commercial agriculture because of steep slopes and a lack of water. As a result, most of the Timorese population relied on swidden agriculture and a barter-based economy.

Agricultural practices during the Indonesian occupation

Timor-Leste was forcibly taken over by Indonesia in 1975, with the main objective of making it a province of Indonesia. Initially, thousands of Timorese who resisted the Indonesian occupation were sent to relocation camps where many starved because they were only allowed to farm small plots near the new camps where it was not possible to rotate the land or the crops. During the Indonesian occupation, there was widespread destruction of forest ecosystems resulting from the unsustainable harvesting of many valuable forests products, including sandalwood (*Santalum album*), teak (*Tectona grandis*) and ebony (*Diospyros* sp.) (Timor-Leste National Action Program, 2008).

The Indonesian government also made Timor-Leste a transmigration destination and relocated thousands of poor Javanese and Balinese farmers to rural areas in Timor where they could construct rice terraces and rice paddies on large tracts of land expropriated from Timorese families (Figure 27-2).

FIGURE 27-2: Rice terraces constructed by migrants on land expropriated from Timorese.

The authors have been unable to identify any citable references related to the construction of upland rice terraces in Timor-Leste. We suspect that most of the existing upland rice terraces were probably constructed by Indonesian transmigrants, most likely by farmers from the island of Bali, where terracing has been part of the agricultural tradition for centuries. Some isolated pockets of rice terraces may date back to Portuguese developments around key sandalwood or administrative centres.

During this period, much of the population was forced into new *campo barro* villages near local towns and highways. Since independence, they have been free to return to their traditional village sites in more remote areas. However, improved infrastructure and economic opportunities remain limited to the 'new villages' (McCarthy, 2015).

To support this effort, the Indonesians developed an agricultural extension service and provided subsidized fertilizers and pesticides along with kerosene for lamps and petrol for imported hand tractors and trucks. Under Indonesian rule, roads were improved and irrigation dams and pumping stations were built to support the production of paddy rice in addition to the traditional staple of maize. Indonesian firms, in partnership with the military, took over key businesses in Timor-Leste, including those that exported sandalwood and coffee (Timor-Leste National Action Program, 2008).

Challenges to forestry and agricultural development after independence

Deforestation

Every year, thousands of hectares of secondary forests are cleared in Timor-Leste for swidden agriculture (Jeus et al., 2012). A number of key factors and conditions

FIGURE 27-3:   Erosion scars a denuded hillside.

have contributed to deforestation, including increased cutting of trees for firewood, over grazing, burning, reduced fallow periods that are too short for soil recovery and increased presence of invasive weeds.

When fallow periods begin after cropping has ended, forest regeneration is slowed by wildfires, unmanaged grazing and degradation of the land by severe soil erosion during monsoon rains (Figure 27-3). Secondary forests in dry areas are largely composed of fire-tolerant species such as palms and *Eucalyptus alba*, but uncontrolled grazing, mainly by goats and sheep, prevents forest regeneration. A modern problem is the fast spread of noxious weeds such as *Lantana camara* and *Chromolaena odorata* on overgrazed lands (McFadyen, 2003). These weeds, which are inedible by livestock, prevent more desirable fallow vegetation from reclaiming former swidden land.

Population density is another key force driving swidden agriculture towards permanent cultivation (de Jong et al., 2001). With the construction of irrigation systems that allow intensive wet-rice cultivation in coastal areas, there is a lesser incentive for farmers to practise extensive swidden agriculture in the uplands. The population is also moving away from rural areas toward the cities, so swidden agriculture is becoming less appropriate and may soon be unable to support a growing population (Molyneux, 2011).

One of the factors contributing to the decline in swidden agriculture throughout Southeast Asia over recent decades has been government classification of swidden practitioners as ethnic minorities, distinguishing them from the dominant, mostly lowland population. This has happened, for example, in the Lao PDR and Vietnam (Lao Swedish Agriculture and Forestry Research Programme, 2004). Other factors that have led to the decline in shifting cultivation include resettlement, the privatization of commercial agriculture and the growing transition from rural to urban livelihoods

(Fox et al., 2009). People who have access to modern amenities such as health care and education in towns are reluctant to return to their traditional areas of upland agriculture. However, high population growth since the conflicts of the Indonesian era may force people to clear and farm less desirable areas in the future. Some people also want to reclaim ancestral lands that were lost during the time of conflict and forced resettlement.

According to Jeus et al. (2012), farmers cut and burn on average less than 1 hectare, and most of them clear upland swiddens for subsistence agriculture rather than to produce crops for sale. These authors further suggest that farmers who do not have fixed garden plots contribute the most to forest destruction.

Firewood harvesting is often blamed for causing deforestation. Firewood is the overwhelming choice for cooking fuel, even in cities, because kerosene and other petrochemicals are not always available and are relatively expensive (Figure 27-4). Industrial users of firewood include bakeries, furniture makers and coconut processing operations. A World Bank study (Shum et al., 2007) estimated that 600,000 metric tons of firewood was consumed per year in Timor-Leste. Urban residents buy firewood, while rural residents collect it themselves. A single species, *Eucalyptus alba*, makes up most of the firewood harvested for market. This species is fire tolerant and may be seen growing in burnt-over areas. It also regenerates well after fires, sometimes growing in monospecific stands. Since almost all firewood cutting takes place in secondary *Eucalyptus alba* stands, it is most likely that most firewood harvesting is being done in secondary, although natural, forests.

The World Bank study concluded that deforestation was unlikely to be the result of firewood cutting. It blamed other pressures on the forests, such as clearing land for agriculture. However, increased firewood harvesting from secondary forests around the capital, Dili, has indeed led to the retreat of the forest. Harvesters report having

FIGURE 27-4:   Firewood harvesting is a big local industry.

to walk further and further each year to gather firewood. But despite the difficulties in finding adequate supplies, there is insufficient economic incentive to establish firewood plantations.

Tara Bandu: A system of Timorese customary law

A bright spot in Timor-Leste's environmental situation is the continued existence in some areas of indigenous systems of land management. *Tara Bandu* is a system of traditional bans or taboos on certain practices that is an important tool for indigenous village leaders in resolving conflicts and managing natural resources at community level. According to the Ministry of Economy and Development (2012), 'within the context of Timor-Leste's history and occupation, the customary law of *Tara Bandu* stands out as a way for traditional societies to regulate both daily social matters and relationships between humans and the environment'.

Tara Bandu varies greatly in different parts of the country. In some communities, it is used to delimit areas of sacred forest. Often, *Tara Bandu* is used to prohibit unsustainable activities such as cutting trees, hunting or overfishing. Normally, the visual representation of *Tara Bandu* is a wooden or bamboo pole in a village with parts of plants or animals attached to it, indicating what forest products or creatures are protected within a specific area (Figure 27-5). Sometimes a *Tara Bandu* pole is accompanied by a pair of water buffalo horns, indicating the number of buffalos that must be sacrificed by someone guilty of violating the prohibitions (Figure 27-6).

The traditional system of *Tara Bandu* was banned during the Indonesian occupation, but since independence it has been actively practised in various parts of the country. Through the renewal of the traditional system, Timor-Leste is investing in local cultural customs and mediation methods to resolve modern-day social conflicts (Ministry of Economy and Development, 2012). However, it remains to

FIGURE 27-5 and 27-6: A *Tara Bandu* pole (left), marking an area where a certain practice is taboo, and the horns of a water buffalo, warning of penalties for violating *Tara Bandu*.

be seen whether future generations, raised in a cash economy, will respect the *Tara Bandu* of their elders.

A more detailed assessment of *Tara Bandu* and how traditional customary law is being used to deal with modern–day social conflicts and resource conservation may provide lessons that can be applied in other developing countries.

National policies and strategies for forest and watershed development

Threats to forest resources

There are currently few activities supported by the Ministry of Agriculture, Forestry and Fisheries in Timor-Leste that aim to rehabilitate areas of severe deforestation. The country's sandalwood resources are nearly exhausted, but there has been some trial production of seedlings and test plantings of sandalwood trees in several rural communities. Following independence, the first Minister of Agriculture, Forestry and Fisheries confiscated harvested sandalwood from a number of rural villages to prevent traders from buying it and selling it to foreign buyers. The Minister said he hoped to identify one or more buyers for a one-time sale of the confiscated sandalwood. The goal was to use the proceeds from the sale to fund a renewed sandalwood propagation and conservation programme (da Silva, 2003). However, such actions alienated local communities and created a disincentive to conserve sandalwood forests. It has been suggested that local communities should be given control and management of sandalwood resources, thus giving them a stake in the regeneration of sandalwood forests (Ora, 2012).

A 2002 survey by the Japan International Cooperation Agency found remnants of valuable forest species still standing. In the central districts of Manatuto, Ainaro and Manufahi the survey found specimens of *Eucalyptus urophylla*, a valuable native timber tree, still standing, but noted that the trees were old and suffering from pest attacks. In the Covalina, Bobonaro, Baucau and Viqueque districts, it found limited numbers of candlenuts (*Aleurites moluccanus*), and in the sparsely populated Lautem district, it found sandalwood and *narra* (*Pterocarpus indicus*). The Ministry of Agriculture, Forestry and Fisheries has begun a forest-resource inventory, but its progress has been hampered by limited resources and staff.

As well as a lack of human and financial resources, swidden farming is another challenge to any forest-rehabilitation programmes. Fifteen areas have been designated as

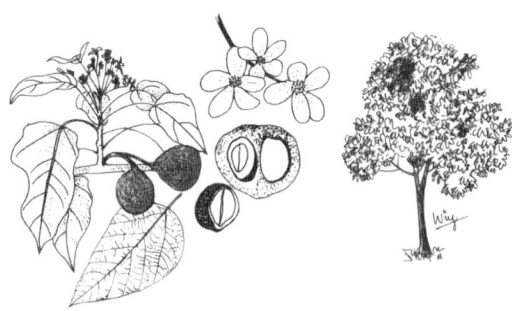

Aleurites moluccanus (L.) Willd.
[Euphorbiaceae]

Candlenuts were among the valuable forest species found in limited numbers in 2002.

national parks or protected natural-resource areas, and although tree cutting, swidden farming and hunting are banned in these areas by law, these activities still proliferate at the hands of families living within the boundaries of these protected areas. Unless a partnership can be forged between rural communities and the government to jointly plan for alternative livelihoods, the problem of agricultural encroachment in forest areas will continue.

A policy and strategy report for the forestry and watershed subsector, released in 2004 by the Ministry of Agriculture, Forestry and Fisheries through its National Directorate of Forest and Water Resources, refers to Timor-Leste's National Development Plan of 2002, which identifies two key goals: enhancing food security and generating rural employment. The two documents form the basis upon which forestry regulations and watershed laws are soon to be formulated. The report suggests that 'the most appropriate way to improve natural resource management is through use of an "integrated watershed approach" along with a focus on community-based management of natural resources'.

Development of a forestry watershed law

The report emphasizes the need to define the legislative framework and enabling mechanisms and regulations necessary for effective management of forests and watersheds. It says that under a well-defined framework, long-term activities can be directed towards participatory, community-based forest management planning with integrated concerns for maintenance of biodiversity and watershed conservation. The report further notes that a sustainable forest and watershed procedure will require a long-term participatory partnership and coordination between the National Directorate of Forest and Water Resources, rural communities, non-governmental organizations, the private sector and various international development agencies involved in forest and watershed conservation.

The National Action Programme to Combat Land Degradation (2008)

The Constitution of Timor-Leste identifies environmental protection as one of the country's key objectives. In Section 61, paragraphs two and three state that 'The state shall recognize the need to preserve and rationalize natural resources and should promote actions aimed at protecting the environment and safeguarding the sustainable development of the economy'.

The projects being conducted within the National Action Programme are vital mechanisms to enable Timor-Leste to meet its obligations as a signatory to the United Nations' Convention to Combat Deforestation. The NAP programmes are focused on (1) minimizing factors that contribute to deforestation, and (2) mitigating and/or alleviating the impacts of land degradation. A series of draft laws on forestry protection and use of fertilizers and pesticides has already been presented for approval to the Council of Ministers.

Chromolaena odorata (L.) R. M. King & H. Rob. [Compositae]

A problematic weed that invades many fallowed upland fields in Timor-Leste.

Despite the government's best intentions, enforcement of environmental-protection and forest-conservation laws has been weak or non-existent. National police rarely follow up complaints with relevant government agencies. A number of local and national non-governmental organizations are involved in sustainable land management activities in different parts of the country. These programmes include a nationwide seed-propagation programme to establish community nurseries and replant forest lands (Timor-Leste National Action Program, 2008). Unfortunately, communications and cooperation between these groups is weak, often resulting in poor utilization of resources.

A recent survey conducted by the Ministry of Agriculture, Forestry and Fisheries through its National Directorate of Forest and Water Resources indicates that most upland families farm land parcels of between 1 and 2 hectares for about three years. After the final crops are harvested and the fields are abandoned, they are commonly dominated by the invasive weed *Chromolaena ordorata* and the rhizomatous grass *Imperata cylindrica*. While the grass is valuable both as a thatch and for grazing, *Choromolaena* is unpalatable to livestock. Both species smother tree seedlings that would otherwise return the fields to secondary forest, and both species burn readily during the dry season.

Fuel-efficient indigenous stoves

As in most tropical countries, food is cooked over open wood fires in Timor-Leste. Some development projects have attempted to introduce improved cooking stoves to reduce the use of wood, improve kitchen environments and reduce labour for women, who must collect the wood before cooking (Figures 27-7 and 27-8). The stoves are, indeed, more fuel efficient, but in many areas scrapwood is so accessible that there is little incentive to try something different. In this sense, it seems that the new stoves are a solution to a problem that hasn't yet been recognized by many people.

FIGURE 27-7 and 27-8: Conventional cooking equipment (left) and improved stoves: A solution for which there is no problem.

Current programmes and strategies

Promoting and maintaining district and community tree nurseries

In many districts, staff of the Ministry of Agriculture, Forestry and Fisheries operate forestry nurseries where seedlings are produced for government-managed reforestation projects and distribution to local farmers. While some of these nurseries do an excellent job of growing seedlings, planting in the field has only limited success because of a lack of follow-up care in a harsh environment where drought, fire and uncontrolled grazing destroy many young trees. Government projects lack the staff to plant trees on anything larger than a demonstration basis.

Small-scale farmers have little incentive to plant forest trees because of uncertain harvests in the distant future. While growing trees for firewood might provide earlier harvests, farmers in one survey conducted by the authors expressed interest in growing higher-value timber trees such as mahogany (*Swietenia macrophylla*) or teak (*Tectona grandis*), rather than the common *Eucalyptus* species for firewood.

Reforestation

Reforestation is constrained by many of the same issues as tree nurseries and out-planting. While the government has proposed legislation related to forest conservation and rehabilitation, implementation remains

Eucalyptus alba Reinw. Ex Blume
[Myrtaceae]

This species provides most of the commercially harvested firewood on Timor-Leste.

a key issue because the government does not have the trained personnel to effectively reforest significant areas.

Summary

A number of 'push-pull' factors and conditions continue to affect swidden agriculture and forest conservation in Timor-Leste. The Portuguese were the colonial rulers of Timor-Leste for more than 400 years, but apart from exploiting sandalwood and introducing coffee as a potential export crop, they had a minimal impact on the countryside and the lives of the most of the people. Therefore, unlike many of their Southeast Asian neighbours, the people of Timor-Leste and their traditional way of life were largely unaffected by the policies and actions of the colonial administration. They were then, and remain today, largely dependent on swidden agriculture for their survival. Farmers open new land only as necessary and because of low population density in rural and upland areas are able to maintain a system of long fallow and field and crop rotation.

These conditions changed drastically during the 24 years of Indonesian occupation, when people were forced into village camps near to towns and roads. This forced relocation aimed to control increasing insurrection and to create a critical population mass needed to sustain an intensive irrigated paddy-rice system. The object was to increase rice production in addition to the traditional staple crop of maize.

In the *campo barro* settlements, villagers were confined to small plots of land adjacent to the camps. There was insufficient land for traditional swidden agriculture. People were not allowed to return to their original village sites or ancestral lands in the hinterlands. The towns exerted a considerable pull away from traditional life, with access to improved infrastructure and urban amenities such as health facilities and schools. In order to maintain this more sedentary lifestyle, the Indonesian government began employing thousands of Timorese in menial support roles to the military and government agencies.

As well as wage income, the local people received subsidized kerosene, fertilizer and petrol for imported hand tractors that were used to prepare the paddy fields. Private Indonesian firms were encouraged to form partnerships with local military leaders to exploit the remaining forest products and non-timber resources for export back to Indonesia.

During the Indonesian occupation, efforts were made to quell the use of and support for the traditional system of bans (*Tara Bandu*) and customary laws. This was seen as giving power to village leaders, something the Indonesians did not want to do. Following independence, *Tara Bandu* re-emerged in many parts of the country. Although the nature of the practice varies from one place to another, *Tara Bandu* once more serves as an important means of dealing with relationships between villagers and the land. Only time will tell how effective the *Tara Bandu* prohibitions will be when young people in rural areas see something to be gained from collecting and selling forest products such as sandalwood and rosewood to eager foreign traders.

Following the end of the Indonesian occupation, people were free to return to their original village sites. Many did so, but the distances to markets and poor communication infrastructure meant that farmers found difficulty in getting their harvests to markets, or to where most of the people now lived. This situation, plus a high birth rate of 3.8% or more in some rural areas, has pushed swidden farming closer to settled agriculture. Increasing numbers of rural youth are also seeing greater opportunities for work in towns and cities and this has resulted in increased rural to urban migration.

Clearly, Timorese society is undergoing significant transitions and these are impacting how people live and support themselves. As a result, the government of Timor-Leste is embarking upon a number of programmes aimed at providing agricultural and agribusiness opportunities in rural areas. As noted by the former director of the Timor Coffee Cooperative's agroforestry programme, Shane McCarthy (2015), government agencies are now being encouraged to 'modernize, technify and intensify'.

References

Brady, C. and Timberman, D. G. (2006) *The Crisis in Timor-Leste: Causes, Consequences, and Options for Conflict Management and Mitigation*, report for USAID Timor-Leste, http://www.apcss.org/core/Library/CSS/CCM/Exercise%201/Timor%20Leste/2006%20Crisis/USAID%20Conflict%20Assessment%20Nov%202006.pdf, accessed 2 May 2015

Chao, S. (2009) 'Democratic Republic of Timor-Leste', brief #7 of 8 in S. Chao (ed.) *National Updates on Agribusiness: Large Scale Land Acquisitions and Human Rights in Southeast Asia*, Forest Peoples Programme, Moreton-in-Marsh, England, pp118-137, http://www.forestpeoples.org/sites/fpp/files/publication/2013/08/lsla-studies.pdf, accessed 2 May 2015

da Silva, Estanislau (2003) personal communication with the former Minister for Agriculture, Forestry and Fisheries (2002-2006)

de Jong, W., van Noordwiik, M., Sirait, M. and Liswanti, N. (2001) 'Farming secondary forests in Indonesia', *Journal of Tropical Forest Science* 13(4), pp705-726, http://www.cifor.org/publications/pdf_files/articles/ADeJong0103.pdf, accessed 2 May 2015

eTN Global Travel Industry News (2015) 'Lessons learned from countries that have experienced conflict', http://www.eturbonews.com/55677/lessons-learned-countries-have-experienced-conflict, accessed 1 May 2015

Fox, J. J. (2003) 'Drawing from the past to prepare for the future: Responding to the challenges of food security in East Timor', in H. da Costa, C. Piggin, C. daCruz and J. J. Fox (eds) *Agriculture: New Directions for a New Nation, East Timor (Timor-Leste)*, Australian Centre for International Agricultural Research proceedings No.113, Canberra, pp105-114

Fox, J., Fujita, Y., Ngidang, D., Peluso, N., Potter, L., Sakuntaladewi, N., Sturgeon, J. and Thomas, D. (2009) 'Policies, political-economy, and swidden in Southeast Asia', *Human Ecology* 37(3), pp305-322, http://www.ncbi.nlm.nih.gov/pmc/articles/PMC2709851/pdf/10745_2009_Article_9240.pdf, accessed 2 May 2015

Jeus, M., Henriques, P., Laranjeira, P., Narciso, V. and Carvalho, M. L. S. (2012) *The Impact of Shifting Cultivation in the Forestry Ecosystems of Timor-Leste*, CEFAGE-UE Working Paper 2012/16, Center for Advanced Studies in Management and Economics, University of Evora, Portugal, http://www.cefage.uevora.pt/en/producao_cientifica/working_papers_serie_cefage_ue/the_impact_of_shifting_cultivation_in_the_forestry_ecosystems_of_timor_leste, accessed 2 May 2015

Lao-Swedish Agriculture and Forestry Research Programme (2004) *Poverty Reduction and Shifting Cultivation Stabilization in the Uplands of Lao PDR: Technologies, Approaches and Methods for*

Improving Upland Livelihoods, Proceedings of a workshop held in Luang Prabang, Lao PDR, January 27-30, 2004

Lopes, M. and Nesbitt, H. (2012) 'Improving food security in Timor-Leste with higher yield crop varieties', paper delivered at the 56th annual conference of the Australian Agricultural and Resource Economics Society, Fremantle, Western Australia, February 7-10, 2012, http://ageconsearch.umn. edu/bitstream/125077/2/2012AC%20Lopes%20CP.pdf, accessed 2 May 2015

McCarthy, S. (2015) personal communication with the former director of the Timor Coffee Cooperative's Agroforestry Programme (2004-2012)

McFadyen, R. C. (2003) '*Chromolaena* in Southeast Asia and the Pacific', in H. da Costa, C. Piggin, C. da Cruz and J. J. Fox (eds) *Agriculture: New Directions for a New Nation, East Timor (Timor-Leste)*, Australian Centre for International Agricultural Research proceedings No.113, Canberra, pp130-134

McWilliam, A. (2003) 'New beginnings in East Timorese forest management', *Journal of Southeast Asian Studies* 34(2), pp307-327

Ministry of Economy and Development (2012) *Sustainable Development in Timor-Leste: National Report to the United Nations Conference on Sustainable Development (UNCSD) on the Run-up to Rio+20*, https://sustainabledevelopment.un.org/content/documents/978timor.pdf, accessed 27 July 2015

Molyneux, N. (2011) 'Looking at crop varieties and farming practices in Timor-Leste is essential to food security in the face of climate change', Issues 94, Control Publications, Australia, http://www. issuesmagazine.com.au/article/issue-march-2011/seeds-life-adapting-food-security.html, accessed 2 May 2015

National Directorate of Forestry and Watershed Resources (2004) *Policy and Strategy Report for the Forestry and Watershed Subsector*, National Directorate of Forestry and Watershed Resources, Ministry of Agriculture, Forestry and Fisheries, Dili, http://www.fao.org/forestry/8973-03afe0e434 fc93beed293a3a19ccfedce.pdf, accessed 2 May 2015

Ora, Y. A. N. R. (2012) 'An integrated management and conservation strategy for sandalwood (*Santalum album* L.) in West Timor, Indonesia', in L. Thompson, C. Padolina, R. Sami, V. Prasad and J. Doran (eds) *Sandalwood Resource Development, Research and Trade in the Pacific and Asia Region*, Proceedings of a regional workshop, November 2010, Port Vila, Vanuatu, Secretariat of the Pacific Community, James Cook University and Australian Centre for International Agricultural Research, pp65-89.

Pannell, S. (2011) 'Struggling geographies: Rethinking livelihood and locality in Timor-Leste', in A. McWilliams and E. G. Traube (eds) *Land and Life in Timor-Leste: Ethnographic Essays*, Australian National University Press, Canberra

Ramos-Horta, J. (2013-2014) *A Brief History of Timor-Leste*, http://ramoshorta.com/about-timor-leste/, accessed 2 May 2015

Shum, S., Terrado, E., Openshaw, K. and Tuntivate, V. (2007) *Timor-Leste Issues and Options in the Household Energy Sector: A Scoping Study*, World Bank and Population Division, United Nations Department of Economic and Social Affairs, https://openknowledge.worldbank.org/ handle/10986/7934, accessed 2 May 2015

Timor-Leste National Action Program (2008) *Revised Draft, Timor-Leste National Action Program to Combat Land Degradation*, Dili, http://www.fao.org/fileadmin/templates/cplpunccd/Biblioteca/ bib_TL_/Timor-Leste_NAP_Revised_Draft.pdf, accessed 2 May 2015

B. Removing the 'shifting' from 'shifting cultivation'

This Agta farmer in the Philippines was reduced to permanent cultivation on a small plot of land. It wasn't state policy that robbed him of his swidden land, but something equally stifling – the eruption of Mt. Pinatubo in 1991.

Sketch based on a photo by Malcolm Cairns.

28

EVOLVING SWIDDEN FARMING PATTERNS IN THE LAO PDR

When policy reverses historically mobile ways of life to impose permanently settled livelihoods

*Laurent Chazée**

Introduction

Until 1991, swidden farmers in the Lao PDR could still hope to maintain their indigenous way of life, based on mobility and a livelihood drawn from shifting cultivation in the country's vast forests. The weight of external drivers had been affecting their mobility since 1950, superseding the internal social forces that governed their way of life. But looming in April 1991 was an external event that was going to change their lives. Beyond their knowledge and understanding, the Lao PDR's Fifth Party Congress began the steady implementation and large-scale strategic planning of a rural-development model that aimed to stabilize agriculture by grouping and permanently resettling itinerant farmers. Policies, strategies, tactics and legal tools were introduced to speed up implementation of this programme, especially after 1993. Supporting this drive was international funding available for developing countries under commitments to environmental and forest protection (following Rio de Janeiro, 1992) and poverty reduction (the Millenium Development Goals, 1995-2015). The realities of the implementation and impact of these policies were documented in Sayabury and Oudomxay provinces, where the author worked for six years (Figure 28-1).[1]

Swidden farmers bore the brunt of these policies, not only in having to adopt new agricultural technologies, but also in finding their entire social, customary and religious beliefs and organization overturned. For the first time in their history, on a national scale, they were no longer in the decisional driving seat of the future production systems upon which their lives depended. Some of them were convinced at the outset that the government's direction was the correct one, and they accepted the challenges.

---

* Dr Laurent Chazée is currently working on the protection of Mediterranean wetlands at Tour du Valat in the south of France, and is involved in publication of books and articles on previous research work in Southeast Asia and Africa (see endnote 1).

Others fled into the remotest forest areas. But the largest proportion were reluctantly swept up in the development programmes at high, and sometimes extreme, social costs.

In 2000, it was still unclear whether this politically driven territorial approach had been socially accepted by swidden farmers, given the unbalanced emphasis given to its territorial and technological dimensions and the relative lack of importance attached to the human dimension. However, since then, social and poverty impact studies have delivered compelling and growing evidence to confirm a rapid and unfortunate degradation of human and social values (Chazée, 1994, 1998; Goudineau, 1997a, 1997b; Epprecht, 1998; Keoketsy et al., 2000; ADB, 2001, 2007; Daviau, 2001; UNDP,

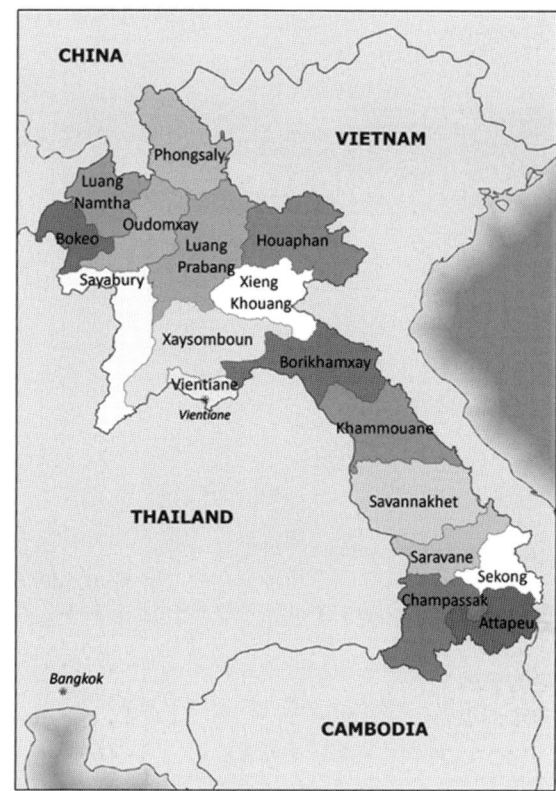

FIGURE 28-1: Lao PDR with provinces mentioned in this chapter.

2001; Moizo, 2004; Evrard, 2004; Baird and Shoemaker, 2005; Chazée and Gehin, 2012, 2017).

Rural development policies and programmes from 1975 to 2000, with a focus on shifting cultivation and swidden communities

Between 1975 and 1990, the priority agenda of the communist party aimed to ensure national security and rice self-sufficiency, as well as 'political consolidation', from central to village levels. In 1986, the country launched the new economic mechanisms. After 1989, the collapse of the Soviet Union and reopening of borders with China and Thailand coincided with widening regional and international cooperation and private-sector development in the Lao PDR, while it maintained a communist political line supported by Vietnam and China.

Up to 1990, the communist party's agenda for rural development was implemented by the government through a series of strategies, tactics and directives, usually influenced by Vietnamese advisers. In rural areas, national security was handled by establishing Vietnamese and Lao camps, and by controlling and resettling swidden

communities in lowland and accessible areas, especially those communities that had been outside the Pathet Lao area of influence during the second Indochina war (the Vietnam war). The political consolidation procedure involved regular central, provincial and district meetings and training sessions. To achieve national self-sufficiency in rice, a collectivist-production or cooperative system was tested in lowland areas between 1976 and 1980. When this failed, all cooperatives were dismantled by 1989. At the same time, the government, assisted by Vietnam, developed irrigated lowland cropping as an incentive for increased production and permanent, settled agriculture as an alternative to swidden farming, which was actively discouraged (Figure 28-2).

FIGURE 28-2: Khmu Khrong swidden farmers from Nahome and Sawang villages were encouraged by local authorities to grow lowland paddy rice, seen here at harvest time. Since the paddy fields were far from their villages, they decided to build granaries on the hillside behind them, to avoid transport problems.

Photo: Laurent Chazée.

Between 1980 and 1995, each district and province was supposed to plan its own rice development programme and report the results to feed national statistics, taking into account the aim of national self-sufficiency and the restrictive policy against swidden farming and cropping. In responding to these central objectives, provincial authorities tended to prepare statistical data based on plans, rather than on field realities, especially during the periods from 1980 to 1987 and 1990 to 1995. The statistical cooking exercise showed rice production, and hence national rice security, increasing from year to year, while the swidden farming area decreased (Table 28-1). In contrast, Forest Department aerial pictures taken from 1982 to 1989 showed an increase in the country's area under swidden.

TABLE 28-1: Evolution of paddy area, yield and production in Lao PDR between 1976 and 1995.

Production data	1976	1980	1985	1986	1987	1988	1989	1990	1991	1992	1993	1994	1995
Total area of paddy (1000ha)	521.8	724.3	655.0	641.6	556.4	544.8	596.2	656.7	556.9	592.5	538.7	600.0	544.3
Swidden area (1000ha)	204.1	297.4	270.4	207.3	256.6	313.5	214.7	260.2	237.1	200.1	188.3	219.1	164.0
Yield (T/ha)	0.99	1.13	1.23	1.33	1.31	1.32	1.55	1.46	1.44	1.47	1.50	1.60	1.54
Production (1000T)	202.0	337.0	345.3	341.0	272.3	282.8	332.9	380.8	337.5	293.6	283.6	341.6	253.1
Permanent rice fields	317.6	426.9	383.1	385.0	349.1	331.3	381.5	396.5	322.8	392.5	350.1	380.8	361.8
Wet season (1000ha)	1.43	1.65	2.11	2.81	2.63	2.07	2.72	2.75	2.61	2.94	2.60	3.10	3.10
Yield (T/ha)	455	705	1023	1082	917	686	1039	1088	842	1153	921	1198	1122
Production (1000T)	2.7	7.7	10.0	10.1	9.7	11.4	10.2	11.4	13.3	15.6	13.0	11.0	18.5
Yield (T/ha)	1.27	1.44	2.31	2.70	2.69	3.02	3.13	3.43	3.28	3.56	3.50	3.40	4.11
Production (1000T)	3.5	11.1	26.5	27.3	26.1	34.5	32.0	39.1	43.7	55.3	45.6	37.8	48
Total national production	661	1053	1395	1450	1216	1003	1404	1508	1223	1502	1250	1577	1423

Notes: Statistical data were incomplete for 1976 and 1980; Rain deficit in 1991 and 1993 led to decrease in national rice production. In 1988, official swidden area rose nearer to reality, in line with better control of statistics.

Source: National Statistics Center (1976–1992, 1993–1995).

In this difficult equilibrium, provincial departments of agriculture were obliged to artificially inflate the yield of cultivated swidden areas and increase the surface area and rice yields in lowland areas, even if lowland areas did not exist. As a consequence, national statistics on rice production in the Lao PDR are far from reliable.

The policies against swidden farming were sometimes translated into simple internal resettlement of swidden farming communities along roads, more as a way of controlling them than developing them. In this period, Laos had no private sector, the prices for several agricultural commodities were controlled by the state, education and health services were very poor and there was a limited budget for development of agriculture and irrigation, so most of the displaced communities did not expect any social and economic advantages from their new mobility. On the contrary, they soon faced serious health problems and food insecurity. Additionally, the psychological shock of resettlement and their vulnerability were enhanced by insecurity related to land property and a feeling of having lost their identity and cultural values. Traditional beliefs led them to fear a negative response from their spiritual guardians to the resettlement and its associated loss of agricultural traditions and customs; to the diminution of their traditional social organization, which was superseded by a new village administration system; and to the loss of their traditional rice varieties because of their failure to adapt to the new conditions. Austro-Asiatic speakers such as the Khmu, Kouene, Bid, Lamet, Nguan, Alak, Souay, Nyaheune and Lovene, who were initially more confident about the government decisions, or at least flexible in their regard for them, ended up being the worst affected. Among Miao-Yao speakers, only the Lantene groups of Luang Namtha accepted the challenge, and suffered high mortality rates as a consequence. The authorities found it more difficult to convince Tibeto-Burman speakers and Hmong to accept resettlement without offers of real compensation.

A series of schemes called 'integrated rural development projects' were launched between 1980 and 1989, with the intention of speeding up the implementation of government plans to resettle swidden-farming communities. The participation of international organizations was welcomed in the form of funding and technical assistance. But international experts were unable to make field surveys, so the projects were poorly planned, usually ill-conceived and then poorly implemented, with limited impact and sometimes high social costs and rates of mortality (Baird and Shoemaker, 2005). Mountain communities were relocated to lowland areas, resulting in several cases of high mortality due to malaria, dysentery and other health problems. These 'top-down' projects were first and foremost designed to serve political goals. Rural communities were regarded as 'beneficiaries' of development initiatives, most of which were conceived in the capital, Vientiane. These projects did not offer many viable options to the swidden farmers, except by chance. They demobilized them, in their already vulnerable condition, with non-productive and community work that they were asked to perform on a 'voluntary' basis. In several cases, their unpaid participation in the construction of irrigation canals, roads, bridges, buildings for rice mills and other jobs requiring their labour was perceived by the swidden farmers more as a time-consuming constraint than as a well-understood and mandatory step towards improving their living conditions. They knew that the infrastructure they were building would belong to the province or district; that they would be called upon annually to maintain it; and, because the price of rice was state-controlled and access to free markets was not possible, they would receive no financial return for their labour. Among these projects, in which swidden farmers were occupied until 1990, the author can recall the Muang Home project in Vientiane and the Dakchung (Xekong) project, both financed by the United Nations Development Programme.

The 1990-2000 period was marked by a series of policy and strategic actions aimed at consolidating permanent agriculture and protecting the natural environment, especially the forest.[2] The rural-development model was a communist-party-driven sector approach based on permanent agriculture and resettlement. It mostly targeted swidden farmers. In continuity of actions already begun in the 1970s, it applied a steady and large-scale strategic leverage to settle and control them. The programme implied a rapid and permanent sedentarization of mountain communities and was institutionalized by an integrated and mutually reinforcing set of policies, laws, decrees, directives and strategies. All of these actions started, or were tested, in early 1990. Implementation increased in 1997, incorporating foreign funding under the umbrella of international commitments to poverty reduction and environmental protection.

In this rural development policy approach, there was no mention of vital issues such as real participatory processes, cultural values, the potential for eco- and ethno-tourism, revival of environmentally friendly indigenous practices, social feasibility and the cost of changes, the value chain of nurturing, valuing, processing and marketing non-timber forest products or the production of handicrafts at community level. The key starting period for implementation of this policy in all parts of Laos was April 1991, with strategic directives given by the Politic Bureau during the Fifth Party Congress and reaffirmed by the Sixth Party Congress in March 1996.[3]

These policies were formalized by a series of instruments such as the Forest Law (No, 96/NA11); the land decree (No. 99, December 1992); and the Land Law (33/PO, May 1997), which were aimed at allocating family plots in mountainous rural areas and attempting to reduce the length of the fallow cycle to three to four years. The village decree (102 PM, July 1993) announced criteria for village consolidation (*chat san ban khongthi*) and encouraged small communities of fewer than 20 nuclear households or 100 persons – mostly swidden farming communities – to join other villages.[4] Several other plans and directives aimed to eradicate shifting cultivation by 2000 (Boupha, 1995).

In 1992, the outcomes of the Rio de Janeiro Summit somehow validated the forest protection policies of the Lao PDR, including the Tropical Forest Action Plan approved in 1991 and the subsequent Forest Law, as well as actions already taken in favour of preserving forest ecosystems. However, in the Lao context, a sectoral approach was being made to forest protection, at least within the territorial boundaries of swidden communities, without much consideration for soil protection, soil management or the livelihoods of the people. Between 1990 and 1995, with the objective of protecting the forest, the narrow sectoral approach adopted in policies and strategies was ignoring well-established foundations of forest sustainability. The intention was to reduce the fallow phase of the shifting cultivation cycle to three years, when agronomists were advising that eight years was a minimum fallow period if sufficient plant biomass and diversity was to be maintained, the soil restored and both a favourable topsoil structure and content of organic matter retained. By reducing the fallow to three to four years, soil erosion began to take a heavy toll on land quality and there was a rapid loss of both fertility and biodiversity. This degradation was already visible in some places in 1995, such as around Luang Prabang town and along main national roads where intensive resettlement of swidden farmers was taking place. The immediate consequences were a fall in swidden rice yields and a rapid increase in annual weed growth across entire farms (Chazée, 1998).

When lost swidden production was not compensated by more beneficial alternatives, both the bottleneck of unavailable labour for weeding and decreasing yields from swiddens created a vicious spiral of food insecurity and household impoverishment (Chazée, 1998).[5] In some cases, 'settled' swidden farmers – for their sheer survival – had no choice but to return to their mountain origins and 'illegally' slash the forest to begin shifting cultivation again (Keoketsy et al., 2000).[6] Shifting cultivators thus lost out on several fronts: fear and psychological shock due to the 'illegality' of what they were doing; a decreased capacity to produce sufficient yields and ensure their food security because of the shortened fallow cycle; and declining labour productivity due to the time needed to travel to and from their remote 'illegal' swiddens, all the while watching their land degrade under the imposed short-cycle fallow and recognizing the probability that their traditional production systems were slipping beyond recovery. At the same time, in an apparent contradiction of the declared intention to protect the forest, large-scale commercial logging began in 1980 at the hands of companies managed by the army, sometimes within the traditional territories of swidden communities. Members of mountain communities were employed to locate stands of valuable commercial trees and non-timber forest products in the Nam Phoui area of Khamouane province, and in Hongsa district, Sayabury province, the domestic elephants of mountain communities were used to haul felled trees from the forest. The operations of the commercial logging companies were particularly extensive in Khamouane, Bolikhamsay and Sayabury provinces. In 1989, despite growing evidence of the social and food-security costs of its anti-swidden-farming policy, as presented to the country's first National Conference on Forestry (sponsored by the World Bank), the government maintained its goal of eliminating swidden agriculture, and maintains it to the present day.

The land decree, land laws and the village decree together were particularly efficient instruments for settling the location of residence and agricultural practices of swidden farmers. After a few pilot projects in 1990 and 1992, the land decree allowed a start to the formalization of household agricultural land and forest allocation (*beng din beng pa*). But there first had to be a 'test period'. This began in Saravane, Sayabury and Luang Prabang provinces in 1993, followed in 1994 by Xieng Khouang, Luang

Namtha, Vientiane and Khamouane provinces. Extension to all provinces began in 1996 and was ongoing in 2005. This programme was supported mostly by the Swedish International Development Agency and the Asian Development Bank. It was aimed mostly at swidden farmers, as 82% of land allocated between 1995 and 2002 was classified as forest land. When it began in 1993, the land decree was used in mountainous areas as a form of leverage to decrease the traditional fallow cycle of 8 to 15 years to 3 to 4 years, and to stop pioneering swidden farming. The process of allocation was based on a standard methodology. In rural areas, the process involved a land-use plan covering the territory of each village and the issue of temporary land-use certificates to all households. According to the Reduction of Shifting Cultivation Extension Centre, by 2004 about 50% of all the households in the country – 350,000 households in more than 5300 villages (40% of all villages) – had been given temporary land-use certificates, but none of these had been converted to permanent ownership (Evrard, 2004). According to a survey on poverty commissioned by the Asian Development Bank, rural people pointed to land allocation as the first cause of impoverishment in the country's northern, eastern and central regions (ADB, 2001). This was confirmed by several other studies (Chazée, 1998; Daviau, 2001; Ducourtieux et al., 2004; Evrard, 2004; Baird and Shoemaker, 2005), which in turn supported earlier observations and studies in the period from 1989 to 1992, which found that swidden farmers were being resettled without enough agricultural land and without sufficient means to intensify or diversify their livelihoods (Chazée, 1998). In this predicament, swidden households were forced to rely on a set of alternatives: practise 'illegal' shifting cultivation in remote and invisible parts of their territory; enter renting or share-cropping arrangements with absentee or retired land owners; or buy or sell land, or lease land.

Between 1993 and 1995, spreading awareness of the measures introduced by the village decree provoked real panic in small swidden communities of fewer than 20 households. Very small minority groups such as the Doy, Kongsat, Bit, Poumone and Poussang in northern provinces and the Atel, Arao, Malang and Phong in central provinces were among the most affected psychologically. On top of being targeted by the decree as small communities that should move to join other villages, they were afraid for their status as minority ethnic groups. Such minority groups, in a way marginalized from the evolving national development process (purposely or not), preferred to remain in single ethnic settlements. Indeed, they felt different from other groups because of their linguistic, religious, customary and cultural features, and preferred not to risk integration for fear of acculturation. The immediate field actions undertaken by local authorities under the village decree obliged swidden households to make quick decisions. Their options ranged from accepting the deal to fleeing into remote and inaccessible areas. Most tried to find their own compromise before local authorities found a solution for them. In 1993 and 1994, I witnessed spontaneous grouping arrangements in several mountain areas of Luang Namtha and Oudomxay provinces, in which people used horses, trucks and even loaded their personal belongings on to their own backs to escape the strictures of the land decree.

They hauled their religious and ritual instruments and their village and household spiritual pillars along with them.

Between 1992 and 1993, provincial and district authorities communicated the measures of the land decree widely to villages and the people. For the first time in their lives, swidden communities realized that the principles of their ancestral agricultural systems were soon to end. The authorities' deadline for eradication of shifting cultivation by 2000 was, at best, unrealistic, and the campaign was still ongoing in 2016. But the strong mobility of that period was not provoked by the so-called arguments of local authorities in favour of permanent agriculture. Rather, it was driven by psychological shock and fear of the future. A rash of migration affected every part of the country, but with greater intensity in the northern provinces, which had the highest density of swidden farming villages. Minority Tai, Austro-Asiatic, Miao–Yao and Tibeto–Burman communities began a race to reach the last flat agricultural territories in 1991. It was a race that was sometimes out of control between 1991 and 1993, as minority groups attempted to move from provinces perceived as being the most authoritarian, such as Xieng Khouang and Houaphan, to more flexible and pragmatic provinces such as Oudomxay and Luang Namtha. Ethnic networks and solidarity movements across provinces were revived to help and inform newcomers, both to meet the village decree criteria concerning the grouping of communities and to find land that was suitable for permanent agriculture. Based on studies among migrating swidden-farming groups in Oudomxay, Luang Namtha and Phongsaly provinces from 1991 to 1993, their main reason for moving was an eagerness to find proper and acceptable land for permanent agriculture before it was too late. However, some small and remote communities tried to escape into even more remote places, while other, larger and better organized villages of Hmong (Nahome area, Oudomxay) and Akha communities (Ano village, Oudomxay) settled into very determined negotiations before moving (Figure 28-3). These migrations continued until at least 2000, but with more control.

In March 1994, a party resolution on rural development established a new strategy based on 'focal sites'(*khetchoutxoum*) (Cabinet of Party Central Committee, 1994). The new approach, which was influenced by Vietnam, had been under discussion in the political arena since 1991. It was based on several lessons, including an evaluation by the communist party and the government of the past poor efficiency of national and international assistance in developing rural areas. A 'focal site' was an area where rural development actions were intensified and integrated with development of access and a comprehensive set of social and economic infrastructures, services and training. The approach suggested that settlement at the site was permanent and that a grouping of villages and their agricultural systems would benefit from development. The focal sites would also have a pilot mandate to test new rural development approaches, including introduction of land reforms. The success of the new approach depended on the sufficiency of national funding for its implementation, and then the capacity of the government to attract further funding from international organizations. To this end, the focal site approach quickly incorporated linkages with national plans for rural development, human development and poverty reduction, all of which were

FIGURE 28-3: Maintaining cultures and traditions is strongly anchored in the livelihood patterns of Hmong communities and clans. Here, a young woman in traditional costume (left) plays a ball game aimed at identifying a future partner during New Year celebrations at Nahome village in 1995, and another youngster displays her traditional headgear.

Photos: Laurent Chazée.

elaborated from the government's socio-economic development plan. A leading Central Committee for Rural Development was established by decree in November 1994, followed by Provincial Rural Development Committees with a mandate and authority to coordinate work at focal sites only (Central Rural Development Committee, 1995). In 1995, provinces and districts were asked to evaluate their land resources in order to develop areas for permanent agriculture, after which they would receive swidden farmers from mountain areas that were unsuitable for permanent agriculture. They were then told to identify a set of appropriate focal sites for priority rural development, based on 'official' criteria, including poor access, the production system, economic poverty, development potential and security. In reality, sites in Oudomxay province that were selected as focal sites were clearly remote areas inhabited by communities of shifting cultivators producing opium. In Luang Namtha, Xieng Khouang, and the northern parts of Sayabury provinces, 'unwritten' selection criteria were used that were related to national security, continuous ethnic control and dismantling of historical opposition networks. In August 1996, a Rural Development Committee was set up under the Prime Minister's Office to speed-up and centralize the decision-making and feedback processes at a high level. Its role and mechanisms for coordination with Provincial Rural Development Committees, chaired by provincial governors, were laid out in detail, to enable rapid communications for immediate implementation (Central Rural Development Committee, 1997).

The implementation of the new rural development programme began in 1996, in 58 focal sites. There were 62 focal sites in 1997 and 87 were foreseen in 2002, to include 1200 villages and 450,000 people, or about 12% of the rural population. It was estimated that about half of the population of the focal sites was made up of ethnic groups displaced from mountain villages (Goudineau, 1997a). However, until the end of 1997, the implementation rate was poor due to limited funding. Between 1996 and 2000, the national budget for rural development was 10 times higher than that in the previous five years and considerable efforts were devoted to increasing this budget through bilateral and international funding agencies and projects. Among these were United Nations agencies such as the United Nations Development Programme, the United Nations Capital Development Fund, the United Nations International Drug Control Programme and the World Food Programme, the European Commission, the World Bank, the Asian Development Bank, the International Fund for Agricultural Development, bilateral agencies from the United States, Germany, France and Luxembourg, and international non-governmental organizations such as World Vision, Quakers Service Lao, Care, Action Contre la Faim, and others. As a consequence, of the 154 million kip (about US$36,636) budgeted in 1998 for the focal site development programme, 83% was from international funding (Government of Lao PDR, 1998 – informal translation of official document).

Most of the budget was channelled to a few priority focal sites – the most sensitive ones, usually involving security and opium issues – to improve access roads, build local rural development centres and begin political training for district and village representatives. The National Rural Development Plan (1998) had foreseen the availability of external funding to speed-up implementation of proposals for infrastructure (41% of the budget); agriculture (28%); education (9%); health (6%); programme management (6%); relocation of villages (6%); income-generating activities (3%); and community development (1%). The author was based in the Provincial Rural Development Office of Oudomxay from 1997 to 1999 for the Strengthening Economic and Social Management Capacity (SESMAC) and Beng Alternative projects, and was able to witness the differing official priorities given to government focal-site governance, on one hand, and internal party-line governance on the other.[7] Within the blurred lines of authority (province and district), priority was clearly given to party-line. Staff of the SESMAC project were expected to follow the government line, in implementing the party line. With differing priorities between party-line (security, relocation and food security) and government-line (socio-economic development), internationally funded projects were backing, either directly or indirectly, cases of involuntary displacement, inequitable access to land, use of improved access roads for commercial logging by the army, political training, and so on. Several other internationally funded projects also faced these dual lines, while usually being aware only of the government line. These included projects funded by the European Union in Luang Prabang and Luang Namtha, a CARE project in Luang Prabang, a CUSO-managed project in Saravane province, and a UNESCO/

New Zealand ecotourism project in Oudomxay province. The degree of party-line enforcement depended on the attitude of local authorities, including their ranking in the party. The governors of Phongsaly, Xieng Khouang and Houaphan were the most active in this regard in the years from 1976 to 1990. In Oudomxay, after a relatively flexible period, a new governor introduced a very tough interpretation of the focal site approach from 1998 to 2000.

Because of this segmentation of directives, it took international organizations some time to realize the dual purpose of the focal site rural-development programme. Because of globalization and commitments to international conventions and agreements that had developed in the decade before 2000, it was not easy for international organizations to act outside the focal sites.[8] However, from 2000 there were increasing reports and growing awareness of this governance duality, and international organizations were able to be more strategic in their development aid. It was nevertheless clear that during the decade from 1990 to 2000, international funding for rural development offered either direct or indirect support for the resettlement of swidden farmers against their wishes. In this period, aid agencies adopted four different attitudes to rural development: some, such as the Asian Development Bank, the Food and Agriculture Organization of the United Nations and the United Nations Children's Fund (UNICEF) provided active and uncritical support for resettlement initiatives. Others, such as World Vision and CUSO International failed to critically analyse the rural situation and maintained an attitude of ignorance, lack of interest and denial. Some organizations, such as Action contre la Faim and the Japan International Cooperation Agency, provided conditional support to resettlement, while a few others, such as the international development agencies of Sweden and Canada, were involved in active resistance to resettlement (Baird and Shoemaker, 2005).

Implementation of policies opposed to swidden farming in focal sites in Sayabury and Oudomxay provinces

Rural development policy in Sayabury focal sites, 1994 to 2000

Following the creation of a Provincial Rural Development Committee and opening of an office in Sayabury province in 1995, 57 rural-development cluster areas were identified as 'focal zones'.[9] The first criterion for the selection of focal sites was that based on national security. Others were remoteness, swidden farming, geomorphologic conditions for agricultural development, proximity between villages to reduce operational costs, development potential for farm and non-farm activities and the issue of opium production (Chazée, 2000).

After several modifications, the number and size of focal sites with development plans covering the years from 1998 to 2003 were scaled down to 14 zones, covering 130 villages, 9269 families and 46,727 villagers (Table 28-2). The programme was supposed to target minority groups in the following percentages: Tai speakers, 58%; Austro-Asiatic speakers, 35%; and Miao-Yao speakers, 7%.

TABLE 28-2: Priority province- and district-managed focal sites in Sayabury province, 1998 to 2003.

Focal site	District	Number of villages and ethnic groups
	Provincially managed focal sites	
1 Muangpa-Nachane	Paklay	6 (Lao Soung, 1; Lao Loum, 5)
2 Phoulane	Sienghone	11 (resettlement of Hmong, Khmu and Plrai communities)
3 Samat-Saysana		26 (Lao Theung)
4 Ban May	Tongmixay	4
	District-managed focal sites	
1 Nalouam	Sayabury	5
2 Naven	Phieng	3 (Lao Loum, 1; Lao Theung, 1; Lao Soung, 1)
3 Bouamlao-Phakeo	Paklay	13 (Lao Loum)
4 Donemen	Kenthao	10 (Lao Loum)
5 Nabornoy	Bortane	6 (Lao Loum)
6 Khokat	Hongsa	8
7 Banethong	Ngern	5 (Lao Theung)
8 Parkpeth	Sienghone	11
9 Panghai	Khorb	6
10 Barngnai	Tongmixay	16

Note: Lao villagers are commonly categorized according to the elevation at which they prefer to live, as Lao Loum (lowland Lao), Lao Theung (midland Lao) and Lao Soung (upland Lao).

Source: Office of the Governor, Sayabury (1998).

Four of the 14 zones – Muangpa-Nachane, Phoulane, Samat-Saysana and Ban May – were given priority and were directly supervised by the provincial authorities. The remaining 10 (Nalouam, Naven, Bouamlao-Phakeo, Donemen, Nabornoy, Khokat, Banethong, Parkpeth, Panghai and Barngnai) were placed under the supervision of district authorities.

Based on implementation of the government's eight national priorities,[10] the provincial authority of Sayabury committed and mobilized human resources and allocated funds from internal and external sources for rural development activities in these focal sites.[11] Actions aimed at the development of ethnic groups included investment in irrigation systems, expansion of ricefields, transportation, allocation of land for farming and management by people focused on settled farming, expansion of dispensaries and schools and water supplies. The largest investments were made in the focal site at Naven, as it had been a pilot project since 1995 and it held a special attraction for foreign assistance because of its geographic location in the Nam Phoui National Biodiversity Conservation Area. However, except for an access road, 'political consolidation' and stabilization of the security situation, there was little successful implementation of sustainable socio-economic activities and the management of the Naven focal site was transferred from the province to Phieng district in April 1998.

The building of access roads, construction of offices, land allocation, land titling and 'political consolidation' were the main activities in other provincial- or district-managed focal sites between 1995 and 1999. Under the food-security programme, the main focus was on new irrigated systems, rearing livestock and new tree plantations. Rice was distributed as food aid in the northern districts, where most resettlement took place, assisted by the European Union and the Italian non-governmental organization CESVI. The programme aimed to eradicate swidden by focusing on support for irrigated agriculture and the allocation and conversion to agriculture of forest land. Early in 1998, this programme had covered 443,905 hectares and involved 455 villages (81% of all rural villages) and 44,326 households. The public infrastructure programme focused on building unpaved roads (175km built between 1994 and 1998), education (27 new schools constructed in focal sites between 1994 and 1998), and health (four new dispensaries built, 29 medical volunteers trained, 37 open wells and one gravity water supply built and five deep wells sunk between 1994 and 1998). Income-generating activities included rice banks, mulberry plantations, silkworm rearing, commercial cropping, weaving and sewing programmes.

All of these activities, in almost all sectors, were largely funded by foreign aid. In 1998, the budget for provincial rural-development planning in focal sites until 2002 was estimated at 46 billion kip (US$10.9 million), with six billion kip coming from the government in Vientiane and 40 billion from external sources. The largest items on the budget were roads, food security, public infrastructure and human development. Between 1991 and 2004, external funds received or committed to the provinces of the Lao PDR amounted to US$54.3 million.

The main donors were:

- In the agriculture and forest sectors: UN Capital Development Fund/UN Development Programme/Dutch bilateral aid, for the rehabilitation of the Nam Than irrigation scheme (1992-1998); Japanese cooperation (1993); Swedish International Development Agency for land allocation planning and conservation of National Biodiversity Conservation Areas (1992-2000); Lao/ International Rice Research Institute for rice experimentation (1996-1997); the international Catholic non-governmental organization CIDSE (1996-1997), AusAid (1995-1997), US non-governmental organization CARE (1996-2002); and UN Development Programme (1997-1999).

- In the industrial sector: Japanese cooperation (1996-1997), Thai cooperation (1996-1997); Asian Development Bank (1997-1998), German cooperation (1992-1994).

- In the communication, transport, post and construction sectors: UN Development Programme/UN Capital Development Fund for road construction between Sayabury, Nam Phoui, Paklay and Kenthao (1992-2000); AusAid (1992-1994);

French cooperation (1995-1999); Lic Nai private Thai Company (1992-1994); International Development Association (World Bank) (1996-1997); UN Development Programme/North American Aerospace Defence Command (1997-1999); Asian Development Bank (1994-1998); German cooperation (1996-1997).

- Education sector: UN Educational, Scientific and Cultural Organization (1994-1998); Lao Red Cross (1996-1997); Japan cooperation (1995-1997); World Bank (1991-1999); European Union education programme (1997-1999).

- Health sector: UN Children's Fund (UNICEF), World Health Organization (WHO) and Japan International Cooperation Agency for vaccination programme (1993-1998); UNICEF for mother and child health care (1993-1998); UNICEF, Save the Children Fund - Australia, CARE for water supply and environment (1992-1998); UNICEF, WHO for disease prevention (1994-1998); Save the Children Fund - Australia and AusAid for health care (1992-1998); Canada, Thailand, Germany, UN Development Programme, UN High Commission for Refugees, Belgian aid, CARE and European Union for funding of several health-related programmes (1993-2000).

- Socio-economic sector: Strengthening Economic and Social Management Capacity Project (UN Development Programme) (1997-2001); income generating activities and UN Development Programme project (1997-2001); UN Capital Development Fund/UN Development Programme microfinance project (1997-2002); International Fund for Agricultural Development/ UN Development Programme northern district development project (1997-2004); Save the Children Fund - Australia, North American Aerospace Defense Command, Swedish International Development Agency, Japan, France, UN High Commission for Refugees, Lao Red Cross, AusAid and the World Food Programme for other related socio-economic development activities.

In the focal sites of Sayabury province, development was less spectacular than the donor list may suggest. At the Muangpa-Nachane focal site, no funding was available until the end of 1998, when the Strengthening Economic and Social Management Capacity Project was able to conduct a participatory rural appraisal and, using a step-by-step approach, define a survey and implementation methodology. At Phoulane, no direct funding was available until 1999, but the area benefited from an access road build between Sienghone and Korb districts. Several resettlements that began in 1997 affected Hmong, Khmu and Plrai swidden communities. At the Ban May focal site, the army was present between 1989 and 1999, but there were no real development activities (Chazée, 2000).

The impact of the focal site approach in 2000

Due to limited budgets and human resources, only five of Sayabury's 14 focal sites received assistance between 1994 and 1998. Most of the activities were concentrated in a single site, at Naven. As well as these limitations, other factors played against the efficiency of focal site implementation. They included a lack of real planning for preparation, implementation and monitoring of the process; poor horizontal and vertical integration of actions with sector institutions; a lack of transparency and openness and a consequent lack of confidence in the process among rural development institutions, sector ministries and donor communities; a relatively 'top-down' approach to the sector development process that created ambiguity and diverging reactions, including fear, fleeing from the focal sites, moving in to the focal sites in hope of benefits, passive participation, and so on. During the period from 1994 to 2000, those of us working in the area were able to witness three main effects of the focal site programme:

1. When a 'critical mass' of development activities took place – as they did in the Naven focal site – rural communities began to perceive the social and economic benefits and external communities were attracted to the focal site as a kind of 'voluntary resettlement'. When several development programmes were launched at Naven in 1996, there was an uncontrolled influx of migrants from outside communities. The sudden demographic surge inflamed social and land conflicts and heightened competition for use of natural resources. Farmers illegally encroached upon the National Biodiversity Conservation Area and the environment suffered from survival mechanisms such as illegal logging and overuse of land. In February 1999 the provincial authorities were forced to impose population controls.
2. When a focal site required the resettlement of swidden communities alongside roads and promises made about development projects were not kept, or were substantially delayed – as happened in Phoulane – the people lost confidence in the project, health problems increased, there was insufficient food, limited land was overused and families became impoverished.
3. When no public development took place at a focal site and there was no resettlement involved, the people usually continued with their traditional life-style and agricultural systems. But this was 'illegal', and from this grew fear of increased controls, the presence of the army, and 'political education'.

Organization and impact of rural development policy in Oudomxay focal sites between 1994 and 2000

The way in which national rural-development policies and related strategies were implemented in Oudomxay province must be understood within the context of the province's recent history. Oudomxay was heavily bombed during the second Indochina War (the Vietnam War). Its geographical position between China, Vietnam and Myanmar led to Oudomxay being settled or transited by several migrating ethnic groups. During the first half of the 1990s the province was managed with a pragmatic approach and a sensibility towards swidden farming by a Hmong governor and a Khmu vice governor. However, the implementation of policy became much more authoritarian when a new governor took over in February 1998.

In 1990–2000, vital lessons had already been learned from resettlement of swidden communities in Oudomxay province. Generally, there had been too little preparation for resettlement and the families involved given too little assistance. This had led to:

- Food insufficiency and impoverishment for the first five years after resettlement because new land had to be cleared and prepared for cropping. The province attempted to help with rice, seeds and fertilizer, but it had only a limited budget.
- Increased mortality among former mountain-dwellers because of a higher incidence of malaria and dysentery, compounded by poor health facilities and little assistance.
- Loss of traditional rice varieties that had been maintained by farming families for many generations. These were replaced by new varieties that were more sensitive to pests and, according to resettled families, much less tasty.
- Assistance promised for irrigated agriculture took too long to arrive, usually because projects were internationally funded. Aid funds came from Vietnamese assistance between 1980 and 1987; Quaker Service Lao between 1982 and 1996; the UN Development Programme/UN Capital Development Fund for a small-scale irrigation project (1991–1995); and the UN Development Programme for a rural development project (1997–2001). As a consequence, some communities that were facing serious food insecurity packed up and returned to their former mountain villages.

After the decision of the Fifth Party Congress in 1991 and implementation of the new strategies and instruments related to villages and settlement of agriculture, the number of villages in rural areas of Oudomxay province decreased significantly, from 1200 in 1990 to 813 in 1996. Small swidden communities had been encouraged to join together as one, or merge into larger communities. One of the options preferred by local authorities was to resettle them along provincial roads built by the Chinese between 1963 and 1975, especially the roads from Xay to Pakbeng (Khmu and Hmong swidden farmers), from Xay to La (Khmu and Akha swidden groups), from Xay to Namo (Khmu, Hmong, Akha and Poussang villagers) and from Xay to Huay Nambak (Hmong and Khmu). Resettlement of swidden communities along these roads took place between 1968 and 1980, especially among Khmu and Hmong groups. For local authorities, this was a preferred option because it not only allowed control of the population, but was also easy and cheap because no new access roads were needed, social, economic and administrative services had easy access to the former shifting cultivators and there was sometimes the potential for permanent agriculture on lower land and in valleys. In 1992, following the Food and Agriculture Organization's Tropical Forest Action Plan and the Convention on Biological Diversity and Sustainable Development, arising from the Rio de Janeiro Summit, a reduction in both swidden farming and opium poppy production were expected, as positive outcomes, from forest regeneration and environmental protection in general. When the Lao PDR's National Poverty Reduction Programme began in

1996 with the establishment of national and provincial committees, Oudomxay province faced a more complex challenge: how to combine environmental protection and the generation of sustainable livelihoods with its existing swidden resettlement programme. Efforts to construct irrigation schemes increased and microfinance projects were launched to support income-generating activities. Most of them were, once again, foreign funded. At the same time, the new strategy to concentrate rural development in focal zones became operational.

In 1996, 10 priority focal zones were defined in Oudomxay province, including three priority zones managed directly by the province and seven others managed by districts[12] (Table 28-3).

Despite the naming of 10 focal sites in Oudomxay, most development actions took place only in the three provincially managed sites, as shown in Table 28-4.

In 2000, the provincially managed focal-site programme had achieved relatively good growth in the area of irrigated rice, access to education and some diversification of agriculture. Yet identification of alternative sources of income to opium remained inconclusive; rearing livestock, vegetable production and fruit-tree orchards were all being trialled, but it was too soon to conclude which of them would have the most beneficial impact on the sustainability of swidden farmers' livelihoods. In 1998, Oudomxay province saw the rise of a new challenge: the rapid expansion of commercial agriculture driven by Chinese demand, including teak and rubber plantations and commercial production of watermelons, sugar cane and other agricultural crops. Along with this trend came chemical fertilizers, pesticides and leasing of land. Given the strength of this agricultural market, it became doubtful how long the initial 'political' model adopted in focal sites would last.

TABLE 28-3: Priority province- and district-managed focal sites in Oudomxay province.

Focal site	District	Number of villages and ethnic groups
Provincially managed focal sites		
1 Namkha	Houn	14 (Lao Theung, 7; Lao Soung, 5; Lao Loum, 2).
2 Nahome	Beng	12 villages, 637 households (6 villages Khmu Khrong; 6 villages Hmong). Almost all swidden farmers.
3 Houanambak	Xay	9 villages, 110 households (Hmong Khao and Khmu Ou groups), all swidden farmers. Five villages displaced.
District-managed focal sites		
1 Namheng	Xay	12
2 Sounvang-Aang	Namo	22
3 Thatmouane	La	9
4 Khone	Beng	15
5 Namphoun	Houn	15
6 Xaysana	Pakbeng	6
7 Houayngouam	Nga	13

Source: Office of the Governor, Oudomxay (1998).

TABLE 28-4: Development actions in the three provincially managed focal sites.

Activities	Focal sites		
	Namkha	Nahome	Houanambak
Agricultural extension	√	√	√
Market construction		√	
Rice bank	√	√	
Dispensary	√	√	√
Primary and medium level schools	√	√	√
Training in hygiene	√	√	√
Institutional capacity building	√	√	√
Irrigation management committee	√	√	Planned
Opium alternative-crop committee	√	√	

Source: Siksidao, 2001.

The case of Nahome focal site

The Nahome area was declared a priority focal site for rural development in 1994, mostly because it was one of Oudomxay's most prominent opium-production areas, revealed after a national survey of Oudomxay by the UN International Drug Control Programme in 1993. The mountainous area of about 63,000ha (Figure 28-4) included 12 villages populated by 637 households comprising 3400 people. All of them were swidden farmers (Coudray, 1999). The six villages of Nahome, Sawang, Nam Long, Nam Ngao, Ban Lak and Poulong where inhabited by 1740 people, most of them of the Khmu Khrong ethnic group of Austro-Asiatic speakers. The remaining six villages of Ban May, Senxi, Senhome, Khokai, Piakeau and Hamnua had a total population of 1660 Hmong Khao people, who were Miao–yao speakers (Chazée, 1999).

The Khmu came from Luang Prabang province and had changed the location of their villages at least six times, always in mountainous areas. In 1972 they were asked by local authorities to settle near the lowland village of Ban Lak, where they were told they would get an access road and an irrigation scheme. A few years later, the authorities divided the villages of Nam Ngao, Nam Long, Nahome and Sawang. There were displacements in several other parts of the province in this period, particularly affecting Khmu Khrong and Khmu Ou people, who agreed to settle along the Beng valley. Most of them did not get access to proper agricultural land or irrigation facilities until the mid-1990s, and faced decades of high mortality as a result of malaria and water-borne diseases, and impoverishment because there was little land available for swiddens and the shorter fallow period led to a rapid decline in crop yields.

The Hmong villagers came from China through Luang Namtha province, or from Vietnam through Houaphan province, escaping rebellion and then seeking suitable

FIGURE 28-4:   The agrarian landscape of the Nahome area in the early 1990s. Swidden farming to produce rice and some associated crops was based on a cropping-and-fallow cycle of about 10 to 14 years. This cycle was reduced by government policy to just three to four years, resulting in a drastic slump in crop yields that was not always compensated by other economic activities.

Photo: Laurent Chazée.

forest land (Figure 28-5). After the long migration, crossing Nga and and Xay districts of Oudomxay province, the first communities arrived at Senhome and Hamnua at the beginning of the 19th century. The others followed, with the last community settling at Ban May in 1978, having been attracted by the prospect of roads, education and health services (Coudray, 1999).

Implementation of the rural development programme began in the Nahome focal site in 1994, with the extension of an irrigation scheme covering 53ha, completion of an all-weather road by 1998,[13] and political and technical training. However, the programme planning was rather 'top-down' (it was ordered from 'the top', with little involvement of people 'at the bottom'). The activities were not integrated and clearly targeted the elimination of shifting cultivation, including opium production. In 1999, the area was one of Oudomxay province's three priority focal sites, and the most advanced in terms of development actions, thanks to international assistance. Several internationally funded projects were focused on Nahome: a small-scale irrigation project in Oudomxay and Luang Namtha provinces (UN Development Programme/UN Capital development Fund, 1991 to 1996); a road project (Rural Development Office/UN International Drug Control programme, 1995 and 1996); an eco-development and irrigation project (social- and land-management survey, training for district management teams and village representatives, UN

FIGURE 28-5: Hmong women are skilled in embroidery. The patterns created by their intricate stitching define ethnic and clanic identity. With the opening of the market economy in the 1990s, embroidery became an income-generating activity. Old women in the Nahome area were sometimes contracted by handicraft trading companies.

Photo: Laurent Chazée.

Development Programme/UN Capital Development Fund, 1998 to 2001); a road construction project (Japan International Cooperation Agency, 1997); another road project (US Aid, 1997 and 1998); the Strengthening Economic and Social Management Capacity (SESMAC) project for institutional capacity building (UN Development Programme, 1997 to 1999); development activities involving irrigation schemes (UN Development Programme/UN International Drug Control Programme, 1998 to 2001);[14] a survey of the assessment and economic programming of the Nahome area's agro-ecosystem (Xishuangbanna Tropical Botanical Garden and Chinese Academy of Sciences, 1997 and 1998); and land allocation (Forest Department and the governor's office, 1998 and 1999).

In 1999, the SESMAC project team, which was housed in the Rural Development Office of Oudomxay province, discovered that the provincial governor was using the international project budgets (for irrigation, health, education, roads, and so on) as 'conditions' with which to abet the involuntary resettlement of Hmong communities in lowland areas of Nahome. 'Political education' had been strengthened in rural areas and some staff from decentralized national ministries, such as agriculture, forestry, health and education, had been ordered to settle in the area to provide local training and assistance. By the end of 1998, the Hmong community had organized a strong protest against the intentions of the provincial administration and Hmong leaders had informed the SESMAC project that their clear preference was to remain in mountainous areas, for social, religious and health reasons. The offices of the UN Development Programme and the UN International Drug Control Programme in Vientiane were informed by SESMAC of the province's intentions to displace the Hmong against their wishes – an action that was contrary to the principles of the United Nations Charter. The affair became a human rights issue. Ultimately, the United Nations failed to address the issue, so the project coordinator of SESMAC and some Lao staff chose to resign rather than validate the political intention to act against the wishes of the swidden farmers – a move that was only possible because of the leverage offered by international aid.

References

ADB (2001) *Participatory Poverty Assessment in Lao PDR*, Asian Development Bank, Manila

ADB (2007) *Participatory Poverty Assessment in Lao PDR II (2006)*, Asian Development Bank, Manila

Baird, G. I. and Shoemaker, B. (2005) *Aiding or Abetting: Internal Resettlement and International Aid Agencies in the Lao PDR*, Probe International, Toronto

Boupha, P. (1995) *Shifting Cultivation Extermination Program up to Year 2000 through Permanent Occupations in Lao PDR*, note for the Leading Central Committee for Rural Development, Vientiane

Cabinet of Party Central Committee (1994) *Resolution about Rural Development Decree issued by the Lao Revolutionary Party*, Government of Lao PDR, Vientiane

Central Rural Development Committee (1995) *Study on Role, Regulation, Duty and RDC Working Methods Appropriate for Provinces*, circular issued by the Central Rural Development Committee on 25 August, Government of Lao PDR, Vientiane

Central Rural Development Committee (1997) *Improvement of RDC Structures in the Provinces, Municipalities and Special Zones*, circular issued by the Central Rural Development Committee on 1 March, Government of Lao PDR, Vientiane

Chazée, L. (1994) 'Shifting cultivation practices in Laos: Present situation and their future', Master's thesis to EPHE, Sorbonne, Paris

Chazée, L. (1998) *Evolution des systèmes de production rurauxen république démocratique populaire du Laos*, L'Harmattan, Bonchamp-Lès-Laval, France

Chazée, L. (1999) *The People of Laos, Rural and Ethnic Diversities*, White Lotus, Bangkok

Chazée, L. (2000) *Xayabury Province. History, Actors, Territories and Resources for Further Socio-economic Development*, photocopy report, Xayabury, Lao PDR

Chazée, L. and Gehin, E. (2012) *Say, femme Poussang. Peuple de la forêt. De la montagne à la plaine, au Laos* (Women from the Poussang minority: People of the Forest, from Mountain to Plain, in Laos), BuchetChastel, Paris

Chazée, L. and Gehin E. (2017) *Peuples, territoires et moyens d'existence du Laos rural, de 1990 à 2005* (Peoples, Territories and Livelihoods of Rural Laos, 1990-2005), draft in progress. Tentative publication years: Vols 1 and 2, 2018; vol 3, 2019; vol 4, 2020.

Coudray, J. (1999) *Study of Production systems in Nahome area, Beng district, Oudomxay province: Recommendations for the Planning and Implementation of an Alternative Development Programme in an Opium Growing Area*, United Nations Development Programme and United National Capital Development Fund, Vientiane

Daviau, S. (2001) *Resettlement in Long District, Louang Namtha Province*, Action contre la faim, Vientiane

Ducourtieux, O., Laffort, J. R. and Stacklokham, S. (2004) 'Land policy and farming practices in Laos', *Development and Change* 36, pp499-526

Epprecht, M. (1998) 'Opium production and consumption and its place in the socio-economic setting of the Akha people of North-western Laos', Master's thesis, Institute of Geography, Faculty of Natural Science, University of Berne, Switzerland

Evrard, O. (2004) *La miseen oeuvre de la réforme foncière au Laos: Impacts sociaux et effets sur les conditions de vie en milieu rural*, Food and Agriculture Organization of the United Nations, Rome

Goudineau, Y. (1997a) *Resetlement and Social Characteristics of New villages: Basic Needs of Resettled Communities in the Lao PDR*, vol. 2, provincial report, United Nations Development Programme, Vientiane

Goudineau, Y. (1997b) *Besoins essentiels pour les communautés rapatriées au Laos: Rapatriement et caractéristiques des nouveaux villages dans six provinces*, vol. 1, rapport principal, Vientiane

Keoketsy, B., Bounthabandid, S. and Noven, J. (2000) *Monitoring and Evaluation of Land Use Planning and Land Allocation Impacts 1998-2000*, Ministry of Agriculture and Forestry/National Agriculture and Forestry Research Institute/Forest Inventory and Planning Division, Lao Swedish Forestry Programme, Vientiane

Moizo, B. (2004) 'Implementation of the Land Allocation Policy in the Lao PDR: Origins, problems, adjustments and local alternatives', in *Shifting Cultivation and Poverty Eradication in the Uplands of the Lao PDR*, National Agriculture and Forestry Research Institute, Vientiane

National Statistics Center (1976-1992), *Annual Statistics from 1976 to 1992*, Committee for Planning and Cooperation, Vientiane

National Statistics Center (1993-1995), *Annual Statistics from 1993 to 1995*, Committee for Planning and Cooperation, Vientiane

Office of the Governor, Oudomxay (1998) *Guidelines for Integrated Rural Development Planning of OudomxayProvince for the Fiscal Year 1998-1999 and for the Period up to 2002*, Provincial Administration of Oudomxay province, Muang Xay, Oudomxay

Office of the Governor, Sayabury (1998) *Guidelines on Rural Development to the year 2003*, Rural Development Office of Sayabury province, Sayabury

SESMAC (1999) *Rural Development Programme in Xayabury province: A Situation Analysis and a Programme for the Strengthening of the Rural Development Office* (Lao/97/002), Strengthening Economic and Social Management Capacity Project, Muang Xay, Oudomxay

Siksidao, P. (2001) *Final report, Rural Development Component*, Strengthening Economic and Social Management Capacity Project (SESMAC), Muang Xay, Oudomxay

UNDP (2001) *National Human Development Report, Lao PDR, 2001: Advancing Rural Development*, United Nations Development Programme, Vientiane

Notes

1. The author spent 11 years in Lao PDR between 1988 and 2002. He was based for six years between Oudomxay, Luang Namtha and Sayabury provinces, but also worked a total of two years in Phongsaly, Luang Prabang, Xieng Khouang, Houaphan, Khammouane, Vientiane, Champassak, Xekong, Saravane and Attapeu provinces. He was implementing projects and missions for the United Nations Development Programme; the United Nations Capital Development Fund; the United Nations Drug Control Programme; the World Bank; the Asian Development Bank; the French Development Agency; the European Union; and various international consultancies in Vientiane, among them the SOFRECO company.

2. These actions included regulation 74/CCM on forest protection; instructions 24/CCM to safeguard the forest and wildlife, including aquatic animals; decree 01/CCM on the protection of specific wood species; decree 185/CCM on the prohibition of wildlife trade; decree 147/CCM related to taxes; decree 117/PCM on the use and management of forest and forest territories; decree 118/PCM on management and protection of wildlife, including aquatic animals; decree 21 PPC (1993) on the responsibility of the Ministry and Forestry (policy of recentralization); and the establishment of penal and forest laws.

3. In its objectives for the second five-year plan (1985 to 1990), the government foresaw a 30% reduction in the area under shifting cultivation. A directive was issued to protect 5 million hectares of productive forest, 9.5 million hectares of protected forest, and 2.5 million hectares of conservation forest. A Tropical Forest Action Plan, financed by international organizations, was approved in 1991, following the country's first national conference on forests, organized in 1989. The action plan foresaw the permanent settlement of 60% of all swidden farmers (900,000 swidden farmers out of 1.5 million), by 2000.

4. In 2004, this decree was updated by raising the minimum size of villages that were encouraged to consolidate. The new minimum sizes were a population of 500 for lowland villages and 200 for upland villages.

5. By reducing the fallow cycle, the impoverishment process was also associated with a loss of diversity of local flora and fauna and a decreasing volume and diversity of non-timber forest products for family use and income. The consequences were lower staple-food security and a need to find second-choice foods and pursue less-preferred economic alternatives, such as longer trips to reach forest products. In some areas, such as around Luang Prabang, Vientiane, Luang Namtha and Pakse,

swidden families began in about 1993 to 'escape' to urban areas, there to begin begging, stealing and even selling their children. Poverty reduction became a key objective in 1995, under the national commitment to the Millennium Development Goals.

6. At this time, production of upland paddy increased by 47% in some studied communities. Other studies showed that upland swidden areas increased in some districts during the period 1997 to 2002.

7. SESMAC, or the Strengthening Economic and Social Management Capacity project, was financed by the United Nations Development Programme, and the Beng Alternative Project was financed by the United Nations International Drug Control Programme.

8. These conventions and agreements included the Alma-Ata Declaration (1978) on primary health care; the Cairo Commitment (1994) on Population and Development; the Beijing Gender-equality Commitment in 1995; The Convention on Biological Diversity and Sustainable Development (Rio de Janeiro, 1992-2002); and The Millennium Development Goals (including poverty reduction, 1995-2015).

9. It should be understood that these zones were selected on the basis of recent historical events that began around 1982, including a conflict with Thailand related to border issues at Nam Lop in the Navene area of Phieng district in 1982; a conflict with Thailand in the Tongmixay area of Paklay district in 1984; border issues that started in the Nabornoy area in 1986 and 1987; conflict with opposition groups at Pongkood and Pongna in Phieng district in 1988 and 1989; and conflict with opposition groups in Hongsa and Sienghone districts in 1992. Until 1999, the western and northern parts of the province and some sensitive strategic areas, particularly Naven, Tongmixay, Samed, Phoulane and Hongsa-Sienghone, were under the control of the army. For the sake of national security, the mandate of the army mission was extended to include some development activities. Beginning in 1993, Naven was the site of a pilot development zone for research on land-distribution methods, with funding assistance from the Swedish International Development Agency. Areas with security issues were given priority for declaration as focal zones, not only for the local people, but also to ensure a permanent presence and a strong measure of control.

10. The priority programmes defined in the Rural Development Plan, 1996-2000, in Sayabury comprised eight objectives: food sufficiency; commercial crop production; eradication of swidden farming and boosting the condition of the environment, water and forest conservation with income-generating activities for communities involved in swidden farming; rural development; human-resource development; infrastructure development; economic relations with foreign countries; and general services.

11. Permanent staff in the Provincial Rural Development Office of Sayabury: four people in 1994-1995; five in 1995-1996, eight in 1996-1997, 11 in 1997-1998 (SESMAC, 1999).

12. Other focal sites were named for later development. The Ano area in La District, which had 13 villages occupied by 700 people of the Akha Ano group (Tibeto-burman speakers), was also considered to be a focal site. The selection of this area was clearly linked to its high-volume opium production and to insecurity. In 1997-1998, after the construction of an access road and school, the province tried to convince the Strengthening Economic and Social Management Capacity (SES-MAC) project, which was financed by US bilateral aid and the UN Development Programme, to fund the area. However, the Ano area came under army control in 1998 and outsiders were forbidden entry.

13. Funded by US Aid, the Japan International Cooperation Agency and the UN International Drug Control Programme.

14. Between 1998 and 2001, 22 small irrigation weirs were built to support fruit-tree orchards, livestock husbandry, gardening, domestic water supplies and fish ponds.

29

'YOUR LAND IS NEEDED'

The fundamental reason behind the sedentarization of shifting cultivators

*Rodolphe De Koninck and Pham Thanh Hai**

Deforestation by swidden cultivators?

> Soil studies clearly indicate that shifting cultivation does not in itself ruin soils and produce destructive erosion (…). There is ample evidence that advanced shifting cultivators attach very high values to the maintenance of an effective vegetative cover in their home territories (Spencer, 1966, p167).

Even before – and even more since – J.E. Spencer published the results of his impressive survey entitled *Shifting Cultivation in Southeastern Asia*, a lot of research had or has been done on the subject. Largely based on a wide variety of case studies, most of the ensuing literature – amply referred to in this volume and in its predecessors – confirms Spencer's very clear conclusions: unless forcefully subjected to a restriction of their territorial range, shifting-cultivation societies have not been environmentally destructive. So why do most Southeast Asian states implement policies which, often originating in colonial times, are basically opposed to swidden agriculture? The two most commonly heard official reasons behind these policies are that shifting-cultivation communities are accused of, first, being the main culprits behind the rapid deforestation occurring in much of the region's marginal lands and, second, of practising a form of agriculture that generates a limited surplus, which would explain their poverty. Consequently, policies – already implemented during the colonial period – to settle and to prevent swiddeners from practising their traditional form of agriculture have, in recent decades, been applied with increasing vigour throughout the entire Southeast Asian region (Padoch et al., 2007).

* Professor Rodolphe De Koninck is holder of the Canada Chair of Asian Research and teaches in the Department of Geography at the University of Montreal; Pham Thanh Hai is a postdoctoral researcher at the Canada Chair of Asian Research.

Of course, other motives are frequently evoked to justify sedentarization policies. Among the more common ones is that they will facilitate the state's attempts to improve the indigenous population's productivity in food-crop cultivation as well as their livelihood (Guérin et al., 2003). That type of argument has been prominent in the contemporary official state discourse of Vietnam, even after being debunked by J. E. Spencer in 1966 and even more by Michael Dove in 1983 and 1985 – and by many others since then, for example Thrupp et al. (1997) and Ducourtieux (2009). As expected, such attempts, even if they were really meant to improve food-crop productivity, have generally failed. This has been the case in Vietnam's Tay Nguyen region – the country's Central Highlands – particularly among the Ede people in Dak Lak province. Starting in the late 1980s, while gradually reducing their production of food crops, a large number of Ede were drawn into the expanding coffee economy because they generally foresaw no improvement in their overall economic situation (Danh, 2005, p203; Doutriaux et al., 2008). The deterioration in their livelihood, along with land dispossession, were in fact the basic reasons behind unrest that erupted in 2001 in Dak Lak and particularly in the provincial capital of Buon Me Thuot.

This led the Vietnamese government, once the unrest had been quelled, to invest massively in the rehabilitation of the local economy, but in such a way that it made the Ede less autonomous and more dependent on the state for their subsistence (Danh, 2005, p209). Throughout the rest of the Central Highlands of Vietnam, notwithstanding recent efforts to intensify market gardening in the Da Lat basin, overall food production has declined steadily since the 1980s, while coffee production has boomed (Fortunel, 2000, 2005; Danh, 2005; Phan and Tran, 2007). Spectacular growth in the production of coffee has of course been directly associated with the planned and relentless settlement of Kinh migrants (Hardy, 2003; Déry, 2004),[1] and along with them, the development of the coffee frontier – which has been mowing down the forest – and the fast-increasing commoditization of the entire regional economy, over which the state has less and less control.

Unsurprisingly, attempts to correlate the map of deforestation and that of traditional swidden agriculture have not been conclusive for a number of reasons, including the actual mobility of shifting agriculture. On the other hand, such attempts have been conclusive when the retreat of the forest has been matched with the expansion of lowlanders' agricultural fronts. An example of this was a 1990s study in Vietnam's Central highlands (De Koninck, 1999), which made it very clear that, contrary to the Vietnamese government's claims, the then ongoing deforestation had very little to do with swidden agriculture practised by indigenous minorities. Rather, it was largely attributable to the expansion of the coffee frontier involving Kinh lowlanders, who were burning the forest down, sometimes by employing the expert services of indigenous people. But unlike that of indigenous tradition, this burning was permanent, without any allowance for fallowing, and with no intention of letting the forest grow back.[2]

That rather simple yet fundamental demonstration led us to start looking more systematically at the relationship between the evolution of land use and population distribution throughout Southeast Asia, particularly since the 1960s. The amount of work required to produce the necessary diachronic maps of both types of distribution at the scale of the entire region has been daunting. But we have come a long way and are now ready to begin analysing and making public the results of our investigations (De Koninck and Pham, 2012). What we present here is provisional and sketchy. First, we look at the case of the Vietnamese Central Highlands, and further demonstrate the link between the expansion of the agricultural frontier and deforestation. We also reveal the ensuing process of cultural dilution, as indigenous groups are gradually losing ground, often becoming minorities within their own territorial realm. Then we briefly document the process of population 'deconcentration' which has been applied throughout most of Southeast Asia over the last half century.

The takeover of Vietnam's Central Highlands

> The permanent association of the state and sedentary agriculture is at the center of this story (…). Modern conceptions of national sovereignty and the resource needs of mature capitalism have brought that final enclosure into view (Scott, 2009, pp9–11).

As mentioned earlier, population expansion of the Kinh into the Central Highlands began well before Vietnam's 1975 postwar unification. In 1936, the Viet or Kinh accounted for only 2% of the total population of Dak Lak province (Do, 1996, p58). By 1944, the Kinh share had increased, but only slightly, to 4%. However, by 1974 it had climbed to 44%, with the proportion accounted for by indigenous minorities – referred to in the literature by the collective term 'highlanders' – dropping from 96% to 56% over that 30-year period (Hardy, 2003, p312).

By the late 1980s and early 1990s, Kinh migrants were increasingly joined by representatives of ethnic minorities moving down from the country's Northern Highlands, where long-brewing and severe environmental degradation had reached crisis level (Rambo et al., 1995; Castella et al., 2006). Since the 1960s, the Northern Highlands had themselves been submitted to in-migration of Kinh settlers from the Red River Delta. This had brought increasing pressure to bear on indigenous swidden communities in the north, resulting in unsustainable agricultural practices and, in turn, deforestation and overall environmental degradation. Gradually, this pressure was transferred to the Central Highlands, towards which an increasing flow of refugee migrants had turned. They included both Kinh and Northern Highlands minority people, particularly Nung and Thai.

An examination of census data from 1979 and 2009 confirms a gradual intensification of the cultural-dilution process throughout the entire Central Highlands (Table 29-1). For example, by 2009, in the provinces of Dak Lak and Dak Nong – carved out of the former Dak Lak in 2004 – the proportion of indigenous people had dropped to as little as 21% and 15% of the total population, respectively. Over three decades, that total population had itself grown nearly fourfold. The acceleration in the

TABLE 29-1: Central Highlands of Vietnam, population and ethnic composition, 1979 and 2009.

Provinces	1979				2009			
			Percentage				*Percentage*	
			Ethnic minorities				*Ethnic minorities*	
	Population	*Kinh*	*Indigenous*	*Other*	*Population*	*Kinh*	*Indigenous*	*Other*
Gia Lai *	595,906	44	55	1	1,274,412	56	41	3
Kon Tum *					430,133	47	49	4
Dak Lak **	490,198	60	37	3	1,733,624	67	21	12
Dak Nong **					489,392	68	15	17
Lam Dong	396,657	70	23	7	1,187,574	76	17	7
Central Highlands	1,482,761	56	40	4	5,115,135	65	27	8

Note:*In 1991, the province of Gia Lai-Kon Tum was divided into Gia Lai and Kon Tum provinces.
** In 2004, the province of Dak Lak was divided into Dak Lak and Dak Nong provinces.
Source: Census 1979 and 2009.

contemporary demographic 'filling in' of the Vietnamese Central Highlands was clearly directly associated with the expansion of the agricultural frontier, in this case, essentially the coffee frontier.

However, mapping that expansion was a difficult task, considering the constantly evolving land-use categories used in the various sources available (Figure 29-1). As already mentioned, much of the problem had to do with the way in which swidden agriculture operated, as well as the way it was defined and represented cartographically.[3] For example, for the 1970 land-cover map produced by the US Central Intelligence Agency (CIA), it is likely that both the 'grasslands' and 'forests' categories included areas devoted to shifting agriculture, while in the 2005 Globcover-project map, it is the 'croplands' and 'grasslands/shrublands' categories that do so (ESA Globcover Project, 2011). Whatever the case, the comparison between the two maps still illustrates convincingly the massive replacement of forests by permanent croplands. The additional comparison between the evolution of land use and population density is equally eloquent, particularly in illustrating the relationship between agricultural expansion and population increase. As settlers move in, the forest recedes and permanent agriculture takes over; the landscape is literally transformed by the intensification of land use and densification of population and settlement patterns. One can hardly find a better demonstration of territorialization by the state, even though other agents are involved, including spontaneous migrants, whether Kinh or not. But these migrants only follow in the footsteps of the state-sponsored ones. The exceptional nature of the demographic 'filling in' of the Central Highlands is clearly illustrated by maps representing very substantial increases in population densities between 1970 and 2004 (Figure 29-1) and even more by another one representing

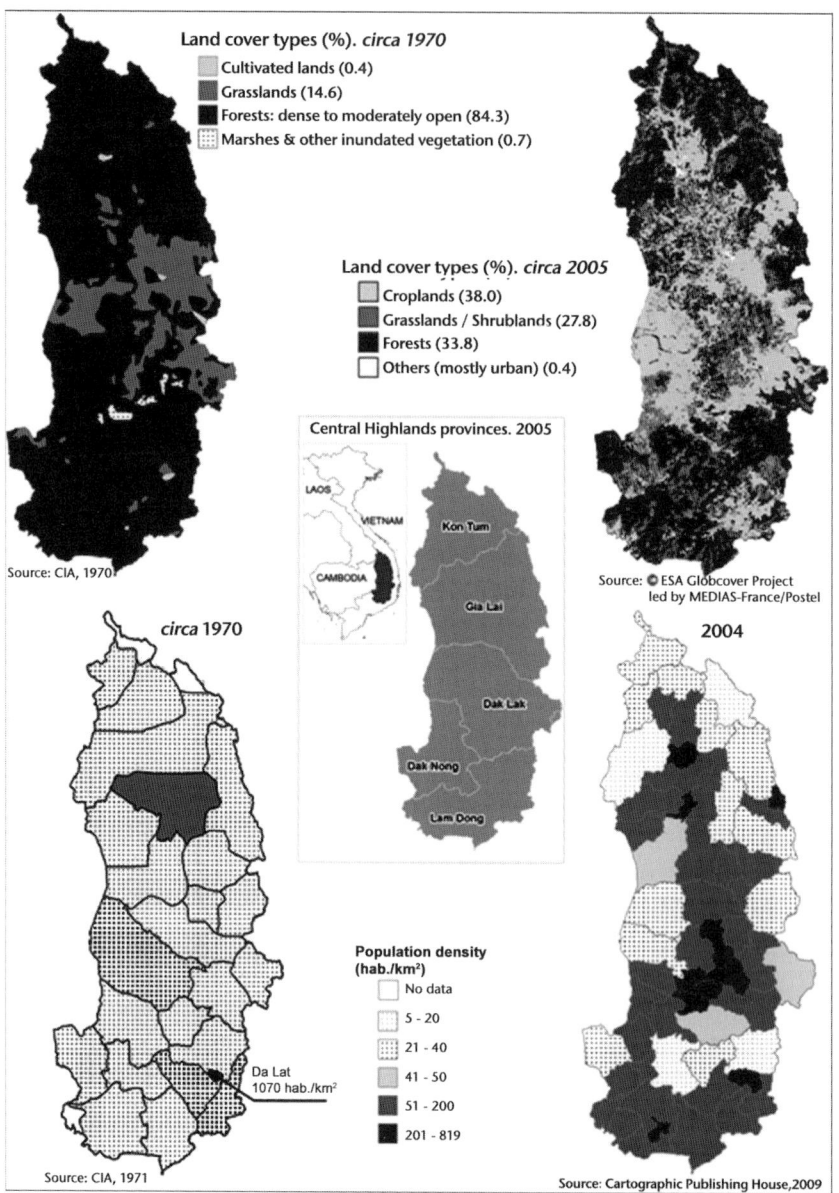

FIGURE 29-1: Central Highlands of Vietnam, land cover and population, 1970 – 2005.

the comparative evolution of population for the whole of Southern Vietnam between 1989 and 1999 (Figure 29-2). The latter map shows clearly that relative population increases have been much more significant in the Central Highlands provinces than in any other throughout the South. The 'last enclosure' (Scott, 2009, p4) of the Central Highlands is well on its way.

Filling in all Southeast Asia's margins

The combined forces of agricultural, forestry, and conservation policies, increased global demands for tropical commodities, and continued population growth are not to be stopped in any part of Southeast Asia for long (Padoch et al., 2007, p41).

Over about 40 years, from 1960 to 2000, the process of population redistribution illustrated in Figure 29-2 has in fact been operational in nearly every country of the region. As is well known, state authorities have, in nearly every country, gradually intensified or simply favoured land pioneering and settlement programmes, particularly since the late 1950s and early 1960s (De Koninck and Déry, 1997). These have always involved the filling in of peripheral lands, frequently borderlands (De Koninck, 2000). The better-known cases are those of Indonesia, the Philippines, Malaysia and Vietnam. What is less well known is that, in all of the region's major countries, except Laos, this has resulted in actual population deconcentration (Table 29-2).[4] In other words, the stated goal of decongesting some of the more densely populated lowlands or islands has in fact been achieved nearly everywhere.[5] And the two countries where the population deconcentration process has been most noticeable are Indonesia – which still has the highest score on the Hoover index – and Vietnam. It so happens that these are also the two Southeast Asian countries where, during the 40-year period under examination, the most elaborate land-development programmes

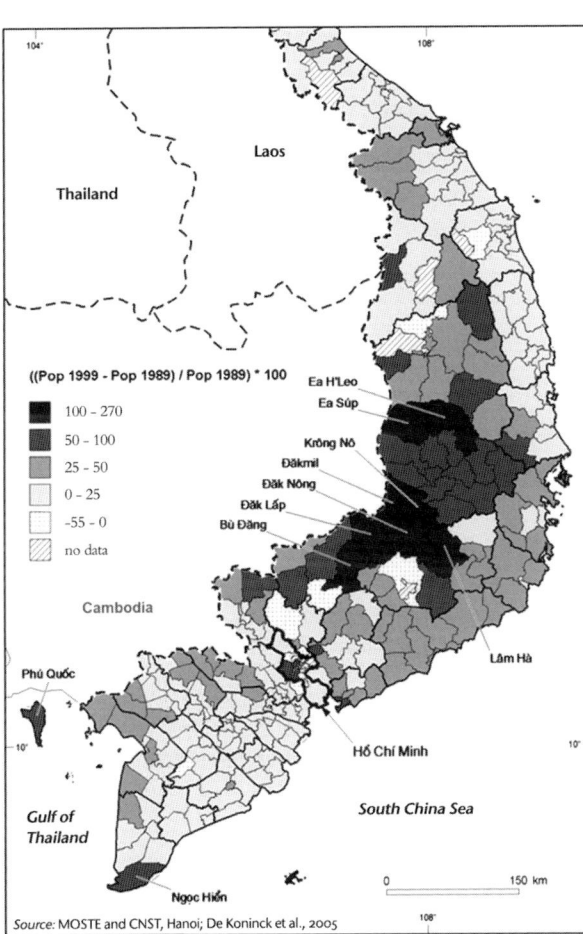

Source: MOSTE and CNST, Hanoi; De Koninck et al., 2005

FIGURE 29-2: Southern Vietnam, population evolution by district and in percentages, 1989-1999.

TABLE 29-2: Southeast Asia: Evolution of population distribution as measured by the Hoover Index, circa 1960; circa 1980; circa 2000.

Countries	Year	Hoover index	No. of territorial units	Countries	Year	Hoover index	No. of territorial units
Burma	1961	38.1	14	Malaysia	1957–60	52.9	52
	1983	40.1			1980	51.2	
	2004	38.4			2000	49.9	
Cambodia	1962	47.3	20	Philippines	1960	34.3	74
	1981	44.0			1980	32.7	
	2008	43.7			2000	33.5	
Indonesia	1961	62.9	28	Thailand	1960	25.9	72
	1985	60.7			1980	24.8	
	2005	57.6			2004	24.3	
Laos	1961	17.7	18	Vietnam	1965	42.3	52
	1985	21.0			1985	39.0	
	2005	23.8			2004	35.7	
				Southeast Asia	1957–65	51.5	329
					1980–85	50.7	
					2000–08	48.2	

Notes: The formula for calculating the Hoover index is as follows:

$$H_t = 0.5 \sum_{i=1}^{n} |P_{it} - a_i| 100$$

Where H_t is the Hoover index at time t, P_{it} is the proportion of the nation's population in area i at time t, and a_i is the proportion of the nation's land area in area i. The index would be 0 if each area had the same share of the nation's population and land; it approaches 100 if everyone were to live in a single locality (Long and Nucci, 1997, p431).

Sources: Department of Statistics Malaysia, 1958, 1987; Jones, 1962a, 1962b; Royaume du Cambodge et al., 1966; Nitisastro, 1970; CIA, 1971; Arumainathan, 1973; Donner, 1978; National Statistical Office of Thailand, 1982, 2004; National Census and Statistics Office, 1985; Hugo, 1987; Taillard, 1989; Sliwinski, 1995; General Statistics Office, 2004; National Statistics Office of the Philippines, 2004; Department of Statistics Brunei, 2005; Durand, 2006; Swee-Hock, 2007; Messerli et al., 2008; Government of the Union of Myanmar, 2008; Cartographic Publishing House, 2009; NCDD, 2009a, 2009b, 2009c through 2009w; Brinkhoff, 2010; Data Statistics Indonesia, 2010; GeoHive, 2010; Geospasial-BNPB, 2010; Revolutionary Government of the Union of Burma (nd).

and population displacements have been conducted, predominantly in upland or peripheral regions (De Koninck, 2006).

Of course, the population of regions such as the very densely populated islands of Java, Luzon and the Red River Delta has continued to grow, but the peripheral and borderland regions of the countries concerned have all seen their own populations grow even more rapidly. The result is that, although the demographic urbanization of Southeast Asia is well underway, and although in each country the central and lowland areas remain much more involved in the overall urbanization process and are becoming even more densely populated, the population of the most marginal areas is growing even faster. Put differently, notwithstanding the intensity of the urbanization and industrialization processes, the largely and increasingly agricultural peripheries – which in several cases could be more broadly termed peripheral resource regions – are still being reined in, with lowland peasantries acting as the territorial spearhead of the state (De Koninck, 1986, 1996). The frontier is still very active, to a point where population deconcentration has become an important and original feature of the region's agrarian transition, with agricultural expansion playing a multifunctional role. The share of agriculture, measured in terms of employment or contribution to exports and to gross domestic product, is declining everywhere (De Koninck and Rousseau, 2012). Yet more and more land is being put under permanent crops, generally at the expense of forest cover, a process which results in a genuine agriculturalization of the landscape, particularly noticeable in Indonesia (De Koninck and Rousseau, 2012).

This apparent paradox is precisely attributable to the very functions of agricultural expansion, whether economic, social or geopolitical. The latter category of functions includes the cultural dilution of the marginal and upland regions, often involving land dispossession and various forms of exclusion (Hall et al., 2011). Consequently, in most instances, indigenous peoples' traditional livelihood sources, including swidden agriculture, are gradually rendered obsolete and land needed and coveted for population redistribution and deconcentration is made available. This land is then predominantly devoted to the cultivation of export crops, through ventures increasingly controlled by large private investors.

References

Arumainathan, P. (1973) *Report on the Census of Population: 1970*, Singapore, Department of Statistics, Singapore

Brinkhoff, T. (2010) *Singapore, City population,* https://www.citypopulation.de/Singapore-Regions. html, accessed 3 May 2017

Cartographic Publishing House (2009) *Vietnam Administrative Atlas,* Cartographic Publishing House, Hanoi

Castella, J. Ch., Boissau, S., Nguyen, H. T. and Novosad, P. (2006) 'Impact of forest land allocation on land use in a mountainous province of Vietnam', *Land Use Policy* 23, pp147-160

CIA (1970) *Indochina Atlas,* Office of Basic and Geographic Intelligence, Central Intelligence Agency, Washington, DC

CIA (1971) *South Vietnam, Provincial Maps,* Population Edition, Office of Basic and Geographic Intelligence, Central Intelligence Agency, Washington, DC

Danh, D. T. (2005) 'Café et agriculture de subsistance; complémentarité ou compétition? Étude de cas d'un village édé au Viêt Nam', in R. De Koninck, F. Durand and F. Fortunel (eds) *Agriculture, environnement et sociétés sur les Hautes terres du Viêt Nam,* Arkuiris and the Institute of Research on Contemporary Southeast Asia (IRASEC), Toulouse and Bangkok, pp191-216

Data Statistics Indonesia (2010) Data Statistik, http://www.datastatistik-indonesia.com/portal/index. php?option=com_tabel&kat=1&idtabel=114&Itemid=165, accessed April 2010

De Koninck, R. (1986) 'La paysannerie comme fer de lance territorial de l'État: le cas de la Malaysia', *Cahiers des Sciences Humaines* (ORSTOM) 22(3-4), pp355-370

De Koninck, R. (1996) 'The Peasantry as the Territorial Spearhead of the State: The Case of Vietnam', *Sojourn: Journal of Social Issues in Southeast Asia* 11(2), pp231-258

De Koninck, R. (1999) *Deforestation in Viet Nam,* International Development Research Centre (IDRC), Ottawa

De Koninck, R. (2000) 'The theory and practice of frontier development: Vietnam's contribution', *Asia Pacific Viewpoint* 41(1), pp7-21

De Koninck, R. (2006) 'On the geopolitics of land colonization: Order and disorder on the frontiers of Vietnam and Indonesia', *Moussons* 9-10, pp33-59

De Koninck, R. and Déry, S. (1997) 'Agricultural expansion in Southeast Asia as a tool of population redistribution', *Journal of Southeast Asian Studies* 28(1), pp1-26

De Koninck, R. and Pham, T. H. (2012) 'Comment mesurer la redistribution de la population en Asie du Sud-Est?' paper delivered to an International Symposium on Measuring Development, GEMDEV and UNESCO, Paris

De Koninck, R. and Rousseau, J. F. (2012) *Gambling with the Land. The Contemporary Evolution of Southeast Asian Agriculture*, NUS Press, Singapore

Department of Statistics Brunei (2005) *Brunei Darussalam Statistical Yearbook, 2002*, Department of Statistics, Negara Brunei Darussalam

Department of Statistics Malaysia (1958) *1957 Population Census of the Federation of Malaya,* Department of Statistics Malaysia, Kuala Lumpur

Department of Statistics Malaysia (1987) *Population Report for Mukims*, Department of Statistics Malaysia, Kuala Lumpur

Déry, S. (2004) *La colonisation agricole au Vietnam*, Presses de l'Université du Québec, Sainte-Foy, Québec

Do, T. D. (1996) 'Impact of increasing population on the natural environment in the Central Highlands region', unpublished PhD thesis, National Economics University, Hanoi

Donner, W. (1978) *The Five Faces of Thailand: An Economic Geography,* C. Hurst, London

Dove, M. (1983) 'Theories of Swidden Agriculture and the Political Economy of Ignorance', *Agroforestry Systems* 1, pp85-99

Dove, M. (1985) *The Agroecological Mythology of the Javanese and the Political Economy of Indonesia,* East West Environment and Policy Institute, Honolulu

Doutriaux, S., Geisler, C. and Shively, G. (2008) 'Competing for coffee space: Development induced displacement in the Central Highlands of Vietnam', *Rural Sociology* 73(4), pp528-554

Ducourtieux, O. (2009) *Du riz et des arbres. L'interdiction de l'agriculture d'abattis-brûlis, une constante politique au Laos,* Institute of Research for Development (IRD) and Karthala, Paris

Durand, F. (2006) *East Timor, a Country at the Crossroads of Asia and the Pacific: A Geo-historical Atlas,* Institute of Research on Contemporary Southeast Asia (IRASEC), Bangkok and Silkworm Books, Chiang Mai

ESA Globcover Project (2011) *Global Land Cover Product (2005-06)*, European Space Agency, http:// due.esrin.esa.int/page_globcover.php, accessed 3 May 2017

Evrard, O. and Goudineau, Y. (2004) 'Planned resettlement, unexpected migrations and cultural trauma in Laos', *Development and Change* 35(5), pp937-962

Fortunel, F. (2000) *Le café au Viêt Nam. De la colonisation à l'essor d'un grand producteur mondial,* L'Harmattan, Paris

Fortunel, F. (2005) 'L'amertume du café dans les plateaux du Centre Viêt Nam. Les structures productives et les autochtones', in R. De Koninck, F. Durand and F. Fortunel (eds) *Agriculture, environnement et sociétés sur les Hautes terres du Viêt Nam,* Institute of Research on Contemporary Southeast Asia (IRASEC) and Arkuiris, Bangkok and Toulouse, pp163-186

General Statistics Office (2004) *Vietnam Statistical Data in the 20th Century*, Statistical Publishing House, Hanoi

GeoHive (2010) *The Democratic Republic of Timor-Leste*, GeoHive, http://www.xist.org/cntry/ timorleste.aspx, accessed May 2010

Geospasial-BNPB (2010) *Administrasi provinsi*, Indonesian National Board for Disaster Management, http://geospasial.bnpb.go.id/category/peta-dasar/provinsi/, accessed April 2010

Government of the Union of Myanmar (2008) *Statistical Yearbook 2006*, Central Statistical Organization, Naypyitaw, Myanmar

Guérin, M., Hardy, A., Nguyen, V. C. and Tan, S. B. H. (2003) *Des montagnards aux minorités ethniques. Quelle intégration nationale pour les habitants des hautes terres du Viêt Nam et du Cambodge?*, L'Harmattan and the Institute of Research on Contemporary Southeast Asia (IRASEC), Paris and Bangkok

Hall, D., Hirsch, P. and Li, T. (2011) *Powers of Exclusion. Land Dilemmas in Southeast Asia*, NUS Press, Singapore

Hardy, A. (2003) *Red Hills: Migrants and the State in the Highlands of Vietnam*, NIAS Press, Copenhagen

Hugo, G. J. (1987) *The Demographic Dimension in Indonesian Development*, Oxford University Press, New York

Jones, L. W. (1962a) *North Borneo: Report on the Census of Population taken on 10th August 1960*, Government Printing Office, Kuching

Jones L. W. (1962b) *Sarawak: Report on the Census of Population taken on 15th June 1960*, Government Printing Office, Kuching

Long, L. and Nucci, A. (1997) 'The Hoover Index of population concentration: A correction and update', *Professional Geographer* 49(4), pp431–440

Messerli, P., Heinimann, A., Epprecht, M., Phonesaly, S., Thiraka, C. and Minot, N. (2008) *Socio-Economic Atlas of the Lao PDR*, National Centers of Competence in Research (NCCR), North-South, University of Bern, Switzerland and Geographica Bernensia, Vientiane

National Census and Statistics Office (1985) *1980 Census of Population and Housing, Philippines*, National Census and Statistics Office, Manila

National Statistics Office of the Philippines (2004) *2000 Census of Population and Housing. Population, Land Area and Density*, National Statistics Office, Manila

National Statistical Office of Thailand (1982) *1980 Population and Housing Census*, National Statistical Office of Thailand, Bangkok

National Statistical Office of Thailand (2004) *Geospatial Statistics*, http://service.nso.go.th/nso/nsopublish/servGis/servGis.html, accessed April and May 2010

NCDD (2009a) *Banteay Meanchey Data Book 2009*, GIS code 1, National Committee for Sub-National Democratic Development, Phnom Penh

NCDD (2009b) *Battambang Data Book 2009*, GIS code 2, National Committee for Sub-National Democratic Development, Phnom Penh

NCDD (2009c) *Kampong Cham Data Book 2009*, GIS code 3, National Committee for Sub-National Democratic Development, Phnom Penh

NCDD. (2009d) *Kampong Chhnang Data Book 2009*, GIS code 4, National Committee for Sub-National Democratic Development, Phnom Penh

NCDD (2009e) *Kampong Speu Data Book 2009*, GIS code 5, National Committee for Sub-National Democratic Development, Phnom Penh

NCDD (2009f) *Kampong Thom Data Book 2009*, GIS code: 6, National Committee for Sub-National Democratic Development, Phnom Penh

NCDD (2009g) *Kampot Data Book 2009*, GIS code 7, National Committee for Sub-National Democratic Development, Phnom Penh

NCDD (2009h) *Kandal Data Book 2009*, GIS code 8, National Committee for Sub-National Democratic Development, Phnom Penh

NCDD (2009i) *Kep Data Book 2009*, GIS code 23, National Committee for Sub-National Democratic Development, Phnom Penh

NCDD (2009j) *Koh Kong Data Book 2009*, GIS code 9, National Committee for Sub-National Democratic Development, Phnom Penh

NCDD (2009k) *Kratie Data Book 2009*, GIS code 10, National Committee for Sub-National Democratic Development, Phnom Penh

NCDD (2009l) *Mondul Kiri Data Book 2009*, GIS code 11, National Committee for Sub-National Democratic Development, Phnom Penh

NCDD (2009m) *Otdar Meanchey Data Book 2009,* GIS code 22, National Committee for Sub-National Democratic Development, Phnom Penh

NCDD (2009n) *Pailin Data Book 2009*, GIS code 24, National Committee for Sub-National Democratic Development, Phnom Penh

NCDD (2009o) *Preah Sihanouk Data Book 2009,* GIS code 18, National Committee for Sub-National Democratic Development, Phnom Penh

NCDD (2009p) *Preah Vihear Data Book 2009*, GIS code 13, National Committee for Sub-National Democratic Development, Phnom Penh

NCDD (2009q) *Prey Veng Data Book 2009*, GIS code 14, National Committee for Sub-National Democratic Development, Phnom Penh

NCDD (2009r) *Pursat Data Book 2009*, GIS code 15, National Committee for Sub-National Democratic Development, Phnom Penh

NCDD (2009s) *Ratanak Kiri Data Book 2009*, GIS code 16, National Committee for Sub-National Democratic Development, Phnom Penh

NCDD (2009t) *Siem Reap Data Book 2009*, GIS code 17, National Committee for Sub-National Democratic Development, Phnom Penh

NCDD (2009u) *Stung Treng Data Book 2009*, GIS code 19, National Committee for Sub-National Democratic Development, Phnom Penh

NCDD (2009v) *Svay Rieng Data Book 2009*, GIS code 20, National Committee for Sub-National Democratic Development, Phnom Penh

NCDD (2009w) *Takeo Data Book 2009,* GIS code 21, National Committee for Sub-National Democratic Development, Phnom Penh

Nitisastro, W. (1970) *Population Trends in Indonesia*, Cornell University Press, Ithaca, NY

Padoch, C., Coffey, K., Mertz, O., Leisz, S. J., Fox, J. and Wadley, R. L. (2007) 'The demise of swidden in Southeast Asia? Local realities and regional ambiguities', *Geografisk Tidsskrift – Danish Journal of Geography* 107(1), pp29-41

Phan, Q. S. and Tran, T. D. (2007) 'The Problems of Shifting Cultivation in the Central Highlands of Vietnam', in M. Cairns (ed.) *Voices from the Forest: Integrating Indigenous Knowledge into Sustainable Upland Farming*, Resources for the Future, Washington, DC, pp705-711

Rambo, A. T., Reed, R. R., Le, T. C. and DiGregorio, M. R. (eds) (1995) *The challenges of highland development in Vietnam*, East-West Center Programe on Environment, Honolulu, Centre for Natural Resources and Environmental Studies, University of Hanoi and Center for Southeast Asian Studies, Berkeley, CA

Revolutionary Government of the Union of Burma (nd) *Statistical Year Book 1967*, Central Statistical and Economics Department, Rangoon

Royaume du Cambodge, Ministère du Plan and Institut National de la Statistique et des Recherches Économiques (1966) *Résultats finaux du recensement général de la population*, Ministère du Plan, Phnom Penh

Scott, J. C. (2009) *The Art of Not Being Governed: An Anarchist History of Upland Southeast Asia*, Yale University Press, New Haven, CT and London

Sliwinski, M. (1995) *Le génocide Khmer rouge: une analyse démographique*, L'Harmattan, Paris

Spencer, J. E. (1966) *Shifting Cultivation in Southeastern Asia*, University of California Press, Berkeley and Los Angeles, CA

Swee-Hock, S. (2007) *The Population of Singapore*, Institute of Southeast Asian Studies, Singapore

Taillard, Ch. (1989) *Le Laos: stratégies d'un État-tampon*, Groupement d'intérêt public Reclus, Montpellier, France

Thrupp, L. A., Hecht, S., and Browder, J. (1997) *The Diversity and Dynamics of Shifting Cultivation: Myths, Realities, and Policy Implications,* World Resources Institute, Washington, DC

Notes

1. The Kinh are the majority ethnic group in Vietnam, representing 86% of the population in 1999.

2. During the years and decades before the massive agricultural-colonization programmes that were launched soon after the 1975 unification, a lot of pressure had fallen on the indigenous minorities' domain, not only because of war operations, but also through slow infringement, whether or not linked to state-sponsored land opening. This often constrained the swiddeners to reduce the duration of the fallow between periods of cropping, a process which had already begun to contribute to the deterioration and degradation of forest cover.

3. On that issue, cf. Padoch et al., 2007. Shifting cultivation is indeed hard to pin down, literally, and to map – a point that tallies with James Scott's (2009) arguments about the difficulty of ruling over upland people.

4. Several methods can be used to measure the evolution of population distribution in a given territory. When population statistics are available for at least two dates, ideally more, and for a sufficient number of territorial units, for example districts or sub-districts, it is possible to measure increasing population concentration or deconcentration. Among the better tools available are the Lorenz curve and the Hoover index. We are currently using both indices for our ongoing investigations (De Koninck and Pham, 2012). Here, we limit ourselves to providing summary results obtained with the Hoover index.

5. The fact that deconcentration has not occurred in Laos is quite understandable. It is the only country, among the eight major countries in Southeast Asia, where the authorities, rather than favouring population transfers towards peripheral and mountainous areas, have instead moved entire communities of swiddeners from such areas to the central lowlands (Evrard and Goudineau, 2004). However, it seems that the process of population concentration that resulted from such population resettlement is currently being reversed. Laos is now favouring new forms of land development in northern borderland areas, largely centered on the cultivation of rubber, practised on behalf of Chinese investors and intended for the neighbouring Chinese market.

C. Opium production: The notorious cousin of shifting cultivation

An Akha Panghok woman harvesting her opium in northern Laos. Probably more than any other issue, opium production illustrates the role of state policies in swidden farming. The success of opium-suppression policies in Southeast Asia in the two decades from the early 1980s to the early years of the New Millennium also has a broader relevance (e.g. state and international efforts to eradicate coca (*Erythroxylum coca*) cultivation in the South American Andes).

Sketch based on a photo by Laurent Chazée.

30

WAS THAILAND'S HIGHLAND POLICY MISDIRECTED?

*Gary Suwannarat**

Introduction

This chapter argues that for decades, conflicting policies in northern Thailand have been shaped by crucial assumptions and beliefs about the nature of forests and forest populations, who has the right to manage forests, and other issues related to forests and people. Moreover, the same assumptions and beliefs continue to shape policy to this day (see Prachatai, 2014). This policy conflict has been further complicated over the years by inconsistent implementation at field level. After more than 30 years, government policy and messages continue to be inconsistent, communication between field and policy levels remains weak, and there is difficulty in monitoring large areas that have become increasingly accessible and attractive to powerful interests, including those from outside the highlands. While the 'opium problem' is largely solved, the opening of the highlands, which began with development projects and has since been spurred by commercial interests, including tourism, has brought new problems.

The stage was set in the profusion of official interventions in the human and natural environments of the northern mountains beginning as far back as the 1960s. The problem was opium; the mechanisms to solve it were replacement crops and parallel enforcement of opium eradication and a crack-down on opium use. These priorities created a disposition in which local customs, traditions, needs and views were largely ignored, including any role that swidden agriculture had played in the highlands. These efforts put into motion large changes among populations that had previously been quite isolated and independent, and which are still playing out today, sometimes in unexpected ways. The introduction of highland projects that prioritized opium eradication complicated the situation faced by highland people:

* DR GARY SUWANNARAT formerly worked on highland development programmes with the United Nations, the United States Agency for International Development (USAID) and the Ford Foundation.

the agricultural options open to them narrowed while the government presence in the highlands increased. Auxiliary government services – principally health and education – were seen as critical to eradicating opium, and these were eventually supported by donor countries and the United Nations. But while health and education addressed real needs among the population (albeit defined by outsiders) the focus on opium relegated other issues in the lives of highland communities to a ranking of lesser importance. Resources for development were provided largely on a basis of external understanding.

Before delving into these themes, it is important to understand the context of the highland development projects. Although the cultivation and use of opium had been both legal and a significant source of revenue for the Thai government until shortly after World War Two, the Single Convention of 1961 defined drug abuse and trafficking as national security threats (UNODC, nd; Crick, 2012). In the post-Vietnam War era, large quantities of illegal drugs flooded Western countries and overwhelmed their prevention capacity, so the international community focused on eliminating drug crops at their source – in places like northern Thailand. The United Nations International Fund for Drug Abuse Control (UNFDAC)[1] and the Government of the United States assisted Thailand in the 1970s, in the first phase of opium replacement. This was narrowly defined as crop replacement only. Subsequently, UNFDAC supported the Thai Office of the Narcotics Control Board (ONCB) in developing a Masterplan for Opium Poppy Cultivating Regions of Thailand, based on ground surveys of population, education, health and agriculture, the physical features of the land and the agricultural systems (ONCB and UNFDAC, 1983). Henceforth, this is referred to as the Masterplan.

In the early 1980s, ongoing development projects provided resources to half of the 74 highland units recognized by the ONCB.[2] These projects were supported by the World Bank, the U.S. Agency for International

Papaver somniferum L.
[Papaveraceae]

The opium poppy: the villain of the piece. It probably originated in the Eastern Mediterranean, but its use predates written history and it now grows in temperate countries around the world. The sap, mostly harvested from slashes in the seed pod, is the source of all opiate drugs, including morphine (and its derivative heroin), thebaine and codeine. The latin name means 'sleep-bringing poppy'.

Development (USAID), and the Norwegian and Netherlands governments. Then, less than a decade after the Vietnam War ended, the opium-focused efforts in the highlands gained significant complexity when Thailand defined its highland problem as one of potential Communist threat. While a security framework was important to both the international community and the host country, the main driver of security concerns probably differed. The international community was focusing increasingly on potential impacts from large-scale heroin addiction in the United States, Australia and other Western countries and less on Communism. However, Thailand focused on a perceived potential Communist threat from highland peoples, and initially had less concern about the impacts of opium addiction and trafficking in the Thai population.

The Social Research Institute of Chiang Mai University was contracted by the UNFDAC to conduct Masterplan-related surveys and, in consultation with the ONCB, to draw up preliminary plans for proposed project areas and two supporting projects: highland rice development and manpower development. The key issues outlined in these preliminary plans included agriculture, health, education and border security, noting that some 1000 hilltribe villages were outside the jurisdiction of any government agency and were 'thus vulnerable to infiltration and subversion' (ONCB and UNFDAC, 1983, vol. 1, p15). My role in this effort began in 1981, when I was asked to edit the report developed by the Chiang Mai University team.

The eight project areas proposed in the Masterplan accounted for about 40% of the 4960 hectares of opium poppies grown in Thailand in the 1982-1983 season. These areas, which can be seen in Figure 30-1, were later administered as four development projects: Sam Muen (areas 1 and 3, Mon Ya and Sam Muen, respectively); Wiang Pha (area 2); Doi Yao Pha Mon (area 4); and Pae Por (area 7). They were managed by the United Nations Project Coordination Office (PCO), which was headed initially by Richard Mann, who was previously a senior adviser to the Highland Agriculture

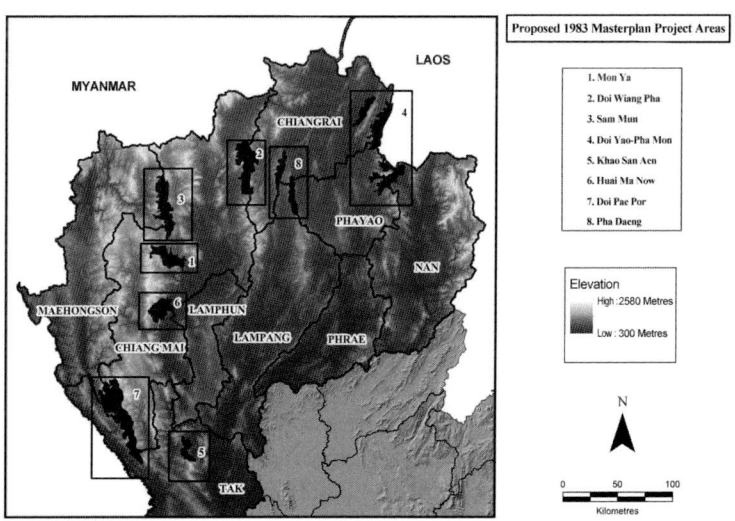

FIGURE 30-1: Highland project areas proposed by the 1983 Masterplan.

and Marketing Project (HAMP). Mann was fluent in the Karen language and well-respected in the highlands, as well as among Thai and international officials.

The Sam Muen, Wieng Pha, Doi Yao Pha Mon and Pae Por projects were each monitored by a Thai national experienced in highland development, but were overseen by the Project Coordination Office, along with other active UNFDAC/ UNDCP projects, until they ended in 1994. The PCO managed projects in cooperation with designated Royal Thai Government project managers, handled UN administrative functions and coordinated activities with the ONCB. Some widely dispersed areas of opium cultivation that were not initially covered by project funding were incorporated in the Integrated Pocket Area Development project, managed independently of the PCO and focusing, as the name suggests, on remaining pockets of opium cultivation.

Basic human needs were high on the agenda for Thailand and the international development community in the 1980s. But, as one Thai reader of highland survey results commented, '*kem mai gradik loey*', (the human-needs status indicator does not rise above zero in the highlands). In other words, highland villages were so far below the standard human needs indicators that a new set of indicators was needed to measure progress. Prior to the project interventions, the government had only a limited presence outside the lowlands. Formal education was offered in only about half of the recognized communities in the hills, and this was mainly limited to primary schools. Some Chinese-language schools were, at the time, suspected of harbouring spies or infiltrators. Satellite settlements outside the 'official' communities rarely had any schools (ONCB and UNFDAC, 1983, vol 1, p39). Infant mortality in highland areas exceeded 70 deaths per 1000 live births, more than double the rate for the Thai lowland population of 33 deaths per 1000 live births (NSO, 1995-1996; Sasiwongsaroj et al., 2008), and consistent with the infant mortality rate found by an internal survey in the Wiang Pha project area (which is no longer available in print or on the Web). Poppy cultivation was widespread; villages relied on the nearest stream or spring for water, and immunizations and basic health care were largely absent.

Access to the project areas was limited, hampering both contact with and services for the villagers and transport to markets of the crops intended to replace opium. Just a few years before the development of the Masterplan, forestry workers assigned to many highland areas had only one means of reaching their yet-to-be-constructed posts: walking from the nearest road, and the roads in those days reached no further than lowland district centres. On one occasion in the early years of the Highland Agriculture and Marketing Project, helicopter transport had to be arranged for a bumper crop of kidney beans.[3] In the mid-1980s, access to villages in the highland units targeted by the Masterplan was largely limited to forestry posts and to major villages, and then, only in the dry season. Earlier Thai and international forays into the northern mountains were mostly confined to logging and mining interests, although the mountains were home to both indigenous groups and ethnic minority migrants from elsewhere in the Mekong subregion. Access to the project areas was improved somewhat by rough dirt tracks that were periodically cleared, and in some cases, surfaced with brick using local labour.

Thailand's broad cooperation with the international community included bilateral efforts with, among others, the governments of Australia, Canada, Denmark, France, Germany, the Netherlands, Norway and the United States. French aid came through the Office de la Recherche Scientifique et Technique Outre-mer (ORSTOM), and from provision of Spot satellite images; Dutch assistance focused on coffee research and development and the Mountain People's Culture and Development Project. Multilateral efforts were carried out by the United Nations, supported principally by Italy and Sweden, and by the World Bank. This vast effort was the product of the ONCB's cooperation with UNFDAC, driven by the view that heroin processed from Thai opium posed a major threat to Western nations, and to Thailand itself. The latter concern became stronger with growing official concern in Thailand for opium trafficking, increasing heroin addiction and related crime.

The highland development crusade began less than a generation after Thailand prohibited the growing, processing and use of opium and opiates in the late 1950s. The government of Field Marshal Sarit Thanarat was then concerned that the role played in the Indochina War by the Montagnards – the indigenous mountain people of central Vietnam – would be replicated by the hilltribes of northern Thailand, and as a consequence, Thailand's security was under threat (Renard, 1986, pp15-16). Thus, a decision that was intended to prevent a problem created many more challenges.

Thailand's original decision to outlaw opium did not come easily. As recently as 1955, the Ministry of Finance opposed any such prohibition, saying that such a decision would be damaging to the country (ONCB, 2007). The Finance Ministry's reluctance probably reflected the historical importance of opium taxes to the State; possibly also impacting the personal income of Ministry staff with interests in opium concessions. Opium was initially imported from China as part of the British-imposed Bowring treaty. It was legally grown within Thailand after 1855, with both imports and domestic production controlled by the Royal Opium Monopoly (McCoy, 1972; Renard, 2001, pp12-13). Combined gambling and opium revenues comprised an estimated 40% of Thai government revenues from 1850 to 1892. From 1905 to 1926, opium accounted for roughly 20% of total state revenues (Ingram, 1955, pp177-179), falling to about 5% in 1950 (Calculated from Ingram, 1955, p185). Opium was important medicinally in remote villages for relief of pain, coughs and dysentery and to suppress the symptoms of malaria. With the near-total eradication of opium, mountain people turned to purchased medications. Although the Mae Chaem District Hospital in Chiang Mai province initiated a project to develop and standardize indigenous herbal remedies, many highland families had already shifted to commercial pharmaceuticals.

Historically, northern highland forests produced teak, controlled through much of the 19th century by British firms and Burmese foresters, with much of the cut timber floated down the Salween river and apparently not entering into Thailand's national accounts (Ingram, 1955, p93). But by 1926, timber accounted for about 4% of government revenues. While timber revenues rose in value over the years, their contribution to state revenues diminished to about 1.3% by 1950 as the size of the economy grew, additional revenue sources were developed and tax collection improved (calculated from Ingram, 1955, p185).

Early projects in the northern Thai highlands focused on opium 'replacement crops', with the United States government refusing to provide funds for anything other than crops in the early 1970s (Morey, 2013, p80). While this situation remained largely unchanged in the 1980s when the Masterplan was developed, the Masterplan incorporated recognition of the broader aspects of development, which were required to address underlying issues. At the same time, the singular focus of the Thai government was arguably fixed on controlling the highlands.

By the early 1990s, the language in the international community relating to highland projects was shifting. The terms 'integrated rural development' and 'alternative development' had gained acceptance, signalling recognition of the reality that a broad scope of development tasks were implicit in any effort to eliminate opium cultivation. On the Thai side, the aims remained unchanged, but these terms were accepted in English. By the mid-1990s, Participatory Land-use Planning (PLP) was pioneered in the Sam Muen Highland Development Project, and this influenced thinking to the extent that this approach was adopted by non-UN projects, including the Thai-German Highland Development Project, which used the term Community-based Land-use Planning (Dirksen, 2001, p90). The Thai-German project also developed community-based approaches to address drug-use problems.

The projects provided a wide range of support to local villagers in addition to extensive agricultural support, including not only seeds and seedlings, but also irrigation. Modest village education centres were built and staffed with teachers hired by the Thai Department of Non-formal Education. These teachers often gained great respect from the children and adults they taught to speak Thai. The children also studied a broader set of subjects consistent with the non-formal education curriculum. However, when UN funding ended, the Ministry of Education closed all the village education centres in favour of boarding schools in larger communities.

Health centres were established in every *tambon* (sub-district). Unlike the village learning centres, these were permanent structures made of reinforced concrete. But 'permanence' was sometimes relative. One such structure, built with materials carried into a remote area of Om Koi district by elephants, fell victim to termites, which reduced its wooden floor to scraps of cellulose in just a year. Villages received assistance to build water tanks and install pipes from the nearest spring or source of clean water, ensuring village water supplies and relieving women and girls of the time-consuming burden of carrying water for household and personal needs. In many instances, implementation of this assistance failed to account for local realities. PVC pipes crossing steep slopes in the intense highland sun delivered hot water and the pipes were trampled by livestock or shattered under vehicular traffic.

Mountain tracks and trails were upgraded, boulders were removed, and repairs were made on an annual basis, if required. While the quality of the roads was nowhere near that of the roads that snake their way through the northern Thai highlands these days, they were nevertheless crucial for farmers to get produce to markets. As rudimentary as they were, the roads brought a new set of developers to the mountains – private-sector middlemen, buying vegetables, fruit and tea, rather than opium.

Regardless of the language used to describe highland projects, their approaches and activities, there was always an outstanding question: was the goal to replace a crop in a given area or raised by specific growers, or was it to replace the economic role that opium played at household level? The implicit definition was that the projects sought to replace a crop grown in a specific area, and to target specific ethnic groups identified as poppy growers. Enforcement authorities learned quickly that eliminating poppies in large fields led the farmers to create handkerchief-sized fields that were harder to spot in aerial photos or in satellite imagery. The authorities then developed more sophisticated reviews of remote images, so the growers interplanted poppies with other crops in an effort to shield the illicit plants from detection. It was a cat-and-mouse game. Enforcement authorities put boots on the ground for ground surveys, so the growers opened new fields outside Thailand's borders and opium remained available within Thailand, partly because of the cross-border production and partly because of the country's existing network linked to international trafficking. In more recent decades, drug consumption within Thailand has exploded – but the product of choice is now *yaa baa* (amphetamines). Mimicking the old opium traffic, much of this comes from across the border and there is little limit on where it can be manufactured. As many who have studied drug control have observed, eliminating illegal drug production is like squeezing a balloon – a squeeze on one part produces a bubble on another.

Both the projects and their donors paid little attention to the economic role of opium poppies among highland households. Field work by Kathleen Gillogly in a Lisu village in the Sam Muen project area provided an insight into the importance of poppy-growing to village families:

> Opium introduced a relatively stable source of wealth, storable and portable either on its own or converted into silver, granting small scale upland peasants a great deal of autonomy regarding when and where to sell their crop. Significantly, opium was ecologically suitable to mountain soils, so it opened up economic and ecological niches that had previously been unusable by humans. It could be grown anywhere; it was very high value per unit of weight; merchants came to the village to buy it. It gave households a cushion of safety for buying food, land and labour when establishing themselves in new locations. In short, opium increased and diversified the household subsistence portfolio, making it more stable (Gillogly, 2008, p684).

Today, opium poppy cultivation is largely gone from Thailand, falling from nearly 9000 hectares in 1984 to a few hundred hectares in 1995, when most UN-supported highland projects ended, and about 200 hectares per year currently (ONCB, 2010; Douglas, 2014). In 2008, ONCB Deputy Secretary-General Pittaya Jinawat attributed this fall in opium production to 'recognition of the overlap of poverty and illicit crops' (Jinawat, 2008). 'Thus,' he said, 'the narcotics crop control programme has been treated as a development issue and integrated into the national social and economic development and highland development plans of the country.' Improved roads now reach far into the hills and government services are more available. More (but not all) people have Thai citizenship and, therefore, access to benefits ranging from freedom of movement to health care, education and greater income-producing alternatives. The health and education status of highland peoples has improved, as has access to markets and integration with the Thai (and global) society and economy.[4]

Nevertheless, misplaced myths have driven, and continue to drive, much of Thailand's policy regarding the highlands and highland peoples. Among these myths are those built upon the following topics:

1.   Who is Thai?
2.   The nature of Thai forests.
3.   Who manages Thai forests?
4.   What characterizes 'ideal' land management in the highlands?

I will now discuss each of these issues before finally reflecting upon the overall opium eradication effort in relation to broader issues of development and the role of the state, the impacts of the projects on dwindling swidden agriculture relative to other agricultural practices, and some larger impacts of highland development.

Who is Thai?

A Bangkok-centric stereotype holds that Thais are all one people. This ignores the great historical, cultural and linguistic diversity within the country's borders. When a group of people obviously does not fit the stereotype, they are often seen as illegal interlopers. The peoples of the northern highlands clearly fall into this category. There are more than a dozen groups, each with its own distinctive language and cultural, clothing and agricultural traditions. The principal ethnic groups include the Karen, Hmong, Mien, Lahu, Lisu, Akha, Khamu, northern Thai and Tai Yai, or Shan. Of these, the Karen or Pawkayoh are documented to have lived in the northern and western forests of Thailand for centuries (Kunstadter, 2015, p137, citing Renard, 1980), and likely lived there for some time prior to any documentation. Lawa or Lua, Thin and Khamu are historical residents of the borderlands of the Mekong subregion. Interviews conducted in 1991 found that most ethnic minorities believed that their ancestors had migrated to Thailand between the late 1800s and the early 20th century, fleeing the chaos of that period in Southwest China. In the same period, many Chinese from other regions arrived in Bangkok by sea. This migration predated the change in the country's name, from Siam to Thailand, and the introduction of democratic government. Others arrived as upheavals related to World War Two impacted neighbouring countries. Smaller numbers of new migrants arrived periodically during the implementation of many of the highland development projects in the 1980s and 1990s, as conditions beyond Thailand's borders drove people out of their homes and farms.

The Hmong and Yao mainly live at the highest altitudes and were among the ethnic groups most heavily involved in poppy cultivation. The Akha, Lisu and Lahu also grew poppies, although perhaps less extensively than the Hmong and Yao. The Karen (who are still said to make up nearly half of the ethnic minority population of northern Thailand), did not historically grow opium poppies and generally lived in the middle zone, below 900 metres, considered inappropriate for poppy cultivation. However, as Karen grew skilled in poppy cultivation as workers in Hmong fields, some began growing poppies themselves.

The involvement of ethnic minorities in what was an illegal activity reinforced the reluctance of government officials to extend them Thai citizenship. Yunnanese Chinese who had served in the Kuomintang (National Revolutionary) Army, on the other hand, were a privileged group with an easy path to citizenship. This cemented the position of some of them in the opium trade, since citizenship conferred ease of movement not available to non-citizens. The respect accorded to former Kuomintang soldiers by many Thai officials also smoothed the path for advantageous dealings with government officials, regardless of the legal status of their business.

From the highlanders' perspective, official reluctance to extend them citizenship reached bewildering extremes. The legendary story of one man resonates because it is so believable: asked by a government official to produce his identification card, he asked "Which one?", and laid out multiple cards of many colours, indicating that he had been included in successive surveys over a number of years, but had never been granted citizenship (Sakboon, 2013; Laungaramsri, 2014).[5] The cultural distinctiveness of the hill people worked against them: as early as 1921, local Thai officials wanted to evict a group of Hmong who resisted Thai conscription, although Renard (2001) says 'calmer voices prevailed'.

The Masterplan itself was silent regarding citizenship status: it addressed village and population characteristics, including population density, growth rates, in-migration and availability of schools. In referring to community groups, it noted that 'more than half of [them] are village development groups established by the District Office' (ONCB and UNFDAC, 1983, vol. 1, p39; see also UNESCO, 2008).[6] Insights regarding indigenous organizations or the citizenship status of the population were absent.

However, United Nations efforts under the Masterplan provided funding for the Ministry of Interior's efforts, through the Department of Local Administration, to achieve household and individual registration with the ostensible aim of expediting citizenship (see Dirksen, 2001, p93).[7] Although a 1956 Cabinet decision called for accelerated extension of citizenship to highland people, risk-averse officials were in no rush to do so and there was little progress in this regard during the 1980-2000 'development project' period. Amanda Flaim (2017) addresses the complexity of the system for considering citizenship applications – a bureaucratic enterprise that she argues has 'contributed as much to the situation of protracted statelessness among highlanders as to the systematic resolution of citizenship claims among them'. Ministry of Interior officials privately justified their reluctance to accelerate the process of extending citizenship on grounds of personal liability, if citizenship were granted to someone involved in illegal activity.

As one Yao man reportedly said in the 1990s: 'Without citizenship, it's hard to cross district boundaries to trade in the lowlands. The only thing I can do is grow opium'.[8] The UNESCO Highland Citizenship and Birth Registration Project noted that lack of citizenship impacted on job and life opportunities (UNESCO, nd). Without citizenship, everyday acts were difficult, if not impossible, including education beyond village schools, legal livelihoods without fear of unscrupulous traders or officials, and the ability to resettle one's family.

While some progress has since been made in extending citizenship to highlanders, more remains to be done. As I recall, an estimated 40% of highland people held citizenship in the mid-1980s, increasing to about half by the year 2000 (Hu and Podhisita, 2008, p19). Currently, about three quarters of highlanders hold citizenship, based on a 2010 UNESCO-Thai Bureau of Social Development and Human Security (BSD) survey. Of the remaining quarter, many meet the legal qualifications but have yet to be granted citizenship. While the UNESCO-BSD (2010) survey is large enough to be statistically significant, scepticism continues regarding highland citizenship numbers (Kongchantuk, 2015).

A fundamental concern in this situation, where hundreds of thousands of highland people are affected, is that citizenship limbo contributes to precisely the problems that governments might seek to avoid. Individuals are more likely to be involved in illegal activities due to a lack of legal choices and they are more likely to be prey for human traffickers. The potential for further immigration from neighbouring countries remains high because of porous borders and known loopholes in the system. Regardless of their eligibility, highland people without citizenship continue to be denied basic rights. Immediate resolution of this situation, by truly expediting the conferral of citizenship on those remaining in limbo, would benefit all stakeholders, including the Thai government.

The nature of Thai forests

The past view of some Thai decision-makers appears to have been that the northern highlands have historically been, and should remain, home to undisturbed forests, preserved without human habitation. These forests are largely tropical hill evergreen, mixed deciduous and dry dipterocarp, and not tropical rainforests as some 'green' non-governmental organizations have suggested. Such an undisturbed state is without historical precedent in Thailand, but it does reflect conditions in many parts of Scandinavia and North America, where Thai academics and leaders studied forestry and environmental science. It also supports other policy objectives related to integration of remote areas and populations. Therefore, it is not too surprising that this view appears to have been incorporated into the national government's watershed classification system, administered by the Royal Forest Department (RFD, and now the Ministry of Natural Resources and Environment). In the mid-1800s, virtually all of Thailand beyond the floodplain was forested (Phongpaichit and Baker, 1995, p7). In 1960, 54% of the country's total land area was forested, but by 1988 this had fallen to 25% (Thomas et al., 2004a, p11). The forest area in northern Thailand has remained consistently higher than in other regions, but it fell from 69% to 43% in the period from 1960 to 1988 (Suraswadi et al., 2005). Northern watersheds are critical to major rivers serving agriculture, households and industry in both the highlands and the valleys of the north, before they flow to the Central Plains and the capital city, Bangkok. Control of these forests and the entirety of northern Thailand has long been a major concern of the central government, whether for opium and teak income, for security reasons during the Vietnam War, for agroforestry managed by the Royal Forest Department, for water supplies, or for the many issues surrounding the contemporary problem of global climate change.

Highland peoples and their land management

Thai and international concepts of appropriate land management in the highlands have long stood at odds with the traditional practices of highland ethnic groups. Slash-and-burn is the pejorative epithet applied by many outsiders to highland farming, ignoring the nuances associated with different systems. This form of land use is mentioned in the 1983 Masterplan: 'With the cultivation of rainfed highland rice and poppies, land is denuded through forest destruction as slash-and-burn agriculture is employed in the cultivation of both these crops' (ONCB and UNFDAC, 1983, Vol. 2, p283). Harold Brookfield (2015, p25) refers to negative attitudes in the international agricultural community about swidden cultivation (specifically that of the Food and Agriculture Organization of the United Nations). This perspective clearly influenced forestry research and thinking in Thailand, as well as the design and interpretation of surveys prior to the development of the UNFDAC/UNDCP projects in the country's northern highlands.

During UN project operations, I observed swidden systems among Karen (see also Trakansuphakon, 2015), Akha, Lisu and Lahu farmers. These were often augmented by raising free-range livestock, particularly among the Karen, with livestock accounting for more than half of household wealth (Kesmanee, 1991). Karen farmers in the Om Koi district of Chiang Mai province acknowledged in 1991 that the swidden cycle, originally around 20 years, had been greatly disrupted. It was then about five years in duration. When asked why, the response was: *luug maag* (lots of children). It is unlikely that this meant the Karen were actually bearing more children. Instead, there were probably fewer children dying in childbirth and the critical early years. Is it possible that a swidden system that had existed for centuries had been upset by improved maternal and child health care? Or were villagers repeating the perceived 'correct' response, based on earlier contacts with the government? The response to this reported change in balance between population and the carrying capacity of the land was two fold: shorten the swidden-cycle period and seek other income alternatives. Karen who had worked in Hmong fields and had learned the skills involved in growing poppies and harvesting the raw opium put those skills to work by growing opium in their own fields. Within a few short years, Om Koi also became a major cash-cropping area – at that time, of tomatoes. More recently, Om Koi and Mae Chaem have seen the development of large areas of contract corn farming, and this has contributed to reduced air and water quality in the Chiang Mai basin.

Historical evidence suggests that the rotational swidden systems of the Karen and the Lua had sustained human populations for hundreds of years, and an early British government official observed that Karen farmers were the only shifting cultivators who had a surplus of rice to sell. In the mid-20th century, a study of established Lua swiddens in Mae Hong Son province concluded they had 'apparently achieved a balance with the conditions of soil, vegetation, geology and climate' (Zinke, 1978). That balance was changing, however, as improved health care and demographic movements fuelled population growth in the highlands.

In the early 1990s, Hmong and Yao farmers were seen as pioneer swiddeners, clearing primary forest and cropping an area until the soil fertility was exhausted, then moving on to open a new area of forest. They were heavily involved in poppy cultivation and often hired labour from other ethnic groups. As this hired labour became sufficiently skilled at growing poppies and harvesting raw opium, some became growers themselves. Thus, there was a need to focus on smaller pockets of poppy cultivation, almost entirely at the hands of Karen growers who had developed poppy-cultivation skills as labourers for Hmong growers. The final UN poppy-eradication project was the Integrated Pocket Area Development Project (IPAD).

The lack of clarity among officials regarding the swidden cultivation cycle was demonstrated in a 1991 IPAD pre-project survey of Om Koi district, in which the author took part. After spending a day visiting villages, the survey team returned to the Om Koi district centre, which was then a village rather than a town, with a market, a few shops, the district offices and a jail. Before leaving the following morning, word came that an elderly Karen couple had been arrested for forest destruction in one of the proposed IPAD project areas. We were told that the old couple and their extended family had made offerings to the spirits of the swidden they were re-opening after some years of fallow. The younger members of the family were sufficiently fleet of foot to escape from an official who arrived on the scene, but the old couple were arrested and locked up in the Om Koi jail. Eventually, the charges were dismissed, but not without lingering misunderstandings. The official made no distinction between old forests and swidden fallows. This was perhaps understandable, since Thai forest policy made no such distinction (Thomas, 2015).

When it came to work in the swiddens, Figure 30-2 makes an important point: most field work was done by women, often with infants tied on their backs and young children accompanying them in the fields. The longest-running of all the externally funded and managed highland projects, the Thai–German Highland Development Project, recognized this in the early 1990s, documenting the fact that women accomplished most of the agricultural work (Dirksen, 2001, p94). Akha women have been found to play a critical role in forest management (Sturgeon, 1998). Would a more nuanced understanding of highland agriculture, focused on swiddening and the roles of women in agriculture, have produced different results? Greater inclusion of women in survey and decision-making processes may potentially have changed the outcomes.

Ideal land management in the highlands

Concerns about the state of the country's forests and impacts on both the quality and quantity of water supplies to the lowlands led to the development of a watershed classification system in Thailand based on the importance of an area as a watershed, the steepness of slopes (presumably reflecting vulnerability to erosion) and the quality of forest cover. Implementation of this system began in northern highland areas in 1982. Class 1A and 1B watersheds were in headwaters surrounded by healthy (1A) or partially degraded forests (1B). Class 2 watersheds were also important as water sources, but were eligible for other uses, such as mining, but not for agroforestry. Class 3 watersheds were of lesser importance as such and appropriate for forestry, mining and tree cropping. Class 4 watersheds were those that had been cleared and

FIGURE 30-2:   A group of Karen women pass through their swidden on their way home after a day's work.

Photo: Gary Suwannarat (1972).

used for field crops. Class 5 included flatlands, valleys and moderate slopes with little or no remaining forest and, therefore, considered appropriate for other uses (Preechapanya, 2015).

While the need for watershed conservation was clear, a number of issues challenged the implementation of the classification system. In a presentation to the Food and Agriculture Organization of the United Nations in 2002, Kaosa-ard and Rutherford raised management and economic issues, including the difference between political boundaries and those for watershed management; fragmentation of forest management among watershed units, national parks and national forests by the Royal Forest Department; the challenges of decentralization to the role of the RFD; and the economic importance of non-timber forest products to highland residents (Kaosa-ard and Rutherford, 2002).

Furthermore, the apparent failure of the watershed classification system to integrate information on human habitation resulted in the investments of highland farmers and, in some cases, of multilateral projects (including those funded by both foreign and Thai governments) being effectively cancelled out. For example, the government accepted international funding for opium-crop replacement in the Pha Daeng area, at the intersection of Lampang, Phayao and Chiang Rai provinces, then a few years after project completion, it evicted the villagers and declared the area a national park. Road networks in the hills have since improved, and the area from which villagers were evicted and told never to return, even to harvest their coffee and lychees, now advertises tourist facilities, like most other Thai national parks. Competition for control of highland land for orchards, plantations, meditation retreats, resorts or other commercial enterprises has increased, and conflicts have arisen over upland water resources as various interests compete for the water that eventually feeds into the Chao Phya river.

Forced evictions, followed by forced relocations to places unsuitable for habitation, destroyed social cohesion and exacerbated weak relations between the Thai government and forest dwellers. Although there were reports of some successful relocations in the 1920s, the record of such actions in the late 20th century is regrettable. Phop Phra, in Tak province (Cultural Survival, 1988) and Pha Daeng village are two such cases: Hmong villagers, in both cases, were evicted with short notice and resettled in areas unfit for agriculture or human life, with no water sources. At the time of these relocations, there was virtually no unoccupied arable land with access to water remaining in Thailand.

As can be seen from the brief description of Watershed Classes above, there is no mention of swidden agriculture as a potentially ecologically friendly land use. Moreover, some foresters expressed quite a hard-line view on ethnic minorities, suggesting to me that they all represented problems, and the American way of putting them on reservations was the only long-term solution. In effect, restrictions on the movement of those without citizenship have had the effect of confining highland minorities to virtual reservations, restricting their movement and leaving them with increasingly constricted agricultural options.

There are alternatives to such draconian measures. Cooperating with academics, the Sam Muen Highland Development Project (SMHDP) introduced three-dimensional mapping as a tool for Participatory Land-use Planning, within the context of the watershed classification system (Tan-Kim-Yong, 1992, pp8-15, 2002, pp4-5). This proved to be extremely useful in developing common understandings among foresters, who by law were tasked with protecting forests and watersheds – and farmers. As a result, villages within the SMHDP project area developed resource management agreements that demarcated protected areas, farming areas and forest areas where local households could obtain timber for minor household repairs and other uses. Perhaps one of the greatest contributions of the three-dimensional maps was that they provided a medium through which people who lacked sophisticated mutual knowledge of each other's language could communicate and develop a greater respect for one another's knowledge and perspective (Thomas et al., 2004b, p16). Perhaps as a consequence, considerable natural forest regeneration occurred during the term of the Sam Muen project.

Drastic enforcement of the ban on opium growing – usually involving the destruction of crops just before the sap was collected from mature pods – and the Royal Forest Department's restrictions on agricultural activities in the highlands, combined with population growth to shorten the swidden cycle as shifting cultivators struggled to meet household needs from less land. Simultaneously, farmers were expected to grow 'replacement' crops, some of which had a lower value than opium and, under normal circumstances, would have required additional land to carry extra crops, in order to meet normal income needs. A failure to recognize fallow forest as old swiddens, rather than 'natural forest', led to deepening conflict between highlanders and the state. Demand for farmland in the highlands continued to be driven by young people who lacked both citizenship and options outside the highlands. Their predicament has been documented by Flaim (2017), who points to demand from outsiders for land in the highlands as an even more problematic influence on their future.

The Royal Forest Department emphasized replanting of degraded forests, which required funding for nurseries, seedlings and manpower. It was inconsistent with indigenous land management, which relied on natural regeneration of forests, including swidden cycles, to provide for regrowth of trees and natural replenishment of soil fertility. During my days with UN highland projects, the UN and donor governments provided funding for reforestation in some project areas, with little noticeable impact. As mentioned earlier, the Sam Muen Project relied successfully on natural regeneration, with forest areas defined by community land-use agreements. In other highland areas managed by the RFD, project funds were used for reforestation with little impact.

A number of lesser issues confronted project workers and villagers alike. Consultations with ethnic-minority villagers were uneven. Language differences and widely differing conceptual frameworks compounded the difficulty of developing common perspectives regarding poppy cultivation, swiddening and other agricultural practices and forest management, among a host of other issues, where developing a common understanding would have been immensely helpful.

In some instances, the RFD's communication with other agencies and highland communities was inadequate. One official of a continuing highland project summed it up: 'The RFD waits until illegal construction in highland resort communities is completed, then presents a warrant for the arrest of the owner, and a years-long legal process begins. They don't sit down and talk with villagers and village leaders to develop constructive dialogue and guidelines' (Anonymous, 2014).

Ethnic-minority villages had long managed their own affairs without a government presence, and they saw little reason to change their traditional practices. That attitude changed as poppy fields ready for harvest were destroyed at an increasing pace, under enforcement programmes that were not linked to the management of development projects – a nuance that no doubt escaped poppy growers. Kathleen Gillogly (2008 and chapter 38 of this volume) summarizes the economic reasons for growing opium and reports on the impoverishment and social impacts imposed on highland families by opium interdiction efforts.

Project areas based on elevation, and therefore potential for opium production, did not follow neat administrative lines. The Sam Muen project, for example, covered areas in both Mae Hong Son and Chiang Mai provinces. A forester accustomed to working within a distinct geographic area found himself dealing intensively with two provinces, several districts, a number of subdistricts and village administrations.

For many government officials, the idea of granting highland villagers a role in forest management was difficult to understand and implement. As one local administration official said after visiting the Sam Muen project in a bid to understand collaboration with highland communities in the development of community forest management plans: 'I'll issue the orders to communities in district X when I return on Monday'. The concepts of collaboration and community engagement had been lost, and the official retreated in bewilderment to his old 'top-down' fixation. He was schooled in action and expected to produce quick results, so he looked for tangible outputs rather than concentrating on the process (Jingsoongnern, 1992).

Misconceptions were more likely to be held by Bangkok-based policy-makers than field staff responsible for project implementation. Therefore, long experience suggests that the disconnection between formal policy and field realities results in a lack of policy clarity, leading to schizoid policy. Cabinet-level policy that is inconsistent with needs on the ground generates actions like forced evictions, relocations and dumping people in unliveable conditions, all of which destroy social cohesion and exacerbate weak relations between the government and forest dwellers. While this is going on, the same government agencies cooperate with bilateral and international agencies to develop agriculture, health services and education facilities in locations previously or later declared to be restricted zones. Agricultural communities are separated from their only known means of livelihood, thus impoverishing and alienating the population.

Unanticipated changes and consequences of development

When I first worked in the highlands, pickup trucks, motorcycles and satellite dishes were largely unknown. While these items are not yet as common as they are in the lowlands, almost every village in northern Thailand now has at least one pickup, one satellite dish and many motorcycles. As local governments have taken over a large part of expenditure for local infrastructure, concrete roads have made their way into once remote areas (Figure 30-3). A huge variety of garden produce, field crops and flowers is trucked daily to Chiang Mai and Bangkok. Tourists flock to once–isolated villages during holidays, and as a result, competition for water in the highlands has increased.

Opium cultivation is down and has stayed down over the past 20 years. A generation of highlanders has now grown to adulthood largely without opium as a crop, an economic resource or a home remedy. Detoxification of highland opium users made limited inroads into the effort to reduce the use of opium during the highland project era. Thailand's most recent and draconian War on Drugs in 2003, which resulted in 2500 deaths, including those of many innocents (Smith, 2007), was largely unrelated to the use of opiates. It failed, once again, to end the curse of addiction or rid the country of drug dealers.

An unanticipated consequence of the drop in the availability of opium in the hills was the flood of heroin trafficked across Thailand's borders. The cheap availability of heroin was promoted by using 'direct sales' models ('get five friends addicted, and your

FIGURE 30-3:   A new concrete road winds its way through forested hills in Mae Wang district, southwest of Chiang Mai city. Recent growth of local budgets has supported a major expansion of road networks penetrating the northern Thai highlands, facilitating delivery of crops to market and increased tourism, property speculation and competition for water and land resources.

Photo: Gary Suwannarat (2015).

supply is free' – with limitations, of course). Easy availability of injection equipment (sometimes shared) and a hit that didn't require long preparation or smoking time made it easier to elude detection. Given that this was happening at the time of the HIV pandemic, the scene was set for a disaster much larger than opium use. Reports still circulate of highland villages populated by large numbers of AIDS widows and their children.

As poppy cultivation declined, one might have expected that the forests of northern Thailand would recover. Instead, conversion of forest to farmland in specific areas of the north has accelerated, abetted by large-scale contract farming, principally of corn (Saengpassa, 2015). Nearly 290 square kilometres were planted in corn in Chiang Mai province in the 2010-2011 season, mainly in Mae Chaem and Om Koi districts. While this was about 1% of the province's total area, it amounted to a quadrupling of the area planted to corn in Chiang Mai province in 10 years (since 68.4sq. km were planted in corn in 2000). This was driven in part by government promotion of 'gasohol' (ethanol) production to reduce consumption of petroleum fuels in motor vehicles. Elsewhere in Thailand's upper north, Nan province grows more than 1000sq. km of corn; an estimated 60% of it on recently cleared forest land (Wangkiat, 2015). Where there is demand, supply will follow, and where there is subsidy, supply will challenge environmental limits.

The environmental impacts of corn extend from farm level to the wider system. Corn is a hungry crop, exhausting soil fertility relatively quickly if grown continuously on the same plots. When grown on steep slopes, the potential for erosion is considerable, resulting in more sediment flowing into the rivers of the north and thence to the country's 'rice bowl' on the Central Plains. However, the most controversy arises from heavy air pollution during the dry season, said to result from the burning of corn wastes. There is perhaps a need to reconsider the contribution to this pollution from corn-waste fires. Lamphun province and the corn-growing areas in Mae Chaem and Om Koi are equidistant from Chiang Mai. Yet Lamphun experiences consistently lower pollution levels than does Chiang Mai. This suggests that motor vehicle exhausts and the burning of municipal and household wastes in a large city with weak regulations are co-culprits in poisoning the atmosphere. Clearly, additional data on a broad set of factors are required, rather than knee-jerk solutions once again penalizing highland farmers.

Foresters are concerned about the spread of agriculture into vulnerable areas of the highlands. Some three decades after the Watershed Classification System was implemented, it is perhaps time to review the country's forest management again. Does Participatory Land-use Planning of the sort pioneered in the Sam Muen project offer an alternate way forward? How do communities deal with outside interests, sometimes protected by political backing outside the hills? By allowing destructive practices (agricultural or otherwise) to proceed and then to take violators to court sets in motion a process that usually takes many years to resolve – often with little or no impact on 'illegal' behaviour. Meanwhile, environmental damage continues, from corn cropping, rampant urban and rural construction and the exhaust fumes of internal combustion engines.

Extending my comments to current problems carries some risks, because my knowledge of contemporary issues is somewhat limited. However, the resources that allowed the government to expand its reach into the northern mountains originated with the highland development projects of the 1980s and 1990s. Those resources also accelerated other changes in the highlands; changes which tied former subsistence households more closely to markets, rather than to their traditional systems, and increased the need for cash income. These changes have also tied households more tightly to Thai administrative systems and services. Yet the children of many ethnic-minority communities have no access to schools in their own villages and must board away from their families and communities for entire school terms. Few of these changes have been favourable to swidden agriculture, and perhaps it is time to consider subsidies for swiddening, rather than for gasohol refined from mountain cash crops.

Accommodating the humanity of ethnic minorities; allowing them the agency to make their own decisions, to negotiate with authorities and to carry full Thai citizenship is essential in order to move forward in a positive manner in Thailand's northern mountains.

References

Anonymous (2014) Interview conducted by the author in confidence, December 2014

Brookfield, H. (2015) 'Shifting cultivators and the landscape: An essay through time', in M. F. Cairns (ed.) *Shifting Cultivation and Environmental Change: Indigenous People, Agriculture and Forest Conservation*, Earthscan by Routledge, London

Crick, E. (2012) 'Drugs as an existential threat: An analysis of the international securitization of drugs', *International Journal of Drug Policy* 23(5), pp407-414, http://dx.doi.org/10.1016/j.drugpo.2012.03.004, accessed 22 December 2015

Cultural Survival (1988) 'Hmong relocated in northern Thailand', *Cultural Survival Quarterly* 12(1), Spring 1988, Cultural Survival Inc, Cambridge, MA, http://www.culturalsurvival.org/publications/cultural-survival-quarterly/thailand/hmong-relocated-northern-thailand

Dirksen, H. (2001) 'Thai-German Highland Development Program (TG-HDP) in Northern Thailand', in *Alternative Development: Sharing Good Practices, Facing Common Problems,* proceedings of a Regional Seminar on Alternative Development for Illicit Crop Eradication Policies, Strategies and Actions, 16-19 July 2001, Taunggyi, Myanmar, UNDCP Regional Centre for East Asia and Pacific, Bangkok

Douglas, J. (2014) 'Opium production soars in Southeast Asia', *Deutsche Welle* 9 December 2014, Bonn, Germany, http://www.dw.de/opium-production-soars-in-southeast-asia/a-18118229, accessed 24 December 2015

Flaim, A. (2017) 'Problems of evidence, evidence of problems: Expanding citizenship and reproducing statelessness among highlanders in northern Thailand', in B. N. Lawrance and J. Stevens (eds) *Citizenship in Question: Evidentiary Encounters with Blood, Birthright, and Bureaucracy*, Duke University Press, Durham, NC, and London, pp147-164

Gillogly, K. A. (2008) 'Opium, power, people: Anthropological understandings of an opium interdiction project', *Contemporary Drug Problems* 35 (Winter), http://www.academia.edu/483852/Opium_Power_People_Anthropological_Understandings_of_an_Opium_Interdiction_Project, accessed 24 December 2015

Hu, J. and Podhisita, C. (2008) 'Differential utilization of healthcare services among ethnic groups on the Thailand-Myanmar border: A case study of Kanchanaburi province, Thailand', *Journal of Population and Social Studies* 17(1)

Ingram, J. C. (1955) *Economic Change in Thailand since 1850*, Stanford University Press, Stanford, CA

Jinawat, P. (2008) 'Thailand's experiences of sustainable alternative development and opium reduction', in *Sustaining Opium Reduction in Southeast Asia: Sharing Experiences on Alternative Development and Beyond*, UNODC Global Partnership on Alternative Development (GLO/I44), Chiang Mai

Jingsoongnern, P. (1992) Personal communication between the author and Pakorn Jingsoongnern, Project Manager of the Sam Muen Highland Development Project

Kaosa-ard, M. and Rutherford, J. (2002) 'Cross-sector linkages in mountain development: The case of northern Thailand,' paper presented to the UN Food and Agriculture Organization, June 2002

Kesmanee, C. (1991) *IPAD Project Area Survey*, the Integrated Pocket Area Development (IPAD) project, no copy available

Kongchantuk, S. (2015) Personal communication between the author and Surapong Kongchantuk of the Thai Lawyers' Association, 24 April 2015

Kunstadter, P. (2015) 'Swiddeners at the End of the Frontier: Fifty years of globalization in northern Thailand, 1963-2013,' in M. F. Cairns (ed.) *Shifting Cultivation and Environmental Change: Indigenous People, Agriculture and Forest Conservation*, Earthscan by Routledge, London

Laungaramsri, P. (2014) 'Contested citizenship: Cards, colors and the culture of identification', in J. Amos (ed.) *Ethnicity, Borders, and the Grassroots Interface with the State: Studies on Mainland Southeast Asia in Honor of Charles F. Keyes*, Silkworm Books, Chiang Mai

McCoy, A. W., with Read, C. B. and Adams, L. P. (1972) *The Politics of Heroin in Southeast Asia: CIA Complicity in the Global Drug Trade*, Harper and Row, New York, http://druglibrary.eu/library/books/McCoy/book/19.htm, accessed 23 December 2015

Morey, R. D. (2013) *The United Nations at Work in Asia: An Envoy's Account of Development in China, Vietnam, Thailand and the South Pacific*, McFarland Publishing, Jefferson, NC

NSO (1995-1996) *Report on the 1995-1996 Survey of Population Change*, National Statistical Office, Government of Thailand, Bangkok, http://web.nso.go.th/eng/en/stat/popchang/popchgt4.htm, accessed 23 December 2015

ONCB (2007) *About Us: Background*, Office of the Narcotics Control Board, Ministry of Justice, Bangkok, http://www.oncb.go.th/EN_ONCB/Pages/background.aspx, accessed on 5 May 2017

ONCB (2010) *Thailand Narcotics Control: Annual Report 2010*, Office of the Narcotics Control Board, Ministry of Justice, Bangkok

ONCB and UNFDAC (1983) *A Masterplan for Development of Opium Poppy Cultivating Regions of Thailand*, Vols I and II, Office of the Royal Thai Narcotics Control Board with the assistance of the U.N. International Fund for Drug Abuse Control, Bangkok

Phongpaichit, P. and Baker, C. (1995) *Thailand: Economy and Politics*, Oxford University Press, Kuala Lumpur

Prachatai (2014) 'Junta's attempt to 'return forest' hurts the poor', *Prachatai English,* Foundation for Community Educational Media (FCEM), Bangkok, available at http://www.prachatai.com/english/node/4441, accessed 22 December 2015

Preechapanya, P. (2015) Personal communication between the author and Dr Pornchai Preechapanya, formerly of the Watershed Management Division, Royal Forest Department, 27 April 2015

Renard, R. D. (1980) 'History of Karen-Tai Relations: Red Karen', PhD dissertation to the University of Hawaii, Honolulu

Renard, R. D. (1986) 'Problems of small farmers in highland areas', in *Data Requirements for Highland Farming Systems Development*, Proceedings of a workshop sponsored by the Food and Agriculture Organization of the United Nations, 21-25 April 1986, Payap University, Chiang Mai

Renard, R. D. (2001) *Opium Reduction in Thailand 1970-2000: A Thirty-Year Journey*, Silkworm Books, Chiang Mai

Saengpassa, C. (2015) 'Private sector must help fight smog blanketing the northern provinces', *The Nation*, Bangkok, 24 March 2015, http://www.nationmultimedia.com/politics/Private-sector-must-help-fight-smog-blanketing-the-30256618.html, accessed 30 December 2015

Sakboon, M. (2013) 'Controlling bad drugs, creating good citizens: Citizenship and social immobility for Thailand's highland ethnic minorities', in B. Coeli (ed.) *Rights to Culture, Heritage, Language, and Community in Thailand*, Silkworm Books, Chiang Mai

Sasiwongsaroj, K., Sethaput, C., Vapattanwong, P. and Ford, K. (2008) 'Child mortality inequality between Thais and hilltribes in Thailand: Study from Population and Housing Census 2000', *Journal of Population and Social Studies* 16(2), pp143-162, http://www.jpss.mahidol.ac.th/index.php?option=com_docman&task=doc_download&gid=120&Itemid, accessed 23 December 2015

Smith, P. (2007) 'Southeast Asia: Most killed in Thailand's 2003 Drug War not involved with drugs, panel finds', *Drug War Chronicle* 152, StoptheDrugWar.org, Washington, DC, http://stopthedrugwar.org/chronicle/2007/nov/30/southeast_asia_most_killed_thail, accessed 30 December 2015

Sturgeon, J. (1998) Personal communication between the author and researcher Dr Janet Sturgeon, in 1998

Suraswadi, P., Thomas, D. E., Pragtong, K., Preechapanya, P. and Weyerhauser, H. (2005) 'Northern Thailand: Changing smallholder land use patterns', in C. Palm, S. A. Vosti, P. A.Sanchez and P. J. Eriksen (eds) *Slash and Burn Agriculture: The Search for Alternatives*, Columbia University Press, New York

Tan-Kim-Yong, U. (1992) *Participatory Land-use Planning for Natural Resource Management in Northern Thailand*, Rural Development Forestry Network Paper 14b, winter, Overseas Development Institute, London

Tan-Kim-Yong, U. (2002) *Constructing Political Process and Reform for Decentralization in Thailand: Three Case Studies in Decentralized Natural Resource Management*, paper delivered to Decentralization and the Environment Conference, 18-22 February 2002, Bellagio, Italy, Institutions and Governance Program, World Resources Institute, Washington, DC

Thomas, D. E. (2015) Personal communication between the author and Dr David Thomas of the World Agroforestry Centre (ICRAF), Chiang Mai, 4 June 2015

Thomas, D. E., Preechapanya, P. and Saipothong, P. (2004a) *Landscape Agroforestry in Northern Thailand: Impacts of Changing Land Use in an Upper Tributary Watershed of Montane Mainland Southeast Asia*, ASB Thailand Synthesis Report 1996-2004, World Agroforestry Centre (ICRAF), Chiang Mai

Thomas, D. E., Preechapanya, P. and Saipothong, P. (2004b) *Developing Science-based Tools for Participatory Watershed Management in Montane Mainland Southeast Asia*, final report to the Rockefeller Foundation, World Agroforestry Centre (ICRAF), Chiang Mai

Trakansuphakon, P. (2015) 'Changing strategies of shifting cultivators to match a changing climate', in M. F. Cairns (ed.) *Shifting Cultivation and Environmental Change: Indigenous People, Agriculture and Forest Conservation*, Earthscan by Routledge, London, pp335-356

UNESCO (2008) *Capacity Building on Birth Registration and Citizenship in Thailand: Citizenship Manual*, Asia and Pacific Regional Bureau for Education, United Nations Educational, Scientific and Cultural Organization (UNESCO), Bangkok, http://unesdoc.unesco.org/images/0016/001621/162153e.pdf, accessed 26 December 2015

UNESCO (nd) *Highland Citizenship and Birth Registration Project*, United Nations Educational, Scientific and Cultural Organization (UNESCO), Bangkok, http://www.unescobk.org/culture/diversity/trafficking-hiv/projects/highland-citizenship-and-birth-registration-project/, accessed 26 December 2015

UNESCO and BSD (2010) *Highland Household Survey: The Impacts of Legal Status on Access to Education and Social Services*, United Nations Educational, Scientific and Cultural Organization (UNESCO) and the Thai Bureau of Social Development and Human Security (BSD), Bangkok, http://www.unescobk.org/culture/diversity/livelihood/surveys/impacts-of-legal-status-on-access-to-education/, accessed 26 December 2015

UNODC (nd) *Single Convention on Narcotic Drugs, 1961*, United Nations Office on Drugs and Crime, Vienna, http://www.unodc.org/unodc/en/treaties/single-convention.html, accessed 22 December 2015

Wangkiat, P. (2015) 'Agro Giants must own up to north haze', *Bangkok Post*, Bangkok, 10 April 2015, http://www.bangkokpost.com/print/524111/, accessed 30 December 2015

Zinke, P. J., Sabhasri, S. and Kunstadter, P. (1978) 'Soil fertility aspects of the Lua forest fallow system of shifting cultivation', in P. Kunstadter, E. C. Chapman and S. Sabhasri (eds) *Farmers in the Forest: Economic Development and Marginal Agriculture in Northern Thailand*, East-West Center, University of Hawaii Press, Honolulu, pp134–159

Notes

1. UNFDAC was first succeeded by the U.N. International Drug Control Programme (UNDCP), and then by the U.N. Office on Drugs and Crime (UNODC).
2. A highland unit was a multi-village, usually contiguous, area growing a significant amount of opium, and targeted for potential development interventions.
3. The U.N. Development Programme later provided funding for the Highland Agriculture and Marketing Project, followed by Norwegian Church Aid as principal donor.
4. Including, among other things, credit, transport (trucks), communications (cell phones), information, and relationships that extend to contract farming.
5. Laungaramsri (2014, pp9-20) shows a variety of civil registration cards issued in Thailand and discusses their role in delineating citizenship.
6. UNESCO (2008) includes a description of the many ethnic categories specifically addressed under Thai nationality and immigration regulations.
7. All highland development projects, regardless of links to the UN effort, likewise supported efforts to register highland peoples in anticipation of them receiving Thai citizenship.
8. As relayed to the author by a project officer in 1994.

31

OPIUM AND SHIFTING CULTIVATION IN LAOS

State discourses and policies

*Paul T. Cohen**

Introduction

The government of Lao PDR recently adopted a hard-line policy towards opium cultivation by upland minorities in the country's northern provinces, with measures including rapid crop eradication and reduction of demand. This prohibitionist policy is linked to a discourse that regards opium as both a form of shifting cultivation and a narcotic drug, and as such, a major cause of poverty. By contrast, the attitude of the French colonial government in Laos was both benign and utilitarian in its regard for opium; it was used as a crucial source of state revenue. Even the communist government of Laos proclaimed the fiscal importance of opium as recently as the 1980s. In this chapter I explain this radical discursive and policy change in terms of the convergence of both foreign and domestic factors and offer a brief comparison with the gradualist policies of neighbouring Thailand. I also examine the immediate impact of the rapid elimination of opium in Laos and assess the longer-term prospects of rubber as the major replacement crop.

Opium and shifting cultivation

Opium has traditionally been grown in the Golden Triangle region of Thailand, Laos and Burma as a form of shifting cultivation (or swiddening), predominantly by ethnic minorities at altitudes about 1000m above sea level, and ideally on limestone-rich soils (Figure 31-1). Other swidden crops cultivated with opium included upland rice as a staple, maize (planted in the opium fields and used mainly as livestock fodder) and vegetables. Some opium cultivators felled and burnt climax forest and used the land continuously for a long period, after which the soil was exhausted, forcing a search for

* ASSOCIATE PROFESSOR PAUL T. COHEN, Department of Anthropology, Macquarie University, New South Wales, Australia.

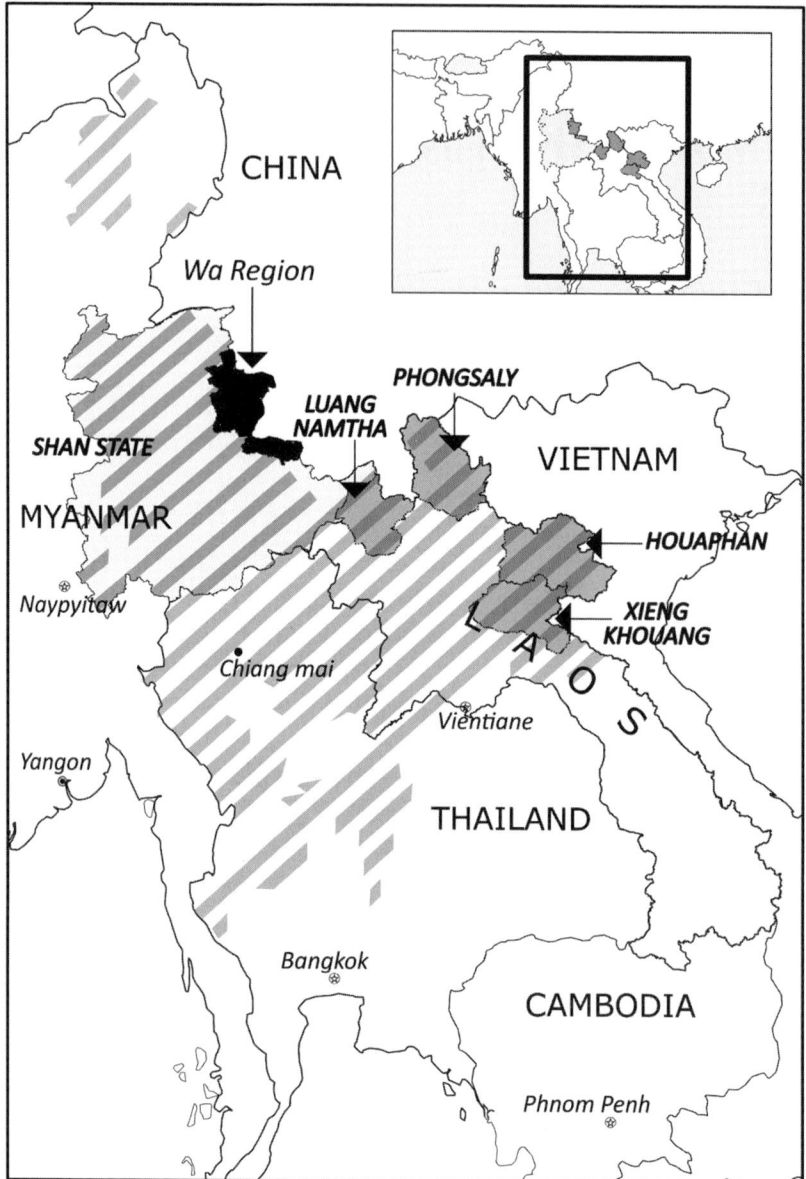

FIGURE 31-1: Areas of past poppy cultivation: The so-called Golden Triangle of Thailand, Laos and Burma.

Source: United Nations Office on Drugs and Crime.

new areas of climax forest that often required migration and village relocation. Thus, Geddes (1970, p4) described the Hmong as 'shifting cultivators in the most extreme sense, who shift not only their cultivation but themselves as well'. However, since about the 1960s, population pressure and rapid deforestation have largely precluded this 'pioneer' type of swiddening. Opium-growing minorities have varied in the

Papaver somniferum L.
[Papaveraceae]

The opium poppy: a beautiful plant with a
bad reputation.

priority given to their production of opium or rice. The Hmong, Yao and Lisu tended to emphasize opium production and to give priority to opium in the selection of land, whereas the primary concern of the Lahu and Akha was the cultivation of rice (Feingold, 1970, p330; Walker, 1980, p148; Alting von Geusau, 1983, pp226–227). In the Muang Sing and Muang Long districts of Luang Namtha province in northern Laos, where I have undertaken periodic research since 1995, the small number of Hmong villages were the only ones to produce substantial quantities of opium,[1] while the Akha, the dominant population of both districts, cultivated opium almost exclusively for household consumption and local trade (Epprecht, 1998, p131).

James Scott has characterized shifting cultivation as a form of 'escape agriculture', writing in 2009 (pp195-196) that swidden and forage crops are 'escape crops' that allow political secession from the state, and that they are 'fiscally sterile' in that their low value-to-weight 'will not pay the efforts of the tax gatherer'. In this regard, opium was exceptional, in that its high value-to-weight ratio and other factors such as storability made it particularly attractive for state appropriation. This was certainly so in the case of the French colonial government, which used opium as a lucrative source of revenue to fund the administration of the Indochina colonies.

French opium monopoly

After the annexation of Saigon in 1862, the French established an opium monopoly and imported opium from India, where the British had been producing it on a large scale for mass markets in China and Southeast Asia. The French imposed a 10% tax and sold the opium to licensed Chinese merchants to supply opium shops and dens. Opium monopolies (*régie*) were later established in French protectorates in Cambodia, central and northern Vietnam and Laos. In 1899 these autonomous opium agencies were united into a single Opium Monopoly (*Régie Générale de l'Opium de l'Indochine*) (Michaud, 2000, p61). For the first time, cheaper opium was imported from Yunnan province in China in order to attract customers among poorer workers and thus expand consumer demand. Between 1898 and 1922 the contribution made by the opium industry to the total gross income of the French Indochina colony fluctuated between 25% and 42% (Michaud, 2000, p62).

In order to increase the production of cheaper opium, French officials forced highland opium growers, in particular the Hmong (Meo), to make deliveries of the drug. Burdensome taxes then led to a prolonged Hmong uprising in northwestern Vietnam and northern Laos (McCoy, 2003, p114). The French colonial government later adopted a more consultative and less exacting system of procuring opium. In the 1930s, the French used Vietnamese, Lao and some Hmong 'brokers' to purchase opium at fixed prices from the Hmong for delivery to the Opium Monopoly. This procurement system (which often required the outbidding of smugglers) persisted during World War Two due to the blocking of shipping and imports of opium from India and the Middle East, and the need to increase local opium cultivation. The French also used prominent Hmong leaders (such as Touby Lyfong) to negotiate prices with Hmong growers (McCoy, 2003, p115).

Post-war policies

After World War Two 'official purchases from indigenous producers decreased significantly and do not appear to have been continued after 1950' (Rapin, 2003, p45). However, during the First Indochina War (1946 to 1954) French intelligence agencies and paramilitary forces exploited the illicit trade in opium to finance covert operations and maintain Hmong loyalty (McCoy, 2003, p131). Opium production in the uplands of Laos increased significantly during the 1950s and 1960s due to protection and distribution by the Royal Lao Government and the growth in the 1960s of the heroin market among United States troops in Vietnam. In 1971, President Richard Nixon launched his global war on drugs – a culmination of the United States' longstanding prohibitionist drug policy, which dated back to the Shanghai International Opium Commission and the Hague Opium Convention of 1909 and 1912, respectively. United States pressure on the Royal Lao Government resulted in opium prohibition and repressive measures against opium consumption and trafficking, and the United States' war against the Pathet Lao resulted in the use of aerial herbicides and bombing to destroy opium fields in Pathet Lao territory (Kuzmarov, 2008, pp360-361). However, after the Pathet Lao seized power in 1975 the new government of the Lao PDR overturned the 1971 ban in order to win the political support of the dissident Hmong, who were the country's largest opium producers. The government set out to control opium production 'in order to earn badly needed foreign exchange by selling opium to Western governments and pharmaceutical companies' (Wekkin, 1982, p190). In a policy that resembled the earlier French *régie*, opium cultivation was officially sanctioned, with a requirement that the Hmong and other uplanders sold to the government at fixed prices (Lee, 1982, p218, Note 33). Opium production in the Lao PDR reached an all-time high of 389 tons in 1989.

The New Economic Mechanism and its aftermath

Economic reforms initiated by the Fourth Party Congress in November 1986 (and labelled the New Economic Mechanism) were the catalyst for change in state policies towards opium. The reforms involved moving from a centrally planned to a market-oriented economy and included the abandonment of collectivized agriculture, the development of private enterprise and liberalization of domestic and foreign trade. The Foreign Investment Law of 1988 allowed foreign investors to enter most economic sectors by way of subcontracting, joint ventures or wholly owned firms (Chantavong, 1996, pp26-41). These economic reforms have since been supported by concessional loans from international lending institutions, including the World Bank, the International Monetary Fund and the Asian Development Bank, and international aid organizations such as the United Nations Development Programme.

The New Economic Mechanism reforms locked the Lao PDR into a global economic system that inevitably made it responsive to pressure from the international system of drug control. In the 1980s, President Ronald Reagan intensified the drug war within the United States and exerted increasing pressure on the United Nations to wage war on drugs around the world. In 1987, the International Conference on Drug Abuse and Illicit Trafficking in Vienna launched a new era of UN drug control. Soon after, the UN General Assembly declared 1991-2000 to be the United Nations Decade against Drug Abuse and, in 1991, established the UN Drug Control Programme (UNDCP). In 1998, a special session of the United Nations General Assembly committed itself to a drug-free world by 2008, heralding a victory for a hard-line prohibitionist approach to drug control (Jelsma, 2003).

Drug prohibition in Laos

In early 1990 Laos embarked upon counter-narcotics cooperation with the United States by signing a memorandum of understanding on Bilateral Cooperation on Narcotics Issues. In 1990 and 1991, the US provided narcotics training for Lao officials and the UNDCP formulated a Comprehensive Drug Control Programme (known as the Masterplan) for the country, to cover the period from 1994 to 2000. In 1996, the Lao government revised Article 135 of the Criminal Code on Drug Trafficking and Possession, thereby formally prohibiting opium production and trafficking. In 2000, Giuseppe (Pino) Arlacchi, the Executive Director of the United Nations Office on Drugs and Crime (UNODC)[2] promised the Vientiane government US$80 million in aid to expedite the elimination of opium in Laos. The agreement was negotiated and signed by a local UNDCP representative who was later described as an 'anti-drug true believer' (Baird and Shoemaker, 2005, p9). This reflected the hard-line prohibitionist and target-oriented policies of the United Nations during the 1990s and its commitment to the imminent goal of a drug-free world. In December 2000 the Prime Minister issued Decree 14, mandating the total elimination of opium cultivation in Laos by 2006 (later altered to 2005).

The UNDCP's Masterplan proposed the 'gradual elimination' of opium cultivation in Laos, over six to 10 years, but the Lao government's Decree 14 imposed a more stringent deadline. Within a short period of five years, the government pursued a punitive campaign combining fines and eradication raids, and reduced opium cultivation in the northern provinces of Laos from 19,052 hectares in 2000 to a mere 1800 hectares in 2005.

Forestry and shifting cultivation

Concurrent with the New Economic Mechanism reforms was an increasing reliance on forestry, especially forest plantations, as a source of state revenue. This spawned discursive and policy attacks on shifting cultivation.

The denigration of shifting cultivation in Laos, as well as in other regions of Southeast Asia, was linked to state forestry after World War Two and the declaration of forests as state land. These 'political forests' were a legacy of 'colonial-era state making' (Peluso and Vandergeest, 2001, p762). Colonial forestry was based on scientific forest management and it served several basic purposes: 'legibility', i.e. forests could be 'read' accurately from tables and maps, and this, in turn, could serve the commercial interests of the 'utilitarian state' (Scott, 1998, p13) and the expansion of the state's jurisdiction (Peluso and Vandergeest, 2001, p 768).

Laos was colonized by the French in 1893. Most of the country's forests and forest lands were subsequently directly and indirectly controlled by the colonial administration (Phimmavong et al., 2009, p505). Traditional forest use (the right of monarchs to use the forest) was restricted by an agreement between the King of Luang Prabang and the colonial administration in 1917 (Evans, 2002, p47). Major changes in Lao forest management and policies in the colonial period included the establishment of forestry institutions to manage the forest and forestland and the introduction of a tax system, property rights, plantations and a regulatory framework to limit the use of natural forests by the local population. For example, exploiting the timber from natural forests was one of the main forest policies in the colonial period. The French administration also encouraged the creation of coffee, rubber and teak plantations, although exports were hampered by

Tectona grandis L. f. [Lamiaceae]

Teak plantations were promoted, along with coffee and rubber, by the French administration in Laos.

poor transport infrastructure, by World War Two and the First Indochina War. From the 1960s through to the 1980s, both the Royal Lao Government and the succeeding Lao PDR government sought to maximize revenue from natural forests. In 1979, 45% of the value of the country's exports came from wood (Phimmavong et al., 2009, pp505, 506). This fell to 40% during the 1990s (Midgley et al., 2012, p161).[3]

Despite the rise in forest revenues, the major impact of the Lao PDR's forest industry on shifting cultivation came earlier, following the initiation of the New Economic Mechanism in 1986, the subsequent internationalization of forest policy and growth of industrial forest plantations. The first national forestry conference was held in 1989 and this led to the Tropical Forest Action Plan, funded by international organizations such the World Bank, the Asian Development Bank and the UN Development Programme. It was followed in 1991 by the National Forestry Action Plan. This aimed to protect remaining natural forests, but also to promote forest plantations, with a preference for fast-growing species. Notably, direct foreign investment in forest plantations increased significantly in 2000 and 2001 due, in part, to 'increasing areas with potential for plantation development as a result of shifting cultivation…' (Phimmavong et al., 2009, p507).

According to Barney (2009, p151), the 'potential' for plantation development in Laos was closely linked to a 'neoliberal-inspired discourse of the Mekong as an untapped resource frontier…' The resource frontier metaphor had been exploited by the Lao state and Asian Development Bank to create the impression that the Lao uplands were an empty space of 'degraded land', i.e. land mostly left fallow by shifting cultivators. Barney concludes: 'The rhetoric of "frontiers" can be understood as a legitimating ideological device, used by state agencies, companies and development banks, in the advertising and promotion of Laos as a "new frontier of corporate investment"' (2009, p147).

Shifting cultivation, opium and poverty eradication

The demonization of shifting cultivation in Laos is not recent. In 1930 the French colonial administration introduced a new forest code (*Code Forestier*) prohibiting shifting cultivation (Phimmavong et al., 2009, p505). It was probably largely ignored given limited state control over the uplands. The Lao, whatever their political orientation, have an entrenched cultural antipathy towards shifting cultivation and a preference for lowland wet-rice agriculture. After seizing power in 1975, the Lao PDR government at first encouraged upland minorities to abandon shifting cultivation, to resettle and move towards collectivized farming of irrigated rice in the lowlands (Wekkin, 1982, p190). The Lao aversion to shifting cultivation was reinforced by a Marxist/Stalinist view that shifting cultivation represented a primitive stage in social evolution (Fox et al., 2009, p307). In April 1978, the Prime Minister Kaysone Phomvihane ordered district party committees to impose a total ban on tree felling to clear land for planting. However, in 1979, Soviet Premier Alexei Kosygin advised the Lao government to postpone resettlement and collectivization (and by implication, the ban on shifting cultivation) so as to prevent further Hmong resistance and migration across the Mekong (Wekkin, 1982, pp191-192).

The government's attack on shifting cultivation resumed and intensified following the Fourth Party Congress of the Lao People's Revolutionary Party in 1986, at which the New Economic Mechanism reforms were written into law and the eventual abandonment of shifting cultivation in favour of wet-rice cultivation was formally announced (Ireson and Ireson, 1991, p930). In 1994, the government decided that shifting cultivation would be eliminated by the year 2000. In that year, the deadline was extended to 2020, but in 2003 it was brought back to 2010 (Ducourtieux et al., 2005, p504).

Coercive enclosures to eliminate shifting cultivation were also imposed under upland tenure policies adopted by the Lao state and supported by large international organizations. The Land and Forest Allocation Programme (LFAP), introduced in the early 1990s, is a land and zoning policy that aims to alleviate poverty by improving access to land for poor farmers through community-based forest management, forest conservation and the 'stabilization' of swidden agriculture (Vandergeest, 2003, p50). Yet, research has revealed that the Land and Forest Allocation Programme has been a major cause of displacement and impoverishment (especially of upland ethnic groups) by zoning village land so that shifting cultivation is permitted only on land without 'secondary' or 'primary' forest cover. This allocation process has forced villagers to shorten fallow cycles, causing soil depletion and low rice yields (Vandergeest, 2003, pp51-52; Ducourtieux et al., 2005, p517). This 'policy-induced' Malthusian squeeze contests the orthodox view that population pressure in the Lao uplands explains the low productivity of shifting cultivation (Rigg, 2006, p126).

Upland resource scarcity, in the form of low yields of upland rice and consequent rice deficits, placed a premium on the cultivation of opium as a means of acquiring rice for household consumption. Epprecht's study of 19 Akha villages in the Muang Sing district of Luang Namtha province revealed a situation in which yields of upland rice were averaging about half the yields of wet-rice in the lowlands. As a result, Akha households produced enough rice for only seven to eight months per year (Epprecht, 1998, pp43-44). All 'productive households' in these villages grew some opium, but most in relatively small amounts. Nevertheless, opium cultivation was critical for household survival and about 65% of all rice purchased or exchanged in the area was paid for directly with opium or cash earned from the sale of opium (Epprecht, 1998, pp66-76).

The vital role of opium as a survival crop in conditions of upland resource scarcity is confirmed by agronomic and anthropological research elsewhere in northern Laos and northern Thailand (Ducourtieux et al., 2008, pp162-164; Cooper, 1984, p62). These findings are at odds with the Lao government's discourse and repeated claims that opium is a major cause of poverty rather than a symptom of poverty in the form of reduced land available for shifting cultivation. The state discourse on opium as a cause of poverty is exemplified in the following declarations in the National

Programme Strategy for the Post Opium Scenario, 2006-2009 (LCDC/UNODC, 2006):

> 'Opium is equal to poverty' (p4) and 'The Seventh Party Congress has identified the elimination of opium and poverty as national priorities. The Government of Laos believes that poverty cannot be eradicated in the northern provinces without the elimination of opium' (p5).

In the eyes of the Lao government, the elimination of shifting cultivation, opium and poverty are inextricably linked. Thus, according to Ducourtieux (2004, p82): 'The goal of eliminating shifting cultivation is motivated by the reasoning that it is one of the main causes of rural poverty... Furthermore, the poverty of families who practise shifting cultivation drives them to grow opium, a source of addiction and therefore increased poverty. The vicious circle is complete and poverty is self-maintained'. The campaign to eradicate opium is an integral part of the National Poverty Eradication Programme, with the aim of enabling the removal of Laos from the list of least-developed countries by 2020 (Ducourtieux, 2004, p81).

'Laoization' and the 'discourse of lack'

The negative symbolism of opium and shifting cultivation has been reinforced by what Pholsena characterizes as the 'remarginalization' of ethnic minorities in Laos since the early 1990s. Under the French colonial administration and the Royal Lao Government, upland minorities (usually referred to as *kha*) were identified as 'primitive' compared to the more 'civilized' Lao and other Tai-speaking groups of the lowlands (Pholsena, 2006, p27). However, when the Lao People's Revolutionary Party seized power in 1975 there was a perceived need to make a radical break with the past and create a new national identity that included ethnic minorities. Such a new national identity would also recognize the debt to upland ethnic groups for their significant contribution to the Pathet Lao victory. The communists envisioned the creation of a 'new socialist man' that accorded

Litchi chinensis Sonn. [Sapindaceae]

Lychees were one of the fruit-tree species that helped farmers overcome the loss of cash income from opium.

equal rights and opportunities to all ethnic groups. Yet, following the introduction of the New Economic Mechanism and increasing integration of Laos into the global capitalist system, the Lao government became 'less dependent on minority support and responsive to minority demands' (Evans, 2002, p212) and promoted policies promoting the 'Laoization' of ethnic minorities (Ireson and Ireson, 1991). 'Laoization' includes the elimination of shifting cultivation, the resettlement of upland ethnic minorities in the lowlands to farm wet-rice and a nationalist discourse that idealizes Lao culture in a way that remarginalizes and stigmatizes ethnic minorities (Pholsena, 2006, pp3, 16, 180ff).

Furthermore, the consolidation of national culture in post-socialist Laos has been 'accompanied by exhortations to economic competitiveness' and characterized by a 'discourse of lack', that is, a perception that Laos is 'backward' compared to other countries in the region and in the world (Pholsena, 2006, pp63-65). The 'national obsession' with economic development corresponds with an obsession with opium as a symbol of primitiveness, backwardness and poverty.

The post-opium scenario

This fast-tracked eradication campaign had a devastating effect on the livelihood of uplanders in northern Laos, due in large measure to inadequate development of alternatives. Even the Lao Commission for Drug Control and Supervision (LCDC) and the UNODC acknowledged that the rapid elimination of opium in Laos had created a crisis: 'Laos is at a critical juncture. …the fact that opium elimination has outpaced the provision of alternative livelihoods has not improved an already difficult situation' (LCDC/UNODC, 2006, p4). By 2005 more than 50% of opium-growing communities in northern Laos did not have the means or the time to develop new cash crops or staple food crops (UNODC, 2006, p1). Consequently uplanders suffered severe rice shortages in the absence of opium to trade for the staple, decapitalization (e.g. by being forced to sell livestock to survive) and migration to lowland areas where there was limited arable land (Ducourtieux et al., 2008, pp163-164; Cohen and Lyttleton 2008,

Pyrus communis L. [Rosaceae]

European pears were among the fruit-tree crops adopted in the highlands as a replacement for opium.

pp136-137). Migration was often spontaneous and uncontrolled. In early 2004, *The Economist* reported 'the displacement of some 25,000 Hmong, Akha and other tribes from their traditional homes in the mountains to the valleys'. In Muang Sing, migration snowballed, resulting in the depopulation of whole upland areas.[4]

The rapid eradication of opium cultivation in the highlands of northern Laos had the effect, intended or unintended, of creating a frontier 'empty' space, open for the development and expansion of forest plantations. The preferred plantation crop was rubber. Rubber plantations have expanded rapidly in the north since 2003, as a result of the convergence and interaction of a number of factors: the desperate need in Laos for a cash crop to replace opium; the symbolic appeal of rubber as a 'modern' crop by contrast with the 'primitiveness and backwardness' of opium; demand for rubber in China coupled with declining yields in traditional rubber-growing areas of Xishuangbanna (Yunnan province); Chinese government investment incentives for Chinese businesses linked to its own opium-replacement policy and schemes; and the perception of northern Laos as an untapped resource frontier (Cohen, 2009, pp426-427). By 2008 there were eight Chinese rubber companies operating in northern Laos (Cohen, 2009, p427). Some were beneficiaries of large concessions that excluded local villages from access to their land, threatened their livelihood and in some areas provoked local resistance (McAllister, 2012). However, while acknowledging that not all villagers have benefitted from the rubber boom, Lagerqvist (2013, p68) argues that research in Muang Sing has found that Chinese 'land grabs' have been offset by thriving smallholder plantations using cross-border social networks. Indeed, the boom has advantaged those villagers with land and capital, but excluded those who were displaced from the uplands by opium eradication and forced into dependence on wage labour in the lowlands (Cohen, 2009, p428). This confirms the claim by Rigg (2006, p131) that in Laos, 'market integration' can be both 'livelihood eroding and livelihood enhancing'. Other problems with rubber plantations as substitutes for opium and other forms of shifting cultivation include destruction of natural forests, soil erosion, reduced biodiversity arising from monoculture, the decline in non-timber forest products and the vagaries of the global rubber market.[5] These problems highlight a fundamental difference between monoculture plantation development as an opium-replacement policy and the alternative development policies of international aid organizations that have emphasized livelihood security, environmental sustainability and poverty reduction through community-based development activities. The Chinese business-oriented model of development has prevailed over the Western aid model, restricting Western international development agencies to a secondary role of monitoring investment activities and analysing the social, economic and ecological impacts of these investments.[6]

Northern Laos and northern Thailand compared

The fast-tracked campaign to eradicate opium in Laos contrasts with the Thai government's gradualist policy. There was a lag of about 25 years between Thailand's

Opium Act of 1959, which prohibited opium cultivation and trafficking, and the beginning of opium eradication in the mid-1980s under pressure from the United States' 'certification policy'.[7] Opium, shifting cultivation and upland minorities were similarly stereotyped in Thailand as the 'hill tribe problem' and denigrated as threats to the Thai nation, but state policy was much more pragmatic and, for a long time, less subject to international pressures. Following the 1967 'Red Meo' revolt and during the 1970s, the government was concerned that opium eradication could alienate 'hill tribe' opium growers and encourage them to join the communist insurgency. Especially influential was the

Diospyros virginiana L. [Ebenaceae]

Persimmons were among the fruit crops used by Hmong farmers to help replace income from opium and to avoid relocation to new village sites.

Opium Replacement Programme, one of King Bhumibol's many Royal Projects, which was launched in 1969, and the king's insistence that opium poppies should not be eradicated until viable alternatives existed. According to Renard (2001, p76): 'The king realized that the radical removal of the hill people's source of income would imperil them'. Implicit in this approach, supported by Thai government agencies and the United Nations, was that opium was not a cause of poverty but, in the absence of alternative cash crops, a vital survival crop for many uplanders. Opium cultivation in the northern uplands of Thailand persisted through the 1970s. Yet, even well before a concerted eradication campaign began in 1984, the cultivation of opium had decreased significantly from 17,920 hectares in 1965-1966 to only 6026 hectares in 1980-1981 (Renard, 2001, p36, Table 3). This can be partly explained by the modest success of crop replacement projects (Renard, 2001, p38) and 'resource scarcity' that led to the degradation of opium land and declining opium yields. The Hmong response was to minimize relocation to new areas and voluntarily invest in permanent agriculture, initially in the form of wet-rice fields and later cabbages, cut flowers and tree crops such as lychees, pears, apples and persimmons (Renard, 2001, pp61-63; Tapp, 1989, pp41-42). Road construction into the uplands in the 1970s and 1980s was initially for the purpose of countering communist insurgency. However, it also facilitated the marketing of cash crops and access to social services.[8]

Conclusions

For many years, colonial and post-colonial governments of Laos exploited opium as a valuable source of state revenue, or at least tolerated opium cultivation, consumption and trafficking despite pressure from the United States and the international drug control system. Yet, quite recently the Lao government embarked upon a campaign to rapidly eliminate opium cultivation within a short period of just five years. I have argued that the Lao PDR government's fast-tracked campaign can be explained in terms of the convergence of several foreign and domestic factors after the mid-1980s.

The Lao government's apparent obsession with meeting international standards of modernity and development in the wake of the New Economic Mechanism resulted in a sense of urgency and a type of deadline-oriented thinking regarding the elimination of opium growing, shifting cultivation and poverty and the attainment of developed-country status.

The economic reforms of the New Economic Mechanism were contemporaneous with intensification of the global war on drugs under US President Ronald Reagan and the establishment of the UNDCP and UNODC, with their hard-line prohibitionist target of a drug-free world by 2008. This international prohibitionist regime was congruent with, and lent authority to, the domestic discourses in Laos in which opium was accused of causing poverty and the uplands were seen as a new 'empty' frontier beckoning plantation development.

The Lao government's hard-line policy and fast-tracked eradication campaign have been in marked contrast with the gradualist policy of the Thai government. While opium, shifting cultivation and upland ethnic minorities in northern Thailand were also denigrated in various ways in national discourses, state policies were more pragmatic and less ideological; opium was never singled out as a cause of poverty. In the 1980s and 1990s, drug-war rhetoric in the United States and the United Nations became more strident and policies more coercive. However, by this time opium cultivation in Thailand was already in decline due to crop replacement projects, road construction and a gradual transition to permanent agriculture. In the post-opium era, state policies in Thailand's northern uplands have aimed at alternative development based on crop diversification, integrated rural development and people's participation – in stark contrast to the potentially precarious and less sustainable future of rubber monoculture in the uplands of northern Laos.

References

Alting von Geusau, L. G. M. (1983) 'The interiorizations of a perennial minority group' in J. G. Taylor and A. Turton (eds) *Sociology of Developing Societies in Southeast Asia*, Macmillan, Basingstoke, pp215-229

Baird, I. G. and Shoemaker, B. (2005) *Aiding or Abetting? Internal Resettlement and International Aid Agencies in Lao PDR*, Probe International, Toronto, Canada

Barney, K. (2009) 'Laos and the making of a "relational" resource frontier', *The Geographical Journal* 175(2), pp146-159

Chantavong, S.(1996) 'Lao-style New Economic Mechanism' in M. Than and L-H. Tan (eds) *Laos'*
Dilemmas and Options: The Challenge of Economic Transition in the 1990s, Institute of Southeast
Asian Studies, Singapore, pp23–58

Cohen, P. T. and Lyttleton, C. (2008) 'The Akha of northwest Laos: Modernity and social suffering', in
P. Leepreecha, D. McCaskill and K. Buadaeng (eds) *Challenging the Limits: Indigenous Peoples of*
the Mekong Region, Mekong Press, Chiang Mai, pp117–142

Cohen, P. T. (2009) 'The post-opium scenario and rubber in northern Laos: Alternative Western and
Chinese models of development', *International Journal of Drug Policy* 20(5), pp424–430

Cooper, R. (1984) *Resource Scarcity and the Hmong Response: Patterns of Settlement and Economy*
in Transition, Singapore University Press, Singapore

Ducourtieux, O. (2004) 'Shifting cultivation and poverty eradication: A complex issue', proceedings
of a workshop Poverty Reduction and Shifting Cultivation Stabilisation in the Uplands of Lao
PDR: Technologies, Approaches and Methods for Improving Upland Livelihoods, 27-30 January
2004, Luang Prabang, Lao PDR

Ducourtieux, O., Laffort, J-R. and Sacklokham, S. (2005) 'Land policy and farming practices in Laos',
Development and Change 36(3), pp499–526

Ducourtieux, O., Doligez, F. and Sacklokham, S. (2008) 'L'Éradication de l'opium au Laos: Les politiques
et leurs effets sur l'économie villageoise' (The eradication of opium in Laos: Policies and their effects
on village economies), *Revue Tiers Monde* 193, pp145–168

Epprecht, M. (1998) 'Opium production and consumption and its place in the socio-economic setting
of the Akha people of north-western Laos: The tears of the poppy as a burden for the community?'
MSc thesis, Institute of Geography, Faculty of Natural Science, University of Berne, Switzerland

Evans, G.A. (2002) *Short History of Laos: The Land in Between*, Silkworm Books, Chiang Mai, Thailand

Feingold, D. (1970) 'Opium and politics in Laos' in N. S. Adams and A. W. McCoy (eds) *Laos: War and*
Revolution, Harper & Row, New York and London, pp322–339

Fox, J., Fujita, Y., Ngidang, D., Peluso, N., Potter, L., Sakuntaladewi, N., Sturgeon, J. and Thomas, D.
(2009) 'Policies, political economy, and swidden in Southeast Asia', *Human Ecology* 37(3), pp305–
322

Geddes, W. R. (1970) 'Opium and the Miao: A study in ecological adjustment', *Oceania* 41(1), pp1–11

GIZ (2012) *Building up Land Concession Inventories: The Case of Lao PDR*, Deutsche Gesellschaft für
Internationale Zusammenarbeit (German Federal Enterprise for International Cooperation), Bonn

Ireson, C. J. and Ireson, W. R. (1991) 'Ethnicity and development in Laos', *Asian Survey* 31(10), pp920–
937

Jelsma, M. (2003) 'Drugs in the UN system: The unwritten history of the 1998 United Nations General
Assembly Special Session on Drugs', *International Journal of Drug Policy* 14, pp181–195

Kuzmarov, J. (2008) 'From counter-insurgency to narco-insurgency: Vietnam and the international war
on drugs', *Journal of Policy History* 20(3), pp344–378

Lagerqvist, Y. F. (2013) 'Imagining the borderlands: Contending stories of a resource frontier in Muang
Sing', *Singapore Journal of Tropical Geography* 34, pp57–69

LCDC/UNODC (2006) *National Programme Strategy for the Post Opium Scenario, 2006–2009*, Lao
National Commission for Drug Control and Supervision and United Nations Office on Drugs and
Crime, Vientiane

Lee, G. Y. (1982) 'Minority policies and the Hmong', in M. Stuart-Fox (ed.) *Contemporary*
Laos: Studies in the Politics and Society of the Lao People's Democratic Republic, University of
Queensland Press, St. Lucia, Brisbane and London, pp215–219

McAllister, K. (2012) 'Rubber, rights and resistance: the evolution of local struggles against a Chinese
rubber concession in Northern Laos', paper presented at the International Conference on Global
Land Grabbing II, 17-19 October, Cornell University, Ithaca, NY

McCoy, A. W. (2003) *The Politics of Heroin: CIA Complicity in the Global Drug Trade*, Lawrence Hill
Books, Chicago

Michaud, J. (2000) 'A historical panorama of the montagnards in northern Vietnam under French rule',
in J. Michaud (ed.) *Turbulent Times and Enduring Peoples: Mountain Minorities in the South-East*
Asian Massif, Routledge, Abingdon and New York, pp51–78

Midgley, S., Bennett, J., Samontry, X., Stevens, P., Mounlamai, K., Midgley, D. and Brown, A. (2012) *Enhancing Livelihoods in Lao PDR through Environmental Services and Planted Timber Products*, ACIAR Technical Reports 81, Australian Centre for International Agricultural Research, Canberra

Peluso, N. L. and Vandergeest, P. (2001) 'Genealogies of the political forest and customary rights in Indonesia, Malaysia, and Thailand', *The Journal of Asian Studies* 60(3), pp761-812

Phimmavong, S., Ozarska, B., Midgley, S. and Keenan, R. (2009) 'Forest and plantation development in Laos: History, development and impact for rural communities', *International Forestry Review* 119(4), pp501-513

Pholsena, V. (2006) *Post-War Laos: The Politics of Culture, History and Identity,* Institute of Southeast Asian Studies, Singapore

Rapin, A-J. (2003) *Ethnic Minorities, Drug Use and Harm in the Highlands of Northern Vietnam: A Contextual Analysis of the Situation in Six Communes from Son La, Lao Chau, and Lai Cai*, United Nations Office on Drugs and Crime (UNODC), Vietnam

Renard, R. (2001) *Opium Reduction in Thailand 1970-2000: A Thirty Year Journey,* Silkworm Press, Chiang Mai, Thailand

Rigg, J. D. (2006) 'Forests, marketization, livelihoods and the poor in the Lao PDR', *Land Degradation and Development* 17, pp123-133

Scott, J. C. (1998) *Seeing Like a State: How Certain Schemes to Improve the Human Condition Have Failed*, Yale University Press, New Haven, CT

Scott, J. C. (2009) *The Art of Not Being Governed: An Anarchist History of Upland Southeast Asia*, Yale University Press, New Haven, CT and London

Tapp, N. (1989) *Sovereignty and Rebellion: The White Hmong of Northern Thailand*, Oxford University Press, Oxford and New York

UNODC (2006) *Strategic Programme Framework, Lao PDR 2006-2009*, January 2006, United Nations Office on Drugs and Crime, Vientiane

UNODC (2013) *Southeast Asia Opium Survey 2013, Lao PDR*, Myanmar, United Nations Office on Drugs and Crime, Vienna

Vandergeest, P. (2003) 'Land to some tillers: Development-induced displacement in Laos', *International Social Science Journal* 55(175), pp47-56

Walker, A. R. (1980) 'The production and use of opium in the Northern Thai uplands: An introduction' *Contemporary Southeast Asia* 2(2), pp135-154

Wekkin, G. D. (1982) 'The rewards of revolution: Pathet Lao policy towards the hill tribes since 1975', in M. Stuart-Fox (ed.) *Contemporary Laos: Studies in the Politics and Society of the Lao People's Democratic Republic*, University of Queensland Press, St. Lucia, Brisbane and London, pp181-198

Notes

1. The Hmong are the most prolific opium growers in Laos, with Hmong villages concentrated in Houaphan, Luang Prabang and Xieng Khouang provinces. Villagers of the four Hmong villages in Muang Sing district of Luang Namtha province migrated there in the early 1990s and district authorities settled them in the lowlands, where there was limited scope for opium cultivation. In the adjacent district of Muang Long there were three Hmong villages which, until the eradication campaign got under way, cultivated opium extensively, some for trade across national borders.

2. The United Nations Office on Drugs and Crime (UNODC) was established in 1997 as an umbrella organization incorporating the UN Drug Control Programme and the UN's Centre for International Crime Prevention.

3. This dropped to 20% in 2002-2003 and 12% in 2006 (Midgley et al., 2012, p16), presumably due to increased revenue from other sources such as hydroelectric power and mining.

4. In Muang Sing, for example, 338 out of 522 households in 13 highland villages migrated to the lowlands between January and April 2003, following visits from officials who ordered the cutting down of opium poppies and imposed fines for non-compliance (Cohen and Lyttleton, 2008, p129).

5. In 2010 the local price of rubber was 15,000 kip (US$1.80) per kilogram. By 2014 the price had dropped to between 7000 and 8000 kip ($0.87 and $0.99). The president of the Hadyao rubber cooperative (which pioneered rubber planting in Luang Namtha province) stated that if the price fell below 5000 kip 'the group may not be able to continue to produce rubber sustainably because of the high cost of cleaning and buying saplings' (*Vientiane Times*, 26 March 2014). The fall in rubber prices and a corresponding rise in opium prices may have contributed to recent increases in opium cultivation in Laos, from 1800ha in 2005 to 7000ha in 2012 and 4000ha in 2013 (UNODC, 2013).

6. For example, the recent study by the German Federal Enterprise for International Cooperation, GIZ, *Building up Land Concession Inventories: The Case of Lao PDR* (2012).

7. 'Certification' provides for aid cuts and trade sanctions against major drug-producing or drug-transit countries unless the United States administration certifies that a country is fully cooperating with its anti-narcotics efforts.

8. Renard notes that in 1970 there were almost no roads in the highlands of northern Thailand, but by 1987 there were approximately 400 kilometres of roads (Renard, 2001, p126).

32

ELIMINATING OPIUM FROM THE LAO PDR

Impoverishment and threat of resumption of poppy cultivation following 'illusory' eradication

*Olivier Ducourtieux, Silinthone Sacklokham and François Doligez**

Introduction

Opium production regularly makes headlines in the world media and is a recurring feature of geopolitical debate. It figures prominently in the most recent annual report of the United Nations Office on Drugs and Crime (UNODC, 2014b)[1] and in its recent report on opium in Afghanistan (UNODC, 2014a), providing an occasion for Western newspapers to lament the expanding cultivation of opium poppies (*Papaver somniferum*)[2] and the failure of international aid to limit opium production in this conflict-ravaged country.

> Either Afghanistan destroys opium or opium will destroy Afghanistan, President Hamid Karzai has warned. As this survey shows, we are coming dangerously close to the second option (UNODC, 2006a, piv).

For several decades now, the United States and the UNODC have promoted an approach to drug control based on supply reduction (Jelsma, 2003; McCoy, 2003). Afghanistan's failure in this respect, so frequently evoked by the media, has incited champions of this policy to draw attention to a more successful example: opium eradication in Laos (UNODC, 2006b).

> Poppy cultivation in Laos declined dramatically in 2005, and this success stands as an unqualified victory for Laos and its international partners, especially the US, in the battle against illicit narcotics (BINLEA, 2006, p273).

* Dr Olivier Ducourtieux, Associate Professor, Comparative Agriculture Research and Training Unit, Agro Paris Tech, Paris; Dr Silinthone Sacklokham, Associate Professor and Vice-Dean of the Faculty of Agriculture, National University of Laos, Vientiane; and Dr François Doligez, Agricultural economist, Institute for Research and Application of Development Methods (IRAM), School of Economic Sciences, University of Rennes 1, Paris.

The success in Laos appears to be well deserved, considering that the country was long vilified as a major producer. For years, official UN and American government publications on drugs and Laos systematically started by recalling the country's position as the world's third-largest producer of opium (a position it continues to hold):

> The remote and mountainous areas of Northern Laos [...] have consistently come in third place as a source of the world's illicit opium and heroin during the last ten years (UNODC, 2003, p1). The Lao People's Democratic Republic (Lao PDR) is the third largest producer of illicit opium worldwide after Afghanistan and Myanmar (UNODC, 2004a, p1). In 2003, more than 90 per cent of the illicit poppy cultivation was concentrated in three countries: Afghanistan, Myanmar and Laos (UNODC, 2004b, p59).

UNODC statistics confirm this position until 2005 (Figure 32-1). But a closer look at the data reveals that although the Lao People's Democratic Republic (Lao PDR) indeed ranked third, the first two countries accounted for 85% to 95% of world production between 1990 and 2013.

While Myanmar was the top producer during the early 1990s, Afghanistan is the current front runner and market trendsetter, producing 93% of the total tonnage in 2013. With production worth little more than US$7 million in 2005, the gross output of Lao farmers was incomparably smaller than that of their Afghan or Burmese counterparts, which was $556 million and $58 million, respectively (UNODC, 2006c). Because trade in opiates is illegal worldwide,[3] profits from processing and commercialization of poppies can be substantial (Chouvy, 2009). But in 2009 the turnover of Lao opium producers was only 0.03% of the total value of the global retail market for illicit drugs (UNODC, 2011a).

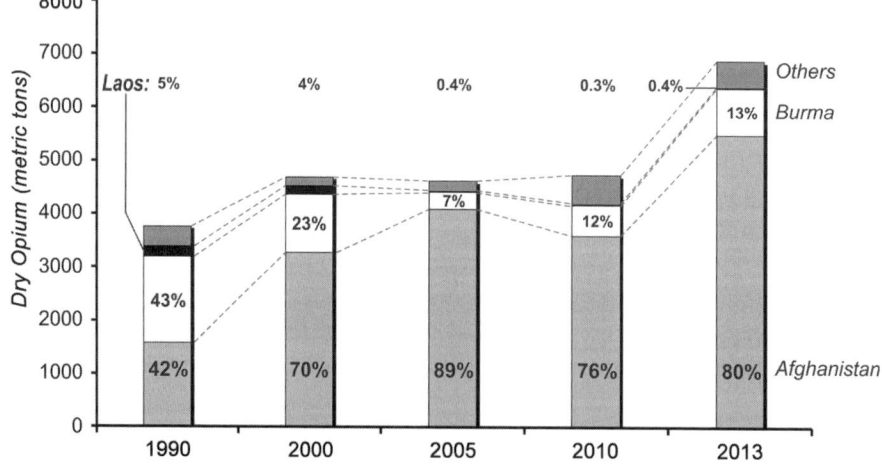

FIGURE 32-1:   Contribution of Laos to world opium production, 1990 to 2013.

Source: UNODC (2014b).

Why and how did Laos go from being demonized as the world's third-largest producer to a country 'freed' from opium (UNODC, 2006b) within just a few short years? It is difficult to understand this process without considering the country's historical position in the Golden Triangle[4] and the geopolitical backdrop of the globalized illegal drugs market and its repression. Although Laos' success in eradicating opium is the result of the Lao government's strong political will, supported by part of the international community, it is noteworthy that the eradication programme was implemented concurrently with a poverty reduction policy, also supported by international aid partners, targeting the ethnic minorities living from shifting cultivation in the mountains of northern Laos.

After reviewing the background and the rise of opium production in northern Laos, we will appraise the impact of the recent eradication policy in the light of results from a field survey in Phongsaly province (Figure 32-2), the primary production region in northern Laos.[5] The survey was designed to track changes to farmers' incomes and answer the following questions: has the livelihood of rural families improved as the social and health burdens of opium decreased? Or has it deteriorated, now that a main source of cash income has been eliminated?

The rise and fall of opium production in northern Laos

Opium production in Laos was declared illegal in the second half of the 1980s. Prior to this, poppy cultivation was legal and often encouraged by local authorities.

A commodity of the colonial era

The term 'Golden Triangle' evokes exotic images of poppy cultivation and opium consumption as an ancient tradition – wrongly, as it turns out: the Golden Triangle is not the birthplace of opium, but a fairly recent creation (Renard, 1996; Booth, 1999; Lintner, 2000). The development of the commercial poppy trade in the region dates back to the early 20th century, and occurred under the influence of French and English colonizers. The term 'Golden Triangle' was coined by the United States State Department in 1972 (McCoy, 2003; Chouvy, 2009).

A source of revenue... or a threat to public health?

France's colonization of Indochina was driven in part by the prospect of reproducing the lucrative trade triangle developed by Great Britain in the late 18th century: manufactured English goods (mainly textiles) were exported to India, Indian opium was exported to China via Macao and Hong Kong, and tea was imported from China (Chouvy, 2009; Lovell, 2011). Before conquering Cochinchina (the southern part of Vietnam) between 1858 and 1867, French marines took part in the second opium war, led by the English against China between 1856 and 1858. China surrendered, opened its borders, and under pressure from the English, legalized the importation of opium (Trocki, 1999). However, colonial hopes to produce great quantities of

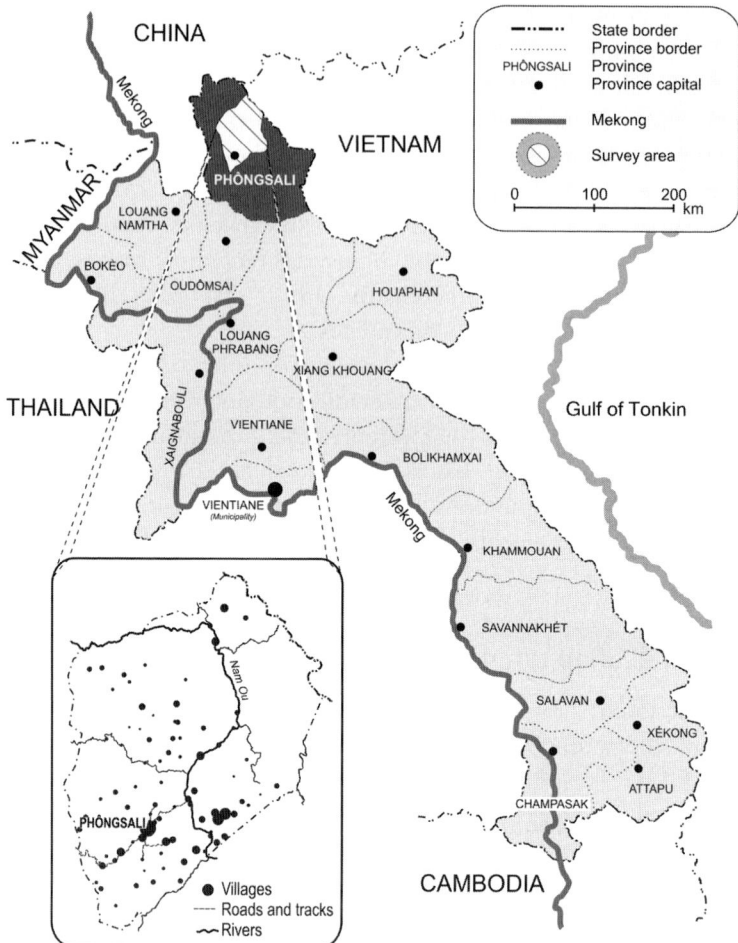

FIGURE 32-2: The Lao PDR and the study area, Phongsaly district, in the province of the same name.

opium in Indochina, as in British-controlled Bengal, were quickly dashed. Climate conditions on the plains were not suitable for poppy production and the northern mountains were too isolated and unsafe (Le Failler, 2001; Rapin, 2008). Moreover, Indochina was far from being the ideal port of entry to China because the Mekong river was not navigable.

These setbacks led the colonial authorities to change their strategy: opium consumption would be controlled and taxed. Although the stated objective of this policy was health-related – the reduction or eradication of opium consumption – economic interests were the main driver: by monopolizing and increasing supplies to smokehouses, the government budget was guaranteed. Between 1899 and 1922, tax revenues from opium represented between 21% and 44% of the General Governorate budget (Descours-Gatin, 1992). The colonial policy led to a rise in opium consumption in Indochina by enlarging the circle of traditional users beyond the Chinese community and mandarins (Booth, 1999). In the words of a French doctor, writing in 1925:

> There are few pages [in our colonial history] that are less worthy of recognition than those in which the residents of Indochina sing the glories of the two-headed monster, alcohol-opium, and where, for purely fiscal reasons, this double poison is supplied to, sometimes even forced upon, the indigenous populations. A collective drug addiction was reinforced by an odious colonial policy, in which the Nation lost the majority of its moral prestige (Legrain, 1925).

The rise in poppy cultivation

The Indochinese Opium Excise Office[6] was originally designed to purchase raw material from India for processing and resale to retailers – the licensed smokehouses (Booth, 1999; Descours-Gatin, 1992). Proximity to China, the world's major consumer between 1830 and 1949, rendered this apparently simple plan complicated. First, the rise in Chinese demand led to local cultivation of poppies in Yunnan province.[7] Second, the vagaries of China's domestic policies in the first half of the 20th century resulted in fluctuations in production of opium and legal importation from Yunnan (Lovell, 2011), despite there being no elasticity in demand, given the millions of addicts. Depending on Chinese markets, the Opium Excise Office was forced to either import opium from Yunnan at the same time as it supported contraband from the same province, or supply contraband networks by exporting to Canton and Shanghai (Le Failler, 2001). Tens of tons of opium passed through northern Tonkin (northern Vietnam) and Laos every year, transported by black-marketers over trade routes used by purveyors of legal goods like tea (Chouvy, 2009). Farmers in these regions already grew poppies for local pharmaceutical use; their location along the opium trade routes simply incited them to increase production as a way to pay taxes and avoid conscripted labour (Gunn, 2003).

Although it was at first ignored by the Indochinese Opium Excise Office, local opium production gradually increased as opium imported from India waned under enactment of international agreements to make opium illegal (discussed below). The first purchases of local opium were made at the end of the 1910s (Rapin, 2008), and by the 1920s, production had taken off (Gunn, 2003). Transactions were official, conducted or ordered by French customs officers on mission for the Opium Excise Office.[8] Colonial authorities passed off purchases as 'health and social measures' designed to protect local populations by controlling supply,[9] but locally produced opium was advantageous for the regime. Not only was it profitable, it served to align the economic interests of minority upland populations with those of the General Governorate (Michaud, 2009; Scott, 2009), thereby reducing the risk of rebellion (Gunn, 2003). During the economic crisis of the 1930s, the colonial administration further developed the Opium Excise Office to compensate for lost fiscal revenues and subsidies from France, rather than raise taxes on a local population that was increasingly receptive to nationalist and communist calls for emancipation (Brown and Zasloff, 1986; Gunn, 1998). The Opium Excise Office's purchases from Lao farmers grew considerably (Descours-Gatin, 1992; Gunn, 2003).

Starting in 1940, the colonial administration in Indochina was first cut off from France, then occupied by Nazis. Public finances became a major problem; budgets were cut and the Opium Excise Office's opium trade was pushed to its limits (Descours-Gatin, 1992; Gunn, 2003). With supplies from India cut off and access to Yunnan province haphazard, customs officials focused on local production, boosted by the attractive prices (Figure 32-3). It was the 'golden age' of opium production for Lao farmers: rising prices, guaranteed markets and public support for a legal crop. By 1944, opium revenues made up more than 40% of the colony's budget (Descours-Gatin, 1992). In the same year, the Viet Minh independence movement, attracted to the opium trade as a way to increase its revenues, joined the fray alongside the Opium Excise Office and Chinese contrabands (Gunn, 2003).

Opium and the Indochina wars

The tumultuous decolonization process in Southeast Asia reinforced the role of opium in upland farming and local economies. During the conflict-riddled three decades from 1945 to 1975, different protagonists financed their war efforts in Laos with opium.[10] The royal government and its French, then American, allies did so illegally and in bad faith (McCoy, 2003); the Pathet Lao openly and vindictively (Vongvichit, 1968).

Northern Laos was one of Indochina's major opium production zones, and confrontations between buyers at harvest time were frequent (Feingold, 1970;

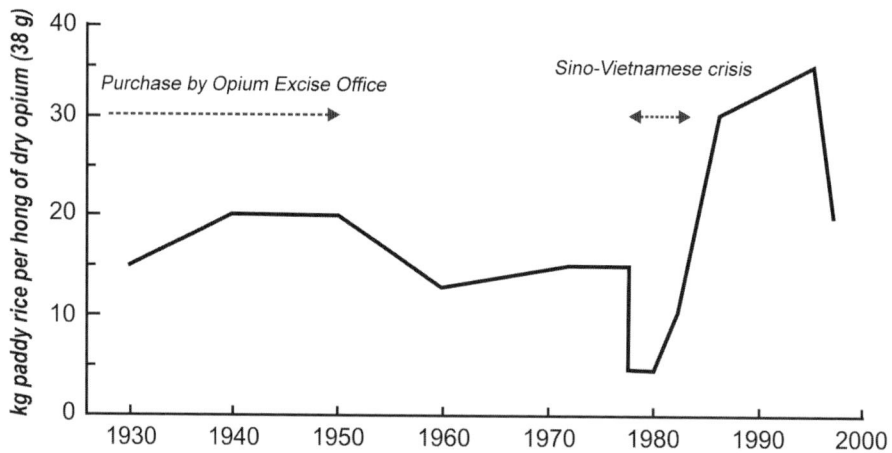

FIGURE 32-3:   Evolution of opium prices paid to villagers in Phongsaly province (northern Laos).

Source: Alexandre and Eberhardt (1998, p85).

McCoy, 2003). Despite Chinese demand having evaporated in the 1950s due to the Communists' rise to power in 1949 (Trocki, 1999; Chouvy, 2009), opium remained an important income source for both the Franco-Lao administration and the Vietnamese-Pathet Lao allies (Feingold, 1970; McCoy, 2003). Upland villagers continued to plant large areas with poppies, despite the drop in prices as competition lessened with the Franco-Lao administration's gradual expulsion from production regions (Figure 32-3).

The Geneva Accords of 1954 marked the end to French colonization in Indochina, but competition to control opium production continued. In regions controlled by Vientiane, the royal government and the United States Central Intelligence Agency (CIA) were actively involved in clandestine, lucrative trafficking, mainly with Hmong minority groups (McCoy, 2003).[11] As the number of regions controlled by the Pathet Lao increased, the new revolutionary organization lobbied highlanders to support the war effort with their labour, rice, and financial resources by vindicating the opium trade. In political tracts from the 1960s and 1970s, poppies are presented as a typical cash crop. For example, Pathet Lao leader Phoumi Vongvichit wrote in 1968:[12]

> With the development of food crop production and other agro-food products, harvesting valuable forest products such as opium, sticklac, cardamom, essential oils, etc. has been encouraged. We have now a major source of exports that can be used for trade of many industrial goods to satisfy a part of the population's immediate needs (Vongvichit, 1968, p162).

After more than half a century of occupation, conflict and political uncertainty, combined with the isolation of northern upland regions, trade with remote farmers began a rapid decline. Subsistence crops replaced some of the commercial crops that had developed during the colonial period, with the notable exception of opium poppies. Under the influence of colonial policy and the cold war, poppies went from being an auxiliary plant for local pharmaceutical purposes to a major contributor to the regional economy (Chouvy, 2009; Steinberg et al., 2004). They became the main 'cash crop' of isolated upland populations, with international market integration. Villagers specialized in poppy production, incited by its comparative advantages (Trocki, 1999).

From trade to trafficking: Prohibition of opium gradually becomes international

Throughout the 20th century, the international community sought to control narcotic consumption and it ultimately became a major issue in diplomatic relations. While the first international conferences organized by China and the United States had little impact (Terry and Pellens, 1928),[13] a series of conventions were passed starting in the 1960s, under the impetus of the United States and China, then the world's largest consumers of illegal narcotics. Drafted within the framework of the

United Nations, these conventions were legally binding on the countries that ratified them: the Single Convention on Narcotic Drugs of 1961, modified by the 1972 Protocol; the Convention on Psychotropic Substances of 1971; the United Nations Convention Against Illicit Traffic in Narcotic Drugs and Psychotropic Substances of 1988, and the unanimous resolution of the United Nations General Assembly Special Session on Drugs in 1998 (Jelsma, 2003). Since its founding in 1945, the General Assembly of the United Nations has adopted 86 resolutions on narcotics, invariably urging their eradication and concerted international efforts for their repression. The Lao PDR ratified the 1961 and 1971 conventions in 1997, and the 1988 convention in 2004.

Growing international pressure

In 1975, the Lao PDR became the official (in this case, meaning legal) supplier of the opium-poppy extract morphine to the pharmaceutical industry of the Council for Mutual Economic Assistance (COMECON) – the organization formed to facilitate and coordinate the economic development of the eastern European countries belonging to the Soviet bloc. Despite this arrangement, a considerable portion of the Lao PDR's opium production was still sold on the international illicit drug market. The fall of the Soviet Union about 15 years later left the poppy growers without a legal market.

During the 1980s, the Lao PDR gradually renewed diplomatic relationships with the West and strengthened its cooperation with agencies of the United Nations (Evans, 2002). In exchange for international aid, influential donor countries repeatedly asked Laos to cease poppy cultivation and opium trafficking. The United States was most active in this process (Pholsena, 2010). Starting in 1990, the United States and Laos began signing annual Letters of Agreement on narcotic drugs control and in 1992, annual cooperation agreements on strengthening anti-drug laws were added (BINLEA, 2006). Between 1989 and 2014, the United States government provided more than $45 million in support to Laos for institutional capacity building and three rural-development projects aimed at eliminating poppy cultivation (BINLEA, 2014, p223). These projects financed rural infrastructure and agricultural development on a small scale. In Phongsaly, for example, roughly 20 villages in two districts were involved.[14] Law enforcement and legal activities, on the other hand, were funded throughout the country.

The UNODC has also been active in efforts to reduce opium production in Laos. Since 1980, the agency has funded rural development projects designed to provide an alternative to poppy farming, with budgets and geographical outreach similar to the US programmes. As with the US aid, funding from the UNODC went to the Government of Laos, while the agency provided technical assistance to reinforce the country's legal and administrative system with the aim of reducing trafficking and opium production. After the UN General Assembly decided in 1998 to set the objective of eliminating the illegal drug problem by 2008, the UNODC signed a high-level agreement with Laos to eradicate poppy crops by 2005, and offered $80 million in financial support (UNDCP, 1999).

This funding was not a firm commitment, but rather the UNODC's estimate of the resources needed to develop alternatives to poppy farming in the villages of northern Laos. The UNODC expected other donors to rally to the cause and encouraged them to include a poppy-eradication component in all of their rural-development activities in northern Laos. However, involvement was limited. With the exception of those funded by the United States and the UNODC, only two projects, funded by the Asian Development Bank and German international cooperation (GTZ/GIZ), explicitly included poppy eradication in their objectives during the period of the initial programme (1999 to 2005). Moreover, the UNODC's resources were cut in 2001 and priority was shifted to Afghanistan. The budget for its 'Opium Eradication in Lao PDR: Alternative Development Module in North Phongsaly' project fell from $4.7 million (UNDCP, 1999) to $3 million (UNODC, 2006f, p15). The deadline for eradicating poppy crops by 2005 went by, unattained. Nevertheless, the UNODC maintained its goal of eradicating opium from Laos in the short term by declaring: 'Strategic objectives: By 2008, to have eliminated or reduced to insignificance cultivation and production of illicit opium and related opium abuse' (UNODC, 2004c, p2). In 2008, with the support of the UNODC, the Lao government published a guiding drug-control strategy document, the 'National Drug Control Master Plan for 2009-2013', based on receiving international funding of $72 million. In 2013, the Lao government extended its master plan to 2015, but as of the end of 2013, only $20 million had been raised by donors (BINLEA, 2014, p223).[15]

A high-stake crop

In quantitative terms, Lao opium production is limited (Figure 32-1), indeed insignificant, in international opiate markets. According to the US Bureau for International Narcotics and Law Enforcement Affairs: 'Illicit transit to the US includes unrefined opium […], but not in sufficient quantities to have a significant effect' (BINLEA, 2006, p268), and in the words of the UNODC: 'We can safely assume that Laos is no longer a supplier of illegal opiates to the world market' (UNODC, 2005b, p1). Therefore, what is behind this international mobilization? The answer is most likely politics. From the first international conferences to the present day, illicit drugs have served political purposes in diplomatic relationships (Trocki, 1999; McCoy, 2003; Chouvy, 2009).

Lao opium production was an excellent opportunity for the supporters of a rigorous anti-drug policy: rapid eradication was possible at a small financial and diplomatic cost in a sparsely populated country of limited geostrategic importance. With a limited budget and some public relations, opium eradication in Laos could become a showcase for the United States State Department and the UNODC:

> Based on USG [US Government] strategic goals […], opium poppy elimination
> in Laos represents a genuine success story (BINLEA, 2005, p327); According to
> US government estimates, between 1998 and 2007, opium poppy cultivation
> decreased by 96 percent due to aggressive government action and international
> cooperation (BINLEA, 2014, p222).

International ambitions to eliminate opium in Laos would have been in vain without
buy-in from the Lao government. Because US authorities had long accused high-
level Laotian military officials of being involved in the opium trade (BINLEA,
2005), it was surprising when, in 1998, the Government of Laos decided to adopt
an eradication policy that apparently contradicted its own interests. The lack of
transparency around decision-making in the Lao People's Revolutionary Party[16]
calls for prudence when it comes to conjecture, and in any case, diplomatic pressure
probably only partially explains why Laos embraced efforts to eradicate opium. Other
explanations may have to do with the political need to mobilize public opinion
around something other than obsolete socialist messages; the opportunity to increase
the control of the lowland-based State over upland ethnic minorities (Cohen, 2013);
and growing problems of addiction outside rural areas. Rising amphetamine use
among urban youth (Lyttleton, 2004) is undoubtedly a cause for concern (BINLEA,
2005, 2006, 2014). Bertrand concludes:

> The political awareness that inability to control drug abuse is a symptom of
> the Party's and other organisations' lack of influence over the population may
> explain their commitment to anti-drug policies as a way to shore up influence
> (Bertrand, 2003, p95).

Another concern may be the threat of new political and economic actors or organized
crime emerging around the illegal trafficking of opium, amphetamines and heroine,
thus putting public security and government stability at risk, as well as the current
regime's prerogatives (historically, this has been a concern of Lao authorities). In sum,
the Lao government's change of heart regarding opium is probably due to a mix of
opportunism, pragmatism and worry.

Opium and poverty: causal relationship or opportunism?

Noting the juxtaposition of poverty and poppy cultivation in Laos, the UNODC has
been quick to assume that correlation equals causality (UNDCP, 1999).

> Opium is equal to poverty (LNCDCS, 2006, p4). The average annual cash
> income of opium producing households was estimated at US$139, while the
> average annual cash income of a non-opium producing household was US$231.
> This discrepancy indicates that opium production is linked to poverty, as opium
> producing households are often characterized by lower productivity due to
> addiction of one or several of their members (UNODC, 2006e, p227).

In the first project document of the UNODC's eradication programme, launched in 1998, the authors emphasized the interdependence of opium and rural poverty; poppy production, they wrote, leads to high rates of addiction in villages, especially among men, who instead of contributing to the household, sap revenues with their drug habit (UNDCP, 1999). This causal relationship was systematically referenced in speeches and documents by the Lao government and organisations supporting opium eradication. Far from being a contributor to villagers' income, the argument goes, poppy cultivation has caused massive addiction in villages, thus debilitating adults and undermining farm production:

Papaver somniferum L.
[Papaveraceae]

Upland farmers and the poorest of the poor have paid a high social cost for eradication of opium (this chapter).

> In northern Laos, it is not really possible to address poverty alleviation adequately without addressing opium issues (UNDCP, 1999, p4). As the Lao government formulated poverty eradication goals, many observers came to identify opium use – and indirectly, opium production – as a major cause of poverty' (UNODC, 2005b, p26).

The rationale is clear: eradicating poppy production is a prerequisite to eliminating poverty, the government's strategic objective. Institutions that want to support the Lao PDR in this priority must contribute to the anti-drug effort and finance the UNODC's programme.

While addiction is a major problem in many villages across northern Laos, causing considerable social and economic hardship (Cohen, 2000; Epprecht, 2000; Cohen and Lyttleton, 2002; Lyttleton, 2004), the argument that opium is a major cause of poverty – rather than a consequence – is based on presumption rather than scientific evidence (Cohen, 2006).

Eradication of opium: National policy, local effects

Recent political will and prohibition of poppy cultivation

The first legal measures to outlaw opium in the Lao PDR were made in 1990 (trafficking) and 1996 (production) (BINLEA, 2005). Both were dead letters for lack of genuine political will and means to enforce them.

In almost all cases where illicit crops have been eliminated, such as by the Chinese in the early-1950s, in Thailand in the 1980s, and with the Taliban in Afghanistan, strong political will was required (UNODC, 2005b, p22).

However, once the Government of Laos and the UNODC signed their agreement in 1998, implementation of the opium eradication policy was swift. The Lao National Commission for Drugs Control and Supervision (LNCDCS) was created under the President's cabinet and directed by a minister. In 2000, the Prime Minister further strengthened the government's position with a decree ordering the elimination of opium production and illegal drug trafficking[17]. Article 135 of the penal code (1990) which had prohibited drug trafficking, but tolerated 'traditional' opium production, was modified in 2001 to ban all production and trade and intensify sanctions (BINLEA, 2006).[18] During its Seventh Congress in 2001, the Lao People's Revolutionary Party reiterated the objectives of its agreement with the UNODC: to eliminate opium production and consumption in Laos by 2006. In 2001, the government launched its 'National Campaign against Drugs' to disseminate new measures among peasant farmers and raise awareness of the repression-based policy among government and party officials. Threats of punishment aimed to compel farmers to voluntarily destroy plantations and to halt new planting, while plantations identified by police or governmental organizations were destroyed.[19] In November 2005, the Office of the Prime Minister ordered provincial governors to complete the process within three months, so as to declare success at the Eighth Party Congress in March 2006 (BINLEA, 2006, pp264-265).

'Opium-free' Laos

During the 1990s, the total area covered by poppies in Laos fluctuated between 20,000 and 30,000 hectares, depending on weather and market conditions. Then, in 1999, surface areas started to decrease by 10% to 20% per cent a year, a trend that picked up pace in 2004 and 2005 (Figure 32-4). By 2005, the total area planted with poppies (1800ha) was only 7% of the 1998 figure; gross opium production tumbled from more than 200 tons in the 1990s to 123 tons in 1998, and to just 14 tons in 2005 – a drop of 93% (UNODC, 2005b).

In 1992, opium poppies were cultivated in more than 2300 villages; in 1998, this number dropped to 2060 (LNCDCS, 2000). The downward trend was slow at first, with a decrease of 5% per year between 1998 and 2003, then it accelerated in 2004 and 2005 as pressure from local authorities was stepped up (Figure 32-5). In 2005, opium was being grown in fewer than 300 villages, i.e. a drop of 87% since the UNODC agreement (UNODC, 2005b).

In the 2005 Laos Opium Survey, just 6% of villages in northern Laos were reported to be cultivating poppies, while one-third of surveyed villages declared that they had ceased production less than two years before (UNODC, 2005b).[20] The official statistics tell us that the government's policy of swift eradication of opium was

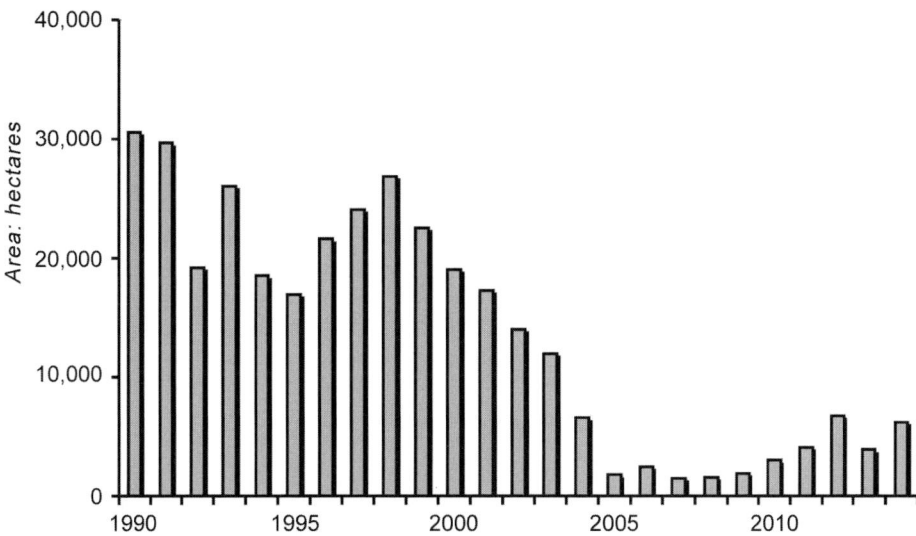

FIGURE 32-4: Evolution of surface area planted with poppies in the Lao PDR (1990 to 2014).

Source: LNCDCS (2000); UNDCP (2001, 2002); UNODC (2003, 2004a, 2005b, 2006d, 2007, 2008, 2009, 2010, 2011b, 2012, 2013, 2014c).

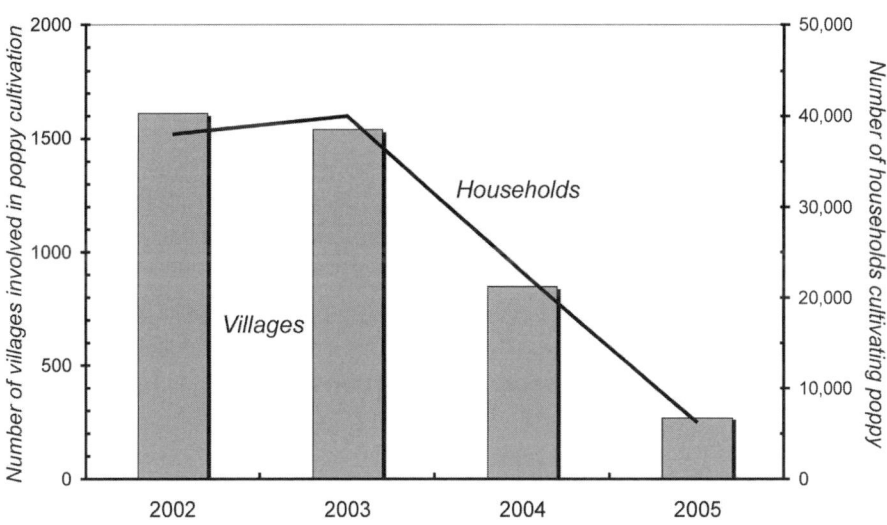

FIGURE 32-5: Number of villages and households involved in poppy cultivation in Laos (2002 to 2005).

Source: UNODC (2005b).

a success: poppy cultivation had almost disappeared from Laos by 2005. Gradually, district and provincial administrations declared themselves 'opium free' and every week, the Lao media reported on the successes stemming from the government's policy.[21]

The reliability of these results is questionable. Political pressure and repression had pushed poppy cultivation into illegality, making it difficult to access reliable information. The UNODC recognized this problem:

> The pressure to overcome the cultivation of illicit crops in Laos has been strong during the 2004-2005 season, raising the risk of receiving unreliable answers from the villagers (UNODC, 2005b, p7).

The UNODC based its 2005 assessment of surface area (1800ha) on an aerial survey by helicopter and estimated a 90% probability that the cultivation area was between 900 and 2900 hectares– a rather large confidence interval.[22] Using remote sensing data, the United States government found a cultivation area of 5500 hectares in the same year (BINLEA, 2006, p265), although this estimate should be viewed with reserve, given technical uncertainties associated with interpreting reflectance in the slope and shadow of mountain zones and the tendency to revise gross estimates based on diplomatic and geopolitical elements (Kramer et al., 2009). Our field experience in villages in Phongsaly revealed the difficulties of conducting opium surveys in Laos. To obtain information, it was necessary to win over the trust of villagers by demonstrating that our work (surveys and development projects) would not involve repressive measures against poppy cultivation. This required months of regular contact. How was it possible for researchers from the UNODC to gather detailed information using rapid appraisals while getting out of their helicopter in the presence of a team directly associated with eradication efforts, including local governmental authorities involved in repression? Villagers may have underestimated opium production in their village. This bias would have suited everyone involved, particularly the government and donors: the underestimation of surface areas would naturally point to the success of the programme.

The effects of eradication in production regions

Positive effects reported

Although it is probably overestimated, the recent and rapid decline of opium production in Laos is real. But is it the direct result of government policy, a coincidence associated with market shifts, or some other evolution in villagers' farming systems? There is consensus about crediting the government for these results. At the time of data collection in Phongsaly (2003 to 2005), farmers declared systematically and spontaneously that they no longer planted poppies under orders of local authorities. During the UNODC 2005 survey (2005b, p22), half of the 181 people interviewed said they had stopped poppy cultivation because it was now illegal.[23]

The UNODC surveys found positive social and economic effects among farmers who had stopped poppy cultivation.The 2003 survey found that opium sales amounted to $90 per farm, accounting for 42% of the average annual household cash income of $215 for poppy-cultivating families (UNODC, 2003, p9).The 2005 survey found that opium sales had fallen to $14 per year, or 10% of the average annual monetary income of $140 of the families involved, while the average annual household cash income of neighbouring non-opium-growing villages was more than $230 dollars (UNODC, 2005b, pp28-29). These gross findings tended to confirm the thesis of supporters of the government's anti-drug policy, that opium was a factor of poverty. A small minority of villages benefitted from assistance to develop alternatives to poppy cultivation, and these villages converted more easily to other agricultural systems (mainly animal husbandry). However, other non-opium-growing villages also saw improvement in their living conditions. According to the UNODC, ceasing poppy production had the following effects:

- Less than 30% of non-opium-growing villages reported rice deficits during the lean season, compared to nearly 60% of villages still growing poppies (UNODC, 2005b, p31);
- Forty-five per cent of respondent households reported a positive livelihood change after ceasing opium cultivation, while only 10% reported a negative change (UNODC, 2005b, p32);[24]
- The situation of women has improved significantly, by eliminating the time-consuming labour required for opium cultivation, thus freeing up time for other agricultural activities (UNODC, 2005b, p38), and reducing the violent behaviour of addicted husbands (thanks to detoxification).

By 2005, opium production and consumption had almost disappeared in Laos; the country no longer fuelled international trafficking and the socio-economic conditions of farmers were improving. The results of the rapid eradication policy therefore deserve enthusiasm. Nevertheless, they are debatable. The effectiveness of repression is widely recognized by field observers: farmers have produced less opium because of government pressure. The debate now concerns the social and economic consequences of rapid eradication on villagers in northern Laos.

A case study in Phongsaly

Dramatic decline in villager income and the latent risk of returning to poppy cultivation

The authors set out to assess the economic effects of the eradication policy on farming families in the district of Phongsaly, in the province of the same name (Figure 32-2). The 25,000 inhabitants of the district's 90 villages practise shifting cultivation, fishing and gathering. Agricultural alternatives are limited by a lack of cultivable lowlands in V-shaped valleys, poor market access and frequent health and pest problems plaguing livestock (Ducourtieux, 2006).

Phongsaly had been a major opium production region since colonial times and only recently had the area under opium poppies diminished. In 1998, about 3600 hectares of Phongsaly were covered with poppies (LNCDCS, 2000) and in 2001 the area was still 3300ha (UNDCP, 2001). The area fell by half, to 1600ha, in 2003 (UNODC, 2005a, p201), but rose again, to 2600ha, by 2014 (UNODC, 2014c). Despite this, Phongsaly was touted as being 'liberated from opium' in the country's National Assembly in October 2005.[25] In 2014, it was still the number one poppy-cultivating province, with 42% of the total area planted with poppies in the Lao PDR.

Poppy cultivation is most prevalent in northern Phongsaly, where farming systems have adapted over time to accommodate the labour-intensive crop (Alexandre and Eberhardt, 1998; Baudran, 2000). Local practices demonstrate that, contrary to the founding principles of UNODC's Balanced Approach to Opium Elimination in Lao PDR (UNDCP, 1999), opium is not intrinsically linked to shifting cultivation. The villages in the northwest of Phongsaly district, characterized by less rugged terrain and greater water-retention capacity, expanded their paddy fields into flooded terraces and gave up shifting cultivation around the end of the 1980s. The rice-cultivation calendar of the paddy fields (May to November) was perfectly complementary to poppy production (November to March). At the same time, the villagers ensured their food security by planting more poppies (Baudran, 2000). In the mid-1990s, opium sales accounted for 90% of annual household cash income, with huge socio-economic differences between households.

We studied the economic impact of a rural-development project in the Phongsaly district within the framework of the project's final evaluation. The objective was to characterize the evolution of household incomes in the region from 1996 to 2005 and identify discrepancies and their causes. Six of the district's 80 villages were selected for their representative characteristics, in terms of ecological and agricultural diversity (Ducourtieux, 2006). In each village, every family was interviewed about their economic activities between 1996 and 2005. After testing a questionnaire in one village, 232 household questionnaires were administered, processed and then statistically analysed to create a model with which to understand the evolution of household incomes. The methodology was based on the following hypotheses and included a number of constraints.

Reconstructing income over a 10-year period based on the narratives of villagers was both audacious and difficult. The quality of information varied between 1996 and 2005, thus limiting the validity of the analysis. In addition, gathering data on income in retrospect probably underestimated regular activities that contributed a little (gardening, gathering and so on), as they stood out less in farmers' memories than occasional large-scale activities like a rice harvest. Numerical results were thus indicative of trends and should not be interpreted as absolute values.

The choice of currency also influenced the results, especially in light of the 1997 regional financial crisis, which drastically depreciated the Lao currency.[26]

Criteria such as land access, access to building materials, use of traditional medicine and so on, were not taken into account. Produce consumed by the household was evaluated at market value (replacement value), which may have differed from the village's typical use value (Ducourtieux, 2006).

Despite these methodological challenges, which were difficult to overcome within the framework of a project evaluation, the study stands out as one of the most precise conducted in the region in recent times. Analysing even simplified data on the living conditions of farmers gives a more reliable idea of how the poppy prohibition has impacted farmers than a necessarily biased study focused entirely on opium production.

In 1996, five of the six villages cultivated poppies; the village closest to the urban provincial centre of Phongsaly was the only one that did not. Seventy-three per cent of surveyed households planted poppies in 1996, but none admitted doing so in 2005. Average annual production of raw opium per family fell from 600 grams to zero in the same period, and average household income for 232 surveyed households fell by 9%, from US$1050 to $960 (Figure 32-6). Average cash income decreased by 40%, to $70 per household per year in 2005. These averages concealed large local disparities. The drop was more dramatic in villages far from the provincial capital, while those closer to the urban centre showed no change, or a slight increase, in income.

Analysis of the different income sources explained these disparities. Farms close to the urban centre cultivated cash crops (vegetables, fruits, tea, and so on), while in remote villages, the absence of poppy cultivation led to a loss (an average of $100 per household per year) as did a decrease in hunting (loss of $66 per household per year). Government intervention was behind both changes: farmers said they had stopped growing poppies under pressure and following threats from local authorities that had intensified after 2001. Hunting was cut back following the confiscation of firearms in 1999.

The district of Phongsaly managed to eradicate poppy cultivation almost completely by 2005, at the cost of increased poverty. Lacking both investment capacity and economic alternatives, the villagers had to decapitalize their farms by selling animals to feed their families. Later, as assistance was neither swift nor substantial, poppy cultivation began to take hold once again.[27]

The UNODC's conclusion that rapid eradication of opium has not had any negative effects does not apply in Phongsaly. Is it possible that independent surveys in other northern provinces would reveal similar findings?

Similar results in other regions of Laos

Phongsaly is not an isolated example: rare studies conducted by actors other than the Lao government, the UNODC or the United States reveal growing impoverishment and food insecurity among the poorest of the poor, in the absence of either opium sales to buy rice or other viable economic alternatives. The consequence is migration of impoverished populations towards the valleys and plains where land is limited (Cohen, 2000, 2006; Cohen and Lyttleton, 2002; Evrard and Goudineau, 2004).

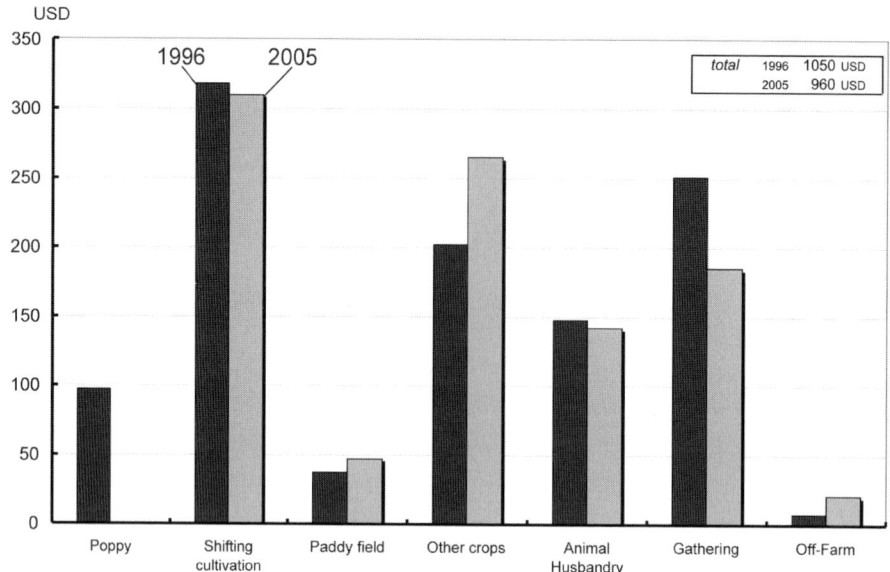

FIGURE 32-6: Evolution of average household income per component in six villages in the Phongsaly district (1996 to 2005).

Notes: The survey involved 232 household interviews. Currency used: US dollars.

Source: Doligez and Sacklokham, unpublished data.

The economic and social effects of forced, swift poppy eradication in Laos are comparable to those observed elsewhere where illegal drug crops have been subject to political repression, such as in Thailand (Crooker, 1988, 2005; Hanks and Hanks, 2001; Lyttleton, 2004; Kramer et al., 2009), Burma (Chouvy, 2009; Kramer et al., 2009; Sarno, 2009), or Latin America (McCoy, 2003; Cohen, 2006). Studies carried out in these countries systematically emphasize the high social cost paid by farmers and the poorest of the poor, as well as the risk of instability in the medium-term.

There appears to be a wide gap between rhetoric and reality. Despite a polished presentation and supporting quantitative data, the rationale for eradicating opium in Laos – that opium is a major cause of poverty – is founded more on socio-cultural prejudices and geopolitical considerations than on actual facts.

Towards a new rise in poppy cultivation in Laos?

The Lao PDR has cleaned up its reputation as a producer of illegal substances, thanks to the swift and repressive eradication policy in the years from 2001 to 2005. Evidence shows that highlanders from ethnic minorities were sorely affected; impoverishment and migration increased, in direct contrast to the stated policy objective of reducing poverty. The impact was made worse by a parallel ban on shifting cultivation. This farming practice was also presented as a cause of poverty.

As villagers were not given viable alternatives to compensate for the impact of poppy eradication, a rebound effect was inevitable; as soon as government pressure lessened, farmers began to replant poppies due to their growing impoverishment (Kramer et al., 2009) (Figure 32-4). This scenario was even more probable in the light of the Lao PDR's growing integration into the world economy (Kramer, 2010), and the likelihood that farmers would seek to specialize their farming systems based on the comparative advantages of their particular ecosystems (Chouvy, 2009; Cohen, 2009). As repression forced production down in Myanmar and Laos, prices for opium tended to increase (Figure 32-7), and this is now contributing to the crop's attractiveness to highland farmers.

By 2006, the UNODC had become somewhat ambivalent. At the same time as it was proclaiming an 'opium-free' Laos, it was making urgent appeals to other international donors, saying they should either mobilize funding for rural development projects to compensate for the drop in income of former poppy farmers or risk a rapid resurgence of poppy (production and consumption) of uncontrollable proportions (UNODC, 2006b):

> Laos is at a critical juncture. The tremendous successes that have been achieved could be reversed if sufficient assistance is not provided (Excerpt from a speech by UNODC's Director-General, Vientiane, 14 February 2006; UNODC, 2006b, p4).

It was a noteworthy change in approach, even a shift in paradigm, from opium as the 'cause of poverty'– eliminate it and all problems will be solved – to opium, 'poverty's consequence':

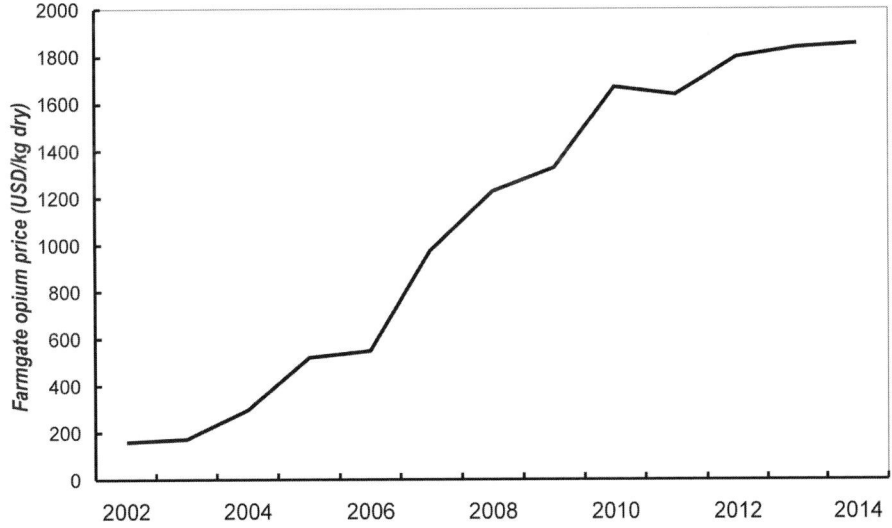

FIGURE 32-7:   Evolution of the farmgate price for opium (2002 to 2014).

Source: UNODC (2003, 2004a, 2005b, 2006d, 2007, 2008, 2009, 2010, 2011b, 2012, 2013, 2014c).

The link between poverty, a lack of alternative livelihood options and the decision to cultivate poppy is clear. The majority of respondents to the 2014 socio-economic survey in Myanmar described the need to provide for basic necessities, such as food, education and housing, as a reason for cultivating opium poppy. The difficulties involved in the cultivation of crops other than opium poppy, and the barriers to transporting goods to market, mean that villagers in remote areas have limited options for earning alternative income. In Lao PDR, no socio-economic survey of poppy-growing villages has been conducted in recent years. However, the data collected during helicopter flights and satellite image analysis indicated that poppy cultivation in 2014 also continued to be a phenomenon linked to villages in peripheral, difficult-to-access locations, far from population and market centres (UNODC, 2014c, p17).

The United States reduced its aid budget, so the UNODC tried without success to mobilize other donors with an urgent – indeed menacing – message: the UNODC has done its job (eradicate opium); if production picks up again, it will be for lack of support from the international community.[28] After liberating Laos from opium in fewer than five years, the UNODC and the Lao government got their desired success, and no one would like to over-stress the relative upturn of opium production in Laos since then.[29]

Eliminating the production of narcotics in a poor country requires long-term effort, sustainable political commitment and massive investments in socio-economic initiatives. The failure is total in Afghanistan (Nathan, 2009). In Thailand, eliminating poppy cultivation took more than 30 years (Renard, 2001) and it still had negative social impacts on the ethnic minorities of that country's northern mountainous regions (Kesmanee, 1994; Francis, 2004; Lyttleton, 2004). To expect similar results after only five years in Laos, where cultivation was more widespread and investment both limited and poorly distributed, was simply an illusion. According to the Global Commission on Drug Policy,[30] 'the global war on drugs has failed, with devastating consequences for individuals and societies around the world' (2011, p2). But that is another story.

References

Alexandre, J-L. and Eberhardt, N. (1998) *Des systèmes agraires de la rive gauche de la Nam Ou* (Farming Systems of the left bank of the Nam Ou), CCL, Paris

Baudran, E. (2000) *Derrière la savane, la forêt* (Behind the Savannah, the Forest), CCL, Paris

Bertrand, D. (2003) 'Le combat contre la drogue en RDP Lao: une analyse à travers la presse 1998-2003' (The Fight against Drugs in Lao PDR: An Analysis through the Press 1998-2003), *Moussons* 7, pp95-114

BINLEA (2005) *International Narcotics Control Strategy Report: Volume I, Drug and Chemical Control*, Bureau for International Narcotics and Law Enforcement Affairs, US Department of State, Washington, DC

BINLEA (2006) *International Narcotics Control Strategy Report: Volume I, Drug and Chemical Control*, Bureau for International Narcotics and Law Enforcement Affairs, US Department of State, Washington, DC

BINLEA (2014) *International Narcotics Control Strategy Report: Volume I, Drug and Chemical Control*, Bureau for International Narcotics and Law Enforcement Affairs, US Department of State, Washington, DC

Booth, M. (1999) *Opium: A History*. St. Martin's Griffin, London

Brown, M. A. and Zasloff, J. (1986) *Apprentice Revolutionaries: The Communist Movement in Laos 1930-1985*, Hoover Press/Stanford University, Stanford, CA

Chouvy, P-A. (2009) *Opium: Uncovering the Politics of the Poppy*, I. B. Tauris, New York

Cohen, P. T. (2000) 'Resettlement, opium and labour dependence: Akha-Tai relations in northern Laos', *Development and Change* 31(1), pp179-200

Cohen, P. T. (2006) 'Help as a threat: Alternative development and the 'War on Drugs' in Bolivia and Laos', *Development Bulletin*, 68-69, pp31-36

Cohen, P. T. (2009) 'The post-opium scenario and rubber in northern Laos: Alternative Western and Chinese models of development', *International Journal of Drug Policy* 20(5), pp424-430

Cohen, P. T. (2013) 'Symbolic dimensions of the anti-opium campaign in Laos', *Australian Journal of Anthropology* 24(2), pp177-192

Cohen, P. T. and Lyttleton, C. (2002) 'Opium-reduction programmes, discourses of addiction and gender in Northwest Laos', *Journal of Social Issues in Southeast Asia* 17(1), pp1-23

Crooker, R. A. (1988) 'Forces of change in the Thailand opium zone', *Geographical Review* 78(3), pp241–256

Crooker, R. A. (2005) 'Life after opium in the hills of Thailand', *Mountain Research and Development* 25(3), pp289-292

Descours-Gatin, C. (1992) *Quand l'opium finançait la colonisation de l'Indochine* (When Opium Financed the Colonization of Indochina), L'Harmattan, Paris

Ducourtieux, O. (2006) 'Is the diversity of shifting cultivation held in high enough esteem?', *Moussons* 9-10, pp61-86

Epprecht, M. (2000) 'The blessings of the poppy: Opium and the Akha people of northern Laos', *Indigenous Affairs* 4, pp16-21

Evans, G. (2002) *A short history of Laos: The land in between*, second edition, Silkworm Books, Chiang Mai

Evrard, O. and Goudineau, Y. (2004) 'Planned resettlement, unexpected migrations and cultural trauma in Laos', *Development and Change* 35(5), pp937-962

Feingold, D. (1970) 'Opium and politics in Laos', in N. S. Adams and A. McCoy (eds) *Laos: War and Revolution*, Harper & Row, New York

Francis, P. (2004) '"Where there is thunder there should be rain": Ethnic minorities and highland development in northern Thailand', *Mountain Research and Development* 24(2), pp119-123

Global Commission on Drug Policy (2011) *War on Drugs: Report*, Global Commission on Drug Policy, Rio de Janeiro, Brazil

Gunn, G. C. (1998) *Theravadins, Colonialists and Commissars in Laos*, White Lotus, Bangkok

Gunn, G. C. (2003) *Rebellion in Laos: Peasant and Politics in a Colonial Backwater*, Second edition, White Lotus, Bangkok

Hanks, J. R. and Hanks, L. M. (2001) *Tribes of the North Thailand Frontier*, Southeast Asia Studies, Yale University, New Haven, CT

Jelsma, M. (2003) 'Drugs in the UN system: The unwritten history of the 1998 United Nations General Assembly Special Session on drugs', *International Journal of Drug Policy* 14(2), pp181-195

Kesmanee, C. (1994) 'Dubious development concepts in the Thai highlands: The Chao Khao in transition', *Law and Society Review* 28(3), pp673-686

Kramer, T., Jelsma, M. and Blickman, T. (2009) *Withdrawal Symptoms in the Golden Triangle: A Drugs Market in Disarray*, Transnational Institute, Amsterdam

Kramer, T. (2010) *An Assessment of the Impact of the Global Financial Crisis on Sustainable Alternative Development: Key Determinant Factors for Opium Poppy Re-cultivation in Southeast Asia*, United Nations Office on Drugs and Crime, Vienna

Legrain (Dr) (1925) *Médecine sociale: Traité de pathologie médicale et de thérapeutique appliquée* (Social Medicine: Treatise of Medical Pathology and Applied Therapy), Maloine, Paris

Le Failler P. (2001), *Monopole et prohibition de l'opium en Indochine: Le pilori des chimères* (Monopoly and Prohibition of Opium in Indochina: The Pillory of Chimeras), L'Harmattan, Paris

Lintner, B. (2000) *The Golden Triangle Opium Trade: An Overview*, Asia Pacific Media Services, Chiang Mai

LNCDCS (2000) *Annual Opium Poppy Survey 1999/2000*, Lao National Commission for Drug Control and Supervision/UNDCP, Vientiane

LNCDCS (2006) *The National Programme Strategy for the Post-opium Scenario (2006-2009)*, Lao National Commission for Drug Control and Supervision/UNODC, Vientiane

Lovell, J. (2011) *The Opium War: Drugs, Dreams and the Making of China*, Picador, London

Lyttleton, C. (2004) 'Relative pleasures: Drugs, development and modern dependencies in Asia's Golden Triangle', *Development and Change* 35(5), pp909-935

McCoy, A. (2003) *The Politics of Heroin*, third edition, Lawrence Hill, Chicago

Michaud, J. (2009) 'Handling mountain minorities in China, Vietnam and Laos: From history to current concerns', *Asian Ethnicity* 10(1), pp25-49

Nathan, J. A. (2009) 'Poppy blues: The collapse of poppy eradication and the road ahead in Afghanistan', *Defense and Security Analysis* 25(4), pp331-353

Pholsena, V. (2010) 'US rapprochement with Laos and Cambodia: A response', *Contemporary Southeast Asia* 32(3), pp460-466

Rapin, A-J. (2008) *Opium et société dans le Laos précolonial et colonial* (Opium and Society in Pre-colonial and Colonial Laos), L'Harmattan, Paris

Renard, R. D. (1996) *The Burmese Connection: Illegal Drugs and the Making of the Golden Triangle*, Lynne Rienner Publishers, Boulder, CO

Renard, R. D. (2001) *Opium Reduction in Thailand 1970-2000: A Thirty-year Journey*, Silkworm, Chiang Mai

Sarno, P. (2009) 'The War on Drugs', *Southeast Asian Affairs* 1, pp223-241

Scott, J. C. (2009) *The Art of Not Being Governed: An Anarchist History of Upland Southeast Asia*, Yale University Press, New Haven, CT

Steinberg, M. K., Hobbs, J. J. and Mathewson, K. (eds)(2004) *Dangerous Harvest: Drug Plants and the Transformation of Indigenous Landscapes*, Oxford University Press, New York

Stuart-Fox, M. (2001) *Historical Dictionary of Laos*, second edition, The Scarecrow Press, Lanham, MD

Terry, C. E. and Pellens, M. (1928) *The Opium Problem*, Committee on Drug Addictions, Washington, DC

Trocki, C. A. (1999) *Opium, Empire and the Global Political Economy*, Routledge, London

UNDCP (1999) *A Balanced Approach to Opium Elimination in Lao PDR: Executive and Strategy Summary*, United Nations Drug Control Programme, Vienna

UNDCP (2001) *Annual Opium Poppy Survey 2001*, United Nations Drug Control Programme, Vienna

UNDCP (2002) *Annual Opium Survey 2002*, United Nations Drug Control Programme, Vienna

UNODC (2003) *Laos Opium Survey 2003*, United Nations Office on Drugs and Crime, Vienna

UNODC (2004a) *Laos Opium Survey 2004*, United Nations Office on Drugs and Crime, Vienna

UNODC (2004b) *World Drug Report 2004, vol. 1: Analysis*, United Nations Office on Drugs and Crime, Vienna

UNODC (2004c) *Strategic Programme Framework, Lao PDR 2004-2007*, United Nations Office on Drugs and Crime, Vienna

UNODC (2005a) *World Drug Report 2005, vol. 2: Statistics*, United Nations Office on Drugs and Crime, Vienna

UNODC (2005b) *Laos Opium Survey 2005*, United Nations Office on Drugs and Crime, Vienna

UNODC (2006a) *Afghanistan Opium Rapid Assessment Survey 2006*, United Nations Office on Drugs and Crime, Vienna

UNODC (2006b) *Towards an Opium-free Lao (PDR)*, newsletter 2, United Nations Office on Drugs and Crime, Vienna

UNODC (2006c) *World Drug Report 2006, vol. 1: Analysis*, United Nations Office on Drugs and Crime, Vienna

UNODC (2006d) *Opium Poppy Cultivation in the Golden Triangle: Lao PDR, Myanmar, Thailand*, United Nations Office on Drugs and Crime, Vienna

UNODC (2006e), *World Drug Report 2006, vol. 2: Statistics*, United Nations Office on Drugs and Crime, Vienna

UNODC (2006f) *Laos: Strategic Programme Framework*, United Nations Office on Drugs and Crime, Vientiane

UNODC (2007) *Opium Poppy Cultivation in South East Asia: Lao PDR, Myanmar, Thailand*, United Nations Office on Drugs and Crime, Vienna

UNODC (2008) *Opium Poppy Cultivation in South East Asia: Lao PDR, Myanmar, Thailand*, United Nations Office on Drugs and Crime, Vienna

UNODC (2009) *Opium Poppy Cultivation in South East Asia: Lao PDR, Myanmar*, United Nations Office on Drugs and Crime, Vienna

UNODC (2010) *South-East Asia Opium Survey 2010: Lao PDR, Myanmar*, United Nations Office on Drugs and Crime, Vienna

UNODC (2011a) *World Drug Report 2011*, United Nations Office on Drugs and Crime, Vienna.

UNODC (2011b) *South-East Asia Opium Survey 2011: Lao PDR, Myanmar*, United Nations Office on Drugs and Crime, Vienna

UNODC (2012) *South-East Asia Opium Survey 2012: Lao PDR, Myanmar*, United Nations Office on Drugs and Crime, Vienna

UNODC (2013) *South-East Asia Opium Survey 2013: Lao PDR, Myanmar*, United Nations Office on Drugs and Crime, Vienna

UNODC (2014a) *Afghanistan Opium Survey 2014*, United Nations Office on Drugs and Crime, Vienna

UNODC (2014b) *World Drug Report 2014*, United Nations Office on Drugs and Crime, Vienna

UNODC (2014c) *South-East Asia Opium Survey 2014: Lao PDR, Myanmar*, United Nations Office on Drugs and Crime, Vienna

Vongvichit, P. (1968) *Le Laos et la lutte victorieuse du peuple Lao contre le néo-colonialisme américain* (Laos and the Lao People's Victorious Struggle against American Neo-colonialism), Neo Lao Haksat, Vientiane

Notes

1. The United Nations Office on Drugs and Crime is the result of a merger in 2002 between the United Nations Office for Drug Control and Crime Prevention (ODCCP) and the United Nations Drug Control Programme (UNDCP). Throughout this chapter, the acronym UNODC will be used when referring to this organization.

2. Poppy cultivation increased by 310% between 1990 and 2014, an expansion matched precisely by the rise in opium production over the same period (UNODC, 2014a).

3. Poppies are cultivated for pharmaceutical morphine in India, Turkey, Australia and France, but in localized and highly controlled circumstances that are theoretically distinct from heroin and opium.

4. An area that encompasses northeastern Myanmar, northern Thailand and northeastern Laos.

5. In 2014, Phongsaly province accounted for 42% of the poppy-growing area in the Lao PDR (UNODC, 2014c).

6. Named by the General Governorate of colonial Indochina as '*Régie de l'Opium*'.

7. McCoy estimates that there were 13.5 million opium addicts in China in 1900, or 3% of the population, and they consumed 38,000 tons of opium, of which 35,000 tons were produced domestically. That amounted to 85% of world production (McCoy, 2003, p5).

8. *Source:* Archives from Indochina (Oversea National Archives, Aix-en-Provence, France), INDO/ GGI/SE/2709 (1937-38) Réglementation foncière en ce qui concerne les populations Rhadé du Darlac (Land regulation concerning Rhade populations of Darlac), Agriculture Department, General Governorate of Indochina, Hanoi.

9. *Source*: Archives from Indochina (Oversea National Archives, Aix-en-Provence, France) INDO/ GGI/42966 (1936) Achat par la Régie indochinoise de l'opium au Laos (Purchases by the Indochinese Opium Excise Office in Laos), Policy Affairs Department, General Governorate of Indochina, Hanoi.

10. Illegal drug trafficking was embedded in local conflicts at this time, but this was not exceptional. Rather, it was the general rule (Chouvy, 2009; McCoy, 2003).

11. Purchasing opium allowed the CIA to generate significant extra-budgetary revenues for special operations and to recruit ethnic mountain minorities for these operations (Feingold, 1970; McCoy, 2003).

12. Phoumi Vongvichit was a member of the political bureau of the Lao Revolutionary Party from 1955 until his retirement in 1991. He served as Secretary-General of the Pathet Lao and vice-Prime Minster after 1975 (Stuart-Fox, 2001).

13. Shanghai, 1907; La Haye, 1911-1913; Geneva (Society of Nations), 1924 and 1925.

14. Personal communication between one of the authors and W. M. Carroll, technical advisor to the project, June 2006.

15. This represented just 28% of funding pledges, five years after the plan was launched.

16. The only political party to hold power since 1975.

17. Decree 14/PM (11/2000).

18. The death penalty applies for production or trade of more than 500g of heroin and more than 3kg of amphetamines.

19. Organizations identifying plantations included youth unions, women's unions and the National Edification Front. In 2005, for example, the government announced that nearly 2600ha of poppies had been destroyed (UNODC, 2005b, p38).

20. An annual study conducted by the UNODC and the Lao National Commission for Drug Control and Supervision.

21. Forty-nine districts out of 80 formerly involved in poppy cultivation were declared 'opium free' by January 2005, i.e. 61% (UNODC, 2005b).

22. In 2008, the estimated area ranged from 600 to 2700ha (UNODC, 2008); in 2011, 2500 to 6000ha (UNODC, 2011b); and in 2014, 3500 to 9000ha (UNODC, 2014c).

23. Six per cent of farmers responded that they had ceased poppy cultivation because their neighbours had ('everyone in the village has already stopped'). Roughly two-thirds of all replies were directly or indirectly linked to the fact that poppy cultivation had been declared illegal.

24. Twenty-five per cent of families who stopped growing poppies reported no significant change to their livelihood and 30% did not respond to the question.

25. *Vientiane Times* (18 November 2005). This is the sole English-language newspaper in Laos, a country that is not known for freedom of the press.

26. Between 1996 and 2005, the exchange rate plunged from 930 to 10,600 Laotian kip for one United States dollar.

27. In the years from 2003 to 2014, poppy cultivation grew by 62% in Phongsaly province (UNODC, 2005a, 2014c).

28. In 2006, the UNODC promoted a three-year, $8.3-million development plan for the 1000 villages most at risk of returning to poppy farming, but offered no funding beyond its two on-going projects, which were scheduled to end in 2007 (BINLEA, 2014; Kramer, 2010; UNODC, 2006f). The message was clear: funding had to come from other international aid agencies.

29. An upturn from 1800ha of poppies in 2005 to 6200ha in 2014, i.e. an increase of 245% (Figure 32-4).

30. A panel of 23 world leaders and intellectuals, including 15 former ministers or heads of state.

D. When shifting cultivators get pushed aside by large agribusiness

The expansion of oil palm plantations, partly in response to high demand for biofuels, is probably the single largest cause of displacement of shifting cultivation by agribusiness in Asia, particularly in Indonesia.

Sketch based on a photo by Lesley Potter.

33

GIVING UP FALLOWS AND INDIGENOUS SWIDDENS IN TIMES OF GLOBAL LAND GRABBING

*Esther Leemann and Brigitte Nikles**

Introduction

Swidden cultivation is important for the livelihoods of between 14 and 34 million people in Southeast Asia (Mertz et al., 2009).[1] However, it is being transformed into other forms of land use at a rapid pace (Schmidt-Vogt et al., 2009). One of the factors contributing to the sudden demise of swidden systems is the expansion of large-scale industrial agriculture, such as oil palm or rubber plantations, in previously peripheral and remote areas (Fox et al., 2009; Baird, 2010). It is noteworthy that 'integral' swidden systems are not only essential to farmers' subsistence, but to the entire way of life of farming communities (Conklin, 1957). Thus, the sudden and enforced loss of land caused by the expansion of plantations has a strong impact not only on the livelihoods of swidden farmers, but also on their social relations, religious beliefs and cultural and political identity. The social and cultural outcomes of plantation-induced disruption of swidden cultivation and fallow management, as essential elements in an entire way of life, have yet to receive much attention. This chapter deals with the case of the Bunong in Bousra commune, Mondulkiri province, northeastern Cambodia (Figure 33-1). It aims to fill this gap and addresses the impact of the sudden decline of swidden farming on people's social and cultural identity.[2]

The Bunong,[3] one of 23 ethnic minorities in Cambodia, are currently confronted with the sudden and rapid expansion of rubber plantations on their customary land. Affected communities most often completely lose access to their swidden sites, cattle

* Dr Esther Leemann received her PhD in social anthropology from the University of Zurich, and is a senior lecturer and researcher in the Department of Social Anthropology at the University of Lucerne. She is currently directing a research project on land grabbing and swidden demise in Cambodia. Brigitte Nikles carried out field research for her MA thesis in social anthropology from the University of Zurich in two Bunong communes. She has been living and working in Mondulkiri, Cambodia since 2007, where she continues her research among the Bunong.

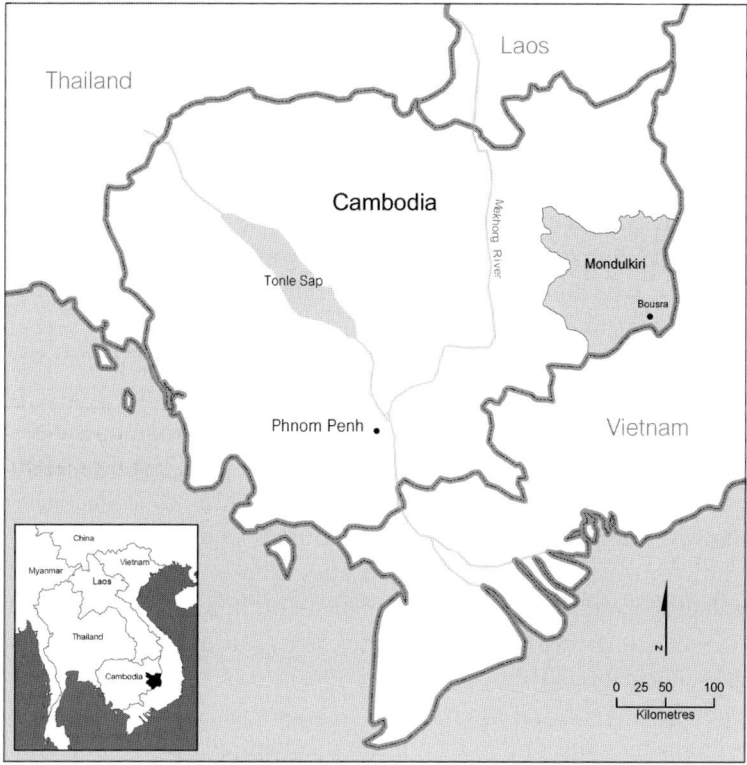

FIGURE 33-1: Location of the study area.

grounds and forest resources. Their traditional livelihood strategies are severely impaired, and in many of their village territories, swidden cultivation with traditional fallow management is no longer feasible. Furthermore, several Bunong communities in Mondulkiri have witnessed the complete destruction of sacred places, such as burial grounds and spirit groves, or at least a drastic reduction in the size of these places. Plantation workers have bulldozed them as if they were any other forest area within the 'economic land concessions' granted to their companies by the Cambodian government.

Land grabbing in Cambodia: 'economic land concessions' and indigenous communities

In Cambodia, the most common means for a company to gain control over vast tracts of land for industrial agriculture is the procurement of economic land concessions from the state. These economic land concessions are discursively justified and legitimized with narratives based on investment, growth, development and modernization, and are said to involve 'idle' and 'unoccupied' lands - depictions common in many contexts at a global level where large-scale land grabs are taking place (Borras et al., 2011; De Schutter, 2011; Li, 2011). A government sub-decree on economic land concessions signed in 2005 notes that an environmental and social impact assessment must be conducted through public consultations with all relevant stakeholders, including villagers and civil society groups, before such concessions are granted. However, this is hardly ever done (Diokno, 2008; Ironside and Nuy, 2010). At the beginning of 2012, economic land concessions covering 2,036,170 hectares had been granted for agro-industrial plantations to grow crops such as rubber, sugar and cassava – an area equivalent to 53% of Cambodia's arable land. According to rights group Licadho, 227 agro-industrial companies were holding government-issued contracts. In addition, 1,900,311 hectares had been granted as mining concessions to companies exploring for minerals such as gold, iron ore, copper and bauxite (Vrieze and Naren, 2012). In Mondulkiri province alone, 94,731 hectares have been granted as concessions for rubber plantations (FIDH, 2011). These areas of Mondulkiri form the basis for the livelihoods of the Bunong, and are managed under local customary tenure systems. Cambodian Land Laws from 2001 provide a legal framework to protect indigenous peoples' rights to collective ownership, thereby allowing the Bunong to formally secure the rights to their land. However, the procedure for obtaining collective land titles is both intricate and costly, and the political will of the authorities to protect the land of the Bunong is minuscule. Their efforts have yielded neither effective collective titles nor interim protection measures (FIDH, 2011; Leemann, 2012).

In Mondulkiri province, several large Cambodian, Vietnamese, Chinese and Malaysian companies and one joint venture involving principals from Luxembourg and Cambodia are currently active (FIDH, 2011), along with an unknown number of private investors controlling small-scale rubber plantations. A rough estimate suggests that 26 villages with more than 3500 families, or about 13,500 Bunong people, are affected by rubber plantations.

People affected by economic land concessions are left with no legal possibility of settling and cultivating in other forest areas of Mondulkiri because most of the province's forests fall under some form of protection. In fact, 80% of the province falls under the protection of either the Forestry Law or Laws on Protected Areas (Diokno, 2008). The Cambodian Land Law of 2001 protects the rights of indigenous communities to those forests that are used for traditional agriculture, including cultivated and fallow fields within shifting-cultivation systems, and the Law on Forestry (2002) states that the grant of a forestry concession or the creation of a protected forest should not interfere with the rights of customary users (NGO Forum on Cambodia, 2009, p22). In practice, however, indigenous people's rights are notoriously disregarded. Interestingly, many economic land concessions are found in the core zone of conservation areas, implying large-scale deforestation in places where forests should be protected (NGO Forum on Cambodia, 2009; IOM, 2009, p24; Vrieze and Naren, 2012).

While transnational rubber corporations use economic land concessions to take control of Bunong communities' customary land, another process of exclusion is simultaneously under way: the rapid alienation of land to migrants from the lowlands.

This involves informal land sales, the simple neglect of villagers' claims, and also incidents of intimate exclusion (Hall et al., 2011). All of this takes place in a climate of intimidation.

Bunong swidden cultivation: The cultural aspects

In some very remote villages in Mondulkiri, where the Bunong are not yet excluded from access to their land, they still practise their traditional system of swidden agriculture. For this reason, we present the following section on the traditional swidden cultivation system of the Bunong in the ethnographic present tense. In reality, when talking about traditional swidden cultivation in the context of Bousra commune, we are referring to the practices of 10 years ago or earlier. At that time, the most important preconditions for an integral swidden system were still met: the availability of large tracts of land and forest and the non-existence of livelihood alternatives.[4] Since the 1960s, when the Bunong at Bousra adopted the cultivation of rain-fed paddies in suitable places on the plains, they have practised what Rambo (1996) called a 'composite swiddening agro-ecosystem', or more specifically, a 'composite paddy-swidden agro-ecosystem' (Rambo, 2007). A decade ago, their main threat came from extensive and unregulated logging operations, which started in the 1990s. But this did not affect the Bunong in a way that forced them to give up swidden farming. The rush of land sales and speculation as rubber plantations began expanding into Bousra did not take off until early 2005, and reached its peak in 2007.

The system of rotational farming

Fallow land

Bunong families have several plots of land lying fallow at any given time. These they refer to as *mir char* (fields that were used before). Farmers say the decision to leave their swiddens fallow is made because of decreasing rice quality, measured by the amount and size of the grains, and increasing competition from grass.[5] Social and spiritual factors may also play a crucial role in a decision to abandon a field. In a case of personal misfortune, such as a death by suicide, the field must be abandoned immediately after the harvest, and no one will ever use that plot again. If a miscarriage or death during delivery occurs on the swidden, purifying ceremonies have to be performed and no one other than the family involved is allowed to cultivate that land again in the future, even if it is left fallow for a long time.

   Even though little or no labour is invested in a fallow, the family who farmed the plot in the cultivation phase continues to claim principal rights to collect vegetables, bananas, tobacco, pineapples and other fruit from the plot, as well as long thatching grass (*Imperata cylindrica*), which is cut for roofing. In the second and third years after cropping, these resources become less abundant, but they are still harvested by the family. Different kinds of medicinal plants can also be found in the fallow plot, some of them for many years. These include *mong leung* (*Chromolaena odorata*) leaves, which are pounded and used against cuts and skin problems; *taem njong* (*Smilax glabra*), a root that is boiled in water, which has revitalizing effects; *taem nggom,* which yields leaves that are used for steam baths or showers during the post-partum period; *taem kroic* (*citrus* sp.), the leaves of which are used for steam baths in the post-partum period after childbirth and against common colds; and *taem dong*, which has bark that is used to produce yeast for rice wine. These medicinal plants can be collected by anybody without restriction and serve as important medical resources for Bunong communities.[6] In addition, during the first two years after cropping, fallows are used as communal grazing areas.

Chromolaena odorata (L.) R. M. King & H. Rob. [Asteraceae /Compositae]

Regarded as an invasive weed in many parts of the world, the Bunong see this as a medicinal plant.

Land is conceived by the Bunong as common property; a community, consisting of a matrilineal descent group, claims a specific area as its land.[7] Families can claim exclusive private rights only to cultivated fields and to plots that have been recently fallowed. As soon as the vegetation returns and a fallowed swidden becomes strongly overgrown, it falls back into common property and consequently can be cleared again by any member of the community. However, the fallow phase lasts for at least six or seven years; the average is around 10 years. If the fallow period is shorter than six years, the land can only be cultivated for three years at the most, because rice yields are adversely affected by thorny scrub and grass that grows back much more rapidly than in fields opened from longer fallows.

Swidden fields

On average, every family cultivates two or three swidden fields in different locations. These are at different stages in the swidden cycle, from a first-year crop to the third or fourth year of cropping. Having several plots increases food security, by reducing the risks of insufficient harvests resulting from natural hazards like insect infestations, droughts and floods. Such events may be explained by villagers as interference by spirits, for similar acts of interference are believed to cause other detrimental consequences, such as sickness, accidents or death.

The most fertile land may be cultivated for as long as six years – an exceptionally long time, in the eyes of the villagers. On average, people use the same swidden for four years; at minimum, for three. The size of swiddens is measured by the number of baskets of rice planted in them each year. In Western parlance, they seldom exceed one hectare. The families usually clear new swiddens directly alongside the old ones.

Conflicts over land issues among kinship groups and villages have reportedly never been a significant problem, as there has always been plentiful land and forest available. The only significant arguments arise when people want to use land that belongs to another community. If two families aim for the same plot, solutions are sought by involving village elders and traditional conflict-resolution mechanisms (Backstrom et al., 2007). In order to avoid conflict, a family has to clarify the tenurial rights of the land before opening a new field for cultivation. Even if it looks like primary forest, most of the land has been used before. If the land belongs to another village, permission must be sought from that community. In any case, whether the land is located within the territory of their own village or belongs to a neighbouring community, a family must research the history of the plot to make sure that no misfortune has taken place at the site and it is not sacred forest. Powerful spirits are believed to inhabit locations designated as sacred forest, and swidden cultivation is strictly prohibited, as is resource extraction. If these rules are ignored, villagers fear punishment by the spirits, such as sickness, accidents and other forms of hardship. However, when the status of the land conforms to the rules, the family puts up a sign with a bamboo pole in the middle of the intended plot to mark the land and to seek approval from the spirits.

Agricultural cycle

When a new plot is opened in dense forest, the agricultural cycle starts around February. First of all, the small trees are cut down, followed by the big ones – a task that is mainly done by the men. Then the tree branches are chopped into pieces. The wood is left to dry for about three weeks and then the plot is burnt. After another week, the remaining wood and plants are piled up for re-burning. When a field that was cropped in the previous year is cleared, work starts in March with the burning of the remaining rice stalks. Then, people start weeding and sweep everything together to burn it again.

In first-year swiddens, bananas and other fruit trees are planted before the rice, along with tobacco, corn and various vegetables (eggplant, chillies, pumpkins, squash, and so on). Some vegetables like cucumbers, pumpkins and gourds are planted together with the rice seeds. People wait for the first rain to fall, and in about June, the rice is sown. In second-year swiddens, the rice is planted earlier – in May or early June. Planting in both new and old swiddens can be postponed to July in cases of delays in preparing fields brought about by a lack of labour or other unforeseen events. To prepare the ground for planting, men make small holes using wooden sticks, followed by women and men sowing the rice. On average, a Bunong farming family will select three varieties of early rice, as well as different varieties of the main rice crop – the so-called mother rice (*ba me*). The early rice varieties, which can be harvested after four months, may be called *ba krong, ba 'yang* and *ba m'et* (sticky rice), among others. These early varieties are planted in small sections of the swidden field, usually in the corners, and take up no more than 30% of the whole swidden. The rest of the swidden is planted with the main *ba me* crop, for which there exist about 20 different varieties, such as *ba wey rahen, ba nggon* and *ba ntol.* The main rice crop is harvested after six months.

Three to four weeks after planting, the field is weeded again, and this is followed by a one-month break from field work. In this time, women dry chillies, tobacco and eggplant; men are busy making baskets and other handicrafts. In late July or August, the people return to the fields to weed the crops again, and for about two weeks in September, they cut firewood. Between September and October, the early rice is ready for harvest and the main rice is ready in late October or November, but harvesting may continue until the end of December.

Work force

Mechanisms for mutual aid are common in Bunong communities. Therefore, families do not work exclusively on their own plots, but help to accomplish the various tasks throughout the agricultural year in the fields of other families. This labour force is either organized into permanent communal working groups or, on occasions of urgent demand, mobilized by summoning a large group of villagers. On average, five and sometimes up to 10 families form a permanent communal working group, called *teum teung ti,* or *teum rap ti.* They are usually related through their matrilineage. From every family, one or two members join the communal working group, which moves from field to field working for one day in each of the swiddens belonging to members of the group. Family members who are not involved in the working group continue their tasks in their own fields. Families forming a communal working group also take part in each other's rice ceremonies.

The most time-consuming and labour-intensive work, such as planting and harvesting, requires plenty of labour at one time, in order to complete the work without delay. People who are invited to help will only work for one full day, even if the work is not yet finished. They are provided with food during the day and with one or several jars of rice wine in the evening. This kind of mutual help is based on reciprocity and is expected to be repaid in labour.

People explain that the communal working groups contribute to strong bonds and social cohesion within villages. The labour force is distributed, as are agricultural outputs, and stability is created within the community as a result. Exchanges of labour between different villages are a means of maintaining or improving existing relationships or forging or consolidating possible future marriage arrangements and alliances.

The spiritual belief system

Ceremonies to the rice spirits

Most of the Bunong are animists and they believe that spirits exist in the natural environment. The relationship with the spirit world forms a fundamental part of their culture and identity; to a certain extent influencing and guiding their behaviour and decision-making processes. In order to maintain good relationships with the natural realm, as well as with ancestral spirits, numerous ceremonies are performed for agricultural, protective, compensative or healing purposes and as an expression of gratitude, acknowledgement or apology.

The performance of various rice ceremonies *(jan ba)* throughout the agricultural year is a manifestation of the interconnection between the Bunong and spiritual forces; of the continuous process of mutual giving and taking. Without the performance of ceremonies for the rice spirits, an abundant harvest is thought to be impossible and the people fear disastrous consequences, either directly or indirectly. The rice ceremonies are related either to a particular time of year or to a specific working step in the agricultural cycle. Women are generally in charge of preparing, organizing and leading the ceremonies, as they are the heads of households, the principal owners of their families' assets and experts at invoking the spirits of ancestors and the natural environment. Basic offerings to the spirits in any ceremony are an animal sacrifice – the preparation of which is the responsibility of a male community member – one or several jars of rice wine, a bowl of uncooked rice and candles. The larger and more numerous the sacrificed animals, the more important the ceremony. The woman who leads the ceremony carefully keeps blood and particular parts of the organs of the sacrificed animal until the recitation of prayers begins. Depending on the significance and meaning of the ceremony, decorative items are crafted or carved, bamboo poles are erected in the swidden field or at home, and elders and relatives are invited from far away. Ceremonies are always important social events at which relatives, friends and neighbours swap news and information about family, agriculture and all other aspects of life, and social bonds are invigorated.

The reasons for the rice ceremonies are manifold. Before sowing, the rice seeds have to become healthy, strong, clean and energetic to prepare them for planting. While the rice is being planted, the spirits are called upon for general support and help throughout the farming period. At the end of June, a ceremony is held to protect the fields and crops from animal pests and to purify the rice, enabling it to flourish. One month later, the rice is considered to be pregnant and in another ceremony, toys are presented to the 'children of the rice' (grains) in order to keep them happy. In September or October the rice is nearly ready to be harvested and a ceremony is indispensable, or villagers will feel unable to proceed with harvesting the main rice crop. The last rice ceremony in the agricultural cycle takes place when everybody in the village has completed their harvest: the rice is brought home and lifted up into the granaries.

Dreams and other signs from the spirits

The whole process of clearing and preparing a new swidden demands careful attention to signs of any kind from the spirits. Dreams and traces of certain animals are recalled or observed with caution. Inauspicious signals may result in the immediate abandonment of a field, even if a lot of work and effort has already been devoted to it. Therefore, on the night after the land is marked out for a new swidden, the adult members of the family will 'listen to their dreams' in order to seek the approval of the spirits to cultivate the land. Good dreams include the appearance of a grave, an anthill, a fish or a snake. The grave and the anthill are symbols of a plentiful harvest and fish are said to bring property or money. The snake symbolizes a cord tethering a buffalo, a cow or an elephant. Warning signs that plans to cultivate the designated plot must be abandoned include the appearance of an elephant. This means that plans must be cancelled. To dream about fire stands for suffering. Broken teeth represent death in the family. To dream about a cow signifies that the field or house might catch fire, and to dream about killing a buffalo or wild animals indicates danger to body and health, and the dreamer may be injured while working on the field. If the family has no dreams at all, it is a good sign and they can start working on the land.

Another test consists of placing seven kernels of rice under a stone where the family has marked the land. If the rice is still in the same form the day after and is not damaged, the family has received the spirits' consent to prepare the field in that location.

After chopping down the small trees – the first task in clearing a swidden – the cultivators listen to their dreams again. If a bad sign appears, the place is instantly abandoned the following day. If a cobra is spotted in the field before the vegetation is burnt, or if a dead cobra or civet cat are discovered in the burnt field, the site is immediately abandoned, and the family starts anew.

The practice of 'listening to their dreams' and observing other warning signs when selecting a new swidden has diminished, and as long as 10 years ago, only a minority of the Bunong still followed this tradition. However, rice ceremonies are commonly seen as indispensable and are still widely performed to date.

Impacts of land shortage and tenure insecurity on fallow management in Bousra

Tenure insecurity and loss of access to land

Bousra commune consists of seven villages with 933 households and a total of 4463 people, of which 86% are Bunong.[8] A growing number of families are facing a serious land shortage because they have lost access to swiddens, fallow fields, grazing areas for cattle and forest to (a) rubber companies that were granted economic land concessions by the government;[9] (b) protected areas and wildlife sanctuaries declared by the state; (c) settlers from lowland provinces who simply claim the villagers' land as theirs; and (d) settlers and absentee landowners who buy land from the intimidated villagers. Exact numbers affected by each 'transaction' category are hard to obtain, but a conservative estimate is that every family in Bousra has lost land in the past few years and about 10% of the Bunong families are now without land, a trend that is increasing. More and more people are now worried about how they will make a living and how they will get through the coming year, as finding new land has become almost impossible:

> To look for land today is not easy, as everywhere, the land actually belongs to someone, even if you go to the forest. And the government says it is their land, it is protected forest. And you go to another village and they say 'it is our land'. And you go to another family and they say 'this is our land'. So nowadays, to look for land is almost impossible (…). Maybe you can ask people if they want to share it. Or if you buy the land, then maybe you can get some. But just to go and find land is not possible any more (Bunong man, 28 years, Pu Cheng village).

Land tenure is no longer secure because the customary land rights of the Bunong are either not recognized or threatened by economic land concessions, forest conservation projects and other state claims. A profound feeling of insecurity is taking hold of the villagers. Their main fears and concerns are that companies, settlers and anybody else will claim their remaining land, especially the fallow land and even the current swiddens.

Giving up fallows

The villagers of Bousra are fully aware of the need to adapt to the current context of highly insecure land tenure and shortage. On top of the actual loss and consequential shortage of land, there is the threat of losing even more of the remaining land, and this has added to a trend of families quickly switching from swidden cultivation to permanent agriculture. In order to demonstrate ownership, fallow fields are cleared as fast as possible and are used permanently. As a consequence, a dramatic shortening of fallow periods can be observed in Bousra commune, directly culminating in the abandonment of swidden cultivation and its related fallow-management system. Villagers say they will not have any fallows in the future, and they will no longer be able to practise swidden cultivation.

> We will not leave our land fallow any more. Because there is no more land available and we want to secure our fields. If we leave it fallow, somebody else might take it and use it (Bunong man, 54 years, Pu Cheng village).

Interestingly, in this transition process from rotational to permanent land use, the very practice of fallow and swidden cultivation appears more and more irrational to the former swidden farmers.

If we leave this land fallow, we waste that place. Because we already put a lot of energy in this field, to clear it and prepare it and if you abandon it again, you waste this energy. Therefore, to grow something else after the rice is much more convenient and less work is involved. It is easy to grow cash crops afterwards. In summary, the main reason for not doing fallows any more is to secure the land and have less work for growing cash crops (Bunong man, 54 years, Pu Cheng village).

There are many reasons why people do not do swidden agriculture any more. One reason is that you cannot do the rotational system any more; you cannot leave the land fallow and find a new place. The police of the forest administration will stop you if you cut new forest. The second reason is that at the moment, if you abandon your land, you look very stupid. Because you already cleared and cleaned this place and it is easy to grow another crop, like cash crops, to get money. Today there is a market. Therefore, if the rice is not good any more, people grow something else on this land and don't allow it to go fallow. A third reason is, if you abandon your land, it is not secure. Even if you say it is your old field, it is not secure, people do not have paperwork, and anybody can take this land.

So these are the three reasons why people stop having fallows. And of course there is less land available, there is just no land any more. This also means that people stop growing rice in these fields. They still grow rice in the paddy fields, if they have some. But in the swidden fields you cannot grow rice any more after four or five years, if you do not fallow it. Either you grow cash crops or rubber (Bunong man, 35 years, Lameh village).

In this new context, both the opportunity and the need to sell cash crops and earn a monetary income prevails over the traditional system where people secured their subsistence needs with their own food production. For the moment, cassava and corn are the most popular cash crops, while a minority of villagers have opted for rubber under a contract-farming scheme with the rubber companies.

Whereas rice yields are known to decline considerably after three or four years of cultivation, villagers perceive cash crops to be capable of yielding a promising output even on soil that they know is no longer suitable for rice. Cash crops are thus seen as a way to make permanent agriculture possible:

People are afraid that it [the land] will be taken by outsiders. They want to secure their land and now they have the opportunity to sell cash crops. So at the moment, I can see people working more actively. They stop growing rice because the land is not good for rice any more and then they directly change to cash crops. But I do not know what people will do when the cash crops are not good any more. Maybe they keep using the land, but it is not clear because right now the land is still good enough to continue to grow cash crops (Bunong man, community land committee member, 45 years, Pu Cheng village).

The villagers realize that cash cropping is prone to risks and uncertainty: the soil fertility will inevitably decline under permanent cultivation, and the crop boom could unexpectedly end with a crash if prices collapse. Nevertheless, villagers are grabbing the opportunity to sell cash crops despite their lack of technical skills, market knowledge and experience.

Cultural implications

Identity

To be a swidden cultivator manifests itself in the social, cultural and political life of the Bunong. The agricultural cycle has a specific rhythm, is marked by ceremonies, entails social responsibilities and sets the pace for the swidden community. Swidden agriculture is one of the Bunong's strongest identity markers. An elementary sense of freedom is associated with the practice of swidden agriculture. Villagers express feelings of independence, contentment and peace, not only when speaking directly of swidden cultivation, but also when referring generally to their forest-based livelihoods, which are only secured through unrestricted access to the surrounding natural environment.

> The Bunong identity is to do upland rice farming, *mir*, and to go everywhere, to have freedom. To go to the forest where we want, to fish and hunt where we want, to collect forest resources and we know the food from the forest and what we can eat and use from the forest. (…) We used to be very free to go everywhere in our forest (Bunong man, 60 years, Lampeu' village).

> The important thing of being Bunong is first our way of living: to do swidden rice fields and paddy fields, to have land, to have our village, people living together. We have our own place, our places to go, like the forest and the land. To practice our spiritual way and to have places to raise our animals, our cattle, elephants, pigs, chickens. Also to have a healthy lifestyle, to have enough rice, enough food to eat and to have a place to go hunting and fishing, to have a good house to stay in and to have good health. All of this is highly important (Bunong man, 40 years, Krong Teh village).

Swidden cultivation is directed at ensuring safety and stability, rather than at accumulating profit and surplus. Safety and stability refers not only to securing subsistence needs, but also to sustaining reciprocal relationships with the spiritual world and, to a certain extent, levelling inequities within the community.

Each Bunong village is living, farming and rotating its fields on an area identified as ancestral land *(bri taem)*, and the Bunong consider themselves strongly rooted in this land. It serves as a point of orientation, from both a practical and a spiritual point of view. On a metaphysical level, land is referred to as mother, *me neh* – a perception of land as the protector and caretaker of the people. Their land means far more to a Bunong community than simply providing an economic basis for life: it is an intensely meaningful place invested with social, cultural and religious activities and imaginings. It is the key to the Bunong definition of themselves as a group.

The sudden and enforced loss of land in Bousra commune has had a tremendous impact on the people, who are left with a perception of emptiness and disorientation. Villagers describe themselves as being in a state of shock, as their familiar surroundings have been transformed by rubber monocultures into vast unrecognizable deserts (Figure 33-2). They feel trapped, betrayed, disrespected and deprived of their identity.

FIGURE 33-2: An elephant passes through a rubber monoculture, which is perceived as a deserted jungle landscape.

Photo: Brigitte Nikles.

'Land or no land'

As mentioned above, Bunong swidden farmers once followed a complex set of rules, which demanded that they 'listen to dreams' and pay attention to signs in nature when selecting a new swidden field. The tide of economic and social change has left this tradition in a state of flux. In today's context of marked land scarcity, villagers cannot afford to turn down a field for spiritual reasons; they feel the need to restrict their considerations to a simple question: 'do we have land, or do we not have land?' However, the inviolability of sacred forest is still highly respected, even among the younger generations, and most Bunong who have converted to Christianity pause to inquire into the history of a potential new swidden, rather than risk cutting trees in sacred places.

> Looking for land was also not easy in the past. Before, we checked the land [to see] if it's good or bad quality and if it is a sacred place. Today, when we are looking for land, we don't care if it is good quality or not, we just need land. Everything else is not important any more. Today is just land or no land. There is no option. Of course, if it is sacred forest, we still do not cultivate this land or if it is really bad quality, for example, if it has too many stones. But if the soil is not of the best quality and overgrown with a lot of grass, we do not care; we take this land. And we do not listen to the dreams any more. Because there is no more land, so there are no other options, you have to take it (Bunong man, 50 years, Pu Cheng village).

Continuance of rice ceremonies

Interestingly, even in today's conditions of land scarcity and tenure insecurity, people still cultivate rice and perform the traditional rice ceremonies (Figure 33-3). Landless families say that they try to borrow land from relatives or neighbours – even if it is a very small plot – so that they can cultivate rice and continue to stage ceremonies for the rice spirits. The people perceive the rice ceremonies as customs

FIGURE 33-3: Rice ceremony for the spirits of the paddy field.

Photo: Brigitte Nikles.

that form a substantial part of their identity; therefore, they cling to these practices. Moreover, their composite paddy-swidden agro-ecosystem seems to allow the majority of Bunong families to continue with the rice ceremonies: around 70% of all Bunong families in Bousra commune have small paddy fields that are located close to the villages and have been spared by the economic land concessions. Even though the cultivation of *ba me* is not possible in the paddy fields and the ceremonies are no longer as detailed as those once held in the swiddens, villagers have obviously retained the larger part of their spiritual practices by adapting them to wet-rice cultivation.

It is said to be most important for the rice ceremonies to do their job of sustaining the cycle of sowing, growing and harvesting – independent of any type of agricultural system. Thus, the rice ceremonies have long been adapted to wet-rice cultivation. The transformations involved in the switch from rotational to permanent agriculture are very recent, and it is still too early to assess or estimate how the Bunong will maintain their bonds with the rice spirits in the future, when swidden cultivation will arguably have disappeared altogether.

Work force: Change and continuity

With the conversion from rotational to permanent cultivation and the transition to other crops, agricultural practices are also changing. Yet the Bunong still rely strongly on communal working groups and continue to work together. They say that sharing workloads, mutual support and an encouraging atmosphere are the main reasons they persevere with communal working groups (Figure 33-4). Interestingly, they affirm that these working arrangements have grown in importance, and that they are linked to the steady influx of new settlers to the commune. Fear and mistrust towards the newcomers create intensified social cohesion.

The communal work has not changed. If you have land – it doesn't matter if you plant rice fields or cash crops – people still work in groups. The Bunong cannot let go of this. Maybe some parts are changing. For example, if people want to have immediate help, they hire people to finish the field work quickly. But people still do communal work, at least in the close family, because we cannot abandon a part of the family. If we work together we can share energy, help each other – the families usually do this. Another reason is that more outsiders are coming now. If you work alone in your field and not in your working group ... I am afraid to go there alone. There might be a lot of outsiders; I wouldn't feel comfortable to go alone to my field. Other people feel the same. As a result, more people need the communal working groups; they prefer to work together. Before, when there was forest all over, we could go to the fields alone. We were only afraid of dangerous animals and some people were afraid of the ghosts on their way. But now, we are not afraid of that any more. Instead, we are more afraid of outsiders who we do not know. Therefore, communal work is a very good way to solve this problem (Bunong woman, 35 years, Pu Char village).

Communal working groups are formed mainly among the closest relatives. Working groups, where social cohesion and solidarity remain strong, consist of only three to five families and no longer reach beyond this close circle. Sometimes, the formation of working groups is not tied to kinship, but is arranged according to spatial and practical considerations. As families who engage in contract farming get their fields allocated by the rubber companies, it is very likely that people from different villages and without kinship relations have plots close to each other. In such cases, communal working groups are set up between unrelated families who happen to work spatially close to each other.

When there is an urgent demand for additional labour, it is common these days for workers to expect monetary payments. This has resulted in this specific communal work mechanism losing its significance because most villagers are unable to afford cash remuneration. In recent years, mutual assistance and dependency, surplus sharing and social cohesion have generally been less extended: they do not encompass the community as a whole, but are increasingly

FIGURE 33-4:   A communal working group that has been contracted as outgrowers in the rubber field of a Bunong family. These collective working arrangements are maintained even in completely changed farming systems.

Photo: Brigitte Nikles.

concentrated on close family networks. The noticeable growth of socio-economic differences among the villagers; the rising competitiveness in the search for land and money and a certain trend towards 'individuality', are arguably challenging the social organization of the Bunong, along with the mechanisms that worked towards social cohesion and security.

Conclusions

The Bunong of Bousra commune have been forced by tenure insecurity and increasing land scarcity to give up fallowing their land – an essential part of the swidden cultivation that was once their sole livelihood. This has arisen as a direct consequence of their ancestral land being allocated to rubber plantations, designated as protected forest by the government, or claimed or purchased by settlers from the lowlands.

The Bunong do not idealize their former, traditional way of life and are frank in acknowledging that their past existence was difficult and burdensome. But the loss of their ancestral land to transactions written remotely and negotiated beyond their knowledge, then disguised behind a narrative of 'development', and the sudden and total disruption of their traditional livelihoods and way of life as swidden farmers, has been altogether traumatic. On top of this has come the destruction of thousands of hectares of agricultural land, sacred forests and burial grounds, leaving them with nothing but despair and emptiness.

Despite the loss of land and the abandonment of very distinct cultural aspects of their lives, most villagers have adopted a pragmatic approach to their future. Ways to improve their livelihoods and standards of living have been identified, and many are struggling against the odds to adapt to the abrupt social and economic change in order to avoid being left behind. Land shortage has put farmers' ability to survive directly at risk and has forced them to make very practical decisions. Some spiritual considerations in their relationship with their environment have been relegated to a lesser role, while their ability to cling to rice ceremonies and communal working arrangements helps to provide a sense of continuity.

It is important to note that the Bunongs' loss of both their land and access to natural resources and their abrupt transition from rotational to permanent agriculture began very recently, and is still happening at this very moment. Not everybody in Bousra commune is able to grasp, much less articulate, either the immediate or the long-term effects of what is happening in their lives. Villagers' reactions are different; some deal with the pressures more easily, while others find themselves shocked and disorientated. One of the most traumatic aspects of the profound changes to the social and cultural identity of the Bunong has been its suddenness, and the extraordinary stresses are already manifesting in widening socio-economic differences and conflicts within the villages.

References

Backstrom, M., Ironside, J., Paterson, G., Padwe, J. and Baird, I. G. (2007) *Indigenous Traditional Legal Systems and Conflict Resolution in Ratanakiri and Mondulkiri Provinces, Cambodia,* United Nations Development Programme, Bangkok

Baird, I. G. (2010) 'Land, rubber and people: Rapid agrarian changes and responses in Southern Laos', *The Journal of Lao Studies* 1, pp1-47

Borras, S. M. jr., Hall, R., Scoones, I., White, B. and Wolford, W. (2011) 'Towards a better understanding of global land grabbing: An editorial introduction', *Journal of Peasant Studies* 38, pp209-216

Condominas, G. (1977) *We Have Eaten the Forest,* Farrar, Strauss and Giroux, New York (translation of *Nous avons Mangé la Forêt de la Pierre Géni Gôo: Chronique de Sar Luk, village Mnong Gar, Mercure de France*, Paris [1957])

Conklin, H. C. (1957) *Hanunoo Agriculture: A Report on an Integral System of Shifting Cultivation in the Philippines*, FAO Forestry Development Paper No. 12, Food and Agriculture Organization of the United Nations, Rome

De Schutter, O. (2011) 'How not to think of land-grabbing: Three critiques of large-scale investments in farmland', *Journal of Peasant Studies* 38, pp249-279

Diokno, M. (2008) *The Importance of Community: Issues and Perceptions of Land Ownership and Future Options in Five Communes in Mondulkiri Province, Cambodia*, Non-Timber Forest Products Exchange Programme and Non-Governmental Organizations Forum on Cambodia, Phnom Penh

Ellis, F. (2000) *Rural Livelihoods and Diversity in Developing Countries,* Oxford University Press, Oxford, UK

FIDH (2011) *Cambodia: Land Cleared for Rubber – Rights Bulldozed. The Impact of Rubber Plantations by Socfin-KCD on Indigenous Communities in Bousra, Mondulkiri*, FIDH (Worldwide Human Rights Movement), Paris

Fox, J. M., Fujita, Y., Ngidang, D., Peluso, N., Potter, L., Sakuntaladewi, N., Sturgeon, J. and Thomas, D. (2009) 'Policies, political-economy and swidden in Southeast Asia', *Human Ecology* 37, pp305-322

Hall, D., Hirsch, P. and Li, T. M. (2011) *Powers of Exclusion: Land Dilemmas in Southeast Asia*, NUS Press, Singapore

IOM (2009) *Mapping Vulnerability to Natural Hazards in Mondulkiri,* International Organization for Migration (IOM), Phnom Penh, Cambodia

Ironside, J. and Nuy, B. (2010) *Development with Identity: Assessment of the Impact of Tenure Security from Legal Entity Registration in Indigenous Communities in Cambodia,* unpublished draft report, Danish International Development Agency (DANIDA), Phnom Penh, Cambodia

Leemann, E. (2012) 'Land grab and the emergence of indigenous peoples' rights and identities: The case of the Bunong in Cambodia', paper presented at a Conference on Intersection of Rights and Laws, 13 January 2012, London

Li, T. M. (2011) 'Centering labor in the land grab debate', *Journal of Peasant Studies* 38, pp281-298

Mertz, O., Leisz, S. J., Heinimann, A., Rerkasem, K., Thiha, Dressler, W., Pham, V. C., Vu, K. C., Schmidt-Vogt, D., Colfer, C. J. P., Epprecht, M., Padoch, C. and Potter, L. (2009) 'Who counts? Demography of swidden cultivators in Southeast Asia', *Human Ecology* 37, pp281-289

NGO Forum on Cambodia (2009) *Fast-Wood Plantations, Economic Concessions, and Local Livelihoods in Cambodia,* Non-government Organisations' Forum on Cambodia, Phnom Penh

Rambo, A. T. (1996) 'The Composite Swiddening Agroecosystem of the Tay Ethnic Minority of the Northwestern Mountains of Vietnam', in B. Rerkasem (ed.) *Montane Mainland Southeast Asia in Transition,* Chiang Mai University, Chiang Mai, pp69-89

Rambo, A. T. (2007) 'Observations on the Role of Improved Fallow Management in Swidden Agricultural Systems', in Malcolm F. Cairns (ed.) *Voices from the Forest: Integrating Indigenous Knowledge into Sustainable Upland Farming,* Resources for the Future Press, Washington, DC, pp780-801

Savajol, N., Vanny, T. and Sam, J. (2011) *Traditional Therapeutic Knowledge of the Bunong People in North-eastern Cambodia,* Nomad RSI Cambodia, Phnom Penh

Schmidt-Vogt, D., Vu, K. C., Truong, M. D., Epprecht, M., Hardiono, M., Heinimann, A., Lacroix, L., Mertz, O., Messerli, P., Thiha and Pham, V. C. (2009) 'An assessment of trends in the extent of swidden in Southeast Asia', *Human Ecology* 37, pp269-280

Vogel, S. (2011) *Aspects de la Culture Traditionnelle des Bunoong du Mondulkiri*, Tuk Tuk Editions, Phnom Penh, Cambodia

Vrieze, P. and Naren, K. (2012) 'Sold. In the race to exploit Cambodia's land and forests, new maps reveal the rapid spread of plantations and mining across the nation', *The Cambodian Daily Weekend*, March 10-11, Phnom Penh, pp4-11

Notes

1. Following Ellis' (2000) framework for the analysis of changing and diversifying rural livelihoods, households are conceived as having a livelihood platform comprising various assets (natural, physical, human, financial and social capital). The access to these assets is shaped by social relations, institutions and organizations in a context of exogenous trends and shocks, thus influencing the choice of livelihoods. Swidden farming is understood to be one element in a larger livelihood system.
2. The Swiss National Science Foundation financed this research. Its support is gratefully acknowledged. We would also like to thank Neth Prak for expanding our understanding of the Bunong's swidden system, and for providing excellent research assistance during data collection and comments and suggestions during the writing process.
3. The Bunong are also referred to as Phnong, Bunoong (see Vogel, 2011) or Mnong (in the seminal work of Condominas, 1977[1957]).
4. Of course, the general negative perception of swidden cultivation as a destructive and unproductive way of farming has been imposing massive constraints and pressures on Cambodia's indigenous population for decades (Guérin, ch 7). However, until very recently, Mondulkiri province has been so remote that these policies have not significantly affected the Bunong.
5. See Rambo (2007) for a discussion on whether swiddens are fallowed because of declining soil fertility or pressure from weeds.
6. For more information on the traditional medicine of the Bunong in Mondulkiri province, see Savajol et al., 2011.
7. A community generally means a village or, in cases where a village consists of two or more matrilineal decent groups, a sub-village. The terms community and village are therefore used synonymously.
8. According to commune census records, 2011.
9. In 2007, the Cambodian Government granted several economic land concessions to two rubber companies, covering a total area of about 12,000ha in Bousra commune. Operations started in 2008 and by now, most of that concession land has been bulldozed and an estimated 80% planted with rubber trees. In addition, several other economic land concessions for rubber production as well as mining concessions have been allotted or are in the process of being granted.

34

THE EFFECTS OF COMMERCIAL AGRICULTURE AND SWIDDEN-FIELD PRIVATIZATION IN SOUTHERN LAOS

*Watcharee Srikham**

Introduction

Swiddening, also known as slash-and-burn farming and shifting cultivation, remains fundamental to local Laotian subsistence. It produces locally grown rice and glutinous rice, and family needs are supplemented by fishing, hunting, livestock husbandry and extensive foraging for non-timber forest products. Shifting cultivators also practise mixed or multiple cropping. Thus, agricultural diversity tends to be higher in shifting cultivation systems than in sedentary, permanent farming systems in lowland areas (Van Gansberghe, 2005). Swidden farming is a diverse, complex and dynamic land-use system that provides for comprehensive landscape management on a long-term basis and incorporates changes in agrarian use.

In Laos, swidden or shifting cultivation is known as *hai*. In the late 1990s, it was estimated that 80% of Laotians lived in rural areas and earned a living from swidden agriculture and forest products (Namura and Inoue, 1998, p23). Shifting cultivation has been blamed for deforestation, soil erosion and degradation of watersheds. Lowlanders and state agencies have long regarded upland groups as primitive, and misunderstandings and conflicts over resource use and management are major problems. In many cases, these misconceptions aggravate and animate the policies and practices of state actors and corporate entrepreneurs in areas that were once – or still are – occupied by swidden-farming communities.

Swiddeners have become increasingly marginalized, both politically and economically, as a result of two forces that were once external to much of their

* ASSISTANT PROFESSOR WATCHAREE SRIKHAM, Lecturer, Department of Social Sciences, Faculty of Liberal Arts, Ubon Ratchathani University, Ubon Ratchathani, Thailand.

Note: This chapter is based on a paper presented at an international conference on 'Revisiting agrarian transformations in Southeast Asia: Empirical, theoretical and applied perspectives,' 13-15 May 2010, Regional Center for Social Science and Sustainable Development, Chiang Mai University, Thailand.

lives: the Lao state and the introduction of capitalism. The state has encroached on to swidden fields and forest fallows, staked claims to forests and promoted the conversion of swiddens to monoculture cash crops, such as coffee and rubber, grown in plantations. Capitalism has brought with it an expansion of market infrastructure and market opportunities, as well as the promotion of industrial agricultural practices. State policies have combined with these pressures by providing market support and facilitating the accumulation of capital by corporations and foreign investors through the expansion and upgrading of roads, electricity and telecommunication networks and offering subsidies for investment (Harvey, 2003; Glassman, 2006). Together, these forces have played a major role in driving the transitions taking place in the practice of shifting cultivation and the livelihoods of communities that depend upon it.

Lao government policy on swidden farming

Most swiddeners in Laos have remained peripheral to state power and are typically portrayed by national decision-makers as 'backward people' over which the state needs to exert control. In Laos, this perspective links shifting cultivation with ethnic minorities, and according to Marxist/Stalinist beliefs, such minorities hold an official position that is considered to be a primitive 'stage' on the social evolutionary ladder (Jamieson, 1991; Rambo, 1995; Sturgeon, 2005; Cramb et al., 2009).

Another key assumption that has underpinned policy formulation on upland development in Laos is that 'shifting cultivation poses a serious threat to the health of the Lao PDR's forests and the welfare of rural farmers, and should be replaced with permanent, commercial agriculture' (Badenoch, 1999, p3). Shifting cultivation is also regarded as a serious impediment to rational forest management and sustainable use of soil and water resources (Phangthalangsy, 1991). Many policy-makers believe that shifting cultivation is one of the main causes of deforestation and plays a central role in accelerating deforestation (Giri et al., 1998; Robichaud et al., 2009, p1).

The Government of Lao PDR has thus denounced swidden as the primary threat to the country's forests, despite the fact that during the 1980s, State forestry enterprises and the military utilized forests for commercial logging. In the late 1980s, the first national forest inventory in Laos distinguished forests into 'scientific' management categories, including conservation, and delineated forest areas on official maps. Subsequently, the Department of Forestry developed new policies and legislation to promote conservation, imposing restrictions on upland peoples' access to their swidden fields and fallows. Under the National Biodiversity Conservation Areas Law of 1993 and Forest Law of 1996, the government implemented its National Land and Forest Allocation Policy, under which village boundaries were established and new areas were demarcated for protection.

Through various means, the national land-use policy aimed to eliminate swidden cultivation by the year 2000 (Fujisaka, 1991). The National Land and Forest Allocation Policy limited the number of plots allocated per household and categorized most fallow land as various types of conservation forest. Use of this land for agriculture was prohibited (Fujita and Phanvilay, 2008). The country's National Socio-economic Development Plan identified the stabilization of shifting cultivators as one of eight national-development priorities and aimed to provide sedentary livelihoods for up to 100,000 families by the year 2000 (Badenoch, 1999, p6). When the implementation of this policy showed that it would not live up to expectations, the Government shifted its objective from 'eliminating' to 'stabilizing' swidden cultivation by the year 2010 (MAF, 2003).

In northern Laos, blame for forest degradation has mainly been directed at swidden farmers, rather than at loggers and industrial forestry developers, who bear a greater degree of responsibility for extensive degradation than do swiddeners (Thapa and Weber, 1991; Thiesenhausen, 1991). While numerous studies have shown that the relative contribution of shifting cultivation to deforestation has been exaggerated, many national policy-makers continue to publicly blame swidden farmers, a position consistent with the prevalent belief among policy-makers that shifting cultivation is the leading cause of forest loss in the Lao PDR (Badenoch, 1999, p6). Many officials also believe that the practice of shifting agriculture will be abandoned once villagers are given better options (Bounthong et al., 2009, p184). From 1975 until the mid-1980s, traditional livelihoods changed rapidly under the influence of policy interventions that sought to directly suppress political insurgency and to 'modernize' agricultural production by discouraging shifting cultivation and encouraging upland people to resettle in the lowlands. After 2005, a national campaign to eradicate opium cultivation in swidden fields also involved moving minority upland peoples to the lowlands (Cohen, 2009). Between 1995 and 2005, villagers in the north of the country were forced to relocate and consolidate, resulting in mountain populations declining in some districts by 30% to 50%. Many of those relocated were resettled along new roads, and began engaging in commercial agriculture or working as agricultural wage-labourers (Fujita et al., 2007).

Commercial agriculture, such as rubber plantations, proved to be beneficial to local smallholders. Manivong and Cramb (2008, pp113–125) conducted a discounted cash-flow analysis of smallholder rubber production in the uplands of northern Laos and found that the expansion of rubber planting in a study village was based on good economic returns, which could be highly profitable. The study suggested that rubber could be considered as having considerable potential for poor upland farmers, in line with the government's policy of stabilizing shifting cultivation and supporting new livelihood options for poverty reduction.

On the other hand, rubber concessions granted to foreign companies had denied local people access to the land on which they once practised swidden cultivation for as long as 50 years. As well as a resulting diminution of the traditional knowledge, wisdom and beliefs of swidden cultivators, their ability to sustain their own existence was washed away, leaving them to rely increasingly on outside resources.

Rubber plantations and commercial agriculture in Laos

Since the 1990s, implementation of government policies aimed at eliminating swidden cultivation policy affected the food security of people in many parts of the country. Therefore, the Lao PDR government changed its focus on subsistence farming in a national socio-economic development plan and opted instead for a market oriented economy. Many government officials at different levels of the administration believed that rubber planting was the best way to reduce poverty, 'stabilize' swidden cultivation and provide jobs for people in rural areas. This perception was clear to see, despite most of these officials lacking knowledge of socio-economic consequences, techniques for rubber production and the global marketing cycle of rubber (Vongkhamhor, 2007, p16).

Rubber was first introduced into Laos in 1930, with the first rubber plantation established in Champasak province, in the far south of the country, by the French during the colonial period (Figure 34-1). There was little other rubber development at this time. In 1992, the Pattana Khet Phudoi company brought the Rubber Research Institute of Malaysia trees from Thailand to plant in Kammouane province, central Laos: 80ha in Thakhaek district and 23ha in Hinboun district. Later, in 1995, the

FIGURE 34-1: The Lao PDR, showing the provinces mentioned in this chapter.

state logging company, Development of Agriculture, Forestry and Industry (DAFI), funded the establishment of a 74-hectare rubber plantation at Houay Tong village, in Bachiangchaleunsouk district, Champasak province.

Between 1994 and 1996, the Hmong village of Hadyao in Luang Namtha province, in the north of the country, established smallholder rubber plantations covering 342ha. Near the capital, Vientiane, rubber trees were planted on 3.5ha in Sungthong village and a 4ha plantation was established at Thapabat village in Bolikhamxay province. In 2003, growers in Luang Namtha and Kammouane provinces began to sell rubber product to Thailand, China and Vietnam. Since then, the area planted to rubber in Laos has increased rapidly, as many individuals, private-sector entities (including foreign companies), and state-sector entities responded to high rubber prices and growth in demand from China (Alton et al., 2005; Vongkhamhor, 2007).

While contract farming was used for rubber development in northern Laos, land concessions for commercial agriculture, especially rubber plantations, were the main method by which shifting cultivation land was privatized in southern Laos. Vietnamese companies were granted five land concessions for rubber plantations, especially in Bachiangchaleunsouk district of Champasak province. In June 2004, the Lao government granted the Vietnam-Laos Rubber Joint Stock Company a 50-year concession to cultivate 10,000ha of rubber in Bachiangchaleunsouk district (Obein, 2007). In 2006, the Kaosouyaotiang Viet-Lao Company (aka Yao Tiang Company), a partnership of three Vietnamese companies, signed a 40-year concession agreement to develop another 10,000ha of rubber in the same district (Champamay, 2008). On July 20, 2004, the provincial governments of Champasak and Salavan approved a 50-year land concession for the Dak Lak Rubber Company to develop 10,000ha of rubber as well as thousands of hectares of coffee, cacao and cashews (Thanh Nien News, 2007). Together, the three Vietnamese companies have a concession area for rubber of 23,653ha in Bachiangchaleunsouk district, representing 30.06% of the district's total land area. Dak Lak Rubber Company expected to plant its entire 10,000ha of rubber by 2010 (Vietnam News Agency, 2008). As well as rubber planted in Champasak, Salavan and Attapeu provinces, it was also planning to build a rubber processing factory, which it expected to be in operation by 2011 (Nhân Dân, 2008).

The extent of land for all three concessions was determined on paper, without identifying it on the ground. Baird (2009, p9) wrote: 'once the concessions were approved, there was pressure on lower-level officials in the district to find "available" land, even when there was not actually much available. This, in turn, led to government officials pressuring village headmen to agree to sign away their land.' Not all of the companies were successful in finding land for rubber plantations, even after getting the concession signed by the central government. In March 2008, two years after its concession was granted, the Kaosouyaotiang Viet-Lao Company reported surveying 2318ha of land and planting only 1305ha with rubber. This was considerably less than expected, and the Vietnamese blamed the delays on the high costs of equipment, materials and fuel in Laos, when compared to those in Vietnam. It added that 'some groups of villagers have not yet cooperated'. (Champamay, 2008).

The Ministry of Agriculture and Forestry has granted a concession to a Vietnamese private company to plant rubber trees on 120,000ha in three provinces of southern Laos – Salavan, Champasak and Attapeu – out of a total of 300,000ha of land approved for such purposes in nine provinces – Bokeo, Luang Namtha, Oudomxay, Bolikhamxay, Khammouane, Savannakhet, Salavan, Champasak and Attapeu (Pongern, 2009). Lao government revenues from the concessions are about US$6 to $9 per hectare per year for good quality land (Phouthonesy, 2007; Vientiane Times, 2008a), which is very low. However, the government's present draft land concession decree provides for land concession fees in areas with little infrastructure of US$100 per hectare per year (Vientiane Times, 2008b). 'Furthermore, this revenue goes to the central government, without guarantees that any share will be provided to the villages being impacted by rubber development' (Baird, 2009, p6).

The effects of commercial agriculture and the privatization of shifting cultivation land in southern Laos, as well as commercial investment from neighbouring countries, especially Vietnam and Thailand, have increased. These newly introduced dynamics have further affected swidden farmers, whose livelihoods were already in transition as a result of domestic policies. The World Rainforest Movement (2008) asserts that investments by foreign companies in commercial tree plantations in the Lao PDR increased sharply between 2004 and 2006. In particular, large-scale plantations were promoted by state land concessions made to foreign investors. The total area for growing rubber throughout the country reached 182,900ha in 2006, after a significant increase. While Chinese companies are the main investors in rubber development in the north of the country through contract farming, Vietnamese and Thai companies have invested extensively in the south of the country, mainly through land concessions for rubber plantations.

In 2007, the Vietnamese army's Military Corps No. 15 completed an irrigation complex in Sekong province for plantation crops, established a coffee plantation in Salavan province and developed plans to set up coffee, rubber and cashew plantations in Attapeu province, as well as building a rubber processing plant in Attapeu with an output of 10,000 tons per year. Attapeu's new rubber plantations cover more than 7000ha. The Lao Government granted the Vietnamese Quang Minh company a licence to establish the first rubber plantation in Attapeu covering 3000ha over a period of 50 years. Further, it allowed Dakruco Company from Vietnam to cultivate rubber on 10,000ha in Attapeu, Champasak and Salavan provinces. By mid-2009, there were five Vietnamese companies in the south of Laos with authority to grow rubber on 42,050ha. Moreover, an estimated 50 Vietnamese enterprises were waiting in the wings, having expressed interest in developing rubber plantations, mainly in the southern provinces (Benge, 2007).

The land given to Vietnamese companies in rubber-plantation concessions has overwhelmingly been the shifting cultivation land and crop fields of local villagers. The result has been that most people living in plantation areas have lost almost all of their swidden land. According to Inthapanya (2009, pp76-79) the swidden fields of one village called Udomsouk, in Champasak province's Bachiangchaleunsouk district, diminished sharply, then disappeared entirely. In 1997-1998, the village's shifting cultivation land covered 35.01ha. It dwindled to 15.47ha in 2003-2004 because of the government's policy of 'stabilizing' swidden farming. The Vietnamese rubber company came in 2005, and by 2007 the village had lost all of its shifting cultivation land. Obein's (2007) study on social and environmental impacts of private rubber plantations on 33 villages in Bachiangchaleunsouk district found that a population of 12,644 shifting cultivators had lost 83% of their land to the Viet-Lao Rubber Company by the end of 2006. Eighteen villages were left with 10% or less of their agricultural land, and four villages had no agriculture land left at all. Obein (2007, p33), then writing for the French Development Agency, made the following recommendation:

> … for which the various households have lost their source of income, the opinion of the Consultant is that resettlement to a new location might be a solution, albeit such relocation is usually a very delicate, expensive and risky process. The option of returning plantations to the villagers is highly preferred.

Baird (2009, pp14-19) identified the main impacts on local people of large-scale rubber development in Bachiangchaleunsouk. These were losing access to communal lands, including forests and grazing areas for cattle and buffaloes; losing agricultural lands; suffering problems of labour restrictions imposed by Vietnamese companies; and rapid transformation pressures on their livelihoods. Villagers were left with only limited paddy fields and village housing areas. The paddy fields were unsuitable for growing rubber, but could not produce enough rice for family consumption. In some cases, entire villages were forced to move out to make way for the rubber plantations of Vietnamese investors.

Some villagers were able to get compensation from the companies for the swidden and crop fields they lost. However, many were unable to claim compensation, in part because they had not officially declared their rights over their land in an attempt to avoid tax and thus did not have official documents with which to make claims.

The livelihoods of villagers who have lost their swidden farmlands to foreign companies have changed drastically. They now rely on work as labourers on the plantations to earn money to buy rice, whereas previously they grew their own rice. But not all family members can work on the land they once owned, because the plantation companies set labour quotas for each village. In some cases, only one family member aged between 18 and 40 can get work. Some villagers manage to get work by falsifying their ages (Inthapanya, 2009, p113). Adding to their impoverishment, workers often have to pay for transport to and from the plantations each month. As a result, some previous landowners have been forced to resettle their families in other provinces where they can find suitable land to grow their rice, since this is a major staple in the diet of nearly all families in Laos. As Phanvilay (2009) noted:

This not only indicates the loss of customary resource-management practices, but a failure of formal resource-management institutions to substitute for those practices and secure long-term access to essential assets such as land and natural resources for less well-off farmers.

Having lost his swidden land to a rubber plantation with less compensation than he expected, one villager I interviewed was employed as a wage labourer alongside his daughter. However, their two daily wages were not enough to buy food for a family of six. In the past, although the family had not generated much cash income, their everyday consumption needs were met by non-timber forest products, fishing and hunting, and vegetables and rice either gathered from the wild or harvested from their swiddens. The family did not have much need for money to buy food. After their swidden lands were taken by the companies, the supply of rice was no longer sufficient for the family. Although some members of the family were able to earn daily wages of 20,000 to 25,000 kip (US$2.29 to $2.86) each from the rubber companies, this did not cover the family's needs, such as buying rice, food and other essential needs. Nowadays, a bowl of noodles costs 12,000 to 15,000 kip. Daily wages for labour can buy only two or three bowls of noodles a day, while rice consumption for four or five family members amounts to one or two kilograms per day, at a cost of 5000 to 14,000 kip per kilogram. Their future looks endlessly bleak.

Many labourers have to buy lunch, instead of making it themselves. When they worked in their swidden fields, lunch was prepared and cooked by family members in the field, or brought from home. But working in a plantation, far from home and with limited time for preparing food in the morning and less time for lunch, they are driven to rely on outside food. The family I interviewed ended up selling their house and migrating to the border area between Laos and Cambodia. Their reasons for moving were their lack of swidden land to cultivate rice and a much-reduced resource from which to find firewood.

In terms of social effects on local people, Inthapanya (2009, pp116-117) found that after land privatization to rubber companies, social relations in Udomsouk village changed from a common regard for others as relatives, offering unconditional help to one another and fully controlling the natural resources in and around their village to relationships based on money. This was because villagers, who were self-sufficient farmers and shared communal lands, were under sudden pressure from the government and foreign rubber companies. They no longer had any control over producing, consuming and managing the resources needed by their village, and this had led to a social imbalance. Old relationships began to dwindle, due to the need for long working hours in plantations; there was no longer time for village traditions, rituals and merit making procedures and some rituals were abandoned, to be replaced by brief Christian observances.

With often-serious consequences, the government's promotion of rubber plantations appears to have been given more importance than the preservation of rice-growing areas and other critical resources used by villagers. Earlier government-sponsored schemes encouraging swidden farmers to switch to permanent agriculture had, in fact, reinforced the practice of shifting cultivation as a significant form of food security in both the uplands and the midlands. Often, 'settled' farmers returned totally or partially to swidden cultivation to compensate for serious shortcomings in various government-sponsored sedentarization projects (Pravongviengkham, 1997). The granting of land concessions to foreign investors has become a major problem in Laos because the lands in question include those that belonged to the people, as well as national conservation forests. This issue has been a major topic of discussion in the Lao National Assembly (Pongern, 2009).

Trying to survive

At least three villages in Bachiangchaleunsouk district, Champasak province – Ban Done, Houay Pheune and Phialath – are now listed as having no remaining agrarian land, following the granting of rubber-plantation concessions (Obein, 2007). A road has been driven through the centre of Ban Done, dividing the community of Ngae or Krieng ethnic groups into Ban Done Tueng (upper Ban Done) and Ban Done Luom (lower Ban Done). People in the plantation project areas of Ban Done have lost all of their swidden land. In 1995, 15% of agricultural income in Bachiangchaleunsouk district was earned by gathering non-timber forest products (NTFP's) (Foppes et al., 1996) as part of the swiddeners' livelihood. However, the official view was that swidden agriculture was destroying both the forests and the livelihoods of villagers, who were very poor as a result. Foreign companies who were establishing rubber plantations would therefore provide permanent jobs for the villagers. In reality, although the Vietnamese company hired at least one member from each ethnic-minority household, the need for labour in the plantations diminished when the rubber trees were about three years old because there was no longer a need for weeding and other labour-intensive jobs. Work contracts were terminated and many people were asked to work reduced hours. The decline in rubber prices exacerbated the problem, and was used as an excuse to lay-off local labourers.

The newly landless villagers of Ban Done struggled for survival. Some families moved to live in a village near the Cambodian border, where they could find suitable agrarian land. Others with family members working for the rubber company were allowed to set up stalls to sell local food and products to tourists at the nearby Pha Suam waterfall. Then, the cassava-growing boom began in the same provinces of southern Laos that were already home to extensive rubber plantations. Cassava has the advantage of being relatively undemanding, and will thrive in poor and even exhausted soils, where few other crops will grow. There was increasing demand for cassava both locally and in overseas markets.

In 2008 it was reported that China's Zhongxing Telecom Equipment (ZTE) had been given approval to grow 50,000ha of cassava in southern Laos, an area that could double to 100,000ha in the future (Vientiane Times, 2008c). Producers of tapioca powder from cassava began expanding both their production facilities and land under cultivation. The county's major tapioca producers are the Lao Indochina Group, with a factory in Pakngeum district, Vientiane prefecture, and KPN Tapioca, with a factory in Champasak province. Two other factories are located in Lao Ngarm district, Salavan province and Meun district, Vientiane province. The Lao Indochina Group reported in 2012 that cassava farming had increased by 76%. The Group sourced its crop from 13,500ha of land in 524 villages, providing employment for 7840 families. In 2011, the group had processed cassava grown on 7689ha, and in doing so had provided employment opportunities for more than 2850 families in 171 villages (Vientiane Times, 2012).

In January 2013, SNV Netherlands Development Organisation began a Cassava Inclusive Business project in Bachiangchaleunsouk and Pathoumphone districts of Champasak province and Lakhonepheng district in Salavan province, promoting the sustainable production of cassava by smallholder farmers. The project connected cassava farmers, banks and government agencies to build trust and understanding among all actors in the value chain and improved planting, harvesting and production techniques through demonstration plots, workshops and exchange visits. Farmers now understand the value of cassava and how to grow it properly. They can profit from cassava-growing, with yields earning as much as US$2500 per hectare. The project is being carried out in partnership with three selected cassava enterprises: Asian Agronomy Company, KPN Tapioca Factory and Sernsay Export/Import Co. Ltd, and long-term contracts with buyers offer market sustainability. Asian Agronomy produced 8000 tonnes of tapioca in 2014. The project facilitated loans from the Agricultural Promotion Bank and Nayoby Bank to expand and improve cassava production. The project is working with 2275 smallholder farming households in 47 villages (SNV, 2014). As well as helping cassava growers, this project also generates income for a group of local middlemen who buy cassava in various villages in the project area. One of these middlemen, Loy, 49, an ethnic Ngae villager from landless Ban Done, said:

> Employees of the Vietnamese rubber company were laid off, and some of them turned into cassava buyers. They go to other villages to buy fresh cassava or sliced and dried cassava and then sell it to bigger buyers. They can earn enough daily income to feed their families.

Cassava is exported from Laos to Vietnam and China. But as well as its traditional role as a food crop, cassava has increased in value as a bio-fuel commodity. Some of the Champasak province crop is now sent across the Mekong river to Thailand's Ubon Ratchathani province, where an ethanol plant with a capacity of 400,000 litres of ethanol per day has been established, using fresh cassava as raw material. There is also ongoing research that aims to increase cassava production by reducing the growing time of cassava crops from 10 to 12 months to between six and eight months.

Conclusion

Privatization of swidden land by rubber plantations in Laos has not only resulted in drastic changes to farming practices, land-scarcity problems and migration of dispossessed villagers, but has also affected social relationships at family and community levels. There is cause for serious concern that traditional technical knowledge related to swidden cultivation will gradually disappear, along with the related rituals that support robust ethnic bonds, comfort hard-working people and preserve ethnic traditions. These important traditions will decrease in value and their functions in local societies will be displaced by the Lao government's policies of 'civilizing the barbarians' and promoting profit over sustainable practices. Furthermore, the future of economic land concessions and large rubber plantations in Laos remains uncertain, so villagers should be given more choices with which to protect the well-being of their families. In order to gain reasonable benefits from the pursuit of commercial agriculture, villagers should be involved in other forms of rubber development, such as contract farming, small-scale rubber plantations or co-operative rubber groups, and receive technology transfers on rubber planting, rubber latexing and marketing.

References

Alton, C., Bluhm, D. and Sannanikone, S. (2005) *Para rubber in Northern Laos: The Case of Luang Namtha*, Lao-German Program for Rural Development in Mountainous Areas of Northern Laos, Vientiane

Badenoch, N. (1999) *Watershed Management and Upland Development in Lao PDR: A Synthesis of Policy Issues*, World Resources Institute, Washington, DC, http://pdf.wri.org/repsi_laowshed.pdf, accessed 2 February 2016

Baird, I. G. (2009) 'Land, rubber and people: Rapid agrarian changes and responses in Southern Laos', *Journal of Lao Studies* 1(1), pp1-47, http://laostudies.org/journal/volume-1-issue-1-jan-2010, accessed 2 February 2016

Benge, M. (2007) 'Vietnam's Tay Tien expansion into Laos and Cambodia', paper presented at a National Conference on Cambodia 2007, 21 October 2007, Washington, DC, Cambodian Americans for Human Rights and Democracy, Cambodian American National Council, Cambodian Association of Florida, Cambodian's Border Committee, Coalition for a Free Cambodia, Khmer Alliance Foundation, Khmer M'Chas Srok, United Cambodian International Council, and O''Bon Khmer Art Studio

Bounthong, B., Raintree, J. and Douangsavanh, L. (2009) *Upland Agriculture and Forestry Research for Improving Livelihoods in Lao PDR*, National Agriculture and Forestry Research Institute, Ministry of Agriculture and Forestry, Vientiane, http://www.unu.edu/env/plec/marginal/proceedings/BounthongCH17.pdf, accessed 2 February 2016

Champamay (2008) '6 deuan pi 2007 bolisath hounsouan phatthana Kaosouyaotiang Viet-Lao pouk yang phala dai 1.305 hectares (In the first six months of 2007, Kaosouyaotiang Viet-Lao plants 1305 hectares of rubber)', *The Champamay* (newspaper), 10-14 March 2008, Pakse district, Champasak city, Lao PDR

Cohen, P. T. (2009) 'The post-opium scenario and rubber in northern Laos: Alternative Western and Chinese models of development', *International Journal of Drug Policy* 20(5), pp424-430, http://hdl.handle.net/1959.14/320152, accessed 2 February 2016

Cramb, R. A., Colfer, C. J. P., Dressler, W. H., Laungaramsri, P., Quang, T. L. and Mulyoutami, E. (2009) 'Swidden transformations and rural livelihoods in Southeast Asia', *Human Ecology* 37(3), pp323-346

Foppes, J., Saypaseuth, T., Ingles, A. and Ketphanh, S. (1996) 'Methods of Selecting Non-Timber Forest Products for action in Champasak Province, Lao PDR', paper presented at a workshop on Research and Planning Methodologies for NTFP-based Conservation and Development Initiatives, 21-23 May 1996, Bogor, Indonesia

Fujisaka, S. (1991) 'A diagnostic survey of shifting cultivation in northern Laos: Targeting research to improve sustainability and productivity', *Agroforestry Systems* 13, pp95-109

Fujita, Y. and Phanvilay, K. (2008) 'Land and forest allocation in Lao People's Democratic Republic: Comparison of case studies from community-based natural resource management research', *Society and Natural Resources* 21, pp2120-2133

Fujita, Y., Thongmanivong, S., Vongvisouk, T., Phanvilay, K. and Chantavong, H. (2007) 'Dynamic land-use change in Sing district, Luang Namtha province, Lao PDR', unpublished report for the International Program for Research on Interactions between Population, Development and the Environment (PRIPODE), Faculty of Forestry, National University of Laos, Vientiane

Giri, C. P., Ofren, R. S., Pradhan, D., Kratzschmar, E. and Shrestha, S. (1998) *Land Use/Land Cover Change in Southeast Asia. Study 1: Oudomxay Province, Lao PDR*, United Nations Environment Programme for Asia and the Pacific, Bangkok

Glassman, J. (2006) 'Primitive accumulation: Accumulation by dispossession, accumulation by extra-economic means', *Progress in Human Geography* 30, pp5608-5625

Harvey, D. (2003) *The New Imperialism*, Oxford University Press, Oxford, UK

Inthapanya, M. (2009) 'The changing of communities' land use for commercial tree plantations: A case study of rubber tree plantation in Udomsouk village, Bachiangchaleunsouk district, Champasak province, Lao PDR', MA thesis to the Faculty of Liberal Arts, Ubon Ratchathani University, Ubon Ratchathani, Thailand

Jamieson, N. L. (1991) *Culture and Development in Vietnam*, East-West Center Working Papers, Indochina Series no. 1, East-West Center, Honolulu

MAF (2003) *Forestry Strategy to the Year 2020*, second draft, Ministry of Agriculture and Forestry, Vientiane

Manivong, V. and Cramb, R. A. (2008) 'Economics of smallholder rubber expansion in Northern Laos', *Agroforestry Systems* 74, pp113-125, http://asiapacific.anu.edu.au/newmandala/wp-content/uploads/2008/09/vongpaphane-and-cramb.pdf, accessed 4 February 2016

Namura, T. and Inoue, M. (1998) 'Land use classification in Laos: Strategy for the establishment of an effective legal system', *Journal of Forest Economics* 44(3), pp23-30

Nhân Dân (2008) 'Vietnamese enterprise plants over 9,000 hectares of industrial plants in Laos', *Nhân Dân*, official newspaper of the Communist Party of Vietnam, 27 October 2008

Obein, F. (2007) *Industrial Rubber Plantation of the Viet-Lao Rubber Company, Bachiang District, Champasack Province: Assessment of the Environmental and Social Impacts Created by the VLRC Industrial Rubber Plantation and Proposed Environmental and Social Plans*, produced for Agence Francaise de Développement, Earth Systems Lao, Vientiane

Phanthalangsy, V. (1991) *Agroforestry Development in Luang Prabang Province, Lao PDR*, Regional Community Forestry Training Center for Asia and the Pacific (RECOFTC). Bangkok

Phanvilay, K. (2009) 'Impacts of land-use and land-cover transitions on people's livelihoods in the uplands of northern Laos: Case studies from Bokeo and Luang Namtha provinces', PhD dissertation to the Department of Geography, University of Hawaii, Honolulu

Phouthonesy, E. (2007) 'Date for resuming land concessions unsure', *Vientiane Times*, 6 July 2007

Pongern, S. (2009) 'Laos to plant rubber trees on 300 thousand hectares in nine provinces', *Voice of America News*, Laos, 19 August 2009

Pravongviengkham, P. P. (1997) 'Local regulatory systems in support of the Lao swidden-based farm economy', in M. Victor, C. Lang and J. Bornemeier (eds) *Community Forestry at a Crossroads: Reflections and Future Directions in the Development of Community Forestry*, Proceedings of an international seminar, 17-17 July 1997, Regional Community Forestry Training Center for Asia and the Pacific (RECOFTC), Bangkok

Rambo, A. T. (1995) 'Defining highland development challenges in Vietnam: Some themes and issues emerging from the conference', in A. T. Rambo, R. R. Reed, Le Trong Cuc and M. R. DiGregorio (eds) *The Challenges of Highland Development in Vietnam*, Program on Environment, East-West Center, Honolulu, ppxi–xxvii

Robichaud, W. G., Sinclair, A. R. E., Odarkor-Lanquaye, N. and Klinkenberg, B. (2009) 'Stable forest cover under increasing populations of swidden cultivators in central Laos: The roles of intrinsic culture and extrinsic wildlife trade', *Ecology and Society* 14(1), http://www. ecologyandsociety. org/vol14/iss1/art33/, accessed 4 February 2016

SNV (2014) *Cassava: Great Crop for Great People*, SNV Netherlands Development Organization, The Hague, Netherlands

Sturgeon, J. C. (2005) *Border Landscapes: The Politics of Akha Land Use in China and Thailand*, University of Washington Press, Seattle, WA

Thanh Nien News (2007) 'Vietnam firm grows industrial crops in Laos, Cambodia', *Thanh Nien News*, 20 March 2007, Ho Chi Minh City

Thapa, G. and Weber, K. (1991) 'Soil erosion in developing countries: A politico-economic explanation', *Environmental Management* 15, pp461-473

Thiesenhausen, W. (1991) 'Implications of the rural land tenure system for the environmental debate: Three scenarios', *Journal of Developing Areas* 24, pp11-24

Van Gansberghe, D. (2005) 'Shifting cultivation systems and practices in the Lao PDR', in *Improving Livelihoods in the Uplands of the Lao PDR*, National Agriculture and Forestry Research Institute, National Agriculture and Forestry Extension Service and the National University of Laos, Vientiane

Vietnam News Agency (2008) 'Dak Lak develops industrial crops in Southern Laos', *Vietnam News Agency*, 6 October, 2008

Vientiane Times (2008a) 'Government to maintain low land concession rates', *Vientiane Times*, 5 August 2008

Vientiane Times (2008b) 'Govt seeks consensus on land concessions', *Vientiane Times*, 9 September 2008

Vientiane Times (2008c) 'Champasak gears up for cassava plantation', *Vientiane Times*, Vientiane.

Vientiane Times (2012) 'New cassava grower puts down roots in Savannakhet', *Vientiane Times*, 29 May 2012

Vongkhamhor, S. (2007) *Sapap karn plook yangpara nai Lao PDR*. (Para-rubber planting situation in Lao PDR), Forest Research Center, National Agriculture and Forestry Research Institute, Vientiane

World Rainforest Movement (2008) 'Rural livelihoods made vulnerable as rubber investments take over land in Laos', based on P. Luangaramsi, R. Leonard and P. Kuaycharoen (2008) 'Socio-economic and ecological implications of large scale industrial plantations in the Lao PDR: A case study on rubber plantation', Faculty of Social Sciences, Chiang Mai University, *World Rainforest Movement Bulletin* 137, http://wrm.org.uy/oldsite/bulletin/137/Laos.html, accessed 1 February 2016

35

FROM TRADITIONAL SUBSISTENCE TO COMMERCIAL AGRICULTURE

A downward trend towards food insecurity in rural Lao PDR

*Paulo N. Pasicolan, Saphangthong Thatheva and Timothy John A. Pasicolan**

The Lao PDR: An economy in transition

The Lao PDR is rich in natural resources, including forests, biodiversity, minerals, water and arable land. These assets are principal capital for the country's economic growth, if they are properly harnessed for optimum effect. Land has recently become a big source of revenue, next to hydropower and minerals, because of the increasing number of land concessions being granted as a result of foreign direct investment (Schoenweger and Üllenberg, 2009; World Bank, 2010). However, despite its vast natural wealth and a low population of about seven million, the Lao PDR is among the least developed countries in the world. The government, in partnership with the international community and the private sector, has taken steps to address rural poverty and continuous degradation of natural resources by introducing policy reforms and development programmes. It envisions that by 2020, the Lao PDR will rise above the United Nation's List of Least-Developed Countries, and will be well on its way to catching up with its neighbours Thailand, Vietnam and China.

To date, the Lao PDR's agricultural and natural-resource sector is characterized by low levels of productivity, poor agricultural and marketing skills, limited market access, less than optimal use of water resources, low technology and lack of domestic capital for investment. The country's level of productivity is considered low by the standards of the Association of Southeast Asian Nations (ASEAN), and it will gain

---

* Dr Paulo Pasicolan, Integrated Water-resource Management and River-basin Planning Specialist, Mekong Subregion, based in the Lao PDR; Dr Saphangthong Thatheva, Director General, Council for Science and Technology, Ministry of Agriculture and Forestry, Vientiane, Lao PDR; and Timothy John A. Pasicolan, Environmental Planner and Human Settlement Specialist, School of Environmental Science and Management, University of the Philippines at Los Baños, Philippines.

national self-sufficiency in rice only if continuous growth in agricultural productivity is ensured.

In response to this great challenge, the Ministry of Agriculture and Forestry launched a Five-Year Development Strategy (2011-2015) with three major points of focus: 1. agricultural commercialization; 2. an emphasis on smallholder production, and 3. application of innovative technologies for high quality and mass production with value-adding processes (MAF, 2010). The aim was to transform the agricultural sector from traditional subsistence production to market-orientation and global competitiveness. This was premised on the assumption that the agricultural sector would remain the primary vehicle for economic take-off, as the country had only recently entered the big playing field of world markets (Pasicolan and Sithammavong, 2012a). A closer look at the transition process brings the rural poor into sharp focus, with the sad reality that they are losing control over their land and face increasing food insecurity as they fail to cope with the trading requirements of the global market.

Food security and commercializing agriculture

Swidden agriculture (rotational farming) has long been regarded by many as a backward agricultural system; an inefficient, laborious and destructive use of upland resources (Cavallo et al., 2009). On the basis of such notions, the Government of Laos, starting in the early 2000s, introduced policies aimed at replacing shifting cultivation with permanent systems of lowland wet rice, cash crops or plantation production. However, traditional farming practices remain prevalent throughout the country, particularly in the poorest districts (Figure 35-1). Government land-allocation procedures have resulted in a gradual transition from swidden agriculture to industrial crop plantations requiring vast tracts of arable land, and this has contributed to acute food shortages in many parts of the country. Poverty and malnutrition continue to be major problems, despite the government's efforts to overcome them (Fullbrook, 2009; Haberecht, 2009). As a signatory to the International Covenant on Economic, Social and Cultural Rights (ICESCR), the Lao PDR is bound to adhere to its provisions, and has created a National Nutrition Policy stating that 'all Lao citizens should be able to avail of the fundamental right to be free from hunger' (Fullbrook, 2009).

Under the government's Agricultural Development Strategy 2011-2020, agricultural commercialization became the banner programme of the Ministry of Agriculture and Forestry in 2011 (MAF, 2010). Its first and foremost aim was to address the long-standing problem of shifting cultivation. The overall intention was to eliminate 'slash-and-burn' cultivation of upland rice and replace it with more sustainable land-use systems at both village and household levels.

Contract farming has since been rationalized as a means of replacing the traditional subsistence-farming patterns of individual farmers (Pasicolan and Sithammavong, 2012b). As a result, there has been a shift from production for household consumption

FIGURE 35-1: Upland rice growing in a recently opened swidden in northern Laos.

Photo: Paulo Pasicolan.

to mass production for sale. Now, in order to be competitive, producers are being compelled to invest in high technology in an effort to meet international trade standards. Moreover, the Government of Laos expects the private sector to invest in improving rural infrastructure (e.g. farm to market roads, bridges, schools, community centres and other social services), as some of the basic requirements for agricultural commercialization. In this way, maximum public spending can be directed to other rural reconstruction projects for countryside development.

Further, contract farming normally requires that inputs, such as improved seed varieties, fertilizers, pesticides, technology and farm mechanization, are provided by the investors (Pasicolan and Sithammavong, 2012b). As a means of assuring product quality, the investors who supply both the production inputs and the technology are also expected to closely monitor farmers' agricultural practices to ensure that international standards are met. Being a new member of the World Trade Organization, the Lao PDR needs this product-quality assurance to comply with international food-safety standards, and this requires high technology. Thus, with a big investment from the private sector, the government is relieved of the need to fund research and extension and even the setting up of a traceability system for phyto-sanitary and animal-quarantine measures.

The agricultural development policies of the Government of Laos are therefore geared towards market-oriented production. And in the most dubious aspect of the government's rationale behind the commercialization of agriculture, rural farmers are compelled to become entrepreneurs.

Land as a capital

With vast land resources and a relatively small population, Laos became an attractive destination for foreign investment in industrial plantations and for interest groups venturing into land-based product-processing industries. They were keen to take advantage of relatively cheap labour and land rent.

Recently, the country emerged as a notable supplier of teak, rubber, *Eucalyptus* and agricultural commodities such as cassava, corn and other high-value crops. Since 2000, there has been a remarkable increase in the granting of large-scale land concessions for industrial tree plantations. A case in point is the dramatic rubber-planting boom since 2007, which began in the northern provinces of Luang Namtha, Oudomxay and Phongsaly (MAF, 2007; Thatheva, 2007; Thatheva and Yasuyuki, 2009) (Figure 35-2).

The Lao Government's open-market policy has given foreign investors considerable privileges, including access to the use of land and other natural resources (Fullbrook, 2007; Hanssen, 2007). This has been a part of its strategy to attract foreign direct investment as a means of generating hard foreign currency to propel its lagging economy. As a result of this generous treatment, control over the land became a strong driver for foreign investors, leading them to apply for lease contracts or even indirectly acquire vast tracts of open land for industrial plantations.

At present, the country's agricultural and natural-resource sector must respond to this overwhelming flood of foreign entrepreneurs wishing to invest in industrial plantations with a regulatory and enforcement framework that is, at best, weak

FIGURE 35-2:   A mountainside in Luang Namtha province, northern Laos, cleared and scraped clean, ready for planting rubber trees.

Photo: Saphangthong Thatheva.

(Schoenweger and Üllenberg, 2009). It lacks coherent and clear policies for governing land concessions and other projects related to foreign direct investment. While this remains the main source of investment in the agricultural and natural-resources sector, the government is eager to manage it as a huge source of foreign earnings. It believes the outcome will benefit a large proportion of the country's poor and dispossessed while providing a source of national revenue.

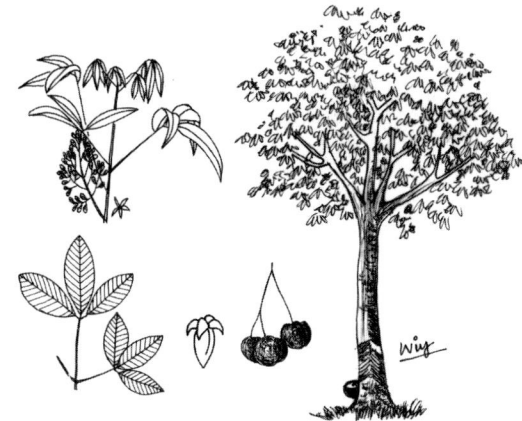

Hevea brasiliensis (Willd. ex A. Juss.) Müll. Arg. [Euphorbiaceae]

Rubber plantations reportedly cover nearly 200,000 hectares in the Lao PDR, many of them on land concessions granted to companies based in China, Vietnam and Thailand. A native of the Amazon rainforest, this tree now grows mainly in the vast rubber plantations of South and Southeast Asia.

Expected gains from contract farming

As the country's economy moves towards greater integration in global markets, demands for a consistent volume of high-quality agricultural products that meet international standards is exerting great pressure on producers.

Contract farming is expected to serve the interests of farmer producers and private firms, as well as the government (Pasicolan and Sithammavong, 2012b). In linking subsistent farmers to bigger playing fields in international markets, contract farming is considered to be a strategic scheme under which the farmers will transition to become entrepreneurs. The following benefits are expected to accrue from the contract farming scheme to farmers, private companies and the government (Baumann, 2000; Hayami, 2003; Patrick, 2004):

The farmers

On the production side, farmers are assured of the following:

1. **Market access.** Small farmers cannot venture into commercial-scale production because they lack sure and immediate markets. With contract farming, a ready market for farm produce is secure.
2. **Increased income.** Farmers can raise non-traditional crops that are sold for higher prices and may be grown without significant extra effort. Past studies have claimed that under a favourable institutional setting, farmers can make better incomes under contract farming.

3. **Reduced risk of price fluctuations.** Increased income is usually associated with lowering price risks. Prices for agricultural crops fluctuate dramatically from region to region and within growing seasons. This is exacerbated by small farmers' inability to access market information. In contract farming, crop prices are predetermined and are generally established during contract negotiations.

4. **Access to production loans.** Small farmers cannot borrow from big commercial or even government banks because they lack collateral with which to redeem the loan in case of default. Firms can provide credit and have the ability to monitor and enforce repayments from a procured crop.

5. **Assured production inputs.** The productivity of smallholders is limited by their lack of non-traditional inputs and production resources, such as improved seeds, fertilizers and equipment. Contracting firms most often provide inputs to ensure timely application and the desired quality of the produce.

6. **Technology transfer.** Smallholders are risk-averse and are often reluctant to adopt new technologies and diversify away from traditional crops. Under contract farming, private firms can provide the inputs and technologies for smallholders to confidently shift from subsistent to market-oriented production.

7. **Reduced production risks.** Contract farming arrangements facilitate risk sharing in the case of crop failure due to poor weather or pest outbreaks. Where production problems are widespread as a result of uncontrollable events, firms can often defer repayment of production advances until the following season.

Private firms

For private companies, the advantages of contract farming are as follows:

1. **Cost efficiency.** Contract farming improves cost efficiency and minimizes risk by overcoming the need to purchase land and hire wage labourers (Hayami, 2003; Patrick, 2004).

2. **Assured supplies of raw materials.** With assured produce from contracted farmers, market risks due to supply and demand fluctuations can be avoided.

3. **Quality consistency.** Since the firm provides production inputs and technology, with close monitoring of farmers' practices, quality assurance is high.

4. **Meeting required trading standards.** Global trade liberalization has resulted in an expanded market base. Multinational agro-business firms can source food and other products from developing countries with rich natural resources and cheap labour. However, food safety and quality have become important attributes to market acceptability in wealthy developed countries. Product traceability and green certification have thus become standard requirements of international trade agreements. These can easily be facilitated under contract farming because the contracting firm has full control over all inputs throughout all stages of production and processing.

Government

The following are the main advantages contract farming offers to the government:

1. **A source of foreign exchange earnings.** An influx of foreign investors increases national revenue in terms of taxes and other fees generated from business transactions.
2. **Promotion of rural development.** With a developed agro-based industry in the countryside, the rural economy is boosted, creating more jobs and improving household incomes and quality of life.
3. **Reduction in public funding of support services.** Government expenditures on research and extension and even setting up a traceability system can be minimized. Similarly, since the contracting firms have to develop complete production strategies that cover plans for product accessibility and overcoming farmers' transport and mobility problems, the government's need to invest in light infrastructure such as farm-to-market roads can be minimized.

Immediate gains from commercializing agriculture

The Ministry of Agriculture and Forestry's Agricultural Development Strategy 2011-2020 is in its fifth year of implementation, so it is too early to conclude its gains and drawbacks. However, from an optimistic standpoint, it can be said that the strategy has continually attracted foreign investments, particularly from Chinese, Vietnamese and Thai business groups.

In the past, about 80% of foreign direct investment in the Lao PDR was derived from hydro-power projects and mineral-resource exploration. However, industrial plantation projects have recently been taking an increasingly significant share of total foreign direct investment (Schoenweger and Üllenberg, 2009). These plantations now occupy between two and three million hectares, or between 10% and 15% of the country's entire land area. Not surprisingly, commercial

Eucalyptus deglupta Blume
[Myrtaceae]

Laos has become a notable producer of timber and pulpwood for paper production, much of it from plantations of so-called 'rainbow eucalyptus', after its multi-coloured bark. This is the sole *Eucalyptus* species with a natural range extending into the northern hemisphere.

tree-crop and industrial plantations are one of the Lao government's priority areas for promotion of foreign investment.

A vast area of under-utilized public land and an increasing portion of land that was once covered by a heavy concentration of shifting cultivation – especially in the northern provinces of Oudomxay, Luang Namtha and Phongsaly – are now being gradually transformed to monoculture cropping for timber, cassava, maize and other high-value crops under contract-growing arrangements (Hanssen, 2007; Fullbrook, 2009; Haberecht, 2009; Thatheva and Yasuyuki, 2009) (Figure 35-3).

This is considered to be one of the immediate positive impacts of agricultural commercialization – the optimized use of land and a rational means of containing the spread of shifting cultivation, especially on steeply sloping land.

The drawbacks of contract farming

While land-concession projects have been continually siphoning hard foreign currency into the Lao PDR economy, the following negative consequences of the rural transformation are believed to outweigh its gains.

Transformation of the natural landscape

A previous study funded by the German technical cooperation agency GTZ found that industrial tree plantations on land concessions were often replacing primary or healthy secondary forest, conservation forest or communal forest used by villagers (MAF, 2005, 2007; Fullbrook, 2007). Some companies had even cleared more land than they were granted. This was the case with many rubber plantations, mostly operated by either Chinese or Vietnamese companies in the northern provinces, which had exceeded

FIGURE 35-3:   A cassava plantation inside a newly cleared protected area in Xaysomboun province, Lao PDR.

Photo: Paulo Pasicolan.

the boundaries of their concessions. Similar cases were found in central and southern provinces such as Bolikhamxay, Khammouane, Savannakhet and Champasak.

The country's transition from purely subsistent farming to an export-oriented economy, as envisioned through the commercialization of agriculture, has placed great pressure on the country's land and natural-resource base. The end result is a mosaic of varied monoculture crop species dominating the landscape and undermining biodiversity, critical watershed functions and sources of livelihood for rural people who are dependent to a large extent on non-timber forest products for food and income (Figure 35-4).

FIGURE 35-4: A maize plantation in Oudomxay province, northern Laos.

Photo: Saphangthong Thatheva.

Breakdown of the traditional rural livelihood

While agricultural commercialization can be seen in terms of high-value crops, new companies, expanded markets and more jobs, it also means loss of traditional livelihoods – shifting cultivation in particular. Most of the shifting-cultivation clearings that used to grow upland rice only a decade ago are now planted to commercial crops. The eradication of shifting cultivation has thus undermined the food security of thousands of upland households.

The shift from subsistence crops to market-oriented high-value agricultural commodities has led to farmers' full transition to wage labourers. It has also meant shifting from diversified cropping to a monoculture cropping system. As a consequence, single crop ventures with longer cultivation periods, such as timber crops, have greatly increased farmers' risks. For example, if market prices fail due to oversupply of products or crops fail because of pest infestation, drought or fire, the shocks are more severe and can steeply increase food insecurity. This leaves farming households increasingly vulnerable amid declining diversity in livelihood activities (Eaton and Shepherd, 2001).

Turning rural farmers into daily wage labourers in industrial plantations results in seasonal out-migration in search of other off-farm jobs during the 'lean season'. According to one official estimate, 300,000 Lao people cross the border to work in Thailand as labourers in various off-farm jobs after working in the plantations (Pasicolan and Sithammavong, 2012b). This phenomenon is the result of leasing their farms to private companies, which later regard them as no more than daily wage earners.

Farmers losing autonomy and self-reliance

Farmers working as daily wage earners complain of losing control and flexibility in the use of their time. Once they are tied to the regimented system of contract farming, they must respond to rigid demands upon their time and efforts so that certain crops can be raised in a given condition.

Monocropping of non-traditional crops also leaves them relying on income from a single cash crop. If the company does not adhere to its contractual obligations, farming households are left helpless, because they no longer grow edible crops and very often have no savings to purchase food (Grosh, 1994).

Most farmers also become entangled in increasing debt. They are locked into a vicious cycle of incurring heavy debt as they find themselves unable to repay their standing debt to the company at harvest time. Much as they want to escape from this chain of indebtedness, they are forced to keep working for the company just to pay the debts of past years. In one contract sugar-growing area in a southern province, about 90% of the contract growers had standing debts to the company averaging 50 million kip, or about US$6250 (Pasicolan and Sithammavong, 2012b).

The undermining of customary norms and social life

The commercialization of agriculture has resulted in a significant decline in the availability of land for smallholders (Moizo, 2005; Hanssen, 2007; Fullbrook, 2007; Pasicolan and Sithammavong, 2012c). As a consequence, smallholder land tenure has also changed dramatically. Under the government's Agricultural Development Strategy 2011-2020, concession projects are seen as an answer to the perceived problem of shifting cultivation and aim to encourage commercial-scale production. This has paved the way for the government to implement a zoning policy for village land, allocating a fixed plot of no more than three hectares to each household (Ducourtieux et al., 2005). This effectively limits the area of land available for cultivation,

Acacia auriculiformis Benth.
[Leguminosae]

Another plantation species popular in Laos, so-called Black Wattle produces wood that is suitable for making paper, furniture and tools. It is also used extensively as firewood and for making charcoal.

encourages commodity production and restricts access to shared land. It abruptly changes the availability of land for subsistence food production, particularly by shifting cultivation.

Many rural farmers and indigenous communities have been bitterly resentful at losing their rights to sufficient land and access to resources because powerful foreign business investors have taken over their land as part of big project concessions. Traditional systems based on subsistence production are now breaking down as the concepts of commercialization and market integration obstruct the established cultural patterns of indigenous communities (Fullbrook, 2007; Haberecht, 2009).

Indigenous farmers claim that they now spend more time working on their farms than they did five years ago, following the conversion of their land to contract monocropping. In the past, they had plenty of time for leisure and attending important out-of-town social obligations. Now, they say their formerly comfortable subsistence is gone and they are reduced to wage labourers. They report less social interaction among neighbours and an increasing incidence of violence and tension within families, and they fear that social life within their communities is breaking down (Haberecht, 2009).

Increased food insecurity

As agricultural commercialization spreads across the uplands of the Lao PDR, an increasing number of rural people are becoming dependent on wage labour. With declining access to land and forests, poorer households need to sell their labour to meet their food requirements (Haberecht, 2009). Skilled workers and traders have better food security than families that depend purely on farming. The most vulnerable, however, are semi-subsistence households that cannot grow or collect enough food to meet their needs, and who perform unskilled tasks to earn cash. This latter group is caught in a vicious cycle, unable to spend enough time on their own plots of land, while being unable to earn a decent wage.

Under all of these circumstances, rural households tend to become consumers rather than producers. A case in point is one area that still has plenty of forest, but industrial plantations are fast consuming the natural landscape. According to one study, an estimated 40% of food consumed in this area came from outside the village. Elsewhere in the uplands, where there is wide-scale conversion to cash crops, many farming households will soon be buying most of their food. As a consequence, vulnerable people and their food security are becoming increasingly dependent on the vagaries of global commodity markets.

Marginalization of farmers

In one case of sugar cane contract farming in Sayboury district, Savannakhet province, growers have complained that the private investor has been able to manipulate them because of the absence of government safeguards to protect their rights and welfare (Pasicolan and Sithammavong, 2012b).

The farmers say that the lack of a government policy stance has resulted in there being no clear, binding agreement between them and the investor and no legal and technical support system to help them. They complain of systematized extortion by the firm, extended working hours, the need for heavy family labour, a marginalized social life, exposure to health risks, and the likelihood of food shortages. They are also locked into a vicious cycle of indebtedness.

Interventions to address these problems may require the formation of farmers' associations, legal assistance, training, organic farming, intercropping, integrated pest management, livelihood diversification, micro-financing, improved farm mechanization and formulation of better policies.

Constraints to successful implementation of agricultural commercialization

At a grassroots level, even though they are integrated into a bigger market, small farmers and subsistent producers in particular are unable to automatically become entrepreneurs. This is due to the following constraints and limitations:

1. Small and subsistent producers lack sufficient capital and technology to compete with bigger interest groups in the open market;
2. Both government and commercial banks lack the confidence to grant loans to these farmers because of their shortage of assets to serve as collateral;
3. Even if the government offers these farmers a package of incentives, such as credit assistance, subsidies or other forms of financial support, the legal requirements and processes are too difficult for them to follow; and
4. They are not strongly organized, so they lack the capacity to function as a distinct legal and viable business entity in order to access funds from donor organizations.

Table 35-1 presents a summary analysis of the directions of current agricultural and natural resources policy in the Lao PDR, in terms of expected gains and perceived drawbacks at both macro and micro levels (Pasicolan and Sithammavong, 2012b).

The Ministry of Agriculture and Forestry's Agricultural Development Strategy 2011-2020 provides direction to decision-makers and planners and serves as a basis for the type of assistance that can be provided by development partners in support of agricultural modernization. In a broad stroke, the strategy describes a clear game plan along with the necessary conditions to attain the government's vision and goals for agricultural development. However, an examination of the interplay between opportunities and global market trends, as well as the constraints for agricultural development as it is envisioned by the Agricultural Development Strategy, leads to recognition of the following perceived shortcomings that must be addressed for successful implementation of the government's agricultural commercialization thrust.

TABLE 35-1: Directions in the Lao PDR's agricultural and natural-resources policies and their implications for economic growth and rural development.

Policy direction	Expected benefits	Perceived drawbacks
1. Land and natural resources as 'engines for economic growth and countryside development'.	• Optimized use of the country's rich natural wealth in generating capital for economic growth and countryside development.	• Rapid degradation of the natural resource base. • Threatens biodiversity and pristine state of the environment.
2. Foreign direct investment.	• Increased national revenue with inflow of foreign capital into the country.	• Over-dependence on cash inflow from foreign earnings and failure to develop internal or local capacity.
3. Attract more foreign investors.	• Creates employment and more jobs. • Increases foreign currency exchange. • Technology transfer. • Improves infrastructure. • Improves local economy.	• Danger of putting too much trust in the hands of the private investors at the expense of established cultural patterns and traditional livelihood systems, and the environment.
4. Turning land into a capital asset by awarding land concessions.	• Generates rent from vast idle lands. • Alternative to shifting cultivation. • Alternative to opium production. • Increases trade with other countries.	• Favours corporate ventures and undermines individual farming effort. • The socio-cultural value of the land is reduced to just an economic asset.
5. Agricultural commercialization: Shift from subsistence to market-oriented production favouring monoculture farming of high-value crops for export.	• Farmers become entrepreneurs. • Farmers become organized to form cooperatives. • Private firms link with farmers' groups. • Favours corporate contract farming. • Economies of scale apply. • Reduced transaction costs.	• Breakdown of traditional livelihood systems. • Threatens agrobiodiversity at the farm level (from diversified to monoculture cropping). • Significant pollution of the environment due to intensive agriculture.
6. Global market integration.	• Laos connects to international markets and competes for a fair share of trade. • Quality and standards of products is improved. • High technology application.	• Big corporate farming groups reap the benefits as they alone have the means to meet international standards. • Poor and subsistent farmers are left behind on the wayside.

A clearly defined transition mechanism

The achievement of agricultural commercialization takes longer and is a more complex process than expected when the move from subsistence to market-oriented production is made in an environment of relatively weak institutional capacity. The Agricultural Development Strategy has identified critical links and nodes that are capable of creating quantum leaps in agricultural productivity, such as the Lao PDR being integrated into regional and global markets. This requires tremendous adjustments, not only in terms of production technology, the scale of operations and global networks, but also in terms of the institutional, social and moral aspects of the transformation. While these latter aspects are of equal importance, they are seldom addressed in many development strategies. The focus generally falls on the build-up of physical infrastructure, fiscal adjustments and investment priorities, organizational and structural reforms and policy measures to effect the desired change. The government's strategy recognizes the need to organize farmers and other grassroots segments into users' or producers' groups as a prerequisite to them becoming market-oriented producers. However, the transformation process requires more than simple social reconfiguration. Value formation and changes in perspective and options, along with government institutional incentives and support systems, such as property-rights reform, the exercise of greater political freedom and other forms of peasant emancipation are basic to effecting social and political change. Before the government's strategy can succeed, there is a need to establish the necessary legal, structural and institutional conditions to enable grassroots villagers and farmers to truly become key components of the global market system, with the same bargaining power as other players.

Transformed institutions that are more functional

A corollary to the need for more transitional mechanisms to affect the desired change is the urgent need for more functional, transformed institutions. There are many line agencies sharing similar sectoral mandates and concerns between the Ministry of Agriculture and Forestry and the Ministry of Natural Resources and Environment. More and more new institutions are being conceived in response to the challenges of agricultural commercialization. At this critical period in the economic transition of the Lao PDR, what is needed most is the strengthening and transforming of existing vital institutions, such as the Ministry of Agriculture and Forestry, rather than continuing to create new agencies and spreading limited resources even more thinly. Institutions related to farmers, producers and grassroots villagers and their interests must take higher priority, as they are currently lagging behind in the process of institutional development. The government has enacted policy reforms establishing social and environmental safeguards to protect the country's rich natural resource base and its cultural heritage (Environmental Impact Assessment Decree, 2010; forest, land and water laws), but implementation of these reforms is weak.

Strong interagency and multi-sectoral coordination

There are many agencies and institutions working in the agriculture and natural resources sector. Yet no concerted or combined efforts are being made to address certain resource-management issues. Every line ministry seeks to maintain its own turf and territorial policy domain. Moreover, each agency tries to interpret the law according to its own vantage point in order to gain complete control over a certain resource base. There should be an institutional mechanism designed to strengthen interagency and multi-sectoral partnerships, coordination and interfaces in order to create greater impact.

Strategic and integrated public and private investment

The three Ministries of Agriculture and Forestry, Natural Resources and Environment, and Planning and Investment must join forces to work out a plan for investment that would build on strategic areas and critical nodes across the entire production continuum (i.e. from seed preparation to harvesting, food processing, marketing, local consumption and exporting). Identification of production bottlenecks, as well as the most promising entry points for intervention within the whole spectrum, could help the agricultural sector to direct its investment priorities to create multiple quantum effects in terms of levels of yield and number of beneficiaries. For instance, the role of farm-to-market roads is critical to the transformation of farmers into entrepreneurs. Likewise, increased profits for farmers would be enhanced by improved post-harvest facilities and the introduction of value-adding processes. All of these should come as one package of incentives to improve farmers' production and incomes.

Improved governance

The Agricultural Development Strategy requires that good public administration and accountability must be ensured. Therefore the Ministry of Agriculture and Forestry has to execute reforms that are effectively implemented in a timely manner at central and decentralized levels. A priority focus on institutional quality is needed for improved Overseas Development Assistance management systems at all levels. Greater transparency and accountability in public decision-making at different administrative levels is necessary to effect change in land-allocation processes. Likewise, there is an urgent need to address the political impasse and conflicting and overlapping policy stands between central and provincial authorities, and among various line agencies involved in the awarding of land concessions. Central to all of this is the need to create a strong and functional 'check and balance' mechanism applying to all levels across the sector to address lack of transparency and accountability in public office.

Human resources requirements

There is a pressing need to build a critical mass of local experts and personnel for the country's agriculture and natural resources sector. International consultants should be hired to work in certain projects and their efforts directed towards strengthening local capacity. There is currently a huge lack of manpower, not only in terms of numbers, but also in skills and competence, relative to the nature of work and outputs demanded from the agriculture and natural resources sector. Research, development and extension services need upgrading and re-focusing.

Increased rights and participation of grassroots villagers and users' groups

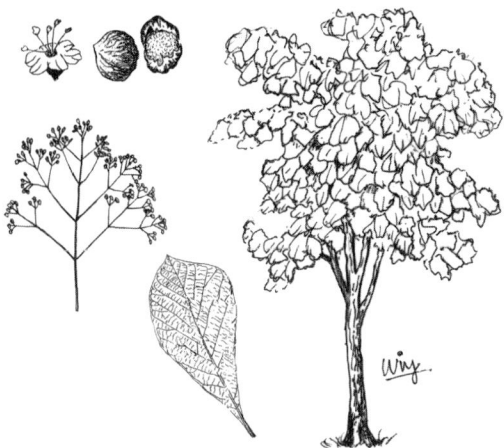

Tectona grandis L. f. [Lamiaceae]

Teak is a tropical hardwood species that grows up to 40 metres tall and is a popular plantation tree in Laos. It is highly valued for its timber, which is extensively used in boat-building and for many decorative purposes. The timber is extremely durable and resistant to termites and other pests.

In response to the government's policy of mainstreaming the participation of farmers and grassroots villagers in the market economy, the Ministry of Agriculture and Forestry envisions incorporating a progressively larger proportion of smallholders into commercial agriculture. To make this happen, there is a need to establish an institutional support system to increase the collective bargaining power of small subsistent farmers in the open market. The formation of cooperatives as a means of organizing farmers into groups of producers, suppliers and consumers requires both training and management. At legislative level, there is no institutional framework for recognizing farmers' groups as legal entities. Thus, farmers' groups lack a legal identity that would enable them to transact official agreements, so they could avail of, or apply for, grants or loans from private, government or donor support institutions.

A greater role and security for the private sector

The private sector plays an important role in transforming the economy. It is the main source of cash flow in the form of business investments coming into the country. There has been an overwhelming number of foreign direct investments and other forms of private business ventures in the Lao PDR. This is presumably because market conditions are currently favourable for foreign investors.

Selective limitation of land concessions and contract-growing projects

The Government of Laos should limit the granting of land concession and commencement of contract-growing ventures in production forests and purely agricultural areas. Critical watersheds in protection forests, protected areas and, in particular, cultural communities should be strictly spared from the widespread growth of massive land concessions in order to preserve the pristine quality and integrity of these areas. There should be a national law to protect the traditional livelihoods and resources of cultural communities.

Conclusions

Based on the above findings and analyses, this chapter makes the following conclusions:

1. Although agricultural commercialization creates new jobs, brings huge amounts of hard foreign currency into the national coffers in the form of foreign direct investments from land-concession projects and improves rural infrastructure, it does not improve food security among rural farmers.

2. While foreign direct investment, particularly through land concessions, provides revenue to the country in terms of foreign exchange earnings, it has a negative bearing on the land rights of individual farmers. Thus, there is a need for the Government of Laos to enact proper social safeguards to benefit rural farmers.

3. Contract farming is a good means of improving product quality and meeting volume requirements through the introduction of high technology and the infusion of huge amounts of private capital. As well as helping the Lao PDR to comply with food-safety requirements and meet the quality standards of the World Trade Organization, it also relieves the need to spend public money on the development of rural infrastructures.

4. However, contract farming converts rural farmers into a labour commodity, resulting in their loss of autonomy. Farmers are tied into production cycles and are compelled to give their full time, attention and effort to producing certain products in a given period.

5. Moreover, when persuaded to enter into agreements with big private companies in order to gain better incomes and stable employment, poor smallholder farmers find themselves subtly losing control over their lands and even access to it.

6. By abandoning their traditional subsistence patterns (i.e. crop diversification for household consumption), in favour of earning daily wages from private industrial farms, the vulnerability of small farmers to market shocks and crop failures increases.

7. In most cases, farmers who enter into contract-growing agreements find themselves entangled in a vicious cycle of indebtedness to the company when crop failures occur, due to their inability to repay loans for seedlings, fertilizers, pesticides and other farming inputs on time.

8. There is substantial promise in land-concession projects and contract-farming arrangements, given the increasing demand for raw materials and agricultural commodities from China, Vietnam, Thailand and other countries in Asia and Europe. However, the trade-offs, if not properly guided and managed, result in local people losing access to their land – their most fundamental economic asset – as well as facing the breakdown of their indigenous communities and their traditional way of life.

9. The rapid transformation of the natural landscape (including the continuous shrinking of natural forest margins and protected areas) into a mosaic of varied monoculture plantations as a result of agricultural commercialization has negative impacts on biodiversity, watershed functions, and the quality of the environment.

10. The complexity and immensity of development issues and concerns arising from the implementation of agricultural commercialization in the Lao PDR calls for a thorough review and evaluation of the programme's gains and losses. This process should fine tune and contextualize its application, so that real benefits are delivered to the majority poor and needy of the countryside. The social, cultural and institutional dynamics of transitioning from subsistence to a market-oriented economy should be given equal importance to economic considerations in order to minimize negative repercussions on the grassroots population and the majority of poor people.

References

Baumann, P. (2000) *Equity and Efficiency in Contract Farming Schemes: The Experience of Agricultural Tree Crops*, Working Paper 139, Overseas Development Institute, London

Cavallo, E., Lawrence, S. and Imhof, A. (2009) *Poverty Reduction in Laos: An Alternative Approach*, International Rivers, Berkeley, CA

Ducourtieux, O., Laffort, J. R. and Sacklokham, S. (2005) 'Land policy and farming practices in Laos: Development and change', in *Poverty Reduction and Shifting Cultivation Stabilisation in the Uplands of Lao PDR: Technologies, Approaches and Methods for Improving Upland Livelihoods*, vol. 2, National Agriculture and Forestry Research Institute, Vientiane

Eaton, C. and Shepherd, A. (2001) *Contract Farming: Partner for Growth*, Agricultural Service Bulletin 145, Food and Agriculture Organization of the United Nations, Rome

Fullbrook, D. (2007) *Contract Farming in Lao PDR: Cases and Questions*, Lao Extension for Agriculture Project, National Agriculture and Forestry Extension Service, Ministry of Agriculture and Forestry, Vientiane

Fullbrook, D. (2009) *Development in Lao PDR: The Food Security Paradox*, Working Paper Series no. 1, Swiss Agency for Development and Cooperation, Lao PDR-Mekong Region, Vientiane

Grosh, B. (1994) 'Contract farming in Africa: An application of the new institutional economics', *Journal of African Economies* 3(2), pp231-261

Haberecht, S. (2009) 'From rice to rubber: Development, transformation and foreign investment in northern Laos – an actor-oriented approach,' PhD dissertation to Bielefeld University, Germany

Hanssen, C. H. (2007) *Lao Land Concessions: Development for the People?*, paper delivered to an International Conference on Poverty Reduction and Forests: Tenure, Market and Policy Reforms, 3-7 September 2007, RECOFTC, Bangkok

Hayami, Y. (2003) *Family Farms and Plantations in Tropical Development*, discussion paper, Series 2003-001, Foundation for Advanced Studies on International Development, Tokyo

MAF (2005) *Report on the Assessment of Forest Cover and Land Use between 1992 and 2002*, Ministry of Agriculture and Forestry, Vientiane

MAF (2007) *Forestry Sector Development Report for 2006. First Stakeholder Consultation on the Forest Strategy 2020 Implementation*, Ministry of Agriculture and Forestry, Vientiane

MAF (2010) *Strategy for Agricultural Development 2011 to 2020*, Ministry of Agriculture and Forestry, Vientiane

Moizo, B. (2005) 'Implementation of the land allocation policy in the Lao PDR: Origins, problems, adjustments and local alternatives', in *Shifting Cultivation and Poverty Eradication in the Uplands of the Lao PDR*, workshop proceedings, National Agriculture and Forestry Research Institute, Vientiane

Pasicolan, P. N. and Sithammavong, T. (2012a) *Policy Trends, Opportunities, Constraints, and Gaps in Relation to MAF 2011-2020 Agricultural Development Strategy*, draft policy working document, Sustainable Natural Resource Management and Productivity Enhancement Project, Department of Planning, Ministry of Agriculture and Forestry, Vientiane

Pasicolan, P. N. and Sithammavong, T. (2012b) *Savannakhet Sugarcane Contract Farming Case Analysis: Towards Designing a Policy Support System for Smallholders*, draft policy working document, Sustainable Natural Resource Management and Productivity Enhancement Project, Department of Planning, Ministry of Agriculture and Forestry, Vientiane

Pasicolan, P. N. and Sithammavong, T. (2012c) *Policy Gap Preliminary Assessment of Farmers' Land Certificate Issuance: Towards Designing a Policy Support System for Smallholders*, draft policy working document, Sustainable Natural Resource Management and Productivity Enhancement Project, Department of Planning, Ministry of Agriculture and Forestry, Vientiane

Patrick, I. (2004) *Contract Farming in Indonesia: Smallholders and Agribusiness Working Together*, ACIAR Technical Reports no. 54, Australian Centre for International Agricultural Research, Canberra

Schoenweger, O. and Üllenberg, A. (2009) *Foreign Direct Investment (FDI) in Land in the Lao PDR*, Deutsche Gesellschaft für Technische Zusammenarbeit (GTZ), Eschborn, Germany

Thatheva, S. (2007) 'Dynamics and sustainability of land use systems in northern Laos', PhD dissertation to Kyoto University, Kyoto, Japan

Thatheva, S. and Yasuyuki, K. (2009) 'Continuity and discontinuity of land use changes: A case study in northern Lao villages', *Southeast Asian Studies* 47(3)

World Bank (2010) 'Natural resource management for sustainable development: Hydropower and mining', in Lao PDR Development Report 2010, Report no. 59005-LA, World Bank, Washington, DC

36

POLICIES, MIGRATION AND COFFEE CULTIVATION IN VIETNAM'S CENTRAL HIGHLANDS

A case study in Dak Lak province

*Truong Hong**

Introduction

In 2014, the area planted to coffee in Vietnam was more than 635,100ha, and of this, more than 580,000ha was in the Central Highlands region (MARD, 2014). Dak Lak province, with more than 203,000ha under coffee, was the biggest coffee producer in the Central Highlands (DARD, 2014). According to their respective agriculture and rural development departments, the provinces of Lam Dong (150,000ha); Dak Nong (140,000ha); Gia Lai (79,000ha); and Kon Tum (12,000ha) followed in order of area committed to coffee production in the Central Highlands.

These areas are already substantially larger than the Vietnam government and the provincial authorities in Dak Lak now say are optimal for sustainable coffee production. However, a combination of official policies led to an uncontrolled stampede to plant coffee in Dak Lak and its neighbouring Central Highland provinces over the past 20 years (Figure 36-1). Many of those caught up in the boom were indigenous shifting cultivators, accused of destroying large areas of forest. But the coffee cultivation to which they turned has caused unprecedented deforestation, with loss of forests keeping pace with growing coffee plantations.

All provinces in the Central Highlands except Kon Tum have ecological conditions that are very suitable for cultivating coffee: the average temperature is between 20 and 23 degrees Celsius; rainfall is between 1600 and 2800mm per year, with two distinguished seasons – a dry season from December to April and a rainy season from May to November; and average humidity is between 75 and 85%. The fertility of the soil is also good, with basaltic soils of volcanic origin covering nearly one-quarter

* Dr Truong Hong, The Western Highlands Agriculture and Forestry Science Institute, Buon Ma Thuot city, Dak Lak province, Vietnam.

FIGURE 36-1:   Coffee and the forest: a scene typical of Vietnam's
Central Highlands.

of the area of the Central Highlands. These soils, with good chemical and physical
properties, result in high coffee yields. More than 97% of the Central Highlands'
coffee crop is *Coffea robusta* (a synonym of *Coffea canephora* Pierre ex A. Froehner).
The rest is *Coffea arabica*.

At the end of the Vietnam War in 1975, the Vietnam government began a policy
of spreading the country's population more evenly, particularly by encouraging
migration from the heavily populated north to the Central Highlands, which was
then thinly populated, with large areas of dense forest. In the past 40 years, about
937,000 people have migrated from the north, and by 2014 they made up 19% of the
Central Highlands' population (Hong, N. L., 2014).

Along with the rapidly spiralling population, the area under coffee spread. Farmers
soon found that coffee was the best crop they had ever cultivated: high yields led to
higher incomes and better living
standards, and this fuelled the
coffee boom. In the 38 years from
1975 to 2013, the area planted
to coffee in Vietnam increased
48.9-fold (Figure 36-2). To claim
land for cultivation, constant
new arrivals slashed back
the forest frontiers and the
cost of burgeoning coffee
production and high migration
soon became apparent. The
Central Highlands lost its dense
forest cover at an alarming pace;

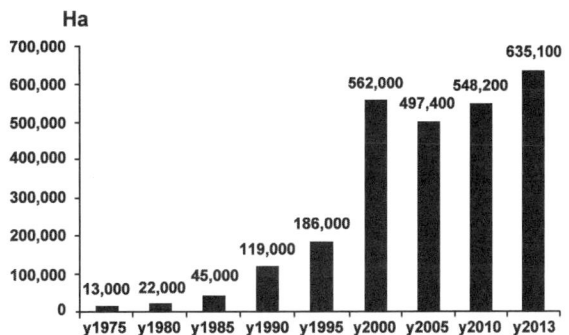

FIGURE 36-2:   Area planted to coffee in Vietnam, 1975
to 2013.

Source: MARD, 2014.

the ecological balance was destroyed, soil erosion began to scar the mountainsides and water resources dwindled (Figure 36-3).

In 2013, the area planted to coffee in Vietnam reached 623,000ha (MARD, 2012a), exceeding a 2008 plan for expansion by about 135,000ha. More recently, growth exceeded a renewed 2014 plan by 35,000ha. There are now concerns that if world coffee prices continue to rise, the area planted to coffee in Vietnam will spiral out of control.

Already, the expansion of coffee cultivation in the Central Highlands is recognized as a primary cause of both forest loss and a decrease in area planted to other, less profitable crops. In addition, forest loss has restricted biodiversity, increased the risk of flooding, reduced the fertility of the soil and increased soil erosion, leading to a general reduction in soil production capacity. Forest loss in the Central Highlands is thus being driven by migration and conversion to agriculture and plantations. Large

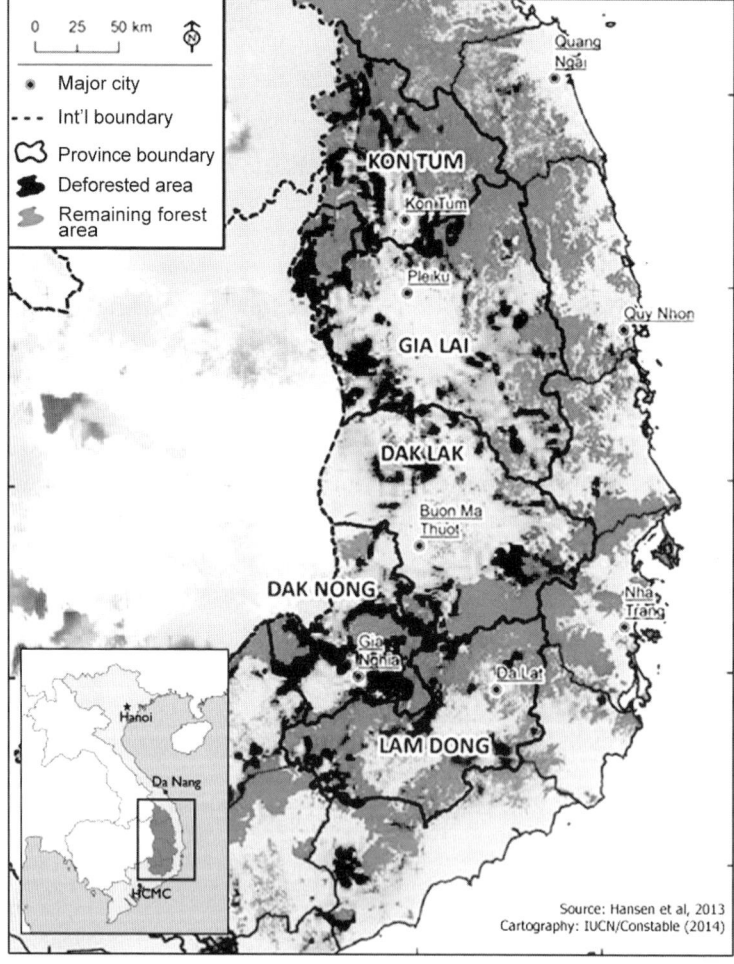

FIGURE 36-3:   Forest loss in Vietnam's Central Highlands, 2000 to 2012.

Source: NIAPP, 2013.

areas along roads and borders have been deforested. By 2012, more than 75% of the forest cover that existed in the Central Highlands in the year 2000 had been lost. However, this calculation did not take regrowth of secondary forests into account, so the loss of forests may be overestimated.

In all of the Central Highlands, the total area of forest converted to coffee plantations is about 500,000ha. In the years from 2000

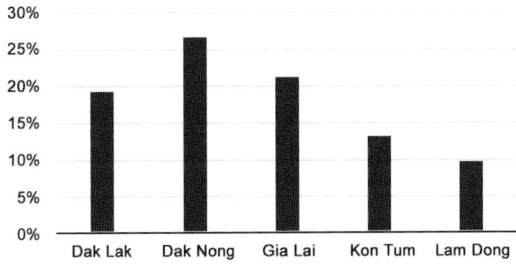

FIGURE 36-4: Forest remaining in 2012, as a percentage of forest cover in 2000, illustrating the rate of forest loss.

Source: Hansen et al. (2013).

to 2014, forest loss was worst in Dak Nong province, followed by Gia Lai, Dak Lak, Kon Tum and Lam Dong provinces (Figure 36-4). Lost forest land has not only been converted to coffee cultivation, but also to rubber plantations, in line with government policy.

Lack of control over the increase of coffee cultivation

The case of Dak Lak province

Dak Lak province has an area of 1,306,201 hectares and a 193km-long border with Cambodia to the west. The average altitude is between 400m and 800m above sea level. It is drained by the Srepok and Ba rivers, and its terrain is very diverse, with a vast plateau at about 600m asl to the west, flat plains around the rivers and sloping, hilly territory in the south. The northwest of the province has a hot, dry climate while the eastern and southern regions are cooler (Table 36-1).

In general, Dak Lak's climate is suitable for many crops, including coffee, black pepper, cacao, cashews and other fruit crops. Lower temperatures in December and January combined with dry conditions are conducive to coffee trees setting flowers to produce a high yield. However, the length of the dry season, from December to April, means coffee crops must be irrigated. The uncontrolled expansion of coffee growing is now causing concerns that planning has been unable to cope with the need for irrigation, and there may be a shortage of water for coffee crops in the near future.

TABLE 36-1: Climatic factors in Dak Lak (2009 to 2013).

Parameters	*Unit*	*Value*
Yearly average temperature	°C	24.0
Yearly average sunshine	hours	2408
Yearly average rainfall	mm	1868
Yearly average humidity	%	83

Source: Dak Lak Statistics Office (2013).

Coffea robusta L. Linden
A synonym of *Coffea canephora* Pierre ex
A. Froehner [Rubiaceae]

Coffee now covers half a million hectares in
Vietnam's Central Highlands. *Coffea robusta*
makes up 97% of this vast crop.

Land resources

Dak Lak has two dominant soil types: grey soil (acrisols) covering 579,309 hectares, or 44.1% of the total land area, and 311,340 hectares of ferrasols, mainly basaltic in origin, covering about 23.7% of the province's land area (Table 36-2).

The qualities of the main soil types, particularly the basaltic soils, make them suitable for cash crops such as coffee, rubber and black pepper, producing high yields with consistently good quality (Table 36-2).

According to the Western Highlands Agriculture and Forestry Science Institute, basaltic soils are suitably fertile to produce high yields of good quality coffee (Table 36-3). The pH is rather acid, but still suitable for coffee; the soil is rich in organic matter and total nitrogen; there are good levels of available phosphorus, but available potassium is a little poor; calcium and magnesium exchange is good. As well, basaltic soil has good physical properties: it is porous and readily absorbs water (Hong, T., 1997).

Dak Lak presently has 537,681ha of agricultural land, of which 203,651ha, or 38.3%, is occupied by coffee. This is roughly equal in area to the amount of land committed to production of annual crops (Table 36-4).

TABLE 36-2: Soil groups in Dak Lak province.

Soil groups	Area (ha)	Percentage of total area
Fluvisols	14,708	1.1
Gleysols	29,350	2.2
Histosols	210	0.01
Luvisols	38,694	3.0
Acrisols	579,309	44.1
Ferrasols	311,340	23.7
Lixisols	146,055	11.1
Phaeozems	22,343	1.7
Planosols	32,980	2.51
Cambisols	23,498	1.7
Leptosols	79,132	6.03
Vertisols	3,794	0.3

Source: DARD (2012).

TABLE 36-3: Fertility of basaltic soils.

Parameters	Unit	Analysed results
pH $_{KCl}$	–	4.60
Organic matter	%	4.10
Nitrogen total	%	0.20
Available phosphorus (P_2O_5)	mg/100 grams of soil	3.50
Available potassium (K_2O)	mg/100 grams of soil	8.95
Calcium exchange (Ca^{2+})	meq/100 grams of soil	3.50
Magnesium exchange (Mg^{2+})	meq/100 grams of soil	2.70

Source: Hong (1997).

TABLE 36-4: Status of agricultural land use in Dak Lak province.

Agricultural land use	Area (ha)	Percentage	
Annual crops	218,388	40.6	
Paddy	60,338	11.2	
Grassland	966	0.2	
Other annual crops	157,084	29.2	
Perennial crops	319,293	59.4	
Coffee	203,561	38.3	
Other perennial crops (rubber, black pepper, avocado, durian, and so on)	115,732	21.1	
Total	537,681	100.0	100.0

Source: Dak Lak Statistics Office (2013).

Table 36-4 shows that the area under coffee area in Dak Lak in 2013 was 203,561ha. This rose to 206,041ha in 2014 (DARD, 2014), an increase of 2480ha in just one year, which, in this case, means that 2480ha of forest was lost in that year.

Perennial crops occupied 59.4% of all agricultural land in Dak Lak province in 2013, covering 319,293ha. Coffee occupied 63.8% of this (DARD, 2013). Although the coffee area was only 38.3% of all agricultural land in the province, income from coffee production represented about 80% of the total value of agricultural production. This underlines the importance of coffee production to the province's economy.

Persea americana Mill. [Lauraceae]

Avocados are another perennial crop thriving in the fertile soils of Vietnam's Central Highlands.

The immigration issue

In almost four decades from 1976, the relative emptiness of Dak Lak province became the destination for a horde of 'free migrants' fleeing the stifling crowds and landless struggle in other parts of Vietnam. In that time, 289,764 new settlers arrived from all parts of the country. They made up 59,488 households. Migration figures show that the spate of newcomers was highest in the two decades from 1976 to 1995, when 49,749 new households settled in Dak Lak. Although the pace slowed significantly, the tide was still flowing at an average rate of about 200 new households per year in the period from 2005 to the middle of 2012.

Migration has brought with it serious problems for the management and protection of forests in Dak Lak. Most of the new households are living on land cleared from forests and converted to agricultural use.

Most of the migrants are of the Kinh ethnic group, which makes up more than 86% of Vietnam's population. They came mainly from northern and central provinces. However, households from northern ethnic minority groups have also been part of the migrant flow since 1980. These groups lived in remote areas at high elevations and had a long tradition of practising shifting cultivation. Most of them were very poor and made the journey to Dak Lak without funds. In their new environment they resumed swidden farming. They cleared areas of forest for coffee cultivation and to grow annual crops. They are accused of negatively changing the natural environment and creating difficulties for local authorities in terms of economy and social security.

The large scale migration and redistribution of Dak Lak's population and its labour force have had a serious effect on the province's original ethnic minorities. The province now has a greater population of newcomers – Kinh people in particular – than original inhabitants. The area of land available to the original groups has diminished considerably, and their living conditions are very poor. They continue to clear the forest for agriculture, and in most cases, the agricultural system they adopt is coffee cultivation.

Table 36-5 shows the relationship between the rate of new arrivals in Dak Lak province and the rate of increase in coffee-growing area.

Although the rate of immigration was slower between 1996 and 2004, the average rate at which the land area under coffee increased was higher because of high coffee prices. The world price for coffee was high in the second half of the 1990s, slumped in 2002 and began a recovery in 2003. The price movements prompted a surge in planting, to average 6800ha of new coffee plantations per year. From 2005 to 2013,

TABLE 36-5: Rate of immigration and coffee-area increase in Dak Lak, 1975 to 2013.

Period	Number of migrant households	Coffee area increase (ha/year)
1975 to 1995	49,749	4872
1996 to 2004	8246	6838
2005 to 2013	1493	3684

Source: DPC (2014).

the flow of migrants diminished, so the increase in new coffee plantations fell to about 3600ha per year.

Economic efficiency of coffee cultivation

Good soil and climatic conditions, along with the application of new technologies for coffee cultivation, have led to very high coffee yields in Dak Lak. The average yield from private plantations is about five to six tons of clean beans per hectare – five times higher than coffee yields around the world.

With high prices for their coffee, the farmers of Dak Lak received a good income and their standard of living improved. The average profit being made by farmers from one hectare of coffee (with an average yield of four tons of clean bean per hectare) was about US$4000 to $5000 per year, at 2013 prices (Table 36-6). This compared with profits of $2000 to $3000 per hectare per year from other crops, such as rubber, cacao, or maize.

As can be seen from Table 36-6, farmers' profits from mature coffee plantations can be very high. In comparison to black pepper, the benefits from coffee are presently lower. However, coffee cultivation is easier, it suits all farmers, investment expenditure is lower than for black pepper and the risks are limited. For this reason, local authorities have favoured the development of coffee cultivation.

World prices and the area under coffee in Dak Lak

In the 24 years from 1990 to 2014, there have been serious fluctuations in world prices for coffee. However, the tendency has been for prices to rise. The highest coffee price was in 1995; the lowest in 2002, and from 2003 the price has been trending higher (Figure 36-5).

There has been a clear relationship between coffee prices and the growth of the area planted to coffee in Dak Lak province. When prices were high, the coffee area increased quickly; when prices fell, the expansion slowed. For example, the price began to increase strongly in 1994 and reached a peak of $2640 for one ton of clean bean in 1995. It was a sign for farmers to expand, and in one year (1999) Dak Lak's

TABLE 36-6: Profits related to prices for four tons of clean bean harvested from one hectare of mature coffee.

Coffee price (US$/ton of clean bean)	Profit (US$/VND/ha)	Value to cost ratio
1500	$3000 (VND62,467,000)	1.0
2000	$5000 (VND104,112,000)	1.7
2500	$7000 (VND145,757,000)	2.3

Note: US$1 = 20,822.538 Vietnamese dong, 30 January 2013.
Source: Author's calculation (2013); www.xe.com/currencytables, accessed 22 May 2015.

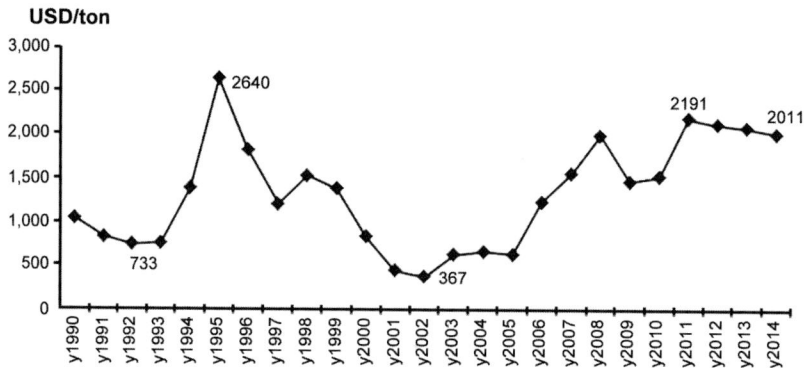

USD/ton

FIGURE 36-5: Coffee-price fluctuations in world market.

Source: MARD (2014).

coffee cultivation area increased by 46,634ha (Hong, N. L., 2014). The new area was mainly cleared from forest, but it also included agricultural land that was converted to coffee from crops of lower value.

From 1975 to the present, the area under coffee cultivation in Dak Lak province has grown 17.6-fold. However, production rose 38.6-fold in that time and yields increased 2.3- to 2.5-fold. In 1975, coffee cultivation covered only 11,503ha in Dak Lak, and the area actually decreased in 1982 when some land reverted to production of food crops. But from 1985 onward, apart from a slight fall in 2004, the increase has been inexorable (Figure 36-6). Dak Lak's coffee-cultivation area reached 206,041ha in 2014.

Piper nigrum L. [Piperaceae]

Farmers can make more money from black pepper, but coffee involves less effort, investment and risk.

Reasons for the uncontrolled growth of coffee cultivation in Dak Lak

The growth of coffee cultivation in Dak Lak has always exceeded official plans. For example, the present agricultural development plan in Dak Lak provides for a coffee-cultivation area of 170,000ha. In reality, coffee already covers more than 206,041ha. Apart from the lure of high prices, there are other major causes for the increase in coffee cultivation:

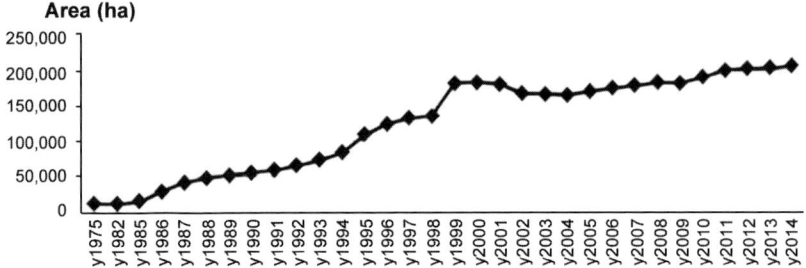

FIGURE 36-6:   Area of coffee cultivation in Dak Lak province.

Source: DARD (2014).

- The government's economic development policy, including the production of agricultural goods to exploit the advantage of the country's natural resources;
- Restructuring of the population and implementation of the migration policy to help agricultural development;
- The land and the climate are perfectly suited to growing coffee with high yields;
- Dak Lak province regards coffee as its major export crop. Coffee contributes more than 60% of the province's total budget and creates jobs for 300,000 direct employees and 100,000 indirect workers. It plays a very important role in Dak Lak's social and economic development, especially in building infrastructure such as roads, hospitals, schools, communications and electrical networks, and so on;
- The coffee plant has high economic value, is easy to grow and care for, and advanced scientific techniques are available to achieve high yields. It is a low-risk, low-cost, high-profit crop;
- Free migration and undisciplined forest management are continuing, with continuous destruction of forest to provide for growth of coffee; and
- Some indigenous ethnic minority households were allocated land to cultivate corn and other food crops, but most have switched to coffee in order to earn higher incomes.

Side effects of increased coffee production

The vast increase in cultivation of coffee in Dak Lak has brought major social and economic benefits to coffee growers and to the province generally. But the race for profits from coffee has meant that the province's land-use plans were ignored and broken, and this has brought major problems that need to be urgently addressed.

The switch to coffee brought about a sharp imbalance in cropping patterns. It meant that other crops, including food crops and short-term industrial crops, such as sugar cane, legumes and cotton, were overlooked and, in some cases, production ceased altogether. The helter–skelter dash to plant more and more coffee not only ignored land-use planning, but in some cases, also ignored agricultural common sense. Much recently developed coffee area expanded onto land with poor soil physical properties and lower fertility. More than 26% of Dak Lak's coffee production now extends over such soils (DARD, 2013). Moreover, 8.5% of the area planted to coffee has no irrigation because water is simply not available. Production from these areas will be very low, even though the cost of inputs has been high, so poor farmers will suffer and households will lose their food security. These issues have escalated the overall risks of coffee production in Dak Lak, and this has impacted on those coffee-production projects that are well planned and responsibly managed. As well as these issues, the sheer extent of Dak Lak's coffee cultivation has now assumed the detractions of extensive monocropping, including loss of biodiversity and increased risks from pests and diseases.

As coffee production grew in popularity, encroachment into forest areas kept pace and the market for good land for coffee growing increased. Some households, particularly of ethnic minority groups, sold their agricultural land to Kinh settlers and joined the ranks of the poor and landless. They moved back into remaining tracts of forest to open swiddens and return to shifting cultivation.

While land-use planners could see a looming scarcity of water supplies to irrigate new coffee plantations in Dak Lak, the new settlers could not, and plans were ignored as the rush to plant coffee continued. According to some recent studies of water needs for agriculture in Dak Lak, more than 65% of the province's coffee area is irrigated by ground water (DARD, 2013). However, ground water levels have receded by as much as three to five metres in some places and the risk to irrigation supplies is very clear. Among the causes of the receding water table are deforestation for planting coffee and the consequent loss of the forest's water-holding capacity.

Another of the major impacts of the expansion of coffee plantations in Dak Lak is the replacement of traditional shifting cultivation, as practiced by the province's indigenous minority people. The diversity of the province's agricultural practices, once a feature of Dak Lak's indigenous communities, has diminished markedly.

In addition to all of these local impacts, the explosion of coffee cultivation in Dak Lak and in the Central Highlands generally has contributed large quantities of coffee to the world's supply, so it has begun to play an important role in governing coffee prices on world markets.

Deforestation

Prior to 1990, the main reason why people cleared the forests of Dak Lak province was to grow food crops, such as rice, maize and cassava, in swiddens. In about 1991, the dual influences of escalating coffee prices and government-sponsored free migration sparked the coffee stampede. According to the Dak Lak Department of Agriculture and Rural Development (DARD), the area planted to coffee increased by 54,000ha between 1995 and 2000, and at its peak was increasing at a rate of about 16,000ha per year (Figure 36-7). The area of forest was diminishing at the same pace. DARD Director Trang Quang Thanh told the author (2013) that during a recent 10-year period, Dak Lak's natural forest area had decreased by more than 74,000ha as forest

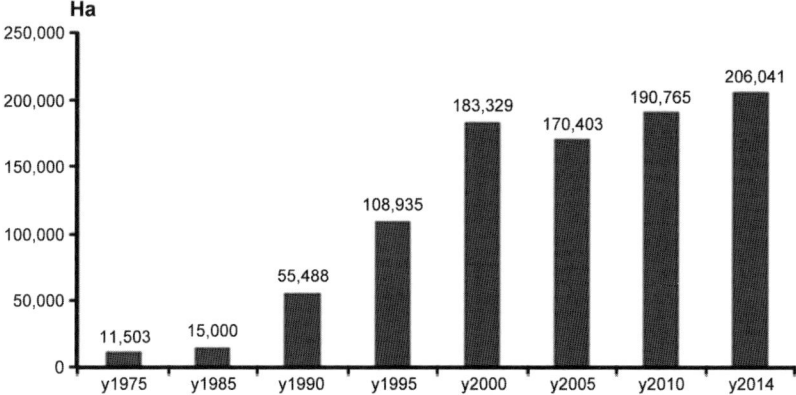

FIGURE 36-7: Area of coffee cultivation in Dak Lak province in five-year periods.

Source: DARD (2014).

frontiers were converted to agriculture and other purposes (Figure 36-8). However, the rate of increase of coffee plantations slowed substantially in the first decade of the New Millennium, to an average rate of about 1600ha per year. At the same time, the number of free migrants also fell, to average about 167 new households, or 842 people per year (DPC, 2014).

The conversion and loss of forest due to coffee cultivation has also deprived local ethnic communities of their traditional access to non-timber forest products, such as wild vegetables and medicinal plants, so their dependence upon coffee or swidden farming has grown.

Continued and uncontrolled deforestation in Dak Lak province is expected to have the following consequences: the biological diversity of forests will diminish sharply; forest cover will be reduced and this will lessen water-retention capacity. The risk of flooding will escalate and, along with it, the risk of soil erosion and soil degradation. Overall, increased impacts from climate change will affect the activities and lives of farmers.

Deforestation is a known contributor to global warming (Miller and Cotter, 2013), and is one of the main causes of increasing greenhouse-gas emissions. Deforestation also allows the soil surface temperature to rise faster, leading to higher atmospheric temperatures. Air temperatures in Vietnam's Central Highlands have risen by about 0.5°C over a recent 10-year period (Hong, T., 2010).

The natural water-circulation cycle is also affected by deforestation. Plants absorb water from the ground and release it into the air. When forests are razed, trees no longer perform this ecosystem function and the climate becomes more arid: both groundwater and air moisture are reduced. In Dak Lak province shortages of water for agriculture have already begun to appear and are becoming more frequent.

Deforestation not only reduces the capacity of a landscape to retain rainwater and store it in underground aquifers, but it also increases surface run off and evaporation. Combined with reduced soil adhesion, this leads to erosion, floods and landslides.

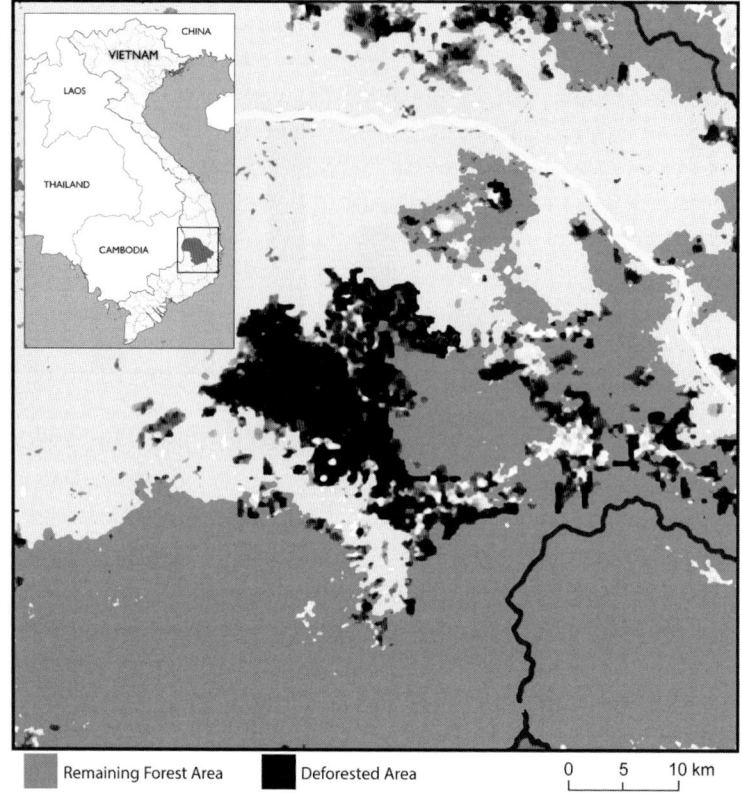

FIGURE 36-8: Forest loss in the eastern part of Dak Lak province from 2000 to 2012.

Source: NIAPP (2013).

In 2013, there was a flash flood in Dak Lak's Ea H'leo district that inflicted serious damage on agricultural production and farmers' livelihoods.

Lack of arable land for indigenous minority groups

Before 1980, Dak Lak province's ethnic-minority families owned both residential and farming land as a result of reclamation, rehabilitation or grants from the government. In those years, large areas of both agricultural and forest land were available with a very low population density. Since then, the availability of land for ethnic-minority households has become much more uncertain, and some have become landless.

Compared to the national rate of population growth, that in ethnic-minority areas has been both rapid and variable, with figures showing a population-growth rate 1.6% higher in ethnic-minority areas than the provincial average. The rapid population growth has the effect of expanding the number of households as children mature and separate from their families. Such new households are most likely to own no productive land, while the forest around them is shrinking. Mired in a poor and difficult lifestyle with an inability to work efficiently, even those households that own land often resort to assigning, pledging or mortgaging what land they own, including their residential land. They are then unable to redeem the land and become landless.

The government's plan for population distribution, especially the free migration scheme, has caused large population fluctuations that have resulted in shortages of both residential and farming land in traditionally ethnic-minority areas. In some cases there have been land disputes between local people and newcomers. Recently, there were still about 6000 households without productive land in Dak Lak province (DPC, 2014).

Restructuring the rural economy of Dak Lak, including labour supply and employment, especially in ethnic-minority areas, has been very slow. In some areas where the population is more than 90% agricultural employees, farming productivity remains low and this leads to increased demand for productive land. The result is a shortage of land.

In a bid to overcome these problems, Dak Lak province has implemented a policy of revoking land titles held by agriculture and forestry companies when the land is not being used efficiently, and allocating it to ethnic-minority households. At the same time, it is mobilizing community support in the search for productive land for landless people. The provincial authorities are also providing support for cattle rearing by poor families, are allocating additional forest land and are finding jobs with agriculture and forestry companies in an effort to help poor ethnic-minority families to break out of the cycle of poverty.

The government's plans to control the growth of coffee cultivation

Immigration policy

From 1980 until the present, the migration of people seeking forested areas and fertile farming land within Vietnam has varied widely. Ultimately, however, there have been two main modes of migration: (1) that organized by state institutions, including the need to populate new economic zones and to recruit labour for state agriculture and forestry companies; and (2) 'free' migration, which allowed households, groups of households, or individuals to seek out new land from which to earn a living. Among these free migrants were ethnic-minority families who were eager to escape difficult and landless circumstances; to find a new farming area to improve their lives.

Negative impacts of immigration

Many aspects of migration within Vietnam have been disordered and uncontrolled. Goals have been broken and local socio-economic development plans ignored. This has led to deforestation and forest clearing in the hunger for productive land and illegal land transfers, particularly for cultivation of coffee.

Free migrant households were often very poor, illiterate, had many children and had low levels of education. Their lives were very difficult, creating additional problems for a province that already had problems enough. As Dak Lak's population increased, so its poverty rate rose.

In order to gradually reduce the negative aspects of free migration, forest clearing for agriculture and the uncontrolled spread of coffee plantations, the Government of Vietnam and authorities of Dak Lak province introduced a number of new policies. These included:

- Decision No. 78/QD-TTg dated November 2008 of the Prime Minister on policies to implement the programme on population distribution (according to Decision No. 193/2006/QD-TTg);
- Decision No. 1776/QD-TTg dated November 2012 on residential arrangements in difficult areas and areas of forest land inhabited by migrants, in order to limit the clearing of forest for agriculture and illegal trading in land;
- People's Committee of Dak Lak province Decision No. 2541/QD-UBND dated December 2006, approving a project for planning the province's population distribution from 2006 to 2010, with a view to conditions in 2015. In particular, this dealt with arrangements for 11,200 scattered ethnic-minority households totalling 52,879 people and 715 households with 3875 people who were living in protection or special forests;
- Decision No. 2763/QD-UBND, dated December 2013 approving an additional project to plan the stabilization of the province's population in the period from 2013 to 2015, with a view to conditions in 2020. This provided for checking, arranging and stabilizing 5479 households of free migrants comprising 27,351 people, with centralized arrangements for 2676 households, intermixed arrangements for 957 households and stabilization of 1846 households in place.

From 2002 to 2012, Dak Lak province has allocated farming land totalling 2771ha to 7737 indigenous ethnic-minority households 'in place' – or where they were living.

The lives of migrants to Dak Lak are now stable. Households were allocated 400sq. m of land on which to build a house and 0.5ha of farmland. Encroachment into forests for agricultural purposes and expansion of coffee-cultivation areas has decreased, but continues to occur.

A number of preventative and restrictive solutions have been implemented, but the best results have come from simple restrictions. Therefore, the need to quickly limit or to stop free migration is a challenge still to be faced by both central and local agencies.

Planning policy

As of May 2012, most provinces in Vietnam had completed land-use plans for the period from 2011 to 2015 and had formulated planning up to 2020. This was in line with implementation of land laws and Resolution 17/2011/QH13 of the National

Assembly, dated November 2011. Normally, land-use planning is carried out five years at a time.

The government has also issued a coffee-area planning policy approving a development plan upto 2020, with a vision to 2030. The plan provides for a total area under coffee in 2020 of 500,000ha and a total area of 479,000ha in 2030. Since the area under coffee in Vietnam was 623,000ha in 2013, the plan envisages a considerable decrease in area, but an increase in productivity, quality and economic efficiency, particularly the addition of value. Dak Lak province, with 206,041ha of coffee already carved from its upland forests, plans to have just 175,000ha in 2020 (DARD, 2012).

Earlier, in an effort to ensure the sustainable development of Vietnam's coffee industry, the central government and Dak Lak province carried out the necessary planning exercises, but management of the plan failed to create sufficient awareness. The area planted to coffee spiralled out of control at the cost of extensive forest loss.

Lately, in a move aimed at limiting the damage being inflicted on Vietnam's forest resources by coffee growers, the government has asked local authorities to complete and implement their land-use planning, forest development and protection plans. Forest-protection contracts will be allocated, with responsibility for afforestation programmes awarded to households (including ethnic-minority households) and communities. Plans will also be set up to protect and develop forests over a five-year period, with annual plans for special, protection and production forests. The management of the forest plans will also be strengthened and forms of forest conversion to agriculture will be prohibited, particularly for the cultivation of coffee.

Sustainable coffee production policy

Coffee is very important to Vietnam's economy, so the government has paid diligent attention to formulating policies for the industry's sustainable development. In 2014, the Ministry of Agriculture and Rural Development decided that by 2020 Vietnam would have a sustainable coffee crop covering about 600,000ha (MARD, 2012b). The goal would be a modern, synchronized and highly competitive coffee industry with high product diversity and quality and high levels of added value. Deforestation for coffee cultivation would be limited by increasing the income of farmers of existing plantations, thereby contributing to the protection of the environment.

In 2008, Dak Lak province identified the coffee industry as one of its key economic sectors. The Dak Lak People's Committee set up a sustainable coffee development project covering the years to 2015, with a view to 2020. This project saw the area of sustainable coffee production in the province as 150,000ha. But by 2014 it had grown to more than 206,000ha. The coffee-area planning policy now seeks an adjustment down to 170,000ha.

The Vietnam government and authorities in Dak Lak province have long been aware of the uncontrolled increase in coffee plantation and its effects on the development of a sustainable coffee industry and the country's forest area. Policies have been issued with the aim of guiding and controlling coffee development, migration, and the protection and development of forests. However, implementation of policies and management of planning has been largely ineffective and controls have met with only limited success. Regulations must now be implemented and the land-use situation in Vietnam's Central Highlands closely monitored if long-term stability is ever to be achieved in the country's coffee industry.

References

Dak Lak Statistics Office (2013) *Yearly Statistics Book of Dak Lak*, Dak Lak Statistics Office, Buon Ma Thuot, Dak Lak

DARD (2012) *Land-use Plan, 2010-2015, Dak Lak Province*, Dak Lak Department of Agriculture and Rural Developmemt, Buon Ma Thuot, Dak Lak

DARD (2013) *Annual Agricultural Production Report*, Dak Lak Department of Agriculture and Rural Development, Buon Ma Thuot, Dak Lak

DARD (2014) *Annual Agricultural Production Report*, Dak Lak Department of Agriculture and Rural Development, Buon Ma Thuot, Dak Lak

DPC (2014) *Report on Status of Free Migrants from the North Provinces to Dak Lak*, Dak Lak People's Committee, Buon Ma Thuot, Dak Lak

Hansen, M. C., Potapov, P. V., Moore, R., Hancher, M., Turubanova, S. A., Tyukavina, A., Thau, D., Stehman, S. V., Goetz, S. J., Loveland, T. R., Kommareddy, A., Egorov, A., Chini, L., Justice, C. O. and Townshend, J. R. G. (2013) 'High-resolution global maps of 21st-century forest cover change', *Science* 342, pp850-853

Hong, N. L. (2014) 'Emigration from the North to the Central Highlands: Why and its after-effects', *Fact and Comment*, October 2014

Hong, T. (1997) 'Fertility of coffee planting soil in the Central Highlands', *Annual Scientific Report*, Vietnam Coffee Research Institute (now the Western Highlands Agriculture and Forestry Science Institute (WASI)), Buon Ma Thuot, Dak Lak

Hong. T. (2010) *Climate Change with Agricultural Production in the Central Highlands,*, Western Highlands Agriculture and Forestry Science Institute (WASI), Buon Ma Thuot, Dak Lak, available at http://wasi.org.vn

MARD (2012a) *Decision No. 1987/QD-BNN-TT, dated 21 May 2012 on Approving Development Planning of Vietnam Coffee Sector to 2020 with a Vision to 2030*, Ministry of Agriculture and Rural Development, Hanoi

MARD (2012b) *Decision No. 3417/QD-BNN-TT, dated 01 August 2012, on Approving Sustainable Coffee Development Project of Vietnam by 2020*, Ministry of Agriculture and Rural Development, Hanoi

MARD (2014) *Document of Crop Restructuring*, Ministry of Agriculture and Rural Development, Hanoi, pp203-204

Miller, C. and Cotter, J. (2013) *Impacts of Deforestation on Weather Patterns and Agriculture,* Greenpeace International, Amsterdam

NIAPP (2013) *Agricultural Plan and Projection for Dak Lak province*, National Institute of Agricultural Planning and Projection, Hanoi

Thanh, T. Q. (2013) Personal communication between Trang Quang Thanh, Director of the Dak Lak Department of Agriculture and Rural Development, and the author.

E. The impacts of policy interventions on the lives of shifting cultivators and the local ecology

Prohibiting shifting cultivation can deprive some of our most vulnerable people of their livelihoods and leave them literally living on sidewalks in nearby urban centres. This mother, apparently raising her baby on the sidewalks of Chiang Mai, in Northern Thailand, may well be a refugee from the country's repressive shifting cultivation policies.

Sketch based on a photo by Malcolm Cairns.

37

THE CHAYANOV LIFE CYCLE IN UPLAND VILLAGES OF LAOS

Socio-economic differentiation driven by state involvement

Olivier Ducourtieux[*]

Introduction

Since the Neolithic era, many human societies have been built on agriculture (Mazoyer and Roudart, 2005). Prestigious vestiges – Angkor, Cuzco, Giza, the Coliseum, and so on – remind us of how powerful farm-based kingdoms or empires used to be, but also how stratified those societies were. The division of labour allowed the states to capture the surplus product from the farmers, and it was used to operate powerful administrations and armies. Historic examples can be found of hydraulic agriculture (Ancient Egypt, Chinese and Khmer kingdoms), or *ager/saltus* (cultivated fields and pasture) agriculture (Roman Empire, European kingdoms), but examples of a strong state built on shifting cultivation are scarce, if they exist at all. Why? Probably because shifting cultivation is not sufficiently productive and does not allow farmers to accumulate surpluses from their labour (Boserup, 1976; Diamond, 1987). Without surpluses, armies and empires are not possible. Another consequence of such a hypothesis is that socio-economic differentiation within societies based on shifting cultivation would remain very limited.

This chapter aims to review the past and current drivers of socio-economic differentiation in the Phunoy villages of Phongsaly, in northern Laos (Figure 37-1) Until very recently, the economy of these communities relied on shifting cultivation. However, the recent implementation of policies aimed at alleviating poverty and banning shifting cultivation have boosted state involvement in the livelihood of Phunoy villagers. What impact has this change had on the socio-economic differentiation in the Phunoy villages?

---

[*] Dr Olivier Ducourtieux is an Associate Professor in the Comparative Agriculture Unit/UMR PRODIG at AgroParisTech, in France. He worked on rural development projects in Laos from 1993 to 2007.

Rationale of shifting cultivation in Phongsaly: an optimal application of workforce

For centuries, Phunoy villagers in Phongsaly have relied on shifting cultivation to meet the needs of their families. Every year, the same story is repeated, over and over. In Samlang village, for example, the cycle begins in November (Figure 37-2), when the council of elders meets to select the area of secondary forest that will be cultivated in the following year (Ducourtieux et al., 2005).[1] In January, the adults from each household clear the plot they will crop. They leave the felled trees lying on the ground to dry for two to three months. At the end of March, the village community organizes the burning, which involves all of the women and most of the men for several days. The elders try to choose windless days, to limit the risk of embers blowing for several kilometres into surrounding forest. There is also a potential danger for villages, where nearly one-third of homes have thatched roofs (Figure

FIGURE 37-1:   Phongsaly province, the northernmost in the Lao PDR, showing Phongsaly district, the capital, also called Phongsaly, and the two Phunoy villages discussed in this chapter.

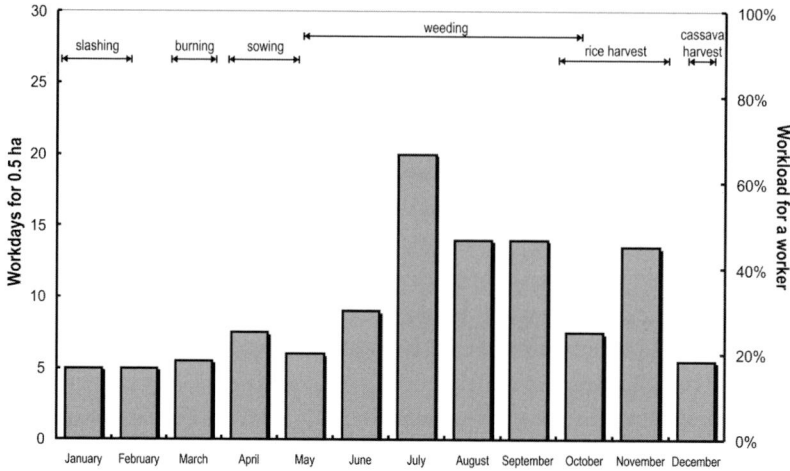

FIGURE 37-2: Shifting cultivation calendar for Phunoy villages in Phongsaly.

37-3). Children or elders who are unable to work in the fields use bamboo ladders and buckets of water to extinguish sparks that land on roofs. When swiddens are close to the village, burning is done at night so that flames can be more easily spotted. In one instance, in March 2003, this precaution was not taken in the Phongsaly village of Chaphou, and all the houses were destroyed by fire in a few tempestuous minutes.[2]

A few days after the fires have died, people start preparing the plot. First, a hut is built with the residue left by the fire. This structure provides shelter at mealtimes or during inclement weather, and temporary storage for tools, seeds and harvested crops. Remaining trunks and branches are then hauled to the edge of the field, from where they are cut and transported to the village for firewood, as needed. These are also used to build a collective fence around all the swiddens on the cleared land, to protect against cattle, pigs, or wild animals. Elephants, however, are not stopped by such trivial human-made obstacles. One of the advantages of grouping the fields at village level becomes apparent when constructing the fence: the total length is less than if every family had to fence its own individual plot. The swidden area for Samlang's 28 households covers 38 hectares; it requires a 2500m-long fence. Fencing each household plot would require 13,000m of fences, implying five times more work.

FIGURE 37-3: The Phunoy village of Samlang, in Phongsaly province.

Photo: Olivier Ducourtieux (2003).

Planting begins with the first rains, expected in April. Sowing occurs in nine steps, spread over more than a month. Shifting cultivation at Samlang combines 15 species of various growth forms, to make optimal use of environmental resources (Brookfield, 2001; Altieri, 2002), maximize light interception for photosynthesis, limit runoff, and so on. Combining crops that have different calendars and react differently to climatic stresses and pathogens helps to secure production (Conklin, 1957). Tubers (sweet potato, red taro) are seeded first. Farmers then sow calabash gourds, followed by cilantro (coriander). Next, comes a mixture of Solanaceae (eggplants and peppers), followed by maize. Rice is mixed in with the pumpkins and cucumbers, which are planted in bunches. Cassava cuttings are then planted on the edge of the plot, and serve to mark the borders of the family field. Finally, sesame and peanuts are sown in May or the beginning of June. The distribution of seedlings and plants is anything but random. Farmers must strike a compromise between (1) the demands and special requirements of each crop; (2) their family's food preferences and commercial opportunities; and (3) differences in the natural environment, including the topography of the field and rainfall. The hoe is the most basic tool for planting, and sowing rice constitutes the bulk of the activities. Households often pool their labour, calling on the extended family to participate in each step. Associating different crops in shifting cultivation not only optimizes photosynthesis, but also limits the farmer's risks. Failure of one or another of the crops for some reason does not jeopardise the family's survival: they can count on the other harvests and activities. The dynamic and evolving allocation of the workforce and the diversification of activities are the two main strategies for limiting risks and maximizing family income. Resources that are scarce, such as workforce, or fragile, such as soil, forests, water and biodiversity, are managed differently, so as to be integrated into these rational strategies in a sustainable way (Brookfield, 2001; Ducourtieux, 2009).

Seeds germinate with the first rains, but so do the weeds. To limit competition for light energy, soil nutrients and water (de Rouw, 1995), farmers commit themselves to the back-breaking and repetitive work of weeding. Between May and September, a family may weed their swidden up to four times, depending on available labour. The first weeding is in mid-May, after the rice plants germinate, requiring about 20 workdays of labour per hectare. The second weeding is in early July (45 workdays/ha), the third in mid-August (35 to 40 workdays/ha) and the last in mid-September (20 workdays/ha). In all, weeding requires 130 to 140 workdays/ha over a period of five months. Timing is crucial: a delay in weeding allows weeds to mature and disperse seeds in the wind; weed growth may explode and become increasingly difficult to control (van Keer, 2003). Because family swiddens are close together, the entire village can be affected by an individual family's weed control, or lack of it. This encourages families to pool their labour. Usually, extended-family members work together, and occasionally villagers will join forces to fill in for a family member incapacitated by illness or death. In an attempt to reduce the workload, some households tried a chemical-based weed remover consisting of herbicide mixed with kitchen salt. However, its effect appeared to be limited, or it required a large and costly volume of herbicide. Whatever, such spraying disappeared within a few years. In other words, a technical innovation became available and farmers immediately tested it for a few crop cycles, only to abandon it upon concluding that it was economically irrelevant. In this way, shifting cultivation practices appear to be more adaptive than stagnant and antiquated.

Harvesting starts in July for calabash gourds and finishes in March of the following year for chillies. The timing of the rice harvest depends on the variety of rice, but the bulk of this work takes place in late October. On average, Samlang villagers harvest about 1300kg of rice per hectare, complemented by the associated crops that double the value of the total harvest. Farmers need about 15 workdays per hectare for harvesting, hauling and storage. It takes roughly a dozen round trips, carrying 60kg loads, to take the harvest back to the village (15 to 20 workdays/ha). This may explain why villagers decided to include only a quarter of their village territory in the rotational swidden system. In order to restore the fertility of the soil after cropping, it makes sense to allow plots to fallow for as long as possible before they are cleared again. If the entire village territory were subject to shifting cultivation, fallow periods would exceed 50 years instead of the current 12 years, but the additional yields would not cover the extra labour required to carry the harvest back to the village, and in the rainy season, every hour spent walking is an hour lost for weeding. Samlang villagers (and most of their neighbouring Phunoy villages) long ago decided not to include in their swidden rotations that forested land that lies within the village domain, but is furthest away from the village.

After the harvest, the forest is allowed to regrow on the village's swiddens for a period of about 12 years (Figure 37-4). These long fallow periods allow for the reconstitution of dense secondary formations and the growth of biomass as a source of fertility for the next slash-and-burn crop cycle (Ramakrishnan, 1992). The fallow land becomes a pasture area for large ruminants; cattle are limited to grazing grassy fallows, whereas water buffaloes graze year round on grassy, shrubby and tree-covered fallows.

Drivers of limited socio-economic differentiation in shifting-cultivation communities

Farmers in Samlang use manual tools exclusively: different sized machetes, hoes, baskets and bags. Everything is made in the village. Few inputs are used: seeds are produced on the farm, while weed and pest management is essentially manual.

In all, farmers dedicate 225 workdays/ha to the annual shifting-cultivation cropping phase, from clearing swiddens to post-harvest processing and storage (Figure 37-5). Weeding is the tricky part of the cropping system: it accounts for more than half of the total workload and is labour-intensive, with a restricted time window. It mobilizes the entire household's labour force in July and August. Even if most of the families were able to clear more land in January to plant in April, they would probably fall short of the labour needed to do a good job of weeding during the

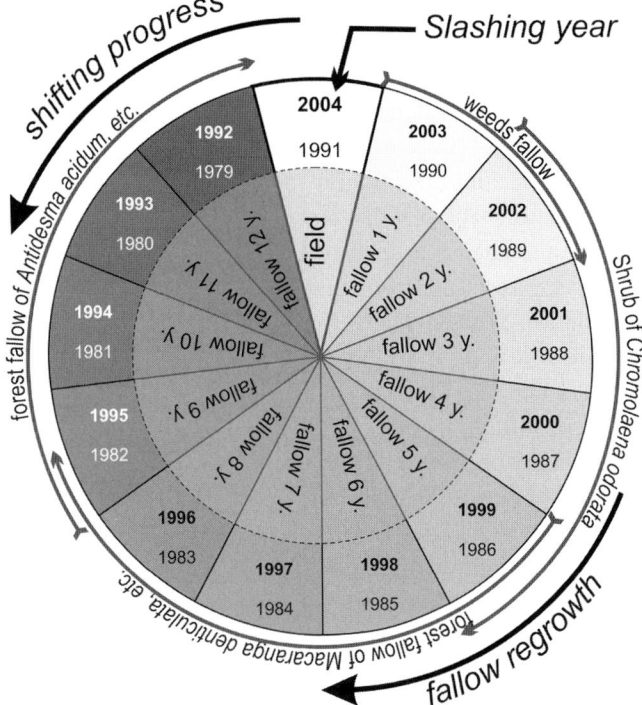

FIGURE 37-4:   Fallow rotation for shifting cultivation in Phongsaly.

Note: The diagram shows the state of a 13-year rotational swidden
system (one year of cultivation and 12 years of fallow) in 2004 and
1991. In this case, the different swidden plots are identified according
to the years in which they were slashed and cultivated. If everything
in the diagram (including the shading) rotates counter-clockwise
EXCEPT the years, the plot that has been fallow for 12 years is slashed
and cultivated in 1992. The plot cultivated in 1991 is in the first year
of fallow, the plot last cultivated in 1990 is in the second year of fallow,
and so on. The years marked in the diagram cover two full rotations,
during which each plot was cultivated twice.

rainy season. Therefore, they avoid wasting their scarce resources (labour, seedlings)
in what would be a foolish enterprise. Weeding is the cropping system's bottleneck: a
farmer cannot weed more than 0.5ha, which is the maximum area of land cultivated
per worker.

Samlang households rely not only on shifting cultivation to secure their livelihood,
but also on many other diverse and complementary activities: animal husbandry
(poultry, pigs, buffaloes and cattle), forest products (hunting, fishing and gathering),
handicrafts (wooden and iron tools, baskets, and so on), and food processing (alcohol
distillation). Forest products provide more than 40% of family incomes in forest
villages.

The diversity of food-production activities confers on the villagers a balanced nutrition, as well as more security over climate, pest and market hazards. Moreover, it allows them to optimize the application of their most limited resource: the family workforce. By combining the calendars of the different activities, Samlang villagers manage to use from 75% to 100% of their time for economic production, without a big discrepancy between months (Figure 37-5). The corollary of the fine-tuning required to mobilize family labour is that any change imposed on household activities may potentially disrupt what is a delicate balance. If, for instance, changes driven by outsiders (projects, government policies, markets and so on) are not thoroughly designed in conjunction with villagers, they may lead to a waste of workforce by either overloading it or leaving it under-employed.

Shifting cultivation takes up much of the workload of the average worker; between 20% and 80% of the monthly capacity of a worker is dedicated to the swidden (Figure 37-5). Peak work periods are July and August, during which the need to weed the swiddens allows for no breaks whatsoever. Economic activities are developed on the basis of constraints and available labour resources. Families aim to distribute the workload evenly throughout the year and maximize labour productivity. Most workers are employed at 70% to 100% of capacity each month. The household economic objective does not appear to be the maximization of profits or rent, but the maximization of labour productivity – a typical strategy in manual-farming economies (Scott, 1985; Mazoyer and Roudart, 2005). Given the way a family labour force is utilized, it does not represent any financial costs, but, as a scarce and limited resource, labour comes with an opportunity cost. A rationale based on opportunity costs is implicitly what drives families to make choices regarding economic activities and their use of labour resources (Dufumier, 2004).

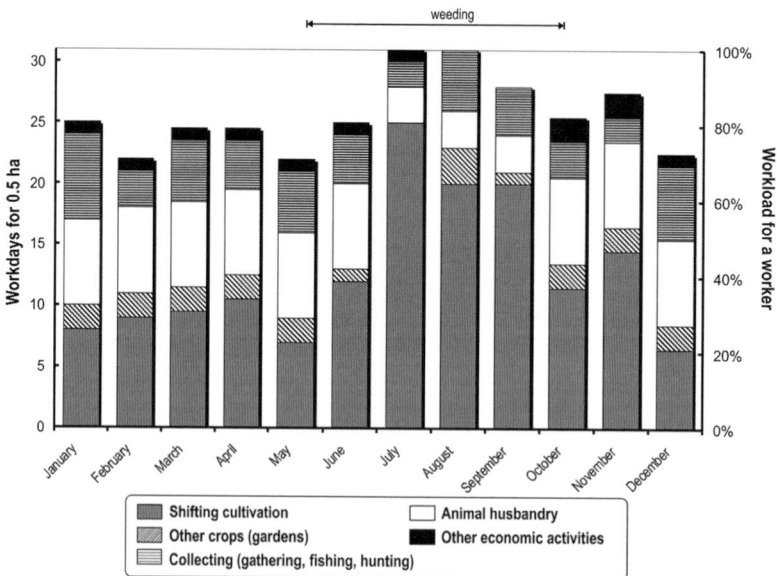

FIGURE 37-5: Workload calendar in Samlang village.

Each household in the village gains access to enough land for cropping (Ducourtieux et al., 2005). Everybody performs almost the same set of activities, with the same set of tools. The range of production per labourer is limited within the village community, and there is only one farming system in Samlang village: shifting cultivation. Half of the households come within ±25% of the average total income per household. This average takes into account self-consumption, computed with the opportunity cost (replacement value) for each product consumed in the household. Despite a differential of one to nine between the lowest and highest total incomes of households in Samlang, the incomes of most of the families are very close, with a differential of less than one to three covering more than 75% of the households. The differences between households come from their capacity to use available family labour over and above that needed to produce the staple food (rice) in shifting cultivation. The surplus labour – if any – is dedicated to forest collecting, which is the main activity creating differentiation between the village's households. Therefore, there is a close correlation between the income of these households and the 'collect' component of their income (Figure 37-6). The economic difference can be exaggerated when a family has access to the only mechanized tools in the village: the automatic firearms of militiamen. A rifle allows them and their relatives to hunt and sell big game (boars, deer and so on). The difference in income can be seen in the case of the household marked (a) in Figure 37-6.

However, the range of income differentiation within the village community is very limited. The Gini index for total household-income distribution in Samlang is as low as 0.24.[3] Economic differentiation is minimal in a farming society such as Samlang. This is due to several factors:

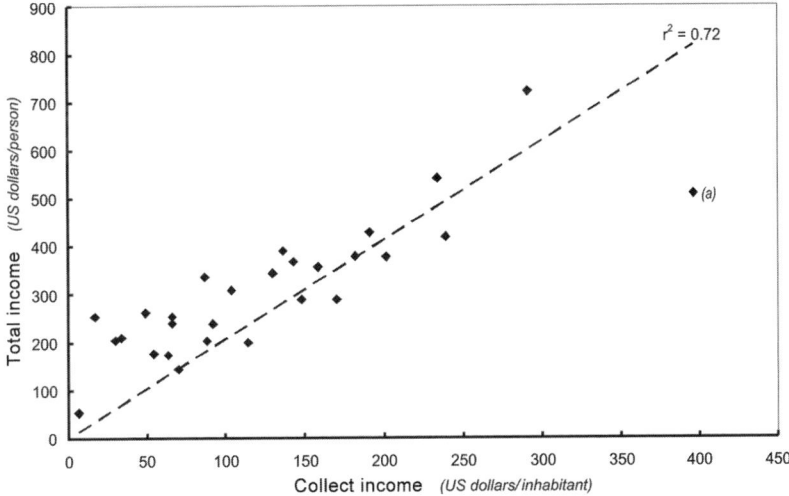

FIGURE 37-6:   Total household income and 'collect' income.

- Land and other natural resources (forests, water) are abundant, and can be accessed equitably (Ducourtieux et al., 2005);
- All households are limited by labour-force constraints;
- All households use the same exclusively manual means of production; and
- Households engage in comparable activities to optimize available labour and secure their livelihoods.

State policies, and unleashed differentiation between village households

The Lao State has managed to become more and more involved in the everyday life of Phongsaly villagers, especially with recent resettlement alongside roads, land reforms, its ban on shifting cultivation and mandatory crop policies. The village of Yapong, 12 kilometres from Samlang in Phongsaly province and previously similar in all respects to Samlang, was used by the author to compare the impact of State policies on Phongsaly villagers. While Samlang's lack of easy access left it mostly outside the policy-implementation process and it was able to continue with its former organization and economy, Yapong received the whole bundle: resettlement in 1996, land allocation in 1999, compulsory cash crops with sugar cane from 1997 to 1998, then tea since 2000; a ban on hunting in 2000; and a ban on shifting cultivation in 2005. Yapong farmers maintained their rice production despite decreasing yields by expanding the area under cultivation within the limits of state land allocation, with two to three successive years of cultivation (compared to one year in Samlang), to the detriment of the fallow duration, which fell to three years instead of 12. With such a short fallow, they were unable to cope with the explosion of weeds. The farmers turned to massive use of herbicides. Production costs rose along with the workload, while yields tumbled. Production per family dropped and rice shortages became the norm in Yapong. Sixty per cent of its families suffered rice shortages for an average of three months every year.

On average, households in villages like Yapong, where the package of reforms has been applied, have lost half of their total income. Earnings from livestock have decreased substantially due to decapitalization: to buy rice, families have sold their animals, including reproductive females. There were 110 head of cattle in Yapong in 1996, but only 85 were left in 2003.

This drop in income does not affect all of the families in a village in the same way. While the highest household income in Samlang is nine times higher than the lowest one, the differential climbs to almost 20 in Yapong, meaning that there is a significant income gap in this community of only 50 families. The Gini index for total household-income distribution in Yapong remains low (0.35), but it has increased by 46% compared to Samlang (Figure 37-7).

Changes induced by state policies have strongly affected the household workload. A drastically shortened fallow period of three years is now the norm in Yapong as a result of land allocation, forest reserves being taken out of the swidden rotation and reduction in the area of fallow land available for swidden cultivation. Rice yields

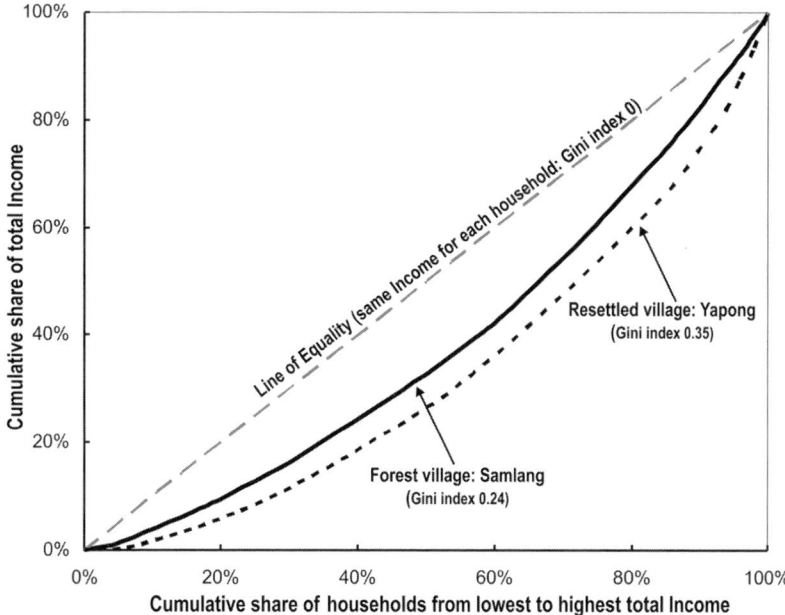

FIGURE 37-7: Lorenz curves of compared income distribution in Samlang and Yapong villages.[4]

have fallen from 1300kg/ha to less than 600kg/ha because of inadequate recovery of fallow fertility and increased weed competition. Covering the basic needs of Yapong households implies the need for more labour in shifting cultivation, to control weed infestation and expand the area cropped. But this workload exceeds the available labour force, so farmers have adopted a number of solutions:

- Reliance on child labour. In Yapong, only 75% of youths under the age of 18 attend schools, compared with 100% in Samlang, despite quicker and easier access to high schools in Phongsaly.
- Pooling of labour and food dependency. Pooling of labour among farm households is less frequent in Yapong than in Samlang, and is limited to relatives (siblings and parents). Another form of labour pooling exists in Yapong, but is based on a relationship of dependency, and is therefore unequal. A family that borrows rice repays through labour, weeding the plot of the lender. Work done outside the household diminishes the amount of time spent on one's own plots, resulting in lower yields; thus begins a chronic cycle of dependency.
- Wage labour. Dependency causes some families to drastically reduce the size of their swiddens in order to work as paid agricultural labourers.

A trend of specialization has emerged in Yapong village. Some households – the poorest – tend to reduce their farm production to devote their time to daily salaried jobs, either in the village or in the provincial town of Phongsaly. At the opposite end of the scale, some households have been able to invest in small businesses and carry out activities on a larger scale than most other households can afford. These include trade (grocery shops), food processing (alcohol distillation, banana trunks) and services (video projection, motorized hulling and taxis). All activities contracted by a third party outside the village are paid in cash. Barter and mutual trade in the village is less systematic than in Samlang.

The difference in average income between Samlang and Yapong is now striking. It is easily explained by the constraints placed on shifting cultivation without provision of sustainable alternatives. However, understanding the increased (and growing) differences in activities and economic results in Yapong, compared to Samlang, is not so clear-cut. Before the trend of reforms initiated in 1996, Yapong households practised a farming system very similar to that in Samlang. There was limited socio-economic differentiation within the Yapong community. Why, then, have some families been able to grab development opportunities brought by road access after resettlement, while many others have suffered from the state's involvement?

A large part of the explanation can be found in research work led in the early years of the 20th century in Russia by the agrarian economist Alexander Chayanov. Let us go back to Samlang village in 2003, when it was not yet affected by the state reforms. The available labour force in a household varies from one year to the next. It is small for young, childless couples who have just set up home, but the labour force gradually increases as children are born and grow. Children start by consuming, and only later contribute their labour. With each additional birth and expanding nutritional needs, food demand grows. Production levels, however, remain the same, since the size of the work force goes unchanged. Once the children become active, agricultural production increases sharply, while consumption remains relatively constant. Over a generation, the household 'consumers/workers' ratio varies. Chayanov described this cycle in the 1920s for families in the Vologda region of northwestern Russia (Chayanov, 1986), practising a farming system involving mostly manual work, and similar in that respect to the system in Samlang. Depending on the number and ages of children, a farming household will either produce a surplus of agricultural products, enabling it to increase sales and capitalize, or will develop a deficit, which will force it to decapitalize to meet its basic food needs. By adjusting some of the assumptions underlying Chayanov's model to give it a basis in the Phongsaly context, it is possible to breakdown the development of a typical household in Samlang into three phases, accumulation, decapitalization and renewed accumulation, as described below. [5]

Accumulation

Rice production potentially exceeds family needs. Rice surplus and by-products can be sold or traded for animals (poultry, pigs), initiating the accumulation process through livestock. Another strategy is to reduce farm labour devoted to shifting agriculture and channel the labour force into other income-generating activities. In

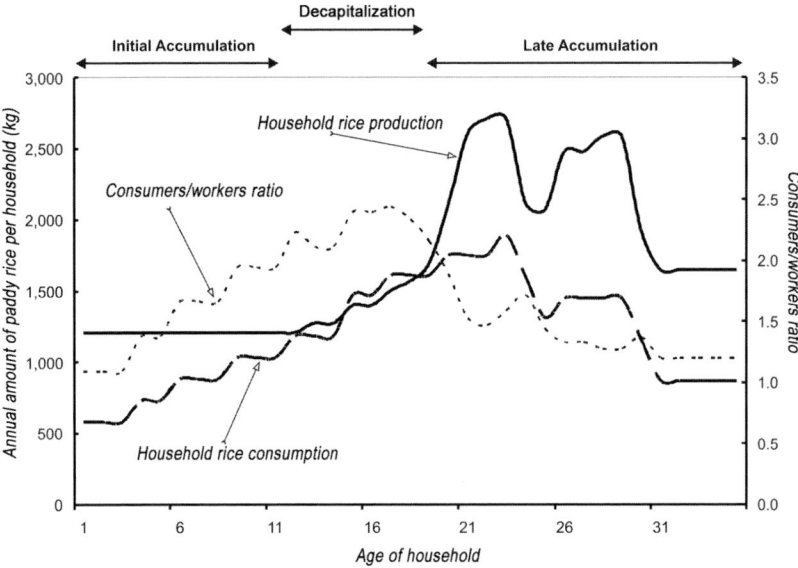

FIGURE 37-8: Successive phases of accumulation and decapitalization of a household in Samlang.

both cases, a 'consumers/workers' ratio of less than 2.3 (Figure 37-8) results in capital accumulation.[6] The capacity to accumulate decreases gradually with the number of children and their age, and cancels out about 13 years after founding the household.

Decapitalization

Rice production falls into deficit. The typical household has two adults and five children, aged 2 to 13. If the 'consumers/workers' ratio is greater than 2.3 (Figure 37-8), agricultural production does not cover the family's basic needs. Food consumption must be reduced, resulting in increased morbidity and the sale of assets such as livestock (pigs and cattle), if the herds are big enough. Although it lasts only five years, this is a precarious time in the life of a household, and will determine its future economic potential. If the family manages to make it through this phase quickly, and without further challenges, it will resume development and will transfer capital to its offspring. However, if the context is unfavourable (poor climate, economic or political upheaval), making it through this phase becomes exceedingly difficult and resources previously accumulated may not be sufficient. The result is impoverishment of both the household and its descendants.

Renewed accumulation

As they grow older, the first children start to contribute to farm work, thereby increasing the cleared area and hence agricultural production. The household can once again start to accumulate (Figure 37-8), in preparation for the older children to settle down and the parents to retire. If the context is favourable during this period, the farm will invest in livestock, which will then enable the parents to give animals to their children, thus quickly initiating the accumulation process for the households of the younger generation.

This sequence of phases, often called the Family Life Cycle (Lee and Kramer 2002; Huaiyin Li, 2005), is a theoretical model (Walker et al., 2002), since in reality, each household has a unique trajectory. In Samlang, the distribution of families according to household age and the 'consumers/workers' ratio is consistent with the model: 14% of families are in the accumulation stage (households under 14 years old); 29% of families are in the precarious phase of decapitalization (11 to 21 years old); and 46% are well established (21 years and over), and once again accumulating assets (Figure 37-9).

The sequence is well known to farmers, who regularly explain it to curious and attentive visitors. Some authors use the 'consumers/workers' ratio as a criterion of economic differentiation for different farming systems in a region (Hammel, 2005; Huaiyin Li, 2005). However, the ratio is constantly changing; the variations are comparable, albeit asynchronous, among families throughout a village. Theoretically, every family goes through the different stages determined by the 'consumers/workers' ratio; it does not constitute a criterion for distinguishing between different types of farm operations. Nevertheless, there is a relationship between the capacity to accumulate over a given period and economic differentiation between families. In a stable environment, each household goes through the phases of accumulation and decapitalization under the same circumstances. But natural hazards are frequent. Moreover, Phongsaly history is marked by frequent changes. These more or less occasional disturbances do not affect households equally; this depends on their assets and ability to accumulate at the time. If a business opportunity presents itself,

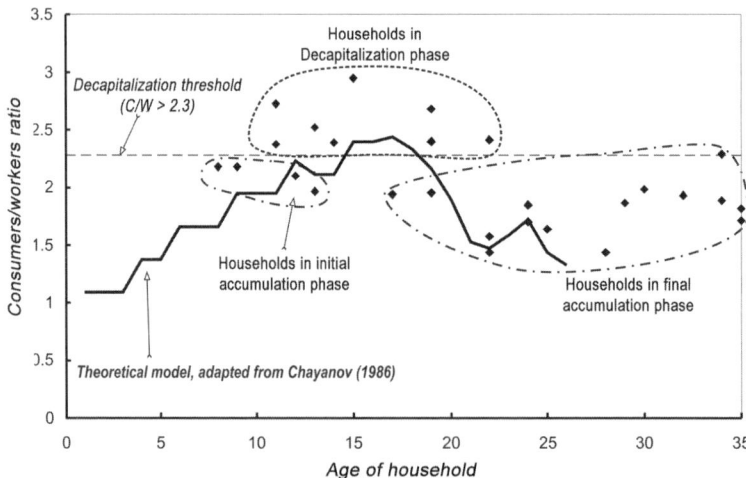

FIGURE 37-9: Evolution of the 'consumers/workers' ratio according to household age in Samlang.

only those households with the available human resources or capital will be able to take advantage of it. Conversely, in a crisis, families with accumulated resources will disinvest in order to wait it out, adapt or migrate. Those with a negative 'consumers/ workers' ratio will sink into trouble. The Chayanov Family Life Cycle is more a driver of socio-economic differentiation than a cause of it.

This explains what happened in Yapong when the State began to implement its reform package between 1996 and 2002. The successive policy measures were 'isotropic'; they were designed to apply to everybody, without any distinction. Nevertheless, those village households in the accumulation phases (with low 'consumers/workers' ratios) were able to sell part of their herds to build spacious and well-equipped houses at resettlement sites along the roadside, and also to invest in market-oriented activities, such as commercial distillation, terraces for irrigated rice, or services (transportation, grocery trade, firewood and timber trade, and so on). These households were able to mobilize their accumulated capital and invest it to catch the opportunities that came with road access. At the same time, those households in the decapitalization phase (with high 'consumers/workers' ratios) had to sell their few animals, if any, to buy rice and cover their basic needs when they were ordered to move. Later, they relied each year on rice credits, arranged under usurious conditions from wealthy households. For a few, the imposition of State policies speeded up their divergence from average status by increasing their income; for most, it meant impoverishment. It initiated a growing process of economic dependence within village households that provided an economic rent to the wealthy among them. The Family Life Cycle does not exist anymore: a few families found a secure escape from the top, while the bulk of the community went down without hope of recovery.

Conclusion

Until recently, the socio-economic differentiation within a Phunoy village community was limited, due to the restricted ability to accumulate capital and the full employment of family labour in shifting cultivation. At any single moment, differences may have appeared to exist, in accordance with the Chayanov concept of Family Life Cycle, but these differences were transitory and each household went through the successive phases of accumulation, decapitalization and renewed accumulation, due to the 'consumers/workers' ratio in the household. Thus, only those social relationships lying within the community power structure explained the limited differentiation within a Phunoy village. Recent state policies to shift from swidden agriculture towards 'modern', market-oriented agriculture have drastically disrupted the former dynamic equilibrium. The political measures crystallized the momentary economic differences from the Family Life Cycle, leading to an increase in the range of income and to the pauperization of the most unprivileged villagers.

The Chayanov Family Life Cycle is a powerful concept with which to appraise the economic differences existing between households involved in manual agriculture (Bernstein, 2009), as well as the processes that families can pursue in order to adapt to an external disturbance, such as a climatic event, a market evolution or a political change. Following the example of Chayanov (1986), some authors analysed the impact of the Soviet revolution on Russian farmers in the 1920s and 1930s (Harrison, 1975; Johnson, 1997; Shanin, 2009). Others extended the concept to historic England (Smith, 2002); to the collectivization and decollectivization in China from the 1950s to the 1980s (Huaiyin Li, 2005); and to the pioneer front in Brazilian Amazonia (Perz and Walker, 2002; Walker et al., 2002). It is also valid for societies based on shifting cultivation, but which, through history, have developed some mechanism to limit the variation of the 'consumers/workers' ratio, notably with kinship organizations (Hammel, 2005).

There may be more in the Chayanov concept than scientific interest. Perhaps it is an effective tool with which to design and appraise state policies. It may allow scientists and administrators to understand the technical, social and economic rationales of shifting cultivators, and to take into account the socio–economic diversity and diverging interests within village communities.

References

Altieri, M. A. (2002) 'Agroecology: The science of natural resource management for poor farmers in marginal environments', *Agriculture, Ecosystems and Environment* 93(1-3), pp1-24

Bernstein, H. (2009) 'V. I. Lenin and A.V. Chayanov: Looking back, looking forward', *Journal of Peasant Studies* 36(1), pp55-81

Boserup, E. (1976) 'Environment, population, and technology in primitive societies', *Population and Development Review* 2(1), pp21-36

Bouté, V. (2007) 'Political hierarchical processes among some highlanders of Laos', in F. Robinne and M. Sadan (eds) *Social Dynamics in the Highlands of Southeast Asia: Reconsidering Political Systems of Highland Burma*, Brill, Leiden, pp187–209

Brookfield, H. C. (2001) *Exploring Agrodiversity*, Columbia University Press, New York

Chayanov, A. (1986) *The Theory of Peasant Economy*, third edition, University of Wisconsin Press, Madison, WI

Conklin, H. C. (1957) *Hanunóo Agriculture: A report on an Integral System of Shifting Cultivation in the Philippines*, Food and Agriculture Organization of the United Nations, Rome

de Rouw, A. (1995) 'The fallow period as a weed-break in shifting cultivation (tropical wet forests)', *Agriculture, Ecosystems and Environment* 54(1-2), pp31-43

Diamond, J. (1987) 'The worst mistake in the history of the human race', *Discover* 8(5), pp64-66

Ducourtieux, O. (2009) *Du riz et des arbres : L'interdiction de l'agriculture d'abattis-brûlis, une constante politique au Laos* (Rice and Trees: The Ban on Slash-and-Burn Agriculture, a Consistent Policy in Laos), Karthala/IRD, Paris

Ducourtieux, O., Laffort, J-R. and Sacklokham, S. (2005) 'Land policy and farming practices in Laos', *Development and Change* 36(3), pp499-526

Dufumier, M. (2004) *Agricultures et paysanneries des Tiers Mondes* (Agriculture and Peasantry of the Third World), Karthala, Paris

Hammel, E.A. (2005) 'Chayanov revisited: A model for the economics of complex kin units', *Proceedings of the National Academy of Sciences* 102(19), pp7043-7046

Harrison, M. (1975) 'Chayanov and the economics of the Russian peasantry', *Journal of Peasant Studies* 2(4), pp389-417

Huaiyin Li (2005) 'Family Life Cycle and peasant income in socialist China: Evidence from Qin village', *Journal of Family History* 30(1), pp121-138

Johnson, R. E. (1997) 'Family Life-Cycles and economic stratification: A case-study in rural Russia', *Journal of Social History* 30(3), pp705-731

Lee, R. D. and Kramer, K. L. (2002) 'Children's economic roles in the Maya Family Life Cycle: Cain, Caldwell, and Chayanov revisited', *Population and Development Review* 28(3), pp475-499

Mazoyer, M. and Roudart, L. (2005) *A History of World Agriculture: from the Neolithic Age to Current Crisis*, Monthly Review Press, New York

Perz, S. G. and Walker, R. T. (2002) 'Household Life Cycles and secondary forest cover among small farm colonists in the Amazon', *World Development* 30(6), pp1009-1027

Ramakrishnan, P. S. (1992) *Shifting Agriculture and Sustainable Development: An Interdisciplinary Study from North-eastern India*, United Nations Educational, Scientific and Cultural Organization, Paris

Scott, J. C. (1985) *Weapons of the Weak: Everyday Forms of Peasant Resistance*, Yale University Press, New Haven, CT

Shanin, T. (2009) 'Chayanov's treble death and tenuous resurrection: An essay about understanding, about roots of plausibility and about rural Russia', *Journal of Peasant Studies* 36(1), pp83-101

Smith, R. M. (ed.) (2002) *Land, Kinship and Life-Cycle*, Cambridge University Press, Cambridge, UK

van Keer, K. (2003) 'On-farm agronomic diagnosis of transitional upland rice swidden cropping systems in northern Thailand', PhD dissertation, Departmnent of Land Management, Faculty of Agricultural and Applied Biological Sciences, Katholieke Universiteit Leuven, Belgium

Walker, R., Perz, S., Caldas, M. and Silva, L. G. T. (2002) 'Land use and land cover change in forest frontiers: The role of household life cycles', *International Regional Science Review* 25(2), pp169-199

Notes

1. The council of elders comprises male villagers who are deemed successful: they have established and maintained a family, keep cordial relationships and participate actively in the community. Membership of the council is informal, and follows an invitation to attend meetings. Generally, the members are heads of household who are more than 50 years old (Bouté, 2007; Ducourtieux, 2009).

2. The elders of Samlang explain that the village of Chaphou had been so badly depopulated that it had lost traditional experience related to events like swidden burning. Chaphou had 59 households in 1980, and only 32 in 2003.

3. The Gini index shows the level of inequality for income distribution within a community. An '0' value means perfect equality, i.e. everybody in the community has the same income; at the opposite end, a '1' value means that one person gathers all of the community income and the others have nothing. For example, the Gini index of 2011 income reads 0.45 for the United States; 0.38 for the Lao PDR; 0.47 for China; 0.31 for Australia; 0.40 for Thailand; 0.63 for South Africa; and 0.30 for the European Union, with an average value of 0.40 for Asia and 0.48 for the world (http://wdi.worldbank.org/table/2.9).

4. The Lorenz curve shows the cumulative income share (vertical) according to the distribution of the population (horizontal axis). If each individual had the same income, or total equality, the income distribution curve would be the straight line in the graph – the line of total equality. Graphically, the Gini index is defined by the area between the Lorenz curve and the line of equality.

5. The following assumptions were used: rice production, 660kg/worker/year (0.5ha/worker); average consumption, 290kg/adult/year and 150kg/child/year (children aged 2 to 12 years). Based on the village survey, it was assumed that there would be five children per household, alternating boys and girls, and that the oldest son would stay on the farm, while the other boys would leave home at age 18 and the girls at 20.

6. Accumulated capital may be reinvested in the farm (livestock, tillage tools, processing equipment, and so on) or into the home (roof, household appliances).

38

POLICY-DRIVEN CHANGES IN LISU SWIDDENING

Social organization as adaptation to a new economy

*Kathleen Gillogly**

Introduction

In the face of fundamental transformations in the political, economic, and environmental conditions of northern Thailand, the people of the upland watersheds have had to change an eco-strategy on which they have depended for more than 150 years. Originally, they were participants in a global market for opium grown by shifting cultivation, but over a relatively short period of 40 years, their lives have been transformed by state policies, including watershed-management laws that alienate them from their land, the enclosure of their land in national parks and reserves, and bans on both swidden farming and opium growing.

The place about which I write in this chapter is a cluster of Lisu villages in the Doi Sam Muen Highland Development Project, which carried out its operations in an area north of Chiang Mai, in Chiang Mai and Mae Hong Son provinces, northern Thailand, in the years from 1987 to 1994 (Figure 38-1). The villages serve as a case study of how policy was put into place, and illustrate the everyday effects of national forestry policy on upland communities. The project was remarkable for its use of a new (at that time) method for bringing about an end to opium cultivation and swiddening. It was called Participatory Land–use Planning. It also followed a specific philosophy of self-contained sustainability. As with other highland development projects, the approach was integrated rural development: introducing new crops, permanent agriculture, marketing strategies, and in general attempting to improve the living conditions of upland peoples.

The People's tactics for adapting to the new economic system were based on changes to their agricultural practices, building relationships with project personnel and other government agents, and undertaking new strategies within their own

* DR KATHLEEN GILLOGLY, Associate Professor, Department of Sociology/Anthropology, University of Wisconsin-Parkside, Kenosha, WI.

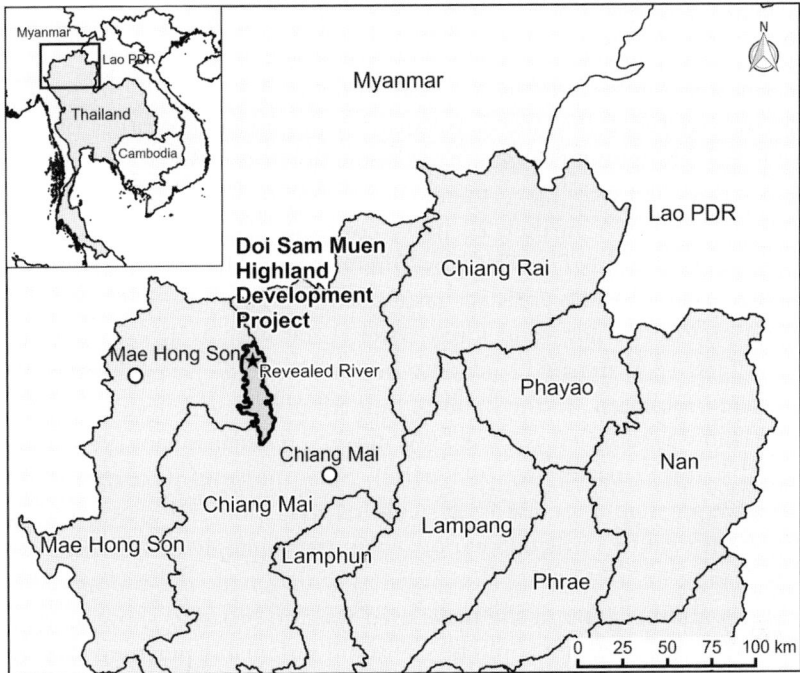

FIGURE 38-1: Northern Thailand, showing the location of the Doi Sam Muen Highland Development Project. Within the project area is the Lisu village that the author called 'Revealed River'.

households by reorganizing the use of labour through shifts in marriage and kinship practices. In short, policies to end swiddening brought about a wide range of shifts in the total socio-economic system, illustrating the intertwining of social organization and economic adaptations. In this case, the changes were policy driven by the Doi Sam Muen Highland Development Project. The extent and meaning of these shifts were often invisible to project personnel, who considered simply the extent to which land was being used for watershed conservation as a result of techniques such as Participatory Land-use Planning.

National policy, upland land use and upland peoples

Swiddening has long been treated by state officials as iconic of everything that is 'wrong' with upland ethnic-minority peoples: peripatetic and thus unreliable, crossing borders and thus a security risk, destroying trees and thus primitive. It is considered a 'primitive' form of agriculture (see Dove, 1986). Swiddening has been criminalized as the cause of forest destruction, the means of opium production, and the facilitator of population mobility, moving swidden farmers beyond control and thus a danger to the state. Thai policy to preserve forests by putting an end to swiddening (and eradicating opium cultivation) has served as a vehicle for expansion of government administration and business (both private and government) into the peripheries of the Thai state, especially since the early 1980s, with the rise of numerous state and bilateral projects in the highlands (Gillogly, 2004).

Of particular significance to this case study is the role of the Royal Forest Department.[1] By the 1990s, this department managed huge swathes of designated watershed land throughout the mountains of the north as national parks and reserves. By 1988, the Royal Forestry Department had claimed 31 wildlife sanctuaries covering 2.47 million hectares. A further 22 parks covering 1.13 million hectares already existed, pending approval. The Thai government's watershed classifications involved in the designation of these areas set restrictions on land according to slope, land form, elevation, soil, geology and forest cover. Land with a 30% slope or over was classified as Class 1A Watershed land, on which settlements and exploitation of forest products were prohibited. In the basin of four tributary rivers in northern Thailand, 27% of the land was classified as Class 1A Watershed (Ganjanaphan, 1987; Wun'Gaeo, 1992; Royal Forest Department, 1993, p182). Such measures were meant to protect the nation's water supply for use in agricultural irrigation and generation of hydro electricity, both vital to Thailand's export economy (Gillogly, 2014). However, little effort was made to correlate the classifications with actual land-use or topographic characteristics, and the scale of the maps used was too broad to show upland valleys and streams where agriculture could be safely and sustainably practised (Weyerhauser, 1993; Phongpaichit and Baker, 1999, p62). As a result, the government was able to declare control over virtually the entirety of the country's northern mountains. This was in keeping with a trend since World War Two of progressive territorialization of rural land through legal mechanisms such as the Forest Act (1941), which declared certain forested lands to be state land regardless of its use by local villagers (Vandergeest and Peluso, 1995, p396). This was a continuation of the process of 'mapping' Siam into a modern nation-state (Winichakul, 1994). As a result of these policies, by the 1990s approximately 10 to 12 million people, equivalent to 25% of the rural population of Thailand, were living on land that was classified as forest reserves, and most upland farmers were classified as illegal settlers (Phongpaichit and Baker, 1999, p63). These upland peoples were all the more vulnerable to harassment given their status as 'immigrants,' or non-citizens of Thailand (see, for example, Lohmann 1999; Toyota, 2007; Network of Indigenous Peoples of Thailand, 2010).

Across the northern uplands, the economic and social consequences of the end of swiddening have been profound. One result has been a widespread decrease in agricultural land available to upland farmers. This is, in essence, a new enclosure (Sturgeon et al., 2013). The land tenure system has been transformed from open to closed access, with consequences for the formation and economic viability of new households. Whereas swidden farming was once limited by available labour, land is now a scarce resource which people must hold close and protect across generations. However, in the case of this cluster of villages, the drastic effects of the end of swiddening were, to some extent, ameliorated by the policies of the Doi Sam Muen Highland Development Project.

The Doi Sam Muen Highland Development Project

The project was one of many programmes that were designed to solve the problems of the highlands through integrated development. It was run by the Watershed Division of the Royal Forest Department with oversight of some programmes by agencies such as the United Nations International Drug Control Program. I conducted research in the cluster of Lisu villages within the project area from 1991 to 1994, and called the cluster 'Thirty Thousand'. I lived in one of the villages, which I named 'Revealed River'. The five separate villages were located in western Chiang Dao district, Chiang Mai province, and had a history of intermarriage, political alliance and cultural exchanges. After a period of time in which the focus was opium replacement, integrated highland development programmes were initiated to fundamentally transform the upland economy. Hoare (chapter 44) discusses the timeline and justifications for this approach. From the perspective of the state, the aim was to reduce poverty and instability by incorporating upland populations into state administration.

The project was considered highly successful, due in large part to the apparent compliance of upland villagers. Other evidence of its success included a decrease in population mobility; a decrease in opium cultivation; a decrease in the area under cultivation; and clear involvement of upland villagers in devising land-use plans for their community and responsibility for maintaining them. Project personnel also pointed proudly to higher numbers of 'resident aliens' holding Thai identity cards, more household registration and higher rates of citizenship among villagers in the project area, although it had been unsuccessful in gaining legal land rights for the people.

The Doi Sam Muen Project was distinctive in that its former head had created a total philosophy that moved the focus away from simple opium replacement, prohibition of swiddening and watershed conservation towards a transformation in the way of life of the people. Specifically, he saw the fundamental problem as being one of producing crops for profit. People opened swiddens to grow opium; opium was a cash crop; and this seeking of profit was an expression of greed. This was a fundamentally Buddhist philosophy rooted in concepts of the 'Middle Path'. From this perspective, replacement of opium with another cash crop only led to economic instability and environmental degradation as farmers expanded their production in order to achieve a base cash income. The project head often told the story of introducing cut flowers as an opium-replacement crop, only to see production expand (with the need to cut more forest) when prices dropped and farmers had to grow more to meet their need for cash. To him, this illustrated the fundamental problem of production for profit. As an alternative, he advocated sustainable self-sufficiency, in which cash crops would be only a small part of the household economy. It was, in a sense, a precursor of the Sufficiency Economy Philosophy now advocated in Thailand (Gillogly, 2014), or at least part of a widespread Thai Buddhist idea that production for markets was negative in that it amplified desires for 'more'; an attachment to the material world. This was also reflective of a romanticized image of rural villages as being insular, harmonious communities. This image was widespread in the culture of the Thai middle classes and among activists in non-governmental organizations (Nartsupha, 1991; Vaddhanaphuti, 2003). Therefore the project head proposed limiting the amount of land any particular household could open to 15 *rai* (2.4ha) and demanding that villagers learn how to live sustainably on this amount of land by planting permanent crops. He also introduced regulations that further controlled villagers' use of trees. Not only was swiddening prohibited, but households also had to ask the project's permission to open any new fields, to cut trees for timber to build houses, collect firewood outside of specific areas, use fallen trees for mushroom production, and so on. While imprisonment was in fact a strong deterrent to opium swiddening (usually the male head of household was jailed, leaving his wife as a de facto widow in the village), the regulation of land use took place on a personal and daily basis. Community organizers were appointed in each village to help develop watershed preservation plans and to encourage people to act appropriately within Thai cultural standards; mountain-village teachers also taught Thai literacy as well as making themselves models of 'appropriate' and respectful behaviour by exercising politeness to superiors[2] and loyalty to the Thai nation. Other project personnel regularly visited each village, haranguing those who broke their rules, and there were regular 'seminars' to educate people in the new agricultural system. While this may sound coercive – and it could be – the project also helped people to obtain expanded identity cards, household registration and even citizenship. Project personnel helped get students into lowland schools, although they were severely critical of young people who returned to the village to resume swiddening. They proudly displayed the mixed farms of a few local farmers, but then threatened people with army violence if they did not obey the rules. The Project set itself up as the guardian of the upland peoples, and in fact attempted to get legal access to land for the villagers. Project officials helped people to obtain household registration, but at the same time withheld their help from others. Therefore, personal relationships were an important part of how the project was implemented, and one of the foundations of these relationships was involvement in Participatory Land-use Planning.

Prunus salicina Lindl. [Rosaceae]

During the opium-crop replacement programme in northern Thailand, Chinese plums became a significant cash crop for the Lisu of Revealed River. Most of the plums sold in Western markets these days are cultivars of this species.

The Project advocated a mixed economy, introducing (first) coffee, then fruit trees and other forms of permanent agriculture. It introduced fish ponds. It tried to protect people from commercial exploitation by vetting all proposals for contract farming, facilitating some and forbidding others. It took over fallow lands and planted these to pine forest. It strongly advocated subsistence farming, and subsistence farming meant growing rice, even though production was very low on permanent mountain fields.

Participatory Land-use Planning

Participatory Land-use Planning (PLP) was, at the time, a procedure unique to the Doi Sam Muen Highland Development Project, and it is worthy of deeper consideration here. Participatory Land-use Planning was developed by Dr Uraivan Tan-Kim-Yong, of Chiang Mai University, and was described as follows:

> [Participatory Land-use Planning is] an operational tool or process which creates conditions of frequent communication and analytical discussions, hence strengthening local organization by generating common understandings and shared rights and responsibilities among project partners who carry out activities that lead to the solving of local forest management problems and other related community problems (Tan–Kim–Yong, 1992).

Uraivan Tan-Kim-Yong was hired in 1986 to study the problem of convincing upland farmers to cooperate with state forest-protection programmes. She redefined the problem as one of different interests (food for farmers, forest for foresters), and recognized that upland peoples had various ways of protecting watersheds that were not recognized by state foresters. Participatory Land-use Planning facilitated communication between foresters and villagers about the specific types of land available in a village's territory as well as existing methods of protection. Using 3D maps built by foresters and villagers working together, they found a vehicle for conversations about what specific areas to protect and how, as well as which lands could be used for farming or other economic activities (Tan-Kim-Yong et al., 1994; Rambaldi and Callosa-Tarr, 2000; Rambaldi, 2010). Out of this arose a common language for land-use planning and personal relationships developed between agents and villagers through frequent communication. Specific land-use plans were developed for each village, with the agreement of villagers as well as foresters. Villagers then took responsibility for enforcing the rules (at least in theory), with the backing of government forestry agents, and foresters could claim that their enforcement of regulations was consistent with plans devised and agreed upon by the villagers. It was an extraordinary technique. Dr Tan-Kim-Yong's goal was to create a 'more level playing field', for equal participation. In this, it was in keeping with a growing trend, both towards democratization within Thailand and participatory development elsewhere in the world. She saw it as a means of communication and conflict resolution, rather than simply as a method for better management of mountain watersheds that would bring an end to swidden farming. Dr Tan-Kim-Yong's use of 3D mapping was a paradigmatic shift, in that it made participatory development both visual and present, and was particularly important in overcoming cultural and language barriers. Participatory 3D mapping made knowledge tangible. It shared knowledge across generations and among people with different worldviews and forms of knowledge (Rambaldi, 2010, pp13-21). While not perfect in its application, PLP certainly provided a greater voice for farmers than any other programme at that time, allowing them to express their views and claim their rights. At the very least, it created clear expectations about land use.

Participatory Land-use Planning became a significant programme and was made the model for watershed conservation and development throughout northern Thailand in 1991, when it was known as RFT (*Raengrat kaan funfuu paa ton nam*), or Accelerated Watershed Development. However, extension of the programme was shaped by the interests of government officials in Bangkok, who sought more 'efficient' (e.g. rapid) application of the policy. The original Participatory Land-use Planning process took more than a year, with a community organizer living in the village, conducting de facto ethnography and building up personal relationships before starting the long process of negotiation to develop a land-use plan. When GIS and satellite photos became more widely available in the mid-1990s, these were used for fine-scale land-use planning rather than consultation with villagers, thereby extending state control over land use in the mountains.

The plan to give mountain minority peoples a stake in the management of mountain lands conflicted with a trend toward centralized state control of forests and watersheds. Therefore, highland development projects such as Doi Sam Muen needed to demonstrate that their policy and practice of keeping upland people in the mountains while introducing new forms of agriculture and participation in the Thai polity was valid to others in the Royal Forest Department. In seeking to demonstrate the effectiveness of what they were doing, project officials needed the help of villagers to show off their farms, involvement in seminars, and so on. This, in itself, gave villagers a bit of power in the system: they were able to define themselves in these settings and officials needed to persuade them to participate, rather than simply lay down the law (Gillogly, 2008). The personnel of the Doi Sam Muen Highland Development Project, therefore, attempted to balance these contradictory national trends of state control and participatory development. And they were successful, but not simply because of a successful development 'technology' (Li, 2007). In Dr Tan-Kim-Yong's opinion, it was the process that was the vital element (Tan-Kim-Yong, 1993).

The dynamics of participation

While the villagers were, indeed, cooperative with the Doi Sam Muen Project, their cooperation cannot be understood simply in terms of a superior development technique (Gillogly, 2008; Gillogly, nd). An example of how cooperation was multi-layered – sometimes a performance, sometimes a strategy and sometimes a shared belief – is illustrated by an interview I conducted at my field village, as part of a team evaluating the effectiveness of Participatory Land-use Planning in highland development projects in northern Thailand under contract with the Projects

Coordinating Office of the United Nations in Thailand.[3] In this interview, three village leaders (heads of their kin groups) sat in one of the few public places in the village, the plaza outside of the Mountain School. Dr Tan-Kim-Yong conducted the interview, but as they answered her questions, they checked reactions from each other and from me. Their answers were pitch-perfect: swiddening and growing opium were bad; commercial gain led to conspicuous consumption and greed; and they saw the value of sustainable agriculture for their future. They were enthusiastic and articulate in their support for the project. But as my colleague looked down to make notes, they whispered to me, 'Is it good, *ajarn* (professor)'? It was not that my presence had altered the interview. They were very experienced in hosting visitors to the project. In fact, I found many people in the village cluster to be vocal supporters of the project in these settings. There was a common intra-village discourse of cooperation in which farmers and village leaders who were in good standing with the project exhorted others to cooperate in the project's plans. As one leader said: 'We must change our ways. We must follow the project's advice. If we don't, we'll lose our land, they won't respect us, they'll think of us as bad people, as uncivilized. We have to progress'. Project officials suspected villagers of acts of resistance, such as setting forest fires, planting opium in project borders, or trading in heroin, but those incidents that did occur were far more prominent in the minds of officials than present in the daily life of villagers. In fact, while villagers might complain about project control, their complaints were often about injustices – taking fallow land for reforestation, disagreeing with officials about appropriate crops to plant, rejection of agricultural contracts, or underpayment for agricultural produce. One particular village was even complicit in a public and highly significant act of cooperation when they 'banned' opium addicts from the village in order to declare it 'drug free'. This meant that a number of elderly parents were ejected from the village, since opium was their main means of managing the pain of illnesses such as arthritis. However, the reward was that Army soldiers were withdrawn from the village. Despite private complaints by villagers, I never saw any direct confrontations between villagers and project officials, even when the officials were aggressively hostile toward villagers about their agricultural practices.

Why, then, was there so much cooperation despite many of the policies causing hardship for families? The villagers were able to do this in part by flexibly defining the project as their ally for survival; they gained access to subsidies for agricultural change, access to education for their children and protection from harassment by the police and land grabs by the wealthy elite. There were instrumental benefits to their cooperation (Gillogly, 2006). The project's emphasis on participatory development gave them access to new micro-level political processes (see Drydyk, 2005; McKinnon, 2007), and this mediated and ameliorated conflict (Steinberg and Clark, 1999). That is, the Lisu villagers and the various project agents evaluated and acted on different standards, but these were at times congruent. By their participation, the Lisu, as former opium swiddeners, validated Thai political culture both within Thailand and on the global stage on which environmental protection was carried out. Project officials

Fragaria x ananassa (Duchesne
ex Weston) Duchesne ex Rozier
[Rosaceae]

Strawberries became an important crop
in the relatively cool climate of upland
northern Thailand following the end of
opium growing. This was despite the need
for labour-intensive harvesting because the
fruit had to ripen on the plants.

needed good outcomes (see Kampe, 1998), and this gave Lisu villagers a certain amount of power (see Tessier et al., 2001). Participation expanded possibilities for political action in an on-going process of engagement (Williams, 2004), so that the Lisu were able to tactically structure some of the effects of land-use planning in their piece of northern Thailand. In addition, the Lisu dealt with the project through their social relationships and exchanges with project agents, who, in turn, leveraged their relationships with individual Lisu villagers to bring about the end results envisioned by the project (see West, 2006). There were inherent contradictions in this: what the Lisu saw as alliances with project officials in an egalitarian sense were seen by the Thai project officials as patronage relationships (older to younger brothers) that required subservience from the Lisu. I heard many complaints on both sides about the kinds of relationships people had; of demands made on villagers by project officials that villagers found burdensome and without sufficient recompense; and of Lisu villagers' unwillingness to accept the authority of project officials. This does not mean that project officials did not have the power of the state behind them. I heard project personnel tell villagers that if they did not cooperate, then the project would be unable to protect them from the army and violence. Nevertheless, all of these actors interacted, negotiated and compromised on a daily basis, with the end result of quite a successful programme.

The actions of the Lisu fitted what de Certeau et al. (1980, p6) referred to as tactics, cunning and self-conscious manipulation of the encompassing system of power and the representatives of that power; opportunities grasped in the moment, makeshift and fragmented (de Certeau et al.,1980, p40). By using tactics such as hosting the 'development picnickers' that visited from all over Thailand, by attending workshops and seminars and participating in national rituals such as Mother's Day, and by developing personal relationships with specific project personnel, the villagers influenced the system to attempt to ensure the economic viability of their households. For the villagers, maintenance of household viability was a fundamental and culturally set goal, particularly because of the severe delimitations on land available to households following the end of swiddening.

Land and social organizations: Adaptations to the new economy

One key result of the ban on swidden farming was that swidden rice was no longer productive. The other was the end of open access to land, on which swiddening had depended. As rational, thinking human beings, the Lisu strategized from their own frame of reference and their own interests. A key feature of their strategies was to acquire and hold on to land for future generations of their households and to provide, at least, a stable income for their households from year to year. In this way, their goals and behaviour differed from the accepted view that participation, in and of itself, would bring about effectiveness and compliance with forestry protection plans.

I never witnessed the ritual sacrifice to the spirits for opening new lands because it was no longer performed. The fallow period was too short to make it necessary. As mentioned earlier, one result of the end of swiddening was a significant decrease in rice production. Only newly opened land, fertilized with the ash of burned vegetation, made rice fields productive (Salzer, 1993, p84; Van Keer et al., 1998, p115). In the past, as rice productivity decreased in a specific field, farmers shifted to opium because its income could be used to buy rice (Durrenberger, 1979). Ironically, the total system of forest protection (ending opium swiddening) made the Lisu ever more dependent on cash crops and the market economy.

While the Project advocated 'sustainable' tree crops such as coffee and fruit trees, farmers experimented with a plethora of cash crops, including taro, strawberries, white potatoes, barley, cabbages and garlic. Some of these, such as barley, were planted under contract. One difficulty was getting the crops to market in decent shape; the other was market competition. Crops faced problems of disease, such as fungal infections in the potato fields as a result of overplanting. In addition, many new cash crops conflicted with the planting schedule for rice. No optimal cropping patterns or cash sources had been developed that were capable of competing with rice, corn and opium (see, for example, Salzer, 1993, p91). Barley was the only cash crop that filled the temporal position left by opium. Crops grown in the off-season required irrigation from March to May. The Lisu did not have much knowledge of irrigation systems, although they certainly worked to learn what they could. However, many lacked the capital to install irrigation; there were risks to installing any permanent irrigation because farmers did not have title to their land and permanent

Hordeum vulgare L.
[Poaceae]

Barley was the only replacement crop grown by upland swidden farmers that fitted the temporal position left by opium. It gained early success in the study villages when grown to supply the brewer of Singha beer.

alteration of the landscape did not fit into watershed planning. They used portable irrigation systems.

Therefore, Lisu farmers seeking ways to keep their household economies viable undertook a variety of economic activities outside of the project's territory, including growing opium across the border in Burma. They found themselves in a newly complex agricultural system that hinged on the management skills of a household leader who could allocate labour across a range of activities (Salzer, 1993). Households sold or pawned their heritage silver (previously used in bride-wealth payments) to buy motor vehicles, allowing them to transport their own goods and to charge others to transport both them and their goods. This was a good economic strategy, but the wear and tear of mountain roads on trucks meant a lot of on-going costs. They also raised cattle in forests not regularly monitored by the project or neighbouring administrative districts.

The new cash crops were far more expensive to produce, and the means of production were not accessible to all because of the labour and capital investments required. The new commercial-farming system seriously widened social and economic disparities among farming households because of their different abilities to invest capital and withstand risk. This is a common socio-economic effect of commercial farming that depends on seeds and technology that have to be purchased (Cuc et al., 1990; Van Keer et al., 1998, p10). Vulnerability increased because of unpredictable production and markets and the need for high investments, and these risks fell on households inequitably. Wealthier households could survive a year of poor production and low market prices because they had their sources to cover the costs of high investments for a year; poorer households could not. Similarly, access to relatively specialized labour, such as driving and maintaining vehicles, knowing how to plant new crops and negotiating in the markets, was not equitably distributed across households. Those households with enough adult or near-adult men had a significant advantage.

By the 1990s, families were using what capital resources they had to 'buy' land in other parts of northern Thailand, bringing about transformations in the spatial patterns of their agriculture and household management of their agricultural economy. Buying land not only allowed households to potentially avoid the project's restrictions on their farming practices, it also gave them land in reserve should they ever lose their right to use land within project-controlled territory. Having other pieces of land diversified the environments and markets available to each household, and it reserved land for future generations.

In short, changes in social organization were a significant means of adapting to the new agricultural economy. Managing their families and households was one of the main domains in which the Lisu had active agency or strategic power to adapt to their new conditions. Open access to land had been a critical feature in Lisu social organization. Migration had been the main means of managing conflict in the village. Open access to land had also affected post-marital settlement patterns: after a period of living with the family of the groom or the bride, a young couple could choose to establish a new household with any relatives (Figure 38-2). Any new household was viable if they had the labour to open land. Thus, settlement, kinship and household developmental cycles were affected by the end of the opium economy.

These shifts were gendered. Weeding requirements increase as fallow time decreases. This is what happens when swiddening is banned, if swiddening is regarded as a long-fallow form of agriculture. The labour of weeding is women's labour. In the post-swiddening economy, the way to increase household productivity was to increase the weeding of existing fields, rather than opening up new fields (Roder, 1997, p1013; Perz and Walker, 2002). Since the work of weeding was women's work, the women bore the brunt of the household's daily work in the fields (Figure 38-3). The irony was that as women's work in fields increased, the value of their work in the household economy fell. In the swiddening economy, the scarce resource is labour. The end of open access to land meant that land was at a higher premium than labour. This had consequences for marriage and kinship. In the swiddening economy, the bride's and groom's families competed to have the young couple live with them for a period of time after marriage. In the post-swiddening economy, young women were told that they had to find a young man whose family had land and to marry out, leaving the limited family land for her brothers. With the enclosure of land, new households could acquire land only through inheritance. Among the Lisu, inheritance is patrilineal, from father to son, so young men must stay with their father and be obedient if they hope to inherit land in the future. While patrilineal kinship groups were dispersed in the swiddening economy – given the ready access of land and management of conflict by leaving – patrilineages have now become dominant in social life. Finally, senior household managers need more and

FIGURE 38-2: A Lisu grandmother and baby. When young couples live with one of their parents, grandparents are a significant source of child care while the young mother works in the fields.

Photo: Malcolm Cairns.

different kinds of labour (wage labour, marketing, driving trucks, herding cattle, and so on), and this is male labour, not female labour. So they put off arranging marriages for their sons year after year. Villagers often expressed concern about there being so many unmarried girls. In fact, age at first marriage for girls has increased from 16 to 25 years (Gillogly, 2006).

Conclusions

The Doi Sam Muen Highland Development Project was successful, if measured by reduction in swiddening and opium cultivation. This success can be attributed to participatory development, particularly Participatory Land-use Planning. But there was more behind this success. The Lisu villagers made use of their relationships with personnel and other resources

FIGURE 38-3: 'Women bore the brunt of the household's daily work in the fields'. A Lisu woman harvesting swidden rice near Huai Thung Choa, north of Chiang Mai. While upland rice was the dominant crop, there were many other crops in the same field.

Photo: Jack D. Ives (1979).

available to them to adapt to their new world. At the same time, they strategized to expand land holdings outside of project territory and diversified their economic activities. Furthermore, they changed marriage practices to manage labour and land allocation, resulting in a shift in their kinship system. That is, adaptation to the new socio-economic policies resulted in changes across Lisu social life. The project created avenues for tactics and strategies in adapting to the radically new economic world in which these Lisu villagers found themselves. But we should not assume that their participation was without its own contradictions and complexities. Development is a type of social relationship, and social relationships are complex and multi-layered, with a range of meanings attached to behaviour. Swiddening was not just a relationship to land or an economic practice; it was, in its way, the basis for a set of social relationships, and these have changed with its demise.

References

Cuc, L.T., Gillogly, K. and Rambo, A.T. (1990) *Agroecosystems of the Midlands of Northern Vietnam: A Report on a Preliminary Human Ecology Field Study of the Three Districts in Vinh Phu Province*, Volume 12, Southeast Asian Universities' Agroecosystem Network (SUAN); the Center for Natural Resources Management and Environment (CRES), Hanoi University; and the Environment and Policy Institute (EAPI), East-West Center, Honolulu

de Certeau, M., Jameson, F. and Lovitt, C. (1980) 'On the oppositional practices of everyday life', *Social Text* 3(Autumn), pp3-43

Dove, M. R. (1986) 'The practical reason of weeds in Indonesia: Peasant vs. state views of *Imperata* and *Chromolaena*', *Human Ecology* 14(2), pp163-190

Drydyk, J. (2005) 'When is development more democratic?', *Journal of Human Development* 6(2), pp247-267

Durrenberger, E. P. (1979) 'Rice production in a Lisu village', *Journal of Southeast Asian Studies* 10(1), pp139-145

Ganjanaphan, A. (1987) 'Land tenure in Thailand', in *Thai Studies in SEAN: State of the Art*, International Thai Studies Conference, Thai Khadi Institute, Thammasart University, Bangkok

Gillogly, K. (2004) 'Developing the "hill tribes" of northern Thailand', in C. Duncan (ed.) *Civilizing the Margins: Southeast Asian Government Policies for the Development of Minorities*, Southeast Asian Studies Program, Cornell University Press, Ithaca, NY

Gillogly, K. (2006) *Transformations of Lisu Social Structure under Opium Control and Watershed Conservation in Northern Thailand*, University of Michigan, Ann Arbor, MI

Gillogly, K. (2008) 'Opium, power, people: Anthropological understandings of an opium interdiction project in Thailand', *Contemporary Drug Problems* 35 (Winter), pp679-715

Gillogly, K. (2014) 'Environmental sustainability policy in Thailand: Global systems, Thai localism', in J. Li (ed.) *Environment and Development in Asia,* volume 4 of *Globalization, Development and Security in Asia*, World Scientific Publishing, Singapore, pp159-179

Gillogly, K. (nd) 'Strategies and tactics in a participatory development project', unpublished paper

Kampe, K. (1998) 'The culture of development in developing indigenous peoples', in D. McCaskill and K. Kampe (eds) *Development or Domestication?*, Silkworm Books, Chiang Mai

Li, T. M. (2007) *The Will to Improve: Governmentality, Development, and the Practice of Politics*, Duke University Press, Durham, NC

Lohmann, L. (1999) 'Forest cleansing: Racial oppression on scientific nature conservation', in *The Corner House Briefing* Paper No. 13, The Corner House, Dorset, UK

McKinnon, K. (2007) 'Post-development, professionalism, and the politics of participation', *Annals of the Association of American Geographers* 97(4), pp772-785

Nartsupha, C. (1991) 'The 'community culture' school of thought', in M. Chitakasem and A. Turton (eds) *Thai Constructions of Knowledge*, School of Oriental and African Studies, University of London and White Lotus Press, London and Bangkok, pp118-141

Network of Indigenous Peoples of Thailand (2010) *Indigenous Peoples of Thailand*, Focus 62, Asia-Pacific Human Rights Information Center, Osaka, Japan, http://www.hurights.or.jp/archives/focus/section2/2010/12/indigenous-peoples-of-thailand.html, accessed 14 December 2015

Perz, S. G. and Walker, R. T. (2002) 'Household lifecycles and secondary forest cover among small farm colonists in the Amazon', *World Development* 30(6), pp1009-1027

Phongpaichit, P. and Baker, C. (1999) *Thailand: Economy and Politics,* Oxford University Press, Kuala Lumpur

Rambaldi, G. (2010) *Participatory Three-dimensional Modelling: Guiding Principles and Applications*, ACP-EU Technical Centre for Agricultural and Rural Cooperation (CTA), Wageningen, the Netherlands

Rambaldi, G. and Callosa-Tarr, J. (2000) *Manual on Participatory 3-Dimensional Modeling for Natural Resource Management*, Volume 17, National Integrated Protected Areas Programme (NIPAP) and Protected Areas and Wildlife Bureau (Biodiversity Management Bureau), Department of Environment and Natural Resources, Quezon City, Philippines

Roder, W. (1997) 'Slash-and-burn rice systems in transition: Challenges for agricultural development in the hills of northern Laos', *Mountain Research and Development* 17(1), pp1-10

Royal Forest Department (1993) *Thai Forestry Sector Master Plan*, volume 5, Ministry of Agriculture and Cooperatives, Bangkok

Salzer, W. (1993) *Eco-economic Assessment of Agricultural Extension in Recommendations for Shifting Cultivators in Northern Thailand*, University of Hohenheim, Stuttgart, Germany

Steinberg, P. E. and Clark, G. E. (1999) 'Troubled water? Acquiescence, conflict, and the politics of place in watershed management', *Political Geography* 18(4), pp477-508

Sturgeon, J. C., Menzies, N. K., Fujita, Y., Lagerqvist, Y., Thomas, D., Ekasingh, B., Lebel, L., Phanvilay, K. and Thongmanivong, S. (2013) 'Enclosing ethnic minorities and forests in the Golden Economic Quadrangle', *Development and Change* 44(1), pp53-79

Tan-Kim-Yong, U. (1992) *Participatory Land-use Planning for Natural Resource Management in Thailand*, Rural Development Forestry Network, Paper 14b (Winter) Overseas Development Institute, London

Tan-Kim-Yong, U. (1993) personal communication between the author and Professor Uraivan Tan-Kim-Yong, Faculty of Social Sciences, Chiang Mai University

Tan-Kim-Yong, U., Limchoowong, S. and Gillogly, K. (1994) *Participatory Land Use Planning: A Method of Implementing Natural Resource Management*, The United Nations International Drug Control Programme (UNDCP) in cooperation with the Office of the Narcotics Control Board, Chiang Mai, Thailand

Tessier, O., Van Keer, K., Nguyen, H. H., Ho, L. G., Quaghebeur, K., Masschelein, J. and Wildemeersch, D. (2001) 'Giving or imposing the opportunity to participate? Reconsidering the meaning of success and failure of participatory approaches', paper presented at a Conference on Participatory Technology Development and Local Knowledge for Sustainable Land Use in Southeast Asia, 6-7 June 2001, Chiang Mai

Toyota, M. (2007) 'Ambivalent categories: Hill tribes and illegal migrants in Thailand', in P. K. Rajaram and C. Grundy-Warr (eds) *Borderscapes: Hidden Geographies and Politics at Territory's Edge*, University of Minnesota Press, Minneapolis:, MN, pp91-116

Vaddhanaphuti, C. (2003) *Discourses on Thai village Community in the Context of Southeast Asian Modernization*, Volume 146, Economic Research Center, Nagoya University, Japan

Van Keer, K., Comtois, J. D., Turkelboom, F. and Ongpraseart, S. (1998) *Options for Soil and Farmer Friendly Agriculture in the Highlands of Northern Thailand*, Tropical Ecology Support Program, F-I/4, Soil Fertility Conservation Project, Deutsche Gesellschaft für Technische Zusammenarbeit (GTZ), Eschborn, Germany

Vandergeest, P. and Peluso, N. L. (1995) 'Territorialization and state power in Thailand', *Theory and Society* 24(3), pp385-426

West, P. (2006) *Conservation is Our Government Now: The Politics of Ecology in Papua New Guinea*, Duke University Press, Durham, NC

Weyerhauser, H. (1993) personal communication between the author and Horst Weyerhauser, who was then a German volunteer with the Doi Sam Muen Highland Development Project

Williams, G. (2004) 'Evaluating participatory development: Tyranny, power and (re)politicisation', *Third World Quarterly* 25(3), pp557-578

Winichakul, T. (1994) *Siam Mapped: A History of the Geo-body of a Nation*, University of Hawaii Press, Honolulu

Wun'Gaeo, S. (1992) *Ecological Movements and Challenges to Sustainable Development in Thailand*, Center for Social Development Studies, Chulalongkorn University, Bangkok

Notes

1. At the time of the Doi Sam Muen Highland Development Project, the Royal Forest Department was part of the Ministry of Agriculture and Cooperatives. In 2002, it was placed under the new Ministry of Natural Resources and Environment. This chapter refers to the period before the establishment of the MNRE.
2. This made no sense to the profoundly egalitarian Lisu.
3. The head of that office was Dr Gary Suwannarat, author of chapter 30.

39

FROM A COMPLEX TO DEGRADED SYSTEM

Laws, customs, market forces and legal pluralism in the Cordillera, northern Philippines

*June Prill-Brett**

Introduction

This chapter focuses on the alarming rate of forest destruction in the Cordillera region of Luzon, in northern Philippines, involving two National Parks and protected areas proclaimed by the Philippine government. It argues that much of the degradation and unsustainable management affecting the National Parks has been brought about by a multiplicity of laws, with their regulatory and executive orders, policies and implementing rules creating competition among state agencies in the protection, conservation and management of resources. This includes the introduction of development programmes and the promotion and protection of the rights of indigenous communities to their ancestral lands and domains. This situation has brought about unintended consequences.

This chapter also looks at the increasing shift among traditional farming communities to commercial vegetable production; a transition that has changed traditional agricultural-resource management. The traditional agricultural practices of inhabitants of the Parks' lower slopes were predominantly shifting cultivation, and this led to the development of a sustainable soil and forest-management system. However, the introduction of market demand for temperate vegetable production opened up an opportunity structure that encouraged farmers to shift from subsistence agriculture to cash-oriented production for the market.

Most current literature, including Robin Broad and John Cavanagh's work *Plundering Paradise: The Struggle for the Environment in the Philippines* (1993), has described how peasant activists, who are considered the main victims of resource degradation, are joining forces to oppose the loggers, mining corporations, industrialists and aquacultural developers who threaten their lands and livelihoods. According to

* PROFESSOR JUNE PRILL-BRETT, Professor Emeritus, College of Social Sciences, University of the Philippines Baguio, Baguio City, the Philippines.

Broad and Cavanagh, these managers of sustainable forest and agriculture, as grass-roots environmentalists, are reclaiming their country from the wealthy landlords and capitalists who have so ruthlessly exploited it. This exploitation and degradation of the environment by powerful 'outsiders' is a general situation that has been documented by a number of studies. However, this scenario does not apply to the case of the Cordillera (see also Jefremovas, 2001), particularly with regard to the two proclaimed National Parks – Mt. Data and Mt. Pulag (Figure 39-1).

In the Cordillera, specifically on Mt. Data and Mt. Pulag, blame for destruction of forest, conversion of land to commercial vegetable farms and illegal logging of pine stands cannot be levelled at large extractive industries such as mining or logging, industrialists or land developers. Rather, these activities have been carried out mainly by people from established indigenous communities inhabiting the lower slopes and valleys surrounding the mountains. Mt. Data and Mt. Pulag have historically been considered as sacred by the indigenous communities, who regard them as mysterious and awesome and perceive them to be inhabited and 'owned' by supernatural beings. Until recently, the mountains were never claimed to be 'owned' by surrounding communities. But now, several municipalities, organized ancestral-land claimants and communities are claiming the two National Parks as part of their territories. This is what Fitzpatrick (2006) described as the tragedy of contested open access. However, I argue elsewhere that this is largely a consequence of plural legal orders and the failure of state management over natural resources (see Prill-Brett, 2007, 2013).

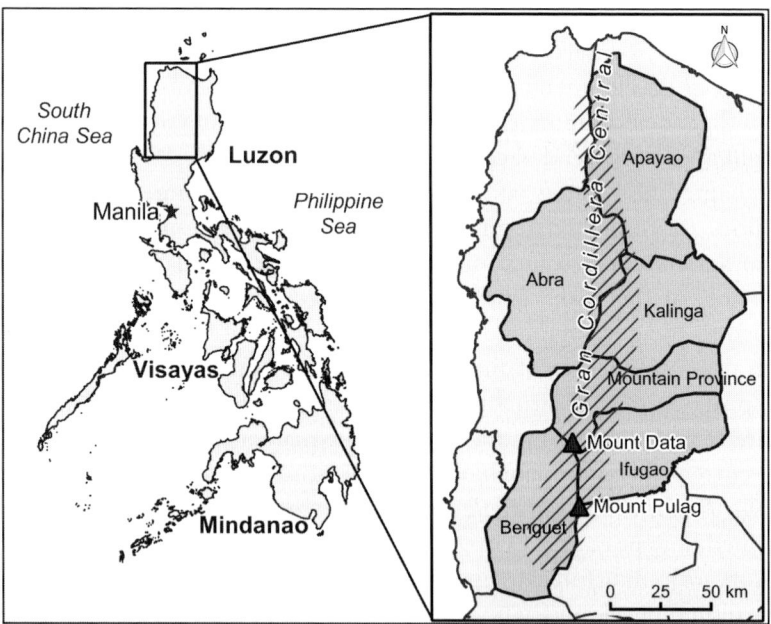

FIGURE 39-1:   The Cordillera Administration Region, Luzon, northern Philippines, showing the location of Mt. Data and Mt. Pulag National Parks.

In order to understand the dynamic role of plural legal orders employed by contesting actors in the Mt. Data and Mt. Pulag arena of resource competition, this chapter applies the concept of legal pluralism, or plurality of legal orders (von Benda-Beckmann, 1983; Griffiths, 1986; Merry, 1988). Legal pluralism is the coexistence and interaction of multiple legal orders within a single social setting or domain. These different legal orders may possibly overlap (e.g. national/state laws and customary, religious, international or project laws, and so on). In situations of legal pluralism, individuals can select and make use of more than one law to rationalize and legitimize their decisions or behaviour. The co-existence of several laws in one domain does not mean that all laws are equal or equally powerful (Meinzen-Dick and Pradhan, 2002). In cases of conflict involving disputes and negotiations, interest groups find it politically or economically advantageous to employ discrepancies and ambiguity between two or more competing jural systems. Von Benda-Beckmann (1983, p5) refers to this as a 'jural jungle'. This situation has been used by claimants in the Mt. Pulag (see Prill-Brett, 2013) and Mt. Data areas. Some interest groups may misrepresent indigenous law to claim property rights and to gain state titles.

The story unfolds with the establishment of laws and regulatory orders by the government to protect the natural scenery and biodiversity of the parks, and later, overlapping laws to promote the development of livelihood activities, such as the introduction of cash crops as an alternative to subsistence crop production. This was followed by the Indigenous Peoples' Rights Act, which gave indigenous people the right to claim title to their ancestral lands and domains.

The two Parks will be described separately, since there are some differences in the evolution of Mt. Data and Mt. Pulag as protected landscapes, and how different agencies with their respective mandates have affected the management of the Parks.

The Mt. Data National Park

The Mt. Data plateau has an estimated area of 3515ha. Half of it is classified as marshy wetlands. Kowal (1966, p396) described the Park as an irregularly shaped, gently rolling 'plateau' about 4km square, with an average elevation of 2100 to 2300 metres above sea level. The earliest mention of the Mt. Data area is found in the records of early German explorers of the Cordillera from 1860 to 1890 (Semper et al., 1975). The area was then uninhabited and described as virgin scrub oak forest and marsh land on the plateau, with pine forest at lower elevations (Delson, 1989). The Kankanaeys, whose communities occupied the lower valleys surrounding the mountain, viewed the plateau as the abode of supernatural beings. The livelihood of the Kankanaeys was based on swidden farming in the pine forests and grassland regions on the slopes of Mt. Data (Kowal, 1966, p400). In 1952, Paktil village, which was located above the Chico River below Mt. Data, moved up from the valley on to the Plateau.

During the American colonial period, government officials saw the need to conserve the natural environment. This led to the issuance of Proclamation no. 217 in February 1929 by Governor-General Henry L. Stimson, establishing the

Central Cordillera Forest Reserve. This covered a wide tract of mountainous land in the provinces of Abra, Ilocos Norte, Pangasinan and Nueva Vizscaya, including the whole Cordillera region. However, this law was not only unrealistic, but it was also not strictly enforced, as people continued to practise their traditional community agriculture and control over resources.

In June 1936, Proclamation no. 65 was issued by then President Manuel L. Quezon, segregating 2398ha of forest land from the Cordillera Forest Reserve as a National Park and a game refuge. This was later amended in October 1940 by Proclamation no. 634, signed by President Quezon, which changed the name of the park to Mt. Data National Park. The area was expanded to 5512ha to include a 200-metre-wide strip of land along both sides of Halsema Highway, from Tublay in Benguet province to Namatec, Sabangan, Mountain Province. The purpose

Pinus insularis Endl. [Pinaceae] a synonym of *Pinus kesiya* var. *langbianensis* (A. Chev.) Gaussen ex Bui

In the words of early explorers, Mt. Data was covered by scrub oak forest and marshland on the plateau, with pine forests at lower altitudes. This was one of the predominant pine species.

was to leave the scenic view unobstructed on both sides of the road, and to conserve the wetlands of Mt. Data as a game refuge. In June 1947, an earlier Act (no. 3915) was amended by Republic Act 122, passed by the Philippine Congress. This law provided for the establishment of national parks, game refuges and bird sanctuaries. However, instead of strictly enforcing the laws that applied to conservation of national parks and wildlife sanctuaries, the government introduced development programmes that encouraged people to farm the wetlands of Mt. Data. These programmes were given higher priority than conservation. In 1950, Executive Order no. 180 directed the Bureau of Lands, Forestry and Soils and the Mountain Province Development Authority to grant Igorot ethnic groups the right to acquire titles to lands they had occupied and cultivated since July 1945. This law encouraged the conversion of swidden farms to permanent cultivation of temperate vegetables. In 1964, Republic Act 782 was amended to become Republic Act 3872, also referred to as the Manahan Amendment. It provided for automatic acquisition of private individual title by indigenous people who had for 30 years or more occupied lands of the public domain suitable for agriculture. The Act was a response to a growing Chinese monopoly of the vegetable industry in Benguet, along the Halsema Highway (Delson, 1989, p7). This Executive Order directed the eviction of alien farmers from the area and the prosecution of Filipino proxies who made claims on their behalf.

Another Executive Order (no. 87), issued at a later date, provided some leeway for farmers to develop some areas of the Mt. Data National Park into farm lots. However, while the intention of the law was good, it was ambiguous, and as a consequence it opened up opportunities for abuses by the farmers themselves (Delson, 1989, p8). It virtually gave licence to people to enter the national park, encouraged most especially by corrupt government officials (Delson, 1989).

After the proclamation of Martial Law by President Ferdinand Marcos, Presidential Decree no. 410 declared that ancestral lands occupied and cultivated by national cultural minorities were alienable and disposable. This decree explicitly provided that 'areas reserved for other public or quasi-public purposes shall not be subject to disposition'. However, this was interpreted differently by unscrupulous people who proceeded to organize associations for the exploitation of the wetlands on the Mt. Data National Park plateau as ancestral land. At the time that Delson (1989) was conducting his fieldwork, a group called the Cordillera Ancestral Lands Manifest, Inc. was claiming portions of the Mt. Data National Park wetlands as ancestral lands. In May 1975, Presidential Decree no. 705, otherwise known as the Forestry Reform Code, was issued. This decree reiterated the position of the government on national parks. In particular, paragraphs 5 and 9 of Section 16 of this decree provided the following descriptions of areas considered inalienable and non-disposable (emphasis added):

> 5) *Ridge tops and plateaus regardless of size found within, or surrounded wholly or partly by forest lands where headwaters emanate;*
>
> 9) Areas needed for other purposes, such as *national parks*, national historical sites, *game refuges, bird sanctuaries*, national shrines…

These provisions of Presidential Decree no. 705 clearly apply to Mt. Data National Park. Its territory cannot be considered alienable and disposable because it is the source of four major rivers, it is a national park, a game refuge and a wildlife sanctuary. However, Section 15 of the same Presidential Decree is inconsistent with the above provisions as it applies to the unique situation of the Mount Data National Park and its peripheral region. Section 15 provides that '… no land of the public domain eighteen per cent (18%) in slope or over shall be classified as alienable and disposable…' There are hardly any areas where slopes are less than 18% in the Cordillera. However, interest groups have interpreted the 18% slope provision to mean that since the plateau is less than 18% in slope, it is therefore open to exploitation. Commercial vegetable farms have sprung up on the plateau, and their expansion has been due to weak enforcement of the law by forest guards, who have allowed people to open up farms for a small fee (Delson, 1989, p11). This inconsistency in the law created open access to farmers, who came mostly from communities in the municipalities of Bauco, in Mountain Province and Buguias and Mankayan, in Benguet province.

The promotion of government programmes seeking to produce temperate vegetables for lowland markets has contributed largely to the death of the Mt. Data plateau wetlands and the oak forest. These vegetables command high market prices and are so attractive that the wetlands and mossy forest on the Mt. Data plateau were a magnet to farmers who moved in to drain the marshlands and use bulldozers to clear-cut the mossy forest for gardens (Figure 39-2).[1] To make matters worse, the government inadvertently encouraged vegetable farming on the plateau by allowing the Philippines-German Seed Potato project to build a potato seed storage facility there. The expansion of vegetable farms into this area dealt a fatal blow to the bird sanctuaries and the game refuge. Another programme encouraged by the government was the raising of shiitake mushrooms (*Lentinula edodes*) that made use of the cut trunks of *kilog* tree (*Lithocarpus luzoniensis*) as a growing medium.[2]

Inconsistencies and defects in the law and non-enforcement of different laws aimed at protecting the park, along with government development programmes, delivered the final blows in the destruction of the Mt. Data wetlands and the mossy forest. Among other contributing factors, the people generally had a lukewarm attitude towards conservation. Swidden farmers readily converted their farms into commercial vegetable gardens because they saw no economic future in shifting cultivation. The prospect of immediate cash payments prompted farmers to choose land-use conversion over conservation of the oak forest and the wetlands. People interviewed by Delson (1989, p16), and by Prill-Brett and Salinas (1991), manifested their preference for cash and gave only half-hearted attention to conservation.

Customary laws and indigenous-knowledge systems governing the sustainable management of resources did not develop in the Mt. Data area because there were no original communities inhabiting the plateau. Thus, the development of property- and resource-management practices among original communities – such as swidden farming, with its ecologically sustainable management – was not applicable (see Figure 39-2, which shows the conversion of mossy forest into vegetable gardens on steep slopes). When people encroached on the Mt. Data forest plateau it was perceived as an open-access movement. The type of market-oriented agricultural production and the technology they used (bulldozers, backhoes and chainsaws), along with

FIGURE 39-2: The conversion of mossy forest into commercial vegetable farms.

Photo: Harley Palangchao (2014).

heavy inputs of chemical fertilizers, were not practices to be expected of subsistence-farming communities (Figure 39-3). Moreover, their source-management practices were equally foreign to indigenous groups.

The proliferation of laws aimed at conserving and protecting the National Park and the simultaneous introduction of development programmes to uplift the lives of community members by exploiting the National

FIGURE 39-3: Part of Mt. Data plateau marshlands, drained and converted into vegetable farms.

Photo: Harley Palangchao (2014).

Park, led to Mt. Data becoming a contested open access landscape. The failure to exclude encroachers was a consequence of the government's inability to protect the forest through weak law enforcement and problems brought about by conflicting legal orders.

The Mt. Pulag National Park

Mt. Pulag is the second highest mountain in the Philippines, reaching an altitude of 2922 metres above sea level. The surrounding National Park covers 11,500 hectares and lies along the Gran Cordillera Central mountain ranges including parts of Benguet, Ifugao and Nueva Vizscaya provinces. It is also a watershed reserve for three large dams.

The summit of Mt. Pulag is covered with short grass and dwarf bamboo plants. Below the summit are small lakes and a cloud forest that supports about 528 species of orchids, ferns, lichens and moss, while pine dominates the lower mossy forest area. Several streams flow from the mossy forest to support multiple rivers on all sides of the mountain that are critical to the water and energy supplies and agriculture of northern Luzon. According to National Integrated Protected Areas Programme assessments, Mt. Pulag is home to 42% of the Philippines' endemic species, including rare, threatened and endangered birds and mammals.

Mt. Pulag National Park is threatened by agricultural activities made possible by technologies such as backhoes, bulldozers and chainsaws to clear areas of forest. This is exacerbated by demands for the building of roads by people living on the slopes of the mountain. Indigenous commercial vegetable farmers see the protection of Mt. Pulag National Park as a threat to the expansion of their farming, while the Department of Environment and Natural Resources sees farming as a threat to the Park, because involved agencies have not yet agreed on guidelines for the Park's protection (Pinel, 2007, p8).

The area covered by the Mt. Pulag National Park is inhabited by the Kalanguya, Ibaloy, Karao and Kankanaey ethnic groups, who share the natural resources found within the Mt. Pulag forest. According to informants, there was no concept of boundaries or exclusionary practices in the area before the introduction of state laws.

Resources are not natural phenomena, but are aspects of the social construction of reality and are socially defined in terms of human values and needs; they have no fixed or permanent definitions (Rambo and Hamilton, 1990, p5). It is therefore important to understand the social factors and processes that influence the ways in which resources are perceived, used and managed (Prill-Brett, 1992, p2). Before the introduction of Western property systems and laws on the management of the forest, the Kalanguya and Ibaloy who inhabit the municipality of Kabayan shared the Mt. Pulag forest and its resources as common property (Prill-Brett, 2013). The people viewed the whole area as open access, in the sense that no one claimed exclusive ownership of resources. There was no concept of boundaries that excluded others because resources were plentiful, such as land for shifting cultivation and wildlife to be hunted. The Kalanguya, who practise swidden farming, inhabit the lower slopes of the mountain, while the Ibaloys of Kabayan, who settled at the foot of Mt. Pulag and the Agno river valley, developed the concept of private ownership of irrigated rice fields (Wiber, 1993). Mt. Pulag was sacred to the Ibaloys of Kabayan, who believed that the spirits of their ancestors dwelt on the heights of the mountain.

The following section looks at state laws that have been introduced by government agencies for the protection and management of the Mt. Pulag National Park.

The National Integrated Protected Areas System (NIPAS) programme

The National Integrated Protected Areas System (NIPAS) programme (Republic Act 7586 of 1992) recognized the right of indigenous cultural communities to ancestral domains within protected areas. It required that livelihood activities within protected areas remain at a subsistence level, based on the assumption that swidden practices governed by indigenous knowledge systems were a model of best practices for sustainable forest management.

The law provided for the establishment of a Protected Area Management Board for each protected area. The 1993 implementing rules authorized these boards to administer protected areas, to demarcate boundaries, buffer zones and ancestral domains, to recognize the rights of indigenous communities and decide matters relating to planning and resource protection in accordance with their general management plans. The boards were also authorized to promulgate rules and regulations, develop biodiversity conservation and local economic projects including tourism, regulate public utility and infrastructure projects, adopt projects and monitor the performance of staff and local government partners.

Under the NIPAS programme, Mt. Pulag National Park was reclassified into one of eight protected-area categories: the category which included natural parks or natural biotic areas established by legal conventions or international agreements to which the Philippines was a signatory. Between 1996 and 2001, the Department of Environment and Natural Resources received funds from both the European Union and the World Bank to establish 18 national parks and protected areas deemed to be the most important in the Philippines for protecting biodiversity.

The Local Government Code and decentralization

The 1991 Philippine Local Government Code (RA.7160) provided for both democratization and devolution of three types of national–government functions and powers to local government units – the *barangays*, municipalities and provinces. The separate and formal powers of *barangay*,[3] municipal and provincial representatives who were members of Protected Area Management Boards were defined by the Code, which decentralized many services and functions by devolving governmental powers and fiscal authority from national agencies to local government units governed by locally elected officials. However, the Code inadvertently created overlapping and somewhat confused areas of authority among local units of government, the Department of Environment and Natural Resources, and indigenous communities. Such was the case with the Protected Areas Act. Although municipalities had resource planning functions and could apply customary laws to manage land and water conflicts, the provinces actually regulated and enforced laws over small watersheds and community forests, which were issued with Environmental Compliance Certificates (Serote, 2004, cited by Pinel, 2007, p58).

The Comprehensive Agrarian Reform Programme (CARP)

The Department of Agrarian Reform was introduced to the Kalanguya communities of Lusod and Tawangan, within Mt. Pulag National Park, via its Comprehensive Agrarian Reform Programme. The intention of the programme was to redistribute agricultural estates to landless farmers and to promote their economic development through crop diversification, enterprise development and the establishment of cooperatives. A memorandum of agreement was drawn up between the Departments of Environment and Natural Resources and Agrarian Reform, which created a programme called 'Operation Highland Wind' (Batcagan, 2007, p42). In this memorandum, signed by the regional office of the Department of Environment and Natural Resources, the Department of Agrarian Reform defined indigenous swidden farms and forests that were recognized in ancestral land claims by the Department of Environment and Natural Resources as being eligible for Certificates of Land Ownership Award, following the same boundaries.

The Department of Agrarian Reform launched its programme in Tawangan in 1996. It instructed residents to identify their agricultural swidden lands in preparation for the issuance of Certificates of Land Ownership Award. Technically, however, the area was not included in the Comprehensive Agrarian Reform Programme because the entirety of *barangay* Tawangan's area was officially classified as inalienable and non-disposable public land, since it was within a protected area. However, the regional office of the Department of Agrarian Reform went ahead and identified Tawangan as an agrarian reform community. This was followed by the creation of other agrarian communities, allowing the extension of agrarian reform services within the Mt. Pulag National Park.

In 1992, based on the Department of Environment and Natural Resources' documentation of ancestral land claims, the Department of Agrarian Reform awarded the 'Lusod Tribal Community' eight 'mother' Certificates of Land Ownership Award covering more than 2000 hectares of land within the protected area of the National Park. This was despite the fact that the agreement between the two departments specifically excluded the National Parks and protected areas from the Agrarian Reform programme. However, the farmers took the opportunity to expand their farms higher up the slopes of Mt. Pulag. The conflict was discovered only when the Agrarian Reform Programme announced plans to construct irrigation facilities and roads within the protected area. This prompted officials of the Department of Environment and Natural Resources to rescind the 1998 memorandum of agreement and dismiss the Certificates of Land Ownership Award as being illegal (Batcagan, 2007, pp54-55). However, the damage had already been done. The Department of Agrarian Reform's programme had encouraged the conversion of traditional swidden farms to commercial vegetable gardens and the forest landscape had changed.

The Indigenous Peoples' Rights Act

The Indigenous Peoples' Rights Act, or Republic Act 8371 of 1997, was crafted under the 1987 Philippine Constitution, which reversed the country's policy of assimilation and committed the government to protecting the rights of indigenous peoples to their ancestral lands and domains, including the upholding of customary laws governing their community life. In all cases of development projects, programmes and activities involving indigenous communities, the law demanded that free prior and informed consent be sought from the communities beforehand. Furthermore, indigenous communities had to be consulted when national plans and policies affected their social, cultural and economic well-being. This policy integrated sustainable natural-resource management with devolution of responsibility. It was a law that enabled indigenous communities to obtain communal titles to their ancestral domains and individuals or clans to claim titles to ancestral lands.

The Indigenous Peoples' Rights Act created the National Commission on Indigenous Peoples to administer these rights. The Commission's implementing regulations specified that existing ancestral-domain claims and ancestral-land claims would be converted to titles of ownership after verification. The Act recognized ancestral-domain title holders as planning and land-management authorities, and stated: 'All protected areas that form part of the ancestral domains and lands shall be maintained with the full participation of indigenous peoples and cultural communities, and conserved in accordance with the sustainable indigenous knowledge systems and practices and customary laws of the indigenous peoples'.

The rules of the National Commission on Indigenous Peoples also asserted self-determination and the right of indigenous peoples to pursue the economic, social and cultural development of their human and natural resources. The application of indigenous knowledge systems and practices could be consolidated into five-year Ancestral Domain Sustainable Development and Protection Plans.

The provisions of the Indigenous Peoples' Rights Act overlapped those of comprehensive land-use and development plans by adding customary practices, laws and traditions and policies for sharing programme benefits among community members. Thus, new governmental or private projects in areas titled 'ancestral' were also subject to the need for 'free, prior and informed consent' by 'the consensus of all indigenous community members in accordance with their customary laws and practices…' However, there remained some ambiguity in the interpretation of laws by implementing agencies, as to what should happen if there was no consensus among communities in the designation of previously proclaimed parks that were initial components of the protected areas system. According to the National Commission on Indigenous Peoples, protected areas were subject to the process of free prior and informed consent because they had to be redesigned under the protected areas law in consultation with indigenous peoples and communities.

The situation was especially unclear when a protected area was not within the exclusive domain of one indigenous community, but rather it crossed multiple ancestral domains and local government boundaries. Mt. Pulag presented a particularly complex problem of stakeholder representation and overlapping land tenure and planning authority. The Park was claimed by two municipalities and at least three *barangays* within the protected area. When the Indigenous Peoples Rights Act promised to award title to indigenous claimants, the main arena for competition shifted from the collaborative efforts of the Protected Areas Management Boards to negotiations regarding boundaries for ancestral domain title. Ancestral domain boundaries were to be self-identified and delineated by the indigenous peoples themselves. Thus, the Kalanguya and Ibaloy representatives to the Protected Areas Management Board presented the Indigenous Peoples' Rights Act as a choice for co-management. As a result, both the Kalanguya Tribal Organization and the municipality of Kabayan passed resolutions claiming exclusive control of the park as their own ancestral domain. Furthermore, some members of the Protected Areas Management Board were also officials of Kabayan municipality, and were unanimously opposed to the very existence of the park (Perez, 2010, p90).

The municipality saw financial interest in increasing and marking boundaries for municipal jurisdiction. There was also a shift of interest in ancestral-domain titles as a mechanism for solidifying boundaries and increasing revenue. This strategy was driven primarily by a Local Government Code financing formula that rewarded local government bodies with large territories. There was competition for revenue, votes and income generated by tourism fees (Pinel, 2007, p192).

The Indigenous Peoples' Rights Act assumed that natural-resource management would be governed by the customary laws and indigenous-knowledge systems of cultural communities. However, the sustainable management and best practices found in swidden communities, supported by indigenous knowledge and governed by customary laws,were no longer applicable when the indigenous farmers shifted to commercial vegetable farming.

As land-use practices changed, there was also an increasing preference for the privatization of land, rather than opting for traditional common-property resource management under customary law. This was partly because tax declarations led to the eventual titling of land. The preference for private titles was also a consequence of the permanent conversion of land to commercial gardens. This entailed heavy investment in permanent improvements, labour costs and cash payments for the bulldozing and terracing of the land, as well as the purchase of chemical fertilizers and pesticides. Furthermore, with land titles in hand, farmers were able to access bank loans to invest in their businesses. This changed the character of both land use and management systems (Prill-Brett, 1992, 2013).

The Mount Pulag National Park gained value when it was made a forest preserve and biodiversity centre. This brought in revenue from tourists and mountain climbers. Then there were unintended consequences. Roads were opened, which allowed the clearing of mossy forest for commercial vegetable farms. This increased the value of the land for cash cropping. The park became a contested open-access natural landscape where natural resources were exploited for their commercial value.

Assessors' records show that some people filed tax declarations on swidden plots in mossy forest areas, then cleared and bulldozed the mountainsides into vegetable terraces, claimed private ownership and sold the land to speculators. This strategy was also observed by Lewis (1992) in a separate study. Tax declarations filed over protected areas and forest preserves have been accepted by municipal governments hungry for revenue. Similar tax declarations have also been filed from swidden plots in the Mt. Data National Park and accepted by the municipality of Bauko, Mountain Province (Prill-Brett and Salinas, 1991). In 1999, a provincial ordinance required all municipal assessors to accept applications for tax declarations in the Cordillera forest preserves because the Indigenous Peoples' Rights Act had declared that areas traditionally occupied and improved by indigenous peoples should be classified 'alienable and disposable' (Carino, 2010).

Conclusions

The proliferation of laws from different state agencies has inadvertently created overlapping boundaries, resource competition and political and legal conflict, and while this conflict has continued, the protected areas of Mt. Pulag and Mt. Data have been vulnerable to contested open access. The situation in both National Parks points to problems of weak enforcement and lack of accountability. As pointed out by Meinzen-Dick and Pradhan (2002, p4), laws are only as strong as the institution or collectivity that stands behind them. This is clearly the case in the Mt. Data and Mt. Pulag National Parks. Furthermore, as stated by Giddens (1979, p69), laws and indigenous norms, rules, and values are invoked and modified by actors in the pursuit of their interests. Sets of rules become structures employed by actors in strategizing, and then are both the medium for, and the outcome of, human action, agency and praxis. Such was the case of traditional swidden farmers whose strategy was to shift to commercial vegetable production as opportunity structures became available. This allowed them to send their children to college, build modern houses, avail of material goods, send family members overseas, pay for medical expenses and other forms of expenditure that were beyond their means as subsistence swidden farmers in communities with ever increasing populations. However, the trade-off is the risk not only of biodiversity loss and destruction of the ecosystem, but the loss of vital watersheds and the growth of health hazards from the use of chemical pesticides and fertilizers. In the Tinoc communities of Ifugao province that converted from traditional *inum-an* shifting cultivation to commercial farming, the areas under rotational farming have diminished and commercial monocrops threaten the existence of 20 varieties of traditional crops. Tinoc's biodiversity resources have also been reduced, including the loss of three species of honey bees and 13 species of mushrooms. Traditional occupations such as hunting have also been threatened, with large areas converted to cabbage and potato growing. Farmers claim that eight kinds of animals and three kinds of birds are no longer seen. Another consequence is the loss of indigenous knowledge, including indigenous terms that define land use and management patterns.

Finally, I conclude that it is not the laws *per se* that have brought about forest destruction, but rather overlapping, ambiguous and often contradictory laws, each with their intended mandates that address the issues and purposes of resource management in the protection of National Parks (Prill-Brett, 2013). When multiple laws are applied over a particular landscape or domain that contains a static range of resources; where lack of coordination and identification of possible contradictions exists among the respective agencies and interest groups; this opens up what von Benda-Beckmann (1983) called a 'jungle of legal pluralism.' In this jungle of legal pluralism it is usually the indigenous elites who are the most successful in grabbing and titling lands, as observed by Jefremovas in her Sagada study (2001). So along with the land conversion we also see increasing differentiation between indigenous elites and ordinary people.

Acknowledgement

I wish to acknowledge the assistance of Dr Joachim Voss, who made some constructive comments on an earlier draft of this chapter.

References

Batcagan, M. A. (2007) 'Issues on management of forest protected areas in the northern Philippines: A case study of the Mt. Pulag National Park', seminar paper submitted in partial fulfilment of requirements for a Master of Arts degree in Social Development Studies, College of Social Sciences, University of the Philippines, Baguio

Broad, R. and Cavanagh, J. (1993) *Plundering Paradise: The Struggle for the Environment in the Philippines*. University of California Press, Berkeley, CA

Carino, D. (2010) 'Benguet's green agenda', *Philippine Daily Inquirer*, 25 March 2010, Manila

Delson, M.T. (1989) 'The death of a wetland in the uplands: The Mt. Data experience', paper presented at a Conference on Wetland Management, 4-9 June 1989, Leiden University, Netherlands

Fitzpatrick, D. (2006) 'Evolution and chaos in property rights systems: The Third World tragedy of contested access', *The Yale Law Journal* 115, pp996-1048

Giddens, A. (1979) *Central Problems in Social Theory*, Macmillan, London

Griffiths, J. (1986) 'What is legal pluralism?', *Journal of Legal Pluralism* 24, pp 1-55

Jefremovas, V. (2001) 'In harmony with nature?: Environmental management and social change amongst the Sagada Igorots of the Cordillera Central, Philippines', in *Towards Understanding Peoples of the Cordillera*, vol. 2, Cordillera Studies Center, University of the Philippines, Baguio, pp30-44

Kowal, N. E. (1966) "Shifting Cultivation, Fire, and Pine Forest in the Cordillera Central, Luzon, Philippines," *Ecological Monographs* 36(4), pp389-419

Lewis, M. W. (1992) *Wagering the Land: Ritual, Capital, and Environmental Degradation of the Cordillera of Northern Luzon, 1900-1986*, University of California Press, Berkeley, CA

Meinzen-Dick, R. and Pradhan, R. (2002) *Legal Pluralism and Dynamic Property Rights*, working paper 22, International Food Policy Research Institute, Washington, DC

Merry, S. E. (1988) 'Legal pluralism', *Law and Society Review* 22(5), pp 869-896

Perez, P. (2010) 'Deep-rooted hopes and green entanglements: Implementing indigenous peoples' rights and nature-conservation in the Philippines and Indonesia', PhD dissertation to Leiden University, Leiden, the Netherlands

Pinel, S. L. (2007) 'Co-management of cultural landscapes: Collaborating to compete at Mt. Pulag National Park, the Philippines', PhD dissertation to the University of Wisconsin-Madison, Madison, WI

Prill-Brett, J. (1992) *Ibaloy Customary Law on Land Resources*, working paper no. 19, Cordillera Studies Center, University of the Philippines, Baguio

Prill -Brett, J. (2007) 'Contested domains: The Indigenous Peoples' Rights Act (IPRA) and legal pluralism in the northern Philippines', in M. Wiber and C. Milley (eds) 'After Recognition: Implementing Special Rights in Natural Resource Management', *The Journal of Legal Pluralism and Unofficial Law*, special issue 55(2007), pp11-36

Prill-Brett, J. (2013) 'Indigenous common property regime, contested open access, the IPRA and legal pluralism in the Cordillera, northern Philippines: The Mt. Pulag case', paper presented at a public forum, September 2013, College of Social Sciences, University of the Philippines, Baguio

Prill-Brett, J. and Salinas, F. (1991) *From Mossy Forest to Cabbage and Potato Farms: The Lamayan Case (Mt. Data)*, Community Profile no. 3, Cordillera Studies Center, University of the Philippines, Baguio

Rambo, A.T. and Hamilton, L. (1990) 'Social trends affecting natural resources management in upland areas of Asia and the Pacific', in *Proceedings of the UNDP Regional Symposium on Cooperation in Asia and the Pacific*, April 1990, East-West Center, Honolulu

Semper, C., von Drasche, R., Meyer, H., Schadenberg, A. and Scheerer, O. (1975) *German Travelers on the Cordillera (1860-1890)*, edited and annotated by W. H. Scott, The Filipiniana Book Guild, Manila

von Benda-Beckmann, F. (1983) 'Why law does not behave: Critical and constructive reflections on the social and scientific perception of the social significance of law', in *Papers of the Symposia on Folk Law and Legal Pluralism*, vol. I, XIth International Congress of Anthropological and Ethnological Sciences, Vancouver, Canada, compiled by Harald W. Finkler, Ottawa.

Wiber, M. G. (1993) *Politics, Property and Law in the Philippine Uplands*, Wilfrid Laurier University Press, Waterloo, Ontario

Notes

1. There were originally 2398 hectares of mossy forest on the Mt. Data plateau. In the 1990s, there were only 89 hectares left. This had been protected as a watershed for surrounding communities. However, this remaining area has been slowly dwindling due to illegal encroachment and is also contested by certain individuals and groups.
2. Personal interviews in 1991, with farmers in the expansion area of Lamayan, Mt. Data.
3. A *barangay* is the smallest unit of government in the Philippines.

40

VIETNAM'S 'RENOVATION' POLICIES

Impacts on upland communities and sustainable forest management

*Tran Thi Thu Ha and Pham Van Dien**

Introduction

The Vietnamese government's 'renovation' policies, known as *doi moi*, have been in place since the late 1980s. These policies enabled a transition from a subsidized centrally planned economy to a market economy, and transferred land-use rights from the state to users. One of the principal goals of *doi moi* in Vietnam's uplands was to reduce deforestation and enhance the sustainability of forests and forest land management. The policy was the result of a combination of top-down and bottom-up forces (Fforde and Goldstone, 1995; Kerkvliet, 2005). From the bottom up there was 'fence breaking' as rural people resisted the central planning system. Farming yields fell on communal land and central authorities were forced to respond to growing food crises. In 1988, Politburo Decree 10 was issued, changing agricultural land from collective to household ownership (GoV, 1988). This allowed individual ownership of production and benefits were quickly apparent. By 1992, Vietnam had changed from being a rice importer to being one of the world's largest rice exporters (World Bank, 1994). The achievements of agricultural reforms in lowland parts of the country gave the government the confidence to introduce a wider set of reforms under the *doi moi* process, including land reforms, social-economic reforms, forestry reforms, and so on.

Since the early 1990s, the *doi moi* policies have been implemented in upland regions with the aim of improving local livelihoods and the management of forest and forest land, among other things. The policies included transferring forest land-use rights from the state to individual households and other users through the Land Law of 1993 (NAoV, 1993). At the same time, the government introduced economic

---

* Dr Tran Thi Thu Ha, Associate Proferssor and Director, Institute of Forestry Research and Development, Thai Nguyen University of Agriculture and Forestry, Thai Nguyen, Vietnam; and Dr Pham Van Dien, Associate Professor and Vice-Rector, Vietnam Forestry University, Hanoi.

and land reforms. The policies sought to achieve optimal use of forests and forest land through the application of appropriate cultivation systems and strategies to combine conservation and development goals in upland areas. It was hoped that appropriate agricultural and forestry production activities would raise local incomes, stop deforestation and ensure that sloping land was managed in a sustainable way. However, the policies have fallen short of achieving the government's objectives (Sikor, 2001; Alther et al., 2002; Zingerli et al., 2002; Tran, 2005; Tran and Hoang, 2013).

This chapter examines the impacts of the government's *doi moi* policies on upland people's livelihoods and forest management.

Research methods

The study was conducted in three provinces in the Northern Mountain Region (Thai Nguyen, Bac Kan and Cao Bang); and two provinces in the Central Highlands (Dak Lak and Dak Nong). The case studies in these provinces covered the three main forest categories: production forest, special-use forest and protection forest. In each province, two or three case-study villages were surveyed intensively. In each case, one was located near a commune centre and close to lowland areas and the others were in remote upland areas (Figure 40-1).

Relevant data and information were obtained for two periods: the early 1990s before the introduction of the policies, and in 2005 and 2014, after the introduction of the policies. Primary data were collected from village meetings and group discussions, field surveys, household surveys and interviews of key informants at different levels by using semi-structured questionnaires and checklists (see Table 40-1).

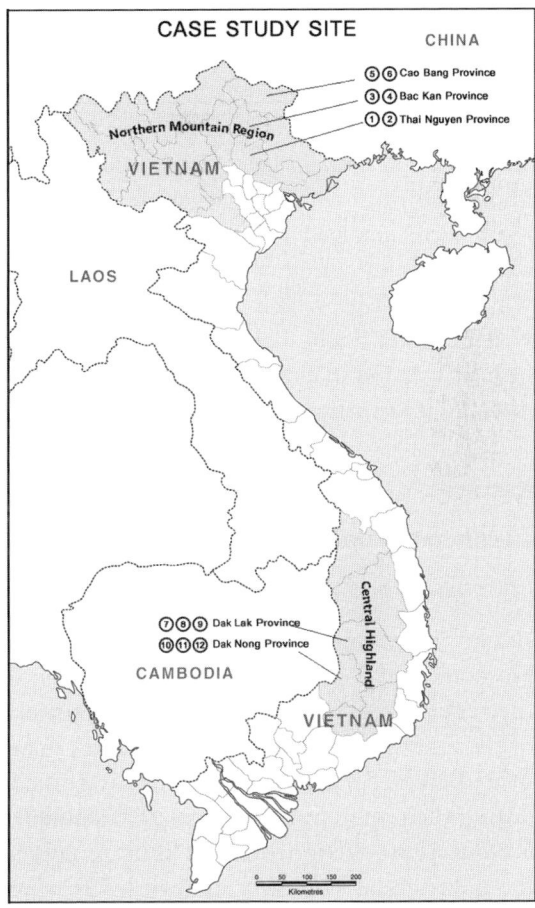

FIGURE 40-1:   Location of the case-study villages.

TABLE 40-1: Summary of the case-study villages.

Province	Case study villages	Principal forest category	Households	Ethnicity
Thai Nguyen	Vau	Production	60	Dao, Nung, Kinh
	Nac		136	Dao, Nung, Kinh, Tay
Bac Kan	Ban Cam	Special use	68	Tay, Kinh
	Khau Qua		41	H'Mong
Cao Bang	Pac Han	Protection	28	Dao
	Khuoi Ken		26	Dao
Dak Lak	Hang Ja	Protection and production	161	M'Nong, Ede, Muong
	Hang Nam		139	M'Nong, Kinh
	Ea Lang		141	H'Mong, Tay, Thai, Kinh, Muong
Dak Nong	Bon Dieng Ngaih	Production	182	M'Nong, Ede, Kinh
	Bon Phung		152	M'Nong, Ede, Kinh
	Bon R'Mnon		266	Ede, M'Nong

A total of 360 households, 145 key informants from the 20 case-study villages, 68 key informants from commune level, and 60 key informants from district and provincial levels were involved in 2005 and 2014. In both field seasons, the households in each case-study village were included in the surveys on the basis of wealth ranking within the village, which was determined during group discussions. Secondary data were collected, including documents and statistics from government agencies. Both quantitative and qualitative analytical methods were used to process and analyse the data and information.

Results

Livelihood of upland communities

Changes in land use

Before the upland reforms in 1993, forest land was managed by the government. However, in the view of interviewed villagers, forest land was then common land that they could freely access for shifting cultivation or to harvest forest products. In most of the case-study villages, forest land was managed by a collective, but the villagers treated it as common land. The government's strategy for food self-sufficiency during the period from 1980 to 1990 resulted in upland people trying to maximize food-crop production through shifting cultivation. In early 1993, introduction of the government's *doi moi* strategy led to changes in cropping systems to promote both greater economic efficiency and respect for environmental values. Cultivation systems in most of the case-study areas changed (Table 40-2).

TABLE 40-2: Changes in major land uses for households in the case study villages,1994-2014.

Categories	Case studies in the Northern Mountain Region		Case studies in the Central Highlands	
	Pre doi moi	*Post* doi moi	*Pre* doi moi	*Post* doi moi
Food crops	– Upland rice dominant. – Maize and cassava dominant. – Paddy rice, one crop only.	– Few upland rice fields. – Maize, cassava dominant in fixed plots. – Paddy rice, two crops.	– Upland rice dominant. – Maize dominant. – Paddy rice, one crop only.	– Few upland rice fields. – Cassava dominant, less maize. – Paddy rice, two crops.
Animal husbandry	– Husbandry free.	– Husbandry intensive.	– Husbandry free.	– Husbandry intensive.
Long term crops		– Fruit trees expanding.		– Fruit trees expanding.
Commercial species		– Tea expanding.		– Primary forests replaced by coffee, rubber.
Plantations		– Plantations expanding (*Acacia* and *Eucalyptus* spp.). Bare land replaced and forests regenerating.		

There have been marked changes in land use since the policy reforms. The area of paddy rice in most study villages has not increased since 1994, but has intensified from one to two crops per year by using irrigation. The area of agricultural land classified as plains for dry food crops has increased in most cases.[1] Upland areas formerly used for dry food crops such as upland rice, maize and cassava have diminished significantly in most cases in the Northern Mountain Region, but market demand has led to expansion of cassava in the Central Highlands villages. In the earlier period from 1980 to 1990, most villages used forest land to grow upland rice and maize to meet their staple dietary needs. Survey data from 2014 (Table 40-2) show that the area used for shifting cultivation in 1994 had, in most villages, reduced by 2005 and was continuing to diminish in 2014 as it was replaced by plantations or commercial crops. However, maize, upland rice and cassava were still being cultivated as the main staple foods by all households in the study villages, although the area under these crops had fallen. Better-off households with enough land for wet rice cultivation were using forest land for commercial crops and plantations, while poor people who lacked both the land for paddy rice and the capacity to develop cash crops continued to grow dry food crops on forest land.

Areas planted to commercial crops and plantations have expanded in most study villages over the past 10 to 15 years. Most areas previously under shifting cultivation were allocated to individual households for reforestation. However, the villagers elected to use these areas to plant commercial crops such as tea, coffee, pepper, fruit trees or plantation crops, which provide higher economic returns than forest trees or naturally regenerating forests.

Income

Before the *doi moi* policies, most households in the study villages lived in similar economic circumstances, typified by temporary houses, no high-value goods, low annual income, food insecurity, no cash savings and little money to invest in production. From the field surveys after *doi moi*, three wealth-ranking groups (poor, average and better-off) were identified and classified on the basis of local criteria shown in Table 40-3. These categories give a more meaningful categorization of local poverty and economic well-being than a strictly cash-based definition.

In the study villages, poor people accounted for between 19.7% and 73.3% of the population. There was a tendency towards a greater proportion of poor people in more remote villages (i.e. Khau Qua, Khuoi Ken, Hang Ja and Hang Nam) and those

TABLE 40-3: Variations for different wealth-ranking groups following *doi moi*.

Poor	*Average*	*Better-off*
Temporary house.	Simple house.	Good house/solid house.
No valuable items in the house.	Some furniture and small appliances.	Good furniture, many valuable items; small appliances (e.g. thresher, TV).
Limited or no paddy rice area; limited or no forest land.	Limited or no paddy rice area; limited forest land.	Large paddy rice area; upland field; large forest land area.
Lack of capital for crop investment.	Lack of capital for crop investment.	Capital for crop investment.
Own 0-1 cow/buffalo.	Own 1-2 cows/buffaloes.	Own 4-6 cows/buffaloes.
Lack food for 4–10 months per year.	Lack food for 1-3 months per year.	Food security, excess food for some months of the year.
Income less than VND400,000 (<US$22) per person per month.	Income between VND400,000 and 600,000 (US$22-$33) per person per month.	Income greater than VND600,000 (>US$33) per person per month.
No savings, in debt from buying food on credit.	No cash savings, in some debt.	Cash savings.
Education to primary school level only.	Education up to secondary school.	Education up to vocational school level.

in the Central Highlands. The proportion of poor households was lower in villages located near a commune or a district centre (i.e. Ban Cam and Bon Phung). Table 40-4 shows the proportion of households in each wealth category.

In most study villages, the income of a large number of people was only slightly above the poverty threshold. Most income generated by the villages was shared among a small number of better-off people. Thus, it was mainly the income of better-off people that had improved as a result of *doi moi*. Before *doi moi*, the differences in income among village people were small. Now, the income gap is much greater.

Inequality in land allocation appears to have been one of the main factors responsible for the increasing gap between rich and poor. In addition, *doi moi* generated new sources of income through market demand, and local people involved in small businesses, services or handicrafts were able to earn a significant income. However, these activities were the prerogative of better-off people. For example, in the study village of Ban Cam, in Bac Kan province, only the better-off people had good boats with which to catch fish on nearby Ba Be lake. Their catches, for household consumption and for sale, were therefore much better than those of poor people who had difficulty in catching enough fish for their own consumption.

At the study sites in the Central Highlands, villagers of the M'Nong and Ede ethnic groups based their pre-*doi moi* livelihoods on shifting cultivation. After the reforms, forest land was allocated for national parks, natural reserve areas and to private companies. Migrants from other parts of the country also claimed land, so there was no more space for shifting cultivation. The people had to switch to fixed or permanent cultivation with new agricultural practices. In these areas, the main income of better-off households came from coffee plantations or small businesses, while the income of poor people still came from maize, cassava and selling their labour. Field observations in 2014 showed that wet rice production was barely able to provide food security. Maize provided a source of cash income for households to

TABLE 40-4: Percentage of households in each of the three wealth-ranking groups in the 12 study villages.

Case study village	Poor	Average	Better-off
Vau	32.7	52.9	14.4
Nac	40.6	50.6	8.8
Ban Cam	21.7	62.0	16.3
Khau Qua	44.4	49.4	6.2
Pac Han	20.2	67.8	12.0
Khuoi Ken	33.3	56.7	6.0
Hang Ja	73.3	22.7	4.0
Hang Nam	63.0	29.6	7.4
Ea Lang	40.2	54.8	5.0
Bon Dieng Ngaih	55.2	40.7	4.1
Bon Phung	34.6	59.5	5.9
Bon R'Mnon	45.0	52.0	3.0

buy rice and other family needs and accounted for about 25% to 35% of their total income. Cassava accounted for a further 10% of total income. The average income from coffee plantations was about 20 to 40 million Vietnamese dong (US$941 to $1882) per hectare per year. But not many households had coffee plantations. The rearing of livestock failed to thrive, with most households having between one and four head of breeding cattle. Off-farm employment (labouring work) was a significant source of income for most poor households. Between 10% and 80% of village households had members employed in the commune or in other districts during the year. These employees were mostly aged between 18 and 45 and included both men and women. Working as hired labourers never occurred in the years before *doi moi*. The people complained that they did not wish to 'work like slaves' for other people, but had no choice. For cash income, they also began harvesting non-timber forest products, such as bamboo sprouts and rattan, in the months from August to October. Poorer people, in particular, turned to selling bamboo shoots in this season to earn cash to buy rice.

Food security

Food security, especially in rice, has always been an issue among Vietnam's upland communities, even before the introduction of the *doi moi* policies. However, in the years before *doi moi*, every household managed to feed itself through shifting cultivation of non-rice crops such as maize and cassava. Following *doi moi*, better-off people were self-sufficient in food production because they had large paddy areas in which to cultivate rice and agricultural land to grow commercial crops to generate income. Poor people, on the other hand, were unable to be self-sufficient because they were left with very small patches of paddy land and other agricultural land. Many poor households had no paddy land at all, and rice shortages became very common among the very poor in all study villages (Table 40-3). Better-off people had cash to overcome shortages by buying rice from markets.

Cash flow, education and employment

In the years since the introduction of *doi moi*, school fees have been a heavy burden on poor families, forcing many of them to borrow money at the start of the school year. Most households are unable to support their children to undertake secondary or high-school education. Thus, following *doi moi*, only better-off families have been able to put their children through high school. This differential access to education is helping to perpetuate inequalities. The children of better-off villagers, who have access to higher levels of education, grow up with greater economic opportunities, while the children of the poor are locked into cycles of poor education and poverty. Commercial cropping, as an important source of cash income in upland areas, is seen as one answer to this problem. But, once again, not every household has the same opportunities, much less the capacity, to earn this income.

Land-use rights

Inequitable access to land rights

Before the *doi moi* policies, every household in upland villages had similar opportunities to access forests and forest land to harvest forest products, cultivate swiddens and gather other benefits. Following *doi moi*, forest land and forests were allocated to real owners or users. In this context, the term 'forest allocation' referred to allocating forests to users for management purposes, while 'forest land allocation' meant allocating forest land for long-term use, including degraded forests and bare land. The new land policies also ruled that forest and forest land allocation differed between the three forest categories. In the production forest category, forest land could be allocated for long-term (50 years) use, with full use rights. In the protection forest category, it could be allocated for protection purposes only, for either a short or long term, while forest land in special-use forests could only be allocated for short-term protection. Therefore, land was allocated to households only in those study villages with forests in the production and protection categories, and not in those with special-use forests. Moreover, in villages where forest land allocation occurred, not everyone received land areas of the same size or quality (Table 40-5).

Poor people consistently own smaller plots of land than the better-off. In Vau, Nac and Pac Han villages, the very poor were allocated on average 1.8ha, 5.3ha and 1.6ha, respectively, while the better-off received on average 7.5ha, 16.8ha and 8.3ha, respectively, despite the households having about the same number of members (Table 40-5). Every household was entitled to apply for an area of land up to a maximum limit, depending on the availability of land resources in the commune.

TABLE 40-5: Forest land distribution and average number of family members in different wealth-ranking groups in Vau, Nac and Pac Han villages.

Categories	Very poor	Poor	Better-off
1. Vau			
Mean land distribution per household (ha)	1.8	3.9	7.5
Mean family membership	4.4	4.2	4.3
Number of households that didn't receive forest land	3		
2. Nac			
Mean land distribution per household (ha)	5.3	11.9	16.8
Mean family membership	5.4	4.7	6.0
Number of households that didn't receive forest land	8		
3. Pac Han			
Mean land distribution per household (ha)	1.6	7.4	8.3
Mean family membership	4.3	5.8	4.8
Number of households that didn't receive forest land	18		

However, in practice, the better-off had money to pay for land surveys, thus acquiring more land than poor people in the land allocation process. More seriously, in the study villages where land allocation had already occurred, not every household had received forest land – contrary to the claims of local government agencies. Clearly, not every household has benefited from the government's land reforms. Moreover, efforts to 'de-collectivize' forestry have suffered frequent policy reversals and are fraught with high levels of tenure insecurity. Many poor households in Nac village have illegally sold their land-use rights for cash.

In study villages that are located in protection and special-use forests, villagers have not received any forest land-use rights at all. So they continue to access land as illegal swidden farmers, and this brings them problematic lives. Allocation of rights to use forest land had not yet occurred in the study villages of the Central Highlands, leaving local households without land-use rights.

Institutional issues

Interviews and field surveys conducted at the study sites revealed that up to 80% of forest land plots marked on land allocation maps were not the true locations in the field. The main cause of this poor quality of land measurement for land allocation was a lack of professional and moral responsibility on the part of government officers and technical staff who carried out the field measurements and the managers who monitored the work. The managers were aware that land measurements were not carried out in the field, but they endorsed false documents in return for 'kick-backs'.

In one case-study province, we also found that local government authorities had changed a forest type from production to protection so that they became eligible to participate in a national reforestation programme for financial gain. The change was made despite the land having already been allocated to households for long-term use. The household land owners were unaware of the change, which could have seriously affected their rights to use the land. Many government officers at provincial, district and commune levels

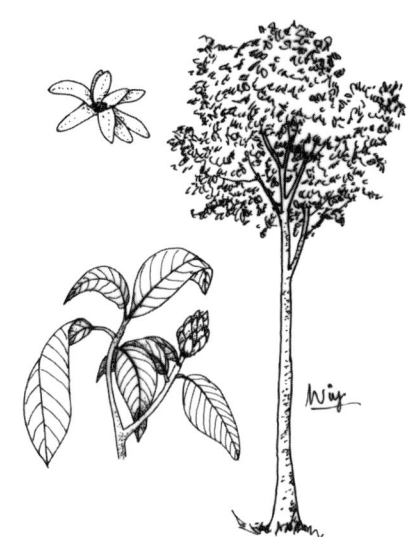

Manglietia glauca Blume [Magnoliaceae] synonym of *Magnolia sumatrana* var. *glauca* (Blume) Figlar & Noot

One of the fast-growing species used by villagers in the Northern Mountain Region in short-rotation timber plantations on land allocated for reforestation. These provide higher economic returns than naturally regenerating forests.

were known to have changed forest land to agricultural land, cleared natural forests for plantations and even supported illegal logging. They got away with these things because of limitations in the monitoring system, lack of policy enforcement and procedures, complicated administrative systems and a lack of adequate understanding of government policies and laws on the part of upland villagers.

In the Central Highlands, forest-land allocation for local households was still a work in progress at the time of the case studies there. However, allocation of forest land was mostly complete in the case of migrant newcomers, including private companies, leading to shortages of cultivable land on which the local people could grow food crops. So the local people cleared more forest. Therefore, there were many conflicts between local people and the authorities, arising from the land allocation process.

Forest management

Forest cover and quality

The *doi moi* policies have brought different impacts on forest management in the Northern Mountain Region and Central Highlands. Before *doi moi*, most forests in the Northern Mountain Region were cleared for shifting cultivation or timber logging (both legal and illegal) and the level of forest cover was very low. Following the policy reforms, the forest cover increased significantly in the period from 1995 to 2013, but the quality of the forests was still very low in all case study provinces. The main reason was that most forests were young and in the process of regeneration, with low biomass. The area of rich and medium-quality forests was very small (2.8% in Bac Kan; 0.6 % in Thai Nguyen; and 0.9 % in Cao Bang). In contrast, the *doi moi* policies led to serious deforestation in the Central Highlands. In the case-study provinces, forest cover diminished significantly in the period from 1995 to 2013, shrinking by 28.2 % in Dak Lak and 16.6% in Dak Nong (Table 40-6). Before the reforms, local people in the Central Highlands practised shifting cultivation in small plots and left the land fallow for five to 10 years before clearing it again for cultivation. After *doi moi*, most primary forests were cleared for commercial crops such as coffee, rubber and pepper. Nowadays, remaining primary forests that are part of national parks or natural reserve areas are threatened with deforestation by illegal loggers and weak institutional management.

In the study villages of Vau, Nac and Khuoi Ken, in the Northern Mountain Region, where land was allocated from production forests, many regenerating and primary forests were replaced by plantations. However, the quality of the plantations was very poor because of the low quality of seedlings and poor silvicultural techniques. More seriously, many hundreds of hectares of good natural forest were cleared for plantations of fast-growing species such as *Acacia mangium*, with a short rotation of only four to six years. This was a consequence of weak implementation of plantation policies. Additionally, many small plots of regenerating forest are still being cut for shifting cultivation by individual households, with a fallow period of five years.

TABLE 40-6: Forest cover changes in different provinces, before and after the *doi moi* policies.

Provinces	Forest cover in different years (%)			Gain or loss over different periods (%)		
	1995	2005	2013	1995-2005	2005-2013	1995-2013
Thai Nguyen	16.1	41.4	47.9	+ 25.3	+6.5	+31.8
Bac Kan	24.0	53.0	70.8	+29.0	+17.8	+46.8
Cao Bang	12.0	47.3	50.1	+35.5	+2.8	+38.3
Dak Lak	62.0	45.5	45.6	– 16.5	-0.1	-16.6
Dak Nong	62.0	56.4	33.8	-5.6	-22.6	– 28.2

Source: GSO, 2000; MARD, 2006, 2014.

In those study villages with protection and special-use forests, the primary limestone mountain forests have many tree species with high economic value, particularly *Nghien* (*Excentrodendron tonkinense*). Some large trees with high value and good form have been felled illegally by different means. For example, a large tree would be selected and girdled by cutting through the bark all the way around the tree, and then it would be left to die or be blown over by wind. At Ban Cam and Khau Qua villages, many plots previously used for swidden cultivation of upland crops were reclaimed and incorporated into Ba Be National Park in 1999 for natural regeneration. In most study villages, households that had been allocated land near to, or within, national parks cleared only small areas of forest every year to avoid unwanted attention from park rangers. Their activities were raised many times at village meetings, but the households ignored requests from other village members and continued to expand their cropping areas.

Illegal logging, by both insiders and outsiders, was still occurring in most of the study villages and the number of large, high-value trees left standing had declined. For example, in Pac Han village, all primary forests were dominated by high value *Nghien*, and illegal logging of these trees was continuing. The timber is used to make chopping boards for sale in

Excentrodendron tonkinense
(A. Chev.) H. T. Chang & R. H.
Miau [Malvaceae] synonym of
Burretiodendron hsienmu W. Y. Chun
& F. C. How

Called *Nghien* in Vietnam, this highly valued natural forest tree continues to be illegally harvested from primary forests allocated to private owners. About 500 to 800 of these trees are felled each year to make chopping boards for sale in China.

markets in China. About 500ha of this primary forest was allocated to 12 households in Pac Han village, with full land-use rights ('Red Books'). Although illegal cutting had declined significantly, the *Nghien* trees were still being harvested by the household-owners, to be made into chopping boards. Since 2002, both villagers and outsiders have adopted the use of chainsaws for illegal logging. Most large trees, in particular those with hard timber like *Nghien* species, have fallen victim to the chainsaws. Some poor households without chainsaws have arranged for others to fell their large trees on the basis of a 50-50 share of the timber. It is estimated that between 300 and 500 *Nghien* trees of 50 to 80cm diameter are being felled every year in these primary forests that have been allocated to private owners. Better-off people, rather than the village poor, are believed to be the main culprits. A key informant in Pac Han village noted: 'Better-off households began to use chainsaws to cut down trees in 2002. It has become a popular means to harvest timber among the wealthy people'.

Similarly, in most study villages in the Central Highlands, villagers still log the forests illegally when they need to build a new house. Others frequently exploit precious timber reserves in national parks, for illicit sales. Every year, at least two to three cubic metres of the highest grade timber is believed to be exploited per household. The average price for timber is about 20 million VND (US$941) per cubic metre. Most village households also gather non-timber forest products, such as rattan shoots and bamboo shoots, for both their own consumption and cash sales.

Lost biodiversity

In 1996, H'Mong people began moving from the Northern Mountain Region to resettle in the Central Highlands. Hunting wild animals soon became a major problem. According to interviews in Ea Lang village, about 40 of the village's 141 households were professional hunters who used firearms in the forest. They shot all kinds of wild animals, big or small, whenever an opportunity arose The prices of wild meat fluctuated according to the species, but wild pig, for example, was worth 200,000 to 300,000VND (US$9.40 to $14.11) per kilogram. The professional hunters reportedly made hundreds of millions of dong per year, spending one or two weeks in the forest at a time, using home-made firearms. They hunted through a national park and sold wild meat at night, to avoid detection. The head of Ea Lang village said a number of villagers began hunting because there was a shortage of land for cultivation. Most of them were habitual hunters.

Environmental management

In the Northern Mountain Region, key informants in most study villages said that problems of soil erosion and lack of water resources had improved in the previous decade. However, many households in most study villages said that the loss of primary forests to illegal logging in watershed areas had led to problems such as soil erosion, flooding and shortages of water for cultivation.

In the Central Highlands, soil erosion, lack of water supplies, loss of biodiversity, flooding and drought, frequent landslides, pollution and other issues have become more serious problems. Most local people say that degradation of forest resources and the over-use of fertilizers and chemical pesticides for commercial crops have resulted in negative impacts on the lives of people. Soils are impoverished and crop productivity is very low despite high investments in inorganic fertilizers. Topsoil is being lost at an average rate of about two to three centimetres per year. In 2009, cassava production was between 12 and 14 tons per hectare. By 2014, this had fallen to seven to eight tons per hectare, even with fertilizers, and the land became exhausted and had to be abandoned after four years of cassava cultivation, adding further to the lack of arable land. Epidemics of insect pests, most commonly a worm that attacks crops, occur regularly and the use of pesticides has caused environmental pollution affecting the air, water and health of communities. Floods and drought result in regular crop failures, poverty and disease. Poverty has had a negative impact on the cultural and spiritual life of villagers, leading to unsustainable land use and social problems among young people.

Acacia mangium Willd.
[Leguminosae]

Hundreds of hectares of good forest have been cleared in Vietnam's Northern Mountain Region for plantations of fast-growing exotic species such as this, with short rotations of four to six years, for use as paper pulp. The land was allocated to private owners from production forests.

Discussion

Income has increased but poverty rate remains high

The change from traditional shifting cultivation to fixed cultivation systems has generally led to an increase in income in the study villages. However, despite the increased income, the research found that the number of poor people in the study villages in 2014 was still very high (up to 73% of households). The income of a large number of people was only slightly higher than the poverty threshold of 400,000 VND (about US$19) per month. Most income in the villages was shared by a small number of better-off families. Thus, the *doi moi* policies have improved the income of only the better-off.

Before *doi moi*, there were small differences in income among villagers, but the income gap is now much greater. The main reason for this is that better-off people have had greater opportunities to access resources to improve their income. As dominant land owners with the capacity to grasp new opportunities from the *doi moi* process, as well as having better access to the free market, the incomes of better-off people have grown very rapidly.

Food security remains a problem

Before *doi moi*, most people in Vietnam had insufficient rice to meet their dietary needs. Since *doi moi*, Vietnam has become the second largest exporter of rice in the world. However, food security in rice is still a serious issue in the upland study villages. In the uplands, the availability of food is closely linked to the production capacity of households and ease of access to the market. While food can be purchased or bartered in markets, many poor upland people lack both cash and goods to barter. Before the *doi moi* policies were introduced, local people managed to feed themselves through shifting cultivation. At present, poor upland people find it difficult to earn enough cash to buy food and have no choice but to return to shifting cultivation even though it is against the law.

Cash flow, employment and education

Access to cash was not a critical issue before *doi moi* because livelihoods were more or less self-sufficient, with welfare and services being provided by the government. Since the reforms were introduced, the government has ceased to subsidize certain welfare services, such as school fees. The needs of upland people are not limited to meeting their food requirements; they also seek to improve their quality of life, including better education for their children. Therefore the demand for cash by upland people is increasing. However, the accumulation of wealth is difficult for poor people.

There is also a strong link between income, employment, education and forest resource

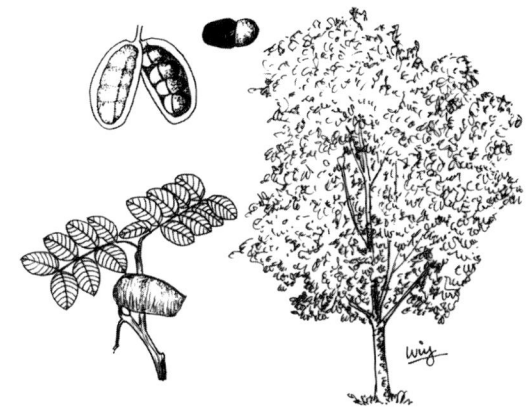

Afzelia xylocarpa (Kurz) Craib
[Leguminosae]

This highly valued and endangered species is found in remaining natural forests in Vietnam. Its timber can be used for ornamental woodturning, carving and making musical instruments. Its seeds and bark are used as a herbal medicine.

management. Most remote upland people who belong to ethnic minorities have low education levels. They have little access to education because there are no secondary schools in remote villages. In addition, without the government subsidizing school fees, only children from better-off families can continue to study. Young people who grow up in villages without a good education find that they cannot compete for jobs in urban areas and have to remain in their villages. This has led to higher village populations and greater pressure on existing forest resources.

The increasing demand for cash income and the lack of jobs are the main reasons why upland people over-exploit forests and encroach on forest land. While in the past upland people cleared forests for shifting cultivation, they are now cutting down trees for sale or clearing regenerating forests for cultivation of commercial crops.

Inequality in land allocation

Several factors contributed to irregularities in land distribution at the study sites. First, land allocation was implemented without the prerequisite land-use planning and field testing to evaluate its impacts. Local government authorities and officers responsible for implementing land allocation did not have a good understanding of the policies. Unlike their better-off neighbours, poor villagers also did not understand the land allocation policies. When land allocation was carried out many of the better-off tried to get as much land as they could, while the poor were reluctant to raise their voices. Second, the government officers who carried out land measurement in the field did not follow the regulations. Some households obtained a large area of forest land – up to three times more than the maximum permitted by the law. These households used different names of family members in their applications, with support from local officers.

Institutional issues

There have been many conflicts and problems associated with implementation of the land allocation policies, including unclear or disputed land boundaries, inequity of land holdings and loss of access. These problems are due to (1) limited human resources and inadequate funding for land-use planning and land allocation; (2) lack of transparency in the evaluation and monitoring process; and (3) provincial authorities who have been concerned with achieving the targets of land allocation, but have ignored issues such as land capacity and suitability and the rights of local people to enjoy sustainable livelihoods. A lack of transparency in the enforcement of the *doi moi* policies by local government agencies has been a serious cause of the policies' low achievements.

Causes of deforestation, before and after doi moi

The causes of deforestation, before and after *doi moi*, are different. Before *doi moi*, the main causes of deforestation were shifting cultivation, logging operations, planting cash crops, infrastructure development and warfare. There was increased clearing of forest land for shifting cultivation because of the government's 'local self-sufficiency in food' policy during the collective period (1976 to 1986). Migration and logging operations by both the government and private organizations also contributed substantially to deforestation in this period. A public perception that the forest was common property – otherwise known as 'the commons problem' – was an important factor that pushed the government to introduce land reforms and transfer the ownership of land to individual households and some local government bodies such as national parks and protection boards. The rate of deforestation declined after *doi moi* came into force, showing that, in this respect, the policy did have some success. However, deforestation still occurs

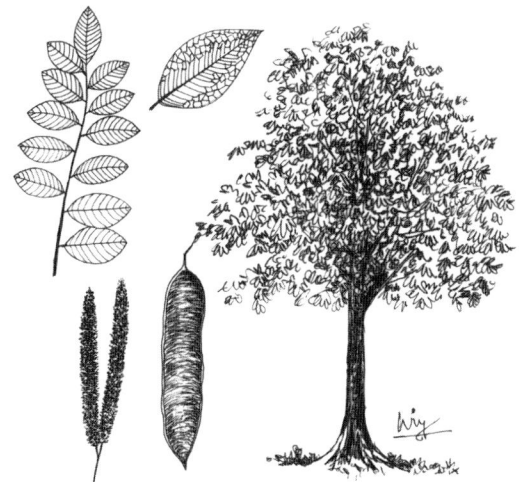

Erythrophleum fordii Oliv.
[Leguminosae]

This species, known as ironwood, is these days found only rarely in natural forests in Vietnam. Its timber is regarded as precious, particularly for furniture and carving. Over-exploitation and deforestation have made it an endangered species, and across the border in China it is under second-class national protection.

in the post-*doi moi* period, in particular in the Central Highlands. But the causes are different from those pertaining to the years before the reforms. In production forest areas that have been allocated to households, large areas of naturally regenerated forest have been cleared to make way for cultivation of cash crops. In protection and special-use forests, deforestation has resulted from illegal logging and encroachment on forest land for crop cultivation because of food insecurity. Poor management of forests in the Central Highlands has resulted in clearing of primary forest for cash crops such as coffee and rubber. Remaining 'rich and medium-quality' forests with high levels of diversity and many highly valued species are constantly subjected to illegal logging and hunting of wildlife. Most high-value timber species in natural conservation forests have been illegally felled. The loss of natural forests and over-exploitation of remaining forest areas mean that biodiversity is steadily being lost.

Conclusion

The *doi moi* process in Vietnam's upland regions is not meeting the government's objectives in improving local livelihoods and achieving sustainable forest management.

The livelihood quality of most better-off villagers has improved due to expanded agricultural and forestry activities, but this has not been the case for poor people. The majority of poor households still face food shortages for many months in every year, and are disadvantaged by their low level of education, high unemployment, poor infrastructure and poor government services. Lack of land for cultivation and lack of experience and knowledge of new crop production techniques have also contributed to high poverty rates. Inequity in both allocation of land-use rights and access to common resources has led to a widening gap between better-off and poor people.

Forests and forest lands have not been managed in a sustainable way. The extent of forests increased significantly in the years following the introduction of *doi moi*, but the quality of both new plantations and existing forests is very low and forests remain under pressure from illegal logging and unsustainable conversion to agricultural uses. In the Central Highlands, forest management has failed, and most primary forests with high economic and environmental value have been destroyed following *doi moi*. Those primary forests that remain are facing ever-greater rates of deforestation. The stage has been reached where the authorities who bear responsibility for this situation must be called upon to answer the question of what will happen if the Central Highlands region is totally stripped of primary forest?

Governance issues, such as ineffective and weak institutions, remain an obstacle to the ability of *doi moi* to achieve sustainable upland forest management. Inadequate services and a lack of transparency in local government activities are major problems for the implementation of *doi moi* policies. As Sayer and Maginnis (2005, p7) noted: 'forest is only well managed when formal institutions are effective and civil society is mobilized to defend the interests of diverse stakeholders'. Therefore, the further decentralization of forest management under the *doi moi* policy reforms is an issue that calls for careful consideration.

References

Alther, C., Castella, J-C., Novosad, P., Rousseau, E. and Tran, T. H. (2002) 'Impact of accessibility on the range of livelihood options available to farm households in mountainous areas of Northern Vietnam', in J-C.Castella and D. Q. Dang (eds) *Doi Moi in the Mountains: Land Use Changes and Farmers' Livelihood Strategies in Bac Kan Province, Vietnam*, The Agricultural Publishing House, Hanoi, pp121-146

Fforde, A. and Goldstone, A. (1995) *Vietnam to 2005: Advancing on all Fronts*, The Economist Intelligence Unit (EIU), London

GoV (1988) *Resolution 10 of the Politburo of the Communist Party of Vietnam*, Central Committee of the Communist Party of Vietnam, National Political Publishing House, Hanoi

GSO (2000) *Statistical Data of Vietnam: Agriculture, forestry and fishery 1975-2000*, General Statistics Office of Vietnam, Statistical Publishing House, Hanoi

Kerkvliet, B. J. T. (2005) *The Power of Every Day Politics: How Vietnamese Peasants Transformed National Policy*, Cornell University Press, Ithaca, New York

MARD (2006) *Decision No. 1970/QD-BNN-KL, dated 06/7/2006 on Announcing the Current Forest Status of Vietnam in 2005*, Ministry of Agriculture and Rural Development, Hanoi

MARD (2014) *Decision No. 3322/QD-BNN-TCLL, dated 28/7/2014 on Announcing the Current Forest Status of Vietnam in 2013*, Ministry of Agriculture and Rural Development, Hanoi

NAoV(1993) *Land Law 1993*, National Assembly of Vietnam, National Political Publishing House, Hanoi

Sayer, J. A. and Maginnis, S. (2005) 'New challenges for forest management', in J. A. Sayer and S. Maginnis (eds) *Forests in Landscapes: Ecosystem Approaches to Sustainability*, Earthscan, London, pp1-16

Sikor, T. (2001) 'The allocation of forestry land in Vietnam: Did it cause the expansion of forests in the northwest?', *Forest Policy and Economics* 2, pp1-11

Tran, N. T. (2005) 'Does devolution really influence local forest institution? Two case studies in the Central Highlands of Vietnam', in P. Cuasay and C. Vaddhanaphuti (eds) *Commonplaces and Comparisons: Remaking Eco-political Spaces in Southeast Asia*, Regional Center for Social Science and Sustainable Development, Chiang Mai, Thailand, pp165-178

Tran, T. T. H. and Hoang, V. C. (2013) 'Climate change and livelihoods of people living in protected areas: A critical perspective', proceedings of an international interdisciplinary dialogue conference *Intergrating Knowledge: The multiple Ways of Knowing Vietnam*, 16-17 December 2013, Thai Nguyen University, University of Hawaii, Vietnam National University and Monash University, pp262-265

World Bank(1994) *Vietnam Changed to Market Economy*, National Political Publishing House, Hanoi

Zingerli, C., Castella, J-C., Pham, H. M. and Pham, V. C. (2002) 'Contesting policies: Rural development versus biodiversity conservation in the Ba Be National Park area, Vietnam', in J-C. Castella and D. Q. Dang (eds) *Doi Moi in the Mountains: Land Use Changes and Farmers' Livelihood Strategies in Bac Kan Province, Vietnam*, The Agricultural Publishing House, Hanoi, pp249-275

Note

1. Agricultural land is characterized by a slope of less than 15 degrees.

41

CHANGES IN SPECIES DISTRIBUTION AND PLANT RESOURCES AFTER THE CESSATION OF SWIDDEN CULTIVATION IN NORTHERN THAILAND

Maki Fukushima*

Introduction

Swidden cultivation, also known as shifting cultivation, is still practised by many tribal people in mountainous northern Thailand. In swidden cultivation systems, fields are left fallow after about one year of cropping and the forest is allowed to regenerate for a period between several years and more than a decade. As a result, village areas typically have a patchy forest cover ranging from undisturbed forest to secondary forest at various stages of regrowth. Young fallow forest stands are usually rich in diversity, because many pioneer species can exist in open, lightened environments. Ridge tops, steep slopes and stream valleys where swiddeners do not cultivate exhibit increased species richness.

Recent studies have reported that wild plants gathered from secondary forests are important sources of food, medicine, firewood, tools and building materials for daily consumption and use (Nawichai, 1999; Johnson and Grivetti, 2002; Delang, 2006). These studies demonstrated that plant species appearing in fallow stands of various ages are utilized by swidden farmers. After being outlawed in Thailand in 1989, swidden cultivation gradually stopped and alternative crops were introduced, beginning around the 1980s (Thomas et al., 2004). When swidden cultivation is no longer practised, the fields are abandoned and forest succession is allowed to progress. Plant composition in older secondary forests changes from that in young fallow forests under the swidden cultivation cycle. This may potentially affect the availability of plant resources in forests, and although many useful plants have been replaced by manufactured goods for daily use, wild forest plants are still important for people living nearby.

---

* MAKI FUKUSHIMA, Lecturer, Tsuru University, Yamanashi, Japan.

In order to clarify the effect of land-use change on species diversity in the forests of northern Thailand, this chapter examines the differences in available plant resources in fallow regrowth ranging in age from the cessation of swidden cultivation to nine years old in a village where swidden cultivation was continuing at the time of the study, and in another village where swidden cultivation had ceased more than 20 years earlier.

Study sites and methods

The study villages were two Karen communities located in the highlands of Chiang Mai province (Figure 41-1). Mae Yot village (Y village) is located in the westernmost part of Chiang Mai province, about 80km northwest of Chiang Mai city. Y village was established by Karen people who came from Myanmar around the middle of the 19th century. In 2006, it had a population of 196 people living in 37 households. They practised swidden cultivation in a cycle involving one year of cultivation followed by 10 years of fallow. The main crop was upland rice, along with various vegetables, herbs and spices. Households opened their fields adjacent to one another to form a large swidden around the village area. They avoided opening swiddens on ridge tops and in valleys near streams. As a result, young fallows were created every year around the village.

Mae Klang Luang (K village) is located in the foothills of Doi Inthanon, within one of Thailand's best known national parks. Karen people probably began settling in the area around K village in the 1890s (Mischung, 1986). K village itself was founded in 1947, when it separated from the founding village. Wet rice cultivation was the dominant form of agriculture, but many households cultivated upland rice in swiddens. The central government proclaimed Doi Inthanon a national park in 1972, and controls over cultivation in the forest area were tightened. Since then, swidden rice fields have been gradually abandoned and shifting cultivation ceased completely by the late 1980s. In 2004, K village had a population of 261 people living in 55 households. They relied mainly on wet rice cultivation. In the village's former swidden cultivation area, secondary forest had recovered. According to older villagers, dry season forest fires occurred frequently in the western part of the area until the 1970s. The frequency of forest fires decreased in the 1980s and disturbances to forest re-growth by forest fires has been insignificant in the past decade (Fukushima et al., 2008).

Ecological censuses were carried out in both Y village and K village. The total number of census plots in Y village was 54 (41 plots in fallow forests, 13 in uncultivated forests) and the number of plots in K village was 30 (23 plots in old abandoned fallow forests, seven in uncultivated forests). Herbs, vines, woody climbers and tree species were surveyed in Y village, but only tree species were surveyed in K village. Therefore, this chapter relates only to tree species. After the tree censuses, I interviewed two villagers (a 47-year-old man in Y village and a 50-year-old man in K village) about the use of all species that were found in the census plots. These

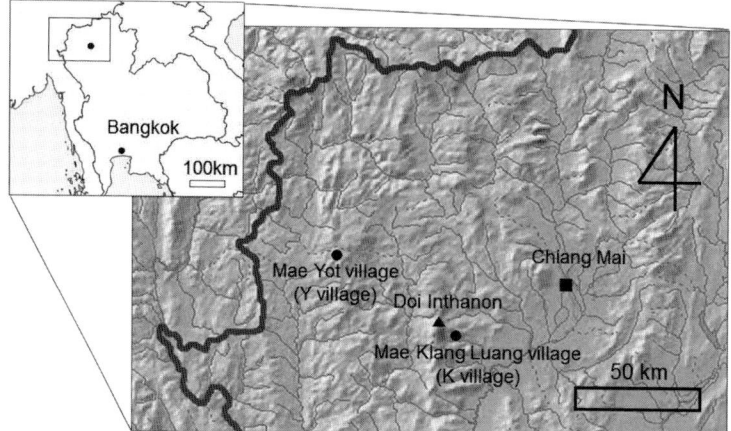

FIGURE 41-1: Location of the study sites in northern Thailand.

villagers also helped during the tree censuses and supplied the local names of the tree species. We also interviewed other villagers, including a 37-year-old woman in Y village and a 34-year-old woman in K village. Plant uses were divided into eight categories: firewood, medicine, food, tools, dyes, fodder and ritual. The latter three categories were not compared between the two villages because fewer than 10 species were collected in these categories. The Jaccard similarity index was calculated as $C/(A + B + C)$, where A and B were the number of species found only in Y village and K village, respectively, and C was the number of species found in both villages.

Results and discussion

Tree species found in both villages

In Y village, 249 tree species were identified from 71 plant families and in K village, 170 species were identified from 58 families. Of these, 165 species were found only in Y village, and these included 19 species of Euphorbiaceae, 13 species of Lauraceae, 11 species of Fagaceae and 11 species of Rubiaceae. Only 86 species were found solely in K village, including 12 species of Lauraceae, eight species of Fagaceae, six species of Theaceae and five species of Euphorbiaceae. There were 84 species in 40 plant families common to both villages. The Jaccard similarity index of species composition between the two villages was 0.25.

Species that appeared with high frequency in the fallow forests of Y village included *Rhus chinensis* (Anacardiaceae), *Glochidion sphaerogynum* (Euphorbiaceae) and *Macaranga denticulata* (Euphorbiaceae) (Table 41-1a). All are short-living pioneer species. Species that appeared with high frequency in uncultivated forests on ridge tops included *Wendlandia tinctoria* (Rubiaceae), *Buchanania glabra* (Anacardiaceae), *Quercus mespilifolia* (Fagaceae) and *Castanopsis argyrophylla* (Fagaceae), while species that appeared with high frequency near valleys with streams included *Xanthophyllum virens* (Polygalaceae), *Knema lenta* (Myristicaceae) and *Tarennoidea*

TABLE 41-1: Species observed with high frequency in fallow stands and uncultivated stands in Y village. Forty-one stands were sampled from fallow forests and 13 stands from uncultivated forests.

Species name		Frequency (%)
(a) Fallow stands		
1 *Rhus chinensis*	Anacardiaceae	80
2 *Macaranga denticulata*	Euphorbiaceae	73
3 *Eurya acuminata* var. *wallichiana*	Theaceae	73
4 *Glochidion sphaerogynum*	Euphorbiaceae	73
5 *Diospyros glandulosa*	Ebenaceae	66
6 *Gluta obovata*	Anacardiaceae	66
7 *Aporosa octandra*	Euphorbiaceae	63
8 *Phoebe lanceolata*	Lauraceae	56
9 *Melastoma malabathricum* ssp. *normale*	Melastomataceae	56
10 *Maesa montana*	Primulaceae	56
An additional 193 species were observed		
(b) Uncultivated stands		
1 *Olea salicifolia*	Oleaceae	62
2 *Gluta obovata*	Anacardiaceae	54
3 *Schima wallichii*	Theaceae	54
4 *Tarennoidea wallichii*	Rubiaceae	54
5 *Phoebe lanceolata*	Lauraceae	46
6 *Nephelium hypoleucum*	Sapindaceae	46
7 *Lithocarpus elegans*	Fagaceae	46
8 *Prismatomeris tetrandra* ssp. *tetrandra*	Rubiaceae	46
9 *Dalbergia cultrata*	Leguminosae	46
10 *Knema lenta*	Myrsticaceae	46
An additional 181 species were observed		
Commonly found in (a) and (b)	130 species	
Total number of species found in Y village	249 species	

wallichii (Rubiaceae) (Table 41-1b). These species are non-pioneers typically found on ridge tops and near stream valleys.

In K village, species that appeared with high frequency in old abandoned fallow forests included *Schima wallichii* (Theaceae), *Machilus bombycina* (Lauraceae), *Wendlandia tinctoria* (Rubiaceae) and *Aporosa villosa* (Euphorbiaceae). The number of pioneer trees was low (Table 41-2a). Fagaceae species such as *Castanopsis diversifolia, Lithocarpus cerifer* and *Castanopsis acuminatissima* appeared with high frequency in uncultivated forests (Table 41-2b). These species were also found in old fallow forests near to uncultivated forest areas.

Species used in both villages

The numbers of species used for firewood, medicine, food, tools and construction materials were 94, 20, 73, 28 and 28, respectively. The total number of species used in Y village was 175, with 72 species having several uses. In K village, the total number of species used was 112, with 58 species having several uses, which was less than in Y village. The number of species used for firewood, medicine, food, tools and construction materials in K village were 83, 18, 42, 9 and 40, respectively. In total, 51 species were used in both villages. More species were used for food and tools in Y village than in K village and more species were used for construction materials in K village than in Y village (Table 41-3).

TABLE 41-2: Species observed with high frequency in fallow stands and uncultivated stands in K village. Twenty-three stands were sampled from fallow forests and seven stands from uncultivated forests.

Species name		Frequency (%)
(a) Fallow stands		
1 *Machilus bombycina*	Lauraceae	91
2 *Schima wallichii*	Theaceae	91
3 *Wendlandia tinctoria*	Rubiaceae	83
4 *Decaspermum parviflorum*	Myrtaceae	65
5 *Eurya acuminata* var. *acuminata*	Theaceae	65
6 *Aporosa villosa*	Euphorbiaceae	65
7 *Styrax benzoides*	Styracaceae	61
8 *Glochidion sphaerogynum*	Euphorbiaceae	48
9 *Symplocos macrophylla*	Symplocaceae	48
10 *Turpinia pomifera*	Staphyleaceae	43
An additional 140 species were observed		
(b) Uncultivated stands		
1 *Schima wallichii*	Theaceae	86
2 *Symplocos macrophylla*	Symplocaceae	86
3 *Machilus bombycina*	Lauraceae	71
4 *Styrax benzoides*	Styracaceae	71
5 *Turpinia pomifera*	Staphyleaceae	71
6 *Lithocarpus cerifer*	Fagaceae	71
7 *Castanopsis acuminatissima*	Fagaceae	71
8 *Myrica esculenta*	Myricaceae	71
9 *Castanopsis calathiformis*	Fagaceae	71
10 *Castanopsis diversifolia*	Fagaceae	57
An additional 83 species were observed		
Commonly found in (a) and (b)	53 species	
Total number of species found in K village	170 species	

TABLE 41-3: Number of tree species used in Y village, where shifting cultivation is still practised, and K village, where shifting cultivation ceased in the 1980s. Some species observed were not used in either village.

Total number of species	Number of species used for					
	Fuel	Medicine	Food	Tools	Building	Total
Observed in Y village (A)	100	28	78	29	31	249
Observed in K village (B)	91	22	42	14	40	170
Observed in both villages (C)	48	15	27	12	18	84
Used in Y village (a)	94	20	73	28	28	175
Used in K village (b)	83	18	42	9	40	112
Used in both villages (c)	34	3	22	6	15	51
Similarity in species (C/A+B+C)	0.34	0.43	0.29	0.39	0.34	0.25
Similarity in use (c/a+b+c)	0.24	0.09	0.24	0.19	0.28	0.22

Of the 84 species that commonly occurred in both villages, 51 were used in both villages, 15 were used only in Y village, six species were used only in K village and 12 were not used. Thus, the similarity index of utilized species (= 0.22) was not much different from the index for the species that occurred in both villages (= 0.25). However, the similarity index for species used for each purpose was variable. The similarity index of species used for medicine was much smaller than the index for the total number of species observed in this category because many common species were not used by either village. On the other hand, the index for species used for food and tools was not much different from the index for species observed in these categories. The differences in similarity indexes for species used for firewood and construction materials were slightly smaller than those for species observed in these categories.

Rhus chinensis Mill. [Anacardiaceae]

Appears with high frequency in the fallow forests of Y village (Mae Yot), where shifting cultivation continues. This short-living pioneer species forms galls when infested with aphids, and the galls are used in Chinese medicine to treat coughs, diarrhoea, night sweats, dysentery and to stop intestinal and uterine bleeding. Compounds from the tree also possess many medicinal qualities.

Discussion

Succession of forest and plant resources

The effects of changes in species composition on available plant resources varied between the two villages. The species used for firewood were similar in Y and K villages, probably because species observed in various successional stages were available for that purpose. On the other hand, the numbers of species utilized for food and tools were higher in Y village than in K village, probably because pioneer species that were able to exist in open, disturbed areas may have had an important role in supplying food and tools in Y village. Young secondary forests maintained in continuous swidden cultivation may be re-evaluated as sources of these kinds of plant resources, as Kunstadter (1978), Schmidt-Vogt (1999), Johnson and Grivetti (2002) and Delang (2006) have pointed out. Species useful for building materials were more common in K village. The recovery of climax species such as those of the Fagaceae and Lauraceae families contributed to the supply of building materials and these species were also an ecologically important element of old-growth forests.

Commonality and differentiation in knowledge of plant resources

The plant resources in secondary vegetation used by people in the two villages were considerably different (similarity indexes ranged from 0.09 to 0.28). These differences were partly due to the differences in flora between the two villages. The total similarity of utilized plants (0.22) was close to the total floral similarity (0.25), indicating that the difference of utilized plants was concordant with the difference of flora occurrence. However, when the focus was changed to medicinal plants in both villages, another factor seemed to be more important.

Commonality and differentiation in the knowledge of plant resources are not only related to differences in the flora available, but also to cultural knowledge. The similarity index of species used for medicine (0.09) was much lower than that of floral similarity in this category (0.43) because many common species were not often utilized in both villages. This suggests that knowledge related to plant resources for medicine might have accumulated independently from community to community. On the other hand, the similarity indexes for species utilized for food and tools were not much different from those for species observed in these categories. This suggests that knowledge about plant resources for food and tools might be closely related to the species available in forests around a community; villagers utilize almost all of them if

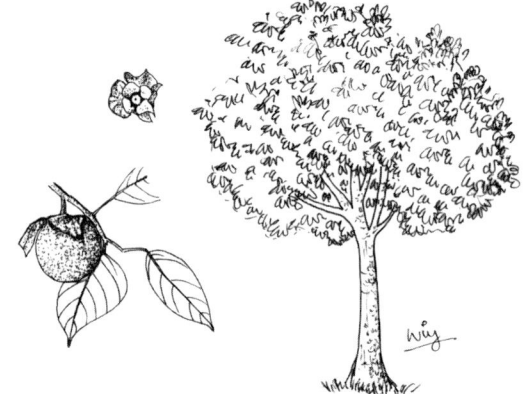

Diospyros glandulosa Lace
[Ebenaceae]

A pioneer species appearing with high frequency in the fallow forests of Y village, where shifting cultivation continues to be practised. It is valued for its timber and as a food source; its fruit are eaten raw when fully ripe. Its fast growth also makes it valuable in reforestation projects.

they are available. Thus, changes in species composition after the cessation of shifting cultivation may erode the knowledge of plant resources for food and tools. If certain species are no longer available, the knowledge about these plant resources may also disappear from a community.

Conclusion

Forest succession after the cessation of swidden cultivation clearly changes the availability of plant resources. The surveys in northern Thailand showed that young forest stands under rotational shifting cultivation were rich in plants used for food and making tools. But as farmers stop burning swiddens and second-growth forests disappear, the use of certain plants may cease along with traditional agricultural practices. In this point of view, rotational shifting cultivation is not only a tool to produce agricultural foods, but also a tool to supply multi-valued forest products.

When the species composition changes, plant resources used for food, tools and medicine may be seriously affected. This study focused on tree species, but many herbs, vines and climbers are also abundant in young fallow forests under shifting cultivation and supply food and medicines to upland communities. Various approaches are needed to the issue of conserving species diversity. One such approach is the planning of plant conservation according to ethnobotanical principles, particularly when villagers' lives are still closely related to forest products.

Conservation of existing old-growth forest is important to protect species that are intolerant of the processes of slashing and burning and will disappear from secondary fallow forest after repeated swidden cultivation cycles. At the same time, swidden cultivation should be re-evaluated as an ecologically adapted agricultural method that provides a basis for the life and culture of people in mountainous areas.

References

Delang, C. O. (2006) 'Indigenous systems of forest classification: Understanding land use patterns and the role of NTFPs in shifting cultivators' subsistence economics', *Environmental Management* 37(4), pp470–486

Fukushima, M., Kanzaki, M., Hara, M., Ohkubo, T., Preechapanya, P. and Choocharoen, C. (2008) 'Secondary forest succession after the cessation of swidden cultivation in the montane forest area in northern Thailand', *Forest Ecology and Management* 255, pp1994–2006

Johnson, N. and Grivetti, L. E. (2002) 'Environmental change in northern Thailand: Impact on wild edible plant availability', *Ecology of Food and Nutrition* 41(5), pp373–399

Kunstadter, P. (1978) 'Sustainable agriculture economics of Lua' and Karen Hill Farmers, Mae Sariang district, northern Thailand', in P. Kunstadter, E. C. Chapman and S. Sabhasri (eds) *Farmers in the Forest: Economic Development and Marginal Agriculture in Northern Thailand*, East-West Center, The University Press of Hawaii, Honolulu

Mischung, R. (1986) *Environmental 'Adaptation' among Upland Peoples of Northern Thailand. A Karen/Hmong Case Study*, final research report to the National Research Council of Thailand, Bangkok

Nawichai, P. (1999) 'Use of wild plants in Karen women's livelihood systems', Master's thesis to the Graduate School of Agriculture, Chiang Mai University, Chiang Mai

Schmidt-Vogt, D. (1999) 'Swidden farming and fallow vegetation in northern Thailand', *Geoecological Research* 8, Franz Steiner Verlag, Stuttgart, Germany

Thomas, D. E., Preechapanya, P. and Saipothong, P. (2004) *Developing Science-Based Tools for Participatory Watershed Management in Montane Mainland Southeast Asia*, final research report to the Rockefeller Foundation, World Agroforestry Centre, Chiang Mai

COLOURED PLATES* ...Part II (Plates 32-69)

In Khonoma village, Nagaland, an Angami Naga research assistant enjoys a moment of mirth, usually at her boss's (the Editor) expense. Zeroniü (Carolin) Meru is a member of the powerful M-khel in Khonoma, and played a large role in the successful completion of the Editor's research in Nagaland.

Sketch based on a photo by Malcolm Cairns (2002)

---

*
Please refer to the regional map, at the beginning of Part I of the Coloured Plates Section, to see where each photo was taken.

Part II picks up where Part I ended, with subject 7. GOVERNMENT POLICY ALWAYS SOUGHT TO REPLACE SHIFTING CULTIVATION WITH SOMETHING ELSE.

B. Plantation crops that enjoy strong market demand. The most significant of these are oil palm and rubber.

32. The steep slope on which this young oil palm is planted was earlier community forest land used for shifting cultivation. It is near West Phaileng, Mamit district, Mizoram in northeast India. The state policy here is to promote and subsidize oil palm production, with the output sold to private companies.

Photo: T. R. Shankar Raman (2014)

The explosive growth of plantation cropping and investment in hitherto remote areas has given powerful interests a motive to 'grab' land that was previously used by shifting cultivators. This is often done with the connivance of state governments.

33. Strong market demand from China has resulted in rubber plantations replacing shifting cultivation in large parts of northern Laos, like this area in Namtha district, Luang Namtha province.

Photo: Vongpaphane Manivong (2011)

While a steady income from rubber is a good thing, the price of rubber latex is volatile, and when it falls, farmers may be unable to afford food from outside. They may be left to regret that 'you can't eat rubber!'

Coffee and tea plantations have also become popular in some upland areas.

34. Terraced hybrid tea was introduced to this Hani (Akha) community at Mengsong, in Xishuangbanna prefecture of China's Yunnan province, as part of a government poverty-alleviation and alternative to shifting cultivation programme.

Photo: Xu Jianchu (2009)

35. Coffee cultivation has transformed Vietnam's Central Highlands. This plantation in Ia Grai district, Gia Lai province, was part of the coffee 'gold rush'.

Photo: Truong Hong (2012)

The explosion of coffee plantations in Vietnam's Central Highlands was so spectacular that it single-handedly depressed world coffee prices (Hong, ch 36).

C. With policy-makers focused on replacing shifting cultivation with something else, there has been a policy vacuum where it comes to conserving the rich germplasm of traditional swidden crops, which humankind can ill afford to lose.

37. Foxtail millet (*Setaria italica* L. Beauv.) is a subsistence crop that is rapidly disappearing. Shifting cultivation is also retreating before the march of plantation crops in the Khagrachari district of Bangladesh's Chittagong Hill Tracts, where this photograph was taken.

Photo: Santua Tripura (2012)

38. Hyacinth beans (*Lablab purpureus* (L.) Sweet) are often planted at the base of trees that had been heavily pruned, providing a trellis for a climbing crop.

Photo: Malcolm Cairns (2001)

36. Asia's shifting cultivators typically plant up to a dozen varieties of upland rice (*Oryza sativa*) in their fields. This makes the swiddens of Asia the 'Noah's Ark' of traditional rice varieties.

Photo: Malcolm Cairns (2001)

As shifting cultivators have been drawn more and more into the cash economy, they have tended to plant the few crops that can earn them the highest returns, and a wide range of subsistence crops have disappeared. Even if farmers wanted to plant them, it is too late. They are gone forever. Some of them may have been genetically suited to degraded soils or survival in the vagaries of climate change. Their loss is a tragic step backwards in the quest for global food security.

D. Some governments have sent their forest departments into shifting cultivators' fields to plant trees and claim the land as state forest.

39. Taking land from farmers, the Royal Forest Department planted pine trees in this field of dryland rice in northern Thailand.

Photo: Malcolm Cairns (1992)

40. Another field in the same village was similarly planted to trees. It is likely that villagers were hired to slash the weeds and care for the trees in a *taungya*-style system.

Photo: Malcolm Cairns (1992)

This Royal Forest Department programme in Thailand appears to take inspiration from the *taungya* system, which originated in Burma. For the first few years after planting the trees, the farmers' routine care for their crops provides simultaneous benefits to the growing trees.

E. Many Asian governments regard grasslands in mountainous areas as undesirable, so in places where grasslands have developed, they rush in to plant trees in the hope of converting it 'back' into forest. However, there may have been no 'unnatural' reason for the area being without forest in the first place.

41. The south slopes of Doi Chang, in Chiang Rai province, northern Thailand, were covered by *Imperata cylindrica* grass after opium production ceased. The grass was used to graze villagers' cattle until the Royal Forest Department planted it with *Pinus kesiya* trees. With their source of grazing thus lost, the cattle then had to be sold.

Photo: Trevor Gibson (1976)

42. A 32-year-old *Pinus kesiya* plantation at Pa Kia, south of Doi Chiang Dao, Chiang Mai province, northern Thailand, showing poor biodiversity. The site was covered by lush herbaceous vegetation before the Royal Forest Department planted the pines.

Photo: Trevor Gibson (2014)

There is believed to be a link between opium cultivation and the development of *Imperata cylindrica* grasslands. Opium poppies thrive in soils no longer fertile enough for cultivating rice, so at the end of rice cropping, farmers would plant poppies. This is believed to have severely depleted soil fertility, to the point where the only plant that would grow was *Imperata* grass, which, in turn, inhibited regeneration of the forest (McKinnon, ch 6).

F. Another 'top-down' policy: for some years, contour hedgerows and grass strips were heavily promoted to counter erosion on sloping lands. The problem was, very few farmers adopted them.

43. This small field of contour hedgerows in Chiang Rai province, northern Thailand, was most notable because it was isolated and none of the neighbours had copied it.

Photo: Malcolm Cairns (1989)

44. Farmers at Claveria, Mindanao, the Philippines, found that they could achieve some of the same benefits as contour hedgerows simply by leaving natural vegetative strips (NVS) across the slope contours.

Photo: Malcolm Cairns (1996)

Farmers' rejection of contour hedgerows raised the fundamental 'top-down' versus 'bottom-up' issue in agricultural development policy. Did soil erosion loom much larger in the eyes of researchers than it did in the concerns of farmers? Did shifting cultivators prefer to use their own techniques for reducing soil erosion? What if, at the outset, we had sought the opinions of farmers?

G. Encouraging shifting cultivators to intensify their land use to permanent cultivation is a common policy objective. Some projects offer draught animals as a first step down that path.

45. Faced with shorter swidden cycles, Jinuo farmers in Jinghong county, Xishuangbanna, in China's Yunnan province, resorted to burning their crop residue and cultivating for longer periods before fallowing the land. Ploughing proved to be the most efficient way of releasing the soil nutrients.

Photo: Xu Jianchu (1996)

46. Farmers in Bukidnon province, Mindanao, the Philippines, struggle with a tree stump to clear a path for their plough. This is typical of lowland farmers who migrate into the uplands and permanently convert forests into agricultural use. A shifting cultivator would happily have dibbled his way around this stump.

Photo: Malcolm Cairns (1996)

The ways that farmers manage their land will depend largely on the scarcity of labour vis–à–vis land.

47. A villager in a newly planted taro garden on the Mamusi Plateau in New Britain, Papua New Guinea. Here, sweet potato has been adopted as a co-staple with taro in response to population growth and increasing pressure on the land. Logs are placed across the slope to reduce soil erosion.

Photo: Mike Bourke (1995)

48. Increasing land-use pressures and greater availability of labour may make it feasible to carve sloping fields into bench terraces, such as these near Baguio, in the Philippines. High-value vegetable crops are grown for urban markets; fertilizers are applied and shifting cultivation has long since disappeared.

Photo: Malcolm Cairns (1993)

Plates 47 and 48 represent polar opposites in land-use intensity. The Papua New Guinea villager (left) is planting crops for his own family's needs, while the dryland terraces near Baguio, in the Philippines (right), are used for market gardening, with high financial returns. A vital precondition for investing in terracing like this is secure property rights.

H. Governments have long encouraged the expansion of wet-rice terraces to reduce reliance on upland rice grown in swiddens.

Photo: Malcolm Cairns (1992)

50. Shifting cultivators often supplement their swiddening with wet-rice terraces in valley bottoms, such as these in Chiang Mai province, northern Thailand.

It seems certain that, despite a long-standing debate on the issue, developing permanent irrigated fields eases pressure to clear forest land.

Photo: Laurent Chazée (1994)

49. Pounoy swidden farmers in Phongsaly province, northern Laos, learnt terraced wet-rice farming from either China or Burma, and it was promoted by the government after 1975. Construction of terraces was considered worthwhile only in clay soils able to retain water and where gravity-fed irrigation was available.

A combination of upland rice on the slopes and paddy rice in the valley bottoms is a wise strategy. On years when one doesn't perform well, the other will probably do well enough to make up the deficit.

I. Bringing shifting cultivators into mainstream development.

51. Providing roads to isolated swidden communities has been a major part of government strategy to bring shifting cultivators into the mainstream economy and give them the benefits of development. Roadwork adjusts the landscape in Chiang Mai province, northern Thailand.

Photo: Malcolm Cairns (1989)

52. Bamboo pipes deliver water to households in a Palaung-Wa village near Kengtung, in Myanmar's Shan state. Such pipelines can stretch for hundreds of metres from water-collection points in the hills.

Photo: Tim Forsyth (2004)

Building roads like this may involve encasing entire mountainsides in concrete to keep them stable. This village may find that the road brings positive influences, such as improved health, education and income. It may also be a way out for young people in search of work, and this may mean a shortage of labour for shifting cultivation.

Bamboo is to upland villagers what PVC pipe is to their lowland counterparts. Clean drinking water is delivered over long distances by bamboo pipes, and development efforts often give high priority to improving village water supplies.

8. STATE AGENCIES OFTEN PREACH 'THE ERROR OF THEIR WAYS' TO SHIFTING CULTIVATORS

53. This poster, with its devastated landscape, shows Akha people complaining to an elderly woman about drought. The old woman responds that the drought has resulted from trees being cut.

Photo: Tim Forsyth (2004)

54. Here, the same old woman is urging Akha farmers to plant trees. The farmers are following her advice, and the poster shows warmer colours and an environment that is already responding, with fish in the stream and new growth.

Photo: Tim Forsyth (2004)

Colour plates 53 to 55, which go together in a series, convey a clear message that the old ways are bad, but the addition of trees is good.

These posters were made by the international non-profit organization Heifer International and placed by the Thai government in an Akha village in Chiang Rai province, northern Thailand. Government agencies and NGOs tend to blame upland agriculture for drought, although long-standing research has cast doubt on this belief (Forsyth, 2015).

55. The third poster is finished in warm colours. It shows mature trees and bodies of water: the results of forest plantation. Ironically, there are now no people in the image. Upland forest plantations have been proposed as a means of restoring water supplies, but the same plantations have been accused of using large volumes of water and taking land away from farmers.

Photo: Tim Forsyth (2004)

A campaign for shifting cultivators to plant preferred trees in their swidden fields (*jhums*) as a kind of assisted fallow, became the main impetus behind the Canadian-supported Nagaland Environmental Protection and Economic Development (NEPED) project in northeastern India.

56. This roadside sign near Kohima, the capital of Nagaland, urges shifting cultivators to plant valued trees in their swiddens, as a kind of assisted fallow regeneration.

Photo: Malcolm Cairns (1995)

9. WHEN PERSUASION DOESN'T WORK, SOME GOVERNMENTS TRY TO LEGISLATE SHIFTING CULTIVATION OUT OF EXISTENCE

57. These Jing Paw women in Myanmar's Kachin state were forcibly relocated in 2011. They must travel long distances to their swiddens, which are increasingly threatened. Some lands have been appropriated by 'crony' companies and other areas are being notified as 'protected'.

Photo: Oliver Springate–Baginski (2015)

58. A mountainside in Myanmar's Chin state shows the difficulties of registering ownership of shifting cultivation land. The large communal areas, called *lopil*, may contain up to 100 temporary individual plots. The huts in the forest mark further communal plots (Ewers Andersen, ch 50).

Photo: Kirsten Ewers Andersen (2013)

Tenure security is a big concern in Myanmar. Shifting cultivators there rarely have formal land titles to fallowed lands and have little protection against private, large-scale land developers or projects taking their land '*in the interests of the state*' (Mertz and Bruun, ch 2).

10. SOME STATES PERMIT SHIFTING CULTIVATION, BUT RESTRICT THE LENGTH OF THE FALLOW

59. Forested mountains on the road between Thimphu and Wangdue Phodrang, in Bhutan. Here, if land is left fallow for more than 12 years, it reverts to state ownership because the farmer apparently doesn't need it.

Photo: Malcolm Cairns (1996)

Longer fallows provide environmental services that should be rewarded, not discouraged (MacDicken, ch 51). Some have called for the setting of minimum fallow lengths, usually around 10 to 20 years.

60. The landscape mosaic created by shifting cultivation in the mountains between Vientiane and Luang Prabang in the Lao P.D.R. In its efforts to curtail swidden farming, the Lao government introduced a policy restricting the length of fallows to three years.

Photo: Gene Hettel for the International Rice Research Institute (IRRI) (1998)

Long fallow periods usually mean that shifting cultivation systems are sustainable. Therefore, state policies that restrict swidden farmers to brief periods of fallow are accused of forcing such systems into decline.

The mountains of Bhutan (left) and Laos (right) are a study in contrast. However, the apparently uninterrupted forests of Bhutan are not necessarily environmentally better. Shifting cultivation creates mosaic landscapes with forests, fallows and fields that may have a positive effect on environmental management.

11. SOME GOVERNMENTS HAVE FORCED SHIFTING CULTIVATORS OUT OF THEIR VILLAGES AND RESETTLED THEM

61. An Atel family's temporary shelter after being forced to abandon their village in Nakai district, Khamouane province, Lao PDR. Poor families could not afford the cost of an obligatory shift to join a larger, permanent village and stayed for months in temporary shelters made of branches and leaves. Many members of such families died.

Photo: Laurent Chazée (2001)

62. The remains of a Khmu Lu village in Oudomxay province, Lao PDR, which was destroyed in less than two hours by a fire that raced through the closely grouped bamboo and thatch houses. The concentrated resettlement of shifting cultivators along roads, as encouraged by the Lao government, increased the impact of such fires.

Photo: Laurent Chazée (2001)

Governments have historically given different justifications for resettling ethnic minorities. During the Second Indochina War (the Vietnam War), it was usually to put an end to 'insurgency', but by the 1980s, it was increasingly a matter of 'conservation'.

Farmers, police face off over reclaimed land

Violence feared after resettled forest 'encroachers' return home

The Nation
Nakhon Ratchasima

LOCAL authorities are preparing for the mass arrest of hundreds of villagers who have defied an order to move from a forest reserve to a resettlement site and have come back to the village they were evacuated from, the provincial forestry chief said yesterday.

Tension ran high yesterday as about 300 people affected by a government-sponsored resettlement programme were gathering at Nong Yai village in Soeng Sang district and about 100 forestry officials and policemen were surrounding them and blocking the passage to the village.

The villagers, who have been living in makeshift shelters at the village since Saturday, said they would not move out and would rather be arrested.

The villagers, members of about 105 families, were ordered to move from their homes in Nong Yai village to Santisuk village in the same district last September under the relocation programme, for which the army is the key involvement, after the land in question was declared part of Thab-lan National Park.

authorities while the others had agreed to resettle at the new place.

He also denied that many villagers who were evacuated from Nong Yai village faced resistance from residents at Santisuk village.

As of yesterday, some elderly people and children among those occupying Nong Yai village were reported sick after being exposed to rain and strong winds.

A highly-placed source at the Second Army Region, which oversees the northeastern region, said authorities planned to drive the villagers out of the disputed village by arresting their leaders first. About 300 soldiers were put on standby to help take part in the arrests, he said.

A village leader, Taeng Chumhrem, said the villagers decided to move from Sra Takhien temple to Nong Yai village after they were repeatedly threatened by certain people.

"From now on, we won't move to any other place. If they arrest us, they will have to take everyone here," he said.

No senior provincial officials had offered to negotiate with the village leaders after their occupation of Nong Yai village on Saturday.

63. Newspapers in Thailand carried frequent stories about official attempts to evict farmers from the forest in the late 1980s and early 1990s. This one relates to Nong Yai Village in Soeng Sang district, Nakhon Ratchasima province.

Photo: Malcolm Cairns (1992)

64. These carefree Karen cousins in northern Thailand were never forced to move. An anthropologist at the Tribal Research Centre in Chiang Mai successfully argued their case with the government by pointing out that the Karen shifting cultivation system conserved the quality of the soil.

Photo: Peter Hinton (1968)

65. Of the various minority groups known in Thailand as 'hill tribes', the Hmong adapted to the market economy better than most. Here, a couple of Hmong women prepare their trinkets for sale at Chiang Mai's Night Bazaar.

Photo: Malcolm Cairns (1992)

Thailand has long disenfranchised its upland minorities by denying them both citizenship and land titles.

There is an interesting juxtaposition between northern Thailand and Nagaland, on either side of northern Myanmar. The difficulty for Thailand's minorities in seeking Thai citizenship is such that they are effectively denied citizenship of the country in which they have lived since birth. In northeast India, the Nagas want to be independent and refuse Indian citizenship. But the aspirations of the Nagas for independence are militarily suppressed by India, and they are told that they have no choice but to be citizens of India.

Dispossessing shifting cultivators of their ancestral lands usually means they also lose their cultural and spiritual milieux. Those who survive are often found living in pitiful conditions in nearby towns and cities.

66. Her story is unknown. But this woman's plight living on the sidewalks of Chiang Mai with her baby could well be the result of heavy-handed Thai government policies, which shattered many shifting cultivation communities.

Photo: Malcolm Cairns (1992)

67. It is likely that many of Thailand's displaced shifting cultivators found their way to slum communities like this one in Chiang Mai.

Photo: Malcolm Cairns (1992)

Displacement of shifting cultivators often means the complete loss of the social safety net that they enjoyed in their home village, where friends and relatives were ready to lend a helping hand.

12. WITH THEIR BACKS AGAINST THE WALL, SHIFTING CULTIVATORS SOMETIMES PROTEST PUBLICLY TO DEMAND BETTER POLICIES

68. Shifting cultivators and other upland farmers protest against the Thailand government's policies outside Chiang Mai's provincial hall.

Photo: Tim Forsyth (1999)

69. It's doubtful that there were shifting cultivators in this protest. With banners claiming justice, freedom and equality and demanding the return of confiscated land, these were permanent farmers in the Tenasserim division of Myanmar's Mon state, protesting about land-grabbing by the country's military rulers. Nineteen of the protesters were jailed on charges including disturbing government officials on duty and cursing. Officials would have regarded shifting cultivators as occupying a lower rung on the social ladder, so the land-grabbing threat in their territory would probably have been worse.

Photo: The Dawei Project (2014)

References

Colfer, C. J. P. (2015) Personal communication between the Editor and Dr Carol Colfer, an anthropologist serving as a senior associate at the Center for International Forestry Research (CIFOR) in Bogor, Indonesia and as a visiting scholar in Cornell University's Southeast Asia Programme, Ithaca, NY

Colfer, C. J. P., Minarchek, R. D., Aier, A., Cairns, M., Doolittle, A., Mashman, V., Odame, H. H., Roberts, M., Robinson, K. and Van Esterik, P. (2015) 'Gender analysis: Shifting cultivation and indigenous people', in M. Cairns (ed.) *Shifting Cultivation and Environmental Change: Indigenous People, Agriculture, and Forest Conservation*, Earthscan, London, pp920-957

Dove, M. R. (2015) 'The view of swidden agriculture by the early naturalists, Linnaeus and Wallace', in M. Cairns (ed.) *Shifting Cultivation and Environmental Change: Indigenous People, Agriculture, and Forest Conservation*, Earthscan, London, pp3-24

Forsyth, T. (2015) Personal communication between the Editor and Dr Tim Forsyth of the Department of International Development, London School of Economics and Political Science, London

Gunasena, H. P. M. and Pushpakumara, D. K. N. G. (2015) '*Chena* cultivation in Sri Lanka: Prospects for agroforestry interventions', in M. Cairns (ed.) *Shifting Cultivation and Environmental Change: Indigenous People, Agriculture and Forest Conservation*, Earthscan, London, pp199-220

Weimarck, G. (1968) *Ulfshult: Investigations Concerning the Use of Soil and Forest in Ulfshult, Parish of Örkened, During the Last 250 Years*, C. W. K. Gleerup, Lund, Sweden

Yaden, T. A. (2015) Personal communication between the Editor and T. Amenba Yaden, former Deputy Conservator of Forests, Social Forestry Division, Nagaland, India and Project Operations Unit member for the Nagaland Environment Protection and Economic Development (NEPED) project

III. POLICY LESSONS THAT WE SHOULD BE LEARNING

Shifting cultivators are usually intimately familiar with their environments and can achieve a comfortable lifestyle by harvesting a wide range of resources. Here, a shifting cultivator in Bukidnon province, the Philippines, completes the construction of fish traps that will add some welcome protein to his family's diet. When state policy tries to integrate them into the mainstream economy, shifting cultivators find themselves needing completely different skills than those learned for survival as traditional forest dwellers.

Sketch based on a photo by Malcolm Cairns.

A. From research by graduate students

Doctoral-degree students are generally highly motivated, and with the support of their universities and advisors, are often highly efficient in the performance of effective field studies.

Sketch based on a photo by Anonymous.

42

TOP-DOWN OR BOTTOM-UP?

The role of the government and local institutions in regulating shifting cultivation in the Upper Siang district, eastern Himalaya, India

*Karthik Teegalapalli and Aparajita Datta**

Introduction

Shifting cultivation has been described as a form of forest farming in which land under crops, located a practicable distance from a farming village, is often rotated annually (Whittlesey, 1937). While several definitions exist in the literature, the general practice is that the land is cultivated for a few years and then left fallow, usually for a longer period of one to two decades, for the soil to recuperate its fertility, to enable another cultivation phase on the same land. Therefore, the practice can be sustainable over long periods, but only with the availability of surrounding secondary successional forests and primary forests in the landscape that can be cleared to extend the cultivable area as the demand for food grows with increasing human population. However, it is important to acknowledge that there is a traditional form of the practice in which farming communities are closely tied to the rotational farming system and the natural resources of their environment for relatively long periods, and a non–traditional form in which the practice has been undertaken more recently (Watters, 1971).

Shifting cultivation has been practised since the Neolithic period, and is still widespread among hill communities in Asia, Africa and Latin America. Until a century ago, it was still practised in Europe (Mazoyer and Roudart, 2006). Shifting cultivation currently provides subsistence livelihoods for between 300 million and 500 million people around the world (Brady, 1996) and is intricately linked to the cultural, ecological and economic aspects of farming communities (Conklin,1961; Ramakrishnan, 1992).

The practice involves the felling of trees for temporary cultivation, so it is blamed for deforestation, soil erosion, loss of biodiversity and contributing to global climate

* KARTHIK TEEGALAPALLI is a research scholar and DR APARAJITA DATTA is a senior scientist, both from the Nature Conservation Foundation, Mysore, India.

change (e.g. WRI, 1985; Bandy et al., 1993; Kotto-Same et al., 1997; Sivakumar and Valentin, 1997; Ranjan and Upadhyay, 1999). Nevertheless, some argue that shifting cultivation is the only viable agricultural practice in the hilly tropics. This group considers the practice ecologically and culturally appropriate and, under certain conditions, the best means of retaining biodiversity in the landscape (Watters, 1971; Gadgil and Guha, 1992; Ramakrishnan, 1992; Fox, 2000). Monoculture plantations, pasture land, large-scale permanent agriculture, logging for timber, oil-palm plantations and other such economic activities have all been shown to affect a landscape in more destructive ways than shifting cultivation (Fox et al., 2000; Seidenberg et al., 2003; Hashim, 2010). Lambin et al. (2001) argued that deforestation was caused by complex global factors often driven not just by shifting cultivation *per se*, but also by commercial factors such as markets and policies, cash-crop production and changing economic conditions.

The two primary factors often cited as being responsible for this age-old sustainable practice now becoming unsustainable are reduction in fallow length (the period between two cultivation phases) due to production pressures arising from human-population increase and commercialization of the practice (e.g. Kotto-Same et al., 1997; Sivakumar and Valentin, 1997; Ranjan and Upadhyay, 1999). However, O'Brien (2002) and Ickowitz (2006) argue that these are largely 'narratives' with hardly any evidence of shifting cultivation being responsible for the deforestation it is accused of causing. Often, these 'narratives' are based on assumptions that the farming communities involved were once upon a time in harmony with the forests around them, but that they have lost this harmony and developmental steps are crucial in order to modify their destructive farming practices and make them nature-friendly once again. By investigating information from several West and Central African study sites published in the first half of the 20th century, Ickowitz (2006) found that the key argument, that fallow lengths had declined due to human population increase, was unproven. In cases where fallow length had declined, Ickowitz further questioned the assumption that this corresponded to a breakdown of the system or led to unsustainability (see also Schmidt-Vogt et al., 2009). The fallow length of a shifting cultivation system often rests upon decision-making at village level that involves several factors other than population increase, such as the ease of clearing secondary forest, the distance from the village to the secondary forest to be cleared, the availability of cultivable land, the area of cultivable land owned by the household and other topographical variables (Coomes et al., 2000; Seidenberg et al., 2003).

Shifting cultivation is commonly known as *jhum* in India, and in the early 1970s it was being practised by about three million people over an area of about one million hectares per year (Agarwal and Narain, 1985, cited by Ninan, 1992). The practice is now prevalent in the eastern and northeastern regions of India. Northeast India comprises eight States: Arunachal Pradesh, Assam, Meghalaya, Manipur, Mizoram, Nagaland, Sikkim and Tripura. About half a million families from more than 100 different tribal groups practise shifting cultivation (Ramakrishnan, 1992) over an area of about 2.7 million ha in the region (NEC, 1974). Northeast India is characterized by high rainfall that fosters rapid forest regeneration and undulating terrain which leaves communities with few alternatives to *jhum* cultivation (Ramakrishnan, 1992). The region is also rich in biodiversity and comprises the eastern part of the Himalayas and the northwestern part of the Indo-Burma biodiversity hotspots (Conservation International, 2005).

The dichotomy in perceptions of shifting cultivation exists in India as well as in other parts of the world. The practice has been termed wasteful, unscientific and destructive of biodiversity by some, who claim that it must be abandoned for the benefit of the people (Borah and Goswami, 1973; Ranjan and Upadhyay, 1999). Others consider the practice to be relatively benign in comparison with other land-uses, such as monoculture plantations, mining and commercial exploitation of forests for timber, and that it is a diversified practice that is customized to local conditions (Ramakrishnan, 1992; Raman, 2000; Malik, 2003; Teegalapalli, 2008). In comparison with other forms of agriculture, shifting cultivation also promotes agrobiodiversity: in the northeast Indian state of Nagaland, a single farming household was shown to cultivate an average of at least 60 different crops in its shifting cultivation fields (Nakro, 2011). While shifting cultivation is usually undertaken on a relatively small spatial scale, land uses such as monoculture plantations and pastures often cover hundreds to thousands of hectares (Finegan and Nasi, 2004; Teegalapalli et al., 2009). For instance, within only three districts of Assam state alone, about half a million ha was under tea cultivation in 2007 (Sharma et al., 2012). Further, shifting cultivation landscapes are often interspersed with secondary and primary or near-primary forests, and therefore they are better able to support biodiversity than other land uses (Fox, 2000).

The 'narrative' of the relatively long fallow periods of the past being reduced to cater for drastic increases in human populations and the crucial need for developmental and policy interventions to keep the practice feasible (Fox, 2000; O'Brien, 2002; Ickowitz, 2006) is also dominant in India (Borah and Goswami, 1973; Ranjan and Upadhyay, 1999; Inter-Ministerial Task Force, 2008). However, Toky and Ramakrishnan (1981) showed that shifting cultivation, even with fallow periods as short as five years, was often more productive in terms of both monetary returns and energy efficiency than permanent forms of agriculture such as terrace and valley cultivation, and was therefore a practice well suited to the local conditions of the farmers (Ramakrishnan, 1984). They also showed that the productivity of shifting cultivation was higher when the rotational cycle lasted for 10 years, rather than five years. Permanent forms of cultivation also need inputs such as fertilizers and pesticides, which can have negative effects on the environment (Lambert, 1996). Singh (1996) indicated that in Mizoram state, evidence that fallow periods were shortening because of an increasing human population was weak. Rather, the farmers in that state chose to cultivate in cycles of 5 to 10 years, even if longer fallows were possible, because the fallow vegetation was easier to clear. Population density could affect fallow lengths, but only after a critical threshold of human population pressure had been reached.

Top-down: A review of Indian policies and schemes regarding shifting cultivation

Government policies, schemes and interventions aimed at shifting cultivation in India over the past century have been based on the paradigm of terminating, controlling or improving the practice. During the colonial period of British rule (between 1827 and 1947), the impetus was on banning shifting cultivation and weaning farmers away from it. However, policies and schemes over the past 60 years have sought to regulate or improve shifting cultivation by bringing about development and technological innovations. A significant part of the northeast Indian region was referred to administratively as the Assam State by the British before the formation of the current northeastern states in the latter half of the 20th century.

In the early days of British rule, shifting cultivation was seen as a primitive and inferior system of cultivation that needed to be abandoned. This ideology also existed prior to British rule, when the Ahoms from upper Burma ruled over Assam state: shifting cultivation was discouraged and wet-rice cultivation was introduced in the non-hilly parts (Sharma et al., 2012). The earliest effort by the British to modify the practice was in 1890 when the German forester Dietrich Brandis, later to become known as the 'father of tropical forestry', introduced the *taungya* system on behalf of the British administration (King, 1979).[1] In this system, while the *jhum* farmers were allowed to cultivate their crops, they were also obliged to nurture seedlings of timber trees in their swiddens. They could continue annual cropping for two to three years, or until the young trees created too much shade, at which stage they shifted the same practice to another site (Malik, 2003). This system had the twin benefits of weaning farmers away from shifting cultivation and employing them as labourers for commercial forestry operations practised in the Assam state until 1969 (Malik, 2003).

In the early 1900s, British policy aimed to prevent long-term forest degradation and to manage forests in India to ensure a constant supply of timber and commercial forest produce to support their campaign in World War One (Malik, 2003). According to the Assam Forest Regulation Act, 1891, the practice of *jhum*, or shifting cultivation, was 'deemed to be a privilege subject to control, restriction and abolition by the State Government, and not a right'. The State Government had the authority to permit or prohibit shifting cultivation, and even if permitted, the Forest Officer could dictate the conditions under which it had to be undertaken. This approach to shifting cultivation was later reflected in the Indian Forest Act, 1927.

The Balipara/Tirap/Sadiya Tract *Jhum* Land Regulation Act, 1947, also enacted by the British Government, vested the District Commissioner or an officer appointed by the District Magistrate with the powers of a 'Land Conservator'. Rents, taxes or other dues had to be paid by shifting cultivators and if they failed to do so they forfeited the right to continue their farming practices. *Jhum* land could also be acquired by the Land Conservator for public purposes. The Land Conservator also had the power to order afforestation of *jhum* land; direct that some part of the *jhum* land be protected; or order that a particular area of *jhum* land not be cultivated for a period of 10 years, in order to prevent soil erosion. Despite the powers held by the Land Conservator, there was an interesting aspect to this law: if a Tribal Council existed in an area with the approval of the governor of that territory, then the Land Conservator would vest his powers in the Tribal Council (Murtem et al., 2008). This was an important step towards recognition of the regulatory roles of Tribal Councils in shifting cultivation communities. A point worth noting here is that, although several Acts and policies were formulated by the British during the colonial period, it is unlikely that they affected the actual practice of shifting cultivation in remote parts of northeast India. This is particularly the case in some districts of Arunachal Pradesh, such as Upper, East and West Siang, then referred to as the Sadiya Frontier Tract, where the British failed to establish relations with the Adis and had a physical presence only, beginning in the early 1900s (Roy, 1997). The British policy for the Siang districts was mostly one of non-intervention and non-administration, but sought nonetheless to facilitate unimpeded trade and communication between the tribes in the hills and the plains (Anonymous, 1995).

Post-independence policies were based on a slightly different idea: food production had to be increased and technology introduced to improve shifting cultivation (Malik, 2003). In a more egalitarian approach, the National Forest Policy, 1952, sought to persuade tribal communities to give up shifting cultivation, rather than to coerce them to do so, and to amalgamate the practice with the *taungya* system, mentioned earlier (Maithani, 2005). With this backdrop, the first pilot project seeking to control shifting cultivation was launched by the Government of India as part of its Fifth Five-Year Plan in the late 1970s (Maithani, 2005). The focus of the project was to encourage 1700 families in the northeast Indian region to shift to settled cultivation by providing a relatively small 2-hectare area per family for paddy cultivation and plantations. The River Basin Scheme was another project that was initiated at the same time by the North-Eastern Council (Maithani, 2005). In 1994, during the Government of India's Eighth Five-Year Plan, the Watershed Development Project in Shifting Cultivation Areas was launched (Tripathi and Barik, 2003). More recently, the New Land Use Policy was introduced in northeast India with the idea of gradually changing the practice of shifting cultivation by introducing a new pattern of land use, commercial use of domestic resources and marketing of agricultural products. In its first five-year stage, the policy envisioned providing sustainable livelihoods to 120,000 families in 750 villages in the state of Mizoram (Chhetri, 2010). In 2012, 4.1 billion rupees (US$81.46 million) was budgeted under this policy to wean 30,000 families away from shifting cultivation.

The common goal of these schemes and policies in northeast India was to introduce rubber, cashew nuts, tea, oranges, black pepper and other such cash crops, or to provide shifting cultivators with settled-cultivation options to improve their economy. An interesting perspective on these central-government schemes and policies was provided by Maithani (2005): although state governments in northeast India did not consider the practice of shifting cultivation a problem, the schemes were nevertheless welcomed as they resulted in a greater inflow of funds from the centre to the states. However expensive these projects and schemes were, there has been no formal assessment of their outcomes. It is not clear if there was any success at all, because only rarely did these interventions actually benefit the shifting cultivators (Tripathi and Barik, 2003; Maithani, 2005). Such schemes and policies were based on the presumption that shifting-cultivation systems had already broken down in all the areas in which they were practised and that the only feasible options were either to terminate them or improve their production by way of technological intervention. At the same time, improper implementation of such policies and schemes affected local customary institutions that had been regulating shifting cultivation for hundreds of years (Singh, 1996).

The bottom-up: regulation of shifting cultivation by local institutions

Institutions, defined by North (1990) as rules that affect human interactions, play an important role in a shifting cultivation system because several factors regarding the practice involve collective, rather than individual, decision-making. Such institutions, which ensure ecologically and economically sustainable production, have been documented around the world (Ramakrishnan, 1992; Alcorn et al., 2003). Factors such as the land area chosen for cultivation, the allotment of individual farming plots, timing of the farming operations and fallow lengths are all regulated by local decentralized institutions that ensure sustainability and prevent long-term land and forest degradation in shifting cultivation landscapes (Singh, 1996; Malik, 2003; Jyotishi, 2006; Saikia and Bhaduri, 2012).

In India, local institutions, their practices and systems of regulation have been documented in many shifting cultivation communities (Singh, 1996; Borang, 1997; Gupta, 2000; Jyotishi, 2006; Saikia and Bhaduri, 2012). In Tripura state, in northeast India, Gupta (2000) reported that traditional shifting cultivators had a strong element of socio-cultural organization that used to regulate their farming practice: they maintained long fallow periods in excess of 20 years, with some patches of primary or climax forest left uncut in the larger landscape (Gupta, 2000). However, the population of the state grew by about 70% between 1941 and 1981 due to influxes of non-tribal immigrants from neighbouring states and from Bangladesh. The immigrants practised non-traditional shifting cultivation with short fallow cycles of three to five years and affected local, traditional shifting cultivators by forcing a reduction in land available for cultivation. The population increase in Tripura coincided with protection of patches of evergreen primary forest by the state administration and several developmental projects, which caused a further breakdown of traditional institutions. Village councils in the Nagaland state in northeast India are empowered by the Nagaland Village and Area Councils Act, 1978 (Saikia and Bhaduri, 2012), and are responsible for regulating land commonly owned by villages for shifting cultivation. Local institutions in Orissa state have been shown to regulate traditional forms of cultivation such as shifting cultivation by shaping property rights and facilitating village-level decision-making, whereas external sources of funding and markets are more important in that state for terrace cultivation (Jyotishi, 2006).

However, the strength of these local institutions has diminished over the last few decades owing to market pressures, the increase in other forms of cultivation, state- and national-level policies (Jyotishi, 2006; Saikia and Bhaduri, 2012), the ingress of government-driven political systems such as the *Panchayati Raj* (Danggen, 2003), and the impact of religious conversion (Malik, 2003; Aisher, 2007). These factors have also brought about changes in the strong social networks and traditional knowledge associated with shifting cultivation (Malik, 2003).

Despite these pressures, systems of self-governance still exist among several tribal communities in the state of Arunachal Pradesh in the eastern Himalaya. These are called *Kebang* among the Adis, *Nogonthun* among the Noctes, *Builyang* among the Apatanis, *Abbala* among the Idu-Mishmis and *Wancho-Wangas* among the Wanchos, to name just a few of them (Baruah, 1960; Roy, 1997; Dollo et al., 2010). In the early 1980s, Arunachal Pradesh had about 50,000 families practising shifting cultivation over an area of about 2100sq. km, representing less than 2.5% of the state's total area (Government of India, 1983). Nowadays, the Adi community remains mostly subsistent on shifting cultivation. Living in the central parts of Arunachal Pradesh, their farming practices are still efficiently regulated by a democratic institution called the *Kebang*, which encapsulates centuries of Adi tradition (Borang, 1997; Murtem et al., 2008; Raj, 2010). The scope of the *Kebang* extends beyond the regulation of shifting cultivation; it is also responsible for resolving intra-village, inter-village and inter- tribal issues (Danggen, 2003). The *Kebang* also administers village matters such as the water supply and sanitation, communal hunting and fishing decisions, maintenance of paths and bridges and decisions on festival dates (Anonymous, 1995). The first mention of the *Kebang* was perhaps made by Wilcox (1825). Krick (1853) attended a *Kebang* meeting of the *Padam* clan of the Adis in Mebo, a village close to Pasighat, the present headquarters of East Siang District in Arunachal Pradesh. In his words, 'each village is self-governing and independent. It has its own administration, both legislative and executive'. Subsequently, several other authors, such as Dalton (1855), Dunbar (1913) and Elwin (1959), have commented on the efficiency of the *Kebang*.

The *Kebang* has judicial, administrative and development functions and is a three- layered institution. The village *Kebang*, at the bottom, administers village affairs and manages and resolves day-to-day problems in the village; the *Bango* or area *Kebang* is formed between several Adi villages and the *Bogum-Bokang Kebang* is the apex body of the Adis in Upper Siang District (Danggen, 2003). The headman of the village (or *head-gam*), is the leader of the *Kebang* and is elected for his oratory and judicial skills. Within a village, the *Kebang* is largely responsible for allotment of land for shifting cultivation, farming operations, settlement of land-ownership disputes between individuals and the length of the fallow period between two cultivation phases (Danggen, 2003). The *Kebang* is administered within the framework of the Assam Frontier (Administration of Justice) Regulation, 1945, with the village *head-gam* acting also as a representative of the administration at village level (Anonymous, 1995).

More recently, however, the authority of the *Kebang* may have been diluted after the introduction of *Panchayati Raj* in the State in 1968, following the recommendations of the Ering Committee, appointed by the Government of India (Danggen, 2003). The *Panchayati Raj* is the oldest form of local government on the Indian subcontinent. This State-sponsored governance has a three-tier structure with the *Gram Panchayat* at village level, the *Anchal Samiti* at block level and the *Zilla Parishad* at district level. The *Anchal Samiti* is responsible for various developmental and welfare activities, for promoting education, improving agricultural methods and increasing production. Therefore, there seems to be a considerable overlap in the roles of the *Kebang* and the *Gram Panchayat.* There also seems to be some disparity in the functional roles of the State-induced *Panchayati Raj* and the *Kebang.* While the *Anchal Samiti* and *Gram Panchayat* are important for the planning and developmental activities of a village, the *Kebang* is crucial in terms of the functioning of the village and in preserving centuries of local traditions and knowledge.

Case Study: description of the shifting cultivation system of Bomdo village

The Adi community in Bomdo village, in the Upper Siang district of central Arunachal Pradesh state, practises subsistence shifting cultivation, with a lesser area under terrace cultivation. The community comprises about 70 families. We interviewed 20 farmers from the village to gather information pertaining to shifting and terrace cultivation. In particular, we sought details of the area owned and used for these two farming practices; the length of fallow periods; benefits and preferences of one practice over the other; the annual rice production from the two practices; the distance of the fields from the village; and any changes that have occurred in the practise of shifting cultivation over the past two decades. Upper Siang communities have been practising systematic shifting cultivation for centuries, with a relatively long fallow period of 10 years. Relatively large tracts of forests still exist in the landscape. Bomdo village is located about 10km from the 483sq. km Mouling National Park (Figure 42-1) and the Dihang-Dibang Biosphere Reserve, which covers 5100sq. km. Fields for shifting cultivation lie within 6km of the village, although the village's territory extends further. Beyond 6km, the farmers find the effort of trekking to their fields too difficult.

Shifting cultivation is undertaken in Bomdo village by several families together, in fields that were earlier referred to as *patats* (Murtem et al., 2008). These *patats*, which are two to four hectares in size, are located within three blocks in the area surrounding the village. Each family usually owns at least one site suitable for shifting cultivation, and often more than one, within each of the three blocks. The entire village undertakes shifting cultivation in a single block each year, and once an area of secondary forest is cleared, it is usually cultivated for at least two years. After spending six years in blocks 1, 2 and 3, the farmers return to different sites in block 1, later clearing different sites in blocks 2 and 3. This allows a fallow period in excess of 10 years for each site. In the case of a family not owning a cultivable site within a block

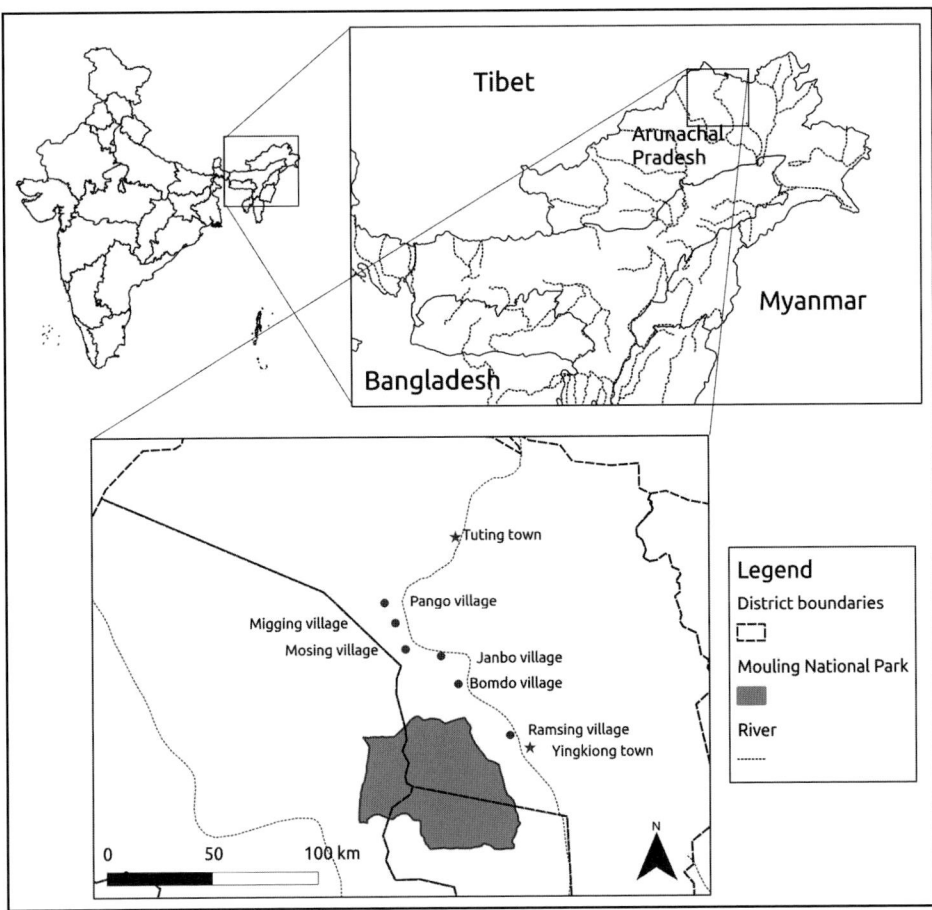

FIGURE 42-1: The location of Bomdo village, in the Upper Siang district of central Arunachal Pradesh.

in a particular year, they may borrow a site that is owned by another family. This arrangement is beneficial to both families if there is a risk that the site may exceed its 10 years of fallow. Sites become increasingly difficult to clear if the vegetation is allowed to regenerate for more than 10 years.

Each of the sites is demarcated by planting *Crinum amoenum*, locally called *riksu sodok*, along the boundaries. This species is fire–resistant and survives for several years, so the site boundaries can be easily identified. On several occasions in Bomdo village, the local *Kebang* has resolved boundary disputes on the basis of the location of the *Crinum amoenum* and the extent of its root, which determines the age of the plant and therefore when it was planted. Interestingly, the farmers also retain the pteridophyte species *Helminthostachys zeylanica*, locally called *Asi Benyé*, in their shifting cultivation fields if it occurs naturally. It is believed that the plant helps to retain soil moisture in the fields, however this has yet to be scientifically investigated. Soil conservation at the field level is undertaken by placing logs perpendicular to the angle of the slope. Instances of families cultivating a site that is not located in the block that the rest of the village is cultivating in a particular year are rare. One of the reasons for this is that additional labour is required for fencing an individual site, whereas fencing an entire cultivated block is a group activity involving the entire village. Crop losses are also mitigated when everyone cultivates together. Granivorous birds and animals feed on rice crops in the months of October and November, before the harvest. To prevent crop losses, the farmers band together to erect scarecrows and assemble bamboo poles that can be pulled to create noise to scare the birds away. The farmers use bow-traps to catch small rodents and set snares around the swiddens for large animals such as monkeys and wild pigs.

Sixteen varieties of rice are cultivated in the shifting- and terrace-cultivation fields at Bomdo village; seven in shifting cultivation and nine in terrace cultivation. Some of the varieties are drought resistant, while others are resistant to high rainfall. These are planted to cover variations in annual rainfall, although the region is

Crinum amoenum Ker Gawl. ex Roxb. [Amaryllidaceae]

Used to mark boundaries between shifting-cultivation sites owned by different households.

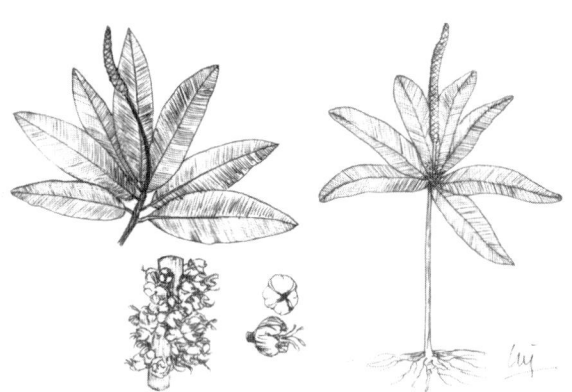

Helminthostachys zeylanica (L.) Hook. [Ophioglossaceae]

Retained in their shifting-cultivation fields by farmers in Bomdo village because they believe the plant results in retention of soil moisture.

more prone to high rainfall. Although it was introduced several decades ago, terrace cultivation arrived in the Bomdo village area of the Siang river valley as recently as 2000, after two canals were built to provide irrigation to the fields. Our interviews in the village indicated that most respondents preferred settled cultivation on the terraces for producing rice because the labour involved was significantly less than that required in shifting cultivation. However, they also said that shifting cultivation had several practical advantages that made the practice indispensable:

- At least 30 species of vegetables that are grown in shifting cultivation fields cannot be grown in terrace-cultivation patches;
- Terrace cultivation involves high initial expenditure, and the use of chemicals makes it relatively more expensive than shifting cultivation, where use of chemicals is negligible;
- Firewood collected during the clearing of swiddens provides essential fuel for warmth and cooking;
- The multi-cropping aspect of shifting cultivation makes it an essential practice because as well as rice, millets and maize are also grown in swiddens;
- Shifting cultivation is a collective activity and is intricately linked to the customs, festivals and traditions of the villagers, while terrace cultivation is an individual activity and has no connections with the village other than individual rice production;
- The terrain of several villages in the district does not allow for terrace cultivation;
- The land-tenure system is suitable for shifting cultivation rather than terrace cultivation. In the latter, land has to be purchased or bartered for and permanently owned, whereas there is 'floating ownership' of land for shifting cultivation;
- Shifting cultivation is better adapted to the local climate and therefore involves less risk than terrace cultivation. For instance, in 2013, the harvest from terrace cultivation was lower because of high rainfall, whereas shifting cultivation was relatively unaffected; and
- The *Kebang* plays a significant role in ensuring that the practice of shifting cultivation remains systematic in Bomdo village.

Conclusion

The National Task Force on Rehabilitation of Shifting Cultivation Areas has reported that there should be 10 to 20 years of fallow after the cultivation phase of shifting cultivation in order to prevent soil erosion and loss of fertility and to allow for forest regeneration. Moreover, it said that these conditions were possible only in areas where the human population density was lower than 20 people per sq. km (Inter-Ministerial Task Force, 2008). The dominant view of the task force was that in regions with high human populations, in the presence of factors such as intrusive markets and alternative economies, shifting cultivation had become ecologically and economically unviable, leading to forest degradation and in some cases, denudation. Our documentation in this chapter shows that in remote areas with poor access to markets and with no practical alternatives, the practice of shifting cultivation is still being undertaken systematically in sites within relatively well forested landscapes.

Research from other villages and the outcome of our research in six other villages in the Upper Siang district (Teegalapalli and Datta, 2016) indicates that the characteristics of shifting cultivation are similar in regions with strong regulatory influence from the local institution, *Kebang* (Borang, 1997; Murtem et al., 2008). The communities, which are almost entirely dependent on shifting cultivation for subsistence, cultivate large patches that are located around their villages. There are relatively large areas of primary forest in the shifting cultivation landscape around most villages. The human population density, at about 5 people per sq. km, is low (Census of India, 2011) and the forest cover is relatively high, at about 79% (Singh et al., 2005), covering a significantly large area of the Dihang-Dibang Biosphere Reserve. Alternative economic opportunities, such as selling ginger, vegetables and firewood, exist only in a few villages that are located close to towns. In villages located further from towns, where the terrain is relatively rugged, shifting cultivation is the mainstay and very few products are sold in markets. We believe that such villages have no practical alternatives to shifting cultivation and any new government policy or scheme should consider the complete dependence of these communities on traditional swiddening. In such areas, strengthening the local institutions can ensure that the practice remains sustainable for many years.

In a proposal referred to as the Shillong Declaration (Kerkhoff and Sharma, 2006), the International Centre for Integrated Mountain Development (ICIMOD) suggested such an intervention in 2004. The significant recommendations made to the Government of India in the declaration were:

- That the practice of shifting cultivation be recognized as an adaptive forest-management practice based on sound scientific and ecological principles;
- That shifting cultivators be able to take part in making decisions about any new policies affecting them; and
- The significance of traditional institutions that regulate shifting cultivation be recognized.

In areas where shifting cultivation being practised by traditional farmers has indeed become unsustainable, the incorporation of composite farming systems can ensure that a landscape remains sustainable and further deforestation can be halted (Cairns and Brookfield, 2011). In this type of farming system – earlier referred to as 'composite swidden-agroforestry systems' by Rambo (1996) – shifting cultivation, settled cultivation, home gardens and tree cultivation may be undertaken simultaneously by a farming community in the landscape surrounding its village (Cairns and Brookfield, 2011). The Nagaland Environmental Protection and Economic Development project (NEPED) initiated a scheme in Nagaland state in northeast India in 1994 to stabilize shifting cultivation through an agroforestry initiative. Farmers in several villages planted alder *(Alnus nepalensis)* trees in shifting cultivation fallows (NEPED-IIRR, 1999). The species has nitrogen-fixing properties, enabling faster recovery of soil fertility and thereby allowing for shorter fallow cycles in an alder-*jhum* system (Nakro, 2011). Trees of a different alder species *(Alnus acuminata)* were also planted in highland pastures in Costa Rica several decades ago for their nitrogen-fixing properties (Holdridge, 1951). Such initiatives and others that have involved amalgamating traditional methods with technological innovations need to be replicated in other sites where shifting cultivation systems have become unsustainable. Such initiatives should be incorporated within government policies (Cairns and Garrity, 1999).

In northeast India – as has been reported in other parts of Southeast Asia – forms of shifting cultivation have been introduced quite recently in some areas by communities that are only partially dependent on the practice. This has been referred to as 'partial shifting cultivation' (Conklin, 1957), and can easily be distinguished from traditional swiddening. The traditional farming system is undertaken for subsistence and is mostly found in a relatively large farm-forest landscape, whereas the partial practice is often located close to an urban fringe and is undertaken for the commercial market (Spencer, 1966; Christanty, 1986). Where such practices are encountered, the only way forward may be to provide the farmers with suitable alternatives, and schemes and policies introduced in India over the past 60 years, aimed at bringing an end to traditional shifting cultivation, may be practically relevant. Similar to variations in the actual practice of shifting cultivation, government interventions should be adapted to the local conditions and to the type of shifting cultivation being undertaken in an area. Well-intentioned, large-scale policies and interventions that set their sights on entire states or regions can cause significant damage at a local level by affecting traditional knowledge and traditional institutions and further marginalizing remote farming communities that have subsisted on shifting cultivation for generations.

Acknowledgements

We thank the Adi community for sharing the information used in preparing this chapter. We also thank the Adit Jain Foundation, the Idea Wild Foundation, the Ashoka Trust for Research in Ecology and Environment (Small Grants for Research in North East India) and the Ravi Sankaran Inlaks Fellowship Programme for funding this research.

References

Agarwal, A. and Narain, S. (1985) *The State of India's Environment 1984-85: A Second Citizens' Report*, Centre for Science and Environment, New Delhi

Aisher, A. (2007) 'Voices of uncertainty: Spirits, humans and forests in upland Arunachal Pradesh, India', *Journal of South Asian Studies* 30, pp479-498

Alcorn, J. B., Bamba, J., Masiun, S., Natalia, I. and Royo, A. (2003) 'Keeping ecological resilience afloat in cross-scale turbulence: An indigenous social movement navigates change in Indonesia', in C. Folkes, F. Berkes and J. Colding (eds) *Navigating Nature's Dynamics,* Cambridge University Press, Cambridge, UK, pp299-327

Anonymous (1995) 'Arunachal Pradesh, East and West Siang district', *Gazetteer of India*, Government of Arunachal Pradesh, Itanagar

Bandy, D. E., Garrity, D. P. and Sanchez, P. A. (1993) 'The worldwide problem of slash-and-burn agriculture', *Agroforestry Today* 5, pp2-6

Baruah, T. K. N. (1960) *The Idu Mishmi*, North-East Frontier Agency, Shillong, India

Borah, D. and Goswami, N. R. (1973) *A Comparative Study of Crop Production under Shifting and Terrace Cultivation (A Case Study in Garo Hills, Meghalaya),* ad hoc study no. 35, Agro-economic Research Centre for North-East India, Jorhat, Assam

Borang, A. (1997) 'Shifting cultivation among the Adi tribes of Arunachal Pradesh', *Journal of Human Ecology* 6, pp145-151

Brady, N. C. (1996) 'Alternatives to slash-and-burn: A global imperative', *Agriculture, Ecosystems and Environment* 58, pp3-11

Cairns, M. F. and Garrity, D. P. (1999) 'Improving shifting cultivation in Southeast Asia by building on indigenous fallow management strategies', *Agroforestry Systems* 47, pp37-48

Cairns, M. F. and Brookfield, H. (2011) 'Composite farming systems in an era of change: Nagaland, Northeast India', *Asia Pacific Viewpoint* 52, pp56-84

Census of India (2011) *Provisional Population Totals*, paper 2 of 2011, Registrar General and Census Commissioner of India, New Delhi, available at http://censusindia.gov.in/, accessed 1 January 2013

Chhetri, P. (2010) 'New Land-Use Policy: Mizoram's Unique Model for development', in *Eastern Panorama*, http://www.easternpanorama.in/index.php?option=com_content&view=article&id=1274:new-land-use-policy&catid=67:february-, accessed 1 January 2013

Christanty, L. (1986) 'Shifting cultivation and tropical soils: Patterns, problems, and possible improvements', in G. G. Marten (ed.) *Traditional Agriculture in Southeast Asia: A Human Ecology Perspective*, Westview Press, Boulder, CO, pp226-240

Conklin, H. C. (1957) *Hanunoo Agriculture: A Report on an Integral System of Shifting Cultivation in the Philippines*, Food and Agriculture Organization of the United Nations, Rome

Conklin, H. C. (1961) 'The study of shifting cultivation', *Current Anthropology* 2, pp27-61

Conservation International (2005) *Biodiversity Hotspots,* Conservation International, Arlington, VA, http://www.conservation.org/where/priority_areas/hotspots/Pages/hotspots_main.aspx, accessed 1 January 2013

Coomes, O. T., Grimard, F. and Burt, G. J. (2000) 'Tropical forests and shifting cultivation: Secondary forest fallow dynamics among traditional farmers of the Peruvian Amazon', *Ecological Economics* 32, pp109-124

Dalton, E. T. (1855) *Correspondence and Journal of Capt. Dalton, Principal Assistant of Luckimpre, of his Progress in a Late Visit to a Clan of Abors on the Dihong River*, Selection from the records of the Bengal Government, No XXII, Calcutta, India

Danggen, B. (2003) *The Kebang: A Unique Indigenous Political Institution of Adis*, Himalayan Publishers, New Delhi

Dollo, M., Gopi, G. V., Teegalapalli, K. and Mazumdar, K. (2010) 'Conservation of the orange-bellied Himalayan squirrel *Dremomys lokriah* using a traditional knowledge system: A case study from Arunachal Pradesh, India', *Oryx* 44(4), pp573-576

Dunbar, G. D. S. (1913) 'Abors and Galongs', *Memoirs of the Asiatic Society of Bengal* 5, pp1-113

Elwin, V. (1959) *A philosophy for NEFA*, 2nd edition, Shillong Publication, Calcutta

Finegan, B. and Nasi, R. (2004) 'The biodiversity and conservation potential of swidden agricultural landscapes', in G. Schroth, G. A. B. da Fonseca and C. A. Harvey (eds) *Agroforestry and Biodiversity Conservation in Tropical Landscapes*, Island Press, Washington, DC

Fox, J. (2000) *Can't See the Forest for the Swidden: Land-Use and Land-Cover Change in Montane Mainland Southeast Asia*, working paper, Environment Series, East-West Center, Honolulu

Fox, J., Truong, D. M., Rambo, A. T., Tuyen, N. P., Cuc, T. and Leisz, S. (2000) 'Shifting cultivation: A new old paradigm for managing tropical forests', *Bioscience* 50, pp521-528

Gadgil, M. and Guha, R. (1992) *The Fissured Land: An Ecological History of India*, Oxford University Press, Melbourne

Government of India (1983) *Task Force on Shifting Cultivation*, Ministry of Agriculture, New Delhi

Gupta, A. K. (2000) 'Shifting cultivation and conservation of biological diversity in Tripura, northeast India', *Human Ecology* 28, pp605-628

Hashim, N. R. (2010) 'Analysing long-term landscape changes in a Bornean forest reserve using aerial photographs', *The Open Geography Journal* 3, pp161-169

Holdridge, L. R. (1951) 'The alder *Alnus acuminata* as a farm timber tree in Costa Rica', *Caribbean Forester* 12, pp47-53

Ickowitz, A. (2006) 'Shifting cultivation and deforestation in tropical Africa: Critical reflections', *Development and Change* 37, pp599-626

Inter-Ministerial Task Force (2008) *Report of the Inter-Ministerial National Task Force on Rehabilitation of Shifting Cultivation Areas*, issued by G. K. Prasad, Additional Director General of Forests and Chairman of the Task Force, Ministry of Environment and Forests, Government of India, New Delhi

Jyotishi, A. (2006) 'Institutional analysis of swidden: The case of swiddeners in Orissa', poster paper presented at the International Association of Agricultural Economists' Conference, 12-18 August 2006, Gold Coast, Australia

Kerkhoff, E. E. and Sharma, E. (2006) *Debating Shifting Cultivation in the Eastern Himalayas: Farmers' Innovations as Lessons for Policy*, International Centre for Integrated Mountain Development (ICIMOD), Kathmandu

King, K. F. S. (1979) 'Agroforestry and the utilisation of fragile ecosystems', *Forest Ecology and Management* 2, pp161-168

Kotto-same, J., Woomer, P. L., Appolinaire, M. and Louis, Z. (1997) 'Carbon dynamics in slash-and-burn agriculture and land use alternatives of the humid forest zone in Cameroon', *Agriculture, Ecosystems and Environment* 809, pp245-256

Krick, N. M. [(1853) 1913] 'An account of an expedition among the Abors in 1853', translated by Rev. Gille, S. J., *Journal and Proceedings of the Asiatic Society of Bengal* 9, pp107-122

Lambert, D. P. (1996) 'Crop diversity and fallow management in a tropical deciduous forest shifting cultivation system', *Human Ecology* 24, pp427-453

Lambin, E. F., Turner, B. L., Geist, H. J., Agbola, S. B., Angelsen, A., Bruce, J. W., Coomes, O. T., Dirzo, R., Fischer, G., Folke, C., George, P. S., Homewood, K., Imbernon, J., Leemans, R., Li, X. B., Moran, E., Mortimore, M., Ramakrishnan, P. S., Richards, J. F., Skanes, H., Steffen, W., Stone, G. D., Svedin, U., Veldkamp, T. A., Vogel, C. and Xu, J. (2001) 'The causes of land-use and land-cover change: Moving beyond the myths', *Global Environmental Change* 11, pp261-269

Maithani, B. P. (2005) *Shifting Cultivation in North-East India: Policy, Issues and Options,* Mittal publications, New Delhi

Malik, B. (2003) 'The "problem" of shifting cultivation in the Garo Hills of North–East India, 1860-1970', *Conservation and Society* 1(2), pp287-315

Mazoyer, M. and Roudart, L. (2006) *A History of World Agriculture: From the Neolithic Age to the Current Crisis*, Earthscan, London

Ninan, K (1992) 'Economics of shifting cultivation in India', *Economic and Political Weekly* 27, ppA2-A6

Murtem, G., Sinha, G. N. and Dopum, J. (2008) '*Jhumias*' view on shifting cultivation in Arunachal Pradesh', *Bulletin of Arunachal Forest Research* 24, pp35-40

Nakro, V. (2011) *Traditional Agricultural Practices and Sustainable Livelihood*, Department of Planning and Coordination, Government of Nagaland, Kohima, Nagaland

NEC (1974) *Basic Statistics of North Eastern Region*, North Eastern Council Secretariat, Shillong, Meghalaya, India

NEPED-IIRR (1999) *Building upon Traditional Agriculture in Nagaland, India*, Nagaland Environmental Protection and Economic Development (NEPED), Kohima and International Institute of Rural Reconstruction (IIRR), Cavite, Philippines.

North, D. C. (1990) *Institutions, Institutional Change and Economic Performance,* Cambridge University Press, New York

O'Brien, W. E. (2002) 'The nature of shifting cultivation: Stories of harmony, degradation and redemption', *Human Ecology* 30, pp483-502

Raj, S. (2010) 'Traditional knowledge, innovation systems and democracy for sustainable agriculture: A case study on Adi tribes of Eastern Himalayas of North-east India', in *Proceedings of a Conference on Innovation and Sustainable Development in Agriculture and Food*, 28-30 June 2010, Montpellier, France, available at http://hal.archives-ouvertes.fr/docs/00/52/33/09/PDF/Raj_Saravanan_Traditional_knowedge.pdf, accessed 19 August 2013

Ramakrishnan, P. S. (1984) 'The science behind rotational bush fallow agricultural systems (*jhum*)', *Proceedings of the Indian Academy of Science* 93, pp379-400

Ramakrishnan, P. S. (1992) *Shifting Agriculture and Sustainable Development: An Interdisciplinary Study from North-Eastern India*, UNESCO Man and the Biosphere Series, Volume 10, United Nations Educational, Scientific and Cultural Organization, Paris

Raman, T. R. S. (2000) '*Jhum*: Shifting Opinions,' in *Environment: Reality and Myth*, Seminar 486 (February), Seminar Publications, New Delhi, pp15-18, available at http://www.india-seminar. com/2000/486/486%20raman.htm, accessed 20 August 2013

Rambo, A.T. (1996) 'The composite swiddening agroecosystem of the Tay ethnic minority of the northwestern mountains of Vietnam', in B. Rerkasem (ed.) *Montane Mainland Southeast Asia in Transition,* Chiang Mai University, Chang Mai, Thailand, pp69-89

Ranjan, R. and Upadhyay, V. P. (1999) 'Ecological problems due to shifting cultivation', *Current Science* 77, pp1246-1250

Roy, S. (1997) *Aspects of Padam-Minyong Culture,* third edition, M/S Purbadesh Mudran, Guwahati, India

Saikia, A. and Bhaduri, S. (2012) 'An institutional analysis of transition to market: The case of shifting cultivation in Mon, Nagaland', *International Journal of Rural Management* 8, pp19-34

Schmidt-Vogt, D., Leisz, S. J., Mertz, O., Heinimann, A., Thiha, Messerli, P., Epprecht, M., Cu, P.V., Chi, V. K., Hardiono, M. and Dao, T. M. (2009) 'An assessment of trends in the extent of swidden in Southeast Asia', *Human Ecology* 37, pp269-280

Seidenberg, C., Mertz, O. and Kias, M. B. (2003) 'Fallow, labour and livelihood in shifting cultivation: Implications for deforestation in northern Lao PDR', *Danish Journal of Geography* 103, pp71-80

Sharma, N., Madhusudan, M. D. and Sinha, A. (2012) 'Socio-economic drivers of forest cover change', *Environment and Political Weekly* 47, pp64-72

Singh, D. (1996) *The Last Frontier: People and Forests in Mizoram,* Tata Energy Research Institute, New Delhi

Singh, S., Singh, T. P. and Srivastava, G. (2005) 'Vegetation cover type mapping in Mouling National Park in Arunachal Pradesh, Eastern Himalayas: An integrated geospatial approach', *Journal of the Indian Society of Remote Sensing* 33(4), pp547-563

Sivakumar, M.V. K. and Valentin, C. (1997) 'Agroecological zones and the assessment of crop production potential', *Philosophical Transactions of the Royal Society B (Biological Sciences)* 352, pp907-916

Spencer, J. E. (1966) *Shifting Cultivation in Southeastern Asia,* University of California Press, Berkeley, CA

Teegalapalli, K. (2008) 'Shifting livelihood options and changing attitudes of communities in the Garo hills, western Meghalaya', *Current Conservation* 2(3), pp18-21

Teegalapalli, K. and Datta, A. (2016) 'Shifting to settled cultivation: Changing practices among the Adis in central Arunachal Pradesh, North-east India', *Ambio*, DOI: 10.1007/s13280-016-0765-x

Teegalapalli, K., Veeraswami, G. G. and Samal, P. K. (2009) 'Forest recovery following shifting cultivation: An overview of existing research', *Tropical Conservation Science* 2, pp374-387

Toky, O. P. and Ramakrishnan, P. S. (1981) 'Cropping and yields in agricultural systems of the northeastern hill region of India', *Agro-Ecosystems* 7, pp11-25

Tripathi, R. S. and Barik, S. K. (2003) 'Shifting cultivation in North East India', in B. P. Bhat, K. M. Bujarbaruah, Y. P. Sharma and Patiram (eds) *Proceedings of a seminar on Approaches for Increasing Agricultural Productivity in Hill and Mountain Ecosystems,* Indian Council of Agricultural Research, Research Complex for NEH Region, Umiam, Meghalaya, pp317-322

Watters, R. F. (1971) *Shifting Cultivation in Latin America,* Food and Agriculture Organization of the United Nations, Rome

Whittlesey, D. (1937) 'Fixation of shifting cultivation', *Economic Geography* 13, pp139-154

Wilcox, R. (1825) 'Member of a survey of Assam and the neighbouring countries, executed in 1825-6-7-8', *Asiatic Researches* 17

WRI (1985) *Tropical Forests: A Call for Action. 1: The Plan,* World Resources Institute, Washington, DC

Note

1. The word *taungya* is of Burmese origin, with *taung* meaning hill and *ya* meaning cultivation.

43

TRANSITIONAL UPLAND RICE CROPPING SYSTEMS IN NORTHERN THAILAND

Priorities for research and development, on the basis of on-farm crop diagnosis

*Koen Van Keer and Guy Trébuil**

Introduction and objectives

In Southeast Asia, swidden cultivation of upland rice (*Oryza sativa* L.) is still a major component of remote highland farming systems (Piggin et al., 1998; Mertz et al., 2008). In northern Thailand, these upland-rice systems have been in transition towards semi-permanent or permanent cropping for several decades (Trébuil et al., 2000, 2006). With no external inputs, grain yields in farmers' fields are typically low (1 to 2.5t/ha), but very variable (Ramakrishnan, 1992; Wang et al., 2010). When cropping intensity is increased by shortening fallow periods, the outcome includes soil fertility depletion, increased soil erosion and increased weed and pest pressure. This is generally believed to exacerbate the problem of low and variable upland-rice yields, but empirical studies documenting and interpreting such multifaceted and transitional upland-rice systems are still relatively scarce (Saito et al., 2006).

In this chapter we revisit the methodology and key findings of a comprehensive empirical on-farm diagnostic survey that was conducted in a remote Red Lahu highland village in Fang district, Chiang Mai province, in the western part of upper northern Thailand (Figure 43-1). The survey covered four successive cropping years (1993 to 1996) and an extensive range of upland-rice farming situations.

Maximum achievable paddy yields of 3.1 and 4.4 tons per hectare respectively for early- and late-maturing local varieties of upland rice were recorded in small monitoring squares under farmer management. Actual yields, however, averaged only 1.1 and 1.3 t/ha respectively. This showed that important yield gaps existed in upland-rice systems based on local cultivars where no external inputs were applied.

* Dr KOEN VAN KEER, former researcher in the Division of Soil and Water Management at KU Leuven (The University of Leuven), Belgium, currently a Sustainability Affairs and Stakeholder Relations Expert at Yara International ASA; Dr GUY TRÉBUIL, researcher and Adjunct Professor, Centre de cooperation internationale en recherche agronomique pour le développement (CIRAD), Montpellier, France.

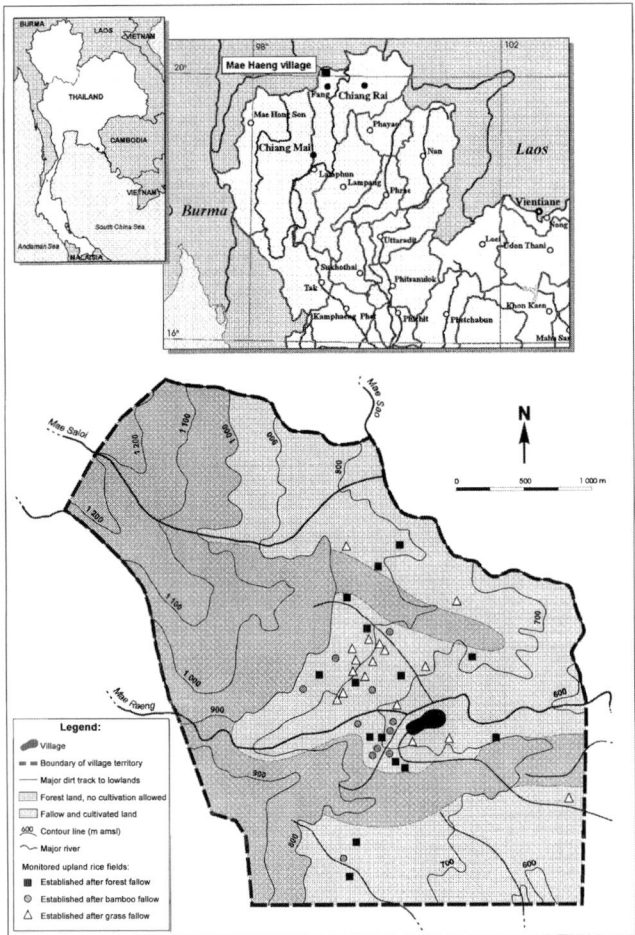

FIGURE 43-1: Location of the study site in Mae Haeng village, Fang district, Chiang Mai province, and the upland-rice fields used for the diagnostic survey.

Hence, we postulate that such gaps can be closed, at least in part, through better resource management and cropping practices. Paddy yields approaching the local potential are seldom reached because of multiple and interacting limiting factors, either permanent or transient, that characterize these diverse and variable upland-rice agroecosystems. Therefore, identifying and ranking these limiting factors is crucial to setting relevant research policies and extension priorities for improving farmers' practices and reducing yield gaps (Doré et al., 1997).

Based on an earlier assessment of the key characteristics of local upland-rice varieties in their actual setting (Van Keer et al., 1998), an interpretative boundary-line model was developed of the successive yield build-up processes of the upland-rice crop. This model was then used to carry out an in-depth on-farm diagnosis of upland-rice yield gaps and their causal factors (Van Keer, 2003).

The objectives of this diagnostic survey were fourfold:

1.   to characterize the crop population, crop environmental conditions and farmers' practices along the upland-rice crop cycle and across several climatic years;
2.   to develop and test a boundary-line approach for on-farm upland-rice yield-gap analysis;
3.   to identify and rank the key periods of yield differentiation, or limitation, in upland rice under actual farmers' conditions and management practices; and
4.   to identify, rank and explain the major on-farm yield-limiting factors affecting upland rice at the study site.

The study site

The village territory of Mae Haeng is located in a mountainous landscape at altitudes between 600 and 900 metres above sea level, near the border of northern Thailand and Burma (Myanmar). It is populated by people of the Red Lahu ethnic minority. Its agroecosystem is a complex mosaic (Figure 43-2) of permanent forests, fallows, lychee orchards and swidden fields with a wide range of weed species, crop diseases and insect and animal pests. The village's fallow vegetation types and weed populations are described in greater detail in Van Keer (2003).

Dendrocalamus membranaceus
Munro [Poaceae]

The most common species growing in bamboo fallows at Mae Haeng village.

FIGURE 43-2:   Overview of the Mae Haeng village agroecosystem showing different types of crops and fallow vegetation.

The dominant soils types on the sloping upland-rice fields in the Mae Haeng area are deep (50cm to more than 200cm) humic acrisols overlying strongly weathered granitic parent material. The topsoil layers are shallow (10cm on average) and have a clay-loamy texture, an average pH of 5.5 and a medium level of soil fertility, except for plant-available phosphorus, which is very low at an average level of only 5mg/kg of soil (Bray-2 P). Such soil types are typical and considered generally suitable for upland-rice cultivation in Southeast Asia. The total annual rainfall measured at the Fang meteorological station (15km south of Mae Haeng at 470masl) varied between 1064 and 1910mm over a period of 16 years (1981 to 1996), with an average of 1482mm. Rainfall distribution follows a slightly bimodal pattern, with frequent dry spells occurring in late June and early July (Figure 43-3). Total rainfall measured on-site during the upland-rice wet season varied significantly across the four cropping years of the study. The 1995 wet season was driest, recording 1319mm of rain after long dry spells at the start and end of the season. The 1994 wet season was wettest at 2000mm, with a short dry spell at the start of the season. The 1993 and 1996 wet seasons had average and well-distributed rainfall levels of 1565mm and 1599mm, respectively.

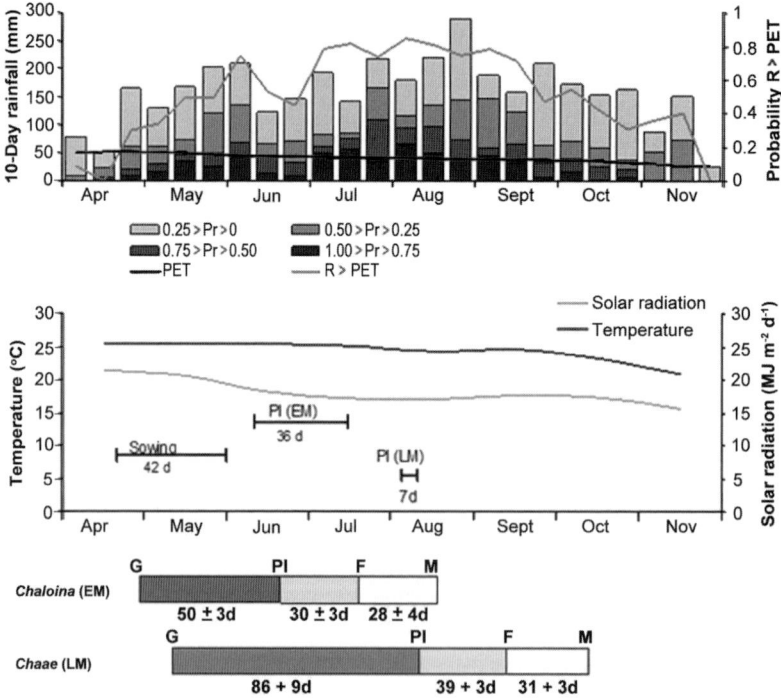

FIGURE 43-3:   Climatic conditions and upland-rice cropping calendars for early (EM) and late (LM) cultivars in Mae Haeng village. Rainfall data are from Fang meteorological station for the period 1981-1996.

Notes: R = rainfall; PET = potential evapotranspiration; G = germination; PI = panical initiation; F = flowering; M = maturity; d = calendar days.

Village history and the farming systems based on upland rice

Ceramic shards discovered near the present settlement of Mae Haeng village were between 500 and 600 year old, suggesting that the area has had a human presence for centuries (Shaw, 1996). However, Mae Haeng village itself was founded as recently as 1976, by seven Red Lahu (*Lahu Nyi*) households immigrating from Myanmar. By 1996, near the end of the study period, the village had 64 households and 355 people, but was not yet accessible by road. Following the eradication of opium poppy cultivation as a cashcrop, local farming systems were still at an early stage of diversification into horticultural crops (mainly small lychee orchards) and integration into the market economy. Livelihood systems were still based mainly on subsistence cultivation of local Lahu varieties of upland rice, similar to those observed in a remote northern Thai Lahu community back in the 1960s (Walker, 1976). The main staple at Mae Haeng village was a late-maturing non-glutinous variety called *Chaae*. It represented nearly 80% of the village's total annual upland-rice crops. Mae Haeng farmers also tended small plots of an early-maturing non-glutinous upland-rice variety called *Chaloina*, which boosted food security while the village households waited for the main harvest, and a late-maturing glutinous variety called *Chanona*, which was used mainly for ceremonial purposes. Mae Haeng farmers still used traditional swidden practices for growing their upland rice, with almost no external inputs. However, their weed-control practices were gradually diversifying following the adoption of hoe tillage and the application of salt (NaCl) as an effective herbicide against weed communities dominated by species belonging to the *Asteraceae* family (more details of this practice can be found in Van Keer et al., 2001). In addition to upland rice, Mae Haeng farmers also grew small plots of local maize (as pig feed) and several minor crops for local usage only. Besides farming, many households were still involved in hunting and gathering activities and also, increasingly, in off-farm wage-earning, mostly during the dry season.

On-farm crop diagnostic survey methodology

Research protocol

The success of agricultural research and development is critically dependent upon a sound diagnosis and prioritization of constraints in actual farmers' fields. The methodological approach for on-farm diagnosis that we adopted draws heavily on the pioneering views and work of Sebillotte (1989) and other French

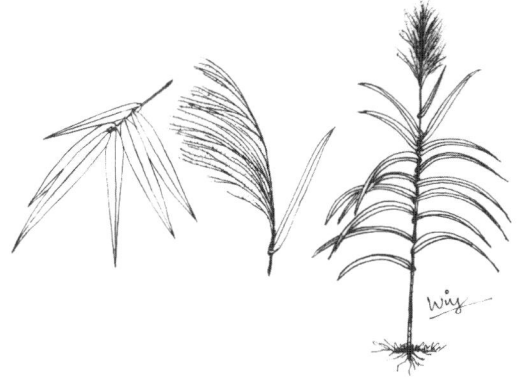

Thysanolaena latifolia (Roxb. ex Hornem.) Honda [Poaceae]

Broom grass was commonly harvested in the dry season at Mae Haeng village as a source of cash income.

agronomists (Doré et al., 1997; Wey et al., 1998). Full details of the survey design, sampling, measurement and data-analysis protocols used are provided in Van Keer (2003). Here, we revisit only the key elements of our methodology.

Survey design

In the course of four successive cropping seasons (1993 to 1996), comprehensive data on crop population status and functioning were collected for two major groups of local upland-rice cultivars: early-maturing (95 to 115 days) and late-maturing (140 to 180 days). Crop environmental conditions and cropping practices were monitored on a two-weekly basis along the entire crop cycle, from a total of 432 intra-field, farmer-managed permanent monitoring squares, each measuring one square metre. These squares (three to six per field) were established when the upland-rice crop emerged in 63 farmers' fields under swidden rotation (Figure 43-4). In the first year, all of the selected plots were in new fields cleared from three different main types of fallow vegetation: grassland, bamboo and young secondary forest. These fields were monitored until they were fallowed again after one, two or three years of successive upland-rice monocropping. Additional newly cleared fields were included throughout the following three years of the survey. The number of fields monitored during the 1993, 1994, 1995 and 1996 wet seasons were 12, 14, 19 and 18,

FIGURE 43-4:   Monitoring squares marked with bamboo sticks in an upland-rice field, with Mae Haeng village in the distance.

respectively. Experimental treatments (not presented in detail here) were superimposed on the main survey axis to conduct additional research on selected cropping practices including burning, weeding, fertilizer application and use of modern cultivars (George et al., 2001; Van Keer, 2003).

Data collection

The genetic variability of the local upland-rice cultivars was assessed by isozyme analysis performed on seed materials (Van Keer et al., 1998; see also Table 43-1). Upland-rice crop status, including canopy height and cover, growth stages, weed cover, incidence of pests and diseases and farmer practices were monitored

Crassocephalum crepidioides (Benth.) S. Moore [Compositae]

A common weed in upland-rice fields at Mae Haeng village that was often used as a green salad vegetable.

and recorded throughout the entire crop cycle, for all 432 permanent monitoring squares. Upland-rice above-ground biomass, grain yield and yield components were measured at harvest from the full 1sq. m of all permanent squares. Vegetative biomass and grain moisture contents were determined from subsamples. All biomass and yield data presented in this chapter are expressed at 0% moisture content. Temperature- and radiation-limited total above-ground biomass potentials were estimated for both main cultivar types, according to the method described by Driessen and Konijn (1992). Detailed measurements were performed on upland-rice rooting systems on selected fields and cultivars, and on topsoil (0-20cm) fertility, weeds and nematodes. Weed measurements included species identification, population densities and above-ground biomass. The dataset for nematodes was incomplete. Of all the variables measured, only those that produced complete and robust datasets for all cropping years (1993 to 1996) were used for statistical Principal Component Analysis with Instrumental Variables (PCAIV)(see full list in Notes below Figure 43-9).

'Boundary line' approach to upland-rice yield-gap analysis

The boundary-line approach (also called the reference- or envelope-curve approach) for the analysis of biological data was first described by Webb (1972). The method has since developed into a useful tool for comparison of cultivars and crop diagnosis purposes (Tasistro, 2012). In an approach comparable with diagnostic studies of other cereal crops (Doré et al.,1997; Wey et al., 1998), the boundary-line method was used to develop an empirical model of the process of the upland-rice yield build-up observed at our study site across four climatically diverse years.

The grain yield of upland rice can be partitioned into five components that are determined successively along the crop growth and development cycle. This can be expressed as: GY = PL x PAPL x SPPA x SFR x W1G, where GY is the grain yield (g/sq. m); PL is the plant population density (no./sq. m); PAPL is the number of panicles per plant; SPPA is the number of spikelets per panicle; SFR is the spikelet (or grain) filling ratio (in %), and W1G is the weight of 1 grain (conventionally and hereafter shown as WTG or the weight of 1000 grains, in g). These yield components are mutually dependent, i.e. the level reached for a given yield component depends on the level reached by the preceding one, for each phase of the crop cycle. Yield components also depend on crop genotype (G) x crop environment (E) x crop management (M) interactions.

The boundary line theoretical model used in this study is presented in Figure 43-5. It is hypothesized that the boundary curve describes the relationship between a

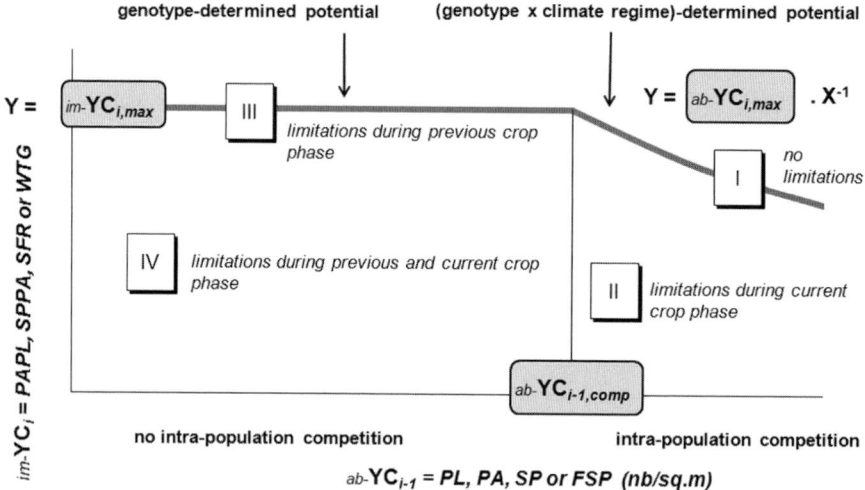

FIGURE 43-5   Boundary line theoretical model used in this study to describe and interpret relationships between successive yield components (YC$_i$ and YC$_{i-1}$) in upland rice.

Notes: YC = yield component; *ab*-YC = 'area-based' (population level) YC; *im*-YC = 'intermediate' (individual plant level) YC; PL = no. of plants; PA = no. of panicles; SP = no. of spikelets; FSP = no. of filled spikelets; PAPL = no. of panicles per plant; SPPA = no. of spikelets per panicle; SFR = percentage of filled spikelets; WTG = 1000–grain weight; GY = grain yield; *ab*-YC$_{comp}$ = intra–population competition threshold level = PL$_{comp}$, PA$_{comp}$, SP$_{comp}$, FSP$_{comp}$; *im*-YC$_{max}$ = PAPL$_{max}$, SPPA$_{max}$, SFR$_{max}$ or WTG$_{max}$; *ab*-YC$_{max}$ = PA$_{max}$, SP$_{max}$, FSP$_{max}$ or GY$_{max}$; Interpretation of individual data points (monitoring squares observations): I = no loss of yield potential has occurred until this crop phase; II = loss of yield potential has occurred during current crop phase; III = loss of yield potential has occurred during previous crop phase; IV = loss of yield potential has occurred during previous and current crop phase.

given upland–rice yield component YC_i and the component YC_{i-1} that preceded it in the crop cycle, for a given cultivar grown under local temperature and solar radiation conditions, when all other crop environmental conditions (water and nutrient supply, pests and diseases) are non-limiting. The model takes into account the existence of intra-population competition effects (also called crowding or compensation effects), which occur when a population yield component exceeds a certain density threshold level (YC_{comp}). It was postulated that on the basis of this model and our survey data scatters, boundary curves and their parametric equations could be fitted for each phase of the yield build-up process and for each upland–rice cultivar. An actual example of such a boundary curve obtained from this study is shown later in this chapter in Figure 43-7, for the panicle determination phase.

In the course of each growth and development phase, a crop can keep or lose part of the yield potential it built up in the previous phase. Data points that deviate from the boundary line reflect field situations where such losses (or yield gaps) emerged due to stress conditions. To identify and rank the periods of yield differentiation (yield gap emergence), a 'phase realization index' (PRI) analysis was performed, based on a method developed by Wey et al. (1998). For each successive growth phase and monitoring square, a PRI value was calculated as the ratio between the simulated (theoretically calculated) potential grain yield achieved at the end of that phase (PGY_i) and the one achieved at the end of the preceding phase (PGY_{i-1}), expressed as $PRI_i = PGY_i / PGY_{i-1}$. This index shows the effects of environmental and management constraints on the functioning of the upland–rice crop during each phase of its crop cycle (PRI = 1 indicates no limitations; PRI < 1 indicates limitations). PRI distributions were analysed to assess the relative importance of key successive growth and development phases on the final yields.

Multivariate statistical analysis

To identify and rank the causes of yield differentiation in the upland–rice crops, a Principal Component Analysis with Instrumental Variables (PCAIV) (Lebreton et al., 1991) was carried out on the complete pooled dataset from 1993 through 1996, as well as for the individual years. PCAIV is a tool used for exploratory descriptive analysis of complex ecosystem datasets containing a mix of qualitative and quantitative data, combining principal component and multivariate regression analysis. For all of the rice field monitoring squares sampled in this study, a simultaneous analysis was made of a dataset of all of the crop response variables, i.e. the grain yield and its yield components against a comprehensive dataset of explanatory variables. These were all of the environmental and management variables that were measured in the course of the study. Relatively high values were obtained for the percentage of inertia in the first two axes and for the pertinence of the PCAIV of both pooled data and that from individual years (see Table 43-3). This endorses the suitability of this statistical method for capturing and explaining the most important information from this complex and diverse dataset.

Results and discussion

Characterization of local upland-rice cultivars

Table 43-1 presents the key physiological and genetic characteristics of local upland-rice cultivars planted at Mae Haeng during the four cropping years from 1993 to 1996. All cultivars were found to be genetically similar and belonging to the panicle-weight type tropical japonicas group (Glaszmann, 1987). However, two physiological groups of cultivars were clearly distinguished, one early-maturing and the other late-maturing, as well as the most common cultivars in each of those groups, namely *Chaloina* and *Chaae,* which were found to be weakly and strongly photoperiod-sensitive, respectively (see Figure 43-3). The total above-ground biomass potentials of early- and late-maturing cultivars were measured to be 9 and 15t/ha, respectively. These values were backed by temperature- and radiation-limited biomass production estimations. The important difference in biomass accumulation between the cultivar groups was consistent with the difference in the length of their growing periods. The canopy architectures of both cultivar types were similar, and characterized by long droopy leaves and intermediate to tall (110 to 140cm) stems, which were not susceptible to lodging except after very strong vegetative growth or late wet season storms. The tillering abilities of both cultivar types were also similar and relatively low, with maxima of five to six tillers per plant at panicle-initiation (PI) stage. These two most common cultivars were used to develop two separate full sets of boundary-line models describing the yield build-up processes in upland rice.

TABLE 43-1: Characteristics of local upland-rice cultivars at Mae Haeng village, 1993 to 1996 cropping years.

Cultivar	Frequency	Varietal group [a]	Photo periodism	Grain type	DD [b] to PI [c]	DD to harvest
Early-maturing types						
Chaloina	Common	Tropical japonica	Weakly sensitive	Non-glutinous	650	1340
Kochokai	Rare				800	1400
Chaloioe	Rare				670	1350
Kochole	Rare				590	1270
Late-maturing types						
Chaae	Dominant	Tropical japonica	Strongly sensitive	Non-glutinous	1060	1900
Chanoko	Common			Glutinous	1060	1910
Chafuma	Common			Non-glutinous	1000	1860
Chazu	Rare			Non-glutinous	1230	2120
Komu	Rare			Non-glutinous	1150	1960
Chanona	Rare			Glutinous	860	1790

Note: [a] Based on isozyme analysis; [b] DD = degree-days (13° Celsius threshold); [c] PI = panicle initiation.

Extent of upland-rice yield variability in farmers' fields

Figure 43-6 shows the distributions of grain yields in the pooled data from the four cropping years for both early- and late-maturing cultivar groups. The figures show an extensive variability in upland-rice productivity, ranging from 0 to 438g/sq. m among the surveyed monitoring squares.

Both types of cultivars produced low yields in a majority of the monitoring squares, with very similar distributions.

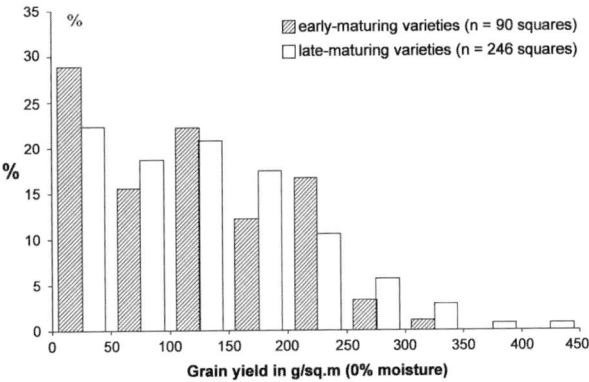

FIGURE 43-6:   Distribution of grain yields for early- and late-maturing upland-rice cultivars in Mae Haeng village. Pooled data for cropping years from 1993 to 1996.

Average yields from pooled data were low and only slightly different at 113 and 130g/sq. m for early- and late-maturing cultivars, respectively. However, the recorded maximum achievable yields were relatively high at 308 and 438g/sq. m for early- and late-maturing cultivars, respectively, and the difference between the two cultivar groups was substantial. The discrepancy between the relatively high achievable yield potentials and the relatively low actual yields observed in many farmers' fields confirmed the existence of important yield gaps.

In such a situation, where there are extensive yield gaps across fields in the same local agroecosystem, it is appropriate to adopt a crop diagnostic approach in order to identify and understand the causes of these gaps.

Establishing upland-rice yield build-up boundary curves and their parameters

Evaluation of all yield component data scatters from farmers' upland-rice fields supported the existence of yield build-up reference curves, similar to those already established for other cereal crops (Van Keer, 2003). For example, Figure 43-7 displays such scatters and reference curves for the relationship between plant densities and the number of panicles per plant for both types of cultivars under study.

On the basis of a set of similar curves obtained for each successive phase of the upland-rice crop cycle, the following three parameters were estimated for each phase and for both types of cultivars (see Figure 43-5 and Table 43-2):

1.   The maximum achievable value of the area-based yield component (ab-YC_{max});
2.   The maximum achievable value of the intermediary yield component (im-YC_{max}); and
3.   The threshold density level of the preceding yield component beyond which crowding effects occur(ab-YC_{comp}).

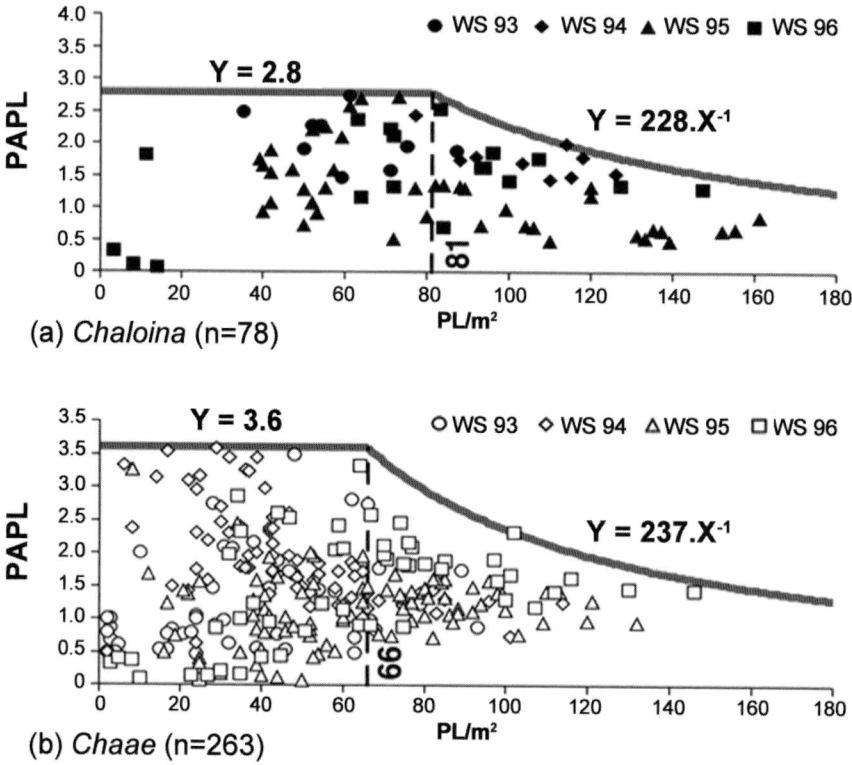

FIGURE 43-7: Yield build-up reference curves on relationships between plant density and number of panicles per plant for *Chaloina* (early) and *Chaae* (late) upland-rice varieties in Mae Haeng village. Pooled data for cropping years from 1993 to 1996. *Note:* WS = wet season.

These parameters were subsequently used to calculate the phase realization indices for each crop monitoring square in order to identify and rank the key periods of yield limitation.

Dating and ranking the periods of upland-rice yield limitation

Figure 43-8 shows the distribution of phase realization indices for each successive yield build-up phase and for both cultivar types. In the vegetative-growth period, it can be seen that determination of final plant population density (i.e. the combined effect of seed germination, seedling emergence and crop establishment), did not limit the yield potential, with nearly all monitoring squares reaching the maximum index value of 1. During this period, poor crop biomass accumulation was found to be playing a more important limiting role than low plant density. Vegetative biomass accumulation (growth vigour) affected panicle and spikelet formation, but not spikelet fertilization nor grain filling (these data are not shown here).

TABLE 43-2: Parameters of the yield build-up reference curves for *Chaloina* (early) and *Chaae* (late) upland-rice cultivars at Mae Haeng village. Pooled data for cropping years from 1993 to 1996.

Cultivar	Chaloina *(early)*	Chaae *(late)*
Spikelet filling		
Maximum grain yield GY_{max} (g/sq. m at 0% H_2O)	308	438
Max. 1000-grain weight WTG_{max} (g at 0% H_2O)	23.2	24.3
Threshold for filled spikelets FSP_{comp} (no./sq. m)	13,300	18,000
Spikelet fertilization		
Max. no. of filled spikelets FSP_{max} (no./sq. m)	16,200	20,600
Max. Grain filling rate SFR_{max} (%)	80	96
Threshold for no. of spikelets SP_{comp} (no./sq. m)	20,100	21,500
Spikelet formation		
Max. no. of spikelets SP_{max} (no./sq. m)	22,100	26,700
Max. no. of spikelets/panicle $SPPA_{max}$ (no.)	144	183
Threshold for no. of panicles PA_{comp} (no./sq. m)	154	146
Panicle formation		
Max. no. of panicles PA_{max} (no./sq. m)	228	237
Max. no. of panicles/plant $PAPL_{max}$ (no.)	2.8	3.6
Threshold for plant density PL_{comp} (no./sq. m)	81	66

In the reproductive growth period, only 60% of squares, or less, did not experience any strong limiting conditions during a given yield build-up phase. Such favourable conditions seemed to happen somewhat more frequently in fields planted to the early-maturing cultivars, for reasons not clearly understood. These distributions of indices also identified the panicle formation and especially the spikelet differentiation phases as key periods of yield limitation during the reproductive period. In the vast majority of the squares, and for both types of cultivars, the yield component build-up process was severely constrained during these phases. In the spikelet fertilization and

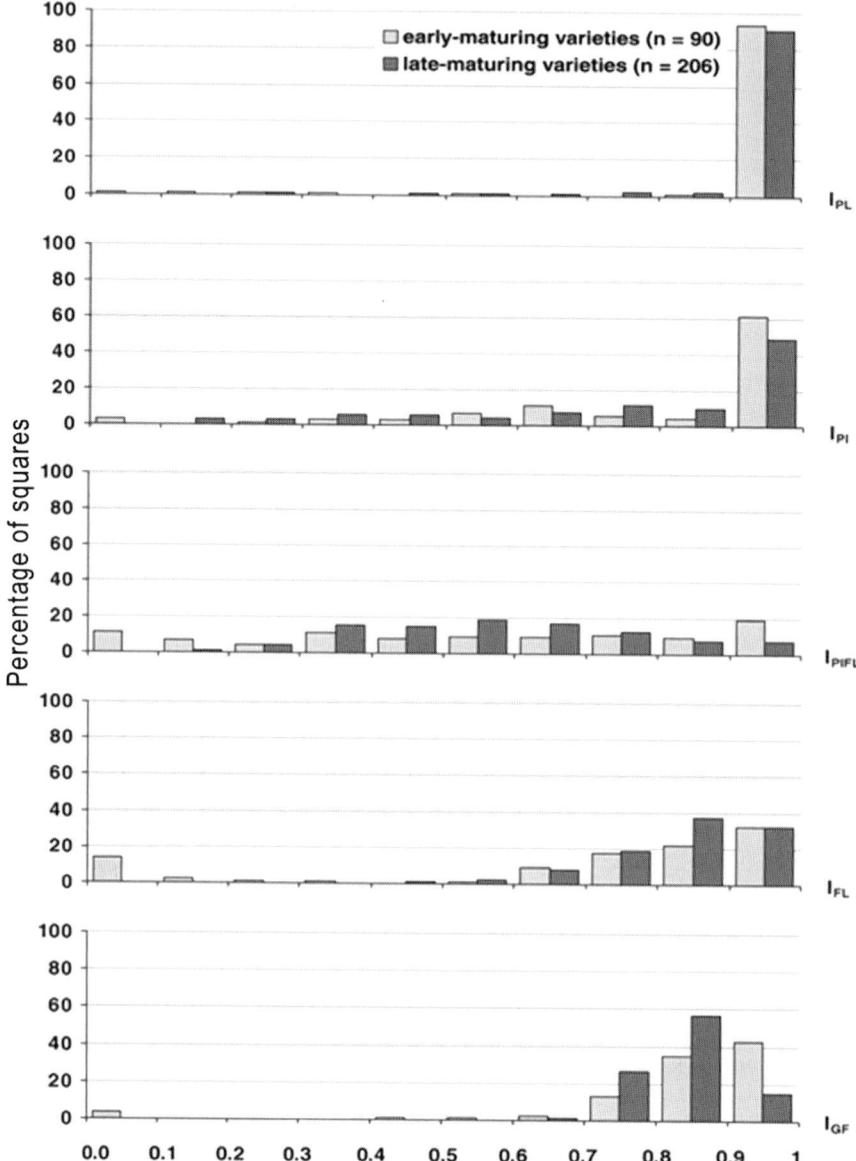

FIGURE 43-8: Distribution of the successive phase realization indices for early- and late-maturing upland-rice cultivars at Mae Haeng village. Pooled data for the cropping years from 1993 to 1996.

Notes: Phase realization index for determination of number of plants per sq. m (I_{PL}); up to panicle determination (I_{PI}); up to spikelet determination (I_{PIFL}); up to flowering/spikelet fertilization (I_{FL}); and during grain filling and maturation (I_{GF}).

grain-filling phases of the crop cycle, a higher proportion of squares registered index values above 0.7, indicating less severe constraints to the yield build-up processes.

Identifying and grading the causes of yield limitation

In the final step of the on-farm crop diagnosis, comprehensive PCAIV statistical analyses were conducted on the complete set of pooled data from the cropping years from 1993 to 1996, and on sub-datasets for the individual years. Figure 43-9 shows the results of the analysis of the pooled data, revealing the relationships between successive yield components and more or less related environmental conditions and management practices.

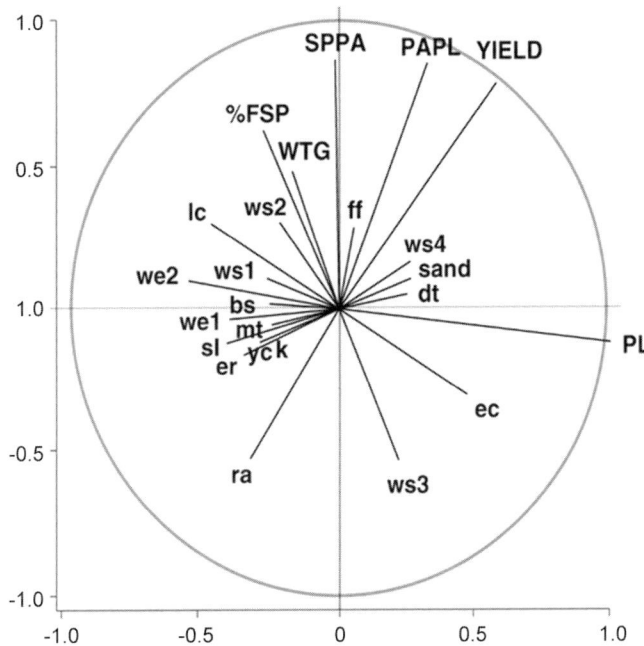

FIGURE 43-9:   Plane of PCAIV first two axes showing the relationships between upland-rice yield, components of yield, crop environmental conditions, and crop management practices in Mae Haeng village. Pooled data for cropping years from 1993 to 1996.

Notes: Crop variables (measured at harvest): PL = no. of plants per sq. m; PAPL = no. of panicles per plant; SPPA = no. of spikelets per panicle; %FSP = percentage of filled spikelets; WTG = 1000-grain weight; YIELD = grain yield. Crop environmental and management variables: ec, lc = early, late cultivars; mt, dt = minimum, deeper tillage; yc = no. of successive years of upland-rice cultivation; ff, bf, gf = forest, bamboo, grass fallows; ws1, ws2, ws3, ws4 = 1993, 1994, 1995, 1996 wet seasons; sl = slope angle; er = soil erosion; sand = percentage of sand; ph = pH; som = soil organic matter; pav = soil available P; camg = soil Ca + Mg content; k = soil K content; we1, we2 = cumulative area under the weed cover curve for 0-60 DAS (days after sowing) and for 60 DAS to harvest, respectively; bs = brown spot infestation; ra = rice-root aphid infestation.

The analysis of the pooled data from the 1993 to 1996 cropping years revealed strongly negative relationships for the following:

a) The number of panicles per plant and the number of spikelets per panicle versus field infestation by rice-root aphids (*Tetraneura nigriabdominalis*). Rice plants affected by these aphids showed severely stunted growth, yellow leaves and reduced tillering (Figure 43-10).
b) Plant density versus extended weed stress (after 60 days) for late-maturing cultivars that were planted at lower sowing densities.
c) The percentage of filled spikelets and 1000-grain weight versus the wet season for late-maturing cultivars in 1995, this being the only wet season suffering drought during the critical reproductive phase of the late-maturing crop of upland rice.

Weaker negative relationships were found between the number of panicles per plant and slope angle, soil erosion, the number of successive upland-rice crops on the same field, minimum tillage and weed competition early in the crop cycle. A weak positive correlation was observed between the older type of forest fallow and panicles per plant, the number of spikelets per panicle and the final grain yield. Brown spot was the only disease with minor importance, mainly as it affected late-maturing cultivars.

The analysis of major and minor limiting factors on upland-rice yields and their components for individual year sub-datasets are shown in Table 43-3. These more detailed analyses diagnosed rice-root aphid infestation as the single major limiting factor, with a strong and consistent effect on final grain yields. Rice-root aphids

FIGURE 43-10: Damage to upland-rice plants caused by rice-root aphids, a major yield-limiting factor in the Mae Haeng village agroecosystem.

Ageratum conyzoides (L.) L. [Compositae]

With seeds drifting on the wind, this was the dominant weed in all monitored fields at Mae Haeng village.

TABLE 43-3: Major and minor factors and conditions influencing upland-rice yield build-up, based on PCAIV analysis for each cropping year in Mae Haeng village. Bold type indicates strongest relationships.

Relationship	1993 WS −	1993 WS +	1994 WS −	1994 WS +	1995 WS −	1995 WS +	1996 WS −	1996 WS +	1993–96 WS −	1993–96 WS +
No. of plants per sq. m.	**lc**, er, we1,we2, mt,sl	**ec**, sand,dt	**we2, lc**, we1, yc,ff	ec	**ra, we2**, we1, er,yc	camg, sl	**we1, we2**, camg, k,yc, lc	ec	**we2**, lc	ec, dt
No. of panicles per plant	**ra**, we1,we2, camg, ph,bf, mt,sl,k	sand, dt	**er**, k, ra,bs, camg, pav		**ra, we2**, gf, er,yc, we1		**er**, **ra**, **sl, mt**, camg, k,yc	**dt**, ff	**ra**, ws3, er,sl, we1,ws3, yc,mt	ff
No. of spikelets per panicle	**ra**, we1,we2, camg,ph, bf,mt,sl	ff	er, k, ra,bs, camg,pav		**ec**, gf,lc, ra,we2	camg, sl	ra, mt, er		**ra**, ws3, ec,er, sl,yc	ec
% of filled spikelets	ec	lc		lc	**ws3**, ec,gf	camg, lc, ec			**ws3**, ec	lc, ws2
1000-grain weight					**ws3**, lc,we1				**ws3**, ec	lc, ws2
Inertia of axes 1–2 (%)	84		87		70		75		84	
Pertinence PCAIV (%)	63		67		52		79		37	

Notes: Crop environmental and management variables: ec, lc = early, late cultivars; mt, dt = minimum, deeper tillage; yc = no. of successive years of upland-rice cultivation; ff, bf, gf = forest, bamboo, grass fallows; ws1, ws2, ws3, ws4 = 1993, 1994, 1995, 1996 wet seasons; sl = slope angle; er = soil erosion; sand = percentage of sand; ph = pH; som = soil organic matter; pav = soil available P; camg = soil Ca + Mg content; k = soil K content; we1, we2 = cumulative area under the weed cover curve for 0–60 DAS (days after sowing) and for 60 DAS to harvest, respectively; bs = brown spot infestation; ra = rice–root aphid infestation.

are commonly reported in upland-rice fields in Southeast Asia, but such a detailed empirical assessment of their impact on crops is rare (Saito et al., 2006). Extended weed stress during the vegetative phase of the crop cycle was found to be the second most important yield-limiting factor (even after the usual manual weeding procedures), followed by soil erosion on steeply sloping fields.

It was only in one out of four cropping seasons that the last two upland-rice yield components (percentage of filled spikelets and 1000-grain weight) were negatively affected by drought, in spite of the fact that the local cultivars were not very deep rooting, with most of their root mass (90%) found in the upper 20cm of the soil.

No very significant fallow effects or soil nutrient-limiting factors were diagnosed. This could be due to the fact that the survey was conducted in an area with relatively homogeneous and fertile soils. Clearer effects of fallow types may also have been further masked by the fact that similarly dense and aggressive weed communities, dominated by the wind-disseminated *Ageratum conyzoides* species, were observed in all monitored fields. It is also possible that the limited soil fertility measurements (only 1 sampling campaign per season, at 20-30 DAS) used in the survey were not adequate to identify relationships with crop population characteristics. Fertilizer trials conducted at the site showed strong vegetative growth responses of the upland-rice crop, suggesting that soil nutrient levels were less than optimal for sustaining adequate yields over time. However, yield gains from fertilizer application were small due to low biomass partitioning to grain (low harvest index) in these local cultivars (George et al., 2001).

Conclusions and research policy implications

Past research and extension policies for remote upland-rice based agroecosystems in Asia have often focused on the introduction of 'high yielding varieties', and on the use of external inputs needed to achieve the yield gains offered by plant genetic improvement. However, the results of this on-farm diagnostic survey show that the genotype-determined yield potentials of local upland-rice cultivars (about 3.1 to 4.4 tons per hectare) are not a major cause of low yields at this site. We observed important yield gaps across fields and cropping years, in both early- and late-maturing cultivar groups. Hence, our analysis suggests that such yield gaps may be narrowed through targeted improvements in resource and crop management (Tittonell and Giller, 2013), while retaining the distinctive advantages of local landraces. Such advantages may include (1) proper adaptation to the difficult conditions in these diverse upland ecologies (Rerkasem, 2008); (2) suitable canopy architecture and swift canopy closure (40 to 60 days), improving weed competiveness and reducing soil erosion (Turkelboom, 1999); and (3) commercial prospects in expanding markets for so-called 'specialty rices' (Chaudhary, 2003).

This study also showed that a well-designed on-farm diagnostic survey, based on in-depth boundary-line analysis, is a pragmatic tool with which to assess and explain yield gaps in upland-rice grown in complex agroecosystems. Yield gaps were found to occur mainly during biomass build-up in the vegetative period and during the early part of the reproductive period (panicle and spikelet formation) of the upland-rice crop. Further analysis of the causes of these yield gaps confirmed the validity of several well-known hypotheses on limiting factors in upland-rice crops in these environmental conditions, such as weed competition, soil erosion and drought stress. However, we also found that closer attention should be paid to management of soil-borne pests (rice-root aphids in particular), possibly associated with increased cropping intensity (shorter fallow periods) and changes in fallow and weed ecologies.

Such in-depth diagnosis of upland-rice cropping enables the definition of a sound agenda for designing and evaluating improved cropping systems. This is not an easy task because crop environmental conditions were found to be far more limiting than farmers' current crop-management practices. Such innovative cropping systems could combine the use of weed-competitive and drought-tolerant upland-rice cultivars in suitable rotations with other field crops, helping to reduce erosion, maintain soil fertility and break the biological cycles of soil-borne pests. For small farmers to find them attractive, new cropping systems should also include less tedious weed-control techniques, allowing a sharp increase in family-labour productivity as well as introducing rotational crops capable of generating critical cash income.

Crop yields must increase substantially over the coming decades to keep pace with global food demand driven by population and income growth. We observe a revived global interest in yield-gap studies (www.yieldgap.org), and in robust diagnostic methods to assess the magnitude of yield gaps and explain their causes in crops grown across a wide range of agro-ecological and market situations. Among such methods, boundary-line analysis of large yield datasets across sites, seasons and management practices is increasingly recognized as a powerful tool for on-farm yield-gap diagnosis (Affholder et al., 2013; van Ittersum et al., 2013). We encourage the broader adoption of on-farm diagnostic surveys by agronomic researchers and policy-makers in order to advance more productive and sustainable food-production systems, especially for smallholder farmers in less favourable agroecosystems. We also foresee that on-farm diagnostics will undergo further methodological improvements (Makowski et al., 2007; Tasistro, 2012), including the adoption of more standardized and robust survey-design and data-analysis protocols (e.g. quantile regression to estimate boundary line parameters), the use of advanced tools to better assess crop status and crop environmental conditions (e.g. on-site and remote sensing technologies), and the use of 'big data'.

References

Affholder, F., Poeydebat, C., Corbeels, M., Scopel, E. and Tittonell, P. (2013) 'The yield gap of major food crops in family agriculture in the tropics: Assessment and analysis through field surveys and modelling', *Field Crops Research* 143, pp106–118

Chaudhary, R. C. (2003) 'Specialty rices of the world: Effect of WTO and IPR on its production trend and marketing', *Food, Agriculture and Environment* 1(2), pp34–41

Doré, T., Sebillotte, M. and Meynard, J. M. (1997) 'A diagnostic method for assessing regional variations in crop yields', *Agricultural Systems* 54, pp169–188

Driessen, P. M. and Konijn, N. T. (1992) *Land-use Systems Analysis*, Department of Soils Science and Geology, Wageningen Agricultural University, The Netherlands

George, T., Magbanua, R., Roder, W., Van Keer, K., Trébuil, G. and Reoma, V. (2001) 'Upland rice response to phosphorus fertilization in Asia', *Agronomy Journal* 93, pp1362-1370

Glaszmann, J. C. (1987) 'Isozymes and classification of Asian rice varieties', *Genome* 30, pp21-30

Lebreton, J. D., Sabatier, R., Banco, G. and Bacou, A. M. (1991) 'Principal component and correspondence analyses with respect to instrumental variables: an overview of their roles in studies of structure-activity and species-environment relationships', in J. Devillers and W. Karcher (eds) *Applied Multivariate Analysis in SAR and Environment Studies*, Kluwer Academic Publishers, Brussels and Luxembourg, pp85-114

Makowski, D., Doré, T. and Monod, H. (2007) 'A new method to analyze relationships between yield components with boundary lines', *Agronomy for Sustainable Development* 27, pp119-128

Mertz, O., Wadley, R. L., Nielsen, U., Bruun, T. B., Colfer, C. J. P., de Neergaard, A., Jepsen, M. R., Martinussen, T., Zhao, Q., Noweg, G. T. and Magid, J. (2008) 'A fresh look at shifting cultivation: Fallow length an uncertain indicator of productivity', *Agricultural Systems* 96, pp75-84

Piggin, C., Courtois, B., George, T., Pandey, S., Lafitte, R., Kirk, G., Kondo, M., Leung, H., Nelson, R., Olofsdotter, M., Prot, J. C., Reversat, G., Roder, W., Schmit, V., Singh, V. P., Trébuil, G., Zeigler, R., Fahrney, K. and Castella, J. C. (1998) *The IRRI Upland Rice Research Program: Directions and Achievements*, International Rice Research Institute, Los Baños, Philippines

Ramakrishnan, P. S. (1992) *Shifting Agriculture and Sustainable Development: An Interdisciplinary Study from North-Eastern India*, Man and the Biosphere Series vol. 10, UNESCO, Paris

Rerkasem, B. (2008) 'Diversity in local rice germplasm and rice farming: A case study in Thailand', *Biodiversity* 9(1 and 2), pp49-51

Saito, K., Linquist, B., Keobualapha, B., Phantaboon, K., Shiraiwa, T. and Horie, T. (2006) 'Cropping intensity and rainfall effects on upland rice yields in northern Laos' *Plant and Soil* 284, pp175-185

Sebillotte, M. (1989) *Approaches of the On-farm Agronomist: Illustrated Methodological Considerations*, Development-oriented Research on Agrarian Systems Project, Kasetsart University, Nakhon Pathom, Thailand

Shaw, J. C. (1996) Personal communication between author Van Keer and Thai ceramics expert and former Chiang Mai University lecturer J. C. Shaw

Tasistro, A. (2012) 'Use of boundary lines in field diagnosis and research for Mexican farmers', *Better Crops* 96(2), pp11-13

Tittonell, P. and Giller, K. E. (2013) 'When yield gaps are poverty traps: The paradigm of ecological intensification in African smallholder agricultures', *Field Crops Research* 143, pp176-190

Trébuil, G., Thong-Ngam, C., Turkelboom, F., Grellet, G. and Kam, S. P. (2000) 'Trends of land use change and interpretation of impacts in the Mae Chan area of northern Thailand', in *Proceedings of the International Symposium II on Montane Mainland Southeast Asia: Governance in the Natural and Cultural Landscapes*, 1-5 July 2000, Chiang Mai, Thailand.

Trébuil, G., Ekasingh, B. and Ekasingh, M. (2006) 'Agricultural commercialisation, diversification, and conservation of renewable resources in northern Thailand highlands', *Moussons* 9/10, pp131-155

Turkelboom, F. (1999) 'On-farm diagnosis of steepland erosion in northern Thailand: integrating spatial scales with household strategies', PhD dissertation to the University of Leuven (KUL), Belgium

van Ittersum, M. K., Cassman, K. G., Grassini, P., Wolf, J., Tittonell, P. and Hochman, Z. (2013) 'Yield gap analysis with local to global relevance: A review', *Field Crops Research* 143, pp.4-17

Van Keer, K. (2003) 'On-farm agronomic diagnosis of transitional upland rice swidden cropping systems in northern Thailand', PhD dissertation to the University of Leuven (KUL), Belgium

Van Keer, K., Trébuil, G., Courtois, B. and Vejpas, C. (1998) 'On-farm characterization of upland rice varieties in north Thailand', *International Rice Research Notes* 23(3), pp21-22

Van Keer, K., Trébuil, G. and Thirathon, A. (2001) 'Farmers' practice of using salt for weed control in upland rice', *IRRI Program Report for 2000*, International Rice Research Institute, Los Baños, Philippines, pp56-59

Walker, A. R. (1976) 'The swidden economy of a Lahu Nyi (Red Lahu) village community in North Thailand', *Folk* 18, pp145-188

Wang, H., Pandey, S., Hu, F., Xu, P., Zhou, J., Li, J., Deng, X., Feng, L., Wen, L., Li, J., Li, Y., Velasco, L. E., Ding, S. and Tao, D. (2010) 'Farmers' adoption of improved upland rice technologies for sustainable mountain development in southern Yunnan', *Mountain Research and Development* 30(4), pp373-380

Webb, R. A. (1972) 'Use of the boundary line in the analysis of biological data', *Journal of Horticultural Science* 47, pp309-319

Wey, J., Oliver, R., Manichon, H. and Siband, P. (1998) 'Analysis of local limitations to maize yield under tropical conditions', *Agronomie* 18, pp545-561

B. From highland development projects

A monthly meeting of Hilltribe Development Division field workers in Mae Hong Son province, northern Thailand.

Sketch based on a photo by Peter Hoare.

44

LESSONS LEARNED IN NORTHERN THAILAND

Twenty years of implementation of highland agricultural development and natural resource management projects

*Peter Hoare**

Introduction

Over the past four decades, major international development projects have usually followed a process in which the rationale, the methodologies and the outcomes were dictated from 'above' – if international aid agencies and their national counterparts are regarded as the apex of a notional project structure, and the beneficiaries, i.e. those who are being 'developed', exist at the bottom. Hence, the common term 'top-down approach', used in development and scientific circles to describe the path down which the methods and objectives of a project commonly flow. The alternative is a 'bottom–up' process in which the needs, opinions and knowledge of the people at the bottom – the intended beneficiaries – carry as much weight in the preparation of project work plans. In terms of achieving successful implementation of a project and widespread adoption of its changes and innovations, involvement of the beneficiaries, using 'bottom-up' approaches, is now widely regarded as the only way to proceed. It was not always like that. For all the successes that have been achieved by international development projects over the past four decades or so, there has been an astounding waste of money and effort on projects doomed to failure because they told the 'beneficiaries' what they were getting and how they should adapt, rather than

* PETER HOARE, Advisor, Faculty of Agriculture, Chiang Mai University, Chiang Mai. Formerly, team leader and researcher on agricultural extension methods for the Thai-Australian Highland Agricultural Project, 1975 to 1980; consultant on extension methods for an Australian Development Assistance Bureau preparation mission for technical assistance to the Thai-World Bank Highland Agricultural and Social Development Project (HASD), 1981; Extension trainer, HASD project, 1981 to 1985; short-term consultant on extension training and smallholder tea with the Thai-German Highland Agricultural Project, 1986; Research Fellow, East-West Center, Honolulu, on agroforestry in the Asia-Pacific Region, 1986; Chief technical advisor for the Thai-Danish Upper Nan Watershed Management Project from 1996 to 2003, with the objective of improved government and community cooperation for sustainable highland natural-resource management.

starting from their perceived agricultural and natural-resource problems and their ideas on how to solve their problems. Those of us who worked on international development projects in that period took the science inherent in such projects and, with considerable effort, made implementing them into an art.

The art arises from implementing such projects so that community participation is accepted as part of the way things should be done. Not only this, but also making community participation a part of the culture of the supervisory agencies; changing their mindset away from the old way of managing projects in what was seen as a logical framework that provided for only limited participation by the people who were the intended beneficiaries. For a consultant dedicated to ensuring that a project benefitted farmers, this meant navigating a tricky pathway through the culture of the agency responsible for the project and the sometimes conflicting attitudes of consulting colleagues. The art of participatory project implementation is to settle on a combination of bottom up and top down processes so that the project work plan is based on needs identified by the villagers that fit the prevailing government and project policies.

I was motivated to write this chapter by the editor, Malcolm Cairns, who wrote, 'You must be one of the last still here who witnessed at first hand the busy period of highland projects in Chiang Mai, so I feel it is absolutely great to have your hindsight on lessons learned during that time.' In this chapter, I will describe how the processes of community participation were field tested and became an integral part of two major agricultural development projects and one natural resource management project in the highlands of northern Thailand. The projects were the Thai-Australian Highland Agricultural Project (TAHAP, 1972 to 1980); the first phase of the Thai-World Bank Highland Agricultural and Social Development Project (HASD, 1979 to 1987); and the Thai-Danish Upper Nan Watershed Management Project (UNWMP, 1997 to 2003).

Northern Thailand in the 1970s

Thailand's policy in the 1970s was to raise the living standards of rural people through improved agricultural production. In 1976, agriculture contributed 26% of Thailand's gross domestic product; 75% of the country's export earnings came from agriculture and the sector employed 76% of the labour force. Northern Thailand had about one tenth of Thailand's population, with 0.6 million urban and 3.8 million rural Thais, most of them living in the lowlands. There were at least 300,000 Karen, Hmong, Yao, Lisu, Lahu, Akha, Lua and Khamu 'hill tribe' ethnic minorities living in the highlands. Their rate of population increase – 2.6% per annum – was above the national level due to natural increase and migration from neighbouring countries (FAO/World Bank, 1978).

Northern Thailand covers an area of 128,480sq. km, or about one quarter of the country. About 75% of it is hilly, steeply sloping and mountainous land. Most of it lies in the watershed of the Chao Phya river, which flows south to provide water for irrigation in the central plains. The altitude of northern Thailand ranges from less than 300 to 2565 metres. The rainy season, with the southwest monsoon, lasts from May to October and annual rainfall ranges from 620mm in the valleys to more than 1600mm on the mountain tops.

In 1969, King Bhumibol Adulyadej founded the Royal Project Foundation to help to solve the problems of deforestation, poverty and opium production in the highlands of northern Thailand. The King and Queen spent two months every year visiting projects in northern Thailand during the cool season. A wide range of highland horticultural crops were developed by the Royal Highland Project over 30 years of publicly funded research under the leadership of Prince Bhisadej Rajani, a senior member of the Thai royal family. In the period from 1969 to 2003, there were 22 government agencies and many international projects supporting research and development in agriculture, education and health for minority groups, as well as opium crop replacement, in the highlands of northern Thailand.

The Australian Development Assistance Bureau (ADAB) – predecessor to the present AusAID – strongly supported the Thai policy of improving agricultural production. Research and staff-development programmes were initiated at three new regional universities of Chiang Mai in the north, Khon Kaen in the northeast and Prince of Songkhla in the south, with funding for post-graduate scholarships. Agricultural development projects were also supported with different Thai agencies in the three regions. There was an urgency to 'get it right' in the 1970s because of communist insurgency in the lower north and border areas of Thailand. Vietnam and Cambodia fell to the communists in April 1975, soon after I arrived in Chiang Mai, and the Pathet Lao took over in neighbouring Laos in the following December.

Field testing participatory agricultural extension processes

The Thai-Australian Highland Agricultural Development Project (TAHAP), 1972 to 1980

The Tribal Research Centre (TRC) was inaugurated in Thailand in 1964, and was later upgraded to become the Tribal Research Institute. The research publications of anthropologists based at the TRC were a useful starting point for agriculturalists who, in the 1970s, needed to understand the culture, economy and farming systems of shifting cultivators. The Hilltribe Development and Welfare Division of the Department of Public Welfare (hereafter referred to as the Hilltribe Division) was made the main coordinating agency for matters related to hill tribes by Cabinet resolution DPW/6/7/1976.

TAHAP was based at the Faculty of Agriculture, Chiang Mai University. The project's Australian implementing agency was the University of Queensland and the Tribal Research Centre was named as a partner. My job included the field testing of new extension methods with highland farmers (Table 44-1).

Participatory agricultural extension methods had been pioneered by Joan Tully of the University of Queensland in her studies of sociological and educational concepts in changing the practices of a group of Australian dairy farmers. She used the 'problem census' technique, which she described as 'a highly structured situation which provides the opportunity for everyone to express their opinions' (Tully, 1966). The 'problem census' and 'problem-solving' techniques were used by Australian agricultural extension officers when working with diverse ethnic groups in shifting agriculture systems in Papua New Guinea, before its independence in 1975. There, the 'problem census' was described as the first part of a process 'to commence an extension programme at the capability and expectation of the community' (Ministry of Agriculture Papua New Guinea, 1966).

Agreement was obtained from senior management at the Hilltribe Division to field test participatory extension methods in hill tribe villages in northern Thailand, with anthropologists from the Tribal Research Centre as observers. The meetings were in Karen, Hmong, Yao and Lahu villages (Lamrock,1978). Village problem census and problem-solving meetings involved field staff of the Hilltribe Division who were based at so-called 'key village centres' and Jack Lamrock from the University of Queensland, a former Divisional Director of Agriculture in Papua New Guinea. A consensus budget meeting was also held to gather data from the farmers on inputs and outputs to calculate gross margins on traditional crops. Later, positive feedback from the observing anthropologists led senior management at the Hilltribe Division to approve the use of these extension methods by their staff. Lamrock reported back to the University of Queensland that the extension method could be written on the back of a postage stamp: 'pc+ps+cb (problem census + problem solving + consensus budget)'.

The need for village water supplies was one of the highest-ranked problems arising from the problem census meetings and the Australian Ambassador's funds were used for pilot projects involving low-budget gravity-fed water supplies constructed in two villages by voluntary labour. As a result, ADAB approved funding for village water supplies as well as Australian technical assistance to the Thai–World Bank Highland Agricultural and Social Development Project, which began concurrently but followed TAHAP. A total of 120 villages benefitted from new water-supply systems over the next decade. The first agricultural credit project for hill tribe farmers with a commercial bank, the Mae Rim branch of Bangkok Bank, was launched to support a potato crop and concluded with 100% repayment by Hmong and Lahu farmers. This project showed that hill tribe farmers

Prunus persica (L.) Batsch [Rosaceae]

The TAHAP project found that growing native peaches for pickling gave much higher returns to family labour than growing upland rice.

TABLE 44-1: Summary of Thai–Australian Highland Agricultural Project objectives, counterpart agencies, location and lessons learned for participatory development.

Project objectives and staff	Counterpart agencies, location	Outputs and lessons learned for farmer-centred research and development
1. Applied research to improve both the livestock industry and subsistence food production of the hill tribe people. 2. Assist in training extension officers of the Hilltribe Division. 3. Assist the Faculty of Agriculture, Chiang Mai University to undertake research in the highlands. Project staff Five Australian agricultural scientists, seven Thai counterparts and many cooperating staff from the Faculty of Agriculture, Chiang Mai University; five from Tribal Research Centre and one from Royal Forest Department (RFD). Staff development Eleven post graduate scholarships to Australian universities after completing research projects with Australian counterparts. My position Team leader and agricultural extension methods.	ADAB funding through the University of Sydney and Queensland University. Faculty of Agriculture, Chiang Mai University as the executing agency; the Tribal Research Centre and Hilltribe Division as partners. Location Field site at Doi Pa Kia, about 80km north of Chiang Mai at an altitude of 1400 metres, in Nikhom Chiang Dao. Surrounding villages were Hmong and Lahu practising shifting agriculture and northern Thai (Khon Muang) native-tea villages producing fermented tea (*Miang*).	1. Extension methods. 1.1. Rapid measurement of existing subsistence farming systems based on one season's recall: location of shifting agricultural fields, field measurement, labour and material inputs and yields on each field, household livestock enterprises and family food consumption. Output: the land area needed for household subsistence and labour available by calendar month for new enterprises. 1.2. Field testing of problem census and problem-solving village meetings with Hilltribe Division staff and Tribal Research Centre ethnic specialists as observers. 1.3. Consensus budgets to obtain gross margins for existing farm enterprises. 2. Lessons learned 2.1. Participatory extension methods work and are approved by Tribal Research Centre and Hilltribe Division. 2.2. Importance of low-budget gravity-fed village water supplies with voluntary labour for improved public health. 2.3. Agricultural credit – first pilot project with Bangkok Bank – hill tribes prove to be credit worthy. 2.4. Village cattle fund with 40 cows in Hmong and northern Thai villages near the field site. 2.5. Returns to family labour for subsistence hill rice are declining and are below the daily wage rate in some villages. 2.6. Tree crops of tea and native peaches give much higher returns to family labour.

Source: University of Queensland (1981).

were creditworthy. In a similar project, ADAB provided credit for the purchase of 40 cows for village cattle funds in villages neighbouring the Pa Kia field site. This project had its genesis in 1976 when research was concentrated on improving pasture and beef production systems.

The Thai government's fourth national development plan (1977 to 1981)

The Thai government's policies for the country's northern region, announced in the fourth National Economic and Social Development Plan included:

- Agricultural production increases where there was an acute shortage of arable land, in order to satisfy the needs of the population and reduce the need for migration.
- Improvements in the economic and social status of minority groups and natives of the highlands, and checks on further destruction of forests.
- Strengthening of farmer groups and cooperatives as a channel for public and private credit and other government services to farmers.
- Increased participation of commercial banks in extending credit to the agricultural sector.
- Protection and reforestation of areas for soil and water conservation (FAO/World Bank, 1978).

Incorporating new extension methods in the Thai-World Bank Highland Agricultural and Social Development Project (HASD), 1979 to 1987

In 1973, the policy of the Royal Forest Department (RFD) was to reforest northern Thailand, and for this purpose it sought financial assistance from the World Bank. The World Bank's mission recommended that in order to reforest the north of the country successfully, shifting cultivators living in the highlands had to be considered. It recommended three subprojects, beginning in 1979. The first was aimed at establishing sustainable farming systems among mainly northern Thai farmers practising shifting agriculture in lowland and upland areas. This subproject was to be executed by the Land Development Department. The second subproject was the Highland Agricultural and Social Development Project (HASD), aimed at highland shifting agriculture with the Hilltribe Division as the Thai executing agency. The third subproject was that for reforestation by the Royal Forest Department.

The Thai-World Bank (HASD) subproject, with technical assistance from Australia and the United Kingdom, was to become the largest and one of the longest-running international development assistance projects during this period in the northern highlands. From the inception of the HASD project through a second phase called the Thai-Australian TA-HASD project (1988 to 1993), the work spread across 14 years. In the first phase, there were eight highland zones in five provinces, with 306 villages and a population of 52,000 northern Thai, Karen, Hmong, Lahu, Yao and Akha people (Figure 44-1). Project implementation involved a staff of 316 people, including 54 who were already employed by various agencies. Additional staff of 44 graduates, 85 extension workers and 69 teaching aides were recruited. This called for continuous on-the-job training in new extension methods for new and existing key-village-centre field staff. The project area for the second phase TA-HASD project increased to six provinces, with an additional 273 villages and 50,000 highland farmers. The teaching aides established schools at key village centres for primary-level education.

The FAO/World Bank Preparation Report recommended that the delivery of project services to highland farmers be 'organized generally along the lines of the training and visit system' that was used by the Department of Agricultural Extension in the lowlands. However, poor road access and widely dispersed farms in the highlands made a training and visit system impracticable. Both satellite villages and shifting agriculture fields were often far from the key village centres of the Hilltribe Division. An extension system also needed to deliver benefits to ethnically diverse communities of shifting cultivators.

In 1981, ADAB agreed to provide technical assistance in the wake of successful field testing of participatory extension methods and the development experience gained with village water supplies under the TAHAP project. It commissioned a project preparation mission through a private consulting firm. I was included in the team to prepare an agricultural extension training programme, together with a sociologist and an engineer with an eye on access roads and small-scale irrigation. We spent a month in the field visiting all the highland project areas in five provinces. After two years of 'top-down' project implementation based on targets and farm models from the FAO/World Bank Investment Centre Report (1978), we found that farmers were not taking part. For example, the FAO/World Bank target was for 440 hectares of Arabica coffee to be planted by year two. Coffee nurseries were established using untrained hired labour in all key village centres. The seedlings, which were mostly of poor quality, were distributed to farmers during the wet season, when they were already fully employed on their subsistence crops. The Hilltribe Division's project monitoring was based on the number of seedlings produced in nurseries, and showed that targets were being met. But after the provision of technical assistance by the UK for development of coffee crops in 1981, fewer than 10 hectares of coffee were found to be successfully established in the field. A similar pattern became obvious in other project components, with monitoring

Coffea arabica L. [Rubiaceae]

'Top-down' targets included planting of 440 hectares of Arabica coffee in year one of the HASD project. Reality fell far short.

data showing targets were being met, but with a lack of participation by intended beneficiaries and poor quality infrastructure on the ground.

I believed there was a compelling case for ADAB to support technical assistance for the introduction of the participatory extension methods that had been field tested and approved by the Hilltribe Division during the preceding TAHAP project. I was therefore shocked when, on the last day for preparation of a report seeking Australian technical assistance from ADAB, the team leader of the private consultancy asked me, 'Where is your extension programme?' I explained for the umpteenth time the 'pc+ps+cb' process to gain farmer participation for preparing the village work plans. He said 'When I was in the South Australian Department of Agriculture we had a calendar for tree crops that showed the activities for farmers for every month of the year.' Fortunately, his fear of delegating control through empowerment of shifting cultivators did not extend to the Thai project management. The Hilltribe Division Project Director wrote:

> At first we used the same approach we had used before – the top-down approach. We set up targets and then went to the people and then tried to persuade the people what we thought was good for them. After we had used this approach we had a big discussion with the Australian advisors and agreed to change the way we approached the project. We switched to a more participatory approach. Then we trained our staff in the problem census approach – to go to the village and help the villagers identify their own problems and make plans to help solve these problems. (Chandraprasert, 1993, p2).

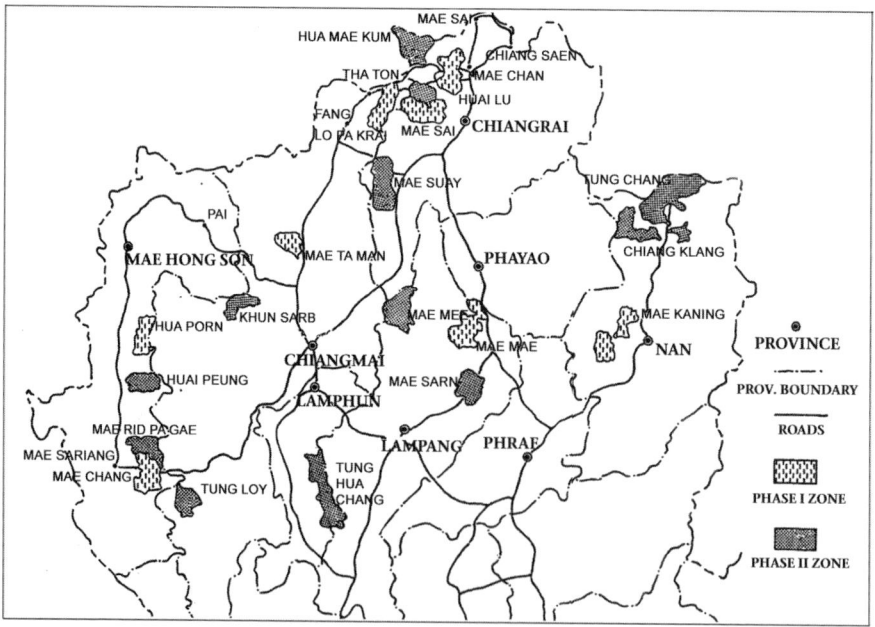

FIGURE 44-1: Northern Thailand showing TA-HASD project areas.

Village problem census meetings

Appointments for the village problem census meetings were arranged through the key-village-centre teams, which consisted of a team leader, a primary health worker, a primary school teacher and a diploma-level agricultural worker from the Hilltribe Division. The teams were usually responsible for three to five highland villages. Most of the villages comprised from 20 to about 70 households, and all households were invited to the meeting. The participants usually sat on the floor, and were asked to form groups of three to seven people in which they felt socially comfortable. There were usually some mixed gender groups. The group chose a discussion leader with the task of ensuring that everyone had a chance to express an opinion. Each group selected a recorder, and a sheet of paper about 1m x 0.80m in size was provided, along with a marking pen.

The meeting participants were then given the frame of reference: 'What are your main agricultural problems?' The groups discussed their problems in their native language and recorded their problems in the central Thai language. If the adults were illiterate, children studying at the primary school were asked to record the problems identified by each group. After about an hour and half, each group listed more than five problems. The meeting facilitators then asked each group to pin their list of problems on the wall and the group leaders read the problems to the meeting and clarified various points if asked to do so. After this process, the facilitators and meeting participants prepared a list of common problems and the number of groups with similar problems. This usually took another hour and a follow-up problem-solving meeting was scheduled within a week.

The most important indicator that the intended beneficiaries were taking part in the project was the number of villages conducting problem census meetings, as shown in Table 44-2. Problem census meetings were conducted in all 96 key village centres and 291 satellite villages, or 65% of the total. The level of female participation in the problem census meetings was highest in Karen villages (40%) and lowest in Hmong villages (14%).

Many villages gave high priority to the need for a village water supply. A total of 120 villages benefitted from this ADAB-funded project component, using voluntary labour from the recipient community. The result was improved public health and labour productivity. Importantly, it was a crucial entry point to increase participation in agricultural production activities.

The problem census is now a widely used participatory process in rural development. The International Institute of Tropical Agriculture has made the problem census and solving process the subject of a research guide to increase farmer participation in research and development in Nigeria (Schulz, 2000). The Bangladesh Department of Agriculture has an online extension manual showing how to structure separate male and female groups in a Muslim community (Department of Agricultural Extension Bangladesh, 1995). The Upper Nan Watershed Management Project used the problem census technique to prepare annual work plans using the frame of reference: 'What are your main natural resource problems?' (DANIDA, 2003).

TABLE 44-2: Number of villages conducting problem census meetings in the HASD and TA-HASD projects.

Province	Total number of villages	Number of village problem census meetings	% of villages with problem census meetings
Chiang Rai	85	74	87.1%
Chiang Mai	207	124	59.9%
Lampang	54	39	72.2%
Lamphun	20	20	100.0%
Mae Hong Son	99	89	89.9%
Nan	78	41	52.6%
Total	541	387	71.3%

Source: Jones (1993, p48).

Village problem-solving meetings

All households were again invited to the problem-solving meetings, and the starting point was attaching the record from the problem census meeting to the wall. After reading the problems and the ranking of problems from the previous meeting, participants were asked, 'How can you solve your own problems?'

Tully (1973), who pioneered the problem-solving meeting process in regularly scheduled bi-monthly meetings with the same group of Australian farmers, considered that there were seven steps to be taken during discussion of the parameters, the inter-relationship of the variables, the causes of the problems and possible solutions. In the HASD project the number of steps in the problem-solving process was reduced to four as regular meetings could not be scheduled with the same groups of farmers. This was because of difficult access during the rainy season, with many farmers spending weeks at distant field huts. The four steps were:

1) recognition of the problem in general terms (problem census);
2) recognition of the causes of the problem;
3) recognition of possible solutions to the problem (the consensus budget); and
4) a 'project proposal' to present to senior management at an annual budget planning workshop seeking budget support for carrying out the chosen solution (TA-HASD, 1988, p36).

Rice was the staple crop. Food consumption data from households of four ethnic groups in 1979 showed that about 80% to 90% of dietary energy and 60% to 80% of protein came from rice. However, this was when it was foot pounded, removing just the husk and leaving most of the outer brown layer of bran. This was before the introduction of village rice mills and the production of white polished rice (University of Queensland, 1981). With an increasing population and shortening fallow periods for hill rice in the 1970s and 1980s, one of the main findings from the problem census meetings was 'not enough rice for subsistence'. At that time, subsistence demanded about 300kg of unhusked rice per adult, or about 2000kg for the average highland farm family, per year, including about half a litre a day for the chickens. The approach to solving this problem changed rapidly between the 1970s and the 1990s, with improved road access, market integration and increasing opportunities for off-farm income in most provinces in northern Thailand, as shown in Table 44-3.

The problem census and problem-solving meetings also created the opportunity for the farmers to modify farm models proposed by the external agencies. The FAO/ World Bank Preparation Report specified the establishment of a coffee (*Coffea arabica*) nursery in each key village centre, with each family terracing and planting 0.2 hectares of coffee. This included the northern Thai growers of *miang* tea. The fermented 'jungle tea', which is chewed as a stimulant, is produced from native tea (*Camellia sinensis* var. *assamica*), an understorey plant, at altitudes of between 900 and 1400 metres. The mature leaves are cut with a small finger knife, tied with bamboo strips into small bundles and steamed for two hours before being stored in pits for anaerobic fermentation. *Miang* has been marketed in northern Thailand and the Shan states of Myanmar for more than a century. A Thai forest researcher, Preechapanya (1998), in a study of indigenous agroforestry systems in northern Thailand, considered *miang* tea grown in an understorey at about 600 trees per hectare in hilly evergreen forest to be the most sustainable highland farming practice in the region, combining farm income with forest conservation and capable of providing harvests over a long period. When problem census meetings were conducted in these villages, the farmers were not interested in planting coffee. Their village development plans proposed the establishment of tea nurseries for high-density terraced tea under evergreen forest

TABLE 44-3: Changes in problem-solving strategies for 'not enough rice for family consumption', with improved road access and market integration.

1970s: Poor road access, limited market access, objective rice food security (1)	1990s: Market integration, road access, resulting in increased agricultural diversification (2)
1) Grow enough rice for family consumption.	1) Farmers buy rice from lowland markets with income from cash crops and off-farm income.
2) Project assistance for small areas of paddy land development.	2) Limited expansion of irrigated rice.
3) Increased yields of hill rice with fertilizer and improved weeding.	3) Decreasing area planted to hill rice.
4) Large hill rice fields on steep slopes, with long slope length. Soil erosion identified as major problem by consultant agronomists.	4) Smaller areas of cash crops planted near villages. Smaller fields of cash crops result in quicker soil cover than rice and decreasing soil erosion.
5) Consultants field test contour grass strips to reduce soil erosion, legumes in cropping systems to maintain soil fertility, and extension recommended to farmers.	5) Increase areas of tree crops and distant swiddens abandoned to natural forest regeneration. Land-use changes in line with government policy for watersheds.
6) Problem of land tenure for agriculture on slopes of more than 35% in Chao Phya watershed being against government policy.	6) Land tenure on slopes over 35% not possible but tree crops believed to provide usufruct rights.

Source: (1) HASD Strategies, 1982 to 1987; (2) Trebuil et al. (1995).

and also the planting of fruit trees. They wanted to diversify and learn tea-nursery technology because they had a ready market for fresh tea leaf at a fixed price at the factory of the nearby Raming Tea Plantation. The factory produced black tea and had offered monthly cash payments. A later World Bank evaluation mission was not impressed and asked 'Why did you let the farmers plant tea? The world market price is low.'

Camellia sinensis (L.) Kuntze [Theaceae]

Farmers growing *miang* tea were happy with their crop and wanted to expand. Project donors thought differently.

Consensus budget meetings

The output of these meetings were analyses of gross margins, with work data and material inputs for traditional crops and livestock enterprises provided by the farmers. The meetings also produced gross margins detailing the inputs needed for new enterprises such as coffee and terraced tea. The farmers participated by giving their estimates of labour inputs required for activities such as land preparation or terracing. A subject-matter specialist detailed the material inputs and cultural practices for new field crops or perennial tree crops, including pruning needs and expected yields. Meetings for field crops usually took an hour, but for coffee and tea, which had a 10-year time frame, they took up to three hours. In the Hmong villages it was interesting to see some farmers using calculators to check data displayed on a board by the meeting facilitator.

In order to prepare village work plans, the key-village-centre staff and farmers were made aware of labour inputs needed by calendar month, the cost of inputs, the availability of credit and expected yields. The gross margins were a starting point for training and the farmers were easily motivated to learn new skills. The consensus budget meetings were also used as field days for annual crops, to discuss the results of on-farm crop trials. Gross margins for the on-farm trials could then be compared with traditional practices.

Annual budget-planning workshops

Annual budget-planning workshops were initiated in 1982 as a crucial new activity in the participatory process. They were intended as a vital link between work plans based on identified village needs, available budget and government policy; ultimately seeking both approval and funding from senior management for village work plans that were in line with project and government policies (Hoare and Crouch, 1988). This new activity was the link between the 'bottom-up' and 'top-down' project processes, and when added to the established extension process, it could still fit on the back of a postage stamp: 'pc+ps+cb+abpw' (problem census+problem solving+concensus budget+ annual budget-planning workshop).

On the first day of the workshop, senior project management informed key-village-centre field staff of the donor and regular Thai funds that were available for the coming year, and priority activities dictated by project and government policies. The key-village-centre teams from the different provinces reported on their achievements in the previous year, and then the village work plans from farmers, arising from the problem census and problem-solving meetings, were presented. In the following two days the village plans were tailored to the available budget and to meet project and government policies. By the end of the four-day workshop, all key-village-centre field staff were familiar with the coming year's draft work plan about two months before the start of the new budget year. Opportunities for corruption were minimized.

The first annual budget-planning workshop was rough on our young Thai counterparts after they reported to senior management that fewer than 10 hectares of quality Arabica coffee had been planted despite monitoring data, which was based on the number of seedlings produced in the nurseries, showing the area as 440 hectares. They were severely criticized in a private meeting for 'showing all the bad things to the *farangs*'(foreigners) and they offered to resign en masse. However, there was a marked change in attitude when management was sufficiently agile to risk change by implementing the first village 'bottom-up' work plans with Australian technical assistance. There was an immediate difference in farmer participation and in subsequent annual budget-planning workshops the attitude of project management changed completely and close attention was paid to the reports of junior field staff who, in their new role as advisors, reported truthfully on what was happening in the field.

The participatory process was dynamic and, over 10 years of project implementation, was decentralized to provincial level and became institutionalized within the Hilltribe Division. The timetable involved in taking the many steps in preparation and implementation of these functional participatory annual work plans are shown in Table 44-4. The timetable fits the Thai government budget year from 1 October to 30 September. Activities for farmers begin with preparations for a dry-season irrigated crop in the cooler season in October. The budget calendar also suits infrastructure activities such as work on access roads, small-scale irrigation schemes, village water supplies and terraces for coffee and tea planting, most of which take place in the dry season between November and April. The planting season for the main rainfed crops begins with the arrival of the southwest monsoon in May.

Detailed extension manuals for training field staff in village water-supply design and construction, the preliminary survey of diversion weirs, extension methods and field and tree crops were compiled from field experience (Thai-Australia Highland Agricultural and Social Development Project, 1988).

TABLE 44-4: The 'functional participatory planning process' institutionalized in the Hilltribe Division: Timetable for Annual Budget Work Plan Preparation, based on the Thai Fiscal Year (1 October to 30 September).

Activity	Completion Date
Plan preparation (12 months in advance)	
Forward budget estimates by HTD	end September
Staff plan for problem census meetings	end January
Village problem census meetings	end February
Prepare project policy guidelines	end February
Project guidelines management meeting	mid-March
Project guidelines provincial meetings	end March
Village problem solving	end April
Family and group interviews	mid-May
Village work plan programmes	mid-May
Zonal work plan programmes	end May
Provincial workshops	end June
Executive workshop	end July
Submit work plan budget to HTD	end August
Plan validation	
Prepare first period training plan	end September
Review first period social plan	mid-September
Thai fiscal year begins October 1st	
Review dry season agricultural plan	mid-October
Prepare second period training plan	mid-January
Prepare second period social plan	mid-January
Review wet season agricultural plan	end March
Prepare third period training plan	mid -May
Review third period social plan	mid-May

Source: Ole (1993).

The Thai–World Bank Highland Agricultural and Social Development Project achieved or exceeded all of the physical and personnel targets laid down by the project appraisal and design teams. The few shortfalls were those requiring line agency cooperation and coordination for schools, access tracks and satellite imagery. In his concluding remarks at the Project Completion Seminar, the Director of the Hilltribe Welfare and Development Division of the Public Welfare Department, Elawat Chandraprasert, said he considered that the functional participatory process was the project's major success:

> I think the major success of the Thai–Australian Highland Agricultural and Social Development Project has been a functional participatory approach for working with the people. …We have listened to hill tribe farmers, we have learned from them, and we have accepted their knowledge. …Our strategy for

transferring new technical knowledge to the farmers …we can advise them, but we must let the villagers make their own decisions on whether to accept the technology or not (Chandraprasert, 1993, p197).

What better endorsement could be found for participatory agricultural development? This was the Project Director who, in the first two years of project implantation, used a top-down work plan based on farm models in the FAO-World Bank Investment Centre Report (1978).

The Upper Nan Watershed Management Project (1996 to 2003)

In Thailand's eighth national plan (1997 to 2001), the policy focus was on holistic, people-centred development and improved natural-resource management. The Local Administration Act of 1993 empowered subdistrict administrations in natural-resource management. The Upper Nan Watershed Management Project was carried out over eight years in a vital watershed of the upper Nan river that provides more than half of the irrigation water for Thailand's agricultural heartland, the central plains. The project was funded under technical cooperation between the Royal Thai and Danish Governments. Danish Cooperation for Environment and Development (DANCED) and later Danish International Development Assistance (DANIDA) were the donor implementing agencies. Its location is illustrated in Figure 44-2 and its outputs and lessons learned are listed in Table 44-5.

The art of project implementation in this case was using the talents of the six Forest Department watershed management unit chiefs and the previous field experience of the 15 community coordinators to achieve the project's objectives. One of the watershed management unit chiefs became an expert in using the problem census and problem-solving techniques for community planning and another led the project geographic information system section. Significantly, during the baseline survey one of the community coordinators documented the existing village regulations for natural-resource management and I realized this could be an important entry point for community participation. This finding, that existing village regulations could be used as an entry point for community participation, is described in detail in chapter 48.

Thailand's Royal Forest Department was known for its 'culture of power', and after a good start, the project struck a crisis when my counterpart Project Director declared: 'Someone is trying to cut the legs off my chair!' The new Project Director didn't like the DANCED funds for village development projects being controlled by the village committees, and refused to approve any more projects. Fortunately, the Danish consultants and DANCED acted quickly and asked their embassy in Bangkok to seek an emergency meeting with the Forest Department's Director-General. The project struggled with a non-communicative Project Director until the end of that year, when he was promoted and departed, and an earlier Project Director took over. By the end of the project, attitudes had changed and all of the Forest

Department project staff were convinced that community participation was necessary for fire management and increasing the forest cover.

Lessons learned: did the project implementation 'get it right'?

The current objectives of the 11th plan of Thailand's National Economic and Social Development Board, covering the years from 2012 to 2016, include a new model of holistic 'people–centred development'. The stated aim is to promote a fair and peaceful society, to increase the potential of all Thais through social institutions, to develop an efficient and sustainable economy and preserve national resources (NESDB, 2012). In this light, Table 44-6 assesses project implementation.

FIGURE 44-2: The area covered by the Upper Nan Watershed Management Project in Nan province, northern Thailand.

Shifting agriculture is no longer a self-contained system; off-farm income needs more attention

With improvements to highland roads, diversification of highland agriculture and rapid integration of highland villages into lowland regional and national economies in the 1990s, shifting agriculture is no longer a self-contained system. A prominent researcher of the Karen 'hilltribe' people, Peter Hinton, wrote:

> Anthropologists are often accused of romanticizing the self-contained community. It seems to me that the TA-HASD project has tended to regard the communities as self-contained units and has tried to build up this self-containment. I would like to suggest some greater attention to those off-farm factors... People go off to town not just out of necessity but to supplement incomes and have some experience of the world outside (Hinton, 1993, p192).

TABLE 44-5: Summary of the Thai-Danish Upper Nan Watershed Management Project: Its objectives, counterpart agencies, locations and lessons learned.

Project objectives and staff	Counterpart agencies, location	Outputs and lessons learned for community and government cooperation in natural resource management
1) Development objective: An organizational framework for sustainable natural-resource management by communities and government agencies on the upper right bank of the Nan river.	Danish Cooperation for Environment and Development (DANCED) and later DANIDA (Danish International Development Assistance).	1) Baseline survey on fires reveals that all villages have rules and regulations and about half of them concern community natural-resource management.
2) Immediate objectives:	Thai government budget for Watershed Management Division of Royal Forest Department, six watershed management units as implementing agency.	2) Land-use surveys using hand-held GPS completed for all agricultural fields in 45 villages to monitor land-use changes.
2.1 GIS-generated participatory tools for land-use planning using 3-D topographical models.		3) Topographical 3-D models prepared for all villages and seven watershed networks to assist land-use planning.
2.2 Income-generating activities initiated by individuals and supported by village committees functioning on a sustainable basis.	Eight subdistrict (Tambon) administrations and two districts in Nan province, as partners	4) A total of 244 income-generation projects appraised and implemented using 7,467,373 baht (about US$181,467)★ with villagers managing grant funds, 1997 to 2000.
2.3 Institutionalize planning and implementation of project activities in village committees, village watershed networks subdistrict and district administrations.	Location: 1007sq. km in Song Kwae and Tha Wang Pha districts of Nan province on the upper right bank of the Nan river including the Pa Nam Yao and Pa Nam Suad National Forest Reserves, covering about 70% of the project area.	5) Survey in 2000 shows that households who used village development funds for paddy development reduced swidden areas by 21%, for tree crops by 40%, and generating income by raising pigs, fish ponds and handicrafts, by 36%. These were mostly distant swiddens on slopes over 35%.
2.4 Increase forest cover on 16% of the project area over seven years by enrichment planting of 156 trees/ha.	Population: 45 villages with 4177 households and 21,561 people from six ethnic groups.	6) Seven village watershed networks formed in 45 villages for improved natural-resource management.
2.5 Increase forest cover on 16% of the project area over seven years by natural forest regeneration.	About 60% of the farmers with no legal land tenure practising shifting agriculture in forest reserves.	7) Boundaries for fire management agreed between Forest Department and village watershed networks over whole project area.
2.6 Implementation of fire controls by villagers and Forest Department staff to reduce burning by 50% in the project area.	Population density about 20 persons per sq. km.	8) Training and equipping village fire volunteers. Community firebreaks reduced area burnt from 21% to less than 3% (Hoare, 2015).
3) Project staff:	Average farm size: 2.1 hectares, small areas of paddy rice and shifting agriculture with crops of hill rice, maize and cotton.	9) Analysis of Landsat images finds forest-cover increased from 28% to 58% between 1996 and 2003 because of enrichment planting and natural regeneration.
3.1 Forest Department project management and staff of six watershed-management units.		
3.2 DANCED/DANIDA staff: Chief Technical Advisor, GIS support staff and trainers and 15 Community Coordinators.	About 80% of cash incomes from off-farm migratory labour.	
The author's position: Chief Technical Advisor.		

Notes: The Royal Forest Department has since been renamed the Department of National Parks, Wildlife and Plant Conservation; ★US$1 had an average value of about 41.15 baht from 1997 to 2000 (US$1=Bt46.89 in December 1997; Bt36.07 in December 1998; Bt38.31 in December 1999; Bt43.33 in December 2000).

TABLE 44-6: Did the project implementation 'get it right'?

Thai government policy	Lessons from project implementation	Recommendations for future projects
People-centred development.	Participatory development project implementation requires a combination of 'bottom-up' and 'top-down' processes.	Project design needs to be flexible to allow 'bottom-up' village planning and project modifications in line with government policy.
Increased agricultural production.	The 'pc+ps+cb+abpw' extension method increased the participation of farmers and increased the flow of project benefits. Improved road access and market integration in the 1990s resulted in smaller areas of cash crops near to villages and off-farm income to buy lowland rice.	Farmers' perceived problems are a key entry point for participation. Small areas of tree crops – 0.2 hectares of coffee, tea or fruit trees and some indigenous agroforestry systems – can generate enough income for self-sufficiency (1) (2). The area planted to subsistence rice is declining, but in Mae Hong Son province, limited market access is still an important issue, with potential for niche markets in organic rice production.
Highland soil conservation.	Practices introduced by projects to slow runoff and minimize erosion involved additional costs to farmers, resulting in limited adoption (3).	Smaller areas of cash crops near to villages and increased planting of tree crops reduces the risk of erosion (4).
Availability of Credit for Agriculture.	Hill tribe farmers proved credit worthy (5). Village-managed development funds reduced swidden areas by 21% for paddy development, 40% for tree crops and 36% for small animal projects and other means of income generation (6).	The availability of credit for village-managed development projects reduces the area of shifting agriculture.
Increased forest cover.	Existing village regulations for natural-resource management are a key entry point for community participation. 3-D topographical models facilitate community land-use planning and participation in natural-resource management.	Village watershed networks are needed, along with agreement between government agencies and local administrations on boundaries for fire management. When less than 3% of a watershed area is burnt, enrichment planting and natural forest regeneration can increase forest cover by up to 50% within five years.

Notes: (1) The farmers' own assessment of the importance of income from new crops of Arabica coffee planted in the TA-HASD project in 1983 was revealed in an article in the Bangkok Post on 9 November 1988, with the headline 'Celebration for Karen's coffee crop'. It was reported that Karen coffee growers would hold a Rab Kwan Kafae (Coffee Blessing) ceremony in Khun Yuam district, Mae Hong Son province;(2) Hoare et al. (2007); (3) Panomtarinichigul (2003); (4) Trebuil et al. (1995); (5) University of Queensland (1981); (6) DANIDA (2003).

Kunstadter (2015) detailed the importance of off-farm factors in Pa Pae, a Lua village in Mae Sariang district, Mae Hong Son province, northwest Thailand, over a period of 50 years. In the 1970s and 1980s many villagers moved out of subsistence agriculture to begin wage work and now half the descendants live in Chiang Mai, even though some of them maintain their household registration in the village. Kunstadter highlighted education as a factor that enabled the children to move away from swidden agriculture.

In the 45 villages involved in the Nan project, about 80% of cash income now comes from off-farm urban migratory labour. I have Karen hill tribe people as my neighbours in Chiang Mai, one as a live-in house help and the other as a handyman managing a dormitory. They send money to their relatives in highland villages, but their children, who are educated in Chiang Mai, are no longer drawn to the isolated and difficult existence in the mountains and want to be a permanent part of Thai lowland society.

Acknowledgements

I would like to acknowledge the leadership of the Thai Project Directors Elawat Chandraprasert, Director of the Hilltribe Development and Welfare Division of the Department of Public Welfare, and Khun Borpit Maneeratana, of the Royal Forest Department, and their staff, who stepped outside their comfort zones to 'risk' the implementation of participatory processes on their projects. Also the Australian expat consultants, especially Peter Jones, who continued the commitment to participatory project implementation after I was sacked in 1995 by AACM Limited when I was forced to take a stand on the integrity of the village planning process on the HASD project. I would also like to thank Kirsten Ewers Andersen, my boss on the Danish-funded Nan watershed project and RAMBOLL and DANCED staff for their professional support. Also thanks to the donors, AusAID (formerly the Australian Development Assistance Bureau) and DANIDA, for the opportunity to work on the development-assistance projects.

References

Bangkok Post (1988) *Celebration for Karen's Coffee Crop*, issue of 9 November, 1988, *Bangkok Post*, Bangkok

Chandraprasert, E. (1993) 'Project introduction', p2, and 'Seminar summary', p197, *Proceedings of Completion Seminar of Thai-Australian Highland Agriculture and Social Development Project*, June 1993, Chiang Mai, Australian International Development Assistance Bureau, Canberra

DANIDA (2003) *Lessons learned from Upper Nan Watershed Management Project 1996–2003*, Thai language with English language summaries, Danish International Development Assistance, Copenhagen

Department of Agricultural Extension Bangladesh (1995) *The Problem Census*, Agricultural Support Sevices Project, Khamarbari Farmgate, Dakha

FAO/World Bank (1978) Preparation Report of the Northern Rural Development Project No 19/78 THA 13, Food and Agriculture Organization of the United Nations, Rome

Hinton, P. (1993) 'Questions and discussion' p192, *Proceedings of Completion Seminar of Thai-Australian Highland Agriculture and Social Development Project*, Chiang Mai, June 1993, Australian International Development Assistance Bureau, Canberra

Hoare, P. (2015) 'Community networking for improved highland fire management', paper presented at the First International Conference on Asian Highland Natural Resources Management, January 2015, Chiang Mai

Hoare, P. W. C. and Crouch, B. R. (1988) 'Required changes to the project management cycle to facilitate participatory rural development', *Agricultural Administration and Extension* 30, pp3-14

Hoare, P., Maneeratana, B. and Songwadhana, W. (2007) 'Ma Kwaen: A jungle spice used in swidden intensification in Northern Thailand', in M. F. Cairns (ed.) *Voices from the Forest: Integrating Indigenous Knowledge into Sustainable Upland Farming*, Resources for the Future, Washington, DC, pp614-619

Jones, P. R. (1993) 'Extension methodology', p48, *Proceedings of Completion Seminar of Thai-Australian Highland Agriculture and Social Development Project*, Chiang Mai, June 1993, Australian International Development Assistance Bureau, Canberra

Kunstadter, P. (2015) 'Swiddeners at the end of the frontier: Fifty years of globalization in Northern Thailand, 1963-2013' in M. F. Cairns (ed.) *Shifting Cultivation and Environmental Change: Indigenous People, Agriculture and Forest Conservation*, Earthscan (Routledge), London, pp134-178

Lamrock, J. (1978) 'Extension methods in Northern Thailand', *Report to Department of Agriculture, University of Queensland*, Thai-Australian Highland Agricultural Project, Brisbane

Ministry of Agriculture Papua New Guinea (1966) *Papua New Guinea Extension Manual*, section DD1, Port Moresby

NESDB (2012) *National Development Plans, Thailand*, fourth plan (1977-1981), eighth plan (1997-2001) and 11th plan (2012-2016), National Economic and Social Development Board, Office of the Prime Minister, Bangkok

Ole, B. T. (1993) 'Planning procedures', P 91, *Proceedings of Completion Seminar of Thai-Australian Highland Agriculture and Social Development Project*, Chiang Mai, June 1993, Australian International Development Assistance Bureau, Canberra

Panomtarinichigul, P. (2003) 'Research on sustainable hill farming in Northern Thailand', paper presented at Department of Water-Atmosphere-Environment Institute, University of Bodenkultur, Vienna

Preechapanya, P. (1998) *Indigenous Highland Agroforestry Systems in Northern Thailand*, Chiang Dao Watershed Research Station, Department of National Parks, Wildlife and Plant Conservation, Chiang Mai

Schulz, S. (2000) *Farmer Participation Research and Development: The Problem Census and Solving Technique*, Research guide 57, International Institute of Tropical Agriculture, Nigeria

Thai-Australia Highland Agricultural and Social Development Project (TA-HASD) (1988) *Extension Manuals for 1) Infrastructure, 2) Extension Methods, 3) Field Crops, 4) Arabica Coffee, 5) Teams*, Department of Public Welfare, Ministry of Interior, Bangkok and Australian International Development Assistance Bureau, Canberra

Trebuil, G., Boch, T., Thong-Ngam, C., Turlkelboom, F. and Begue, A. (1995) *Market Integration, Agricultural Diversification, and Erosion Risk in Northern Thailand*, study funded by the International Rice Research Institute, Philippines and CIRAD, Montpellier, France

Tully, J. (1966) 'Changing practices: A case study', *Journal of Cooperative Extension*, Fall edition, pp143-152

Tully, J. (1973) 'Aims and methods in extension', *Journal of the Australian Institute of Agricultural Science*, March edition, pp53-55

University of Queensland (1981) *Final Report on Agricultural Research and Development in the Highlands of Northern Thailand*, prepared by University of Queensland, Brisbane and the Faculty of Agriculture, Chiang Mai University for the Australian Development Assistance Bureau, Canberra

45

PUTTING UPLAND AGRICULTURE ON THE MAP

The TABI experience in Laos

Andreas Heinimann, Chris Flint, Rasso Bernhard and Cornelia Hett[*]

Introduction

For decades, shifting cultivation has been the dominant land-use system in the uplands of the Lao People's Democratic Republic (Lao PDR; hereafter referred to as Laos), securing the livelihoods of a large percentage of the rural population in this mountainous region. Despite the pull factor of regional economic integration, which favours other forms of land use, and the push factor of government policies aiming to eradicate the practice, recent studies show that shifting cultivation remains widespread in the Lao uplands, and in some regions, it is even increasing (Heinimann et al., 2013; Hurni et al., 2013; Liao et al., 2015).

Since the 1990s, the Lao government has sought to reduce the area under shifting cultivation based on the perception that it is a backward and underdeveloped form of land use (Mertz et al., 2009). While this view is largely flawed, it is nevertheless rooted in the beliefs of many policy-makers and political elites. In this oversimplified view, shifting cultivation represents a 'poverty trap' (Bounthong et al., 2003; Ducourtieux, 2006) and is the main cause of ongoing deforestation (Lawrence et al., 2010) and forest degradation (Fox et al., 2000; Thongmanivong et al., 2009).

Based on this rationale, the Lao government passed various pieces of legislation to promote forest conservation, which impose restrictions on upland peoples' access to their fields and fallows (Fox et al., 2009). At a technical level, the perception of shifting cultivation as a disadvantageous or damaging practice led to its neglect in

[*] Dr Andreas Heinimann, Institute of Geography and Centre for Development and Environment, University of Bern, Switzerland, and The Agro-Biodiversity Initiative of the Lao PDR (TABI), Vientiane, Lao PDR; Chris Flint, The Agro-Biodiversity Initiative of the Lao PDR (TABI), Vientiane; Rasso Bernhard, the Centre for Development and Environment, University of Bern, Switzerland, and The Agro-Biodiversity Initiative of the Lao PDR (TABI), Vientiane; and Dr Cornelia Hett, the Centre for Development and Environment, University of Bern, Switzerland, and The Agro-Biodiversity Initiative of the Lao PDR (TABI), Vientiane.

Laos' first national forest inventory, conducted at the turn of the millennium. In this inventory, the entire upland/fallow system of shifting cultivation was classified as 'unstocked forest' – wrongly implying the absence of agricultural activities. This classification created serious problems for evidence-based natural resource planning. It also led to two problematic consequences for shifting cultivation communities with regard to access to land. First, it supported the idea that large-scale agribusiness was the best way to promote economic integration in rural areas, thus opening the door to the granting of seemingly idle land to large-scale investors (e.g. commercial tree plantations such as rubber). Second, it created an inaccurate understanding of the country's forest cover: fallow areas were characterized as 'forests' in which villagers had only restricted access to land (Schmidt-Vogt et al., 2009). This characterization contributed to the national policy goal of reaching 70% forest cover by 2020. The subsequent creation of three main national forest management zones formalized this limited access to land for shifting cultivation communities (Lestrelin, 2010). The policy created the following three main classes of forest: 'conservation forests' where human activities are prohibited in order to preserve fauna, flora, biodiversity and areas of cultural, educational or scientific interest; 'protection forests' where human activities are prohibited in order to prevent soil erosion and associated natural disasters as well as to protect water sources and maintain national defence areas; and 'production forests' where controlled logging and the limited collection of forest products are permitted.

This forest categorization became the main instrument for area-based planning through the Land-use Planning and Land Allocation Programme (Rigg, 2005). This programme has largely favoured forest conservation over the allocation of agricultural land for smallholder agriculture. For instance, it limited the number of agricultural plots to three plots per household while allocating most fallow land to the three forest categories listed above (Fujita and Phanvilay, 2008). The Land-use Planning and Land Allocation Programme also reduced the permitted fallow length to two years, thus rendering shifting cultivation unsustainable (Bourgoin, 2012). Despite the slight increase in local participation as a result of an updated Participatory Land Use Planning (PLUP) approach, and the endorsement of a manual for this approach (PLUP manual), the policies and regulations affecting shifting cultivation continued to ignore the local realities of upland villages and limited community access to land by fostering a mentality of 'shifting cultivation eradication'.

Due to the expansion of large-scale land acquisitions for agriculture, tree plantations and hydropower since the mid-2000s (Fenton et al., 2011; Schoenweger et al., 2012), many upland communities have been resettled and have lost access to their land. Their new land – when considered in the light of the three-plot policy mentioned earlier – is frequently insufficient to sustain livelihoods. Consequently, a substantial number of resettled households have been forced to 'illegally' return to their former upland plots, often more than a day's walk from their new settlements. In some cases the loss of land has left villagers with no choice but to encroach on previously forested areas for agricultural purposes, leading to an increase in primary shifting cultivation is some areas (Heinimann et al., 2013).

Despite the strong policy-led push factors and market-led pull factors, shifting cultivation remains the main type of agriculture practised in the uplands of Laos (Heinimann et al. 2013). Shifting cultivation communities are marginalized by existing policies, but have no viable alternative cropping practices to help them cope with the drastic changes that the policies demand (Lestrelin, 2010; Vongvisouk et al., 2014). Consequently, there is a large gap between the goals of land-related policies and regulations, on one hand, and their frequently negative impacts on local livelihoods and socio-economic development in the uplands, on the other (Lestrelin et al., 2012). This is largely due to the fact that a number of these land policies and regulations – i.e. the national forest classification, forest management zones and the 'three plot policy'– are not based on the reality of the country's upland agriculture.

In this context, The Agro-Biodiversity Initiative of Lao PDR (TABI) has advanced a multi-step strategy and developed concrete approaches and tools to contribute to more evidence-based, realistic, inclusive and effective resource planning and monitoring of forest and land use in the uplands. TABI is a long-term initiative (2008 to 2020) funded by the Swiss Agency for Development and Cooperation (SDC) and led by the Ministry of Agriculture and Forestry (MAF) of Laos in partnership with multiple stakeholders at various administrative levels. It aims to promote local empowerment and improve secure access to land and forest resources for improved, agrobiodiversity-based livelihoods. This approach includes the following elements:

1. To capture reality in local land-use planning activities and place shifting cultivation areas, including necessary fallows, on the map.
2. To plan realistically, by including shifting cultivation in village land-use plans.
3. To test, monitor and adapt village land-use plans.
4. To facilitate communal land titles for upland areas that provide a legal basis for compensating communities in case of expropriation by the state.
5. To bridge the gap between availability and accessibility by making data and information accessible in order to foster transparency regarding village land use and community aspirations.
6. To promote the benefits of development, as well as planning, by identifying concrete development opportunities in upland communities and providing villagers with development support.
7. To effect changes to national policy. Based on the success of local level land-use planning, to alter the national land-use classification scheme to include upland agriculture and fallows as an official class under agriculture, and to update the three national forest categories accordingly,

In the following sections, we outline the different elements of this strategy, as well as the lessons learned from this approach.

1. Capturing reality in local land-use planning activities and placing shifting cultivation areas, including necessary fallows, on the map

The initial aim of land-use planning must be to gain an accurate and up-to-date assessment of the current resource endowment. Previous approaches have not taken this into account, but rather have applied the current national policy – which ignores shifting cultivation as a basis for the livelihoods of upland villages – to the local level. As a result of quick planning and limited participation from villagers, a large portion of rotational shifting cultivation land was allocated to one of the three national forest management zones without considering the actual land uses and livelihoods that depended upon it. The Lao government's Participatory Land Use Planning (PLUP) manual (MAF and NLMA, 2009) and other approaches developed subsequently (e.g. NAFRI, 2012) have addressed some of these problems, such as the lack of participation. However, the problematic doctrine of the three forest classes remained, especially in the case of the 'PLUP process' of participatory land-use planning according to the PLUP manual.

Using a multi-stage approach, TABI and its partners have developed an efficient and accurate approach for assessing the current land-use reality and livelihoods of village populations as the basis for future land use planning and zonation (MAF-TABI, 2014).

The following three stages are involved in the process of capturing reality in local land-use planning activities and placing shifting cultivation areas, including necessary fallows, on the map:

Stage 1: Baseline assessment of natural resources and socio-economic conditions

In order to reach a proper assessment of the current land-use situation in a village, it is critical that the landscape elements be accurately assessed in the field. First, a topographic and a satellite-image map based on available Government of Laos orthophotos,[1] or high-resolution satellite imagery such as RapidEye, are produced and taken to the village as A0 format (841mm x 1189mm) hard-copy maps. In a series of field visits that include the participation of villagers, as well as office-based GIS work, the geographic features of a village are mapped and named. Our experience has shown that, with a good moderator in the field, villagers are generally able to orient themselves well on the printed maps and provide detailed information regarding geographical features. Using these features, they provide information about the location, extent and types of current land use.

Stage 2: Determining and mapping village boundaries

In close collaboration with villagers, landscape features such as rivers and creeks are mapped and their local names recorded. At the same time, the first draft of a village boundary is delineated. This process is the reason why land-use planning should, whenever possible, be done simultaneously in all the villages that are part of a village-cluster, or *kumban* – the lowest administrative unit in Laos – made up of four to 10 villages. The village borders are negotiated in a series of meetings with participants from all the villages of a *kumban*, as well as district officials. After this negotiation process, the final village boundaries are mapped and formalized in the land-use plan, as well as in the village agreement. This is the first step towards formally claiming village land for its inhabitants (i.e. the local people).

Stage 3: Assessment of current land use

Current land use is assessed and mapped – with a high level of detail, both spatially and in terms of classification – by a technical team consisting of project members and provincial and district government representatives, with the participation of local authorities and villagers. It is crucial to have a classification scheme that allows the process to capture the complex reality of upland landscape mosaics (Figures 45-1 and 45-2). The process explicitly focuses on the fallow areas of shifting cultivation and different uses of forest areas (e.g. for the collection of non-timber forest products) that have not been cleared for long periods (20 years or more), and are therefore not considered part of the rotational system. This map of current land uses serves as a critical input for the next step in the process: land-use planning and the delineation and definition of management zones.

FIGURE 45-1:   Examples of current land use (left), and a forest and land-use management plan (right) for a village in northern Laos (maps simplified for the purposes of this chapter).

2. Plan realistically: include shifting cultivation in the village land-use plan

The negative socio-ecological impacts of shifting cultivation 'eradication' policies in the uplands of Laos have been taken up by many scholars (e.g. Fox et al., 2009; Lestrelin, 2010; Vongvisouk et al., 2014). The issue is also recognized by a number of development initiatives, local government officials (because they are close to the reality) and some reformers in the national government.

However, until a few years ago, there was no approach to land-use planning at village level that was not driven by mistaken policies; that captured complex local realities, and that included participatory land-use planning, zoning and monitoring. TABI has taken on this task, from a strategic perspective, since 2010. The only way to change problematic policies is to provide local government officials and national government reform actors with concrete, bottom-up tools and approaches with which to more meaningfully deal with shifting cultivation in planning and zoning. Other important initiatives (e.g. such as NAFRI, 2012) have worked to further develop the initial participatory land-use planning (PLUP) approach of 2009 (MAF and NLMA, 2009), with a focus on how to enhance village-level participation in the planning and zoning processes (Lestrelin et al., 2011; Bourgoin, 2012; Bourgoin et al., 2012) and on the negotiation process regarding the ecosystem outcomes of different planning scenarios (Bourgoin and Castella, 2011).

TABI's approach begins with a strategy to stabilize, and not to eradicate, shifting cultivation. The process is based on a thorough assessment of the current situation in a village. This includes an assessment of current land use, as outlined above, as well as a number of household-level socio-economic and agricultural surveys (including the use of non-timber forest products, aquatic resources and wood) aimed at understanding the development opportunities and constraints related to land. Through an iterative and participatory process, this information is used to translate village aspirations into a future land-use plan (for more details see MAF-TABI, 2014). This includes a precise delineation of different land-use zones as well as the development of corresponding management plans. A crucial aspect of this approach is the clustering of households' upland fields for all subsequent years (see Figure 45-1). From a soil fertility and weed infestation perspective, delineated cultivation and fallow zones allow for fallow periods with a minimum duration of five years (Seidenberg et al., 2003; Tanaka et al., 2005). This takes into account the results of the socio-economic surveys with regard to land demand, the productivity of different areas and anticipated future developments. In certain ethnic groups, the clearing of land *as a village* has already been practised for centuries. This allows for easier fire control and – critically for our approach and objectives – a land-use plan that looks much more 'tidy' than household-level fields and fallows scattered across the landscape. This may seem like an obscure argument, but experience has shown that it is crucial for convincing local authorities to approve a plan that includes shifting cultivation beyond the official three-plot policy. The final output of the process is a draft land-use planning map and corresponding management plan.

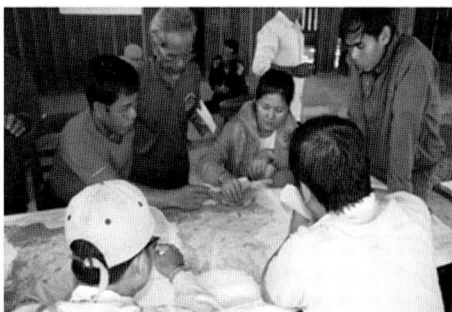

FIGURE 45-2:   Initial village–border delineation and participatory land-use planning.

3. Test and adapt: monitoring and adapting the village land-use plan

Unlike the Lao government's Participatory Land-use Planning (PLUP) Manual (MAF and NLMA, 2009), which is still officially valid, the approach developed by TABI and its partners does not assume that a formal village agreement will be developed and established immediately after the management zone borders are 'drawn on the map'. In addition to the previously-mentioned shortcomings of the participatory land-use planning approach according to the PLUP manual (use of the three forest classes, lack of current land-use or other resource assessment, and limited local participation) this timeline represents a fourth major shortcoming. Any local land-use planning endeavour should be tested for one to two years, while adjustments are made, before it becomes official; otherwise it will be useless or potentially problematic for the local people. Consequently, TABI's approach supports villages in monitoring the implementation of the draft plan for at least one full year, and facilitates discussions and adaptations towards a final plan or zonation. Only then is it transformed into a formal agreement signed by village and district officials.

Village agreements can be a tool for increasing secure access to land, especially for shifting cultivators. While they do not provide formal tenure security – or even the right to compensation if authorities give the land away for other purposes (e.g. land concessions) – they acknowledge, for the first time, the local population's land claims and formalize the existence of shifting cultivation beyond the three-plot policy.

4. Formalizing: communal title for shifting cultivation land

Communal land titling is still in its infancy in Laos, where the first communal land titles were issued in 2011 (Baird, 2013). Unlike in neighbouring Cambodia, where the link between communal land titles and indigenous status has been extensively studied (e.g. Baird, 2013; Dwyer, 2015), this link does not exist in Laos (Hirsch and Scurrah, 2015). While the rules concerning communal land are not yet formally established (for example, the National Land Policy has not yet been endorsed) it is likely that communal land titles would allow villages to receive adequate (admittedly, a rather vague term) compensation if the state decided to seize collective land.

This is also why TABI included communal land titling in its approach. The idea is that, by giving out communal land titles for shifting cultivation and fallow zones in the land-use plan, the local people will gain access to a larger part of the village territory – namely, all current upland fields and fallow zones as defined in the land-use plan. Pilot programmes are currently being carried out to develop the most effective technical approaches (case studies are forthcoming).

5. From availability to accessibility: making data and information accessible to foster transparency regarding village land use and aspirations

In spite of intense land-use planning activities by government agencies and numerous development projects, there is still no systematic storage and management of village land-use planning data and agreements in Laos – at district, provincial or national levels. This is in part because specific mandates regarding land-use planning, and for particular land classes, exist within different ministries and departments and they often overlap. Furthermore, participatory land-use planning in Laos is largely driven by different development actors, each with its own agenda, finite duration and limited financial resources. As a result, even when thorough village-level land-use planning takes place and agreements are made, the information is not necessarily accessible either to the government or to the general public. Driven primarily by development projects, and in the absence of central coordination by the government, there are frequently several projects doing land-use planning in the same village. This is both inefficient in terms of resource use and unbearable for local people.

TABI took action to mitigate this situation by following a two-fold strategy. First, the long-term strategy is to develop and maintain a clear and consistent database and process-management system (e.g. provincial-national data exchanges or clear guidelines for development projects regarding data standards) for village-level land-use planning within key agencies of the government – specifically, but not solely, the Ministry of Agriculture and Forestry's Department of Agricultural Land Management. A key objective of this strategy is long-term and continuous capacity development (e.g. geographic information systems (GIS), database management, participatory processes) in key government offices. These activities have been going on for the past few years, with progress made towards a harmonization of approaches and coordination across administrative levels. As is to be expected, progress is slow and at times hampered by the particular objectives of certain development projects over the larger common goal.

Second, in the short term, an online pilot platform was developed to share locations and data (e.g. maps, GIS data, agreements) of village land-use planning activities.[2] The platform was designed as a crowd-sourced system where land-use planning data, maps and documents of any project or government agency could be uploaded. Since no government agency has yet developed the capacity to manage and coordinate data management, this platform provides a concrete tool for all actors in village-level land-use planning to see who is engaged in land-use planning and where, in order to avoid unnecessary or redundant activities, as well as to make data and information accessible to the public. This allows agencies and experts to cross-check data, and potentially allows civil society organizations to use the information for advocacy purposes, e.g. in the case of a village losing access to its land because of a land concession. In the medium term, we are hopeful that this will become the one-stop shop on land-use planning for Lao government agencies.

6. From planning to development impact: agrobiodiversity-based livelihood improvement

The land-use planning process and its outputs should be linked to the identification of concrete development opportunities for a village, as well as offering support for those opportunities through extension and/or development projects (NAFRI, 2012). The land-use plan and corresponding management plan represent the villagers' vision for their village based on a thorough assessment of their current situation. However, support is needed in planning, guiding and implementing the intended activities. Local development interventions (be it through development projects or investments) should be embedded in land-use planning activities as described above. This allows for a much better matching of local needs with technical or financial support through development interventions within a long-term planning horizon for the village.

Within TABI, development activities that support the development of sustainable, agrobiodiversity-based livelihoods in upland villages are linked to, or even based on, village-level land-use planning. This is done through a sub-project system that includes a wide range of activities, from domestication and enhanced supply chains for non-timber forest products to educational activities, improved fallow management and general improvement of agricultural practices.[3]

7. Going national: altering the national land classification system and forest management zonation to be more evidence-based and to incorporate shifting cultivation

In addition to the abovementioned increase in local interventions, TABI aims to enhance current selected policies and planning systems that tend to ignore or even marginalize shifting cultivation practices. This neglect of shifting cultivation is based on the ideology that seeks to eradicate shifting cultivation, misguided policy goals and the use of flawed or insufficient information, or a complete lack of data, to support polices. As a consequence, there is a large gap between policy goals and planning systems on one hand and the reality in the field on the other.

Most development projects state that they aim to influence policies (e.g. to make them more evidence-based), but due to a number of factors that lie beyond the scope of this chapter, this is rarely achieved in reality. TABI is trying to make a difference at a national level by demonstrating what works at a local level (e.g. through the land-use planning processes described above) and offering feasible, concrete approaches and tools to enhance systems, regulations and policies to better reflect the current situation in the uplands.

7.1 Putting shifting cultivation in the national land classification system

As mentioned earlier, the current official national land cover/land use classification system in Laos is highly problematic when it comes to shifting cultivation practices or land planning in general. In the current classification system, the entire fallow areas of rotational shifting cultivation systems are classified as 'unstocked forest', thereby wrongly indicating that there are no agricultural activities in these areas. One grave consequence of this categorization is that fallow areas may be regarded as 'idle land', hence opening the door for the 'development' of these areas, e.g. by large-scale commercial investments such as tree plantations. This in turn results in a loss of land for upland communities, with grave consequences for their livelihoods and for the ecosystem services that these landscapes provide.

Based on its success in getting local authorities (both district and provincial) to include shifting cultivation in village-level land-use planning, TABI began working towards a new, meaningful national land classification system that recognizes the entire shifting cultivation complex as agricultural land. As all attempts over the past 15 years to convince medium-and high-level decision-makers (e.g. department directors-general and vice-ministers, respectively) to alter the classification system have failed to bring about change, TABI and its partners pursued a different strategy.

With a technical staff from different departments of the two main ministries related to land – the Ministry of Agriculture and Forestry and the Ministry of Natural Resource and Environment – and partnering with the main development actors in the land sector, a new land classification was developed over a period of nearly three years. This process guaranteed bottom-up ownership without 'political ideological' issues. The technical staff within the different departments were able to communicate any problematic issues and possible solutions higher up within their respective ministries, leading to a series of national-level workshops on the new classification with high-level government participation.

The new land classification includes current land cover and land use, as well as land-use planning and management zonation. It consists of three tiers, ranging from a small set of broad land-cover classes at the first tier[4] to a highly detailed set of land-use classes at the third tier. It is important to note that more categories are constantly being added to the third tier as different projects venture into new areas to carry out land-use planning. Additionally, there is an ongoing debate among experts as to whether it would be wise to create an 'agroforestry' land-use category at the first tier, and to move all classes related to shifting cultivation at tiers two and three into this new category. As previously mentioned, the shifting cultivation system, including upland fields and all fallow stages, is now categorized within the main agriculture class at the first tier. This marks the first time that upland agricultural areas are made explicit at the national policy level and are not categorized as idle land (or 'unstocked' forest, as previously classified). Rather, they are seen as part of an agricultural system that is the basis for the livelihood (and bio- and agro-diversity) of the population of the Lao uplands.

This 'bottom-up' approach to altering the land classification has been successful and the new classification is now used by a wide array of government agencies and development projects in village-level land-use planning and in land-cover assessment based on remote sensing at higher levels of aggregation. While the government has not yet officially ratified the new national land classification, we expect this to take place in the near future.

In addition to its efforts to make national land classification more realistic and meaningful, TABI has been working over the past two years to improve the three national forest management categories, regulations and corresponding zones.

7.2 Reality-based national forest zonation that acknowledges shifting cultivation

As discussed earlier, the ideology of eradicating shifting cultivation and the target of reaching 70% forest cover by 2020 led to the creation of the three national forest management zones (conservation, protection and production). The use of these zones in village-level land-use planning (e.g. the Land-Use Planning and Land Allocation Programme and even participatory land-use planning (PLUP)) effectively formalized limited legal access to land for shifting cultivation communities. Moreover, these three classes were 'mapped' at a national level by creating a data set with formal zones. However, this 'mapping' was done without any participation or reference to meaningful sources, leading to large and vaguely delineated zones, especially for protected forest. As a result, the protected forest class covers large parts of upland Laos; an area with almost 3000 villages and a population of about one million people with limited access to land.

Despite the inaccuracy of the state forest zones, they are still used – or at least 'quoted' – for national, provincial and district planning and management purposes. The country's Five-year Social and Economic Development Plans are based on them; revised drafts of both the Forest Law and the Land Law refer to the three categories; management responsibility is allocated to government agencies according to the three categories; and legal limits on land uses are based on them. This has led to: (1) ineffective and inefficient land management due to incorrect macro-level forest land zoning; (2) misguided development planning at national, provincial and district levels due to the fact that state forest land bears no relation to actual 'forest-cover';[5] and (3) unrealistic and confusing village land-use planning due to higher-level legal constraints imposed by the three state forest zones.

There is growing consensus in Laos that if state forest management is to contribute to the sustainable social and economic development of the country, forests must first be properly delineated and mapped. The National Assembly recognizes this and has ordered the Ministry of Agriculture and Forestry and the Ministry of Natural Resource and Environment to review and re-survey the state forest categories nationwide. Based on its village-level land-use planning approach, TABI was able to step in and support the development and implementation of a pilot programme for Luang Prabang province (Figure 45-3).

TABI and its partners developed a participatory approach to re-delineate the three forest management zones. In addition to recent high-resolution satellite imagery and orthophotos (already available to the government), TABI used a set of auxiliary maps to assist the process of re-delineating the zones (Figure 45-4). These included 30-metre

FIGURE 45-3: The Lao PDR showing Luang Prabang province, one of the provinces where TABI is developing and implementing land-use planning and delineation of national forest categories.

FIGURE 45-4:   Participatory re-delineation of the three national forest management zones.

resolution deforestation data from 2000 to 2014 (Hansen et al., 2013) to obtain information on the level of human forest disturbance; a map showing the three old forest management zones; and a map with additional land-use information (e.g. land concessions). Based on district-level data and maps, the current extent of old forest (excluding secondary forest on fallow land that was still within a rotational system) was delineated as part of a process involving various stakeholder groups, including local and national government officials, civil society conservation organizations and village populations. The participants determined which areas should belong to which forest management zone. Subsequently, the borders were accurately delineated, including descriptions of landscape features as border elements. In this setting, villager participation was to some extent limited. Nonetheless, the newly delineated forest management zones will provide a macro-level spatial zoning framework within which village land-use planning (as described earlier) will be revised and finalized with the full participation of villagers. In the pilot programme for Luang Prabang province, this approach led to a reduction from an illusory 71% down to a realistic 41% of the area covered by one of the three forest management categories.

The pilot programme was well received by high-level decision-makers and the process to scale it up to national level is currently underway. There is hope that meaningless national forest management zones may – after more than 15 years – finally be replaced by an evidence-based and participatory multi-level system, thereby ending the de facto 'criminalization' and marginalization of shifting cultivation in large parts of the uplands of Laos.

Conclusions

For more than two decades, the 'shifting cultivation eradication paradigm' has dominated upland development and forest conservation policy in Laos. It stood in stark contrast to a reality in which shifting cultivation was widespread and supported the livelihoods of a large proportion of the rural upland population. Many of the country's agriculture and forestry policies and corresponding spatial planning rules and regulations are not based on reliable data and approaches. As a result they neglect the real situation on the ground and promote unrealistic goals (e.g. of reaching 70% forest cover by 2020). The TABI project and its partners have addressed this issue and have bridged the 'policy-reality gap' by developing concrete approaches to land-use planning at various levels. This accurate representation of the current situation creates a starting point for future participatory planning. The ultimate aim of this approach is to empower upland communities and improve secure access to land and forest resources for improved livelihoods based on agrobiodiversity.

At a local level, the TABI project developed a participatory land-use planning approach that began with an accurate assessment of the current socio-ecological system in a village. This approach was initially developed and tested in a small number of villages in northern Laos, starting in 2010. Today a number of provinces have officially requested the use of this approach, which is constantly being developed and refined. So far, it has been applied in more than 300 villages and its reach is rapidly expanding. Our experience has shown that if this initial assessment of land use is not done with the utmost care, any further land-use and development planning will be meaningless. In other words, it will become just another piece of paper that is impossible for villagers to implement. Even today, there are many village-level land-use planning endeavours in Laos that do not take this issue seriously and apply flawed official approaches, causing harm to upland villages (e.g. by formalizing limited access to land). They are often driven by development projects that are under pressure due to their visibility and output expectations.

A detailed assessment is vital, not only for meaningful village planning, but also to encourage inclusion and create trust among government officials. If these officials feel a sense of 'ownership' of the data and the approach, they are more likely to 'dare' to be part of, and officially accept, a village-level land-use plan that ignores current policies and regulations, for example by planning for shifting cultivation beyond the 'three-plot policy' or ignoring meaningless macro-level forest management zones. This bottom-up 'ownership', coupled with work to revise the national land classification system and re-delineate the three national forest categories, has proven effective as a first step towards reversing two decades of discrimination against shifting cultivation in national planning systems. While not all of the changes are officially endorsed and ratified, we are hopeful that this will soon be the case. What is certain is that the ship has left the harbour: there have been positive steps towards official recognition of shifting cultivation practices and achievement of secure access to land for upland communities.

In order to continue moving towards a more effective and transparent inclusion of shifting cultivation in land-use planning and policy, we make the following recommendations:

- Any kind of 'rapid village-level land-use planning' exercise is a waste of time and can potentially be problematic for a village – for example, if inaccurate, top-down approaches based on the three forest categories are 'formalized' in a village map or agreement, as was largely the case with the initial participatory land-use planning (PLUP) approach of 2009. These outputs are harmful to upland villages as they formalize limited access to land and erode both the livelihood base and development possibilities.

- Any village-level land-use plan and related agreement (e.g. communal land titles) should be shared publicly (e.g. within a land information system). Given current commercial pressures on land, this transparency enables the public or advocacy groups to intervene if a village loses its land without its consent or without adequate compensation.
- Local-level development interventions should always be explicitly linked to a sound (i.e. evidence-based and participatory) local land-use plan, allowing for a better matching of local aspirations with potential technical or financial support through development interventions within a long-term planning horizon. This helps to avoid the problems that arise from the short-term planning horizon of many development projects, and sometimes fragmented or uncoordinated interventions.
- In tandem with advocacy and policy work, sound and concrete approaches should be developed that demonstrate how shifting cultivation can be stabilized and integrated into a spatial planning system at local and national levels.
- Create a sense of 'ownership and trust' among government officials regarding sensitive issues, as a strategy for changing long-standing policies and regulations.

References

Baird, I. G. (2013) '"Indigenous peoples" and land: Comparing communal land titling and its implications in Cambodia and Laos', *Asia Pacific Viewpoint* 54, pp269–281, DOI: 10.1111/apv.12034

Bounthong, B., Raintree, J. and Douangsavanh, L. (2003) 'Upland agriculture and forestry research for improving livelihoods in Lao PDR', in K. G. Saxena, Liang Luohui, Kono Yasuyuki and Miyata Satoru (eds) *Small-scale Livelihoods and Natural Resource Management in Marginal Areas of Monsoon Asia*, Dehra Dun Bishen Singh Mahendra Pal Singh, New Delhi

Bourgoin, J. (2012) 'Sharpening the understanding of socio-ecological landscapes in participatory land use: A case study in Lao PDR', *Applied Geography* 34, pp99–110

Bourgoin, J. and Castella, J-C. (2011) '"PLUP fiction": Landscape simulation for participatory land use planning in northern Lao PDR', *Mountain Research and Development* 31(2), pp78–88

Bourgoin, J., Castella, J-C., Pullar, D., Lestrelin, G. and Bouahom, B. (2012) 'Toward a land zoning negotiation support platform: "Tips and tricks" for participatory land use planning in Laos', *Landscape and Urban Planning* 104, pp270–278

Ducourtieux, O. (2006) 'Is the diversity of shifting cultivation held in high enough esteem in Lao PDR?', *Moussons* 9-10, pp61–86

Dwyer, M. B. (2015) 'The formalization fix? Land titling, land concessions and the politics of spatial transparency in Cambodia', *Journal of Peasant Studies* 42, pp903–928

Fenton, N., Lindelow, M., Heinimann, A. and Thomas, I. (2011) *The Socio-geography of Mining and Hydro in Lao PDR: Analysis Combining GIS Information with Socioeconomic Data*, The World Bank, Washington, DC, http://boris.unibe.ch/6734/, accessed 27 March 2016

Fox, J., Truong, D. M., Rambo, A. T., Tuyen, N. P., Cuc, L. T. and Leisz, S. (2000) 'Shifting cultivation: A new old paradigm for managing tropical forests', *Bioscience* 50, pp521–528

Fox, J., Fujita, Y., Ngidang, D., Peluso, N., Potter, L., Sakuntaladewi, N., Sturgeon, J. and Thomas, D. (2009) 'Policies, political-economy, and swidden in Southeast Asia', *Human Ecology* 37, pp305-322

Fujita, Y. and Phanvilay, K. (2008) 'Land and forest allocation in Lao People's Democratic Republic: Comparison of case studies from community-based natural resource management research', *Society and Natural Resources* 21, pp120-133

Hansen, M. C., Potapov, P. V., Moore, R., Hancher, M., Turubanova, S. A., Tyukavina, A., Thau, D., Stehman, S. V., Goetz, S. J., Loveland, T. R., Kommareddy, A., Egorov, A., Chini, L., Justice, C. O. and Townshend, J. R. G. (2013) 'High-resolution global maps of 21st-century forest cover change', *Science* 342, pp850-853

Heinimann, A., Hett, C., Hurni, K., Messerli, P., Epprecht, M., Jørgensen, L. and Breu, T. (2013) 'Socio-economic perspectives on shifting cultivation landscapes in northern Laos', *Human Ecology* 41, pp51-62

Hirsch, P. and Scurrah, N. (2015) *The Political Economy of Land Governance in Lao PDR*, BRICS Initiatives in Critical Agrarian Studies, Mosaic Research Project and the Land Deal Politics Initiative, the International Institute of Social Studies, Rotterdam; the Transnational Institute, Amsterdam; and the Regional Centre for Social Studies and Sustainable Development, Chiang Mai University, Chiang Mai

Hurni, K., Hett, C., Heinimann, A., Messerli, P. and Wiesmann, U. (2013) 'Dynamics of shifting cultivation landscapes in northern Lao PDR between 2000 and 2009, based on an analysis of MODIS time series and Landsat images', *Human Ecology* 41, pp21-36

Lawrence, D., Radel, C., Tully, K., Schmook, B. and Schneider, L. (2010) 'Untangling a decline in tropical forest resilience: Constraints on the sustainability of shifting cultivation across the globe', *Biotropica* 42, pp21-30

Lestrelin, G. (2010) 'Land degradation in the Lao PDR: Discourses and policy', *Land Use Policy* 27, pp424-439

Lestrelin, G., Bourgoin, J., Bouahom, B. and Castella, J-C. (2011) 'Measuring participation: Case studies on village land use planning in northern Lao PDR', *Applied Geography* 31, pp950-958

Lestrelin, G., Castella, J-C. and Bourgoin, J. (2012) 'Territorialising sustainable development: The politics of land-use planning in Laos', *Journal of Contemporary Asia* 42(4), pp581-602

Liao, C., Feng, Z., Li, P. and Zhang, J. (2015) 'Monitoring the spatio-temporal dynamics of swidden agriculture and fallow vegetation recovery using Landsat imagery in northern Laos', *Journal of Geographical Sciences* 25, pp1218-1234

MAF and NLMA (2009) *Manual on Participatory Agriculture and Forest Land Use Planning at Village and Village Cluster Levels*, Ministry of Agriculture and Forestry and National Land Management Authority, Vientiane

MAF-TABI (2014) *Manual: Participatory Forest and Land Use Planning, Allocation and Management*, Ministry of Agriculture and Forestry and The Agro-biodiversity Initiative of the Lao PDR, Vientiane

Mertz, O., Padoch, C., Fox, J., Cramb, R. A., Leisz, S. J., Lam, N. T. and Vien, T. D. (2009) 'Swidden change in Southeast Asia: Understanding causes and consequences', *Human Ecology* 37, pp259-264

NAFRI (2012) *Handbook on Participatory Land Use Planning. Methods and Tools Developed and Tested in Viengkham district, Luang Prabang province*, National Agriculture and Forestry Research Institute (NAFRI), L'Institut de recherche pour le développement (IRD) and Center for International Forestry Research (CIFOR), Vientiane, http://participatorygis.blogspot.com/2015/08/handbook-on-participatory-land-use.html, accessed 27 March 2016

Rigg, J. (2005) *Living with Transition in Laos: Market Integration in Southeast Asia*, Routledge, London and New York

Schmidt-Vogt, D., Leisz, S. J., Mertz, O., Heinimann, A., Thiha, T., Messerli, P., Epprecht, M., Cu, P. V., Chi, V. K., Hardiono, M. and Dao, T. M. (2009) 'An assessment of trends in the extent of swidden in Southeast Asia', *Human Ecology* 37, pp269-280

Schoenweger, O., Heinimann, A., Epprecht, M., Lu, J. and Thalongsengchanh, P. (2012) *Concessions and Leases in the Lao PDR: Taking Stock of Land Investments*, Geographica Bernensia, University of Bern, Switzerland, http://boris.unibe.ch/17591/, accessed 27 March 2016

Seidenberg, C., Mertz, O. and Kias, M. B. (2003) 'Fallow, labour and livelihood in shifting cultivation: Implications for deforestation in northern Lao PDR', *Geografisk Tidsskrift, Danish Journal of Geography* 103(2), pp71-80

Tanaka, S., Kendawang, J. J., Yoshida, N., Shibata, K., Jee, A., Tanaka, K., Ninomiya, I. and Sakurai, K. (2005) 'Effects of shifting cultivation on soil ecosystems in Sarawak, Malaysia: IV. Chemical properties of the soils and runoff water at Niah and Bakam experimental sites', *Soil Science and Plant Nutrition* 51, pp525-533

Thongmanivong, S., Fujita, Y., Phanvilay, K. and Vongvisouk, T. (2009) 'Agrarian land use transformation in northern Laos: From swidden to rubber', *Southeast Asian Studies* 47(3), pp340-347

Vongvisouk, T., Mertz, O., Thongmanivong, S., Heinimann, A. and Phanvilay, K. (2014) 'Shifting cultivation stability and change: Contrasting pathways of land use and livelihood change in Laos', *Applied Geography* 46, pp1-10

Notes

1. An orthophoto is an aerial photograph that has been geometrically corrected or 'ortho-rectified' such that the scale of the photograph is uniform and may be utilized in the same manner as a map.
2. www.landuseplanning.la
3. A list of TABI's sub-projects is available at http://www.tabi.la/index.php/en/tabi-sub-project/overview-of-subproject
4. These are the land-cover classes mentioned in the Lao Land Policy, which relate to Land and Forestry Laws.
5. Only about 50% to 60% of the area currently covered by the three state forest categories has any forest cover left.

C. From efforts to mediate the role of farmers in the Indonesian forest

The damar agroforests of Krui, Indonesia, are one of the most compelling examples of farmers transforming their shifting cultivation system into a permanent forest. Nevertheless, their initiative was, at first, not recognized by the Indonesian government.

Sketch based on a photo by Hubert de Foresta.

46

NEGOTIATING FOR COMMUNITY FORESTRY POLICY

The recognition of damar agroforests in Indonesia

Tuti Herawati, Hubert de Foresta, Dede Rohadi, Mani Ram Banjade and Chip Fay *

Introduction

This chapter recounts the development of a forestry policy innovation designed specifically to protect the customary rights of local people after their agroforests were subsumed into an area of state forest. In the 1990s such actions were a common problem in Indonesia, where local people had developed various kinds of agroforests covering large areas. Rubber agroforests alone covered more than 2.5 million hectares in the late 1980s (de Foresta, 1992). Regrettably, community-created agroforests were ignored in the designation of state-forest boundaries outside Java in the late 1980s, and in the adoption of provincial Forest Land-Use Master Plans by Consensus (*Tata Guna Hutan Kesepakatan*, or TGHK). The Krui region, in West Lampung district, on the southwestern tip of Sumatra, was seriously affected. More than half of the area covered by damar agroforests that had been planted and managed by local people for generations was designated as state forest in 1991 and officially gazetted as such in 1996 (Michon et al., 2000, 2007).

The damar agroforests are a successional agroforestry system developed and managed by local farmers without any external support. When mature, they form an impressive forest-like cover dominated by tall cultivated resin-producing *Shorea javanica* trees, locally called 'damar' (de Foresta and Boer, 2000), and fruit trees such as durian (*Durio zibethinus*) and *duku* (*Lansium domesticum*). Damar collection, in which the *S. javanica* trees are usually tapped for resin once or twice a month,

* Dr Tuti Herawati, Seconded Scientist, Forests and Governance, Center for International Forestry Research (CIFOR), Bogor, Indonesia; Dr Hubert De Foresta, Editor, *Forests, Trees and Livelihoods*, Senior Scientist, Institut de Recherche pour le Développement (IRD) UMR AMAP, Montpellier, France; Dr Dede Rohadi, Seconded Scientist, Forests and Livelihoods, CIFOR, Bogor, Indonesia; Dr Mani Ram Banjade, Post-Doctoral Research Fellow, Forests and Governance, CIFOR, Bogor, Indonesia; and Chip Fay, Advisor, the Climate and Land Use Alliance, Indonesia.

forms the backbone of all agricultural activities in the agroforests. In 1993, resin produced by damar farmers was estimated to be worth US$3.25 million. The first damar agroforests were planted in the mid-19th century by pioneer farmers. They were soon joined by others, so that by 1994, the agroforests had spread over more than 50,000ha in the Krui region between what is now known as the Bukit Barisan Selatan National Park and the Indian ocean (Michon et al., 2000, 2007) (Figure 46-1).

The damar agroforests, their history, the processes of their establishment and their ecological, economic and social functioning, as well as the various conflicts in which damar farmers have found themselves embroiled, have been described and analysed in detail by Michon et al. (2000, 2007). This chapter is a follow-up to those articles and the authors assume that the elements described and analysed in them are known to readers, who may refer to them if and when needed.

In the conclusion of their 2007 article, Michon et al. wrote on page 560: '…the agroforest situation does not fit any of the existing legal forest categories, and a new legal status needs to be devised to suit the needs of damar farmers and to ensure a future for damar agroforests.'

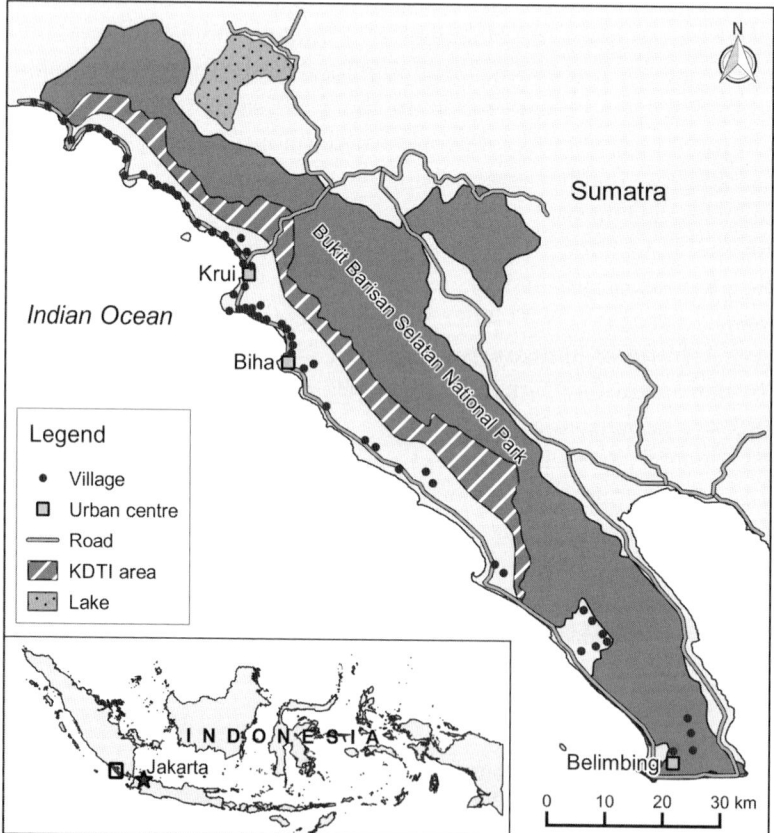

FIGURE 46-1:   The area around Krui in Sumatra, showing the KDTI area and the Bukit Barisan Selatan National Park.

This is precisely what the Minister of Forestry, Djamaludin Suryohadikusumo, did in January 1998, when he issued a decree (Minister of Forestry Decree Number 47/Kpts-II/1998) establishing a new legal status for the state-forest area already occupied by damar agroforests. The new category for the Krui agroforests was called *Kawasan Dengan Tujuan Istimewa* (KDTI) – a Forest Zone with a Special Purpose. The Krui case had been chosen by the Minister to test a new forestry policy, and the outcome was a significant policy innovation. For the first time in Indonesia, the decree gave legal recognition to the rights of local people to control, maintain, develop and pass on to the next generation their (agro)forest management systems, all within the state-forest zone (Fay and de Foresta, 2001).

This chapter focuses specifically on the negotiation process and the institutional context that surrounded the issuance of the KDTI decree. Based on both written and oral sources, it benefits from the experience of co-authors de Foresta and Fay, both of whom played an active role in the negotiations leading up to the KDTI decree. Our main sources were documents related to research projects carried out since the early 1990s by IRD[1] (ex-ORSTOM) and the World Agroforestry Center (ICRAF)[2] with their research and NGO partners (CIFOR,[3] WATALA,[4] FORDA,[5] LATIN[6] and P3AE-UI[7]). These documents include reports, correspondences, research publications and minutes of meetings. Oral information came from interviews with several key informants in local communities and local government. We first describe the context and the process that led to the KDTI decree. Then we present the main reactions to the decree, and conclude this chapter by drawing some policy lessons.

The background of the KDTI decree

A unique research context

There are many examples of agroforestry systems developed by local farmers in Indonesia (Michon et al., 2000, 2007), but none of them has been researched with such intensity as the damar agroforests of the Krui region. We believe that the depth of understanding of these agroforests, most particularly the evidence the research provided that they were indeed planted and not 'natural' forests, was a critical element in the Minister of Forestry's decision to choose Krui as a test case.

Damar agroforests were first discovered by the outside world through a Dutch forester who visited the Krui region in 1936. He reported the existence of stands of *Shorea javanica* trees that had been planted by local farmers around Krui for producing damar resin. He estimated the area covered to be around 70ha, with some plantations being more than 50 years old (Rappard, 1937). The local farmers' 'invention' of damar agroforests could thus be dated to the 1880s, at least.

The man-made damar plantations then fell into oblivion for 40 years, until their re-discovery in 1979. After a long and difficult journey across the southern tip of Sumatra, a group of master's students from the University of Montpellier in France, along with their botany professor (F. Hallé), and a researcher from the Southeast Asian Regional Centre for Tropical Biology (Seameo-Biotrop) (Y. Laumonier), arrived in

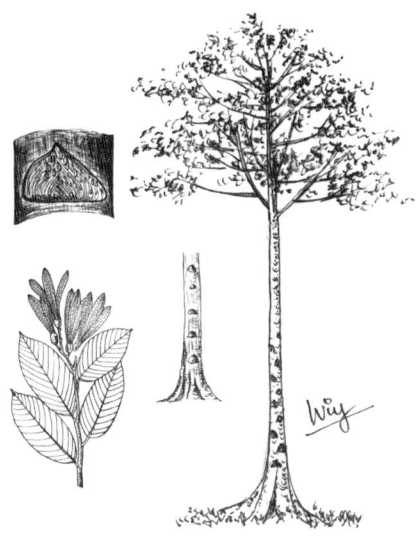

Shorea javanica Koord. &
Valeton [Dipterocarpaceae]

Soaring, straight-boled trees with
dome-like canopies, the producers
of damar resin have long been
planted and nurtured by local
people in the Krui agroforests of
Sumatra. The resin is collected from
permanent wounds cut into the
trunks of the trees.

Krui for a one-week field trip. Their aim was to study the local primary rainforest for a vegetation map of Sumatra (Laumonier, 1997). Tropical deforestation was already making newspaper headlines, and the group walked for hours in a deforestation pioneer front to arrive in what they decided was a 'pristine' forest, high in the mountains. On their way, however, the group spent even more hours walking below a majestic canopy dominated by a species of resin-producing trees locally called 'damar'. After questioning local villagers, it became obvious that this majestic forest cover was not 'natural', but the result of a cultivation process. The group was deeply impressed by the reforestation success they had unexpectedly encountered. It was a success that appeared as an island of hope in a desolate ocean of deforestation; a success that had to be studied and widely reported.

One of the students in the group returned to Indonesia a few years later (in 1982) as a scientist for the Regional Centre for Tropical Biology (Biotrop), and undertook the first study of the composition and structure of this man-made dipterocarp forest (Torquebiau, 1984). Another student from the group (the co-author of this chapter, H. de Foresta) returned to Krui in 1983 and 1984, and assisted a group of PhD students, also from the University of Montpellier, in field data collection. This group was engaged in a research project devoted to the ecology and socio-economy of 'agroforests' – a word first coined by the same group. One example was the damar agroforests (Mary and Michon, 1987). Between 1989 and 1994, two of these early participants (de Foresta and Michon) were posted as scientists to Biotrop by Orstom (the French government research organization, later known as IRD). They launched a new series of research studies on the damar agroforest system, including its biodiversity, long-term monitoring of tree stands and associated anthropology. Both researchers joined the World Agroforestry Center (ICRAF)'s Southeast Asia regional office as IRD scientists in 1994. De Foresta continued working at Krui quite intensively until 1999, and more occasionally afterwards. His most recent field visit was in 2014.

In 1992, local government authorities in Krui asked the IRD scientists to help them by providing data and scientific information on damar agroforests to counter

FIGURE 46-2: T-shirts designed and produced by IRD scientists to help popularize damar agroforests and their environmental benefits. The T-shirts were proudly worn by Krui farmers when the Minister of Forestry visited the area in 1993.

a plan by the Ministry of Forestry to establish monoculture *Acacia* plantations in the area. This came as a shock to the scientists, who realized that the damar agroforests were largely unknown in Indonesia, even though the international scientific and forestry community considered them a model of indigenous sustainable forest management. The IRD scientists decided to launch a campaign to popularize the main results of their research, including translation into the Indonesian language and dissemination of publications, promotional T-shirts (Figure 46-2) and illustrated calendars.

This effort soon resulted in local newspaper articles, and most importantly in a new awareness in the provincial office of the Ministry of Forestry that led to the abandonment of the *Acacia* plans and a visit to Krui by the Minister of Forestry in 1993. Deeply impressed by the success of the damar agroforest farmers, the Minister decided to press for new research on damar agroforests by the Forestry Research and Development Agency (FORDA) and P3AE-UI, a research organization based at the University of Indonesia. The IRD scientists' efforts also attracted the interest of two environmental non-governmental organizations (NGOs), LATIN and WATALA, whose members decided to launch research and support programmes in Krui.

A new wave of research on damar agroforests thus began in 1994. By then, it was deeply rooted not only in the international research framework (IRD, ICRAF and CIFOR), but also in the national research and NGO context (FORDA, LATIN, P3AE-UI and WATALA).

It is not common for an indigenous agroforestry system to be the subject of long-term research endeavour, encompassing a wide array of disciplines from plant ecology and forestry to socio-economy and anthropology. Thanks to numerous scientific publications and communications, the damar agroforests soon became famous around the world. Collaboration between scientists and NGOs that began in 1994 resulted in efficient dissemination of information and awareness among the Indonesian public. It was perhaps not surprising, therefore, that the reputation of the damar agroforests, established by a substantial knowledge base accumulated over more than 20 years of research, created a unique setting in 1997 and 1998 for the Krui region to become a test area for a new category of state forest in Indonesia (Figure 46-3).

FIGURE 46-3: A damar agroforest at Krui, showing the dominant *Shorea javanica* trees in a dense forest setting, their boles marked by permanent wounds from which the damar resin is regularly collected.

Sketch: Dr Genevieve Michon.

The Krui access-rights context: From recognition to negation[8]

To understand the conflict related to land access that existed between the state and damar farmers in the time before the KDTI decree, we must refer back to colonial times. The forest in the Krui area was divided into a clan-forest zone (*hutan marga*) and a reserved-forest zone by the Dutch colonial administration in 1937, following a consultation process with the clans (*marga*). After this, even though local people at times entered the reserved forest and even established some damar agroforests there, they were always aware of its borders, which they called (and still call) 'BW' (for *boschwesen*, meaning 'forest reserve' in Dutch). Most of the damar agroforests, which now cover more than 50,000ha, were gradually established in the clan-forest zone, between the borders of the reserved-forest zone and the coast of the Indian Ocean. The reserved-forest zone became the Bukit Barisan Selatan National Park in 1991. Most of the damar farmers did not have legal title to their agroforestry land. However, they believed their tenure was secure because they had abided by the rules of local customary access. They saw these as the only valid and legitimate rules because they had been officially recognized by the colonial power in 1937, with the delineation of their clan-forest zone.

However, another story was being devised in Jakarta, far from Krui, under the New Order regime of President Soeharto. In the 1970s, the international development paradigm was based on a massive short-term liquidation of forest resources to provide capital for the industrialization of tropical countries (Pretzsch, 2005). The Indonesian government began encouraging the development of its forest sector for the sake of national development. In Krui in 1972, this national effort was instrumental in the granting to a logging company of 'timber harvesting rights' (converted to 'forest concession rights' in 1981) over 52,000 hectares of forest. Almost all of the damar agroforests were included in the concession area. However, the company halted its activities in 1991, after having harvested timber from rainforest remnants in the extreme north and south. It did not enter the damar agroforests, so the farmers never knew that their agroforests were under threat of being 'legally' clear cut, as had already happened in neighbouring Bengkulu province.

In 1991, the Minister of Forestry issued a decree formalizing the Forest Land Use Master Plan by Consensus for Lampung province. The associated map did not mention the existence of man-made damar agroforests in the Krui area, and the decree created a new strip of state forest covering about 42,000ha between the border of the national park and the coast, covering what had, in fact, been clan-forest land since before independence. This new state-forest area was designated mainly as production forest, with small pieces of protection forest distributed thinly along the western border of the national park.

The local people only gradually became aware of this change. Forestry Services delineated the new state-forest borders in the field between 1992 and 1996. Forestry regulations stipulated that local people should be informed of the localization of planned borders and any claims to land that was owned and managed within the state-forest area had to be taken into account and the border modified accordingly. However, according to village heads in the Krui area, field delineation officers never informed village authorities of their mission (as required by law), nor of the potential consequences for local people of what they presented as a simple field exercise. As a result, the border of the state forest in the Krui area was delineated in full accordance with the Ministry of Forestry map. The new and official production forest included more than half of the area covered by the damar agroforests, legally depriving the farmers of their previous ownership rights.

The context leading to the KDTI decree: Researchers and NGOs as mediators of local communities' concerns

With the emergence in 1994 of new research and support programmes in Krui, researchers, NGOs and representatives of local communities began to meet regularly in Bogor, West Java. The informal group – soon known as 'Team Krui' (Suporahardjo and Wodicka, 2003; Kusters et al., 2007) – held regular meetings through 1999. Its initial aim was to coordinate various programmes and share information. However, escalating social tensions caused by two oil-palm plantations (Michon et al., 2000, 2007) and the inclusion of damar agroforests in the state-forest area lifted the matter of securing the rights of local communities over their damar agroforests to prominence on Team Krui's agenda.

In October 1995 the group decided to form a new research and support programme. It was funded by the Ford Foundation and involved ICRAF, LATIN and WATALA, with H. de Foresta (IRD, on behalf of ICRAF) as programme coordinator. The sole aim of the CBFSM-Krui programme (standing for Community Based Forest System Management – Krui) was to help secure local people's rights over their agroforests by developing new models for relationships between the government and local forest communities, with the damar agroforests of Krui as an example. Although the

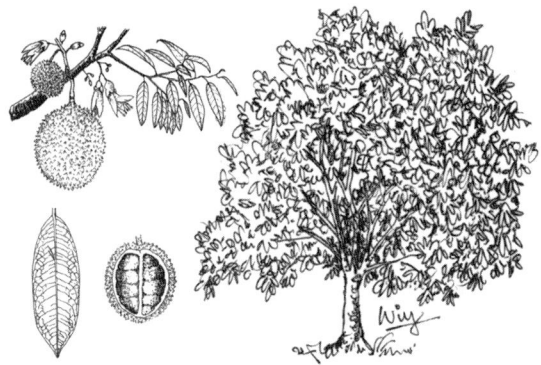

Durio zibethinus L. [Malvaceae]

A native of Southeast Asia, durian grew from seed and was protected and nurtured by the farmers of the damar agroforests. Its sweet flavour and pungent odour have now made it a popular fruit in many Asian markets.

programme was in line with the Ministry of Forestry's policy discourse, it faced significant resistance from within the Ministry and its services, so that the programme could not be a joint exercise with forestry services, as originally planned. However, it was able to conduct various activities to pave the way for official recognition of the rights of damar agroforestry farmers who were inside the state-forest zone. These included:

- Participatory mapping in eight villages and land-use mapping covering the whole region;
- Supporting the development of two local institutions: the Krui Customary Communities' Association and the Damar Agroforest Farmers' Community Association;
- Supporting local communities in expressing their concerns and forwarding them to various authorities, such as the Minister of Forestry;
- Providing complementary data on the ecological impact of damar agroforest management, in particular on the impact of local timber-harvesting practices;
- Informing various local government authorities about the damar agroforests and the problems faced by the farmers;
- Proposing that the Krui customary communities be nominated as recipients of the 'Kalpataru' National Environment Award in 1997 (all institutions involved in Team Krui sent official forms to the Ministry of Environment, based on drafts prepared by the programme).

Resistance within the Ministry, mentioned earlier, did not come from the Minister himself. Following his visit to Krui in 1993, he said he was strongly convinced that damar agroforests were a model of sustainable forest management by local people. The Ministry's support was later decisive, not only in recognizing damar farmers' rights within the state-forest zone, but also in protecting damar farmers from the plans of oil-palm companies.

In November 1996 two ICRAF members of Team Krui (C. Fay and G. Michon) were able to discuss the concerns of Krui communities directly with the Minister of Forestry, particularly their fears related to the development of two oil-palm plantations, which were the most pressing issue at that time. The Minister acted as he promised: within a few weeks, the company operating in the south of the state-forest zone halted its activities. The company planning to establish oil-palm plantations in the north halted its land-measuring activities and abandoned its project, while the Governor of Lampung issued a decree protecting damar trees.

In March 1997, at a meeting with the Minister and his high-level staff, two ICRAF scientists (C. Fay and H. de Foresta) raised concerns about the impact on the management of damar agroforests from the recent gazetting of a major part of the Krui damar agroforests as state forest. In response, the Minister asked ICRAF to collaborate with the Directorate General for Reforestation and Land Rehabilitation (RRL) in assessing the problems and devising potential solutions. He suggested the organization of a 'discussion panel'.

The discussion panel was held in June 1997, a few days after the Kalpataru National Environment Award had been presented by the President of Indonesia to a representative of the customary Krui communities for the environmental contribution made by their damar agroforests. The synchrony could not have been better. The Kalpataru Award sat prominently on the table of the communities' representative for the entire duration of the discussion panel, as a symbol of the communities' success in attracting national attention to their case. Three days later, the panel's main conclusions were reported to the Minister and his high-level staff by C. Fay (ICRAF), H. de Foresta (IRD-ICRAF) and N. Gintings (FORDA). ICRAF and FORDA listed six demands made by the Krui communities:

1. That the markers delineating the border of the state-forest zone (Figure 46-4) be moved back to the border of the Bukit Barisan Selatan National Park;
2. That products from the damar agroforests not be taxed as forest products;
3. That damar agroforest farmers be allowed to continue to implement their agroforestry system;
4. That harvesting and marketing of timber from the damar agroforest be unrestricted;
5. That all rights over the damar agroforest be inheritable; and
6. That the government formally recognize the damar agroforest management system.

FIGURE 46-4: Signs such as this one were posted in damar agroforests by Forestry Services between 1992 and 1996 to mark the boundary of the new state-forest zone. The signs created a feeling of unsecure tenure in damar farmers' communities.

The Minister said he was unable to meet the primary demand of the Krui communities, that the markers delineating the border of the state-forest zone be moved back to where they were before 1991. He said the best solution 'for the time being' would be the creation of a special-use zone within the state-forest zone for the Krui agroforests. He cited as an example the State Forestry Corporation in Java, which classifies forests in and surrounding graveyards as *Lahan dengan Tujuan Istimewa* (land with a special purpose). He asked ICRAF to work with the Directorate General of RRL and the Ministry's Legal Bureau on a decree that would similarly classify the Krui agroforests as *Kawasan dengan Tujuan Istimewa*. He instructed the group to ensure that the basic demands of the farmers were met through such an approach.

The KDTI decree-drafting process

The Minister of Forestry signed the KDTI decree on January 23, 1998. His action ended the seven-month process of developing the decree – a process that was far longer than expected and not without complications. But a comparison of the final document and the first draft produced by mid-level RRL staff reveals enormous improvements in content and indicates that, for all involved, an important learning experience had taken place. For the sake of simplifying a complex chain of events, the process can be divided into three phases:

1. An internal Ministry process with limited outside involvement (June to November 1997);

2. An official working group with increased participation by ICRAF and LATIN (November and December 1997); and
3. High-level Ministry discussions and consultations with other relevant ministries (January 1998).

Phase 1: RRL/Legal Bureau/ICRAF-LATIN

After the discussion panel, N. Gintings from FORDA and the Director of Social Forestry and Reforestation (a sub-section of RRL) provided the Minister with details of the discussion panel in separate reports. While the FORDA report closely resembled ICRAF's account, the RRL view of what transpired in Liwa was more selective. There was no mention of the fundamental dispute over land and the RRL staff clearly saw the *Hutan Kemasyarakatan* (HKm) scheme as the only option for addressing the presence of damar agroforests inside the state-forest zone.[9] This position set the tone for discussions between RRL and ICRAF/LATIN for the next several months.[10]

On 2 July 1997, the Minister of Forestry sent a letter to the Legal Bureau with copies to all Directors General, instructing that:

1. All damar agroforests within the state-forest zone would be classified as a *Kawasan dengan Tujuan Istimewa* (KDTI);
2. Future generations of damar farmers would be able to inherit the gardens from their ancestors; and
3. Damar agroforests outside the state-forest zone should receive certificates of private ownership in coordination with the National Lands Agency.

In mid-August, the Ministry of Forestry's directorate responsible for forest-boundary delineation and mapping, INTAG (*Inventarisasi dan Tata Guna*), called a meeting to discuss the delineation of the Krui agroforests. This meeting never took place. It appeared that the Legal Bureau let it be known that they were responsible for calling meetings that concerned Krui. From then on, INTAG took part in the process only with great reluctance. This began a period of about two months in which ICRAF and LATIN waited for the Legal Bureau to present a first draft.

In this period, LATIN took the initiative and drafted its own version of the decree. It centered on making the entire Krui area a buffer zone for the adjacent national park. LATIN also met the head of the HKm programme to discuss its draft.

It soon became apparent that RRL staff involved in drafting the decree preferred to keep ICRAF and LATIN in a purely advisory position. The draft, based only on what was allowed within the restrictive HKm framework, also bypassed the Legal Bureau, a breach of procedure that we learned was common in the competitive environment of the Ministry of Forestry. It was also sent on the Minister's stationery in the hope that he would give it a cursory review and quickly sign it. ICRAF staff learned of this directly from the RRL lawyer who drafted the decree and was orchestrating its signing. Since the draft was already on the Minister's desk, he was pleased to provide a copy. ICRAF staff who reviewed the draft had many serious concerns, the major one being that if the RRL approach was signed, there was little prospect of achieving anything in the field apart from further antagonizing the Krui farmers. ICRAF then spoke to staff at the Legal Bureau, and discovered that although the RRL draft had reached the Minister's office, a Legal Bureau ally 'at the gate' had noted the lack of Legal Bureau endorsement. He then sent it back for endorsement. While unhappy at having been bypassed, the Legal Bureau staff were concerned that they might be accused of delaying the process (the Minister had complained twice about the slow pace). They quickly signed off on the draft and sent it directly back to the Minister.

ICRAF staff then prepared a preliminary report to the Minister. It once again detailed the main concerns of the Krui farmers and made recommendations on the overall size of the area in question (RRL had stuck with a figure of 7000ha, while ICRAF believed it to be more than 20,000ha). It also urged that there be no time limits on land rights provided by the Ministry. Having developed a positive working relationship with a key staff person in the Legal Bureau, ICRAF staff sought his guidance before sending the report to the Minister. He was genuinely surprised to learn that ICRAF had not been able to review the draft before it was sent, and he confessed that the draft had left the Legal Bureau with a note saying that ICRAF had endorsed it. The Legal Bureau contact then called an ally in the Minister's office and, for a second time, managed to have the draft plucked from the Minister's in tray. He recommended that, rather than ICRAF sending its report to the Minister, he would rewrite the Legal Bureau's cover letter saying that ICRAF had not yet endorsed the draft and had concerns about it. He then included much of the text of ICRAF's preliminary report in the Bureau's cover letter and sent the draft to the Minister.

The Minister commented immediately, expressing substantial dissatisfaction with the draft prepared by RRL. He went as far as to ask who had mustered the courage to send him such a 'half-baked' concept. Then, for the first time, he outlined how he expected the decree to become a prototype for securing the rights of isolated communities whose traditional lands were within the state-forest zone. He sent it back to the Director General of RRL for revision. He also requested that the boundary-delineation and mapping directorate, INTAG, prepare a map of the area in question. ICRAF staff assisted by supplying a satellite image and a preliminary map of the area, which showed that more than 29,000ha of damar agroforests had been included in the state-forest zone. The process of creating the map was drawn out and, at the time, contentious. In defiance of the minister's direct order, INTAG staff resisted supplying a map that estimated the size of the area. Ultimately, however, they had to comply.

Phase 2: The RRL Official Working Group

The final two months of 1997 may be considered the second phase of the Krui policy development process. The Minister assigned the Secretary of the Directorate

General of RRL to coordinate the process and a working group was formed. The first meeting, on 24 November was attended by forestry officials from Lampung and from the Ministry headquarters, along with ICRAF and LATIN staff. The RRL's draft was discussed and revised over several days. The main issues were the definition of the agroforestry system, the estimated size of the area, and the nature of the agreement – or, indeed, whether there should be an agreement at all. Of these, the shape of the agreement generated the most discussion. Over the course of developing several new drafts, the duration of tenure rights in the agreement became a point of contention. ICRAF and LATIN recommended that it be open ended, and not bound to a specific

Lansium domesticum Corrêa
[Meliaceae]
a synonym of *Lansium parasiticum*
(Osbeck) K. C. Sahni & Bennet.

The fruit known as Langsat or Lanzones originated in western parts of Southeast Asia and was a common component of the damar agroforests. The tree, coming from the Mahogany family, now supplies fruit to markets throughout the region.

period. RRL staff preferred a 20-year period that could be renewed. ICRAF argued successfully that a limited period created a potential disincentive for agroforestry farmers to continue to invest in their systems and created uncertainty as the end of the agreement period approached. After some 'back and forth' on this issue, the Minister's staff accepted an open-ended agreement and agreed that a five-year evaluation process would give the Ministry an opportunity to review the agreement if the community was violating the Ministry's requirements.

Another aspect of the agreement that received much attention was whether it was awarding rights to individuals or to groups. Following a lengthy explanation of the social organization in Krui by LATIN, RRL staff agreed that individual contracts were unrealistic and that agreements with each social group or clan (*marga*) were preferable. Over several meetings, much discussion focused on whether the Ministry could enter in 'agreements' with local people. The Legal Bureau said it could not, and that strictly speaking, the Ministry awarded rights over given areas of the state forest zone. It was eventually agreed that rights over clan areas that were covered by damar agroforests would be awarded through a decision letter issued by the head of the Ministry's office in Lampung province to the head of the clan, in the name of the community. The clan, in turn, would sign a declaration letter stating that it agreed to the rights and responsibilities as outlined in the decision letter.

Another important point of discussion concerned agroforestry activities in areas classified as protection forest. From the outset, RRL staff opposed timber extraction from these areas. However, ICRAF and LATIN staff analysed existing regulations and argued that cutting trees was not prohibited in protection forests. However, there were heavy sanctions for those who *disturbed the ecological functions* of these forests. The Minister agreed that Krui farmers could continue with limited extraction of timber from agroforests located inside protection forests, but with restrictions aimed at protecting the functions of these forests. ICRAF staff assisted in refining these restrictions so they were compatible with existing farmer practices.

Only ICRAF and LATIN representatives and two or three RRL staff attended the final drafting meetings of this phase. During this time, the Minister became concerned about clearly defining the legal basis upon which he could make a forest classification such as was being proposed, and requested that a RRL lawyer prepare a memorandum on the issue. The resulting analysis provides an important definition of the legal boundaries within which a Minister of Forestry can operate and has implications for other policies, particularly those concerning the recognition of *adat* (customary) areas within the state-forest zones.

Phase 3: High-level discussions, consultations with other relevant ministries and signing of the decree

During the latter part of December 1997 and most of January 1998, the Minister hosted several meetings with his senior staff concerning the final draft of the KDTI policy. It was clear that the Minister saw such an approach in Krui as a major initiative towards securing the rights of all *adat* (customary) communities living inside state-forest areas. At one point, he asked the Legal Bureau for an opinion on whether he should consult the President. However, he chose to make sure that other relevant Ministries were consulted, but planned to report the ground-breaking policy to the President after the fact. In early January, staff from the Legal Bureau held meetings with legal staff of the Ministry of Interior and the National Lands Agency. After reporting to the Minister that there was no opposition to the KDTI initiative from these meetings, the Minister of Forestry signed the decree on January 23, 1998.

Responses to the KDTI decree

The KDTI decree did not respond to the primary demand of damar farmers, that they be given back full ownership rights and that the border of the state forest be returned to match the boundary of the Bukit Barisan Selatan National Park. However, the decree was unprecedented in Indonesia at that time, in that it:

- Recognized a community-based natural-resource management system as the official management regime within an area of the state-forest zone;
- Devolved the management responsibility for state-forest lands to a traditional community governing structure; and
- Provided for inheritable rights that were without a time limit (Fay et al., 1998).

The decree was regarded as a major forestry-policy breakthrough by almost all experts dealing with social and community forestry programmes and concepts in Indonesia (Fay and de Foresta, 2001). However, it received criticism from some human rights NGOs that focused (correctly) on the fact that damar agroforests remained in the state-forest zone and on the lack of participation by the intended beneficiaries in the development process.

In Krui, perceptions of the decree were nuanced. Before the seismic political events that culminated in the fall of President Soeharto and assumption of power by the 'reformasi' government, Team Krui organized a series of meetings in villages around Krui to present the decree and measure the reactions of local villagers (Figure 46-5). In general, there was disappointment on one hand, because the land was still gazetted as state forest and damar agroforests were not explicitly recognized as the result of the hard work of local people. There was also local concern about restrictions such as the need for both the continuous existence of the customary community and its official recognition, about the monitoring and sanctioning role of the Ministry of Forestry and about the complexity of the formal process that was needed to get their rights officially recognized. On the other hand, there was satisfaction in learning that their rights had been legally recognized, not only for themselves but also for their children.

Following the fall of Soeharto in June 1998, the attitude of local communities changed to total rejection of the KDTI as a solution to their problems. The KDTI, which would have been acceptable under policy conditions prevailing when it was issued, was no longer considered acceptable. Basically, the Krui communities stuck to their claim that gazetting of the land they had developed into damar agroforests and other land-use systems was a violation of their private property and basic rights.

On the other side, the Forestry Services whose job it was to implement the KDTI decree had shown their resistance to the decree on many occasions and had no

FIGURE 46-5: The cover page of a booklet prepared for presenting the decree to local communities in Krui. In 1999, hundreds of booklets were distributed by Team Krui in villages bordering the KDTI area.

reason to be enthusiastic about it because they saw it as curtailing the power they held over land that they felt was under their full authority.

With both sides reluctant to implement the decree, and given the major political upheaval that occurred after its signing, it is perhaps no surprise that over the past 17 years, no attempt has ever been made to bring the decree into force: the Ministry of Forestry has not undertaken the delineation of the KDTI and no customary-community head has ever applied for a Damar Concession Right.

In 2005, Kusters et al. (2007) reviewed the impacts of the KDTI decree. They found that although it had never been implemented, the KDTI decree had been successful in reaching its objectives, i.e. improving security and maintaining the agroforest area under community management. Importantly, they noted that it had been 'instrumental in stopping outsiders' attempts to appropriate damar agroforests', and that damar farmers had continued to manage and reap the full benefits of the agroforests located inside the state-forest zone, with no restrictions from the Forestry Services (Kusters et al., 2007).

However, the production forest within the KDTI area (24,835ha) was designated as a *Hutan Tanaman Rakyat* (community-based timber plantation) project area by the Minister of Forestry in 2010.[11] Up to mid-2014, eight cooperatives had obtained permits and 865ha of land had been planted with various fast-growing timber species such as *jabon* (*Anthocephalus cadamba*), acacia (*Acacia mangium*), and *cempaka* (*Michelia champaca*) (Herawati, 2013).

This is the beginning of another story. Although no damar agroforests (*repong damar*) have yet been converted, the new designation, which re-asserts the power of the Ministry of Forestry over state forest land at Krui, and the random process by which the project has been implemented so far, are both worrying and worth monitoring. These factors could well lead to conflict within the Krui communities, between those who wish to continue to manage their damar agroforests and those who wish to pursue the community-based timber plantation scheme.

Conclusion

Important lessons about development policy in general and forestry policy in particular may be drawn from the experiences recounted in this chapter, surrounding the KDTI decree and the processes that led to its issuance.

In detailing the processes leading up to the decree, this chapter reiterates the commonly held concern that local communities have a limited capacity to present matters of disputation to policy-makers and seek external support in terms of facilitation, funding, information and technology. The experiences presented here also show that key findings and recommendations provided by researchers, academicians and practitioners need to be reinforced by advocacy efforts in order that they are properly understood and most likely to generate positive responses from policy-makers. This important process of advocacy, i.e. bridging science to the policy-development process, is overlooked in many policy studies, so that recommendations often go no further than scientific publications.

Seven main lessons may be drawn from experiences in the process of developing and drafting the KDTI decree:

1. Forestry policies in Indonesia emerge from competing interests both inside and outside of the Ministry of Forestry. Formal procedures are often bypassed and conceptual development of new policy is largely done quietly, if not in secret.

2. Non-governmental organizations and academics, including international institutions, may provide critical support for policy reform by 'entering the process from within', and providing decision-makers with technical information that they need as well as honestly analysing the possible or even likely implications of various decisions.

3. The success of research and advisory institutions 'entering the policy process from within' is heavily dependent on them having a mandate from at least one faction within the Ministry of Forestry. In the Krui case, the ability of ICRAF and LATIN to provide recommendations benefited significantly from having a mandate from the Minister himself.

4. Success is also dependent on being present. The haphazard nature of policy development requires those involved to be prepared to respond at a moment's notice and, at times, to be proactive. For example, some of the most important meetings during the development of the KDTI decree took place spontaneously, following a chance encounter in an elevator at the Ministry's headquarters. A proactive approach can also create such 'spontaneous' encounters, or go one step further and 'keep the ball rolling' by assisting in organizing meetings or moving information.

5. To be successful, institutions working from within must be well informed and have a solid basis for their recommendations. This often means 'having one foot in the field' and maintaining close relationships with groups working with local communities, as well as with advocacy groups. These latter groups join with local communities to 'enter the process from the front', often projecting analysis and recommendations on the basis of a political agenda focused on promotion of justice and human rights. Pressure from these groups can often create the political space within government institutions that reform-minded officials and policy-advisory groups need to achieve change. Local communities and advocacy groups 'entering from the front', can create the need for a response and empower those from within to shape that response in a way that meets the needs of the local communities and the environment.

6. The process of developing a policy such as the KDTI is a process of negotiation, and one that does not end with the signing of a decree. The Krui case illustrates this clearly. In the end, the local communities reviewed the government's response to their concerns and decided that it did not go far enough. It is important to note that prior to the May 1998 change in government, it appeared that the Krui farmers and heads of the *margas* (clans) would accept the KDTI,

albeit not without opposition from within. It seems that the perception of greater political space and opportunities for reform from the new government weighed heavily in the community's decision to stand by their initial demand of moving the production-forest boundary back to the national park. Significantly, the Krui communities have faced no serious outside threats to the integrity of the overall damar agroforestry area following the issuance of the KDTI decree. This has potentially lessened the urgency to follow through with moves to secure their agroforestry boundaries.

7. Policy development in Indonesia is heavily dependent upon precedent. A policy breakthrough applying to a specific geographic area can have impacts beyond the immediate location. This is why the process of developing the KDTI decree was so difficult and why its signing was so important. For a while, the decree became another tool for local people to use in their efforts to gain tenure rights over traditional lands. It also became a tool for those working to develop a national policy that creates a process for recognizing traditional rights, along with the forest creation and management capacities of communities whose lands have been classified as state forest.

An important lesson may also be drawn from reactions to the KDTI decree, pointing to the identification of its domain of relevance.

The evolution of local communities' reactions to the decree reveals that it would have been acceptable only as a temporary solution; in other words, it was judged better than the previous situation but still far from what local communities considered fair: an unequivocal official recognition of their land-ownership rights. In similar situations elsewhere, it is probable that a solution such as the KDTI decree would have been found acceptable only with the emergence of serious outside threats to the integrity of the overall area.

It is important to note that when land that is already covered with agroforests or other agricultural land uses is gazetted as state-forest land it means that the field-delineation process has not been conducted according to national laws and regulations and the ownership rights of local communities have been at best ignored or at worst violated. Local communities, if they can gain access to official documents, would thus have strong legal arguments to contest the gazetting of their agricultural land as state-forest land. In such situations, KDTI-type solutions will only temporarily soothe the wounds resulting from flawed implementation of national laws and regulations. Permanent solutions have to be found through revisions of the delineation processes, in a much-needed attempt to establish state-forest zones that are really 'clear and clean'.

In fact, therefore, KDTI-type solutions cannot be considered as permanent or acceptable rulings in state-forest areas where agricultural activities were developed before gazetting. If they are to be regarded as permanent, KDTI-type solutions will only fit those natural forest areas that may be claimed equally – or almost equally – by local communities on one hand and the government on the other. In such cases, the rights of local communities to manage their immediate environment and the rights and responsibility of the government to preserve forests and develop forest utilization in a sustainable way for the benefits of the whole nation may be equally defensible claims.

Dedication

This chapter is dedicated first to the late head of Pahmungan village, Pak Rusba Toha, who almost fainted when, after being shown the new official land-use map of the West Lampung district, he realized that the independent government of Indonesia had negated local people's rights that were recognized even by the colonial government. The map showed no damar agroforests in the whole Krui region; they had been swallowed within the new boundaries of the state forest.

And second, to the late Restu Ahmaliadi, from LATIN, who devoted years of living in Krui to helping in community organization and participatory mapping, and who, along with the three other 'musketeers' (C. Fay, H. de Foresta and M. Sirait), struggled through the KDTI drafting process to ensure that the rights of damar agroforest farmers were recognized and that damar agroforests could continue to be managed sustainably.

Last but not least, it is also dedicated to the Global Comparative Study on Tenure (GCS Tenure), an ongoing CIFOR research project aimed at securing tenure rights for forest-dependent communities. It is funded by the International Fund for Agricultural Development (IFAD) and by the Global Environment Fund in association with the Food and Agriculture Organization of the United Nations.

References

de Foresta, H. (1992) 'Botany contribution to the understanding of smallholder rubber plantations in Indonesia: An example from Sumatra', in *Proceedings of the International Symposium, Sumatera: Environment and Development, its Past, Present and Future*, Seameo-Biotrop APHI/MPI, Bogor, Indonesia, pp363-368

de Foresta, H. and Boer, E. (2000) '*Shorea javanica* Koord. and Valeton', in E. Boer and A. B. Ella (eds) *Plant Resources of Southeast Asia, no. 18: Plants producing exudates*, Backhuys Publishers, Leyden, pp105-109

Fay, C., de Foresta, H., Sirait, M. and Tomich, T.P. (1998) 'A policy breakthrough for Indonesian farmers in the Krui damar agroforestry', *Agroforestry Today* 10(2), pp25-26

Fay, C. and de Foresta, H. (2001) 'Progress towards recognizing the rights and management potentials of local communities in Indonesian State-defined forest areas', in B.Vira and R. Jeffery (eds) *Analytical Issues in Participatory Natural Resource Management*, Palgrave Macmillan (Global Issues Series), London and New York, pp185-207

Herawati, T. (2013) 'Studi ekonomi dan standar harga produk hutan tanaman rakyat di provinsi Lampung' (Economic studies and standard product pricing for plantation forests in Lampung province), *Project for Strengthening the Capacity of Stakeholders for the Development of Community Based Plantation Forests at Three Selected Areas in Indonesia*, Reports of the International Tropical Timber Organization, Jakarta

Kusters, K., de Foresta, H., Ekadinata, A. and van Noordwijk, M. (2007) 'Towards solutions for state vs. local community conflicts over forestland: The impact of formal recognition of user rights in Krui, Sumatra, Indonesia', *Human Ecology* 35(4), pp427-438

Laumonier, Y. (1997) 'The vegetation and physiography of Sumatra', *Geobotany* 22, Kluwer Academic Publishers, Dordrecht, The Netherlands

Mary, F. and Michon, G. (1987) 'When agroforests drive back natural forests: A socio-economic analysis of a rice-agroforest system in Sumatra', *Agroforestry Systems* 5(1), pp27-55

Michon, G., de Foresta, H., Kusworo, A. and Levang, P. (2000) 'The damar agroforests of Krui: Justice for forest farmers', in C. Zerner (ed.) *People, Plants and Justice: The Politics of Nature Conservation*, Cambridge University Press, New York

Michon, G., de Foresta, H., Kusworo, A. and Levang, P. (2007) 'The damar agroforests of Krui, Indonesia: Justice for forest farmers', in M. F. Cairns (ed.) *Voices From the Forest: Integrating Indigenous Knowledge into Sustainable Upland Farming*, Resources for the Future Press, Washington, DC

Pretzsch, J. (2005) 'Forest related rural livelihood strategies in national and global development', *Forests, Trees and Livelihoods* 15(2), pp115-127

Rappard, F. W. (1937) 'De damar van Bengkoelen', *Tectona* D1XXX, pp897-915 (in French)

Suporahardjo and Wodicka, S. (2003) 'Conflicts over community-based "repong" resource management in Pesisir Krui region, Lampung Province, Indonesia', in A. P. Castro and E. Nielsen (eds), *Natural Resource Conflict Management Case Study: An Analysis of Power, Participation and Protected Areas*, Food and Agriculture Organization of the United Nations, Rome

Torquebiau, E. (1984) 'Man-made dipterocarp forest in Sumatra', *Agroforestry Systems* 2(2), pp103-127

Notes

1. IRD (Institut de Recherche pour le Développement, ex ORSTOM) is a French government research organization. See https://en.ird.fr/ird.fr.
2. ICRAF (World Agroforestry Centre) is a CGIAR consortium research centre. ICRAF headquarters are in Nairobi, Kenya, with five regional offices located in Cameroon, India, Indonesia, Kenya and Peru. See http://www.worldagroforestry.org.
3. CIFOR (The Center for International Forestry Research) is a non-profit research centre of the CGIAR consortium that conducts research on the most pressing challenges of forest and landscapes management worldwide. Its headquarters are in Bogor, Indonesia. See http://www.cifor.org/.
4. WATALA (Friend for Nature Environment) is a non-governmental organization based in Lampung. See http://www.watala.org.
5. FORDA (Forestry Research and Development Agency) is the research and development unit of the Indonesian Ministry of Forestry. See http://www.forda-mof.org/.
6. LATIN (Lembaga Alam Tropika Indonesia) is a non-governmental organization based in Bogor. See http://www.latin.or.id.
7. P3AE-UI (Program Penelitian dan Pengembangan Antropologi Ekologi Universitas Indonesia) Department of Ecological Anthropology, University of Indonesia.
8. This section is based primarily on Michon et al., 2000 and 2007.
9. The *Hutan Kemasyarakatan* (community forestry in a state-forest zone) scheme was formalized in a decree from the Minister of Forestry in 1995. This scheme was conceived in order to involve communities in the rehabilitation of degraded forest land, but was strongly restrictive in terms of community rights. For the Ministry of Forestry, it became a solution for all problems related to the presence of local communities on state forest land. Its basic flaw, not only in the case of Krui, but also for numerous other cases in Indonesia, was that it did not take into account the possibility that local communities had access rights to land before it was gazetted as state forest land.
10. Because of its in-depth knowledge about land-access issues at Krui acquired over recent years by LATIN field staff, ICRAF asked the Ministry to include the NGO in the decree-drafting process. The demand was accepted.
11. According to the Minister of Forestry's Decree No. 47/Menhut-II/2010.

47

LAND LAW AND SWIDDEN CULTIVATION

Indonesian *adat* communities and the struggle for statutory rights

*Rebakah Daro Minarchek**

Introduction

The chapters in this volume detail the myriad ways in which national-level policies can affect and, in turn, be affected by swidden agricultural systems. In this chapter, I explore the possibilities for such policy impacts in Indonesia, following a landmark ruling by the country's Constitutional Court in May 2013, the major aspect of which was a declaration that indigenous peoples' customary forests should no longer be classed as state forest areas. I use the word 'possibilities' because, although Constitutional Court Ruling 35/PUU-X/2012 (hereafter referred to as MK35) has been made, there has not yet been any implementation, in a legal sense, by the Indonesian government at national level.[1] Therefore, I will expound upon the steps taken by the government and local communities as they explore the possibilities for implementation on their own terms in the coming years. In particular, I focus on swidden agricultural practices among indigenous communities that are impacted by MK35 and the importance of recognizing their shifting cultivation systems in future implementation of the ruling.[2]

Swidden agricultural systems in Indonesia must always be discussed within the context of the sheer expanse of forest land owned by the central government, as many swidden-dependent communities live in forested areas and use primary and secondary forest land for swidden production. Currently, the Indonesian central government claims ownership of more than 96% of the country's forest area (RRI, 2015). This claim was debated during the MK35 court proceedings, between March 2012 and May 2013. The case was brought before the Court by the Indigenous Peoples' Alliance of the Archipelago (AMAN), along with representatives of

* Rebakah Daro Minarchek is Managing Director of the Southeast Asia Center in the Henry M. Jackson School of International Studies at the University of Washington, Seattle, WA, and a PhD candidate in the Department of Development Sociology at Cornell University, Ithaca, NY

indigenous peoples from Kenegerian Kuntu, in Riau Province, and Kaesepuhan Cisitu, in Banten Province.[3] They sought a judicial review of the Forestry Law of 1999 on the grounds that the government had used this law for more than 10 years to 'take over the rights of indigenous people over their customary forest areas to become state forest, which then on behalf of the state were given/or handed over to capital owners, through various licensing schemes to be exploited without consideration to the rights and local wisdom of indigenous peoples in the region' (Constitutional Court of Indonesia, 2013). The Court decided in favour of AMAN and the indigenous representatives, removing the word 'state' from the following passage of the Forestry Law: '*adat* forest is state forest which lies within the lands of *adat* communities' (Government of Indonesia, 1999).[4]

This decision creates the possibility of returning nearly 40% of state-owned forests to indigenous communities throughout Indonesia. Although there has been no framework set in place for enforcing the Constitutional Court ruling, the decision set local communities, civil society organizations and government ministries in motion. All actors are now preparing evidence of their land claims in readiness for future implementation of the ruling. While civil society organizations such as AMAN played a significant role in successfully arguing the MK35 case, its role in that process is not the focus of this chapter. Rather, my emphasis lies with the impact that process will have on the swidden systems of indigenous communities.

Historically, there has been a divide over swidden systems between academics, swidden practitioners and civil society organizations on one hand and governments on the other. The feeling among many academics, practitioners and international-level organizations is that swidden agriculture is more sustainable than previously thought, with studies showing that swiddening is a rational and potentially beneficial agricultural system in tropical climates (Conklin, 1957; Geertz, 1963; Dove, 1983; de Jong, 1997; Mertz et al., 2008). However, governments often see swiddening as a destructive practice that leads to forest degradation, and those within Southeast Asia have been less than tolerant of traditional shifting cultivation. Indeed, the common approach for governments has been to discourage swidden farming by means of regulation and ideological campaigning in the assumption that the practice degrades forest landscapes. Duncan (2004, p88) criticized this view by arguing that 'these environmental concerns about swidden can be considered a red herring, because often the Indonesian government opened up the forests for logging, mining, or transmigration as soon as forest-dwelling communities were resettled'. Forms of regulation imposed in Indonesia have included outright prohibition; prohibition without a permit; prohibition on certain classes of land; and tenure-related disincentives (Cramb et al., 2009). As well as this there is ideological persuasion, by which governments act to coerce or otherwise convince swiddening communities that their agricultural practices are outdated, backward and in need of conversion. Of all of these possibilities, I believe that ideological persuasion is the most dangerous and the most convincing among the indigenous communities involved in this study. Swidden farming cannot be assumed to be either a sustainable or an unsustainable practice by academics on one hand or governments on the other unless a variety of factors is first taken into consideration. Generally, the dependent variable cited for sustainability is the length of time for which formerly cultivated swiddens are left fallow. This will be discussed below. While there are various claims about the number of years of fallow that should be considered 'best practice', Boomgaard (2007) gives eight years as a minimum for sustainable practices. This chapter will not debate the sustainability of swidden practices, but rather show that despite academic or government debates, many forest-dwelling communities – including those at the heart of MK35 – continue the practice and it still forms an integral part of their everyday lives.

In order for implementation of MK35 to be successful, I contend that the central government must recognize that swidden farming is occurring and is an important aspect of the lives of certain segments of their population, and that indigenous groups must push for its recognition as a central component of their *adat* traditions. These steps would (a) recognize the connection between *adat* and forest use in such a way that the ability of indigenous communities to manage their own land would be respected; (b) increase community mapping activities in areas where swidden agriculture is still practised and possibly enable communities to capture swidden land use on maps where government maps have failed to do so; (c) protect both community forests and swidden agriculture in areas where privatization, concessions and conservation zones are common; and (d) recognize community ownership of land and the right of indigenous communities to parcel out swidden areas for use by community members.

Research for this study was carried out between June and October in 2014 and June and December, 2015. The trips were similar in research focus, but differed in the scale at which data was gathered. Both trips were focused on the MK35 issue, indigenous mapping efforts and conflicting interests in land currently occupied by members of the Kaesepuhan community in Indonesia's West Java and Banten provinces. However, the 2014 research focused primarily on community-level research and interviews, whereas the 2015 research focused on the activities of government and civil-society actors at provincial and national levels, along with community responses to these activities. Interviews, surveys and/or focus groups were conducted in four Kaesepuhan communities, most extensively within the Ciptagelar community in Sukabumi, West Java. This location serves as the study site for this chapter. Of particular relevance were the following activities for gathering quantitative data: 57 (10 male and 47 female) activity diaries and food diaries; 61 (23 male and 38 female) participatory mapping exercises; 89 (55 male and 34 female) surveys of community members; two multi-day national-level meetings on MK35 (one in Rangkasbitung, Banten, and another in Jakarta); two strategic planning meetings coordinated across civil society organizations (Bogor and Jakarta); five Kaesepuhan-level meetings of ruling councils (one each in Cisungsang, Bogor and Sinar Resmi and two in Ciptagelar); 12 focus-group meetings with community members and civil-society staff; 308 formal and semi-formal interviews with community members and leaders; 28 personal interviews with staff; and attending meetings of four primary civil-society or non-governmental organizations, along with countless hours of participant observation with government workers, community members and civil-society staff.

This chapter analyses the extent of swidden practices among indigenous households in Kaesepuhan communites in West Java, in order to justify the recognition of these practices in the course of the Indonesian central government's implementation of MK35. First, I provide information on the Gunung Halimun Salak National Park study site and the Kaesepuhan indigenous group. Then I provide a brief history of policies in Indonesia related to *adat*, national parks, conservation and swidden agriculture. While this is not an exhaustive history, it uses the most significant and impactful of the Indonesian government's statutory laws to illustrate the policy trajectory that has led to the current state of swidden agriculture for many forest-dwelling communities. Finally, I expound upon the practice of swidden agriculture and the connection of the Kaesepuhan to the forests and the land in order to show the relationship that exists between community members, *adat*, and the essence of life. The chapter concludes with a discussion on the future of MK35 and Kaesepuhan swidden agriculture.

The study site

Gunung Halimun Salak National Park (hereafter the National Park) lies in three districts on the island of Java – Bogor and Sukabumi districts in West Java province and Lebak district in Banten province (Figure 47-1). The site of village-level research, Ciptagelar, lies in the Sukabumi district, but it is very close to the border with Lebak district and villagers often have the use of land parcels in both districts and provinces. The Park has areas of colline, montane and submontane forests covering more than 100,000 hectares at elevations ranging from 500 to 1929 metres above sea level (Harada, 2003). The National Park is significant because it is the largest area of remaining forested land with the highest levels of biodiversity on the island of Java (Takahashi, 2006; Kubo and Supriyanto, 2010).

The Gunung Halimun area was first designated as a protected forest in 1932, based on the hydrological function it provided to the Bogor-Jakarta area (Harada, 2004). As Galudra et al. pointed out in a 2008 study, the protected forest status was gazetted and delineation begun under the Dutch administration between 1906 and 1938. In first gazetting the area, the Dutch rendered invisible the communities living in and around it. These communities protested the erasure of their existence in the 1920s, and in 1922 more than 3000 swidden farmers from the area were imprisoned for protesting the inclusion of their lands within the state's 'protected forest'. In response, the Dutch colonial government ruled that a new gazetting was needed, in order to take the local people into consideration. However, World War Two began and the process never took place. After Independence at the end of World War Two, the Indonesian government used the old Dutch maps as proof of an 'empty space' when it created the National Park. The government also claimed that the space could not

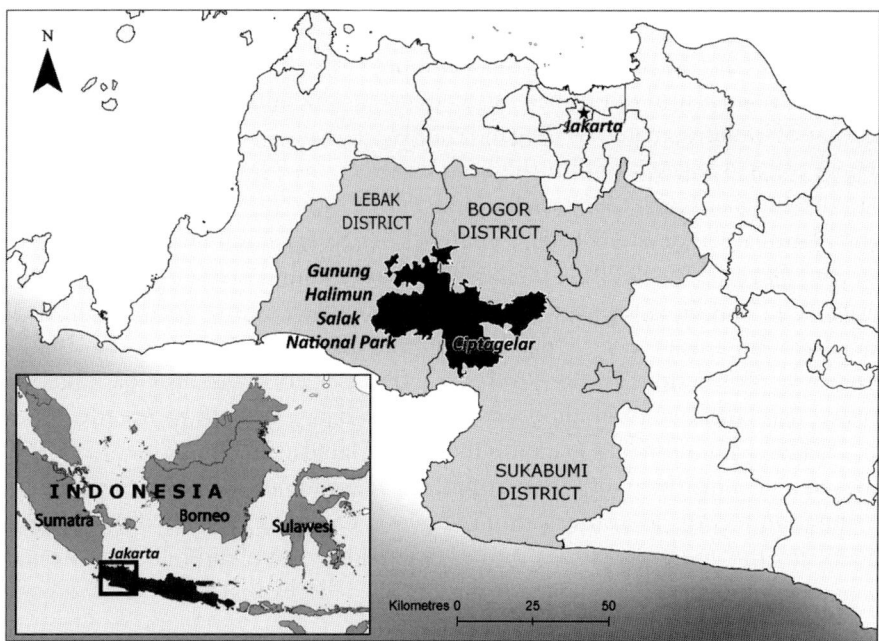

FIGURE 47-1: The research site in West Java, Indonesia, in relation to Gunung Halimun Salak National Park.

be re-gazetted to incorporate local communities, as the Dutch colonial government had planned (Galudra et al., 2008).

As the rest of Java succumbed to deforestation and to the land needs of a rapidly growing population, Gunung Halimun's isolation and lack of transportation infrastructure soon helped it to develop a higher level of biodiversity than the rest of the island. In recognition of its ecological importance, the Indonesian government changed the area's status to a nature reserve in 1979 (Decree No 40/1979), using the details of the old Dutch gazette notice to define the area included in the reserve (Harada, 2003). These designations were largely confined to paper. The local people continued to live on their land, often without knowing that it was now considered state or national park land (Kubo, 2008).

Under reserve status, overlapping land claims between the state and villages were dealt with by using a profit-sharing model: the communities could stay and farm (swidden or irrigated) within the park boundaries, but they were required to give 25% of their profits or harvests to the forest authority. In 1992, the area's status was changed once more (with much protest from local communities) into a national park. Under a stricter core zoning system, the profit-sharing model was officially discontinued. However, the Ciptagelar community, perhaps in an effort to legitimize its occupation of park land, continues to this day to pay harvest taxes. In an event known under *adat* as *tatali*, villagers report their rice harvests at a central building and pay a portion to the village council. The council then divides the rice into portions for payments to the government, to village resources (which can be resold to villagers), tributes to the community leader, and for use in ceremonies. Kaesepuhan communities that now fall within the National Park's boundaries have complained about limitations placed upon their agricultural production, firewood collection and gathering of building materials (Suganda, 2009).

There are 314 villages within the boundaries of the National Park, with around 100,000 residents (Kubo, 2008; Kubo and Supriyanto, 2010). Villages are classified by the National Park system as adjacent, enclaves or encroachments (Harada, 2003). Adjacent villages are just outside the National Park's boundaries. Enclaves are villages or plantations that lie within the boundaries, but outside of the Park's administrative jurisdiction. Encroachment villages are within the boundaries of the Park and subject to its jurisdiction. Projects to resettle the villagers outside the Park have been discussed, but community members are understandably resistant to resettlement and their villages remain within park boundaries (Harada, 2003). Geisler (2003, p69) defines people who 'involuntarily part with their livelihood claims in places set aside for natural protection' as 'conservation refugees'. The tenure rights of encroachment villagers are the most precarious and they are the group most likely to become conservation refugees. These are not necessarily villages that have 'encroached' on National Park land since its original designation (Harada, 2003, 2004). Many were in the area before formulation of the National Park's management plan, but for one reason or another were never zoned as 'enclaves'.

While this is not always the case, enclaves and encroachment villages tend to be labelled according to their claims to 'indigenous' status. Encroachment villages are more likely than those in other categories to be populated by rural subsistence farmers who, at sometime in the past, moved to the area from elsewhere in Java. While they may be 'indigenous' to other regions of the island, they are not considered as such in the region of the National Park. The village of Cibedug is an example of an encroachment village. Harada (2003) describes its people as 'politically disadvantaged' and notes that they have less of a claim than others to land ownership in the National Park and little expectation of ever gaining it. MK35 does not apply to rural farmers within the National Park. As Hall et al. (2011) note, 'inclusion' always involves an aspect of 'exclusion', and in this case the inclusion of 'indigenous' peoples in MK35 inherently means the exclusion of small-scale, rural farmers that cannot claim indigeneity to the region in which they live.

According to their oral history, the Kaesepuhan people are descendants of the Pajajaran-Bogor or Sunda kingdom, which existed in West Java between the 10th and 16th centuries. They are led by an *Abah* (father), which is a hereditary position. Members of the Kaesepuhan *adat* community number around 16,000 (Suganda, 2009). There are 13 main Kaesepuhan villages, most of them within the boundaries of the National Park. However, there are many more followers scattered throughout other villages within and around the Park. Ciptagelar is the central village for followers of *Abah* Ugi Sugriana Rakasiwi. There are conflicting stories about the legitimacy of various *Abahs*, but most are from the same family and Ciptagelar is the largest of the Kaesepuhan communities. The village lies in the corner of the National

Park's southwestern boundaries and is home to about 300 households. It is the most recent settlement of the Kaesepuhan of *Abah* Ugi's line. The village was relocated to its present site in 2000, after *Abah* Ugi's father, *Abah* Anom, was instructed by his ancestors to shift the village. The technical problem with this spiritual instruction is that Ciptagelar now lies within the National Park's boundaries, whereas its earlier site did not, and it was formerly listed as an adjacent village. Stories about the semi-nomadic lifestyle of the Kaesepuhan vary, but two common themes relate to spiritual beliefs and territory. One belief is that the Kaesepuhan must periodically gather the most devout and relocate, to start anew. Another is the need to continually identify and maintain control over customary lands that are scattered across three districts (Suganda, 2009). When the swidden practices of the Kaesepuhan are considered, the need to relocate every decade or so may also be a practical way of finding new and possibly healthier lands for agricultural purposes.

On average, a community lives in a village for about 10 to 15 years before relocating. Ciptagelar's people believe that *Abah* Ugi is spiritually connected to the ancestors and is able to communicate with them. They have lived at the present site since 2000, and 'perhaps soon' the ancestors will instruct *Abah* Ugi to move the village again. One villager stated that *adat* required them to move, and if the ancestors instructed it, they would follow their *Abah*. They are, nevertheless, fully aware of likely official reaction to the ritual migration of 300 Ciptagelar households through the territory of the National Park. One said, without hesitation, that he and others would take up arms if necessary to protect their right to go where the ancestors instructed them. They claim to be searching for *ugakebakcawane* (the 'promised land').

Government and forests in Indonesia

Government policies within Southeast Asia have long served to discourage shifting cultivation and the number of swidden practitioners in the region has been in steady decline as a consequence, particularly in Indonesia. Perhaps the heaviest impact in Indonesia has come from the territorialization of large tracts of forest land within the country, both before and after independence. Regardless of the human occupants, the central Indonesian government has claimed more than 96% of the country's forested land. A large proportion of this has been declared protected areas and national parks. As previously explained, the study communities for this chapter live within the Gunung Halimun Salak National Park, so much of the following discussion of the history of Indonesian policy affecting swidden agriculturalists will be linked to national parks, conservation, land tenure and *adat*. In addition, since the MK35 ruling has only recently been delivered and has not yet been implemented, the discussion of statutory and *adat* land laws and their connections is intended to shed light on the historical trajectory leading up to the Court's ruling.

Starting in 1830, the Dutch colonial government introduced an agricultural regime known as *cultuurstelselor* 'the cultivation system'. This arrangement marked Indonesia's point of entry into the modern world economy. The cultivation system was feudal in nature and required the native population to contribute one-fifth of their land to growing export crops for the colonial administration, or if they had no land, to contribute their labour (Szczepanski, 2002). This system was more strictly enforced on Java than on any other island. It led to widespread poverty, as it forced peasants to devalue their own labour in order to meet production quotas (Bernstein, 2010). It also brought famine as peasants found themselves with little time or energy to devote to rice production. A tax system effectively became a forced labour system.

Elettaria cardamomum (L.) Maton [Zingiberaceae]

Green or True Cardamom appears commonly in Kaesepuhan swiddens, where it is known as *kapulaga*. Its seeds provide the spice, cardamom.

In 1870, the Dutch turned away from 'the cultivation system' (which had largely collectivized land) and implemented *Agrarische Wet* or the Agrarian Law. This law was meant to help establish more private businesses in Indonesia by granting 75-year leases to foreigners. It was also intended to formalize land holdings with titles for the local population. In reality, large numbers of people who had recognized land-use titles, but no formal title, were dispossessed of their land rights. This included forest lands, *adat* land, village communal land, private agricultural land and native lands under use rights (Djalins, 2012). In a classic colonial model, the Dutch seized 'any useful lands or resources from the local people, without compensation' (Szczepanski, 2002, p234). The effects of this programme are still being felt within Indonesia. Much of the land claimed by the colonial government more than 140 years ago, especially indigenous lands, now forms the country's system of protected areas, and this has become the basis of the present Indonesian government's land claims.

With the advent of World War Two, Dutch colonial rule gave way to the Japanese occupation in 1942 and 1943. The two administrations had vastly different approaches to land and conservation. In Java, the Dutch were concerned with forest preservation for production reasons, while adding an element of conservation. The Japanese, perhaps understanding the fleeting nature of their rule while feeling the financial pressure of waging a wide-ranging war, focused on agricultural production and logging (Peluso, 1992; Galudra et al., 2008). This helped to finance the war for the Japanese and brought about widespread deforestation. Many villagers hurried to expand their agricultural fields during this period as the laws and regulations relating to protected forest were declared invalid (Harada, 2003).

Although the Dutch pressed to regain control of Indonesia after the war, Indonesia resisted and drafted its first Constitution in 1945 as the people prepared for independence. After the widespread deforestation, destruction and violence suffered during the Japanese occupation, the newly independent Indonesian government idealized the Dutch model to some degree and reverted to the Dutch approach to forests (Peluso, 1992; Szczepanski, 2002; Kubo, 2008). Picking up the Dutch legacy, it continued to write into statutory law the idea that 'the preservation narrative carried with it the assumption that only the state (in this case, the government), which had the right to access and control, was capable of protecting the uniqueness of nature or wildlife species' (Galudra et al., 2008, p7). Consequently, the maps created during the colonial regime took on a definitive status, regardless of whether they were originally disputed, as they were in the Gunung Halimun region.

In the early years of the Republic, land struggles were aligned largely according to class. The rural poor were somewhat united in calling for land reform and the return to lower-class citizens of large tracts of land held by native and foreign elites. In an unfortunate game of politics, rural farmers were caught up in the struggle between Indonesia's first president, Sukarno and the then Major-General Suharto, who was to become Indonesia's second president in 1967. Any small-scale rural farmer calling for land reform was labelled a 'communist' and was likely to be swept up in purges that killed an estimated 500,000 to one million people. The mass killings effectively silenced class-based land struggles and continue, even now, to influence the views of activist farmers. Li (2003) raised concerns that class-based struggles in Indonesia were being forgotten in the rise of campaigns based on indigeneity and waged by groups like AMAN.

The overthrow of the Suharto regime in 1998 and the beginning of a transitional period known in Indonesia as *reformasi* opened up a space in which civil society actors could focus on indigeneity as a motivation for land reform (Kubo, 2008). Earlier, forestry officials did not have to consider the opinions and practices of local villagers in implementing management plans for national parks. But as Kubo (2008, p85) noted: 'the state forest bureaucracy now faces difficulties in implementing state policies without consideration of local interests due to the transformation of the state-civil society relationship that took place upon the commencement of *reformasi*'. Prior to that, there had been a widespread homogenization of society, reinforced by *pancasila* – the philosophical foundation of the Indonesian state, which calls for unity in diversity through five unifying principles concerning the "beliefs" of Indonesian citizens (Li, 2003). Although introduced by Sukarno, Indonesia's first President, Suharto took *pancasila* to a new level by attempting to create a homogenous Indonesia in which there were little or no rights for being 'indigenous'.

To this day, Indonesia's government is hesitant to legally recognize *masyarakat adat* as 'indigenous people.' To do so would not only grant them rights that differ from those of the rest of the population, but also give them international protection, and the government favours neither of these outcomes. It has used a number of terms to refer to *masyarakat adat*, but has maintained the discourse that all people in

Piper betle L. [Piperaceae]

The 'betel' in betel nut, the mild stimulant that plays a role in ceremonies prior to planting of the Kaesepuhan swiddens (below).

Indonesia (apart from ethnic Chinese) are indigenous and none deserve rights over and above any other group (Szczepanski, 2002; Li, 2003). However, the government did attempt to incorporate *adat* communities (rather than indigenous communities) into some of the legal framework concerning protected areas.

The Basic Agrarian Law of 1960, along with the Forestry Laws of 1967 and 1999, had a significant impact on *adat* communities in Indonesia. These laws were offered up by the Indonesian government as its inclusion of *adat* into national laws. However, with striking similarity to the *Agrarische Wet* of the Dutch administration, these laws ignored customary rights to land since rights claimants did not possess legal title issued by the government. Once more, the rights of *adat* communities gave way to government land grabbing in the name of national interest. After the Forestry Law of 1967 was passed, 75% of land in the country that was not held under official private title passed into state control (Li, 2003). To this day, the government wins most land disputes because, 'in the majority of cases under review, land-holders have never registered their lands, because of both the cost and the bureaucratic procedures involved. Thus, the only proof of ownership or cultivation rights is the length of time they have been cultivating the land and their payment of all financial obligations' (Lucas, 1992, p84).

After the ousting of Suharto in 1998, civil society called for the revision of many laws written during Suharto's New Order regime. One of these was the Forestry Law of 1967, which claimed to take *adat* into consideration, but in reality was contested by *adat* communities. According to Li (2003, p397) Indonesia's Forestry Laws of 1967 and 1999 both had the same mission: 'to recognize peoples' presence in forested areas while conceding nothing on the issue of rights, and to enmesh them more securely in state regulatory regimes'. The 1967 law in particular was contested because it did not allow ownership of forest land, and this affected the land rights of between 40 and 60 million Indonesians, most of whom were indigenous (Szczepanski, 2002). The 1999 revision of the law allowed private ownership, but power ultimately remained in the hands of the state because of the wording: '*adat* forest is state forest which lies within the lands of *adat* communities' (Government of Indonesia, 1999). This is precisely the phrase that brought about the Constitutional Court review that resulted in the MK35 ruling.

There have been other statutory laws in which the government claimed to be incorporating *adat* and *adat* agricultural practices. Imitating an early European model of conservation, it implemented a 'fence-and-fine' approach or exclusion model, in which 'restrictive regulations are enforced because a human presence in tropical forests is incompatible with the conservation of biodiversity' (Kubo and Supriyanto, 2010, p1786). However, a ministerial decree in 2004 (P.19-Menhut-II/2004) changed who could legally become a partner in conservation efforts (Mulyana et al., 2010). Whereas the idea of forest management earlier ended with government actors, the decree stipulated that local people and non-governmental organizations could also be formal partners in conservation efforts in Indonesia. In 2005, the Ministry of Forestry officially switched to a 'participatory' approach.

Following the decree, provision was made to implement co-management plans in national parks in Indonesia. The government, however, retains ownership rights and overall authority. Most importantly, communities can still be ordered to leave the land at any time if it is seen to be in the country's best interests. The co-management approach assumes that equal power is held by the co-managers and each is respected equally within the process (Kubo, 2008). However, reality has already proven that this is often not the case within Indonesian national parks.

A pilot project for co-management of the Gunung Halimun Salak National Park was set up in 2005, in order to experiment with the local capacity for park management (Kubo, 2008). A 'team' of national park officers and donor-agency representatives was asked to determine the best course of action for gaining the trust of the local people. It may have been easier to accomplish this if representatives of the local communities had been included on the co-management 'team', but they were not. Kubo (2008, p90) later pointed out that 'without critical reflection, there is a risk that the co-management process simply co-opts rural people for [the] state bureaucracy'.

In addition, a ministerial decree in 2006 amended park zoning to allow for a zone in which livelihoods based on customary tenure were recognized. Park officers were to cooperate with and educate local communities about the change in zoning as part of a participatory approach. In reality, the programme was understaffed and underfinanced and park officers found it difficult to change their approach to training, which was built on the colonial legacy (Kubo and Supriyanto, 2010). Nevertheless, the participatory approach is gaining support, but it is still very much about 'training' the local people to 'understand' the importance of conservation in the forests surrounding their newly tolerated swiddens, rice paddies and forest gardens.

Most recently, the Indonesian government has attempted to prepare for the implementation of MK35 by hosting a number of indigenous-community forums to discuss land conflicts. The meetings, organized by the National Commission on Human Rights (*Komisi Nasional Hak Asasi Manusia*), were held from September to December 2014 on the islands of Bali, Sumatra, Sulewasi, Kalimantan, Maluku and Java. A culmination meeting in Jakarta at the end of the process highlighted gender issues that had been raised in the earlier forums. The meeting in Java was held in Rangkas Bitung, the capital of Banten province. The focus was primarily on the Kaesepuhan and Baduy communities and their land disputes with the Gunung Halimun Salak National Park office and the Ministry of Forestry. Forest resources and swidden cultivation were important topics for the Kaesepuhan, with one participant remarking that he was 'tired of being made to feel like a thief on his own land'. The government stated that the meetings would form the basis for its future implementation of MK35 in all parts of Indonesia, although all actors remain confused about how MK35 will apply to national parks and other conservation zones.

The Kaesepuhan and forests in Indonesia

The Kaesepuhan community claims it can trace its history in West Java back for nearly 700 years. In that time it has existed with only limited contact with the outside world. The region's status as a protected area over the past century has limited the growth of infrastructure and led to a lack of public transport and a dearth of stores, traders and markets. Aspects of *adat* have also contributed to the Kaesepuhan's self-imposed isolation. For example, community members are not allowed to buy or sell rice, so markets have been slow to arrive. However, isolation is relative and there are community members who argue that the Kaesepuhan have never been all that isolated because of gold mining, tourism, logging, plantation agriculture and trade networks.

Throughout the centuries, the Kaesepuhan have relied on swidden rice production, although they have also grown irrigated rice for more than one hundred years. They plant and harvest only once in every year, whether the crop is in swiddens or irrigated paddies. Spiritually, the two methods of rice production are very different. Both are controlled by *adat* ceremonies, but the only connection with the ancestors is made through swidden production. In fact, the Kaesepuhan resisted irrigated-rice cultivation up until the late 1800s. One factor contributing to its eventual adoption was the impact from the colonial cultivation system in the mid-1800s that forced the native population to devote a certain percentage of their land or

Solanum americanum Mill.
[Solanaceae]

This wide-ranging nightshade species is called *leunca* by the Kaesepuhan and is grown along with rice in their swiddens.

labour to growing export crops. This system forced the conversion of large swathes of Kaesepuhan land into state-owned plantations. The community responded by slowly converting small areas to irrigated-rice production, which generally produces higher grain yields than swidden production and requires less land.

Of the four Kaesepuhan villages involved in research for this chapter, all had some residents who still practised swidden agriculture, although it was a much more vibrant practice in some villages than in others. Those villages that were closer to market access, main roads, National Park offices or gold mining operations tended to have fewer swidden farmers. However, the practice was particularly strong in Ciptagelar, where residents proclaimed that wet-rice agriculture was for production purposes and was introduced by the Dutch, whereas rice grown in a *huma* (swidden) was for *adat* and for 'life'; it was how the Kaesepuhan had traditionally grown their rice. Indeed, those Kaesepuhan communities that still practised swidden farming were those mostly closely linked with *adat* power and perseverance.

Kaesepuhan swidden agriculture usually occurs on a seven-month cycle, with planting in October and harvest in April or May, depending on the rice variety. It follows the cycle of *nyacar, ngaduruk, ngaseuk, ngoreddanpanen* (cleaning, burning, planting, weeding and harvesting). These names are not merely agricultural activities within the annual cycle, but are synonymous with *adat* ceremonies that are performed at each stage. No activity in the field, garden or forest can take place without the proper ceremony. For example, during the *ngaseuk* stage of the cycle, *Abah* Ugi and members of the community will perform ceremonies that seek permission from the ancestors to proceed with planting, give thanks for the use of the land and seek guidance about planting times, efforts and locations.

On the day of planting, the main ceremony involves the community's spiritual leader covering his head with a white cloth and meditating and praying over the basket or baskets of seeds to be planted. The basket of seeds sits inside a square of bamboo slats, covered by a white cloth. A small triangle of woven palm fronds is staked into the ground nearby. The basket of seeds also holds a variety of ritual objects, including a small mirror and a comb for *Dewi Sri*, the rice goddess; betel nut packets; an antique comb; incense; a small ceremonial dagger; and essential oils. The spiritual leader chews betel nut during the process and then sprays the juice on to his open palms. He then presses them down onto the seeds in the basket. An *adat* elder then analyses the spread of seeds stuck to the spiritual leader's hands and chooses the two that are most similar in position, size and shape. These are then placed in a small white cloth to be buried in the earth beneath the basket of seeds.

The ceremony takes place in front of all participating community members, who will later help to plant the seeds in the swidden. It will be replicated at household-level by Kaesepuhan families throughout the following few weeks. Fields are generally planted on the day of the week on which a prominent household member was born. At the end of the *adat* ceremony, community members gather around the ceremonial baskets of seeds to take handfuls for planting. Men usually walk first around the swidden plot with dibble sticks, while the women follow behind, planting three or four seeds in each hole poked into the soil by their counterparts. However, these roles are not static and depend on the availability of labour. Some women will help with dibble sticks and some men (usually younger men or adolescents) will help to plant the seeds.

Some of the women also plant a variety of other seeds and small seedlings to be intercropped with the swidden rice. The swiddens of the Kaesepuhan are often intercropped with chillies, garlic, beans, *leunca* (*Solanum americanum* Mill.), terubuk (*Saccharum spontaneum* var. *edulis* (Hassk.) K. Schum.) and shallots among the rice. In this way, the biodiversity in the swidden

Saccharum spontaneum var. *edulis* (Hassk.) K. Schum. [Poaceae]

Called *terubuk* by the Kaesepuhan, the unopened flowering heads of this close relative of sugar cane are a popular vegetable.

fields is often quite high, compared to the rice monocrops in paddy cultivation (Styger et al., 2007). Small saplings of various tree species are also left around the swidden to start its regrowth into secondary forest once cropping ends in a few years. Larger trees are harvested for building materials and firewood in the early stages of field preparation.

With regard to land tenure – a vital issue in the implementation of MK35 – there are two general swidden categories. First, there are 'community' swiddens that are worked by communal labour for the benefit of the *Abah*, his family, the workers and guests. Then there are family plots that are worked by family units for their own benefit. These are often on land to which community members have use rights. Being located in a national park means there are no official certificates or ownership rights, either communal or individual, but most community members hold rights to use plots of land. These rights are only occasionally disputed as the *Abah* and *adat* leaders allocate all rights. Families sometimes trade and sell use–rights letters that detail a plot's location, but these trades must still be approved by the *Abah* and the *Baris Kolat* (*adat* leaders).

The *Abah* and members of the *Baris Kolat* have taken part in regional- and national-level discussions concerning their land-use rights in the National Park area, including the government's MK35 meetings. However, these activities are often focused on the use rights in general, rather than on the activities taking place on the land where the use rights are invoked. Another common topic of community discussion is forest use and the gathering of forest resources. Some people are more forthcoming than others about their use of the land within the boundaries of the National Park. For example, a community member may deny both the existence of swidden farming and any knowledge of swidden practices by other members of the community, but then will be seen taking part in *adat* ceremonies for swidden farming. Other members will openly discuss the preference of park officers for irrigated rice agriculture, but confess to be puzzled by the preference, because in their understanding the two practices have both benefits and drawbacks for the environment. Throughout the research, swidden agriculture was a common practice and community members would openly discuss its connection and importance to *adat*. However, it was rarely discussed in relation to land-use policies, the National Park's zones or rules, or MK35. This could leave the Kaesepuhan at a severe disadvantage in coming negotiations on use and ownership of the land if it is given over to community ownership through the MK35 decision, even though swidden farming is not included as an 'approved' use of the land.

Conclusions

In this chapter I have presented the historical connections between Indonesian national policies and *adat* (customary law), conservation and land use in order to highlight the practise of swidden farming by indigenous communities. Primarily, I have focused on the recent Constitutional Court ruling known as MK35, as implementation into statutory law seems imminent and could significantly impact the swidden practices of many communities. As Indonesia's historical policy trajectory shows, statutory laws have long influenced the land tenure and farming practices of the Kaesepuhan and it seems reasonable to conclude that implementation of MK35 will continue that trend.

This study has also shown that swidden agriculture is an important component of the Kaesepuhan's food security, land-use cycle and forest resource management. Swidden practices, ceremonies and land use are integral components of the customary laws of Kaesepuhan communities and are impossible to separate from the cycles and customs dictating their everyday life. These practices must be recognized in the process of implementing MK35 in the territories occupied by the Kaesepuhan if there is to be an effective transition to greater communal land holdings for indigenous groups in Indonesia. The Kaesepuhan are not alone in their swidden practices, and while this chapter does not intend to speak for indigenous groups generally, it does seek to raise awareness of swidden farming practices in order to encourage the incorporation of these practices into the implementation of MK35 in coming years.

Rather than placing the burden of inclusion on the Indonesian government, I contend that indigenous communities must work on both national and regional levels to ensure that locally relevant land-use policies are incorporated into the implementation of MK35. For the Kaesepuhan community of West Java, that includes swidden rice production. By acknowledging the practice and using the many scientific studies that support the appropriateness of swiddening in tropical climates, the Kaesepuhan have an opportunity to raise awareness of local land-use traditions and potentially influence the formulation of statutory laws related to MK35, as well as other legislation affecting *adat* communities in Indonesia.

Acknowledgements

Thanks to Max Pfeffer and Carol Colfer for their excellent guidance and analysis throughout this project. Thank you to Matthew Minarchek for careful editing and revision suggestions and thanks also to Malcolm Cairns for the opportunity to contribute to this volume. Of course, I owe gratitude to the civil-society organizations within Indonesia that allowed me to document their activities and discuss my research, especially RMI (Bogor), which organized a preliminary presentation of this chapter and provided feedback. Also, thanks to all of the Kaesepuhan community members who took part in my research with such excitement that they often excited me in the process! Finally, thank you to the Fulbright-Hays Doctoral Dissertation Research program, the American Institute for Indonesian Studies and the Borlaug Graduate Research fellowship program, all of which provided funding to support this research.

References

Bernstein, H. (2010) *Class Dynamics of Agrarian Change*, Fernwood, Halifax, Nova Scotia

Boomgaard, P. (2007) *Southeast Asia: An Environmental History*, ABC Clio, Santa Barbara, CA

Conklin, H. C. (1957) *Hanunóo Agriculture: A Report on an Integral System of Shifting Cultivation in the Philippines*, Food and Agriculture Organization of the United Nations, Rome

Constitutional Court of Indonesia (2013) Ruling 35/PUU-X/2012, Constitutional Court, Jakarta

Cramb, R. A., Colfer, C., Dressler, W., Pinkaew Laungaramsri, Quang Trang Le, Mulyoutami, E., Peluso, N. and Wadley, R. (2009) 'Swidden transformations and rural livelihoods in Southeast Asia', *Human Ecology* 37, pp323-346

de Jong, W. (1997) 'Developing swidden agriculture and the threat of biodiversity loss', *Agriculture, Ecosystems and Environment* 62, pp187-197.

Djalins, U. (2012) 'Subject, lawmaking and land rights: Agrarian regime and state formation in late-colonial Netherlands East Indies', PhD dissertation to Cornell University, Ithaca, NY

Dove, M. R. (1983) 'Theories of swidden agriculture and the political economy of ignorance', *Agroforestry Systems* 1, pp85-99

Duncan, C. (2004) 'From development to empowerment: Changing Indonesian government policies toward indigenous minorities', in C. R. Duncan (ed.) *Civilizing the Margins: Southeast Asian Government Policies for the Development of Minorities*, Cornell University Press, Ithaca, NY

Galudra, G., Nurhawan, R., Aprianto, A., Sunarya, Y and Engkus (2008) *The Last Remnants of Mega Biodiversity in West Java and Banten: An In-depth Exploration of RaTA (Rapid Land Tenure Assessment) in Mount Halimun-Salak National Park, Indonesia*, Working paper no.69, Center for International Forestry Research (CIFOR), Bogor, Indonesia

Geertz, C. (1963) *Agricultural Involution: The Process of Ecological Change in Indonesia*, University of California Press, Berkeley, CA

Geisler, C. (2003) 'A new kind of trouble: Evictions in Eden', *International Social Science Journal* 55 (1)

Government of Indonesia (1999) *Regulation No. 41/1999 regarding Forestry*, Chapter 1, Part 1, Article 1, Number 6, Government of Indonesia, Jakarta

Hall, D., Hirsch, P. and Li, T. (2011) *Powers of Exclusion: Land Dilemmas in Southeast Asia*, University of Hawai'i Press, Honolulu

Harada, K. (2003) 'Dependency of local people on the forests of Gunung Halimun National Park, West Java, Indonesia', *Tropics* 13(3), pp161-185

Harada, K. (2004) 'Attitudes of local people towards conservation and Gunung Halimun National Park in West Java, Indonesia', *Journal of Forest Research* 8(4), pp271-282

Kubo, H. (2008) 'Diffusion of policy discourse into rural spheres through co-management of state forestlands: Two cases from West Java, Indonesia', *Environmental Management* 42(1), pp80-92

Kubo, H. and Supriyanto, B.(2010) "From fence-and-fine to participatory conservation: Mechanisms of transformation in conservation governance at the Gunung Halimun-Salak National Park, Indonesia', *Biodiversity and Conservation* 19(6), pp1785-1803

Li, T. M. (2003) '*Masyarakat adat*, difference, and the limits of recognition in Indonesia's forest zone', in D. Moore, J. Kosek and A. Pandian (eds), *Race, Nature and the Politics of Difference*, Duke University Press, Durham, NC, pp380-406

Lucas, A. (1992) 'Land disputes in Indonesia: Some current perspectives', *Indonesia* 53, pp79-92

Mertz, O., Wadley, R., Nielsen, U., Bruun, T. B., Colfer, C. J. P., de Neergaard, A., Jepsen, M., Martinussen, T., Zhao, Q., Nowed, G. and Magid, J. (2008) 'A fresh look at shifting cultivation: Fallow length an uncertain indicator of productivity', *Agricultural Systems* 96(1-3), pp75-84

Mulyana, A., Moeliono, M., Minnigh, P., Indriatmoko, Y. and Limberg, G. (2010) *Establishing Special Use Zones in National Parks: Can it Break the Conservation Deadlock in Indonesia?*, Brief No. 10, Center for International Forestry Research (CIFOR), Bogor, Indonesia

Peluso, N. L. (1992) *Rich Forests, Poor People: Resource Control and Resistance in Java*, University of California Press, Berkeley, CA

RRI (2015) *Protected Areas and the Land Rights of Indigenous Peoples and Local Communities: Current Issues and Future Agenda*, Rights and Resources Initiative, Washington, DC

Styger, E., Rakotondramasy, H., Pfeffer, M. J., Fernandes, E. and Bates, D. (2007) 'Influence of slash-and-burn farming practices on fallow succession and land degradation in the rainforest region of Madagascar', *Agriculture, Ecosystems and Environment* 119, pp257-269

Suganda, K. U. (2009) 'The Ciptagelar Kasepuhan indigenous community, West Java', in *Forests for the Future: Indigenous Forest Management in a Changing World*, Indigenous Peoples' Alliance of the Archipelago (AMAN), Bogor, Indonesia, http://www.downtoearth-indonesia.org/story/forests-future-indigenous-forest-managament-changing-world, accessed 20 July 2015

Szczepanski, K. (2002) 'Land policy and *adat* law in Indonesia's forests', *Pacific Rim Law and Policy Journal* 11(1), pp231-255

Takahashi, S. (2006) 'Field-based research and biodiversity conservation: Review and prospects of the biodiversity conservation project in Indonesia', *Tropics* 15(3), pp259-265

Notes

1. Timelines vary according to those asked, but primary actors within civil society organizations expect a law to be in place within one to two years.
2. I use the term 'swidden' to describe the agricultural system employed by my case-study population as it most closely aligns with their processes, which include annual burning and shifting to new fields every three to five years.
3. Kaesepuhan Cisitu is a part of the same network of Kaesepuhan communities as that featured in this study. Focus groups and interviews were conducted in Cisitu, but are not a part of this chapter.
4. *Adat* is a Bahasa Indonesia term that can be translated as 'customary law'. It is also used to refer to communities that still follow customary law, rather than, or along with, statutory law. Indigenous communities in the country may also be referred to as *masyarakat adat*.

D. From the merger of national and local

If the communal land-tenure systems of shifting cultivators cannot be mapped and codified into national laws and legal systems, then fallowed lands will continue to be at great risk from 'land grabbers'. A system of cadastral registration of shifting cultivators' communal lands is needed as a step towards providing title to these lands. Pictured here, an Angami Naga research assistant works with a village leader in Khonoma, Nagaland, to mark a waypoint on her global positioning system in the course of mapping Khonoma's swidden blocks.

Sketch based on a photo by Malcolm Cairns.

48

EXISTING VILLAGE REGULATIONS FOR NATURAL RESOURCE MANAGEMENT

A key entry point for community participation in sustainable management

Peter Hoare[*]

Introduction

The Upper Nan Watershed Management Project was launched in 1997. It set out to achieve sustainable management of natural resources in an area of 1007 square kilometres of mountainous forested land that is one of Thailand's most important watersheds. The project area had 45 villages of six ethnic-minority groups, with a population of 21,561, and its objective was to develop processes for cooperation between these communities and government agencies in order to institutionalize cooperation for improved natural resource management.

One of the earliest tasks was a baseline survey conducted by 15 Community Coordinators employed by the project. They made an extraordinary discovery: all 45 villages not only had their own village rules and regulations, developed over generations of social interaction, but these also covered natural resource management and were in general agreement with project objectives and government policy (DANIDA, 2003). Therefore, the existing village regulations provided a logical starting point for improved community and government cooperation in the sustainable management of natural resources.

---

[*] PETER HOARE, Advisor, Faculty of Agriculture, Chiang Mai University, Chiang Mai. Former team leader and researcher on agricultural extension methods for the Thai-Australian Highland Agricultural Project, 1975 to 1980; consultant on extension methods for an Australian Development Assistance Bureau preparation mission for technical assistance to the Thai-World Bank Highland Agricultural and Social Development Project (HASD), 1981; Extension trainer, HASD project, 1981 to 1985; short-term consultant on extension training and smallholder tea with the Thai-German Highland Agricultural Project, 1986; Research Fellow. East-West Center, Honolulu, on agroforestry in the Asia-Pacific Region, 1986; Chief technical advisor for the Thai-Danish Upper Nan Watershed Management Project from 1996 to 2003, with the objective of improved government and community cooperation for sustainable highland natural resource management.

This chapter examines the detail of village rules and regulations listed by communities of shifting cultivators in the Upper Nan Watershed Management Project area. This will illustrate the suitability of customary laws as an entry point for achieving broader national goals in the management of natural resources. Such an approach does have international precedent. In Nepal, customary village regulations have been used in all community forestry arrangements when establishing collective action on natural resource management, especially when linked to rights to resources (Andersen, 2011). However, the Independent Evaluation Group of the World Bank reported limited success in a high proportion of development projects in Nepal because they suffered from time and cost overruns, and government commitment at the policy and implementation level was inadequate (IEG, 2012).

Thailand's eighth National Development Plan (1997-2001)

In the 1990s, the focus of highland development assistance in northern Thailand moved from agriculture to improved natural resource management. The Local Administration Act of 1993 empowered subdistrict (*tambon*) administrations in natural resource management, and Thailand's Constitution of 1997 encouraged and supported local participation in development. The eighth National Economic and Social Development Plan (1997-2001) called for 'people-centred development' to replace the previous 'growth-oriented' model and the greater participation of local organizations in the management of natural resources.

The Thai-Danish Upper Nan Watershed Management Project (1997-2003)

The Upper Nan Watershed Management Project (UNWMP) built on participatory processes (Participatory Land-use Planning) pioneered by the Doi Sam Muen Highland Development Project (1987-1994). The Danish government provided development assistance through the Danish Cooperation for Environment and Development (DANCED) agency, which in turn contracted the project to the Danish consulting firm, RAMBOLL. The Thai executing agency was the Royal Forest Department (RFD) through six Watershed Management units.[1] The local administrations of the Tha Wang Pha and Song Kwae districts of Nan province, with eight subdistricts (*tambon*) were partners.

The project area was 1007sq. km (Figure 48-1), and it contained 45 villages with a total population of 21,561, living in 4177 households. They were of northern Thai, Thai Lue, Khamu, H'tin, Hmong and Yao ethnic groups. The project area covered the vital Nan river watershed, which provides more than half of the irrigation water for the 'Rice Bowl' on Thailand's Central Plains. The project's development objective was the sustained management of natural resources through community and government cooperation. Immediate objectives included livelihood enhancement through credit provided for village-managed projects, a reduction of 35% in the area under shifting cultivation over seven years, and a reduction of 50% in the area burnt every year

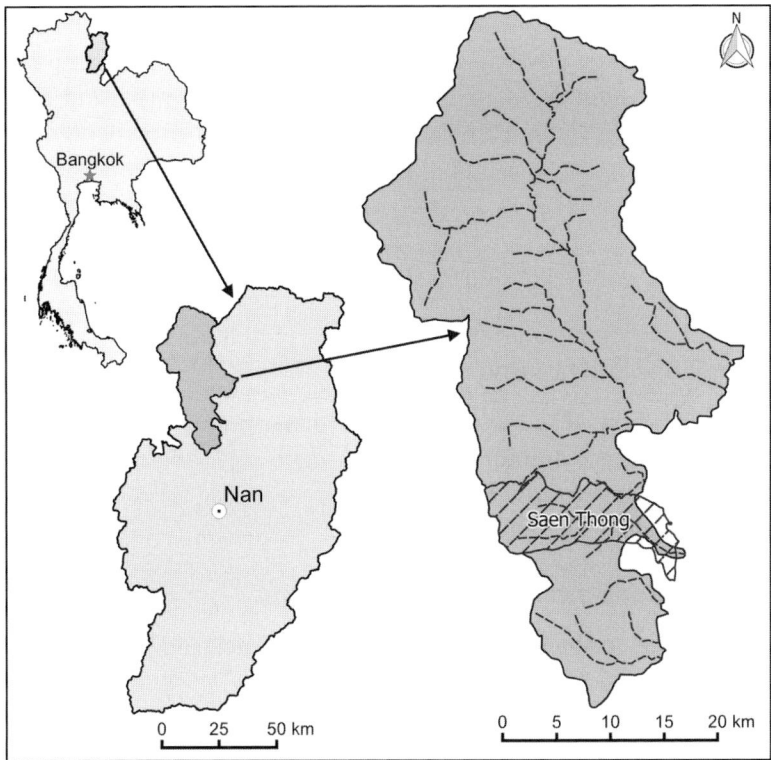

FIGURE 48-1: Locations of Nan province and the Upper Nan Watershed Management Project, showing the Saen Thong subdistrict, which was the first to institutionalize its local Nam Rim Village Watershed Network (below).

within the project area. With the establishment of effective fire management, the project's forestry objectives were an increase of 16% in the forest area by enrichment planting of 156 trees per hectare by the RFD and an additional increase by natural forest regeneration on a further 16% of the project area.

Discovery of existing village rules and regulations

The project employed 15 so-called Community Coordinators, and this group began its duties in 1997 by conducting a baseline survey of the communities with which the project would be involved. Seventeen of the project villages were northern Thai and Thai Lue and the other 28 were Khamu, Lue, H'tin, Yao and Hmong. Twenty seven of the villages were located in the Pa Nam Yao and Pa Nam Suad National Forest Reserves, but most of them had been established before the reserves were declared. One village had been established for just three years, but most of those that could be regarded as 'recent' had been in existence for about 20 years. A northern Thai village was the oldest of the communities, having been established for more than 300 years. About 60% of the farmers in the project area had no legal land tenure, but practised shifting cultivation. It was soon discovered that all 45 villages had their own rules and regulations governing social behaviour. These had been passed down from their ancestors and modified and adjusted over many years, and because the entire culture of these people was shaped by their forest environment, many of the regulations related directly to natural resource management. Surprisingly, there was often a common thread to these village regulations, despite the variety of ethnic backgrounds.

The natural resource management issues covered by the regulations in the 45 project villages included:

The use of firearms and hunting

- Shooting is prohibited in the village, with higher fines for government officers, headmen and village committee members.
- Shooting is prohibited at night.
- Hunting of wildlife is prohibited in reserves.

Aquatic resources and fishing in rivers and streams

- Use of explosives, poisons and electric current to catch and kill fish is prohibited.
- Fishing in a fish sanctuary is prohibited.
- Established rules for appointment of a water regulator in an irrigation area.
- Throwing chemical containers and rubbish in rivers or streams is prohibited.

Watershed land use

- Agriculture in designated watersheds is prohibited.
- Agriculture close to streams or rivers is prohibited.
- When farming sloping land, contour-planting of fruit trees or vetiver grass (*Chrysopogon zizanioides*) is required.
- Fields that are not used for two years are reallocated to other farmers.

Forest conservation

- Cutting trees in forest reserves is prohibited.
- Cutting trees for sale to a middleman is prohibited.
- Cutting trees in village cemeteries and community forests is prohibited.
- Transport of wood through the village is prohibited.
- Cutting bamboo shoots and rattan is prohibited.

Forest fire management

- Allowing fires to spread from agricultural land to damage forests is prohibited.
- Burning swiddens without a firebreak is prohibited.
- Burning swiddens without informing the village committee in advance is prohibited.

- Burning for hunting is prohibited.
- Fines are imposed for fireguards who desert their duty.

Overall, about half of the village rules and regulations concerned natural resource management (Table 48-1). Moreover, they were in general agreement with the objectives of the project and government policies concerning natural resource management and land use. For serious offences such as cutting trees in national forest reserves or fishing with explosives, offenders could even be sent by the village committees for prosecution under Thai national forest and fisheries laws. These baseline data showed that using these village rules as an entry point was a logical strategy for reaching the project's objective of sustained management of natural resources through community and government cooperation.

TABLE 48-1: Frequency of similar village rules and sanctions in 45 villages, Upper Nan Watershed Management Project.

Village rule or regulation	Fines in Thai baht $US1=38 baht*	Applied in how many villages out of 45
Cutting trees in forest reserves, community forests.	Bt500 to Bt10,000; court action possible; three times the value of timber cut.	36
Shooting in the village.	For an officer, Bt1000 per shot; headman, Bt500; committee member, Bt300; villager, Bt100.	31
Fishing with explosives, electric current, poison in rivers.	Mostly Bt500 to Bt5000; outside court action possible.	28
Theft.	Mostly Bt300 to Bt5000, part going to the whistle-blower; pig, Bt1000; chicken, Bt300; vegetables and fruit, up to Bt500.	26
Brawling, disturbing the peace.	Bt500 to Bt1500; making a loud noise after 11pm, Bt250.	21
Allowing a swidden fire to spread to damage forest.	Bt500 to Bt5000; pay damages; outside court action possible.	21
Burning a swidden without informing committee in advance.	Bt300 to Bt3000; Bt1000 baht per rai.[#]	16
Allowing domestic animals to damage crops.	Bt50 to Bt500, headman and committee member pays more; Bt1 per plant; offending animals can be shot, with the exception of a pregnant pig.	16

TABLE 48-1 (cont.): Frequency of similar village rules and sanctions in 45 villages, Upper Nan Watershed Management Project.

Village rule or regulation	Fines in Thai baht $US1=38 baht*	Applied in how many villages out of 45
Failing to assist in village development work.	Bt20 to Bt100, headman and committee member pays more; offender can be expelled from village after third failure to assist.	16
Fishing in river conservation zone.	Bt500 to Bt5000; Bt1000 per fish.	16
Failure to attend village meetings.	Bt20 to Bt100, higher fines for headman and committee member; offender can be expelled from village after third failure to attend.	15
Burning a swidden without a firebreak.	Bt500 to Bt5000; Bt2000 per rai.#	14
Selling illegal merchandise or drugs in village.	Bt500 to Bt10,000; five times the value of the merchandise; possible outside court action.	13
Burning forest for hunting.	Bt500 to Bt5000.	9

Notes: # 1 hectare=6.25 rai; *The value of the Thai baht fell dramatically in 1997, from 25.61 baht to the US dollar on 6 January to 38.04 on 20 October and 45.54 on 29 December.

1. The fines were different in all villages. Those listed in the Table 48-1 represent the most common fines. Where more than one fine is mentioned, only one fine applied per village.

2. In 1997, the farm wage was about Bt180 per day, so the fines were significant as they represented from two days' to a month's income for violations of the many natural resource management regulations.

3. The importance given to prohibiting the use of explosives, electric current and poison in fishing in rivers and streams, and the creation of fish sanctuaries, was due to the work of the non-governmental organization Huk Muang Nan, which promoted fish sanctuaries.

Source: UNWMP Baseline Survey, 1997

These rules and regulations could be enforced by village committees if transgressions occurred within their traditional village boundaries. However, existing village regulations for fire management could not be enforced if the fires started outside a village. Fire management plans for risk reduction in the 1998 fire season were prepared by a fire consultant with a staff of six Royal Forest Department Watershed Units and the 45 villages. Videos on the environmental damage caused by uncontrolled forest fires were shown in each village, followed by discussion groups, and volunteer fire guards were trained and equipped with knapsack sprayers and rakes for fire-fighting in each village. Firebreaks were prepared by the RFD to reinforce community firebreaks in order to protect reforestation areas. Despite these measures, analysis of Landsat satellite images showed that in 1998 the area burnt within the project increased from the 1997 baseline of 5% to 21%, with considerable damage to orchards and private property.

Cajanus cajan (L.) Millsp. [Leguminosae]

One of the world's most useful crops, Pigeon pea is a major source of protein, a popular forage and cover crop, a soil-enriching legume, and is recognized in this chapter as a species used in hedgerows for erosion control on sloping land.

Village Watershed Networks strengthen existing natural resource management rules and regulations

In the 1998 fire season many fires spread from neighbouring villages and in these cases the existing village rules shown in Table 48-1 could not be enforced because there was no agreement between neighbouring jurisdictions. The government's forest rules concerning fires were also difficult to enforce. They specified fines of up to 50,000 baht and up to five years in jail, and were counterproductive if used against rural offenders (DNP, 2016). Conflict-resolution meetings between the 45 villages concerning the fire damage resulting from uncontrolled fires were organized by the project's Community Coordinators and the chiefs of the RFD Watershed Management Units. This led to the formation of seven Village Watershed Networks each with six to eight villages. After two or three meetings for the resolution of conflicts over damage inflicted on orchards and properties in the 1998 fire season, the village representatives agreed with the six RFD Watershed Management Units to establish boundaries for fire and natural resource management. These boundaries covered 100% of the project area, including areas where villagers had no legal land tenure, but accepted responsibility for fire management as a trade-off for practising shifting cultivation in national forest reserves and the right to harvest non-timber forest products.

In many villages, households were assigned responsibility by village committees for fire management in all of the micro-watersheds. After the 1998 fire season, everyone knew who was responsible for a fire, even if it did not start on their swidden. Over the following two years, the different laws in each of the six to eight villages in each Village Watershed Network were rationalized in public hearings, and agreement was reached. Table 48-2 shows a summary of this process, beginning with conflict resolution and creating Village Watershed Networks.

After successful implementation and modification of the village rules, the area burnt within the project fell to below 3% for three successive years (Hoare, 2015). In some subdistrict (*tambon*) administrations, the Village Watershed Network rules were institutionalized.

TABLE 48-2: Summary of the process of forming Village Watershed Networks to strengthen and institutionalize village regulations for natural resource management (NRM).

Stakeholders	Description of process	Project inputs
Representatives from 45 villages. Staff of Royal Forest Department Watershed Management Units.	Meetings facilitated by Community Coordinators of the six to eight villages to discuss causes of NRM conflicts. After four to six meetings, agreement is reached between villages and the RFD on boundaries for NRM issues, including fire management. Uniform regulations and sanctions are established.	Community Coordinators are trained in facilitating village meetings. Budget for food and meeting venue.
Representatives of the 4177 households attend public hearings in all project villages.	Public hearings in all villages in the Village Watershed Networks to endorse boundaries, uniform regulations and sanctions for NRM.	3D topographical models, scale 1: 12,500, to facilitate village land-use planning and marking of fire management boundaries (1)
Village Watershed Network (VWN) village representatives. Subdistrict (*tambon*) administrations.	Implementation of the NRM regulations and modification for two to three years, with changes where necessary. Institutionalizing the VWN regulations for NRM under the Subdistrict (*tambon*) Administration Act of 1993.	Regular meetings between VWN representatives and sub-district administration. 3D topographical models to include all villages in each Village Watershed Network, with local administration boundaries.

Notes: (1) The 3D topographic models, created in the context of participatory watershed management, had their origin in the Doi Sam Muen Project (1987 to 1994). The experience in using 3D models as a tool with which to facilitate village land-use planning and achieve agreement on NRM boundaries in the Upper Nan project was documented in *ASEAN Biodiversity* (Hoare et al., 2002).

Village development funds for livelihood enhancement: A factor contributing to improved natural resource management

The availability of credit for livelihood-enhancement ventures was an important component of the Upper Nan project. A total of 244 of these schemes, submitted from all 45 villages, were appraised and implemented, using a grant of 7,467,373 baht provided by DANCED and Danish International Development Assistance (DANIDA) between 1997 and 2000.[2,3] They included paddy-land development, wet-rice yield increase, tree-crop planting, pig raising, fish ponds, handicrafts and other income-generating projects. The funds were deposited in bank accounts controlled by Village Development Committees. Training in bookkeeping and the management of the funds was provided and the accounts were audited annually.

A survey in 2000, after three years of project implementation, showed that farmers who used the village development funds for paddy-land development had reduced the area of land they had under shifting cultivation by 27%. Those who planted fruit trees reduced their swiddens by 40% and those undertaking other income-generating projects by 36%. Land-use surveys showed that the most distant swiddens were the first to be abandoned to allow natural forest regeneration in the forest reserves. This also reduced the risk of fire spreading from the distant swiddens to burn areas of forest (UNWMP, 1999a).

Institutionalizing Village Watershed Networks and *tambon* rules on natural resource management

Zanthoxylum limonella (Dennst.) Alston [Rutaceae]

Ma Kwaen – as it is called in northern Thailand – is a 'jungle spice' with increasing popularity, both as a food additive and an alternative crop in swidden fallows, particularly in the study area for this chapter. It is a non-timber tree closely related to the tree that produces Sichuan pepper.

Over a period of two years, the seven Village Watershed Networks covering the project area discussed and adopted rules and regulations related to their functions and jurisdiction. Each of them achieved this in the course of eight regular meetings with subdistrict (*tambon*) administrations in the districts of Tha Wang Pha and Song Kwae, in Nan province. The Saen Thong subdistrict administration was the first, in 1999, to institutionalize the local Nam Rim Village Watershed Network of eight villages. The rules related to natural resource management adopted by Tambon Saen Thong are listed below:

> With provision under Section 67 (7) of the Tambon Administration Act (1993), the National Environmental Quality Enhancement and Maintenance Act (1991), and Section 14 of the National Reserved Forests Act (1963), the committee of the Tambon Administration Organization Council of Saen Thong and the Tha Wang Pha District Office proclaim the Tambon Saen Thong rules, as follows:
>
> 1. This set of rules shall be called the 'Tambon Rules' on natural resource management in Saen Thong Tambon Administration and Organization Watershed Network.
>
> 2. This set of rules shall be applied within the jurisdiction of the Tambon Administration Organization of Saen Tong and shall come into force 10 days after the announcement is posted in the public area of the Office of the Tambon Administration Organization.
>
> 3. In this act of Tambon rules 'the committee' means the Network Committee of Tambon Saen Thong's Administration Organization.
>
> 4. Natural resources means forest, wildlife, aquatic resources, soils, rocks, minerals. The Network Committee of the Tambon Administration of Saen Thong shall have a direct duty to take responsibility for the said resources together in the form of a committee.

5. The Chairman of the Committee shall coordinate works amongst tambons, government agencies and other involved agencies.

6. Land use matters

6.1   The Committee shall oversee the land for farming and monitor the results of land use.

6.2   Soil erosion control systems must be established for sloping land by means of conservation practices like hedgerow planting of *Leucaena*, vetiver grass or pigeon pea.

6.3   Clearing land for farming in watershed areas is prohibited. Violators shall be penalized with a fine of no more than 500 baht per rai [0.16ha] and the cleared area confiscated. The violator shall be sent for legal procedures under the stipulation of section 14 of the National Forest Act (1963) and section 100 of the National Environment Quality Enhancement and Maintenance Act (1991).

7. Forest fire control

7.1   Burning to hunt wildlife is prohibited. Violators shall be fined 500 baht per rai. In the case where damage occurs to an individual's property, farm field, or orchard, the owner of such properties shall be the one who seeks compensation and sends the offender for legal procedures as provided by Section 14 of the National Reserved Forests Act (1963).

7.2   If field burning will be practiced the Committee must be informed at least five days before burning.

7.3   Before burning a field a firebreak must be made every time, at least eight metres wide.

7.4   If fire spreads into other areas the fire lighter must take responsibility.

7.5   If a fire takes place and no guilty party can be identified, the clearer of that field for farming must assume all responsibilities and proceed to find the culprit for further legal action.

7.6   Whoever finds a person burning forest shall inform the Committee. The informant shall receive 50% of what has been paid as a fine and the name of informant shall not be disclosed.

8. Control of chemical use

8.1   Chemical containers after being emptied must be carefully and entirely buried underground.

8.2   Cleaning or disposal of chemical containers in the neighbourhood of, or in a water source is prohibited. Violators shall be fined from 100 to 500 baht per bottle.

9. Control over timber utilization

9.1   Processing of wood outside the village is prohibited. If the committee finds the case the village shall receive a cut in budget or receive no consideration for its budget. In the case where the village committee informs the Network Committee or officer, the budget shall not be cut. In the case where the Village Committee neglects to inform the Committee or officer, the village budget shall be cut or receive no consideration.

9.2 Cutting wood and damaging forest in conservation zones of the village and Tambon are prohibited. Violators shall be subject to penalty according to village laws and shall be sent for legal procedures as provided by section 14 of the National Reserved Forests Act (1963) and section 100 of the National Environmental Quality Enhancement Act.

10. Control over wildlife hunting and aquatic life harvesting

10.1 Fishing and hunting in a reserved zone is prohibited, particularly for protected species. The violator shall be fined 500 baht and sent for legal action according to the National Reserved Forests Act (1963). No one shall hunt animals in other village's territory. The violator shall be sent for legal action according to village rules.

10.2 No one shall harvest fish by means of electrification, poisoning, and explosive in a water source. The violator shall be fined 500 baht per time per person and sent for legal action under the stipulations of Fishery Laws.

11. Villagers in the Tambon Administration Organization of Saen Thong Watershed Network can report the incidence of violation of forest, soil and water destruction to the Committee and officer of the Tambon Administration Organization of Saen Thong Watershed Network.

12. The Tambon Administrative Organization of Saen Thong Watershed Network Committee shall meet 12 times a year to follow up on work performed by the Network Committee.

13. This set of rules shall receive revision, addition, and modification to fit the existing circumstances and conditions.

14. The Tambon Administration Organization of the Saen Thong Watershed Network Committee shall safeguard this set of village rules

Chairman of the Committee.........................Date..................

Source: UNWMP (1999b).

Important points arising from the Saen Thong *Tambon* Administration Organization's rules on natural resource management

1. Existing village regulations are recognized in control over timber utilization and aquatic resources, where violators are subject to penalties according to village rules and may also be prosecuted under government forest, environment and fishery laws.

2. The offence of illegally cutting trees, if not reported to the Village Watershed Network committee, can result in punitive action against the village by cutting its budget. The violator(s) can also be sent for prosecution under national forest and environment laws.

3. The Saen Thong Watershed Network Committee meets 12 times a year to follow up on work and issues.

4. The *Tambon* fines of 500 baht and 500 baht per rai are much less than those under most of the villages' original regulations. However, for serious offences, violators can be sent for prosecution under government laws.

Following the formation of Village Watershed Networks and agreement on boundaries for fire management, the area burnt per year fell in the years from 1999 to 2003 to less than 3% of the project area. With effective fire management, the objectives of natural forest regeneration and enrichment planting on 32% of the project area were achieved (Hoare, 2015).

National level endorsement of the local rules

At a meeting with the Tha Wang Pha district staff to mark the completion the first phase of the Upper Nan Watershed Management Project, one of the Royal Forest Department's Watershed Management chiefs asked his boss, the RFD's Deputy Director: 'We now have two sets of rules for fire management: one from the RFD at the national level and now the local subdistrict approved Village Watershed Network rules. Which ones should we use?' He replied: 'Use the local rules. But if a village resident does not abide by those rules, then use the stronger national forest rules.'

This exchange showed how the project had succeeded in strengthening cooperation between government agencies and highland communities for improved participation in natural resource management. The villagers' rules and regulations for natural resource management, institutionalized by the subdistrict administration, were recognized by the Royal Forest Department as being more enforceable. In a recent communication, the former Director of Danish Cooperation for Environment and Development wrote: 'I still refer to the Nan Project as a special DANCED success' (Dhyr-Nielsen, 2014).

Chrysopogon zizanioides (L.)
Roberty [Poaceae]

Vetiver grass is famous for its prodigious roots. Instead of spreading, they grow down as far as four metres in the course of one year, making this tough species a valuable defence against erosion. Local rules in the study area require the planting of species such as Vetiver along contours on sloping land.

Conclusions

Development projects should henceforth include the documentation of village rules and regulations in a baseline survey as these may provide entry points for community participation in line with project objectives. To recap the Upper Nan project's experiences in this regard, all 45 villages of six ethnic groups in the project area had village rules and regulations that provided for substantial fines for violators. About half of these rules and regulations concerned natural resource management and were in general agreement with government policy and the development objectives of the project. Under the local rules, serious offenders could be sent for prosecution under Thai law.

The design of the Upper Nan Watershed Management Project was appropriate for a merger of national and local, to achieve the project's development objective of sustainable natural resource management through cooperation between highland communities and government agencies.

Acknowledgements

I would like to acknowledge the participation of the Royal Forest Department's Project Director, Borpit Maneeratana, his six Watershed Management Chiefs, and the 15 Community Coordinators and training staff, all of whom accepted that improved natural resource management and achievement of the project's forestry objectives could only be achieved with the participation of farmers living in the forest. I am grateful for the unwavering support of the Director of Danish Cooperation for Environment and Development, Dr Mogens Dyhr-Nielsen and his staff, and for the support of Kirsten Ewers Andersen of RAMBOLL during the testing times of project implementation.

I would also like to acknowledge the encouragement of the Editor of this volume, Dr Malcolm Cairns, who wrote to me: 'you are the only one who has pointed out that villagers have themselves traditionally enforced their own sets of regulations, and they would provide a logical starting point in trying to develop a national policy' (Cairns, 2015).

References

Andersen, K. E. (2011) *Communal Tenure and Governance of Common Property Resources*, Land tenure working paper number 20, Food and Agriculture Organization of the United Nations, Rome

Cairns, M. F. (2015) personal communication between the author and Dr Malcolm Cairns, the editor of this volume

DANIDA (2003) *Lessons learned from the Upper Nan Watershed Management Project 1996-2003*, Thai language with English-language summaries, Danish International Development Assistance, Copenhagen

DNP (2016) *Policies and Legislation*, Forest Fire Control Division, Department of National Parks, Wildlife, and Plant Conservation, Bangkok

Dhyr-Nielsen, M. (2014) personal communication between the author and the Director of Danish Cooperation for Environment and Development concerning the Upper Nan Watershed Management Project

Hoare, P. (2015) 'Community networking for improved highland-management', paper presented at the First International Conference on Highland Natural Resource Management, Faculty of Agriculture, Chiang Mai University, Thailand

Hoare, P., Maneeratana, B., Songwadhana, W., Suwanmanee, A. and Sricharoen, Y. (2002) 'Relief models, a multipurpose tool for improved natural resource management', *ASEAN Biodiversity*, January–March 2002, Department of Environment and Natural Resources, Philippines, pp11-16

IEG (2012) *Natural Resource Management in Nepal: An Analysis of Project, Policies and Institutional Reforms*, Independent Evaluation Group, World Bank, Washington, DC

UNWMP (1997) *Upper Nan Watershed Management Project: Village Rules and Regulations for 45 Villages in six Watershed Management Units* (English translation), Royal Forest Department, Nan, Thailand

UNWMP (1999a) *Completion Report Executive Summary Phase I*, Upper Nan Watershed Management Project, Danish Cooperation for Environment and Development and Royal Forest Department, Nan, Thailand

UNWMP (1999b) *Village Watershed Network Rules and Regulations, Upper Nan Watershed Management Project* (English translation), Danish Cooperation for Environment and Development and Royal Forest Department, Nan, Thailand

Notes

1. The Royal Forest Department has since been renamed the Department of National Parks, Wildlife and Plant Conservation (DNP).
2. The Thai baht-US dollar exchange rate in this period averaged about 38.69 baht to the dollar, making this sum equivalent to about US$193,000.
3. Danish Cooperation for Environment and Development (DANCED) was created in 1944 following the UN Rio de Janiero Environment Conference as part of Denmark's assistance to developing countries under the Ministry of Environment. In 1999, existing DANCED projects were incorporated into the Danish International Development Agency (DANIDA), of the Department of Foreign Affairs.

49

POLICIES THAT TRANSFORM SHIFTING CULTIVATION PRACTICES

Linking multi-stakeholder and participatory processes with knowledge and innovations

*Madhav B. Karki**

Introduction

In this chapter, I reflect upon and analyse available information on shifting cultivation in the highlands of Asia, from the perspectives of evolving policy debate, implementation experiences and ongoing local innovations (Inter-Ministerial Task Force, 2008; Grogan et al., 2012; Leduc and Choudhury, 2012). I argue that overall, the legislation, policies and programmes of national governments have either ignored traditional shifting cultivation or harmed or damaged both the farming system and its practitioners. I will give examples of policies that have enabled farmers to innovate and adapt to ongoing socio-economic and environmental changes, as well as citing cases in which interventions that have not worked at all have been imposed on shifting cultivation systems that were already working well. I argue that there is a need for national and subnational policies that are not only focused on shifting cultivation, but are also informed, inspired and influenced by local realities and propose processes that are sufficiently dynamic to accommodate ongoing changes and issues.

Shifting cultivation is a common traditional farming system in the Asian tropics, subtropics and highlands that is socio-culturally, economically and environmentally challenging. It has attracted the attention of policy-makers, researchers and development practitioners around the world, to debate its many facets and dimensions; to laud its virtues or scorn its detractions (Cairns and Garrity, 1999; Erni, 2005; Fox et al., 2009; Leduc and Choudhury, 2012). It is estimated that between 300 and 500 million of the world's farmers – more than 200 million of them in Asia alone – still practise shifting cultivation on about 300 million hectares (ha) of land, most of it forest land (Erni, 2009; Singh et al., 2010; Dev et al., 2013). They are responsible for about 8% of

* Dr Madhav Karki, Advisor, IDS-Nepal; South Asia Chair, IUCN/Commission on Ecosystem Management (CEM); Task Force and Expert Member, Indigenous and Local Knowledge and Policy Guide and Catalogue Support Group, IPBES/UN – Nepal.

global food production (Inter-Ministerial Task Force, 2008). In northeast India, shifting cultivation occupies between six and seven million ha of land, to meet the subsistence needs of poor indigenous communities (Tripathi and Barik, 2003; Inter-Ministerial Task Force, 2008). In the countries of mainland Southeast Asia, between 14 and 34 million people, mostly indigenous communities and poor farmers belonging to ethnic minorities, are believed to be cultivating around five million hectares of swidden land (Fujisaka, 1991; FAO, 1995; Hossain, 2001; Rerkasem et al., 2009a; Dev et al., 2013). Shifting cultivation therefore remains a dominant traditional farming system practised largely by poor, marginalized and indigenous people living on hilly and mountainous slopes and the forested plains of Eastern Himalaya, the Mekong basin, the Southeast Asian highlands and tropical and subtropical regions of Asia-Pacific (Sharma, 1976; Arifin, 1998; Tripathi and Barik, 2003; Rerkasem et al., 2009a).

There is no universally agreed definition of shifting cultivation, but most agree that it has three main elements:

Setaria italica (L.) P. Beauv. [Poaceae]

Foxtail millet is a common subsistence crop grown by shifting cultivators. It is the second most commonly grown millet and has the longest history of cultivation, having been grown in China since sometime in the sixth millennium BC. Reaching maturity in less than 90 days, its efficient use of available water makes it suitable for dry areas.

(1) an alternation between a brief period of cultivation and a relatively long duration of fallow; (2) in most cases, cyclical shifting of fields; and (3) clearing of natural vegetation, normally using fire (FAO, 1995; Erni, 2005; Scott, 2009). Although fallow periods and levels of productivity constantly change, one common characteristic of shifting cultivation is its critical role as a source of subsistence livelihood for a large number of poor and marginalized farmers, although some of the larger practitioners have moved or are moving from being subsistent to early consumer-oriented or even semi-commercial farmers (Brady, 1996; Leduc and Choudhury, 2012). Generally, there are four types of shifting cultivation: (1) traditional swiddening: this normally follows a cycle of clearing the forest with fire, a short period of cultivation, and abandonment to allow forest recovery; (2) rotational agroforestry: farmers are on the path of transformation from traditional practices to different intensification or diversification options; (3) composite shifting cultivation: farmers maintain permanent wet-rice farms in the lowlands and shifting cultivation fields in rainfed

uplands; and (4) temporary farming: usually practised by nomadic hurders and pastoralists who cultivate a patch of land for either one season or a year, then move on (Inter-Ministerial Task Force, 2008; Erni, 2009). Whatever shape or form it takes, shifting cultivation is clearly the livelihood mainstay for millions of poor, vulnerable and marginalized people (Kerkhoff and Sharma, 2006).

Many attempts have been made and initiatives taken to address issues related to both the rights of shifting cultivators and the environmental consequences of their farming. A process of constant ecological and social change has added pressure to these moves (Darlong et al., 2008). Traditional-rights groups have criticized what they see as government apathy towards shifting cultivation issues. Particular criticism has been levelled at policies and laws that compromise or diminish the value of communal land tenure, indigenous forest management, traditional biodiversity conservation, fallow-management systems and the right of individuals to pursue their chosen options, including the alternative of permanent agriculture (Rerkasem and Rerkasem, 1995; Cramb et al., 2009; Fox et al., 2009; Mertz et al., 2009b; Erni, 2009). In recent years, while preparing for implementation of the Reducing Emissions from Deforestation and Forest Degradation (REDD/REDD+) programme, which was introduced by the United Nations Framework Convention on Climate Change to combat greenhouse-gas emissions, many government institutions have pointed to shifting cultivation as one of the drivers of deforestation and degradation (Griffiths, 2008; AIPP/IWGIA, 2011). Supporters of swidden farming reject these claims as 'simply shifting the blame' (Erni, 2009). Nevertheless, the land area under shifting cultivation and the food production from swiddens is generally believed to be decreasing.[1] At the same time, the relevance of the traditional system and its importance in providing livelihoods to poor and indigenous people shows no signs of diminishing (AIPP/IWGIA/IKAP, 2009; Choudhury and Sundriyal, 2003). With rapid demographic, socio-economic and environmental changes occurring all around them, shifting cultivators comprehend the concerns of the global community and are working in virtual isolation to innovate and embrace various changes in their farming systems. The current state of shifting cultivation will therefore be analysed in this chapter and the impact of policy interventions on issues related to traditional vocational and property rights will be discussed. This will refer especially to conditions in Bangladesh, Bhutan, India, Lao PDR and Nepal (Figure 49-1). Policy recommendations will be made on the basis of proven scientific and indigenous knowledge.

When it comes to debate on shifting cultivation practices and policies in Asia, there are basically two distinct schools of thought, or opposing paradigms (Rasul and Thapa, 2003; Inter-Ministerial Task Force, 2008; Leduc and Choudhury, 2012). Supporters and practitioners of shifting cultivation argue that it is a complex and rich mixture of different traditional and culturally ingrained agricultural practices. They say that poor and indigenous communities choose shifting cultivation as a way of life, to earn their livelihoods by managing natural resources in a socially acceptable, environmentally sound and economically rationalized manner (AIPP/IWGIA, 2011; Erni, 2009). Moreover, they stress that traditional shifting cultivation has a

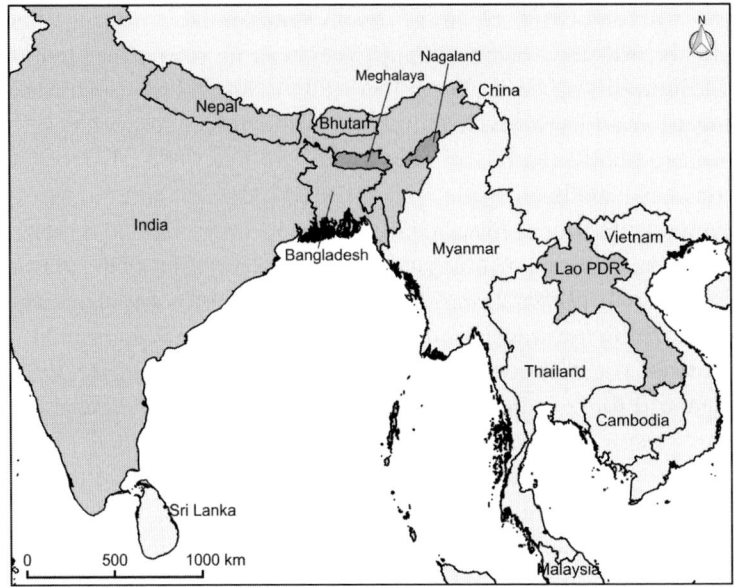

FIGURE 49-1: The countries in which existing conditions for shifting cultivation and the state of property rights are discussed in this chapter.

unique system of access that ensures universal tenurial security to land and, in turn, to food security. Traditional institutions regulate the practice through an equitable governance system managed by village elders or elected community leaders (Inter-Ministerial Task Force, 2008; Leduc and Choudhury, 2012). In fact, the tenurial frameworks of shifting cultivation are sufficiently robust to provide a social safety net to subsistence farmers, and their egalitarian concepts have helped to bind tribal societies together. This has played a powerful role in building strong social capital among shifting cultivation communities (Choudhury, 2012).

Despite its many positive attributes, shifting cultivation has been the target of government decisions to alter the communal land-management practices of upland communities in favour of a command-and-control system of forest-land management or permanent agriculture (Kerkhoff and Erni, 2005). Bureaucrats and professionals generally make sweeping judgements against shifting cultivation, mostly without concrete evidence or information (Laungaramsri, 2005; Kerkhoff and Sharma, 2006). They accuse shifting cultivators of deforestation, degrading forests and causing biodiversity loss by quoting national data on annual loss of forest cover. These antagonists of shifting cultivation argue that it is a primitive and wasteful form of agriculture, the evolution of which dates back to the hunting and gathering era of primitive societies, and claim that its time has now passed (Inter-Ministerial Task Force, 2008). They argue that swidden farmers must gradually move toward 'sedentarized' or permanent agriculture, a move already made by many of their compatriots in the tropical and subtropical highlands of Asia (Rerkasem et al., 2009b). Some countries, such as Bhutan, the Lao PDR, Thailand and Indonesia, have attempted to eradicate shifting cultivation, either by using existing forest laws or issuing new policies and acts that declare shifting cultivation illegal (Romagny, 2004; Namgyel et al., 2008; Erni, 2009; Choudhury, 2012). In the case of the Lao PDR, where the law now bans shifting cultivation, cultivators have been switching to settled agriculture in the same or new locations (Karki et al., 2003; Moizo, 2004; Ducourtieux, 2005; Ducourtieux et al., 2006).

While views supporting shifting cultivation have generally come from ecologists, social scientists and non-governmental and civil-society organizations, negative opinions have been expressed by forest-land administrators and managers and development economists. The Food and Agriculture Organization of the United Nations has termed shifting cultivation 'the greatest obstacle … to the immediate increase of agricultural production', and 'also to the conservation of the production potential for the future' (FAO, 1995). These two viewpoints are fundamentally at odds with each other, and do little to give policy-makers a clear picture of the complex problems involved in finding sustainable solutions for maintaining and improving the livelihoods of more than 200 million Asian farmers currently dependent on shifting cultivation. Therefore, there is an urgent need to find systematically analysed and clearly defined policy options that put the whole debate in a dynamic context, within national and local frameworks and perspectives. Definitely, a one-size-fits-all solution has not worked in the past and will not work in the future, and a country-driven process of policy reform will be necessary to find progressive and flexible solutions. This is the main rationale for seeking to identify a process by which policies can be developed with a clear view to implementation. This is the line this chapter takes.

The changing context of policy debate and dialogue

The historical evolution of shifting cultivation is based on the practice of a primitive system of 'slash-and-burn', or swidden agriculture (Spencer, 1966; Gilruth et al., 1995; Rasul et al., 2004; Garrity, 2007; Singh et al., 2010). Over the years, shifting cultivation has undergone constant changes through innovative practices and evolving knowledge. However, in all these evolutionary processes, policy has existed as a dominant factor in making changes in shifting cultivation practices at local and national levels (Rasul and Thapa, 2003). Over the years, various internally and externally driven innovations have contributed to its gradual change – in both positive and negative directions. These have led to different forms of agroforestry and/or mixed cropping systems, which in some cases have achieved sustainability (van Vliet et al., 2012). Better fallow-management practices in some areas have led to the maintenance of both biodiversity and productivity (Kerkhoff and Erni, 2005; Schmidt-Vogt et al., 2009; Aryal et al., 2010). In other areas, various distortions are cropping up, such as replacing shifting cultivation with plantations, commercial and exotic crops, private-sector managed tourism resorts and even urban settlements, and these disrupt the most important tenets of shifting cultivation – equitable access to land, protection of customary rights of the cultivators and maintenance of minimum fallow periods (Choudhury and Sundriyal, 2003). In some cases, especially where long fallow periods have been maintained and traditional practices followed, shifting cultivation has helped to maintain and improve forest cover and conserve agrobiodiversity. However, where plantation crops have expanded and the size of common lands has shrunk, fallow periods have been diminishing as farmers attempt to produce more crops from less land. In such communities, maintaining household food security and conserving biodiversity has not always been possible (Garrity, 2007; Lawrence and Lam, 2009).

There is widespread recognition that these traditional systems were sustainable, within a certain range of fallow length and demographic conditions (Ramakrishnan, 1992). Proponents of this view argue that these systems were basically sustainable under an extensive system of management. However, as the population dependent on traditional agriculture kept growing, intensification processes began to erode soil nutrients, biodiversity conservation and ecosystem integrity, gradually leading to productivity and biodiversity decline (Gafur, 2001; Gafur et al., 2003; Borggaard et al., 2003). Bahuchet (1990) and Brady (1996) argued that fallow periods of 10 to 20 years were needed to prevent soil erosion, loss of

Alnus nepalensis D. Don [Betulaceae]

One of the most beneficial species associated with shifting cultivation from the eastern Himalayas to the highlands of southwest China. Symbiotic bacteria in root nodules fix nitrogen to rejuvenate depleted or degraded soils, enabling a shorter period of fallow. It is used to stabilize eroding slopes and is interplanted with crops to provide nutrients and shade. It also provides valuable firewood.

fertility, disruption of the water balance and weak forest regeneration. Ramakrishnan (1992) recommended a fallow period of 10 to 15 years in the context of northestern India. However, Choudhury and Sundriyal (2003) proposed that top-down policies seeking to convert land use were more critical in reducing fallow periods than mere increases in population density. Such policies sought to convert shifting cultivation land into protected areas, commercial plantations, and so on, resulting in increased pressures on available land. In some countries, settled agriculture, expansion of human settlements and land acquisition for building roads and power transmission lines, as well as for activities related to national security, have distorted traditional land-allocation systems for shifting cultivation (Spencer, 1966; Choudhury and Sundriyal, 2003). But in other places, population density related to carrying capacity has remained an important variable determining the sustainability of shifting cultivation (Inter-Ministerial Task Force, 2008). An ideal carrying capacity that allows more than 10 years of fallow is agreed by most researchers to be appropriate (Ramakrishnan, 1992), and in general, an optimum carrying capacity varies from 6 to 20 people per sq. km (FAO, 1995; Brown, 2006). Diaw et al. (2008), quoting Jian Ge and Xi Sheng (1976) reported that in ancient China, fallow periods as long as 42 years sustained small settlements of about 300 forest-dependent people. In this case the local policy domain played a major role because land-use decisions were made by local leaders. More recently, the Inter-Ministerial Task Force (2008) recognized that traditional shifting cultivation was sustainable, given 10 to 20 years of fallow and low demographic pressure (20 people per sq. km).

Key policy issues and challenges

When we talk about transforming shifting cultivation into other land-management systems, the critical policy issues include protection of customary property rights or ancestral domains; rights to practice traditional vocations; and social equity and justice in gaining access to common property resources (Aryal and Kerkhoff, 2008). There are many forces exerting pressure on shifting cultivators to change their traditional practices, and national-government agencies are the main ones. However, governments have an obligation to either provide viable alternatives to shifting cultivation or allow them to continue their vocation according to traditional norms and informal regulatory frameworks. Therefore, finding equitable solutions that are able to be implemented for the welfare of shifting cultivators requires that policy-makers face the key policy, legal and institutional issue: the choice between continuation or transformation of shifting cultivation (Choudhury et al., 2008; AIPP/ IWGIA/IKAP, 2009; Jamir, 2012).

According to Sedon (2011), the main issue in shifting cultivation lies in 'cultivation' and not in 'shifting'. The shifting refers to a type of rotation which has scientific merit and is acceptable. But the way cultivation is done, by removing the forest cover and burning the biomass, leads to erosion of top soil and loss of nutrients. The fire used to prepare land for cultivation emits carbon dioxide and other greenhouse gases into the atmosphere and also removes primary or climax vegetation, replacing it with grasses and secondary-growth forests. Another vexing issue is the growing influence of market forces on shifting cultivators. These have both the potential to benefit farmers and pitfalls that may bring them harm if adequate safeguards are not taken. For example, unregulated markets have resulted in unsustainable extraction from forests of commercially important medicinal, aromatic, dye-producing and food plants and other non-timber forest products, including bamboo, rattan and small timber species. This often results in shifting cultivators losing important sources of livelihood and governments losing potential sources of revenue. Management of these market-related issues needs enabling national and local policies, especially providing shifting cultivators with effective controls over the management, extraction, local processing and transporting of non-timber forest products.[2]

Some of the major policy issues confronting shifting cultivators include: (1) productive and conservation-oriented fallow management; (2) creating incentives for practising traditional shifting cultivation; (3) ecosystem restoration and landscape conservation; (4) conversion to organic-production systems with affordable certification tools such as a Participatory Guarantee System; (5) strengthening of traditional institutions to ensure equitable resource governance; and (6) co-development of acceptable alternative options, including rotational agro-forestry and conservation farming, to combine the wisdom of shifting cultivators and the scientific concepts of smart agriculture.

Old and new drivers of policy change

Climate change has been one of the major new drivers of land-use and land-cover change in Asia (Rasul and Karki, 2007; van Vliet et al., 2012). It is therefore also associated with shifting cultivation (Griffiths, 2008; Kafle et al., 2009). Equally important for its even greater impact on shifting cultivation is the role of globalization and socio-economic change (Lawrence et al., 2007). However, in the case of climate change, shifting cultivators – unlike other land-based sectors – have been cast as both victims and villains (Eaton and Lawrence, 2009; Carling et al., 2009; Erni, 2009; Mertz et al., 2009a). Traditional use of fire to clear land releases carbon dioxide and therefore contributes to global warming. There are policies in several countries directly and indirectly prescribing alternative methods of land clearing or banning uncontrolled fires to clear forests. Supporters of shifting cultivation argue that traditional use of fire is controlled and that traditional farmers are simply misunderstood and unfairly blamed for accelerating climate change (Fox, 2000). Erni (2009) stressed that it was non-traditional slash-and-burn agriculture, and not traditional shifting cultivation, that produced harmful greenhouse-gas emissions. Traditional systems are managed under well-regulated conditions, with institutional oversight. In fact, well-managed traditional shifting cultivation systems store much higher carbon stocks than any form of monoculture or conventional agriculture (Erni, 2009). There are a number of local innovations, such as those practised by the *Angami* and *Apatani* peoples in northeast India, which do not use fire for clearing forest biomass (Ramakrishnan, 1992; Saikia and Jogamaya, 2003). 'Fireless shifting cultivation' systems are also found in Papua New Guinea, on some Micronesian islands and among the *Tangkhuls* of Manipur, India (Singh, 2009). Decades of research on virtually every aspect of shifting cultivation have generated much evidence to prove that shifting cultivation has both positive and negative aspects that need to be considered in any policy-reform process (Choudhury et al., 2003; Kerkhoff and Erni, 2005; Leduc and Choudhury, 2012). The often generalized comments by policy-makers and forest-land administrators, on the other hand, might be based on a lack of knowledge and information about realities in the field.

Rapid socio-economic changes, especially those driven by globalization, human migration and market penetration, are also affecting shifting cultivation, particularly by spreading the influence of consumerism (Geist and Lambin, 2001, 2002). Yet even in the interests of greater participation in global markets, the use of fire to clear large tracts of forest cannot be defended scientifically or politically, especially when temperature extremes and droughts are occurring more frequently and mitigation programmes such as REDD+ are being launched with the potential to actually pay farmers not to 'slash-and-burn' or degrade the forest landscape. It is a basic reality that in a globalized market economy and a changing socio-economic milieu, all rural households need a cash income. Shifting cultivators are no exception. They are already looking for favourable government policies, fiscal and institutional incentives and negotiation skills to help them participate in a market-led economy. However, they lack the skills to benefit from selling surpluses when they must deal with increasingly intrusive and exploitative market forces. Changing the traditionally entrenched subsistence nature of swidden farming into a semi-commercial production system will not be easy, particularly while retaining the basic values of a traditional culture that prides itself in being egalitarian. Maintaining social and gender equity and environmental sustainability at the same time is yet another challenge (Choudhury et al., 2008).

Given growing international demand for natural and organic products, especially natural food, herbal medicine and nutritional supplements, shifting cultivators are becoming prime targets as suppliers of raw materials for manufacturers of these products. Swidden farmers must be empowered with the capacity to enter into market-led ventures. They must also be cautioned against unduly exposing land that is common property to privatization or conversion by private entrepreneurs, leading to disfranchisement of lands set aside for shifting cultivation and creation in its place of private plantation cropping, commercial farms, fruit orchards and tourist resorts (Inter-Ministerial Task Force, 2008). Such pitfalls may further reduce fallow periods by depriving shifting cultivators of access to their native lands. They could even destroy the very basic tenets of age-old shifting cultivation tradition: the upholding of the principles of equity and justice and creation of an egalitarian society.

Policy-making challenges and shifting domains

The main challenge faced by traditional shifting cultivation communities has been an unfavourable, unclear and complex policy environment (Voss, 2007). This is becoming even more challenging in the context of climate change and rapid socio-economic transformation. The policy environment in most Asian countries is generally neither enabling nor positive in its regard for shifting cultivators (Thomas et al., 2003; Voss, 2007, Fox et al., 2009). Contradictory policies held by different government departments involved in land management (such as those responsible for forests, agriculture, rural and local development and social welfare) have not only obscured overall policy direction, but have also led to incorrect interpretations and implementation problems. This has created an increasingly narrow space for shifting cultivators to either continue or transform their practices. In Nepal, the government has, on one hand, ratified International Labour

Organization conventions 111 and 169, granting rights to indigenous communities to practise shifting cultivation, while on the other hand it has passed its Wildlife Act, 1973, and Forest Act, 1993, making shifting cultivation on forest land both illegal and punishable (Aryal et al., 2010; Choudhury, 2012). In Bhutan, despite a ban on shifting cultivation, farmers in the remote east of the country continue to practise it because they have been offered no viable alternatives (Moktan, 2012). Bhutan's Land Act, 2007, has made shifting cultivation illegal and prescribed the conversion of shifting cultivation land to other uses, such as dry-land pasture, wet-land cultivation areas and perennial horticulture orchards (Moktan, 2012). In Bangladesh, the government has, in some cases, provided tenurial rights to shifting cultivators or *jhumias*, while at the same time it has promoted land-use conversion and land allocation to outside settlers for other uses, as well as allowing the establishment of permanent-cultivation systems by outsiders, such as horticulture farming, without considering the interests of the *jhumias* (Gain, 2000). The most complex, ambiguous and heterogeneous situation is found in India. While shifting cultivation is discouraged and has almost been replaced by settled agriculture in most parts of the country, a constitutional protection of customary rights and traditional practices under the Sixth Schedule of the Constitution provides a legal provision that allows shifting cultivators to continue their traditional practices. As a result, shifting cultivation still prevails in large parts of northeastern India (Choudhury, 2012), where many states have their respective *jhum*-control regulations and Acts adopted from the colonial era (Darlong, 2002). Jamir (2012) argues that prevailing policies have failed to recognize two distinct realities: (1) particular geophysical and ecological contexts; and (2) distinct and diverse socio-cultural beliefs and widely varying social norms and values held by different tribes and villages. Acceptance of these shortcomings and the development of more location-specific and flexible policies could promote deeper involvement of key actors at all levels. The persistence of traditional shifting cultivation is due not only to a lack of policy support, but also to the absence of viable alternatives. Policy failure is often wrongly attributed to the reluctance of shifting cultivators to change the status quo (Leduc and Choudhury, 2012). Given the contradictions in various policies and laws, shifting cultivators have, in most cases, had no choice but to continue doing their best to earn their livelihoods, and this has created social and political conflicts with competing land users. Moreover, within shifting cultivation communities, those with traditional clan-based entitlements have been converting land set aside for shifting cultivation to permanent plantation crops such as tea, coffee, pineapples or rubber, taking advantage of land-registration provisions in government land regulations. To make matters worse, new and powerful groups such as tourism entrepreneurs, the 'timber mafia', new settlers and commercial-plantation entrepreneurs have, under one pretext or another, been encroaching upon shifting cultivation land (Inter-Ministerial Task Force, 2008).

Given the ambiguous policy environment, swidden farmers in many countries have, of their own accord, been participating in state-sponsored programmes promoting transformation away from shifting cultivation. These programmes have

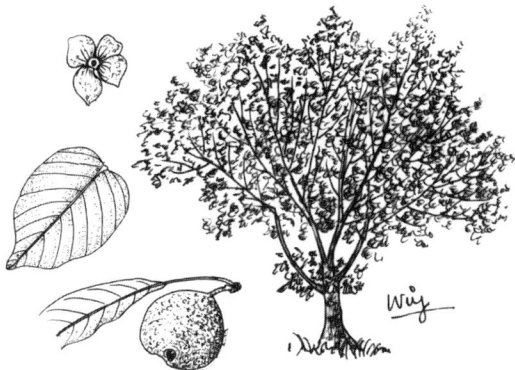

Pyrus Pashia Buch.-Ham. ex D. Don [Rosaceae]

The wild Himalayan Pear is a small- to medium-sized tree that is commonly grown in home gardens at mid altitudes, where it stands out in spring with its abundant white flowers. The fruit is edible but astringent and is usually eaten only when it is overripe.

been introduced mainly by government agencies. Some incentive-based programmes have helped to improve tenurial security and provided subsidies for farming inputs as well as offering micro-credit and other facilities aimed at 'weaning the cultivators away' from their age-old practices and convincing them to adopt alternative livelihood options. However, a sizeable group of stakeholders regards these policies as unsustainable, and posits the view that shifting cultivation is better able to embrace changes such as biodiversity conservation at a landscape level than either settled agriculture or managed forests (Naughton-Treves and Salafsky, 2003; Rerkasem et al., 2009b). This group proposes that where conditions are favourable, shifting cultivators should be allowed to continue their farming practices through enabling policy support and incentive mechanisms (Erni, 2009; Leduc and Choudhury, 2012). They also argue that extractive shifting cultivation can be transformed into optimum land-use systems as long as new policies and regulations ensure property rights, provide investment in infrastructure development and arrange suitable extension and credit services. Therefore, the challenges facing shifting cultivators and their advocates is how to use local perspectives and innovative examples to influence national and sub-national policies in their favour, so that they can preserve the positive aspects and minimize the negative impacts of policies while helping farmers to adapt to climatic and socio-economic changes, globalization and the market economy. Traditional cultivators also need to engage in eco-friendly practices, such as those found in Mizoram, India, where farmers practise strictly controlled burning (Darlong, 2008). There are other practices they can learn from and adapt to, such as alder-based *jhum* cultivation, as practised by the Angami and Tangkhul peoples (Singh, 2009; Ramakrishnan, 1992), and home gardens such as those developed by the Angami in Nagaland. These are classic examples of how farmers have been able to conserve and maintain rich and complex biodiversity while meeting a family's nutritional and cash needs (Majumdar, 1992; Mathur, 1992; Godbole, 1997; Orapankal, 1999).

Finding converging points in policy dialogue

In debating shifting cultivation, one point at which both supporters and detractors seem to converge is that shifting cultivation is the source of livelihood security and socio-cultural benefits for a sizeable population of poor, marginal and disadvantaged communities in Asia (Erni, 2005, 2009; Leduc and Choudhury, 2012). They also seem to agree that the traditional practice has to continually adapt to the changes and challenges that constantly confront both farmers and ecosystems (AIPP/IWGIA/ IKAP, 2009; Choudhury, 2012).[3] Therefore, policy-makers, researchers and land managers throughout Asia must seize these opportunities and use their common understanding to help shifting cultivators find appropriate and adaptive strategies, options and approaches that will enable them to transform their shifting cultivation practices while holding on to traditional socio-cultural tenets and norms that are much more equitable in delivering resource governance. This point is acknowledged by the Inter-Ministerial Task Force (2008) in its report to the Government of India, which said '... [the] level of diversity is a striking testimony to human ingenuity and may provide rare insights for reversing the environmental, social and economic problems resulting from destabilising shifting cultivation systems'. Here, the role of enabling policies and institutional frameworks becomes of paramount importance in transforming shifting cultivation. If policy-makers can recognize the fact that it is not a forest-management issue alone, but a complex combination of issues related to land, natural resources and livelihood management, then perhaps they can find a middle-ground solution. Addressing the real problems of shifting cultivators requires a holistic and multi-stakeholder approach, which is so far missing in the policy-making processes of Asian countries. There is an urgent need for a pragmatic approach to dealing with both the livelihood of the cultivators and the sound health of ecosystems or landscapes, so that shifting cultivators can co-exist with broader society in a harmonious manner, within a defined management framework (Ramakrishnan, 1992; Leduc and Choudhury, 2012). Top-down approaches will not work, but what can help them to improve their productivity to meet not only their household needs, but also to serve local and distant markets, is a gradual movement into improved fallow management, conservation farming, value-chain development, marketing and community-based enterprise-development activities, supported by organic-certification schemes. This has been demonstrated by some recent initiatives in northeast India. With technical, financial and institutional support provided by two projects – Nagaland Environmental Promotion and Economic Development (NEPED) and the North East Region Community Management Project (NERCOMP) – in Nagaland and Meghalaya, shifting farmers have been able to develop intensive fallow management, community-based management of natural resources, revitalized traditional institutions and diversified and improved livelihood opportunities (Leduc and Choudhury, 2012).

A number of authors have called for policy incentives and institutional support for innovative cultivators so that they can gain secure tenure, equitable benefit-sharing mechanisms and easy access to support services, enabling them to develop viable alternatives such as organic farming and fruit and vegetable cultivation (Brady, 1996;

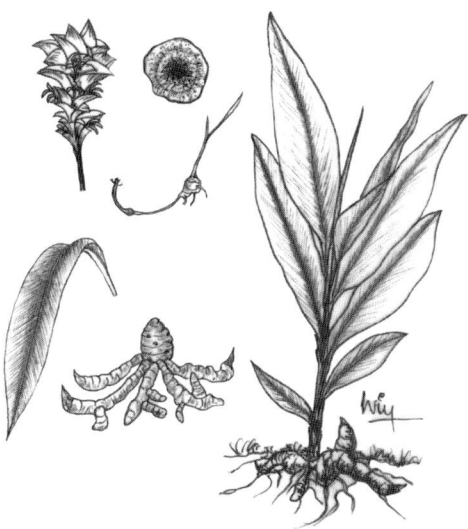

Curcuma longa L. [Zingiberaceae]

A popular crop among alternatives to shifting cultivation in northeast India. The rhizomes of turmeric are a principal ingredient in many Asian and Middle Eastern cuisines. It has long been used as a dye, and is under study for its antifungal and antibacterial properties as well as its potential to treat human ailments, including Alzheimer's disease and diabetes.

Khan and Khisa, 2000; Knudsen and Khan, 2002; Rasul, 2005; Tripathi et al., 2006; Rasul and Karki, 2007; Bala et al., 2008; Leduc and Choudhury, 2012). The alternatives have to demonstrate clear improvements, not only in farm production but also in meeting the increasing cash needs of shifting cultivators. In Nagaland, reforestation with both timber and non-timber species was promoted under the NEPED programme. However, once a market surplus was created, marketing issues became a clear bottleneck to sustaining the popularity of a drive to expand agroforestry as an alternative to shifting cultivation. The NERCOMP project, on the other hand, focused on both locally consumable products such as food grains and niche and high-demand cash crops such as ginger, turmeric, medicinal plants and other non-perishable non-timber forest products that seem to work better (Karki, 2003a, 2003b; Rerkasem et al., 2009b; Schmidt-Vogt et al., 2009; Leduc and Choudhury, 2012).

Learning from policy successes and failures

It is clear that shifting cultivation is governed by a range of policies and legal frameworks both originating from and covering multiple sectors and sub-sectors. Improved and integrated natural-resource management and biodiversity conservation are key requirements for successful long-term economic growth. Appropriate land- and water-management practices, as well as controls on siltation and the loss of other physical resources, are needed to prevent the degradation of forest land due to uncontrolled grazing and burning. Intensification of agriculture is needed to raise the living standards of farming communities in countries like Bhutan. All of these measures are needed so that opportunities can be created for faster economic development. This route to prosperity has been followed by many countries, and has involved the transfer of shifting cultivation land from subsistence agricultural production to productive forest or watershed uses, in the hope of maximizing income from minimum use of land and labour, and diverting part of the rural labour force into other profitable activities. The well-being of a majority of farmers can best be served by pursuing a simple approach that increases labour productivity and farm output. There are many prerequisites for this type of transformation, but the most important of them are appropriate technology, adequate financial resources at farm level, a marketing infrastructure, agricultural inputs and agricultural extension, education and training, and most importantly, willingness on the part of the farmers. In the biophysical setting of a typical shifting cultivation area, intensification of agriculture will also demand higher inputs of labour, and this means that unless cash crops are included in the cropping system, intensive agriculture will not necessarily result in improved livelihoods. In the absence of appropriate land-use and equitable land-tenure policies, this may push poorer sections of the farming community off economically profitable agricultural land and into more marginal areas.

The role of local innovations as evidence for policy-making

There is wide and growing attention being paid to local innovations in shifting cultivation (Cairns and Garrity, 1999; Choudhury et al., 2003; Regmi et al., 2005; Kerkhoff and Sharma, 2006; Lopez-Casiro and Bhuju, 2009) and to work aimed at improving fallow management (Cairns, 2007). Using these innovations as evidence of the need to change policies and practices would be the most appropriate way of developing transformation pathways that would address the multiple issues of shifting cultivation in the Asian highlands (Boxes 49-1 and 49-2). Policy-makers should initiate a process of designing and developing a consultative and participatory policy-making process that accommodates the views of different stakeholders and enables them to adapt to inevitable changes taking place both within and outside shifting

---

Box 49-1: Local innovation leads to successful transformation in Meghalaya, India (Choudhury, 2012)

The key feature of this IFAD-supported project was forming and building the capacity of Natural-Resources Management Groups and Self Help Groups. Subsequently, the Natural-Resources Management Groups joined associations with other such groups and the Self-Help Groups joined federations and formed partnerships with non-governmental organizations. The project first sought to improve incomes from forest products, to provide incentives for the groups. The Natural-Resources Management Groups were very effective in improving forest- and biodiversity-resource management by enforcing rules and regulations formulated in a participatory manner. The groups were also registered with the government to make them legal entities, but their activities were decided by the farmers. The major achievements were: (1) a sustainable supply of forest resources to meet livelihood needs; (2) an increase in household incomes through processing and sales of non-timber forest products; and (3) the adoption of cluster-based collection and marketing of broom grass and bamboo to ensure income. The key outcomes were: (1) inclusiveness: both male and female members from each household were allowed to join; (2) proper institutionalization: formal links to the Village Development Council were found to complement, rather than compete with traditional institutions; and (3) high standards of transparency, accountability and empowerment were maintained by the project.

Box 49-2: Rehabilitation of degraded shifting cultivation land in Lao PDR using community-based natural-resource management (Moizo, 2004)

A major hydropower project resulted in displacement of a large number of shifting cultivation families, due to inundation as well as dam construction. As a part of a compensation and rehabilitation plan, the project rehabilitated the shifting cultivation community in areas of degraded forest. Since upland rice was the most critical component of the traditional system, the alternative sought to maintain or even increase household rice production in permanent systems. The project-evaluation study found that while the area under shifting cultivation in the highlands dropped from 84% to 48%, the percentage of agricultural land use in the highlands increased from 12.7% to 23.6%. The farmers switched from shifting cultivation to permanent systems. Richer farmers adopted the permanent land-use pattern, growing paddy rice and cash crops, raising animals and collecting and marketing non-timber forest products. The results after two to three years indicated that while the upland-rice yield fell by half, the lowland paddy yield almost doubled. Overall, 94% of the farmers felt that the changes had led to an improvement in their livelihoods. The negative impacts were the introduction of chemicals, exotic fruit varieties and tree crops, affecting native biodiversity. Moreover, regeneration of primary-forest vegetation did not happen as expected. The positive factors included the development of local leadership and use of local wisdom in adjusting to the new farming system and increased cash income.

---

cultivation systems. Local innovative practices can provide convincing evidence to policy-makers of the need to craft policies capable of stabilizing shifting cultivation systems while maintaining socio-cultural and biological diversity. Three well-known systems can be cited here: (1) The *Apatani* system of integrated farming practised by the *Apatani* people of Arunachal Pradesh, in India. Although this is not a shifting cultivation system, it uses similar traditional wisdom and practices in integrated land management, maintaining undisturbed clan forests around catchments and seven different categories of land uses, to meet all household needs (Ramakrishnan, 1992); (2) Similarly, the *Chakhesang* tribes of Nagaland practise a unique integrated land-management system called *zabu*, wherein forests, farms and animal species grow together and are managed interdependently in a terrace-based farming system. This is best known for its soil- and water-conservation properties, which protect catchments and water resources along hill slopes (Cairns et al., 2007); and (3) Hedgerow intercropping technology that attempts to mimic shifting cultivation. This has been practised in the Philippines and modified versions have been tried in the eastern Himalayas. Due to its high labour requirement it has not been socially acceptable (LSUAFRP, 2005; Saxena et al., 2007), but it has potential as an alternative to shifting cultivation, and it provides evidence of the need to devise good policies for stabilizing the system (Leduc and Choudhury, 2012). Many locally successful innovations have the potential to transform shifting cultivation, and should be carefully examined. They may enable and empower farmers to move into 'next generation' shifting cultivation, based on community-based processes and sustainable practices in which customary rights are recognized, guaranteeing inalienable access to land for genuine shifting cultivators while recognizing its limits in fragile ecosystems.

Policy options for shifting cultivation

Improving shifting cultivation policies means creating incentives, institutions and instruments that build on farmers' innovative good practices and embrace emerging technological, socio-economic and environmental knowledge and innovations. Such policy solutions must be developed in a participatory and consultative discourse, according to local and national needs (Rasul and Karki, 2007; Choudhury et al., 2008). These efforts may or may not lead to replacement of shifting cultivation with alternative land-management practices. However, they gather local ownership and buy-ins, and in the process, farmers tend to move in the direction of adaptive transformation.

A number of local innovations have shown good potential for transforming shifting cultivation practices in South and Southeast Asia into locally viable livelihood options (Sanchez et al., 1993; NEPED and IIRR, 1999; Cairns et al., 2007; Leduc and Choudhury, 2012; IFAD, 2013). Development research on these innovations has strengthened the argument that transformation of shifting cultivation has to start from within the swidden-farming community, by first understanding the local change process from the perspective of the farmers. For example, scientific testing has indicated that indigenous weeding practices in northeast India are ecologically sound (Ramakrishnan, 1992). Similarly, community-based fire-management tools employed in Mizoram, India are much more effective and cheaper than those prescribed by the government, due to strict enforcement of customary rules (Darlong, 2002). New farming technologies and practices, such as different agroforestry systems, are willingly adopted by farmers as long as the basic tenets of shifting cultivation, including equitable access to communal land, are upheld by village councils (Kerkhoff and Sharma, 2006). There seems to be no resistance to adoption of outside technologies provided they are piloted and extended in a participatory and empowering manner. Examples from Lao PDR and Nagaland can be cited as models to follow (Cairns et al., 2007). One common thread found in these packages is insurance of household food needs and provision of incentives in terms of secured land tenure, together with other development assistance and market access.

Many government agencies, non-governmental organizations and scientists are gradually appreciating the value of local innovations (Leduc and Choudhury, 2012). The starting point was the Shillong Declaration (ICIMOD, 2006), in which stakeholders agreed that any new processes of change in shifting cultivation had to be based on more appreciative and participatory approaches. Thus, the process of incorporating local innovations now involves: (1) documentation and evaluation of farmers' innovations; and (2) sharing these innovations in regional-level policy dialogue for cross fertilization of knowledge and cross regional learning.

Formulating inclusive and enabling shifting cultivation policies

One of the recognized strengths of shifting cultivation is its ability to conserve and maintain agrobiodiversity as well as supporting productive forests. One can argue that due to its huge agricultural diversity and its ability to stimulate socio-cultural cohesion, shifting cultivation has an inherent capacity to adapt to ongoing change. It follows that with proper incentives, shifting cultivation has the potential to transform itself into more resilient mixed farming or agroforestry land uses that will withstand unpredictable and extreme weather patterns. Given that shifting cultivation basically regards nature as the source of its inputs, it has good potential to play an important role in agricultural adaptation, climate-change mitigation and poverty reduction. Therefore, in transforming shifting cultivation into an adaptive form of agriculture, the basic concept of the land–use intensification must be built upon the principles of diversity, resilience and social capital, as its core values. We can draw important lessons from traditional shifting cultivation and its local innovations, by: (1) conserving a rich agrobiodiversity within shifting cultivation areas to meet the requirements of diverse and climate-smart agriculture; (2) designing cropping systems based on microclimatic conditions, to provide harvests throughout the year and avoid climatic risks; and (3) planting a wide variety of crops (and various landraces within each crop) to minimize the risk of pest and disease damage, thus providing added insurance against crop failure due to adverse weather. Ensuring wide crop diversity has long protected shifting cultivators from the risk of complete crop loss and possible starvation. Within the broader agricultural sector, shifting cultivation can also build-in sub-sectoral diversity by including diverse edibles, meat and medicines in cropping patterns, as well as several utility products, the most important of which is firewood, and non-timber forest products such as medicinal, aromatic and dye-producing plants that can generate cash income. This is possible if shifting cultivation can be transformed by a community-based natural-resources management regime, such as those operating successfully in some of the study-site examples.

Panax pseudoginseng Wall.
[Araliaceae]

This species is commonly known as wild ginseng in Nagaland. It is an example of a non-timber forest product that was over-harvested from Nagaland's forests after it attracted a high market price. Angami Nagas report using it as a medicinal plant for purification of the blood.

Based on the examples given in this chapter, the premise of developing successful policies based on the use of local innovations and adaptive community practices as the drivers of change seems to be a practical one. However, the manner in which some policies have attempted to replace shifting cultivation with settled agriculture or other permanent land-based activities has seen them fall short of full acceptance by shifting cultivators and fail to achieve positive results. A substantial body of evidence suggests that policies and regulations that are introduced with adequate consultation and are implemented in a participatory manner and in full understanding of local practices are likely to succeed in promoting both conservation and traditional-knowledge-based development.

The need to strengthen institutions

Adaptive management powered by local innovations in shifting cultivation depends to a large extent on the existence or development of dynamic, resilient and high-capacity institutions (Diaw et al., 2008). Evidence indicates that the robustness of traditional institutions depends on their adherence to the principles of fair access and secure tenurial rights, as these principles affect those who have inherited the land, based on either genealogical or reproductive processes of tenure establishment (Jamir, 2012). This is an important point in crafting resilient institutions that embrace both modern concepts of property rights and age-old traditional norms and good practices, so that they can become a vehicle for institutional reform. This is a common approach to implementing community-based forest conservation and management (Pandit, 2001; Rasul and Karki, 2007; Aryal and Kerkhoff, 2008). In Nepal and Bhutan, for example, new forestry laws (of 1993 and 1995, respectively) initiated an era of *de jure* community forestry that made provision for the devolution of management decisions to community-forestry user groups representing the rural population most dependent on forests for their livelihoods. In the context of shifting cultivation, traditional institutions have been found to be vibrant and dynamic and contribute to the maintenance of good governance and social responsibility in general. However, many of these institutions have been either relegated to ceremonial roles or have suffered neglect and degeneration. The strengthening of these traditional institutions must be an integral part of transforming shifting cultivation and bringing it under good governance, by building their capacity and resilience.

Discussion and analysis of policies aiming to transform shifting cultivation

Some of the best policies in environmental and natural-resources management have been informed, inspired and influenced by the perspectives of multiple stakeholders; by scientific and traditional knowledge and by farmers' innovations and best practices. Some or all of these elements are sometimes bundled under the term 'evidence-based' or 'tested-knowledge' solutions (Pattanayak et al., 2010). One characteristic of good policy-making is the use of systematically collected and validated information to strengthen and enrich the foundation for dialogue linking knowledge to policies aimed at transformational change. Increasingly, best practices in natural-resources policy-making are believed to involve providing both policy-makers and stakeholders with sound scientific and practical information, insights and knowledge related to suggested policy options, together with an assessment of the likely implications for resources (Guldin et al., 2005). Therefore, it is argued that appropriate policies aimed at transforming shifting cultivation will be those that involve multi-stakeholder processes in their formulation, are developed with the active participation of cultivators, and focus on improving their livelihoods with a sensitive regard for their socio-cultural norms and values. Since shifting cultivation traditionally attempts to address social, environmental and economic issues all within one framework, policies should also be based on interdisciplinary and integrative knowledge and practices related to these three pillars. The following steps are suggested in order to reform and refine policy-development processes and paradigms:

1. Create a win-win policy and institutional framework and a collaborative environment that accommodates the norms and values of traditional shifting cultivation while embracing scientific knowledge–based solutions aimed at transforming cultivation systems using approaches to land and natural-resources management that are both flexible and technically supportive. These should include alternative systems such as rotational agroforestry, farm forestry, agro-pastoral systems and leasehold forestry practices (Gautam et al., 2003). Government agencies should provide simple technologies and marketing support to help shifting cultivation communities address their growing socio-economic and environmental challenges. This will call for a multi-sectoral approach to the policy-development process involving all land-based sectors such as forestry, agriculture, watershed, pasture and irrigation. Officials and development agencies will have to work with

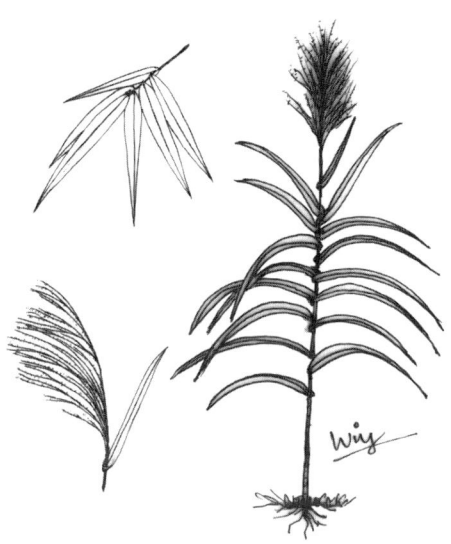

Thysanolaena latifolia (Roxb. ex Hornem.) Honda [Poaceae]

Broom grass forms dense thickets, often covering entire hillsides. Its inflorescences are an important product harvested from the wild throughout South and Southeast Asia to make brooms. It is also cultivated as a crop to provide cash income from regular harvests.

shifting cultivation community leaders to create tailor-made production-to-consumption systems that benefit from rapidly emerging science, technologies and market-information resources. This will enable farmers to improve their sustainable livelihood systems and take care of their environment.

2. Ensure land and tree tenure for shifting cultivators by strengthening the property-rights regime and establish a good-governance system to ensure equity and fairness. This will mean the recognition by governments of cultivators' customary rights to own and cultivate plots of land big enough to meet the food-security needs of their households. Given the need to oppose the growing trend of privatization of commons land in shifting cultivation areas, there is a need to address those land-tenure and governance issues for which the role and capacity of traditional institutions have to be strengthened. This can be achieved by technical and institutional capacity building, incorporating traditional norms into legal codes and reconciling the differences between customary and statuary laws.

3. Develop a better understanding among supporters and detractors of shifting cultivation in order to develop a truly empowering, learning and sharing environment in favour of shifting cultivation communities to help them adapt to gradual change.

4. Since the size of the shifting cultivation population is shrinking, the issue of scale and economic viability of market-oriented operations will be critical. Regional marketing chains or value chains will have to be developed within which producers' associations will be able to develop common products with a common brand using a common organic-production certification system (Karki, 2003b). This will minimize production and processing costs as well as cut unnecessary competition among the cultivators. The Participatory Guarantee System (PGS) or group organic certification (Karki, 2003b; Darlong, 2008) would be suitable models to aim for. Policy-level initiation will be needed as the process requires local, national and regional sharing of experiences and cross-border learning, for which suitable knowledge platforms will have to be created, learning and sharing workshops organized and cross-border exchange visits and peer-learning exercises planned.

5. There will also be a need to create an effective convergence within and between different government- and donor-funded programmes and initiatives, especially those related to livelihood forestry, such as joint forest management, leasehold forestry, non-timber forest products development, community-based forest management and others. Ways in which the communities can be rewarded for good environmental stewardship should be found in order to help the process – perhaps through a system such as Payments for Ecosystem Services.

6. Since shifting cultivation is driven basically by livelihood pursuits, developing sustainable-livelihood strategies to ensure household and community needs throughout a transformation process becomes critical. Therefore, policies must promote sustainable livelihoods, household cash and non-cash generation

and measures for building human capital. Among the tools available for this, promotion of high-value products and services has become quite successful in promoting forest-based livelihoods (Karki et al., 2003).

7. Finally, recognizing the increasing role of knowledge in achieving sustainable development and the importance of research in generating knowledge, governments and development agencies must invest in research-based knowledge production to support the process by which shifting cultivation transforms. Generating location-specific knowledge solutions and policy options is a continuous process, particularly when decision-makers are faced with competing claims and complex social and political relationships within shifting cultivation societies. Research, knowledge and innovations should be recognized as crucial elements for sound economic and social development.

Conclusion and recommendations

The sustainable transformation of shifting cultivation requires policies that strengthen property rights, especially land tenure, governance and social and gender equity, in order to ensure livelihoods and food security. These needs are clear from a review of research literature and findings, clues taken from farmers' innovations, lessons learned from dialogues between stakeholders in South and Southeast Asia, attention paid to the views of various government agencies and realization of the need to address the growing impacts of climate and socio-economic change. There are many differences in context, status and scenario among the region's diverse shifting cultivation communities, landscapes and countries. However, issues common to them all demand integrated and holistic solutions, especially dynamic and inclusive policies and legal frameworks at regional, national and local levels. The major issues are food and livelihood security, biodiversity conservation and addressing the impacts of climate change. But these vital issues should not diminish the importance of regional market development for high-value products and services, such as medicinal and aromatic products, organic and nutritional foods and eco-tourism services. Traditional knowledge, innovative practices and successful pilot projects should be used to generate evidence, not only for developing good policies and practices, but also for peer learning and knowledge sharing. Given the complex nature of the social and gender dimensions, alternative options to shifting cultivation must be gender- and culture-sensitive. Participatory-action research tools should be used to co-develop knowledge and generate evidence of new and emerging issues, especially gaps, in the implementation of policies. This will ensure that stakeholders can inform and influence policy-reform process whenever they face new challenges and opportunities. Experience gathered in Nagaland and Meghalaya, in India, and while implementing leasehold forestry polices in Nepal, can provide important insights. The key messages and lessons that can be distilled from several good practices are as follows:

- Consolidate learning from past policy designs and implementation processes and use this to scale-up and scale-out good practices;
- Use participatory-action research as a tool to co-develop innovation-based knowledge;
- Utilize opportunities and programmes arising from climate change, such as Reducing Emissions from Deforestation and Forest Degradation (REDD+), Clean Development Mechanisms (CDM) and Payments for Ecosystem Services (PES) to achieve triple dividends in terms of adaptation, mitigation and poverty reduction;
- Promote an innovative and integrated approach to diversify livelihoods and minimize environmental damage; and
- Mainstream local innovations, farmers' learning and traditional institutions in the shifting cultivation transformation process.

References

AIPP/IWGIA/IKAP (2009) *Shifting Cultivation and Climate Change*, briefing paper for intersessional meeting of the United Nations Framework Convention on Climate Change, the Asia Indigenous People's Pact (AIPP), the International Work Group for Indigenous Affairs (IWGIA) and Indigenous Knowledge and People's Network (IKAP), Bangkok

AIPP/IWGIA (2011) *Drivers of Deforestation: Facts to be Considered Regarding the Impact of Shifting Cultivation in Asia*, submission to the Subsidiary Body for Scientific and Technological Advice, United Nations Framework Convention on Climate Change, by the Asia Indigenous People's Pact (AIPP) and the International Work Group for Indigenous Affairs (IWGIA), http://unfccc.int/resource/docs/2012/smsn/ngo/235.pdf, accessed 21 September 2013

Arifin, B. (1998) 'Does shifting cultivation really cause deforestation? Lessons from a communal forest area in Sumatra, Indonesia', paper presented at *Crossing Boundaries*, the Seventh Annual Conference of the International Association for the Study of Common Property, 10-14 June 1998, Vancouver, Canada, http://hdl.handle.net/10535/1729, accessed 22 September 2013

Aryal, K. P. and Kerkhoff, E. E. (2008) *The Right to Practice Shifting Cultivation as a Traditional Occupation in Nepal*, a case study to apply ILO Conventions 111 (Employment and Occupation) and 169 (Indigenous and Tribal Peoples), International Labour Office, Kathmandu, Nepal

Aryal, K. P., Kerkhoff, E. E., Maskey, N. and Sherchan, R. (2010) *Shifting Cultivation in the Sacred Himalayan Landscape: A Case Study in the Kanchenjunga Conservation Area*, World Wide Fund for Nature, Kathmandu, Nepal

Bahuchet, S. (1990) 'Cultivating the forest: A historical background of cultivated plants in Central Africa', in C. M. Hladik, S. Bahuchet and I. de Garine (eds) *Food and Nutrition in the African Rain Forest*, United Nations Educational, Scientific and Cultural Organization, Man and the Biosphere Programme, Paris, pp28-31

Bala, B. K., Haque, M. A., Hossain, M. A., Hossain, S. M. A. and Majumdar, S. (2008) *Management of Agricultural Systems of the Upland of Chittagong Hill Tracts for Sustainable Food Security*, National Food Policy Capacity Strengthening Programme, Final Report PR #1/08, Government of Bangladesh, USAID, European Union and Food and Agriculture Oranisation of the United Nations, Dhaka, Bangladesh, http://ifs-b.org/images/com_adsmanager/files/532290247ea77cd96f6cbcbbdf76167d.pdf, accessed 20 May 2017

Borggaard, O. K., Gafur, A. and Petersen, L. (2003) 'Sustainability appraisal of shifting cultivation in the Chittagong Hill Tracts of Bangladesh', *Ambio* 32, pp118-123

Brady, N. C. (1996) 'Alternatives to slash-and-burn: A global imperative', *Agriculture, Ecosystems and Environment* 58, pp3-11

Brown, D. R. (2006) 'Personal preferences and intensification of land use: Their impact on southern Cameroonian slash-and-burn agroforestry systems', *Agroforestry Systems* 68, pp53–67

Cairns, M. F. (2007) 'Conceptualizing indigenous approaches to fallow management: A roadmap to this volume', in M. F. Cairns (ed.) *Voices from the Forest: Integrating Indigenous Knowledge into Sustainable Upland Farming*, Resources for the Future Press, Washington, DC, pp16–36

Cairns, M. F. and Garrity, D. P. (1999) 'Improving shifting cultivation in Southeast Asia by building on indigenous fallow management strategies', *Agroforestry Systems* 47(1), pp37–48

Cairns, M. F., Keitzar, S. and Yaden, A. T. (2007) 'Shifting forests in northeast India: Management of *Alnus nepalensis* as an improved fallow in Nagaland', in M. F. Cairns (ed.) *Voices from the Forest: Integrating Indigenous Knowledge into Sustainable Upland Farming*, Resources for the Future Press, Washington, DC, pp341–378

Carling, J., Erni, C. and Trakansuphakon, P. (2009) *Shifting Cultivation and Climate Change,* briefing paper for an intersessional meeting of the United Nations Framework Convention on Climate Change, International Work Group for Indigenous Affairs, Asia Indigenous People's Pact Foundation and Indigenous Knowledge and People's Network, Bangkok, http://ccmin.aippnet.org/pdfs/bp_sc_final.PDF, accessed 22 September 2013

Choudhury, D. (2012) 'Why do *jhumias jhum*? Managing change in shifting cultivation areas in the uplands of northeastern India', in Sumi Krishnan (ed.) *Agriculture in a Changing Environment: Perspectives on Northeastern India*, Routledge, New Delhi, pp78–100

Choudhury, D. and Sundriyal, R. C. (2003) 'Issues and options for improving livelihoods of marginal farmers in shifting cultivation areas of northeast India', *Outlook in Agriculture* 32, pp17–28

Choudhury, D., Ingty, D. and Jamir, S. (2003) 'Managing marginalisation in shifting cultivation areas of northeast India: Examples of community innovations and initiatives', in Tang Ya and Pradeep Tulachan (eds) *Mountain Agriculture in the Hindu Kush Himalayan Region*, International Centre for Integrated Mountain Development (ICIMOD), Kathmandu, pp207–212

Choudhury, D., Jodha, N. S. and Sharma, E. (2008) *Policy Approaches in Management of Shifting Cultivation: Compromising Equitable Access, Food Security and Common Property Institutions,* International Centre for Integrated Mountain Development (ICIMOD), Kathmandu

Cramb, R. A., Pierce Colfer, C. J., Dressler, W., Laungaramsri, P., Trang, Q., Muloutami, E., Peluso, N. L. and Wadley, R. L. (2009) 'Swidden transformations and rural livelihoods in Southeast Asia', *Human Ecology* 37, pp323–346, DOI: 10.1007/s10745-009-9241-6

Darlong, V. T. (2002) 'Traditional community-based fire management among the Mizo shifting cultivators of Mizoram in northeast India', in *Communities in Flames,* proceedings of an international conference on Community Involvement in Fire Management, FAO Regional Office for Asia and the Pacific, RAP Publications, Bangkok

Darlong, V. (2008) 'Harmonizing *jhum* (shifting cultivation) with PGS organic standards in Northeast India: Key features and characteristics of *jhum* for process harmonization', poster at Cultivating the Future Based on Science, 2nd conference of the International Society of Organic Agriculture Research (ISOFAR), Modena, Italy, 18–20 June 2008, http://orgprints.org/view/projects/conference.html, accessed 22 September 2013

Darlong, V., Jamir, A., Barik, S. K., Tiwari, B. K., Choudhury, D. and Nakro, V. (2008) *Harmonizing* Jhum *in Northeast India with PGS Organic Standards,* North Eastern Region Community Resource Management Project (NERCORMP) and the Regional Centre of the National Afforestation and Ecodevelopment Board, Shillong, Meghalaya

Dev, S., Lynrah, M. M. and Tiwari, B. K. (2013) 'Technological innovations in shifting agricultural practices by three tribal farming communities of Meghalaya, northeast India', *Tropical Ecology* 54(2), pp133–148

Diaw, C., Oyono, P. R., Lescuyer, G., Ribot, J. C., White, A. and Diaw, M. C. (2008) 'The public domain, conservation and decentralization in Francophone Africa', paper presented to a workshop on Forest Governance and Decentralization in Africa, 8–11 April 2008, co-organized by the governments of South Africa and Switzerland, Durban, South Africa

Ducourtieux, O. (2005) 'Shifting cultivation: The Phunoy traditional management system', in *Improving Livelihoods in the Uplands of the Lao PDR*, National Agriculture and Forestry Research Institute, Vientiane, Laos, pp71-77

Ducourtieux, O., Visonnavong, P. and Rossard, J. (2006) 'Introducing cash crops in shifting cultivation regions: The experience with cardamom in Laos', *Agroforestry Systems* 66, pp65-76, DOI 10.1007/s10457-005-6645-1

Eaton, J. M. and Lawrence, D. (2009) 'Loss of carbon sequestration potential after several decades of shifting cultivation in the Southern Yucatan', *Forest Ecology and Management* 258, pp949-958

Erni, C. (2005) 'Shifting cultivation and indigenous peoples in Asia', in *Indigenous Affairs* 2/05, International Work Group for Indigenous Affairs, Copenhagen, http://www.iwgia.org/iwgia_files_publications_files/IA_2-05.pdf, accessed 23 September 2013

Erni, C. (2009) 'Shifting the blame? Southeast Asia's indigenous peoples and shifting cultivation in the age of climate change', paper presented at a seminar on *Adivasi*/ST Communities in India: Development and Change, 27-29 August 2009, International Work Group for Indigenous Affairs (IWGIA), Delhi

FAO (1995) 'Shifting cultivation', *Unasylva* 11(1), Forestry Department, Food and Agriculture Organization of the United Nations, Rome

Fox, J. (2000) 'How blaming "slash and burn" farmers is deforesting mainland Southeast Asia', *Asia-Pacific Issues*, East-West Center, Honolulu

Fox, J., Fujita, Y., Ngidang, D., Peluso, N., Potter, L., Sakuntaladevi, N., Sturgeon, J. and Thomas, D. (2009) 'Policies, political-economy and swidden in Southeast Asia', *Human Ecology* 37: pp305-322, DOI 10.1007/s10745-009-9240-7

Fujisaka, S. (1991) 'A diagnostic survey of shifting cultivation in northern Laos: Targeting research to improve sustainability and productivity', *Agroforestry Systems* 13(2), pp 95-109

Gafur, A. (2001) 'Effects of shifting cultivation on soil properties, erosion, nutrient depletion and hydrological responses in a small watershed of Chittagong Hill Tracts', unpublished PhD dissertation, Chemistry Department, The Royal Veterinary and Agricultural University, Copenhagen

Gafur, A., Borggaard, O. K., Jensen, J. R. and Peterson, L. (2003) 'Run off and losses of soil and nutrients from watershed under shifting cultivation (*jhum*) in the Chittagong Hill Tracts of Bangladesh', *Journal of Hydrology* 279, pp293-309

Gain, P. (2000) 'The Chittagong Hill Tracts: Life and nature at risk', in *Enhancing Access of the Poor to Land and Common Property Resources,* Asian NGO Coalition for Agrarian Reform and Rural Development (ANGOC) and the International Land Coalition (ILC), published by the Society for Environment and Human Development (SEHD), Dhaka, Bangladesh

Garrity, D. P. (2007) 'Challenges for research and development in improving shifting cultivation systems', in M. F. Cairns (ed.) *Voices from the Forest: Integrating Indigenous Knowledge into Sustainable Upland Farming.* Resources for the Future Press, Washington, DC, pp3-7

Gautam, M. K., Roberts, F. H. and Singh, B. K. (2003) 'Community based leasehold approach and agroforestry technology for restoring degraded hill forests and improving rural livelihoods in Nepal', paper presented at an International Conference on Rural Livelihoods, Forests and Biodiversity, 19-23 May 2003, Bonn, Germany

Geist, H. J. and Lambin, E. F. (2001) *What Drives Tropical Deforestation? A Meta-analysis of Proximate and Underlying Causes of Deforestation based on Subnational Case Study Evidence,* LUCC Report Series no 4, Land Use and Cover Change Project, University of Louvain, Louvain-la-Neuve, Belgium

Geist, H. J. and Lambin, E. F. (2002) 'Proximate causes and underlying driving forces of tropical deforestation', *BioScience* 52(2), pp143-150

Gilruth, P. T., Marsh, S. E. and Itami, R. (1995) 'A dynamic spatial model of shifting cultivation in the highlands of Guinea, West Africa', *Ecological Modelling* 79, pp179-197

Godbole, A. (1997) 'Angami home gardens', in A. Rastogi et coll. (eds) *Applied Ethnobotany in Natural Resource Management, Traditional Home Gardens*, International Centre for Integrated Mountain Development (ICIMOD), Kathmandu, pp65-71

Griffiths, T. (2008) '*Seeing 'REDD'? Forests, Climate Change Mitigation and the Rights of Indigenous Peoples and Local Communities'*, report prepared for the 14th Conference of the Parties, United

Nations Framework Convention on Climate Change, 1-12 December 2008, Poznan, Poland, http://www.forestpeoples.org/sites/fpp/files/publication/2010/08/seeingreddupdatedraft3dec08eng.pdf, accessed 20 May 2017

Grogan, P., Lalnunmawia, F. and Tripathi, S. K. (2012) 'Shifting cultivation in steeply sloped regions: A review of management options and research priorities for Mizoram state, Northeast India', *Agroforestry Systems* 84, pp163-177, DOI 10.1007/s10457-011-9469-1

Guldin, R. W., Parrotta, J. A. and Hellström, E. (2005) *Working Effectively at the Interface of Forest Science and Forest Policy: Guidance for Scientists and Research Organizations,* IUFRO Occasional Paper no. 17, Task Force on the Forest Science-Policy Interface, International Union of Forest Research Organizations, Vienna

Hossain, M. A. (2001) 'An overview on shifting cultivation with reference to Bangladesh', *Scientific Research and Essays* 6(31), pp6509-6514, http://www.academicjournals.org/sre/pdf/pdf2011/16Dec/Hossain.pdf, accessed 23 September 2013, DOI: 10.5897/SRE11.1282

ICIMOD (2006) *Debating Shifting Cultivation in the Eastern Himalayas: Farmers' Innovations as Lessons for Policy,* International Centre for Integrated Mountain Development (ICIMOD), Kathmandu

IFAD (2013) *Report to the 12th Session of the UN Permanent Forum on Indigenous Issues,* International Fund for Agricultural Development, Rome, www.un.org/esa/socdev/unpfii/documents/2013/agencies/2013_IFAD.pdf, accessed 22 September 2013

Inter-Ministerial Task Force (2008) *Report of the Inter-Ministerial National Task Force on Rehabilitation of Shifting Cultivation Areas,* issued by G. K. Prasad, Additional Director General of Forests and Chairman of the Task Force, Ministry of Environment and Forests, Government of India, New Delhi

Jamir, A. (2012) *Science and Policy Disconnect in Himalayas,* Mountainvoice TV M-20 Campaign, Climate Himalaya, Rudraprayag, Uttarakhand, India, available at http://chimalaya.org/2012/06/02/science-and-policy-disconnect-in-himalayas/, accessed 22 September 2013

Kafle, G., Limbu, P., Pradhan, B. and Fang, J. (2009) *Piloting Ecohealth Approach for addressing Land Use Transition, Climate Change and Human Health Issues,* NGO Group Bulletin on Climate Change, Local Initiatives for Biodiversity, Research and Development (LIBIRD), Pokhara, Nepal

Karki, M. (2003a) 'The organic production of medicinal and aromatic plants: A strategy for improved value-addition and marketing of products from the Himalayas', in Y. A. Thomas, M. Karki, K. Gurung and D. Parajuli (eds) *Proceedings of Wise Practices and Experimental Learning in Conservation and Management of Himalayan Medicinal Plants,* Ministry of Forests and Soil Conservation, Kathmandu, Nepal, pp56-69

Karki, M. K. (2003b) 'Organic conversion and certification: A strategy for improved value-addition and marketing of medicinal plants products in the Himalayas', *The Indian Forester* 129(1), Indian Medicinal Plants Special - 1, pp. 130-142

Karki, M. K., Tiwari, B. K., Badoni, A. K. and Bhattarai, N. K. (2003) 'Creating livelihoods and enhancing biodiversity-rich production systems based on medicinal and aromatic plants: Preliminary lessons from South Asia', oral paper presented at the Third World Congress on Medicinal and Aromatic Plants for Human Welfare, 3-7 February 2003, Chiang Mai, Thailand, IHCS, Netherlands, http://lib.icimod.org/record/11208/files/1332.pdf, accessed 22 September 2013

Kerkhoff, E. and Erni, C. (eds) (2005) 'Shifting cultivation and wildlife conservation: A debate', *Indigenous Affairs* 2/05, International Work Group for Indigenous Affairs, Copenhagen, pp22-29

Kerkhoff, E. and Sharma, E. (2006) *Debating Shifting Cultivation in the Eastern Himalayas: Farmers' Innovations as Lessons for Policy,* International Centre for Integrated Mountain Development (ICIMOD), Kathmandu

Khan, N. A. and Khisa, S. K. (2000) 'Sustainable land management with rubber-based agroforestry: a Bangladeshi example of upland community development', *Sustainable Development* 8, pp1-10

Knudsen, J. L. and Khan, N. A. (2002) 'An exploration of the problems and prospects of integrated watershed development in the CHT', in N. A. Khan, M. K. Alam, S. K. Khisa and M. Millat-e-Mustafa (eds) *Farming Practices and Sustainable Development in the Chittagong Hill Tracts,* Chittagong Hill Tracts Development Board nd VFFP-IC, Chittagong, Bangladesh, pp165-180

Laungaramsri, P. (2005) 'Swidden agriculture in Thailand: myths, realities and challenges', in C. Erni (ed.) *Shifting Cultivation and Indigenous Peoples in Asia*, http://www.iwgia.org/iwgia_files_publications_files/IA_2-05.pdf, accessed 23 September 2013

Lawrence, D. and Lam, N. T. (2009) 'Environmental consequences of the demise of swidden cultivation in montane mainland Southeast Asia: Hydrology and geomorphology', *Human Ecology* 37, pp361-373, DOI 10.1007/s10745-009-9258-x

Lawrence, D., D'Odorico, P., Diekmann, L., DeLonge, M., Das, R. and Eaton, J. (2007) 'Ecological feedback following deforestation creates the potential for a catastrophic ecosystem shift in tropical dry forest', *Proceedings of the National Academy of Sciences* 104, pp20696-20701, DOI:10.1073/pnas.0705005104

Leduc, B. and Choudhury, D. (2012) 'Agricultural transformations in shifting cultivation areas of Northeast India: Implications for land management, gender and institutions', in D. Nathan and V. Xaxa (eds) *Social Exclusion and Adverse Inclusion: Development and Deprivation of Adivasis in India*, Oxford University Press, New Delhi, pp237-258

Lopez-Casiro, F. and Bhuju, U. R. (2009) *From Shifting Cultivation to Sustainable Livelihood Creation: Strengthening Marginalized Communities through Institutional Development and Micro-finance for Agroforestry and Energy-efficient Technologies*, Institute of Global Environment and Society, Rockville, MD

LSUAFRP (2005) *Poverty Reduction and Shifting Cultivation Stabilisation in the Uplands of Lao PDR: Technologies, Approaches and Methods for Improving Upland Livelihoods*, proceedings of a workshop held in Luang Prabang, 27-30 January 2004, Lao-Swedish Upland Agriculture and Forestry Research Program, Vientiane, Lao PDR

Majumdar, K. (1992) 'An Angami village', in S. M. Channa (ed.) *Nagaland: A Contemporary Ethnography*, Cosmo Publishers, New Delhi, pp49-74

Mathur, N. (1992) 'Religious ethos of the Angami Naga' in S. M. Channa (ed.) *Nagaland: A Contemporary Ethnography*, Cosmo Publishers, New Delhi, pp121-146

Mertz, O., Padoch, C., Fox, J., Cramb, R. A., Leisz, S. J., Lam, N. T. and Vien, T. D. (2009a) 'Swidden change in Southeast Asia: Understanding causes and consequences', *Human Ecology* 37(3), pp259-264, DOI 10.1007/s10745-009-9245-2

Mertz, O., Leisz, S., Heinimann, A., Rerkasem, K., Thiha., Dressler, W., Pham, V. C., Vu, K. C., Schmidt-Vogt, D., Pierce Colfer, C. J., Epprecht, M., Padoch, C. and Potter, L. (2009b) 'Who counts? The demography of swidden cultivators', *Human Ecology* 37(3), pp281-289, DOI 10.1007/s10745-009-9245-2

Moizo, B. (2004) 'Implementation of the land allocation policy in the Lao PDR: Origins, problems, adjustments and local alternatives', in *Shifting Cultivation and Poverty Eradication in the Uplands of the Lao PDR*, National Agriculture and Forestry Research Institute, Vientiane, http://lad.nafri.org.la/fulltext/LAD010320040562.pdf, accessed on 20 May 2017

Moktan, M. (2012) 'Policy issues on shifting cultivation in Bhutan', in S. Chaudhary and K. Thapa (eds) *Proceedings of the Regional Policy Dialogue Workshop on Shifting Cultivation*, 22-23 March 2012, Kathmandu, Nepal, International Development Research Centre (IDRC – Canada) and International Centre for Integrated Mountain Development (ICIMOD), Kathmandu, http://www.icimod.org/resource/7130, accessed 27 September 2013

Namgyel, U., Siebert, S. F. and Wang, S. (2008) 'Shifting cultivation and biodiversity conservation in Bhutan', *Conservation Biology* 22(5), pp1349-1351, DOI: 10.1111/j.1523-1739.2008.01019.x

Nathan, D. and Xaxa, V. (2012) *Social Exclusion and Adverse Inclusion: Development and Deprivation of Adivasis in India*, Oxford University Press, India

Naughton-Treves, L. and Salafsky, N. (2003) 'Wildlife conservation in agroforestry buffer zones: Opportunities and conflict', in G. Schroth (ed.) *Agroforestry and Biodiversity Conservation in Tropical Landscapes*, Columbia University Press, New York

NEPED and IIRR (1999) *Building Upon Traditional Agriculture in Nagaland, India,* Nagaland Environmental Protection and Economic Development, Nagaland, India and International Institute of Rural Reconstruction, Silang, Cavite, Philippines, pp27-30

Orapankal, A. (1999) 'The Roman Catholic Church in a leadership paradox: A case study of an Angami parish in Nagaland, India', PhD dissertation in ETD Collection for Fordham University, New York, http://fordham.bepress.com/dissertations/AAI9938914, accessed 23 September 2013

Pandit, B. H. (2001) 'Non-timber forest products on shifting cultivation plots (*khorya*): A means of improving livelihood of Chepang Rural Hill Tribe of Nepal', *Asia-Pacific Journal of Rural Development* 11, pp1-14

Pattanayak, S. K., Seven Wunder, and Ferraro, P. J. (2010) 'Show me the money: Do payments supply environmental services in developing countries?' *Review of Environmental Economics and Policy* 4(2), pp254-274, doi:10.1093/reep/req00

Ramakrishnan, P. S. (1992) *Shifting Agriculture and Sustainable Development: An Interdisciplinary Study from Northeast India,* United Nations Educational, Scientific and Cultural Organization, Man and the Biosphere series, Parthenon Publishers, Paris and Carnforth, Lancashire, UK, republished by Oxford University Press, New Delhi, 1993

Rasul, G. (2005) *State Policies and Land Use in the Chittagong Hill Tracts of Bangladesh*, IIED Gatekeeper Series No. 119, International Institute for Environment and Development, London

Rasul, G. and Thapa, G. B. (2003) 'Shifting cultivation in the mountains of South and Southeast Asia: Regional patterns and factors influencing change', *Land Degradation and Development* 14, pp495-508

Rasul G. and Karki, M. (2007) *Participatory Forest Management in South Asia: A Comparative Analysis of Policies, Institutions and Approaches*, Talking Point series 5/7, International Centre for Integrated Mountain Development (ICIMOD), Kathmandu

Rasul, G., Thapa, G. B. and Zoebisch, M. A. (2004) 'Determinants of land-use changes in the Chittagong Hill Tracts of Bangladesh', *Applied Geography* 24(3), pp217-240

Regmi, B. R., Subedi, A., Aryal, K. P. and Tamang, B. B. (2005) *Shifting Cultivation Systems and Innovations in Nepal*, unpublished report, Local Initiatives for Biodiversity, Research and Development (LIBIRD), Pokhara, Nepal

Rerkasem, K. and Rerkasem, B. (1995) 'Montane mainland Southeast Asia: Agroecosystems in transition', *Global Environmental Change* 5, pp313-322, DOI 10.1016/0959-3780(95)00065-V

Rerkasem, K., Yimyam, N. and Rerkasem, B. (2009a) 'Land use transformation in the mountainous mainland Southeast Asia region and the role of indigenous knowledge and skills in forest management', *Forest Ecology and Management* 257, pp2035-2043

Rerkasem, K., Lawrence, D., Padoch, C., Schmidt-Vogt, D., Zeigler, A. D. and Bruun, T. B. (2009b) 'Consequences of swidden transitions for crop and fallow biodiversity in Southeast Asia', *Human Ecology* 37(3), pp347-360, DOI: 10.1007/s10745-009-9250-5

Romagny, L. (2004) 'Resettlement: an alternative for uplands development?', paper presented at a workshop *Poverty Reduction and Shifting Cultivation Stabilization in the Uplands of Lao PDR: Technologies, Approaches and Methods for Improving Upland Livelihoods*, 27-30 January 2004, Luang Prabang, Lao PDR, National Agriculture and Forestry Research Institute and Lao-Swedish Upland Agriculture and Forestry Research Program, Vientiane

Saikia and Jogamaya, K. K. (2003) 'Religion: A factor of culture change amongst the Angamis of Nagaland', in S. Sengupta (ed.) *Tribes of North-East India: Biological and Cultural Perspectives*, Gyan Publishers, New Delhi, pp220-228

Sanchez, P., Garrity, D. and Bandy, D. (1993) *Sustainable Alternatives to Slash and Burn Agriculture and the Reclamation of Degraded Lands in the Humid Tropics,* International Centre for Research in Agroforestry (ICRAF), Nairobi, Kenya

Saxena, K. G., Liangan, L. and Rerkasem, K. (2007) *Shifting Agriculture in Asia: Implications for Environmental Conservation and Sustainable Livelihood,* UNU/NIRO, Bishen Singh, Mahendra Pal Singh, Dehradun, India

Schmidt-Vogt, D., Leisz, S. J., Mertz, O., Heinimann, A., Thiha, T., Messerli, P., Epprecht, M., Pham, V. C., Vu, K. C., Hardiono, M. and Dao, T. M. (2009) 'An assessment of trends in the extent of swidden in Southeast Asia', *Human Ecology* 37(3), pp269-280, DOI 10.1007/s10745-009-9239-0

Scott, J. C. (2009) *The Art of Not Being Governed*, Yale University Press, New Haven, CT

Sedon, D. (2011) Unpublished e-conference report on constraints and challenges for sustainable mountain development, International Centre for Integrated Mountain Development (ICIMOD), Kathmandu

Sharma, T. C. (1976) *The Pre-historic Background of Shifting Cultivation in North-East India*, North-East Indian Council for Social Science Research, Shillong, Meghalaya, India

Singh, J., Borah, I. P., Barua, A. and Barua, K. N. (2010) *Shifting Cultivation in Northeast India: An Overview*, Rain Forest Research Institute, Indian Council of Forest Research and Education, Jorhat, Assam, India, www.icfre.org/UserFiles/File/rfri/rpap23.htm, accessed 15 September 2013

Singh, L. J. (2009) 'A case study of shifting cultivation practices among the Tangkhuls of Ukhrul district, Manipur', PhD dissertation submitted to the Department of Ecology and Environmental Sciences, Assam University, Silchar, Assam, India

Spencer, J. E. (1966) *Shifting Cultivation in Southeastern Asia*, University of California Publications in Geography, Berkeley, CA, reprinted by Bishen Singh Mahendra Pal Singh, Dehradun, India

Thomas, D., Bouahom, B., Vonghachack, S., Manivong, K., Sophathilath, P., Tat, V. and Philakone, P. (2003) *A Brief Review of Upland Agricultural Development in the Context of Livelihoods, Watersheds and Governance for Area-Based Development Projects in the Lao PDR*, ICRAF report for the International Fund for Agricultural Development (IFAD)

Tripathi, R. S. and Barik, S. K. (2003) 'Shifting cultivation in Northeast India', in B. P. Bhatt, K. M. Bujarbaruah, Y. P. Sharma and Patiram (eds) *Approaches for Increasing Agricultural Productivity in Hill and Mountain Ecosystems*, Indian Council for Agricultural Research, Northeast Region, Shillong, Meghalaya, pp317-322

Tripathi, R. S., Pandey, H. N., Barik, S. K. and Kumar, A. (2006) *Meghalaya State of Environment 2005*, Ministry of Environment and Forests, New Delhi

van Vliet, N., Mertz, O., Heinimann, A., Langanke, T., Adams, C., Messerli, P., Leisz, S., Pascual, U., Schmook, B., Schmidt-Vogt, D., Castella, J-C., Jørgensen, L., Birch-Thomsen, T., Hett, C., Bech-Bruun, T., Ickowitz, A., Vu, K. C., Yasuyuki, K., Fox, J., Dressler, W., Padoch, C. and Ziegler, A. (2012) 'Trends, drivers and impacts of changes in swidden cultivation in tropical forest-agriculture frontiers: A global assessment', *Global Environmental Change* 22(2), pp418-429

Voss, J. (2007) 'Foreword' in M. F. Cairns (ed.) *Voices from the Forest: Integrating Indigenous Knowledge into Sustainable Upland Farming*, Resources for the Future Press, Washington, DC

Notes

1. Although the diminution of shifting cultivation is generally agreed, finding authentic data is difficult. Schmidt-Vogt et al. (2009), in *Human Ecology* and Leduc and Choudhury (2012), in Nathan and Xaxa (2012), provide more information.
2. Related to this issue, government agencies in India question the special protection and rights available to shifting cultivators under the VI Schedule Act of the Government of India.
3. For example, the Inter-Ministerial National Task Force on Rehabilitation of Shifting Cultivation Areas, set up by the Government of India, concluded that 'the appropriate way to tackle this issue ... is to support a hybrid technology which combines the traditional wisdoms and innovations as well as technological advances' (Inter-Ministerial Task Force, 2008).

50

CODIFICATION OF CUSTOMARY COMMUNAL TENURE OF UPLAND SHIFTING CULTIVATION COMMUNITIES IN MYANMAR

*Kirsten Ewers Andersen**

Introduction: The political economy and regulatory framework of shifting cultivation in Southeast Asia

The Southeast Asian uplands are host to a multitude of ethnic groups, each with its own culture, kinship system and language, all practising shifting cultivation, often combined with irrigated terraces in mountain valleys and grazing lands on mountaintops and in forests. They consider themselves different from the lowland wet-rice cultivating 'hydraulic civilizations' of the Bamar in Myanmar, or the Thai, Lao or Kinh in neighbouring countries to the east (Wittfogel, 1957). Shifting cultivators in the region have, over centuries, managed their village lands communally, without private ownership of land. This applies to communities in territories stretching from Meghalaya in northeastern India to the Central Highlands of Vietnam and the Southeast Asian Massif, an area that more or less coincides with the *Zomia* described by James Scott (2009) and van Schendel (2005) (see also Michaud, 2010). The historical severance of these ethnic groups from valley civilizations has been explained by Scott (2009) as the active decision of these groups to spread uphill to avoid the hegemony of the state, its sedentary agriculture and demands for tribute and labour. In the hills, they have lived by their own rules and culture for centuries.

Nowadays, these groups are structurally included in nation states by the respective governments of Southeast Asia, which basically want these groups to become 'civilized', like the majority of their populations living in the lowlands. In the Lao PDR, many upland communities have been forced to resettle at lower elevations and merge into multi-ethnic villages, where their access to land is very limited. Many communities have, thus, suffered livelihood and cultural deprivation. Most governments in Southeast Asia have adopted strong policies against shifting cultivation, arguing that

---

* KIRSTEN EWERS ANDERSEN, anthropologist, researcher on indigenous peoples' issues and land and forest tenure in Southeast Asia. Consultant to the Land Core Group, Myanmar.

it is a primitive system of low productivity compared to commercial agriculture, and that it causes deforestation and land degradation. The same governments close their eyes to illegal timber logging, other extractive industries, dam and road construction and, in particular, the expansion of commercial farming, which causes widespread forest degradation and deforestation. Attempts by governments to promote privatization and commoditization of land – or 'turning land into capital' as the Lao PDR government puts it – have had impacts on upland communities' fallow land, leading to its appropriation by outsiders. In the Central Highlands of Vietnam, the sustainable land management of indigenous communities practising rotational fallow farming was completely destroyed by an influx of large numbers of Kinh people from the lowlands and the conversion of the communities' land into coffee plantations run by provincial parastatals (World Bank, 2009).

Shifting cultivators in Asia have never had title to their land, with the exception of a number of communities in Cambodia and the Philippines. These two countries have legislation in place for cadastral registration of the communal shifting cultivation lands of indigenous communities. In these cases the land parcels that make up the registered common property are composed of the land cultivated in the year of the cadastral survey as well as the fallow land claimed and used by the community in other years.[1] All land parcels are registered in the name of a legally incorporated community. In Cambodia, the legal incorporation of the community is based on its own statutes or by-laws, which are combined with the community's customary Internal Rules for managing and sharing the land in a way that gives everyone in the village access to land.

Communities in the Lao PDR, Cambodia and Myanmar that have no title to their shifting cultivation areas are at high risk of land grabbing by influential parties, or of having their fallow land included in the demarcation of National Parks and protected areas by the state. In eastern Lao PDR and Cambodia, indigenous communities have lost large areas of land to Vietnamese rubber magnates who were granted land concessions by the Lao or Cambodian governments, which declared the communities' fallow land 'idle and free' (Global Witness, 2013). Local villages impacted by rubber concessions owned by or affiliated with these companies have lost vast tracts of land and forests. As a result, households are facing impoverishment, while their spirit forests and burial grounds have been destroyed. By the end of 2012, the Cambodian government had granted concessions covering 2.6 million hectares – equivalent to 73% of the country's arable land. This has affected 400,000 people in 12 of the country's 25 provinces alone (Global Witness, 2013). Similarly, Kachin communities in northern Myanmar have lost land because it was given by the Myanmar military government to the Chinese to carry out a so-called opium-crop substitution programme, which in reality became concessions to the Chinese for rubber plantations (Kramer and Woods, 2012).

In Myanmar the rural population makes up almost 70% of the total population. Around 42% live in upland areas. Studies by Myanmar scholars conclude that about 155,607 square kilometres is under shifting cultivation by forest-dependent communities totalling 30 million people (San Win, nd). At least half of these forest-dependent communities are upland villages practising shifting cultivation. These upland communities consist of different ethnic groups which since independence have engaged in civil wars with the rulers of Myanmar, particularly with the military

FIGURE 50-1: Myanmar, showing the author's study sites of Hakha, in Chin state, and Lashio, in Shan state.

Source: United Nations Department of Field Support.

junta of General Than Shwe, who 'retired' in 2011, handing over the reins of power to a seemingly reformist president, Thein Sein. The period since 2012 has seen a gradual opening up, which has allowed for studies of the customary land use of ethnic upland communities for engagement with the government over land rights by the Land Core Group (LCG), a group of national and international non-governmental organizations (NGOs) in Myanmar. Against this background, the LCG supported the author to carry out a rapid anthropological land-governance study and action research among upland communities in northern Chin state (Hakha township) and northern Shan state (Lashio township) in 2013 and 2014, in order to understand the communal tenure of their shifting cultivation areas and permanent farming plots (Figure 50-1). The action research aimed at recommending procedures to the government for cadastral registration of villages' shifting cultivation lands, with the ultimate aim of protecting this land against alienation. Two villages in each of northern Chin state and northern Shan state joined in the action research.

Land grievances in Myanmar

The Ministry of Agriculture and Irrigation's Master Plan for the Agricultural Sector, 2000-2030

The Ministry of Agriculture and Irrigation has promised to convert 10 million acres (4.05 million hectares) of 'wastelands' to commercial agricultural production. Based on official statistics, by May 2013 a total of 377 domestic companies had been allocated 2.3 million acres of 'vacant, fallow and virgin land' and 822 companies or individuals had been allocated a total of 0.8 million acres (324,000ha) of forest land (outside of Mon state) (Byerlee et al., 2014). The Ministry's data indicate that land under concessions increased by at least 0.3 million acres (121,500ha) between 2010 and 2013. By far the largest areas have been allocated for planting rubber, oil palm, rice and jatropha, followed by rice, sugarcane and cassava, prompting Byerlee et al. (2014) to note: 'Given ongoing granting of concessions, a major priority is to protect the land rights of traditional land users operating under customary tenure in extensive long fallow farming systems'.

Pinus wallichiana A. B. Jacks.
[Pinaceae]

One of the pine species cleared for high-altitude swiddens in Chin state, it regenerates quickly on fallow land. Native to the Himalayas, it is also known as blue pine.

During military rule, the army (*tatmadaw*) expropriated the land of many communities for its own use. This caused a backlash of grievances when the new government came into power in 2011-2012. A Committee for Land Allocation and Scrutiny was set up to hear grievances, but handing the land back to communities is very slow and subject to corruption. As recently as June 2014 it was confirmed that 'battalions and military units under the Ministry of Defence did transfer the land to the General Administrative Department and the Settlement and Land Records Department. However, the battalions, the military units and the Settlement and Land Records Department in the Ayeyarwaddy region have conspired to sell the land to businessmen, rather than ensure its return to the rightful owners' (Asian Human Rights Commission, 2014).

Members of Parliament are now warning that land-grab tensions could trigger uprisings. A publication entitled *Guns, Cronies and Crops* by Global Witness in March 2015 claimed that military and local government officials had grabbed the agricultural land of a number of Shan communities in northern Shan state in the first decade of the 2000s (Myanmar Times, 2015). In this case, the *tatmadaw* is accused of confiscating nearly 6000 acres (2430ha) of farmland during military rule under the regional command of the general who is now the Minister of Agriculture and Irrigation (Myanmar Times, 2015). Elsewhere in Myanmar, a group of human rights activists is planning to take land-grab charges to the International Criminal Court in a special case of crimes against humanity.

Farmers might have some ability to withstand the land grabbing if they had title to their land. But only 15% of farmers in Myanmar as a whole have land-use certificates. Many have annual agricultural-tax receipts, but these do not count as legal proof of land ownership. Most land in Myanmar, in particular that in the uplands, is not titled. In 2012, the new government issued two laws related to land: the Farmland Law and its associated Rules and the Vacant, Fallow and Virgin Land Management Law and its associated Rules. In the latter, 'vacant and fallow' land is often defined as 'wasteland', with the result that many areas of land that are fallow in shifting cultivation systems fall under the 'vacant and fallow' classification and are therefore eligible for

Panicum miliaceum L.
[Poaceae]

Proso or broomcorn millet is the main crop in many Chin swiddens. A crop well suited to dryland and poor soils, this species is known to have been cultivated in many parts of the world for at least 7000 years.

granting as land concessions to business investors by the Ministry of Agriculture and Irrigation.

The Farmland Law consists of 13 chapters and 43 articles.[2] Implementation procedures for the law are found in the associated Farmland Law Rules of 2012. This law was innovative, in the sense that it provided for individual farmers or organizations to have their land surveyed and registered by the Settlement and Land Records Department, after which they would be issued with land-use certificates and would have the right to sell the land (Chao, 2013). They did not previously have such rights.[3] In the Farmland Law, 'farmland' is defined as paddy land, *ya* land (dry land not suitable for rice), *kaing* land (alluvial land along rivers in dry season), perennial plants land, *taungya* land, *dhani* land (nipa palm land in the Delta region), garden land for vegetables and flowers and alluvial land (near the sea). *Taungya* land refers to permanent upland fields and not to rotational fallow farming fields or shifting cultivation, which would be named *shwe pyaung taungya*. Shifting cultivation land is mentioned in the Farmland Law only under its associated Rules, which stipulate in Article 116 that the government desires to eradicate shifting cultivation.

The Vacant, Fallow and Virgin Land Law (VFV) has 34 articles.[4] Like the Farmland Law, the VFV Law also has Rules to guide its implementation. It defines the role of the Ministry of Agriculture and Irrigation's Central Committee for the Management of Vacant, Fallow and Virgin Lands and provides methods and procedures for the granting of concessions of VFV lands for business development. The VFV Law defines vacant, fallow and virgin land in Articles 2 and 3. The first two are defined as land areas that were used or tenanted in the past, but are currently without any tenant cultivating them. Thus, many areas of land that are in use by farmers and community groups practising traditional or customary shifting cultivation would be classified as 'vacant and fallow'. The concessions can vary in size according to the purpose, crops grown and number of years worked. The chairman of the Central Committee for the Management of Vacant, Fallow and Virgin Lands is the Minister of Agriculture and Irrigation. The committee can grant rights to the use of VFV land for 30 years or more, covering areas from 5000 acres up to 50,000 acres (2025ha to 20,250ha) for the growing of perennial plants and/or industrial crops.[5]

Taungya land classification

The word *taungya* is used in everyday parlance in Myanmar to describe rain-fed permanent upland agriculture, e.g. for growing corn. However, the same word is also used globally to identify an agroforestry system, for example in Ghana or Mexico, where trees are planted and surrounding farming communities are allowed to grow crops in between them for several years, until the expanding tree canopy precludes further cropping. This system was introduced in Myanmar 150 years ago by Dr Dietrich Brandis, a German forester who joined the British service in 1856 as superintendent of teak forests in Pegu division, eastern Burma. Teak trees were planted on shifting cultivators' lands and the local communities were allowed to plant crops in between them. When the trees grew too large, new land was taken and the process was repeated. This was a forester's way of initiating new plantations. The villagers provided labour for clearing, planting and weeding in exchange for the right to cultivate crops.

In contrast to the global meaning given to the word *taungya*, its modern use in Myanmar does not identify the Brandis system. It means primarily upland cultivation, and as a permanent use of the uplands, it often refers to monocropping. This contrasts with shifting cultivation, which is locally called *shwe pyaung taungya*. The Rules and by-laws accompanying the Farmland Law mention *shwe pyaung taungya*, or shifting cultivation, only in the last article but one, in order to differentiate shifting cultivation from permanent *taungya*. Article 116 states: 'The central farmland management committee shall encourage effectively for the vanishing of slash and burn cultivation and to introduce terrace cultivation on high land for the environment conservation, [so as] not to spoil the watershed area [of] the forest, not to spoil [the] topsoil, and to regulate the climate.'[6] In this way the Myanmar government, represented by the Ministry of Agriculture and Irrigation, shows its desire to do away with shifting cultivation in much the same manner as the governments of the Lao PDR and Thailand, without ever having examined its biological, economic, social or cultural characteristics.

In colonial times, the British Frontier Administration was interested in exploiting lowland areas that it regarded as 'wastelands' but it recognized the agricultural systems of the uplands, including the traditional land-tenure systems of ethnic communities. The 1935 Government of Burma Act confirmed the application of different laws to different ethnic groups and geographic areas in the 'Frontier Regions', and left upland communities alone.

The policy expressed in the 2012 Farmland Law Rules for the eradication of shifting cultivation, on one hand, and the risk of land grabbing created under the VFV Law on the other, leave Myanmar's shifting cultivators at considerable risk. A large number of upland ethnic communities still practise shifting cultivation, and as mentioned earlier, they do so with no rights or title to the land they use. Solving land issues, including attention to all of the grievances that have arisen in recent decades,

Phaseolus vulgaris L.
[Leguminosae]

The common bean is a highly variable species popular in the lopil swiddens of the Chin. This species is widely grown for dried beans or for immature pods, eaten as green beans.

is one of the prime contentious issues and sticking points in the country's ongoing peace process. Recognizing this, and seeking peace with the ethnic nationalities in the uplands, a new Land Use Policy was drafted by a committee set up in the Ministry of Environment, Conservation and Forestry in 2014. It was circulated for consultation among stakeholders, including civil society, at the start of 2015. The final policy draft was subsequently issued in mid-2015. In contrast to the Farmland Law of 2012, it recognizes the customary land tenure and land use of ethnic communities, and accepts shifting cultivation as subsistence agriculture. The policy was yet to be endorsed by the end of 2015. It was expected that a new government, led by the National League for Democracy, would take over in 2016.

The committee in charge of preparing the new Land Use Policy and Land Resources Law is chaired by the Ministry of Environment, Conservation and Forestry, which has a special interest in shifting cultivation because much of the uplands, where shifting cultivators live, is also Myanmar's forested area. Some of this forest is natural and undisturbed, while some has regenerated on fallow land. The Ministry of Environment, Conservation and Forestry holds that shifting cultivation is a cause of forest degradation. However, it recognizes that it is a cultural practice and way of life that has evolved in consonance with the physiographic set up. In cooperation with other sectors, the Forest Department hopes to change shifting cultivators' land into community forests. It plans to focus on a number of activities:

1. Community forestry based on agroforestry systems;
2. Provision of improved technologies to complement traditional forest-related local knowledge;
3. Recruiting shifting cultivators into routine forestry operations such as plantation establishment;
4. Enhancing income-generating opportunities; and
5. Provision of awareness-raising campaigns and extension services.

In a 2013 study of shifting cultivation and community forestry in Myanmar for the non-governmental organization Pyoe Pin, Oliver Springate-Baginsky suggested ways in which swidden cultivation practices could be legitimized and tenure security established by using the government's Community Forestry Instructions of 1995 to hand over the land as community forests and thus protect it against land grabbing (Springate-Baginski, 2013). However, a closer analysis of existing customary tenure characteristics will show advice against turning shifting cultivation land into community forests, and that only unused fallows should be thus converted. Community Forestry Certificates are temporary and are not the same as full communal cadastral registration of a village's common shifting cultivation parcels. The Community Forestry Instructions of 1995 have also allowed for the capture of communal land by the elite, as they allow a group in a village, and not the whole village, to hold the land rights. The Community Forestry Instructions are therefore not supportive of the equity of access and benefit-sharing principles that exemplify customary systems of shifting cultivation management.

FIGURE 50-2: Ancestors of the shifting cultivators of Chin state. Members of the ceaselessly warring tribes of western Burma, as they were encountered by British officials in the late 19th century. Haka women and a typical Haka house (top) and members of the Siyin tribe (bottom), from their chiefs (left) to their 'modes of coiffure' (right).

Source: Carey and Tuck (1896).

The cultural construction of a Chin swidden landscape

Customary communal tenure in Myanmar is village-based. A village is a suitable unit for managing land because of low communication costs, allowing management of the land to follow a village's Internal Rules. The territory of a village generally has clear boundaries with the land of neighbouring villages. Inside the village territory there are forests, grazing land and agricultural shifting cultivation land as well as plots of permanent agricultural land such as terraces, orchards and gardens. The boundaries of the shifting cultivation areas are known. Each year new fallow land is opened up and land from which crops have already been harvested is allowed to return to fallow. In some parts of Southeast Asia, the swidden plots are scattered here and there in the forest, as the author found among the Pwo Karen in Western Thailand in the 1970s, or they are joined to make up a large block of adjacent swiddens on a mountainside, such as with the Lua of Nan province in Thailand, the Chin in Myanmar and the Nagas in northeastern India. The plots making up such large mountain tracts are shared among the families of a village, according to the village's customary Rules.

In shifting cultivation communities, the identity of the villagers is clearly linked to a dense network of particular places, each having different cultural and material values and containing a mosaic of resources. There is a central connection between history, identity and land and, in former times at least, there was a strong village-based relationship to the fertility-granting spirits of the land, and these spirits would receive ritual offerings of the first fruits of labour. This old link between land rights and residence in the village is often referred to as the *dama ucha* principle in Myanmar. According to this principle, the residents of a village and its headman are the descendants of the 'first founder of the domain' who wielded the machete (*dama ucha*) to clear the land and establish the relationship with the fertility-granting spirits – the owners of the land (see Tannenbaum and Kammerer, 2003; Chit Hlaing, 2003).

This earlier and very explicit relationship with the spirits of the land explains one of the principles still held today by both Christian Chin and Buddhist Shan villages: that access to the agricultural land of the village is dependent upon physical residence in the village. A newcomer would also be granted access to the land, but only if he or she was living in the village. This rule can be interpreted as an articulation of the old belief that territorial spirits guarantee the fertility of the agricultural land for those physically living on the village's land. Nowadays, the Chin who have converted to Christianity may no longer provide such an explanation for this tradition, but the rule still exists as an internal rule of the village (Carey and Tuck, 1896).[7]

In Chin state, northwest Myanmar, communities practise shifting cultivation on slopes at high elevations (over 2000m); their adjacent plots making up mountain tracts of swidden cultivation with a 10-year fallow period. The land areas in Chin are fairly vast; the landscape is made up of forested mountains. In contrast, in Shan state, eastern Myanmar, bordering China and Thailand, the undulating agricultural land is located at lower elevations. The size of land parcels held by villages is smaller and there is more permanent cultivation in the landscape, interspersed with small areas of shifting cultivation.

In Chin state an area of shifting cultivation making up a tract of land with clear physical boundaries on a mountain side is called a *lopil*. One

Canavalia ensiformis (L.) DC.
[Leguminosae]

Known as sulphur beans in the *lopils* of Chin state, but more commonly known as jack bean. The beans are mildly toxic, but are used for both animal fodder and human food when carefully prepared.

lopil may contain 30 to 100 swidden plots (*lo*), each about three acres (1.2ha) in size, and with boundaries marked by stones on the ground (Figure 50-3). One village may have 12 to 18 *lopil*, depending on the size of the individual *lopil*. Some *lopil* are located at cold elevations and some at warm elevations, with cropping patterns adjusted accordingly. The Chin landscape is characterized by steep slopes and narrow valleys, and the *lopil* – as a bounded landscape category – contain not only land fit for agriculture, but also some terraces or orchards, as well as rocky outposts unfit for agriculture. The uppermost cold *lopil* may also contain grazing lands. Each year, around September, the village elders, the headman and other village VIPs discuss which of the fallow *lopil* to open in the following year.

The villages are commonly located near a road, but uphill, close to the *lopil*. The crops cultivated are part of a subsistence economy; there is a low degree of commercialization. Food crops grown include grains, pulses, roots and vegetables. Grains comprise millet, Job's tears, maize, sorghum and rice. Pulses include gram, peas, small beans, dhal (pigeon peas), and sulphur beans or jack beans. Root crops include sweet potatoes, yams, turmeric and ginger, and vegetables include pumpkins, cucumbers, onions, chillies, eggplant and wild varieties of spinach. Rice is often a luxury food, but when a community gains access to lowland or valley-bottom paddy fields or terraced fields at lower elevations, rice cultivation gradually becomes part of the farming calendar. Rice may also be imported from the lowlands by the more well-to-do families. The number of households working in *lopils* increases along with population growth, unless a village establishes terraced paddy fields and available labour is directed elsewhere. Some young men also travel to Mizoram in India each

FIGURE 50-3: Mountainside *lopil* in Chin state. Small huts are visible both within the *lopil* and just outside them, for temporary residence of families cultivating individual plots, earlier assigned by lottery.

Source: Kirsten Ewers Andersen (2013).

year for temporary work, and villages may keep a number of *lopils* under forest if the households do not have enough labour to cultivate them.

Communal tenure

Research on the characteristics of customary communal tenure in shifting cultivation areas in Myanmar has applied theoretical principles of collective action that are typical of common property institutions. These analytical principles are derived from the theory of common property developed since the 1980s by the Workshop in Political Theory and Policy Analysis at Indiana University, under the leadership of the late Elinor Ostrom (National Research Council (of the US), 1986; Ostrom, 1990). Ostrom set the framework for analysing the evolution of institutions for collective action and contributed greatly to knowledge about original indigenous common property systems for land, grazing land, fisheries and forest. Her work also covered the crafting of new 'induced' institutions, where none existed before.

Communal tenure refers to a situation where a group holds secure and exclusive collective rights to own, manage or use land and natural resources, referred to as common pool resources, including agricultural lands, grazing lands, forests, trees, fisheries, wetlands or irrigation waters. In common property or common pool resources theory, communal tenure can be defined as self-governing forms of collective action by a group of people, most often a village (Andersen, 2011). Such communal tenure can be observed today in the indigenous communities of eastern Cambodia, among ethnic groups in the southern and eastern parts of the Lao PDR, in Nagaland in India (Cairns, 2007), and in Myanmar. This has been the traditional system for a large part of Asia, where communal tenure has been the norm and land has never been a commodity. The view among Western jurists in the 19th century was that the origin of the concept of property was the occupation of land by a single proprietor and his family, but the English jurist Sir Henry Sumner Maine published his *Ancient Law* in 1861, from which we learnt that 'it is more than likely that joint ownership, and not separate ownership, is the really archaic institution, and that the forms of property that will

Coix lacryma-jobi L. [Poaceae]

A native of Southeast Asia, Job's Tears is grown by the Chin at higher altitudes where rice and corn do not perform well. It produces food grain and also beads for necklaces.

afford us instruction will be those that are associated with the rights of families and of groups of kindred'.

For most traditional shifting cultivation communities in upland areas of Southeast Asia, the village is the unit for managing its own affairs, including communal land rights (Cupet and Tips, 1998) and, in former times, its warfare (Carey and Tuck, 1896).[8] A village territory is sacrosanct when rituals for the spirits of the land are performed. Village sacrifices to the spirits often require fencing or erecting gates to prohibit strangers from crossing the boundaries of the village and disturbing the relationship between the villagers and the spirits of the land (Figure 50-4).

The Elinor Ostrom principles that characterize common property are as follows:

1. Clearly defined boundaries (effective exclusion of external un-entitled parties);
2. Rules regarding the appropriation and provision of common resources that are adapted to local conditions (in Cambodia and Myanmar, a village's Internal Rules for sharing the land and defining who can take what, where, when and how);
3. Collective-choice arrangements that allow most resource appropriators to participate in the decision-making process (e.g. when making the village a legal entity that may own land, creating the 'statutes' that define governance of the village as a legally incorporated body);
4. Effective monitoring by people who are drawn from the appropriators;
5. A scale of graduated sanctions for resource appropriators who violate community Rules;
6. Mechanisms for conflict resolution that are cheap and easy to access;
7. The community's self-determination is recognized by higher-level authorities; and
8. In the case of larger common-pool resources, organization in the form of multiple layers of nested enterprises, with small local common-property regimes at the base level (Ostrom, 1990).

In shifting cultivation areas, boundaries to different categories of land and the tenure rules applying to each are well known. All families in a shifting cultivation community have the right to access land, but they may, in some areas, have to make a tribute, tithe or payment to certain aristocratic high-status clans that control the land. In a democratic communal tenure system for shifting cultivation land, a person living in a village has an enforceable right recognized by the village's social system. Internally, some families can hold individual permanent or temporary rights to particular resource niches within the common property. These may include a standing crop or the source of a seasonal non-timber forest product, such as resin or malva nuts in the Lao PDR and Cambodia; a piece of land based on an ancestral claim as in Chin state; part of a lake or fish harvested by a specific technology as in Cambodia.

FIGURE 50-4:   A village gate erected at an Akha village in Luang Namtha province, Lao PDR.

Photo: Kirsten Ewers Andersen.

There are four basic and important kinds of rights in the analytical framework of common property theory. These are the right of access to land; the right to withdraw and appropriate produce from common property; the right to take part in decision-making; and the right to exclude outsiders. Investigating these kinds of rights among the ethnic groups of Myanmar enables an analysis of the customary communal tenure systems for shifting cultivation land.

Customary communal management of shifting cultivation land in Chin state

In a northern Chin village, the Internal Rules stipulate how the land plots within a specific *lopil* or mountain block will be allocated each year by lottery. This is the village's democratic system for sharing shifting cultivation land. Claims of individual rights over ancestral plots in a *lopil* are still recognized and claimant families have priority in making use of such plots. However, if in a given *lopil* in a particular year a family with ancestral claims to six plots has sufficient labour to work only two plots, the remaining four plots become part of the common pool allocated through lottery and the family with the ancestral claims does not receive any payment in compensation. In the following year, if a new *lopil* is opened and the same family has no ancestral claims, it will still benefit from land allocation by lottery. The system in northern Chin is one where no one gains at the expense of others and every family resident in the village is guaranteed land.[9] No one can sell or give land to outsiders, but if an outside family comes to live in the village, it is guaranteed land.

In his *Study on the Evolution of the Farming Systems and Livelihood Dynamics in Northern Chin State*, for the French international non-governmental organization GRET, San Thein (2012) notes that 'Zathal village practised communal land tenure with no ownership. The village chief, with the assistance of the village committee, allocates land to all households. About three family members (Village chief and his committee members) and those who have no terrace fields are considered to have priority in selecting fields to their preference in the *lopil* and the remaining fields are then equally assigned to all remaining households'. Perhaps Zathal village does not have a lottery system, but many villages encountered by the author do have such a system. In Zathal there are also no private claims to land, except possibly to terraces.

The customary tenure existing in the shifting cultivation landscape of northern Chin fulfils all the criteria of common property: clearly defined boundaries (effective exclusion of external un-entitled parties); Rules regarding the appropriation and provision of common resources (who can take what, where, when and how); and collective-choice arrangements (the elders, headman and village assembly discuss and decide on which *lopil* to open in following years). The collective choice arrangements also allow the villagers to participate in the decision-making process. The Internal Rules recorded in the 2013 research for managing shifting cultivation land have a strong similarity to the Rules recorded by the British officer Stevenson, who lived in the town of Falam in northern Chin in

Amorphophallus paeoniifolius (Dennst.) Nicolson [Araceae]

The deciduous Elephant's Foot yam has a regular place in Chin swiddens. Often known as Stink Lilly, it produces a single flower smelling of decaying flesh. The underground corm can weigh up to 8kg and is used as human food.

the 1930s. His studies of the customary tenure 80 years ago detail the social and ritual attributes linked to the sharing of shifting cultivation land and show how this sharing is, on one hand, fully democratic, but on the other, linked to other aspects of village life and to the status of democratic and aristocratic clans and warfare, altogether representing the cultural construction of spatial realities (Stevenson, 1937, 1943).

In the 1930s, Stevenson observed that the *klang ram* – although 'communal' land – was subject to many internal and individual ancestral claims to different plots in different *lopils*, and that the *bul ram* – although 'private' land – was subject to many rights of the community as a whole, so it was not exclusively privately owned. In the 1930s, the village headman had the mandate to distribute the land, but within certain bounds. He could not sell it or rent it out or lend it to someone who was not a resident of the village. His first duty as headman was to see that every village resident had sufficient land to cultivate. Stevenson commented about the right of sons to inherit ancestral claims, in accordance with the strong patrilineal kinship system of the Chin. Land over which such hereditary cultivation titles existed was known as *saihrem nam* and *sumhmui* among the Shimhrin and Hualngo Chin. The first term applied to plots where the cultivation titles were based on the rights of the original founders of the village, and the second term to land over which 'first-clearing' rights were established by the cutting of virgin jungle by later immigrants to the village. The distinction emphasized descent from the oldest families in the village, but did not imply major differences in access to land for those living in the village. Stevenson warned against privatization as long ago as the 1930s, and held that privatization at the expense of communal ownership was bad, because it led to 'the evil of absentee ownership' and 'landless rent payers' – an observation that rings true 80 years later.

In comparison to the Chin, the Shan communities studied in northern Shan state had limited shifting cultivation. Their village land was mostly under permanent cultivation. However, they had no cadastral land titles and they still managed the land communally. The Shan mode of communal tenure did not employ lotteries in the annual distribution of plots. The Shan considered all agricultural land to belong to the village and were very afraid of land grabbing by outsiders. Even informal private claims on irrigated paddy land in the village could not be sold to an outsider, and if a family left the village its land would return to the common pool to be redistributed. The Shan villages opted to join the ongoing research because it sought ways and means of acquiring future communal cadastral land registration. They were aware that under the Farmland Law they would have the option to register their land individually, but they did not want to take this path.

The way towards future cadastral registration of communal land

Communal land registration, in the way that it applies in Cambodia, is desired by both the northern Chin and northern Shan villages studied for this chapter. The people took part in the action research with free prior and informed consent, knowing that the Land Core Group was seeking ways and means of lobbying the government to legally recognize the land rights of the upland communities to their shifting cultivation land and their permanently cultivated fields under customary communal tenure. The research aimed to provide evidence-based recommendations for the cadastral registration of these lands and to devise ways and means of lobbying the government to institutionalize legal procedures for cadastral land registration. The study made use of lessons learnt from Cambodia and the Philippines and prepared recommendations based on an innovative [construed] reading of the regulatory framework of the Farmland Law of 2012, later drawing support from the draft Land Use Policy of mid-2015.

Land titles are legal documents that can serve to safeguard the villagers' land against being swept up in land concessions. Cadastral registration of communal land requires at least three things: identification of the rights-holder as a legal entity in whose name the common property parcel titles are to be issued; recording of the community's customary Internal Rules for managing and sharing the land to ensure sustainability of the principles of equity; and clear demarcation of the boundaries of the land parcels, carried out through surveying and mapping. In eastern Cambodia, the first two steps have been carried out in a number of indigenous communities with support from non-governmental organizations as well as the involved ministries. The NGOs have also supported a form of 'preliminary mapping' of the indigenous communities' lands. The final cadastral survey must be performed by the government's competent authority – in Myanmar, the Settlement and Land Records Department, which, by the end of 2015, had been renamed the Agricultural Land Management and Statistics Department (ALMSD).

In practise, the first step in identifying the rights-holder is historically simple under customary tenure as everyone physically living in the village has rights of access to land. However, when it comes to recognizing a rights-holder that is a community, and not a private person, the Myanmar government needs new procedures and a legal framework to see the community as a legal entity owning land. The solution reached in the action research was to suggest the use of Article 6(b) of the Farmland Law, which allows an 'organization' or 'association' to be a rights-holder of a land-use certificate. Article 6 specifies the entities that 'have [a] right to farming', and besides individual persons or farmers, these include an organization, a government department, a government organization, a non-governmental organization or a company.

In order to use this article, the community would have to incorporate legally as an association and create statutes or by-laws defining its membership and governance structure. Thus, as part of the research, draft statutes were developed and discussed with the communities. The statutes included the mechanisms for election of a land caretaker/management committee; defined the committee's role; and described decision-making arrangements, in which a general assembly of all villagers had the final say, including a say in changes to both the village's Internal Rules or the statutes themselves. The statutes, likewise, referred to a dated list of all members of the legal entity and to clear entry-and-exit Rules of that body. The dated list would be kept in the village as well as the village tract office and would be updated when new households were established in the village. The idea was that the statutes would only outline the objectives and the governance structure used to set the village up as a legal entity, and in this form it would apply to the General Administration Department for status as an association under the new Association Law and its forthcoming Rules. The statutes were written and agreed upon by the villages in 2015, but the communities had yet to apply for status as associations because the necessary implementing Rules had not then been published for the new Association Law, which meant that the operational mechanisms for implementing the Association Law were not yet in place.

The statutes developed for the villages in Chin and Shan states are very similar, and it seems that statutes may look the same for many other villages. But the Internal Rules for land management and land sharing, which are the expression of customary tenure, differ from village to village. They depend on the village's land resource endowment, the size of its total land area, population density, ethnic group identity and culture, clan identity and kinship, women's status, private land claims within the common property, inheritance, bride prices, and transactions to borrow or lend land between village members. The statutes themselves will refer to the village's Internal Rules, and thus avoid establishing new forms of authority and power.

The Internal Rules articulate the particular customary communal tenure of different individual villages. These Rules specify that no one can sell land to

Pinus kesiya Royle ex Gordon [Pinaceae]

Commonly found at high altitudes in Chin state. This is one of the most widely distributed pine species in Asia and a popular source of multi-purpose timber.

outsiders. They also specify that no one can sell land internally. Transactions among village residents are limited to lending or borrowing land with payment of money or tithe, but for one year only. This prevents any influential village family from accumulating permanent claims on shifting cultivation land. In Chin state the Rules also specify the rights of women, inheritance, and the conditions under which a family can ask to establish a private terrace in the common property of a *lopil*. There are also Rules for whether or not a person may dig rocks or sand in his assigned plot to sell to road construction companies. And there may be Rules – such as exist in Nagaland – involving individual withdrawal rights to cut an old planted tree on ancestral land, but after which the coppice re-growth will revert back to the common property of the entire clan (Cairns, 2007).

In Chin state the research team recorded the Internal Rules during village meetings by drawing maps of the landscape on flip charts. The meetings in Chin state were mainly attended by men, but in southern Chin state women also took part (Figures 50-5 and 50-6). In the end, however, the villages in southern Chin state declined to join the action research because the traditional influential 'land owners' did not want communal land registration despite the purported risk to 'their' untitled land from the Vacant, Fallow and Virgin Land (VFV) Law. These 'landowners' received payments each year from other families in the village wanting to use a plot in the *Kho Khmang* (swidden block) that was opened in the current year.

The third step in the process towards cadastral registration of communal land is the measurement of the land parcels that make up the full shifting cultivation system. This is a precondition for registration and provides proof that villagers can hold in their hands for tenure security. It requires land surveys and mapping and could be a stumbling block to the entire process, because it is a very difficult and lengthy process – literally, because of the terrain and figuratively because of the effort required. First, if the swidden land is adjacent to the land of another village, the exact boundaries must be located in agreement with the neighbouring village. This requires negotiations, but can eventually be solved. The time-consuming part is that all parcels making up the village's system of shifting cultivation must be identified and surveyed by global positioning system (GPS), assuming that the government will require cadastral maps to accompany the issuance of titles. In Cambodia such mapping was funded by

FIGURES 50-5 and 50-6:   Drawing village maps of land use and tenure (left) and pipe-smoking, tattooed village women take part in meetings to discuss the communal tenure issue.

Photo: Kirsten Ewers Andersen.

international donors in the belief that the government had neither the capacity nor the desire to set aside money for communal land surveys.

The map in Figure 50-7 shows the common property parcels of a Cambodian indigenous village. The land survey established 388 GPS points for this cadastral index map of the village's common property. Ongoing communal land registration in Cambodia often covers more than 50 parcels for one village. Each of the 50 parcels would have its own cadastral identity, but in the Registry Book the ownership of each would be recorded in the name of the village.

A first attempt at the mapping of shifting cultivation land in Chin state in Myanmar was supported by the Land Core Group as part of the research. It revealed the problems inherent in accomplishing precision when dealing with large areas of shifting cultivation land. The shifting cultivation landscape is managed with reference to the *lopil*, and each year maybe two *lopil* are opened up. But the issue is that the community does not use all of the land in a *lopil* and the land use for the same *lopil* may differ between years. Therefore, the question emerges: which precise land area should be surveyed for the preparation of cadastral index maps? Only that land currently under cultivation, or all cultivable land including fallows? The area under cultivation in any given *lopil* at the start of a new cycle may be smaller than in previous years if, say, half the young men of a village go to Mizoram to work for wages. Figure 50-8 is an illustration of this problem: the broken lines on the map show the boundaries of a *lopil* and Google Earth shows the parcels currently under shifting cultivation. They make up just one-quarter of the *lopil*. Additionally, mapping land parcels for registration in 10-year-old forested fallows is difficult and requires that the surveyor is constantly guided by villagers as to the location of boundaries.

While the demarcation of shifting cultivation land for the purpose of land registration in Cambodia was relatively easy despite being covered in vegetation in many places, in Chin state, because of the landscape, the preliminary mapping has opened up a new range of issues that need to be addressed and solved in consultation

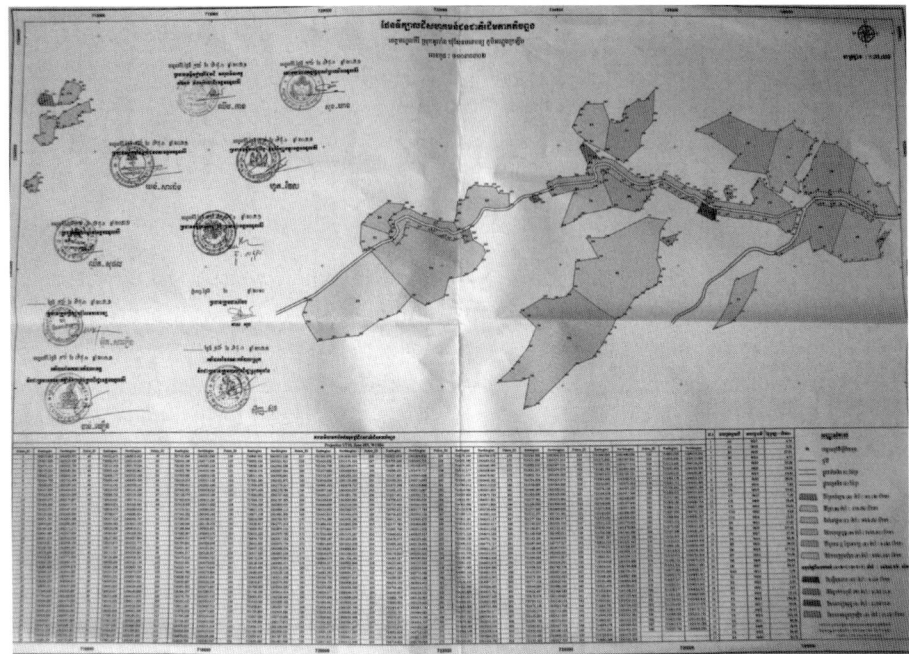

FIGURE 50-7: Cadastral Registration Index Map of the customary communal tenure of an indigenous village in Mondulkiri province, Cambodia. The lighter and darker areas signify shifting cultivation land under cultivation and fallow respectively.

Source: World Conservation Society and the Ministry of Land Management,
Urban Planning and Construction, Cambodia.

with villagers. The maximum size of the common shifting cultivation property of a Cambodian village would be around 1300 hectares, with many villages having only 800 hectares. The total area of arable shifting cultivation land inside, say, 18 *lopil* belonging to a Chin village would be more than 11,000ha. Of this only about 5500ha would be used as communal shifting cultivation land. The rest would be orchards, terraces and grazing lands or rocky land, and these should not be included in the shifting cultivation parcels receiving farmland title under Farmland Law Article 6b. This means that all parcels making up the notional 5000ha would need to be surveyed, if a proper registration of all land claims were to take place. This is a large area and it remains to be seen whether the government would allocate this amount of land for registration. However, 5000ha is in fact not much more than what a community of 300 households would need if each used a three-acre (1.2ha) plot per year in a 10-year rotation.

It is not possible at the moment to carry out measurements by hand-held GPS devices, but it may in the future be possible to use drones, if the government can afford them and someone is ready to pay for the drones to be used in mapping the land of upland ethnic communities. In the meantime, the Myanmar government should issue what in Cambodia are called 'Interim Protection Measures'. In Cambodia, once an

FIGURE 50-8: A satellite image shows a *lopil* bounded by dashed lines. The current cultivation can be seen occupying about one-quarter of the cultivable land.

Source: Google Earth.

indigenous community incorporates legally with the Ministry of the Interior based on their statutes, the provincial governor issues an 'Interim Protection Measure' to safeguard the community's land against land concessions. However, such safeguards have not always worked. The Phnom Penh government, through its Ministry of Agriculture, Forest and Fisheries, has continued to grant land concessions without scrutinizing the uses of the land.

In Cambodia, preliminary land-use mapping, conducted with the support of non-governmental organizations, was used by indigenous communities to substantiate their claims against encroaching Vietnamese companies. Having the maps made them feel empowered. Preliminary mapping is also useful for future support of actual cadastral land registration, because the government needs preliminary maps, be they digital or orthophotos,[10] when conducting surveys for communal or private land registration. Preliminary maps are one of the forms of evidence needed to substantiate a claim over land. The process of preparing preliminary maps of the village territory and its various kinds of land also helps identify the community as a single body.

During 2015 and 2016, funding from non-governmental organizations and donors may be used for preliminary land-use mapping in Myanmar. Such mapping should establish forward linkages with tenure security for indigenous land, as part of the ongoing drafting of land law and other legislation. However, it should not be assumed that preparation of maps will lead to recognition of land rights as such, unless other legal and procedural instruments are in place.

Draft of the new Land Use Policy of mid-2015

The rapid research carried out during 2013 and 2014 showed that the two Chin and two Shan villages wished to obtain communal titles in the name of the villages (as registered associations), with statutes, and to base their land management on customary Internal Rules. When the research began in 2013, the new draft Land Use Policy of 2015 was not yet formulated, so we turned to Article 6b of the Farmland Law, 2012. While this may be used to allow an association to request land titles for its farmland, this approach may strike a significant obstacle: Article 116 of the Rules related to the Farmland Law prescribe the eradication of shifting cultivation. This would therefore obstruct the registration of shifting cultivation land as farmland. By the first half of 2015, we were able to find several articles in the new draft Land Use Policy that were supportive of the customary land use of ethnic groups, including shifting cultivation. Article 51 of the draft policy states: 'Rotating and shifting cultivation shall be considered as subsistence agriculture, and the rate of land tax shall not be more than the maximum rate collected for ordinary smallholder farmer or smallholder household'.

Myanmar has a critical lack of accessible data on land use, land cover and land tenure on which to base land registration, and the government has called for the development of a new 'Onemap' open-access platform with a database that is made up of data from various sources. This is aimed at creating a land governance system. However, the system will initially be simply descriptive, as the procedures for a real cadastral land registration are not in place, least of all for communal land registration. The data in this platform will add to the information base, but it is no more than a first step toward legally valid land registration.

In developing the new Land Resource Law in 2016, it is hoped that the acceptance of shifting cultivation in the sixth version of the National Land Use Policy will ensure full recognition of the positive side of rotational fallow *taungya*, or shifting cultivation (see Regional Shifting Cultivation Policy Dialogue Workshop for the Eastern Himalayas, 2004). These positive issues include:

1. Rotational fallow *taungya*, or shifting cultivation, is practised under a broad range of community-based tenure regimes that regulate the sharing of the land and natural resources. Social and food security is one of the main functions of the local customary tenure institutions of upland shifting cultivators.

2. Farmers practising shifting cultivation conserve more forests on their land than any other farmers, and make it productive at the same time. Techniques used in these systems are generally appropriate for their agro-ecological contexts (although they are not 'modern'), and cultivators often have complex and comprehensive knowledge about resources, land use and the surrounding environment.

3. In comparison to sedentary or permanent cultivation, shifting cultivation has a lower environmental impact: the agricultural ecology is far greater, especially during the fallow period when forest regrowth provides diverse habitats.

4. Virtually no toxic external inputs are used, such as pesticides, herbicides and synthetic fertilizers that are damaging to the soil and water systems.

5. The institutional mechanisms ingrained in traditional rotating fallow *taungya* systems ensure access to productive resources for every member of the community.

Continued research into operational procedures for registration of customary land is warranted. This is in line with Article 82 of Myanmar's draft Land Use Policy, which calls for pilot research on participatory land-use planning and zoning, which – for instance through the Onemap initiative – could be seen as preliminary to cadastral land registration. In the end – with any luck – it will feed into the new Land Law and its associated Rules, and these should contain procedures for shifting cultivators on how to develop statutes, record their Internal Rules, undertake preliminary mapping, and gather all that is required for verification of a land registration claim. There is long-term work ahead in Myanmar, until the land rights of shifting cultivators for sustainable land use can be formally registered through cadastral registration of communal tenure.

References

Andersen, K. E. (2011) *Communal Tenure and the Governance of Common Property Resources in Asia: Lessons from Experiences in Selected Countries*, FAO Land Tenure Working Paper 20, Food and Agriculture Organization of the United Nations, Rome, http://www.fao.org/docrep/014/am658e/am658e00.pdf, accessed 9 December 2015

Asian Human Rights Commission (2014) *Burma: Army Prosecutes Farmers for Trespassing on Their Own Land*, Asian Human Rights Commission, Hong Kong, http://www.humanrights.asia/news/urgent-appeals/AHRC-UAC-090-2014, accessed 9 December 2015

Byerlee, D., Kyaw, D., San Thein, and Kham, L. S. (2014) *Agribusiness Models for Inclusive Growth in Myanmar: Diagnosis and Ways Forward*, MSU International Development Working Paper 133, Michigan State University, East Lansing, MI, http://ageconsearch.umn.edu/bitstream/189109/2/idwp133.pdf, accessed 9 December 2015

Cairns, M. F. (2007) 'The alder managers: The cultural ecology of a village in Nagaland, N. E. India', PhD dissertation to the Research School for Pacific and Asian Studies, Australian National University, Canberra, http://himalaya.socanth.cam.ac.uk/collections/rarebooks/downloads/Malcolm_Cairns_Thesis.pdf, accessed 9 December 2015

Carey, B. S. and Tuck, H. N. (1896) *The Chin Hills: A History of the People, Our Dealings with them, their Customs, Manners, and a Gazetteer of their Country*, vol. I & II, Government Printing House, Rangoon

Chao, S. (ed.) (2013) *National Updates on Agribusiness: Large Scale Land Acquisitions in Southeast Asia, Brief no 8: the Union of Burma*, Forest Peoples Programme, Moreton-in-Marsh, UK, http://www.forestpeoples.org/topics/agribusiness/publication/2013/agribusiness-large-scale-land-acquisitions-and-human-rights-sou, accessed 6 December 2015

Congress of the Philippines (1997) *Philippines' Republic Act No 8371: An Act to recognize, protect and promote the rights of indigenous cultural communities/indigenous peoples, creating a national commission on indigenous peoples, establishing implementing mechanisms, appropriating funds therefore, and for other purposes*, Republic of the Philippines, Manila, http://bmb.gov.ph/index.php?option=com_docman&task=doc_download&gid=195&Itemid=156, accessed 9 December 2015

Chit Hlaing (F. K. Lehman) (2003) 'The relevance of the founders' cult for understanding the political systems of the peoples of northern Southeast Asia and its Chinese borderlands', in N. Tannenbaum and C. A. Kammerer (eds) *Founders' Cults in Southeast Asia: Ancestors, Polity, and Identity*, Monograph 52, Southeast Asia Studies, Yale University, New Haven, CT, pp15-39

Cupet, P. and Tips, W. E. J. (1998) *Among the Tribes of Southern Vietnam and Laos: 'Wild' Tribes and French Politics on the Siamese Border (1891)*, White Lotus Press, Bangkok

Global Witness (2013) *Rubber Barons*, Global Witness, London, https://www.globalwitness.org/campaigns/land-deals/rubberbarons/, accessed 9 December 2015

Kramer, T. and Woods, K. (2012) *Financing Dispossession: China's Opium Substitution Programme in Northern Myanmar*, Transnational Institute, Amsterdam, http://www.burmalibrary.org/docs13/tni-financing_dispossesion.pdf, accessed 9 December 2015

Maine, H. S. (1861) *Ancient Law: Its Connection with the Early History of Society, and its Relation to Modern Ideas*, John Murray, London

Michaud, J. (2010) 'Editorial – Zomia and beyond', *Journal of Global History* 5(2), pp187-214

Myanmar Times (2015) 'Minister faces land grab accusations', http://www.mmtimes.com/index.php/national-news/13773-minister-faces-land-grab-accusations.html, accessed 6 December 2015

National Research Council (of the US) (1986) *Proceedings of the Conference on Common Property Resource Management*, April 21-26, 1985, Washington, DC, Panel on Common Property Resource Management, Board on Science and Technology, National Research Council, National Academy Press

Ostrom, E. (1990) *Governing the Commons: The Evolution of Institutions for Collective Action*, Cambridge University Press, Cambridge, UK and New York

Regional Shifting Cultivation Policy Dialogue Workshop for the Eastern Himalayas (2004) *The Shillong Declaration*, International Centre for Integrated Mountain Development (ICIMOD), Kathmandu, http://www.icimod.org/resource/1250, accessed 8 December 2015

San Thein (2012) *Study on the Evolution of the Farming Systems and Livelihood Dynamics in Northern Chin State*, Group of Research and Exchange of Technologies (GRET), available at http://www.burmalibrary.org/docs17/Evolution_of_Farming_Systems-Chin_State.pdf, accessed 8 December 2015

San Win (n.d.) '*East and Southeast Asia Subregional Workshop for GEF Focal Points: National GEF Priority Setting*, Global Environment Facility, Washington, DC, www.thegef.org/gef/sites/thegef.org/files/gefcs/docs/61.ppt, accessed 9 December 2015

Scott, J. C. (2009) *The Art of Not Being Governed*, Yale University Press, New Haven, CT

Springate-Baginski, O. (2013) *Rethinking Swidden Cultivation in Myanmar: Policies for sustainable upland livelihoods and food security*, University of East Anglia/Pyoe Pin, http://www.burmalibrary.org/docs20/Springate-Baginski-2013-Rethinking_Swidden_Cultivation_in_Myanmar-en-red.pdf, accessed 6 December 2015

Stevenson, H. N. C. (1937) 'Land tenure in the Central Chin Hills of Burma', *Man* 37, pp44-49

Stevenson, H. N. C. (1943) *The Economics of the Central Chin Tribes*, Times of India Press, Bombay

Tannenbaum, N. and Kammerer, C. A. (eds) (2003) *Founders' Cults in Southeast Asia: Ancestors, Polity, and Identity*, Monograph 52, Southeast Asia Studies, Yale University, New Haven, CT

van Schendel, W. (2005) 'Geographies of knowing, geographies of ignorance: Jumping scale in Southeast Asia' in P. Kratoska, R. Raben and H. Nordholt (eds) *Locating Southeast Asia: Geographies of Knowledge and Politics of Space*, Singapore University Press, Singapore

Wittfogel, K. A. (1957) *Oriental Despotism*, Yale University Press, New Haven, CT

World Bank (2009) *Country Social Analysis, Ethnicity and Development in Vietnam*, World Bank, Washington, DC, http://siteresources.worldbank.org/INTEAPREGTOPSOCDEV/Resources/499760ESW0Whit1C10VietnamSummary1LR.pdf, accessed 9 December 2015

Notes

1. Philippines' Republic Act No 8371 (Congress of the Philippines, 1997): 'An Act to recognize, protect and promote the rights of indigenous cultural communities/indigenous peoples, creating a national commission on indigenous peoples, establishing implementing mechanisms, appropriating funds therefore, and for other purposes', and Cambodia's Land Law (2001).
2. An official translation of the Laws and an unofficial translation of their associated rules by UN-Habitat were used in preparing this chapter.
3. Individual villagers who wish to register a land title under the system introduced in 2012 may face high expense, institutional corruption and a complicated system of passage through Farmland Administration Boards at different levels. The Central Farmland Management Body, which operates under the Ministry of Agriculture and Irrigation, is independent of the judicial system and there is no procedure for independent hearing of any grievances.
4. The translation of the law used in this chapter was a regular translation, while details of the Rules were based on UN-Habitat's informal translation.
5. The duration of concessions granted under the VFV Law can contrast sharply with those granted under the Foreign Investment Law 2012, which regulates foreign investment and land-use concessions and allows land grants of up to 70 years.
6. Unofficial translation of the Farmland Law rules found in the Ministry of Agriculture and Irrigation's Notification No 62/2012 14 *waxing wagaung* 1374 (31 August 2012), on promulgation of rules with reference to section 42, subsection (a) of Farm Land Law, with the approval of *Pyidaungsu* Government. Although the clauses in the Farmland Law related to *taungya* were inserted by the *Pyidaungsu Hluttaw* (Assembly of the Union) after lobbying by the Land Core Group and other civil society organizations, the rules were drafted by the Ministry of Agriculture and Irrigation without public consultation or parliamentary oversight.
7. Carey and Tuck write: 'Falam chiefs have a sacred grove near the place where the suspension bridge is built [and] we cut some trees for bridging material; the Chins were much alarmed and prophesied disaster and, when an accidental fire shortly afterwards destroyed a large portion of their capital, they at once exclaimed that the disaster was the result of our inconsiderate conduct to the spirit of the village, who had naturally avenged the insult not on us, but on them, as they are under his control' (Carey and Tuck, 1896, Vol. I, p198).
8. During colonial rule, French researchers on the border between Cambodia and Vietnam in Indochina noted in the 1890s that 'there is no land in the Moi country [Central Highlands] without an owner, but that most of it was collectively owned'. They continue to explain that 'For the duration of its use the *ray* [swidden field] is the private property of the one who has cleared it; once abandoned, it reverts to the community, and the person who had cultivated it retains no rights to use it'.
9. Northern Chin villages are different from those in southern Chin, where claims by (aristocratic) ancestral clans prevail and individual families who are descendants of high status clans can claim ownership of whole mountainsides, i.e. what in northern Chin is a *lopil*, but in southern Chin is called *kho (k)hmang*.
10. An orthophoto, orthophotograph or orthoimage is an aerial photograph geometrically corrected so that the scale is uniform: the photo has the same lack of distortion as a map.

IV. CONCLUDING SECTION

It was while passing through slums on the outskirts of Thimphu, in Bhutan, that the Editor photographed this child and captured her quizzical expression. If given the opportunity to voice her opinion, she might well have expressed the hope for wiser policy decisions, so that her generation might inherit a healthier planet.

Sketch based on a photo by Malcolm Cairns.

51

SHIFTING CULTIVATION POLICY DECISIONS THAT COUNT

*Kenneth G. MacDicken**

Introduction

Policy discussions often begin with the assumption that policy is effective in shaping land management. Yet we live in a world where policy is not often translated into effective regulation, much less enforcement, and the desired outcomes are left behind. We are surrounded by policies that have not always worked in other sectors: arms, narcotics, pollution control, hunger and poverty alleviation, to mention just a few. Land-use policy often contains the right words, but perhaps more often than not, it fails to deliver in the field, where it ought to matter most.

The fact that policies – and the regulations that seek to put them into practice – are often vague, ambiguous and frequently changed makes one question how important they are in shaping the lives of shifting cultivators. Chazée (ch 8) documents 24 key drivers of swidden farmer mobility in the Lao PDR from 1830 to 2000, many of which changed in the past 30 years. Often, policy and regulatory ambiguity itself tends to leave so much room for interpretation that those with access to power can, in essence, 'make the rules' related to shifting cultivation, local people's rights and access to land.

Falvey (ch 3) states clearly that policy must be as dynamic as the environment to which it applies, which means that there can be no single 'correct' policy related to shifting cultivation. He notes that policies that affect shifting cultivators have not been directed at their welfare or at the environments in which they live.

This and the following concluding chapter cite works in this volume with the name of the relevant author(s) and a chapter number, for example, Mertz and Bruun (ch 2). Citations of works in this volume are not listed in References at the end of this or the following chapter. However, all other citations are listed in References.

* Dr Kenneth MacDicken was formerly the Senior Forestry Officer, Global Forest Assessment and Reporting, Forestry Department, Food and Agriculture Organization of the United Nations, Rome.

This chapter seeks to extract a few policy lessons from the vast experience reported in this volume and its predecessor, with the intention of highlighting key policy choices and what has, and has not, worked.[1] Experience and a measure of common sense indicate that there are several key policy questions that directly affect shifting cultivation, and there are no doubt many more policies, regulations and market drivers that indirectly influence the extent and nature of shifting agriculture. This chapter is limited to those policies and regulations that directly affect shifting cultivators and shifting cultivation.

The key questions faced by governments include:

- Should shifting cultivation be allowed or prohibited?
- If allowed, what does the government need to regulate?
- If allowed, what does the government need to do to support it?
- If it is prohibited, what are shifting cultivators to do?

Should shifting cultivation be allowed or prohibited?

This is the most fundamental issue in shifting cultivation policy: should the practice be allowed to exist? Everything else that can be done by the public sector really relates to this fundamental choice. It is a choice about both a farming system and, in many cases, a way of life for traditional shifting cultivators. It is clear from many of the chapters in this volume and its predecessor that those who think deeply about shifting cultivators see them either as positive, useful citizens who use a diverse, sustainable set of agricultural practices or as primitive farmers who destroy the environment and are resistant to change. There are few views between these extremes.

It is also clear that perhaps the first policy decision that governments make regarding shifting cultivation, consciously or not, is whether to allow it or to prohibit it.

Allowing shifting cultivation requires the ability to perceive this farming system as an option that, just like sedentary agriculture, has advantages and disadvantages compared to alternative land uses. Prohibiting swidden agriculture rejects the idea that this set of practices has sufficient value to allow them to exist. Kunstadter (2015) makes the case that shifting cultivators are still the group that is blamed for most environmental problems in the uplands of the developing world. This remains true throughout Asia where soil erosion, floods, fires, deforestation and forest degradation are often said to be caused by shifting cultivators – regardless of how conservative their farming practices are or how untrue these claims may be.

Many of the chapters in this volume and its predecessor point to the lack of evidence-based decision-making when it comes to swidden farmers and the practices of shifting cultivation. Shaoting Yin (2015) provides a clearly stated example from China, where academic and social perspectives are at times erroneous and biased in ways heard many times over the past two centuries. Facts are clearly vital for making just, effective policies that improve lives and protect the environment. Unfortunately, there are few examples cited where facts have been used in setting policies and regulations related to shifting cultivation.

Karki (ch 49) makes a powerful review of policies that can help transform shifting cultivation to meet modern day realities. Because the local circumstances of shifting cultivators, their surrounding populations and governance realities vary so much, he recommends local learning as a key step in setting policy. This includes a serious local effort to learn from past policies and to use these lessons to modify those that have not worked, or had negative results, and to scale up those that contain positive practices. He also notes opportunities, on one hand to use participatory-action research to innovate, and on the other to tap external financing for the environmental services provided by longer fallows.

However, the discrepancy between reality and rule is an all too frequent characteristic of shifting cultivation policy. Van der Ploeg and Persoon (ch 1) note that this is the case in Indonesia and the Philippines. They also note the limitations of an overly complex legal framework and an ineffective judicial system, combined with political interference, institutional conflicts, low staff morale and rent-seeking as inhibitions to the implementation of shifting cultivation policies.

Shifting cultivators themselves seldom have access to governments in ways that help them to be heard as a group. They generally have too little influence, power or connections to make their case. Guérin (ch 7) observes that the modern-day expropriation of land used by shifting cultivators is strikingly similar to the processes used during European colonial expansion, which were often characterized by uncompensated taking of land. Minarchek (ch 47) points out that a critical court decision that potentially reverts some 40% of Indonesia's state-owned forest to indigenous communities leaves open the question of the legal existence of shifting cultivation. The link between indigenous community ownership of forests and the right to use swidden farming practices on those lands seems obvious on one hand – but is not yet recognized by the government on the other.

Governments that have policies to ban shifting cultivation have generally been unsuccessful in implementing those policies. In Bhutan, Dukpa (ch 22) points out that while shifting cultivation is no longer recognized by law, it is still practised because the ban was never strictly enforced. He further notes that the ban did not work because no alternatives existed. The lack of options for shifting cultivators, or the lack of government resources to either enforce land tenure regulations or to help shifting cultivators make the transition to other livelihoods, makes it practically impossible to stop shifting cultivation without major violations of human rights or further impoverishment of farming families who are already resource-poor.

In Vietnam, shifting cultivation is formally banned, with responsibility for land allocation moved from traditional leaders to new village-based forest-management boards (Bayrak et al., 2015). This system seems to have increased food shortages, disadvantaged the poorest villagers and greatly reduced the spiritual values that were traditionally a crucial part of forest management.

Some changes are taking place to at least put an end to discouragement of shifting cultivation. The Shillong Declaration of 2004 recommended the revision of policy to shift it away from disincentives for shifting cultivators and move it towards benefiting them, including land-tenure security, research and extension, market development and availability of credit (Teegalapalli and Datta, ch 42).

If allowed, what does the government need to regulate?

Land tenure

The lack of secure land tenure is by far the most frequently cited constraint to evolution of shifting cultivation into more sedentary forms of agriculture. Policy has generally failed to lead to laws and regulations that effectively give shifting cultivators secure rights to land. While there are some who argue for implementation of customary land tenure for shifting cultivators, it can be argued that while this may be desirable, governments are generally uninterested or unable to empower a set of diverse customary land-tenure arrangements.

Examples from China, Vietnam, Thailand and India all note that governments have too often failed in land reallocation policies, resulting in the destruction of traditional decision-making that has helped to sustain shifting cultivation for centuries. There are too few examples of successful land-tenure improvements that help swidden farmers, although there are many academic voices calling for such reforms. Even the availability of new sources of external finance such as REDD+ are likely to fail unless land-tenure issues are first resolved (Catacutan et al., ch 25; de Royer et al., ch 12).

Fallow length

Although sedentary agriculture is not often regulated in terms of soil erosion rates, soil carbon or fallow length, some have called for the setting of minimum periods for which swiddens are left fallow to allow forest recovery and soil rejuvenation. These minimum periods generally range between 10 and 20 years, although this is a range that needs careful evaluation based on ecozone, fallow composition, topography, climate and soil type. Several authors in this volume have observed that government policies in the Lao PDR are reducing the fallow period, resulting in increased invasion by herbaceous plants, lower yields and increased site degradation (Ducourtieux, ch 15, ch 37; Srikham, ch 34; Heinimann et al., ch 45).

Yet there are costs involved in longer fallows. Additional land is required, larger areas need to be managed or protected, time must be spent in moving to more distant sites, and so on. Ducourtieux (ch 15) notes that longer fallows also provide environmental services that should be rewarded, not discouraged. Payment for carbon, water and biodiversity services provided by longer fallows fits with current thinking about paying some of the costs of providing these services, yet shifting cultivators seem to have been largely forgotten as providers. The recommendation for paying for longer fallow periods is a sound one that cries out for action in international and domestic schemes that provide payments for environmental services.

Environmental restrictions

Shifting cultivation most likely needs regulation when it is practised on excessively steep slopes, adjacent to water courses or in sensitive watersheds. Useful restrictions applying to fallow length may also be dependent on local site conditions. In this regard, restrictions commonly used for commercial forest harvesting may be useful; for example, no shifting cultivation within 100m of a watercourse or on slopes of greater than 35%.

The advent of climate change mitigation through Reduced Emissions from Deforestation and Degradation (REDD+) poses a new challenge for shifting cultivation policy. Shifting cultivation is perceived by some as a form of forest degradation that contributes to anthropogenic emission of greenhouse gases (Fox et al., 2009; van Vliet et al., 2012). There are many greenhouse-gas emission uncertainties related to land-cover and land-use transitions (Ziegler et al., ch 11) that need to be resolved quickly as REDD+ moves forward. Ziegler and colleagues note some of the many questions that relate to REDD+ impacts on shifting cultivators, including how forest farmers might be compelled to adopt new crop technologies that ostensibly reduce land degradation. De Royer et al. (ch 12) take the argument further and suggest that if land rights and boundary issues are underestimated and left unattended they will have a strong negative effect on the success of REDD+. Given the potential size of REDD+ investments, it is vital that shifting cultivation practices and land degradation be evaluated carefully – not at a broad global or national scale, but at the 'cultivation practice' scale.

If allowed, what does a government need to do to support it?

Regulations have value only when they are well-designed and enforceable. How can a government help to make shifting cultivation more successful, in terms of productivity and profitability? A number of chapters in this volume and its predecessor demonstrate how education, credit, access to markets and extension can help.

It has also been suggested that governments need to support local long-term resource-use systems, as opposed to non-local shifting cultivation, which appears to be less sustainable. The idea that traditional cultivators have a longer-term view and a higher capacity to sustain their farming systems than those who are either new to the practice or are less familiar with local environments, is not a new one. The suggestion that governments should give regulatory support to local cultivators and discourage non-local farmers is both bold and difficult, even though it may make perfect sense from a long-term sustainability perspective.

The pace of policy-related change is often overlooked as an important factor in the success or failure of shifting cultivation policy. Cohen (ch 31) notes that despite the many limitations of policy and regulation in Thailand, there has been a more gradual transition to crop diversification, integrated rural development and people's participation than in the Lao PDR. While this example relates to the transition from opium cultivation to more permanent cultivation of legal crops, the reality is that planned changes in traditional farming practices and livelihoods often take time – and that 'quick fixes' often do not work. Leemann and Nikles (ch 33) note that the abrupt loss of ancestral lands and disruption of the traditional livelihoods of the Bunong people in northeastern Cambodia has been traumatic. They note that one of the most traumatic aspects of these changes is the suddenness of change and the exceptional stress this brings.

Many authors in this volume point out the fact that policies evolve through trial and error, with perhaps a longer record of error than of success. Recognizing this reality is critical for those who wish to influence, to create, or to modify policies related to shifting cultivation.

Education

For decades, education systems have emphasized the negative environmental effects of shifting cultivation. For shifting cultivators to be able to make progress in improving their practices, it is important for the public to understand that there is possibly a place for alternative cropping systems such as shifting cultivation. For practitioners, it is also important to know how to adapt to new opportunities. These may include new varieties, crop rotations or species mixtures. Making shifting cultivation a part of elementary- or secondary-education curricula would increase awareness of both the opportunities and risks associated with shifting cultivation and the importance of these for traditional shifting cultivators.

Credit

In some cases, additional capital can help farmers to obtain better inputs, such as planting materials and fertilizers, to improve product quality or introduce processes for adding value. Market access can also require additional funds for transport and storage. Credit can be a valuable tool for those who are good fiscal managers, yet shifting cultivators seldom have access to credit because they have neither secure land tenure nor a history of financial management. Policies that support access to credit for shifting cultivators could provide new opportunities for improved earning capacity.

Fallow improvement

This volume, its predecessor, and Cairns (2007) describe a wide range of improved fallows. These include herbaceous, shrub and woody fallows of various compositions and duration. Government policies and regulations that encourage improved fallows may help to improve crop yields, diversify production and help to achieve shorter fallow lengths without environmental degradation.

If shifting cultivation is prohibited, how can policies help swidden farmers find positive alternatives?

While government policies and regulations may prohibit the practice of shifting cultivation, they seldom provide good alternatives for the practitioners. Most often the expectation is that swidden farmers will find positive options on their own. However, there are some good examples of government support that has helped shifting cultivators to move to other livelihoods.

Transition to other farming systems

Darlong (ch 18) notes that in the shifting cultivation communities of Nagaland (India) only about 8% of young men and women would prefer to practice shifting cultivation if there were other alternatives. The same chapter also describes successful transitions from shifting cultivation into agroforestry mixtures, cash cropping and horticultural plantations. Similarly, Gunasena and Pushpakumara (2015) indicate that swidden farming is no longer profitable or sustainable in Sri Lanka and that the government needs to assist in transitions to other forms of agroforestry, including raising livestock.

A series of policy changes in Vietnam since the early 1960s has moved shifting cultivators into sedentary agricultural systems such as permanent estate crops, terraced rice fields or permanent cultivation with no fallow period (Catacutan et al., ch 25). These practices have not always been successful, in some cases forcing cultivators to revert to shifting cultivation in order to survive. Hong (ch 36) notes that while shifting cultivators in Vietnam were often accused of destroying large areas of forest, a shift to widespread coffee cultivation has caused unprecedented deforestation.

Financial assistance through the Reducing Emissions from Deforestation and Degradation (REDD+) programme is mentioned as a possible future means of helping farmers with the transition to more sedentary cultivation systems. This shift is welcomed by some, but to others it is a threat to cultural values and stability. A number of authors suggest caution in rushing to REDD+ due to a lack of understanding of degradation, tenure and social or economic impacts associated with large-scale land-use changes that might be required to enhance carbon stocks.

These transitions can be a real risk to communities, particularly as they affect lifestyle aspects other than economic development. Yin et al. (2015) explain how the introduction of large-scale rubber farming in China's Yunnan province has resulted in a steady loss of biodiversity, genetic resources and indigenous knowledge. The loss of cultural heritage as a result of transition to introduced farming practices is a serious threat that disrupts crucial links between the present and the past. The trade-offs between retention of indigenous ways and economic development are at least sensitive issues that should be decided locally, where conscious decisions can be made that will affect future generations.

Conclusions

Formulating, implementing and enforcing the right policies and regulations that directly affect shifting cultivation and its practitioners are towering tasks. When policy and regulation allow shifting cultivation, critical decisions must be made on how to help farmers ensure or improve long-term sustainability in a changing world. Many chapters in this volume and its predecessor note that local decision-making is vitally important in shaping upland farming practices. Most communities have not been encouraged to find their own balance of economic development and cultural identity. Some, such as the Karen in northern Thailand, have had the advantage of a well-developed sense of their cultural heritage and, as a consequence, have had a stronger voice than other groups in deciding their own future.

Forsyth and Walker (ch 9) suggest that the process of 'getting policies right' often means justification of positions and social roles based on differing perspectives rather than more dispassionate analyses. They argue convincingly that equitable representation of shifting cultivators requires use of diverse evidence and examination of the uncertainties of the impacts of their practices on the environment and livelihoods. Getting policy right is probably too simplistic, unless it allows for analysis of evidence and respect for the perspectives of swidden farmers. Most swidden-dependent communities have had neither the freedom nor the resources to help them to make choices about their future. These choices affect both their future economic options and the fate of their cultural identities. A key recommendation made repeatedly in these volumes is that governments should aim for policies and regulations that encourage local decision-making.

There is some hope that shifting cultivation may one day receive due recognition as a sustainable land use. The World Heritage Committee is considering the recognition of Karen shifting cultivation – an act that would increase the positive visibility of sustainable shifting cultivation and encourage the Karen and others to carry on. The Government of Thailand has recently recognized the value of the Karen way of life, which was described by a national-level committee as a form of 'sufficiency economy'. This is a substantial step away from decades of government efforts to eliminate shifting cultivation in northern Thailand.

Government policies can help to recognize the value of diverse farming systems – including shifting cultivation – and help those who practise sustainable forms of swidden agriculture.

References

Bayrak, M. M., Tran, N. T. and Burgers, P. (2015) 'Formal and indigenous forest-management systems in central Vietnam: Implications and challenges for REDD+', in M. F. Cairns (ed.) *Shifting Cultivation and Environmental Change: Indigenous People, Agriculture and Forest Conservation*, Earthscan, London, pp319-334

Cairns, M. F. (ed.) (2007) *Voices from the Forest: Integrating Indigenous Knowledge into Sustainable Upland Farming*, Resources for the Future Press, Washington DC

Fox, J., Fujita, Y., Ngidang, D., Peluso, N., Potter, L., Sakuntaladewi, N., Stugeon, J. and Thomas, D. (2009) 'Policies, political economy and swidden in Southeast Asia', *Human Ecology* 37, pp305-322

Gunasena, H. P. M. and Pushpakumara, D. K. N. G. (2015) '*Chena* cultivation in Sri Lanka: Prospects for agroforestry interventions', in M. F. Cairns (ed.) *Shifting Cultivation and Environmental Change: Indigenous People, Agriculture and Forest Conservation*, Earthscan, London, pp199-220

Kunstadter, P. (2015) 'Swiddeners at the end of the frontier: Fifty years of globalization in northern Thailand, 1963-2013', in M. F. Cairns (ed.) *Shifting Cultivation and Environmental Change: Indigenous People, Agriculture and Forest Conservation*, Earthscan, London, pp134-177

Shaoting Yin (2015) 'Shifting cultivation and its changes in Yunnan province, China', in M. F. Cairns (ed.) *Shifting Cultivation and Environmental Change: Indigenous People, Agriculture and Forest Conservation*, Earthscan, London, pp134-177, pp122-133

van Vliet, N., Mertz, O., Heinimann, A., Langanke, T., Pascual, U., Schmook, B., Adams, C., Schmidt-Vogt, D., Messerli, P., Leisz, S., Castella, J-C., Jørgensen, L., Birch-Thomsen, T., Hett, C., Bruun, T. B., Ickowitz, A., Vu, K. C., Fox, J., Cramb, R. A., Padoch, C., Dressler, W. and Ziegler, A. (2012) 'Trends, drivers and impacts of changes in swidden cultivation in tropical forest-agriculture frontiers: A global assessment', *Global Environmental Change* 22, pp418-429

Yin, L., Xue, D. and Wang, J. (2015) 'Rubber plantation, swidden agriculture and indigenous knowledge: a case study of a Bulang village in Xishuangbanna, China', in M. F. Cairns (ed.) *Shifting Cultivation and Environmental Change: Indigenous People, Agriculture and Forest Conservation*, Earthscan, London, pp811-825

Note

1. This volume is the second in a trilogy about shifting cultivation in the Asia-Pacific region. The first – the predecessor to this volume – was *Shifting Cultivation and Environmental Change: Indigenous People, Agriculture and Forest Conservation* (M. F. Cairns (ed.) (2015) Earthscan, London). The third volume (forthcoming) has the working title *Farmer Innovations and Best Practices by Shifting Cultivators in Asia-Pacific*. It was in production when this volume went to press. Production of the trilogy followed the publication in 2007 of the landmark volume *Voices from the Forest: Integrating Indigenous Knowledge into Sustainable Upland Farming* (M. F. Cairns (ed.), Resources for the Future Press, Washington, DC).

52

LESSONS LEARNED FROM THE IDENTIFICATION AND IMPLEMENTATION OF POLICIES AFFECTING SHIFTING CULTIVATION IN THE ASIA-PACIFIC REGION

A summary

William C. Found[*]

Introduction

How can one summarize, synthesize, or draw conclusions from the 64 detailed and highly informative chapters in this book and its electronic 'Addendum'? It is doubtful that any concluding chapter could do justice to the immense quantity and quality of information presented in this volume. Fortunately, other parts of the volume pose the book's major questions, summarize important commonalities in several of the chapters and describe some of the latest thinking concerning the general issue of policies for shifting cultivation in the Asia–Pacific region. In particular, the reader is referred to Colfer (Foreword), Cairns (Preface), Mertz and Bruun (ch 2), Falvey (ch 3), Oughton (ch 4), Karki (ch 49) and MacDicken (ch 51). Reading these sections will give the reader an excellent overview of the book's central issues. But they barely begin, of course, to reflect the important content for specific locations, detailed in the other contributions.

My approach to this chapter will be to summarize the volume's major themes and cite relevant chapters to document my generalizations. Then I will analyse policies affecting shifting cultivation in the region from the standpoint of their implementation. How have various policies, devised over centuries, been (or not been) implemented? For whom are the policies intended? Who are the beneficiaries? What resources, incentives, administrative structures and mechanisms have been used in their

[*] PROFESSOR EMERITUS WILLIAM C. FOUND, Department of Geography and Faculty of Environmental Studies, York University, Toronto, Canada. Much of Professor Found's research and related teaching over several decades has been international in scope, with major projects centred in a variety of Caribbean locations, East Africa and Indonesia. His work has included visiting research/teaching appointments at Harvard, Umeå (Sweden), the University of Toronto, and the Academies of Sciences in Cuba and the Soviet Union.

implementation? This sort of analysis will provide insights into the identification of policies, into attempts to put them into place, and into the fundamental questions of 'what works?' and 'what lessons should be learned?'.

Many contributors to the volume tend to be pessimistic about the fate of shifting cultivators in the region. While the volume is rich in analyses of policies and programmes related to swidden agriculture, it documents very few successes in addressing its problems in a changing world. Yet there are indications that the region might be at a point of change, and that more successes may lie around the corner.

A brief summary of previous chapters

This book concerns shifting cultivation in the Asia–Pacific region, a practice now restricted almost exclusively to highland – and often marginalized – parts of the area. It is practised most commonly in countries within which there are areas of poor socio-economic status, at least as measured by conventional indicators. The main body of the volume is divided into sections that are designed to introduce the major themes, conditions in specific parts of the region and various aspects of the modification of shifting cultivation under pressure from national policies, migration, changes in population density, growth in commercialization, cultivation of new crops, pressure from powerful companies, moves to at least partial sedentary agriculture, and non-governmental organizations. Most chapters provide detailed accounts for specific countries or sub-regions, with several papers each describing conditions within the Lao PDR, Thailand, India, Indonesia, Vietnam and Bhutan. Issues in Bangladesh, Cambodia, China, Japan, Myanmar, Nepal, the Philippines and Timor-Leste are considered in one or two chapters each. Several chapters address conditions throughout the region, and others deal with specific sub-regional issues: historic French Indochina; Eastern Himalayan countries; the Mekong River Valley; and areas affected by the UN Programme 'Reducing Emissions from Deforestation and Forest Degradation in Developing Countries' (UN-REDD and REDD+). Shifting cultivation in Malaysia (largely restricted to the states of Sarawak and Sabah) and Papua New Guinea is described within chapters that concern the entire region.

Many of the chapters address similar topics, not because the authors were required to do so, but because of a high degree of commonality in the authors' views. Several, including Cairns (Preface), ask 'How is shifting (or swidden) cultivation defined?' The question arises because the classic stereotype of small groups living in isolation, clearing and burning plots within an apparently infinite forest and then moving on to new plots as soil nutrients diminish, is nowadays rarely if ever found. This image of humans perfectly integrated within local ecosystems, occupying only a small portion of the forest, and returning to former swiddens only after many years of forest regrowth, forms, at best, one end of a spectrum. The other end of the spectrum includes sedentary or settled farmers who use fire to clear only a small plot or two of previously cultivated land in order to supplement their agricultural enterprises. These same farmer households may access employment beyond agriculture, and may be thoroughly immersed in a modern, commercial economy. Several authors present classifications of shifting cultivation in an effort to describe real contemporary conditions. The description by Karki (ch 49), who estimates that 200 million people in Asia are involved with shifting cultivation on 300 million hectares of land, is a good example:

> (1) traditional swiddening: this normally follows a cycle of clearing the forest with fire, a short period of cultivation, and abandonment to allow forest recovery; (2) rotational agroforestry: farmers are on the path of transformation from traditional practices to different intensification or diversification options; (3) composite shifting cultivation: farmers maintain permanent wet-rice farms in the lowlands and shifting cultivation fields in rainfed uplands; and (4) temporary farming: usually practised by nomadic herders and pastoralists who cultivate a patch of land for either one season or a year, then move on.

Others (e.g. Forsyth and Walker, ch 9) make an important distinction between cultivators who clear primary forest in a moving frontier, and those who carefully rotate their clearings among sites that are allowed to regenerate forest before they are cleared and cultivated again. They illustrate the point by contrasting Thailand's Karen people ('forestry guardians', who rotate their cultivation sites) with the Hmong ('pioneer shifting cultivators'). The former are viewed positively in local political rhetoric, and the latter are vilified for their ruthless invasion of the forest. In this case, the classification of shifting cultivation type is enmeshed in politicized environmental narrative, tied to questions of unwelcomed migration and the larger-scale politics of the region.

Several chapters begin with an historical survey of governmental policies towards shifting cultivation, going back to the colonial period when the practice was legally forbidden by the colonial powers. An 1874 quote attributed to Baden-Powell, a British forestry officer in India, is repeated in two of the chapters (Darlong, ch 18; Saikia and Bhaduri, 2017). Even in the following edited and reduced version, the vehemence and misunderstanding of colonial officials is clear:

> [T]he system is so wasteful that somehow or the other it must be put a stop to ... It consists in destroying a large and valuable capital to produce a miserable and temporary return. To put a stop to it is only to anticipate by a few years the natural determination of the system, which will happen if the system continues long enough, because there will be no more forest to cut down and burn. ... Efforts should be made to change people to permanent agriculture.

Ducourtieux (ch 15) and McKinnon (ch 6) go further by attributing racism and ignorance to colonial officials, and Ducourtieux adds 'greed' to the list. In his Preface, Cairns refers to the 'cultural genocide' of shifting cultivators. The descriptions of colonial policy towards shifting cultivation form some of the most detailed and well-documented portions of several chapters. There is general agreement, explicit or implicit, that colonial governments did not understand the detailed workings of shifting cultivation, nor the complex relationships between the agricultural practices and the cultures of shifting cultivators. However, they did understand that acquiring and using the land of shifting cultivators would lead to commercial gain by private investors and enrichment of government tax coffers.

Few of the chapters describe in detail the inner workings of either the technology or the cultures of shifting cultivators, during historical or present times. Saikia and Bhaduri (2017) come closest to describing how traditional shifting cultivators reflect a radically different culture – even a way of thinking – from those coming from 'modern' commercial traditions. The agricultural and cultural systems of shifting cultivators are dynamic, changing as conditions require; their practices have evolved from millennia of trial-and-error experience rather than short-term experiments based on cause-and-effect formulations. Some of the patterns of cultivation and society that have emerged from the past may not appear to reflect conventional conscious choice. Rather, they may be attributed to behavioural patterns best perceived as a mix of pragmatic and ritual undertakings (Found, 1971, pp124-134). The kind of learning reflected in these practices is at odds with the so-called 'rational' mind of European conquerors, and beyond the understanding of Westerners except for the few with the time, patience, curiosity, communication skills, training and open-mindedness to learn (e.g. Conklin, 1957 and 1975; contributors to Cairns, 2007 and Cairns, 2015).

A few chapters provide concrete descriptions of important aspects of the shifting cultivation process. These are very important contributions since a major objective of the book is to determine what policies 'work' for shifting cultivators, and some of the answers may lie partially hidden in their current practices. Oughton (ch 4), Guérin (ch 7), Chazée (ch 8), Springate-Baginski (ch 13), Riba (ch 19), Tripathi et al. (ch 20), Catacutan et al. (ch 25), McArthur et al. (ch 27), Leemann and Nikles (ch 33), Ducourtieux (ch 37), Gillogly (ch 38), Fukushima (ch 41), Hoare (ch 48) and Andersen (ch 50) provide particularly significant analyses of both the technical and cultural aspects of shifting cultivation in a variety of settings. The observations have a general similarity, but the most important lesson learned from them is that the details of crop selection, the sizes of the areas cleared and burned, the length of fallow, the amount of human labour, the degree of participation in commercial agriculture, the presence or absence of plots of permanent cultivation, the use or non-use of domesticated animals, the decision-makers who decide when and where to clear new plots for cultivation, the adoption of 'Green Revolution' crops and technology, and myriad cultural practices (e.g. marriages timed to shifting cultivation cycles) vary greatly from one place to another. Even within a village, shifting cultivation practices can vary, depending on local elevations, soil and climatic conditions, land tenure, population density, use of exotic species (e.g. teak, coffee, oil palm, rubber, cacao, opium), and pressure from governments or local leaders. The authors avoid treating shifting cultivators as romantic stereotypes, somehow living in perfect harmony with nature. Rather, they are described as people constantly challenged by changing contexts that require adjustments and survival skills, usually in the face of governments (colonial or modern) seeking to eradicate their traditional practices. Their mistakes or weaknesses are also acknowledged (e.g. different communities are often compared, and the relative successes of some cultivators in adapting are gauged and compared to less successful groups). Examples of conflict between groups of shifting cultivators are cited, as is the duplicitous behaviour of some village members in seeking to take advantage of commercial opportunities. One can only describe the commonalities of shifting cultivators at a highly generalized level, although almost all examples in the book refer to marginalized communities in comparatively remote, hilly locations. The greatest anomaly from the general pattern is the study by Takahashi et al. (ch 10) of the cultivation of grasses through periodic shifting and burning in a park setting in southern Japan. All other cases involve the use of forest land. The preservation of a fire-based traditional land use in the Japanese grasslands is held up as a principle that could be extended to shifting cultivation more widely in the region, with the proposition that if Asia's most advanced economy sees benefit in preserving such a land-use system, then other, far less developed nations in the region should not be so quick to label similar systems as 'primitive', and having no place in a modern economy.

Positive views of shifting cultivation, but recognition that it is disappearing

Virtually all of the authors express very positive views regarding shifting cultivation, along with the realization that it is gradually disappearing. Their enthusiasm for the practice reflects much more than empathy for the large numbers of people whose cultures are seriously threatened by government policies, loss of land and contemporary commercial life. They reiterate the ecological advantages of swidden agriculture over sedentary monocropping, over the use of 'Green Revolution' varieties and technologies, its provision of medicinal plants and its comparative assurance of food security in sloping uplands. These features have been well documented in several recent publications (e.g. the first volume in this trilogy (Cairns, 2015), and its predecessor (Cairns, 2007)). They explain that the practice of burning reflects millennia of experience; that it not only provides the only feasible method available to traditional upland forest-dwellers for clearing land for cultivation, but also offers a measure of control over unwanted botanical and animal pests, provides high levels of potassium from ashes and access to nutritious nitrogen and phosphorus in the freshly exposed soil layers, and does it all with convenient access to nearby forest resources. It is also resilient to climate change (Springate-Baginski, ch 13). It can be viewed as a form of rotational agroforestry (Phuntsho et al., ch 17), capable of sustaining populations of modest density, within the bounds of available labour. They also assert that many of the secrets of successful swidden agriculture are invisible to the uninformed eye, and that these practices may make important contributions to new forms of land management as well as to the maintenance of cultural diversity. It is unfair to very large numbers of people, ecologically irresponsible, and wasteful of important traditional knowledge to recklessly terminate a practice that has provided a dependable livelihood to countless peoples over the ages.

At the same time, the authors recognize that the demographic, technological, political and economic contexts in the Asia-Pacific region have been rapidly changing, and that the current focus needs to be on finding alternative and viable forms of livelihood for shifting cultivators. Most of the book's chapters tell of how such alternatives are being pursued throughout the region, along with an appeal that change should occur slowly enough for proper involvement of shifting cultivators in the decision processes, and for proper testing of alternative forms of land use. Mertz and Bruun (ch 2) point out that cultivators would pursue alternatives if they existed and were accessible, but shifting cultivation tends to be terminated without available and acceptable alternatives. Even terracing is inappropriate for steeper lands (Oughton, ch 4). Fukushima (ch 41) argues that the botanical and ecological advantages of shifting cultivation are so great that the practice should simply be allowed to continue.

Importance of land tenure

Land Tenure and the mapping of property boundaries are important themes in several chapters. Weinstock (ch 5), Ironside et al. (ch 14), Kumar (ch 21), Leisz (ch 26), De Koninck and Thanh (ch 29), Herawati et al. (ch 46), Andersen (ch 50) and Minarchek (ch 47) offer compelling analyses of the importance of the legal description and the legal rights associated with property. Traditional shifting cultivation in the region involved communal land tenure, in which the community established and defended a territory, and within which rights might be held by individuals, households or kin groups. When population densities were low, forest land was abundant, and although conflicts over property rights between villages were not unknown (particularly when tribal groups from other areas migrated into an area), the legality of land tenure was seldom an important issue. Decisions regarding when and where to clear, burn and cultivate new plots were made communally, often by village elders. With the coming of European colonial powers, with their historic understanding of private and state-owned property, shifting cultivators found themselves forced into a legal system that was alien to their experience. This was critical, of course, when the new powers wanted to take the cultivators' traditional land, and when they and the newly independent states that followed them outlawed the burning of forest land. More recently, some states have tried, reluctantly, to acknowledge historical communal rights to land, although this legal recognition of communal tenure sometimes obscures individual property rights that are part of customary systems. The 1998 KDTI decree by the Government of Indonesia was a breakthrough (Herawati et al., ch 46), and the May 2013 decision by Indonesia's Constitutional Court is potentially the most important legal land-tenure shift in the region (Minarchek, ch 47). If the decision is upheld, and its full implications materialize, 40% of the land in Indonesia could become community-owned, subject to the traditional *adat* rules and customs of the inhabitants.

Many of the authors argue that the features and advantages of shifting cultivation systems and cultures will only flourish with the granting of communal land tenure. This is a question of both ecological integrity and human rights. Setting aside land as state forests or parks, even if swiddeners are granted community rights within them, has not proven to be a satisfactory solution. Cultivators are all too aware of the property boundaries, which seldom encompass all of the land that they once used for rotational forest agriculture. Of greater concern, burning of the forest is now forbidden in many government 'protected lands', a position that has been exacerbated by global environmental concerns. As noted, few of the chapters analyse in detail what actual 'management systems' are used in swiddening (the previous volume (Cairns, 2015) dealt with this issue), but most authors conclude that shifting cultivation is best served if cultivators are granted legal protection over the ownership and use of their traditional lands. The Shillong Declaration (Karki, ch 49; Teegalapalli and Datta, ch 42), drafted at a Shifting Cultivation Regional Policy Dialogue Workshop for the Eastern Himalayas in 2004 and published as a wide-ranging recommendation by the International Centre for Integrated Mountain Development (ICIMOD) in 2006, epitomizes belief in the great value of shifting cultivation, and management by traditional cultivators:

> That the practice of shifting cultivation be recognized as an adaptive forest-management practice based on sound scientific and ecological principles;
>
> That shifting cultivators be able to take part in making decisions about any new policies affecting them; and
>
> The significance of traditional institutions that regulate shifting cultivation be recognized (ICIMOD, 2006).

Mapping and the categorization of forests into different types are important to questions of land tenure. When governments undertake cadastral surveys using Western conventions, boundaries are imposed on landscapes where they may have been unknown in pre-colonial times. They can be used by governments to claim territory that will be taken up by modern commercial interests, yielding wealth to entrepreneurs and taxes to the government. Some recent projects have tried to involve shifting cultivators in determining appropriate boundaries – a kind of participatory mapping. This is just one of the innovations that led several authors to conclude that, despite a long history of exploitation and neglect of shifting cultivators, the situation is in a state of flux, with a real possibility for improvement.

Unique insights and participatory methodology

Most contributions to the volume describe conditions in particular locations without reference to complex theories or methodology, but Ducourtieux (ch 37) is an important exception. His application of the Chayanov family life-cycle model to upland villages in the Lao PDR provides unique insights into the dynamics of shifting cultivation and its relation to local cultures. By outlining the complexity of the system, he illustrates the great challenge of changing the system through modern interventions without creating high levels of differentiation in the economy and social position of specific families, and thus undermining the government's societal objectives. The other contemporary concept that underlies the work of a handful of authors is participatory methodology, including participatory action research, participatory rural appraisal and various forms of participatory consultation and decision-making. Ducourtieux (ch 37), Gillogly (ch 38), Hoare (ch 44) and Xu et al. (2017) all address the importance – and practical advantages – of involving shifting cultivators in the planning and implementation of changes designed either to protect their traditional livelihoods or to improve their lives. The successes of recent participatory processes mark one of the few bright spots in the adaptation and transformation of swidden livelihoods and culture.

Impact of sketches, drawings and maps

Sketches, drawings and maps greatly enhance the impact of the book's content. Paradorn's wonderful charcoal sketches of shifting cultivation scenes add a high degree of realism to the volume. He manages to highlight important features that might be difficult to spot clearly in traditional photographs. Similarly, Wiyono's ink drawings of botanical species that are often found in cultivated plots, distributed strategically throughout the text, provide images of vitally important plants that are clearer than might be achieved by mere photographs. The locational maps, which ensure that the location of each study area is clearly identified, are indispensable, providing a geographical context for each chapter.

Surprisingly little attention to a few topics

Climate change, an issue very high on the international agenda, has important implications for shifting cultivation, particularly since the burning of forest can appear to fly in the face of efforts to reduce the production of carbon dioxide. Ziegler et al. (ch 11), de Royer et al. (ch 12), and Bayrak and Marafa (2017) analyse how implementation of the UN's REDD+ programme, designed to combat climate change, can adversely affect shifting cultivators, even though traditional swidden practices may well sequester more carbon than alternatives (e.g. land clearing for intensive permanent agriculture). They argue that burning in swidden agriculture is normally well managed, and limited to only a small portion of forest land. This point, they say, is misunderstood by those not familiar with the details of the practice. Accepting payments to stop shifting cultivation may provide useful cash for governments or cultivators, but it may also strengthen the general perception that swiddening contributes to climate change and is not a sustainable system for people's livelihoods. Karki (ch 49) acknowledges that climate change is one of the most important international issues driving policy formation, and that shifting cultivation appears as an easy target. He also draws attention to adaptations that change the production of carbon dioxide, such as 'fireless shifting cultivation'.

A second issue receiving little attention is gender. Van der Ploeg and Persoon (ch 1), Suwannarat (ch 30) and Gillogly (ch 38) acknowledge that, as in all traditional societies, men, women and children play important and differentiated roles in the daily lives of shifting cultivation groups. Besides contributing to labour, women are typically the custodians of the household's seeds and manage the ritual and practical aspects of sowing and harvesting. Men tend to form the community decision-making groups, and take responsibility for particular aspects of the clearing and burning processes. In some communities, the full impacts of change in swidden cultivation are different for women and men; and sometimes women suffer most from adjustments to traditional systems. Some authors document recent successes in involving shifting cultivators in decisions over government programmes, and it is hoped that special efforts are made to involve both women and men in those discussions.

Politics: an overarching theme

Politics emerges as an over-arching theme in several of the chapters. Indeed, it cannot be avoided when dealing with large numbers of disadvantaged people practising shifting cultivation 'illegally', without formal land ownership and without access to political power. Formal policies are almost always the purview of central governments, not local groups, and marginalized communities of swidden farmers are normally targeted for attack and change. The vehemence of these attacks is intensified if the cultivators are viewed with suspicion. For example, are they migrants who may not mix well with local groups, or who may support threatening political movements (Wulandari, 2017)? Do the governments in power fear shifting cultivators' dispersed control of the rugged parts of the countryside; or do swiddeners' practices make the collection of government taxes infeasible (Kumar, ch 21)? While governments might try to administer policies and programmes with a degree of equity throughout their jurisdictions, they encounter groups who cooperate and those with whom they cannot share political goals, nor make progress in effecting change (Chazée, ch 28). Chazée also raises the question of human rights in his discussion of 'voluntary resettlement' in the Lao PDR. Ducourtieux (ch 15) describes the 'civilizing mission' assumed by colonial governments, their paternalistic attitude towards local peoples, and their use of 'divide and rule' tactics. He also explores the role of more contemporary political movements: the international environmental movement of the 1980s that led to greater restrictions on those who burn forests; and the aftermath of the 1992 Rio Earth Summit, which intensified such political action. Hoare (ch 44) and Teegalapalli and Datta (ch 42) consider the top-down/bottom-up balance in policy implementation and acknowledge that the primarily top-down practices of the past can be much improved in Thailand and India with a sharing of responsibility and power at different levels of the political hierarchy. One of the clearest examples of politics affecting swidden farmers is the cultivation of opium, which was once encouraged – and then banned – because of political pressures exerted by very powerful countries far removed from the Asia-Pacific region (Suwannarat, ch 30; Cohen, ch 31; Ducourtieux et al., ch 32).

Policy implementation in the Asia-Pacific region: A framework for analysis

While the authors of this volume present a wealth of important information about shifting cultivation policies in the Asia-Pacific region, the studies do not follow a common format and it is challenging to compare how different policies have – or have not – been implemented. In this section, I present a framework for analysing the implementation of policies, programmes and projects. I use this structure, which has proven useful in several other contexts, to draw together the main findings of the volume by examining the success of implementation.

The framework

Figure 52-1 shows the schematic for a framework that I have developed over the years, beginning with my collaboration with Donald Warwick (Harvard) in 1986, and amplified recently through work with Helen Hambly (2000), Rita Lindayati (2003) and Suzanne Hurley (2012). The framework allows one to identify the various components of the implementation process, to visualize how they interconnect, and to compare successes and failures for implementation in different settings. The overall implementation process is dynamic and assumes that changes will occur through successive evaluations of outcomes. The framework reflects much of the theoretical and practical work of several innovators in the field (e.g. Pressman and Wildavsky, 1973; Van Meter and Van Horn, 1975; Bardach, 1977; Grindle, 1980; Warwick, 1982; Mazmanian and Sabatier, 1983; Elmore, 1985; Matland, 1995; Najam, 1995; Brinkerhoff, 1996; and Crosby, 1996). Earlier published versions of this framework can be found in Found (1992 and 1999).

The implementation gap

A key starting point in analysing a policy's implementation success is to identify the gap between what was planned and what actually happened. This volume abounds with examples of shifting cultivation policies, programmes or projects where intended outcomes were never reached, and where unintended consequences were all too common. This is a trend that straddles the entire study period, from the early days of colonial rule through independence to the present. This is not to assert that the objectives of all of the policies have been inappropriate. But the number of failed implementations far outnumbers the successes. Van der Ploeg and Persoon (ch 1) provide the first clear example. A contemporary young woman is involved with shifting cultivation even though the practice has been outlawed since colonial times and her family has no legal right to the property being cultivated. What an apt introduction to a volume crammed with examples of failed policies and programmes!

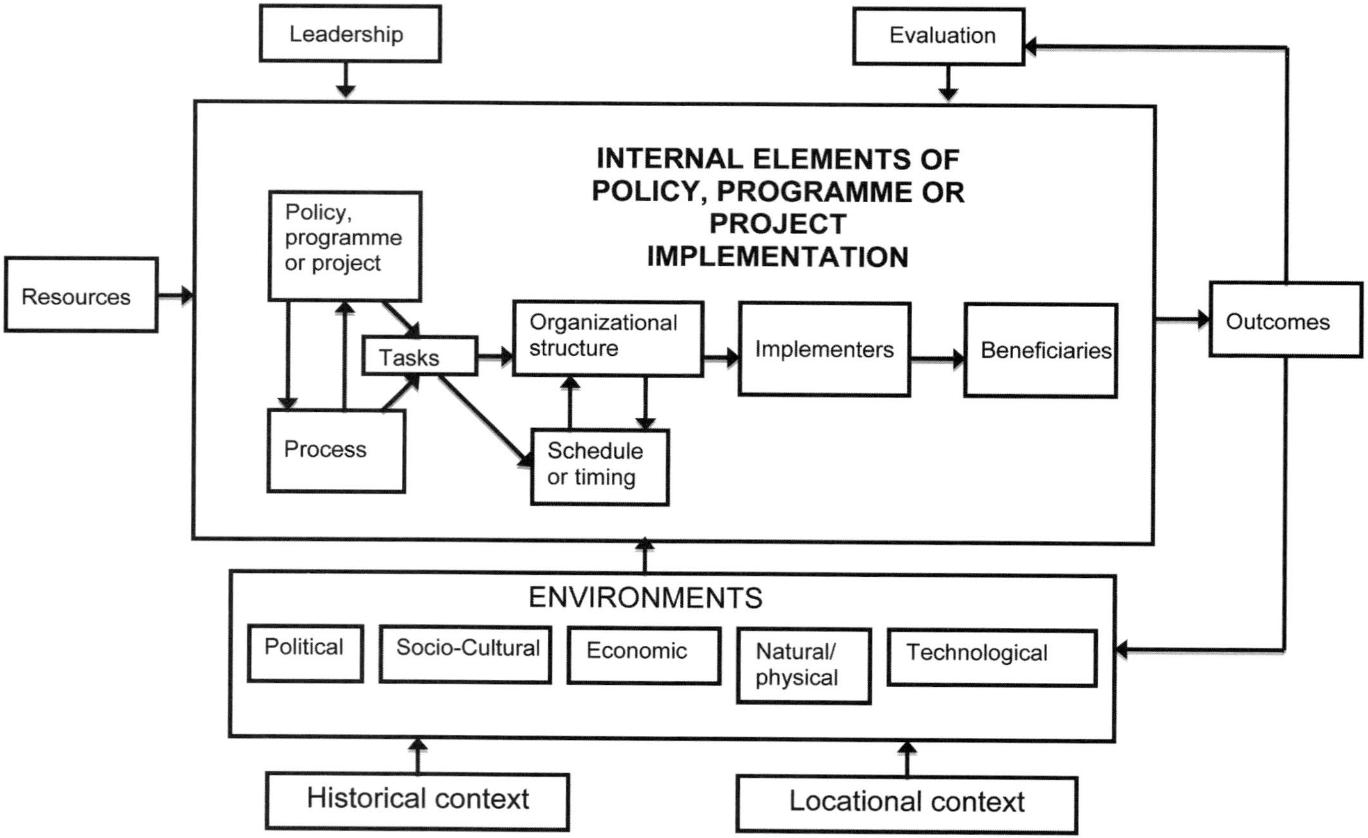

FIGURE 52-1: A framework for policy, programme and project implementation.

Source: William C. Found.

This creates a significant challenge if we are to find success stories or best practices for swidden agriculture policy. From the standpoint of shifting cultivators, of course, failed implementation of laws forbidding the burning of the forest represents a highly desirable outcome. On the other hand, as Ironside et al. reveal (ch 14), failure to implement Article 23 of Cambodia's 2001 Land Law, which supports shifting cultivation through the registration of traditional indigenous communities, would be a severe blow to swidden agriculture and its practitioners. This would represent an example of poor implementation of a good policy. The same question is raised by Herawati et al. (ch 46) and Minarchek (ch 47), concerning the possibility of the Indonesian Government recognizing and following through on the 2013 court decision to grant land titles to traditional communities.

Historical and locational contexts

As indicated at the bottom of Figure 52-1, we begin an analysis of any particular policy by recognizing the historical and locational contexts involved. These basic parameters provide the setting within which particular political, socio-cultural, economic, natural/physical and technological environments emerge. Each of these can have a profound effect on policy and the chances of successful implementation. The various chapters illustrate how policies within different political regimes have differed in detail and implementation, although it is noteworthy that governments of all periods and ideologies have tended to target shifting cultivation for eradication. Suwannarat (ch 30) relates the Thailand government's treatment of highland agriculturalists to its fear that they once represented a political threat.

Many chapters compare policy implementation in different socio-cultural environments. Hoare (ch 48) analyses the lack of agreement among village jurisdictions, and Forsyth and Walker (ch 9) contrast shifting cultivation undertaken by the Karen and the Hmong peoples. Differences in economic environments have a significant effect on the specifics of policies and their implementation, although it is sometimes difficult to distinguish clearly between political and economic contexts. Colonial governments had a primary concern with the business and wealth of citizens and corporations from the home country, which helped to justify their denial of property rights for local indigenous populations, and their use of government resources for armed forces and administrative systems to manage their new colonies (Ducourtieux, ch 15). The construction of roads, ports and railways to provide access to colonial resources helped to force shifting cultivators into marginal locations. Independent governments tended to proclaim support for poor people in the countryside, and they provided funding for a broad range of programmes designed to increase agricultural production and farm incomes. As described in many of the book's chapters, swidden farmers as a whole did not benefit from these initiatives, although some of those who were prepared to switch to at least some sedentary agriculture, adopt new crop varieties, use Green Revolution inputs, enter the commercial economy, or take jobs in towns did advance their material welfare (see Ziegler et al., ch 11; Springate-

Baginski, ch 13; Tripathi et al., ch 20; Chazée, ch 28; Pasicolan et al., ch 35; Hong, ch 36; Prill-Brett, ch 39; Tran and Pham, ch 40; Wulandari, 2017; Jyotishi and Manjula, 2017; Saikia and Bhaduri, 2017).

Karki (ch 49) analyses the impact of globalization on local economic environments, and the extension of commercial agriculture into traditional areas of shifting cultivation. His chapter makes the link between political and economic environments absolutely explicit. Similarly, Suwannarat (ch 30), Cohen (ch 31), and Ducourtieux et al. (ch 32) describe how opium cultivation was first encouraged – and then banned – by governments. Falvey (ch 3) points out that policies towards shifting cultivation tend to reflect 'world views'; and also makes clear that swidden agriculture is a reflection of the local natural or physical environment.

Many authors indicate how the plant varieties used by shifting cultivators – and recently introduced commercial exotic crops – relate to micro-geographic conditions of soil, drainage, slope, elevation and climate. The technological environment is also a key determinant of shifting cultivation policy, in part because it has such an important impact on government programmes to intensify agriculture. Most programmes introduced since the 1960s tend to emphasize increases in rural production through use of Green Revolution technology, including new crop varieties, chemical inputs, irrigation and motorized machinery. As noted in many chapters, few of these innovations have benefitted shifting cultivators in upland locations, and shifting cultivation as a practice and a culture has been negatively impacted by technological change (e.g. see Mertz and Bruun (ch 2), Pasicolan et al. (ch 35) and Hong (ch 36)). On the other hand, those cultivators who have adopted Green Revolution crops in permanent fields (e.g. lowland irrigated rice), or who use small, motorized machinery (e.g. chain saws, rice-milling machines), have benefitted significantly from technological change.

Internal elements of policy implementation

The large box at the centre of Figure 52-1 represents the *internal elements of policy, programme or project implementation*. The emphasis in this volume is on policies, although some of the more recent policies in specific countries are expected to be implemented through programmes with limited time horizons (see details in Heinimann et al. (ch 45) of the Agro-Biodiversity Initiative of the Lao PDR).

The implementation *process*, as presented in Figure 52-1, refers to the methods and procedures used to implement the policy. Few of the chapters provide details concerning the implementation process for specific policies. However, it can be assumed that implementation during colonial times was carried out in an authoritarian, top-down process with general policies determined in the seats of colonial power and implemented by a combination of government, military and corporate action. No doubt local administrators or other public servants had some influence on policy and its implementation, particularly when typical colonial bureaucracies delegated power through administrative hierarchies. One might expect these people, closely attuned to conditions on the ground, to play an important and well-informed role in policy development.

Several chapters call for much more collaboration between government and local communities in contemporary settings (e.g. Mertz and Bruun, ch 2; Gillogly, ch 38; and all chapters from 44 to 52). Hoare (ch 44 and 48) and Teegalapalli and Datta (ch 42) specifically address the problem of finding the proper balance between top-down and bottom-up planning and implementation – the much-studied issue of centralization or decentralization. The last few chapters of the book document a trend towards participatory processes in planning and implementation, partly because of a sense of fairness, but primarily because these processes can lead to a much better understanding of how shifting cultivation works, and how it can be usefully protected and enhanced.

The *tasks* box refers to specific actions that need to be undertaken in order to implement a policy. Recent examples include mapping of property boundaries, registering of land titles, provision of seeds or seedlings for new varieties of crops, and payment of subsidies for undertaking desired changes in land use. Implementation also involves the creation of an appropriate *organizational structure*, which usually extends beyond the existing governmental public service, and a *schedule or timing* for carrying out the policy. Administrative details are seldom mentioned in these chapters and the time horizon for most policies described here appears to be undefined.

The *implementers* are the individuals and local organizations at the interface between those intended to adopt the policy and the more central administration; those tasked with ensuring that the 'beneficiaries' of the policy comply with its requirements. These would include, for example, the persons or groups responsible for ensuring that shifting cultivators stop burning the forests – the official policy across the Asia-Pacific region throughout most colonial and post-colonial times. The authors make little mention of the actual implementers of policy, although Dukpa (ch 22) is an important exception. He mentions that some local officials in Bhutan did not enforce the ban on shifting cultivation because of their 'humanity'. This revealing observation echoes the findings of many who work in the policy-implementation field; who find that those who work most closely with beneficiaries are in a unique position to see the implications of a policy and sometimes take the initiative to act accordingly (e.g. Pressman and Wildavsky, 1973). Policy creation and implementation can be greatly enhanced if implementers are part of the planning process. Takahashi et al. (ch 10) mention that volunteers are used to burn grassland in a shifting cultivation park in southern Japan, making them implementers at the ground level. Hoare (ch 48) notes that villagers in the Upper Nan watershed of Thailand had existing regulations that turned out to meet the government's development objectives, making them policy implementers by happenstance.

Beneficiaries are, of course, central to policy implementation, not just because they are the people expected to benefit from the policy, but because they can have a profound effect on whether and how implementation takes place. Asking the simple question: 'Who are the intended beneficiaries?' is an essential step in analysing the effectiveness of policy, and a useful exercise in considering shifting cultivation in this volume. This argument is implicit in the chapters that describe the benefits of participatory methods, pointing out that involving shifting cultivation beneficiaries in policy formulation and implementation works to everyone's advantage (e.g. Ironside et al., ch 14; Hoare, ch 44 and 48; Karki, ch 49). As it happens, in only a few cases are the policies analysed in this volume intended to benefit shifting cultivators. The normal definition of 'beneficiaries' common to development projects does not really apply in many of the chapters. Who were the beneficiaries when policies banned shifting cultivation? Certainly not the swiddeners. Forestry departments or governments might be seen as beneficiaries if destroying an agricultural system and its practitioners allowed them to effectively control territory. Large corporations and governments might benefit if the land formerly occupied by swiddeners was leased or sold to businesses. Environmentalists who are not aware of the intricacies of swidden agriculture might take satisfaction from the elimination of the practice. But in only a few cases are the policies analysed in this volume intended to benefit shifting cultivators, although governments could argue that recent policies to intensify agriculture, raise production and help cultivators to adopt sedentary practices are actually for the benefit of society in general and, in some cases, shifting cultivators in particular.

Leadership

As Figure 52-1 indicates, *leadership* is identified as a critical role that affects the entire implementation process, and several implementation scholars have documented how leaders can be responsible for policy success or failure (e.g. Bardach, 1977; Warwick, 1982). The studies in this volume seldom identify particular leaders, although Gillogly (ch 38) observes that success in Thailand's Doi Sam Muen Highland Development Project was highly dependent on the actions of a particular leader. Several non-governmental organizations were seen as providing ideas, motivation and leadership (see, e.g. chapters 44, 47, 50 and Jyotishi and Manjula, 2017), a feature that has become increasingly important in more recent times. In line with participatory methodology, leadership can be enhanced when leaders work closely with implementers and beneficiaries.

Resources

Resources are identified on the left side of Figure 52-1. Whether they be funding, human labour, management, new technology, transportation, communication or other inputs, they are a requirement for implementation success. No doubt the failure of policies to eradicate shifting cultivation was partially due to a lack of resources to bring this about – a fortunate outcome from the standpoint of swidden agriculture. In more recent programmes to support agricultural intensification and in relation to REDD+, many resources have been available, primarily from governments. In some cases (e.g. Kumar, ch 21), funding has been abundantly available, but wasted, as many modern programmes have failed to deliver the desired outcomes. Much funding has been available from international sources through aid programmes (e.g. numerous bilateral programmes designed to increase agricultural production and rural incomes), bringing with it the tendency for donors to have a large influence on programme design and implementation (see Mertz and Bruun, ch 2). The United States provided much funding to support the eradication of opium production, requiring radical and often painful adjustments for shifting cultivators (ch 30, 31 and 32).

Evaluation

Periodic evaluation is an essential part of successful policy implementation, and this ideal is incorporated in Figure 52-1. Policy outcomes are evaluated and compared with planned outcomes, and then alterations are made in an effort to improve implementation. This is the kind of flexible policy development that one would like to envision in the real world. The book's chapters involve a great deal of evaluation by their authors, and a few point out the need for flexibility in evaluating conditions in different contexts – at different times and in different locations, and over time to see how policies and their implementation are changing. Saikia and Bhaduri (2017) underscore how shifting cultivation itself is constantly changing, and is anything but static! McKinnon (ch 6) refers to 'landscape plasticity', a term used to describe dynamic changes in shifting cultivation among Akha groups in northern Thailand and southern China. Several other authors show how selected areas or villages have changed in response to government policies and programmes (e.g. Ducourtieux, ch 37; Tran and Pham, ch 40; Jyotishi and Manjula, 2017). Karki (ch 49) makes a plea for pragmatic, context-specific policies for shifting cultivation, particularly given the current array of unclear, complex policies.

Although almost all of the authors evaluate policies and programmes, they do not use a common methodology and their analyses tend to be historical narratives. This is understandable, given the paucity of information about the specifics of either impacts or implementation over much of the past 500 years, and given the different points of focus of the authors' interests. Further, all of the chapters concern the impacts of policies on shifting cultivation and shifting cultivators, even though almost all of the evaluated policies and programmes were designed for quite different 'beneficiaries' – colonial empires, large companies, governments, national or regional populations as a whole, or international organizations interested only in environmental management or suppression of specific practices (e.g. burning of the forest, cultivation of opium poppies). Only recently does one encounter policies that actually support shifting cultivators (e.g. Karki, ch 49; Andersen, ch 50; Minarchek, ch 47). All of the authors' evaluations are valuable contributions to an understanding of the key issues, in anticipation of closer links among beneficiaries, implementers, leaders and creators of policy.

Conclusions

This volume presents a comprehensive analysis of the various policies, developed since the early days of colonialism, that affect shifting cultivation and shifting cultivators. The authors are sympathetic to both the practice and the people of swiddening, and for the most part, lament their fate in the face of policies that ignore their status, wishes and needs. Shifting cultivation has been misunderstood by everyone from governments to major world bodies (e.g. the Food and Agriculture Organization of the United Nations, the World Bank). Given the right circumstances, swiddening is a sustainable and wise method of resource management, probably less problematic for climate change or fragile ecosystems than many modern alternatives. It can provide a rich and dependable livelihood for traditional people who live in and manage their environments somewhat communally. The human rights of those communities have been trampled for centuries, with only a few signs of change.

Shifting cultivation has survived for centuries despite punitive policies, although the forms and specific practices vary greatly from one location to another. Both the practice and the cultivators are adjusting, sometimes rapidly, to changed contexts; particularly modernization, which brings with it an emphasis on commerce, new crop and livestock varieties (most imported from elsewhere), new technologies, very different cultures, permanent settlement and more comprehensive integration with the non-farm economy. The authors understand that these changes are inevitable and in the best long-term interests of traditional communities. But they want the changes to occur at a pace that allows the communities to experiment with change, to decide what directions make the most sense, and to work with governments, non-governmental organizations and other interested parties to develop the most sensible and pragmatic policies. They also want the best practices of shifting cultivators to be identified and preserved, often because they can serve as a basis for modern techniques that can best serve national and local interests.

In his Preface, Cairns calls for the identification and analysis of those best practices, along with examples of policies that have worked to serve the real interests of swidden agriculture and its cultivators. Although this book presents a few examples of such policies, most of the chapters decry the failure of policies, both ancient and modern, to contribute positively. Important case analyses of the details of shifting cultivation practices are presented, and a handful of policies that have been useful are described, but the authors tend to argue that a great deal of analysis and policy development has yet to be done. At the same time, they point to useful developments that hold promise for the future. They also argue that the best methodologies for analysis and policy development are participatory, with shifting cultivation communities playing a central role.

I have presented a framework for analysing policy implementation, and have used it to assess the observations of the various authors. The framework includes the components identified by recent researchers in the field of policy and programme analysis, and throws light on the authors' contributions. It identifies key findings, gaps and lack of clarity in our understanding of policies related to shifting cultivation and provides a guide for future work. The analysis concludes that we need an essential focus on the *processes* of shifting cultivation, policy creation, and policy implementation. Similarly, *the development and use of future policies should be process-oriented, with ongoing evaluation, and with defined roles for those engaged in swidden agriculture as well as for those designing and implementing policies.* The studies in this volume provide a highly important backdrop for that challenge.

As noted at the beginning of this chapter, this volume includes a number of chapters that themselves provide very useful overviews of shifting cultivation and related policies in the Asia–Pacific Region (see, in particular, Mertz and Bruun, ch 2; Falvey, ch 3; Oughton, ch 4; and Karki, ch 49). The reader is urged to revisit those chapters. They, like most of the other contributions, document very well the problems involved in shifting cultivation. They also include a measure of optimism that the future can be better for traditional communities if people and organizations learn from the past and collaborate effectively. This book certainly makes a worthy contribution in that direction.

Acknowledgements

Several reviewers examined an early draft of this chapter and provided a wealth of insights, corrections and suggestions. I am very grateful to them and to the editor for seeking the input of such a knowledgeable collection of individuals. Their comments contributed greatly to revisions.

References

Bardach, E. (1977) *The Implementation Game: What Happens after a Bill becomes a Law*, MIT Press, Cambridge, MA

Bayrak, M. M. and Marafa, L. M. (2017) 'The role of sacred forests and traditional livelihoods in REDD+: Two case studies in Vietnam's Central Highlands', available at http://www.cabi.org/openresources/91797

Brinkerhoff, D. (1996) 'Coordination issues in policy implementation networks: An illustration from Madagascar's Environmental Action Plan', *World Development* 24(9), pp1497–1510

Cairns, M. F. (ed.) (2007) *Voices from the Forest: Integrating Indigenous Knowledge into Sustainable Upland Farming*. Resources for the Future Press, Washington, DC

Cairns, M. F. (ed.) (2015) *Shifting Cultivation and Environmental Change: Indigenous People, Agriculture and Forest Conservation*, Earthscan, London

Conklin, H. (1957 and 1975) *Hanunoo Agriculture: A Report on an Integral System of Shifting Cultivation in the Philippines*, Forestry Development Papers, No 12 (vol. 2), Food and Agriculture Organization of the United Nations, Rome

Crosby, B. (1996) 'Policy implementation: The organizational challenge', *World Development* 24(9), pp1403–1415

Elmore, R. (1985) 'Forward and backward mapping: Reversible logic in the analysis of public policy', in K. Hanf and T. Toonen (eds.) *Policy Implementation in Federal and Unitary Systems: Questions of Analysis and Design*, Martinus Nijhoff, Dordrecht, the Netherlands

Found, W. (1971) *A Theoretical Approach to Rural Land-Use Patterns*, Edward Arnold, London

Found, W. (1992) 'Implementing environmental-management programs: A general framework for analysis', in F. Carden, W. Found and R. Amir (eds.) *Program and Policy Implementation: Proceedings of a Workshop at the Bandung Institute of Technology, November 1991*, Research Paper no. 42, University Consortium on the Environment, Bandung, Indonesia, pp1–7

Found, W. (1999) *Techniques of Project Planning and Implementation: Ten Steps to Success*, Institute for Leadership Development, Toronto

Grindle, M. (ed.) (1980) *Politics and Policy Implementation in the Third World*, Princeton University Press, Princeton, NJ

Hambly, H. (2000) 'Implementation and institutionalization of agroforestry in Western Kenya: Gender and agency analysis', PhD dissertation to the Faculty of Environmental Studies, York University, Toronto

Hurley, S. (2012) 'Women's rights, culture, and conflict: Implementing gender policy in Amboko Refugee Camp, Chad', PhD dissertation to the Faculty of Environmental Studies, York University, Toronto

ICIMOD (2006) *The Shillong Declaration*, International Fund for Agricultural Development and Natural Resources Management, International Centre for Integrated Mountain Development, Kathmandu, www.icimod.org/resource/1250, accessed 29 September 2016

Jyotishi, A. and Manjula, M. (2017) 'Revisiting statutory laws and customary norms governing swidden agricultural systems: A study based on swidden farmers in southern Odisha', available at http://www.cabi.org/openresources/91797

Lindayati, R. (2003) 'Ideas and policy change: Indonesia's locally-based forest management policy', PhD dissertation to the Faculty of Environmental Studies, York University, Toronto

Matland, R. (1995) 'Synthesizing the implementation literature: The ambiguity-conflict model of policy implementation', *Journal of Public Administration Research and Theory* 5(2), pp145-174

Mazmanian, D. and Sabatier, P. (1983) *Implementation and Public Policy*. Scott Foresman, Glenville, IL

Najam, A. (1995) *Learning from the Literature on Policy Implementation: A Synthesis Perspective*, International institute for Applied Systems Analysis, Laxenburg, Austria

Pressman, J. and Wildavsky, A. (1973) *Implementation: How Great Expectations in Washington Are Dashed In Oakland*, University of California Press, Berkeley, CA

Saikia, A. and Bhaduri, S. (2017) 'Institutional design and occupational "opportunity": The case of shifting cultivators in Nagaland, northeast India', available at http://www.cabi.org/openresources/91797

Van Meter, D. and Van Horn, C. (1975) 'The policy implementation process: A conceptual framework', *Administration and Society* 6(4), pp445-488

Warwick, D. (1982) *Bitter Pills: Population Policies and Their Implementation in Eight Developing Countries*, Cambridge University Press, Cambridge, UK

Wulandari, C. (2017) 'Policies that transform shifting cultivation and encourage community-based forest management in Lampung province, Indonesia', available at http://www.cabi.org/openresources/91797

Xu, J. C., McLellan, T. and Hiwasaki, L. (2017) 'Integrating swidden agricultural-knowledge systems into sustainable intensification in the Central Mekong region', available at http://www.cabi.org/openresources/91797

BOTANICAL INDEX

Note: bold page numbers indicate figures and tables. Numbers in brackets preceded by *n* are chapter endnote numbers.

ETHNIC GROUP INDEX

Note: bold page numbers indicate figures and tables. Numbers in brackets preceded by *n* are chapter endnote numbers.

GENERAL SUBJECT INDEX

Note: bold page numbers indicate figures and tables; numbers in brackets preceded by *n* are chapter endnote numbers.